STRIKING IRON

Striking Iron

THE ART OF AFRICAN BLACKSMITHS

Edited by Allen F. Roberts
Tom Joyce
and Marla C. Berns

with William J. Dewey
Henry J. Drewal
and Candice Goucher

With essays and contributions by Rowland Abiodun, Gassia Armenian, Marla C. Berns, Shadreck Chirikure, William J. Dewey, Isaie Dougnon, Henry J. Drewal, Candice Goucher, Manuel Jordán, Tom Joyce, Colleen E. Kriger, E. C. Krupp, Scott MacEachern, Patrick McNaughton, Philip M. Peek, Allen F. Roberts, Raymond Silverman, and John Watson

FOWLER MUSEUM AT UCLA † LOS ANGELES

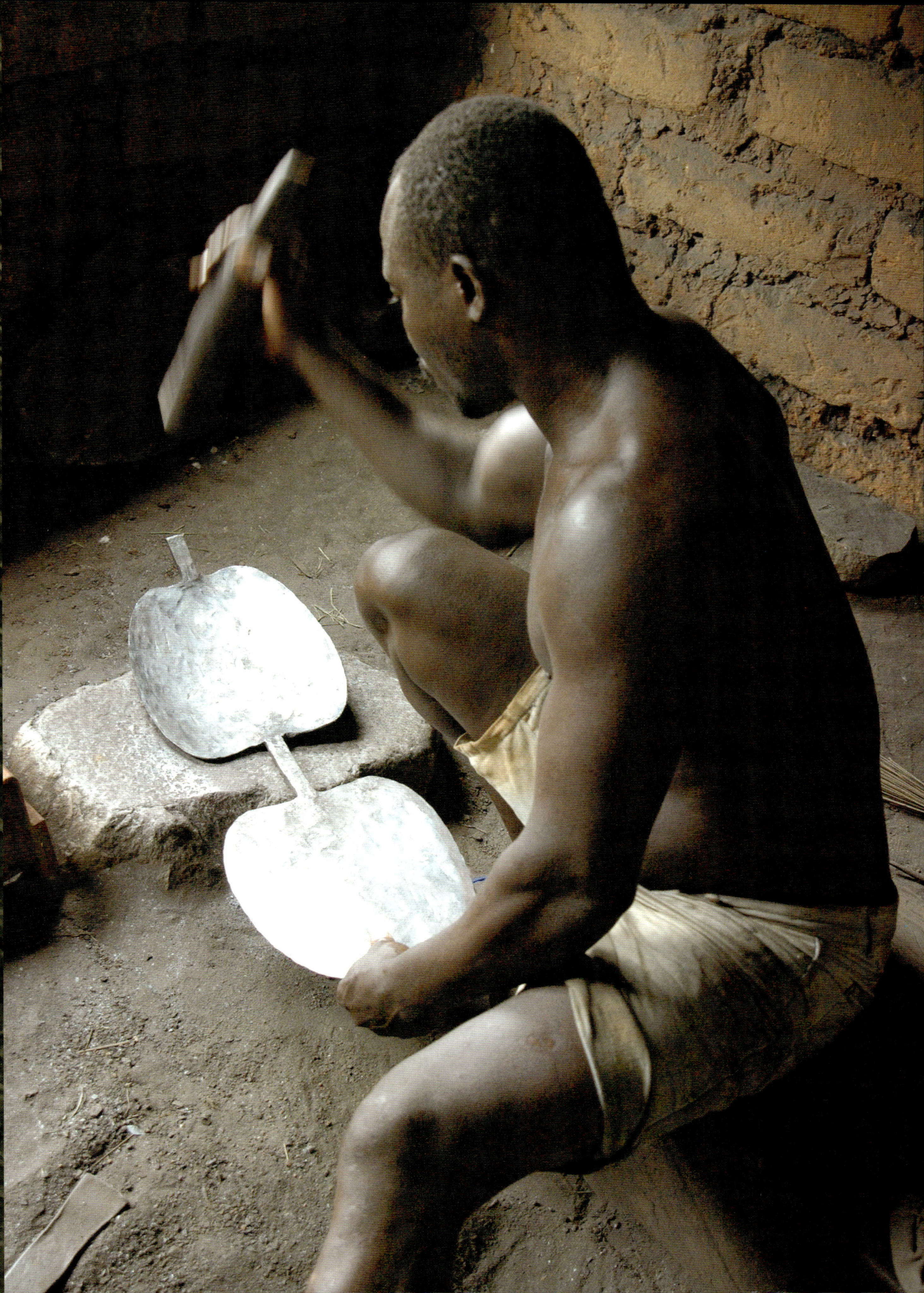

Striking Iron is made possible by major funding from the National Endowment for the Humanities* and by an award from the National Endowment for the Arts.

Lead sponsorship for the publication is provided by the Carl & Marilynn Thoma Art Foundation, with additional support from the Ahmanson Foundation on the recommendation of the late Foundation Trustee Emeritus, Lloyd E. Cotsen, and the Ethnic Arts Council of Los Angeles.

Generous support of the exhibition is also provided by the Martha and Avrum Bluming Exhibition Fund with additional funding from the Fowler Exhibition Fund, Cindy Miscikowski, Lee Bronson, Andrew Adelson, Richard Scheller and Susan McConnell, and Richard and Susan Ulevitch.

Education programs are made possible in part by The Ralph M. Parsons Foundation.

*Any views, findings, conclusions, or recommendations expressed in this book do not necessarily represent those of the National Endowment for the Humanities.

Lenders to the Exhibition

Individuals

Jan Baptist Bedaux
Mina and Samir Borro
Najwa Borro
Jean-Paul Chazal
Mark Clayton
Joel and Zachary Cooner
Charles and Kent Davis
Jo De Buck
Nicole and John Dintenfass
Bernard Dulon
Jan Elsen
The Felix Collection
The Ghysels Collection
Donald and Adele Hall
Mr. and Mrs. Hansen
The Itzikovitz Collection
Tom Joyce
Amy and Elliot Lawrence
Claudie and Alain Lebas
Henry Lu
Shaun Mirams
Claude-Henri Pirat
Private collections
Amyas Naegele
Richard H. Scheller
Mrs. Jerome (Ellen) Stern
Susan and Richard Ulevitch
Agnès Woliner

Institutions

Afrika Museum, Berg en Dal, the Netherlands
American Museum of Natural History, New York
Barbier-Mueller Museum, Geneva
Brooklyn Museum
Cleveland Museum of Art
Dallas Museum of Art
Field Museum of Natural History, Chicago
Fine Arts Museums of San Francisco
Fowler Museum at UCLA
Iris & B. Gerald Cantor Arts Center for Visual Arts at Stanford University
Mel Fisher Maritime Museum, Key West, Florida
The Menil Collection, Houston
Musée Africain de Namur, Belgium
Musée du quai Branly - Jacques Chirac, Paris
Museum aan de Stroom, Antwerp
Musical Instrument Museum, Phoenix
National Museum of African Art, Smithsonian Institution, Washington, D.C.
Neuberger Museum of Art, Purchase College, State University of New York
New Orleans Museum of Art
Samuel P. Harn Museum of Art, University of Florida, Gainesville
Virginia Museum of Fine Arts, Richmond
Yale University Art Gallery, New Haven

Exhibition Venues

Fowler Museum at UCLA
Los Angeles, California
June 3–December 30, 2018

National Museum of African Art, Smithsonian Institution
Washington, D.C.
April 17–October 20, 2019

Musée du quai Branly–Jacques Chirac
Paris
November 19, 2019–March 29, 2020

PREVIOUS PAGE (FROM PG 144), OPPOSITE PAGE (FROM PG 126): PHOTOGRAPHS BY TOM JOYCE, TCHARÉ, TOGO, 2010.

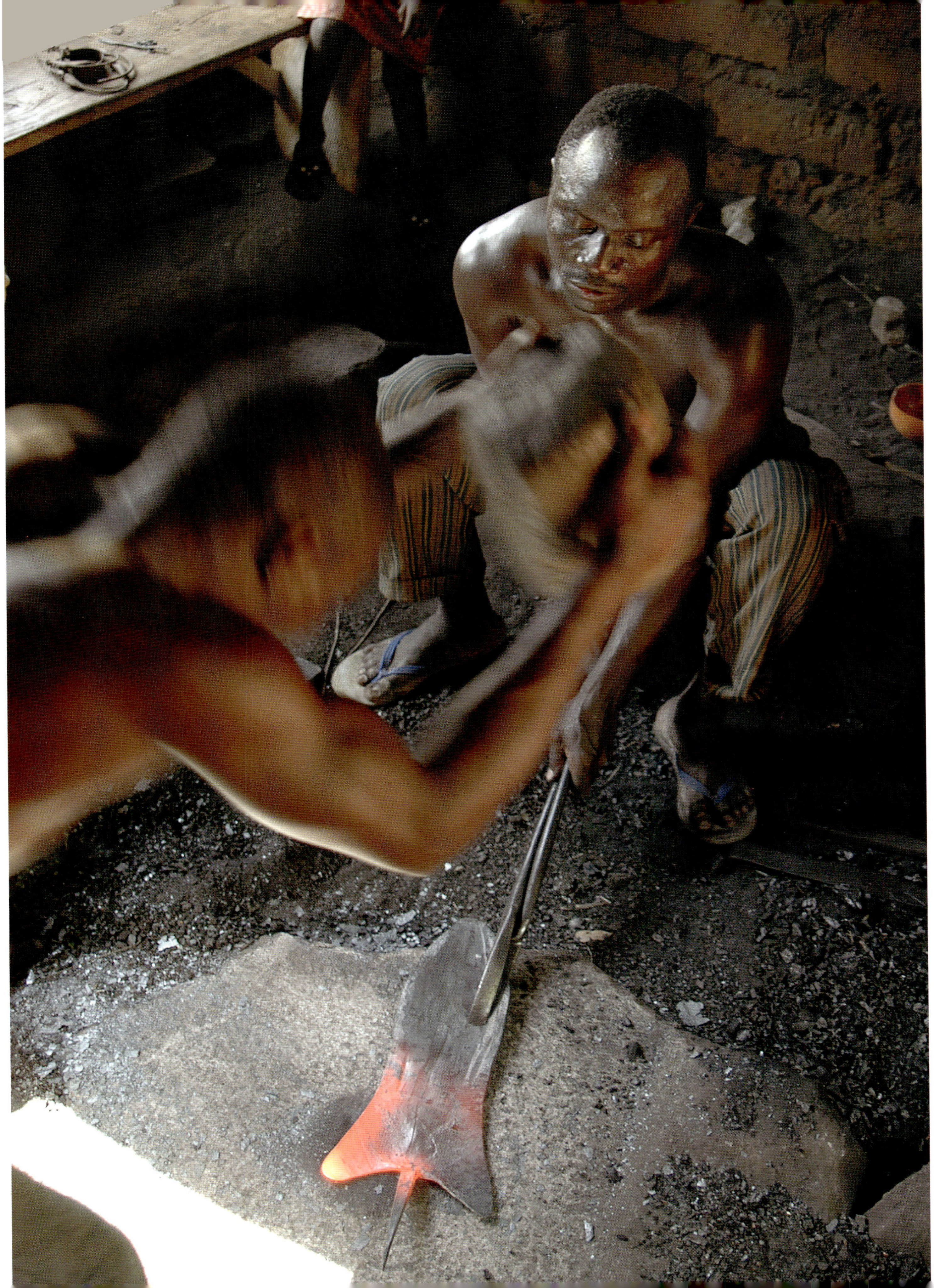

CONTENTS

DIRECTOR'S FOREWORD

Marla C. Berns

Striking Iron: The Art of African Blacksmiths continues the Fowler Museum's ambition to organize exhibitions and produce publications that tackle subjects of great depth and breadth, drawing on the original research and perspectives of scholars and specialists from many disciplines. Over the past three decades at the Fowler, the lineage of Africa-based exhibitions that share *Striking Iron*'s impressive scope and singular focus include: *The Essential Gourd: African Art from the Obvious to the Ingenious* (1986); *Elephant: The Animal and Its Ivory in African Cultures* (1992–93); *Sleeping Beauties: African Headrests and Other Highlights from the Jerome L. Joss Collection at* UCLA (1993–94); *Crowning Achievements: African Arts of Dressing the Head* (1995); *Music in the Life of Africa* (1999–2000); *Inscribing Meaning: Writing and Graphic Systems in African Art* (2007–08); *Mami Wata: Arts for Water Spirits in Africa and Its Diaspora* (2008); *and African-Print Fashion Now!: A Story of Taste, Globalization, and Style* (2017). The realization of *Striking Iron* has been among the most challenging, with the aspiration of encompassing the vast geographical expanse of Africa south of the Sahara and the 2500 years, at least, since the start of the African Iron Age. The selection of objects represents myriad genres of forged iron produced by blacksmiths belonging to well over one hundred ethnic groups living in nineteen countries. In selecting these works, more than 225 in all and dating from the seventeenth century to recent times, the curatorial team and its advisors strove to demonstrate how, through technical and conceptual sophistication, blacksmiths have transformed one of Earth's most plentiful natural resources into objects of great cultural importance and meaning (figs. 1, 2). Iron revolutionized Africa and forever altered human civilization practically and symbolically. Perhaps nowhere else has the work of smiths' been as deeply embedded in the lives of communities as in sub-Saharan Africa, even today.

Striking Iron was conceived by a team of co-curators led by artist Tom Joyce, a MacArthur Fellow trained in the art of forging iron, along with Allen F. Roberts, Professor of World Arts and Culture/Dance at UCLA; Henry J. Drewal, Evjue-Bascom Professor of Art History and Afro-American Studies at the University of Wisconsin–Madison; and William J. Dewey, Director of the African Studies Program and Associate Professor of Art History at Pennsylvania State University. Tom, Al, Hank, and Bill each brought deep historical, ethnographic, and art historical insight to the project, based on decades of research in particular countries and with specific peoples. They initially met with the Fowler's staff in 2011, during the Triennial Symposium held by the Arts Council of the African Studies Association (ACASA) in Los Angeles, where they had convened three multidisciplinary panels of scholars presenting papers on topics ranging from the archaeological beginnings of the African Iron Age to diverse genres of forged iron objects. Many of the participants in the symposium became project advisors to *Striking Iron* and authors of essays in this book.

We at the Fowler are indebted to the team of co-curators for working with us over a seven-year period to focus the exhibition's themes, objects, and interpretive strategies, and for their participation in this book as editors and authors. Their collective contributions have been immense from start to finish, and offered generously despite each having a demanding "day job." As the project progressed, I had the honor of joining them as a co-curator, and I have learned so much along the way.

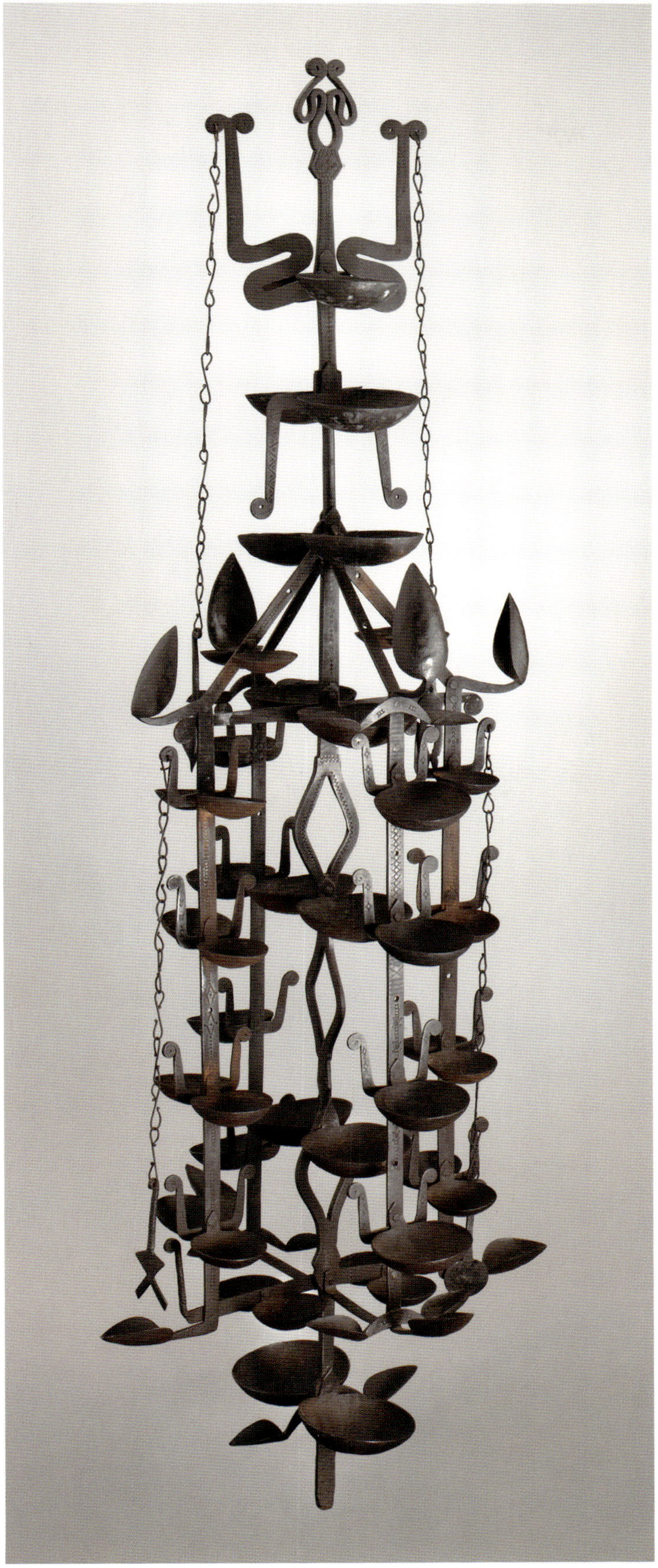

1
Artist unknown (Sorko, Bozo peoples, Mali)
Oil lamp (*ftula*), 19th century
Iron
H: 142.2 cm
YALE UNIVERSITY ART GALLERY, 2005.19.1; GIFT OF LABELLE PRUSSIN, PHD, 1973
PHOTOGRAPH BY DON WHAPLES, 2005. © YALE UNIVERSITY ART GALLERY

This lavish oil lamp, forged by a blacksmith of virtuosic skill, resembles the "Tree of Life" or "Tree of Light" known in Jewish, Christian, and Islamic traditions. Its forty-six cups are organized in numerical combinations linked to the sacred numerology of African origin myths, Islamic Sufism, and Judaic Kabbalah: three, four, seven, ten, fourteen, and twenty-two. More symbolic references are contained in the lamp's embellishments. Its plant- or serpent-like tendrils indicate growth and renewal, while its scrolled finials recall ram's horns and perhaps ritual animal sacrifice, associated with ancestor veneration in Islamic and Judaic traditions. Such sacred objects were planted in the ground at important private and public gatherings, from weddings and funerals to initiations and sporting events. This particular lamp was lit annually for the Tabaski Feast (feast of Abraham), observed during the last month of the Muslim lunar year, when each of its cups was filled with oil or animal fat and lit with a cotton wick. Brightness was adjusted using the small tools—shaped like a blacksmith's tongs—that dangle from the sides.

2
Artist unknown (Songye peoples, Democratic Republic of the Congo)
Figurative staff (detail), late 19th century
Iron, copper
H: 90.8 cm
FELIX COLLECTION
PHOTOGRAPH BY CHARLES MEUR, 2012. COURTESY OF FELIX COLLECTION.

Songye blacksmiths are renowned for their masterful modeling of the human face in forged iron or copper. A blacksmith sculpts iron like clay, but instead of fingers manipulating the medium, the metal is struck upon with hand-held, blunt-ended tools that forcibly move the material into the desired form. This staff would have been first roughed out hot with a hammer, and the finer details then cold-worked with custom-made punches and chisels.

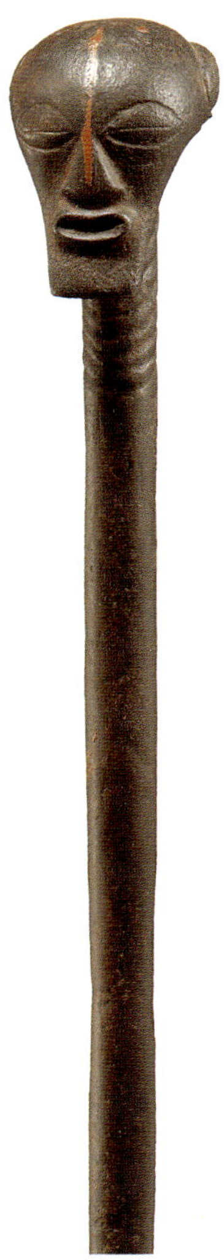

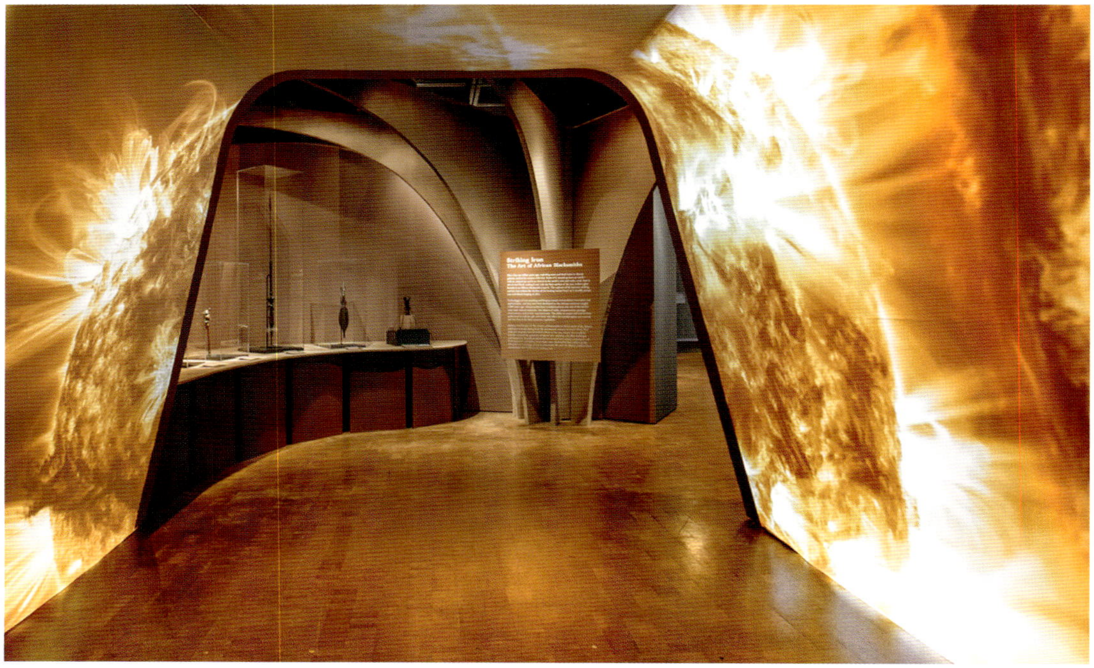

3 Installation view, entry tunnel, *Striking Iron: The Art of African Blacksmiths*, Fowler Museum at UCLA, 2018.
PHOTOGRAPH BY JOSHUA WHITE.

When the curatorial team first presented an outline of the comprehensive exhibition they envisioned, our Fowler exhibition development team was impressed by the extensive typology of ironworks made across Africa as well as the extraordinary beauty and complexity of these sculptural forms, which sometimes involve other metals and organic materials such as carved wood. We learned about how the processes of smelting and forging iron set up metaphors of transformation that have guided peoples' understanding of social change and transition and of the potencies of spiritual transactions. These narratives made for a highly persuasive argument in favor of organizing an exhibition that could convey the profound and multivalent significance of blacksmiths and their art across the continent. For the diverse audiences who attend the Fowler—UCLA students, faculty, and staff; museum members (whether adults or families); and visitors from throughout Southern California and beyond—we recognized that *Striking Iron* offered a fertile teaching opportunity.

The bibliography included in this volume gives some sense of the extensive literature on the subject. The project owes a special debt to Africanist historian and metallurgy specialist Eugenia W. Herbert, whose pioneering volume *Iron, Gender, and Power: Rituals of Transformation in African Societies* raised awareness of how the mastery of ironworking's transformative processes "offers a precious window into African cosmologies and a model for other technologies with similar cosmological grounding" (1993, 3). *Striking Iron* has also gained from previous exhibitions on African iron arts, especially *Iron, Master of Them All* (co-curated by William J. Dewey and Allen F. Roberts; University of Iowa Museum of Art, 1993) and *Life Force at the Anvil: The Blacksmith's Art from Africa* (curated by Tom Joyce; University of North Carolina at Asheville, 1998; National Metal Museum, Memphis, 1999).

Also significant to the Fowler's commitment to this effort was Tom Joyce's compelling slideshow, compiled to introduce audiences to the critical importance of iron to our existence through images of the cosmos, where the element originates in the furnaces of stars; of planet Earth, its crust glowing reddish orange from lateritic soils and deposits; and of the red hemoglobin cells that oxygenate our blood and keep us alive. We began to consider how the exhibition's design might likewise open with a vivid display of iron's potency in the lives of all humans—a wonder that most of us take for granted—to offer a dramatic and meaningful segue into a gallery filled with African artistry (fig. 3).

In contemporary America, iron as a metaphor for strength and perseverance lingers in common phrases such as "ruling with an iron fist," "strike while the iron is hot," and "too many irons in the fire," but few of us know much about the North American continent's history of blacksmiths, except perhaps for their work shoeing horses. This was not always the case, for as Tom Joyce reminded us, well past the American colonial era the occupation was highly regarded and specialized in many other categories of professional engagement beyond the work of farriers. In Colson Whitehead's 2016 Pulitzer Prize–winning historical novel, *The Underground Railroad*, about a young enslaved woman's journey towards freedom and violent pursuance by "slave catcher" Arnold Ridgeway, I encountered this eloquent passage:

> Arnold Ridgeway's father was a blacksmith. The sunset glow of molten iron bewitched him, the way the color emerged in the stock slow and then fast, overtaking it like an emotion, the sudden pliability and restless writhing of the thing as it waited for purpose. His forge was a window into the primitive energies of the world.
>
> He had a saloon partner named Tom Bird, a half-breed who took a sentimental turn when lubricated by whiskey. On nights when Tom Bird felt separate from his life's design, he shared stories of the Great Spirit. The Great Spirit lived in all things—the earth, the sky, the animals and the forests—flowing through and connecting them in a divine thread. Although Ridgeway's father scorned religious talk, Tom Bird's testimony on the Great Spirit reminded him of how he felt about iron. He bent to no god save the glowing iron he tended in his forge. He'd read about the great volcanoes, the lost city of Pompeii destroyed by fire that poured out of mountains from deep below. Liquid fire was the very blood of the earth. It was his mission to upset, mash, and draw out the metal into the useful things that made society operate: nails, horseshoes, plows, knives, guns. Chains. Working the spirit, he called it. [2016, 73–74]

During the transatlantic slave trade, African smiths were hunted down, captured, and brought aboard ships to plantations in the Americas to exploit their expertise (see chapter 16 and interleaf L, this volume). The respect and often fearful reverence long commanded by African blacksmiths in their own communities (which their European captors attested to) is conveyed throughout the present publication. The "Great Spirit" of Native American cosmology in Whitehead's narrative runs parallel to what Yorùbá of Nigeria call *àṣẹ*, which Henry J. Drewal describes as being "alive with spirit"; *àṣẹ* inheres in iron itself as well as the blacksmith's ability to unleash the metal's "performative power" to make things happen.

Allen F. Roberts has served as the *Striking Iron* publication's chief editor, and we are enormously grateful to him for more than two-and-a-half years of incisive commentary on drafts of its essays as well as helpful bibliographic suggestions, thoughtful refinements to the table of contents, and patience with the rounds of copyediting required to finalize texts, notes, and illustration selections and captions, working closely with the Fowler's Managing Editor, Deirdre O'Dwyer. Al's commitment to excellence is evident in every chapter and interleaf of this volume. In addition to introducing the volume with an elegant summary of arguments offered by scholars across several disciplines about the practical and philosophical aspects of "making," he wrote, co-wrote, or translated five chapters and two interleafs. Indeed, this publication is infused with his substantial mastery of the literature and theoretical perspectives on African blacksmiths, the things they make, and the profound meanings these things hold, in particular for the peoples of central Africa among whom he conducted lengthy field research. He was joined in the editing process by the project's co-curators along with project advisor Candice Goucher. I also thank Al for his generous help in writing drafts of the grant narratives that won us major support and for co-authoring the exhibition preview article that appeared in *African Arts* (2018, 51:1, 55–85).

4 Installation view, "Sustenance from the Anvil," *Striking Iron: The Art of African Blacksmiths*, Fowler Museum at UCLA, 2018.
PHOTOGRAPH BY JOSHUA WHITE.

Striking Iron has additionally relied on the critical input of seven advising scholars whose diverse fields of study have extended the project's interdisciplinary and geographical reach. Rowland Abiodun, Shadreck Chirikure, Candice Goucher, Colleen E. Kriger, Scott MacEachern, and Patrick McNaughton joined the co-curators and Fowler staff in an intensive two-day planning session to discuss, expand, and clarify the thematic sections and narrative flow of the exhibition as well as the publication's table of contents; after that session, Manuel Jordán joined the esteemed cohort, who collectively answered a constant stream of questions large and small about the articulation of general concepts and details of specific objects. We are grateful to them for their largesse, patience, and perseverance over the years of *Striking Iron*'s planning and implementation. Each advisor also wrote an essay, sometimes more, for this book, as did Isaie Dougnon, E.C. Krupp, Philip M. Peek, Raymond Silverman, John Watson, and the exhibition's co-curators. The Fowler acknowledges these authors for enabling us to tell many if not all of the rich and compelling stories of iron's importance to African communities, and for making possible what we anticipate will stand as the definitive resource on the subject.

Tom Joyce's leadership as a practitioner exploring both the art and science of forging has lent the project formidable authority. He offers in-depth perspectives on the wisdom and finesse of making—as is evidenced in his essay for this volume on the specificities of blacksmiths' tools, methods, and places of work—opening our eyes to the divine gift of creating with iron. Tom drew on his exhaustive knowledge of the literature on ironworking and the holdings of public and private collections worldwide to select most of the artworks included in *Striking Iron*. His personal relationships with numerous lenders certainly helped facilitate the museum's ability to borrow the finest forgings by African blacksmiths. Listening to Tom explain the subtleties of particular objects and the ways that virtuosity enhances efficacy led us to include his commentary on video "tour" stations within the exhibition. He was the final voice in attending to the myriad details of producing *Striking Iron*, reviewing and helping to craft grant narratives and exhibition texts, making selections of photographs and archival footage, and reading, as technical editor, all of the publication's chapters. One of the great pleasures of this project was working closely with Tom on final refinements to the exhibition checklist, to the flow of the installation's sections, and to the details of positioning and mounting every object. We thank him warmly for his passionate dedication to bringing *Striking Iron* to fruition.

Gassia Armenian, the Fowler's Curatorial and Research Associate, has provided unwavering support for *Striking Iron* since its inception, and no challenge large or small was beyond her reach. In my role as project director, I relied on her to keep track of the object list as it developed over the years, and she worked with Bridget DuLong, the Fowler's Project Manager, Exhibitions, on maintaining and updating information about individual objects and lenders as changes were made. Gassia was tenacious in searching for mystery loans and lenders, seeking information from colleagues and communicating to lenders the Fowler's well-regarded reputation for organizing international exhibitions of this scope and complexity. Gassia also worked with *Striking Iron*'s co-curators, advisors, authors, and Managing Editor, Deirdre O'Dwyer, on securing permission to reproduce archival and field photographs in the exhibition and publication—no easy task. Above all, her passion for and dedication to the project have been inspiring to us all.

An untold number of colleagues have helped answer questions and supply information, photographs, and film and video footage to assist in meeting the demands of such an expansive topic. We warmly thank them all (and apologize to anyone whom we may have inadvertently overlooked in this listing): Codjovi Étienne Joseph Adande; Rob Aft; Susan Allison; David Amster; Ramona Austen; Gasim Badri; Viviane Baeke; Alex Baradel; Randi Barndon; Philip de Barros; Jennifer Belt; Francis Benyure; Paul Berliner; David J. Binkley; Sandra Blakely; Huib Blom; Anne-Marie Bouttiaux; Peter Breunig; An Cardoen; Brigitte Cavanagh; S. Terry Childs; Cinematek–Brussels; Herbert J. Cole; Els Cornelissen; Maarten Couttenier; Nicholas Crowe; Jenny Daly; Nicholas David; Yolanda Davidson; Luc De Boe; Regina De Martelaere; Els De Palmenaer; Frédéric Dehaen; Paul Delucco; David

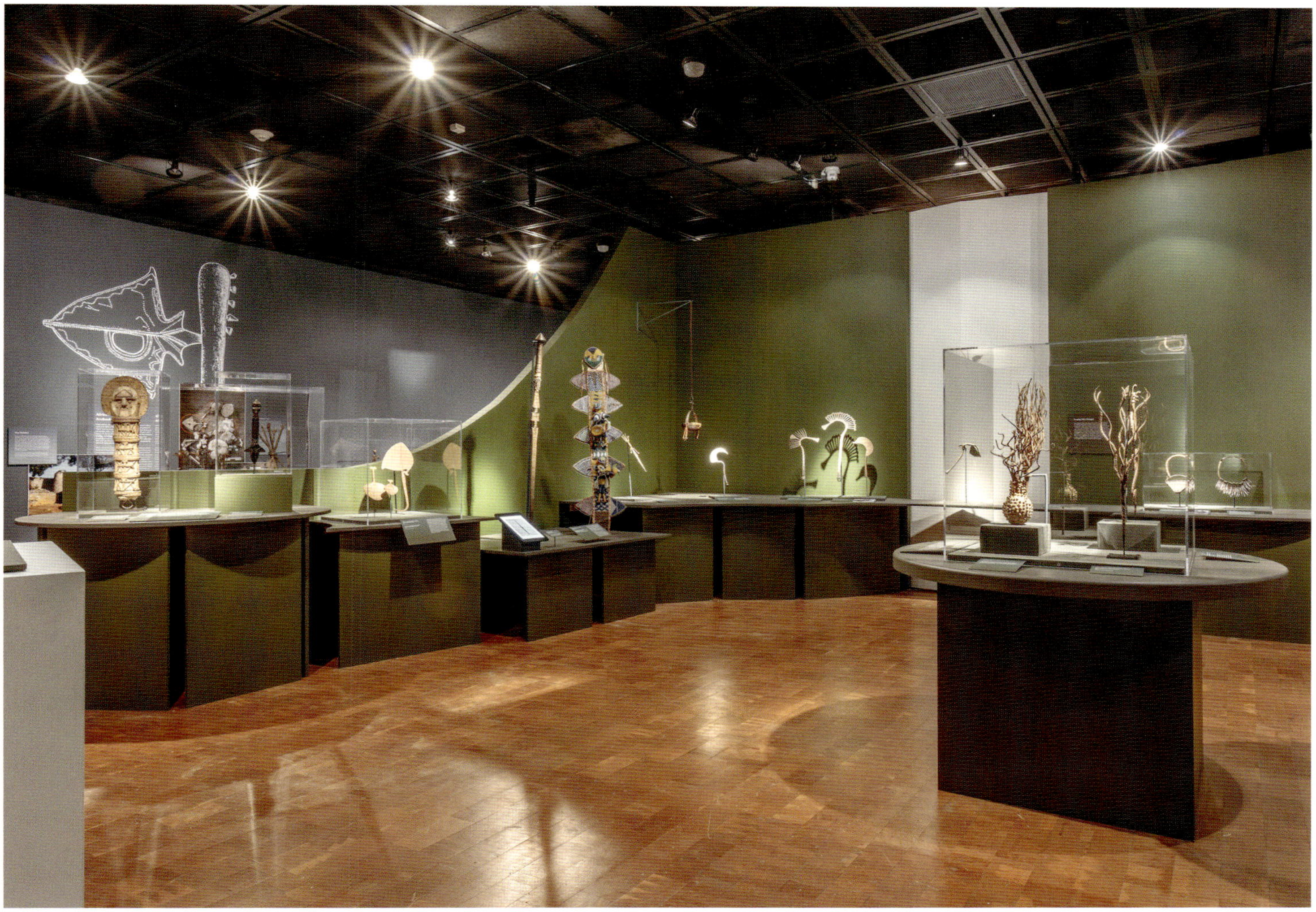

Doris; Ingeborg Eggink; Stefan Eisenhofer; Aliaa El Sandouby; Jan Elsen; Alan Etzel; Richard Fardon; Hamza El Fasiki; Paul Feindt; Marc L. Felix; Sandra Ferracuti; Gabriele Franke; Susan E. Gagliardi; Berhard Gardi; Eric Gent; Abdelhamid Ghandi; Marc Ghysels; Michelle Gilbert; Paula D. Girshick; James Green; Christiane Gruber; Karin Guggeis; Sondra Hale; Lauren Hancock; Keith Hart; Salah Hassan; Branwen Hennig; Dunja Hersak; Thomas Hinnebusch; Abdelouahed Hmadou; Jane Humphris; Rémy Jadinon; John Janzen; Bogumil Jewsiewicki; Abdurahman Juma; Kizumina Kabulo; Sidney Kasfir; Jane Katoke-Labani; Eline Kevenaar; David J. Killick; Michaela Koller; Hans-Joachim Koloss; Andrea Kuprecht; Chapurukha M. Kusimba; Carol Labani; Alisa LaGamma; Adria LaViolette; Talia Lieber; John Mack; Casey Mallinckrodt; Pierre de Maret; Bárbaro Martinez-Ruiz; Zivanai Masango; Evan M. Maurer; Jim McDougall; Wyatt MacGaffey; James A. Miller; James A. Moore; Gift Mugwidi; Maria E. Murguia; Vincenzo Negro; Karel Nel; Fallou Ngom; Bully Nimaga; Stephanie Noland; Molly O'Guiness Carlson; Kolawole Oshitola; Monica Park; Daryl Peavy; David Phillipson; Len Pole; Katie Prichard; Angela Rackham; Jeremiah Ra Richards; Mary Kujawski Roberts; Mary (Polly) Nooter Roberts; Peter Rohowsky; Doran H. Ross; Christopher D. Roy; Asma Sayeed; John Scarlett; Katja Scharff; Peter R. Schmidt; Janet Siroto; Rebecca Skinner Green; Pamela Jane Smith; Neil Sobania; Christopher B. Spring; Amy J. Staples; Judith Sterner; Zoë Strother; Elaine Sullivan; Katarina Thorlin; Robin Truesdale; Steven Tucker; Wouter Van Beek; Gary Van Wyk; Hein Vanhee; Jan Vansina; Kathelijne Verlinden; Honoré Vinck; Alexandre Vral; Pierre Wachholder; Nathalie Wagemans; Rudolph Ware; Sarah Walpole; Brad Weiss; John Watson; and Nicola Woods.

5 Installation view, "Sounding Forms," *Striking Iron: The Art of African Blacksmiths*, Fowler Museum at UCLA, 2018.
PHOTOGRAPH BY JOSHUA WHITE.

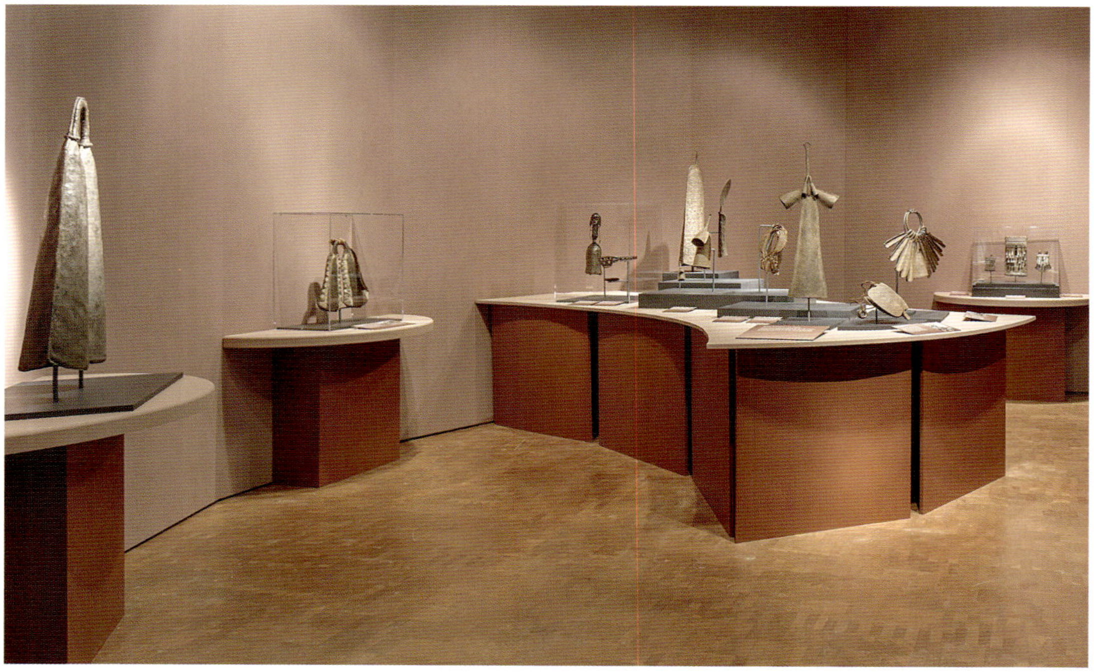

The exhibition's international travel resulted from the extraordinary generosity of forty-nine private and institutional lenders. We extend our gratitude to the following private lenders in the United States: Mark Clayton; Joel and Zachary Cooner; Charles and Kent Davis; Nicole and John Dintenfass; Donald and Adele Hall; Tom Joyce; Amy and Elliot Lawrence; Amyas Naegele; Richard H. Scheller; Mrs. Jerome (Ellen) Stern; Susan and Richard Ulevitch; and anonymous individuals. Additional gratitude goes to the following private lenders in Europe and Asia: Jan Baptist Bedaux; Najwa Borro; Mina and Samir Borro; Jean-Paul Chazal; Jo De Buck; Bernard Dulon; Jan Elsen; the Felix Collection; the Ghysels Collection; Mr. and Mrs. Hansen; the Itzikovitz Collection; Claudie and Alain Lebas; Henry Lu; Shaun Mirams; Claude-Henri Pirat; Agnès Woliner; and anonymous individuals. We thank these institutional lenders in the United States (and their staffs): American Museum of Natural History (Ellen V. Futter, Alex de Voogt, Kristen Mable); Brooklyn Museum (Anne Pasternak, Nancy Spector, Elizabeth Largi); Cleveland Museum of Art (William Griswold, Heather Lemonedes, Ugochukwu-Smooth Nzewi, Gretchen Shie Miller, Rachel Beamer); Dallas Museum of Art (Augustin Arteaga, Roslyn Adele Walker, Tricia Dixon); Field Museum (Richard Larivière, Angie Morrow, Lauren Hancock); de Young Museum, Fine Arts Museums of San Francisco (Max Hollein, Christina Hellmich, Deanna Griffin, Sue Grinols); Harn Museum of Art (Rebecca Nagy, Susan Cooksey, Jessica Uelsmann); Iris & B. Gerald Cantor Arts Center for Visual Arts at Stanford University (Susan Dackerman, Elizabeth K. Mitchell, Tammy Fortin, Amanda M. Maples, Katie Clifford, Jennifer Daly); Mel Fisher Maritime Museum (Melissa Kendrick, Dylan Kibler); The Menil Collection (Rebecca Rabinow, Susan Slepka Anderson); Musical Instrument Museum (April Salomon, Manuel Jordán, Linda Kirsh, Katarina Thorlin); National Museum of African Art, Smithsonian Institution (Augustus Casley-Hayford, Christine Mullen Kreamer, Kevin D. Dumouchelle, Clarissa Fostel, Laura Hauvenac, Amy Staples); Neuberger Museum of Art, Purchase College, State University of New York (Tracy Fitzpatrick, Helaine Posner, Patricia Magnani); New Orleans Museum of Art (Susan Taylor, Lisa Rotondo-McCord, Beth Bahls, Jennifer Ickes, Marie-Page Phelps, Laura Povinelli); Virginia Museum of Fine Arts (Alex Nyerges, Richard B. Woodward, Casey Mallinckrodt, Nancy Nichols); and Yale University Art Gallery (Jock Reynolds, L. Lynne Addison). We thank these institutional lenders in Europe (and their staffs): Afrika Museum (Stijn Schoonderwoerd, Annette Schmidt, Marit Jacobs, Ingeborg Eggink); Barbier-Mueller Museum (Laurence Mattet, Anne-Joëlle Nardin); Musée Africain de Namur (François Poncelet); Musée du quai Branly – Jacques Chirac (Stéphane Martin, Fred Chih-Chia Chung, Laurence Dubaut, Agathe Moroval, Sylvia Linard); and Museum aan de Stroom (Marieke van Bommel, Els De Palmenaer, Kathelijne Verlinden).

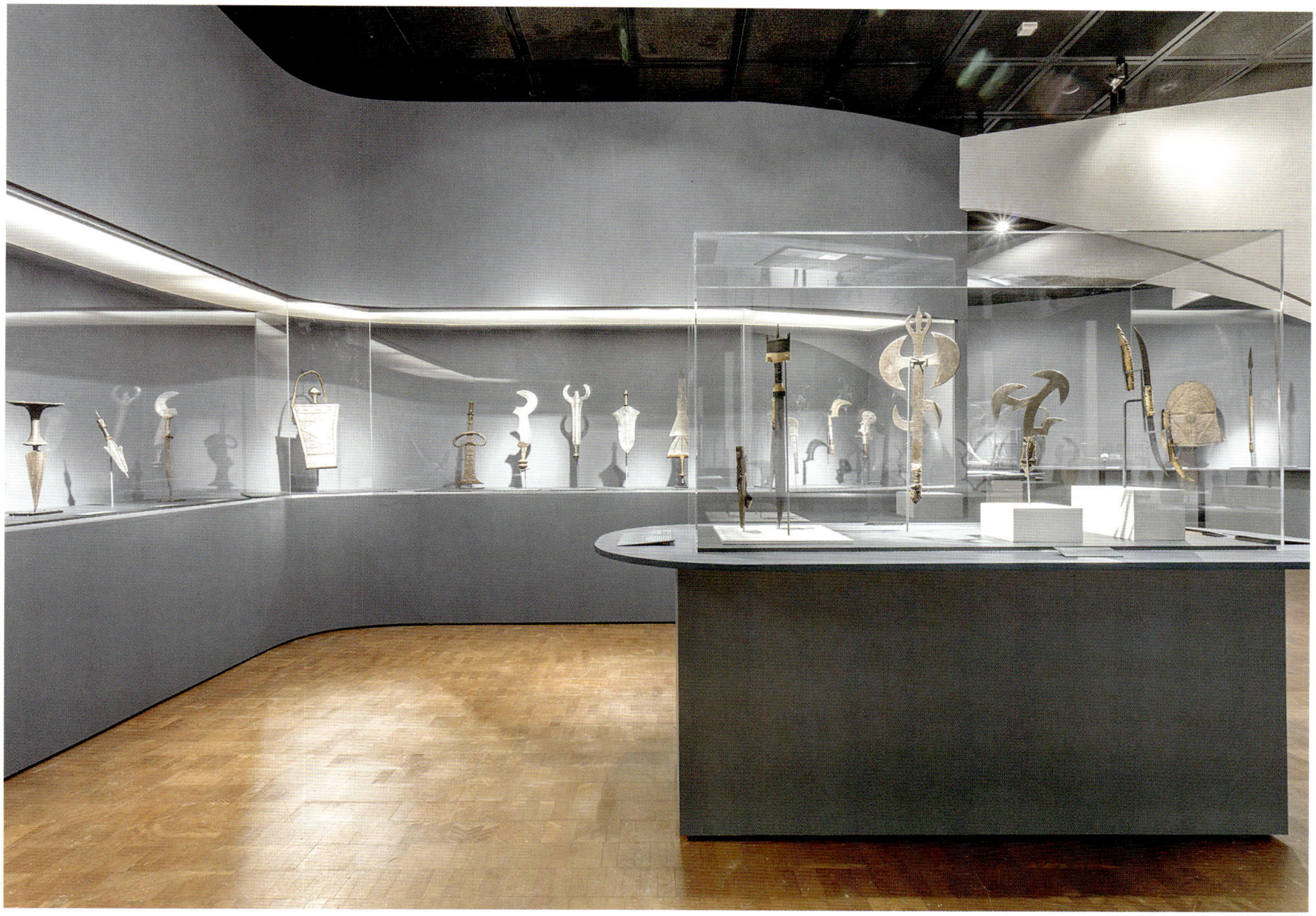

We are very pleased by *Striking Iron*'s travel to two additional museums, both of which have long-standing and mutually beneficial partnerships with the Fowler, and we thank our colleagues at both. At the National Museum of African Art, Smithsonian Institution, Washington, D.C., we thank: Christine Mullen Kreamer, Kevin Dumouchelle, Clarissa Fostel, Douglas Johnston, and Lisa Vann. At the Musée du quai Branly – Jacques Chirac, Paris, we thank: Christine Drouin, Stéphane Martin, Fred Chih-Chia Chung, Théo Ciora, Marine Martineau, and Sophie Chambonnière.

No project of this scale and scope is possible without substantial funding from a number of quarters. Major support came from the National Endowment for the Humanities, which provided both Planning and Implementation grants, and from the National Endowment for the Arts. The support of both agencies lent significant weight to the project at the same time as it acknowledged the endorsement of our peers. Generous exhibition support was also provided by the Martha and Avrum Bluming Exhibition Fund, with additional funding from the Fowler Exhibition Fund, Cindy Miscikowski, Lee Bronson, Andrew Adelson, Richard Scheller and Susan McConnell, and Richard and Susan Ulevitch. Lead sponsorship for the publication was provided by the Carl & Marilynn Thoma Art Foundation, with additional support from the Ahmanson Foundation on the recommendation of the late Foundation Trustee Emeritus Lloyd E. Cotsen, and the Ethnic Arts Council of Los Angeles. Education programs were made possible in part by The Ralph M. Parsons Foundation. Our gratitude goes to all who supported our vision for this project.

6 Installation view, "Blades of Power and Prestige," *Striking Iron: The Art of African Blacksmiths*, Fowler Museum at UCLA, 2018.
PHOTOGRAPH BY JOSHUA WHITE.

The entire Fowler Museum staff is to be recognized for its sustained effort in making *Striking Iron* a reality. Drawing on years of experience and a shared dedication to excellence, every member of our team can share credit for having helped produce a beautiful, innovative, and critically acclaimed traveling exhibition; a handsome accompanying scholarly publication; and a roster of diverse programs for visitors of all ages. All staff members are listed at the end of this volume, and I recognize individuals here for their key roles in the project. David Blair, Deputy Director, joined me in the task of administering this project with his usual financial and organizational acumen. Sebastian M. Clough, Director of Exhibitions, demonstrated once again how his effective immersion in the subject matter leads to inspired installation design. In this case, he conceived a platform for *Striking Iron*'s ideas that was both gorgeous and sympathetic to the objects on view (figs. 4–7). His exhibition crew, led by Armando G. Cortes, labored long hours to build the set and safely install over 225 objects. Derek Dubler, Exhibition Graphic Designer, developed a "striking" graphic identity for the exhibition that captured its essence. Bridget DuLong, Project Manager, Exhibitions, ran a tight ship and kept close track of logistical issues as the exhibition developed, and she oversaw the details of the tour with her usual quiet competency and equanimity. Matthew H. Robb, Chief Curator, was a helpful sounding board in the final stages of the project and, along with the Fowler's other curators, supported our efforts throughout to implement the project. Rachel Raynor, Director of Registration and Collections Management, adeptly oversaw the complexities of borrowing and shipping works from forty-nine lenders internationally. She was ably assisted by her Collections group, who were tasked with showing project curators (more than once) iron objects within the Fowler's collection and with receiving and monitoring loans as they entered the museum and then the gallery. Christian de Brer, Head of Conversation, oversaw the labor-intensive job of cleaning and treating the Fowler's objects and other loans for installation.

Terri Geis, Director of Education and Interpretation, joined me in writing the exhibition script and refining the exhibition's other methods of sharing information with visitors, and oversaw her education staff's creative programming and outreach efforts to reach adult visitors and K–12 students and their teachers. Betsy D. Quick, the Fowler's previous Director of Education and Curatorial Affairs, participated in the project's planning process and identified Adam's Forge in Los Angeles as an especially promising programmatic partner. Terri later pursued this collaboration, through which audiences of all ages partook in demonstrations of the skills required to work iron. Erin Connors, Director of Communications and Engagement, assisted with various strategies for promoting the exhibition and its many programs. We also thank Peter Kirby of Media Art Services in Los Angeles for filming and/or editing all of the videos in the exhibition. He was enormously helpful to us in determining how many, how long, and how they should be presented.

Fundraising for this project stretched over several years. Lynne Brodhead Clark, former Fowler Director of Development, wrote the successful government grants that provided substantial funding for the exhibition. Current Director of Development, Kris Lewis, stepped in to oversee the last stages of marshaling support, especially for the publication.

Any volume of this breadth and intricacy has required prolonged and keen attention by its producers. Allen Roberts has already been mentioned for his fortitude as chief content editor and for writing many of the book's key chapters. He worked closely with Deirdre O'Dwyer, the Fowler's new Managing Editor, who began her tenure in June 2017 with one of the Museum's most substantial challenges. She has handled the task with supreme patience and intelligence, and has been an effective and resilient collaborator. Don Cole, Director of Photography, took beautiful images of objects whose intricate details and reflective surfaces are not easy to capture. Danny Brauer, Director of Publications, is to be congratulated for yet another elegant book design and for managing the lengthy and complex course of its production. There are others who work behind the scenes to keep the Fowler running smoothly, and among them is Sophia Livsey, my Executive Assistant, who excels at organizing my days with efficiency and humor. She has been stalwart in helping others complete aspects of the exhibition and publication from its inception.

Lastly, my thanks go again to the team of co-curators who entrusted the Fowler with producing *Striking Iron*. They have been generous partners on our collective journey. Yet, it is to the artistry and sophistication of sub-Saharan Africa's countless blacksmiths that we offer final words of gratitude. The opportunity to share their magnificent and unparalleled achievements has been the Fowler Museum's great privilege and honor.

Marla C. Berns
SHIRLEY & RALPH SHAPIRO DIRECTOR, FOWLER MUSEUM AT UCLA

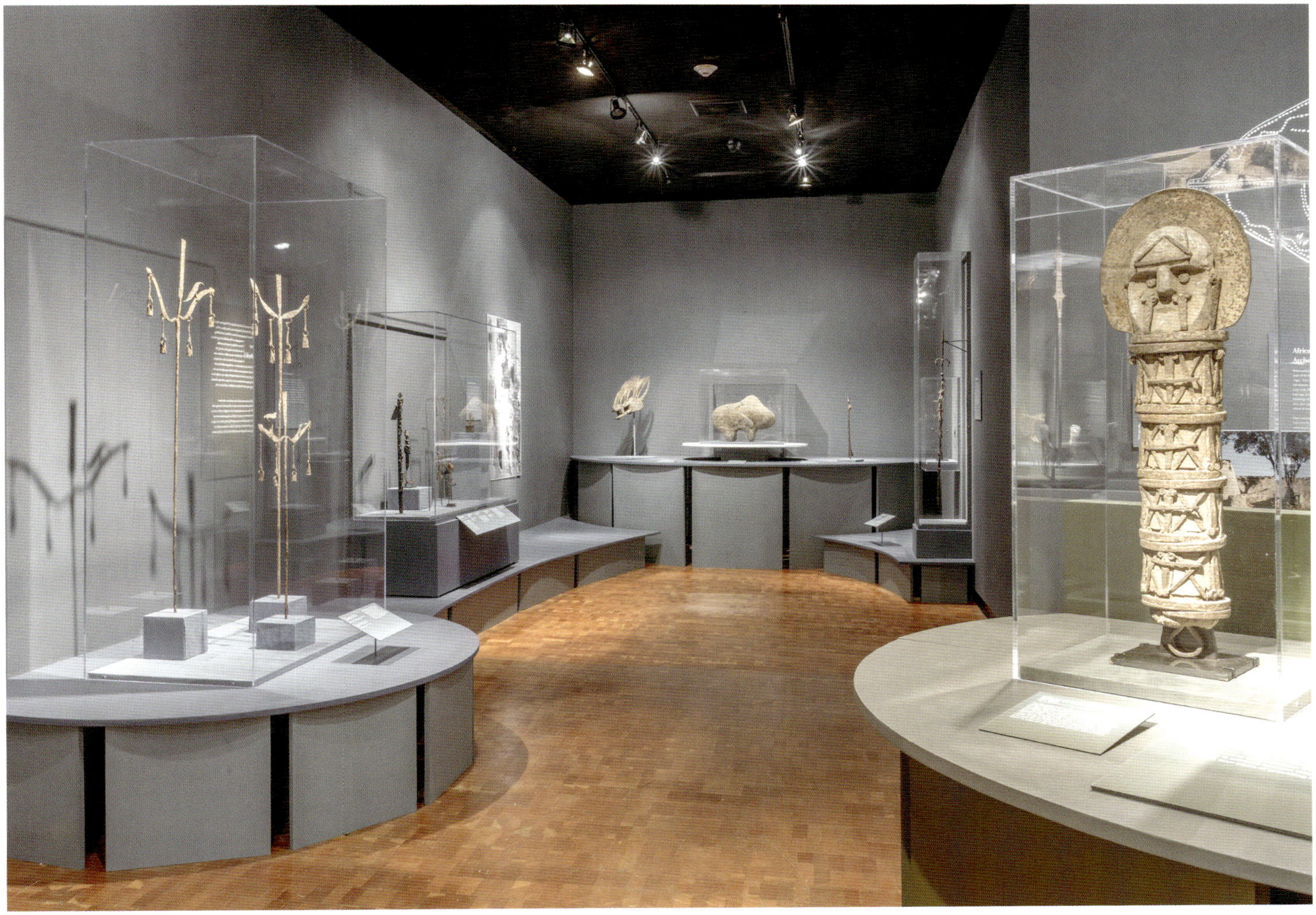

7 Installation view, *Striking Iron: The Art of African Blacksmiths*, Fowler Museum at UCLA, 2018.
PHOTOGRAPH BY JOSHUA WHITE.

CURATORS' ACKNOWLEDGMENTS

Tom Joyce *Striking Iron: The Art of African Blacksmiths* is dedicated to African creators of all things iron—blacksmiths who continue to shape the world through the objects they launch into service. It is with respect and gratitude that I thank the many practitioners I met and spent time with in Ghana, Burkina Faso, Mali, Togo, and the Republic of Bénin during five research trips in preparation for this exhibition. Their professional embrace was welcoming and generous. The intonation of their hammers, as they passed on expertise to another generation, meant that embers were carried from one fire to the next everywhere I visited, as young people picked up the tools of their families' trade.

A sincere appreciation goes out to my traveling collaborators, Codjovi Étienne Joseph Adande, Anne-Marie Bouttiaux, Henry J. Drewal, Steven Feld, Sarah K. Khan, and Charlie Piot, for their reliable and joyful companionship on the road in west Africa. While I pursued a nearly full-time, parallel life away from the studio and home front, my efforts were supported from afar by Helen Colton, Caleb Kullman, and my daughters, Kate and Irene Joyce. This project, in its present form, would never have been completed were it not for the ironclad support and encouragement of Marla C. Berns and Anne-Marie Bouttiaux. Their levity and patience every step of the way created the conditions for generating enough steam to keep moving ahead while juggling intercontinental travel, home, and my studio practice.

I would like to acknowledge the following blacksmiths and their families who graciously received me in their homes and workshops and allowed me to observe, interview, film, and photograph them at work: Damien Akati and family, Latevi Alboubou, Galbert Atakpa, Kokou Atiogbe, Koumedjina Atiogbe, Yao Atiogbe, Sanou Babadjibi and family, Zan Bagayoko, Henri Simonet Biokou, Soumaila Diabate, Adechian Adjadi Djoko II, Damien Dossougbete, Gobert Dossougbete, Herve Dossougbete, Kwashie Atwe Fosu and family, Moqadar Gidan, Leon Hounsounougan, Djidji Silon Hountondji, Jean Abiala Hountondji, Essozimna Ide and family, Patrice Idekambi, Sagara de Idielina, Jouli Kao, Kossi Kao, Ibelu Kasogue, Capko Pascal Kokossou, Amourou Kwomi Kondo, Salif Kone, Fode Konte, Moussa Konte, Sedu Konte, Tamba Konte, Djayo Kpadja, Essozimna Kwando, Hodenou Noglo, Mensah Sowah Okpti, Yom Pyim, Amadou Sagara, Musa Samassakou, Souleymane Samassakou, Mohammad Samasse, Kpatakpe Sinhuto, Mamadou Sounkoro, Manuel Takyi, Samuel Takyi, Armand Tessi, Amadou Traore, Bienvenue Yekedo, Rogier Yekedo, William Dieudom Yekedo, Yawovi Yovo, and Togbui Raphaël Kougblenou Yovo-Agosso.

I also thank the following anthropologists, archaeologists, art historians, curators, ethnomusicologists, metallurgists, scientists, and museum staff who keenly addressed inquiries, facilitated introductions, opened archives, and in myriad ways assisted my research: Alexis Adande, Kofi Agorsah, Elise Alexander, Luc André, Mary Jo Arnoldi, Viviane Baeke, Michael Barrett, Philip de Barros, Edna G. Bay, Harrie Bazunu, Kathleen E. Bickford Berzock, David J. Binkley, Eric Blinman, Marsha C. Bol, Nichole Bridges, Marie-Thérèse Brincard, Temesgen Burka, An Cardoen, Georges R. Celis, Christa J. Clarke, Frédéric Cloth, Herbert J. Cole, Johnnetta B. Cole, Donald J. Cosentino, Els De Palmenaer, Boureima Diamitani, David Doris, Kevin D. Dumouchelle, Jan Elsen, Pamela I. Eze-Uzomaka, William A. Fagaly, Christiane Falgayrettes-Leveau, Anna M. Fariello, Till Förster, Jung Ran Annachiara Forte, Susan E. Gagliardi, Christraud Geary, Marc Ghysels, Colette Gounou, James Green, Jan-Lodewijk

Grootaers, Guido Gryseels, Sedou Guindo, Jane I. Guyer, Eugenia W. Herbert, Joe Horsecapture, Ellen Howe, Julie Hudson, Eric Huysecom, Nicholas K. Ivor, Philippe Jespers, Manuel Jordán, Nestor Kahoun, Ivan Karp, Sidney Kasfir, David J. Killick, Sandra Klopper, Sandra E. Knudsen, Corrine A. Kratz, Christine Mullen Kreamer, Chapurukha M. Kusimba, Alisa LaGamma, Frederick J. Lamp, Yves Le Fur, John Mack, Tim Maggs, Pierre de Maret, Stéphane Martin, Laurence Mattet, Evan M. Maurer, Pamela McClusky, Patrick McNaughton, Wayne Modest, Bossman Murey, Didier N'Dah, Ken Kelountang Ndiaye, Steven D. Nelson, Ugochukwu-Smooth Nzewi, Constantine Petridis, John Picton, Charles Piot, Barbara Plankensteiner, François Poncelet, Robin E. Poynor, Lydia Puccinelli, Valentino Ouedraogo, Betsy D. Quick, Polly Richards, Doran H. Ross, Victoria L. Rovine, Christopher D. Roy, Alimata Sawadogo, Enid Schildkrout, Peter R. Schmidt, Klaus Schneider, Anna Seiderer, Thomas K. Seligman, Samuel Sidibe, Lowery Stokes Sims, Christopher B. Spring, Zoe S. Strother, Barbara Thompson, Carol Thompson, Kristina Van Dyck, Guy Van Rijn, Sophie de Ville, Khristaan D. Villela, Monica Blackmun Visona, Susan M. Vogel, Julien Volper, Alex de Voogt, Roslyn Adele Walker, Boris Wastiau, Philip Yayson, and the Museum of New Mexico Board of Regents.

The following collectors and gallerists hospitably opened their homes, galleries, storage, and photographic archives to accommodate my comparative search for artifacts in iron: Alex Arthur; Eric De Backer; Jean-Baptiste Bacquart; Adolfo Bartolomucci; Jan Baptist Bedaux; Carlo Bella; Stephen Benjamin; Jacques Billen; Mina and Samir Borro; Alain Bovis; Martial Bronsin; Olivier Castellano; Walter Cecil; Jean-Paul Chazal; Didier Claes; Mark Clayton; Kevin Conru; Joel Cooner; John Crawford; Sandy and Taylor Dale; Pierre Dartevelle; Charles and Kent Davis; Douglas Dawson; Jo De Buck; Peter De Groote; Michel De Mol; Seth Dei; Lin, Emile, and Charles Deletaille; Nicole and John Dintenfass; Bernard Dulon; Maine Durieu; Frank Eerhart; Jan Elsen; Lance Entwistle; Christophe Evers; Marc L. Felix; Yann Ferrandin; Natalie Fitz-Gerald; Edith, Roland, and Julien Flak; Patrik Fröhlich; Mona Gavigan; Charles Gerhardt; Jacques Germain; Colette and Jean-Pierre Ghysels; Patrick Girard; Nadia and Philip Gould; Bernard de Grunne; Philippe Guimiot; Alain Guisson; Wallace B. Gusler; Ursula and Herbert Hansen; Owen Hargreaves; Skip Holbrook; Udo and Wally Horstman; Richard Hunt; Norman Hurst; Max Itzikovitz; Auriane and Baudouin Jolly; Corine Kerkhoven; Reynold C. Kerr; Joseph Knopfelmacher; Fritz Koenig; Greg LaChapelle; Philipe Laeremans; Jay Last; Amy and Elliot Lawrence; Serge Le Guennan; Claudie and Alain Lebas; Alain Lecompte; Luc Lefebvre; Michel Leveau; Johann Levy; Brian Leyden; Pierre Loos; Henry Lu; Gianni Mantovani; Angel Martín; Kip McKesson; Thomas McNemar; Ondine and Patrick Mestdagh; Zaira and Marcel Mis; Alain de Monbrison; Bernd Muhlack; Thomas Murray; Amyas Naegele; Emmanuel Ndé Feukam; Karel Nel; Michael Oliver; Charles J. Opitz; Joaquin Pecci; Claude-Henri Pirat; Hazel and Klausdieter Pollak; Dominique Rabier; François Rabier; Michael Rhodes; Eric Robertson; Franck Robichez; Catherine and Patrick Sargos; Richard Scheller; Adrian Schlag; Fatou and Joseph Schmidt; Serge Schoffel; David Serra; William Siegal; Mert Simpson; Ellen and Jerome Stern; Louis de Strycker; Joanne and Arnold Syrop; Nathalie Tarquini; Susan and Richard Ulevitch; Rut Van Caelenbergh; Anne Van Coutsem; Caroline and Michel Van Den Dries; Anne and Michel Vandenkerkcove; René Vanderstraete; André Vanhecke; Renaud Vanuxem; Caroline and Ignacio Villarreal; Jerome M. Vogel; Robert T. Wall; Thomas G. B. Wheelock; Lin Chen-Willis and James Willis; Agnès Woliner; Maureen Zarember; Marie-Cecile, Lionel, and Christine Zinsou; and Manfred and Rainer Zirngibl.

Lastly, I would like to express my deep appreciation to the following individuals who helped in various ways in Africa while I conducted fieldwork during 2008–12: Théophile Ajimon Djavi Adande, Georges Adéagbo, Yao Adjankli, Hunyo Agblagon, Martin Agblagon, Ahomlanto Agbowadan, Honorat Aguessy, Nii Otoo Annan, Oloude Georges Anselme, Kossi Assou, Kérékou Atakpai, Djoua Atchawenon, Tété Azankpo, Ogunin de Babalola Iclona, Kouwènam and Tikenawé Basseliki, Jean-Dominique Burton, Sidiki Camara, Pierre Dagba, Agonsan Deguenon, Mamadou Djenapo, Eugenie Drakes, Glele Euloge, Zantigue Fane, François Vianou Godonou, Théophile Gomez, Dominique A. D. Hazoume, Ameyovi Homawou, Stephan Koehler, Garba Kokou Koufomal, Pakétam Kourakoma, François Magliore, Affokpe Mahounon, Matthias Masewa, Joe Nkrumah, Viten Noel, Nii Noi Nortey, Nii Yemo Nunu, Vida Obodai, Ahmed Oketane, Virginia Ryan, Filomena and Symphorien Santos, Kofi Setordji, Dagounon Tcheokon, Dinah Tiwah, Adomou Tonassin, Ahmed Toure, Kodjo Volta-Tineh, Mensah Volta-Tineh, Kondi-mani Waké, William Dieudonne Yekedo, and David Yemadje.

Thank you, one and all!

Allen F. Roberts African ironworking has been a fervent interest of mine for over fifty years. Far too many friends have been generous with their intellect and hospitality to be mentioned here, but I hope all are aware of my enduring gratitude. Sincere thanks are extended to the colleagues who have collaborated on the present volume as well as to Codjovi Étienne Joseph Adande, Andoche Adjovi, Rob Aft, Ahouanvoedo Augustin, Claude Ardouin, Ramona Austen, Gasim Badri, Randi Barndon, Paul Berliner, Betol Jean, Sandra Blakely, Degenhart Brown, Charlene Cerny, S. Terry Childs, Herbert J. Cole, Donald Cosentino, Maarten Couttenier, Christopher Davis, Avery Davis-Roberts, Paul Delucco, Marie-Claude and Georges Dupré, Nicole Échard, Aliaa El Sandouby, Alan Etzel, Marc L. Felix, Iroko Félix, William Gallois, James Green, Christiane Gruber, Andrew Gurstelle, Jane Guyer, Sondra Hale, Keith Hart, Salah Hassan, Eugenia Herbert, Dunja Hersak, Luc de Heusch, Thomas Hinnebusch, Félix Iroko, John Janzen, Bogumil Jewsiewicki, Kizumina Kabulo, Sidney Kasfir, S. Lloyd Kaufman, David J. Killick, Christine Mullen Kreamer, Alisa LaGamma, Laondigui Georges, Talia Lieber, Wyatt MacGaffey, John Mack, Mairo Joseph, Pierre de Maret, Bárbaro Martinez-Ruiz, Evan M. Maurer, Jim McDougall, James A. Moore, Moroyouang Paul, Sultani Mpala Kaloko, Jean-Claude Muller, Karel Nel, Fallou Ngom, Len Pole, Mary Kujawski Roberts, Mary Nooter Roberts, Doran H. Ross, Christopher D. Roy, Dana Rush, Alison Saar, Asma Sayeed, Peter R. Schmidt, Suzanne Seriff, Janet Siroto, Tohoessou Sourou, Dexter Story, Elaine Sullivan, Hein Vanhee, Jan Vansina, Rosalind Walker, Ernie Wolfe III, Gary Van Wyk, Rudolph Ware, Brad Weiss, Marcia Wright, and Dominique Zahan.

William J. Dewey This project has been many years in development, and a number of people deserve thanks for helping bring it to fruition. In addition to the many African blacksmiths who demonstrated techniques and shared knowledge, countless colleagues worldwide have helped answer questions, directed me to archaeological sites, showed me collections (both public and private), and supplied information. In addition to my archaeologist collaborators Shadreck Chirikure and Scott MacEachern, there are many from the Society of Africanist Archaeologists (SAfA) who were immensely helpful. Among those individuals and institutions deserving special thanks in all categories are: Philip de Barros, Peter Breunig, S. Terry Childs, Els Cornelissen, Gordon Crawford, Nicholas David, Els De Palmenaer, Stefan Eisenhofer, Jan Elsen, Bill Fagaly, Marc L. Felix, Rebecca Skinner Green, Eugenia Herbert, Udo Horstmann, Jane Humphris, Abdurahman Juma, David J. Killick, Hans-Joachim Koloss, Chapurukha M. Kusimba, Adria LaViolette, John Mack, Pierre de Maret, Evan M. Maurer, George Mvenge, the National Archives of Zimbabwe, the National Museums of Swaziland, the National Museums of Zimbabwe, Karel Nel, David Phillipson, Nicolle Rupp, Peter R. Schmidt, Christopher B. Spring, and Gary Van Wyk.

Henry J. Drewal Many friends and colleagues have helped me on this journey of learning about African blacksmiths and their incredible artistry and virtuosity. With Tom Joyce in the Republic of Bénin in 2010, when we discussed the plan to do *Striking Iron*, it was Codjovi Étienne Joseph Adande who assisted us; in 2012, when Sarah K. Khan and I traveled with Tom and Anne-Marie Bouttiaux to Mali and Burkina Faso, it was Anne-Marie's friend and colleague Daouda Keita. In the United States, it was my brother-from-another-mother, John Scarlett, a former student of Tom's who went on to become a renowned smith (and haiku poet) in upstate New York. In Morocco, there were these wonderfully generous blacksmiths: Muhammed and Hassan, Abdel and Ahmed, and Allal (Fez); Muhammed (Meknès); and el Haj (Risanni). I thank, too, the scholar and architect Abdelhamid Ghandi and his daughter Meriam; Hamza El Fasiki, craftsman extraordinaire; David Amster, Director of the Arabic Language Institute in Fez (ALIF); James A. Miller, former Executive Secretary of the Moroccan American Commission for Educational and Cultural Exchange (MACECE) and historian of Sijilmasa; Kay and Arnauld Dastorg; Youssef Ouadderrou; and especially Abdelouahed Hmadou, my most excellent research assistant. I am grateful for a Fulbright Senior Scholar award that funded my stay in Morocco. The joy of this trip was due to the love, energy, and presence of my partner, Sarah Khan, and sustained by the warmth of our adopted Fez family of Abdou, Najia, and Ghita.

Blessings and balance to all these folks, and to the blacksmiths who remind us to "strike while the iron is hot," for life is brief and filled with constant challenges. May we travel well and leave positive, indelible marks behind, like those of master blacksmiths, who sustain, invigorate, and beautify culture.

FROM PG 120. PHOTOGRAPH BY TOM JOYCE, OUIDAH, REPUBLIC OF BÉNIN, 2010. © TOM JOYCE

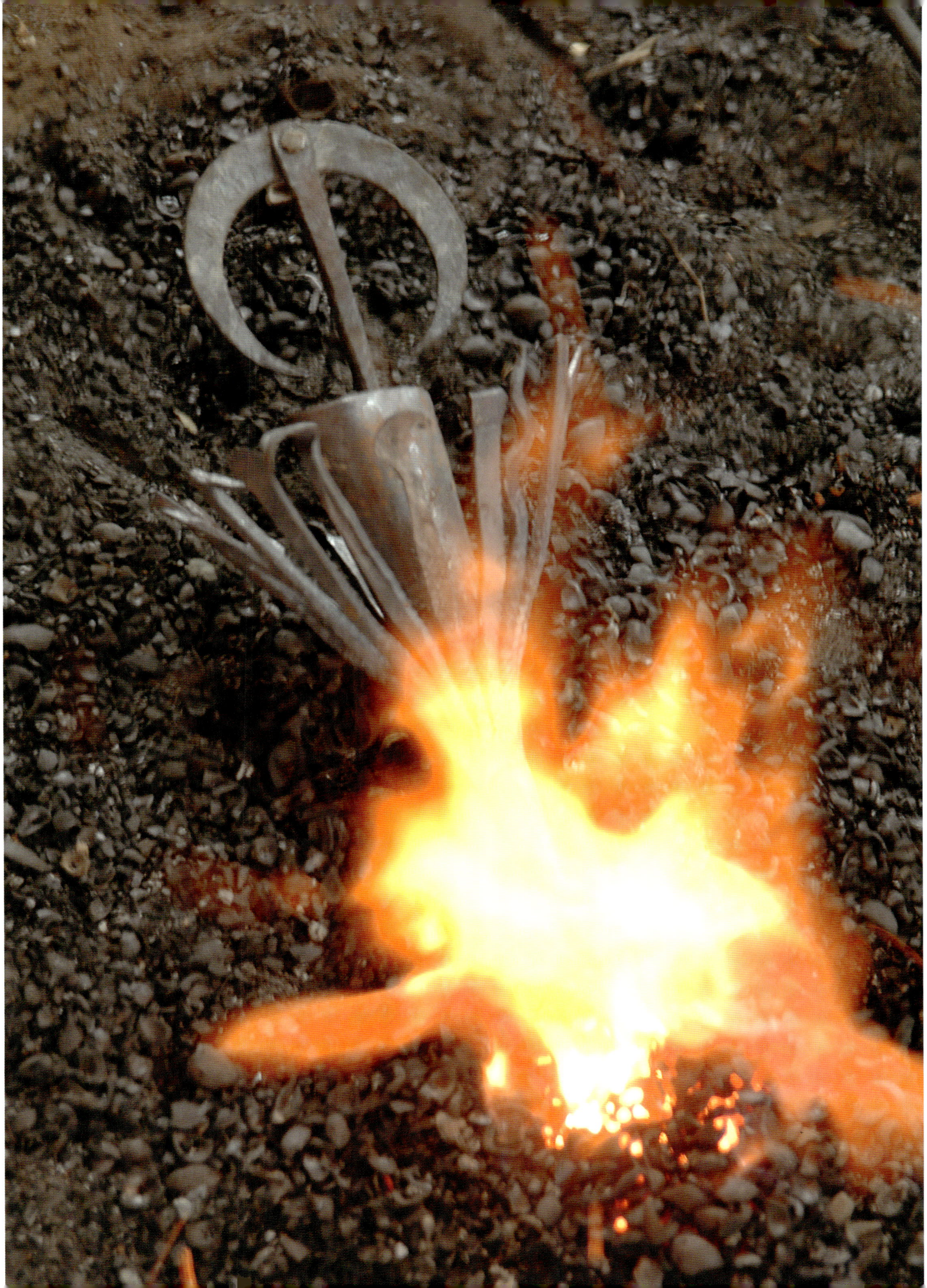

PREFACE 1 *LINGUA FERRUM*: SOLILOQUIES OF SMITHS

Tom Joyce

The language necessary for communicating with iron by the blacksmiths who forge it is universal. Remarkably, this ancient "dialect," spoken in melodious rhythms by hammers ringing on anvils where hot iron is shaped, has not been lost or modified much since it was first articulated some three thousand years ago as the Iron Age dawned in different parts of the world. In essence, "colloquial etymologies" expressed through local, regional, and global forging technologies continue to infuse the profession with innovative solutions as vital in the present as they were in the past. In many African communities, "fluency" in the art of forging iron is more pronounced and practiced today than where mechanization has replaced handwork; where access to skilled practitioners is not just around the corner; where ferrous material, once highly regarded, is discarded or shipped elsewhere for recycling rather than repurposed on the spot to extend its usefulness to society. Wherever blacksmiths are casually observed working, neighbors will possess at least rudimentary understanding of what iron means within the surrounding region and perhaps within one's life—how to handle it, sharpen it, shape it, share it, protect it, respect it, and pass it along (figs. P1.1, P1.2). Sites of integrated blacksmithing activities allow for seamless appreciation among citizens for the specialized goods and services they provide—objects of beauty and purpose that benefit human beings, and upon which human beings rely.

By contrast, when I was learning to forge in America nearly fifty years ago, this scenario was no longer common. The profession was drastically waning and all but a few of the elder smiths who might have been mentors had already laid down their blacksmiths' aprons by the time I picked up a hammer. After acquiring basic skills through an informal apprenticeship as a teen in the rural community where I grew up, I began studying objects that my predecessors had left behind for practical clues, and for pockets of instruction to inform my own practice. I sought out places in the world where blacksmiths were still needed and perhaps even indispensable within their communities. This search led consistently to sub-Saharan Africa. To this day, few places hold as rich a connection to iron—as a material, and one charged with metaphorical substance and power.

The first forging from Africa that I encountered as a teenager was a currency bar attributed to Mumuye peoples of Nigeria (fig. P1.3).[1] From the existing literature, I gleaned that it was a forge-welded consolidation of worn-out hoes accumulated by a farming family, and it would have been offered during matrimonial transactions in exchange for the woman's hand in marriage (see Eyo 1979, 49–54; chapter, 5, this volume). The many layers of this folded form were testimony to the young groom's training in the field, providing tangible proof of his ability to deliver sustenance for his bride and the family they would likely raise together. This balanced shape of stored material could become anything her family desired from a smith, and was therefore acknowledged as "pregnant" with symbolic potential, as in the promised union of this couple (Quiggin 1949, 87). The swelling at the center is expressive of the melding of families, of ancestral bloodlines extending to future generations, just as iron—

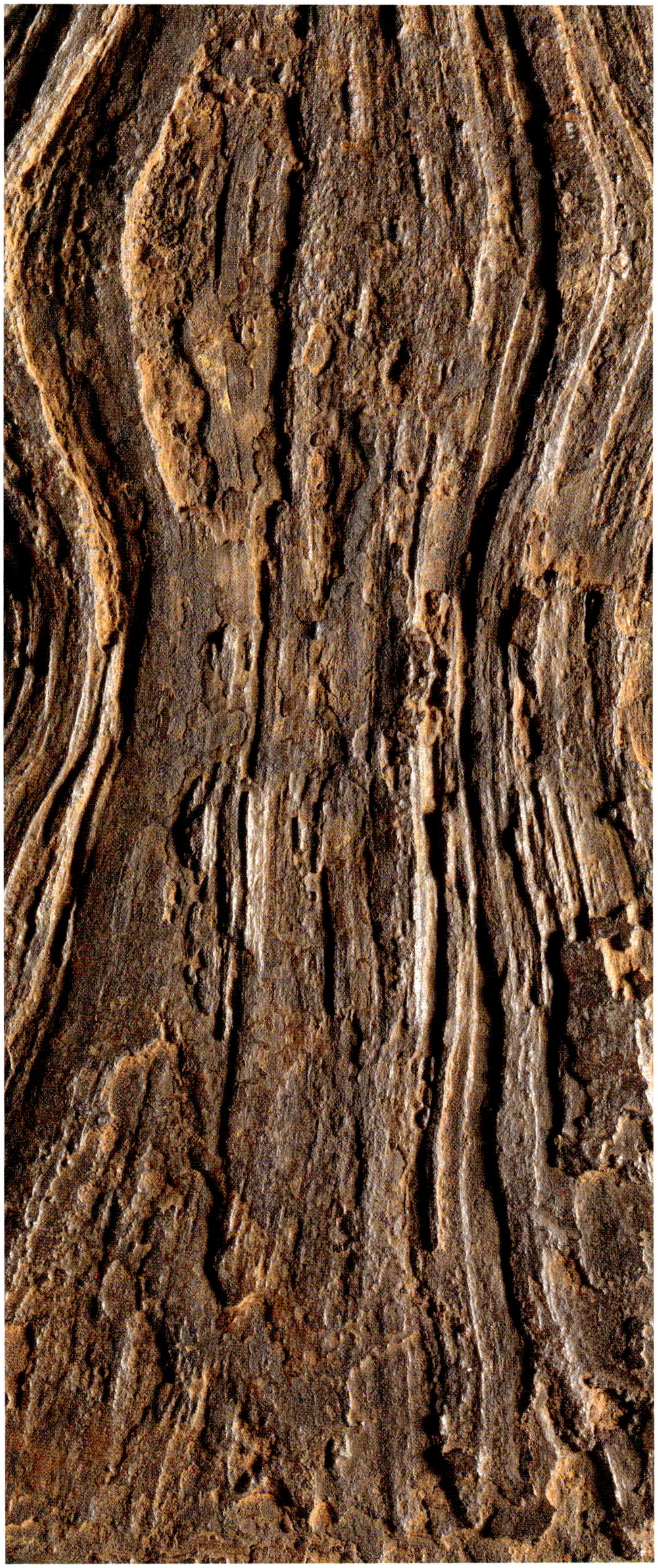

P1.1 Detail of oxidized bloomery iron showing the inherent fibrous grain structure in its forged layers.

considered by some to be the cultural equivalent of blood from the earth—tills fertile ground and affords *pro*creative possibilities. The blacksmiths who conceived this and the extraordinary objects illustrated throughout this volume have played central roles in the many different communities where forge fires are lit, as readers will learn in forthcoming chapters.

African blacksmiths are inextricably connected to a long and substantial legacy that to this day is intact and flourishing across the continent (fig. P1.4). It may come as a surprise to some readers that blacksmiths are still indispensable in facilitating certain tasks in Africa. But blacksmiths are in fact operating throughout the world, in virtually everyone's backyard—if often undetected. The mechanized world as we know it, stimulated by economic conditions that thrive only when things people need or desire are being made, traded, and purchased, would quickly collapse if blacksmiths active today, at this very moment, were to vanish from their workplaces. Cutting-edge forging activities, whether large or small in scale, remain critical to supporting lives in Africa as in every nation. The blacksmiths who conduct business in global spheres work in industry and are simply out of public view, performing astonishing technological innovations inside industrial forging facilities that are closed to outsiders. Their ability to move hot iron is no keener in its inventiveness nor more revolutionary culturally than the skills developed in Africa long before science deciphered the wondrous processes indigenous metallurgists perfected through trial and error.

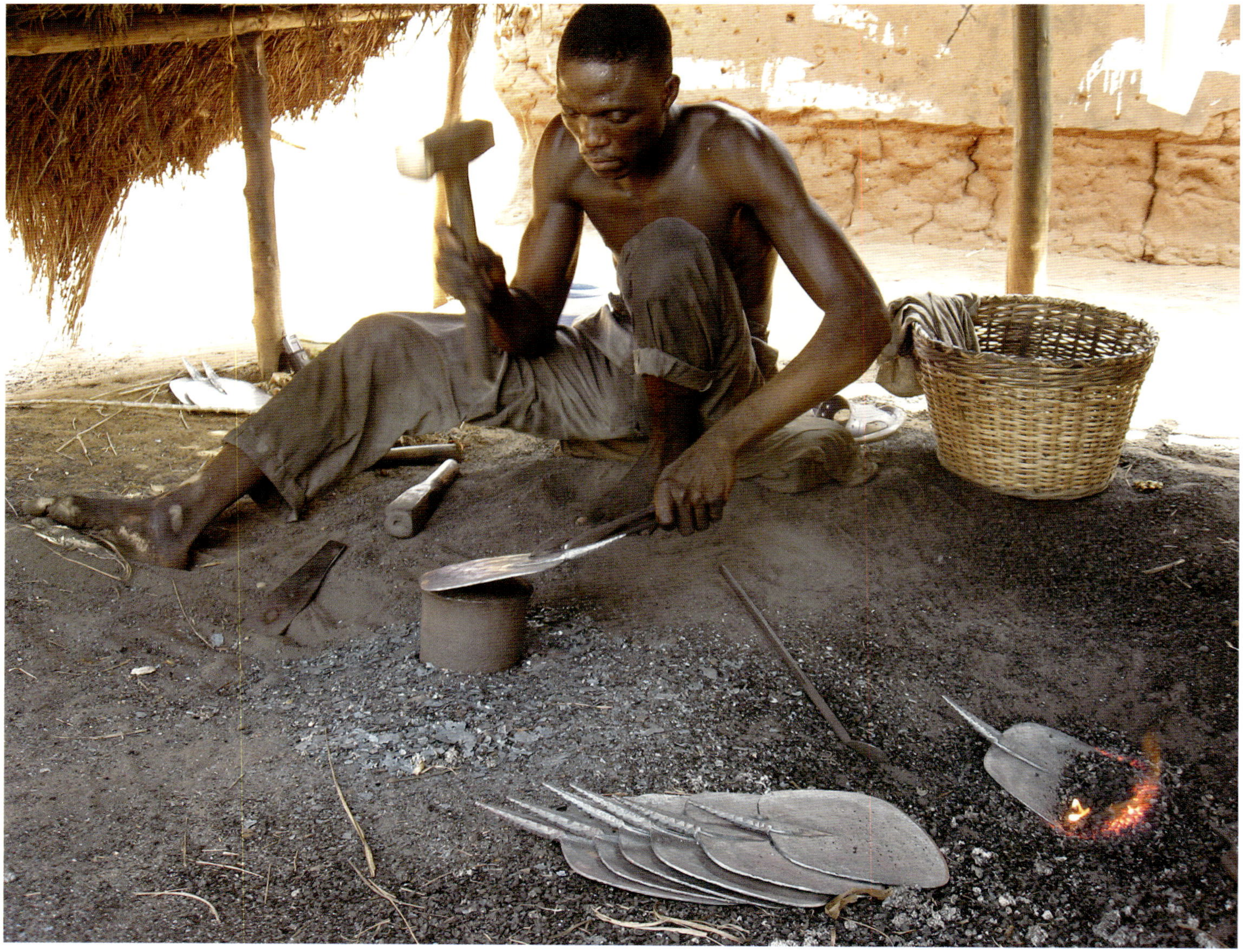

P1.2 Ewe blacksmith Yawovi Yovo forges a batch of farming hoes (*kodjowe*) at his workshop in Yohonou, Togo.
PHOTOGRAPH BY TOM JOYCE, 2008. © TOM JOYCE

P1.3A–B
Artist unknown (Mumuye peoples, Nigeria)
Currency bar (*tajere*, *saka*), early 20th century
Iron
L: 56 cm
PRIVATE COLLECTION
(NOT IN EXHIBITION)

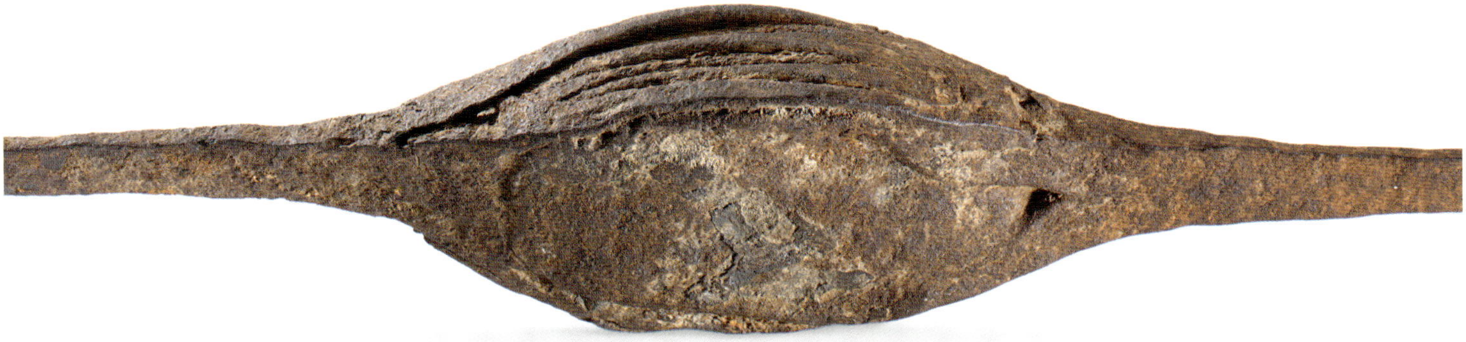

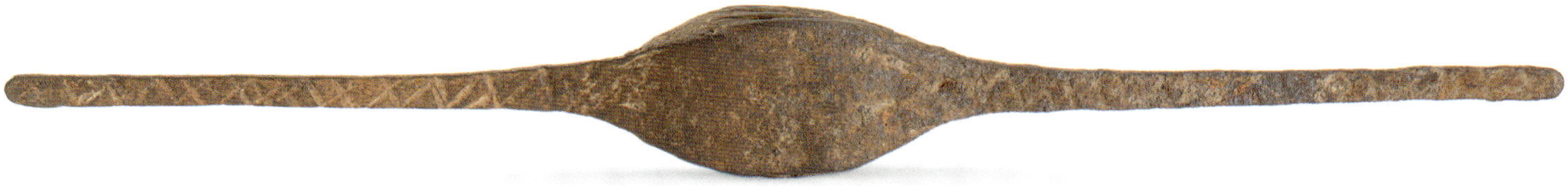

Pioneering metallurgist and scientific historian Cyril Stanley Smith reinforced this point: "An immense and important body of knowledge on useful alloys was discovered long before anything approaching modern science existed. And even today theoretical metallurgists are principally engaged in explaining phenomena that has been known for thousands of years" (Smith 1961, 357). He explained further that artists who experiment with different materials to select the best for their purposes are as much contributors to the development of science as the observational astronomer, or the classifying mineralogist, or the naturalist: "Before good [scientific] theory has been formulated, intelligent observation alone is king" (ibid.). African observational sciences inform nearly every page of this book.

It was instructive for my curatorial inquiry to approach the tools, forging and smelting procedures, and the objects made from them through the lens of current conditions surrounding manufacturing that require forged components to operate efficiently. The parallels are striking, and in contemplating the cultural importance of items made close to home, one can relate these necessities to practical equivalents in Africa. With this proportional association in mind, one can begin to take stock of the splendid as well as the staggeringly massive forged mechanisms churning away at the heart of making almost everything we depend on. These machines produce energy; cultivate and process food; move earth and extract mineral resources; smelt and mill metals; fabricate contraptions that fly, float, dive, and drive; pulp wood into paper; weave plant and synthetic fibers into clothes; and transform monocrystalline silica into semiconductors, LEDs, lasers, and solar cells. Parts blacksmiths forge are deployed in both offensive and defensive strategies to protect homes globally. Iron components they have worked hot allow access to the deepest seas as well as space exploration, drifting toward planets and even galaxies far beyond our own. In all of this and more, communities worldwide require the expertise of blacksmiths working furtively around the clock.

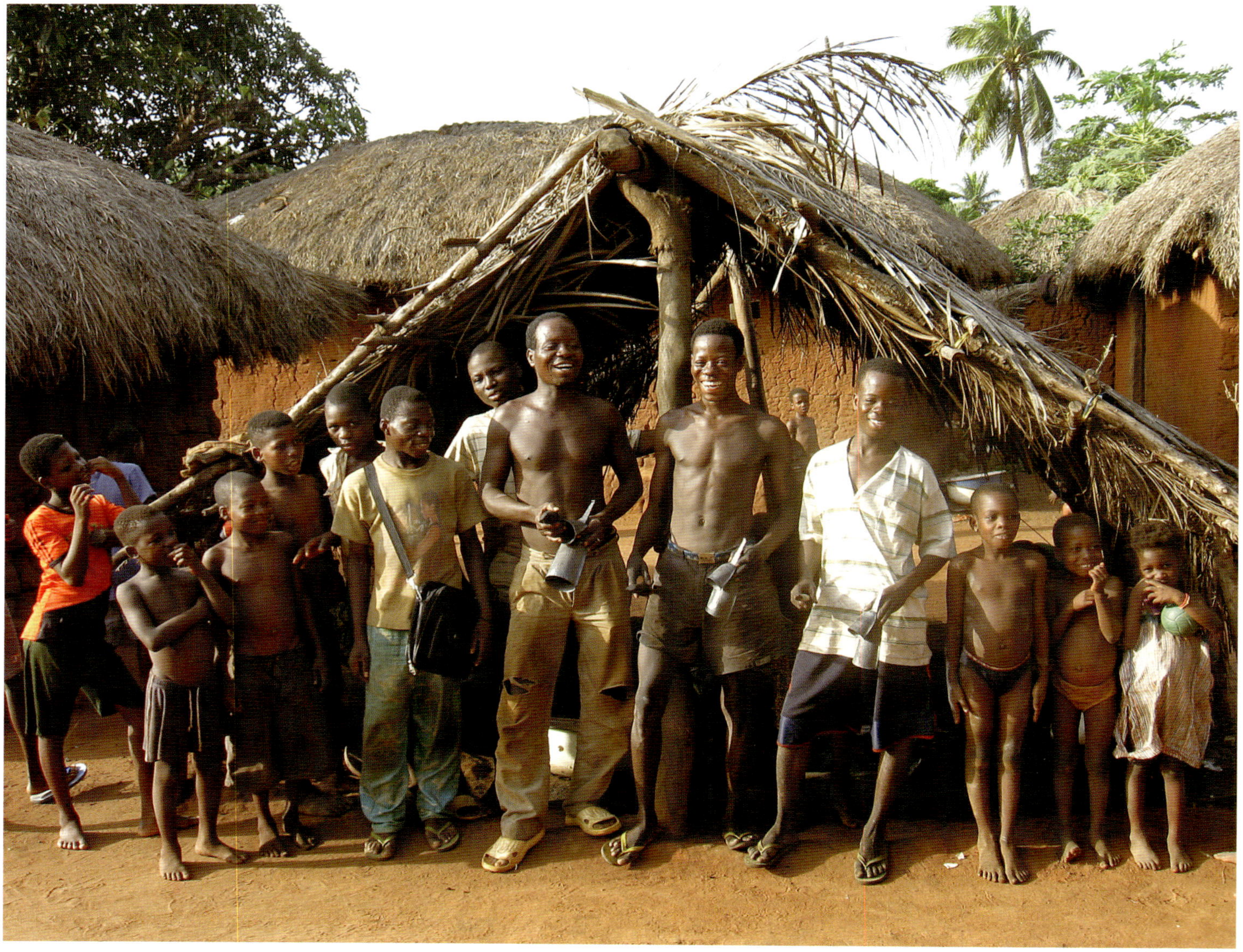

P1.4 Koumedjina Atiogbe and his sons Yao and Kokou (center, left to right) holding bells (*gangokui*) in front of their forge at the end of the workday with younger siblings and children from the neighborhood.
PHOTOGRAPH BY TOM JOYCE, YOHONOU, TOGO, 2008.

Though the tools, methods, and scales of production may be different from workshop to workshop, any place iron is forged offers a glimpse inside a microcosm that affects its encircling environment. Sometimes the results of a blacksmith's output are of massive consequence, as in times of conflict and war, while at others the quantifiable effects are more difficult to calculate. Harnessing and projecting the potency of the forge is not lost on blacksmiths wielding hammers for a living. Whatever the purpose, blacksmiths' actions have implications in the world, and for the most part, they bear responsibility for what they make. Innovations are conceived, material is worked, intentions are manifest, and objects are released into the hands or operation of others. Our debt to these blacksmith technicians and problem-solvers—in theory the descendants of the earliest tool makers, artists, and inventors—is as unparalleled in our time as it was in the time of our ancestors.

Keeping one's finger on the pulse of the political, social, economic, and environmental conditions that drive this profession, whether in Africa or elsewhere, is helpful in positing the importance of forging capabilities harnessed by human beings since the early Iron Age. Each of the objects selected for the *Striking Iron* exhibition and illustrated in this volume has played such a role. Removed from circulation, they are appreciated now out of the contexts of their manufacture, and we must remember that these artfully forged implements of groundbreaking originality fulfilled and maintained transformations of great cultural significance.

As lead curator of the *Striking Iron* exhibition, I was committed to including a broad and exceptional range of African blacksmiths' achievements, and conducted comparative research in 196 public and private collections in the U.S. and abroad over the last twelve years. From the thousands of objects viewed, handled, sketched, and photographed, the curatorial team and I distilled over 225 exemplary works of art and utility to tell the stories of how ironworking has shaped African cultures in practical, intelligent, and aesthetic ways. Now, as in times past, blacksmiths in Africa and throughout the world are translating for our benefit the rhythmic linguistics of *lingua ferrum*, a resonantly nuanced language that not long ago we could all hear and interpret on our own.

PREFACE 2 HEAVY METAL: IRON IN THE UNIVERSE

E.C. Krupp

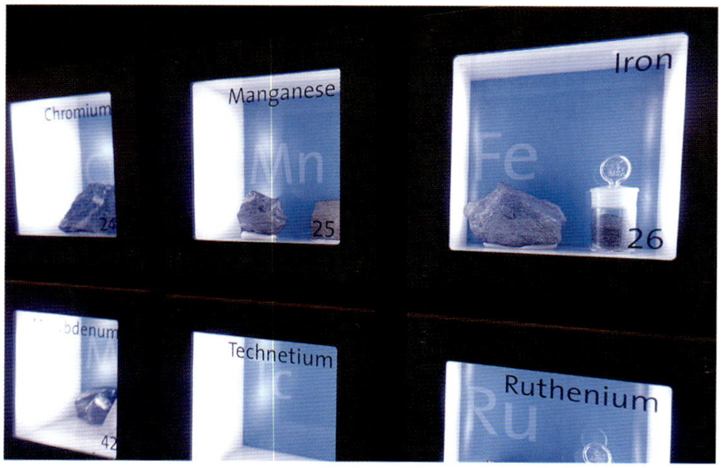

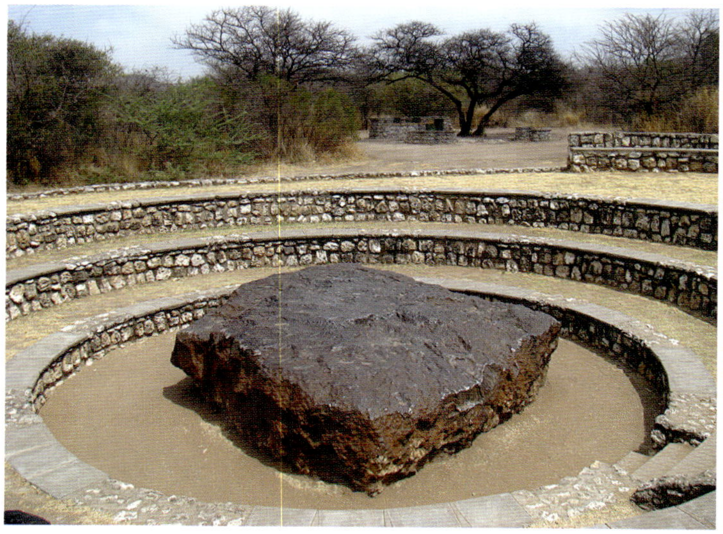

There was no iron in the fire of the universe's Big Bang kick start 13.8 billion years ago. There is now, however, plenty of iron in the world. Over time, the iron hand of gravity manufactured the massive stars it takes to make iron, and some of it wound up here on earth—with us, around us, in us (fig. P2.1). Iron is in our blood. Our hearts are constantly pumping iron, which is part of the hemoglobin protein molecule that makes blood cells red (Emsley 2002, 206).

Compared to hydrogen and helium, which comprise about 98 percent of the atoms in the universe, iron is scarce. It ranks a distant tenth in abundance in our solar system, but that is enough to fortify the earth with iron. The earth's iron core is solid at its center, with a molten iron shell, and there's iron ore in the earth's rocky crust. Altogether, a little more than a third of the planet is iron. Most of it is more than 2900 kilometers below the surface, but over the last few thousand years we've contrived tools, weapons, machinery, and monuments from iron extracted from ore in the crust (Emsley 2002, 205–11).

As an element in the cultural landscape, iron is an emblem of power that colorfully activates the vocabulary of might and will: "Iron Lady" (U.K. Prime Minister Margaret Thatcher), "iron horse" (Victorian phrase for a steam locomotive), "Iron Duke" (First Duke of Wellington and a Pontiac automobile engine), "iron lung" (medical ventilator for artificial breathing), "Old Ironsides" (nickname

P2.1
Iron, with twenty-six protons in its atomic nucleus, is number 26 in the Periodic Table of the Elements. The majority of iron atoms also have thirty neutrons. The symbol for iron, Fe, originates with *ferrum*, the Latin name of the metal. As an active metal, it readily combines with nonmetals, particularly oxygen and sulfur, and so is rarely found naturally as a pure metal, except in meteorites. Iron is represented in the element collection at Griffith Observatory, Los Angeles, by a sample of iron ore and pure iron in an air-tight container.
PHOTOGRAPH BY E. C. KRUPP, 2017.

for the U.S. frigate *Constitution*, with oak strong enough to repel enemy fire), "Iron Curtain" (the Cold War boundary between the West and the Soviet bloc), "iron fist" (ruthless severity), and "Iron Butterfly" (American hard rock band). In French, the term *chemin de fer*, for the railroad, means "iron way."

In the modern industrial age, we depend on iron in structures, equipment, and technology. Indeed, as we slip into the silicon-leveraged Information Age, iron is so much a part of everyday life that we take it for granted. But long before Tony Stark, the industrial entrepreneur and inventive genius in Marvel comic books and movies (see Lee 1963), fabricated the high-tech armor that turns him into the superhero Iron Man™, Assyrian armies equipped with iron spears, iron swords, iron helmets, iron shields, and iron-plated tunics turned their region, roughly 2700 years ago, into an Iron Age empire. Their gods understood the metal's power. In one omen, Sin, the Assyrian moon god, declared "he would bestow evil death through the flashing iron dagger" on those who conspire against the king (Saggs 1984, 245).

Iron was known and valued as far back as the third millennium BCE, for although there is no native metallic iron on the earth's surface (Delmonte 1985, 238), it sometimes falls to earth from outer space. On average, the earth intercepts over two hundred tons of interplanetary debris every day, a small percentage of which survives incandescent passage through the atmosphere to reach the ground. A few of those meteorites—about 7 percent—are pieces of iron, alloyed with a small amount of nickel, which is not found in terrestrial iron (McSween 1991; fig. P2.2).

The oldest known worked iron is meteoritic and not smelted from ore. Beads fabricated in approximately 3300 BCE from an iron meteorite were recovered from Predynastic tombs at Gerzeh, about seventy kilometers south of Cairo (Johnson et al. 2013). In the fourteenth century BCE, a dagger of meteoritic iron was placed in the wrappings of the famous Egyptian pharaoh Tutankhamun. Accompanying the gold and other treasures in his tomb, the dagger indicates the value assigned to the metal by early Egyptians, who did not possess ironworking technology at the time (Deatrick 2016; Cornelli et al. 2016). Iron remained rare in Egypt until the Roman period, though by then Egyptians had long been making small symbolic iron instruments for resurrection-of-the-dead ceremonies (Scheel 1989, 17–18). By 1200 BCE, some of these implements were made of terrestrial iron, but a few ancient texts imply that such iron was imported from the east. Meteoritic iron was worked even earlier in Iran at Tepe Sialk (Bevan and de Laeter 2002).

Meteoritic iron's use for tools occurred as early as 2500 BCE; a scrap was collected from grave 580 in the Royal Cemetery of Ur, in Iraq (Bjorkman 1973, 125). Meteoritic iron was cold-forged into blades fitted into bronze handles by a Zhou dynasty metalsmith in China about three thousand years ago. These ceremonial weapons, now at the Smithsonian Institution's Freer Gallery of Art in Washington, D.C., include a broad ax and a dagger ax with an iron point. Use of iron in ritual instruments like these suggests that rarity and functionality made the iron valuable (Gettens 1971).

In North America, other peoples found iron meteorites, made use of them, and treated them ritually. By H. H. Nininger's account, there were more than a dozen indigenous prehistoric and historic uses of meteoritic iron on the continent (1952, 7–8). Beads of meteoritic iron discovered in a prehistoric mound in Illinois were placed there by Havana Hopewell people around 350 BCE. The iron, from a meteorite that had fallen in Minnesota at a time unknown, had been traded into their lands in the Illinois and Mississippi River Valleys (McCoy et al. 2008; McCoy 2016). Near Camp Verde in northern Arizona, an iron meteorite was discovered in a room of the Wingfield Mesa Ruin, ritually deposited there by prehistoric Sinagua people (Nininger 1952, 80). At Casas Grandes (or Paquimé), in the present-day Mexican state of Chihuahua, peoples of the Mogollon culture of the ancient Southwest wrapped a two-and-a-half-ton iron meteorite in linen and left it in one of the site's two thousand rooms, most likely sometime in the fourteenth or fifteenth century CE (Nininger 1952, 8).

P2.2
Iron meteorites also contain nickel, a telltale sign of extraterrestrial origin. After exposure to the earth's atmosphere, fallen iron meteorites look dark on the oxidized surface. Weighing more than sixty tons, the Hoba West meteorite is the largest known single mass of meteoritic material on the planet. It remains where it fell to earth, perhaps eighty thousand years ago.
PHOTOGRAPH BY E. C. KRUPP, HOBA WEST FARM, NAMIBIA, 2011.

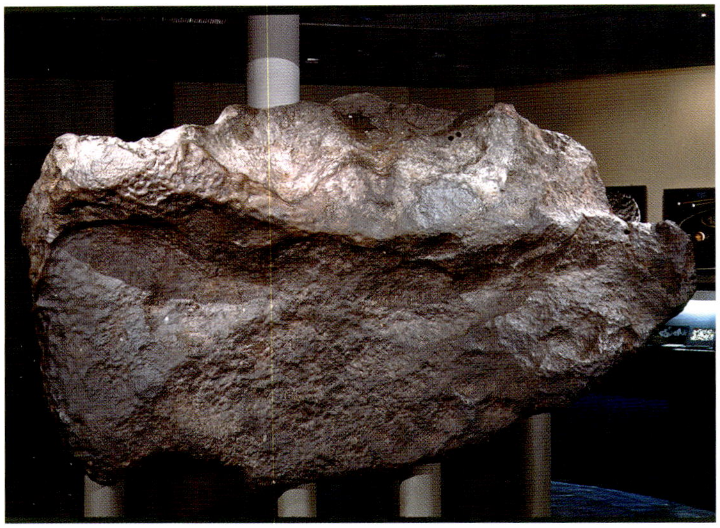

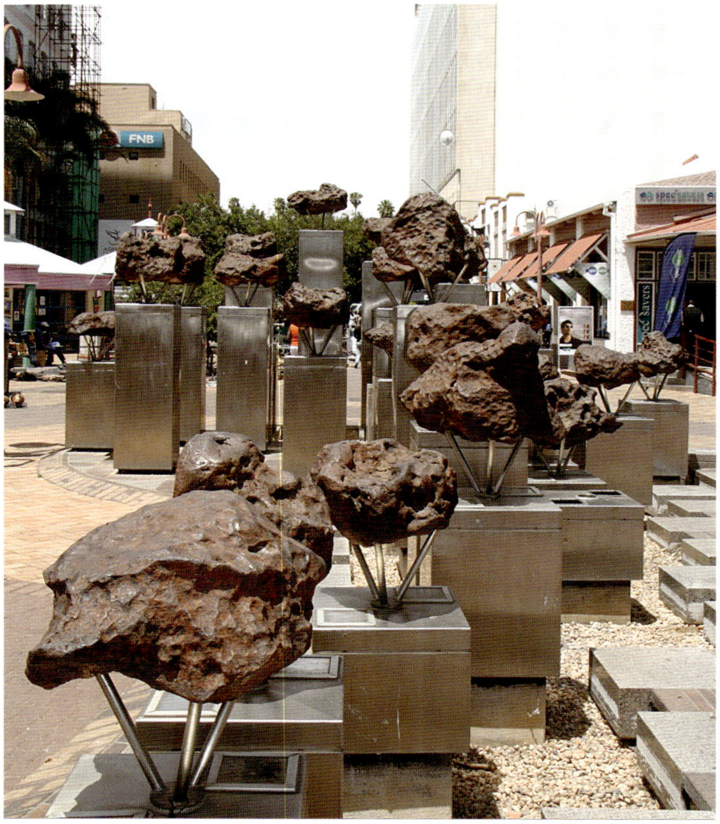

In yet another environment, to make cold-forged blades for harpoons and other tools, Inuit hunters, probably for centuries chipped fragments off massive pieces of an iron meteorite in Greenland, estimated to have fallen to earth ten thousand years ago (Delmonte 1985, 238). Arctic explorers first heard about the fragmented meteorite, known as the Cape York, in the early nineteenth century; Robert E. Peary located it in 1894. The most massive of its three pieces is on public display in New York's American Museum of Natural History and called Ahnighito, meaning "tent" (Aldersey-Williams 2011, 45; fig. P2.3).

Instances of the opportunity-driven use of meteoritic iron are reported in sub-Saharan Africa, although such accounts have not been firmly substantiated. A meteorite found along the Senegal River in 1716 is said to have been wrought into weapon points by local peoples. Nama peoples of present-day Namibia are thought to have fashioned small spear points from Gibeon meteorites, from which early European settlers may have cold-forged a few farming implements (Rickard 1941, 55; fig. P2.4). In the 1940s, a colonial administrator in what was then British Cameroon (now along the southern border of Cameroon and Nigeria) was shown a large boulder assayed to be of unusual ferrous chemistry and thought to be a meteorite, from which Bamenda peoples "took iron for tool-making" (Jeffries 1955, 167). Such reports, however, seem to be more based in conventional wisdom about what "must" have happened than proven through direct observation, archaeology, or collecting. Although little chemical analysis of early iron implements has been undertaken to date, archaeologist David Killick predicts that use of new analytic technologies will reveal that the earliest iron implements yet found in the Sahara and northern Sahel—where meteorites are relatively easy to spot, revealed by wind-blown sand—are of meteoritic origins.[1]

Furthermore, since the late nineteenth century, some Egyptologists have asserted that Egyptians sometimes imagined the

P2.3
Meteoritic iron has been fashioned into tools, jewelry, and ritual objects. As early as 1818, arctic explorers reported that Inuit hunters had blades and tools of worked iron. The source of the metal was finally discovered by Robert E. Peary in meteoritic fragments near Cape York, Greenland. Inuit named the most massive piece, at thirty-four tons, Ahnighito. Peary managed to move it to New York and sold it to the American Museum of Natural History there, where it is still on display.
CAPE YORK METEORITE, AHNIGHITO SECTION, HALL OF METEORITES (1975–1990). COLLECTION OF HISTORIC HALLS OF THE AMERICAN MUSEUM OF NATURAL HISTORY, NEW YORK, K14835. PHOTOGRAPH BY P. HOLLEMBECK. © AMNH

sky to be made of iron; as W.M. Flinders Petrie noted, "The walls of heaven are described as iron, and presumably meteorites were regarded as fragments of the heavenly city" (1940, 94). Egypt's iron sky is not directly documented in texts, but instead appears to be an extrapolation from an Egyptian word for iron, *ba-en-pet*. Hittites of highland Anatolia, who first capitalized on the smelting of terrestrial iron, are said to have similarly imagined the sky to be made of iron (Johnson et al. 2013, 1003). A ritual text mentions "the black iron of heaven they brought from heaven," and several Hittite experts have stated that "black iron" is technical vocabulary for meteoritic iron (Bjorkman 1973, 110).

In ancient Mesopotamia, Akkadian and Sumerian words with a root that can mean "sky" may have also referred to iron (Oppenheim 1978, 663). An Old Babylonian hymn refers to the "fall of iron to the ground," perhaps a hint that some ancient eyewitness actually saw a meteorite hit the earth (Bjorkman 1973, 113; see also Oppenheim 1964, 322). The chances, however, for recovery of a meteorite actually observed to fall were, and still are, remote. The iron meteorites of antiquity almost certainly were found by chance long after they had struck the earth (Krupp 1991, 109). The iron, however, was recognized for the unique and distinctive material it was.

Other explanations for iron's celestial affiliation are possible. Iron is strong and hard. The notion of an iron sky could simply reflect the structural integrity required to support a heavenly vault. Such reasoning is found in the cosmologies of sub-Saharan Africa, notably among Tabwa peoples living on the peripheries of the Congolese Copper Belt, who once regarded the vault as formed of copper (Roberts 2012, 194). Iron may have just had the "right stuff" for celestial architecture. After all, lightning strikes are known to fuse rock. A chance encounter with what appeared to be a piece of once-melted metal might well have inspired the notion of iron's celestial and fiery origin (Krupp 1991, 108–10).

Almost all of the iron known to the ancients before the middle of the second millennium BCE was likely meteoritic, but hematite, a common iron ore that has been widely used for red pigment since the Paleolithic era in sub-Saharan Africa and many other parts of the world, could have left bits of terrestrial iron in the ashes of a fire. Limited smelting seems to have occurred in Mesopotamia and Syria in the first half of the third millennium, for instance, perhaps the result of the chance presence of iron-bearing minerals on the site of an intense fire. The earliest pieces of terrestrial iron known in Europe are a 3600-year-old ring from the Kakovatos tombs on the west coast of Greece, and pieces from Vaphio, near Sparta (Buchwald 2005, 74).

In the middle of the second millennium BCE, Anatolian ironworkers, known as Kizwadana by their Hittite rulers in what is now eastern Turkey, discovered how to smelt iron by creating a high-temperature furnace, heating the ore, and removing the oxygen bonded to the metal. The Hittites obtained this iron technology and used it to alter the landscape by means of new tools for land clearing and agriculture (Muhly 1995, 1517). Many of their axes, scythes, sickles, and ploughs are similar to the African blades illustrating the present volume. The development and eventual diffusion of smelting terrestrial iron were primary factors in the transformation of societies around the Mediterranean as of 1200 BCE.

The Kizwadana seem to have been the same people the ancient Greeks knew as the Chalybes, whose territory was Pontus, along the Black Sea coast of northeastern Turkey (Gurney 1954, 82–84). In the fourth century BCE, the historian Xenophon spotlighted the ironworking reputation of the Chalybes of Asia Minor in his personal account of the military contest for the Persian throne (Anthon 1847, 334). Evidence of smelting dating to the second millennium BCE has been found in east-central Niger (Paris et al., 1992; see also chapter 8, this volume). Clearly, more comparative research is needed, and especially through metallurgical archaeology in Africa.

P2.4
Nearly three dozen iron meteorites collected from the area of Gibeon, in southern Namibia, are on outdoor public display in the Post Street Mall in Windhoek, Namibia's capital.
PHOTOGRAPH BY E.C. KRUPP, 2011.

In the eighth century BCE, the ancient Greek poet Hesiod allied his era with the revolutionary introduction of iron and called the Greeks of his time "a race of iron" (1973, 64). So, the ancients identified meteoritic iron as a rare and valuable commodity and made special use of the few samples they found. They understood the metal's durability and strength and also learned to make it on earth with furnace and forge. As they hammered iron into tools and weapons, they transformed daily life and power politics.

All of the iron we know of today—the iron in meteorites, the iron that makes blood cells red, the iron in the crust and core of the earth, the iron in the spinach that adds instant mass and strength to the muscles of Popeye the Sailor Man, and the iron that makes the planet Mars glow red in the night sky—originated in stars that were already gone when the sun and the solar system were condensing from a primordial nebula of interstellar gas and dust more than four and a half billion years ago (Trimble 1996; fig. P2.5).

The ancients had no idea how iron was hammered into atoms by generations of massive stars, but astronomers have since teased this information out of starlight and the laws of physics. Although stars have been forming, evolving, and in some cases catastrophically bursting since the universe emerged as an explosive expansion of space and time, there was no iron in the beginning—neither we nor the earth were possible until iron was forged in the furnaces of stars.

Hydrogen and helium produced in the Big Bang in time collapsed into stars that shine by generating energy through thermonuclear fusion in their cores. At the extraordinary pressure and temperature deep within a star, the atomic nuclei of the lightest elements have enough energy to overwhelm the strong force of electrostatic repulsion that keeps atomic nuclei apart. They instead merge, through atomic fusion, to create new and heavier elements (fig. P2.6). The temperature in the core of the star is key, and central temperature is directly linked to the star's mass. The weight of the overlaying material of the star creates high central pressure, which lifts the temperature. When a star's core is heated enough for thermonuclear reactions, matter is converted into energy. The energy keeps the atomic particles hot enough to create a pressure that holds up the interior against gravity's power to pull it down. We've known how this works since 1929, when two physicists at the University of Göttingen in Germany, Fritz Houtermans and Robert Atkinson, showed how hydrogen nuclei can fuse together and release the energy by which a star shines (Time-Life Books 1988, 40–69).

In the simplest process, four hydrogen nuclei successively join to produce a helium nucleus, very slightly less heavy than four times any one of the hydrogen particles. Einstein's emblematic equation, $E=mc^2$, turns that small amount of lost matter into a lot of energy. As hydrogen turns into helium, the composition of the star gradually changes. This is, in fact, what takes place in the core of the sun, where the temperature is about fifteen million degrees Celsius. Iron demands a much higher temperature. Nonetheless, analysis of sunlight tells us there is iron in the sun—just a trace, which must have come from somewhere else.

As most stars like the sun (and those that are up to four times more massive than the sun) evolve and age, the temperature in the core rises extravagantly and helium is turned into carbon, but that's the end of their cosmic alchemy. The heavier elements are products of a thermonuclear process that requires a temperature of about one hundred million degrees Celsius. Stars more than four times the mass of the sun achieve temperatures that allow them to go beyond carbon. They burn carbon into oxygen, oxygen into silicon, and silicon into iron (Trimble 1996). At that point, those old, massive stars are unstable, and in time explode, collapse, and shed most of the new atoms they've created into space, where they are

P2.5
The ancients knew Mars as the "red planet." To the unaided eye, it looks red in the sky. In images acquired by sophisticated telescopes, it looks red-orange because its surface is loaded with iron oxide. Mars, like the earth, has iron, but much less of it sank to the core, and we can observe iron-ore dust covering much of the planet.
IMAGE COURTESY OF HUBBLE SPACE TELESCOPE, NASA, ESA, AND THE HUBBLE HERITAGE TEAM (SPACE TELESCOPE SCIENCE INSTITUTE/AURA)

P2.6
The universe blossomed like a skyrocket in an incredibly hot, primordial explosion 13.8 billion years ago. After the first three minutes, the cosmos had cooled enough to allow subatomic particles to bond into atomic nuclei—the first elements. Only the simplest and lightest atoms were forged by the Big Bang: hydrogen, helium, and lithium. Iron had to wait until massive stars formed, evolved, and then dispersed the heavier elements they had manufactured in their cores.
ILLUSTRATION BY DON DIXON. COURTESY OF GRIFFITH OBSERVATORY, LOS ANGELES

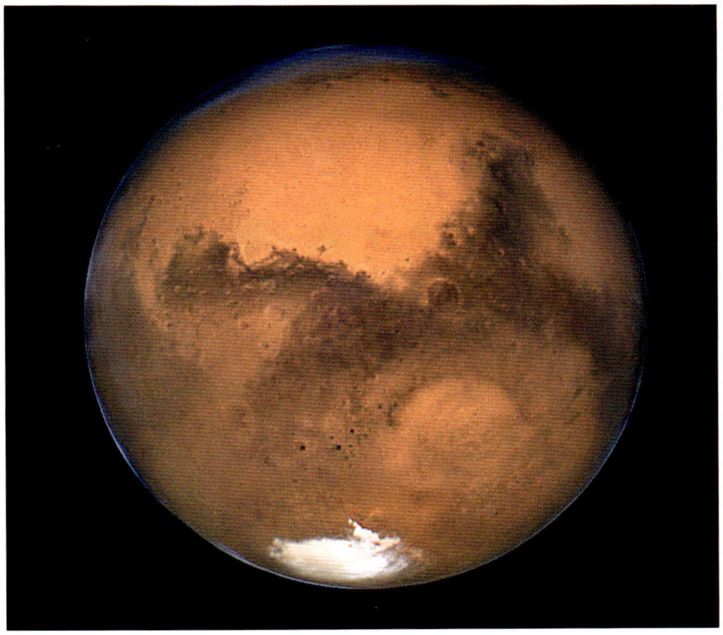

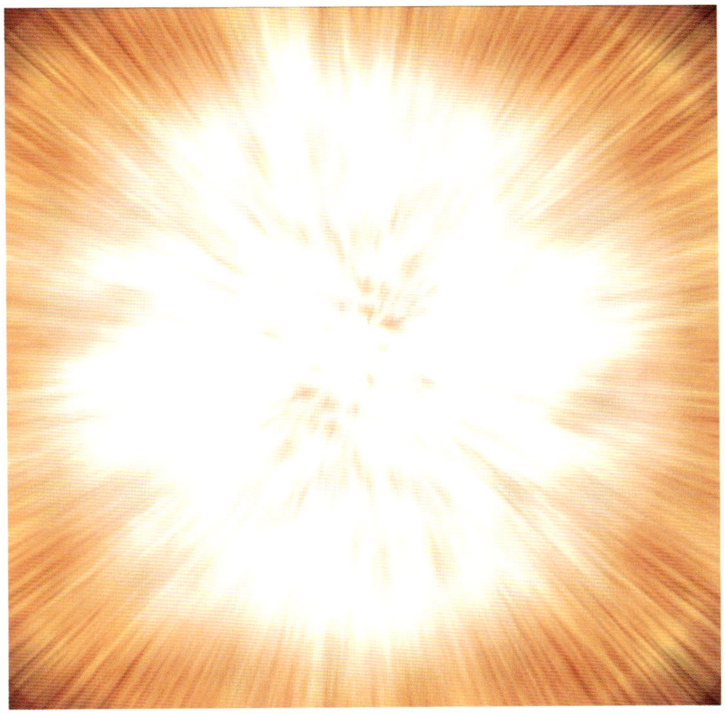

gathered into interstellar clouds, from which emerge new generations of stars, enriched with heavier elements including iron.

Material ejected from a supernova about five billion years ago may have been what induced the collapse of the cloud that became our solar system (Time-Life Books 1988, 89–101). This is where we get our iron—not from raisins, spinach, or supplements, but from archaic generations of massive and fast-evolving stars that began making iron and dispersing it everywhere, like Johnny Appleseed in supernova fury, soon after the Big Bang.

As a heavy metal, iron is kind of a leftover enriching our solar system. For every ton of material in that original solar cloud, there were two pounds of iron. Planetesimal objects—small primordial aggregates that would eventually collide, grow hot, merge, and become the world we know—formed in the disk of gas and dust that rotation created around the infant sun. Iron, turned solid at high temperatures in the disk, combined with oxygen to build tiny grains that were gathered into the growing planetesimals, where they melted and were segregated from the more rocky material. Gravity kept more of the heavier material closer to the bosom of the sun, and much of it was collected into the solar system's inner planets: Mercury, Venus, Earth, and Mars. The denser metal melted and sank early and fast to the center of each of these worlds (Wood, 1999).

Mercury, the nearest planet to the sun, is 70 percent iron, all concentrated in a core that rises three-quarters of the way to the planet's cratered surface (Strom 1987, 66–80). Venus, the second planet from the sun, is nearly the same size as the earth, and its core is thought to be iron and similar to the earth's. About half again as far from the sun as we, Mars is about half the size of the earth. It, too, seems to have an iron core, but small and cooled solid. The jury is still out on our moon, which may have a small metallic core or perhaps none at all (Head III 1999; fig. P2.7).

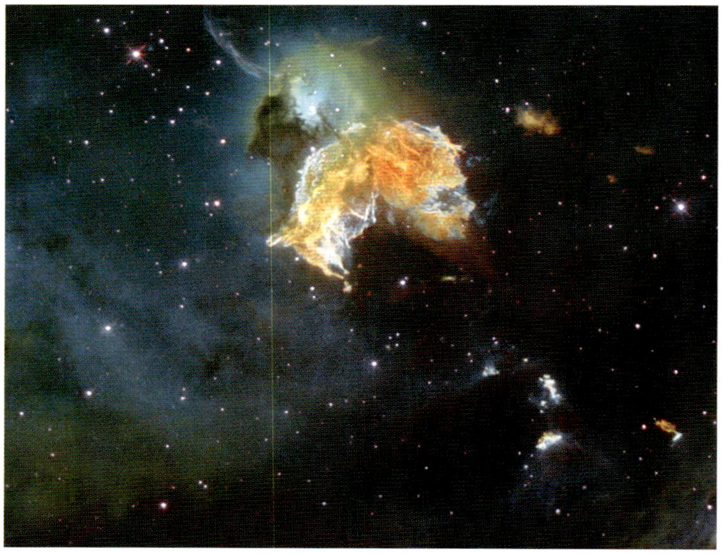

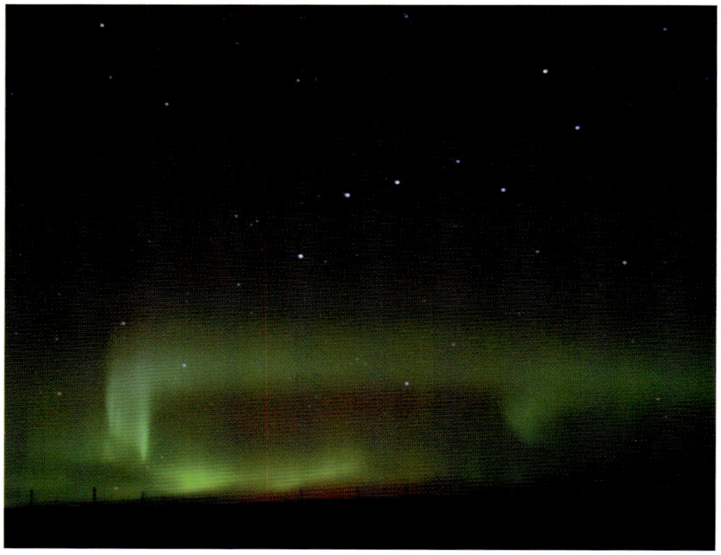

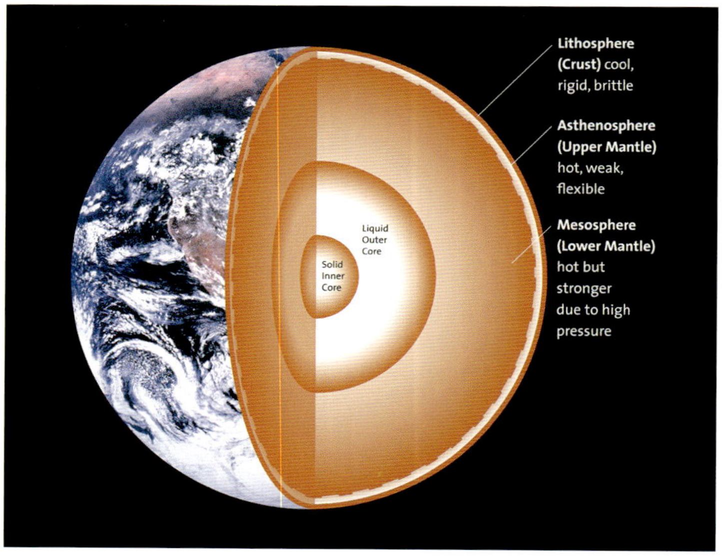
Lithosphere (Crust) cool, rigid, brittle
Asthenosphere (Upper Mantle) hot, weak, flexible
Mesosphere (Lower Mantle) hot but stronger due to high pressure
Liquid Outer Core
Solid Inner Core

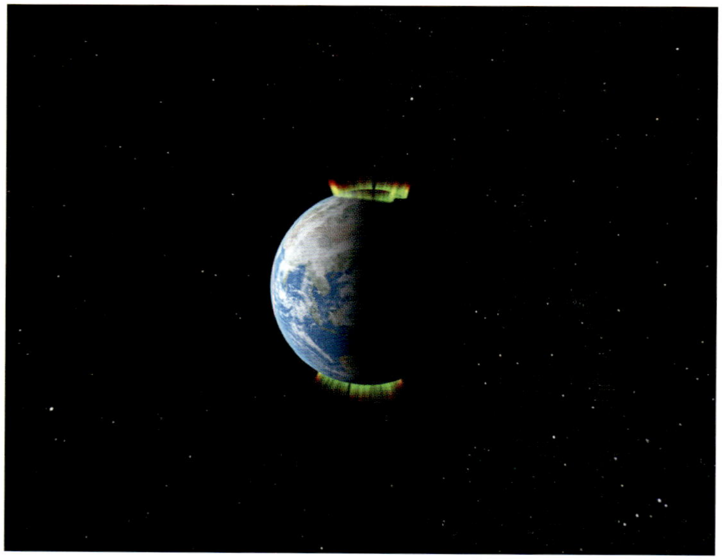

Meteoritic iron comes from planetesimals that were not captured by the developing planets but remained asteroids, the small, rocky objects with orbits between Mars and Jupiter. Fragmentation, largely through impact, liberated pieces of interplanetary iron, some of which over time intercepted the earth, passed through the atmosphere, and landed on earth (McSween 1999).

Although native iron is the most abundant element on the planet, only 4 percent of the earth's crust is iron, and there it is oxidized and available only as rocky ore (Emsley 2002, 209). All of the iron we have ever touched and fabricated and used comes from meteorites, the iron of which was smelted from cosmic "ore" through aggregation, melting, and differentiation, or from the iron ore that floated and froze into the earth's crust to be mined and smelted and forged by us. We depend on that iron—not just for tools and technologies—but to carry oxygen on iron-red corpuscles. Because iron atoms are so chemically attracted to oxygen, our iron-clad blood cells transport it from our lungs and through our blood vessels to redirect it to the body's cells, which use it to burn the fuel we eat to live.

We also rely on that nearly pure iron in the core of the earth to keep us alive (fig. P2.8). Molten iron inside the earth is hot enough to be electrically charged. Moved by the earth's rotation and by internal convection, it runs like an electrical current and turns the earth into a magnet (figs. P2.9, P2.10). The iron core of the earth generates the magnetic field we all experience with a compass in hand (Davis 1992, 93–116). That magnetic field is not just an entertaining planetary accessory. It is a necessity for life. It diverts and continuously captures powerful atomic particles from space—from the sun and the stars beyond—and provides shelter from a storm that would otherwise eradicate life from the surface of the earth. We are by hot iron bound and lucky to be so.

P2.7
The dense knot of gas in supernova remnant N63A was created about two thousand years ago in the Large Magellanic Cloud, a galaxy neighboring the Milky Way. As massive stars age, thermonuclear fusion fabricates heavier elements in a sequence of shells centered on a core of iron, which has the most massive atomic nucleus the star can craft until it collapses and explodes catastrophically as a supernova. Remnants of supernovae are propelled into other clouds of interstellar gas to be scattered or compressed into new cradles of star and planet formation, now seasoned with iron and other heavy elements.
IMAGE COURTESY OF NASA, ESA, HEIC, AND THE HUBBLE HERITAGE TEAM (SPACE TELESCOPE SCIENCE INSTITUTE/AURA, Y.-H. CHU AND R. M. WILLIAMS, UIUC)

P2.8
The size of the earth and its mass imply an average density significantly higher than rock. This suggests that much of the earth's interior is iron. Seismic waves indicate a solid inner core and molten outer core.
ILLUSTRATION COURTESY OF GRIFFITH OBSERVATORY, LOS ANGELES

P2.9
The earth's magnetic field guides highly energetic electrically charged particles from outer space toward the planet's magnetic poles, where they collide with atoms and molecules in the upper atmosphere. Momentarily energized, the atmosphere then re-emits the energy as the aurora.
PHOTOGRAPH BY E. C. KRUPP, ICELAND, 2010.

P2.10
Iron makes the earth a magnet. Cosmic rays flow along the earth's lines of magnetic force and generate two auroral rings: one at the far north, the other over Antarctica.
ILLUSTRATION BY PATRICK SO AND MATTHEW BURLANDO, 2010. COURTESY OF GRIFFITH OBSERVATORY, LOS ANGELES

PREFACE 3 ÒGÚN, THE *ÒRÌṢÀ* WHOSE FAME IS WORLDWIDE

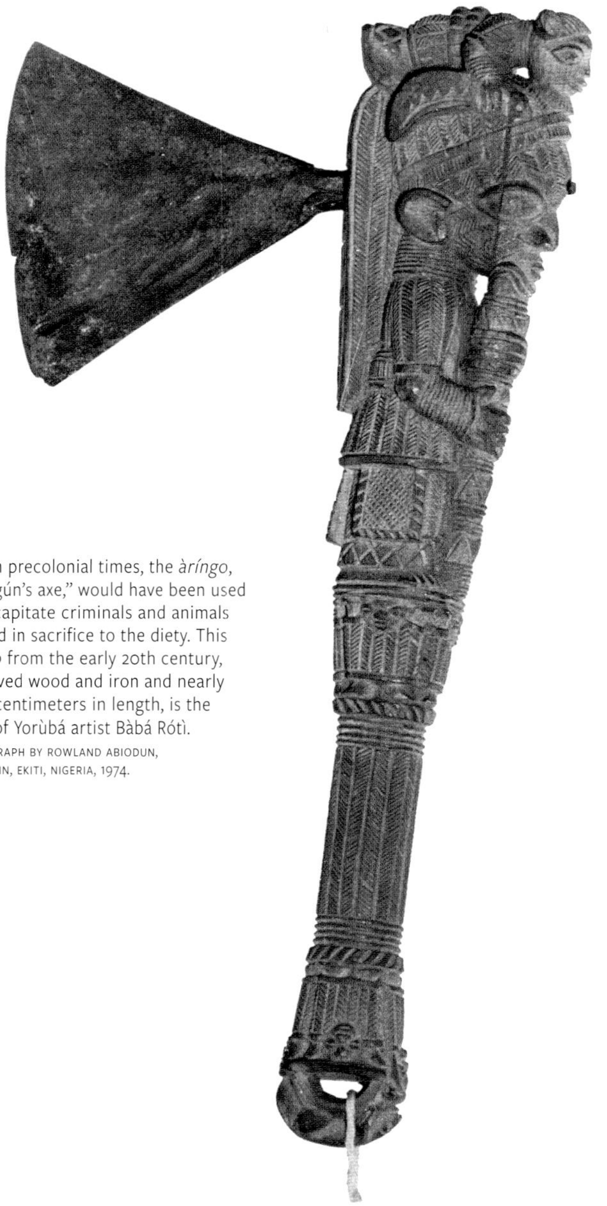

P3.1 In precolonial times, the *àríngo*, or "Ògún's axe," would have been used to decapitate criminals and animals offered in sacrifice to the diety. This *àríngo* from the early 20th century, of carved wood and iron and nearly forty centimeters in length, is the work of Yorùbá artist Bàbá Rótì.
PHOTOGRAPH BY ROWLAND ABIODUN, OSÍ-ILỌRIN, EKITI, NIGERIA, 1974.

Rowland Abiodun

> The lion never allows anybody to play with his cub.
> Ògún will never allow his child to be punished.
> Ògún do not reject me!
> Does the woman who spins ever reject the spindle?
> Does the woman who dyes ever reject a cloth?
> Does the eye that sees ever reject a sight?
> Ògún, do not reject me.
>
> FROM AN *ORÍKÍ* (BEIER 2002, 43)

By virtue of being born into a family of Ògún devotees in Ìjẹ̀bú-Ọ̀wọ̀, in western Nigeria, where the Ọjọmọ, our local sovereign, is aptly praised as "Ọjọmọ Olúdà, Agada mínmín yé rékùn ẹ̀jẹ̀" (Possessor of the great sword, the sharp sword that draws blood), I will be called one of Ògún's children—"cubs." Ògún, the Yorùbá *òrìṣà* (deity) of iron, hunting, and warfare—metaphorically called "the lion" in the opening *oríki* (citation poetry)—"will never allow his child to be punished" (fig. P3.1), and such guaranteed protection is necessary because Ògún also possesses an opposite and much dreaded personality:

> Ògún is god of frenzy
> who will ask questions after seven hundred and eighty years.
> Ògún have pity on me:
> whether I can reply, or whether I cannot reply:
> Ògún don't ask me anything! [Beier 2002, 43]

In virtually all African societies where iron smelting technologies have been recorded, there are oral sources that attest to the metal's indispensability in peoples' lives. Specifically, there are myths pointing to the existence of primordial iron smelters and blacksmiths, who are revered as culture heroes or deities. Most of these narratives are valuable because of their antiquity, which indirectly supports the two-millennia-plus scientific dating archaeologists have assigned to artifacts of the ubiquitous metal. Let us consider here the example of Yorùbá peoples of west Africa, including myself, who refer to Ògún as deity and iron interchangeably, and recognize him as one of the very first *òrìṣà* to descend, with his iron chain, at the dawn of creation, to this world from the otherworld.

With his cutlass/machete, Ògún cleared virgin forests, enabling *òrìṣà* and the first humans to settle on earth. Appropriately, the most important annual festival in Ilé-Ifẹ̀, the cradle of Yorùbá civilization, is the Ọlọ́jọ́, which specifically honors Ògún. It is marked by the presence of the Ọọ̀ni as the sole representative of Oòduà, progenitor of Yorùbá peoples. For this festival, the Ọọ̀ni dons the *àṣẹ-* (power-) impregnated Arè (a crown that conceals his face) and affirms his *òrìṣà* status. Rituals performed during the Ọlọ́jọ́ festival enliven iron and Ògún as the patron divinity of ironworkers; both material and *òrìṣà* are entreated to bless the community. Ògún is acknowledged in the rituals and ceremonies of other Yorùbá deities as well—of herbal medicine (Ọ̀sanyìn) (see fig. 12.7, 12.8), of divination (Ọ̀rúnmìlà) (see fig. 12.1), of the rustic and nature (Òrìṣà-Oko) (see fig. 12.10), and of creativity (Òrìṣàńlá/Ọ̀bàtálá), to mention only a few—who would not be able to perform their duties efficiently without his cooperation.

It is also important to mention, however briefly, the prominent and irreplaceable role of iron sound-making instruments to signify authority, power, and leadership at meetings and gatherings involving revered persons. The typically cone-shaped bell generally called the *agogo* (fig. P3.2) in Yorùbá is known as the *ọ̀mọ̀* or *ọmà* (see figs. 12.4, 12.5) in Ọ̀wọ̀ and Ife among other western Nigerian towns. Significantly, the instrument's *oríkì* is "*ọ̀mọ̀*, it is, that knows the way, no one dares to block or cross *ọ̀mọ̀'s* path." The *lágháḷogho*, a flange-welded iron bell, is the favored musical instrument—rather than drums—during the Igogo festival in Ọ̀wọ̀, when it is used to signal the beginning of the new year.

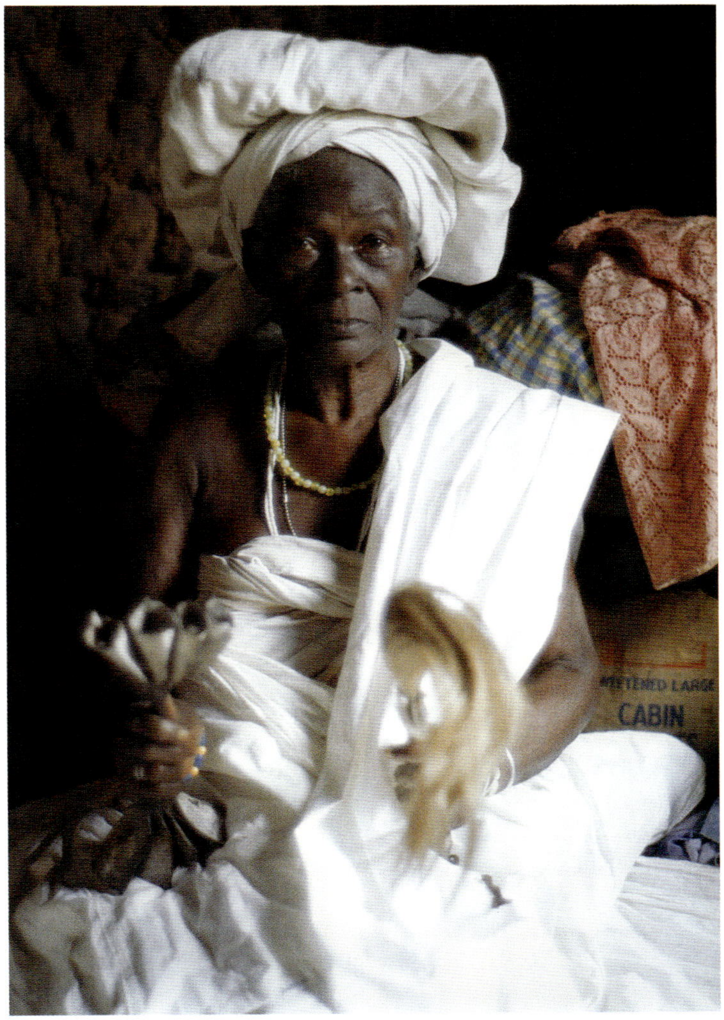

P3.2 To call the Yorùbá goddess Odua, a priestess strikes an iron bell called an *agogo* (the name itself is an onomatopoeia). *Agogo* are also struck at the Igogo festival in Ọ̀wọ̀, where they are called *lághálogho*, when orchestras of thirty or more announce the beginning of a new year. Such music "clears the way" to resumption of ordinary life through reference to the order of political authority and the harmonies of the universe. The Yorùbá deity Ògún is present at such moments, and the conical shape of an *agogo* suggests that Ògún is king and possesses "final authority" through analogy to the conical beaded crown of an *ọba* (king). PHOTOGRAPH BY HENRY J. DREWAL, ILARO, NIGERIA, 1975.

Metaphorically as well in real life, Ògún should be understood as representing the preeminent leader, trailblazer, innovator, and artist in the world, and the *oríkì* of Ògún give rich glimpses of the ways African workers and users of iron interact with and conceive Ògún ritually, socially, and aesthetically. They can also help us reconstruct the meaning and place of iron in precolonial oral-literate African societies. Take the *oríkì* "Ògún, with the bloody cap, let me see the red of your eye" (ibid., 42). Metaphorically speaking, it not only references the red-hot glow of iron being forged, but also Ògún as the patron of all users of iron, and especially warriors and hunters "who, after storing water in abundance in his house, then proceeded to have a bath of blood" (Babalola 1989, 164). Taken a little further, the language connotes the red laterite crust of the earth where iron oxides are found (Herbert 1993, 8). I believe that this *oríkì*'s application extends beyond Yorùbáland, as we can learn from another praise name of Ògún, Lákáayé, meaning "He whose fame is worldwide." Lákáayé is a universal presence who recognizes no borders (Babalola 1989, 169, n. 8).

In the following *oríkì*, not just the weight and size of iron are captured but also its sharpness and quickness. These could be challenging characteristics for those who use and interact with the unpredictable element but don't fully understand it. Iron's unpredictability contributes to Ògún's image as a temperamental *òrìṣà* to be treated with the utmost respect. Abundant rewards await those who are knowledgeable about the powers of Ògún and who possess, to the benefit of society, the rare skills of iron smelting and blacksmithing:

> Ògún, possessor of a massive stock of iron, suffices me as
> a divinity to give worship to.
> Ògún, when I placed you in the roof's interior, you made
> a gash in it.
> Ògún, when I put you in a scabbard you split the scabbard in
> a jiffy. [Ibid., 159]

In precolonial times, expertise in the working of iron enabled men to become not only famous but also wealthy.

> Ògún has plenty of money in his house
> Ògún acquires plenty of money on the road
> Ògún's dwelling is a polyhedral house in Ọrun
> [the otherworld]...
> Supporter and provider for any orphaned youngster.
> [Ibid., 164]

Among many African peoples, iron was used as currency. Suffice it to say, the knowledge of iron was pivotal in the creation, control, and expansion of numerous African kingdoms. It was only natural, therefore, that ironworkers belonged to a special class of their own even as they determined the future and fortunes of most African societies.

Ògún's overwhelming influence and force notwithstanding, there is a female-related efficacy that challenges and may even negate the male-dominated control of iron smelting and blacksmithing.[1] In my own research in Ọ̀wọ̀, I have found that the red tail feathers of the African gray parrot, understood to allude to the powers of women, are strictly forbidden in the blacksmith's forge. This prohibition apparently hinges on the belief that these tail feathers have the ability to change the chemical properties of metals, posing a threat in a sphere of male supremacy. Echoes of this rivalry occur in Ọ̀wọ̀ when an artist is commissioned to carve, using super-sharp knives, the important *àkó*, or second burial effigy. The carver must be not only attentive but sober, and he must abstain totally from sex during the period of its execution (Abiodun 2014, 93, 188).

Artistically, not only those who work in iron owe their success to Ògún but also wood sculptors. Without iron tools, they could never attain such formidable heights of creativity, as is directly and indirectly alluded to in the *oríkì* of the foremost Yorùbá carvers. It is enlightening to recall a few lines from some well-known Yorùbá artists, beginning with Àrè Làgbàyí, Aŕá Ọ̀jọ̀wọ̀n, alive during the early part of the nineteenth century:

> The itinerant carver Làgbàyí, who changed a wooden object into a human being;
> You, my father, inscribed beautiful and delicate *kẹ́kẹ́* facial marks on a wooden object in the city of Ọ̀jọ̀wọ̀n,
> Làgbàyí, the itinerant artist who carved wooden objects resembling real persons during the reign of Abíọ́dún.
> On the day Làgbàyí, the itinerant artist, was returning from one of his carving expeditions,
> The *ap*á trees collided with one another in trepidation,
> The *ìrókò* trees[2] collided with the palm trees in great fear;
> They wondered where next Làgbàyí would place his load of carving tools. [Abimbola 1994, 139]

Without "his load of carving tools," Làgbàyí would not have been able to accomplish his legendary monumental commissions. Note also the mention of "*kẹ́kẹ́* facial marks," an important consideration in the aesthetic ranking of Yorùbá figural wood carvings. Those made on real human faces and bodies require expert handling of iron blades.

Next, the *oríkì* of Ọlọ́wẹ̀ of Ìsẹ̀, active in the early twentieth century and arguably one of the greatest sculptors of the past several, who was commissioned by many a Yorùbá sovereign to carve palace veranda posts (fig. P3.3), doors, thrones, and stools:

> One who carves the hard wood of the *ìrókò* tree as though it were as soft as a calabash.
> One who achieves fame with the proceeds of his carving [...]
> My lord, I bow down to you.
> Leader of all carvers. [Walker 1994, 100–01]

Ògún is acknowledged implicity as enabling Ọlọ́wẹ̀ to perform wonders with wood using iron tools. The *oríkì* of Táíwò, of Òrè's compound Ìlá Ọ̀ràngún, who carved several veranda posts during the restoration program of the sovereign Ọ̀ràngún Oyinlọlá Arọ́jọ́joyè (1924–1936), offers much insight into his manual dexterity. It further suggests that the king was so pleased with his work that he made the pronouncement that Táíwò's "family shall be known for their carving skills . . . [and] shall continue to prosper for generations" (Pemberton 1994, 127).

> He carved the wood, and it was perfect.
> The wood had eyes, a mouth and a powerful chest.
> It had a capped penis.
> This carving was perfect, perfect in the eyes,
> Perfect in the mouth, perfect in the chest,
> To crown it, he depicted perfectly the penis with the cap. [Ibid., 127]

This last line is especially informative because it tells us something about Táíwò's extraordinary ability to use the knife to accomplish the rare feat of rendering that small but delicate part of the male anatomy without breaking it. Indeed, we have reason to believe that circumcision experts who use the iron blade on male infants have also been highly regarded as artists.

Ògún, as the *òrìṣà* of blacksmiths, is explicitly acknowledged in the *oríkì* of another famous carver, Dàda Aréògùn (1880–1954), from the northern Ekiti town of Ọ̀pìn:

> Dàda, who has Ògún's money to spend [...]
> The expert, whose sculptures dazzle the beholder
> Ìrèmògún [Ògún] who turns wood into money [...]
> Son of those who came early [whose ancestors were early settlers], Natives of Ìlágbẹ̀dẹ [home of blacksmiths]
> Ìrèmògún, native of the town where people are well fed. [Ibid., 121–22]

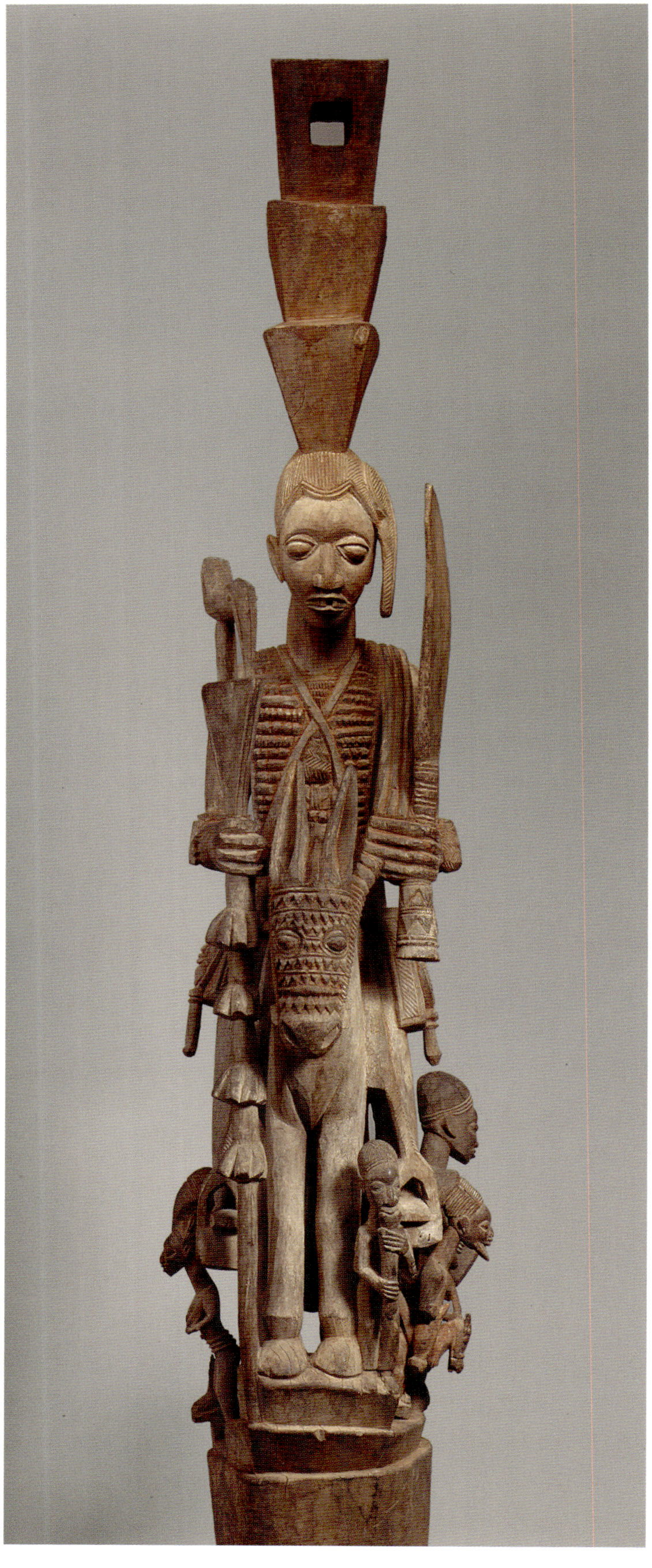

That it is not uncommon for an artist to be both blacksmith and wood sculptor was exemplified by my paternal great uncle, Ọlọ́gán Ògúnlẹ́yẹ, a renowned carver of *àkó* in Ọ̀wọ̀. His last name is an abbreviated *oríkì*, for *Ògún-ní ẹ́yẹ* means, in the dictionary of Yorùbá names compiled by Adeboye Babalola and Alabi Olugboyega, "This family-head, a devotee, has attained a position of distinction."[3] Another name connecting Ògún with creativity is Ògúnbọ́nà; *Ògún-bá-ọnà* translates to "This child from the *òrìṣà* of iron has found artistry and craftsmanship [in the home]."

There are also the names of people who are not artists but whose lineages are traceable to the families of Ògún devotees or practitioners of blacksmithing and iron smelting, including: Ògúngbénlé (*Ògún-gbémi lé*[*-òkè*]: "The *òrìṣà* of iron has exalted me / has made me prominent"); Ògúngbilẹ̀ (*Ògúngbilẹ̀*: "The *òrìṣà* of iron has gained ground / has become widespread"; Ògúnfọlájì ([*Ọmọ-tí-*] *Ògún-fi-ọlá-jì*: "Child whom the *òrìṣà* of iron has endowed with nobility"); and Ògúnjẹ́milá (*Ògún-jẹ́-mi-lá*: "The *òrìṣà* of iron has made me to become wealthy"). In praise of the formidable role of Ògún in war and hunting are names such as Ògúnlàdá (*Ògún-ni- àdá*: "The *òrìṣà* of iron and cutlass are the same: The cutlass [or machete] is a veritable image of the *òrìṣà* of iron. The real executive divinity or causative agent is the *òrìṣà* of iron."). Note how the following Yorùbá names also allude to Ògún's power: Ògúnmọ́dẹdé (*Ògún-mú-ọdẹ-dé*: "The *òrìṣà* of iron and warfare has brought along this [new] hunter"); Ògúnòmọjú (*Ògún-ò-mọ-ojú*: "The *òrìṣà* of iron is no respecter of persons"); Ògúnṣoro (*Ògún-ṣoro*: "The *òrìṣà* of iron has displayed his fierceness"); Ògúntẹ̀lúwò (*Ògún-tẹ-ìlú-wò*: "This family-head, an Ògún devotee, made an attempt at founding a new town"); Ògúntówọn (*Ògún-tó-wọn*: "The *òrìṣà* of iron is a match for them [my enemies] / is equal to them"); and Ògúnwà (*Ògún–wà*: "The *òrìṣà* of iron and warfare is alive").

The importance of Ògún is neither threatened nor diminished by the advent of new technologies, from the automobile or the aviation industry. Indeed, it is no exaggeration to say that technology as we conceive it today, in all its forms and manifestations, would be impossible without Ògún. All too often Yorùbá tout the prowess of Ògún as equal to the technical wonders they see from the West. As the hard-edged path-breaker, Ògún cuts through (neo)colonial pretenses. He conquers, possesses, and redirects introduced metal objects to meet local ends. Yorùbá agency is thus preserved and furthered by Ògún's intervention as a culture broker. We get this message from these Yorùbá names: Ògúntóyìnbó (*Ògún-tó-òyìnbó*: "The *òrìṣà* of iron is as great as the whiteskin people"); Ògúntówò (*Ògún-tóó-wò*: "The *òrìṣà* of iron suffices as the one I should look up to"); Ògúndiwúrà (*Ògún-di-wúrà*: "The support of the *òrìṣà* of iron has become [for me] as valuable as gold / has become a veritable goldmine"); and Ògúnjẹ́mitán (*Ògún-jẹ́-mi-tán*: "The *òrìṣà* of iron has been absolutely responsive to me"). That these names are given to Yorùbá offspring to this day suggests that, as for most African peoples, myths and sciences surrounding iron remain intertwined in life and thought, inseparable even into the twenty-first century.

P3.3
Ọlọ́wẹ̀ of Ìsẹ̀ (active early 20th century; Yorùbá peoples, Nigeria)
Veranda post, before 1938
Wood, pigment
H: 212 cm
MUSEUM FÜNF KONTINENTE, MUNICH
(NOT IN EXHIBITION)
PHOTOGRAPH BY SWANTJE AUTRUM-MULZER, 1999.

Carved by the highly revered sculptor Ọlọ́wẹ̀ of Ìsẹ̀, this veranda post featuring a mounted figure once graced the Onyimon family compound in Egako quarter, Ọgbagi, in the Akoko district of Yorùbáland. The figure holds prestige objects connoting iron and Ògún—a staff adorned with palm fronds, a sword. His equestrian status also suggests power and vitality, as rulers and warriors are thus represented in Yorùbá art.

INTRODUCTION

Allen F. Roberts

Everyone who practices a craft is a poet.

PLATO, *SYMPOSIUM* (AS CITED IN SUSAN STEWART 2011, 14)

Let the tuyère eat rapidly, *bangati banga*;
Let the tuyère eat rapidly, *bangati banga*;
A python comes with its friends, *bangati banga*.

LYRICS FROM A CADENCED SONG DEEMED NECESSARY TO SMELT IRON SUCCESSFULLY WHILE EXHORTING EFFICIENTBELLOWS WORK BY RUKUBA IRONWORKERS IN CENTRALNIGERIA, LEST ENEMIES DISRUPT THEIR CONCERTED ("PYTHONS") EFFORTS (AS CITED IN JEAN-CLAUDE MULLER 1974, 104)

Life has an edge, sometimes keen, sometimes blunt. Enter iron—digging, cutting, piercing, shining, rusting, disintegrating. Iron is "born" when smelting releases it from rock, and in a sense the metal possesses a life of its own. Iron transforms circumstances when forged implements extend hands to realize all manner of essential endeavors. How did sub-Saharan Africans affect such complex processes, fashion furnaces by hand, and create malleable iron to change their worlds? What observational sciences were mastered, what dramas performed? How did people know which ores to select, which wood to cut, and how to stoke and control fires of the astounding intensity required to smelt iron, at 1200 degrees Celsius (2200° Fahrenheit)? What tools were required, and with what uncanny strengths could smelters and smiths wield them, often through exhaustingly sustained labor? Who knew what to do, how to hammer, when to quench, and what to reheat? Why this shape for a forged weapon or emblem of status and not that, and how was and is perfection achieved and recognized? Why were "enchanted technologies" (Gell 1992) involving symbolic substances and carefully choreographed movement practices deemed necessary to ironworking, and how did "the craft of ritual make faith physical" (Sennett 2008, 12) as well as practical? Above all, what did ironworking reveal of the intellect and abilities of the individuals and communities undertaking such crucial activities, as well as the "enduring, basic human impulse ... to do a job well for its own sake"?[1]

Forged iron objects from sub-Saharan Africa are often striking in design, execution, and utilization, and a primary purpose of this book is to celebrate such creative, adaptive artistry (fig. INTRO.1).[2] Yet as Huston Baker suggests, "there are no arrested *things*, only myriad relations in the making" (1994, 186). Works of iron may seem static, yet as the essays to follow will demonstrate, they never sit still. Instead, they bear the "wisdom" of materials and histories of processes that inform the value people recognize in them, and they engender human and spiritual exchanges of many sorts as they serve their varied purposes. As Jane Bennett suggests, "metal is always metallurgical, always an alloy of the endeavors of many bodies, always something worked on by geological, biological, and often human agencies." Indeed, "metalworkers are themselves emergent effects of the vital materiality they work.... The desire of the craftsperson to see what a metal can *do* ... enables him to discern a life in metal and thus, eventually, to collaborate more productively with it" (2010, 60). "Lives in metal" are a primary focus of this book, and notice the word *collaborate*: iron offers its own efficacies as it is worked, its own "incipient tendencies and propensities" that lead to its "vibrant materiality ... that is *itself* heterogeneous, itself a differential of intensities, itself a life" (ibid., 56–57).

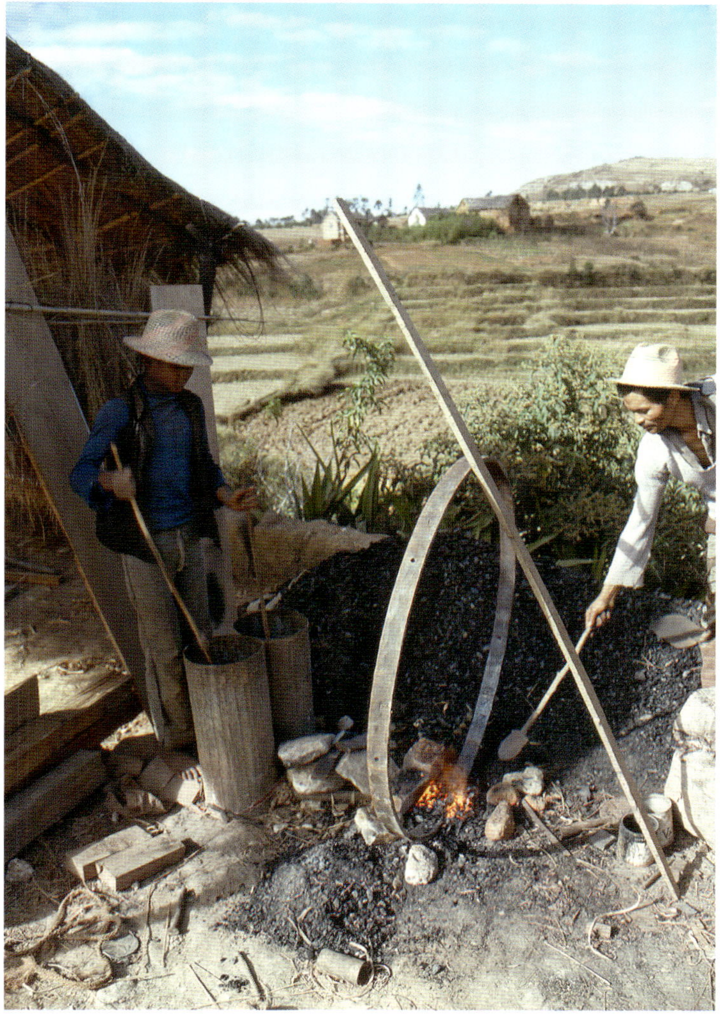

INTRO.1A–B Methods, means, and end products of sub-Saharan ironworking vary considerably throughout the continent and its islands. Here, Malagasy blacksmiths stoke a forge fire with piston bellows to weld an iron rim for the wheel of an ox cart.
PHOTOGRAPHS BY WILLIAM J. DEWEY, NEAR ANTANANARIVO, MADAGASCAR, 1993.

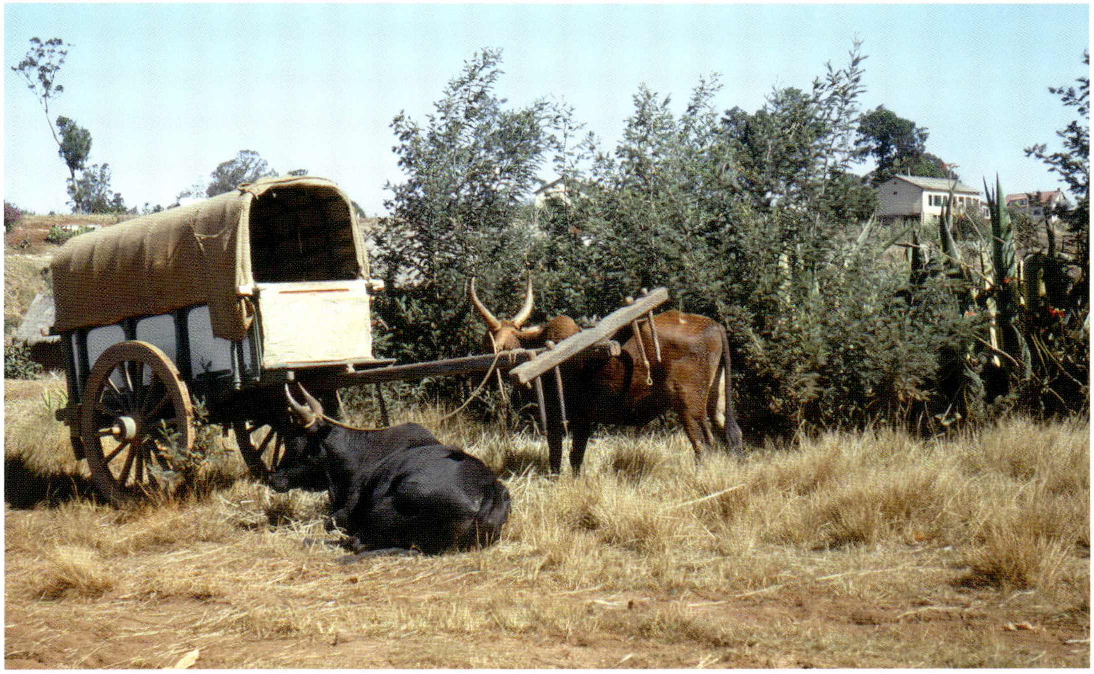

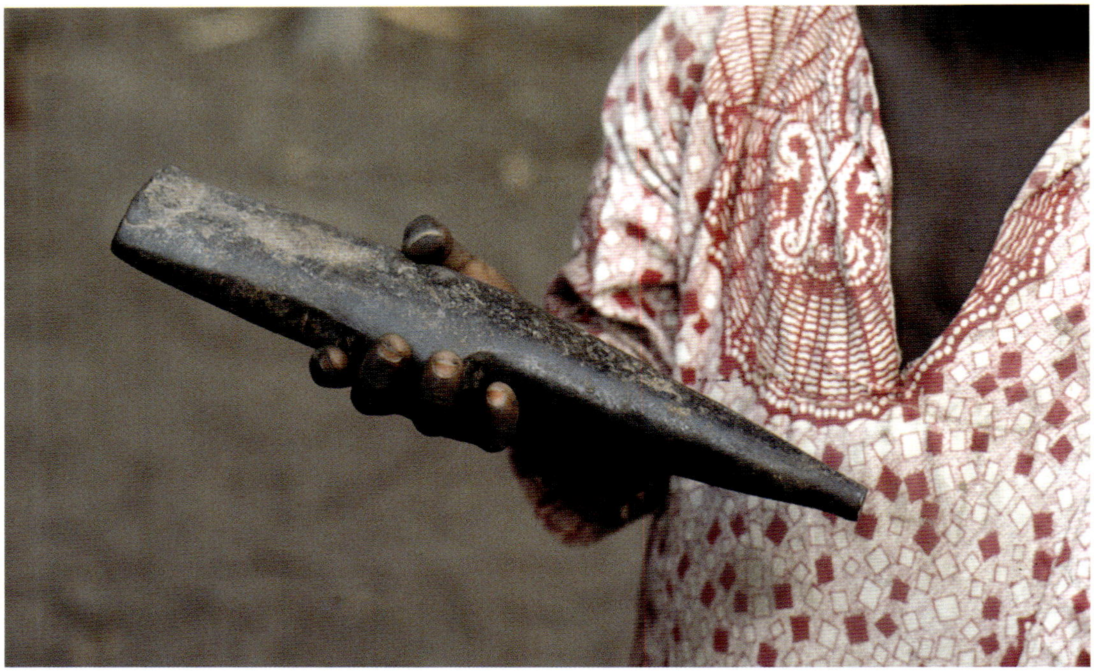

INTRO.2 Tonja Najombe of Nangbani, a village near Banjeli, Togo, holds an heirloom hammer from his family of ironsmiths.
PHOTOGRAPH BY CANDICE GOUCHER, NANGBANI, BASSARI REGION, TOGO, 1985.

Then there is *rust*. "Fickle iron" meets its slow but inexorable demise as oxygen (mostly via water) confronts perdurance, and one realizes that through corrosion "iron is not stable enough to be identical with itself."[3] Rusting characterizes all ferrous processes and products. Like a shadow, an echo, or an evocative aroma, rust is an inevitable if elusive impermanence. In Swahili, one says that iron has "sprouted" rust (*chuma kimeota kutu*), and the same verb *kuota* describes life bursting forth from seeds or inspiration emerging from dreams (Johnson 1988, 232, 356). In a sense, iron's vibrant materiality is organic, but associations among germination of seeds, dreams, and rust may seem improbable until one understands how often transcendent processes are instigated and influenced by spectral presence. Many central African peoples understand that beloved ancestors inform dreams and may be reincarnated (Stefaniszyn 1954). As is found in philosophies around the world, ends beget beginnings, and that the "life" of iron is similarly cyclical must have informed senses of its ontological value.

In earlier days, the training and mystical practices of African smelters and smiths were informed by processual and relational histories of iron as a "living" material that contributed to the status of such persons as well as the implements they forged (Muller 2001; Rowlands and Warnier 1996; Baroin and Moñino 1991). Sub-Saharan smiths were often wealthy, influential men (Childs 2000; Kriger 1999), and again, one must assume that over and above benefits from the exchange of objects they forged, their prestige was due to associations with the complex stories of the materials they worked (Guyer 1993, 254–55). Among Mande peoples of Mali and adjacent countries, for instance, ironworking "was a collection of techniques and a body of expertise considered professional secrets (*gundow*) that were hard-earned and proprietary, and that provided the economic and social foundation for a special way of life" (Patrick McNaughton pers. comm., 2018).

Once forged, implements sometimes begin careers of their own. Examples abound. Oromo people of Ethiopia wear iron beads for protection from spiritual or human aggression, with their moments of effectiveness long remembered, one can assume (Kassam and Megerssa 2004; Burka 2008–09; cf. Blier 1984). In the late nineteenth century, Glele, a king of Dahomey whom divination revealed to be mystically associated with "audacious knives," commissioned the blacksmith Akati Akpele Kendo to forge Gubasa swords immense in size, iconographical complexity, and divine

and exported, even as "voyage iron" became a ballast of choice for European sailing ships because it could be traded along the coast for enslaved African persons and other valued commodities. These dire histories remain little known, but are essential to understanding politico-economic and cultural dynamics of the Black Atlantic. Three interleafs complement the chapter. A first, by Colleen Kriger on voyage iron, presents the complexities of centuries of coastal west African commerce. Henry Drewal offers a second, on artistic achievements of Moroccan blacksmiths descended from Malians who had been enslaved and brought across the Sahara. And a last, by art historian Raymond Silverman, permits reflection upon how Christian crosses, and so faith itself, have long been forged in the Horn of Africa.

The final two essays capture joyously boisterous aspects of contemporary African life as cadenced by "striking iron." In chapter 17, art historian Manuel Jordán discusses sounding forms. For centuries, iron bells, rattles, rasps, and resonators have made music across the continent and on into the Black Atlantic, where they produce the backbeats of many sacred and festive events to this day. Ironworking itself produces a fugue of hammering and bellows work, with bells and other instruments often contributing their complementary rhythms. Marla Berns offers an interleaf on iron instruments and ritual music among Chamba peoples of Nigeria.

Anthropologist Isaie Dougnon, in chapter 18, takes us to marketplaces in and around Bamako, Mali, to visit blacksmiths with whom he shares Dogon ethnic heritage. The dynamism of technological adaptation is at play, with "technology" including both practical skills and "enchanted" energies. A final portfolio by Tom Joyce and Henry Drewal, with photographs by Joyce of blacksmiths working in Bamako, permits one to "sense the vibrations, the sound of hundreds of hammers striking iron, a hundred rhythms of different body-minds pounding out their metallic songs—different pitches, different timbres, different tones at different paces—reflecting and counterpointing the beating hearts, sweating brows, and pulsing lungs of these engaged and focused workers, these musician-blacksmiths" (Henry Drewal pers. comm., 2017).

> Chief ironworker, handsome ironworker.
> All ironworkers are chiefs.
> The beautiful woman, the handsome boy, the pretty house,
> All are the work of the ironworker.
> Strike the iron, but do not strike the man.
> If you are satisfied, thank the ironworker.
> The ironworker is equivalent to the remedy against hunger,
> The end of war, the antidote against disease.
>
> BARIBA PRAISE SINGER DAMA GUERRA OROU OF NORTHEASTERN REPUBLIC OF BÉNIN (AS CITED IN J. VOGEL 2000, V)

> Thus at the flaming forge of life
> Our fortunes must be wrought;
> Thus on its sounding anvil shaped
> Each burning deed and thought!
>
> HENRY WADSWORTH LONGFELLOW, "THE VILLAGE BLACKSMITH" ([1840] 1992, 7–8)

...

Allen F. Roberts dedicates all his contributions to this volume and the "Striking Iron" project more generally to the memory of Dr. Mary "Polly" Nooter Roberts (1959–2018), his spouse. Polly's long and profound art historical research with Luba friends of the Democratic Republic of the Congo informs and inspires many of these pages and the exhibition they complement. May her everlasting radiance guide us all onward.

PART I

IRON'S ESSENCE AND EFFICACIES

CHAPTER 1

THE PLACE OF IRON IN AFRICAN COSMOLOGIES

Allen F. Roberts

> Dreamed hypotheses . . . [are] at the root of even the clearest of technical processes.
>
> GASTON BACHELARD, "IMAGES OF THE SMITHY" (1991, 133)

African cosmologies tell truths, reveal wonders, explain perplexities, facilitate politics. Origin stories "relate how one state of affairs became another: how an unpeopled world became populated; how chaos became cosmos; how the seasons came to replace a climate without seasons; how the original unity of mankind became a plurality of tribes or nations; how androgynous beings became men and women" (Turner 1968, 568). Cosmologies and the religious practices built from them can lead to "the formulation, by means of symbols, of an image of such genuine order of the world which will account for, and even celebrate, the perceived ambiguities, puzzles and paradoxes in human existence." By doing so, we overcome "the strange opacity of certain empirical events, the dumb senselessness of intense or inexorable pain, and the enigmatic unaccountability of gross iniquity" (Geertz 1973, 107–08). Cosmologies are of deep, enduring significance, then, and little wonder that such narratives are always told in the present, to accommodate contemporary needs and circumstances even as inspiration is sought from persons, events, and divine interventions of the past.

Africans smelting iron from local ores and forging purposeful things from it must be considered in light of cosmological narratives and the purposes, practices, and "moral imagination" they explain and promote (Barndon 2012, 43). Indeed, smelting technology is so stunning and smithing so striking that surely such genius is a gift of divinity (see fig. 1.10). Are we ourselves wrought from raw matter, then "forged" to become what we may and realize what we can? How do we triumph if not in the hands of the Master Blacksmith, God of—and as—transcendent practice?

As art historian Sidney Kasfir has noted, all too often scholars have concentrated on one aspect of ironworking to the exclusion of others, thereby missing opportunities to understand African technologies as *lived* logic and praxis formulated through cosmologies (2007, 135). Furthermore, like people everywhere, Africans are reflective: they ponder why things happen the ways they do. Through their own, mostly oral idioms, they have developed what the late Kenyan savant Henry Odera Oruka termed "African sage philosophies" (1990). That is, in African societies as in all others, not everyone is given to deep reflection, nor do all have the intellect or the desire to be so engaged; yet some do, and they are looked to for guidance by their communities.[1] For example, Germaine Dieterlen wrote of Dogon peoples of Mali that "whether it concerns a ceremony or familial event, the analysis of ritual text or a simple inquiry, everyone . . . seems to have a keen taste for speculations. . . . Those Dogon who have penetrated deeply into their traditions become scholars in their own fashion. They learn to manipulate exegeses . . . that their own customs, beliefs, and arts have inspired among their people. This is their work, and the reason why . . . they are . . . sought out when a problem arises in their community" (1989, 43). That many such sub-Saharan "sages," "scholars," and "masters of knowledge" (Dieterlen 1964, 14) have been and are ironsmiths is central to the essays of the present volume.

With an eye to cultural dynamism, problem-solving, and capacities to accommodate social change, this chapter will examine the significance of iron and ironworking through cosmologies of Dogon of Mali and Tabwa and Luba peoples of the present-day Democratic Republic of the Congo. We shall seek to permit Africans to speak for themselves to the degree possible without dismissing their sagacity

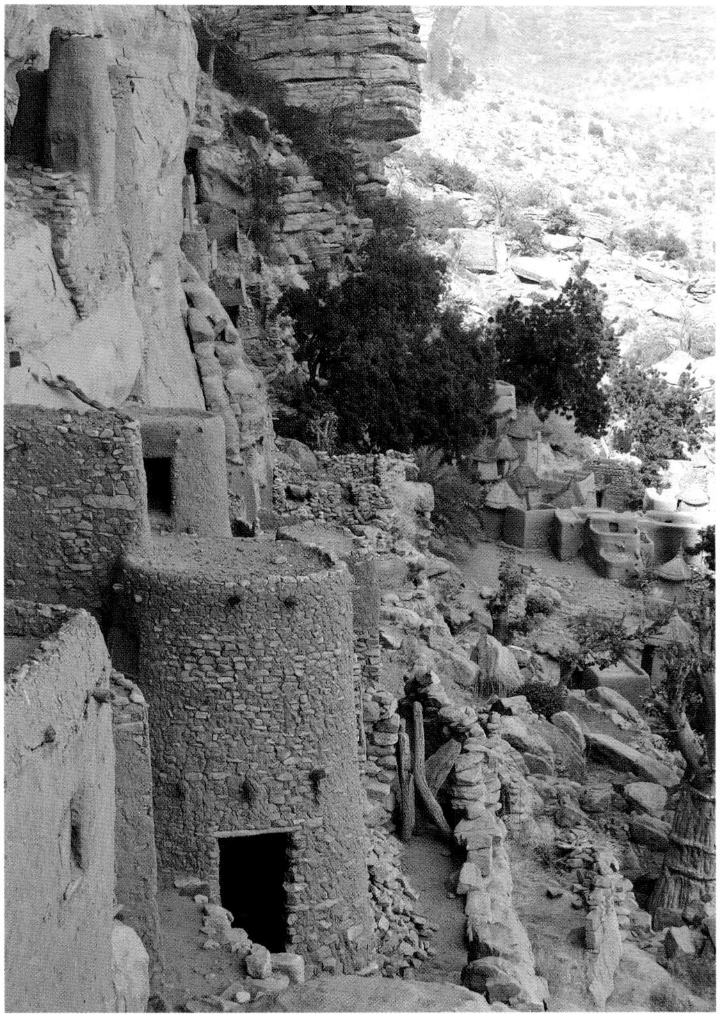

1.1 House of Dogon founding ancestor. Dogon homes and granaries with cap-like thatched roofs may be clustered near cliff walls providing niches for storage, burials, and protection from aggressors.
PHOTOGRAPH BY HUIB BLOM, PEGUE TOULOU, TORO, MALI, 2006. © HUIB BLOM, SWITZERLAND

as "myth" and "magic" as opposed to truth, knowledge, and observational science (see Wiredu 2002; Sousa Santos 2016).

STEALING A PIECE OF THE SUN

Dogon live on the southern edge of the Sahara where farming is precarious because soils are so sandy and rains so scarce. Long ago, Dogon chose to live in such remote, rugged terrain to avoid the depredations of neighbors (fig. 1.1), and they understand the purposes and perils of life through complex narratives explaining how, in the beginning, the Supreme Being, Amma, lived "above" with proto-human beings called Nommo. In the mid-twentieth century, the Dogon sage Ogotemmeli recounted such origin stories in a remarkable intellectual exchange with the French ethnographer Marcel Griaule.[2]

Ogotemmeli explained how the Seventh Nommo noticed that there was no fire on earth for use by those who would become human. Wishing to be a willful "maker of civilization" (Thompson 1973a, 191), he slipped into the forge of the Great Nommo, who, as the blacksmith of Amma, had mastered the fires of the sun itself. The Seventh Nommo stole a piece of the sun using a carved wooden instrument (figs. 1.2, 1.3) and fled Amma's wrath by climbing atop a granary as an ark bearing within it "the world system, set in place and classified into categories of creatures," measures, celestial bodies, and all other beings, objects, and phenomena (fig. 1.4).[3]

The Seventh Nommo provocatively shouted *Guyo!*—"Stolen!"—which is the word still used by Dogon to refer to their granaries, for "without the fire of the smithy and the iron of hoes there would be no crops to store" in such structures (Griaule and Dieterlen 1975, 43). He then directed the ark downward by following the curve of a rainbow, standing with his blacksmith's hammer in hand and a heavy anvil in a sling over his back. The Nommo already resembled a human except that his arms and legs remained sinuous like those of other Nommo, inchoate beings all. Wrathful Amma furiously hurled lightning bolts, so accelerating the ark that it crashed to earth, scattering—and so distributing—the beings, plants, objects, and related phenomena it contained (fig. 1.5). The impact was such that the Nommo's hammer and anvil broke his arms and legs at the elbows, knees, and other points of articulation necessary to undertake human labor. As humanity's first ancestor, the Nommo divided the land about him into eighty square farming plots to distribute to the families that would descend from him as Dogon. He then created a forge where, as the first blacksmith, he would create iron implements for agricultural life and other important pursuits (ibid., 44, 194; fig. 1.6).

Since those earliest of times, Dogon blacksmiths have been consulted for judgment of fraught matters within their communities. To resolve a point, a smith strikes the ground with his hammer; all parties then know that further debate is bootless (Calame-Griaule 1965, 275). Indeed, so central to ordinary pursuits are fire and, secondarily, ironworking skills, that the Nommo's originary theft is celebrated to this day by a society of ritual thieves called Yona, who bear elaborately sculpted wooden crooks (*yo domolo*) over their shoulders (see figs. 1.2, 1.3, 1.7–1.9).[4] Yona are selected from each Dogon clan, and their principal activity is to reenact the Seventh Nommo's theft of fire at the death of one of their members. *Yo domolo* feature open-mouthed zoomorphic heads with which to seize a piece of the sun. The Nommo's descent is depicted down their handles to where they are grasped by Yona as they dance the great cosmogonic drama that permits them to receive divine energies. *Yo domolo* are themselves complex works: while each is different in its particularities and iconographical emphases, all bear sufficient "semantic equivalence" to be "read" by Yona as they compose clan histories tailored to meet circumstantial needs.[5]

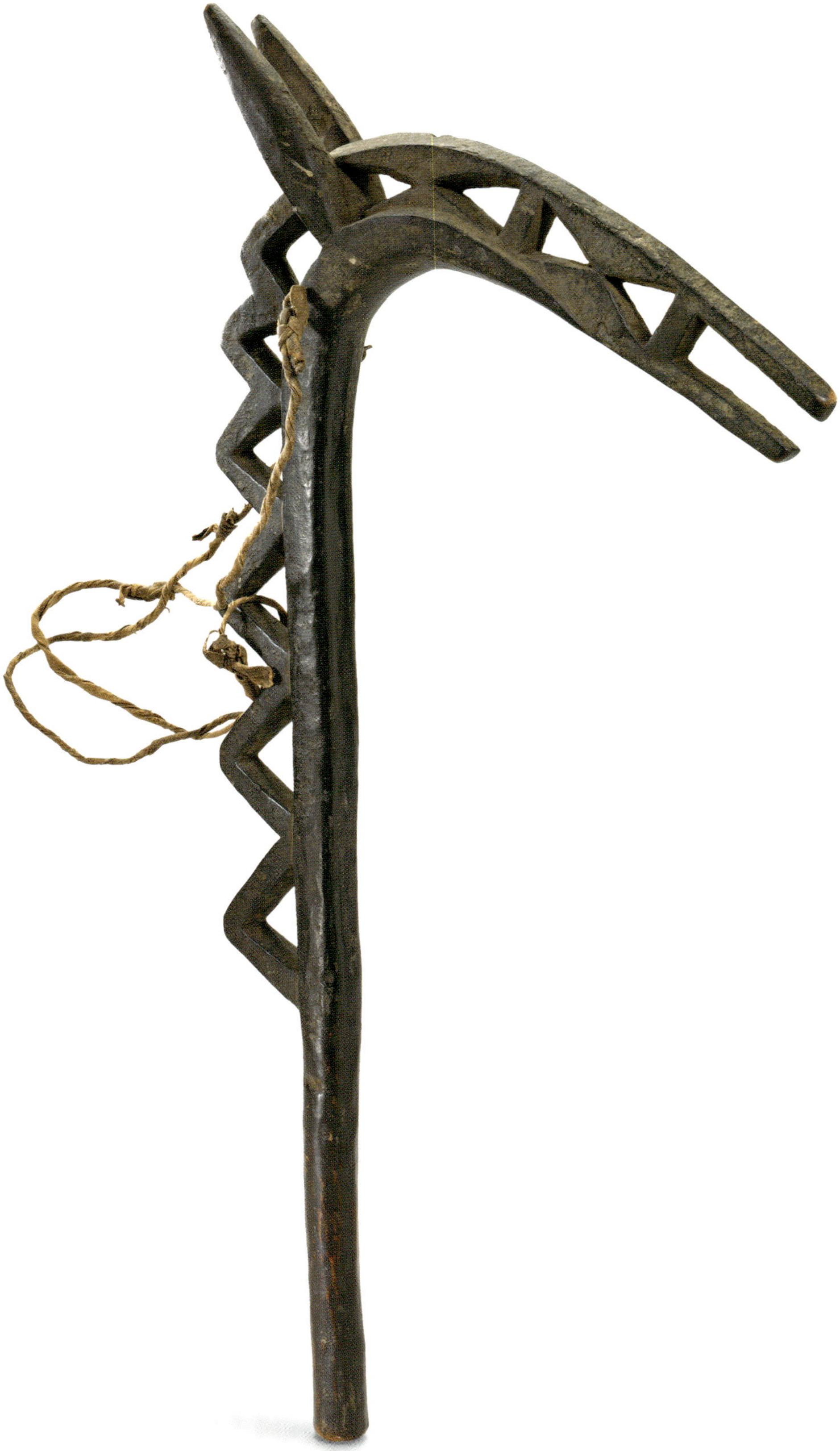

1.2
Artist unknown (Dogon peoples, Mali)
Staff (*yo domolo*), late 19th
to mid-20th century
Wood, leather
H: 69.8 cm
FOWLER MUSEUM AT UCLA, X2009.6.3; ANONYMOUS GIFT
PROVENANCE: PURCHASED BY DONOR IN BAMAKO, MALI, 1986
(NOT IN EXHIBITION)

The open maw of a *yo domolo* replicates where the Seventh Nommo carried a piece of the sun. The ears probably refer to several animals at once, and the zigzag down the haft brings audacious powers to the man gripping the end of the handle in dance.

1.3 As a member of the Yona society of "ritual thieves," a Dogon elder named Wagu Seru Dolo wears a *yo domolo* wooden crook over his shoulder. The sculpture commemorates and is used to reenact the theft of a piece of the sun by the Seventh Nommo. With such a device the Nommo brought fire from the heavens to earth, where it was needed for ironworking and therefore essential to human culture.
PHOTOGRAPH BY ALLEN F. ROBERTS, SANGHA, MALI, 1986.

1.4 Dogon granaries conserve various foodstuffs in intricate architecture designed to convey order and symbolic importance to what is so preserved. Following a Dogon origin story, the Seventh Nommo founded human civilization when his granary-as-ark crashed, dispersing the plants, animals, people, and landmark places of the now-known world.
PHOTOGRAPH BY ALLEN F. ROBERTS, SANGHA, MALI, 1986.

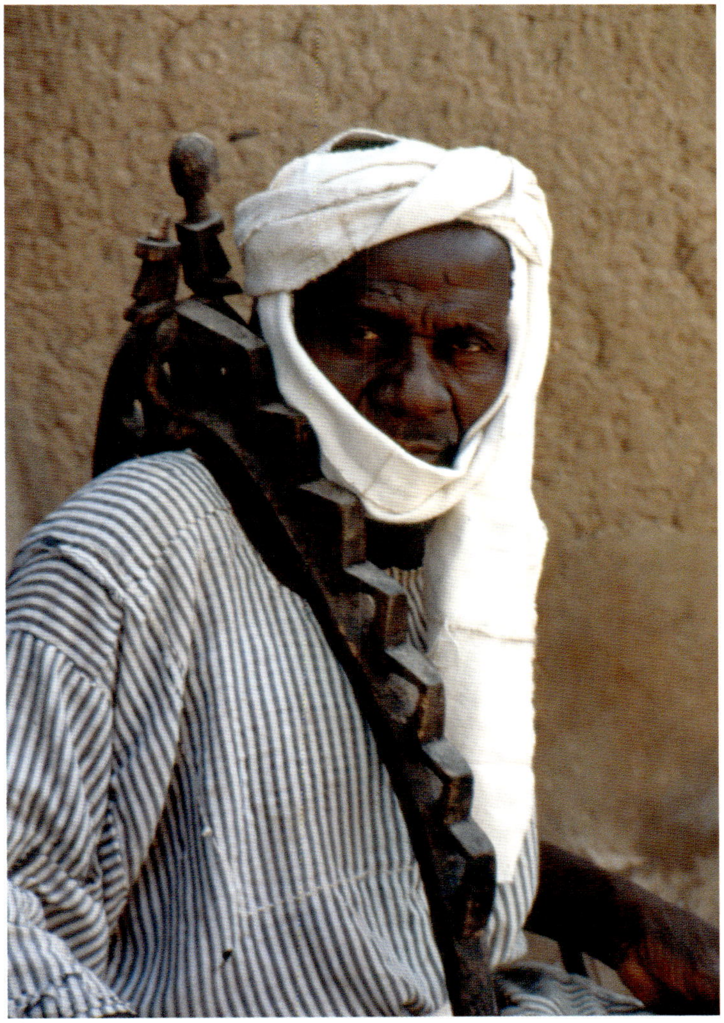

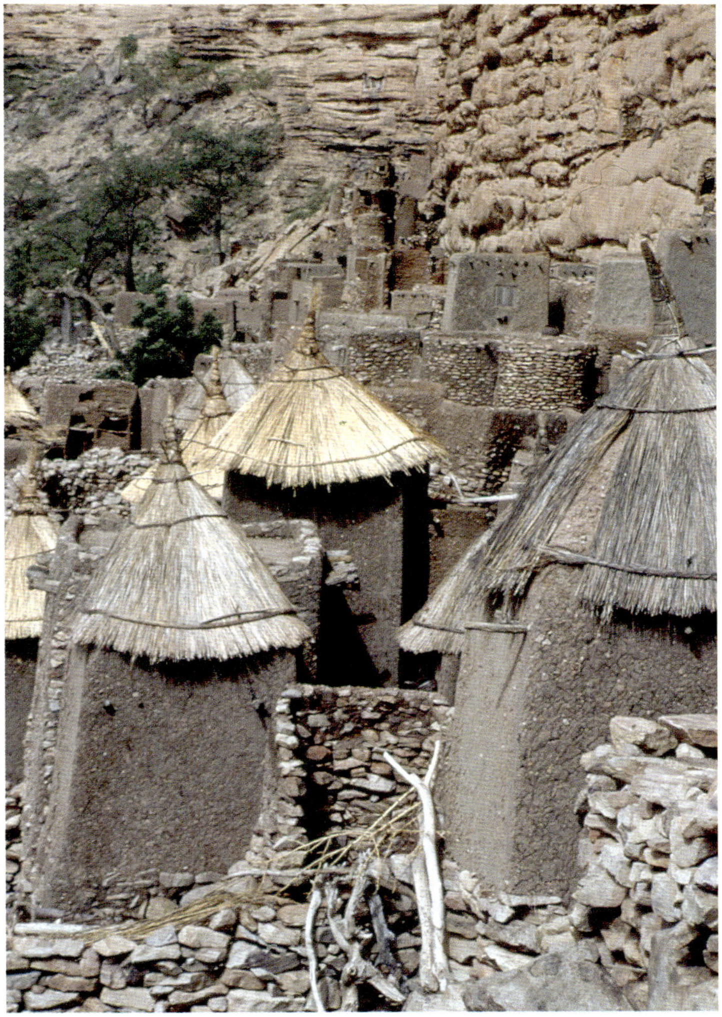

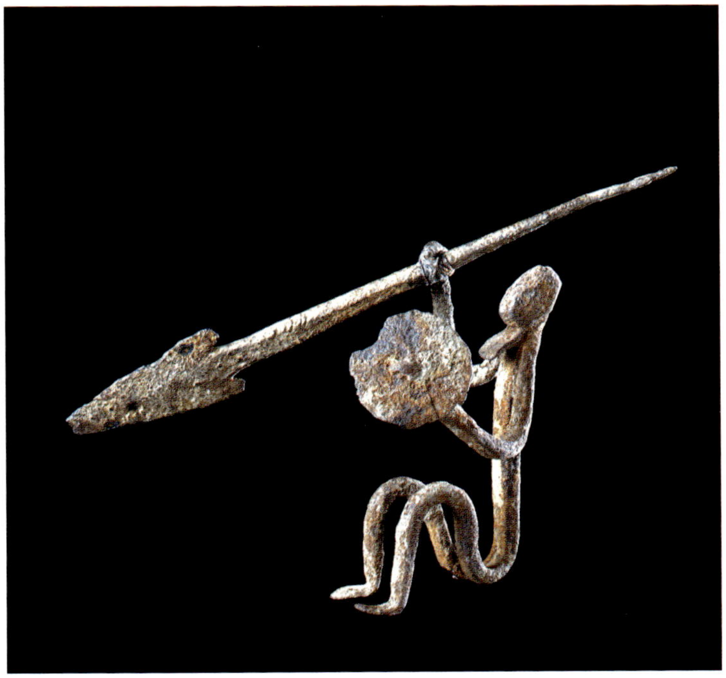

1.5
Artist unknown (Dogon peoples, Mali)
Warrior figure with spear and shield, 19th century
Iron
L: 14 cm
BARBIER-MUELLER COLLECTION, GENEVA, 1004-133
IMAGE COURTESY OF THE BARBIER-MUELLER MUSEUM, GENEVA. © STUDIO FERRAZZINI-BOUCHET

In earlier times, Dogon (or those who preceded them, often called Telem) sought refuge from predatory neighbors in the craggy cliffs of the long Bandiagara Escarpment of central Mali. Lineage-based social organization meant that more often than not, if Dogon men were bellicose, it was in defense of families and foodstuffs. Was this poignant figure of a warrior armed with a spear and holding a buckler forged for a shrine where spiritual assistance might be sought to protect and bring success to valorous individuals? Or might the work have instead referred to the flight of the Seventh Nommo protecting himself from Amma's lightning bolts with his shield (Blandin 1992, 85; cf. Ezra 1988, 80)?

1.6A–B
Artist unknown (Dogon peoples, Mali)
Ritual sickle, 19th century
Iron, wood
H: 38 cm
COLLECTION JEAN-PAUL CHAZAL, PARIS
PHOTOGRAPH BY BRIGITTE CAVANAGH, 2017.
COURTESY OF JEAN-PAUL CHAZAL, PARIS

The figural haft of this ritual sickle may portray a Nommo or perhaps a kneeling female ancestor wearing an iron necklace bearing a pendant summoning transformative energy. The forms of sickles and other useful implements bearing iron blades are often graced with—and so empowered by—agents of divine blessing by blacksmiths among Dogon and many other African peoples.

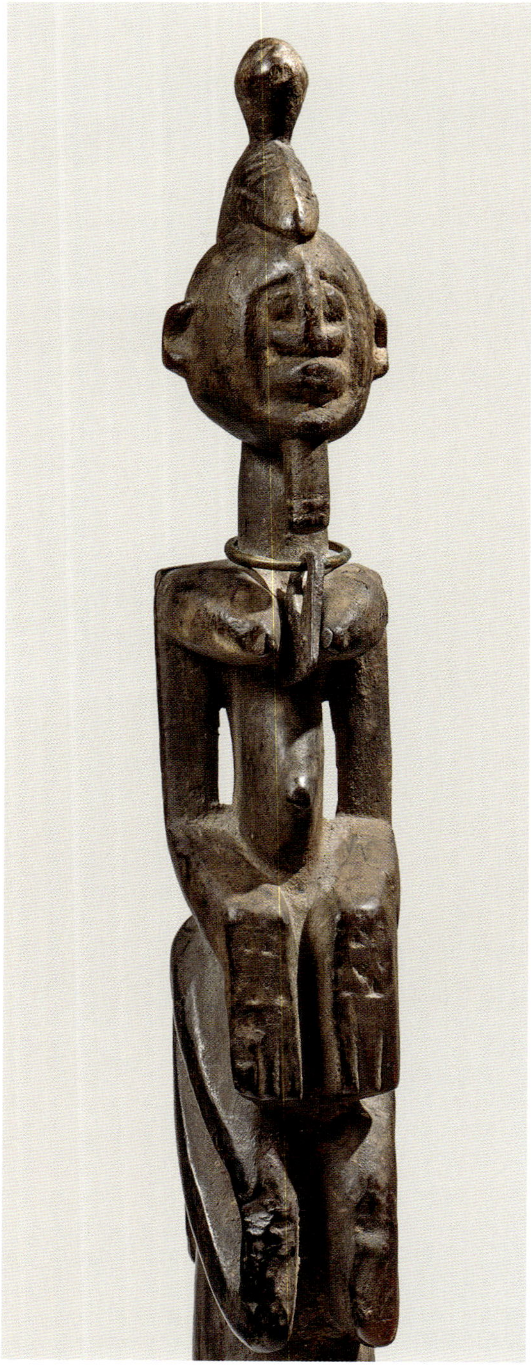

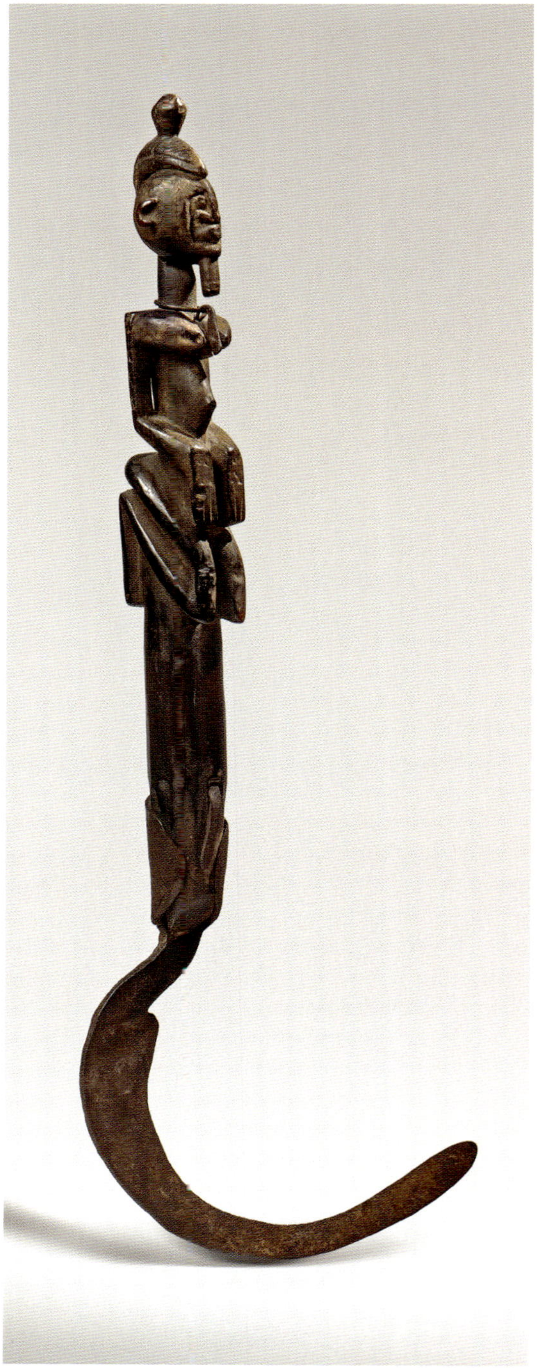

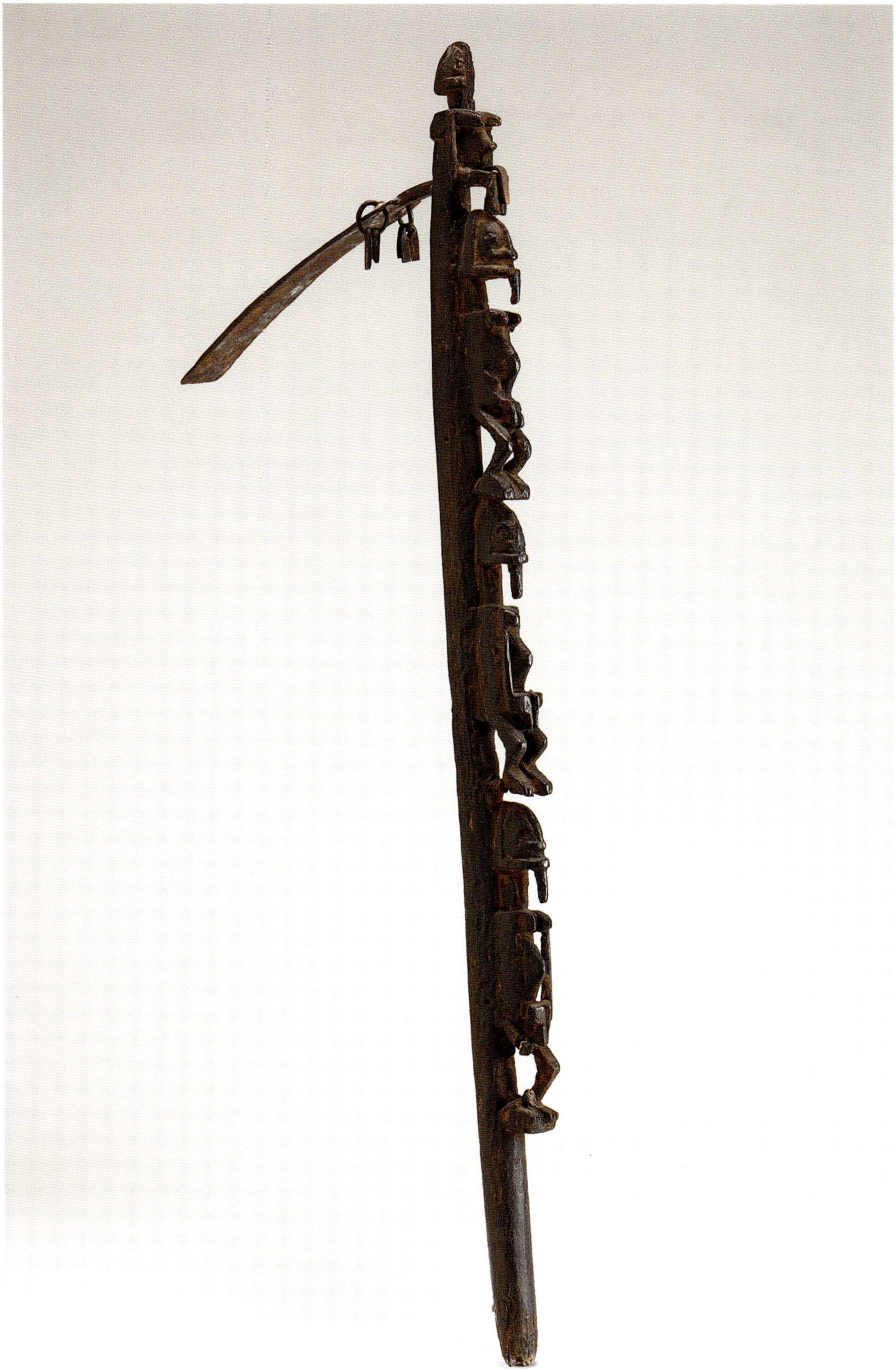

1.7
Artist unknown (Dogon peoples, Mali)
Ceremonial adze, early 20th century
Iron, wood
H: 72.1 cm
FINE ARTS MUSEUM OF SAN FRANCISCO; GIFT OF THE ERLE LORAN FAMILY COLLECTION, 2008.38.48
IMAGE COURTESY OF THE FINE ARTS MUSEUMS OF SAN FRANCISCO

Like the handles of *yo domolo* crooks, this instrument brings the creative energies of iron down the haft to the hand of a dancer. Beginning at the finial, a cascade of four Nommos down the shaft to where it is gripped brings cosmology to bear upon the needs of a particular performance.

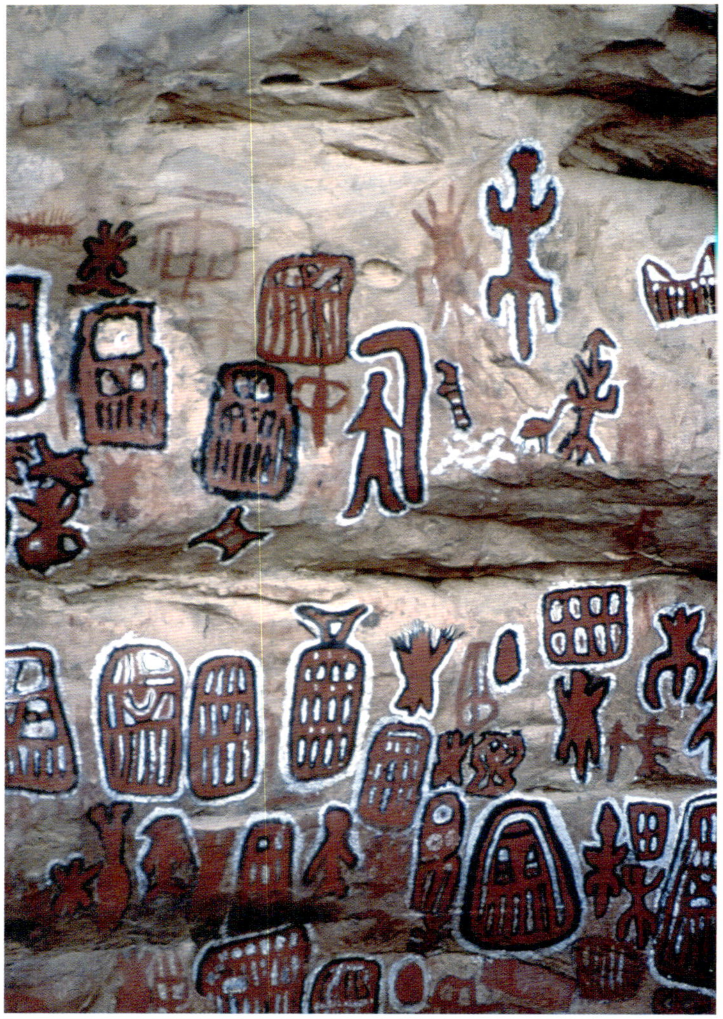

1.8 Dogon paintings on the wall of a rock-shelter largely dedicated to the primordial blacksmith, where large stones for seating recall the organization of a forge. At the shelter young men are initiated into Dogon lore as led by blacksmiths (Dieterlen 1964, 21). At the center of the image, a figure holds what is said to be a larger-than-life *yo domolo* (see figs. 1.2, 1.3).
PHOTOGRAPH BY ALLEN F. ROBERTS, SONGO, MALI, 1987.

1.9 The wall painting on this Dogon Binu shrine features, to the left, what may be two leopards and a *yo domolo* shoulder crook. The painted grid on the facade is an elemental Dogon symbol of imposed order, as when the Nommo designated the first human farms as squares of land. The shrine receives sacrifices on its roof and its sanctum is a place of spiritual power directed to promoting rain and fertility.
PHOTOGRAPH BY ALLEN F. ROBERTS, SANGHA, CENTRAL MALI, 1987.

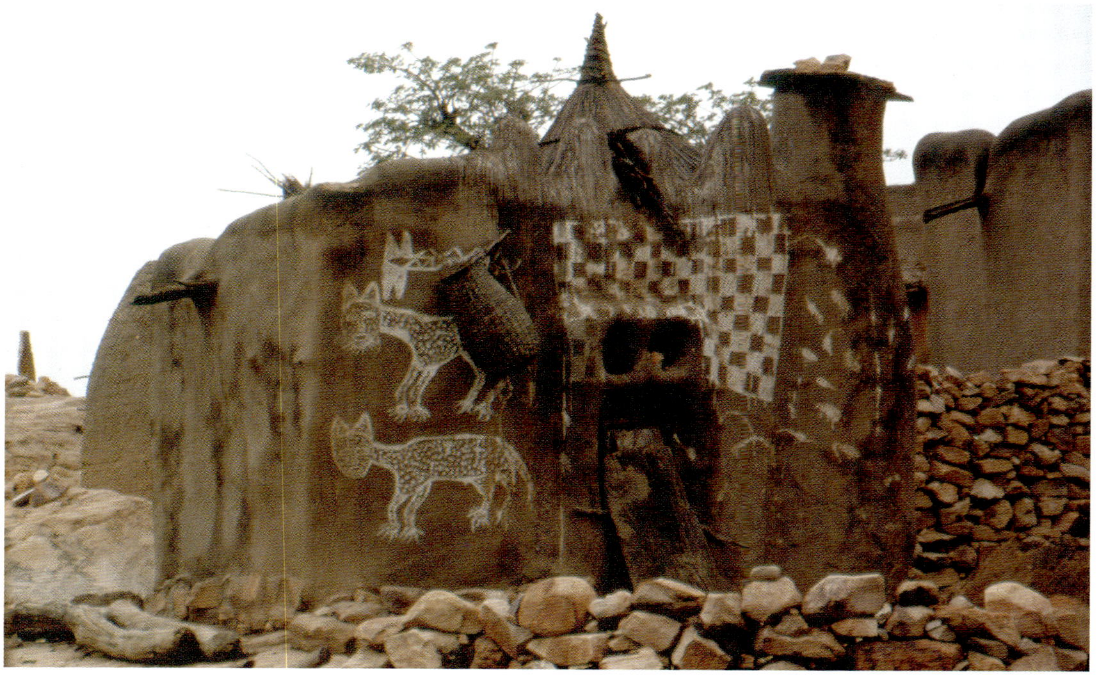

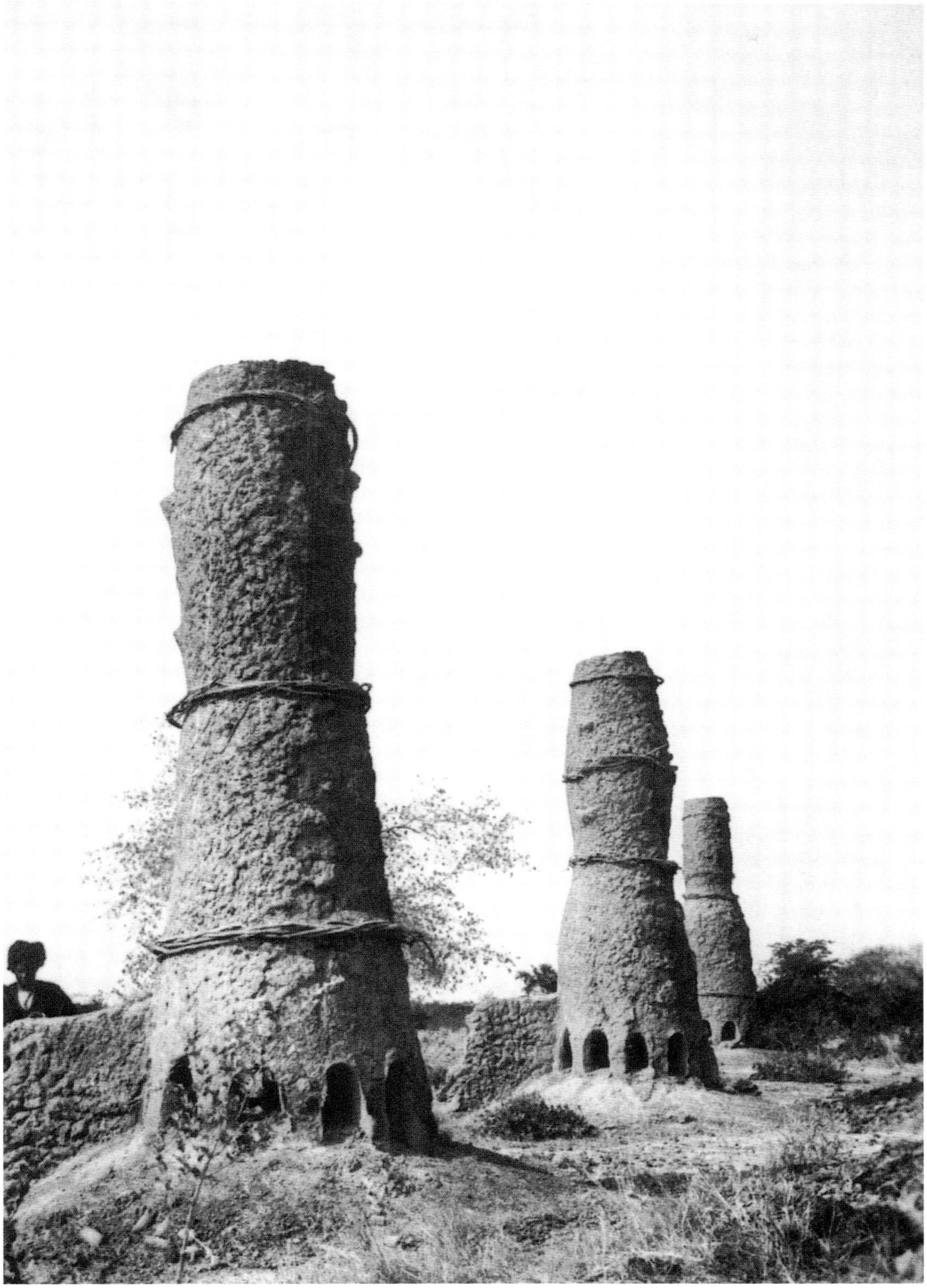

1.10 Earthen smelting furnaces in the Seno plain below Segue, Burkina Faso, 1957.
PHOTOGRAPHER UNKNOWN, PUBLISHED IN *DOGON STATUARY* (STRASBOURG: EDITIONS AMEZ, 1994).

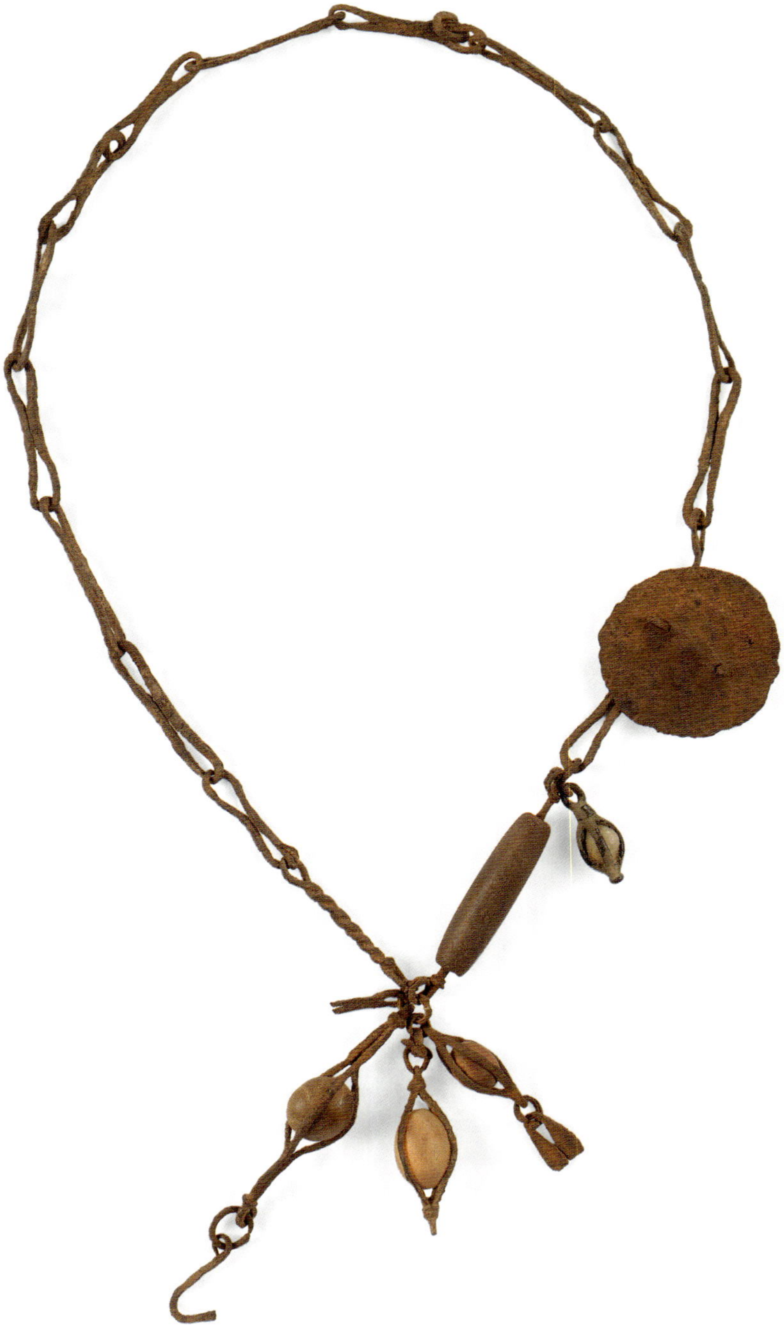

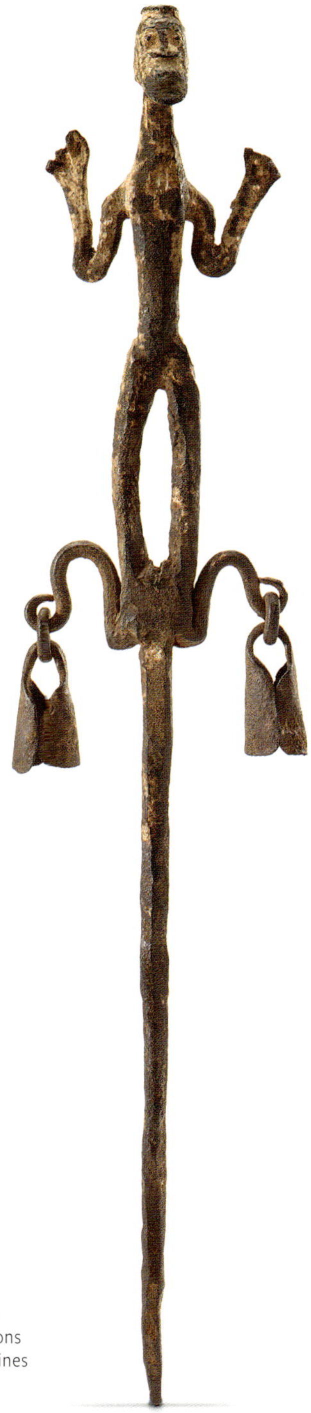

1.11
Artist unknown (Dogon peoples, Mali)
Pendant necklace (*dugo*, *duge*), late 19th to early 20th century
Iron, quartz, copper alloy
L: 58.4 cm
PRIVATE COLLECTION

As pendants, this complex necklace bears three *duge* stones, a long wooden bead, and a shield-like disk that may have made reference to the need for defense from aggressive neighbors, or to the device with which the Seventh Nommo protected himself from Amma's thunderbolts after stealing a piece of the sun. If so, some hold that the circle may also refer to the chamber of a bellows (Ezra 1988, 80).

1.12
Artist unknown (Dogon peoples, Mali)
Ritual staff, 19th century
Iron, pigment
H: 33 cm
PRIVATE COLLECTION

Blacksmiths forge figures like this at the behest of Dogon priests and other persons for placement in sanctuaries and on shrines where they "attract rain, grain, and life force" (Ezra 1988, 81).

1.13
Artist unknown (Dogon peoples, Mali)
Ritual figure, 19th century
Iron
H: 25.4 cm
DR. JAN BAPTISTE BEDAUX
PHOTOGRAPH BY PETER ROTHENGATTER, 2010. COURTESY OF DR. JAN BAPTISTE BEDAUX

This remarkably attenuated form, fashioned by a blacksmith, possesses upstretched arms, perhaps in supplication for rain. Are its two pendant devices earrings, perhaps suggesting how acutely one listens for an advancing storm? Sadly, no information exists about this work's commission, iconography, placement, or purposes, although it is known that Dogon smiths forge figures placed on shrines or stuck in the ground at key points where masked dancers pause to execute complex steps and so recognize the "cosmic connections" of iron.

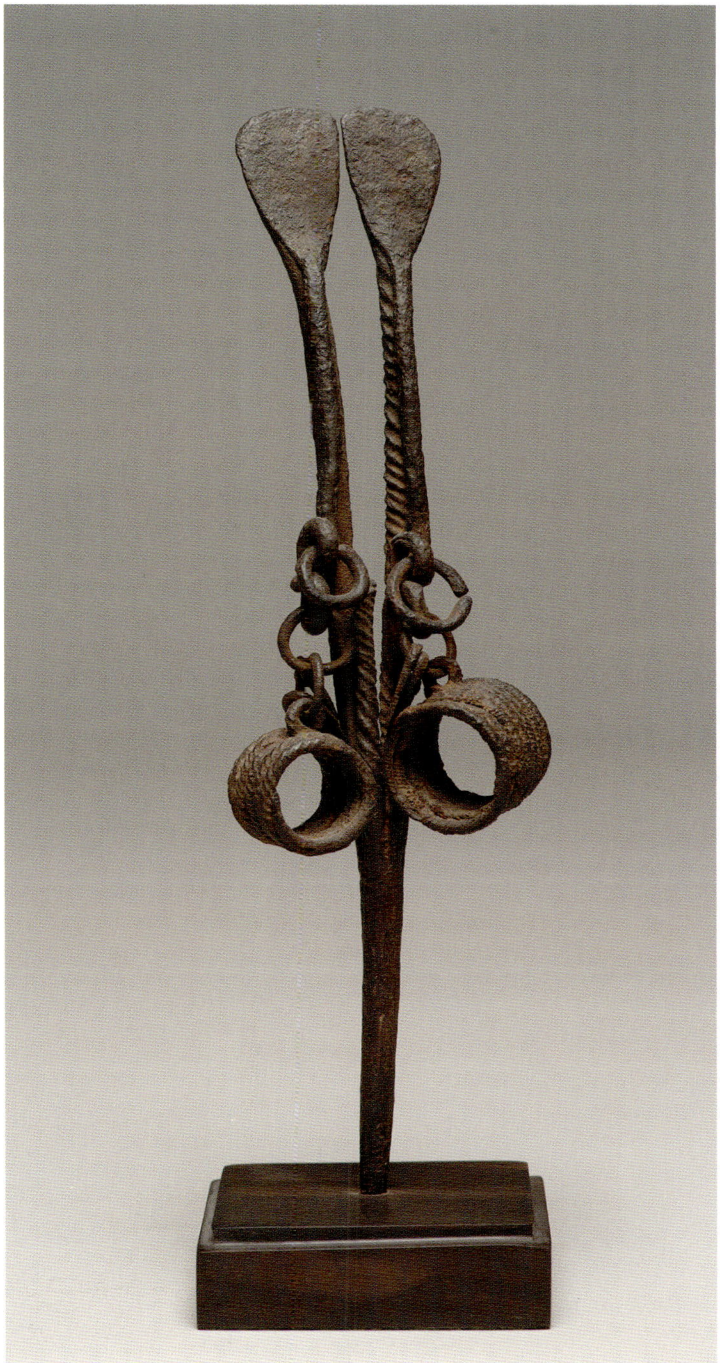

Knowledge of smelting (fig. 1.10), ironworking, and the beginnings and passages of human culture is activated through other Dogon ritual processes as well. A necklace of iron links bearing a suspended stone called a *binu duge*, understood as "Amma's egg" or "the Nommo's ark" (figs. 1.11, 3.7), was once worn by a priest who presided over Binu, a ritual practice that seeks to resolve "the rupture between the visible and invisible worlds caused by death" (Zahan 1988, 56–57; cf. Ezra 1988, 19, 22). Any person privileged to possess such a potent device "wears the world around his neck" as wrapped in wrought iron. Binu shrines are remarkable anthropomorphic structures, and their eye-like niches have been said to represent Sirius and its invisible companion star as important elements of complex cosmology (Dieterlen 1964, 6–7; Roberts 1986). Animals sacrificed at Binu shrines offer their life force to assure the fertility of crops, livestock, and people; pleas for rain are made, pregnancies facilitated, and newborns named (see fig. 1.9).

In their harsh environment, worries about sufficient rain are so acute among Dogon that incessant prayers and sacrifices must be addressed to the heavens. Blacksmiths may make such purposes and practices manifest when they forge devotional sculptures with arms raised in poignant supplication (figs. 1.12, 1.13). Some Dogon hold that blacksmiths bear "Nommo-like powers, most notably the ability to bring rain, said to be Nommo's semen," and that the figures they forge are placed in sanctuaries and on shrines to "attract rain, grain, and life force" (Ezra 1988, 81). Bellows that blacksmiths carve for themselves sometimes suggest a Nommo (or perhaps another spiritual being) with arms similarly raised (Blandin 1992, 50). Iron shrine staffs generically called *gobo* also gesture upward (fig. 1.14); what seem to be their arms may also be "cloud hooks" pulling in rain-heavy weather, and the tinkling of clapperless bells hanging from them may attract divine energies (Duchâteau 2013, 61, fig. 20). The elaborate fragility of such staffs may make poetic reference to people's insecurities, but *gobo* nonetheless assert themselves as "a source of growth, a promise."[6]

Rather than simple designs, such creations are a profound yet "virtual writing in iron" of worldly entreaties, as Armand Duchâteau terms it (2013, 57). His reference to writing is not casually poetic, for Dogon maintain a complex system of signs and glyphs painted on ritual buildings and rock shelters (Griaule and Dieterlen [1965] 2015). Overall, then, practicalities of everyday life as well as pressing needs of particular occasions are often met by Dogon blacksmiths, but as framed by deep thought expressed through cosmological narratives and related visual and performance arts (see Apter 2005).

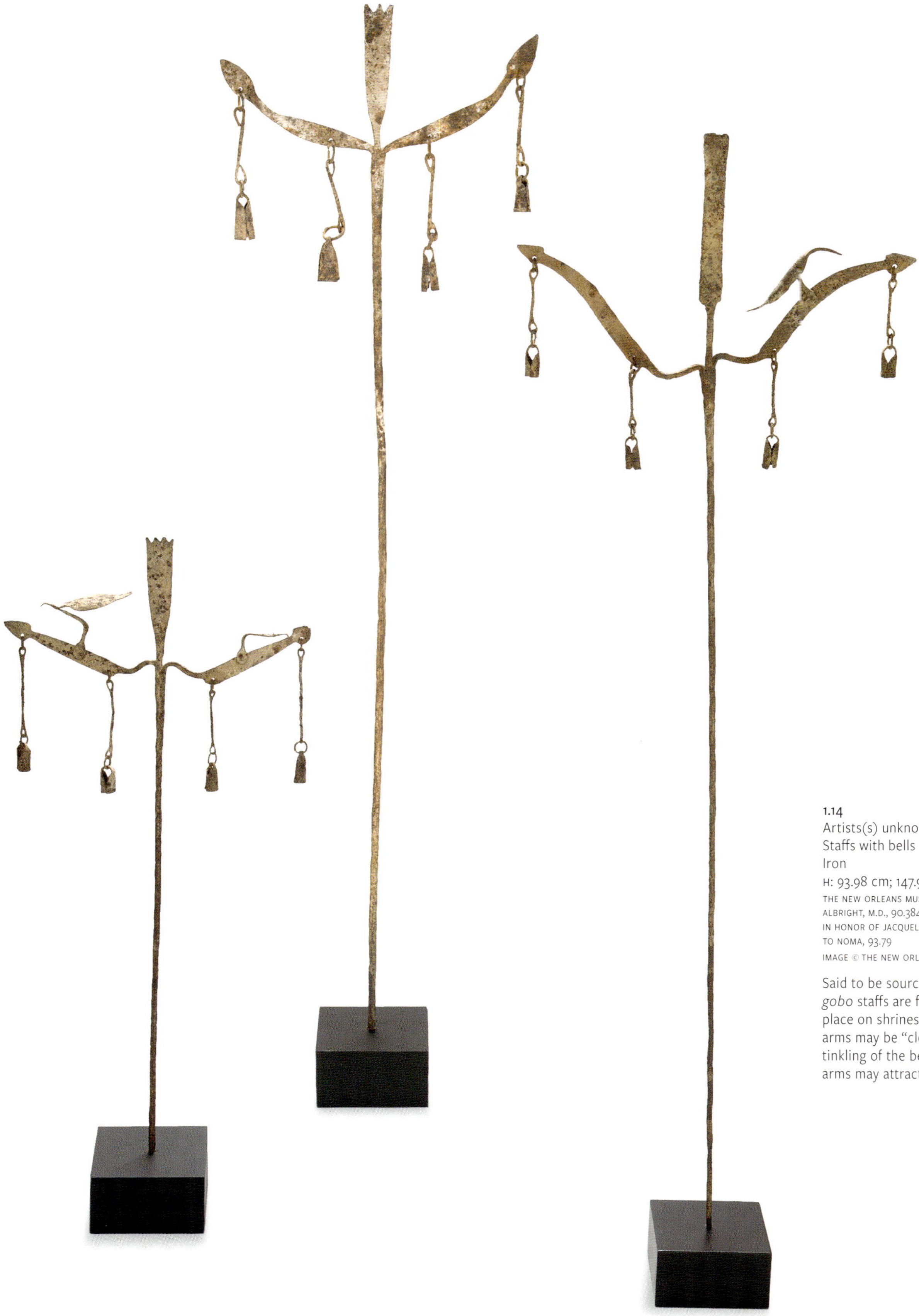

1.14
Artists(s) unknown (Dogon peoples, Mali)
Staffs with bells (*gobo*), 18th to 19th century
Iron
H: 93.98 cm; 147.95 cm; 144.78 cm
THE NEW ORLEANS MUSEUM OF ART: GIFT OF H. RUSSELL ALBRIGHT, M.D., 90.384, 90.385, GIFT OF E. JOHN BULLARD IN HONOR OF JACQUELINE SULLIVAN'S 20 YEARS OF SERVICE TO NOMA, 93.79
IMAGE © THE NEW ORLEANS MUSEUM OF ART

Said to be sources of growth and promise, *gobo* staffs are forged by blacksmiths to place on shrines for rain. Their uplifted arms may be "cloud hooks," while the tinkling of the bells that hang from their arms may attract spirits.

THE RED AND THE BLACK

Let us turn to cosmological explanations of iron and ironworking among two closely related Congolese peoples: Tabwa, living southwest of Lake Tanganyika in scattered, lineage-based communities sometimes composed of only a family or two, and to their west, Luba, who developed celebrated kingdoms along the Lualaba River. Distinct social organizations and the natural resources upon which they have been based have presented different intellectual and political opportunities and needs. Luba developed mythical charters for royal prerogatives unnecessary for most Tabwa, yet both Luba and Tabwa cosmologies are based upon the same logical structures and key metaphors, and ethnic identities often merge into each other in regions where they meet.[7]

Tabwa lands are marked by jutting peaks, ancient volcanism, ongoing seismic activity, and striking rock formations. The carrying capacity of soils is poor throughout the region, and population density relatively low. In the late nineteenth century, a European visitor found Tabwa lands to be "bestrewn with . . . countless small villages, each nearly independent from the others, and without political cohesion" (Storms ca. 1884). Toward the end of the same century, several Tabwa chiefs sought to consolidate their political authority through participation in the east African ivory-and-slave trade (see Roberts 2013a), but even during my research in the mid-1970s, most Tabwa communities remained modest in size and led by lineage-based chiefs.[8] All the more remarkable, then, that smelting and forging of iron should have been so highly developed among these same people in precolonial times, as was the intense regional trade so implied.

Several sources of high-grade iron ore exist in Tabwa lands, such as "a very important magnetite mine" on Mount Kalolo that assayed in Europe at 75 to 80 percent iron, placing the specimens "among the richest of known minerals" (Schmitz 1903, 40). The site and all related activities were the province of an *ngulu* earth spirit called Mother of the Furnace, to which an "ecological cult" was addressed that helped assure success in smelting. The *ngulu* inhered in two immense lodestones adjacent to each other on Mount Kalolo, one male, the other female. Should a person with evil intentions dare pass between them, his or her iron arms would be seized and held in place by the spirit.[9]

Hematite (Fe_2O_3) is far more common than magnetite (Fe_3O_4) and found "everywhere" in Tabwa lands (Colle 1903). Tabwa differentiated between the two, calling black magnetite *malamba* and reddish hematite *mutapo*.[10] They bore different symbolic associations important to Tabwa cosmology through their colors, for in this and other circumstances the "red" generative "heat" of women must be tempered by the cooling "black" qualities of men (Roberts 1993b; cf. Barndon 2012, 41, 43). Historically, when the two ores were smelted together (Stuer 1903), hematite was probably loaded into the furnace first to produce the slag necessary to consolidate bloomery iron, followed by richer magnetite (S. Terry Childs pers. comm., 1993). The remarkable result, in the estimation of an early European visitor, was "a soft steel comparable to that of Sweden" (Schmitz 1903, 40). Another noted that when polished, such metal had "the gleam of silver, and it rusts less quickly than that of Europe" (Morisseau 1910, 15). In 1879, Joseph Thomson opined that Tabwa hoes forged of such iron were "certainly not equaled in workmanship by any other tribe" that he had encountered in his central African wanderings (1968, II, 43).

In the late 1800s, a number of Tabwa villages were entirely populated by smelters, blacksmiths, and their families (IAAPB 1892, 63), and every sizeable community had several smiths working in open-sided buildings also used by men to gather and gossip and where adolescent boys spent their nights (Scheymans 1903). Such ateliers fostered humane endeavors "where labor and life mixed face-to-face" (Sennett 2008, 53) as Tabwa blacksmiths (*nsendwe*) forged hoes, axes, awls, gouges, hammer-anvils (*manyundo*), tongs, nails, knives, swords, arrow and spear points, bow-stands, anklets and bracelets (fig. 1.15), torques, razors, hairpins, beads, flasks for snuff, clapperless bells large and small, keys for plucked idiophones ("thumb pianos"), fire-strikers, pottery-scrapers, cigar-shaped ingots, and undoubtedly still other tools, weapons, gadgets, and religious paraphernalia.[11]

Ironworking was strikingly significant to community life, local political economy, and regional trade, while smelting and smithing technologies were grounded in Tabwa cosmology. Origin stories recorded at the turn of the twentieth century explain how, in the beginning, the world was dark, cold, and silent, and a place where only Mutumbi, an anthropomorphic aardvark, hunted with his dogs. One day, the pack pursued a cane rat down a seemingly endless tunnel. Mutumbi followed and suddenly encountered the realm of the Supreme Being, Leza, radiant with light. Mutumbi told Leza how lonely the earth's surface was, and Leza gave him a man, a woman, and a covered basket. Mutumbi led his charges back up the tunnel, and once they had arrived and felt the frigid air, the man made a fire. When the woman opened her basket so that the sun, moon, and stars could escape to provide diurnal light, Mutumbi fled forevermore—as aardvarks do when dawn approaches.[12]

1.15
Artist unknown (Tabwa peoples, Democratic Republic of the Congo)
Bracelet, 19th century
Iron
W: 1.4 cm
COLLECTION ROYAL MUSEUM OF CENTRAL AFRICA, 31813
(NOT IN EXHIBITION)
IMAGE © ROYAL MUSEUM FOR CENTRAL AFRICA, TERVUREN

Tabwa blacksmiths forged a great many different sorts of tools, weapons, accoutrements, and gadgets, and judging from those few known in museum collections, many were incised or stippled with a triangular motif called *balamwezi*, meaning "the rising of the new moon." Referring to hope and promise as key to Tabwa ontology, the same design enhanced all manner of material culture and body arts. Such a collective sensitivity realized across media recognized and imposed human order on nature.

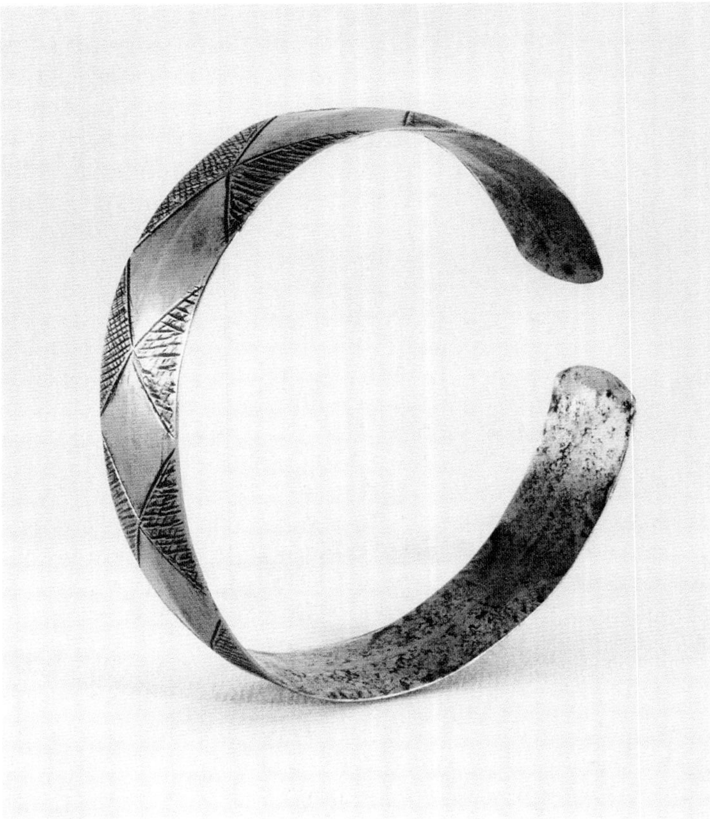

Few knew of Mutumbi by the time of my research among Tabwa in the mid-1970s, but the structure of such tales remained important. By then, a man named Kyomba had replaced Mutumbi in the paradigmatic role of culture hero. Being a named human protagonist, political use of origin stories became possible via a narrative idiom called *milandu*, through which the status of lineage chiefs, rights to natural resources, and other prerogatives were argued during disputes about relative seniority (Roberts 2013a, 50–75). One of the basic purposes of *milandu* is to establish political hierarchy with reference to descent from Kyomba. Associated with the moon and understood to have had exceptionally dark skin, Kyomba is endowed with the profundities and secrets of that which is "black," following Tabwa color symbolism. He is also "father of all clans," and because Tabwa observe matrilineal descent, his wives are the founding ancestresses through whom kinship is traced (fig. 1.16). A Tabwa elder named Kizumina Kabulo—a "man of memory" able to recount *milandu* histories of lineages and much sought out by local people to talk about "the old days"—told me that Kyomba led migrations of various Tabwa clans to the lands they presently occupy, introduced "all symbols and modes of chiefship," brought fire to humanity, and carried the seeds of agriculture in his complex coiffure.[13]

Kyomba was also the first to smelt iron and forge it into useful implements. Proving this relationship, I was told, was the use of the word *kyomba* to refer to a Tabwa smelting furnace.[14] *Nungu* and *lungu*, as more common terms for the same structures, are from the Bantu root *ung*, "to join together," as when ores coalesce to form a *butale*, "bloom." *Ung* may also refer to sexual union or more generally mean "to create." The Congolese comparative linguist Clémentine Faïk-Nzuji has studied the root in thirty Central Bantu languages and summarizes its sense as relating to "mysterious and unfathomable realities found in the permanent aspirations of man without being directly accessible to him, except through the intermediary of symbols" (1992, 47–50). Such references through words for "furnace" suggest the level of conceptual and spiritual power necessary to assure transformation of nature's raw matter into critically useful tools, weapons, and articles of value.

1.16
Artist unknown (Tabwa peoples, Democratic Republic of the Congo)
Ancestor figures, two females and one male in Manda style, ca. 1920
Wood
H: 26.7 cm
UNIVERSITY OF MICHIGAN MUSEUM OF ART, MUSEUM PURCHASE MADE POSSIBLE BY A GIFT FROM HELMUT STERN, 1987/1.157.1–3 (NOT IN EXHIBITION)
PHOTOGRAPH BY CHARLIE EDWARDS, 2018. © UNIVERSITY OF MICHIGAN MUSEUM OF ART

This trio of Tabwa ancestral figures—a male, in the center, and two females—once graced the shrine in a small outbuilding of a Tabwa lineage elder. They may be a presentation of the perplexities of matrilineal descent, through which a man's loyalties are divided between the children of his sisters, whom he loves, teaches, disciplines, and sees as his descendants, and the children of his wives, with whom he shares warm home life and yet who look to their own maternal uncles for social identity.

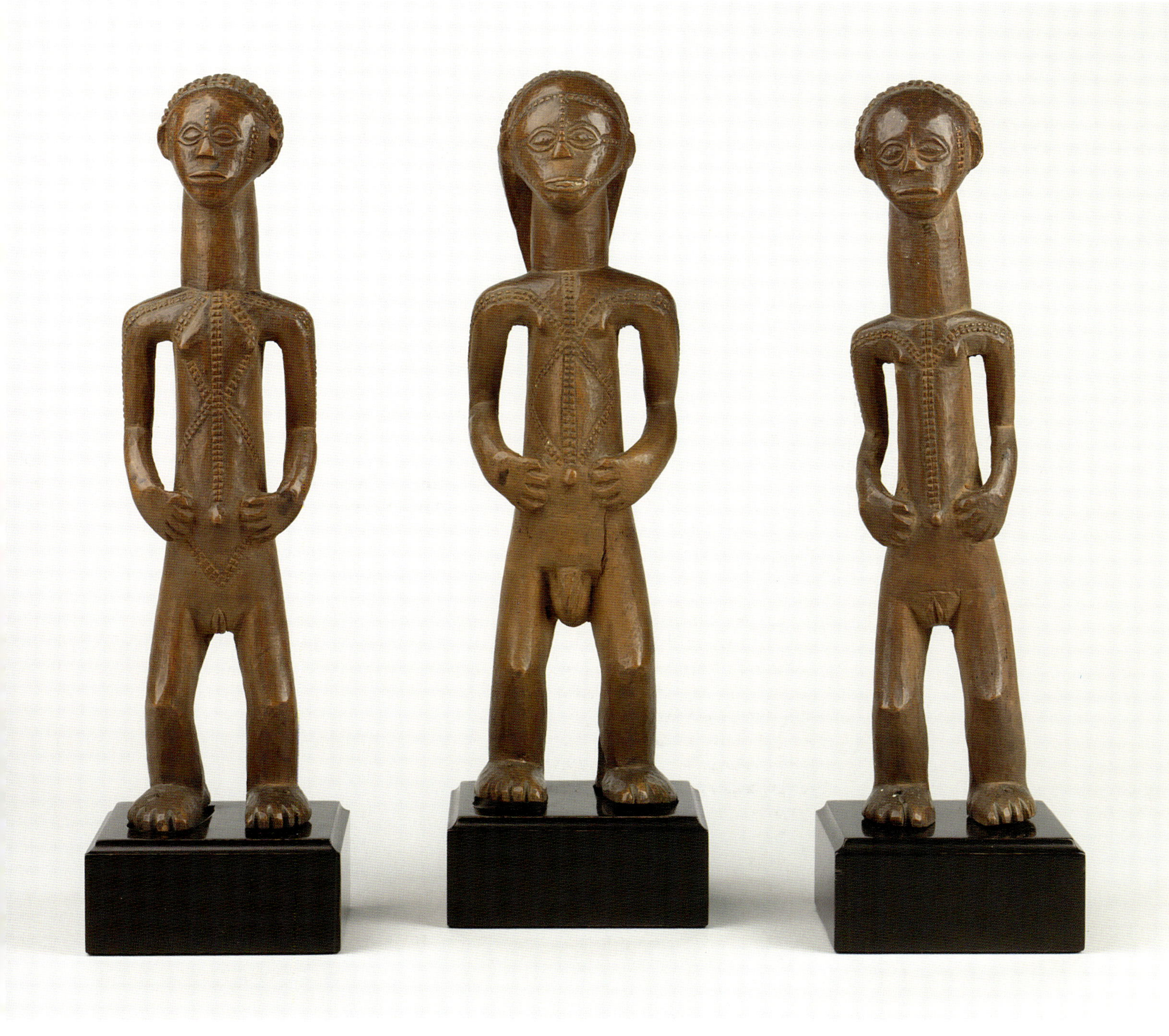

1.17
Artist unknown (Luba peoples, Democratic Republic of the Congo)
Ceremonial adze, 19th century
Wood, iron
H: 34.3 cm
COLLECTION NEUBERGER MUSEUM OF ART, PURCHASE COLLEGE, STATE UNIVERSITY OF NEW YORK, 1999.06.112; GIFT OF LAWRENCE GUSSMAN IN MEMORY OF DR. ALBERT SCHWEITZER
IMAGE COURTESY OF THE METROPOLITAN MUSEUM OF ART, NEW YORK

The spirit heads on either end of the handle of this adze bear miniature hammer-anvil-shaped pins in their coiffures, "closing" esoteric knowledge and purpose from aggressive intervention. They also served as points of creative transformation, for like the hammer-anvils that Luba blacksmiths once used, they brought about social change. The heads look in opposite directions so that the dignitary who wore the adze over his shoulder or danced with it in hand was graced with ancestral wisdom and foresight.

The elderly men with whom I conferred about precolonial ironworking called a master smelter an *ngezia* or a *sisinungu*, "father of the furnace," and understood him to control invisible forces and generative processes that Tabwa associate with women's fertility and procreation. Smelting iron was like giving birth and a furnace like a woman's body, cared for and promoted by master smelters and their assistants (as was also true among Luba [Burton 1961, 119]). The *ez* root of *ngezia* refers to the strength and power to do, make, or accomplish, and appears in the most common praise name for Leza as the Supreme Being. Accordingly, an *ngezia* was the embodiment of Kyomba, "father of all clans," the first to till the ground with the tools of iron he had smelted and forged, and the first to plant the seeds he carried in his elaborate coiffure (Colle 1913, 519, 203).

In many of these traits, Kyomba is the structural equivalent of the Luba culture hero Mbidi Kiluwe. Indeed, the two are explicitly equated in certain narratives, suggesting shared culture and imbricated political histories. Mbidi was a quiet, self-possessed, handsome man with glowing dark skin. His secondary name, Kiluwe, places him in a genre of central African heroes linked with the moon, rain, fertility, and arcana within the symbolic category "black," as is Kyomba for Tabwa. Coming from the east (where Tabwa now live), Mbidi encountered the primal ruler Nkongolo Mwamba. As avatar of excess given to all manner of debauchery and deceit, Nkongolo is a member of a central African genre of immense serpents sometimes but not always personified. They are associated with the sun, breathe forth the rainbow, and cause desiccation, the dry season, sterility, and other violent transformations associated with "red" (Heusch 1991; Roberts 1991, 1993).

Mbidi and Nkongolo's sister bore a son together named Kalala Ilunga, who, after epic struggles, vanquished Nkongolo to become the model of and for sophisticated, assertive royal behavior. As the Luba social theorist V. Y. Mudimbe has noted, however, Kalala possessed the "conflictual inheritances" of the rough-hewn, "red" transformative energies of Nkongolo as well as the "black" esoterica of Mbidi. These dialectical capacities have been "rearticulated and recomposed" over generations, "realizing and assuming . . . the opposing expectation of each social formation" as the ambiguous basis of local-level politics (1996, 245–47). Yet they are, as well, the essence of iron as material, of symbolic aspects of ironworking technologies, and of the efficacies of forged, double-edged implements. For example, hematite and magnetite were mixed in Tabwa furnaces not only for their metallurgical properties but also because tension between "red" and "black" forces was understood by master smelters to be the essence of creative potential (see Roberts 2012).

Kalala Ilunga is praised by Luba as "the one who forges suns." Significantly, when Luba kings are enthroned in Kalala Ilunga's image, they are "forged" through a ritual process called "the striking of the anvil" (*ku komena nyundo*). Their knees are "hammered" with dignitaries' fists as though with smiths' hammer-anvils (see interleaf E and chapter 15, this volume), thus transforming mortals into sacred beings who transcend ordinary human limitations and factional interests. Harold Womersley, an evangelical missionary in the region during the 1920s and 30s, offered a detailed account of how, as the king-to-be's knees were "hammered," he was reminded that his "'forefather Kalala introduced ironworking to this land. He was a wise man. Whether for weapons of war or tools of peace, whether for arrows and spears or for axes and hoes, the anvil is the secret of power and progress. Remember your people,'" the king was exhorted, and "'do not be satisfied merely to take their tribute, [but instead] give them your wisdom and your protection and the success of your kingdom will be assured.'"[15] Material and performance arts further attested to the centrality of ironworking to symbolic life, as when a ceremonial adze with miniature hammer-anvils "closing" the heads of the spirit figures at either end of its handle was danced in a majestic choreography to reenact epic tales of Kalala and Nkongolo (fig. 1.17).

Again, Tabwa did not develop complex kingdoms as did their Luba neighbors. It seems they had no need to differentiate between Mbidi and Kalala as the archetypal father and son of a society observing patrilineal descent; instead, Kyomba subsumed both roles. Tabwa do have their equivalent of Nkongolo, called Nfwimina, but rather than being understood as a person, Nwifimina is an immense solar serpent whose rainbow breath, emitted from the deep recesses of termite mounds, is toxic and brings dire illness to those encountering it. Tabwa say that men sometimes suffered from Nfwimina's noxious fumes during smelting processes, and today they may encounter them while working in deep-shaft mines of the Copperbelt (cf. Blakely 2006, 156). Indeed, Nfwimina is "the king of the below," I was told by a Tabwa healer, and gold and other precious minerals mined with such avidity are Nfwimina's "excrement," for "isn't Nfwimina the mother of wealth?"[16]

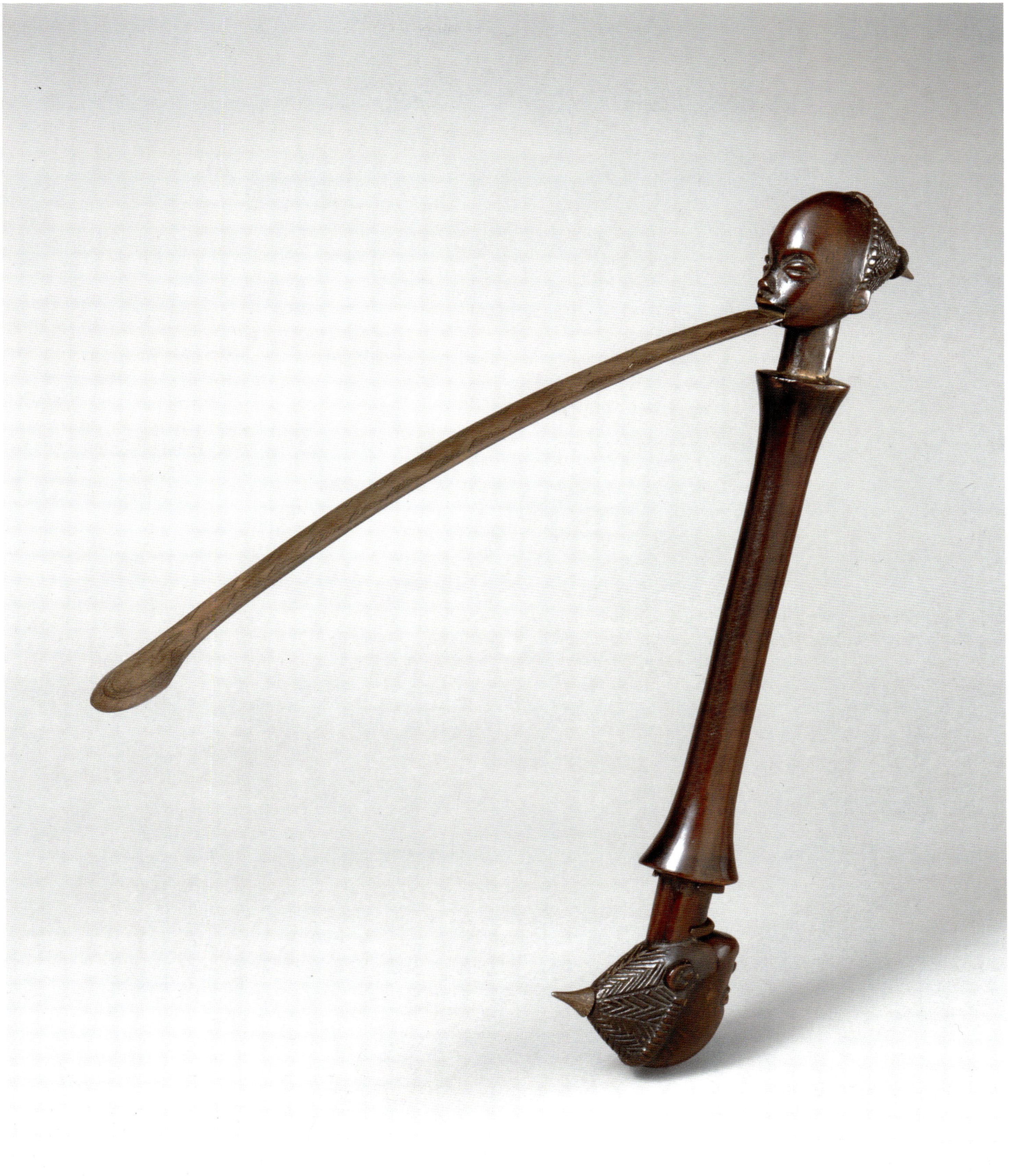

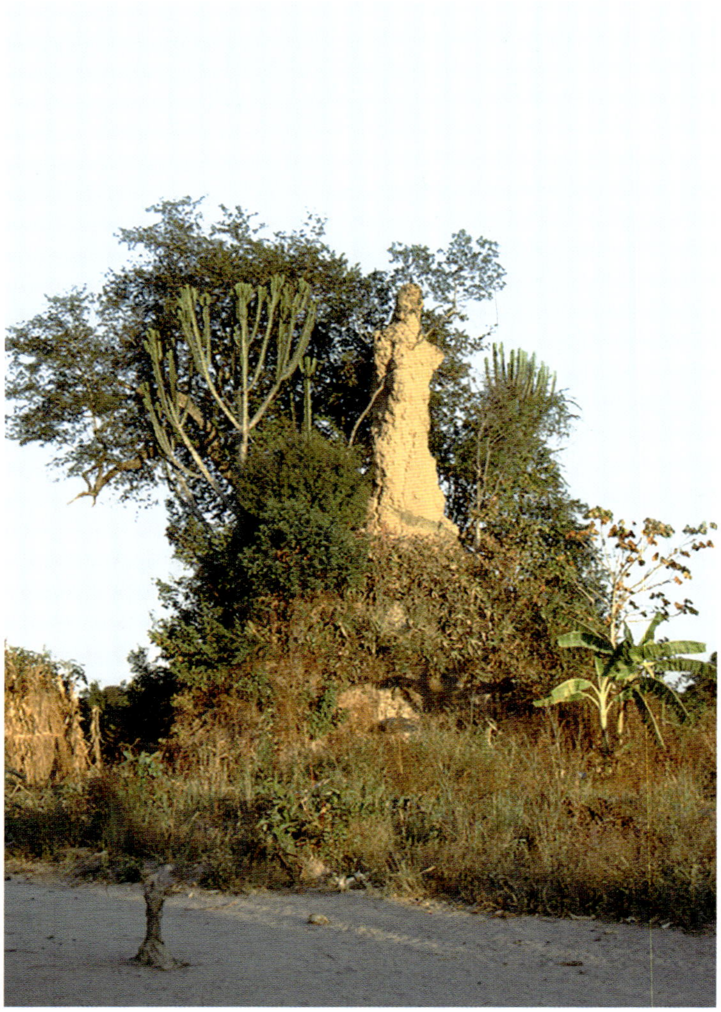

1.18 The vegetation around this tall termitary is especially luxuriant because of nitrogen produced by termites. From such plants Tabwa healers confect fertility medicines (*viabiko*), once essential to iron smelting.
PHOTOGRAPH BY MANUEL JORDÁN IN CHITOFU, ZAMBIA, 1992.

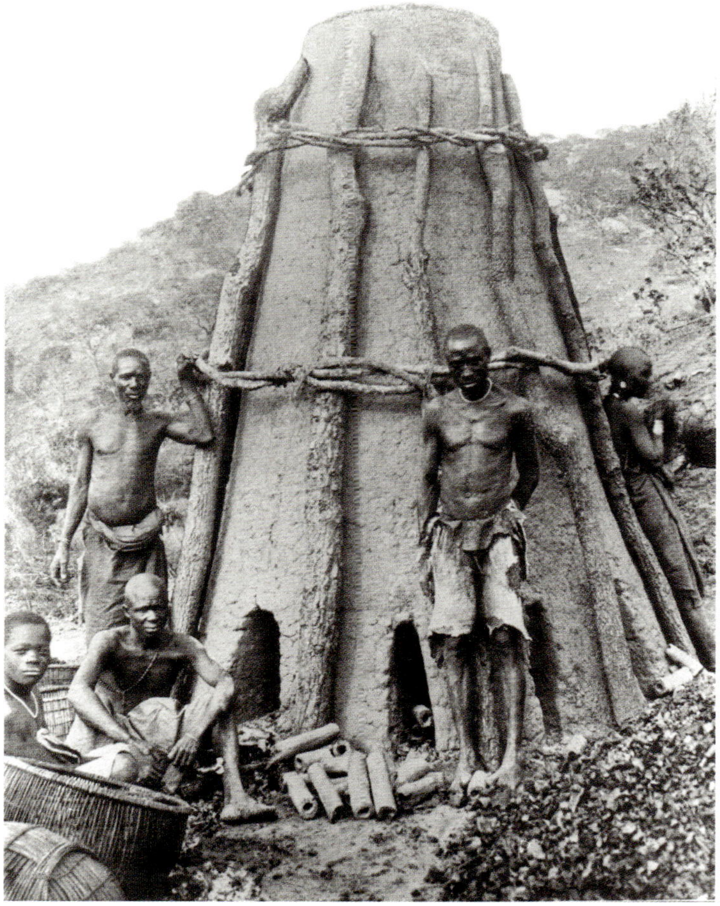

1.19 Natural-draft furnace of a model shared by Tabwa and Fipa peoples living along southwestern and southeastern shores of Lake Tanganyika. They are closely related through political, religious, and technological histories, and both developed remarkable iron industries, although to date, archaeology has only been undertaken among Fipa.
PHOTOGRAPH PUBLISHED IN ADOLPHE LECHAPTOIS, *AUX RIVES DU TANGANYIKA* (MAISON-CARRÉE, ALGIERS: PÈRES BLANCS, 1913).

Nfwimina is also associated with termitaries, from which Tabwa smelters sought the red soil they used to sculpt their several sorts of furnace (figs. 1.18, 1.19), with the largest standing two-to-three-meters tall and the smallest for refining bloomery iron (Childs 1991). The unusual, termite-assisted chemical characteristics of such earth resists cracking in intense heat (Davison and Mosley 1988, 75), but the same substances and their origin also realize symbolic purposes. Nfwimina's refuge lies in the depths of what is understood as a female earth: Tabwa say that "one's whole body is earth" (*mwili yote ni bulongo*), and within a woman's abdomen Nfwimina has its counterpart in a serpent-like creature called Kisumi that is responsible for fertility. Menstruation, called "the little dry season" among other euphemisms, is caused by Kisumi's biting a woman, just as Nfwimina's rainbow breath brings forth the dry season in nature; and Kisumi capriciously decides whether or not to retain a husband's "water" (*maji* in Swahili) imparted through coitus, thus permitting or denying conception (Roberts 1980, 254–57).

Tabwa deploy important herbal and symbolic medicines (*viabiko*) to regulate human fertility and birth, specifically by interacting with and calming Kisumi. The most potent *viabiko* to treat infertility or complications of birth are composed of tree leaves, bark, and roots gathered atop termitaries because of their exceptional potency—that scientists have attributed to nitrogen produced by termites (see Pevenasse 1920). Among their symbolic activating agents (*vizimba*), *viabiko* may include bloomery iron from earlier days. Conversely, medicines placed in the floors of Tabwa smelting furnaces included ingredients of *viabiko* so that iron would coalesce in a process referred to as *kupatana* or *kukamatana*—verbs that describe fetus-formation from female blood and male "water." Significantly, among the most potent of the *vizimba* activating agents for smelting was a relic of Nfwimina to assure "conception" and produce a bloom, as was found among Fipa smelters as well. Mastering so powerful an element, in combination with others chosen to create a "little world" (*kadunia*) of protection within and around the furnace, "closed" the smelting from the meddling of evil adversaries (*balozi*) seeking to cause accidents or the failure of the bloom in order to bring injury, economic disappointment, and loss of face to the *ngezia* and his assistants.[17]

Different medicinal bundles were placed at the openings at the base of the tall *nungu* natural-draft furnace, with several ingredients recalled in the 1970s because they were still used to increase a lover's ardor. An *ngezia* and his assistants abstained from sexual activities while smelting, lest the iron fail to coalesce into a successful bloom; instead, the furnace was their collective "spouse," and a "mother of twins" at that, with reference to women whom Tabwa celebrate for their unusual fecundity (Roberts 2010). Might the thinking have been, as among Chishinga (Brelsford 1949, 28), that because the furnace was their "spouse," coitus would be "adultery" and so bring failure of the bloom as illicit sex does when one's wife is pregnant? As among Fipa and neighboring Pangwa (Barndon 2012), might protecting the welfare of a "pregnant" furnace and that of a spouse have been equated in moral reflection?

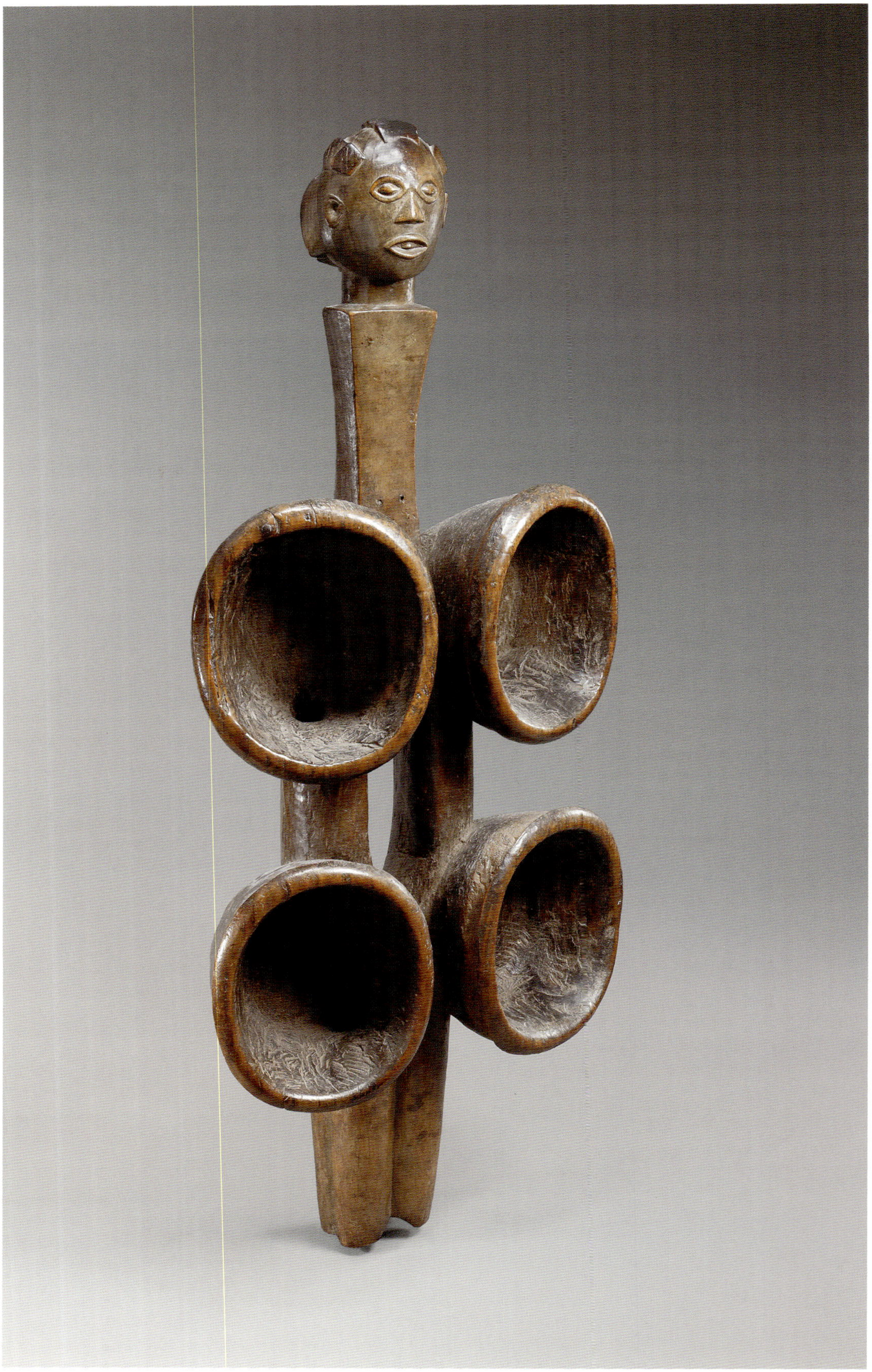

1.20
Artist unknown (Luba peoples, Democratic Republic of the Congo)
Bellows, 19th century
Wood
H: 106.6 cm
BERNARD DULON GALLERY
PHOTOGRAPH BY VINCENT GIRIER DUFOURNIER, 2012.

Like any bellows used in smelting or smithing iron, this one would have been fitted with leather sacks attached to its chambers that were either pumped by grasping the bags directly or, more probably, by thrusting sticks attached to the sacks up and down, in this case as managed by two assistants. The nozzle of the bellows would have been fitted into a tuyère of fired clay, and the latter was then inserted into a smelting furnace or the edge of a forge fire to provide oxygen and intensify the heat. The head of a spirit appears on this bellows to oversee and assure the safety of the work, her tongue protruding slightly as a complex sign of sexuality.

Apprentice smelters (*baswaswaiba*, an onomatopoeia) pumped double-chamber wooden bellows inserted into four entrances understood to provide cardinal winds to the transformative process within the furnace.[18] Such "conscious naming of furnace attributes constituted a complex message to the spirit world and to the non-smelting members of the culture" (paraphrasing Childs 1991, 347). Like bellows carved by artists among other central African peoples (see figs. 1.20, 4.26, 4.27), some Tabwa examples have spirit faces in the saddle between the pumping "pots." More frequently, subtle allusion was made to anatomical aesthetics through a raised, serrated line that referred to scarification along the human body midline, as commonly practiced in earlier times to highlight the balance of right/left and other dichotomous ontological principles (see Roberts 1988c). Bellows provided a phallic complement to the gynecomorphic furnace, permitting "conception" and "birth" of a bloom to be realized (cf. Jacobson-Widding 1990, 68; Lanning 1954).

Medicines were not the only elements instrumental to the "enchanted technologies" of iron smelting.[19] An *ngezia* led his men through elaborate baths of medicinal smoke and oversaw ritual aspects of all phases of iron production including imposition of prohibitions (*vizila*) and curfews. Once the furnace was charged with wood and ore and then fired, an *ngezia* tied a skirt of symbolically significant animal pelts around his waist that was further empowered with medicinal bundles (*erizi*) to bring their efficacies to bear upon the event. Next, he danced around the structure and sang songs, quite possibly celebrating the mothers of twins (Roberts 2010). As he did, he rhythmically tapped the tuyères with an iron staff to judge the progress of smelting, but in a rhythm accompanied by *baswaswaiba* working their bellows and singing refrains, as "the two ritual modalities—dance and medicine—articulate[d] the metonymy between birth and metallurgy" (Blakely 2006, 78). "Smelting magic" was clearly "a technology in its own right," then, and following such logic, the smelting furnace was an "actor" with its own efficacies to achieve practical and symbolic tasks (Échard 1983, 220–21). Indeed, following such logic, a Tabwa furnace as an empowered and empowering device was an *nkisi* (see chapter 15 and interleaf K, this volume).

An *ngezia's* performance would have been a "danced causality," to borrow an apt phrase from Aby Warburg, for without such practices undertaken carefully and correctly, how could one be sure the process would come to fruition, and safely so?[20] As R. Wise wrote of Fipa people, one must recognize "the overwhelming importance of the rituals, magical and ceremonial, without which the kiln [furnace] and all the smith's [and/or smelter's] work would fail. For the successful working of the kiln, the living and the dead, the spirits of the air, wood and rock must work together" (1958a, 232). "Ethical behavior was implicated in the technical work" of an *ngezia* (Sennett 2008, 64), and any "attempt to consider the smelter's magic separately from his technology ignores the fact that in the indigenous perspective, these two are inextricably combined" (Blakely 2006, 69). As Merwe and Avery would add, materialist authors who dismiss the "dreamed technologies" of ironworking, as per this chapter's epigraph, are "simply missing the point of African ideologies" (1987, 144, slightly paraphrased).

When smelting was successfully completed, the men returned to their homes in triumph. An elderly *ngezia* nicknamed Kapalamoto remembered a Tabwa song sung then: "I can even sleep by a termitary [furnace], and those passing will dust me with white flour." The *ngezia* had assured the moral behavior and purpose of his men, who had abstained from coitus and conducted themselves appropriately while working their furnace. They were bringing wealth and prosperity home with them as bloomery iron, and they were to be blessed and cheered with white powder in acknowledgment of their debt to ancestors, earth spirits, and other spiritual forces. The same song accompanied an *ngezia*'s funeral, Kapalamoto added—leaving one to hope that it did upon his own passing soon after I met him in 1977.

In sum, cosmologies and the *performance* of technologies—both practical and mystical—have long informed and confirmed one another among various sub-Saharan peoples. Ironworking in particular has been anchored in systems of thought and morality directed toward mundane and lofty purposes (Barndon 2012), and such poetic senses continue to empower people as they contribute mightily to the meaning and value of forged iron implements today.

CHAPTER
2

SUPERTRANSFORMERS: IRON SMELTING AND GENDER

Candice Goucher

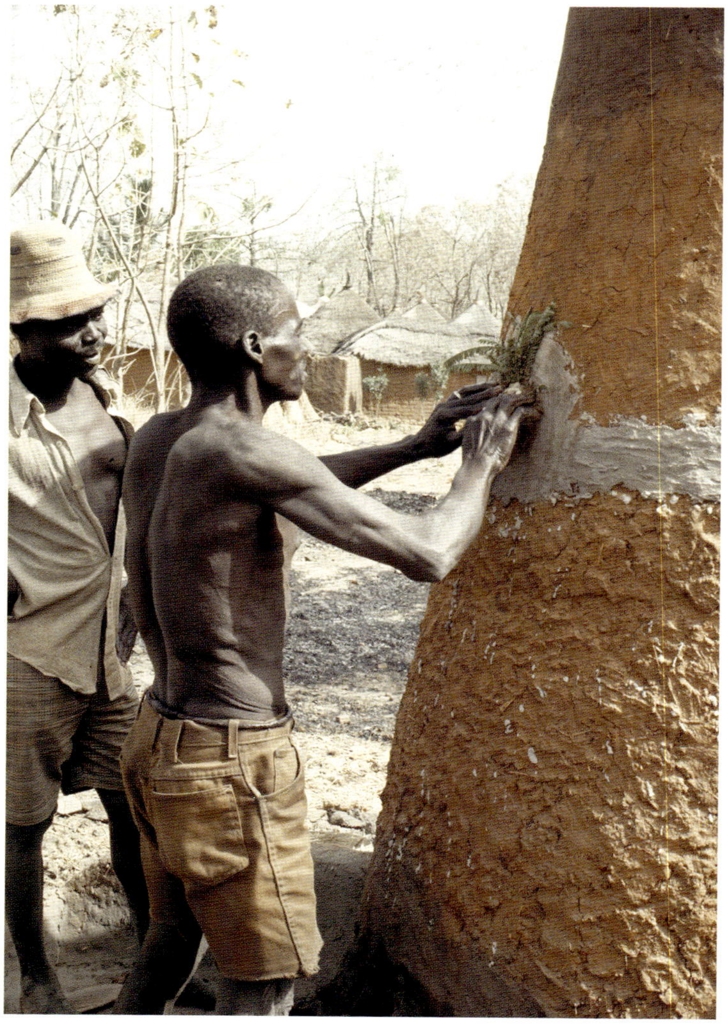

Give birth to good iron. Give birth to good iron.

BELAM DIYAMBO, FURNACE INVOCATION AT NANGBANI, TOGO, 1985

The blood of the wolves, mingled together...
The tools of the workman...

HYMN FROM ERETRIA, CA. FOURTH CENTURY BCE (BLAKELY 2006, 85)

The master smelter Belam Diyambo stood in the center of a sacred space specially marked on red earth. The circle would bring to life a furnace newly constructed, in 1985, for the first attempt to smelt iron in Banjeli, Togo, in more than fifty years (fig. 2.1). Belam Diyambo called to the ancestors and spirits to come build the furnace and make "good iron." Embedded in the walls of the new furnace would be pieces of an earlier one, insistently symbolic of the genealogical memory of iron technology reaching back more than eight centuries in Belam Diyambo's family history. Turning rocks dug from the earth into useable metal relied on technical expertise passed down from one generation to the next, as well as the spiritual blessings of the ancients.[1] Through ritual, the furnace would become a woman, who would "breathe," "eat," and "give birth" to iron. Transmission of knowledge benefited from the repetition that remembered rituals evoked. At the heart of every transformation from ore to metal rested the very definition of "civilization" in the age of iron and steel. Successful smelting of iron enabled yam mounds to be dug with iron hoes and the weapons of armies to succeed in battle. Nothing was more critical or more enduring than the making of iron.

2.1 Belam Diyambo, master smelter, prepares to feed the furnace at Banjeli, Togo.
PHOTOGRAPH BY CANDICE GOUCHER, 1985.

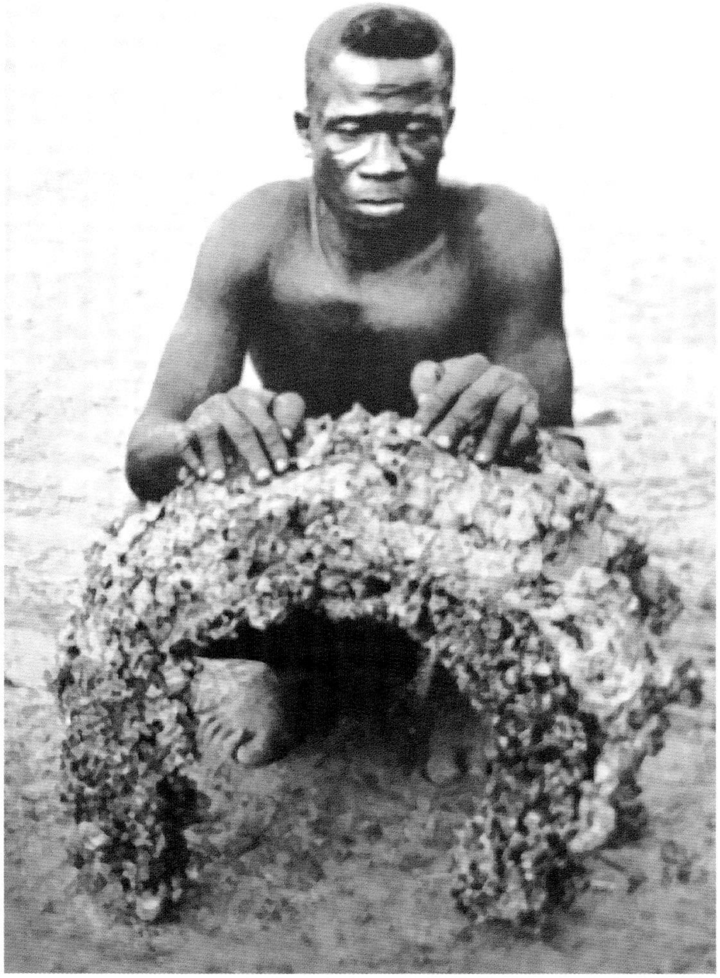

2.2 After tapping the slag containing unwanted impurities from the ore, a smelter holds the iron bloom produced in his smelting furnace in Banjeli.
STILL FROM THE FILM *IM DEUTSCHEN SUDAN* (1914), BY HANS SCHOMBURGK.

BEYOND RITUAL PRACTICE

The metallurgy of iron needed more than a scaffolding of beliefs and a framework of rituals. The African Iron Age, constituting one of the world's oldest metallurgies, was built on the elaboration and expansion of smelting technologies, the complex processes of chemical alteration required to extract workable metal from rocky iron ores found in the earth and necessary to separate the metal from unwanted waste hidden in the ores. Unlike the characteristic green of malachite or the other flamboyant colors of copper deposits, iron ores usually present relatively few visual clues to their ultimate metallic content and potential. Both observational sciences and experimentation with high-temperature processes would have been necessary to obtain high-quality ores from common iron compounds, unlike copper, which was sometimes cold-forged. Exceptionally, the rare *ba-en-pet* (literally, "metal of heaven") iron was directly harvested from meteors by ancient Egyptians. The source's pure iron and high nickel and cobalt content as well as its qualities of trace elements and distinctive crystalline structure permitted its working without smelting first. The nine meteoric beads that were skillfully heated and hammered by Egyptian artisans some time around 3300 BCE were discovered at the site of el-Gerzeh in Lower Egypt, and have been dated much earlier than the iron smelting technology the Egyptians eventually adopted around the sixth century BCE (Rehren et al. 2013). They are humankind's oldest iron artifacts, but the iron had arrived from the heavens essentially ready for the smith's hammer.

According to historical linguists and archaeologists, the African Iron Age likely began outside of the Nile Valley well before Egyptian ironworking technology. Some archaeologists are pushing back the beginning of African iron production to at least the middle of the second millennium BCE (Alpern 2005; see chapter 8, this volume). While dating the origins of ironworking has proven to be difficult and contentious, documenting its determined spread across the African continent during and after the first millennium BCE remains the subject of ongoing research by archaeologists, historians, ethnographers, and linguists. The African continent's wealth of ethnographic and ethnoarchaeological information has been gleaned from interviews with practitioners and observations of living smelting contexts. Such recent studies have helped inform the understanding of diverse technologies around the world, including those of ancient Greece (Blakely 2006). This chapter's epigraph from the ancient Greek city of Eretria signals the widespread pattern of sexuality and ritual transformation embedded in ferrous metallurgies and the victories of male over female. Among other factors, living traditions have enabled researchers to explore human dimensions of technology and change.

African smelting sites offer abundant archaeological residue; they are important sites of memory, having enabled the expansion of subsistence strategies that supported larger polities and construction of new social and political identities. Across the continent, blacksmiths obtained their workable iron from smelted "blooms" (the spongy mass of iron particles, separated from the flow of unwanted "slag") that could be hammered and shaped into tools, weapons, and other items (fig. 2.2). Iron blooms, iron templates, and iron currencies became efficient means for the valuation of iron, facilitating its movement through interregional trading networks. Currency forms sometimes mimicked tool shapes and assumed a value against which other goods could be measured (see chapter 5, in this volume). Seeking regular supplies of iron for village blacksmiths meant gaining access to techniques and knowledge required for highly specialized smelting furnaces as well as long-distance trade in their products.

In archaeological contexts, the heaps of unwanted slag left behind with debris from abandoned furnaces attest to the achievements of African smelting technologies—often more so than the metals produced. Iron could be cast (that is, a high-carbon product retrieved directly from the furnace), as in the down-draft bloomery of Mafa of Cameroon. At Banjeli, iron in this state was subsequently decarburized by a crucible step and returned to a product capable of being forged. Typically, the African preference was for iron to be forged rather than cast. The flow slag tapped from some furnaces—as at Bassari, Togo, or in the bowl furnaces of Lopanzo, Democratic Republic of the Congo—during smelting processes contained more than unwanted minerals from the original ore. Slags are well-preserved witnesses to the smelting processes that took place centuries before.[2] Their study can reveal ore quality and source, temperatures, and other conditions inside furnaces, and even some of the technical decisions made by expert smelters working at furnace sites. While archaeologists have debated the likelihood of an independent invention of iron smelting technology in sub-Saharan Africa, the abundance and variety of available iron ore amidst a staggering variability of technological features (from bellows types to furnace shapes) testifies to the remarkable range of iron technological achievements and constant innovation in the African past regardless of origins.

ECOLOGICAL PARAMETERS: TRANSFORMING AFRICAN LANDSCAPES

To achieve the goal of producing metallic iron, smelting required construction of furnaces and the achievement of reducing conditions (expert control over air flow) under extremely high temperatures. Based on the analysis of slag, it is estimated that optimal temperatures as high as 1100–1250 degrees Celsius were reached regularly, with failed smelts being rare. Smelting husbanded massive amounts of natural resources: ores mined from the earth; charcoal fuel distilled from special hardwoods; limestone used as fluxes to remove unwanted matter from ore; and clay and other materials for the construction of crucibles, tuyères (clay pipes), and the various furnace types (from lined, shallow bowl pits to tall, upright structures). Critical to achieving the necessary high temperatures was the selection of relatively slow-growing tree species that could provide dense wood, often with high silica and alkali contents, for the distillation of long-burning charcoal.[3] Preference for trees with challengingly lengthy life cycles resulted in gradual deforestation everywhere that smelting took place, in turn creating landscapes not easy to defend against conquest by invading armies since the grasses that tended to replace forests permitted increased sightlines and easier access by raiders. Thus, strategic control over iron and resources figured mightily in the political organization of early African communities.

Smelting required both skilled and unskilled labor for mining and haulage of ore, clay, fluxes, and wood for charcoal. Knowledge of the technical parameters of the smelting processes often remained proprietary knowledge, controlled by hereditary membership in guilds, clans, and lineages as well as by restrictions on marriage. Further, smelting was a dangerous activity, owing to its use of intense heat and to its environmental impact. Large-scale industries not only systematically deforested many vulnerable parts of the African continent, they also polluted its landscapes: as the scale of iron smelting increased, dense concentrations of toxic debris spewed forth from hundreds of tall furnaces. In the nineteenth century, European travelers described polluted air, toxic soils, degraded landscapes, and stunted vegetation (dust-covered leaves were preventing photosynthesis). Given their potential for environmental devastation, it is little wonder that smelters were viewed with some degree of suspicion and fear.

Smelters' arcane knowledge unlocked the secrets of metal sealed inside of seemingly ordinary rocks, not in the least bit shiny. Secrecy and hidden meanings and metaphors served more than a proprietary purpose of protecting valuable knowledge. The transcendent metaphors moved societies to focus on the meaning of the creative act that gave birth to iron. As Sandra Blakely has noted, given the diversity of African smelting, a surprisingly singular "African system" that holds transformation to be the "dominant metaphor that joins smelting and gestation" finds wide distribution (2009, 229). Such transformative knowledge enabled smelters to hold positions as healers, diviners, and ritual specialists, from Mande and Mandara to Haya and Barongo peoples.

GENDER AND IRON SMELTING

Smelting iron was, as a rule, everywhere a male activity. Men, as ironworkers, would have smelted the raw ores and worked the blooms produced from smelting furnaces—sometimes performing both tasks. Though occasionally groups relegated bellows operation and ore preparation to women and children, or women might have assisted in the exterior construction of the furnace, they are not

2.3
Artist unknown (Guro peoples, Côte d'Ivoire)
Bar currencies (*bro*), late 19th to early 20th century
Iron
H (each bar): approx. 27 cm
PRIVATE COLLECTION

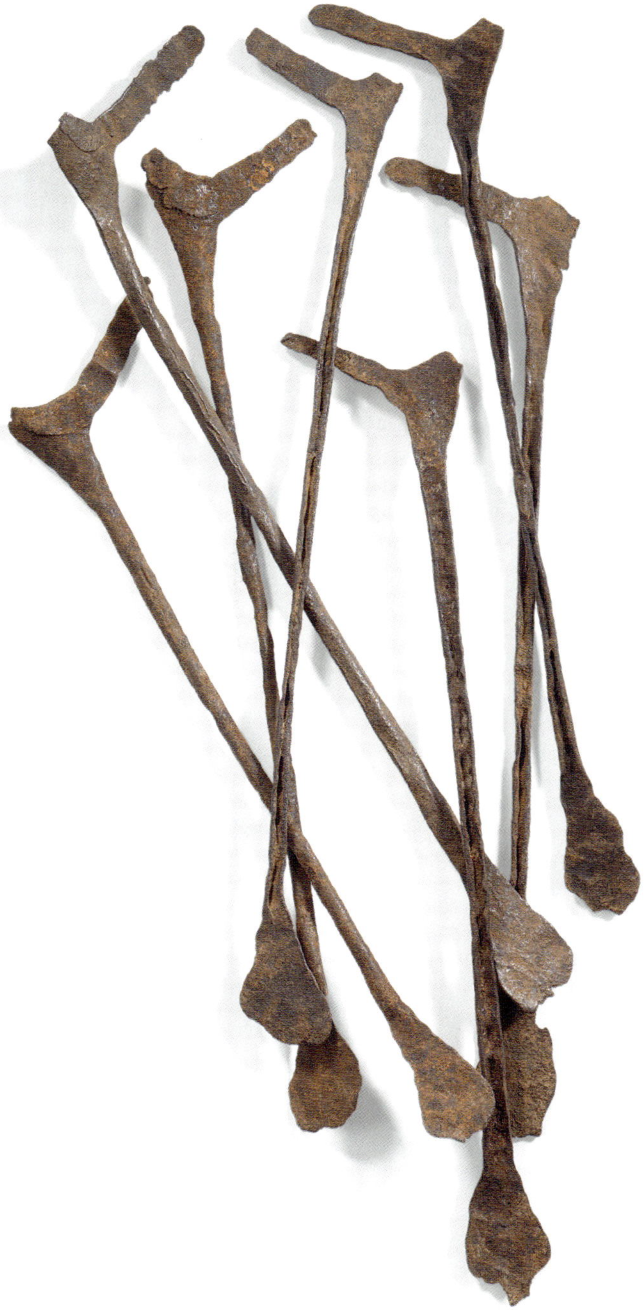

known to have come into contact with the interior spaces and the "enchanted technologies" executed there (Gell 1992). While technologies produced social outcomes, technical processes also had the power to shape how the world was seen. Metallurgies reinforced the control of men over both the technical and symbolic aspects of smelting. Women, particularly those of childbearing age, were systematically excluded from any proximity to the processes of furnace operation, including its rituals. Not only was the expertise and labor used for producing iron restricted by gender, but iron's transformative metaphors across the African continent were also highly gendered, empowering reflections on the understanding of spiritual, material, and political realms of existence. Louise Iles has reported the use of vastly different technologies in relative proximity, including the combining of two ingredients, one deemed male, the other female—perhaps a hard, black manganiferous flux combined in an "embrace" with the reddish, soft hematite (2017).[4]

Hammering blooms removed impurities left behind by smelting and provided blacksmiths with iron to forge into small metal trade bars or hoe-shaped templates for subsequent crafting of tools and weapons on anvils of iron and stone (figs. 2.3–2.5). Bundles of iron bars and miniature and minimally shaped objects took on symbolic importance, suggesting metaphors of the fertile womb of the furnace when given as bridewealth (Guyer 1985, 15–16) and as "material praise" for women's work (see chapter 5, this volume). Iron bars carried sacred connotations direct from the mouth of the furnace, as a reference to a sacred source of knowledge. Among Owerri Igbo smiths, the earth deity Ani used an iron bar as a footrest. Smelted iron fed a range of secondary industries. Products of the smelting furnace were conveyed directly to neighboring smiths, whose specialized knowledge and skills permitted them to create useable metals *and* metaphors. In this way, the working of iron empowered both those who toiled at smelting furnaces and those who forged.

Technological and ritual practices were not distinct, and both combined to bring esteem to their practitioners. The complexity of the technical parameters needed to change ores to metal helped assure master smelters positions of power and authority in many African societies. The meanings assigned to the smelting performance help explain its power as a supermetaphor for transformation itself. After all, the skills and metallurgical expertise of the iron smelter turned a rock into the spears and swords of battle and the hoes of the farm.

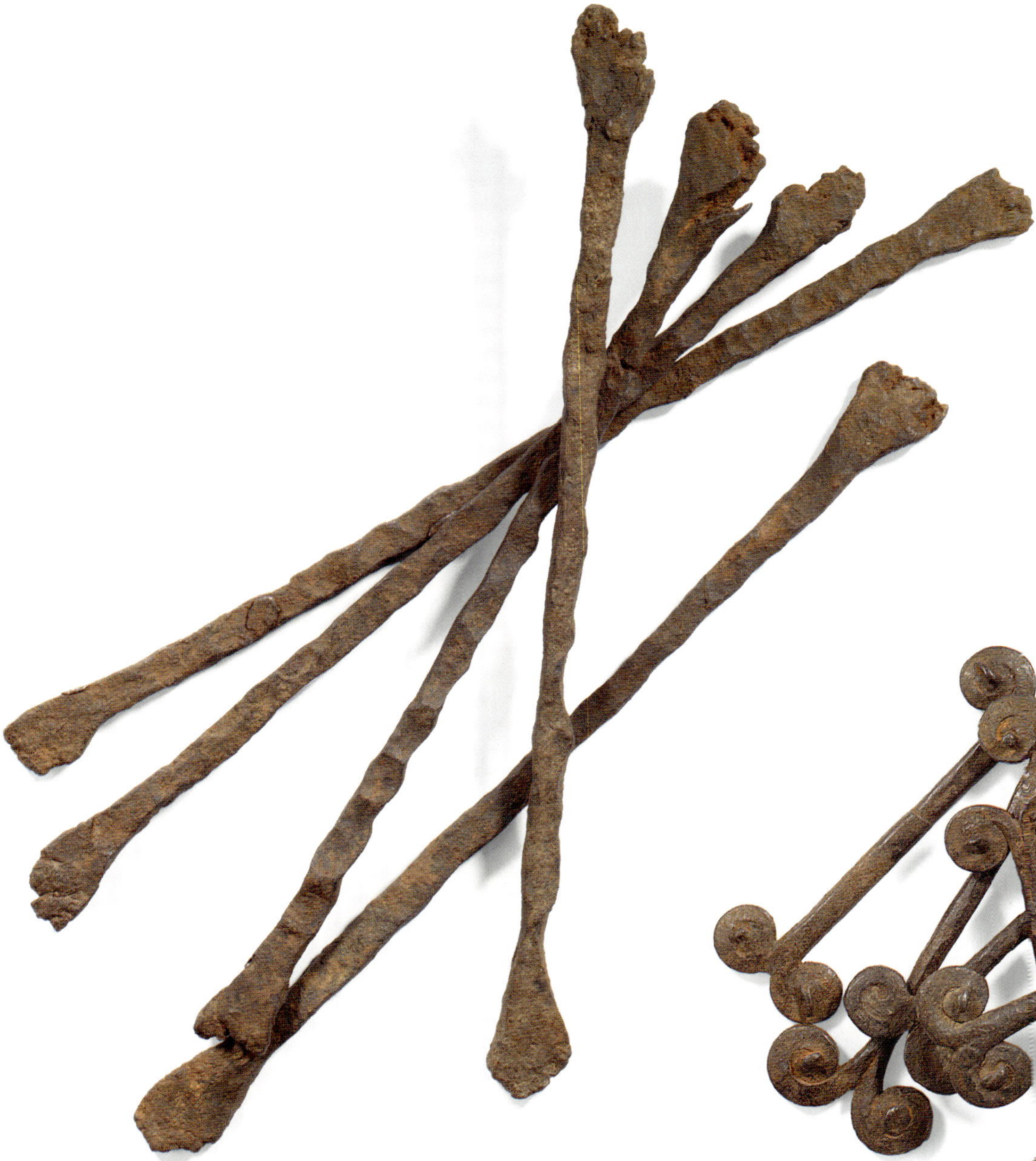

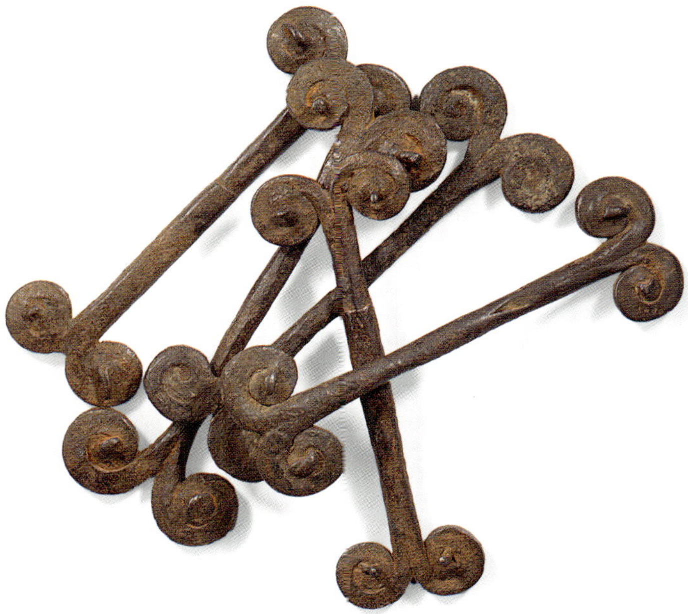

2.4
Artist unknown (Northeastern Nigeria)
Bar currencies, late 19th to early 20th century
Iron
H (each bar): approx. 42 cm
PRIVATE COLLECTION

2.5
Artist unknown (Mafa peoples, Cameroon and Nigeria)
Bar currencies, late 19th to early 20th century
Iron
H (each bar): approx. 17 cm
PRIVATE COLLECTION

Once cool enough to handle, iron bloom was broken apart into more easily managed pieces. These were evenly heated in a forge and hammered at the anvil into uniformly sized bars. Individually produced, some were raw with exposed rough ends to show the quality of the bloom. Others were meticulously shaped for use as symbolic currencies in local exchange and regional trade; these were relatively easy to store, distribute, and forge into useful objects. As bridewealth, locally produced iron bundles represented the intended fecundity of the daughter and wife.

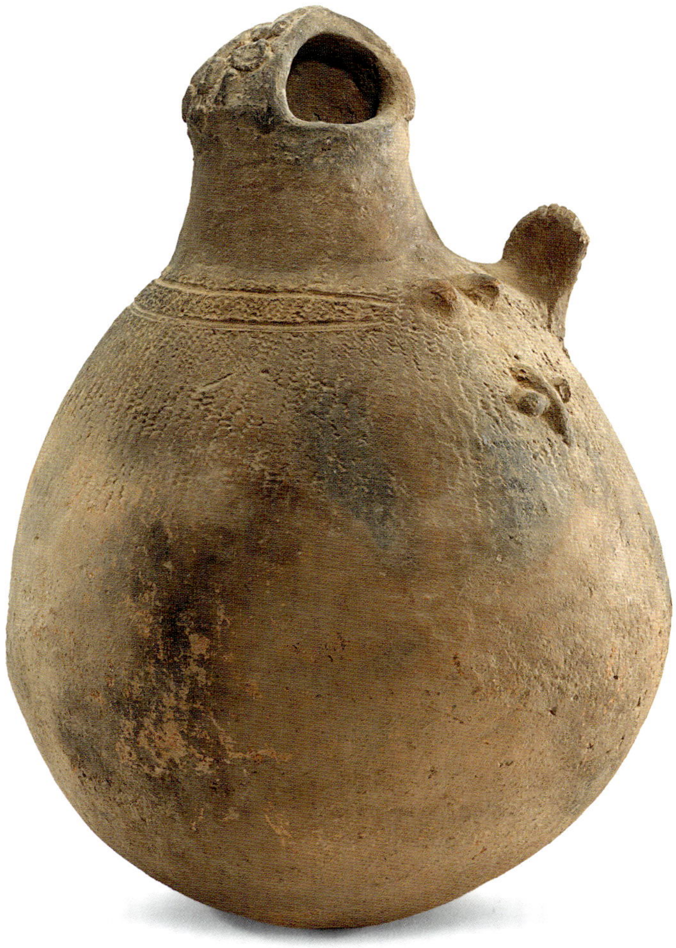

2.6
Artist unknown (Mafa, Bulahay peoples, Cameroon)
Spirit pot (Zhígílé), mid-20th century
Ceramic
H: 25.5 cm
FOWLER MUSEUM AT UCLA, X2014.36.5; MUSEUM PURCHASE

Although Mafa women were excluded from the iron smelting process, wives of smelters made clay crucibles, conical tubes for furnace tuyères, pots for bellows, and ritual vessels that represented ancestors and claimed prominent places at smelting sites. The agencies of this Mafa spirit pot ensured a successful smelt.

Smelting was related to other pyro-technologies. In Cameroon, Mafa clay pots employed by master smelters were anthropomorphic representations of the god Zhígílé with iconography referring to the spiked roofs of the smelting furnaces (fig. 2.6; see also chapter 13, this volume). Such a pot made by a smelter's wife easily became the focal point of rituals and sacrifices of smelting processes. Everywhere iron was smelted, the technology continued to be categorized in familiar human terms, for smelting was a social activity, not simply a technological practice. Sometimes (as in the opening example from Banjeli, and among Mandara traditions) the furnace was understood to be a woman and the smelter was her husband, affirming that:

> In all major enterprises, success depends not only on human skills but also on the intervention of ancestors and spirits who are a part of the natural world. At the same time, this world is humanized: acts of transformation are expressed in terms of . . . the complementary but often asymmetrical interplay of male and female. Iron smelting reflects both positive and negative views of femaleness simultaneously: creativity and danger. By insisting on the exclusion of women from it, the male smelter reaffirms his control of a fundamental source of power and wealth. [Herbert and Goucher 1987, 13]

The iron bloom was born from a smelter's control over his furnace, as his fertile bride. The furnace gave birth to glowing red iron, the blood of life. Thus, distinctly gynecomorphic furnace shapes and associated rituals recalled African smelters' employment of the ultimate creative act—giving birth—as an expression of iron technologies and their material and spiritual empowerment.

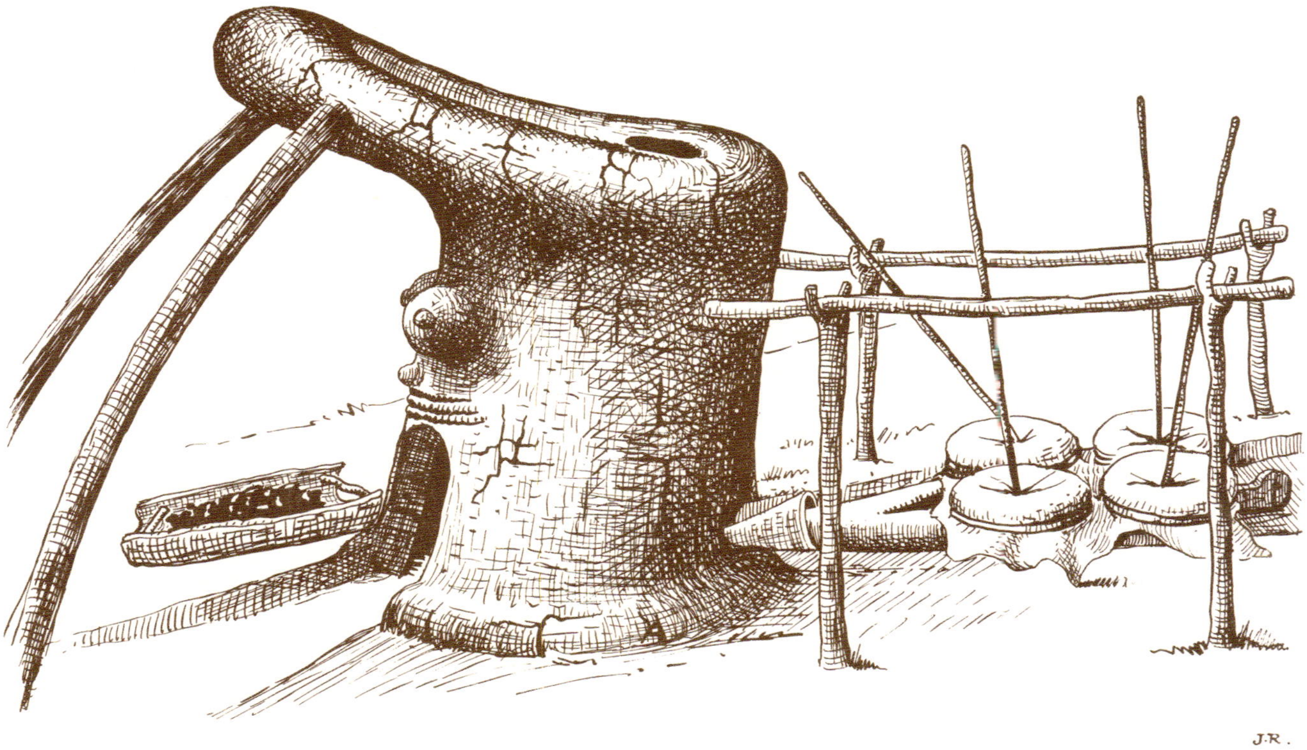

2.7 Illustration of a Chokwe anthropomorphic smelting furnace with parts including a set of double bellows. Furnace imagery depicts breasts, navel, and female scarification, while terminology refers to the legs (extending from either side of the ditch for slag collection), arms, and other human features.
PUBLISHED IN JOSÉ REDINHA, *CAMPANHA ETNOGRÁFICA AO TCHIBOCO* (LISBON: DIAMANG, DUNDO MUSEUM, 1953).

Bassari smelters observed in the last century constructed their tall, natural-draft, female furnaces in a series of rituals that eventually brought them to life (Goucher and Herbert 1996). Gradually the furnace became a woman, eating and breathing and giving birth to iron. The parts of her body were named: mouth, nose, lungs, belly, and womb. Yet while recognized, they were metaphorical body parts, not constructed out of clay to resemble their human counterparts (as was sometimes affected in other sub-Saharan societies). Similarly, among Fipa peoples of southwestern Tanzania, tall clay furnaces were not decorated with physical depictions of the female bodies they recalled and represented; among Pangwa (Tanzania), Shona (Zimbabwe), Chokwe (Angola), and Oromo and Chara-speakers (Ethiopia), as well as other peoples, smelters took the time to construct furnaces with visible female breasts and genitalia (figs. 2.7–2.9).[5] During the smelting process, when men conducted rituals, women, particularly menstruating women, were banned from all such sites, leaving no doubt that a male smelter was believed to impregnate the furnace and claim his control over the dangerous technology. In a related manner, men seeing to iron smelting often observed sexual abstenance. A Pangwa master smelter observed by Randi Barndon in 1990 washed his furnace

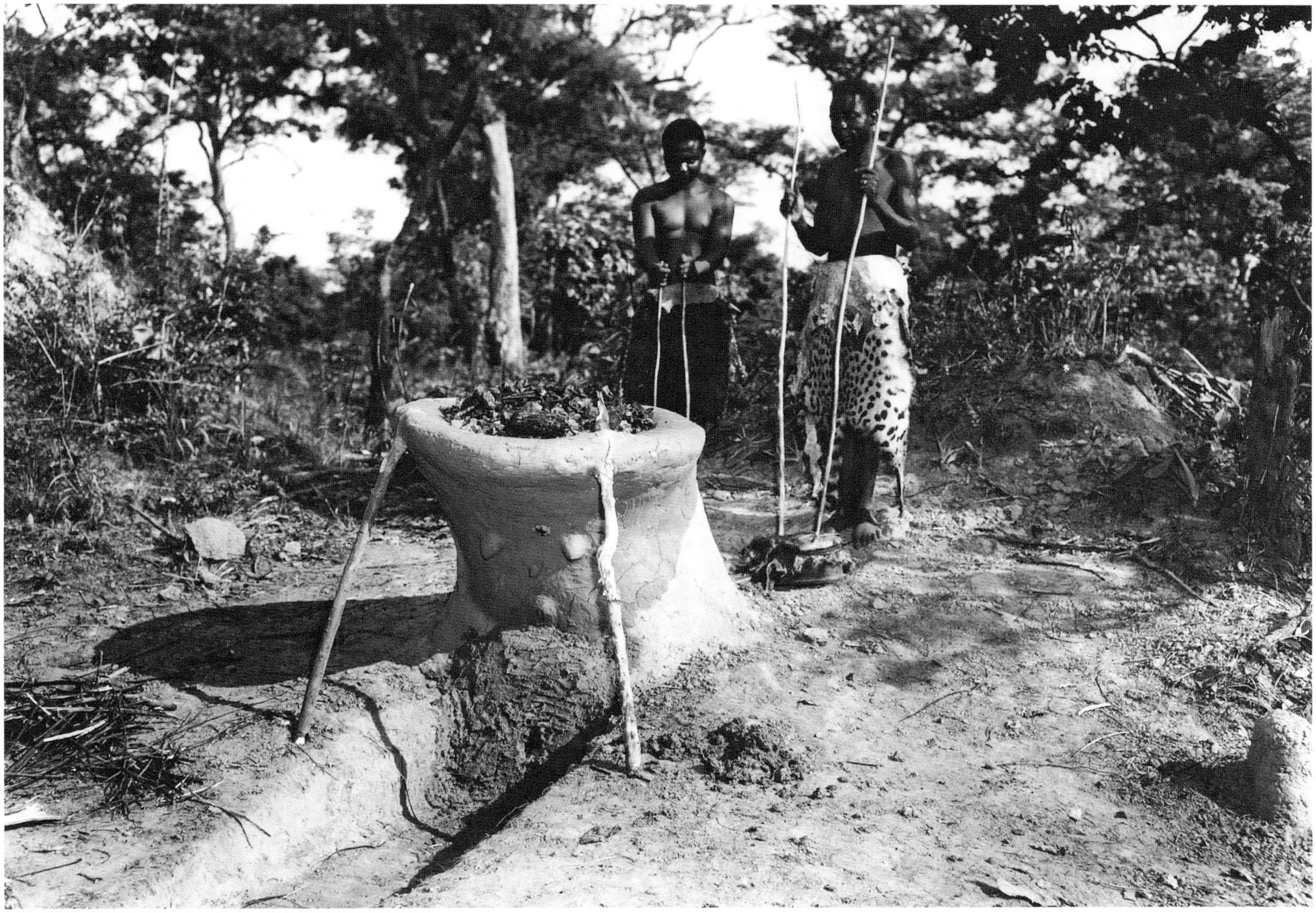

before the delivery of the bloom in order to cool it down, a gesture likened to purifying a pregnant woman with a water bath prior to childbirth (2012).

A different smelting system observed among Ekonda smelters in central Africa consisted only of a bowl-shaped hearth in the form of a pit with thatched roof (Celis 1987; Ackerman et al. 1999). Their operations extended the actions of the male smelter into the furnace by means of the clay pipe or tuyère supplying air. Smelting relied on a charcoal fire combined with a tuyère (*boongo*, meaning "vagina") and bellows to achieve the reducing conditions of a proper furnace. Its characteristics and identifiably female sexuality also emerged in the process of smelting iron ore. The parts of bellows were further anthropomorphized as possessing chest, nose, and tail. The pole, or handle, of the tuyère was called *mpoló*, for the male sexual organ (a slightly different tonal pronunciation, *mpolo*, can refer to the vagina, thus suggesting gender dialectics at play). Molten slag that runs from the Ekonda or Banjeli furnace was likened to the menses, and the required ceremonies to respond to concepts of pollution included sexual taboos and prescriptive medicines applied prior to delivery of a successful bloom. Elsewhere it was characterized as the gush of waters that precede childbirth.

2.8 The anthropomorphic form of this Chokwe smelting furnace would have suggested a gendered agency, undoubtedly enhanced by medicines and ritual practices to assure the "birth" of bloomery iron from its "womb." In this view, iron is conceived and life promoted through the objects it will become.
PHOTOGRAPH BY C. LAMOTE (INFORCONGO), MID-20TH CENTURY. COLLECTION ROYAL MUSEUM OF CENTRAL AFRICA, TERVUREN

The gynecomorphic characteristics of furnaces received various degrees of elaboration in west, east, central, and southern African societies. Furnaces were sometimes dressed with women's waist beads, marked with scarification to connote female genitalia or the body arts deployed to accentuate them, or provided with food and medicines that protect women in childbirth. At other times, as among Bassari smelting operations, references were verbal and female human identity was called out by the smelter himself. Bassari and Fipa smelters characterized their standing furnaces as living, breathing female brides regardless of whether their anatomy was reproduced in clay. Shona and Chokwe used scarification and breasts to signal female identities for their furnaces, with "legs" spread away from the slag hole through which birth occurred. Their phallic tuyères were inserted into the genital area (vulva) of the furnace, which then might produce slag (the "placenta") and give birth to an iron bloom of metal (specifically, a son). Elsewhere, the slag could be associated with that which was characterized as unwanted: feces, placenta, menses. The typically male associations with the tuyère and bellows might even be elaborated with named parts that were complementarily female and male.

Like giving birth, cooking is a transformative act associated with women and the transforming heat of fire. Not surprisingly, in other highly gendered examples of African technology, the smelting furnaces were sometimes envisioned as cooking pots in which civilization itself was brought into being by the master smelter/chemical chef. In the settlement of Sirak, a small community of Mefele-speakers of northeastern Nigeria and north Cameroon, the ambiguous status of ironworkers (literally smelters, but also smiths and potters) derived from a myth about eating something "bad" in the distant past, an act which could seemingly explain how their power and status vis-à-vis farmers was curtailed (David and Stern 2012, 91). Yet they were also transformers, changing—with the help of ancestors and spirits—ore into metal, clay into healing (food) pots, fetuses into human beings. Among Hausa workers of northern Nigeria, the two pyro-technical processes (cooking and smithing) were held in tandem. Hausa trace the origins of their ironworking clan to a woman, Lola, who fell from the sky. Holding a ladle of hot food in one hand and a smith's hammer in the other, Lola displayed her supernatural powers by handling both the heat of the cooking fire and the flames of the forging hearth. Further, Hausa of Niger describe the sequence of smelting in food terms: the process of ore reduction occurs as the furnace "eats" the iron (Echard 1983).

Ntaatakwa, Fipa founder of the royal lineage of modern-day Tanzania, is another cultural hero who dropped from the sky with a metallurgical toolkit to empower lineage members and their descendants (Willis 1981). Still other African traditions describe the emergence of iron technology from the earth itself. At the early Iron Age site of Nsesrekeseso (Ghana), local Akan understood that the first human emerged from a sacred hole in the ground with an iron hammer. (Archaeologists assume this to explain the origins of local iron technology, see Merrick Posnansky pers. comm., 1978.) Yorùbá speakers of southwestern Nigeria and southeastern Republic of Bénin maintain that iron existed at the beginning of time, since their deities climbed down from the heavens on an iron chain to create the world from a gourd filled with sand (Drewal et al. 1989). Such cosmological beliefs enabled the act of smelting iron—as a secretive, hidden transfer of sexual energies—to be foundational to the worldly roles of men and women in building and transforming civilizations.

SMELTING AND TECHNOLOGICAL CHANGE

Smelting systems cultivated complicated and ever-changing social, political, and economic networks across vast landscapes; and smelting had a serious impact on the natural world through deforestation. Not only ores reacted to the conditions of furnace, flux, and fuel—smelters themselves were compelled to respond to the constant challenge of new environments that required technical adaptation and constant innovation (Goucher 1981; Goucher 1984; see also chapter 16, this volume). Their smelting systems were responsive to networks of semantic, gendered, and political associations, accessed through myth and ritual and expressed at both the individual level of decision-making and group practice (Rehren et al. 2007; Ngonadi and Eze-Uzomaka 2015).

Over millennia of making iron innovations in the technical parameters reflected depleted ore and fuel supplies, mobility of smiths and smelters, increasing demand for iron implements, environmental limitations on the seasons of production, and challenging constraints on labor. The use of certain fluxes, tuyères, bellows, furnace models, fuel, and crucible technologies could and did dramatically affect the efficiency of ironmaking processes and the ultimate quality of products. While archaeological investigations can yield the physical evidence of technological practices across a deeper past, more recent ethnographic observations point us toward

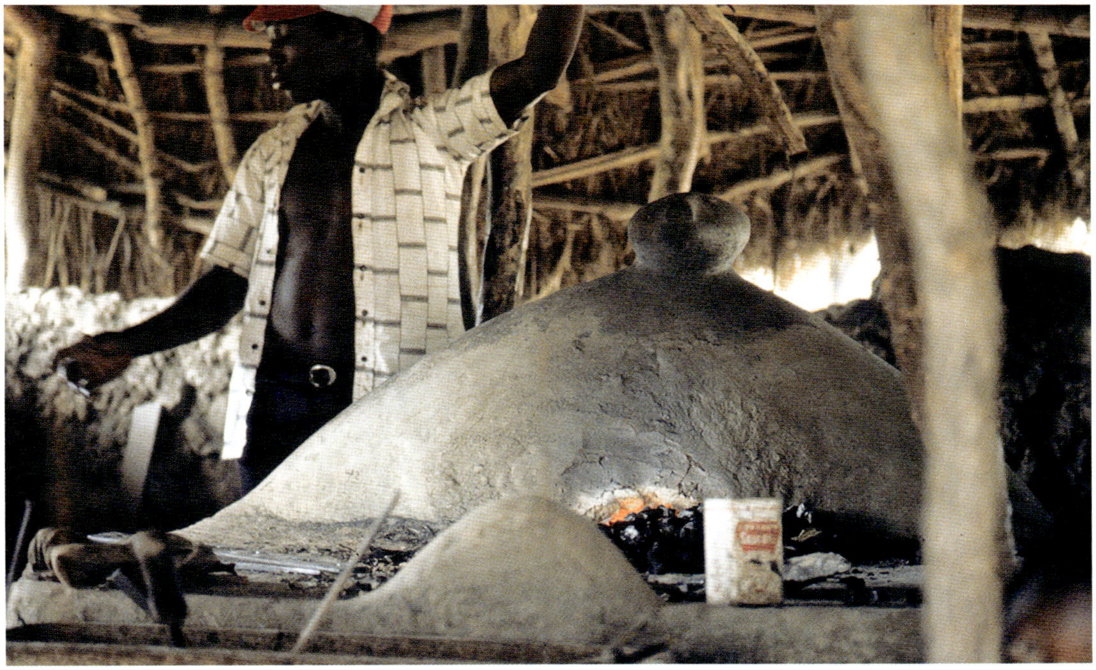

2.9 A European-style (raised) smithing furnace shaped to resemble the female form (its wall and opening are referred to as "vulva" and "vagina"), employed to repair guns in the late 1970s.
PHOTOGRAPH BY CANDICE GOUCHER, BRAWHANI, GHANA, 1979.

understandings of meanings and values in the metaphysical associations that inform the work of blacksmiths. Smelting knowledge and rituals may have served to perpetuate successful practices through repetition and transmission of successful formulas for making iron, but they could also have been used to contain opportunities for innovation and variability. The culmination of iron production in the era of the Atlantic slave trade was advantageous to the transformation of people into commodities with the aid of furnaces that gave birth to things. Persistence of smelting into the twentieth century, despite colonial policies that attempted to eradicate local competition with metropolitan economies, is testimony to the resiliency of African technological and sociopolitical achievements.

CONCLUSION

Both scientific and magical knowledge were needed to make iron. Framed in this way, African iron production was both a technological and supernatural event capable of transformations beyond the natural laws of the seen world. Understanding the practice of smelting iron as supertransformative act does not entirely explain why acquisition of its technical and ritual secrets resulted in personal empowerment or the formation of state structures that managed technological wealth, whether in the form of intellectual or material culture. Did proprietary knowledge yield economic and material accumulation and result in altered social status? Was it the restriction of proprietary knowledge that gave advantages to certain individuals and groups? Structured by gender, iron-making certainly provided men with the essential building blocks of economic power and political and patriarchal authority. It might be argued that the exclusion of women of childbearing age served another purpose, perhaps based on observation and experience: to avoid the harmful contaminants of the toxic smelting furnace and thereby allow for the reproduction of society. Are the origins of the beliefs in the fecund furnaces themselves visible attempts to "marry," or conjoin, opposites, to define or control a restricted male space where gender dialectics were made material? Regardless of their origins, common conceptualizations of iron smelting informed the meaning of materials in the work of African blacksmiths, understood as the fundamental male mastery of female creative forces and life-sustaining efforts to transform the world.

CHAPTER
3

IRON'S EMPOWERING PRESENCE

Henry J. Drewal

When a satellite flies over Africa, what does it see? The reddish tinge of lateritic, iron-rich soil. This earth might well remind us of the presence and power of iron in the making of African civilizations. One of the earliest was Nok, in central Nigeria between 900 BCE and the cusp of the Common Era, a major site of iron smelting and forging in association with one of the world's finest traditions of ceramic sculpture.[1] Nok was named for a village where clay artifacts were initially collected and researched by Bernard Fagg beginning in the 1940s (see Fagg 1968; 1969). Pottery figurines were found within and around furnaces, as was the case at thirteen furnaces found by Fagg in the valley of Taruga, fifty-five kilometers southeast of Abuja. Such sculptures, for example two of women holding their breasts and seated on stools (fig. 3.2), likely indicate that the transformative processes of firing clay, smelting, and ironworking were understood as acts involving supernatural forces (see also figs. 9.1, 9.2).[2] Iron is a *cosmic connector*.

As of 2016, researchers have uncovered in the area of Nok iron implements (mostly simple forgings—spear points, bracelets, small knives; fig. 3.3) together with a variety of stone tools, confirming that Nok culture used both at the same time (Breunig and Rupp 2016). Investigations of Nok add further archeological evidence to support the view that African civilizations proceeded directly from Stone Age to Iron Age cultures without an intervening Copper or Bronze Age, in contrast to other parts of the world (see chapter 9, this volume).[3] The presence and accessibility of high-content iron ore in the lateritic soils of west and central Africa may help to explain this phenomenon and technological transition. But there may be other cultural and philosophical reasons as well.

In sub-Saharan Africa, the world—plants, animals, soils, rivers, humans, words, gestures, songs, dances—is alive with spirit. Rooted in ancient, indigenous knowledge and philosophical systems, such understanding has served local needs and purposes for millennia. Everything possesses spirit as much as consumable usefulness. Rather than thinking of an arbitrary divide between "animate" and "inanimate" things, or "sacred" and "secular" matters, a continuum of things and actions exist *because* they are alive with spirit, whether created by a supreme source or a host of divinities. Molded by deified ancestors, celebrated cultural heroes, and others, objects and people alike are active agents (see Abram 1996; Pollan 2013). In this worldview, iron figures prominently as a medium imbued with special spirit, effective as an *activator of spiritual power* that ensures the efficacy of acts both sacred and utilitarian. Because the smith is master of the transformative processes of making iron from the earth into culture's tools, and because he is considered able, through his awesome power, to mediate the forces of life and death, he is often charged with making the very objects used to invoke, represent, and mediate deities and other mediums of supernatural agency.

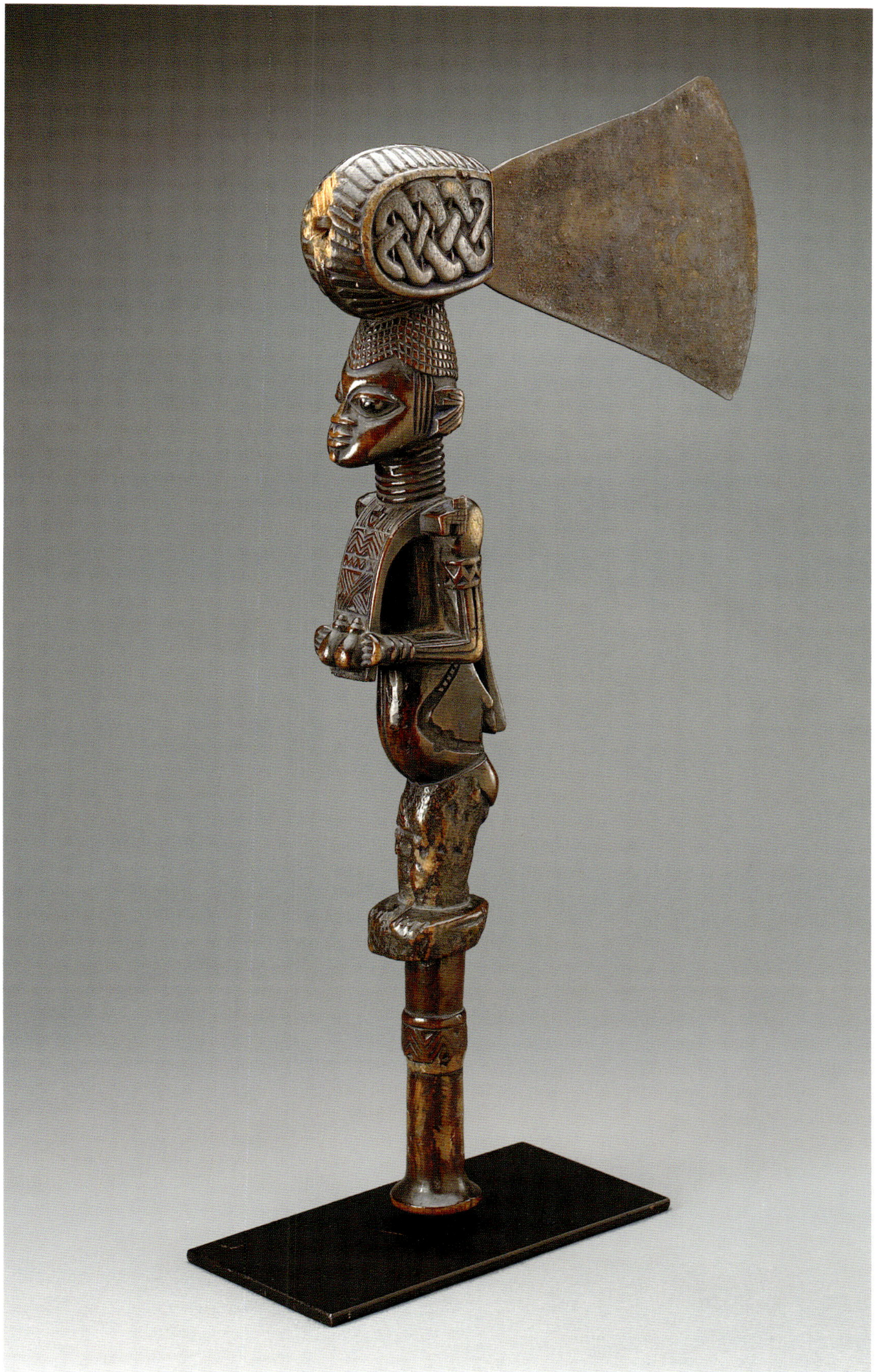

3.1
Artist unknown (Yorùbá peoples, Nigeria)
Ritual axe, late 19th to early 20th century
Wood, iron, pigment
H: 50.8 cm
SAMUEL P. HARN MUSEUM OF ART, UNIVERSITY OF FLORIDA, GIFT OF RODNEY D. MCGALLIARD IN HONOR OF DR. ROBIN POYNOR
PHOTOGRAPH BY RANDI BATISTA, 2009. COURTESY OF SAMUEL P. HARN MUSEUM OF ART

This Yorùbá axe with carved handle, probably a favorite tool of the artist, possesses the life force, or "performative power," of the maker/artist, embodying his knowledge, wisdom, sweat, toil, and prowess.

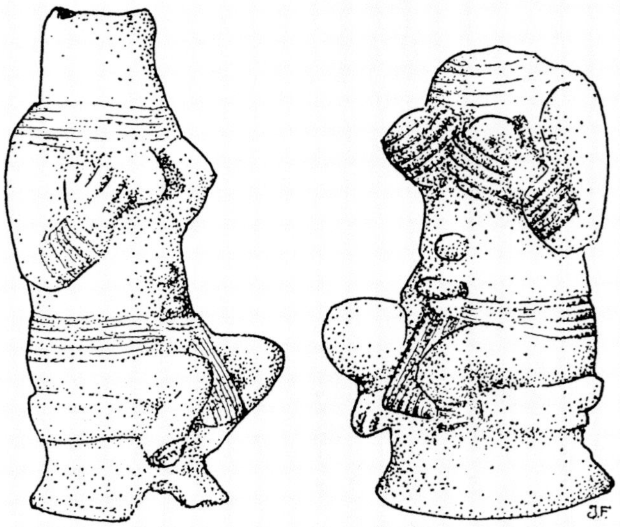

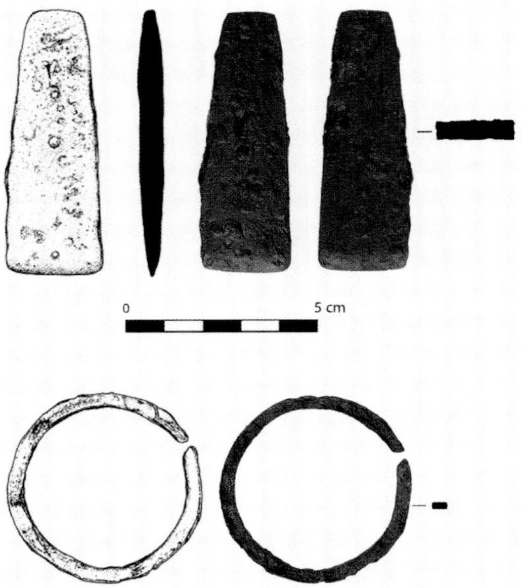

3.2 Terra-cotta figures made by Nok sculptors have been found near to iron smelting sites in Nigeria at Taruga and, recently, at Janjala, suggesting that ironworking and pottery sculpting, two fire-based processes, were viewed as acts involving supernatural forces.
RENDERING BY JOHN FAGG, SON OF BERNARD FAGG, 1968.

3.3 Examples of forged iron found in association with stone tools at Nok sites document a transition to iron technology from stone without an intervening Bronze or Copper Age.
IMAGE © GOETHE-UNIVERSITÄT FRANKFURT, INSTITUT FÜR ARCHÄOLOGISCHE WISSENSCHAFTEN, ARCHÄOLOGIE UND ARCHÄOBOTANIK AFRIKAS, GERMANY

THE POWER OF THE BLACKSMITH'S TOOLS

What comes to the mind's eye when we think of African art? Mostly objects made of wood and carved. What we tend to forget are the means by which they were created, that is, the *iron tools* of the blacksmith and the *iron blades* of the sculptor. It is the sharpened blade of an adze that penetrates wood to reveal forms within, materializing the immaterial. Iron is an ever-present yet often neglected necessity for the creation of most African art (see Drewal 1980). Indeed, those who fashion the tools of creation—blacksmiths—are also sculptors. Blacksmiths, with their intimate body-mind knowledge of working with iron, are best prepared to wield iron tools and carve wood. Just as a sculptor transforms the trunk of a tree into images of humans and animals and sacred entities, so too the blacksmith transforms hot iron into tools that make and shape cultural worlds, creating a myriad array of objects serving both worldly and otherworldly needs (see figs. 3.1, 3.4).

Over a lifetime of use, the blacksmith's tools (hammer, anvil, poker, tongs) come to possess the life force, or "performative power," of the maker himself—his toil, touch, breath, sweat. When they are passed on to the next generation of smiths, they are revered as entities alive with spirit and fearsome potential. The same is true for the carver's tools (axe, adze, knife). All these tools are understood to embody the smith's knowledge, wisdom, and virtuosity, and on an array of remarkable objects from different west African cultures they appear as embellishments (figs. 3.5, 3.6). Without the smith's tools, none of this would be possible.

On a Dan spirit-mask (see fig. 3.5), miniature forged iron implements—among them hairpins and blades—are embedded along the crest of the head. Dan peoples live along the forested mountainous border between Côte d'Ivoire and Liberia, a region of iron-rich ore. Dan posit two distinct but interpenetrating realms: the forest of potentially dangerous spirits, and the village of tamed (that is, socialized) human behavior (Fischer and Himmelheber 1984; Fischer 1978; Siegmann n.d.). Masks, often envisioned in dreams sent to humans by forest spirits, are material embodiments of such spirits. Humans use spirit-masks to harness the forces of the forest for the benefit of the village. This mediating role is evident in the anthropomorphic form of the mask that also possesses the spirit-soul (*du*) of a forest being. The mask is *alive with spirit*, crossing the liminal space between forest and village, nature and culture.[4]

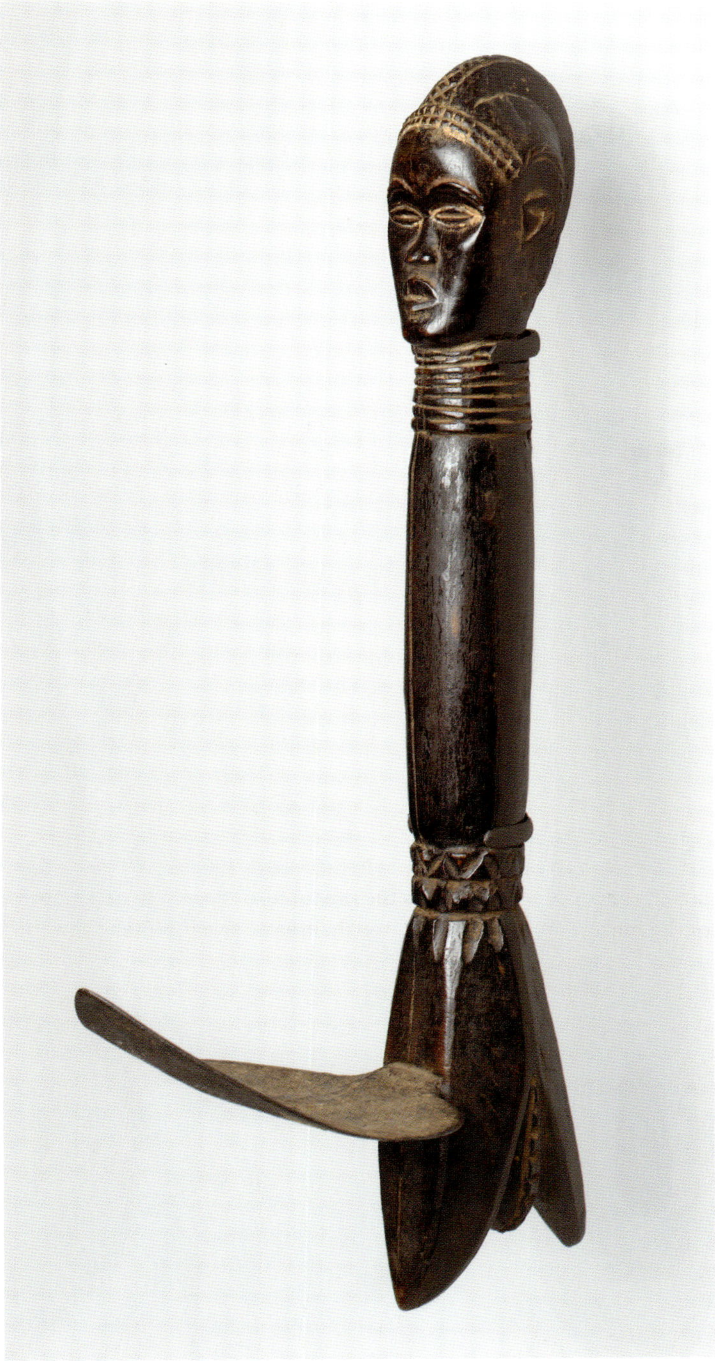

3.4
Feia Tomekpa (active 1940s to early 1950s, Dan peoples, Liberia and Côte d'Ivoire)
Ceremonial hoe, 20th century
Wood, iron
H: 35.9 cm
BROOKLYN MUSEUM, GIFT OF MR. AND MRS. BRIAN S. LEYDEN, 87.216.1
IMAGE COURTESY OF BROOKLYN MUSEUM

The blacksmith Feia Tomekpa may have both forged this hoe's beautiful blade and carved its exquisite wood handle featuring a Dan woman with elegant coiffure. Dan women work in cooperative groups using short-handled hoes for rice cultivation, and each community chooses a group leader based on her reputation for energy and oversight. This ceremonial hoe was a symbol of a leader's status. Although the intricately detailed head could be considered a portrait, it is more likely expressive of Dan ideals of beauty.

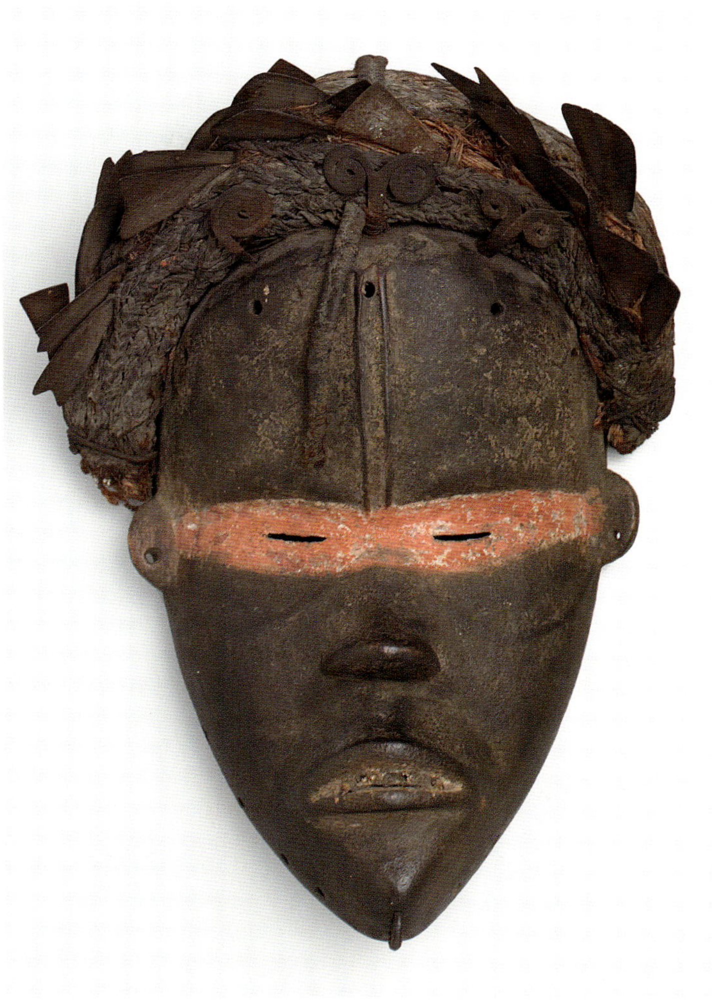

3.5
Artist unknown (Dan peoples, Liberia and Ivory Coast)
Mask, late 19th to early 20th century
Iron, wood, copper alloy, fiber, pigment
H: 28.6 cm
YALE UNIVERSITY ART GALLERY 1954.28.7; GIFT OF MR. AND MRS. JAMES M. OSBORN FOR THE LINTON COLLECTION OF AFRICAN ART
PHOTOGRAPH BY DON WHAPLES, 2004. © YALE UNIVERSITY ART GALLERY

The presence of iron—embedded in the crest of this face mask as miniature forged implements including hairpins and blades—amplifies the sacred potential of the forest spirit and enables a masquerader to mediate between the village and beyond. Oral traditions among Dan are filled with battles and wars, matters in which blacksmiths, as weapons makers, would have been intimately engaged. The band of red that surrounds the thin eyes of the spirit-mask may evoke associations of aggression and violence as well as of spiritual insight. While a spirit-mask may change its identities and purposes over time, we may be looking at Bugle, the warriors' spirit, provider of sacred protection and courage in battle. The mask may also be an expression of *tin*, a Dan concept concerning the display of status and success, reputation and renown.

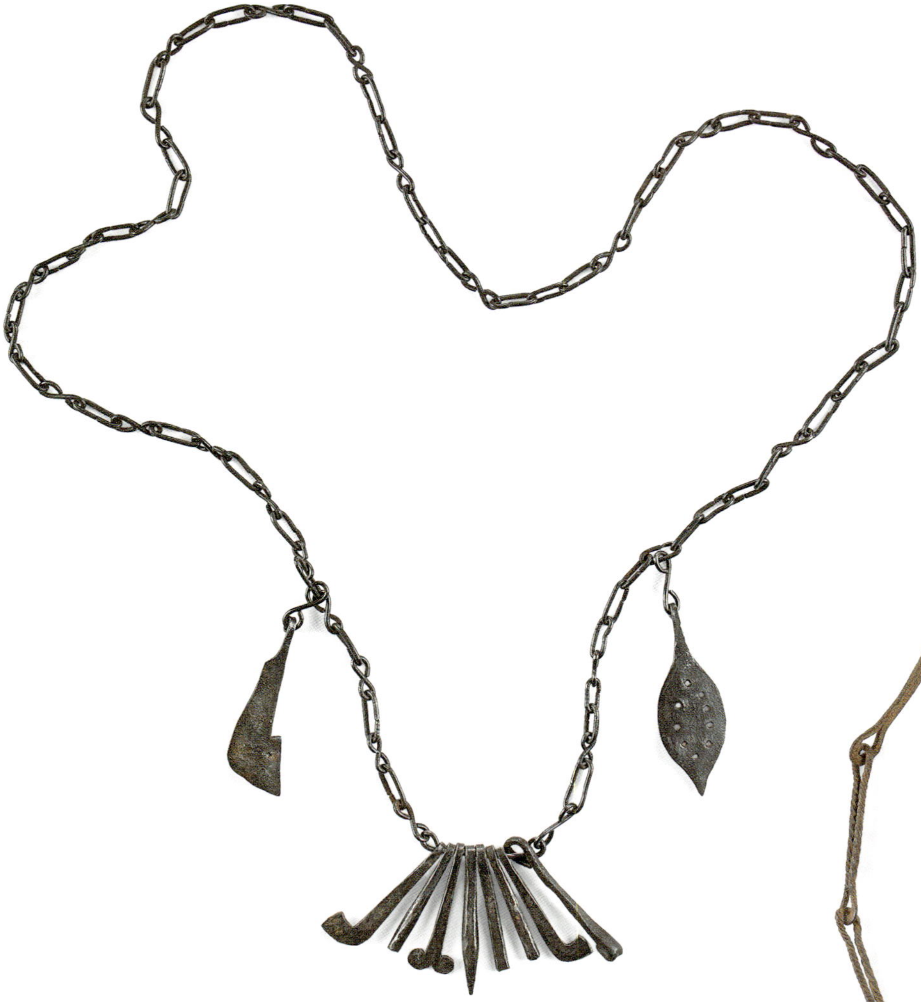

3.6
Artist unknown (Edo peoples, Nigeria)
Necklace, late 19th to early 20th century
Iron
L: 62 cm
FOWLER MUSEUM AT UCLA, X65.9220; GIFT OF THE WELLCOME TRUST
PROVENANCE: MR. H. NEVINS, DISTRICT OFFICER, BENIN DIVISION, 1927

Blade forms adorn an Edo necklace, flanking pendants of blacksmith's tools. The *ada* (far left; see fig. 3.16) represents a functional weapon used to build and expand the Benin Kingdom; the *eben* (far right; see fig. 3.19) is a ceremonial dance implement wielded to celebrate achievement and affirm loyalty to established royal order. Assembled together as a virtual tool kit of diplomatic possibilities, they symbolize the making and sustaining of culture by balancing decisive action and filial reflection, both embodied in iron.

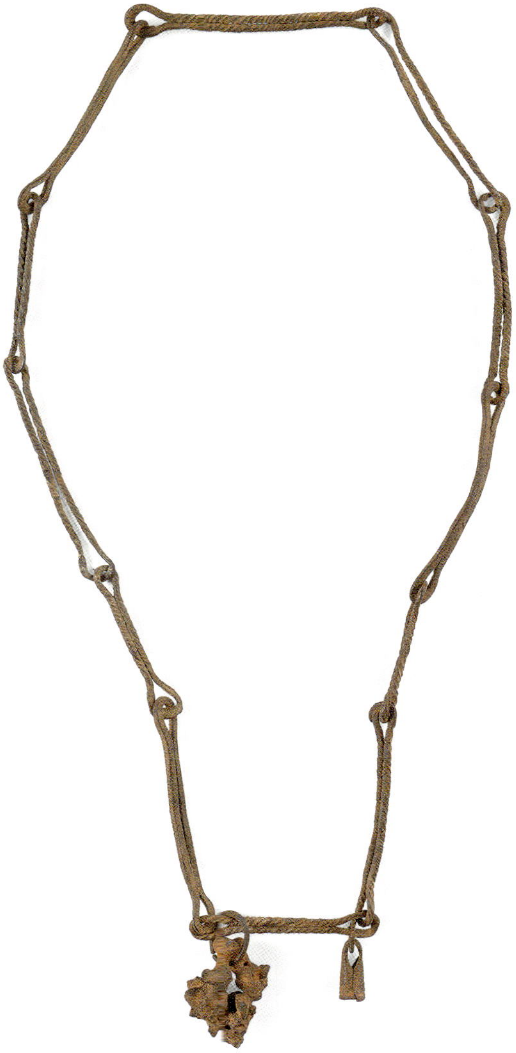

3.7
Artist unknown (Dogon peoples, Mali)
Pendant necklace (*dugo*, *duge*), late 19th to early 20th century
Iron, stone
L: 54 cm
PRIVATE COLLECTION

Dogon pendant necklaces are made of forged iron links from which stones, glass beads, and forged iron rings, miniature bells, shields, and tools can be suspended. The necklaces identify the wearer's status or role in society, whether priest or blacksmith, and also visualize the fundamental principle that an unbroken "chain of being" unites every living Dogon person with ancestors. Exhibited here is a version featuring a small piece of iron bloom, suggesting it may have belonged to a blacksmith, and another with four *duge* stones, understood to be cosmic eggs. He who dons so potent a necklace "wears the world around his neck."

In the absence of a detailed provenance for the mask illustrated here, we can only speculate about its former uses, identities, and meanings; these may have changed over time as it was inherited and used by custodians over several generations.[5] Despite these challenges, the presence of embedded iron objects offers some possibilities. They may have been inserted to amplify the sacred powers of a forest spirit, but also to invoke the transformative powers of the smith in allowing the masquerader wearing this mask to take on a forest force. Or, they may indicate the mask-owner's dangerous occupation as a blacksmith, and his liminal position in Dan society as one who possesses the awesome power to transform earth into iron and iron into tools to create, among other things, weapons used in warfare.

The transformative potential of iron and blacksmiths' tools can be redirected to many more purposes. Among Dogon peoples of Mali and Edo peoples of Nigeria, amulets in the form of miniature forging tools strung on chains instill in those who wear them the same courage and strength the blacksmith commands when working with fire and forging iron. But what is an amulet? It is a portable protective and empowering device that contains and conveys spiritual force (see Mack 2007; Stewart 1993). Its medium is its sacred medicine. It embodies sacred efficacy because its innate energy has been enhanced. It is "fed" and "prepared" with many substances, prayers, and ritual actions to have "performative power," or the "power to accomplish things"—a vital life force called *nyama* by Dogon, *du* among Dan, *sunsum* among Akan, *àṣẹ* by Yorùbá and Edo peoples, *se* among Fon, and glossed as *simbi* among Bantu-speaking peoples of central, eastern, and southern Africa (Barbaro Martinez-Ruiz pers. comm., 2018).

The presence of blacksmithing tools empowers, heals, and protects because they possess the transformative power of the maker; his abilities are invested with the transformative essence of iron itself. Certain forged iron pendant necklaces made by Dogon peoples are called *dugo* or *duge* (fig. 3.7; see also fig. 1.11). Many are made with stones, glass beads, iron rings, miniature bells, and/or tools. They represent "signs of allegiance" between the wearer and a spiritual power, a close or mythic ancestor (Dieterlen 1938; see chapter 1, this volume). Worn close to the beating heart, the necklace activates the bond between the living descendant and the departed spirit. As Kate Ezra explains, "every element is a link connecting the wearer to the spiritual powers that determine his or her identity as an individual, and as a member of a family, village, and clan" (1988, 104–05). A pendant necklace displays its wearer's status or role in society, whether priest or blacksmith, and also visualizes a fundamental Dogon principle—an unbroken "chain of being" that unites every living Dogon person with ancestors, whether recently or distantly departed. Family ancestors are known as "those who are far away" (*vageu*), departed but not "dead" or unresponsive. The more mythic, elevated, and revered clan ancestors are called "those who went away and returned" (*binu* or *binu yay*), suggesting a more remote past and a long journey of return and connection (ibid., 18).

As necklaces possessing *nyama*, *dugo* work for their wearers as embodiments of a past acting in the present—they honor ancestors and bring health, fertility, and prosperity. A personal shrine (*kutogolo*) for a person's *nyama* might also contain iron in the form of hooks embedded in a ball of earth along with pottery and occasionally, on the shrine of a blacksmith or hunter, wooden figures. Such a shrine honors a person's head (*ku*), where thought and spirit reside (Dieterlen 1965, 15; Musee Guimet 1959, 116).

BLACKSMITHS' ROLES IN SOCIETY

In the region where Mali, Burkina Faso, and Cote d'Ivoire come together, Senufo peoples live in towns and villages consisting of distinctive matrilineages (*katioula*) that reflect economic and social status within the community, for instance Kule (blacksmiths), Kpeene (carvers), and Tyeli (brass casters).[6] With the spiritual presence, authority, and guidance of ancestors and myriad spirits, certain societies—Poro, Sandogo, Wambele and, among Fodonon Senufo, Tyekpa—govern cultural life, preserve genealogies and histories, educate youth, and maintain social laws. To guide the actions of individuals and communities, these powerful associations use a wide range of ritual objects, inspiring diverse and dynamic art forms created with blacksmiths' iron tools and the blades of carvers. Divinatory spirits and the sculptures created for them are often referred to as *ndebele*, *madebele*, or *tugubele* (s. *ndeo*, *madeo*, or *tugu*; see Gagliardi 2010). That the names for spirits and their materializations are the same is indicative of how Senufo understand these divinatory objects—as imbued with spirit.

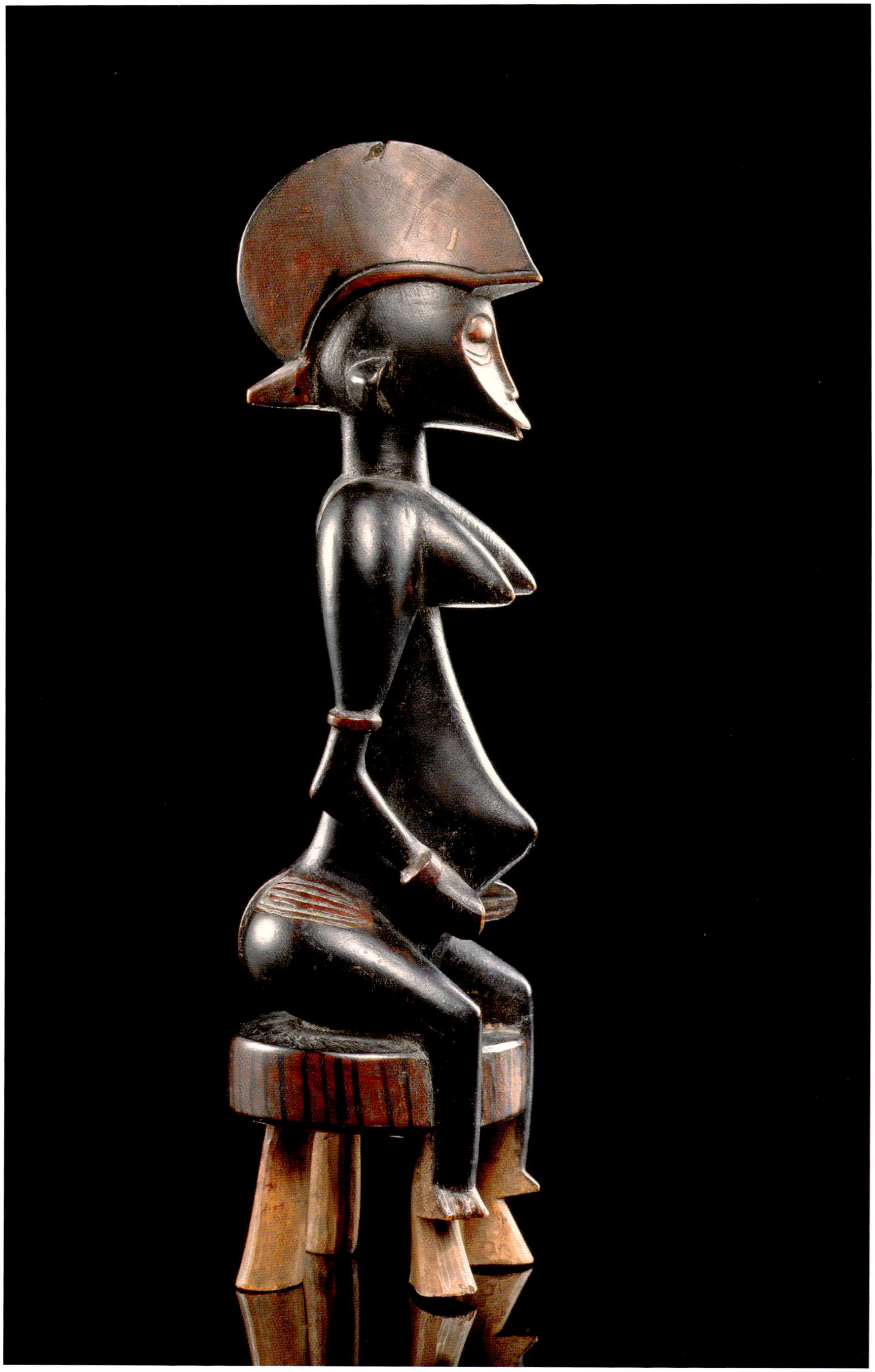

3.8
Artist unknown (Senufo peoples, Côte d'Ivoire)
Female figure, before 1912
Wood
H: 31 cm
BARBIER-MUELLER COLLECTION, GENEVA, 1006-13
PROVENANCE: "KOROKO" (KORHOGO), CÔTE D'IVOIRE, 1912 (PER LABEL ON BASE); JOSEF MUELLER, SOLOTHURN, SWITZERLAND, BEFORE 1939
(NOT IN EXHIBITION)

Senufo diviners, mostly women and members of the Sandogo association, commission sculptures to appeal to capricious spirits, seeking their support and good will. Advanced practitioners seek wooden sculptures like this one for their consultation rooms. This elegantly refined work of an unrecorded Senufo master epitomizes a physical and spiritual vision of female beauty, admired in diviners and characterized as both alert and balanced.

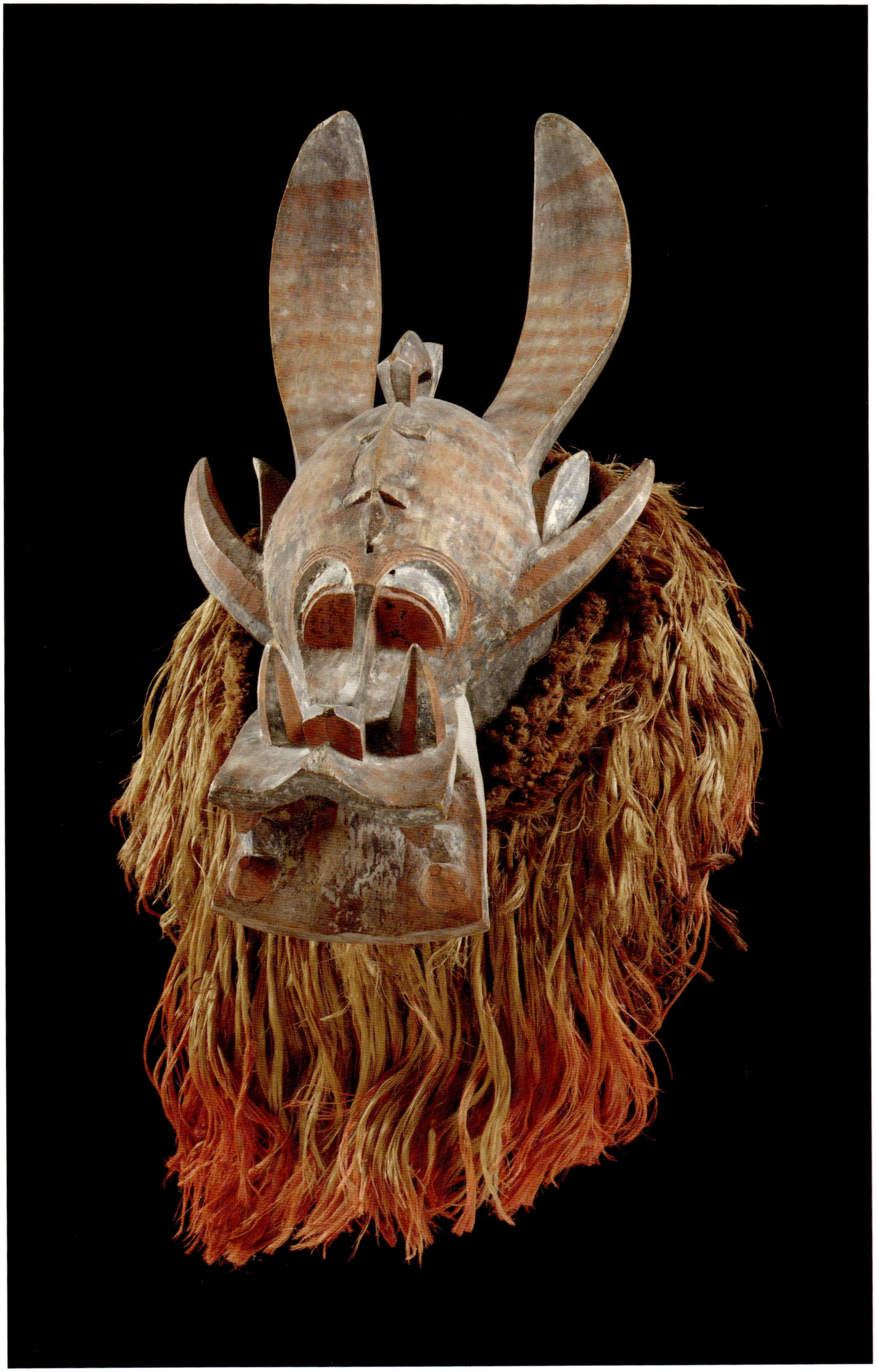

3.9
Artist unknown (Senufo peoples, Côte d'Ivoire)
Helmet mask (*kponyugu*), early 20th century
MUSEUM RIETBERG ZURICH, GIFT OF RAHN & BODMER BANQUIERS, 1 RAF 321
(NOT IN EXHIBITION)
PHOTOGRAPH BY RAINER WOLFSBERGER.

A Poro power object, created with blacksmiths' iron tools and carvers' blades, guides the actions of individuals and communities. Masks like this are used in initiations to teach positive social and moral behavior and sustain community. Such composite masks feature animals symbolic of both positive and negative traits to be emulated or avoided, for example, the mindless aggression of crocodiles or warthogs, or the slow, thoughtful reflection and action of chameleons.

3.10
Artist unknown (Tuareg, Niger)
Dagger, 19th to 20th century
Iron, wood, copper, leather, pigment
H: 54 cm
MUSÉE DU QUAI BRANLY-JACQUES CHIRAC, 71.1934.192.2.1–2
(NOT IN EXHIBITION)

Tuareg blacksmiths engage in "mystical" work as they control the fiery transformation of iron. The virtuosic finishing on this forged and engraved dagger blade, fused with a refined copper handle and sheathed in embossed leather, embodies the widespread warrior renown of Tuareg.

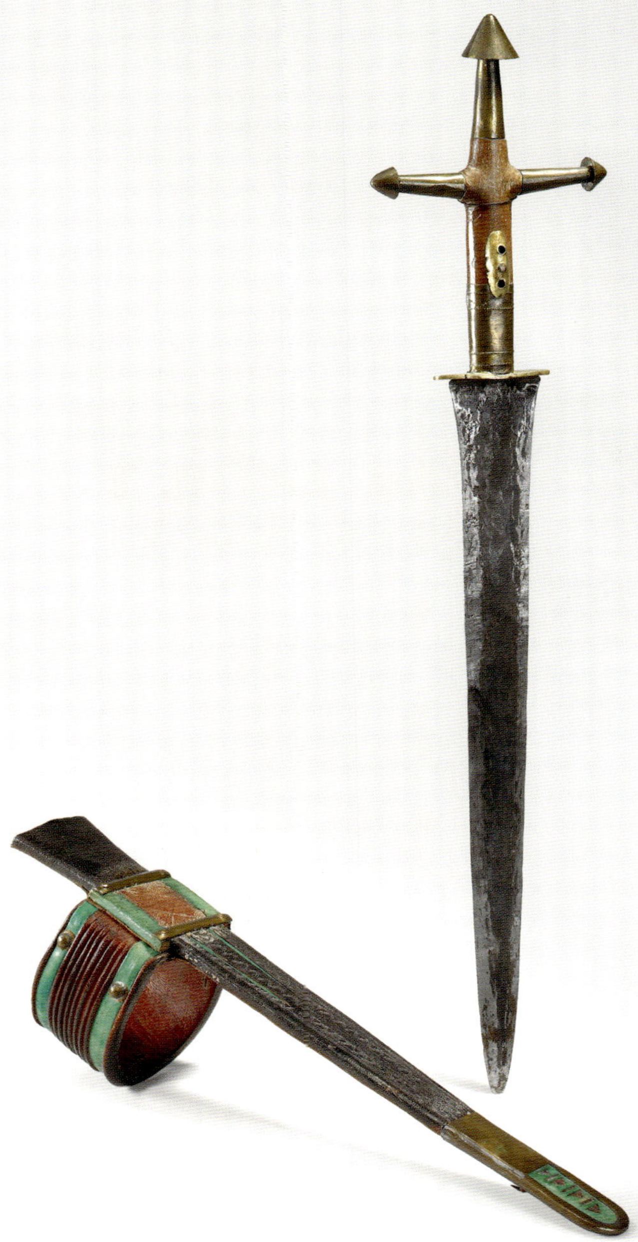

Similar to Dan, Senufo peoples venerate ancestors and a host of spirits. Diviners are consulted to appeal to spirits associated with unknown, unpredictable places beyond human settlement (such as forests and uncultivated lands), and they commission sculptures to appeal to these capricious spirits, to seek their support and good will (fig. 3.8). Many of the most distinguished diviners in Senufo communities in northern Côte d'Ivoire belong to the Sandogo association, and most initiates are women, with membership typically passed through the mother's line. New diviners often acquire miniature sculptures cast in brass in the form of figurative twins, while, for their consultation rooms, more advanced practitioners seek wooden sculptures shaped using iron tools.

The Poro society is primarily for men, although young girls and postmenopausal women are permitted to join. Its main function is to guarantee good relations between the living and the ancestors. Poro, too, benefits from the empowering presence of iron. Figures and masks of Senufo diviners and Poro members are wooden armatures carved by sculptor-blacksmiths (fig. 3.9). Forging, shaping, and sharpening iron adze blades for this momentous work, blacksmiths are essential to society as masters of their craft, creating spiritually powerful objects for associations that ultimately forge individuals, society, culture, and history.

As cosmic connectors, blacksmiths and their transformative powers are often feared and at the same time respected and admired for their differentness, their extraordinary body-mind power (see Kasfir 2007; see also chapter 11, this volume). Blacksmiths among Taureg or Kel Tamajaq ("speakers of Tamasheq") of northern Niger and Mali are regarded as "superhuman" (Rasmussen 2013). Their rituals and social roles and the significances of their artistry help to explain their special status and the tensions they provoke among customers. One origin myth that incorporates foundational stories of Islam claims that the art of smelting and the blacksmith's tools of hammer, tongs, and anvil were "handed down to them by Father Adam and the Prophet David or *Sidna Daouda*" (Morgan 2010).

Tuareg society is stratified among nobles, clerics, artisans, and the "unfree" taken in raids. Many metalsmiths (*inadan*) come from this last, historically enslaved group (Heath 2010; Prasse 1995; see also interleaf M, this volume).[7] Despite their low status, they are

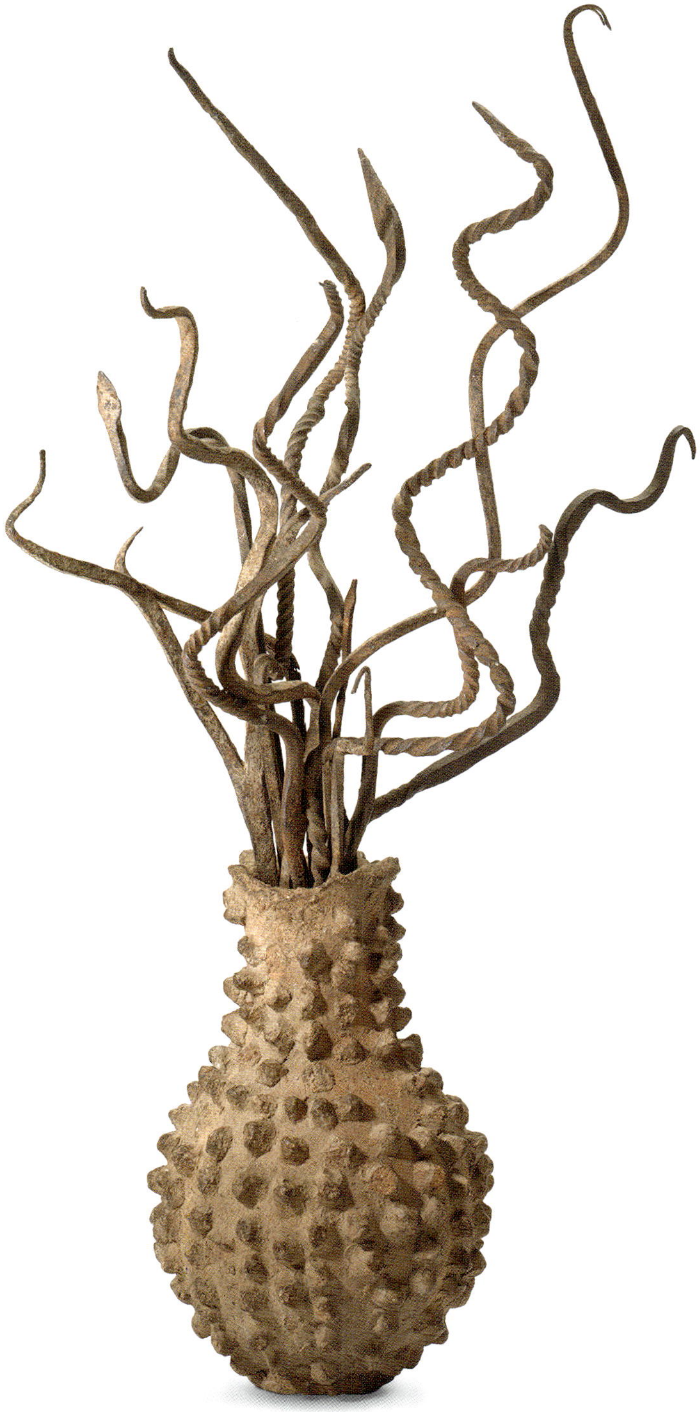

3.11
Artist unknown (Mumuye peoples[?], Nigeria)
Vessel with rainmaking wands, mid-20th century
Ceramic, iron
H: 44.4 cm
FOWLER MUSEUM AT UCLA, X2008.32.3; MUSEUM PURCHASE, 2008
PROVENANCE: PACE PRIMITIVE, NEW YORK; LEE LORENZ COLLECTION, NEW YORK; AMYAS NAEGELE COLLECTION, NEW YORK

To become vertical, these iron rainmaking wands were likely impaled in soft clay inside the ceramic vessel before it was fired. The bump-covered surface of the pot and its fire-born contents signal ritual purpose and potency.

essential "makers" for society, needed because they create or repair items for survival in the harsh and unforgiving Sahara desert. At the same time, they are feared, and their interactions with clients often involve feelings of envy and blame as well as accusations of "sorcery," probably because of the medium they work, for Tuareg believe that both iron and gold possess dangerous, negative spirit energy. Traditionally silver was a most acceptable metal in Tuareg jewelry, considered "pure, clean, holy, luminous and warm, like the heart of a true believer," while gold was proscribed by the Qur'an, feared for its malicious properties and seen as ostentatious (Abigail Balbale pers. comm., 2017). Those who control the fiery transformation of iron from ore into wondrous swords, knives, lances, and daggers (fig. 3.10)—weapons that earned Tuareg their reputation as fierce warriors—are engaged in "mystical" work touched by spirits, or jinn. The "othering" of Tuareg blacksmiths may also stem from their use of a special language, called Tenet, to communicate exclusive knowledge among themselves in the presence of nonartisans (Morgan 2010). Though set apart and feared, blacksmiths are at the same time respected. Their ritual knowledge has profound impact on Tuareg society and religious life, contributing to a complex and dynamic cultural system.

IRON, FERTILITY, AND RENEWAL

Iron has a particular potency as offerings are made to secure the seasonal rains that ensure the fertility of crops and the survival of all living beings. Mumuye peoples of central Nigeria consider rainmakers among the most powerful protectors of community survival. Their ritual supplications require iron rods forged in the form of zigzag-shaped staffs (figs. 3.11–3.13) to represent the flash of lightning in the sky and the sudden movement of slithering snakes, thought to presage rain. Blacksmiths empower rainmakers by producing staffs that can be secured in the ground to marshal the earth's life force. Like metal lightning rods on the roofs of homes that are grounded deep in the earth to save us from conflagration, Mumuye rainmaking rods are visual petitions for protection, meant to sustain society. These exquisite forms also resonate with anticipated rivulets of water on parched land. Such devices illustrate a key message of African iron art: virtuosity and efficacy are interrelated.

3.12
Artist unknown (Mumuye peoples, Nigeria)
Rain wand, mid-20th century
Iron
H: 70 cm
PRIVATE COLLECTION

3.13
Artist unknown (Mumuye peoples, Nigeria)
Rain wand, mid-20th century
Iron
H: 60 cm
COLLECTION OF MARK CLAYTON

Sinuous iron rods used by rainmakers evoke rivulets of running water, the flash of lightning, and the movements of snakes that forecast the coming of rain.

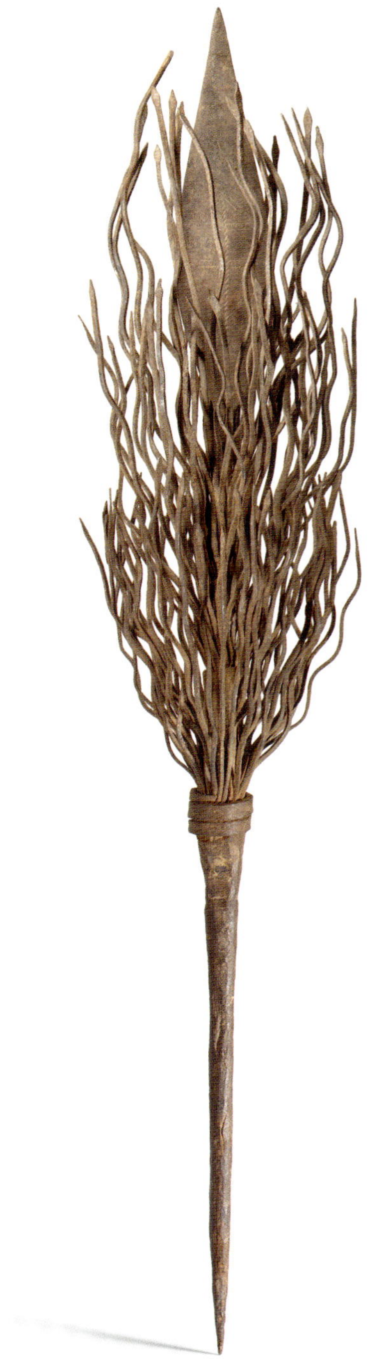

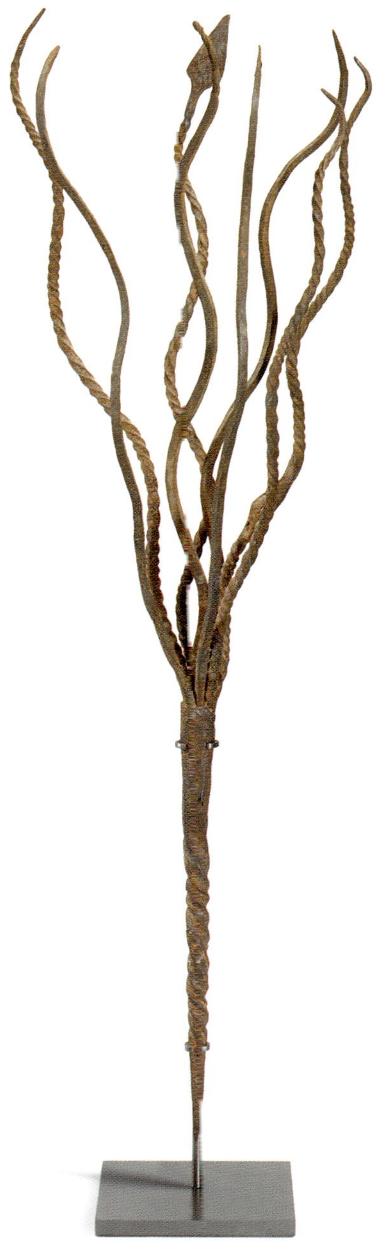

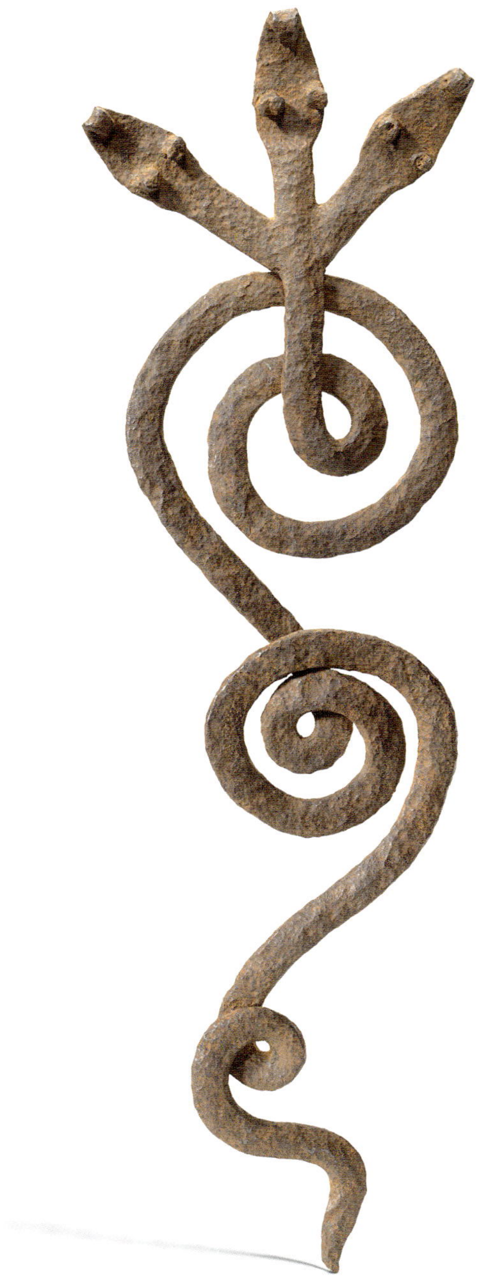

3.14
Artist unknown (Gan peoples, Burkina Faso)
Figurative staff, 19th to early 20th century
Iron
H: 54 cm
PRIVATE COLLECTION

This remarkable spiraling composition of a three-headed python was planted upright in the ground to summon rains and protect people from lightning. Among Gan, snakes inspire reverence and awe, and they are seen as mediators of earth and sky and of the rains that fall to sustain life. Such iron serpentine forms were placed on altars to honor the spirits of the land whom Gan call *thil*.

A remarkable forging from Gan, Lobi-related peoples of Burkina Faso and northern Ghana, features a spiraling snake with three heads, used to call down the rains and protect people from lightning (fig. 3.14). For Gan, snakes are profound, quick, fearless beings that move in mysterious ways. Some burrow underground, presumably "dead and buried," only to emerge "reborn" with new skin. All such startling attributes must contribute to the reverence and awe with which Gan/Lobi regard snakes. They are mediators of the world and the otherworld, earth and sky, the rains that fall from above to sustain life below. Fashioned into rain staffs and iron currencies, serpentine forms are placed on altars in honor of *thil*, spirits of the land. Perhaps blacksmiths' intimate interactions with earth, heat, fire, air (via bellows), and water (to quench hot iron) link them to some of the sensory environmental elements associated with the coming of a storm: the building heat and humidity, blinding flashes of light, gusty winds, and then downpour. Blacksmiths live with and manipulate natural forces that make them cosmic connectors who can bring life-giving rains.

IRON, MIGHT, AND MAJESTY

South-central Ghana is the land of Akan peoples, and long the scene of conflict and competition for the riches that gave the region its colonial name, the Gold Coast. The wars that were fought there over past centuries required weapons of iron. These still appear, but now as symbols of leadership, past victories, and dominion over vast trade in gold. Akan smiths' awesome physical and spiritual abilities made them candidates for transformative work as priests and healers (see Konadu forthcoming).

The special status of Akan blacksmiths has produced extraordinary ferrous finery, given prominence at the Akuapem Odwira festival, an annual new year/new yam festival of renewal and purification. Founded in 1733, Akuapem (northeast of Accra) celebrates histories of military might with arrays of swords, spears, arrows, knives, and guns. During the festivities, sensory elements assemble and build to proclaim power from wealth in gold and from glory achieved with iron weaponry. In the reports of guns, shouts, songs, bells, drums, and laughter; in the smells and tastes of feasts from first harvests; and in the exquisite motions of riders, dancers, and acrobats, Akan peoples recall battles of the past, celebrate victories in the present, and anticipate struggles to come.

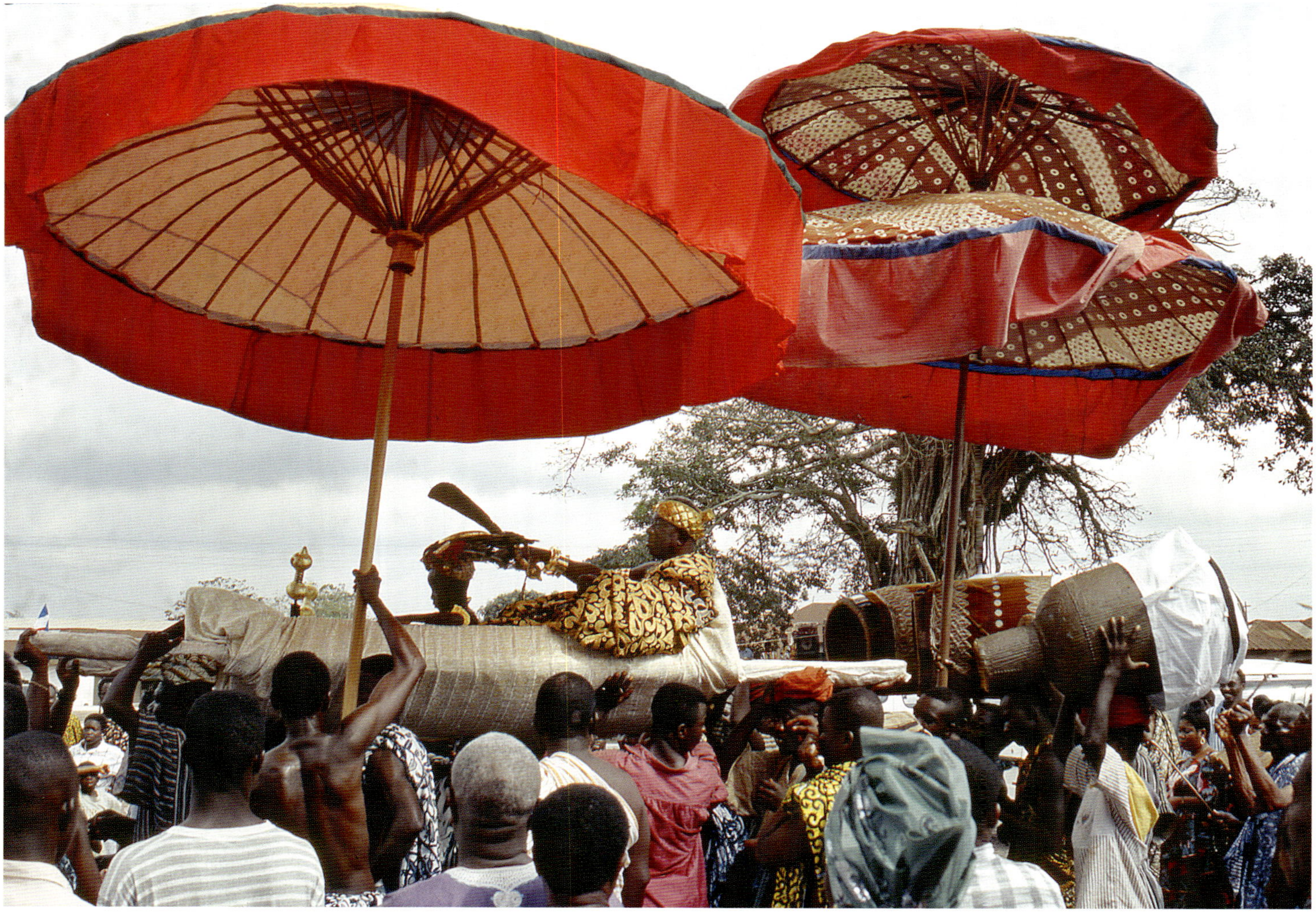

3.15 During the annual Odwira festival, King Nana Addo Dankwa III of Akuapem brandishes a sword with an elaborately embellished blade of iron. It is the sword of authority, an extension of his sacred status. He dances with it while carried in palanquin: it shows he is going to war. This sword is also used by chiefs when they swear their loyalty to the king.
PHOTOGRAPH BY MICHELLE GILBERT, AKROPONG, GHANA, CA. 1977. © MICHELLE GILBERT

The king, whose "well-being is linked to that of his people," is sacred (fig. 3.15). During Odwira he is "anointed with the grave dust of his royal ancestors on his forehead for glory, his chest for bravery, and the small of his back for fertility" (Gilbert 1996, 1, 4). Regalia such as state swords and stools embodying the souls of the king's royal ancestors must be washed and fed like humans because they, too, are alive with spirit. They are offered new yam, regarded as the first and only "real" food. Historically, warrior ancestors' stools were "anointed originally with human blood, gunpowder (associated with men and war), and a spider's web gathered from the kitchen (symbolizing the wisdom of women and the trickster Ananse)" (ibid., 3, 6). Substances from the battlefield such as blood and gunpowder (which might propel a bullet through an *iron* barrel) add to such wisdom their accumulated energies (*sunsum*) to empower and protect.

On a certain night during the festival, executioners (*abrafo*) come and close all roads and shutter everyone in homes. They carry iron swords, spears, and bells on chains to execute the dangerous nocturnal task of collecting the remains of Odusu, the king's most powerful deity, captured from his enemies during Akuapem's 1826

3.16
Artist unknown (Edo peoples, Nigeria)
Ceremonial sword (*ada*), before 1897
Iron, brass, coral, plant fiber
H: 88 cm
DUMAS-EGERTON TRUST LOAN TO PITT RIVERS FOR 100 YEARS, 1991.13.17.1–2
(NOT IN EXHIBITION)

Though its sheath is made of coral (associated with the sea god Olókun) and not removable, this Edo state sword is nonetheless a powerful object. It embodies a history of warfare and conquest, and possesses, through the *àṣẹ* of its iron blade, a mystical power to make any proclamation "come to pass." The iron *ada* was a key symbol of the early rulers known as Ogiso, "Rulers of the Sky."

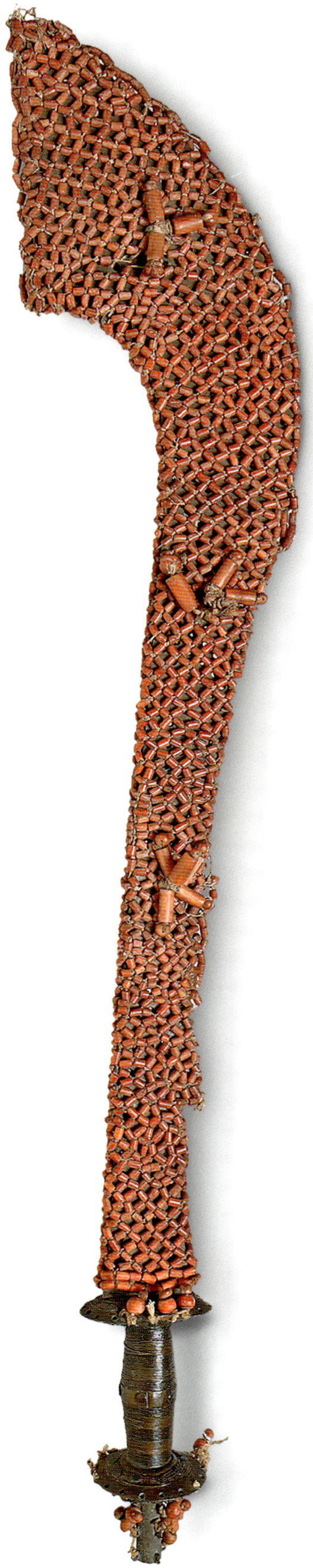

war with Asante. Odusu is an object too powerful to display openly at the start of Odwira, for it embodies the violence of a war-filled past and the souls of dead enemies. The executioners whistle and hum (normally prohibited) to call and remove the spirits of those enemies sacrificed in the past who might seek vengeance, for it is said that they may lurk about for years afterwards (ibid., 9).

The tools (weapons) of executioners and warriors symbolize matters of life and death. Their cutting blades act decisively. They therefore are most appropriate for taking an oath, swearing an allegiance, or proclaiming a truth. Among Yorùbá, persons swear on iron, kissing it to affirm the truthfulness of their words. Sourced from earth that is the abode of ancestors as the givers of law, iron is witness to truths and falsehoods (see chapter 12, this volume). Iron is judgment. Similar ideas and practices appear among the Asante, Akan-speaking peoples with their center in the city of Kumasi, in southern Ghana, for whom, "after the royal stools, the state swords are the most important regalia items. They are categorized in a complex hierarchy for a variety of ritual functions, the most important of which is swearing an oath of allegiance to the paramount chief" (Ross 1996, 18).

Blades of war and conquest also figure prominently in the Kingdom of Benin, founded by Edo people. The early formation of the kingdom (1000–1300 CE; in present-day Nigeria) was the period of rulers known as Ogiso, "Rulers of the Sky." One of their symbols was a scimitar-bladed iron sword called an *ada* (figs. 3.16, 3.17). Like the state swords of the Akan, the *ada* associated with Edo state formation embodies histories of warfare and conquest that eventually led to a centralized kingdom with its epicenter in Benin City and its locus of power in the person of the divine ruler (*ọba*). The *ada* sword symbolizes a king's power to take human life because "the Edo believe that iron has the mystical power of *àṣẹ* to ensure that whatever proclamations are made will come to pass" (Ben-Amos 1995, 25). This consolidation of sacred and political power was the accomplishment of King Ewedo, who reigned in the latter part of the fourteenth century. He declared that as a principle insignia of office the *ada* should be his alone (Ezra 1992, 9). During the age of "warrior kings," in the fifteenth and sixteenth centuries, the realm was expanded greatly, as far west as Ouidah (in present-day Republic of Bénin); northward to Owo, Ekiti, and Ishan; and eastward to the Niger River. In such times, an *ada* was sent to a vassal lord as an empowering emblem of authority, with iron the supreme sign and source of power over life and death.

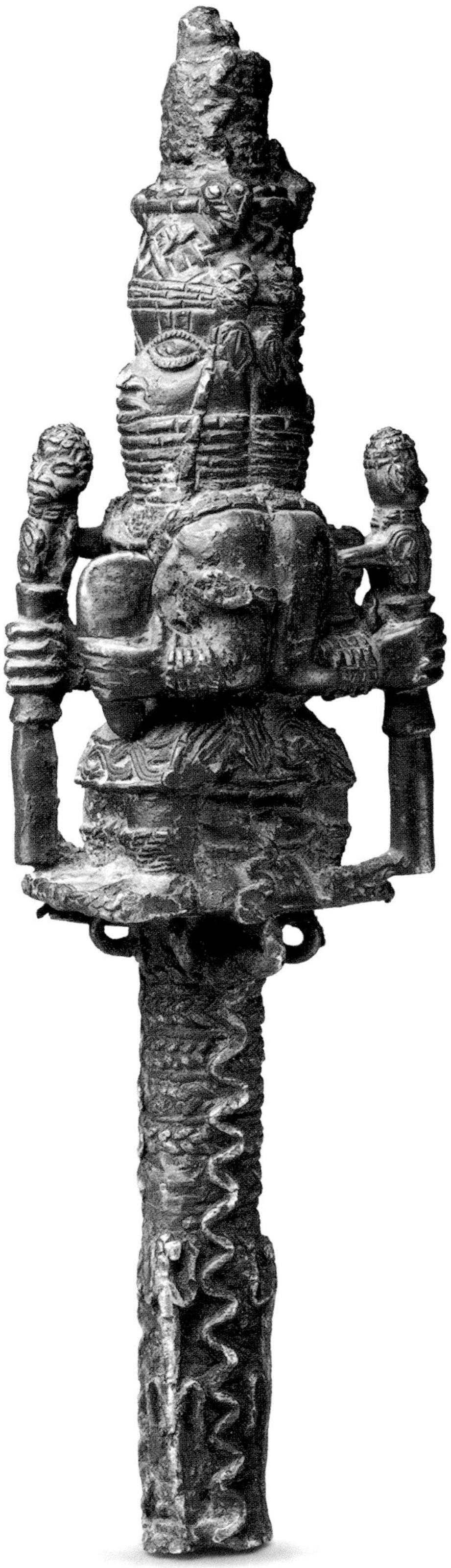

3.17
Artist unknown (Edo peoples, Nigeria)
Metal sculpture, 18th to 19th century
Brass, iron
H: 34.3 cm
THE METROPOLITAN MUSEUM OF ART, NEW YORK, 1991.162.8
GIFT OF MR. AND MRS. KLAUS G. PERLS, 1991
(NOT IN EXHIBITION)
IMAGE COURTESY OF THE THE METROPOLITAN MUSEUM OF ART

Iron's presence activates this royal staff of office. Cast in brass over a core of iron, it usually depicts the divine ruler (*ọ́ba*), a high official, or a spiritual force that must be possessed or confronted. The staff fuses the iron-related powers of Ògún, the medical interventions of Osun, and the beautiful and fearsome red brass of divine royalty. Like the *ada* sword (see fig. 3.16), such a staff carries authority to make proclamations and pronounce judgements with sacred power (àṣẹ).

The forgers of iron ceremonial swords were based outside Benin City, in the Edo village of Igueben, a location echoing the placement of blacksmiths in transitional spaces—between centers and peripheries—in other west African cultures (among them Bamana and Yorùbá). The blacksmiths' guild was one of sixty-eight under royal patronage; these guilds included brass casters, ivory carvers, and wood carvers. The Edo adage "smiths and woodworkers will never suffer from poverty" expressed the benefits of royal support, for such important craftsmen were "servants of the king" (Ben-Amos 1995, 15). The artistry of Benin bronze casters and ivory carvers has received the most scholarly attention, probably due to Eurocentric attitudes toward so-called "noble" media, but it should be remembered that ivory and wood carvers were dependent on blacksmiths' iron tools (fig. 3.18).[8]

Certain Edo blacksmiths belonged to a village guild dedicated to creating ceremonial dance implements known as *eben* for the king and his high-ranking chiefs (fig. 3.19). An *eben* consists of a large iron blade pierced with elaborate openwork designs and a handle that allows the blade to whirl around when tossed in the air. The shape of the blade puns upon several important royal symbols: broad *ewere* leaves used in title-taking rites to signify peace, prosperity, and the king's role as the supreme "ritual specialist" of Edoland; the fish form, possibly a reference to Olókun, god of the sea and linked spiritually to the divine king; and royal fans wielded by court attendants to cool the king (Curnow 1996, 17).

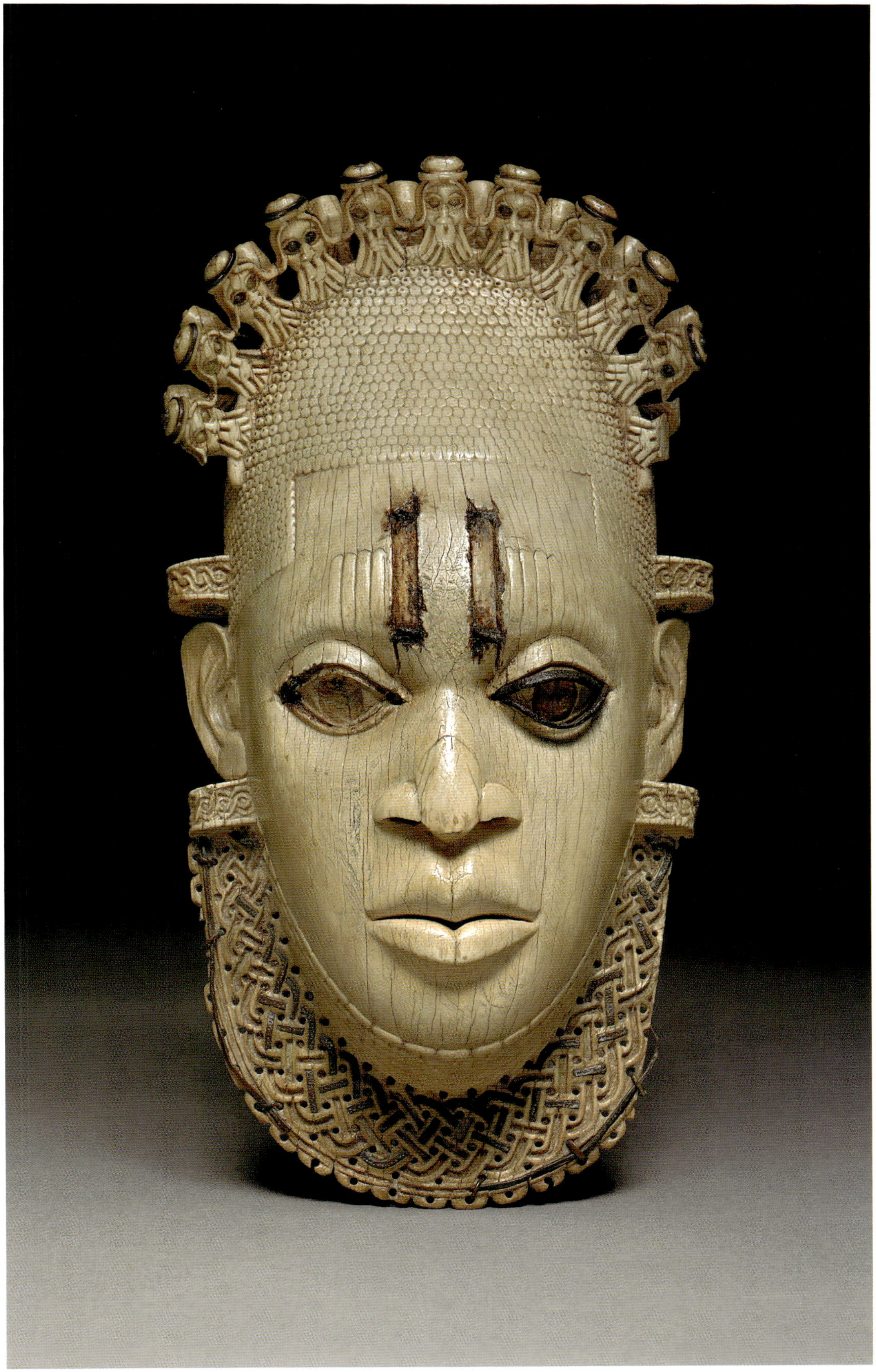

3.18
Artist unknown (Edo peoples, Benin City, Nigeria)
Mask pendant, 16th century
Ivory, iron, bronze
H: 24.5 cm
BRITISH MUSEUM, AF1910,0513.1
(NOT IN EXHIBITION)

Waist pendants of carved ivory faces with iron eyelids and pupils as well as forehead scarifications may have been developed to achieve drama, aura, and efficacy in order to convey the omnipresence and omnipotence of royal ancestors. Polished iron pupils in the eyes of royalty suggest a deified person with awesome powers to protect or destroy. The presence of iron eyes, like iron in the core of brass royal staffs of office, or commemorative heads on ancestral altars, offered a kind of spiritual "charge" (*àṣẹ*), activating such objects to connect living royals with departed ancestors.

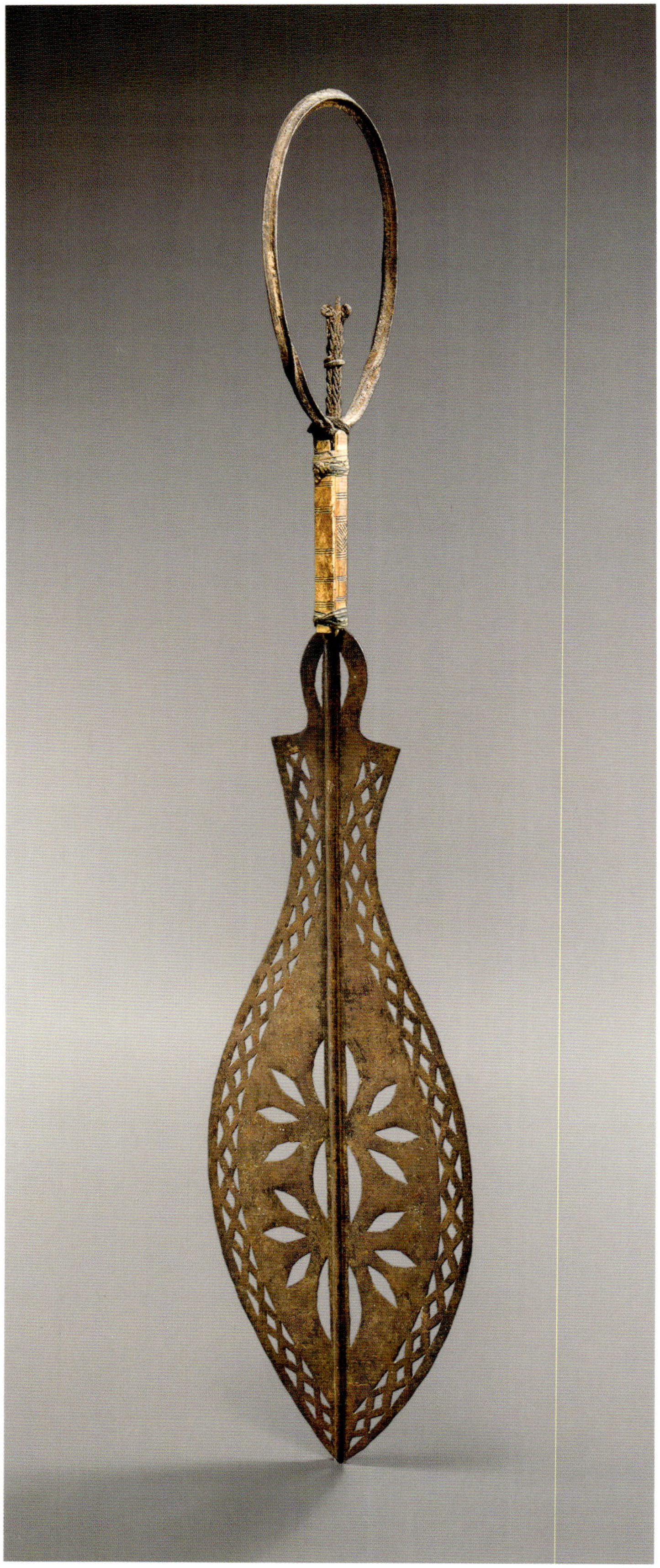

3.19
Artist unknown (Edo peoples, Benin Kingdom, Nigeria)
Ceremonial sword (*eben*), mid-20th century
Iron, ivory, cotton
H: 105 cm
NICOLE AND JOHN DINTENFASS
PHOTOGRAPH BY VINCENT GIRIER DUFOURNIER, 2017.

Along the edge of the *eben* illustrated here is an interlace pattern associated with royalty. In the blade's center are three designs. The triad appears often in Benin arts as a reference to the divine king (at the apex of hierarchical power) and mutual reliance (the support on both sides of attendants and society at large). An *eben*'s handle allows it to be whirled and tossed in the air to perform dramatic gestures of allegiance, fealty, praise, and prayer.

The significance of the *eben* as a *dance* implement involves the senses of sight, motion, and touch. The edges of some *eben* are fitted with iron rings that jingle. In performance settings, an *eben* is raised aloft and sometimes tossed in the air, its blade twirling, by those who have achieved high chiefly rank. In this dramatic gesture of allegiance, fealty, praise, and prayer, which is also a demonstration of agility and endurance, the whirling iron blade flashes in the sun,[9] asserting a chief's rights and privileges, but also his debt, devotion, and loyalty to the king (fig. 3.20). At the same time, the motion creates a cooling breeze of honor, respect, blessing, praise, and protection. Skillfully throwing and catching the *eben* in the presence of the king raises a chief's status, but many avoid the risk, since failure is met with fines and disgrace (Curnow 1996, 16–17).

The *ada* and the *eben* together embody multiple significances and purposes (see fig. 3.6). The *ada* recalls and enacts the power of life and death, a king's sanctioned authority to take a life. The *eben*, in its shape, evokes ideas of peace and prosperity, and in its danced performance, the ideals of loyalty and dedication to a divine monarch and to sustaining Edo society. In both, iron is the medium and the message.[10]

CONCLUSION

Iron permeates the lateritic soils of Africa as well as the beliefs, practices, and histories of the continent's myriad cultures. Taken from earth, ore containing iron must be heated and its impurities separated before the iron itself is worked tirelessly to be transformed into matter useful for culture. Ironworkers—smelters and blacksmiths—are those possessing the extraordinary bodily knowledge and wisdom to make momentous transformation possible. They, together with the medium they transform, are alive with spirit. This presence makes them cosmic connectors capable of transforming nature to create culture.

Iron makes worlds, not only through those forms used to forge the shape and substance of cultures and histories, but also those objects that connect to and activate otherworldly entities—deities, ancestors, and spirits. The artistic virtuosity of blacksmiths imbues objects with performative power, called by different names in many cultures. It is this powerful sacred presence, embodied by iron in its countless forms, that makes all things possible.

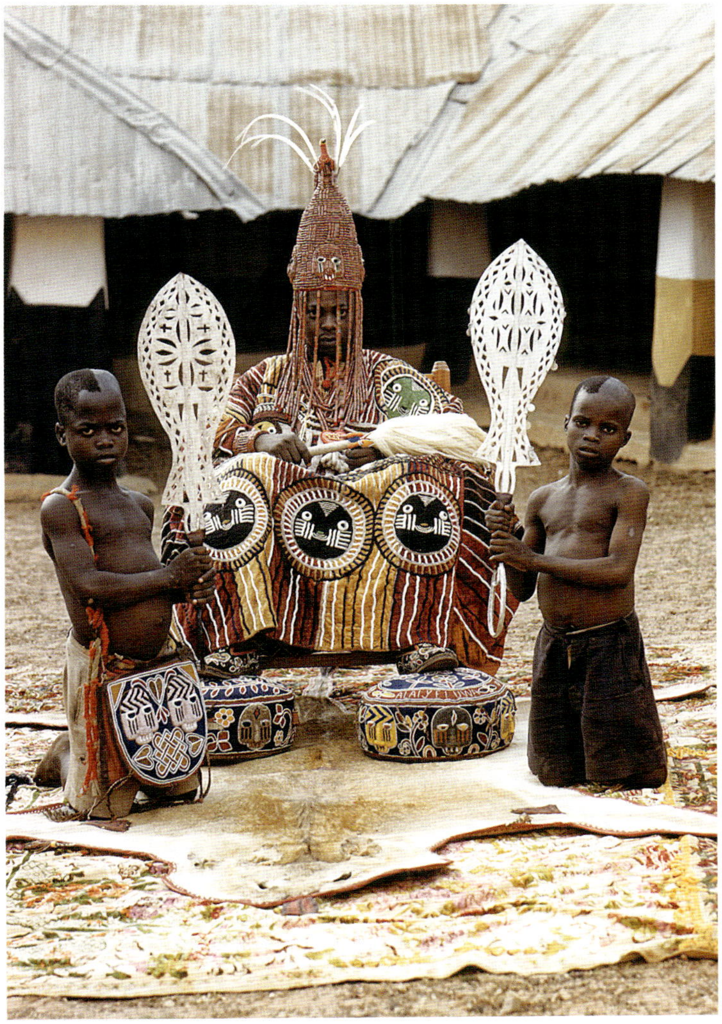

3.20 Ọba Ademuwagun Adesida II, the Deji of Akure, seated on his throne in the courtyard of his palace flanked by attendants carrying *eben*. The Deji, a Yorùbá ruler whose realm has a long history of interaction with the Benin Kingdom, has incorporated the Edo *eben* as part of his royal regalia.

PHOTOGRAPH BY ELIOT ELISOFON, AKURE, NIGERIA, 1959. ELIOT ELISOFON PHOTOGRAPHIC ARCHIVES, EEPA EECL 2074, NATIONAL MUSEUM OF AFRICAN ART, SMITHSONIAN INSTITUTION

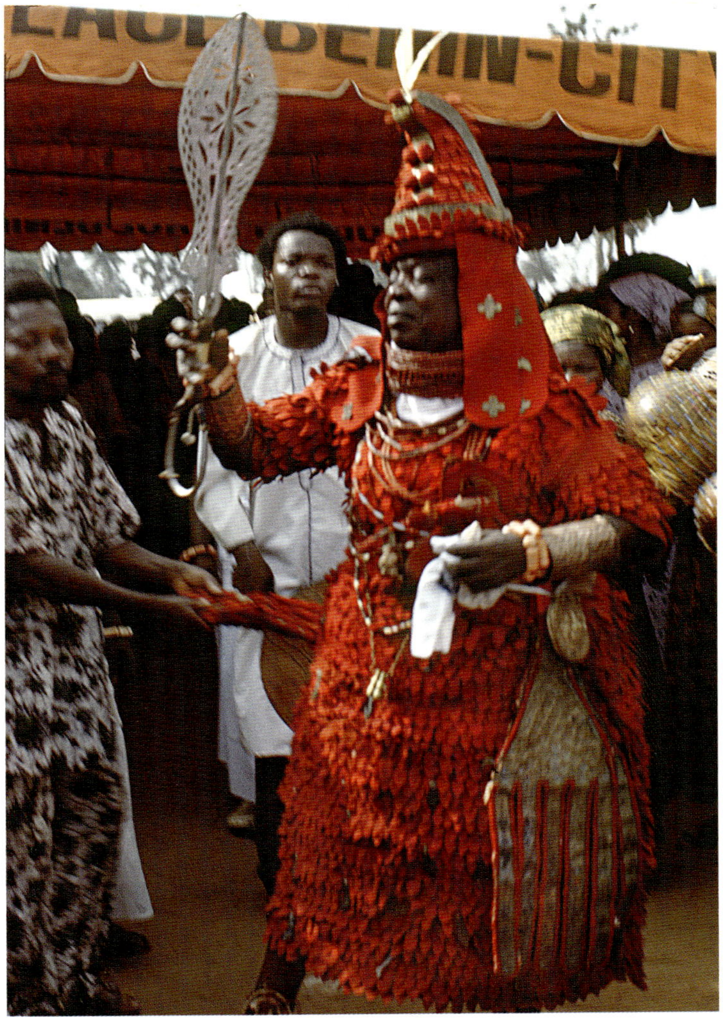

3.21 After extensive and costly rites, a newly installed chief is encouraged and instructed by more experienced colleagues to manipulate his *eben* in several ways. Here, Chief Iyamu, the Esogban of Benin, first salutes the monarch by tipping the point of the sword to the ground several times to acknowledge royal ancestors and the king's divine authority. He then uses the *eben* to recognize his family, supporters, and other chiefs. And in response, well-wishers shout "Iyare!" ("May you go and return safely!").

PHOTOGRAPH BY PAULA GIRSHICK, ROYAL PALACE, BENIN KINGDOM, 1981.

INTERLEAF

A

GẸ̀LẸ̀DẸ́ MASK HONORING BLACKSMITHS

Tom Joyce   Henry J. Drewal

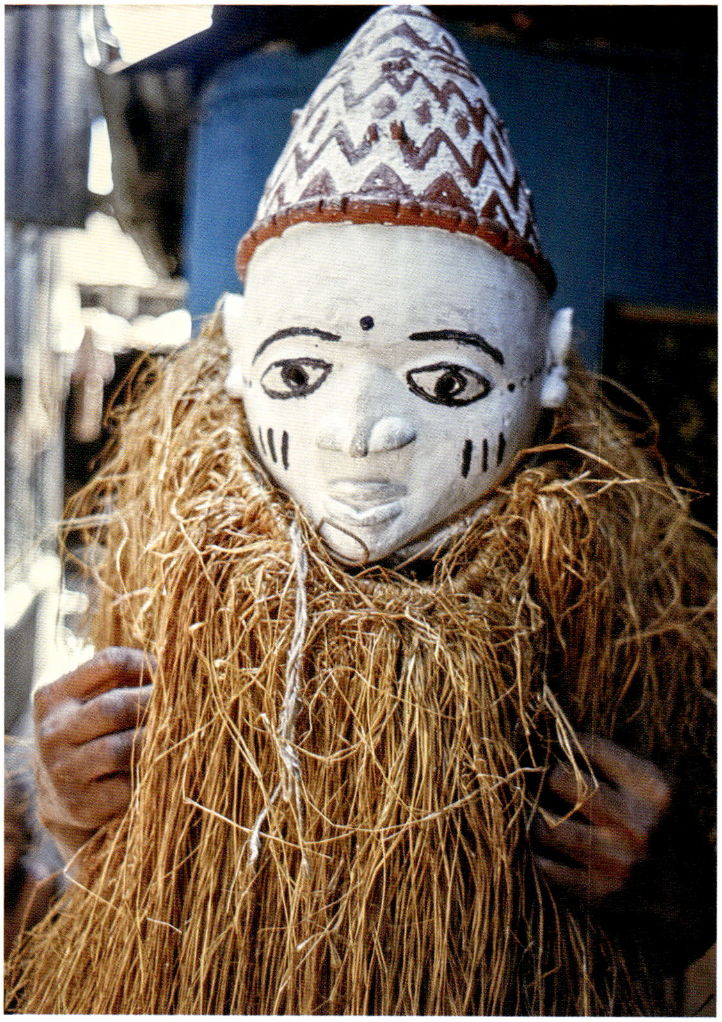

Gẹ̀lẹ̀dẹ́ masquerades celebrate female spiritual powers, especially those of goddesses and elderly women affectionately known among Yorùbá as “our mothers” (*awọn iyá wa*), and reflect gender and social dynamics among community members, considered “the children of our mothers.” Whether by satirizing antisocial behavior or praising dutiful individuals and groups, Gẹ̀lẹ̀dẹ́ masquerades teach, critique, and entertain while shaping public opinion about current events (fig. A.1). A Gẹ̀lẹ̀dẹ́ mask created by Alaiye Adeisa Etuobe (fig. A.2) suggests that the artist may have been a blacksmith or a member of a blacksmith’s family, given his acute interpretation of the workings of the forge.

A.1
Arabi Ajigbale, representing Ògún, the Yorùbá god of iron and patron of blacksmiths, opens the nocturnal performance of Oro Ẹ̀fẹ̀ that precedes the afternoon Gẹ̀lẹ̀dẹ́ dances. His costume of palm fibers (*mariwo*) “sweeps” clean and sanctifies the performance space.
PHOTOGRAPH BY HENRY J. DREWAL, LAGOS, NIGERIA, 1978.

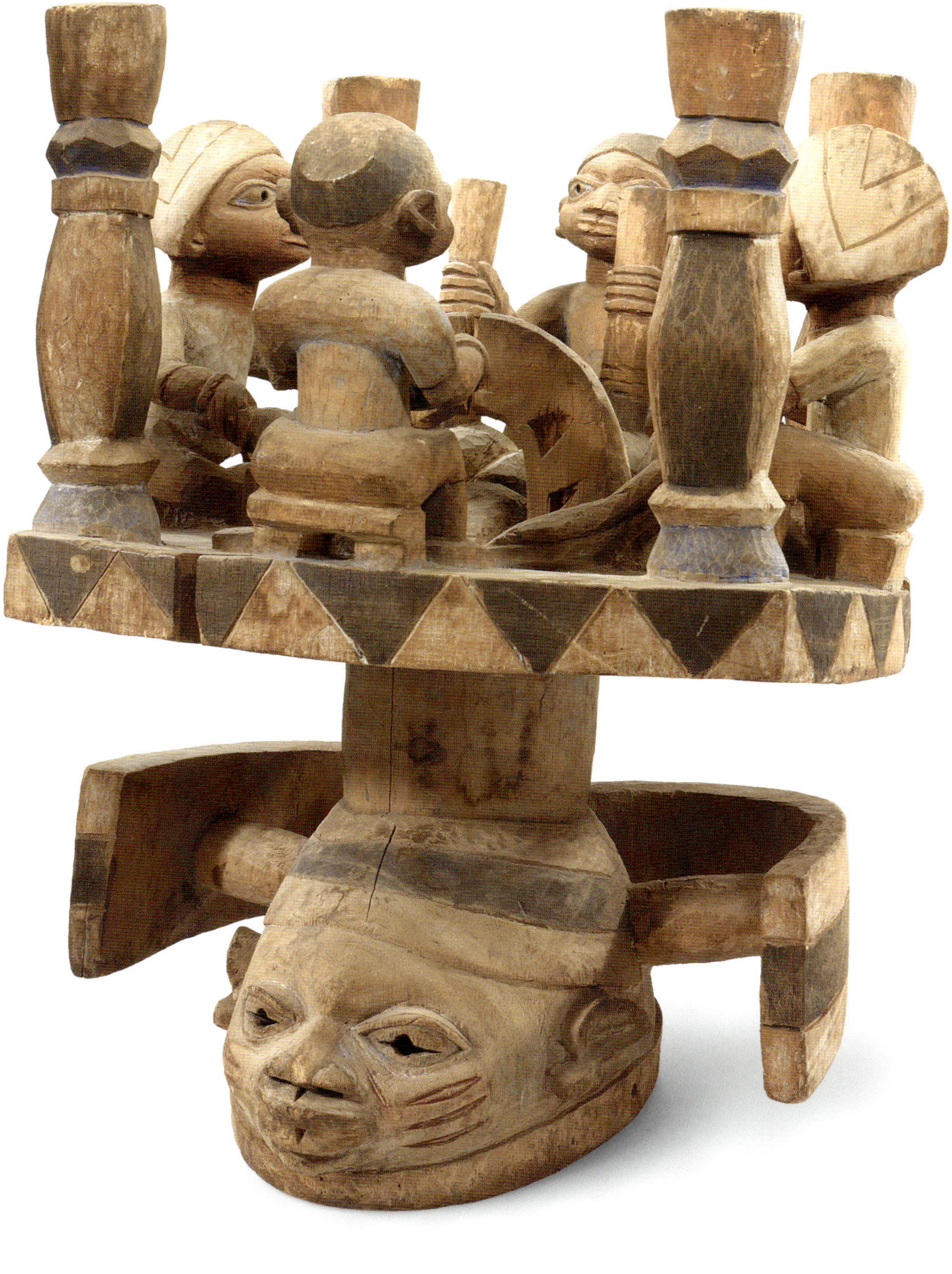

A.2A
Attributed to Alaiye Adeisa Etuobe (active mid-20th century; Yorùbá peoples, Nigeria)
Gẹ̀lẹ̀dẹ́ mask, mid-20th century
Wood, laundry bluing, pigment
H: 62 cm
FOWLER MUSEUM AT UCLA, X70.990; GIFT OF MR. AND MRS. HARRY HUGHES

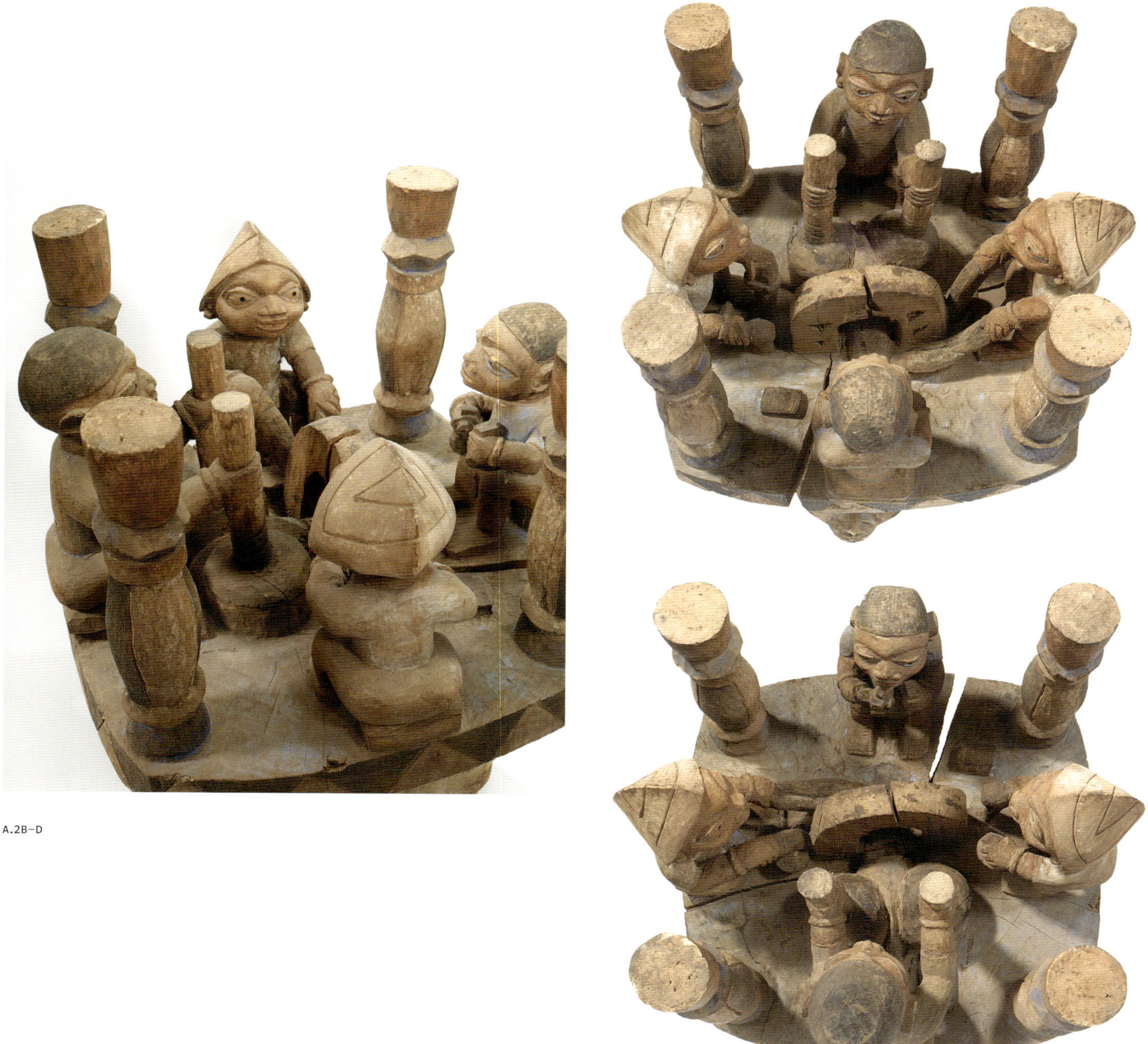

A.2B–D

Atop the mask is a scene featuring coworkers in a blacksmith shop, each poised to execute a specific task when fashioning objects of purpose and beauty. The workshop is directed by the lead blacksmith, sitting with a stake anvil beneath his left hand and a haftless hammer in his right. Opposite him, wearing the same dog-eared caps (*fila abeti aja*) signifying seniority, is the fire tender, with tongs and poker for safely managing hot iron in the fire—a task never given to novices. A well-trained smith has intimate knowledge of every aspect of forging and anticipates each step in the process so that iron is ready and in sync with the tempo set by the lead smith's hammering. Sometimes the iron being heated is more valuable than the labor required to forge it; therefore, the lead smith and the fire tender likely shoulder most of the workshop's financial risk. To the lead smith's right is the striker, perhaps an apprentice, with a handled hammer in his right hand for working hot iron upon the anvil embedded in the ground to his left; following in quick succession every blow directed by the master smith, his efforts more than double that of his mentor in efficiently bringing an object to final form. The bellows operator, to the lead smith's left and shielded behind an arching heat deflector, pumps at a pace in step with production, to move the work forward while mitigating fatigue for the team. The attentive gaze shared by the participants, who have their eyes on each other and on the work at hand, indicates the carver's awareness of their strategic collaboration. The symmetry and balance of the superstructure, and indeed, this focused alertness—a kind of "coolness under fire"—echo the composure of the female face below, adorned with a floating head tie.

When this mask was worn at Gẹ̀lẹ̀dẹ́ performances, gathered community members might have imagined the sounds of the forge through the rattle of masqueraders' iron *ìkù* anklets (fig. A.3). According to one origin myth of Gẹ̀lẹ̀dẹ́, negative forces are dispelled by *ìkù* accompanied by the verbal and rhythmic phrases (*eka*) of the drum ensemble, protecting the dancer and all present (Drewal and Drewal 1983, 226–28). When the mask by Alaiye Adeisa Etuobe was worn, striking iron was evoked from head to foot, from the forging activities above to the forged leg rattles below, extolling the makers of society: blacksmiths and our mothers.

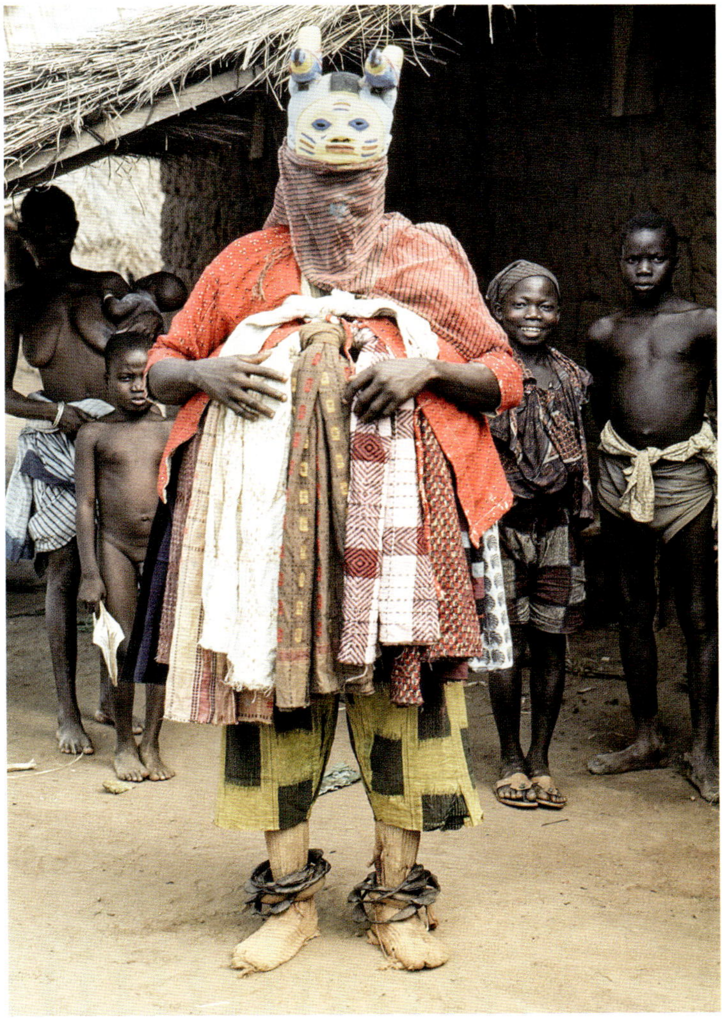

A.3
Gẹ̀lẹ̀dẹ́ masquerader fully costumed with carved mask, cloth head ties borrowed from women, and iron anklet rattles (*ìkù*) worn to accompany the verbal rhythms of the drum ensemble.
PHOTOGRAPH BY HENRY J. DREWAL, IDOFA, NIGERIA, 1971.

CHAPTER
4

POTENCY CONVEYED WITH A HAMMER'S BLOW

Tom Joyce

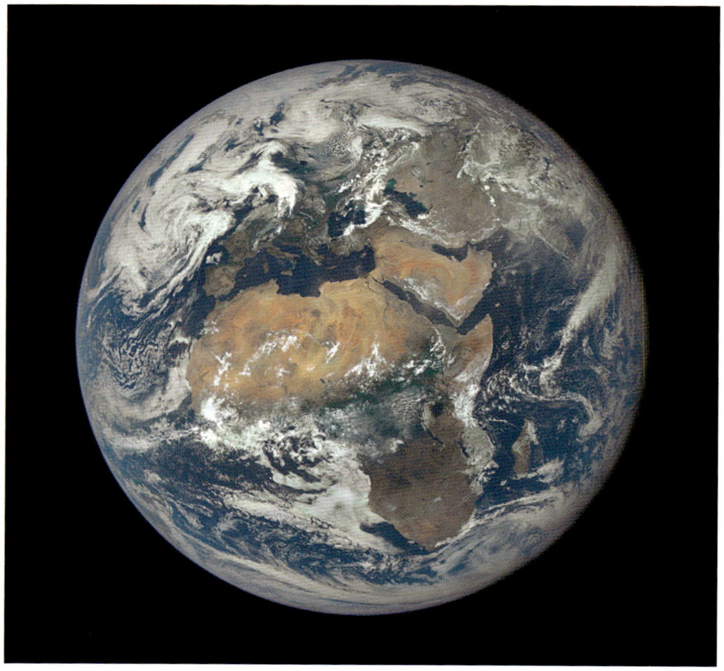

4.1 Taken on July 6, 2015, from a vantage point one million miles from earth, this image shows Africa front and center. The continent's surface appears rust-colored due to lateritic deposits rich in iron oxide.
PHOTOGRAPH COURTESY OF NASA'S EARTH POLYCHROMATIC IMAGING CAMERA (EPIC) ON (DSCOVR)

> Iron doesn't council the hammer.
>
> HAYA PROVERB, TANZANIA (CABAKULU 2004, 42)

> Art is not a mirror held up to society but a hammer with which to shape it.
>
> BERTOLT BRECHT (1924)

IRON ON EARTH, AND IN AFRICA

Before iron passed into the hands of blacksmiths, it made a long and circuitous journey: from the macro to the micro extremes of both celestial and terrestrial spectrums; from its explosive origins in dying suns to its introduction as nanoparticles in soil, plants, animals, and all living things; from microscopic bacteria to the largest mammals inhabiting the earth. Iron in blood and the oxygenating properties it facilitates not only sustain life as we know it, they also acutely affect many other aspects of human health: from the body's immune system to brain development; from declarative learning to elasticity of memory; from regulating metabolism to maturation of motor skills.[1] Blacksmiths in Africa, as elsewhere in the world, owe much to the hefty role iron plays in the hand and in the flesh, balancing cognitive ability and manual dexterity.

Earth's existence, too, is dependent on the life-affirming effects of iron at its core, where a solid sphere, nearly half the size of our moon and hotter than the sun, churns in dynamic movement within the planet's molten "hearth." Four and a half billion years ago, this inner core of iron began generating electromagnetic currents that spun our young earth with sufficient force to slowly cool from its fiery consolidation into a planetary mass, allowing water vapor in its gaseous atmosphere to condense into clouds. While the earth's surface was still too hot for the moisture clouds released to touch the ground as rain, geoscientists believe it took some three hundred

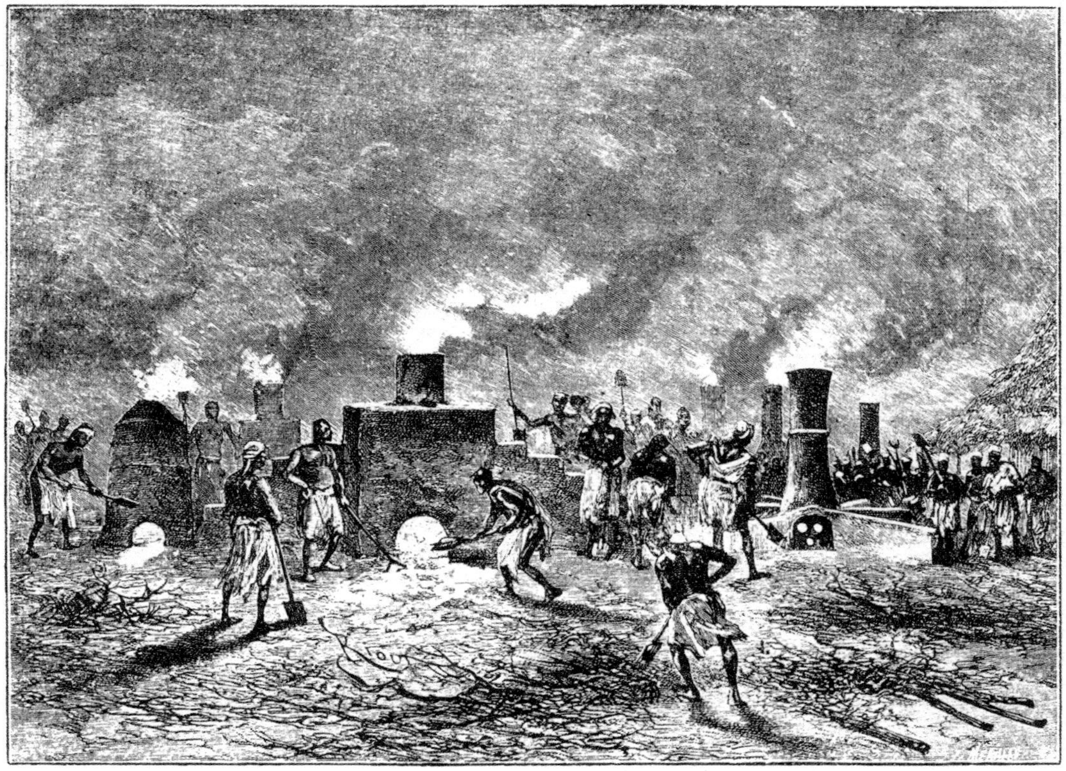

4.2 This engraving illustrates smelting operations circa 1888 at Oumalokho village, in a Senufo and Mande region south of Sikasso in Mali. Captain Louis-Gustave Binger, who claimed Côte d'Ivoire as a French colony in 1893, reported in the travel log that he had observed fifteen tall draft furnaces manned by coordinated teams of capable smiths reducing high-quality iron blooms. REPRODUCED IN *LOUIS-GUSTAVE BINGER, DU NIGER AU GOLFE DE GUINÉE PAR LE PAYS DE KONG ET LE MOSSI* (PARIS: LIBRAIRIE HACHETTE ET CIE, 1892).

million years for evaporative temperatures to recede enough for a global downpour to commence (Logan 1995, 10–13). Once begun, the deluge fell unceasingly for tens of thousands of years, and then on and off again for millions more.[2] Microscopic iron particles embedded in the earth's mantle were loosened from the rain's steady abrasion, enriching early bodies of water. The earliest land masses to emerge during the Archaean Era (four to two and a half billion years ago) existed only in crusty shades of black and gray (Flint 1977). Because oxygen was entirely absent from earth's atmosphere, the natural browns, yellows, reds, purples, and oranges characteristic of geologic features today did not yet exist (Vermaas 2001, 1–7).

Oxygen arrived with the photosynthetic processes of the first single-cell microorganisms, blue-green algae (cyanobacteria) that thrive on carbon dioxide in the air and iron sulfide in the sea (Frei et al. 2009, 250–53). Over eight hundred million years, as oxygen levels steadily rose, floating specks of iron oxidized into heavier ferric compounds that precipitated out of solution and sank to the ocean floor (Jacquot 2009). Settling in colorful layers of magnetite (Fe_3O_4) and hematite (Fe_2O_3) this sedimentary crust, known as the Banded Iron Formation, was in some places forced to the surface by geologic upheaval and now provides all of the 150 million tons of iron ore mined throughout the world each year.[3] Though the earth is composed of 35 percent iron by mass, less than 5 percent resides in the outer mantle and far less is found at depths that are economically viable to mine for processing into a workable medium.[4]

Clearly visible in satellite images from outer space, the surface of Africa is rich in rust-colored lateritic deposits (fig. 4.1). This ready availability of raw material certainly advanced African Iron Age technologies, but even with high concentrations of ferric oxide, it took remarkable ingenuity for smelters and blacksmiths to coax something useful out of this resource—to reduce suitable ore into forgeable material, ushering in path-breaking cultural developments to forever change civilizations (see, for example, Haaland 1985, 50–72; Kriger 1999, 57–85).

As the noted metallurgical historian Cyril Stanley Smith reminded us, the big leaps in metallurgy have resulted less from linear-driven scientific research than from a combination of "inspiration, intelligent observation, deduction, and empirical experiment" (1968, 365). Although innovation builds upon foundations of the past, it has always been apparent that science is more important in explaining technologies long in use, rather than in directly promoting an entirely new invention.

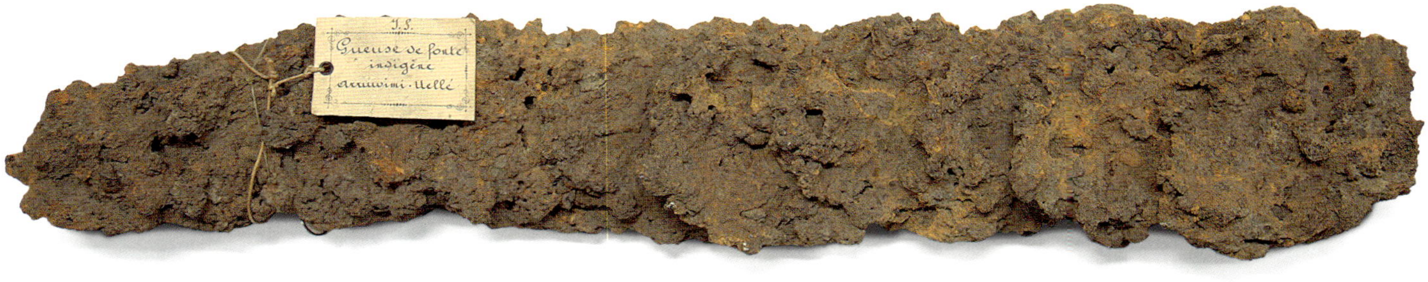

4.3 This iron bloom, sixty-two centimeters in length, was smelted by Boa blacksmiths in the Aruwimi/Uele district of what is now the Democratic Republic of the Congo. It was collected in 1910 by Belgian Lieutenant Mahan and is presently housed at the Royal Museum for Central Africa, Tervuren.
PHOTOGRAPH © TOM JOYCE, 2018

The world's debt to African metalworkers for their contributions over thousands of years is increasingly being recognized, in part due to archaeological excavations that continue pushing the earliest dates of iron smelting farther back than previously estimated (see chapters 2 and 8–10, this volume). Excavations have uncovered iron artifacts along with forge and furnace fragments indicating a diversity of designs that could only have arrived through blacksmith's acute and repeated analysis of every step in ironworking procedures. Constant reevaluation of material qualities, ore peculiarities, air-delivery systems, fuel efficiencies, construction materials, and methodologies have propelled incremental success that in some places resulted in iron of the highest quality manufactured anywhere in the world (figs. 4.2, 4.3).[5]

IRON IN THE HANDS OF AFRICAN BLACKSMITHS

African blacksmiths' activities are enormous in scope, and they command decisive influence in their communities. The staggering orbit of objects emanating from a blacksmith's workshop includes works of art and utility that assist in growing food and harvesting crops, hunting animals and arming warriors, adorning bodies, outfitting rites of passage, managing disputes, making music, summoning ancestors, honoring the deceased, recognizing leaders, symbolizing authority, praising deities, and more (figs. 4.4, 4.5).[6] This essay's commentary is based on observations made through the eyes and hands of a maker trained in the art of forging iron, and it offers an approach to reading layers of meaning in the objects African blacksmiths produce and pass into the hands of others (fig. 4.6). The processes and tools used to create such objects, I would argue, are inseparable intermediaries that enhance an artist's intention by investing the work with "value inherent in its making" (Fariello 2005, 169–70).

In communities throughout Africa, where workshops pass from fathers to sons, it is often customary that every young member of a blacksmithing lineage learn at least rudimentary skills at the forge, even if eventually other professional paths draw him away from the family business (fig. 4.7).[7] While this arduous yet edifying vocation is not for everyone in the lineage, it is understood that the education received is impactful, providing, in a Jungian sense, an archetypically *systematic* model from which to launch one's life (Chen et al. 1999, 44). Blacksmiths are often helped by relatives who begin as apprentices pumping bellows, or who are trained as strikers, wielding a separate, usually larger hammer than the one held by the master smith. Wherever the lead smith strikes, for example, the striker follows, listening and watching for nonverbal signals that the smith communicates through tapping, turning, or some other agreed-upon method.

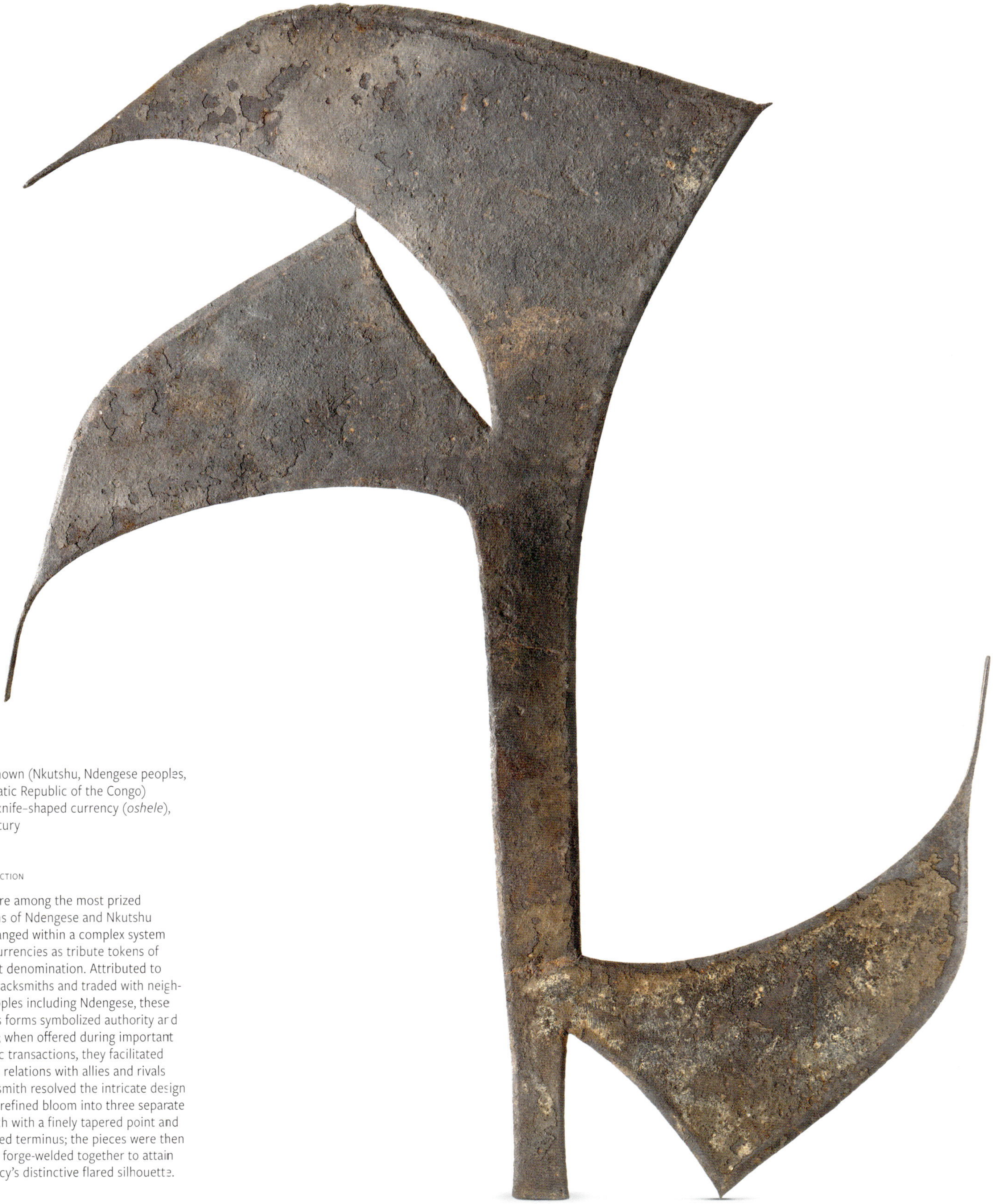

4.4
Artist unknown (Nkutshu, Ndengese peoples, Democratic Republic of the Congo)
Throwing knife-shaped currency (*oshele*), 19th century
Iron
H: 80.7 cm
PRIVATE COLLECTION

Oshele were among the most prized possessions of Ndengese and Nkutshu elite, exchanged within a complex system of metal currencies as tribute tokens of the highest denomination. Attributed to Nkutshu blacksmiths and traded with neighboring peoples including Ndengese, these prestigious forms symbolized authority and leadership; when offered during important or strategic transactions, they facilitated diplomatic relations with allies and rivals alike. The smith resolved the intricate design by forging refined bloom into three separate pieces, each with a finely tapered point and an elongated terminus; the pieces were then seamlessly forge-welded together to attain the currency's distinctive flared silhouette.

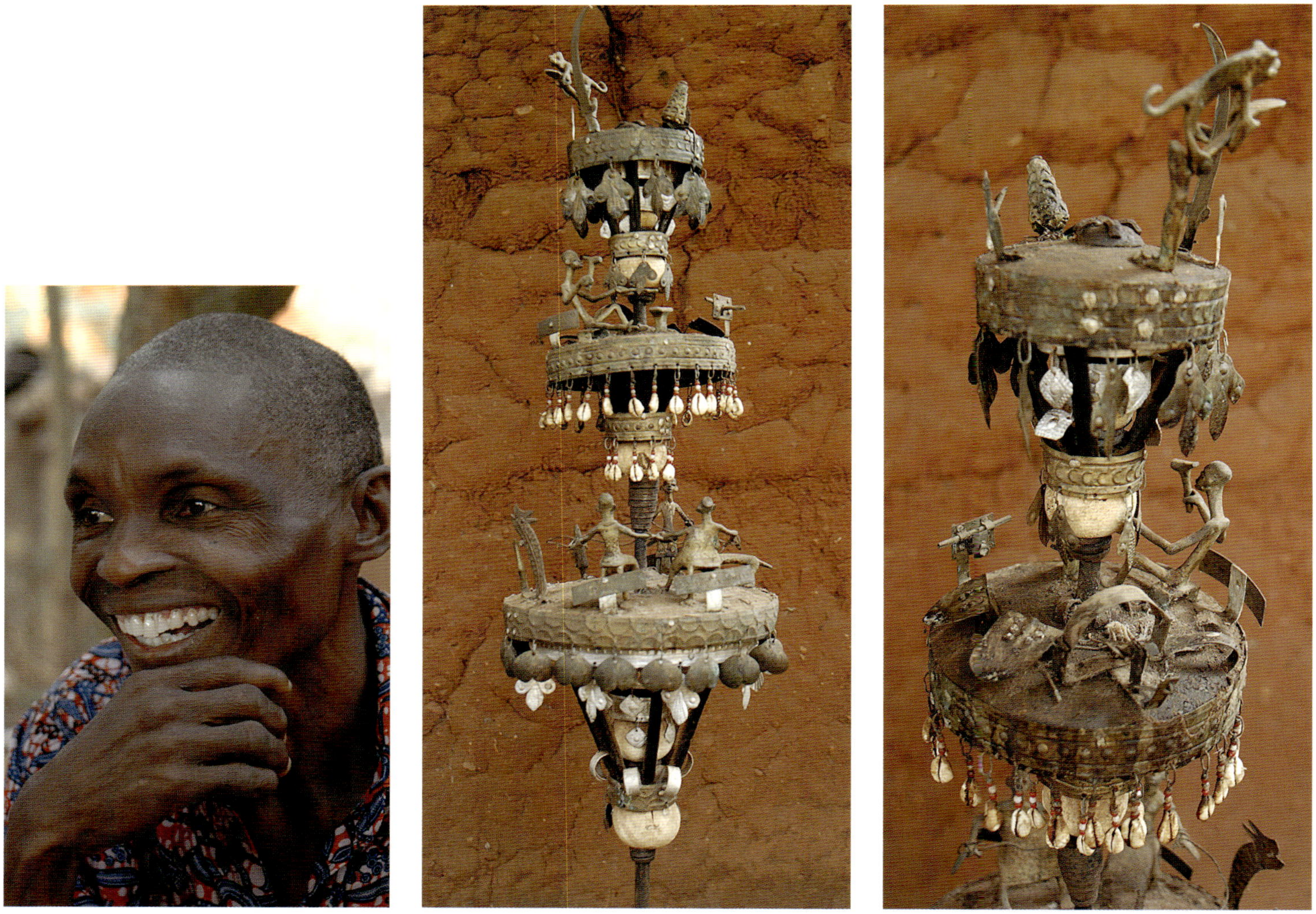

4.5A–C Fon blacksmith Leon Hounsounougan forges in two workshops: one is open to the public for general commissions; the other, in a cloistered space behind closed doors and high walls, is where he produces funerary staffs (*asen*) for clients petitioning to honor deceased relatives (see also figs. 12.18–12.21). Discretion is born out of respect for the mourning families, and to keep secret the stories of loved ones, entrusted to the smith for allegorical tableau made of iron, tin, copper, wood, and cast bronze. Leon Hounsounougan made the three-tiered composition pictured here for himself. It is embellished with family members, trade tools, miniature objects of his own design, and a likeness of him with hammer in hand.

PHOTOGRAPHS BY TOM JOYCE, ABOMEY, REPUBLIC OF BÉNIN, 2010.

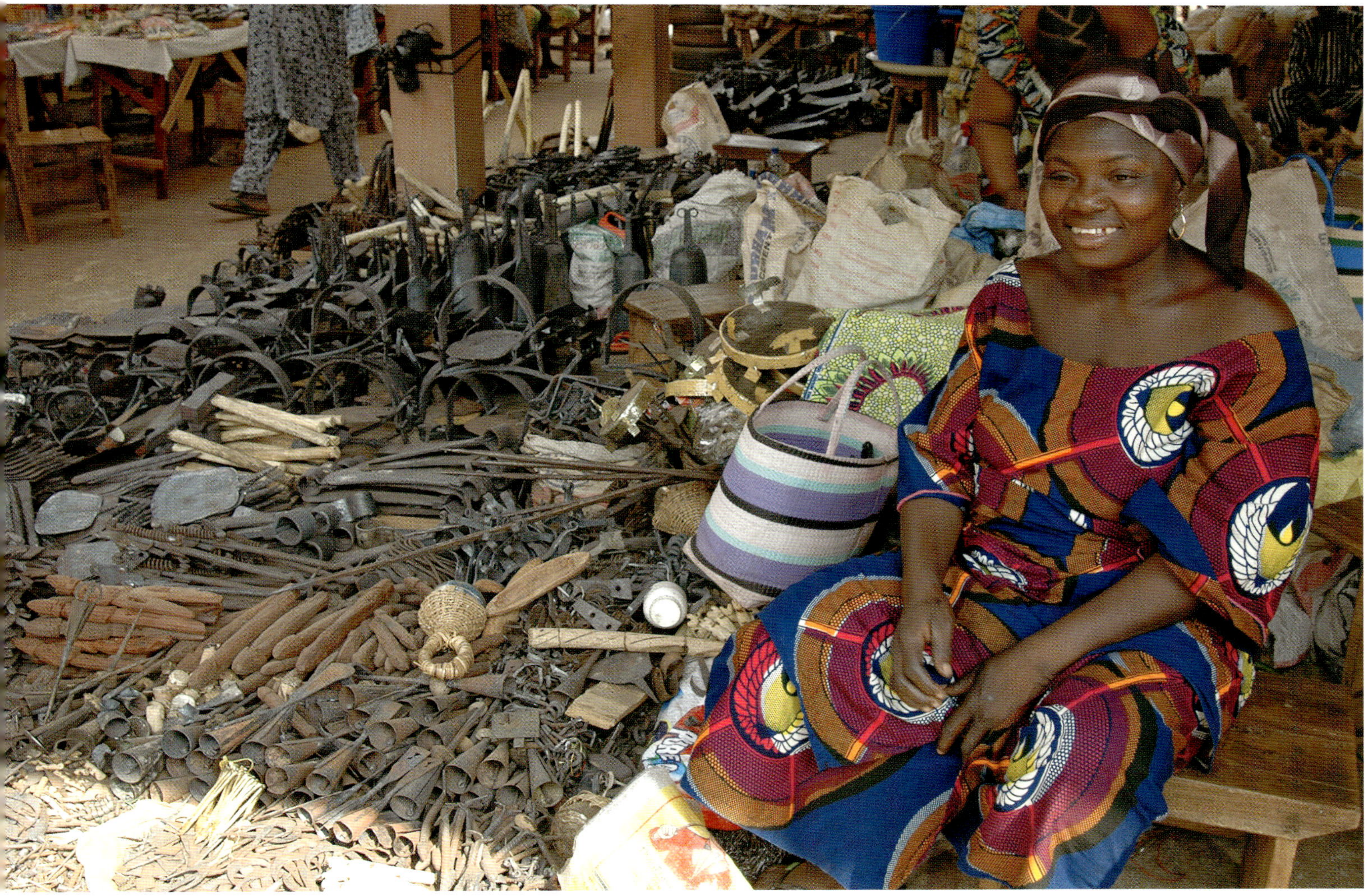

4.6A–C Market vendors in rural and urban areas may be members of a blacksmith's family, or selling may be a cooperative venture where professional market women gather items from neighboring workshops to offer more variety than a single family could provide. At this stall in Abomey, Republic of Bénin, the inventory of articles for sale include: hoes, shovels, sickles, rakes, cultivators, plows, axes, picks, hammers, crowbars, hinges, nails, latches, locks, hooks, chains, animal traps, bells, rattles, knives, machetes, jewelry, raw material, cooking utensils, pots, pans, divination objects, miniature *asen*, and staffs honoring Fon deities. PHOTOGRAPHS BY TOM JOYCE, 2010. © TOM JOYCE

4.6A

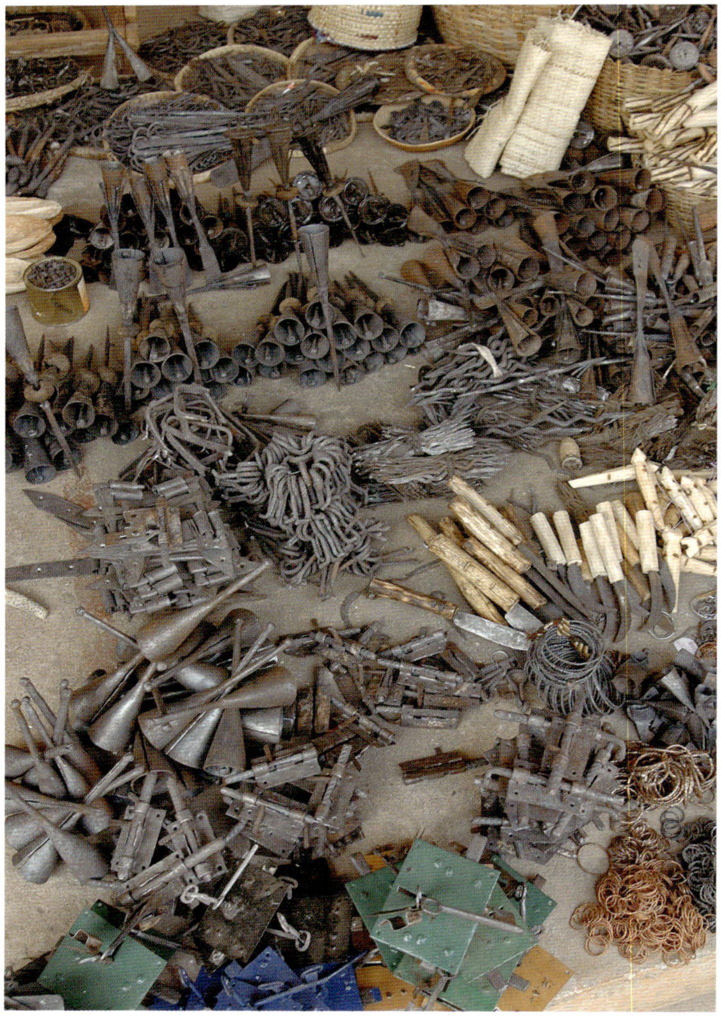

4.6B

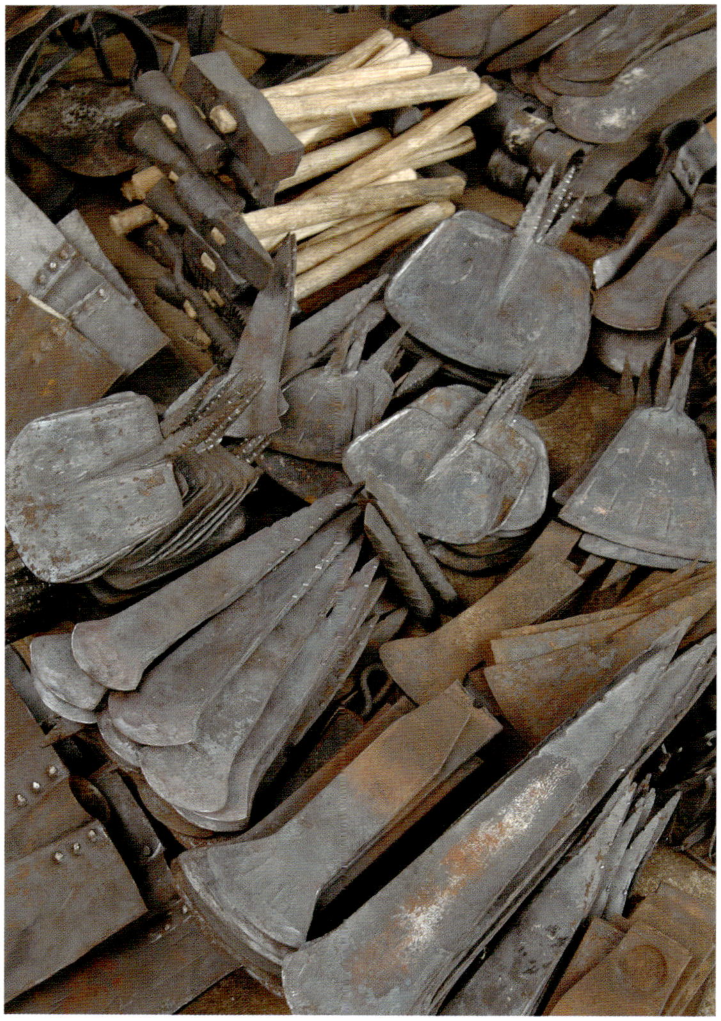

4.6C

4.7A–G The Babadjibi family has worked for generations under the shaded canopy of a cottonwood tree, near to where fields meet the neighborhood of Kunima in Bobo-Dioulasso, Burkina Faso. The patriarch of the family, Sanou Babadjibi (fig. 4.7A, standing, third from left), has trained his relatives to carry out commissions at the forge, providing farming tools and musical instruments, for example, for customers within a fifty kilometer radius of the shop. A portion of the day's production is hand carried to neighboring fields where custom orders are delivered and farmers are also able to buy direct (fig. 4.7B).

Several blacksmiths work from the same hearth in rhythms monitored by the bellows operator. Two strikers wielding French-patterned sledgehammers alternate blows to move iron quickly atop a mushroom-domed stake-anvil. In just six heats, they shape a *daba*, the most popular cultivating hoe they make (fig. 4.7C–D).

Sons are introduced early to the cadence of the forge while tools for pyro-engraving handles heat next to hoe blades made from salvaged angle iron (fig. 4.7E). At this full-service workshop, wooden handles for all the various tools produced are adzed from well seasoned, light-colored hardwood that receives signature designs pyro-engraved to signify the family forge (fig. 4.7F–G).

PHOTOGRAPHS BY TOM JOYCE, 2010. © TOM JOYCE

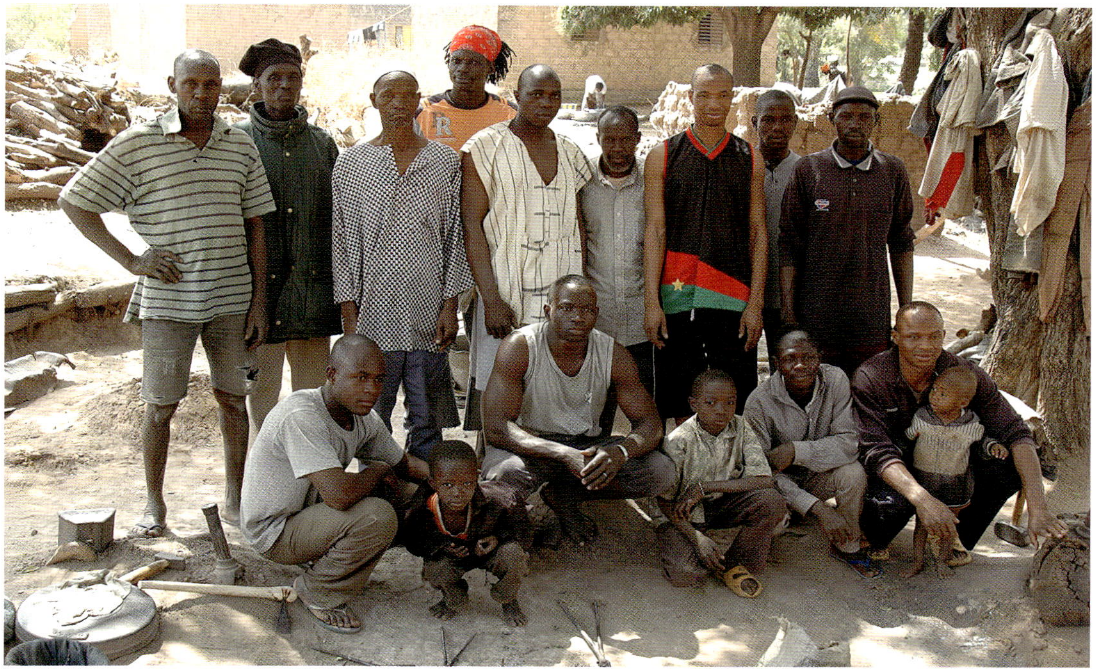

4.7A

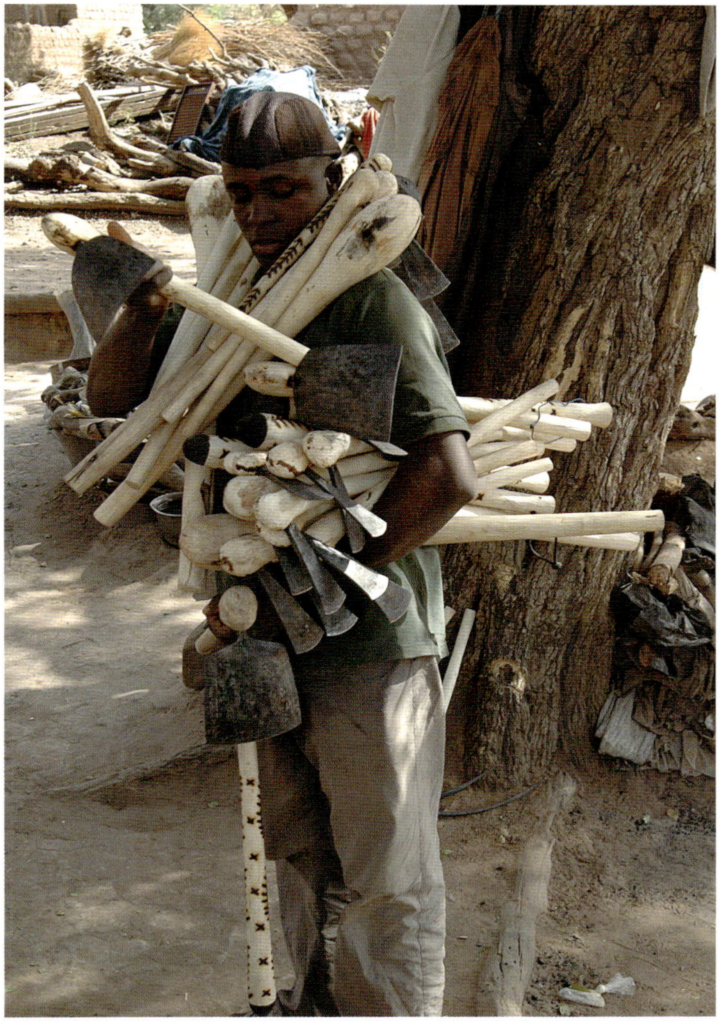

4.7B

4.7C

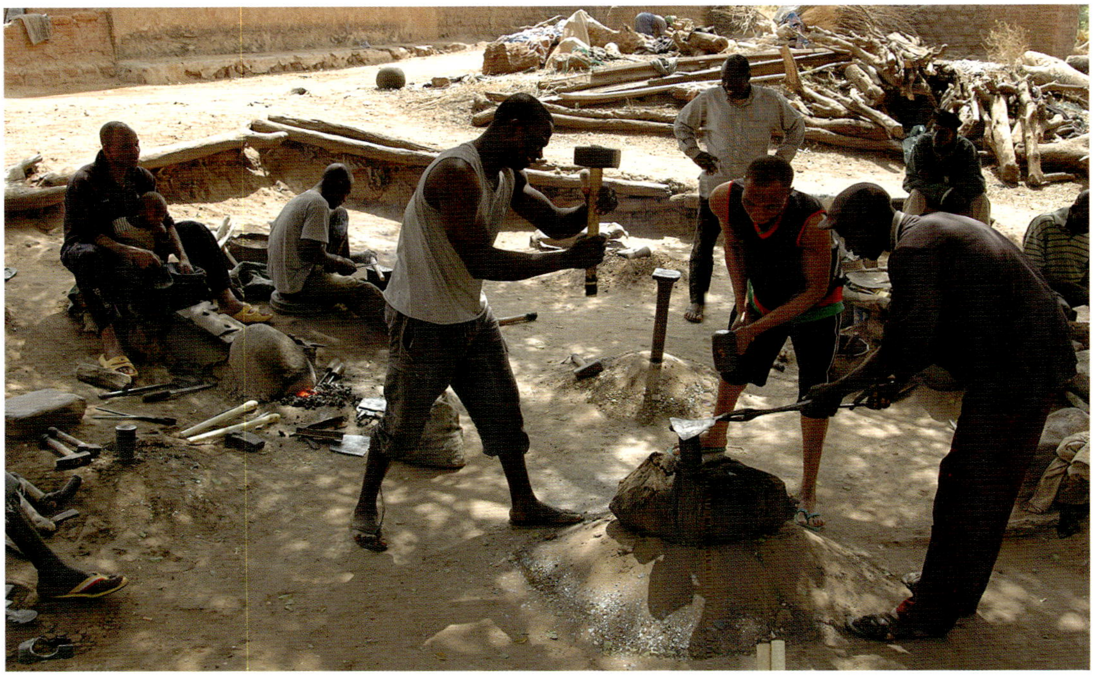

4.7D–E

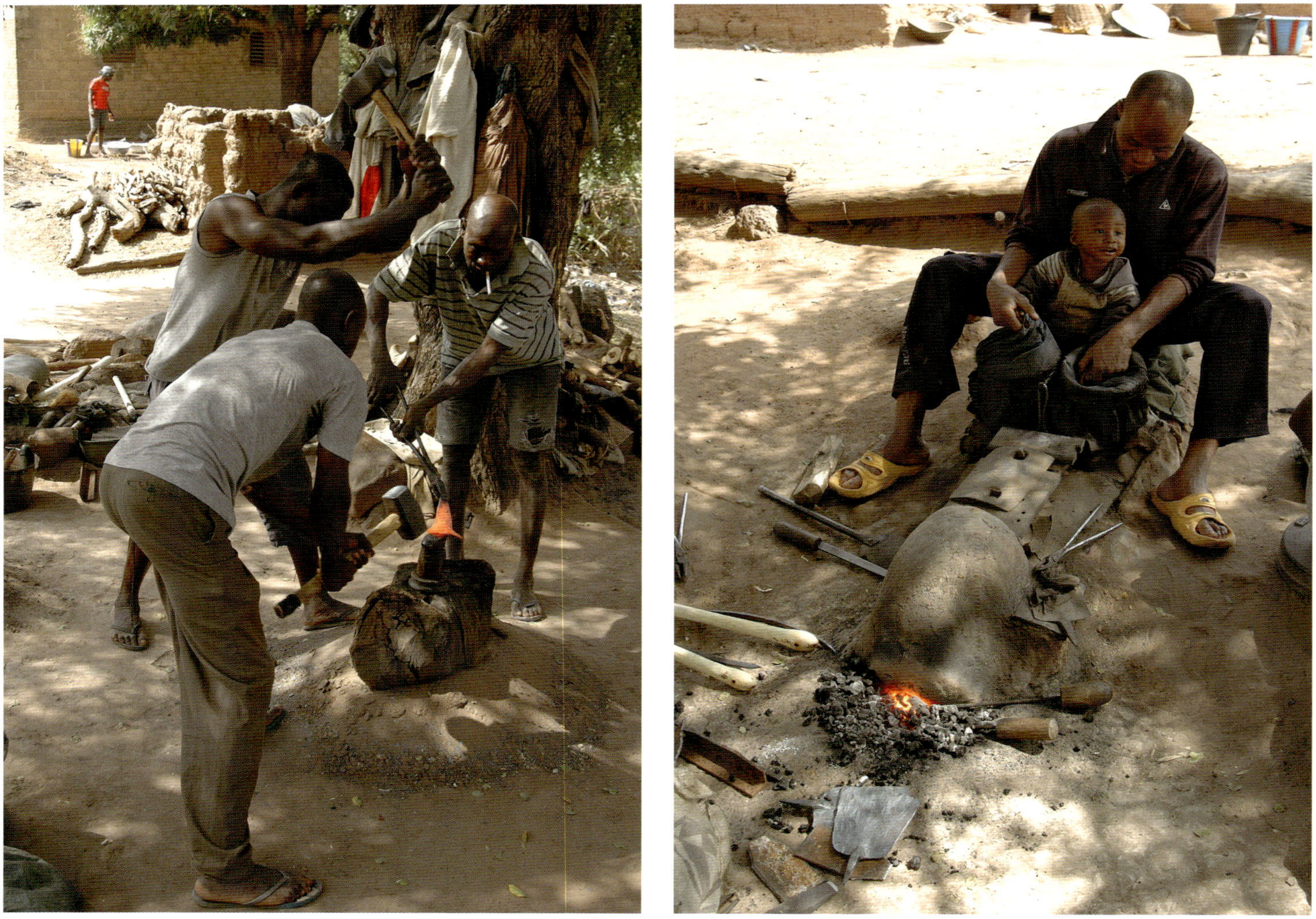

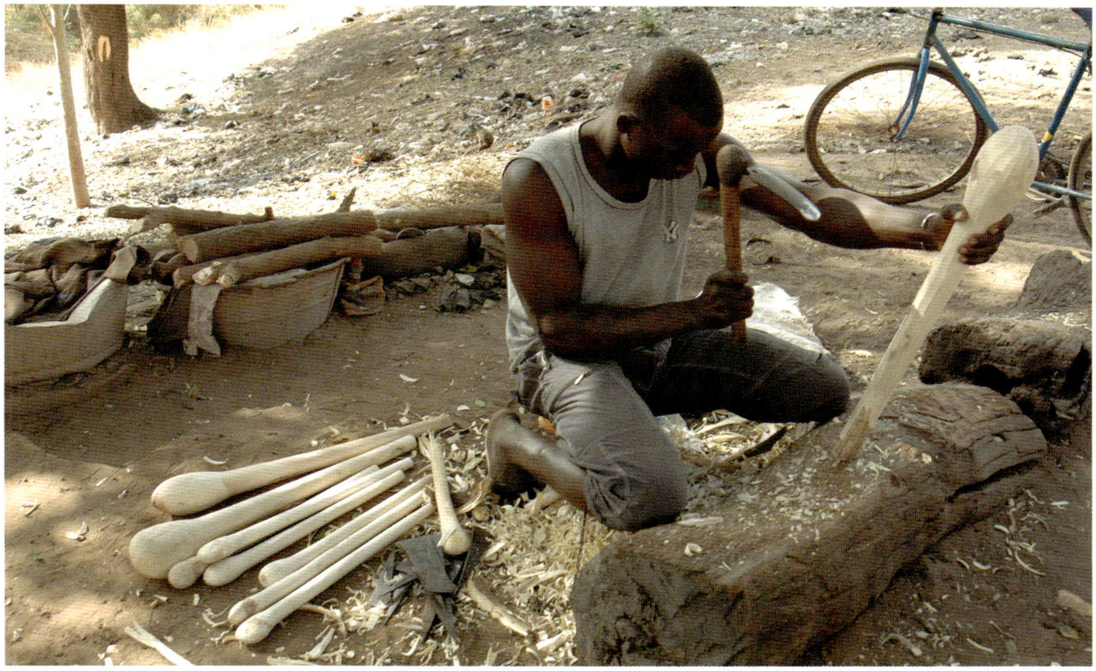

4.7F

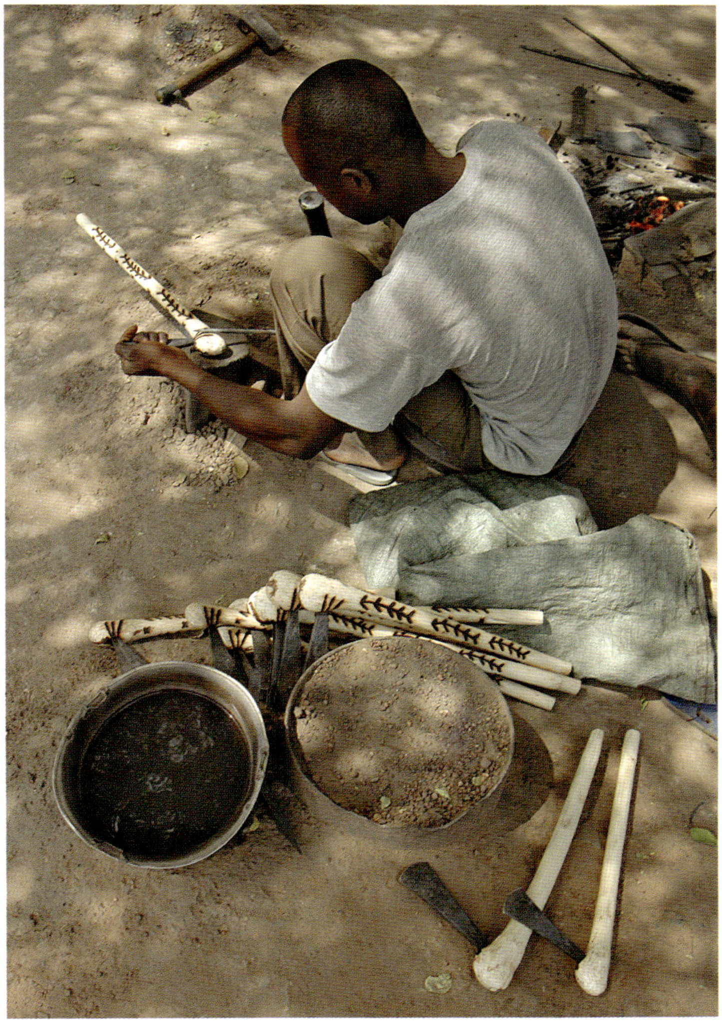

4.7G

4.8A–B Fon blacksmith Herve Dossougbete works at his family's forge, where a steady stream of commissions are produced from dawn till dusk. While he and his brothers are accomplished in forging a vast range of tools as well as musical instruments, they specialize in devotional objects such as ritual knives (fig. 4.8A, lower left) and staffs for family and community shrines (fig. 4.8B).
PHOTOGRAPHS BY TOM JOYCE, OUIDAH, REPUBLIC OF BÉNIN, 2010. © TOM JOYCE

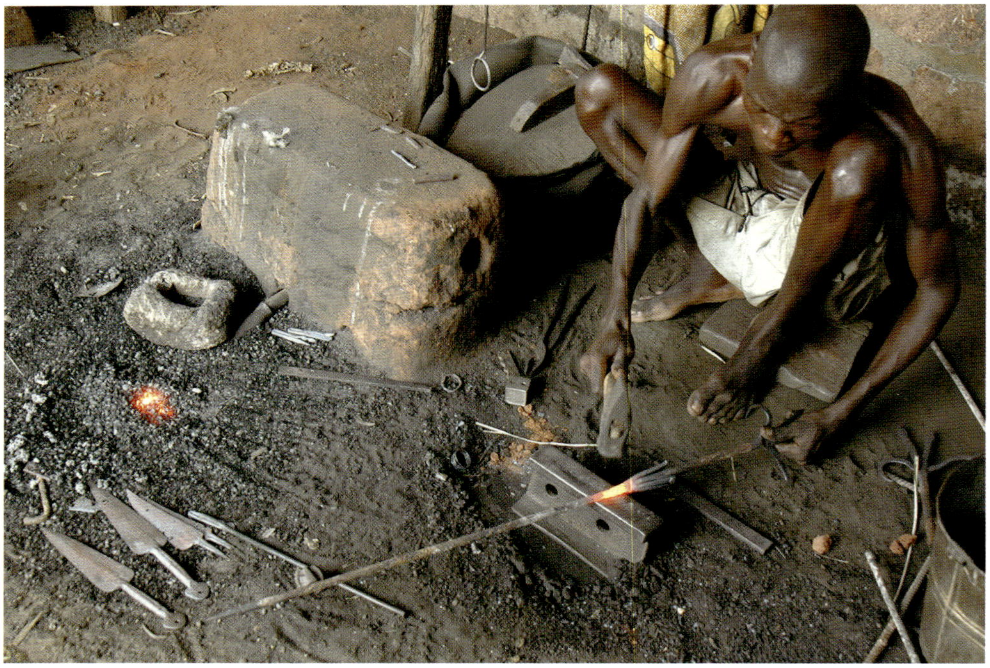

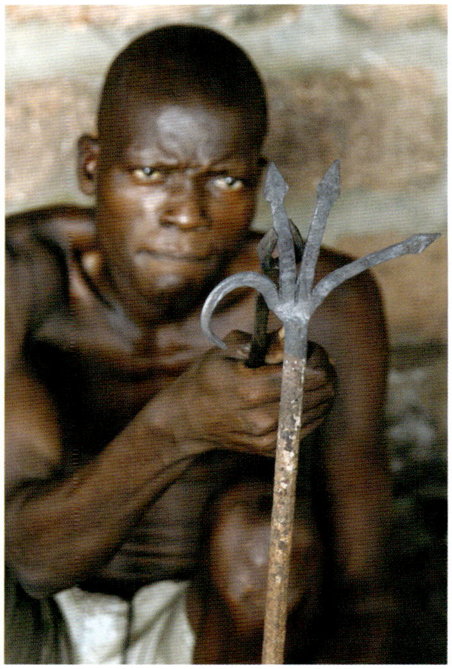

Depending on the young person's interest and the willingness of the mentor to take on a student, intensive training may begin during preteen years. Able to observe a mentor's methods at close range, the assistant, if geared towards the work, assimilates what he sees. This proximity is the quintessential blackboard in the lecture hall, and the mentee's capacity to embody information and other physically demanding responsibilities at a formative age allows instruction to be deeply absorbed along with haptic expertise (fig. 4.8). Metallurgical engineers often speak of "metal memory" as a force to contend with during manufacturing operations (Otsuka and Wayman 1998). Like long-term muscle memory in humans, which enhances motor skills through repeated physical exercise, dedicated practice, day after day, is a blacksmith's only recourse to excel beyond mere proficiency. This is the primary reason apprenticeships and other formal training are administered over many years and usually managed and monitored closely within blacksmithing lineages. There is simply more to learn than meets the eye, and tacit knowledge may be a family's legacy—a priceless inheritance to be invested in future generations (fig. 4.9).

As Richard Sennett writes: "Making physical things provides insight into the techniques of experience that can *shape* our dealings with others. Material challenges like working with resistance or managing ambiguity are instructive in understanding the resistances people harbor to one another or the uncertain boundaries between people. Social consequences are built into the structure and the functioning of the human body; as in the workings of the human hand." He argues that "the capacities our bodies have to *shape* physical things are the same capacities we draw on in social relations" (Sennett 2008, 289–90). It is no wonder that the blacksmith's tasks at the forge offer many symbolic references applicable to decisive action in governance, religion, and warfare, as represented in objects leaders wear or hold, or that accompany them in rituals, politics, or diplomacy.

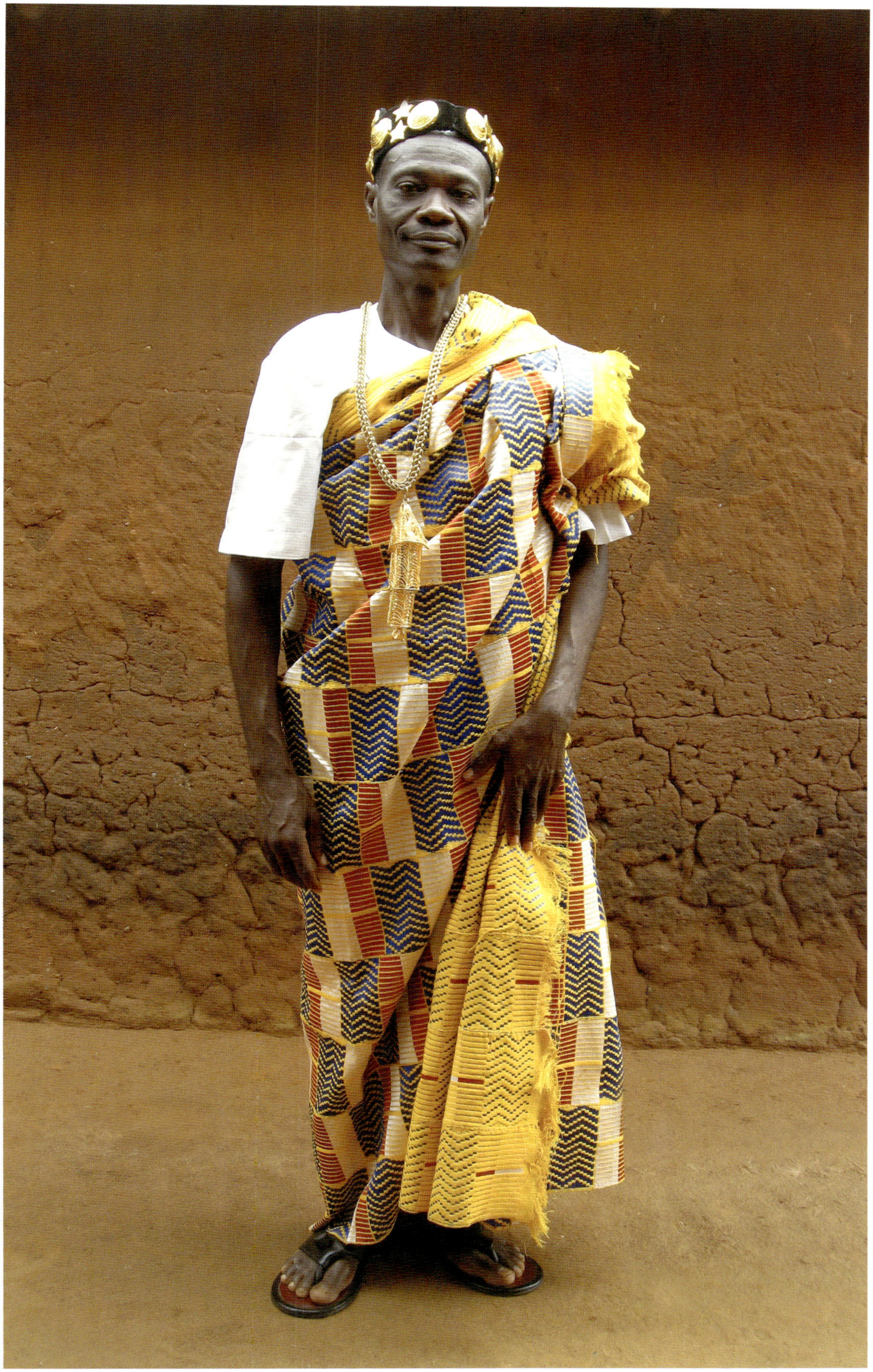

4.9 Togbui Raphael Yovo Agosso is the chief of Yohonou, Togo; his ancestor Togbui Kankoe founded this ironworking center in the eighteenth century. With approximately six thousand inhabitants in several satellite districts, this community is regarded as the cradle of the blacksmithing profession in southern Togo, and its name is a composite of two words: *yoho*, meaning "forge," and *nou*, "entrance." Though professional lineage is inherited, higher education is encouraged, and recent infrastructure beyond the junior secondary school has been implemented to offer children other choices of livelihood. Togbui Agosso is quick to add, however, that his education at the forge has helped him navigate the complexities of governance and diplomacy. As a Senufo proverb has it, "When a blacksmith hammers near the fire, he's careful of the sparks around him" (Garrard 2001, n.p.).

PHOTOGRAPH BY TOM JOYCE, 2010. © TOM JOYCE

4.10 More than half the Ewe population in Yohonou, Togo are blacksmiths. Open-air conditions under shade provide ample ventilation for forging in this tropical environment.
PHOTOGRAPHS BY TOM JOYCE, 2008. © TOM JOYCE

4.11 Pictured here is a small selection of hammers, tongs, and an anvil. This triad along with bellows comprise a blacksmith's basic toolkit. Though most workshops are owned and operated by a single family, every male trained within the lineage possesses his own set of tools scaled to his physical proportions, to be held comfortably in the hand.
PHOTOGRAPH BY TOM JOYCE, 2008, YOHONOU, TOGO. © TOM JOYCE

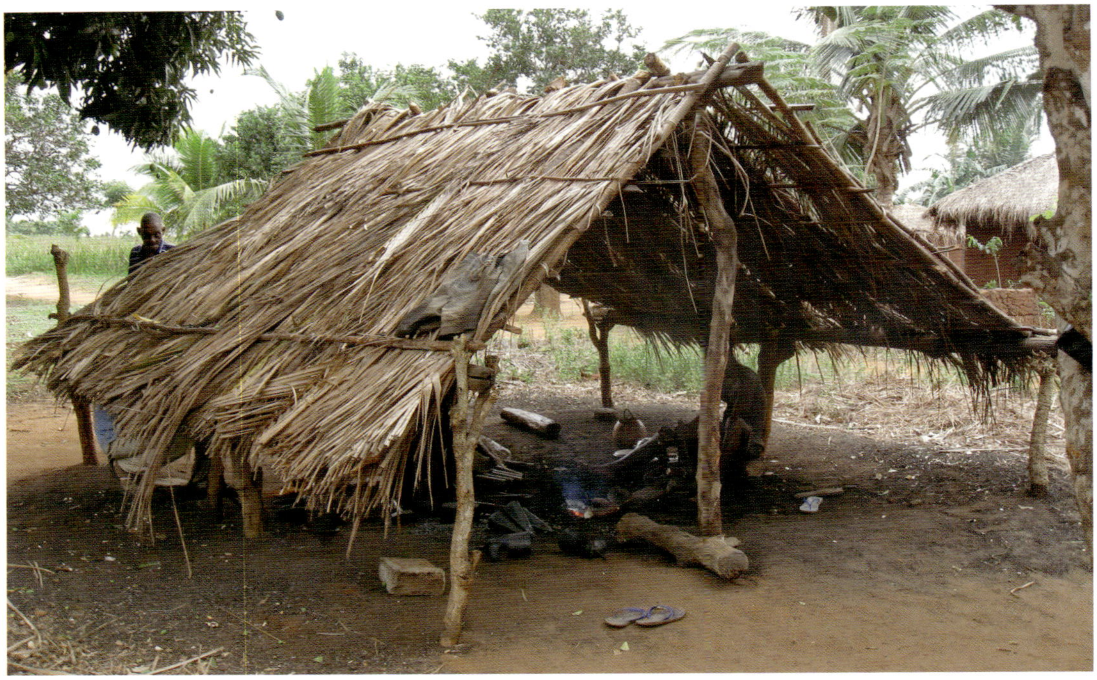

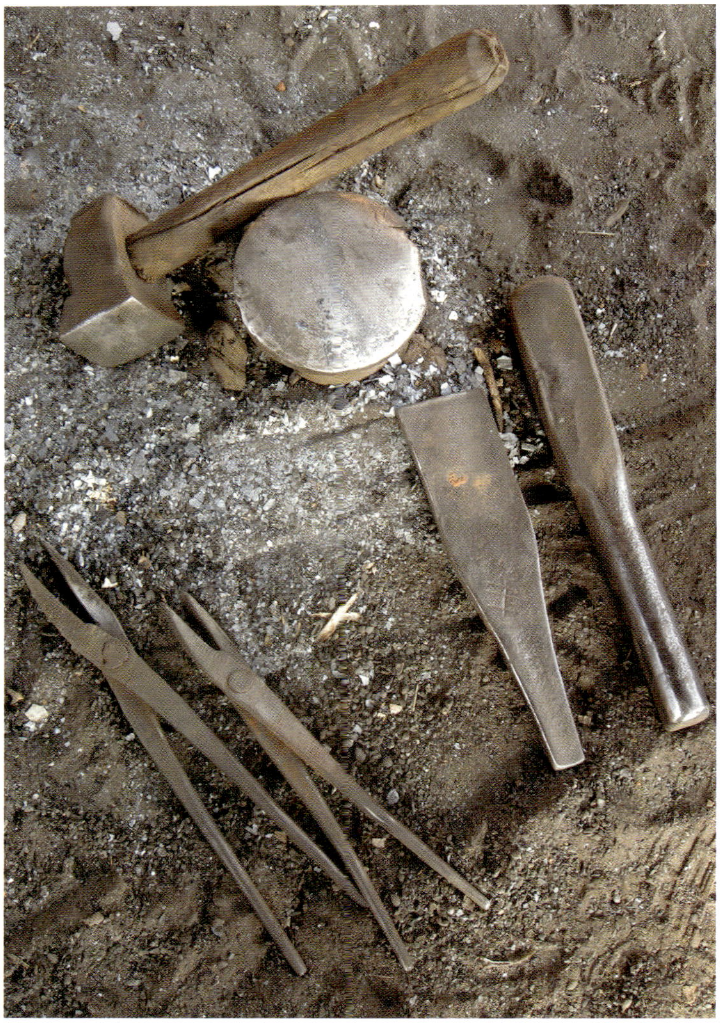

The blacksmith, whether wielding a hammer or manipulating tongs, knows that there is no part for chance, luck, or wishful thinking to play in achieving a successful result. Whenever a forging is made, he relies on his experience and skills with the material he works, the fuel he uses, the tools he has made, and the preparations he has taken. These are manageable factors within his control, but even so, nothing is assured. The designer David Pye writes about the "workmanship of risk," an operative concept for blacksmiths, where at any step along the manufacturing path the object being worked can be ruined—whether through inattention, inexperience, or accident ([1968] 2015, 9, 20). When delicate operations such as welding in the fire are conducted, even weather conditions and barometric pressure can wreak havoc, or a breeze ventilating the workspace, moisture content in the ground, distractions in the neighborhood, trouble at home, ejected scale[8] burning the skin, an insect in the eye—all these unforeseeable factors and more may affect the outcome. Well-trained blacksmiths confidently pivot around the fiery hearth at the uneasy center of a dynamic, ever-changing, and unpredictable environment. The comfort zones are fleeting and elusive at best, and no academic laurels can be rested upon in lieu of stepping up to the anvil, time and time again, to demonstrate one's worth to a community called home. In his workshop, a blacksmith's guard can never be lowered, for the risks on almost every level are present from start to finish, from the youthful trial to the hammer's last strike.

THE WORKSHOP

A blacksmith in Africa, as elsewhere in the world, employs in his workshop (the physical space casually called a smithy) four basic tools and a fire in a hearth (figs. 4.10, 4.11). The coordination of hammer, anvil, tongs, and bellows is essential to modeling iron safely and effectively, as a sculptor might work clay, at near molten

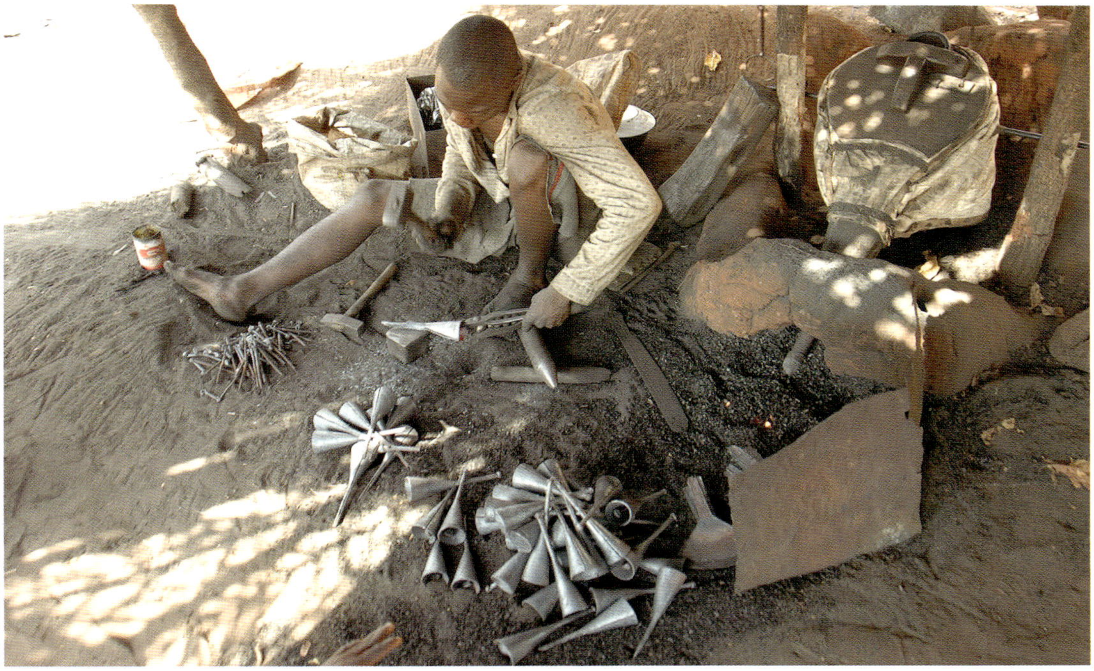

4.12A–B His adoption of European-style two-chambered bellows allows blacksmith Armand Tessi (fig. 4.12A) to work alone when forging small clappered bells (*adja* or *alingle*) used in ancestor worship among Fon peoples in the Republic of Bénin, Togo, Ghana, and Nigeria. Such bells are played by a priestess (fig. 4.12B) wearing a headdress (*lègbaza*) festooned with wooden phalluses and an entourage of attendants who clear the path of malevolent spirits in preparation to honor the most powerful of Vodun divinities, Legba, considered the "guardian at the door of spirits" (Codjovi Étienne Joseph Adande pers. comm., 2010).

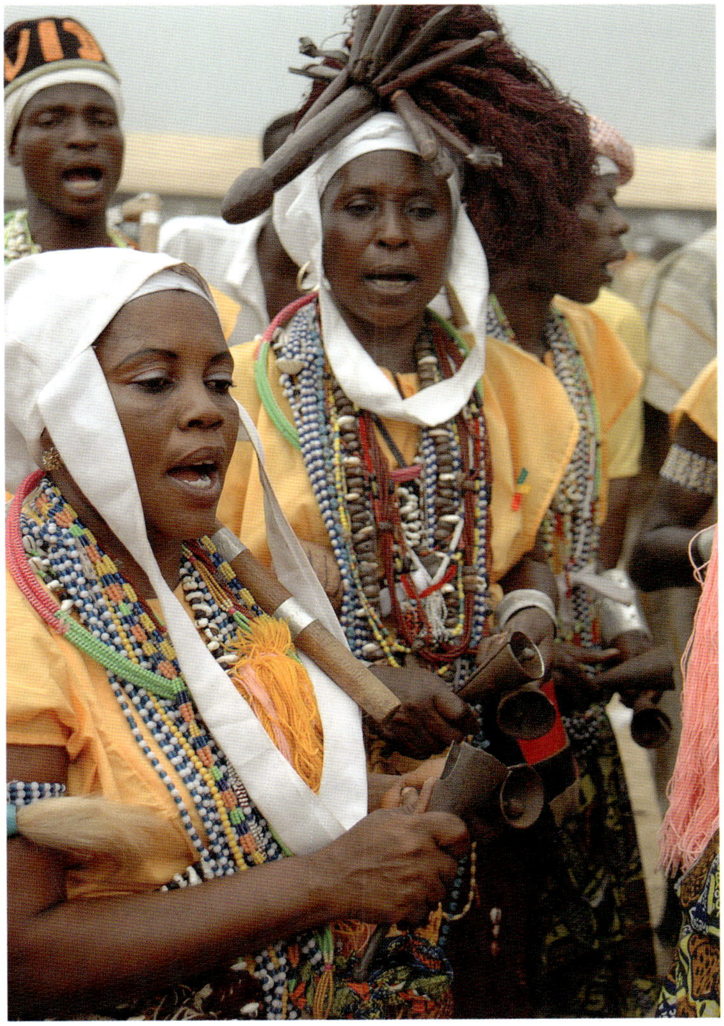

temperatures. These tools have been refined over generations into pragmatic and efficient implements, and their designs are often a seamless combination of form, function, versatility, and ergonomic comfort. Further, the innovations evident in certain tools of the trade may be unique to a specific community and found nowhere else.

A workshop is laid out with consideration of the spatial relationships of the workers and their tasks. Though it is possible to operate a smithy as a solo venture, with the same person forging and pumping the bellows (fig. 4.12), most working scenarios involve at least two individuals. While the tasks of smithing and bellows operation may be periodically rotated among similarly trained members of the workshop, it is typical to have a collaborative division of labor based on seniority and proficiency. If the owner is an elder, for instance, he may have trained another smith, perhaps a relative, to perform heavier or more demanding work, saving the more specialized or delicate operations for himself.

African smiths often work while sitting in front of their anvils, keeping the basic tools within the arc of an outstretched arm (fig. 4.13). Most smithing tools are made in a personal assortment of shapes, weights, and sizes to fit one's own hand and enable skills at coaxing hot iron into desired forms. Sitting at a comfortable striking distance, the smith, whether left- or right-handed, is a quarter-turn from his toolkit in one direction and a quarter-turn within reach of the forge fire facing the bellows operator in the other. In the direction of the hearth, iron stock cut to length and ready to forge is stored alongside a sack or basket of charcoal fuel and a container of water. No shop is complete without the latter, drawn from a basin to selectively cool iron as it is being worked, or to quench handheld tools that absorb heat, such as tongs. Additional fire-tending tools are kept near the smith: on the one side, a long-handled poker, rake, and spoon-like paddle for moving and tamping charcoal; on the other, hammers, tongs, chisels, punches, mandrels, knives, and files.

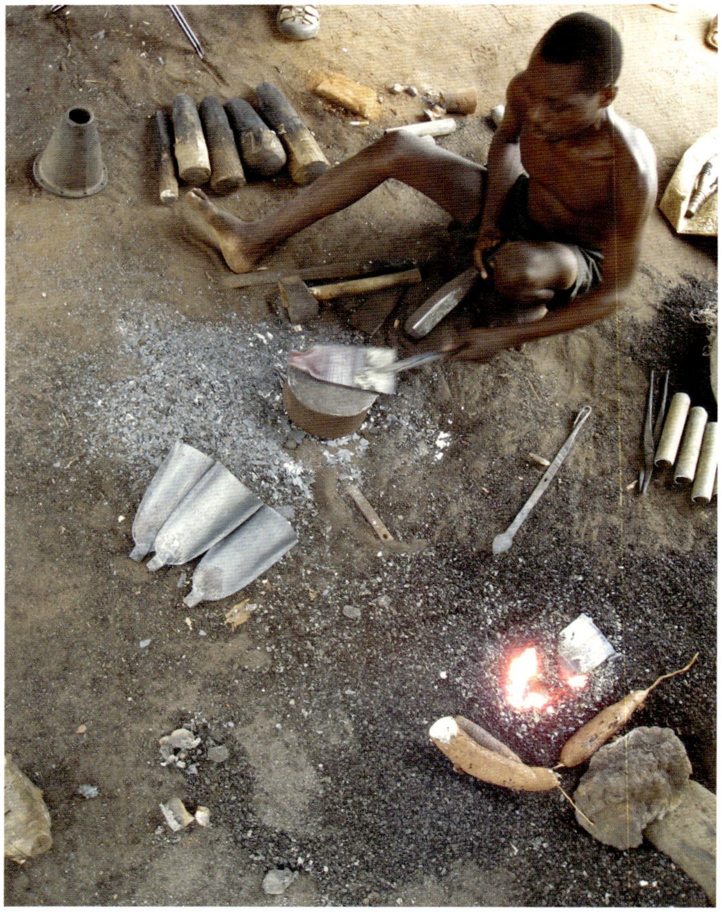

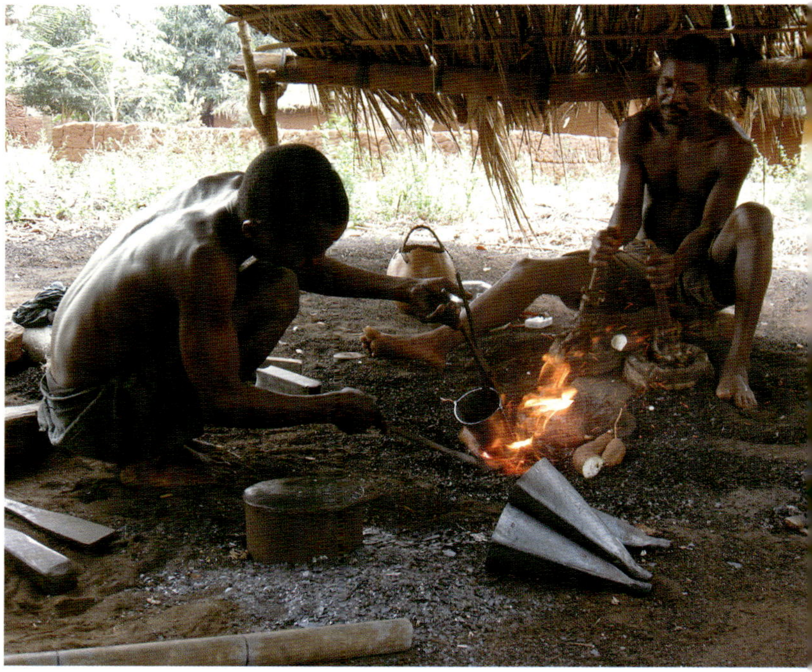

THE FORGE

The radiant epicenter of a blacksmith's environment is the forge, an earthen or clay hearth strategically placed near the middle of the workspace in a shallow, conical depression in the floor. Effectively insulated by the earth around it, heat from a charcoal fire is concentrated toward the center of burning embers, increasing the reflected field of conductivity necessary to reach—with the least amount of effort, time, and fuel—the high temperatures required for forging. Though in heavily forested areas trees may be harvested to make charcoal (see chapter 16, this volume), other more environmentally savvy means have been devised recently, especially in agricultural areas, to process by-products from harvested crops into fuel for the forge.[9]

The earthen or clay hearth in an African smithy's floor is the perfect habitat to encourage the exact conditions necessary for forge-welding, an ancient assembling technique widely adopted by skilled blacksmiths in Africa. Developed alongside iron smelting, it established a metallic bond between iron atoms through localized fusion of ferrous material. The process requires that an oxygen-starved pocket of heat surround the iron parts so that scale is reduced and a coating of molten flux is allowed to "wash" their surfaces of particulate contaminants just before the smith brings the incandescent pieces to the anvil to hammer them together (figs. 4.14, 4.17). Naturally occurring grains of silica sand along the margins of the hearth melt to create a self-fluxing environment conducive for joining pieces. Prior to welding, the surfaces are prepared with forged contours so that they can receive each other in a coupling embrace. This predetermined configuration locks their position, preventing them from sliding as the hammer repeatedly strikes the slippery, near-molten surfaces with enough pressure to bond them forever.[10] This procedure is so demanding that forge-welding is often carried out in secret, with perhaps only family members present to call on an ancestor of the forge for a guiding hand.[11]

4.13A–B
Ewe blacksmith Galbert Atakpa and bellows operator Hodenou Noglo forge small production runs of musical instruments and agricultural implements during the dry season. Efficient production to supply local demand during the months of harvest and preparation for the next growing season is dependent on a well-arranged workspace. Here, the lead smith sits in front of his anvil, close enough to tend the fire. Within reach of his right hand are hammers, cone mandrels made of wood and iron, files, and salvaged pipe to be used as starting stock (from which to cold-chisel sections for forging). On his left is material already cut to length, a fire poker, tongs, and charcoal fuel. On the ground before him appear single bells in-process with an accumulation of fresh scale (flakes of oxidized iron) around the anvil.

PHOTOGRAPHS BY TOM JOYCE, YOHONOU, TOGO, 2008. © TOM JOYCE

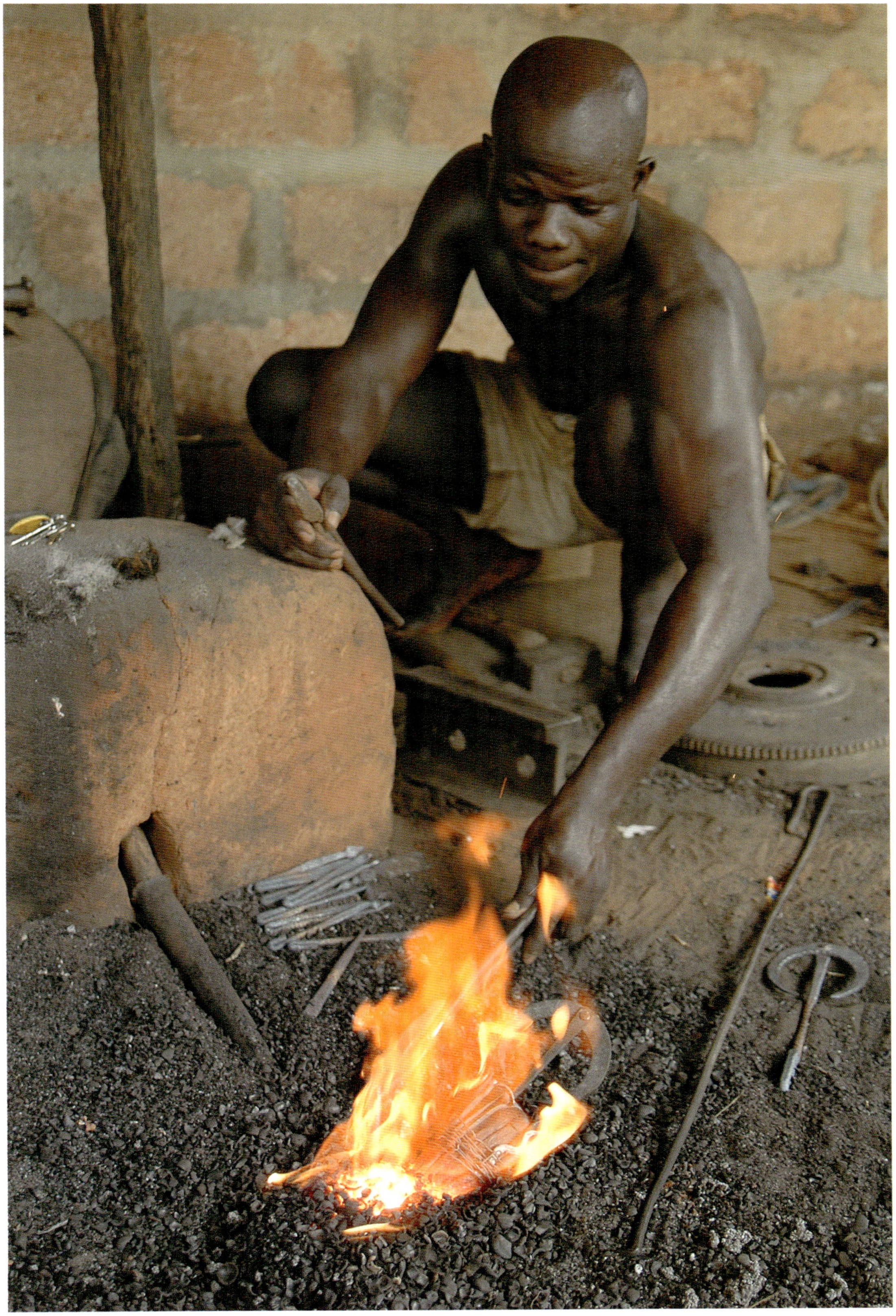

4.14A–C In a workshop situated next door to a temple built to honor Hevioso, Fon god of thunder, lightning, and rain, blacksmith Damien Dossougbete forge-welds a staff to be used for ritual offerings on a shrine. Below its crescent-shaped finial are fourteen miniature tool, weapon, and serpent forms encircling a representation of Hevioso's axe (*sosivi*). Once the cluster is uniformly heated to semimolten temperature, it is gently forged together into a radiating core that projects the power of Gu, god of iron, in all directions simultaneously.

PHOTOGRAPHS BY TOM JOYCE, OUIDAH, REPUBLIC OF BÉNIN, 2010. © TOM JOYCE

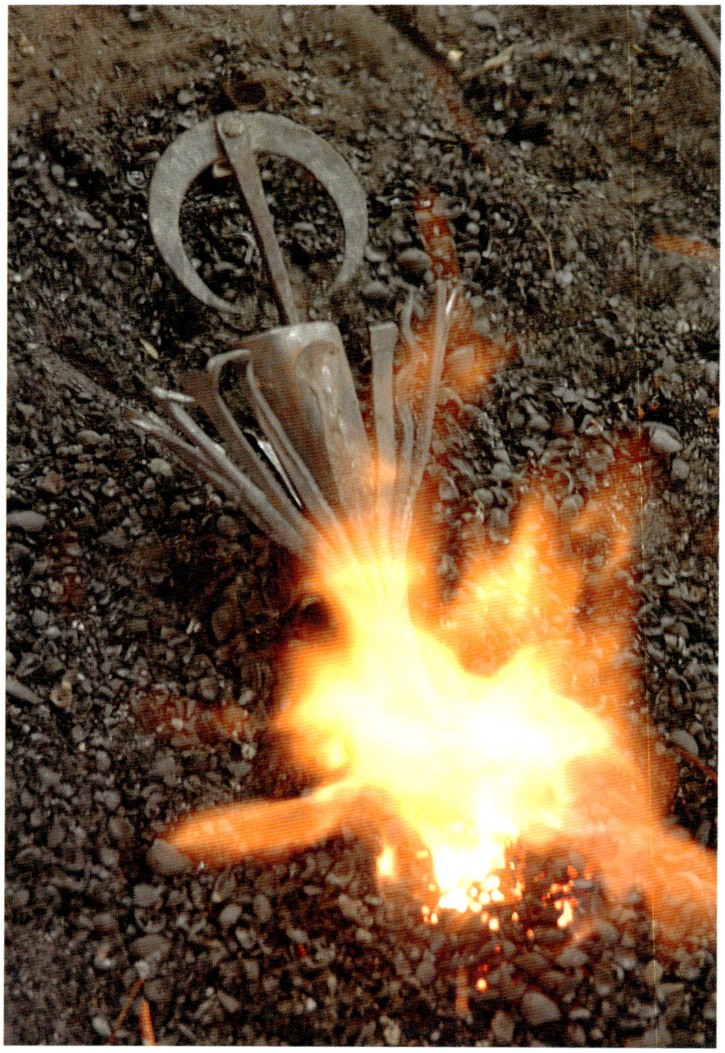

4.14B

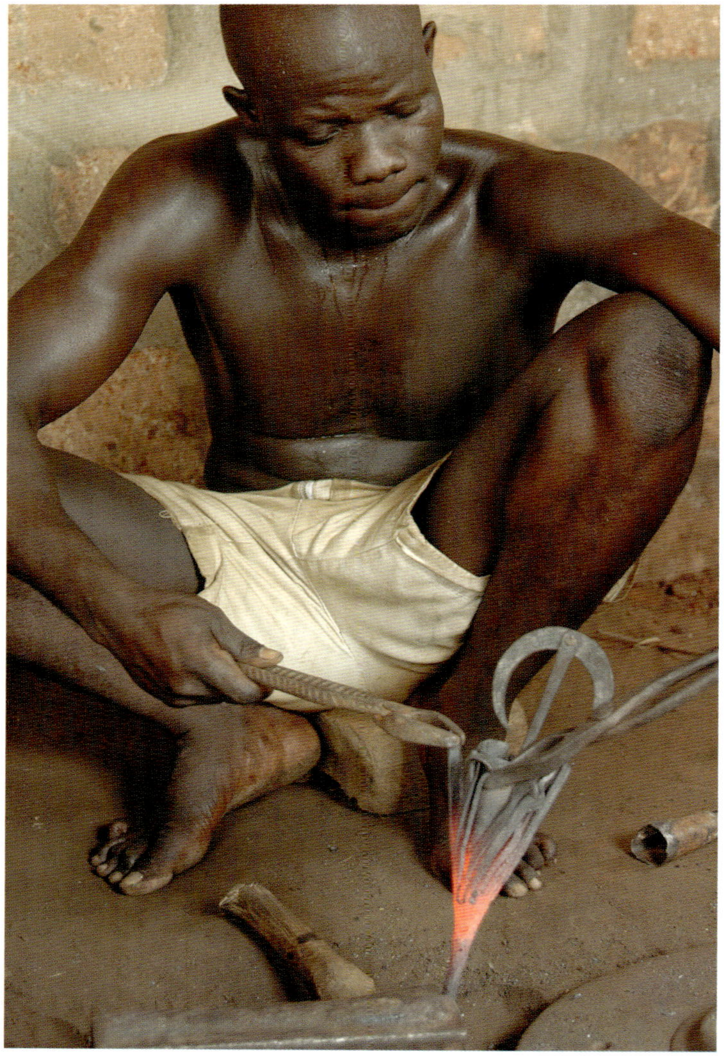

4.14C

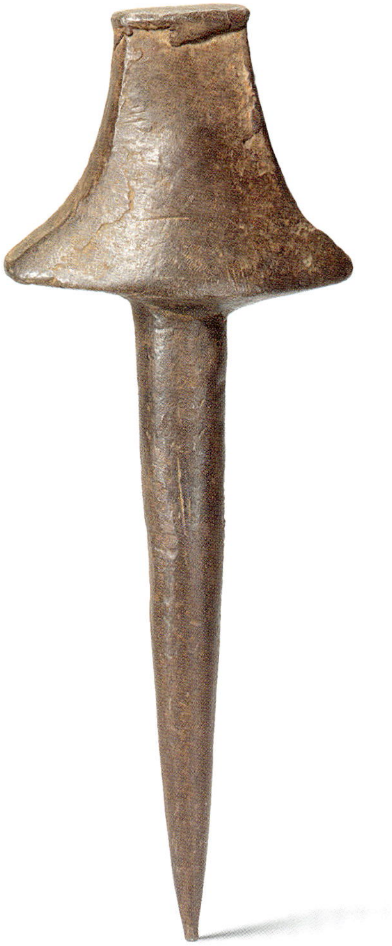

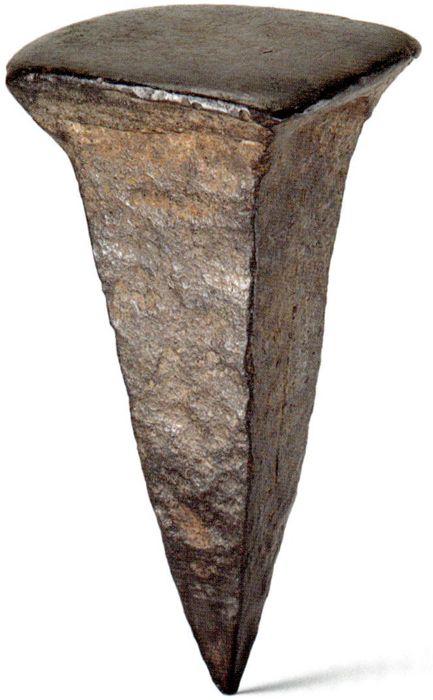

Especially when working outside, the forge fire must be shielded from direct sun, so the smith can monitor by eye the color range of the iron as it grows hotter. This color-differentiated spectrum of heat indicates a specific reach of malleablity enabling iron to move in distinct ways when worked at that precise temperature. As it heats, iron changes from charcoal gray to dull red, ascending rapidly through cherry red, bright orange, straw yellow, white hot, and finally, incandescent-white. Though casual observers commonly describe iron in a blacksmith's forge as "red hot," this witnessed temperature actually places near the bottom of the malleable range for moving iron without fatigue. "White hot" approaches the top end, where iron is at its softest state; this is therefore the most desirable temperature at which to begin shaping an object. The very hottest, however, is the incandescent-white that signifies the near-to-melting temperatures required to forge-weld multiple pieces together. (Iron in this state is so hot that if it is looked at directly, one's retina could be spot-seared, as if exposed to the sun without protective lens.) This heat is the most precarious blacksmiths manage, and irreversible failure lies on either side of its narrow forging range: too cool, and the process fails because the pieces are not hot enough to fuse; too hot, and the iron burns, becoming permanently destroyed for further forging.[12]

4.15
Artist unknown (Mongo peoples, Democratic Republic of the Congo)
Anvil (*njondo*), 19th century
Iron
H: 35.3 cm
PRIVATE COLLECTION

4.16
Artist unknown (Soninke peoples, Gambia)
Anvil (*simbo*), late 19th to early 20th century
Iron
H: 23.5 cm
PRIVATE COLLECTION

4.17 In forge-welding on the anvil, the hammer's impact showers molten flux and scale in all directions as multiple parts are fused together. It is a mesmerizing process to observe, as the smith's attention and purpose in deploying his hammer gives metaphoric potency to every object he creates. Here, Yawovi Yovo forges a four-rattle instrument.

PHOTOGRAPH BY TOM JOYCE, YOHONOU, TOGO, 2008.

4.18
Artist unknown (Angas peoples, Nigeria)
Hoe-blade-shaped currencies, late 19th to early 20th century
Iron
H: 102 cm; 72.5 cm
PRIVATE COLLECTION

These substantial hoe-blade-shaped currencies are constructed from a solid patchwork of dozens of iron fragments forge-welded together by several blacksmiths working collectively. Imprinted across the surfaces are blow-by-blow accounts left by hefty hammers and the coarse texture of a stone anvil. Planted upright on their tangs in the ground to line a matrimonial pathway as bridewealth, these commanding forms offered an abundant store of iron for future use by the bride's family.

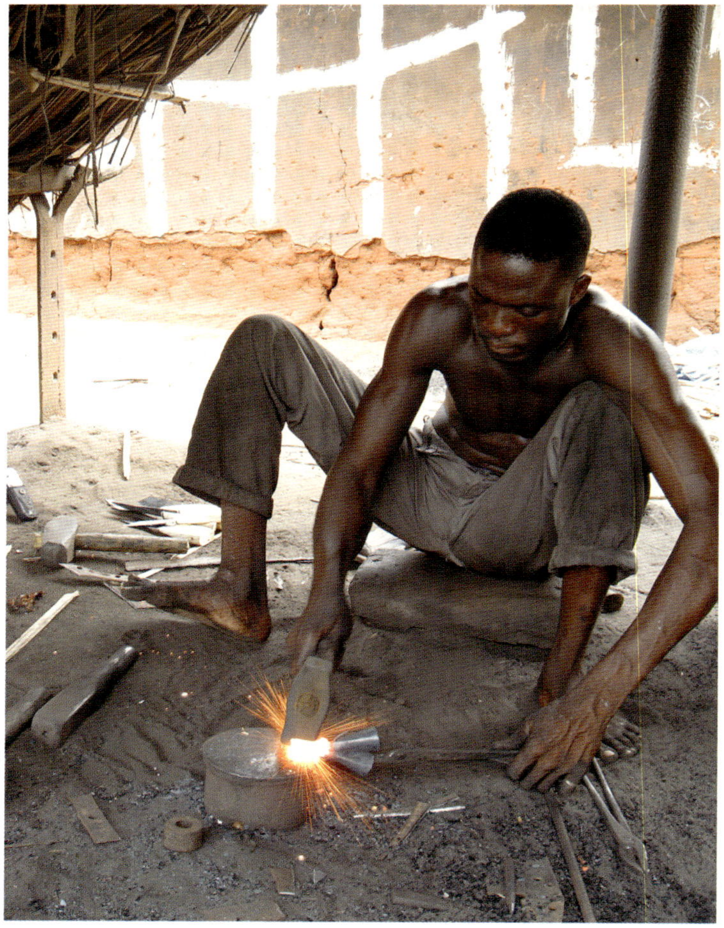

THE ANVIL

Firmly planted in every smithy, the anvil is the still point at the creative nucleus of the blacksmith's realm, and the stage where iron is prepared for the myriad roles it will play. Typically fashioned from stone or iron, this vital tool—available in an astounding diversity of shapes, sizes, and alignments (figs. 4.15, 4.16)—provides foundational support for all hot work. It is to the blacksmith what the keyboard is to the writer, standing for boundless creative potential while representing unsettling practicality, for it is balanced between the smith's imagined intent and an uncertain outcome. All of the blacksmith's accumulated experience is on the line every time his hammer is raised above the glow of hot iron illuminating the working arena below.

Though the anvil's presence is stationary, once the compressive force stored in its solid grounding is released by the awakening impact of the hammer's blow, what appeared cold and static in disuse is instantly activated by the penetrating physics stimulated in the collision zone (fig. 4.17).[13] The momentum of a squarely placed blow directs high-velocity influence into the heat-softened work that rebounds off the anvil in a burst of kinetic energy. This recoiling "fuel" then powers subsequent blows in succession. The compressive stress induced at the point of contact drives a shockwave deep into the work with "post-collision velocity" (Goldsmith 1960), generating heat, sound waves, and percussive vibrations that move the metal and permeate the senses with "transient physical excitation" (Toropov 1998, 315–19).[14]

When forging on an anvil, blacksmiths are displacing mass to manipulate form. On a subatomic level, iron is being compressed and redistributed as its structural integrity is relied on to perform with characteristic traits (Beer et al. 1992). The procedural choices a blacksmith wagers when fashioning any item realign the relative position of every atom in the forging, and he is mindful to forge in a linear direction to elongate iron's fibrous composition, thereby maintaining and enhancing its strength (fig. 4.18). Iron is unforgiving in this respect, and no amount of surface enhancement, applied as decorative embellishment or elaborate twists and turns, can deceive or cover for improper handling of basic forging procedures (fig. 4.19). It usually takes minimal sleuthing to trace individual steps of problematic manufacture backwards and disclose what went wrong, why, and at what stage in the process.

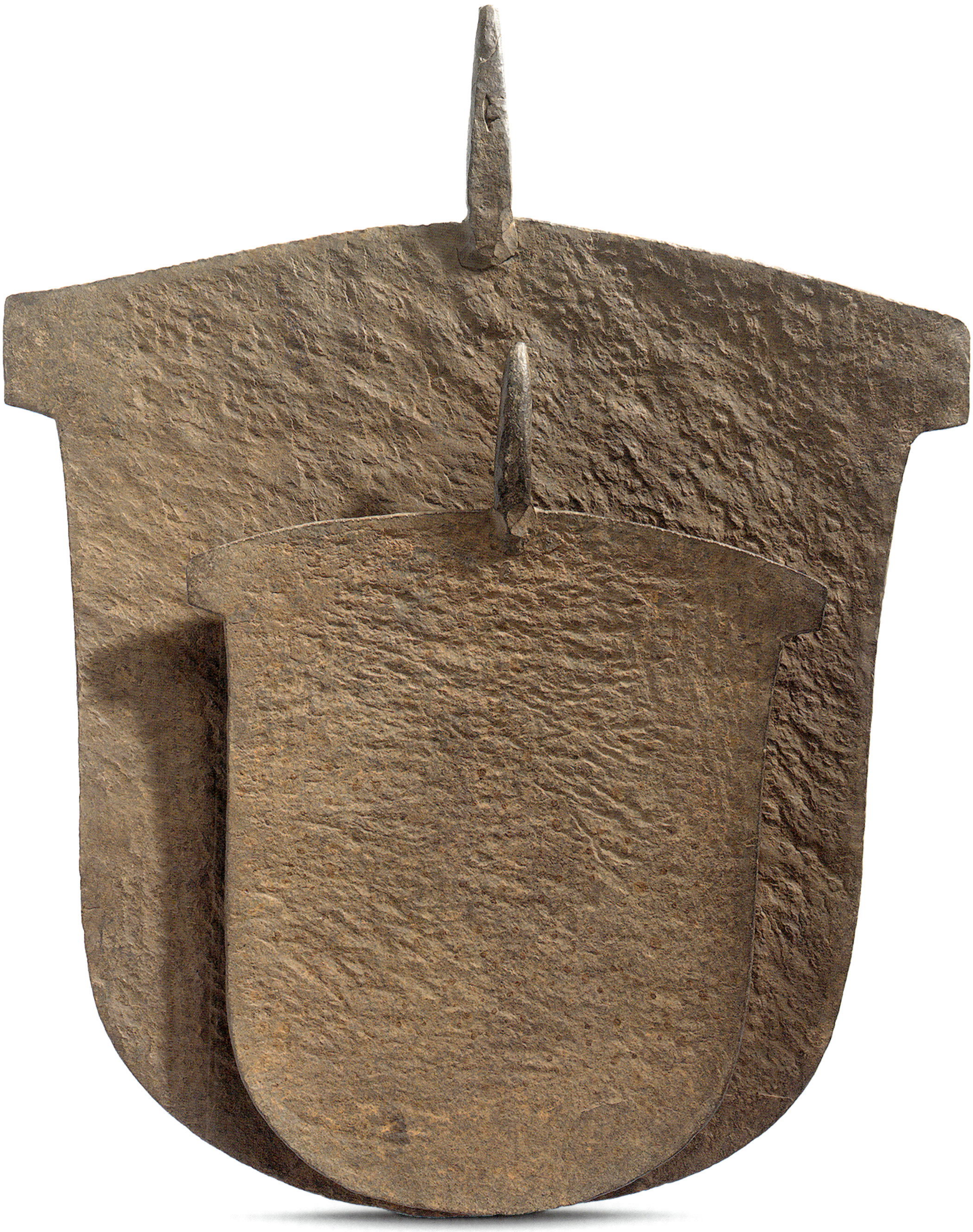

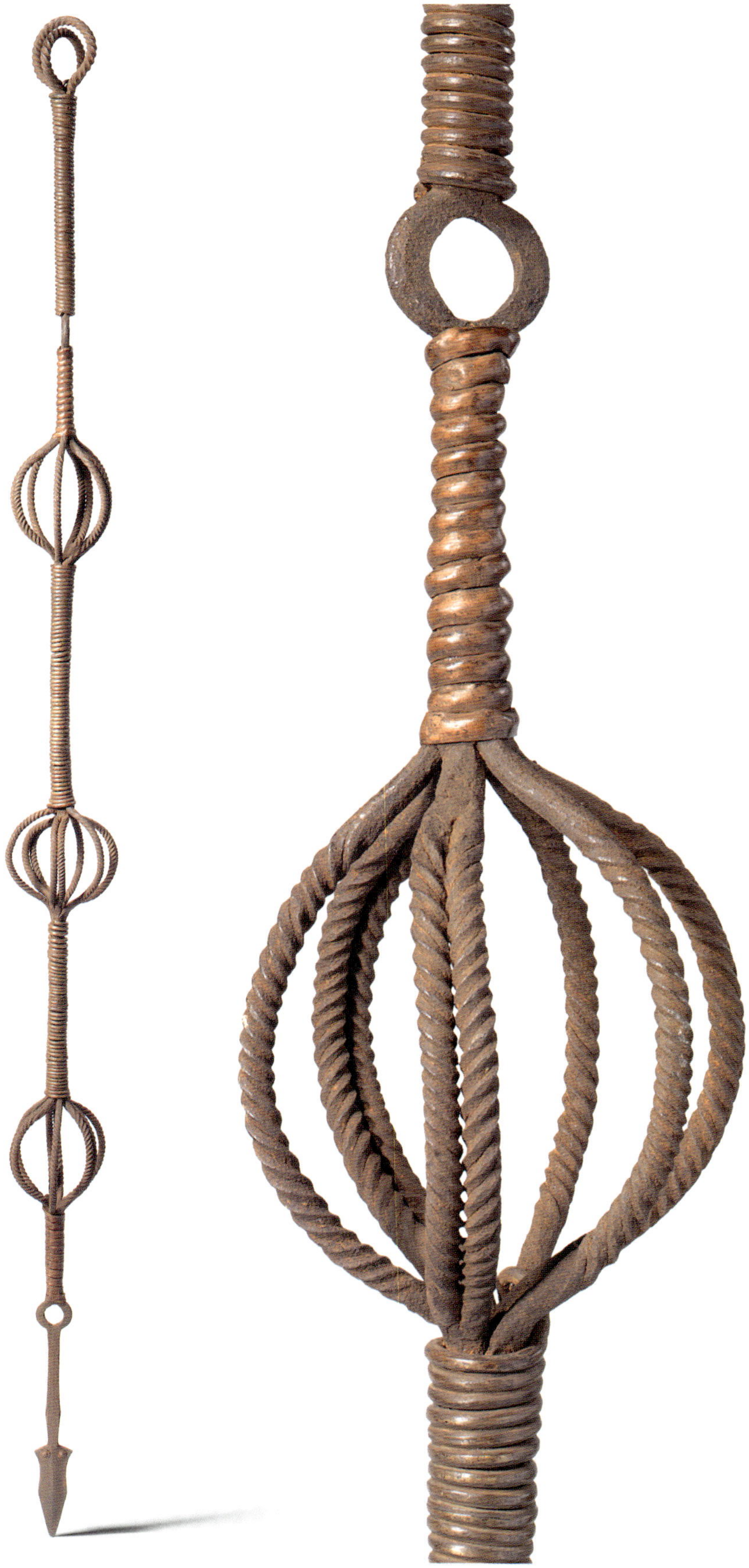

4.19A–B
Artist unknown (Igbo peoples, Nigeria)
Staff, mid-20th century
Iron, copper, brass
H: 174 cm
PRIVATE COLLECTION

This spear-like staff was held by a titled elder of an Igbo community to reflect his distinguished status. Title staffs typically have a single "basket" comprised of multiple twisted strands forge-welded together in the center; however, this example combines a series of three such spherical enclosures, each made from pairs of alternating twisted rods and three single twisted rods. The exceptional variation in this composition is further enhanced by differently contoured strips of copper alloy that coil around most of the staff's length.

4.20 Titled Igbo man with staffs in Ezeagulu Aguleri, Nigeria.
PHOTOGRAPH BY HERBERT M. COLE, 1983.
COURTESY OF THE PHOTOGRAPHER

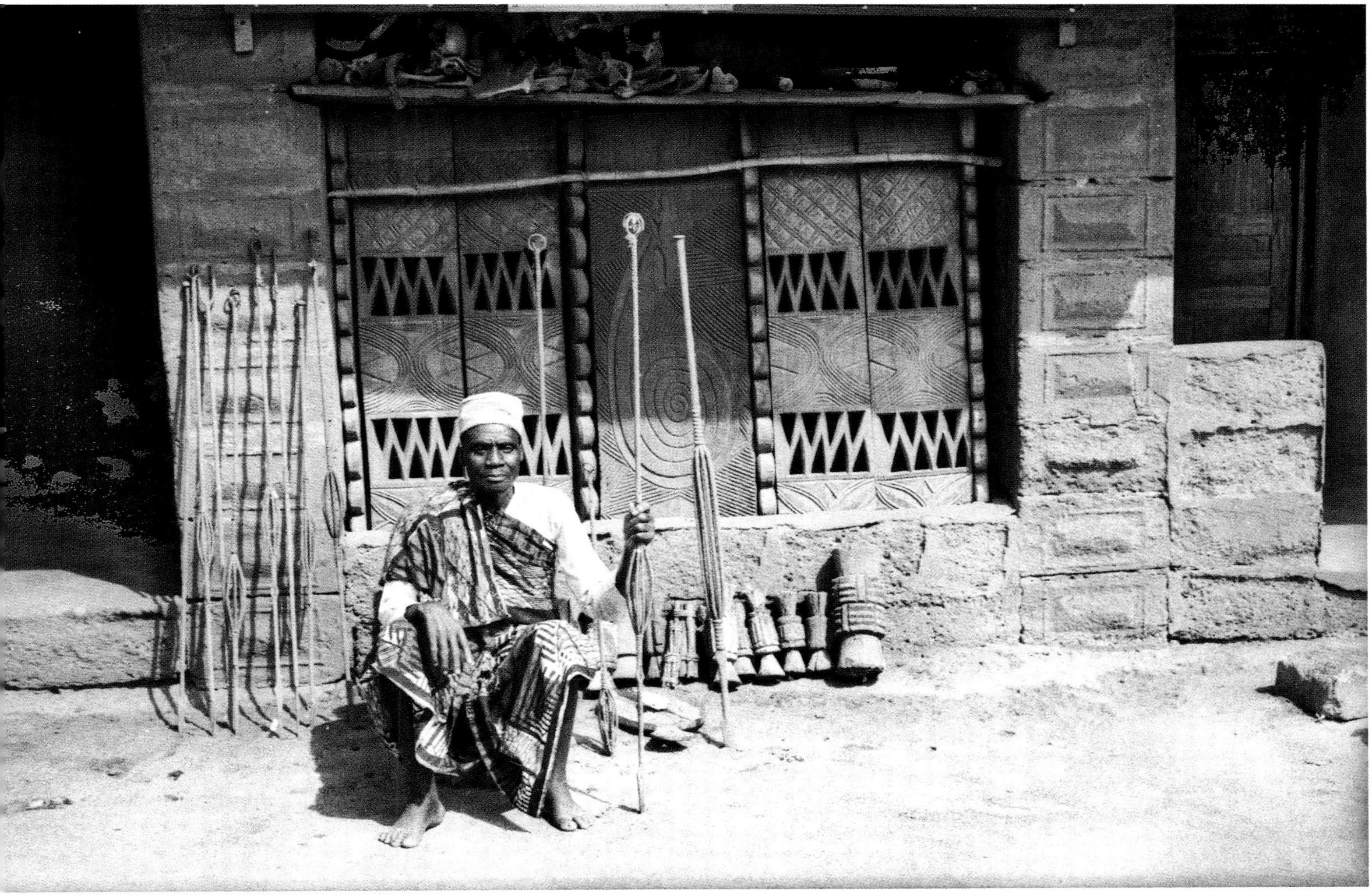

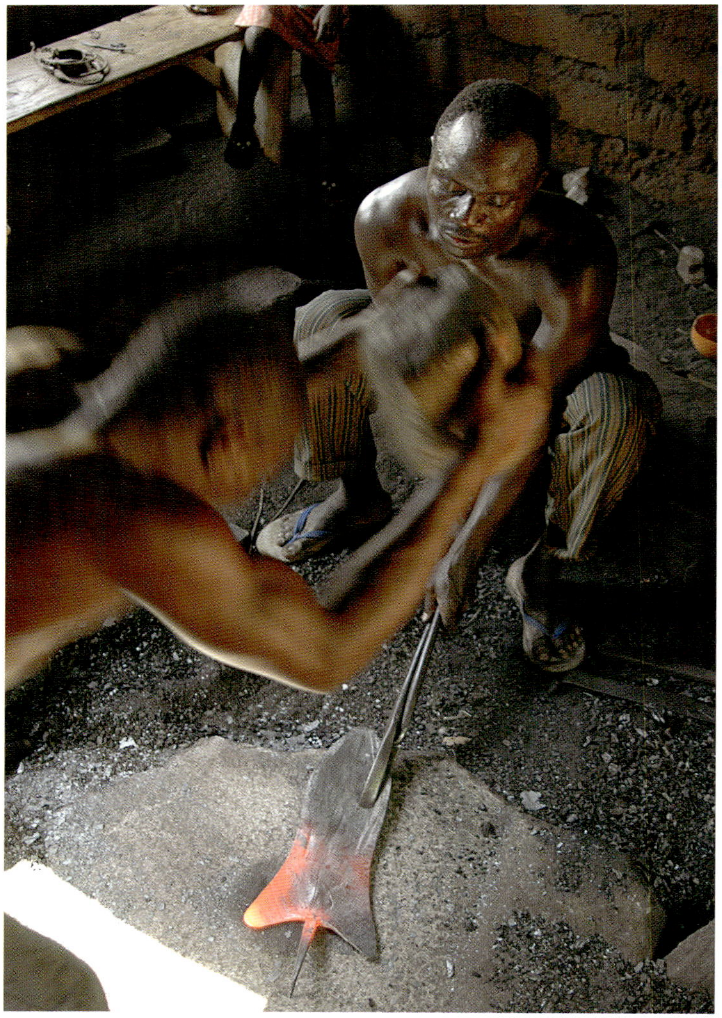

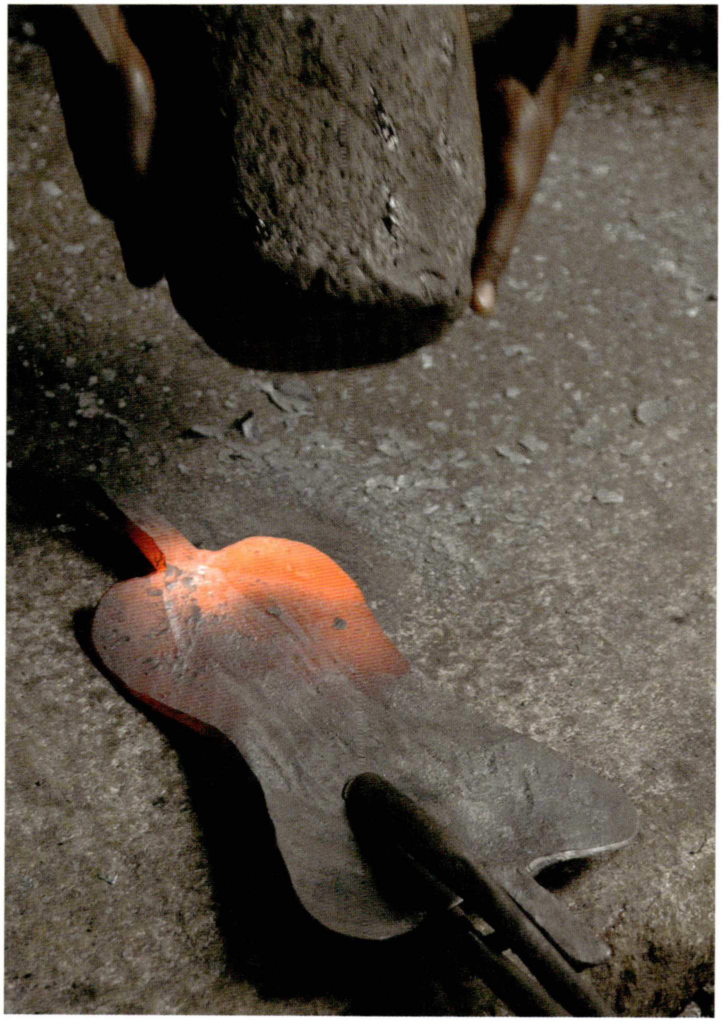

After roughing out a forging at high temperatures (940–1200 degrees Celsius), lighter hammering follows at cooler temperatures (540–620 degrees Celsius) as the iron object is finished on the anvil (fig. 4.21). The blacksmith steers the multidirectional deformation of material to make the piece thinner, wider, or longer. Like any trained material scientist or mechanical engineer, he is constantly evaluating vectors of change and progress, and recalibrating his approach as necessary. Visualizing a completed shape even within a basic repertoire of objects and teasing the most efficient sequence of steps from beginning to end—this is essential business for blacksmiths.[15] When iron is completely cool, the effects of manufacture are frozen in structural tension.

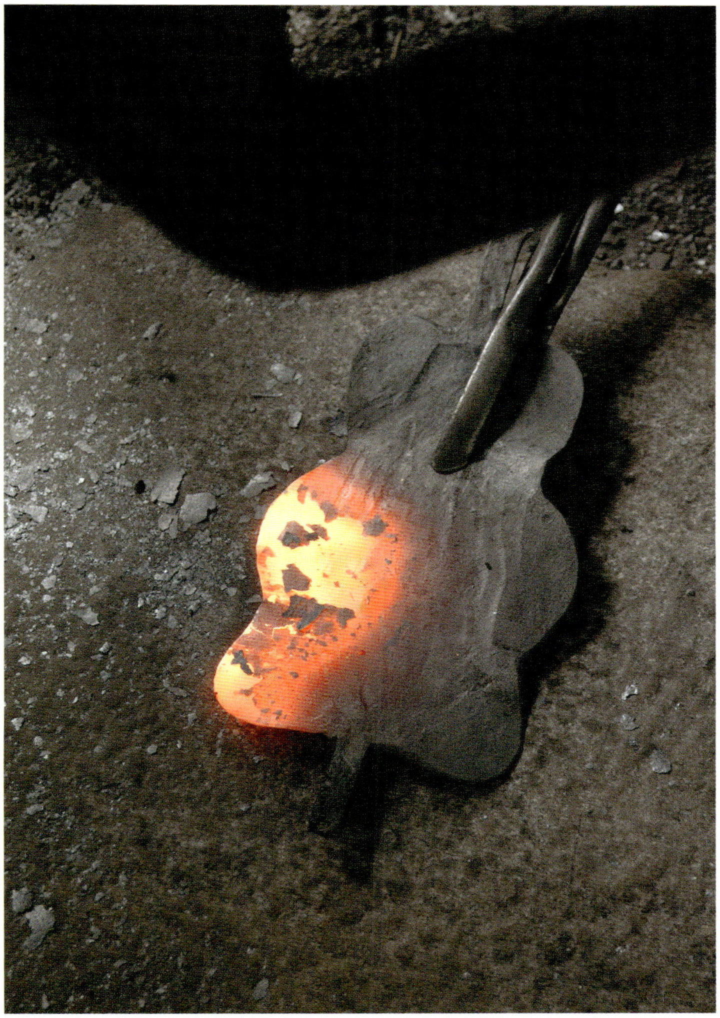

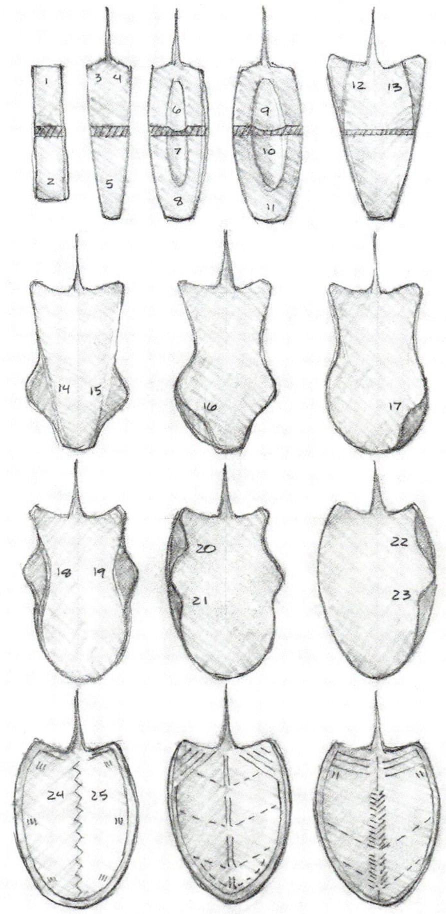

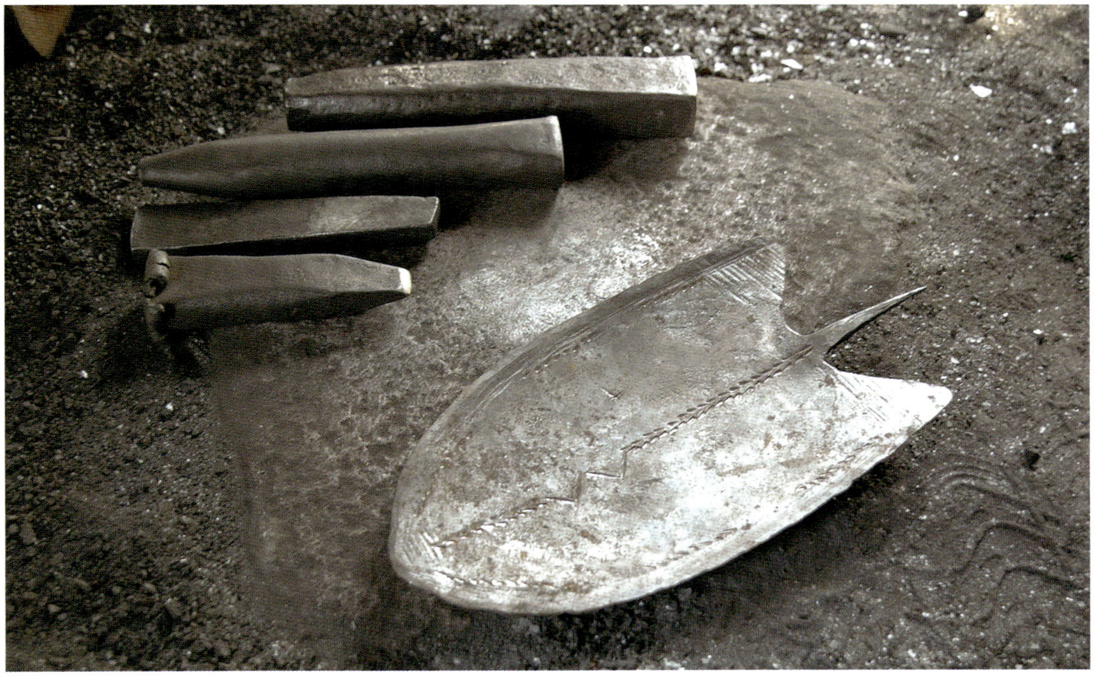

4.21A–E This distinctive Kabre yam-hoe blade (*aku*; H: 55 cm) was forged by blacksmith Kossi Kao and striker Essozimna Ide in a sequentially precise redistribution of material that took full advantage of the basalt hammer's fifteen kilograms of falling weight. Each hoe is ornamented with cold-chiseled designs that embellish the tool while delineating the maker's signature. Recognizable logograms allow village merchants to fairly compensate every blacksmith or household for products gathered for sale at local and regional markets. The incised patterns are often the same as scarification marks identifying family lineages on the cheeks of Kabre men and women.

Forging an iron Kabre hoe over twenty-five heats begins with the chiseled section of a recycled truck-wheel hub for roughing out into the starting blank. The sequence advances until the tool is finished, signed, and ready for insertion into a handle.

PHOTOGRAPHS AND DRAWING BY TOM JOYCE, TCHARÉ TOGO, 2010. © TOM JOYCE

4.22A–B
At Bamana blacksmith Zan Bagayoko's workshop in Konko, Mali, tongs (*balan*) for handling hot iron rest on a stone mortar used as a quench basin. Though a single pair of tongs can be adapted to fit a range of different-sized materials, several graduated sizes are necessary in most workshops to accommodate the variety of works produced.
PHOTOGRAPHS BY TOM JOYCE, 2011. © TOM JOYCE

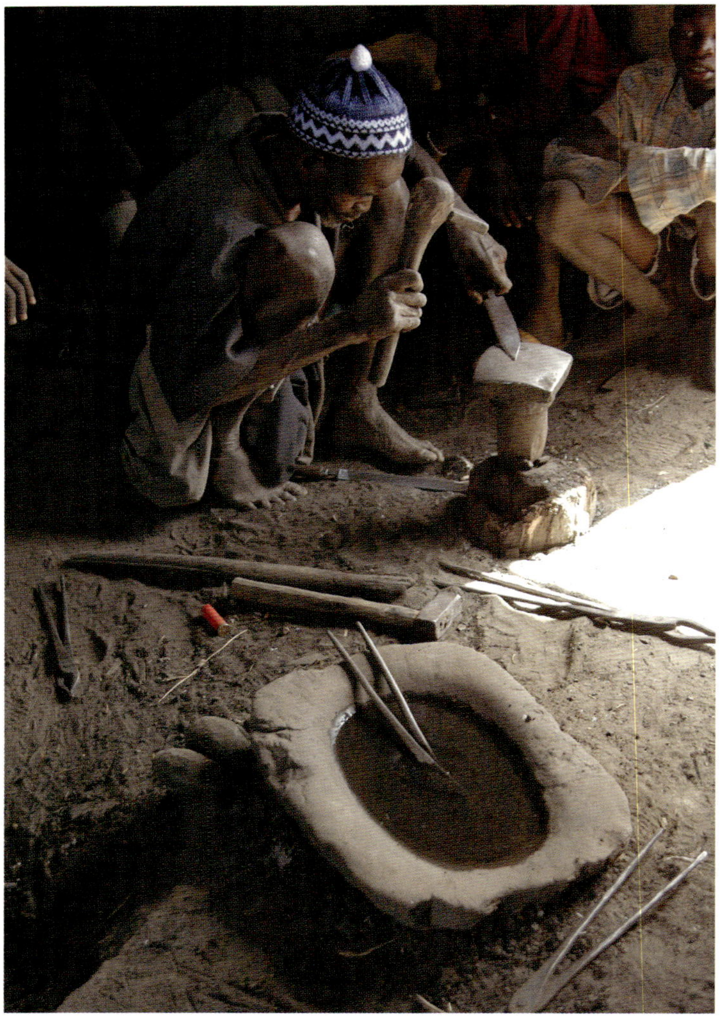

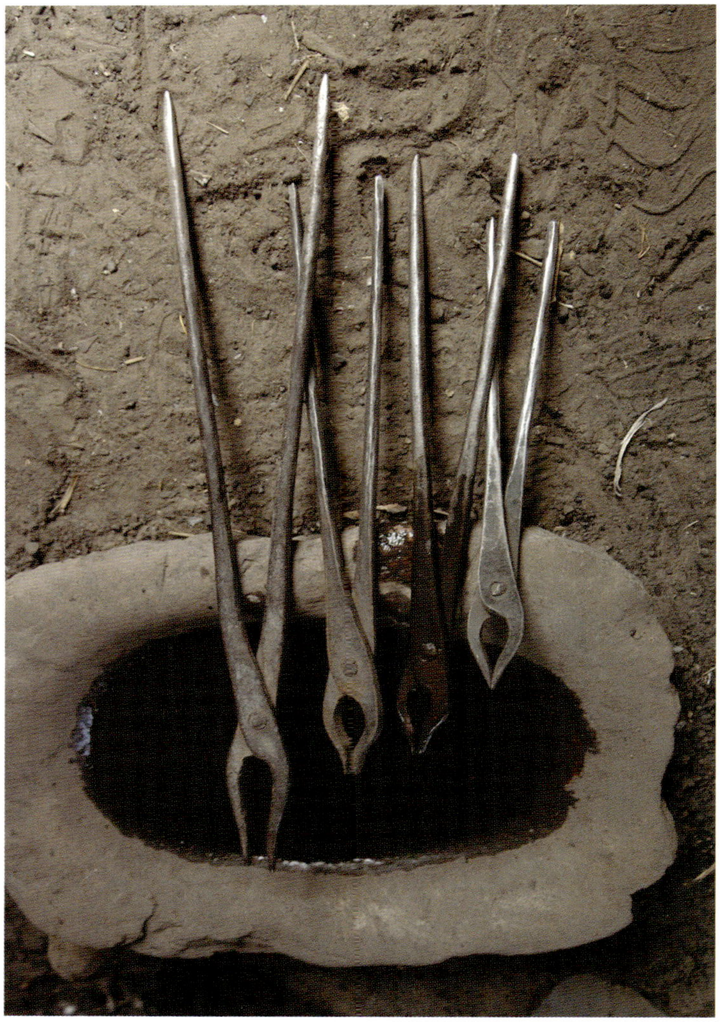

THE TONGS

As intermediaries between hot iron and the hand, tongs are requisite tools in a blacksmith's kit, for they safely enable one of the most dangerous aspects of work at the forge (fig. 4.23). Engineered to clamp hot iron, every pair of tongs is appropriately scaled for single-handed manipulation by the owner. Though lightweight and springy in use, tongs can surpass their modest stature when used to comfortably handle larger and heavier material.

In operation, tongs stimulate sensory receptors in the skin, muscles, and bones of the hand, sending signals received through touch along neural pathways into the body's somatosensory system (Robles De La Torre 2001, 44). Acute haptic perception is key to proficient forging because tongs—much like divining rods held in the hand when dousing for water—communicate to the blacksmith the precise positioning of the iron piece held in its jaws, and they relay with pinpoint accuracy all distinguishing characteristics of individual hammer blows when each strikes its intended target.[16] Though an observer's gaze tends to gravitate toward the unfolding action of the hammering hand when watching a blacksmith work, it is in the subtle movement of the hand holding the tongs that a smith's deftness and skill can be most appreciated.

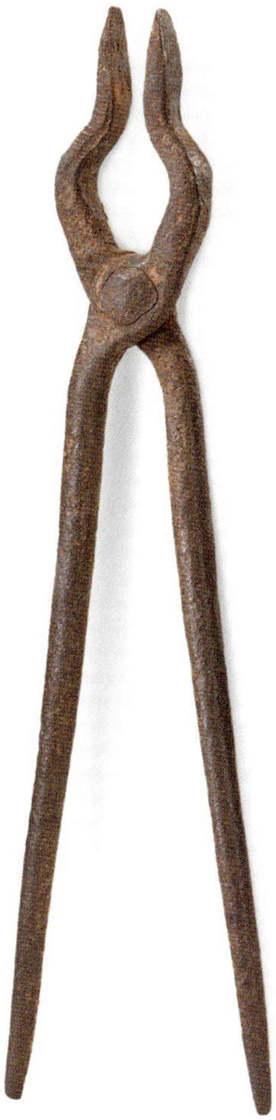

4.23
Artist unknown (Soninke peoples, Gamɔia)
Tongs (*kampa*)
20th century
Iron
H: 37 cm
PRIVATE COLLECTION

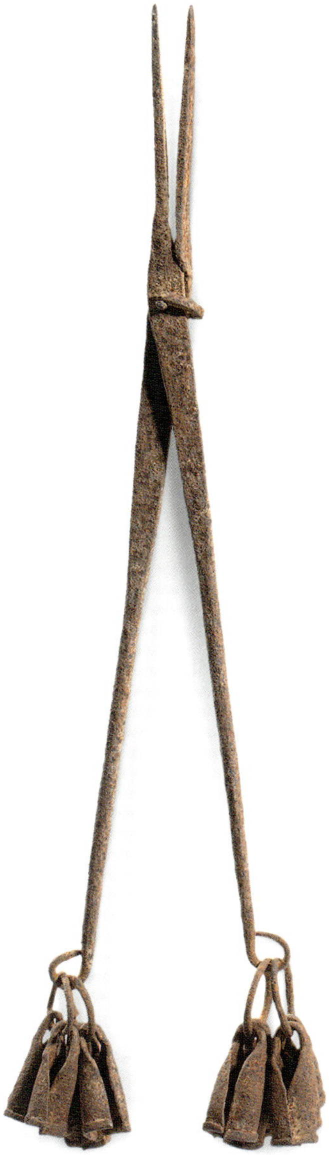

4.24
Artist unknown (Dogon peoples, Mali)
Ritual tongs, 19th century
Iron
H: 45 cm
PRIVATE COLLECTION
(NOT IN EXHIBITION)

So critical for safely managing iron in the fire, tongs are also considered symbolic intermediaries and sometimes perform outside the workshop as ritual instruments such as this Dogon pair of ember tongs with rattles attached, used in ceremonial contexts.

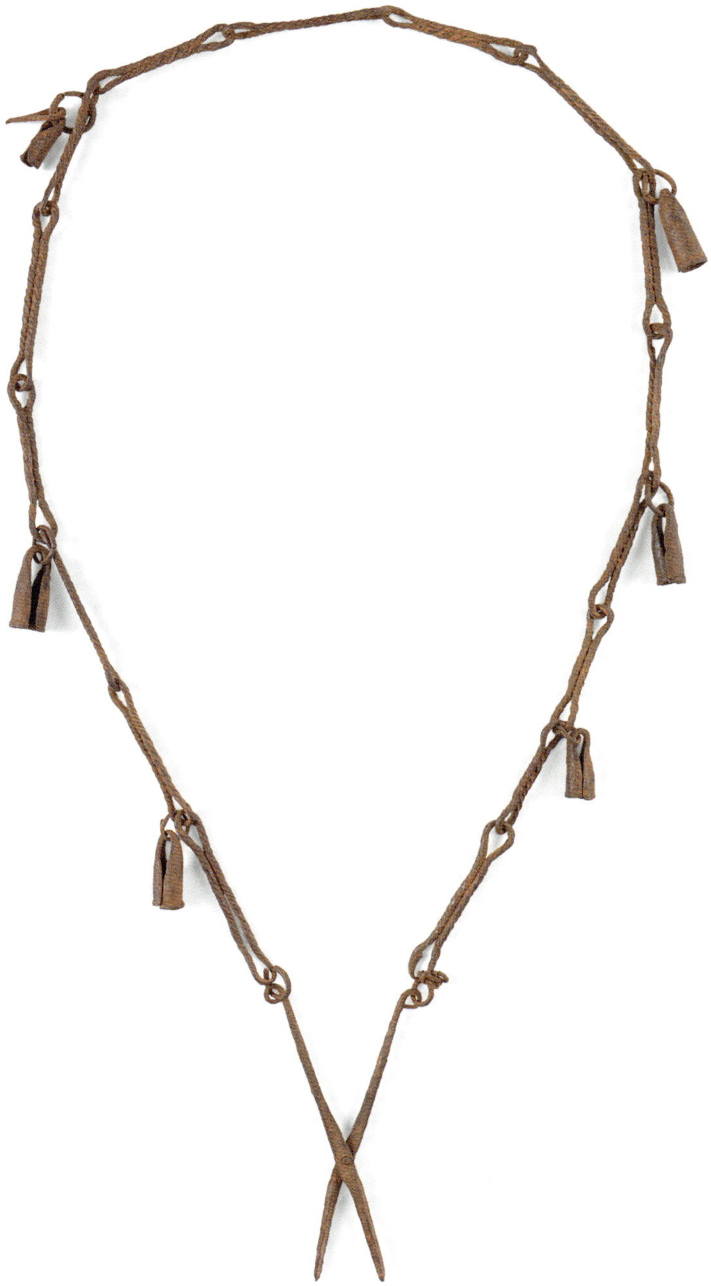

4.25
Artist unknown (Dogon peoples, Mali)
Necklace (*binu duge*), 20th century
Iron
L: 73.7 cm
PRIVATE COLLECTION

Pendant necklaces include signs of allegiance between the wearer and a spiritual power, a close or mythical ancestor. Hanging from this forged iron necklace, miniature blacksmith's tongs may have activated such a bond. This particular symbol is sometimes worn by Dogon women before and during childbirth, to signal the child's safe emergence from the "hearth" of the womb.

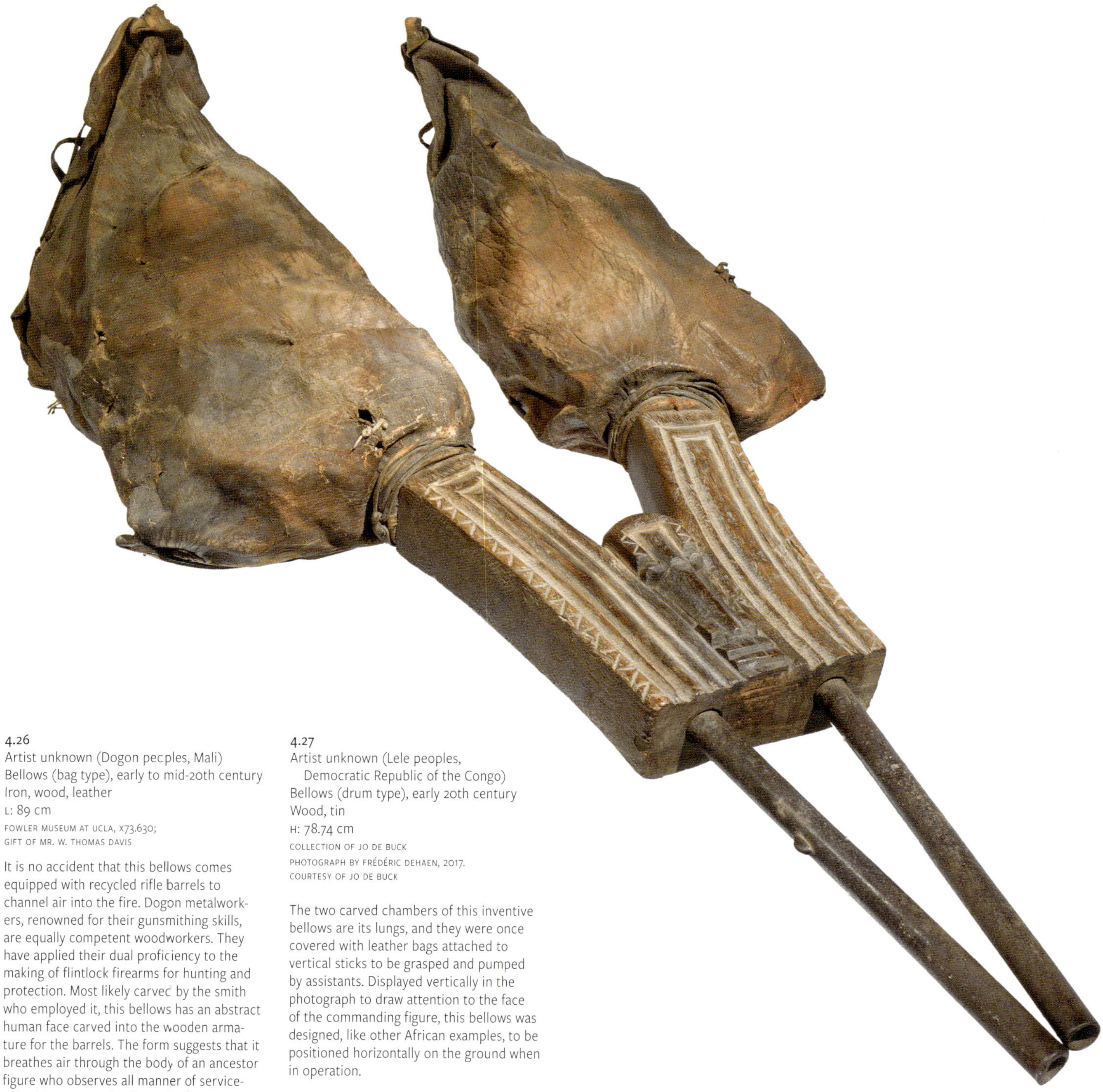

4.26
Artist unknown (Dogon peoples, Mali)
Bellows (bag type), early to mid-20th century
Iron, wood, leather
L: 89 cm
FOWLER MUSEUM AT UCLA, X73.630;
GIFT OF MR. W. THOMAS DAVIS

It is no accident that this bellows comes equipped with recycled rifle barrels to channel air into the fire. Dogon metalworkers, renowned for their gunsmithing skills, are equally competent woodworkers. They have applied their dual proficiency to the making of flintlock firearms for hunting and protection. Most likely carved by the smith who employed it, this bellows has an abstract human face carved into the wooden armature for the barrels. The form suggests that it breathes air through the body of an ancestor figure who observes all manner of serviceable objects emerging from the flames.

4.27
Artist unknown (Lele peoples, Democratic Republic of the Congo)
Bellows (drum type), early 20th century
Wood, tin
H: 78.74 cm
COLLECTION OF JO DE BUCK
PHOTOGRAPH BY FRÉDÉRIC DEHAEN, 2017.
COURTESY OF JO DE BUCK

The two carved chambers of this inventive bellows are its lungs, and they were once covered with leather bags attached to vertical sticks to be grasped and pumped by assistants. Displayed vertically in the photograph to draw attention to the face of the commanding figure, this bellows was designed, like other African examples, to be positioned horizontally on the ground when in operation.

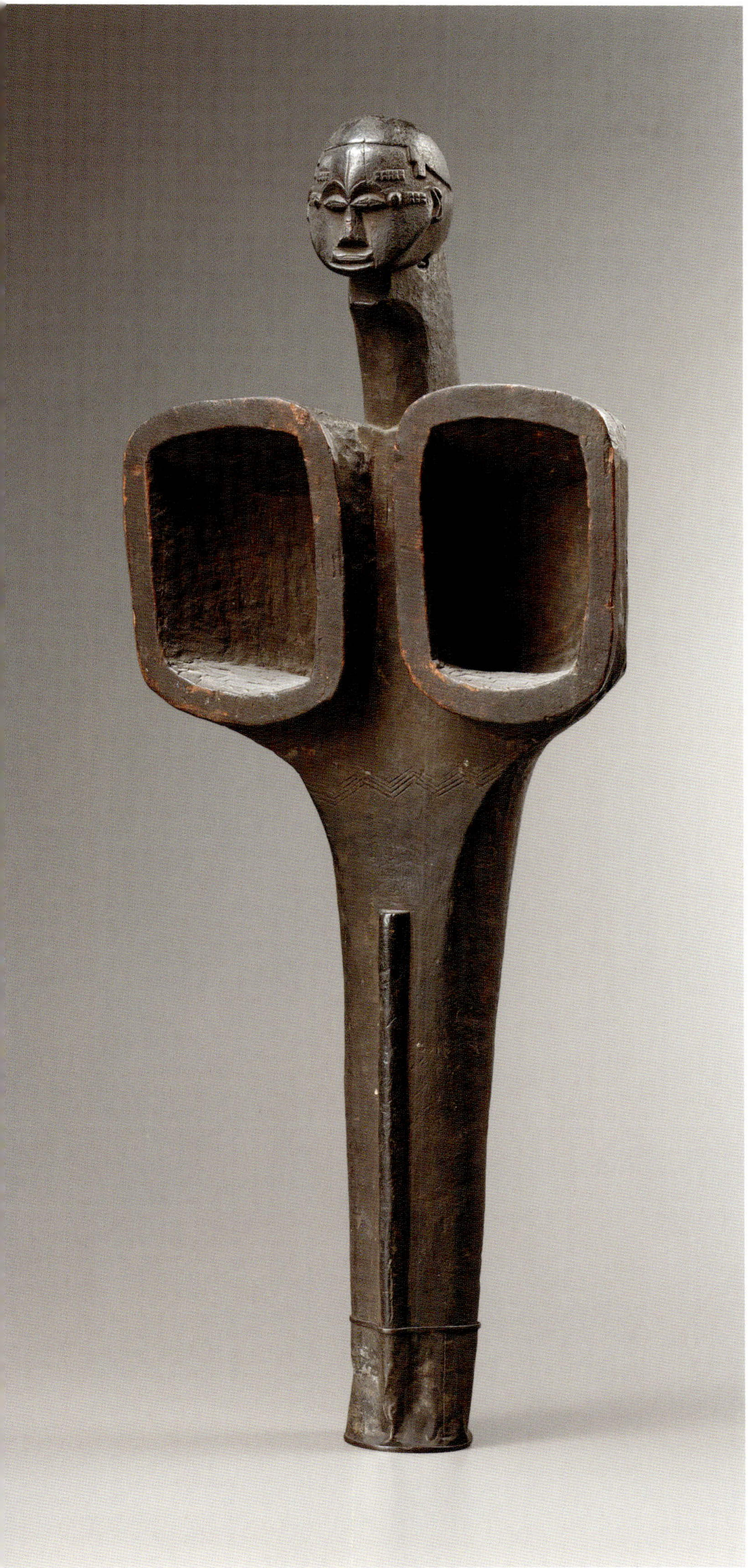

THE BELLOWS

Bellows provide a sophisticated air-delivery system devised to introduce bursts of oxygen into a blacksmith's fire. African designs offer astonishingly beautiful examples of blacksmiths' artistic ingenuity in solving this metallurgical hurdle (figs. 4.26, 4.27; see also fig. 1.20). The invention of compact, mellifluous, multichambered bellows that preheat air and initiate the conditions of atmospheric reduction in the forge and furnace proliferated in Africa centuries before the rest of the world understood the mechanical problems thus solved.[17] This equipment served to advance metalsmithing efficiencies, lighten the labor load, reduce fuel consumption, produce better products in less time, and meet increased local demand for goods and services. Perhaps most significantly, implementation of the invention within iron smelting contexts accelerated pivotal shifts in production methods, enabling workers to manufacture consistent, high-quality bloom at expanded scales of operation, commensurate with burgeoning trade networks building momentum during the height of Africa's Iron Age.[18]

The three basic bellows types developed in Africa—"bag" (see fig. 4.26), "drum" (fig. 4.28), and "concertina" (fig. 4.29)—have culturally distinct variations across the continent (Cline 1937, 102–08; Friede and Steel 1986, 12–16). Usually constructed of wood, fired clay, sun-baked earth, or animal skin, they have two or more air-trapping bladders that inject air into the forge as they are deflated. The drum-like bellows type is most common, and its chambered reservoirs are covered by a thin, pliable diaphragm of leather, plant fiber, cloth, or rubber, sealed tightly to envelope an adequate volume of air for the scale of work conducted. Rhythmically pumped with both hands by a blacksmith's helper, air compressed beneath the membrane is discharged in measured amounts into a pair of tubular passages to regulate the charcoal-fueled flames. The lung-like device inhales warm air near the fire on the helper's up-stroke, which then mixes with cool air on the down-stroke to introduce a calibrated blend of preheated air and swiftly attain forging temperatures. The air-emitting nozzle at the end of the bellows is closest to the flames and can be protected by a conical- or bowl-shaped clay tuyère that funnels exhaled air to the base of the fire at the lowest point of entry (fig. 4.30). This encourages more thorough combustion to ensure a reducing atmosphere that consumes oxygen before it is able to reach any iron being heated at the upper level of the charcoal mound.[19]

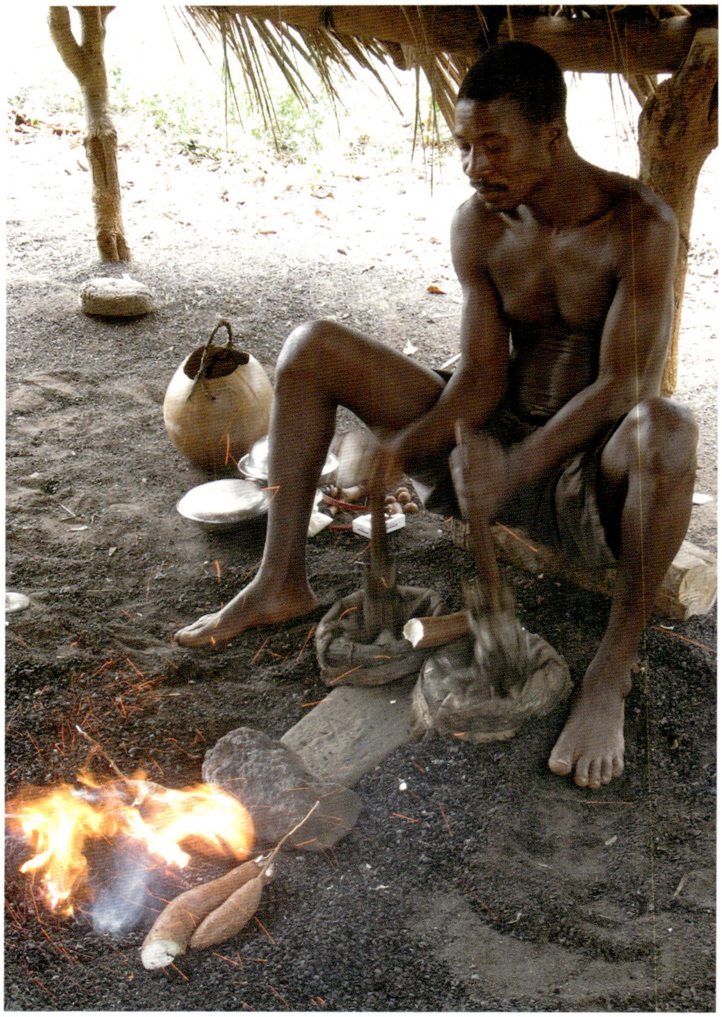

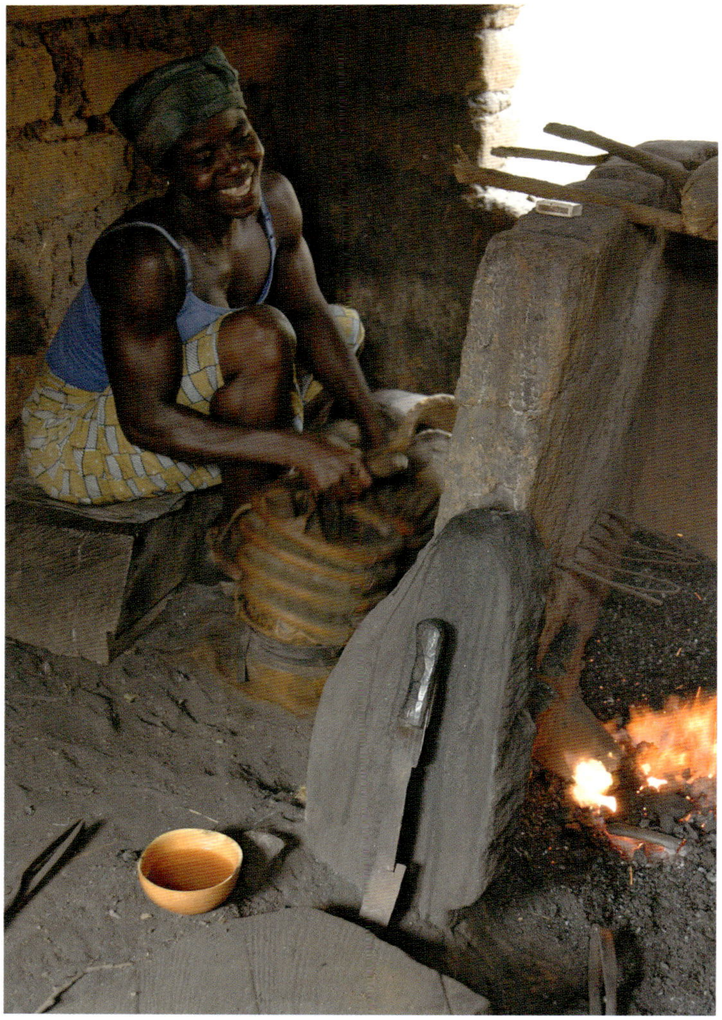

4.28 In Yohonou, Togo, Ewe blacksmith Hodenou Noglo operates goat-skin-clad bellows that ingeniously preheat air to increase combustion temperatures when bringing iron up to heat in the forge. The flat stone resting on the nose of the bellows is a heat deflector to keep the flames from catching the wooden tool on fire.

Yams roast beside the charcoal flames to accompany lunch, served in covered dishes with a handled gourd of fresh water to drink.

PHOTOGRAPH BY TOM JOYCE, 2008. © TOM JOYCE

4.29 In Kossi Kao's workshop in northern Togo, his wife Jouli pumps "concertina"-style bellows, widely used throughout the Arab world and likely introduced into northern Africa with the spread of Islam beginning in the seventh century CE. Their expandable corrugated bladder envelops a larger volume of air than other African models and is protected from heat by an earthen wall surrounding the forge on three sides.

A daily libation of millet beer (in the gourd on the ground) is offered atop a stone stele, an ancestor's anvil once embedded in the workshop's floor and now venerated as a shrine in front of the forge.

PHOTOGRAPH BY TOM JOYCE, TCHARÉ, TOGO, 2010. © TOM JOYCE

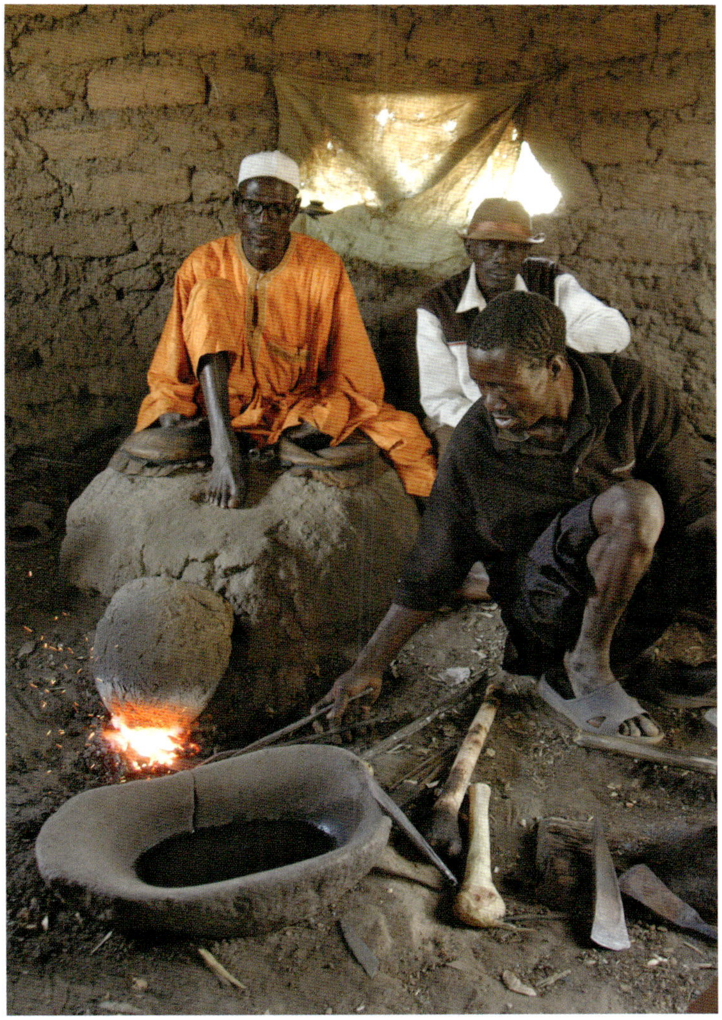

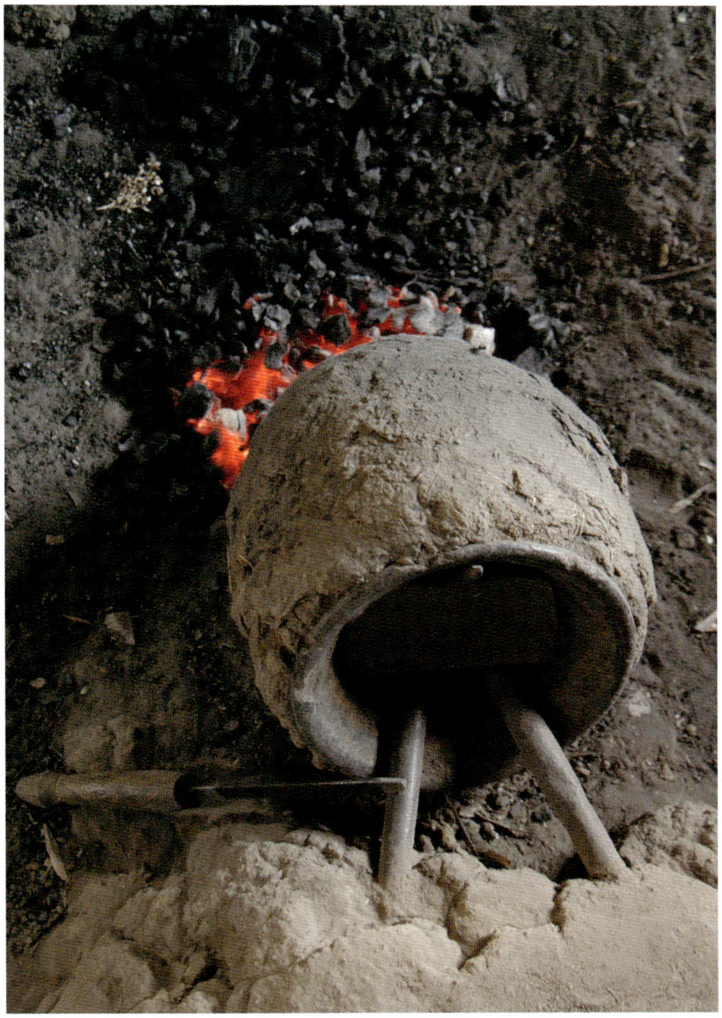

4.30A–B In their workshop in Kansamala, Mali, Bamana blacksmith Tamba Konte heats an axe while his brother Moussa Konte looks on and another brother, Sedu, operates the bellows (*fandaga*, an amalgam of *fan*, for "forge," and *daga*, "cannon"). Three fired ceramic pots have been made to construct the chambers and tuyère (*fanburu*): two are buried side-by-side in a raised earthen plinth (*bambe*) that functions also as a bench, elevating the bellows operator above the forge's heat; the third ceramic pot (fig. 4.30B) is clad in unfired clay to insulate it from the incandescent charcoal, and to regulate the temperature of the air being sucked into the bellows and then returned to the fire in a preheated blast.

PHOTOGRAPHS BY TOM JOYCE, 2012. © TOM JOYCE

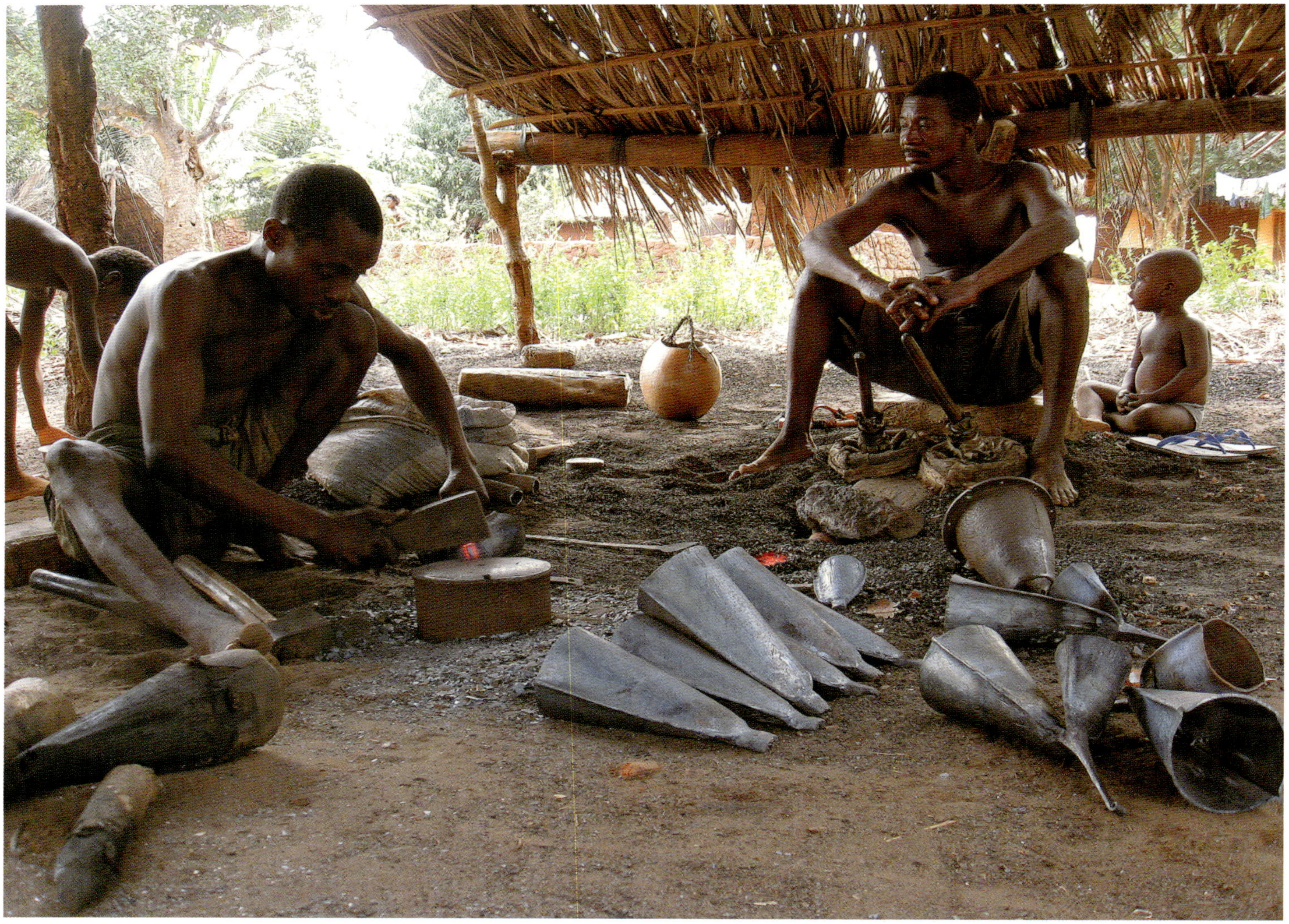

4.31A–B Ewe smiths and accomplished percussionists Galbert Atakpa and Hodenou Noglo forge several sizes of single, double, and triple bells known as *gangokui* (see fig. 17.15). Famous in the region for performances and ceremonial use, the instrument is also renowned in music ensembles internationally for its resonant tonal range. When the smiths tune the instruments (fig. 4.31B), similar rhythms are invoked as the bellows operator plays with two sticks similar in size to those used to pump the bellows, while the blacksmith plays the base beat with one stick (like a hammer) serving as a metronome for the soloist's lead melody.

PHOTOGRAPHS BY TOM JOYCE, YOHONOU, TOGO, 2010.

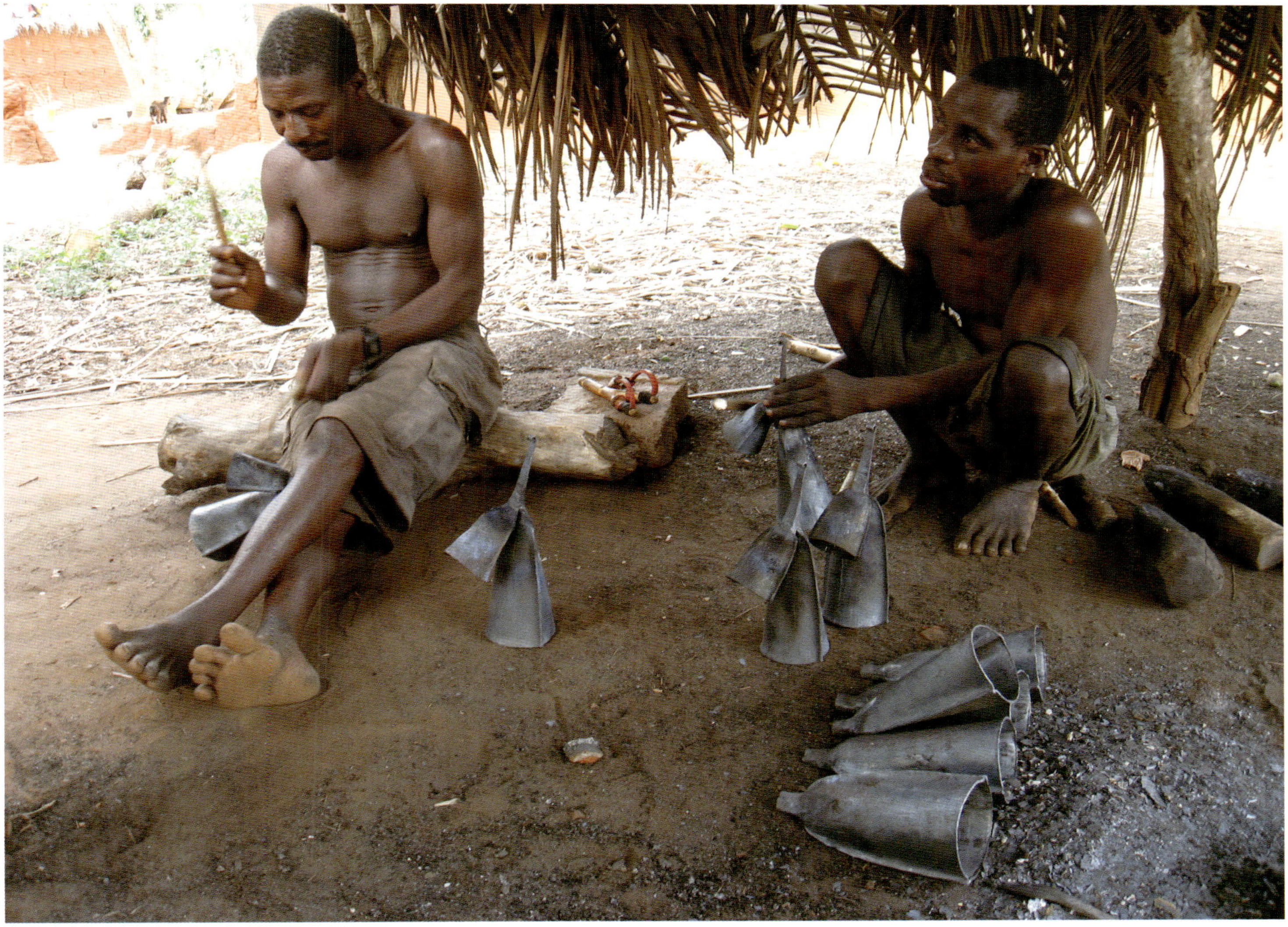

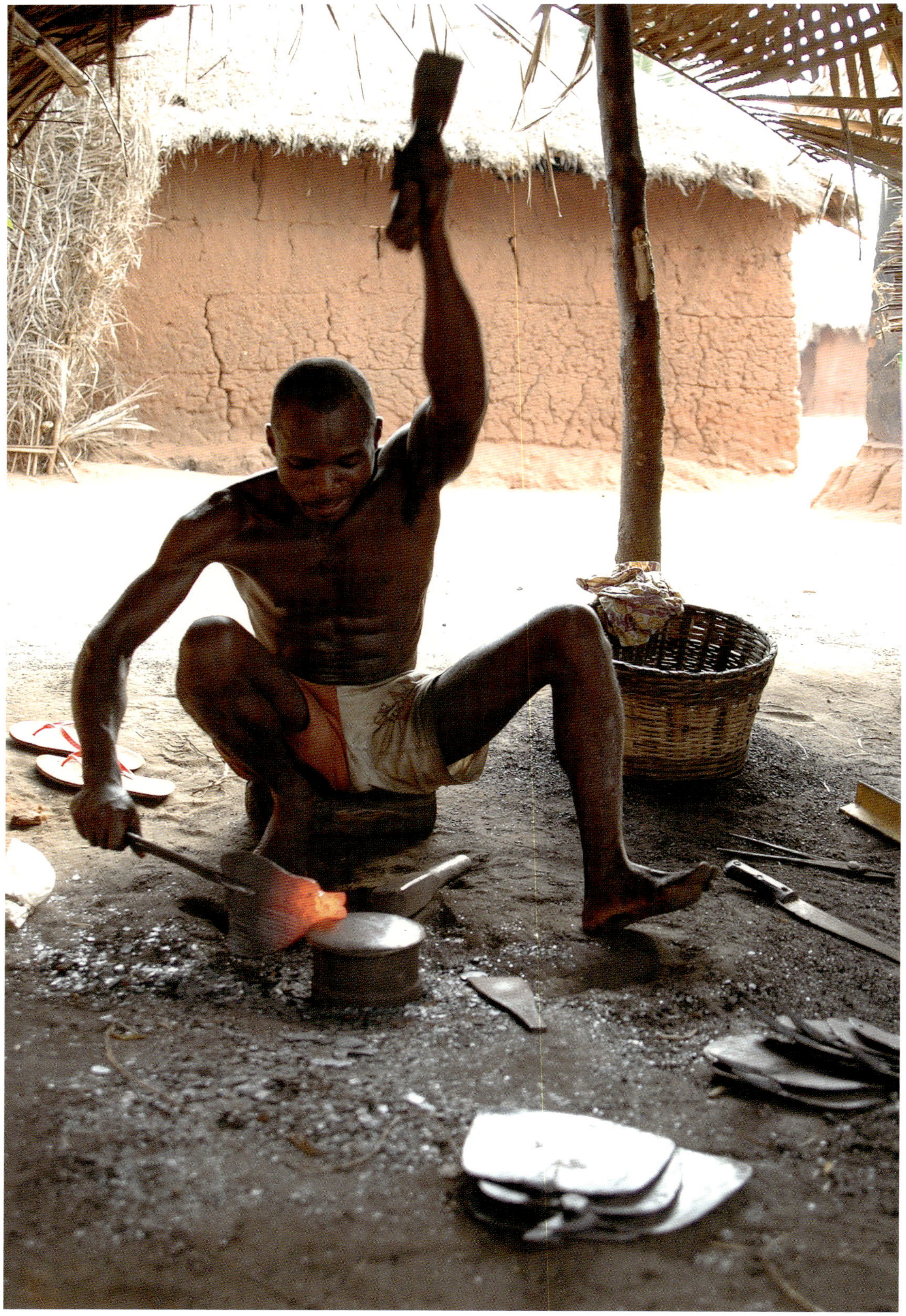

4.32 Whether a blacksmith is left- or right-handed, his center of gravity is most effective when the transfer of energy is vertically oriented along the spine to direct a blow with maximum power. Here, the smith is shouldering material to draw out the tang on a hoe blade (*kodjowe*) like those stacked up in the foreground.

PHOTOGRAPH BY TOM JOYCE, YOHONOU, TOGO, 2010.

The pumping rhythm and cadences of the bellows are carefully calculated to deliver blasts of hot air at a modulation and rate of delivery determined according to the worked object's thickness, length, and width. This assignment becomes ever more formidable for the bellows operator when several pieces of iron are being worked in the fire simultaneously. The helper must proactively respond at the right tempo to move the work along at the blacksmith's pace—anticipating the smith's every move is imperative. For example, it can be a stressful timing exercise to have the next piece in the stack evenly heated and ready to be pulled at precisely the moment the smith slips the previous work underneath it for reheating over the fire. The correlation of the bellows pulse and the percussive beat of the hammer serve to signal, without words, the necessary interval. This sonic interaction—the melodious interweaving of the sounds of the forge—is remarkably consistent wherever these tools are in use, more often than not musically coordinated in a 6/4 time sequence (see also chapter 17, this volume). It seems quite probable that this deliberate working tempo spawned some long-established drumming rhythms of African performing arts, today heard throughout the continent and abroad in international music ensembles (fig. 4.31).[20]

THE HAMMER

In a blacksmith's hand, the hammer functions largely within a vertically oriented space defined by the outer reach of the arm, as it hinges at the elbow to pivot downward toward hot iron on the anvil. Though this didactic explanation sounds quite obvious, what is unseen by the observer and matters most to the smith is the inertia behind the hammer blow—what he visualizes as extending *beyond* his reach, *beyond* contact with solid surfaces. It is the force of energy that he propels, that passes *through* iron, rather than simply landing upon it. Without this mode of delivery, the blow would move only the surface of the material; it would lack the follow-through that produces sufficient power to move the center and drive friction-generated heat back into the work, allowing it to be forged longer (and thus brought to shape more quickly). At the vertex of the parabolic gesture, the hammer is released ever so slightly from the blacksmith's grip, allowing the muscles of the hand and arm a moment's rest. In this split-second pause, a weightlessness is induced, a breath taken, and stamina replenished. Then the smith partners with gravity to amplify the tool's falling weight (fig. 4.32).

A legion of hammer blows is necessary to bring any iron form to life and each is precisely calibrated to respond to the effects of the preceding blow. That is to say, they are delivered in a progression that varies in strength and cadence. Rarely does the hammer strike in exactly the same place; its use is woven into an improvisational sequence orchestrated through intensive training to manipulate ferrous material in its plastic state. African blacksmiths inventively fashion hammers to take full advantage of all physical factors available to get the job done (figs. 4.33–4.39). The size and type of work performed determines the most suitable hammer for the task—whether stone or iron, whether with attached handles or without, whether wielded by the smith alone or in synchrony with a striker as a forging partner.[21]

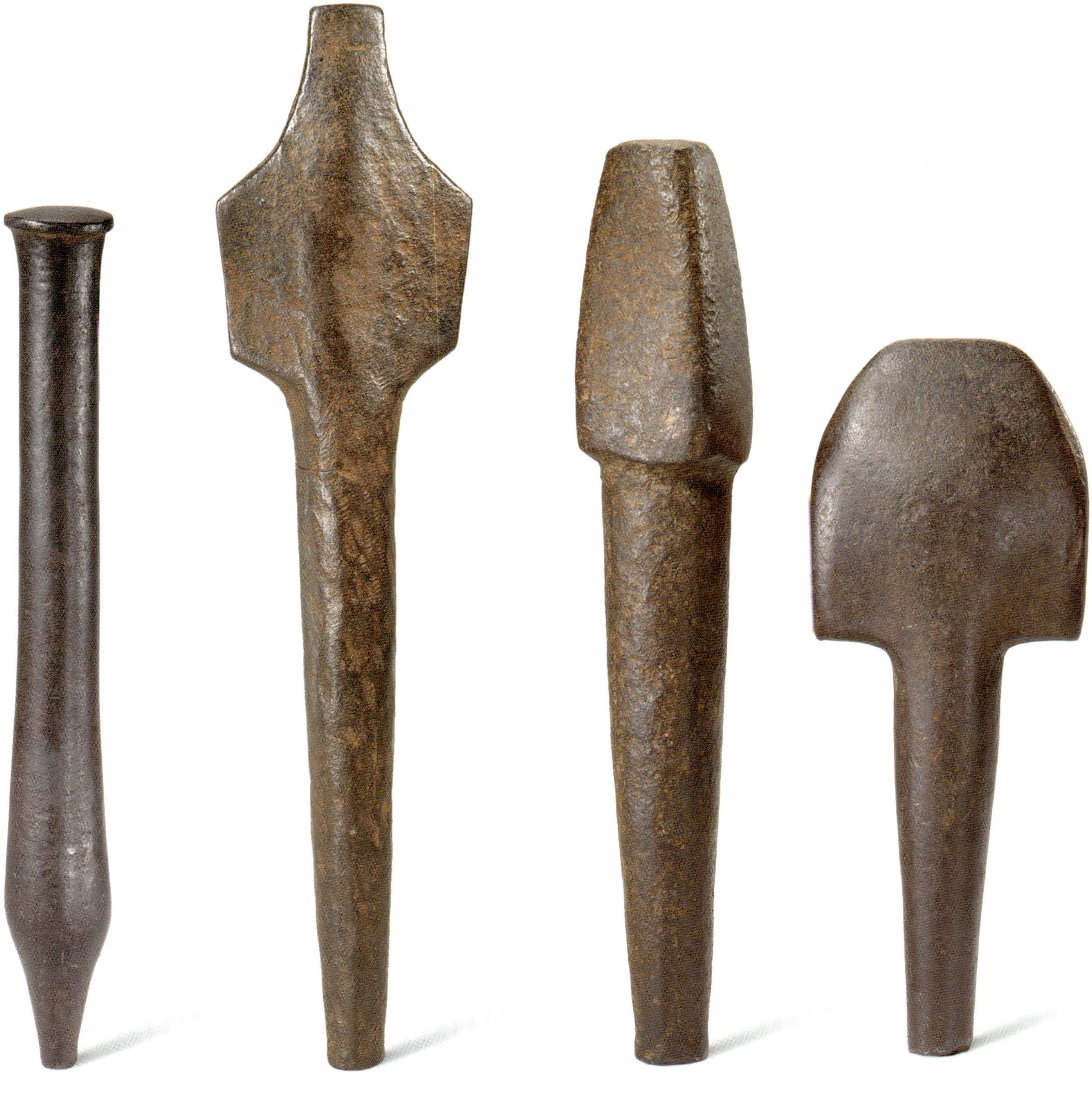

4.33
Artist unknown (Sonike peoples, Gambia)
Hammer (*taneh*), 20th century
Iron
H: 31.5 cm
PRIVATE COLLECTION

4.34
Artist unknown (Kota peoples, Gabon)
Hammer, 19th century
Iron
H: 38.1 cm
PRIVATE COLLECTION

4.35
LEFT: Artist unknown (Yorùbá peoples, Nigeria)
Hammer (*owú*), late 19th to early 20th century
Iron
H: 33.5 cm
PRIVATE COLLECTION

RIGHT: Artist unknown (Igbo peoples, Nigeria)
Hammer (*otutu*), 20th century
Iron
H: 26.5 cm
PRIVATE COLLECTION

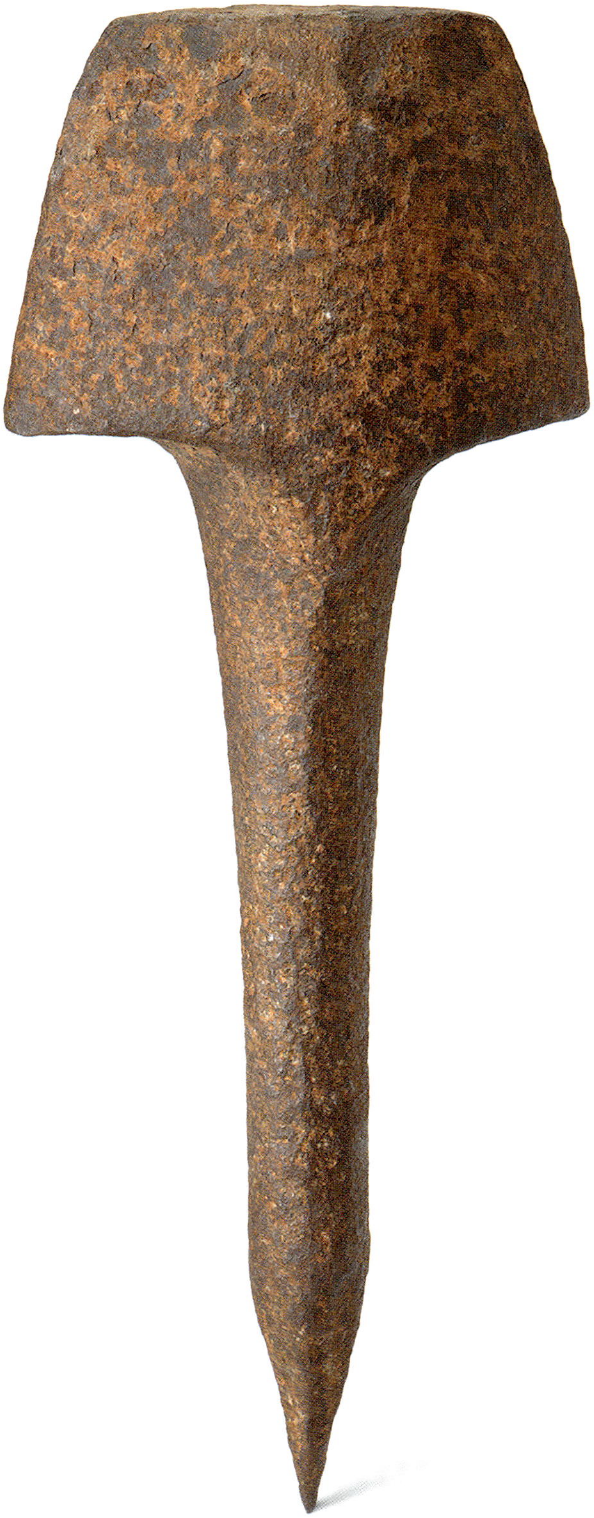

4.36
Artist unknown (Mfumte peoples, Cameroon)
Hammer (*mala*)
Iron
H: 41 cm
PRIVATE COLLECTION

This hammer exemplifies a tool shaped for a specific purpose, for it is perfectly contoured to fit the curvature of a distinctive hoe-blade-shaped currency called *chu'npo* (see figs. 4.37, 4.38), exchanged among Mfumte peoples until the mid-1970s. Broad and heavy enough to draw large material quickly, this hammer facilitated making one of the region's most important matrimonial offerings.

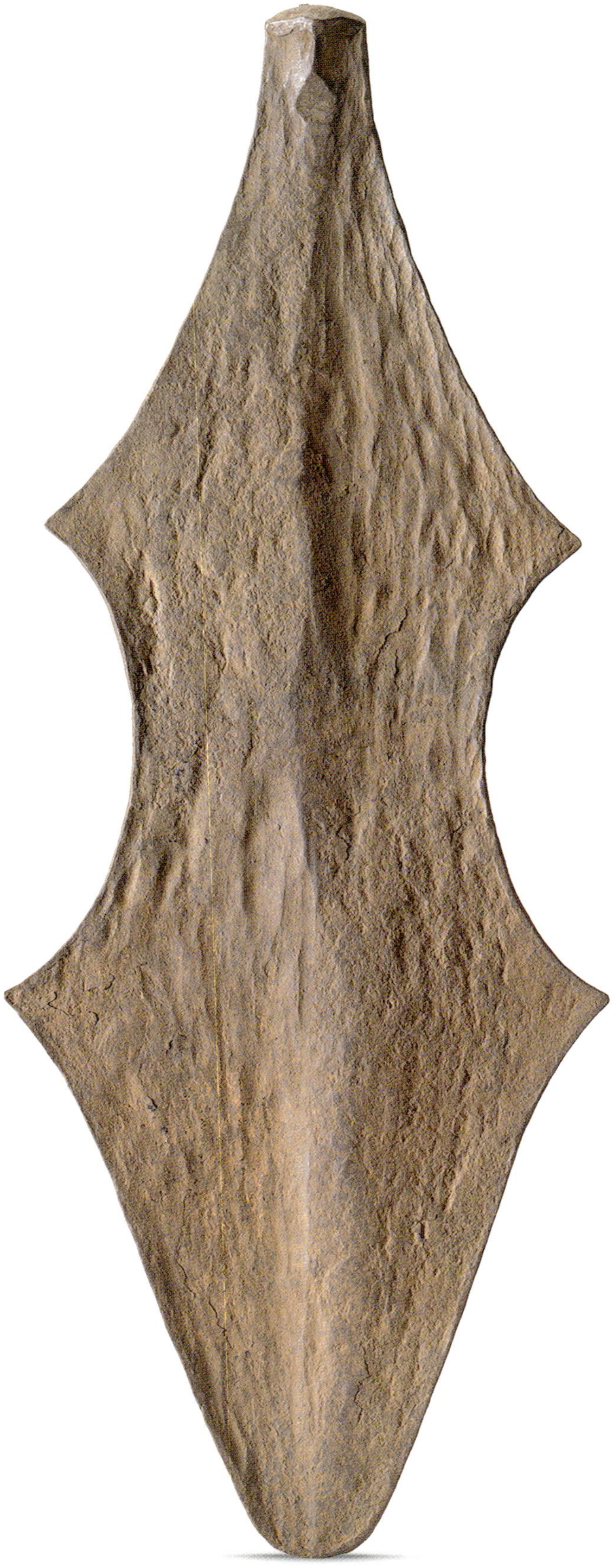

4.37
Artist unknown (Mfumte peoples, Cameroon)
Hoe-blade-shaped currency (*chu'npo*), late 19th century
Iron
H: 69 cm
PRIVATE COLLECTION

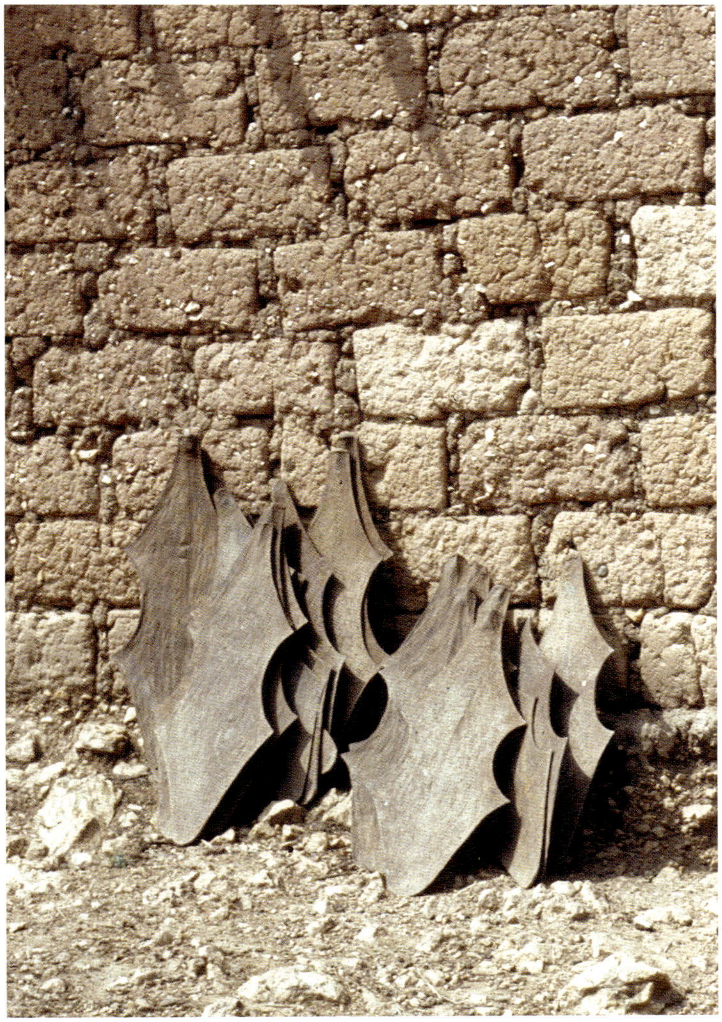

4.38 *Chu'npo* currencies stored behind a home in a Wuli village of the Mfumte community, a consolidation of several ethnic groups as an administrative federation during the colonial era. In a remarkable cooperative effort, three neighboring villages worked together in a joint venture to produce large quantities of *chu'npo* for regional trade. Miners from Bibdji (Yulara peoples) excavated high-grade iron ore for smelters to process and forge into bars in the village of Lus (Wuli peoples); these were transferred to Kwadja (Ungpo peoples), where blacksmiths forged them into final shape for distribution as currencies. Respected for their excellent iron, Mambila communities nearby considered *chu'npo* the highest standard bridewealth a groom could offer as compensation (Viviane Baeke pers. comm., 2018).

PHOTOGRAPH BY VIVIANE BAEKE, LUS, CAMEROON, 1980.

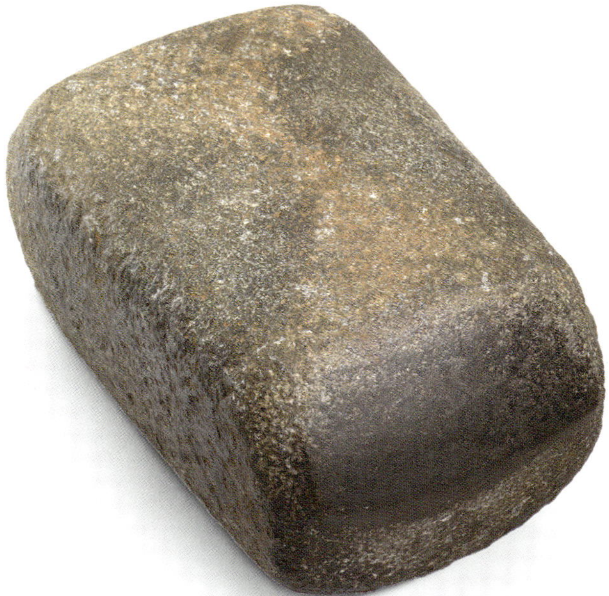

4.39
Kossi Kao (b. 1978, Kabre peoples, Togo)
Hammer (*kima biye*), 1999
Basalt
H: 25 cm
PRIVATE COLLECTION

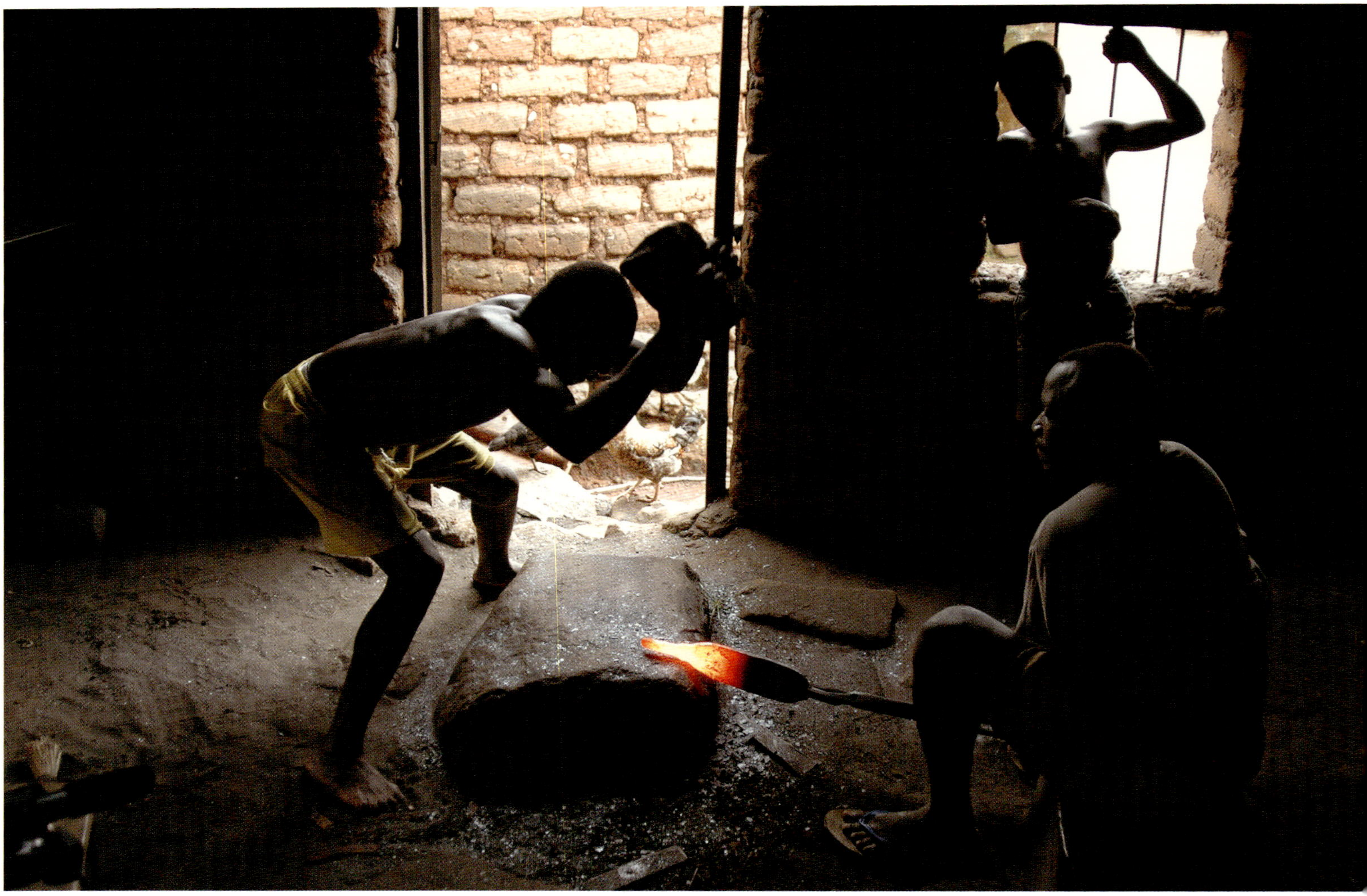

4.40 In the hands of skilled practitioners such as Kossi Kao and Essozimna Ide, pictured here, stone tools are efficient and ergonomically appropriate for the tasks they perform. Basalt hammers (*kima biye*) and anvils (*taata biye*) are standard equipment among Kabre blacksmiths. The dense volcanic stone doesn't absorb heat as readily as steel, and because the iron can be worked longer, the number of heats needed to finish a shape is reduced.
PHOTOGRAPHS BY TOM JOYCE, TCHARÉ, TOGO, 2010.

Widely used throughout the continent, single-piece iron hammers are made so that every face, edge, and terminus serves as a forging plane (Cline 1937, 91–101; Maquet 1965, 5–8). Some are ingeniously designed to be driven into the ground at their tapered shank so that they may temporarily serve as stake-like anvils for light work. Haftless hammers are forged in an astonishing range of shapes, varying in form even among neighboring peoples, though many tend toward stylistically phallic profiles that proclaim the procreative powers blacksmiths garner when "birthing" objects into a community (see Herbert 1993, 99; see also chapter 2, this volume).

To wield a hammer with precision and relay the required force and amplitude of movement, a blacksmith must master two distinct hand grips: the "power grip" (using the palm as a buttress) and the "precision grip" (thumb in opposition to the fingers). Research on human biomechanical anatomy has revealed that the hand, as an integral part of the entire arm, is in effect "a specialized termination of a crane-like structure suspended from the neck and the upper chest" (Wilson 1998, 8). The pull of tendons and muscles near the elbow, articulated by the hand, defines a margin of control that is neurologically registered through tasks that inform the skillful gripping of tools. An intricate balance of visually stimulated and haptic reflexes is essential to developing and administering the grip necessary to make *anything* with *any* tool held in the hand. The manual dexterity and kinesthetic memory required to instantly process and react to changes caused by a succession of hammer blows on hot iron exponentially antes up the equation, even when enhanced proprioceptors[22] are rallied into action. Hundreds, if not thousands, of hammer blows—directed at an iron target an arm's length away, sometimes glowing an incandescent 1232 degrees Celsius—are typically delivered to translate a conceptual representation into a completed work (see figs. 4.40, 4.41).

This tool that a blacksmith grips his entire working life is so significant and emblematic that, upon death, it may be buried with him or enshrined on altars dedicated to ancestors.[23] The hammer he used most frequently is believed to possess a life force incrementally achieved through effort, and it can be venerated by successive generations of family members; after all, it is through the deceased person's labors that the living prosper in the present. On the other end of life's trajectory, when an apprentice is ready to light a fire in his own forge, it is customary among some mentors to fashion a hammer for him (Fowler 1990, 232). The new tool often incorporates, forge-welded into its working surface, a small fragment chipped from the master's own hammer. When the young smith strikes hot iron with his hammer, the blow is guided by his teacher's vitality as well as that of his teacher's teacher. Hard-won mastery is sourced from countless generations of blacksmiths—this, the hammer's call-and-response, the "ferrous DNA" of apprenticeship, is the same the world over, whether at the university or in the field.

CONCLUSION

> To know must therefore be to interpret . . . from the visible mark to that which is being said by it and which, without that mark, would lie like unspoken speech, dormant within things.
>
> MICHEL FOUCAULT ([1966] 1994, 32)

The hammer and its collaborators—anvils, tongs, and bellows—have much to teach about interpreting kindred alliances between a blacksmith's *work* and the *work* a blacksmith makes.[24] No matter where one stands when approaching any object forged in iron, it is imperative to look in many directions simultaneously—for the most obvious of clues as well as the most hidden of affinities.[25]

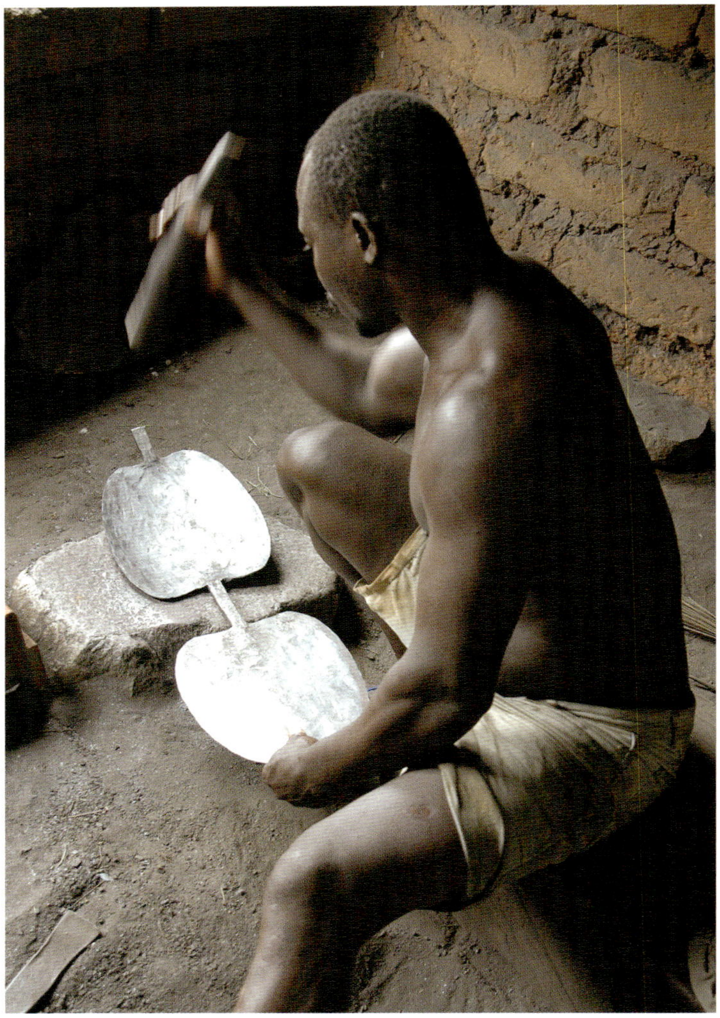

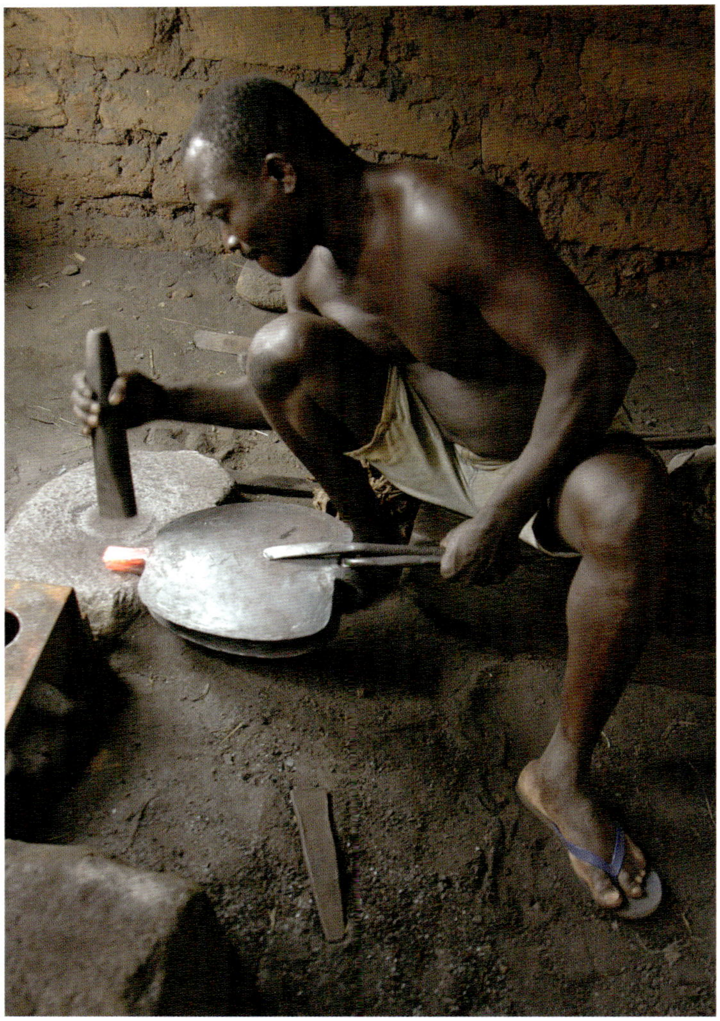

4.41A–B With a hammer (*sire kimabiye*) shaped for both vertical and horizontal forging, blacksmith Kossi Kao refines the concave halves of a Kabre initiation instrument (*ekpande*) with a hammer-induced smoothing technique known as planishing. Cold forging like this work-hardens surfaces, in this case to enhance the bell's ringing quality prior to bending it into its distinctive clam-shell form (see fig. 17.18).
PHOTOGRAPHS BY TOM JOYCE, TCHARÉ, TOGO, 2010.

At the heart of all works in iron lives the *entire* life cycle of iron. Not on the periphery of its existence, but at its subatomic nucleus—from its supernova origins to its meander through bloodlines, from its most mundane use to its most sublime purpose. Iron exists in a perpetual state of transformation and blacksmiths in Africa, as elsewhere, have been momentarily privileged with the immense responsibility of strategizing for its use before it is released, yet again, into the hands of other humans.

> As the *hammer* strikes, so too are the objects it makes called to decisive action.
>
> As the *anvil* supports and deflects, so too do the objects forged upon it absorb and rebound from foundation-shaking impact.
>
> As *tongs* grip, so too do objects they have secured safely manage difficult or dangerous situations.
>
> As *bellows* breath, so too do objects thus nourished with warmth encourage life through things birthed into the world.

Each of these practical devices function as poignant intermediaries, in close alliance with the smith whenever iron is shaped. Without a blacksmith accompanying them in the workshop where they perform, each is but a useless lump of wood, iron, or stone. A tool's proximity to use or disuse, purpose or obsolescence runs parallel to requested needs or desires met, just as the longevity of a tool's relevance in the hands of a maker runs parallel to forging skills learned or lost (fig. 4.42).

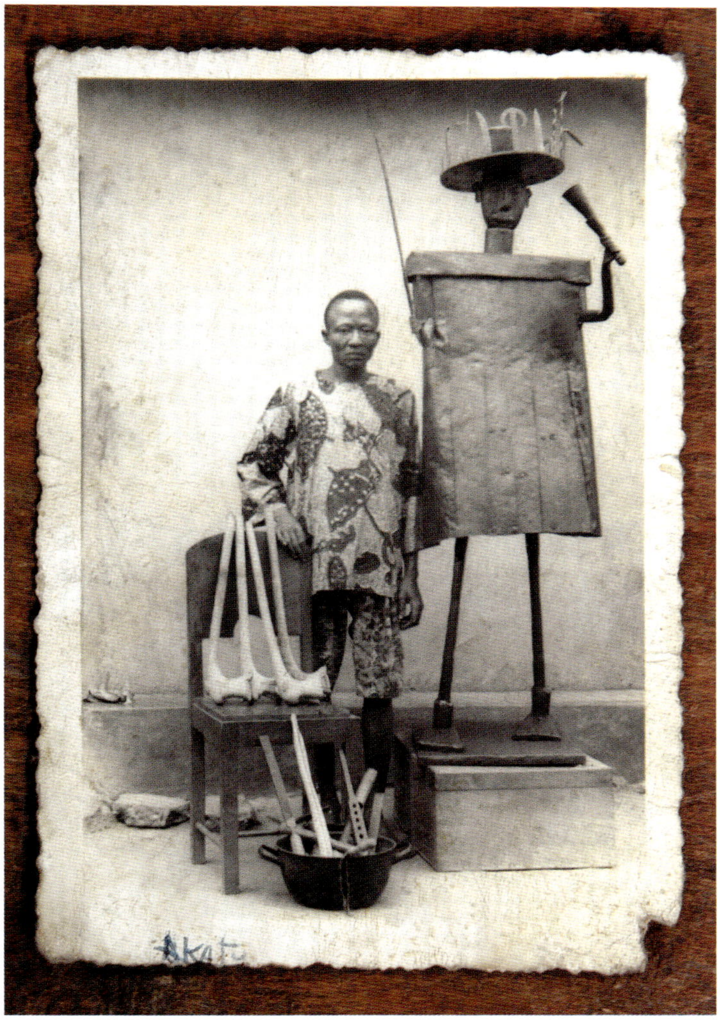

4.42 In Abomey, Republic of Bénin, a family legacy is proudly carried on at the forge of Simon Gounon Akati, pictured here circa 1950 with a representation of Gu, the Fon god of iron. His father Ekplekendo Zomabodo Glenegbe Akati (active 19th century) created another commanding representation of Gu, now on permanent display at the Louvre in Paris (see interleaf H, this volume). Today Simon's sons and grandsons conduct work in the smithy, employing patterns and hand tools used by their ancestors, as well as an anvil made from a repurposed cannon fragment. PHOTOGRAPH BY UNKNOWN PHOTOGRAPHER, REPHOTOGRAPHED BY TOM JOYCE WITH PERMISSION OF THE AKATI FAMILY, 2010.

INTERLEAF

B

BLACKSMITHS ONSTAGE

Tom Joyce

In photographic archives worldwide, researchers can find images of African blacksmiths working in open-air scenes like those illustrated here as figs. B.1–B.5 (see also fig. 17.2). Though a large body of scholarship has already deliberated on the camera's historic role in constructing and reinforcing stereotypes that present troublesome "taxonomies of otherness" (Geary 2007, 76), to my knowledge images of blacksmiths have not been specifically analyzed through this particular lens. Historical photographs of "working" environments in Africa, taken since the late nineteenth century by missionaries, colonial officers, travelers, filmmakers, ethnographers, and anthropologists, include portraits of blacksmiths from which we can gain some knowledge, but the viewer should be cautioned about the veracity of such documentation. These staged tableaus often misrepresent blacksmiths' forges, technological innovations, and the interactions of a smith, his assistants, and their tools.

Early photographers often overcame the lack of light in darkened workshops by directing forging teams and their equipment outside, into ambient natural light. Though necessary for adequate film exposure, such a handling positions the smith in a makeshift setting, rarely clarified, leading to potential misinterpretations in "reading" images related to African blacksmiths and their forges. For rather obvious reasons, blacksmiths avoid working without a protective roof or other shielding from bright sun or inclement weather. Without sufficient shade, a smith cannot accurately gauge forging temperatures, a process that requires acute observation of iron's gradient color as it heats. Also, full sun on the smith's iron tools risks rendering them too hot to handle, while on cloudy days, a sprinkling of rain would halt work, lest the moisture extinguish the fire, quench hot iron, and damage leather- or fiber-clad bellows. Were work to be conducted under public scrutiny in open-air scenarios, the secrecy and discretion required of a blacksmith in creating certain ritual objects, performing delicate operations, or realizing client commissions would be compromised.

Photographers have favored uncluttered views featuring narrow selections of tools, and they have typically positioned smiths and their assistants in expedient arrangements relative to the camera's lens—for instance, with workers facing in the same direction, seated on the "wrong" side of the fire relevant to the hammering hand of the smith, and too close or too far from the hearth for "real" work to occur (see figs. B.1, B.2, B.4, B.5). Another misleading setup depicts the anvil as only temporarily secured to a surface. When installed with permanency in a workshop, an anvil might be attached to a partially submerged stump or have a dovetail-shaped wooden base wedged into the ground for stability, rather than resting loose on top of soil (see figs. B.2, B.4, B.5). Alternately, anvils with long, tapered shanks would typically be anchored directly into a shop's floor; these images often present anvils driven too shallowly by the tang, thus cushioning a hammer's blow and allowing the tool to wobble unsteadily when struck (see figs. B.1, foreground of B.4). Likewise, the area around an anvil is not, realistically, clean. Highly indicative of a staged scene is a lack of oxidized iron scale, which accumulates in a scattering of thin flakes around an anvil's base in quantities commensurate with the amount of work performed. This naturally occurring by-product of forging gathers swiftly because iron loses approximately 3 percent of its thickness to flaking every time it is heated.

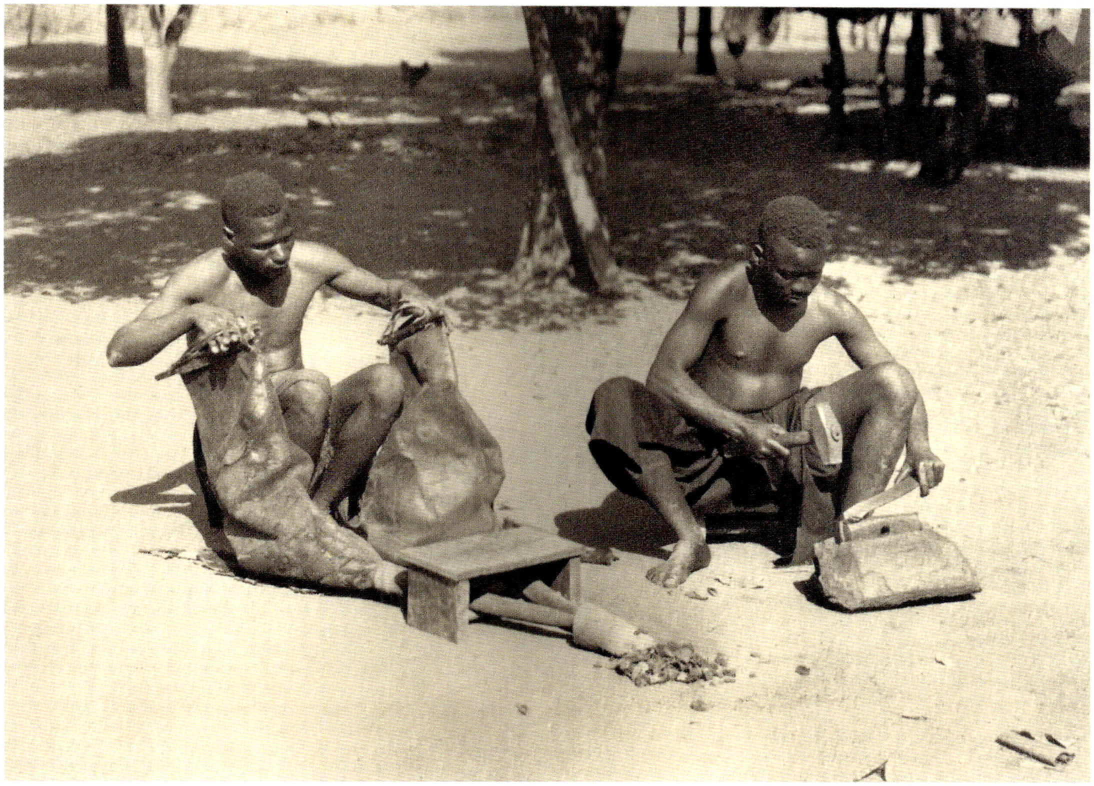

B.1 Chopi blacksmiths at Zavala, Portuguese East Africa (present-day Republic of Mozambique), 1933.
IMAGE COURTESY OF THE DUGGAN-CRONIN COLLECTION, MCGREGOR MUSEUM, KIMBERLY, SOUTH AFRICA

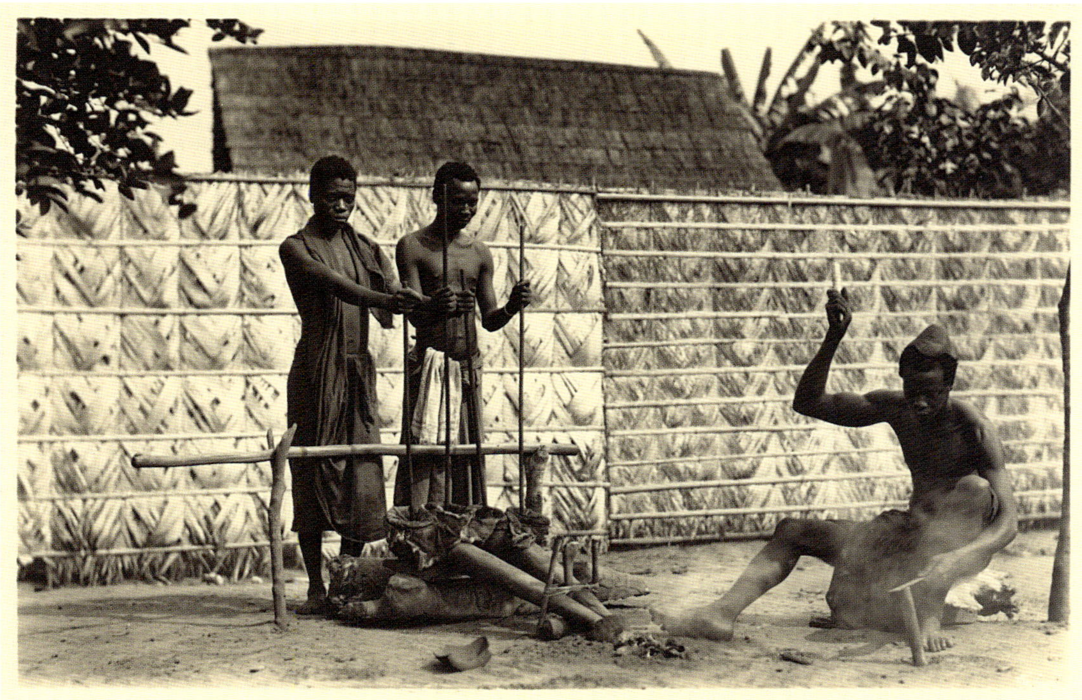

B.2 Kuba blacksmiths (present-day Democratic Republic of the Congo).
CASIMIR ZAGOURSKI AFRICAN POSTCARDS, 1924–41 (INCLUSIVE), MANUSCRIPTS & ARCHIVES, YALE UNIVERSITY. IMAGE © C. ZAGOURSKI

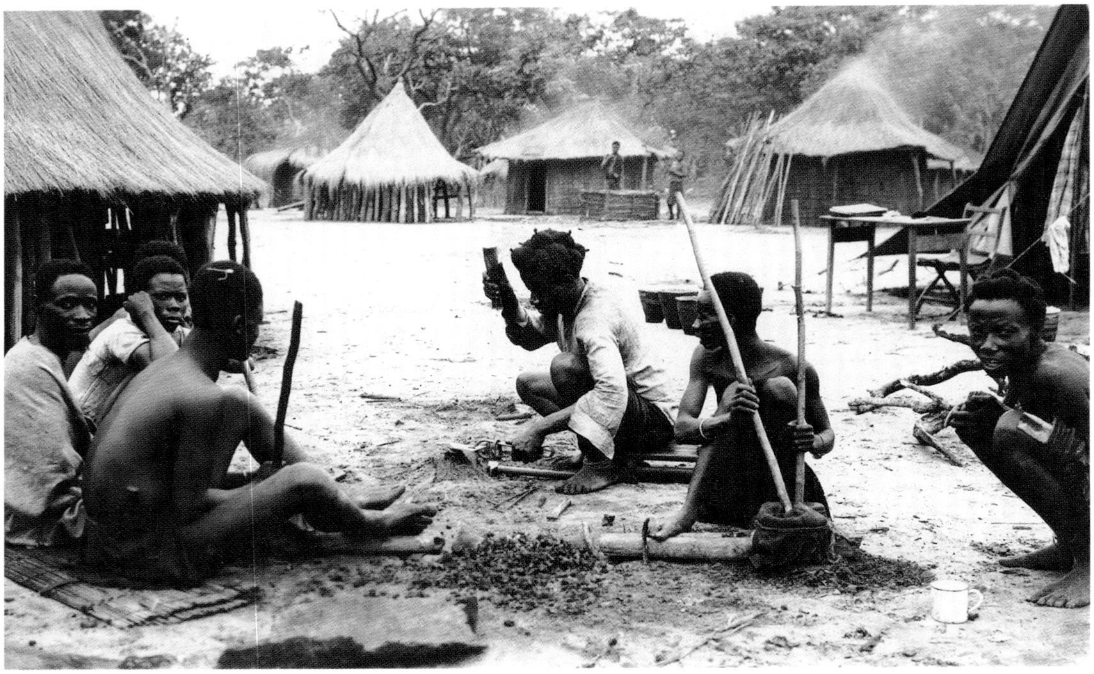

B.3 Chokwe blacksmiths, Katanga Province (present-day Democratic Republic of the Congo).
PHOTOGRAPH BY E. STEPPÉ, CA. 1930S. COURTESY OF THE ROYAL MUSEUM OF CENTRAL AFRICA, TERVUREN

Images showing fires lit on level ground should also be received with some skepticism. Bowl-shaped hearths dug as depressions in the ground proliferate in workshops precisely because they insulate glowing embers, saving fuel as charcoal is shielded from inflammatory breezes that would speed combustion without adding heat to the iron. A submerged hearth also protects the smith and his helpers from full exposure to the charcoal's radiant intensity and reduces the chance of sparks igniting other flammables within the work space.

In some stagings, bellows operators appear in erroneous positions: with their backs to the blacksmith, off to the side rather than in line with the tool, or pumping opposite one another in a crossfire that defeats efficiency (heating the same charcoal through tuyère nozzles placed too close together, as in fig. B.3). They sometimes seem to struggle with unstable, rocking equipment, necessitating the grip of feet or legs to hold the bellows down—an implausible solution and potentially dangerous (again, see fig. B.3). While smiths do apply weights and various tie-down methods, they do so with considerable forethought and not in the haphazard configurations of cumbersome rocks on bellows improvised in some photographs. Implying that such methods and working conditions are usual insinuates a lack of common sense on the part of the smith rather than implicating the scenographer behind the camera.

Though academic field work and professional documentary procedures are, in large measure, handled differently today—and we are, indeed, fortunate to have these photographic archives available—imprecisely arranged compositions institute an overly simplistic record of the methodologies employed by skilled technicians and artists.[1] Wherever blacksmiths have not been a part of one's daily life, a public's assumptions about them tend to conjure up popular clichés even from the most "worldly" and highly educated practitioners in the field (Clifford 1988, 21–54). As Belgian anthropologist Anne-Marie Bouttiaux writes: "It behooves us, therefore, to state at the outset that what we don't know exceeds by far what we do know, and that our claims to 'expertise' are always suspect" (2010, 22–25).

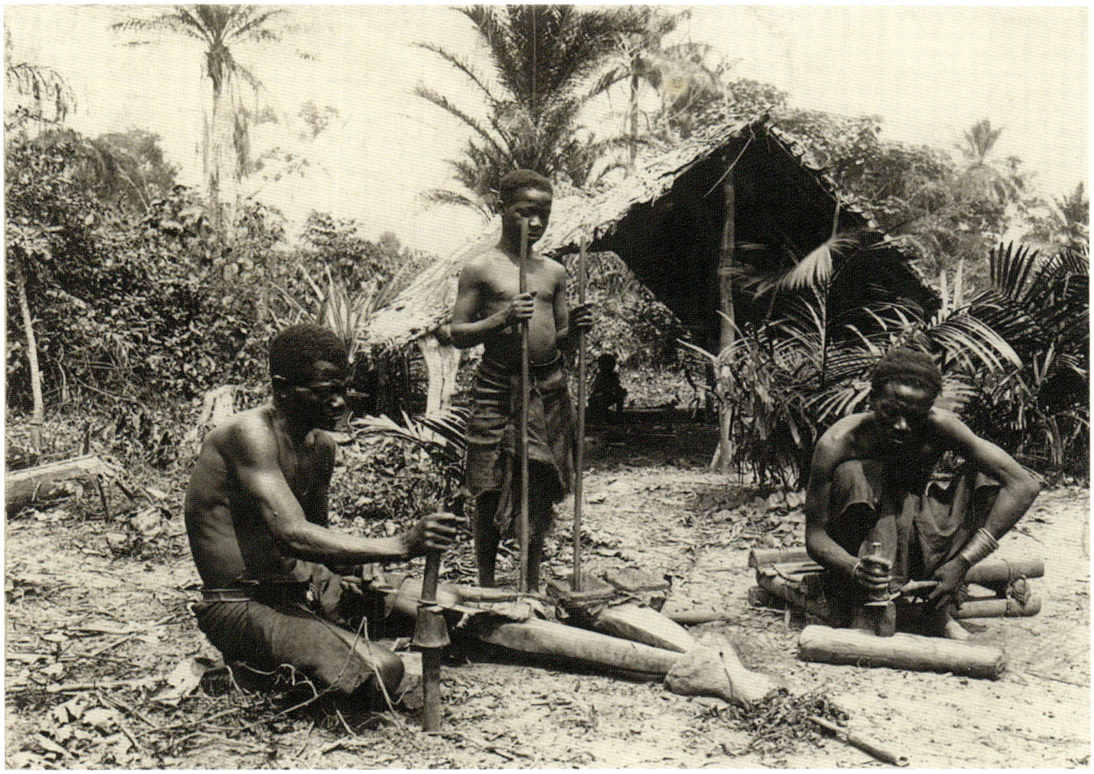

B.4 Kumu-Kuru blacksmiths in the village of d'Lpamu (present-day Republic of the Congo).

PHOTOGRAPH BY R. P. DELAERE, CA. 1922–26. COURTESY OF THE ROYAL MUSEUM OF CENTRAL AFRICA, TERVUREN

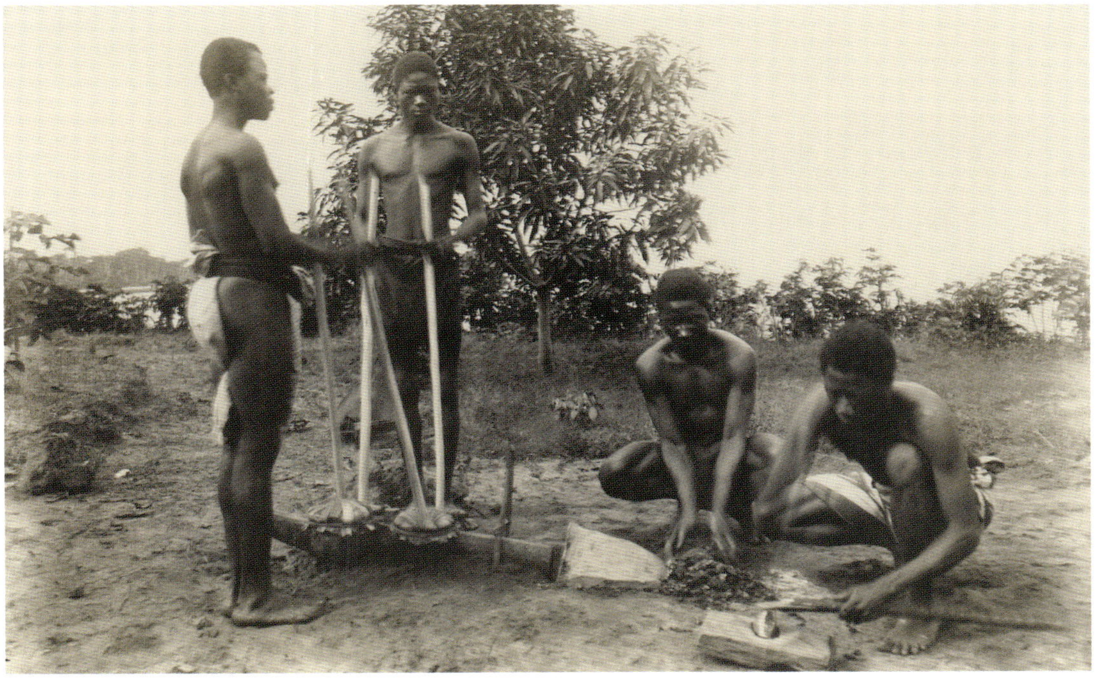

B.5 Knundo (Mongo) blacksmiths near Lake Mai-Ndombe, northeast of the Congolese city of Bandundu (present-day Republic of the Congo).

PHOTOGRAPH BY L. VAN DEN BROECK, EARLY 20TH CENTURY. COURTESY OF THE ROYAL MUSEUM OF CENTRAL AFRICA, TERVUREN

CHAPTER
5

PRECOLONIAL CURRENCIES: VALUE, POWER, AND PRESTIGE

Allen F. Roberts

For a responsive creature, to exist is to evaluate.

BARBARA HERRNSTEIN SMITH (1988, 42–43)[1]

Among Akan-speaking peoples of Ghana, the philosopher Kwasi Wiredu tells us, "'thinking'. . . means 'measuring something'"—determining its dimensions both literally and figuratively, that is, to decide its value and eventual purpose for use or exchange.[2] As thoughtful measures of value, iron currency tokens are among the most compelling and virtuosic of earlier sub-Saharan African blacksmiths' designs (fig. 5.1; see also fig. 4.4). The economic anthropologist Jane Guyer holds that "the 'bewildering jungle'" of tokens forged in such wondrously different shapes and sizes "constitutes one of the most complex examples of non-capitalist monies in the world" (1993, 243, citing Quiggen 1949, 2).[3] Furthermore, different styles produced by otherwise closely related peoples affirmed particular social identities and aesthetics, even as they served to articulate senses of self and community.

How can *these* things be "money," though? Then again, what *is* money to begin with? Even as general-purpose currency seems necessary to contemporary life in Africa and everywhere else in the world, "money as such is nothing in and of itself" (Frisby 2011, VIII), despite how fervently we all hope "that the money we live by has a secure objective foundation." Instead, as Keith Hart asserts, money is "an instrument of collective memory" that "conveys meanings, and these tell us a lot about the way human beings make communities" that "exist by virtue of their members' ability to exchange meanings that are substantially shared between them."[4] Significant here is Georg Simmel's century-old assertion that money is "'a symbol of our interdependence, locating its value in the trust that comes from membership in society.'" Money, then, "with language, [is] the most important vehicle for. . . collective sharing" (Hart 2005, citing Simmel [1900] 1978).

Memories of interdependence, collective sharing, and, one can add, histories of mutual indebtedness (see Thomas 1991) are the essence of *society* as spelled out in culturally specific cosmologies and philosophies; yet any such "systems of connectedness" also lead to an individual's awareness of self as "predicated on a system of reciprocal exchanges in the visible world." Looking back to Marcel Mauss's theory of the gift (1925), David Napier notes that "individuals 'know' themselves by actually exchanging with others those objects by which they are 'identified'. . . since much of what is worth knowing is quite literally self-evident. The self, in other words, becomes evident through a visible demonstration of its connectedness" (1992, 7–8).

Mauss's "'spirit' of the gift, meaning that the giver remains always part of the thing given," should be understood as active capacity, however, for it can contribute to rather than impede adaptation to changing senses of self and community (Mines and Weiss 1997, 161). Indeed, "money is principally a vehicle for the expression of human wishes. . . trapped for a moment, frozen in money transactions that allow us to meet others in society who are capable of satisfying them. . . . It follows from this that money considerably expands the capacity of individuals to stabilize their own personal identity by holding something durable that embodies the desires and wealth of all the other members of society" (Hart 2001, 256, 259, reflecting on Buchan 1997, 269). At the same time, *forgetting* interdependence, sharing, and indebtedness may be as important to community vitality as remembering such relationships when circumstances and needs evolve (Weiss 1997).

5.1
Artist unknown (Mbole peoples, Democratic Republic of the Congo)
Spear-shaped currency (detail), late 19th century
Iron, wood
H: 198.1 cm
NICOLE AND JOHN DINTENFASS
PHOTOGRAPH BY VINCENT GIRIER DUFOURNIER, 2017.

Tall presentation objects like this one as well as smaller versions of similar design are attributed to Mbole and Topoke peoples of east-central DRC and may have been traded and used by nearby regional communities as well. Nothing is known of their particular uses, but the combination of basic blade shapes with exquisitely forged points and adorning circlets suggests references to related aesthetics.

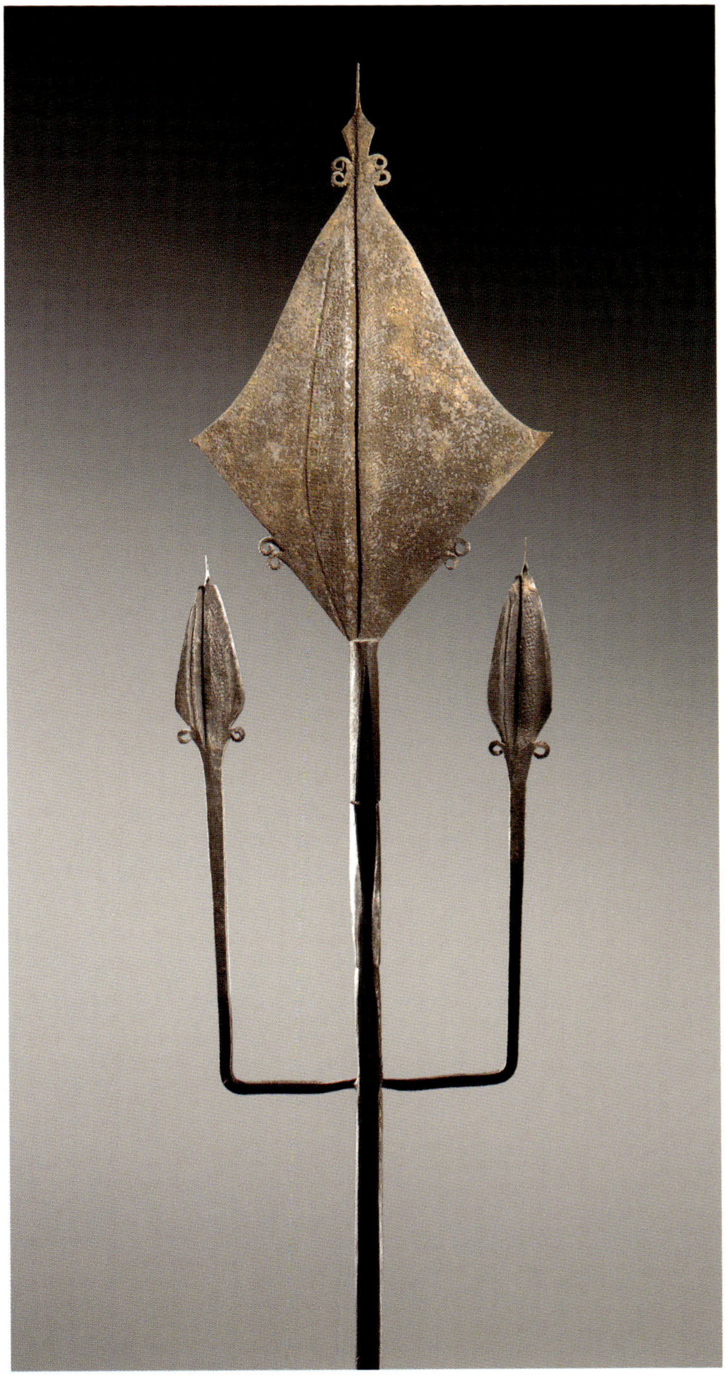

In many precolonial African societies, such dynamic relationships were made material through forged iron tokens exchanged in moments deemed particularly fraught or otherwise especially significant. Such currencies were multireferential—ambiguous "things of inbetweenness," as Paul Basu might put it. "Inbetweenness is a middle ground, a contact zone, a borderland," he writes. Accordingly, precolonial iron currencies helped constitute and remember relationships, and indeed, such "things *were* their relationships," with all the complexity so implied (Basu 2017, 8, 11; cf. Ingold 2011). Here we shall briefly ponder what constitute the materials and values of currency tokens, and then look to case studies of these intriguing items as once created by Sara peoples of the Republic of Chad and Kwele of the Gabon/Congo-Brazzaville borderlands.

IRON AS MATERIAL

Even as blacksmiths "participate in the world of materials, the properties of materials are not attributes but histories" that are "processual and relational." Indeed, as Tim Ingold writes, "every property is a condensed story." Such recognition leads him to propose a "radical . . . ontology that assigns primacy to the processes of formation as against . . . final products"—such as the remarkable objects illustrating the present book. That is, iron and the ways things are forged from it bear stories of their own that "give shape to the forms of thought" as often as "the mind imposes its forms on material objects."[5] Such processual and relational histories contribute to the value of iron as a medium.

Before it is worked, iron as a substance offers its potential to become what a blacksmith can and will forge of it. As Jane Bennett suggests, the materiality of iron is *creative* insofar as it bears "incipient tendencies and propensities" recognized and exploited by a blacksmith (2010, 56). Looking to an African way of understanding these complexities, elements of Yorùbá philosophy suggest that potential is among the qualities of *iwa*, the essence of something. Such an inchoate sense is informed by *cultural entailments* that Yorùbá would consider to be further aspects of iron's *iwa*, as constituted in the experiences, "condensed stories," and arcane knowledge required for smelting, that in turn inform what a blacksmith creates.[6] All these intersections, all these inputs, constitute what sculptor Alison Saar has termed the "wisdom" of materials: where they have been, what they have "seen," what has happened in their presence, and the memories they hold and instigate that then contribute to what is made from them next.[7]

To gain a sense of the wisdom of iron—hence local understandings of its value—let us recall that African cosmologies contribute to processual histories of iron as a material as well as its smelting and smithing. Among Tabwa people of the Democratic Republic of the Congo, for example, rich magnetite ore was gathered on Mount Kalolo under the aegis of an Earth spirit called Mother of the Furnace, and minerals and other underground resources are the province of the rainbow-breathing serpent Nfwimina (see chapter 1, this volume).

A master smelter choreographed the creative process, and his "danced causality" (Warburg in Kosky 2013, 32) addressed the solar serpent to assure transformation of raw ores into malleable metal. Nfwimina is also responsible for human fertility and conception via Kisumi, its double within every woman's body; and so the "enchanted technologies" (Gell 1992) of ironworking implicated embodiment. *Vizimba* activating agents associated with the "super-chthonic" serpent were among fertility medicines placed in furnace floors and portals, and the remarkable way the greatest healers of yore obtained such *vizimba* added to the "processual and relational" efficacies of iron and to its value. So did other uses of the same *vizimba* in medicines to assist hunting the most dangerous of beasts or to provoke or prevent rainfall. Any such application of Nfwimina's transformative abilities metonymically informed all others. So did environmental factors contributing to ironworking, such as weather and terrain; contingent materials like the soil taken from termitaries to sculpt smelting furnaces and tuyères; and the varieties of wood selected to carve bellows, stoke furnaces, and create the best charcoal for smiths' forges—and all the acute observational sciences so implied.

Humane dimensions of ironworking also added to the value of the metal as it was forged into currency tokens. For example, in writing of smelting among Gbaya peoples of the Central African Republic, Yves Moñino was struck by the joy men took in their work (fig. 5.2), as reflected in accompanying rhythms, "encouragements they shouted to each other, or their punning on the analogies of smelting, hunting, and sexuality" (1983, 307). These activities reflected "the pleasure of creating" itself, he surmised. As Jane Guyer might add, beyond the brawn required for such hard work, "there was potential for self-valuation, for partial authorship" among those assisting the master smelter, as well as benefits of association with a prestigious man often closely associated with political authority, wealth, human healing, and procreation (1993, 254; also see Échard 1983, Muller 1974). Such "pride of men" (Kriger 1999) surely obtained among Tabwa smelters and smiths, who felt themselves descended from Kyomba, their culture hero.

The value of objects forged of iron consists of entangled stories of cosmologies and materials, then, as well as traces of the "taskscapes" of their *making* (Ingold 2010, 97) and practical outcomes of their use. Just as "condensed stories" informing the wisdom of iron should be taken seriously in assessing what it means to work and deploy the metal, so should the realization that currency tokens "hold the forces of heart, tongue, hand, and eye that were involved in their own making" (Stewart 2011, 30). For example, able Yòrúba smiths benefit from and generate *àṣẹ*—"performative power"—as what they forge is invested with the energies of Ògún, god of iron and "Lord of the Cutting Edge." That is, iron is an "actant" that possesses a "vital materiality" for Yorùbá, a capacity "to act as a quasi agent or force with trajectories, propensities, and tendencies of its own" that impact human circumstances even as humans put iron to their own purposes.[8]

Furthermore, although a blacksmith's work may seem routine, what he does produce is informed by what he might have but did not, leaving an "*otherwise* that remains unrealized" as potentiality, a "might make" to be executed another time (Stewart 2011, 16–17). A completed work "comes to dominate as all of the possible 'otherwises' in the artist's choices fall away and the work emerges as what it is as being as well as happening—the outcome of a series of decisions that are dense with the possibilities of what was not chosen, what was not actualized," but is still possessed in the potential of crafted objects (ibid., 118). Such capacity must

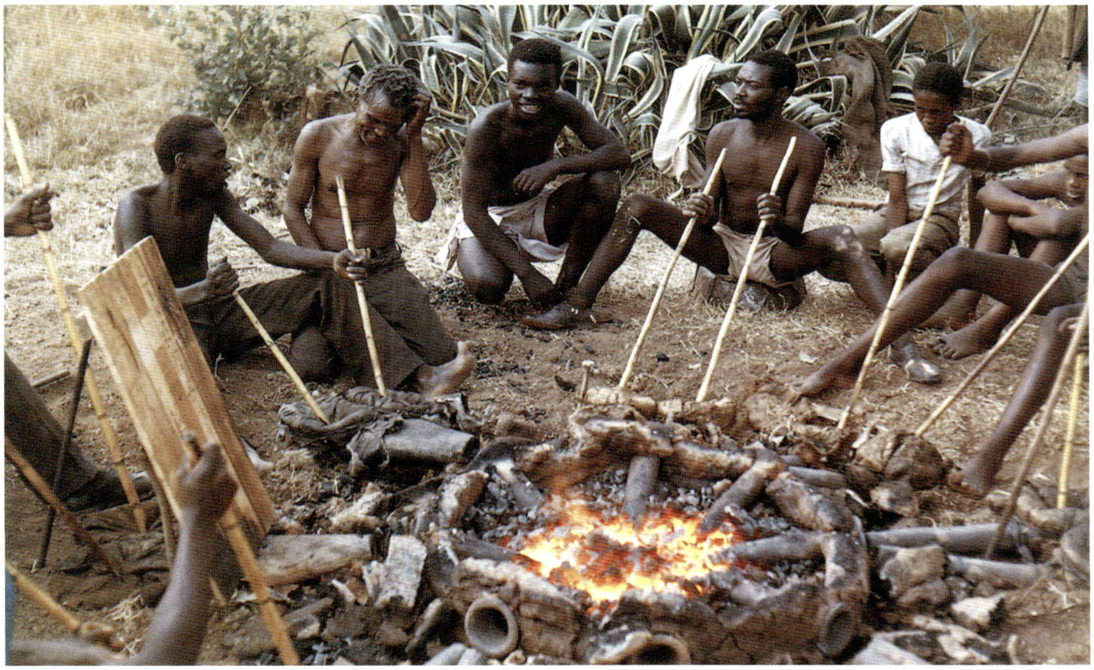

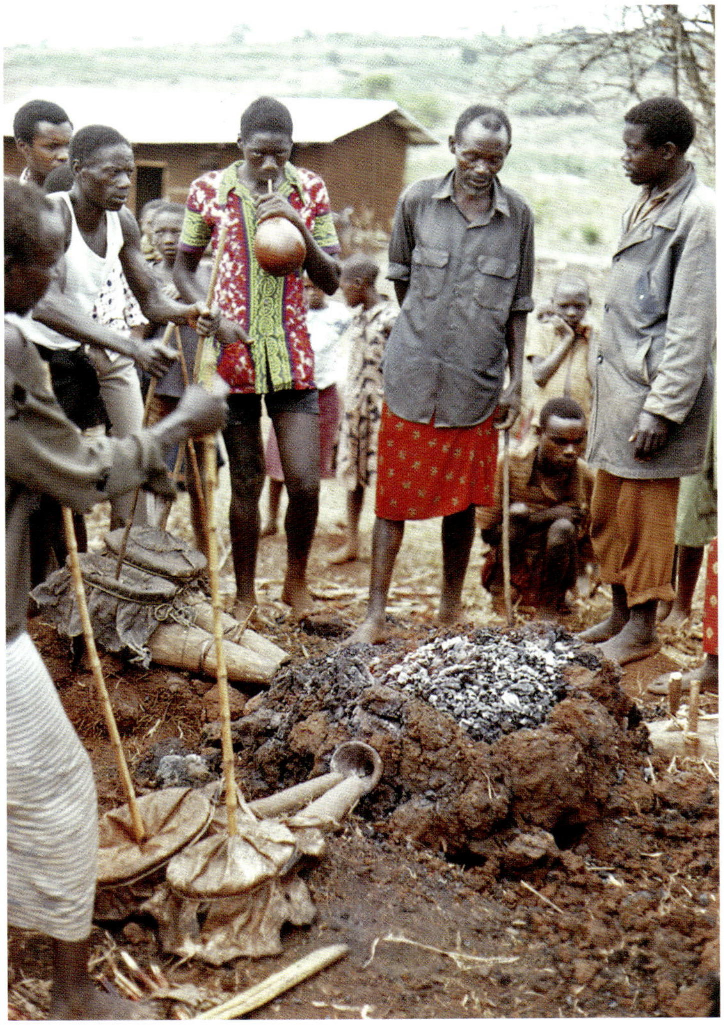

5.2A–B
Hutu men of Rwanda and Burundi smelted iron in the old days and forge iron still, for their own communities and those of Tutsi and other regional peoples. The bellowsmen are stoking what may be a forge fire but that takes the form of earlier low furnaces (*bas-fourneaux*) locally used for smelting iron.

Hutu smelted locally gathered iron ore in low furnaces of the sort seen here, that were dug a few feet into the ground and then mounded over with earth. Steady working of bellows created the intense temperatures required, but of significance here are social aspects of such taskscapes, as men and boys gather around to observe and learn from the labor but also to swap stories, drink locally brewed beer (as one can assume the young gentleman is from the gourd he holds), and otherwise enjoy convivial circumstances of communal endeavor.

PHOTOGRAPHS BY GEORGE CELIS, CA. 1989.

depend upon and create its own *àṣẹ* for Yorùbá, following the logic presented above.

Shadow energies of the sort were (and are) also recognized among mystically inclined African Muslims, for whom what is tangible and evident (*zahir*) is locked in dialectic with that which is hidden and potentially powerful (*batin*) because divinely inspired. Having memorized the Noble Qur'an as children, most African Muslims know stories of how Allah gave the Prophet-King David knowledge of ironworking and his son Solomon control of winds (as in bellows), and they are also aware of Solomon's mastery of jinn (Qu'ran 34:10, 14). In some African Muslim societies, such as Mandinka and Joola of the Casamance (Senegal) and adjacent lands, blacksmiths are associated with jinn as beings created by Allah from "smokeless fire" and closely associated with wind, high heat, and flame (Fallou Ngom pers. comm., 2018; El Zein 2009, 27–28). Some feel that as "divers," following a Qur'anic epithet, jinn have concealed and can reveal ores and other subterranean resources, and Zara people of southwestern Burkina Faso perform masquerades dramatizing the dangers and boons of relations with "hot" yet wondrous jinn who, in a sense, personify the ambiguous essence of iron itself (Bravmann 1977; Homann 2011). In such societies, the wisdom of iron and the value of exchange expressed through iron currency tokens are informed by divine associations like these.

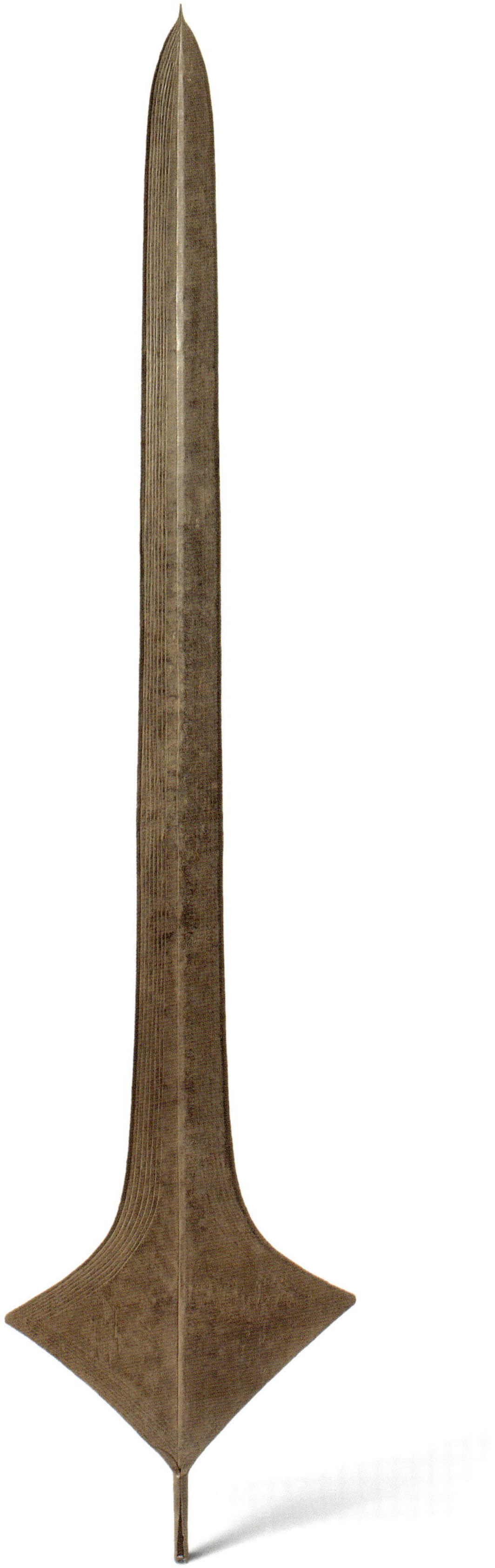

TOKENS OF MATERIAL PRAISE

Iron bars, as a first stage of forging after smelting, were widely exchanged through regional networks, for the material itself and the potential to create useful implements from it made such bars highly fungible. Indeed, in central Africa they "passed through cultural and economic boundaries with ease" (Kriger 1999, 96; see also chapter 14, this volume), as did other commodities, including locally produced salt, palm oil, tobacco, and smoked fish as well as foreign trade items like cowrie shells and glass beads (cf. Appadurai 1986). As a next step in refined workmanship and aesthetic design, iron currency tokens usually served particular purposes limited to a community's most significant exchanges, such as bridewealth (figs. 5.3–5.5), deferential gifts to political authorities, ransom of battle captives, payment of ritual fines, or trade for horses or cattle.[9]

In some societies, iron tokens were used to mediate wider transactions, and that they could be reforged into useful tools and weapons increased their distribution. In a few cases, "polyvalent money" (Maret 1981, 120; 1995) was developed to meet the needs of complex economies, as was the case among Sara-Laka people of Chad, who produced four different sorts of iron currencies, the smallest of which were bound in counted bundles (Roberts 1988b). In even more particular instances, as among Beti of southern Cameroon, tokens became "open value receptors" subject to inflation, useful to regional as well as local exchange, and merging with general-purpose money introduced by colonial authorities (Guyer 1993, 245–46). Such currencies became so numerous that producing them removed substantial amounts of iron from practical use (Guyer 1985, 4; 1986; 1995).

5.3
Artist unknown (Topoke, Lokele peoples, Democratic Republic of the Congo)
Spear-blade-shaped currency (*liganda doa*), 19th century
Iron
H: 182.9 cm
PRIVATE COLLECTION

A masterfully produced currency token such as this—taller than a man and produced by a team of smiths—represented community wealth and well-being. As a young man prepared for marriage, his family would negotiate with his bride's to decide bridewealth that could include impressive devices like this. Such "material praise" would acknowledge the love, nurturing, household management, and labor that the woman would bring to her husband's family. Her own family would then deploy the same tokens to secure marriages of their own young men.

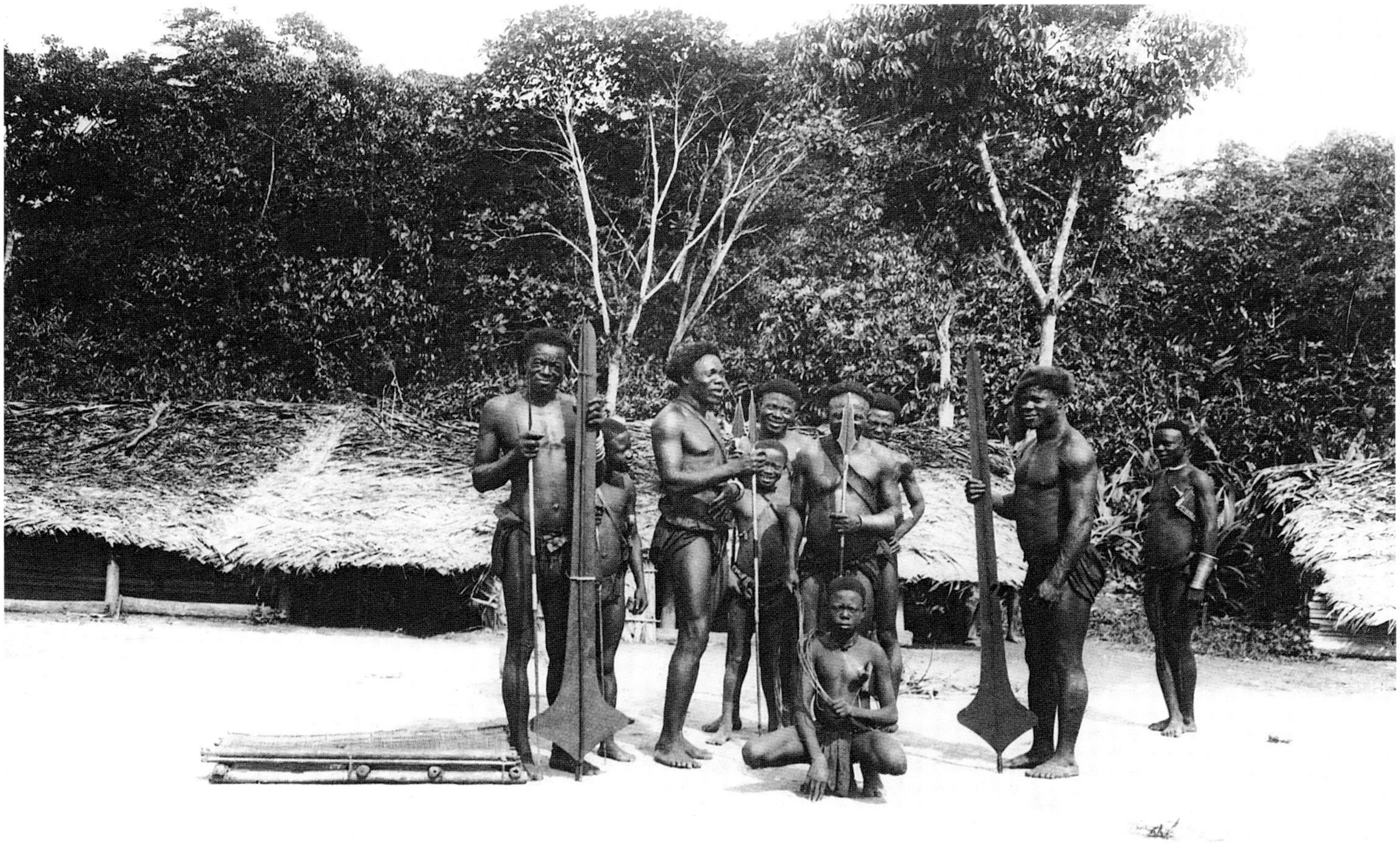

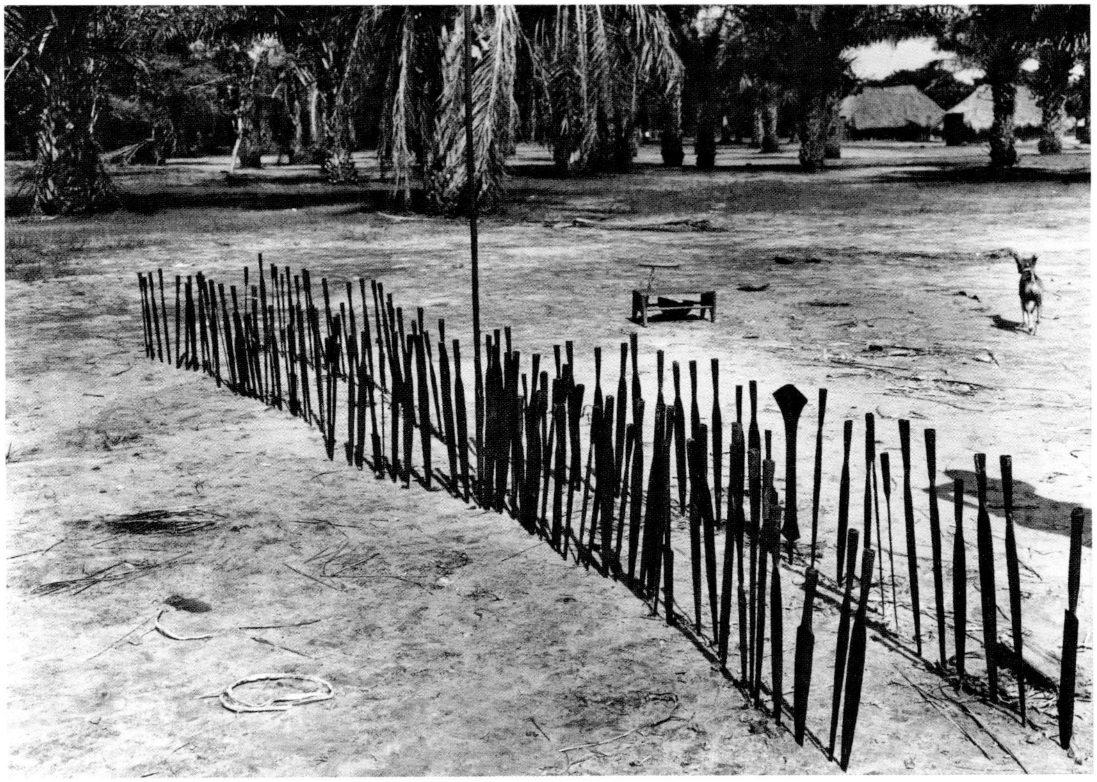

5.4 Topoke men holding large spear-blade-shaped currency tokens representing wealth and authority and used in important exchanges like bridewealth, ransoming captives taken in battle, and offering deference to political superiors. The tokens were banned by the Belgian government in the 1920s as Congolese people were obliged to earn colonial francs to pay taxes, thus boosting the colonial economy.

PHOTOGRAPH BY EMIL TORDAY, BRITISH MUSEUM CONGO EXPEDITION, ORIENTALE PROVINCE, 1907–09. © COLLECTION ROYAL MUSEUM OF CENTRAL AFRICA, TERVUREN

5.5 Some two hundred Zande and Mobenge currency blades planted in the ground during a marriage negotiation. Such a display marked a significant moment, when the agreed exchange was realized publicly, making the husband's family's wealth and determination apparent. Sadly, we have no explanation of what seems to be a pathway between two rows of tokens—might the bride have passed through as an aspect of wedding ritual?

PHOTOGRAPH BY H. GOLDSTEIN (INFORCONGO), 1952. COLLECTION ROYAL MUSEUM OF CENTRAL AFRICA, TERVUREN. IMAGE © 2018 ARTISTS RIGHTS SOCIETY (ARS), NEW YORK / SOFAM, BRUSSELS

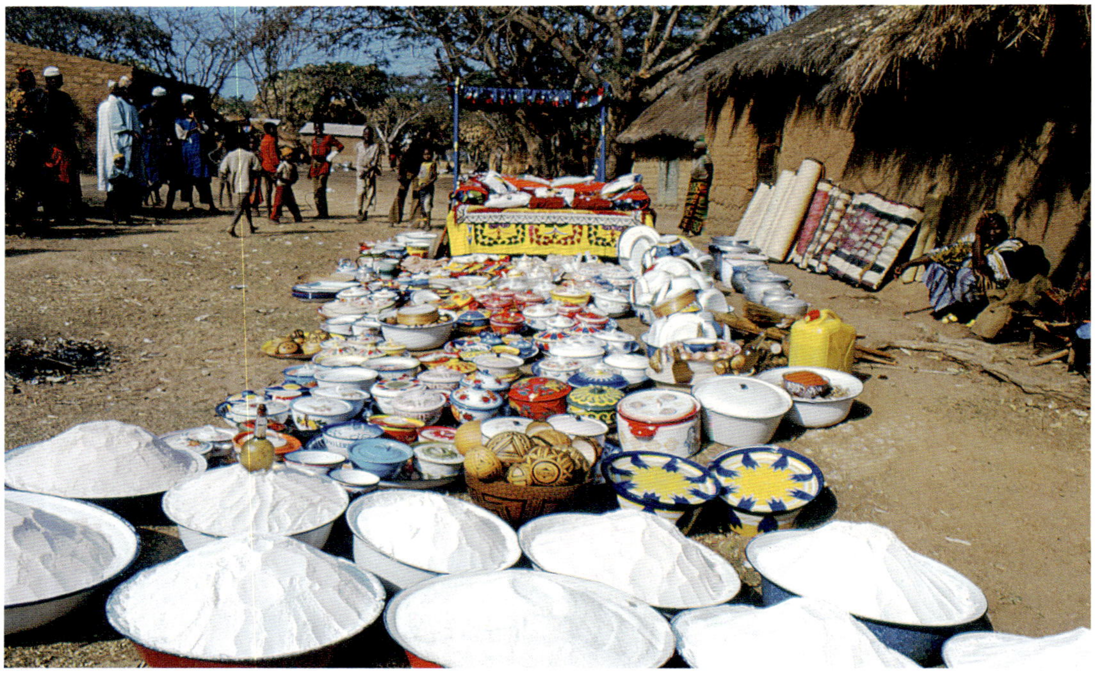

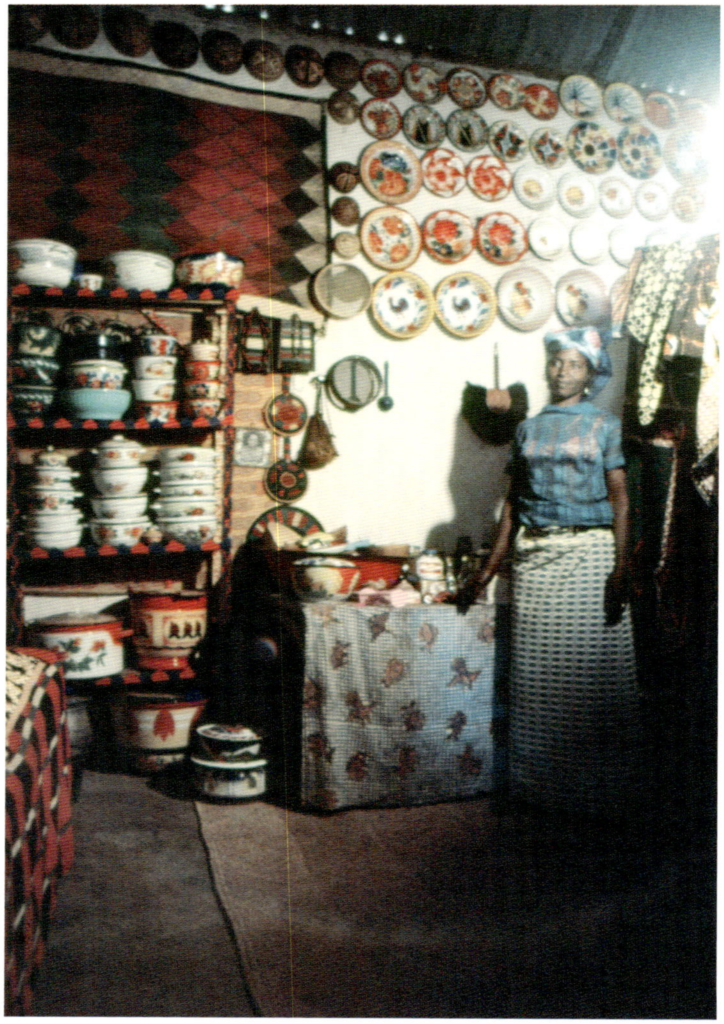

5.6A–B As a distinguished Hausa man married his third wife, he offered, following Muslim custom, a vast display of gifts. Above, his wealth as a long-distance merchant possessing a large truck was manifest in foodstuffs, festively decorated enamelware pots, pyro-engraved gourds, rolled mats, and a large bed. Some items would be kept by the new spouse, others distributed to her family. An enormous feast soon followed.

Below, in the chambers of the first wife of the same Hausa man, durables and lengths of cloth that she had received during her own marriage and accumulated since are displayed with great, if private, pride. Because the pots and pans had been removed from use, they stood as "material praise" for all the woman had brought to her husband's ever-extending family.

PHOTOGRAPHS BY ALLEN F. ROBERTS, BAÏBOKOUM, CHAD, 1973.

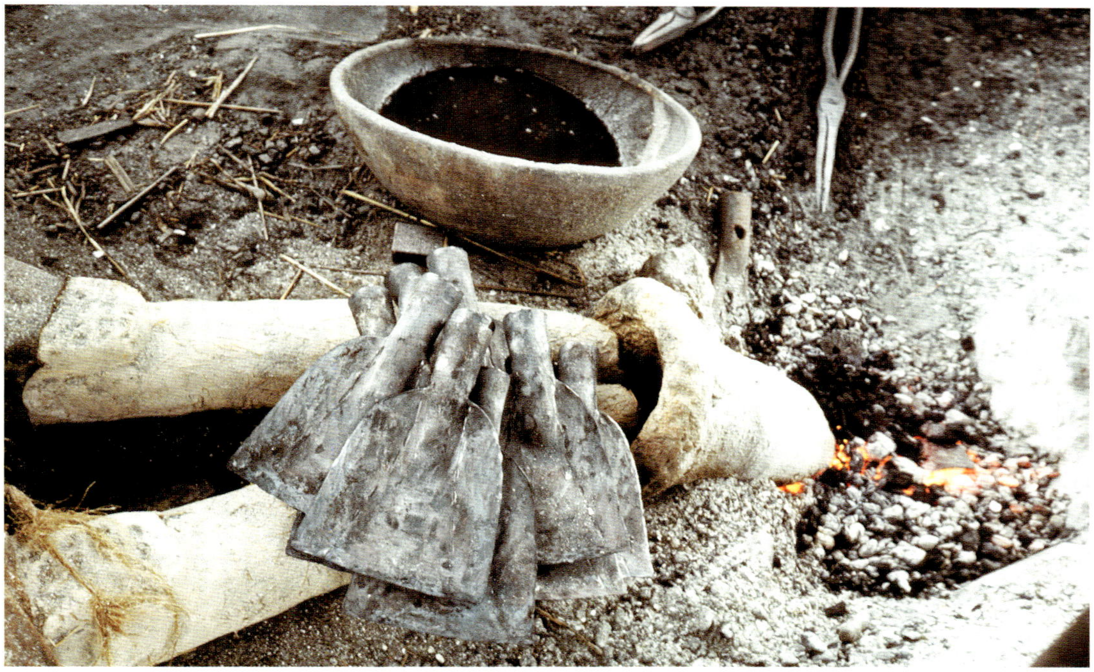

5.7 A Mossi blacksmith's forge fire stoked by a bellows extending to the left, which pumped air into a clay tuyère inserted into burning charcoal. Tools and water for quenching were kept handy to the work space, where (at the request of the photographer) Mr. Kientega proudly displayed hoe blades (*daba*) as objects of value that he forged and repaired as needed. As he explained, in the old days Mossi smiths forged and hafted *daba* to give to farmers in exchange for food; with French colonization, hoes had to be sold so that blacksmiths could pay taxes and purchase food.

PHOTOGRAPH BY CHRISTOPHER D. ROY, YOKO VILLAGE, BURKINA FASO, 1977.

Creation and exchange of precolonial iron currency tokens applied senses of processual value to particular human circumstances, based upon a widely shared sense of "wealth in people." In many parts of Africa, "'a wealthy man increased productivity by... aggregating human dependents' in a context characterized by the 'centrality of human life... but also its fragility.'"[10] Significantly, accumulation of wealth in people was complemented by the intellectual endowments of those so assembled, for individuals were "repositories of a differentiated and expanding repertoire of knowledge" (Guyer and Eno Belinga 1995, 120). Attracting people through marriage, fictive kinship, access to natural resources, or other means was an important gauge of a leader's status, as was the case for earlier Tabwa lineage chiefs who encouraged smelters, smiths, and other specialists to join and remain in their communities. Precolonial currency tokens facilitated such interdependency, with bridewealth their most frequent application.

Bridewealth was and often still is the "glue" of small-scale sub-Saharan societies, for it depends upon solidarity among a prospective husband's kin as it does among a wife's, even as it creates a bond between the two parties.[11] After negotiations between two families to decide the terms of bridewealth, a young man turns to his parents and they network with kin to accumulate the goods required. Once designated gifts are amassed, presented, and, often enough, festively displayed as testimony to the status of bride and groom (fig. 5.6), they are usually redistributed so that young men of the bride's lineage may marry. All parties have vested interest in a marriage's success, for over and above the wish for conjugal and community harmony, the wife's family will not want to lose face by being obliged to reassemble already distributed goods to return them to the husband's kin. Nor would the husband's lineage wish to lose the wife's love, nurturance, children, and the labor she and they will provide.

In earlier days, the forms of many currency tokens used in bridewealth bespoke such social relations. Hoe-shaped currencies, for example, stood for the hard work, harvests, and bounty a bride's family hoped to give and a husband's hoped to receive (fig. 5.7). Occasionally, currency tokens were derived from the shapes of other implements, as seen in tokens illustrated here. More poetically, though, in their symbolic removal from practical work—one could not farm with frail or under- or oversized iron objects—tokens used in bridewealth constituted *material praise* of the woman as an individual and of her life-sustaining capacities. Such currencies stood for affection, work, and family welfare, that is, as reinforced by value derived from the entailments of iron as a substance and the "enchanted technologies" required to smelt and forge tokens. As Susan Stewart notes, "praise is... a kind of expense... of spirit and something offered" that is "meant to be circulated" (2011, 36, 41). As such, exchange of currency tokens in bridewealth added the wisdom and material vitality of iron as well as complex processual and relational histories of smelting and smithing to community cohesion and the greater economies of household and lineage for both husband's and wife's families.

5.8
Artist unknown (Sara-Kaba peoples, Chad)
Axe-blade-shaped currencies (*kokio*),
20th century
Iron
H (each): approx. 6.5 cm
FOWLER MUSEUM AT UCLA, X2014.40.21, X2014.40.22, X2014.40.23; GIFT OF ANONYMOUS
PROVENANCE: PURCHASED BY THE DONOR IN N'DJAMENA, CHAD, 1985

5.9
Artist unknown (Sara-Madjingaye peoples, Chad)
Throwing knife-shaped currencies (*kuur*)
late 19th to early 20th century
Iron
H: 6.5 cm; 8.5 cm
FOWLER MUSEUM AT UCLA, X2014.40.25, X2014.40.26; GIFT OF ANONYMOUS
PROVENANCE: PURCHASED BY THE DONOR IN N'DJAMENA, CHAD, 1985

Sara-Madjingaye smiths forged currencies of a form derived from the sacred throwing knife called Miya-bo, possessed by the Supreme Being, who used it to summon rains to an arid land. The value of these fragile, even human-like forms derived from cosmology and manly pursuits, specifically throwing knives deployed in self-defense and hunting.

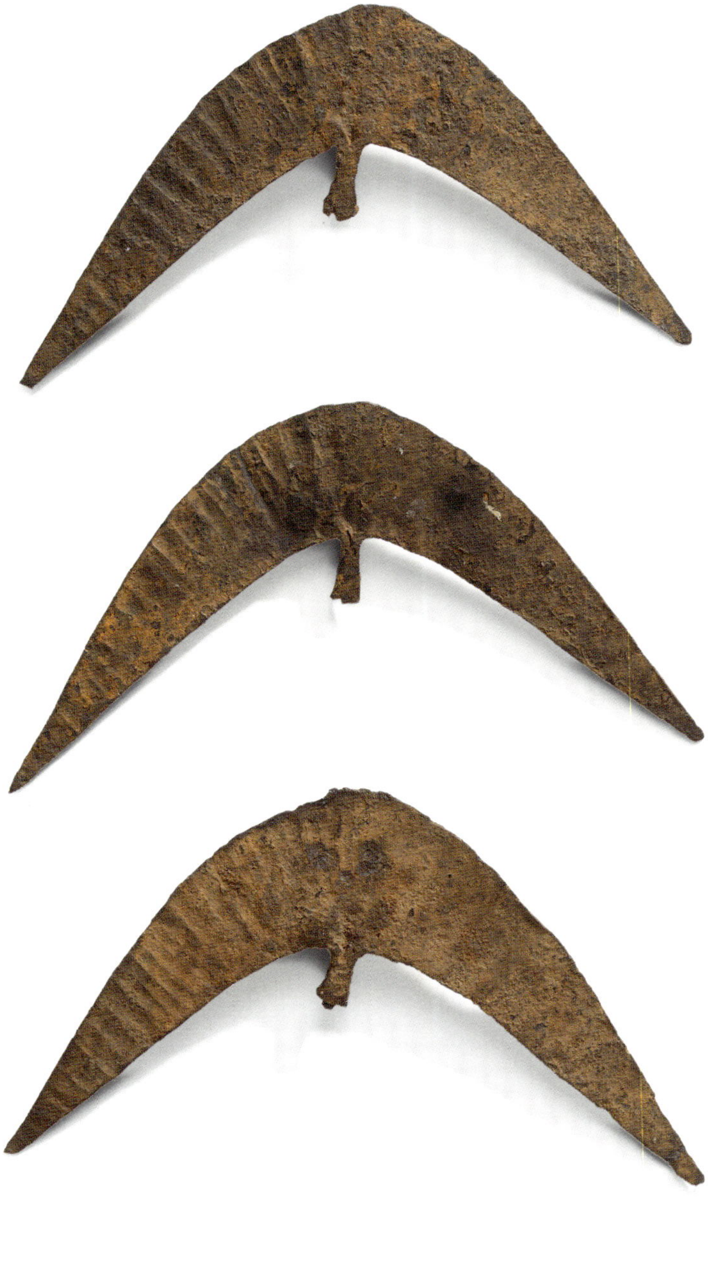

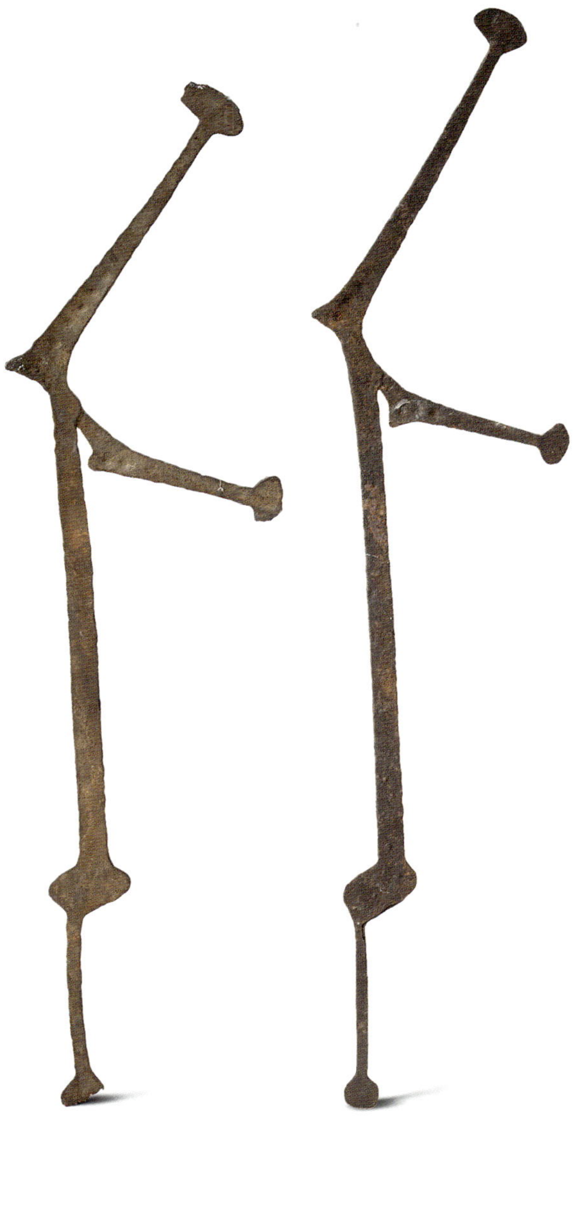

IRON CURRENCIES OF SARA PEOPLES OF THE REPUBLIC OF CHAD

With these thoughts in mind, let us turn to two brief case studies of precolonial iron currency tokens. The first concerns a congeries of peoples called Sara mostly living in southern Chad.[12] In earlier times, Sara-Laka used a mystical ability called *muum* to locate iron deposits from which ore was smelted. *Muum* also permitted someone to become a butterfly in order to approach hunted prey, or to speak with animals and gain their guidance in finding herbal medicines (Roberts 1988b, 101). *Muum* was a sharing of "matter-energy" (Bennett 2010, 57) among humans, iron, and animals, then, and a most potent entailment of the metal itself.

Blacksmiths produced several sorts of iron currencies from locally smelted iron, including *lar ndul* and *rom* that were still used as heirlooms in Sara-Laka bridewealth in the 1960s (Roberts 1988). Hoe-shaped currencies called *kokio* were forged by Sara-Kaba (fig. 5.8), who featured them in bridewealth and to empower healing by priests of the land, and throwing-knife-shaped currencies known as *kuur* were crafted by Sara-Madjingaye smiths (fig. 5.9).[13] Rather than agriculture, however, the value of these fragile, attenuated forms was derived from cosmology and references to manly pursuits. In particular, Sara men wielding throwing knives demonstrated their valor in subsistence hunting and in battle with fierce equestrian raiders from the Bagirmi kingdom, northwest of their territories, who were bent on "predatory accumulation" at the expense of vulnerable populations like Sara.[14]

Sara-Madjingaye hold that the Supreme Being possessed a sacred throwing knife called Miya-bo that produced thunder, and that He commanded the trickster hero Suu to bring the device down to earth to become the "palladium of royalty" of the King of Bedaya. Miya-bo was carefully hidden away and fed daily as an animated presence who could transform itself into a python during the day and a beautiful woman at night. The blade was only exhibited in ritual, as when Miya-bo initiated the rainy season, and it was celebrated with sacrifices at harvests and the inauguration of river-fishing seasons.[15] Because they were derived from the sacred form of Miya-bo, then, when throwing-knife-shaped *kuur* currencies were offered in bridewealth or other important exchanges, "condensed stories" (Ingold 2007, 14–15) of cosmology and mystical practice informed the tokens' value, connotations, and transformative efficacies, as well as expectations of what such transactions would afford (fig. 5.10).

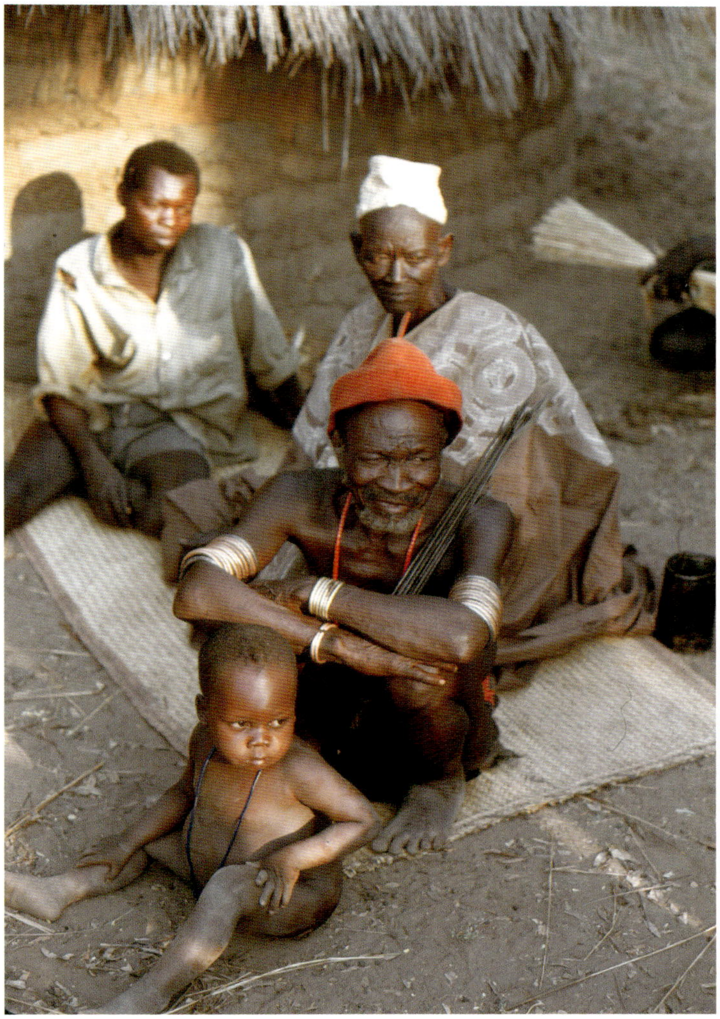

5.10 A Sara-Laka earth priest bears a bundle of throwing knives over his left shoulder, their number suggesting symbolic as well as practical value. The man's many bracelets and strand of red glass beads speak to ritual purposes, while his jaunty felt cap is an heirloom from French colonial days. A Hausa man in a beautifully embroidered robe shares the shade on a hot afternoon, as do a toddler and a young man. Just off-camera, someone is fashioning a straw broom of the sort used to fastidiously sweep homes and courtyards.
PHOTOGRAPH BY ALLEN F. ROBERTS, NEAR BAÏBOKOUM, CHAD, 1969.

5.11
Artist unknown (Kwele peoples, Gabon and Republic of the Congo)
Currency token (*zong*, pl. *mezong*), 19th century
Iron
H: 53 cm
PRIVATE COLLECTION

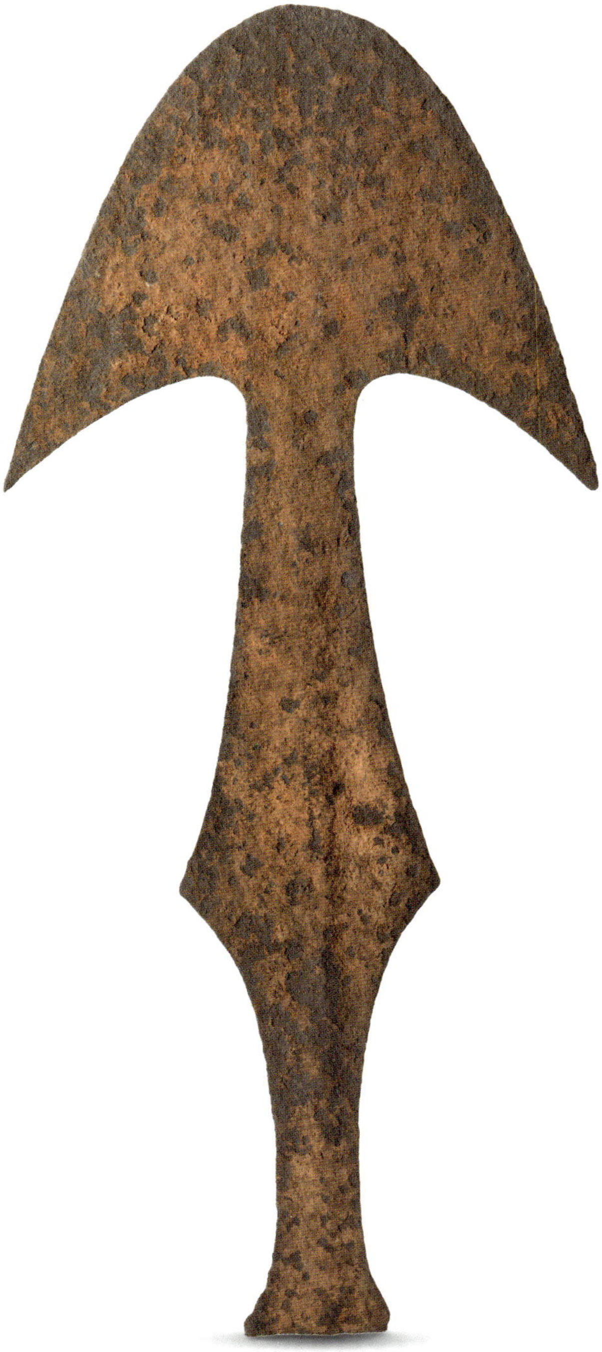

THE UNCOMMON CURRENCY OF MASKS

A second case study highlights aesthetic issues and relational histories of precolonial currency tokens produced by Kwele peoples of the Gabon/Congo-Brazzaville borderlands. Kwele did not smelt iron but traded for it with neighboring groups who did, thus participating in regional commodity-commerce. Tokens called *mezong* (s. *zong*; fig. 5.11) were used in bridewealth and other key exchanges, and their form refers to several tools useful to life in the rainforest (Dupré 1995, 86–88).

Most Kwele still live in small communities without apparent political interconnections. Marriage between persons of widespread lineage-based hamlets provides societal cohesion as articulated in earlier times with forged iron tokens, themselves among object types learned directly from the Supreme Being. Indeed, God's gift of such knowledge initiated human society as Kwele came to know it, for primordial ambiguities including incest were overcome and formal marriage—hence kinship—established. "From then onwards, humans had," as anthropologist Georges Dupré writes of such tokens, "the means and instrument for life in society." From then onwards, we might add, such cosmological associations informed the "inbetweenness" of iron tokens used as marital prestations binding families and wider communities together, such as when a bride visited her kin after moving to her husband's village, holding *mezong* in each hand to signal her new status and reminding people of their having agreed to mutual obligations.[16]

In the 1890s, French colonizers established concessionary companies in the region. As elsewhere in heavily forested central Africa—and most notoriously in what was then the Congo Free State—exploitation of wild rubber quickly became of huge economic significance as vulcanizing technology permitted bicycles and automobiles to be fitted with rubber tires. Discovering how significant *mezong* currencies were to Kwele political economy, traders imported European copies to barter for raw rubber (see Dupré 1995, 79, 91). Such tokens (fig. 5.12) were based upon the indigenous model, and Kwele smiths reforged earlier *mezong* into tools and weapons as replicas manufactured in Europe became the standard for bridewealth (see also chapter 16, this volume). Colonial officials decried use of imported *mezong*, however, for they sought to oblige use of francs that they could control and collect as taxes in order to enmesh Africans in capitalist economy. Kwele territory was sufficiently remote that such admonitions were ignored, for barter permitted French merchants to manage exchange rates for rubber to their advantage (ibid., 79–80).

5.12
Maker unknown (Europe)
"Kwele-like" currency token (*mondjo*, pl. *mondjos*), early 20th century
Iron
L: 52.5 cm
COLLECTION OF THE MAS, ANTWERP, BELGIUM (AE. 1951.0001.0156), BEQUEST J.M. SIEREN, ANTWERP, 1951
PHOTOGRAPH BY MICHEL WUYTS, 2017. © MAS | MUSEUM AAN DE STROOM, ANTWERP

In order to participate in—and hoping to dominate—local exchange in raw rubber and other commodities precious to international markets, French traders imported industrially produced facsimiles of Kwele currency tokens to remote borderlands between Gabon and Congo-Brazzaville. Such objects were tauted as "better" than locally produced ones because of their origins deemed "exotic" to local people; their introduction was meant to subvert local circles of production and exchange to the advantage of European merchants.

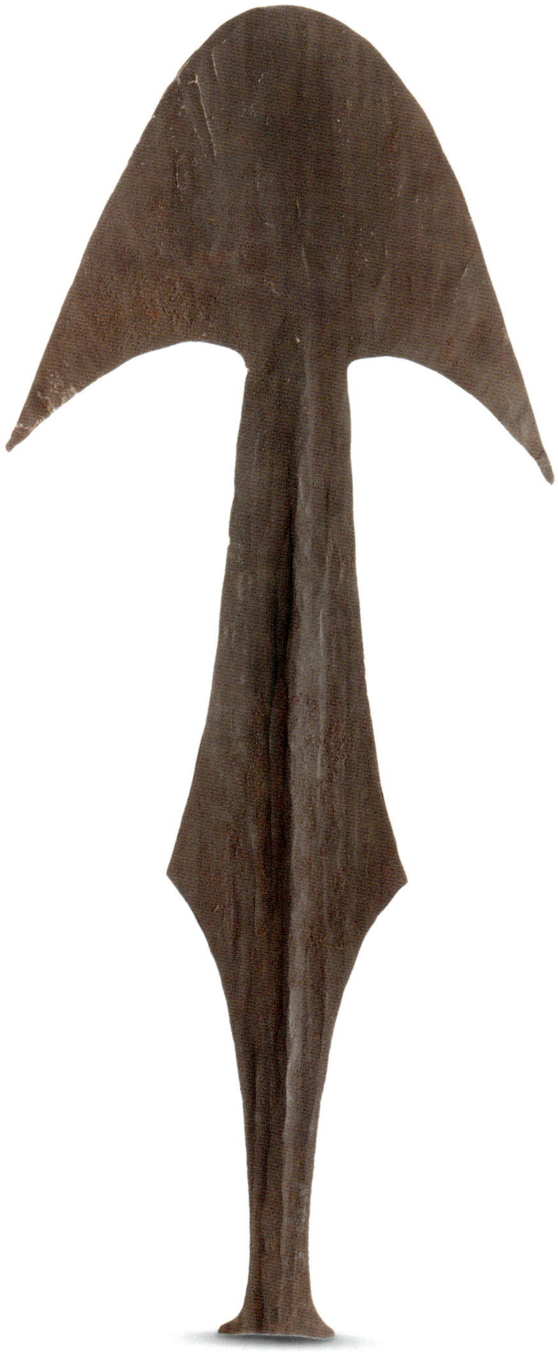

Early industrial *mezong* were soon replaced with imported ones cut from sheet metal to reduce manufacturing costs. These were wrought in a shape derived from earlier models but now modified, possibly to resemble an anchor as a symbol of colonial power. Quite remarkably, though, these tokens were in turn quickly replaced by Kwele smiths who altered their form to create what are now known as *mandjong* (s. *ndjong*; fig. 5.13). "In so doing, they acted politically," Dupré suggests, for "Kwele society gave itself new monetary norms" reflecting colonial realities, even as blacksmiths answered local aesthetics (1995, 91, 93). It is difficult to imagine a more clear-cut example of what Nicholas Thomas has termed "the promiscuity of objects" in colonial contexts (1991, 27–30). "Rather than possessing identities fixed in their structure and form, it is the 'mutability of things in recontextualization' that impresses" as the source of new "inbetween forms."[17]

As a last reflection, let us consider what "promiscuous" cultural entailments may have influenced Kwele smiths in this regard. *Mandjong* were hand-forged, are slightly asymmetrical, and they possess flanges at the ends of their appendages as aesthetic "corrections" to the simpler French design. Kwele assigned anatomical terms to the "revised" tokens, with the upper point the "head"; below that the "chest," "body," and "tail"; and the two lateral features the "ears" (Dupré 1995, 87). Looking to a well-known genre of Kwele wooden masks called Ekuk (fig. 5.14), a striking resonance with *mandjong* tokens may be perceived.[18]

Through their combination of anthropomorphic and zoomorphic features, Ekuk depict forest spirits. Their long, downward-curving horns, which sometimes frame the face of the mask in "radiating and retracting visual play" (M.N. Roberts 2017, 61 63), refer to antelopes of several species hunted by Kwele, and the masks mediate between forces of nature and village.[19] Ekuk masks were performed and displayed by members of the Beete society that assured community social harmony. Their white pigmentation suggested and conveyed light and clarity, but also an ability to transcend boundaries between the living and the dead. Among reasons for Kwele smiths' adaptation of *mandjong*, might there have been a desire to create tokens as "material praise" of intercourse between human and spirit realms realized in masquerade, which could further inform marital exchanges through bridewealth in changing colonial times? In other words, was there a related *currency* between Kwele iron tokens and forest-spirit masks via iconographic commonalities deemed pertinent to senses of self and community through memories of earlier days, but as updated to cope with contemporary circumstances?

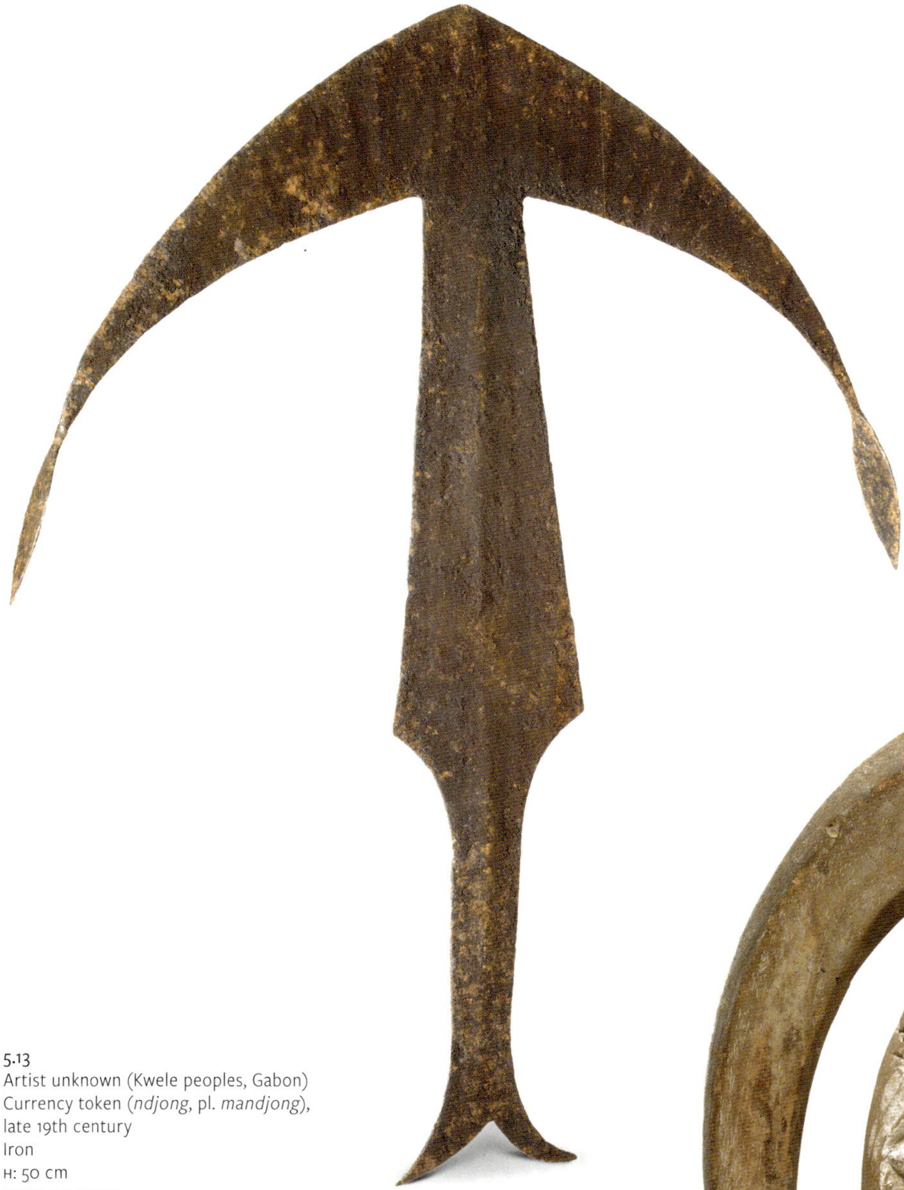

5.13
Artist unknown (Kwele peoples, Gabon)
Currency token (*ndjong*, pl. *mandjong*), late 19th century
Iron
H: 50 cm
PRIVATE COLLECTION

Currency tokens manufactured in and imported from Europe replaced locally forged ones, yet Kwele blacksmiths soon adapted the tokens or forged their own, making them slightly asymmetrical and adding features understood through anthropomorphic terms: the upper point is the "head"; below that are the "chest," "body," and "tail"; and the two flanges extending out on either side are "ears." A hypothesis advanced here is that in so doing, the smiths achieved deliberate aesthetic resonance with an important genre of Kwele masks, perhaps as "material praise" of exchanges between human and spirit realms as realized through masquerade performances. If so, the currencies were lent otherworldly currency.

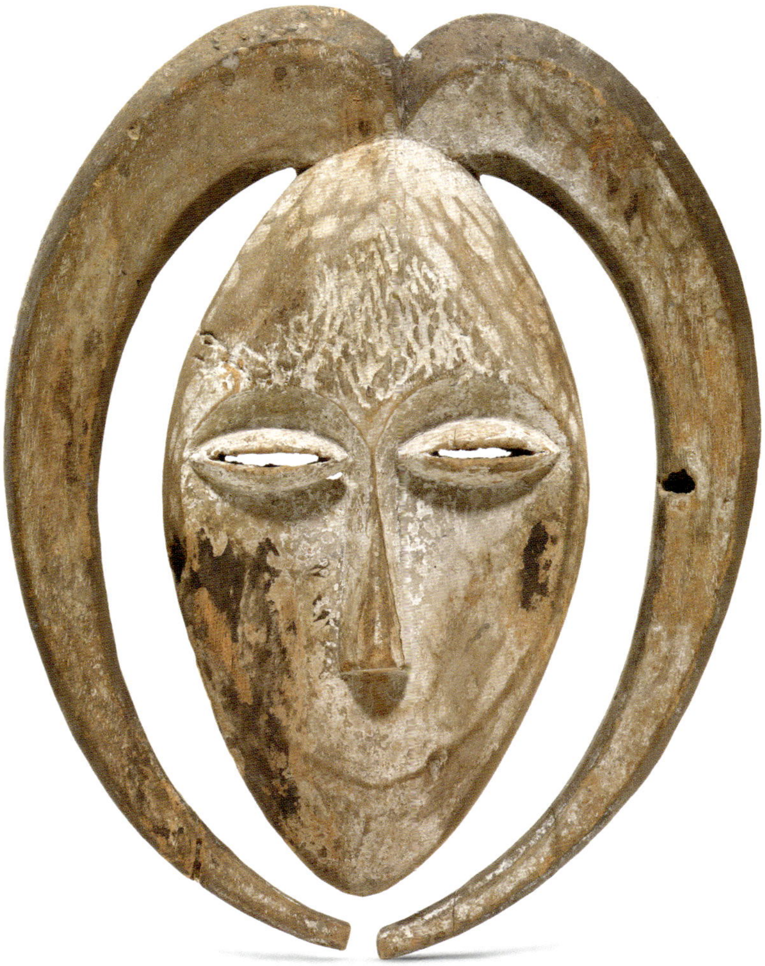

5.14
Artist unknown (Kwele peoples, Gabon)
Mask
Wood, pigment
H: 38.2 cm
FOWLER MUSEUM AT UCLA, X87.1505; THE JEROME L. JOSS COLLECTION
(NOT IN EXHIBITION)

Ekuk masks like this one combine anthropo- and zoomorphic features to depict forest spirits who presumably reside within the masks as made manifest in performances of the Beete society that assured community harmony. The long horns framing the face may refer to several forest animals hunted by Kwele. The balance of forms resonates with that of *mandjong* currency tokens, suggesting that the latter were forged to refer to and participate in the efficacies of such masks and the social circumstances they helped articulate.

That *mandjong* bear anthropomorphic and spiritual references leaves one wondering what other earlier iron currencies might have as well. A similar case may be proposed for so-called "Kissi pennies" forged by Toma and related coastal peoples of Sierra Leone and Liberia. Such tokens have "ears" perpendicular to one end of a thin, spiraling shaft, and a flattened "nose" at the other, as features reminiscent of the medial lines of flat masks worn on top of performers' heads in rituals and festivals. Indeed, packets of Kissi pennies were sometimes bound to such masks. Furthermore, Kissi pennies are small and light in weight, and so may have "traveled" easily (fig. 5.15). A Kongo *nkisi nkondi* power figure created in far-off coastal central Africa and now in a Los Angeles private collection bears several Kissi pennies among the iron shards piercing its body, for instance, leaving one to ponder what histories of exchange brought them from west African to Kongo hands, and then whether their association with an *nkisi nkondi* bore wisdom beyond iron as their essential material (see interleaf K, this volume). Surely, the currency of these currencies was based upon "inbetweenness" extending to and from masks and other vehicles of mediation.[20]

In conclusion, then, the essence and entailments of iron as a material, the taskscapes of its smelters and smiths, and the "wisdom" and "inbetweenness" of currency tokens and other devices forged from iron all implicate intricately interconnected histories. *These* are the determinants of iron's value, as negotiated over time and across differences of ethnic identity, and as vital potential always at the ready to further goals of power, wealth, and prestige. As we shall see in the following chapter, the same can be said of the panoply of blades long in the repertoires of African blacksmiths.

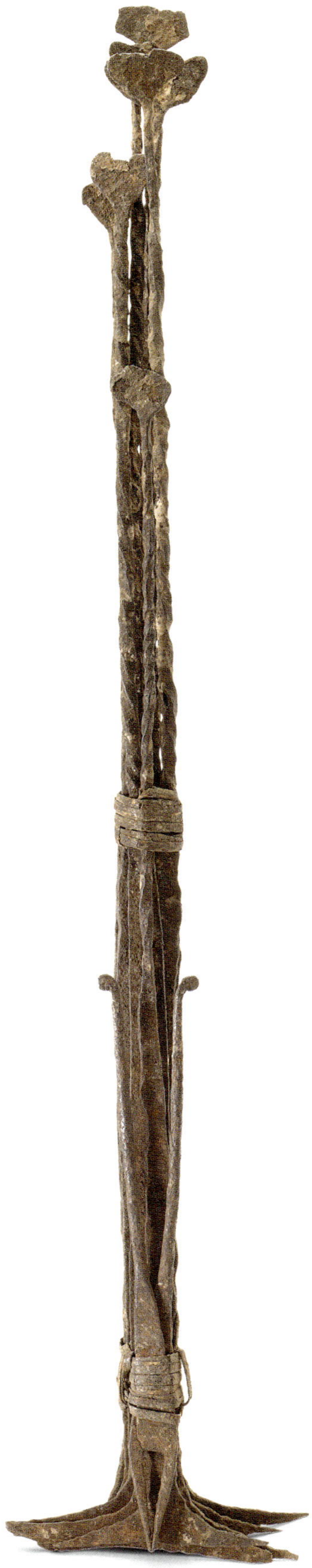

5.15
Artist unknown (Kissi, Toma peoples, Sierra Leone, Guinea, and Liberia)
Currency bar bundle, early 20th century
Iron
H: 75 cm
PRIVATE COLLECTION

This tightly bound bundle of "Kissi pennies" with miniature forging tongs attached may have been offered as bridewealth to satisfy a marriage contract between a blacksmith and his bride's family. The agreed-upon objects of value would have been given to her family to acknowledge the love, nurturance, and hard work the woman would bring to the husband's family. Of interest here is how anthropomorphic details were assigned to such tokens, so that the broadened flange was the "nose," the split and widened points at the opposite end the "ears." Might an analogy have been drawn between these forms and the unusually abstract features of masks danced by peoples of the region?

CHAPTER

6

BLADES: OF HARD WORK AND HONOR

Allen F. Roberts

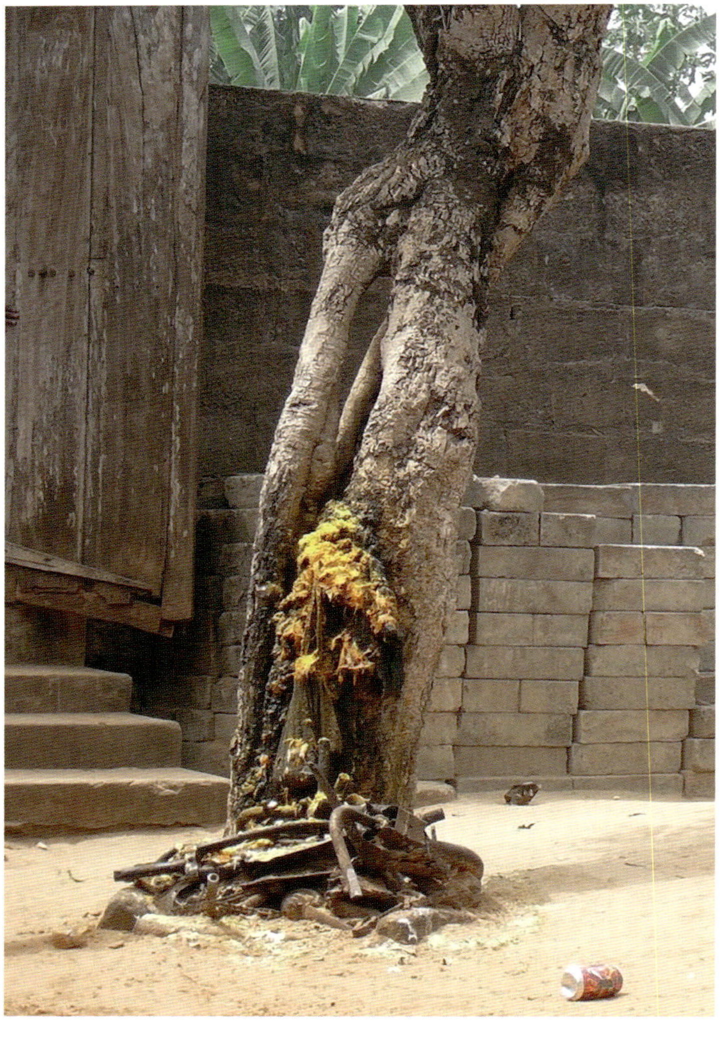

. . . make me a side-arm to take on the earth,
A suitable tool for digging and grubbing the ground,
Lightsome and pleasant to lean on or cut with or lift,
Tastily finished and trim and right for the hand.

SEAMUS HEANEY, "POET TO BLACKSMITH" (2007, 27)

Over the centuries, iron as a substance of value and agent of change has helped Africans to forage and hunt, till the soil, and assure protection and prosperity. Iron blades sustain lives, but take them too. Forged knives, hoes, iron-tipped digging sticks, sickles, machetes, axes, and adzes have long assured clever, efficient, and fulfilling management of household and agricultural chores, while spears, arrows, daggers, swords, scepters, ceremonial axes and adzes, and other blades of an astounding range of designs, sizes, and purposes have promoted kings' ambitions and warriors' fortunes. This is, after all, "the ambivalence of iron, giving as it does to man [and woman] both tool and sword" (Bachelard 1991, 130). Furthermore, iron as a master metaphor was and is more than metal, keener than cutting. To this day, shrines and graves sometimes bear blades in recognition of the practical genius and "enchanted technologies" (Gell 1992) through which they were produced, and of their resulting "wisdom"—so central to human vitality, health, power, and prestige (fig. 6.1).

6.1 A shrine to Gu in Ouidah, Republic of Bénin, installed in a family compound by a Gunon priest to Gu as the god of iron among Fon and related peoples. Composed of broken blades, bicycle and car parts beyond repair, and other metal debris from the household, as well as pertinent additions by the Gunon, the shrine is regularly fed with yellow palm oil and other offerings to assure Gu's energetic blessings.
PHOTOGRAPH BY ANDOCHE CELESTO ADJOVI, OUIDAH, REPUBLIC OF BÉNIN, 2017.

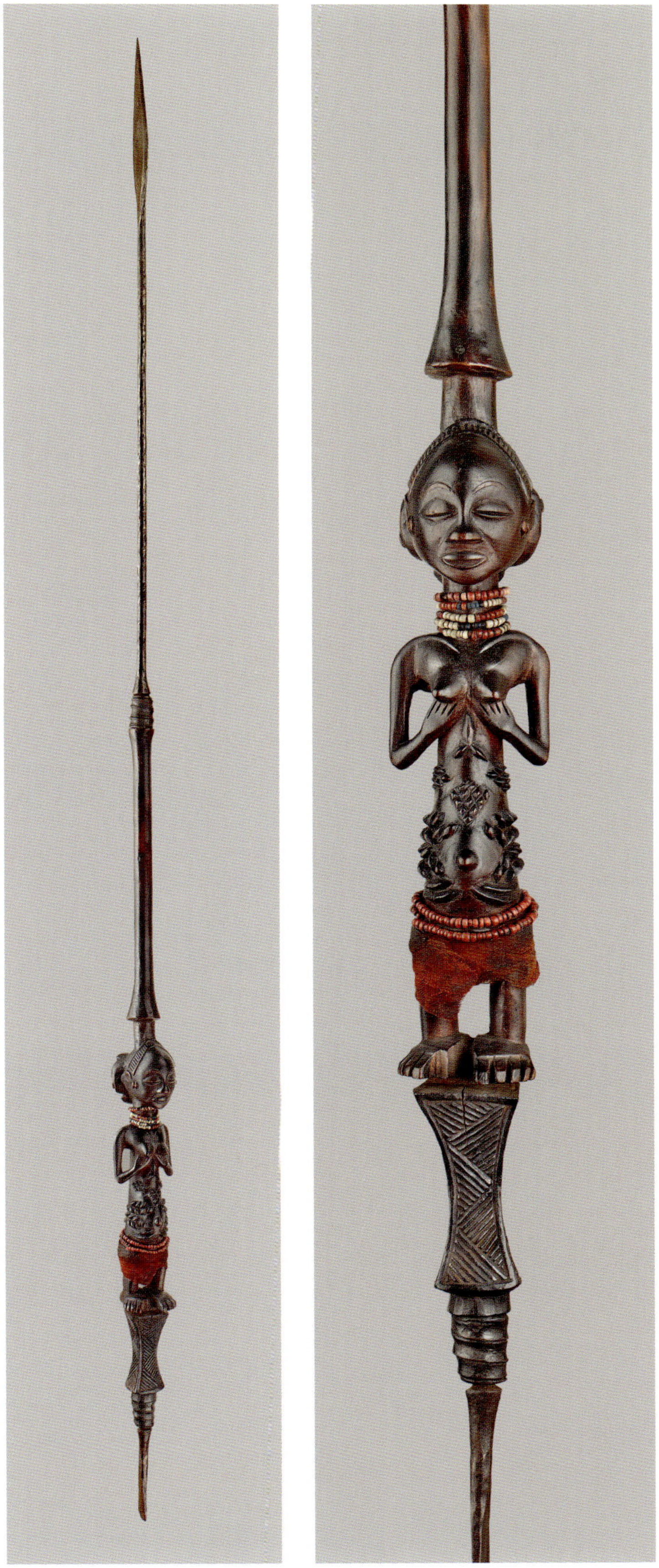

6.2a–b
Artist unknown (Luba peoples, Democratic Republic of the Congo)
Royal spear, 19th century
Iron, wood, glass, shell, cloth, resin
THE FIELD MUSEUM, CHICAGO, 210462
PHOTOGRAPH BY JOHN WEINSTEIN, 2018. © THE FIELD MUSEUM

This stunning spear was among the most compelling components of a Luba king's insignia. Its wooden shaft features the beautifully carved figure of a female protective spirit whose presence is safeguarded by defensive iron at either end. She is a reminder that, for Luba, only the bodies of women are powerful enough to hold the secrets of political authority. Such instruments played an important role in royal investiture, when a spear kept close to the king-to-be as he underwent ritual transformation was then brandished as he danced to affirm his new calling. Royal spears' formal features recall histories of political importance including how Luba kings, following the culture hero who was the first blacksmith, were "forged" through ritual processes and recognized as semidivine.

Here we shall consider two intersecting principles. First, blades extend the hands of those holding or propelling them, and so enhance labor and symbolic work of many sorts. As this chapter's epigraph from Seamus Heaney suggests, there is something ineffably encouraging and simply *good* about a "suitable tool" just "right for the hand." Indeed, in investigating ways that sub-Saharan Africans have understood and wielded bladed implements, we may find a sense of "tool-being," which, to draw from philosopher Graham Harman's development of this concept, "exceed[s] every relation into which they might enter."[1]

In many central African societies, implements could and can be prosthetic extensions of a person—and vice versa.[2] In writing of Kongo peoples, for instance, Wyatt MacGaffey suggests that *nkisi* is a personalized force from the invisible land of the dead" that "has chosen, or been induced, to submit itself to some degree of human control effected through ritual performances" (1991, 4; 2000, 43). Such a force is manifest in iron implements, medicinal devices, movement practices, and sacred places, but especially through figural sculptures called *minkisi*, some genres of which have been studded with iron shards through votive practice, making them archives of intentionality (see interleaf K, this volume). Kongo "chiefs relied heavily on major *minkisi*, objects that were themselves 'chiefs.' In turn, chiefs were also *minkisi*, treated at times as though they were objects" (MacGaffey 1998, 229). As Paul Basu might add, "things create people as much as people create things" (2017, 3; cf. Hardin and Arnoldi 1996, 1).

Following the logic of "tool-being," forged implements may stand in for those possessing them. For example, some Luba royal spears were given the honorific name Fundanshi, or "Earth Stabber" (Womersley 1975, 82; fig. 6.2). They are often virtuosic sculptures depicting tutelary spirits and terminating in finely forged and inscribed iron points. Such instruments played important roles in royal investiture, when a spear kept close to the king-to-be as he underwent ritual transformation was then flourished as he danced to affirm his new calling. The same spear—understood to be that of the Luba kingdom's founding ancestor Kalala Ilunga, whose spirit endowed the blade—was planted in the ground next to the royal stool during audiences and used in performances reenacting the origins of Luba society (see chapter 1, this volume). Like the king's powers, iron is stable, durable, and resistant; and it also underscores the indigenous nature of his rule, for like iron, the ruler is not foreign to his lands and his authority is "rooted deep in the dynastic soil that nurtured Luba royal history" (Nooter 1991, 166). Such a spear was an intimate extension of the king that he dispatched as his "ambassador," for he was "there" in the spear to oversee affairs of his realm.[3]

The second principle to be underscored is that in earlier days, and often still today, blades have contributed to African senses of virtue and valor as tools used skillfully in workaday circumstances and weapons deployed in perilous pursuits like big-game hunting, protection of community and self, and aggressive battle. As John Iliffe asserts, "honor was the chief ideological motivation of African behavior" in earlier times, as it is now, and "it has remained a powerful motivation even for those who accepted world religions" (2005, 1). As a primary manifestation of such deeply felt values, "manliness demands capacity to sustain and defend a household, to maintain personal autonomy, to avenge insult or violence, and perhaps to demonstrate hospitality, integrity, wit, and style" (ibid., 1, 3 slightly paraphrased; see also Forni and Ross 2017).

"Womanliness" contributed to these same senses of purpose, of course, and such qualities cannot be more evocatively portrayed than by a late thirteenth- to fourteenth-century Gwan figure from Bamana peoples of Mali (fig. 6.3). Such works have been carved by blacksmiths. "'Gwan' alludes to ardent will and supernatural powers and was the name also given to the tall natural-draft furnaces in which iron was smelted and 'born' to provide raw metal forged into essential tools and weapons" (M. N. Roberts 2017, 67). Furthermore, "Gwan celebrates both childbirth and the metaphoric 'birthing' of iron from the furnace," and is the name given to a women's association and the society of blacksmiths.[4] The passage of centuries has by no means diminished the pride of motherhood that the figure conveys: with serene and stately bearing, the woman (and/or female spirit) turns her gaze ever-so-slightly downward toward the infant she nurses (now mostly lost). The figure speaks to *badenya*, "the Bamana concept of 'motherchildness'" linked to "centripetal social forces: stability, unity, cooperation, those traits that pull a person back into the group, toward the hearth and the mother" (Cole 2017, 142); yet on her left upper arm, the mother wears a prominent dagger to fend off any indignity. Her protective nurturance is to be taken very seriously indeed.

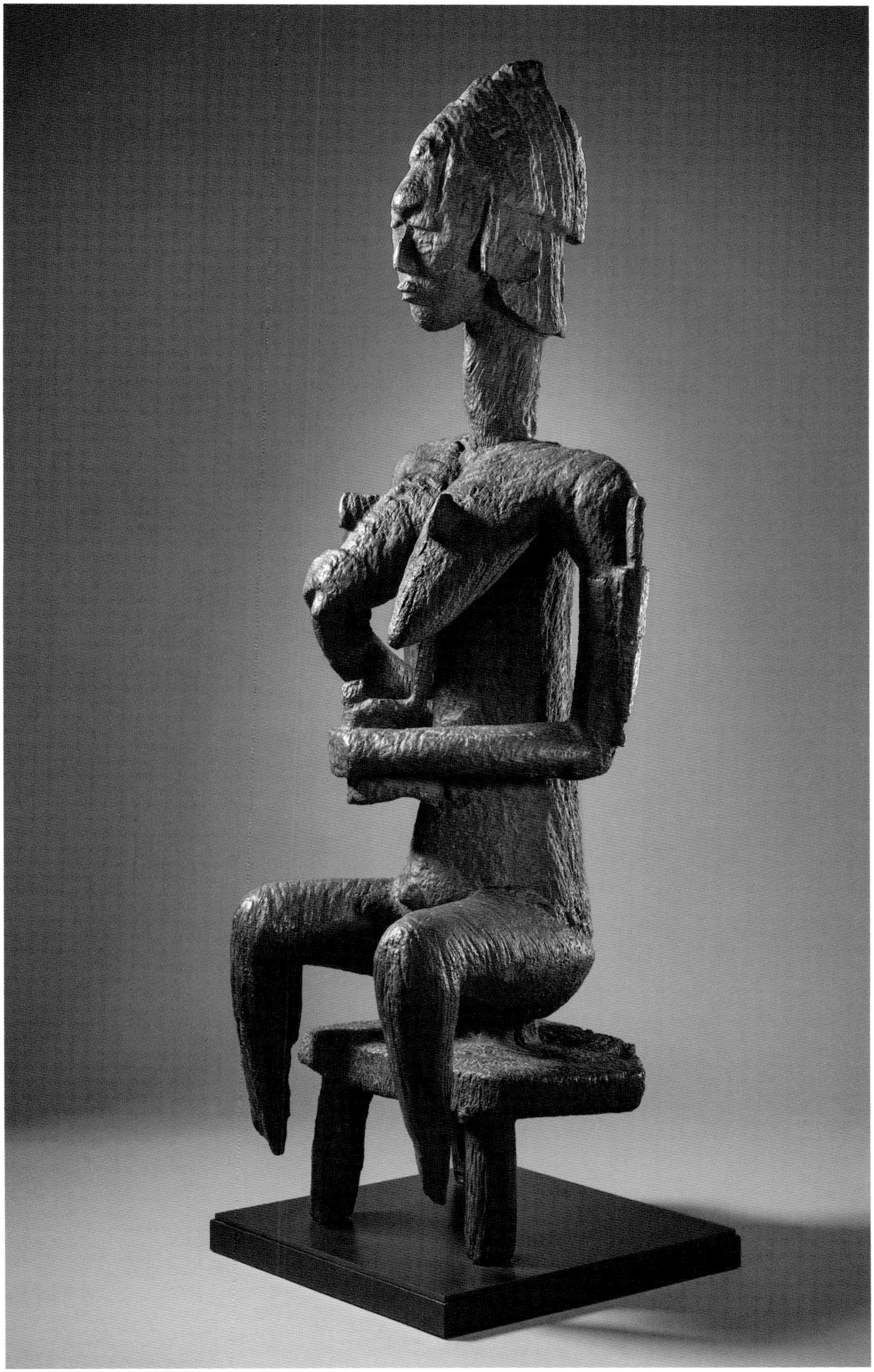

6.3
Artist unknown (Bamana peoples, Mali)
Mother and child figure for the Gwan Association, ca. 1279–1395
Wood
H: 96.5 cm
LOS ANGELES COUNTY MUSEUM OF ART, GIFT OF THE 2013 COLLECTORS COMMITTEE, WITH ADDITIONAL FUNDS PROVIDED BY KELVIN DAVIS AND BOBBY KOTICK (M.2014.19)
(NOT IN EXHIBITION)

"Gwan" is the name given to monumental figures like this one that speak to fertility and nurturance, but also to tall natural-draft smelting furnaces with which earlier Bamana ironsmiths "gave birth" to iron derived from local ores. In this most ancient sculpture, a mother's pride is depicted as she nurses her infant (now mostly lost to time), while the prominent dagger strapped to her upper arm is at the ready to protect her precious charge.

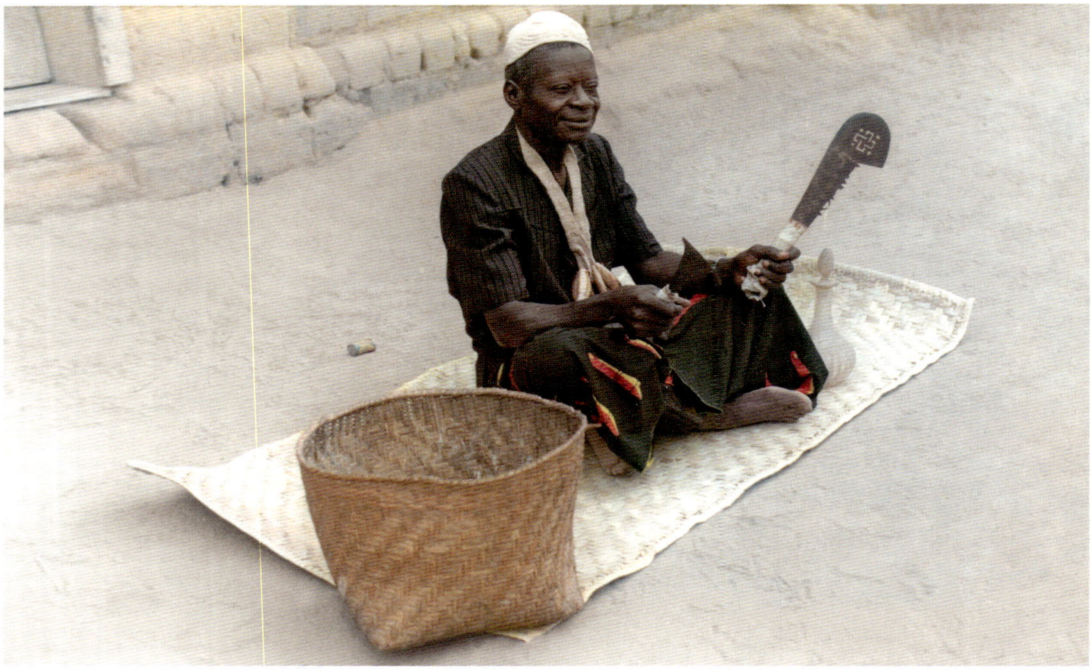

6.4 In a Woyo village on the Atlantic coast of the Republic of the Congo, a chief proudly displays heirlooms, holding a worn but still elaborate "knife of power" (*mbele a lulendo*) in his left hand and a smaller hooked knife in his right. The larger blade's motifs "offered a vocabulary to the chief, who would interpret them with proverbs, according to the needs of the moment" (Vanhee 2013, 113). The gentleman wears a sheaf of what may be medicinal or prayer packets around his neck, and an elegant stoppered bottle sits to his left, material evidence of a family history but possibly an *nkisi* power object as well.
PHOTOGRAPH BY DUNJA V. HERSAK, 1975.

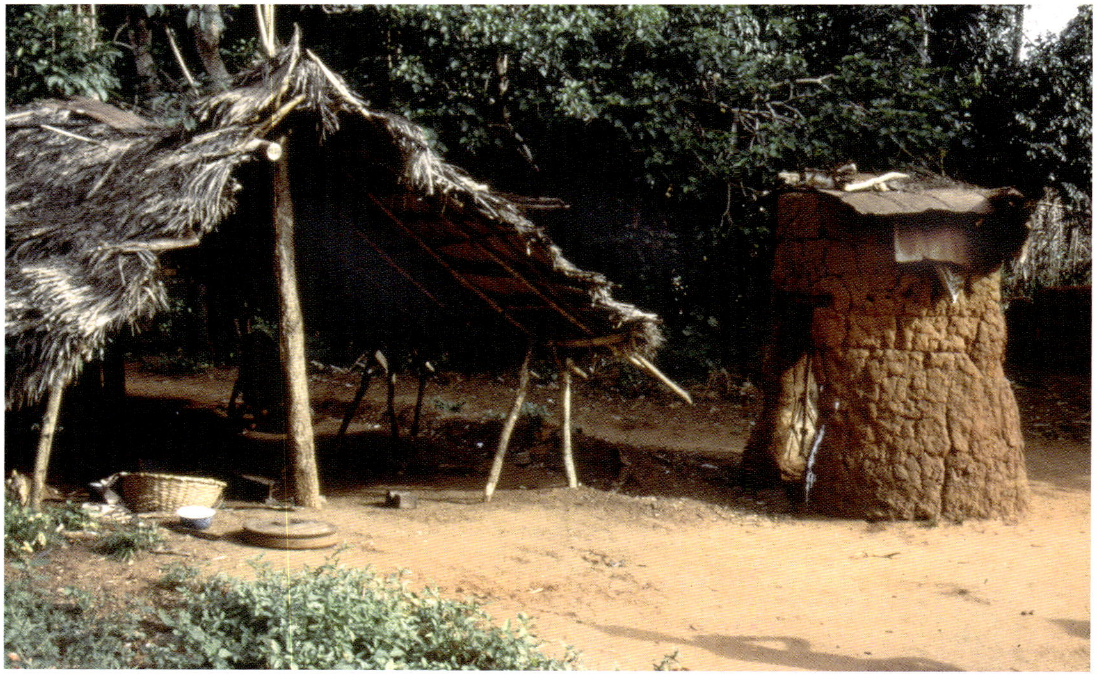

6.5A–B The Toehoessou family of southeastern Republic of Bénin has a long heritage of ironworking. In the mid-1990s they forged and spot-soldered *assanyin* (elsewhere called *asen*) and other devices for ancestral altars from recycled materials. To the right of a sheltered workshop is a shrine to Grandfather Toehoessou, who founded the household; museum-like, it holds his smithing tools and other heirlooms. *Assanyin* to Grandfather (opposite page) bear and convey his protective and promotional energies.
PHOTOGRAPHS BY ALLEN F. ROBERTS WITH C. JOSEPH ADANDE, REPUBLIC OF BÉNIN, 1994.

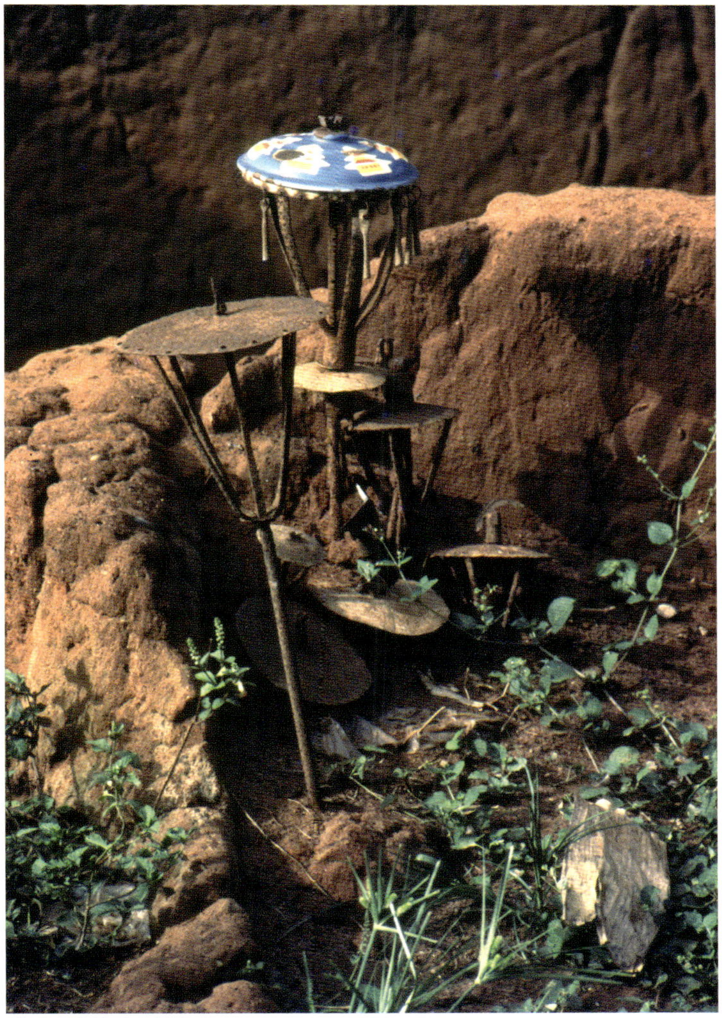

their haft and heft that my hand grips what my father's once held. That African shrines so often include blades identified with ancestors bespeaks such relationships of hands extended through tools to realize intentions and achieve hopes. In the compound of the blacksmith Sourou Tohoessou, in southeastern Republic of Bénin, for example, the crafting of recycled materials for local marketplaces is pursued in the presence Mr. Tohoessou's grandfather, long deceased but ardently remembered as founder of the family forge. In the corner of the plot where Grandfather's home stood, rusting *assanyin* commemorative sculptures mark where the old man died—they keep his spirit close (fig. 6.5).[5] The platform of one is made from the colorful enamelware top on an imported cooking pot, conveying *ara* (inventiveness) as a gauge of ancestral creativity through Ògún, god of iron and patron of smithing. A small building stands where Grandfather's atelier once did, and within it are several of his ironworking tools, his walking stick, and more *assanyin*, all assuring that his ongoing inspiration will be as "lasting and enduring as metal" itself, as a Yorùbá praise-poem has it.[6]

Pastoral people equip themselves to facilitate their on-the-go lifestyles, often by trading with sedentary blacksmiths for bladed implements. Most households in rural African towns and villages are equipped with knives, hoes, axes, and machetes; in rainforests and other ecologies more specialized tools are required.[7] Indeed, shapes of hoe blades and designs of their wooden handles vary widely from culture to culture depending upon climate, terrain, soil type, and crops preferred (figs. 6.6–6.10). Most hoes are forged with tangs inserted into wooden handles, although socketed blades fitted over the ends are produced in Ethiopia and a few other places (Simoons 1958; Blandin 1992, 140; Kriger 1999, 95). Blacksmiths often see to the hafting of hoe blades (fig. 6.11), and handles can be quite ingeniously ergonomic, directing strong muscles of backs, shoulders, and arms to the rigors of farming, or helping direct strokes efficiently for careful gardening.

Axes are used for chopping down trees, pruning, splitting firewood, and other rough tasks, while turning the blade ninety degrees so that it is perpendicular rather than parallel to the handle gives one an adze for more precise work.[8] Adzes are used to chip-carve mortars, pestles, headrests, stools, doors, and many other useful household and ritual items. Some blacksmiths forge adzes of different blade widths and lengths that are tang-set into straight handles. Other adzes are mantle-set with their blades ending in an iron sleeve clamped over a V-shaped handle to direct tiny blows with great accomplishment (Burssens 2003, 21).

BLADES OF HOUSEHOLD, FIELD, AND FOREST

Forged iron blades bespeak the will to work diligently using knives, hoes, axes, sickles, and a host of other tools and gadgets. Most African blacksmiths' daily labor is concentrated on fulfilling such community needs by producing and repairing mundane tools to accomplish ordinary chores, that is, simple implements forged of scrap that can be afforded by most. Even so, individuals often inscribe the blades, pyro-engrave or wrap wire around their wooden handles, or otherwise personalize reliable tools—as do working people the world around. There may be practical reasons to differentiate one's implements from someone else's, yet something far more *empathic* is suggested by ornamentation and the "sensual pleasures of shape, patina, and surface" it confers (Kasfir 2007, 101). Ornament is a form of mediation (Grabar 1992) providing an "inbetweenness" (Basu 2017) rather like the hyphen of "tool-being." Any such decorative acts contribute to tools being "right for the hand": personal or cultural signatures are imparted and purpose is *shared* between hand and tool, tool and hand.

In many ways, the more mundane the blade, the more "wisdom" it bears, and so all the more reason to cherish the tools of lost loved ones (fig. 6.4). I have several from my own family, knowing as I test

6.6
Artist unknown (Senufo peoples, Côte d'Ivoire)
Short-handled hoe, ca. 2010
Iron, wood
L: 54 cm
PRIVATE COLLECTION

The ergonomic design of this short-handled, wide-bladed hoe is perfect for working gardens of soft or sandy soil. A man or woman farming with such a tool stoops low to the ground so that the power of back and arm muscles is translated to comfortable yet efficient thrust-and-drag digging and mound-building for root crops such as yams, manioc, and peanuts.

6.7
Artist unknown (Tutsi, Hutu peoples, Rwanda and Burundi)
Long-handled billhook (*mugishu, nmuhoro*), 20th century
Iron, wood
H: 62.5 cm
FOWLER MUSEUM AT UCLA, X65.3132; GIFT OF MR. W. THOMAS DAVIS

Billhooks serve banana farming in Rwanda, Burundi, and adjacent countries with tropical ecologies. Many varieties of bananas and plantains are prepared in regional cuisines similar to staple dishes of maize, manioc, and millet enjoyed elsewhere in sub-Saharan Africa. Carefully forged to provide edges and ergonomic design, billhooks are so useful that occasionally—as among Lega people of east-central Democratic Republic of the Congo—their form has been adapted to symbolic purposes.

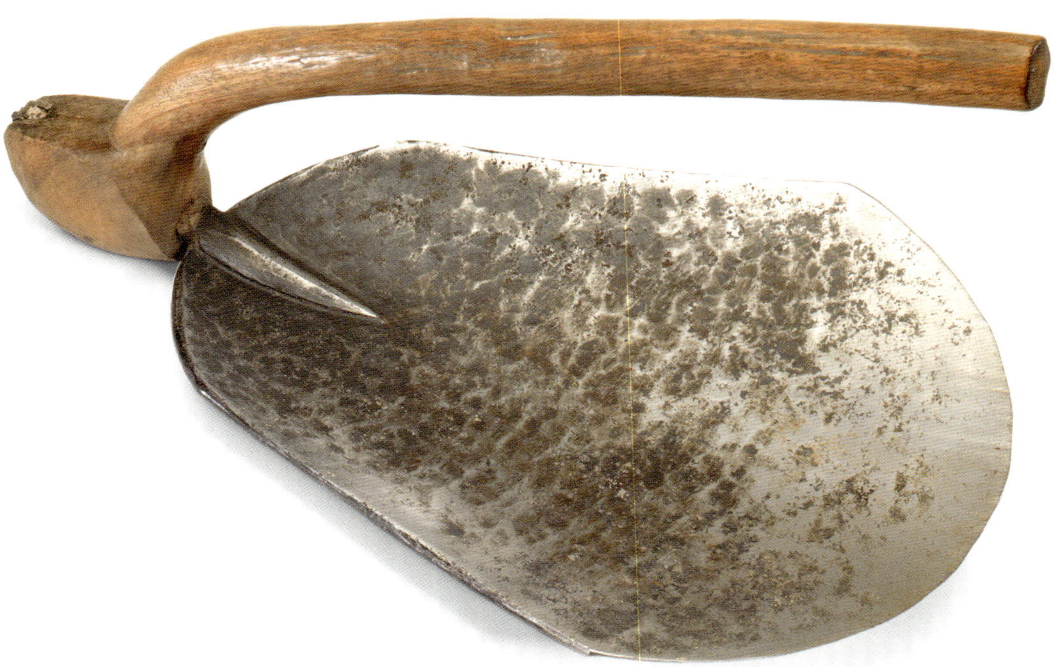

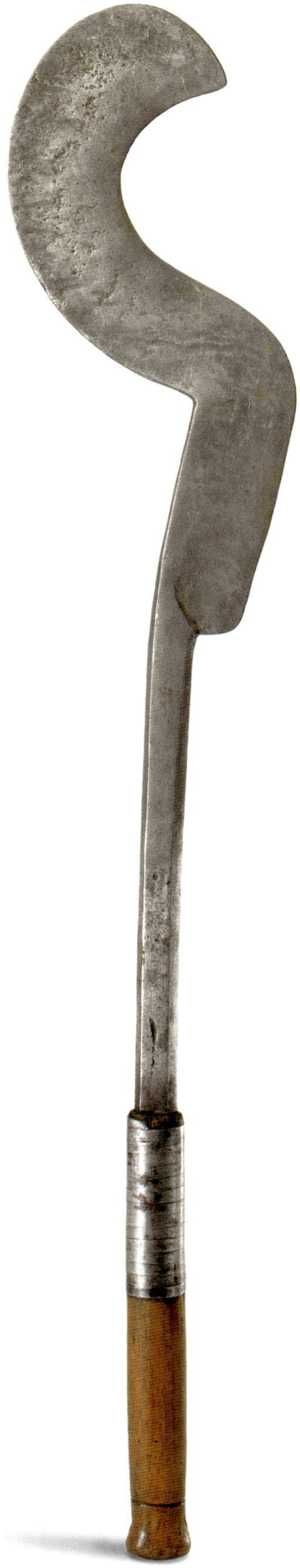

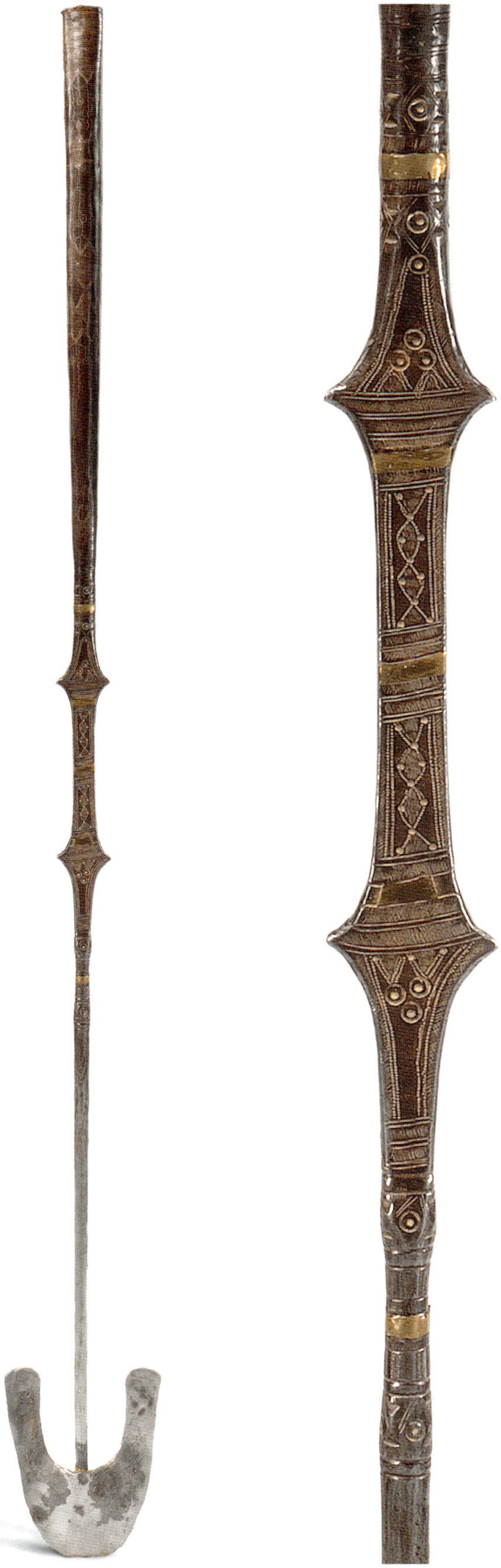

6.8A–B
Artist unknown (Djerma-Songhai peoples, Niger)
Long-handled hoe (*haywa*), 19th to early 20th century
Iron, copper alloy
H: 154 cm
PRIVATE COLLECTION

In contrast to the previous design, this long-handled tool is as much a spade as a hoe, and is used as a farmer stands thrusting and cutting through tough soil, weeds, and roots as cultivation is begun and gardening sustained. That its haft as well as its blade are of forged iron may increase the force with which it can strike the ground or pry clods or obstructions upward. The same implement is used to weed gardens in sandy soil, easily cutting through roots and lifting plants (Celis 1991, 70).

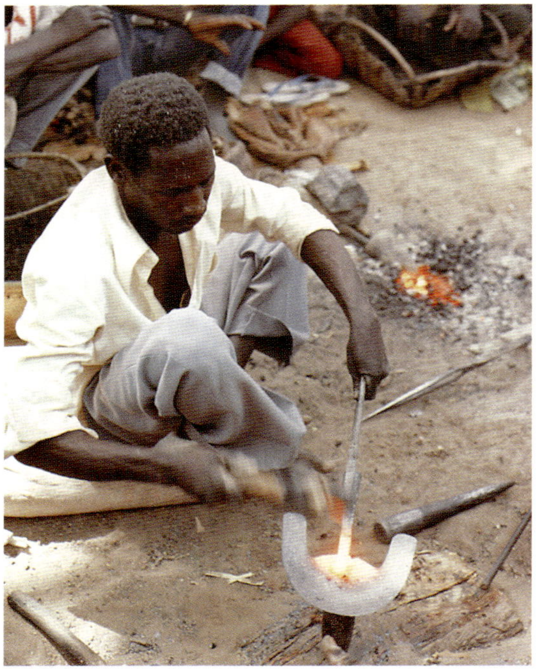

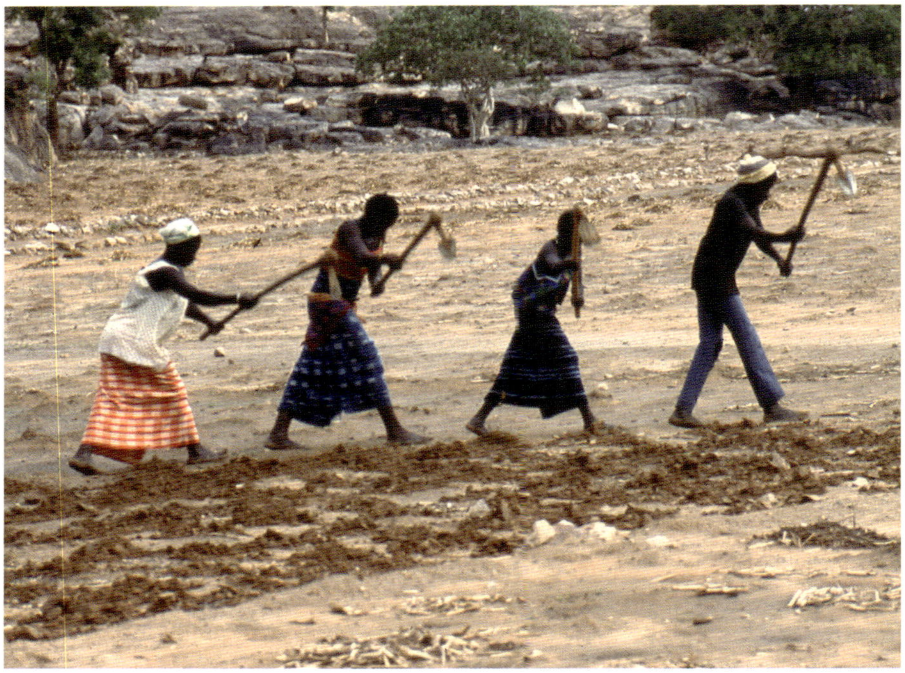

6.9 An unidentified Djerma blacksmith in the village of Yelwani, Republic of Niger, uses two different types of hammer to forge a long-handled hoe (see fig. 6.8) from iron just smelted under the direction of Abdou Modi. The elderly gentleman had engaged in such work into the 1950s, and the furnace and technologies were reconstituted for documentation by Georges Celis (1991, 56–70).

PHOTOGRAPH BY GEORGES CELIS, 1986. PUBLISHED IN GEORGE CELIS, *EISENHUTTEN IN AFRIKA: BESCHREIBUNG EINES TRADITIONELLEN HANDWERKS* (FRANKFURT AM MAIN: MUSEUM FÜR VÖLKERKUNDE, 1991).

6.10 Soon after the first rain of the season, Dogon young people till the sandy soil of a field, incorporating natural fertilizer. A man takes the lead in a call-and-response song with three women, as steps are paced and hoe thrusts coordinated. The resulting grid where seeds will be planted replicates the pattern seen in Dogon arts, bespeaking human organization as initiated by the culture hero who stole a piece of the sun and became the first ironsmith (see fig. 1.9).

PHOTOGRAPH BY ALLEN F. ROBERTS, NEAR SANGHA, MALI, 1986.

6.11 An elderly blacksmith fashions the handle of an axe or adze for a blade he has forged. He uses an adze to chip-carve the head through which a hole will be burned to accommodate and hold fast the tang of the blade.

PHOTOGRAPH BY CHRISTOPHER D. ROY, PELIGNAN VILLAGE, BURKINA FASO, 2003.

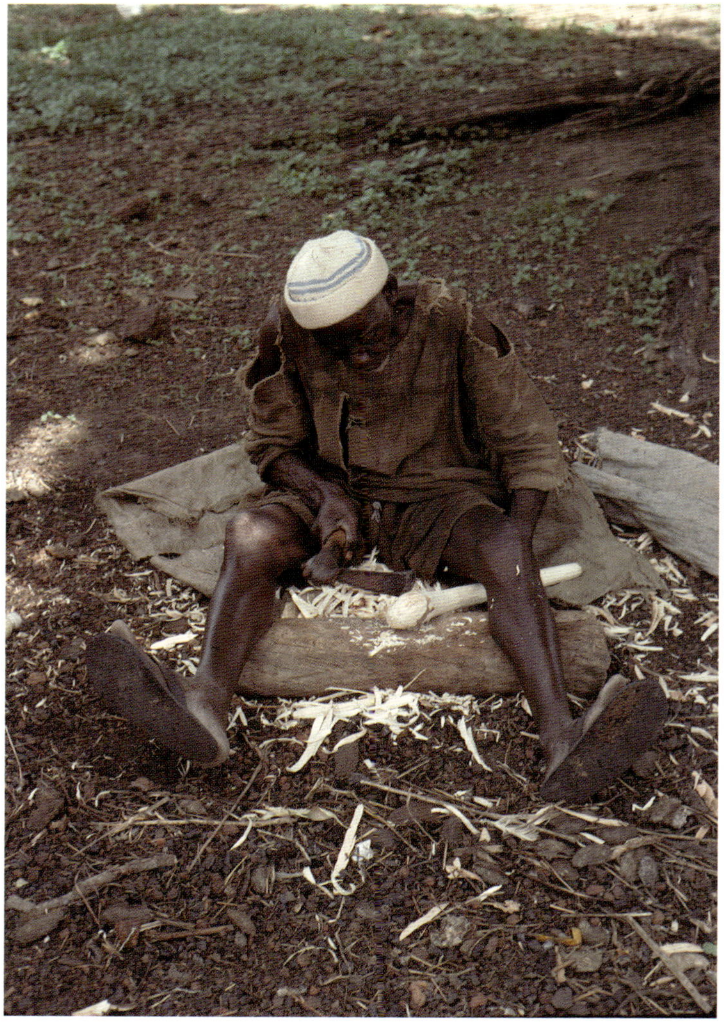

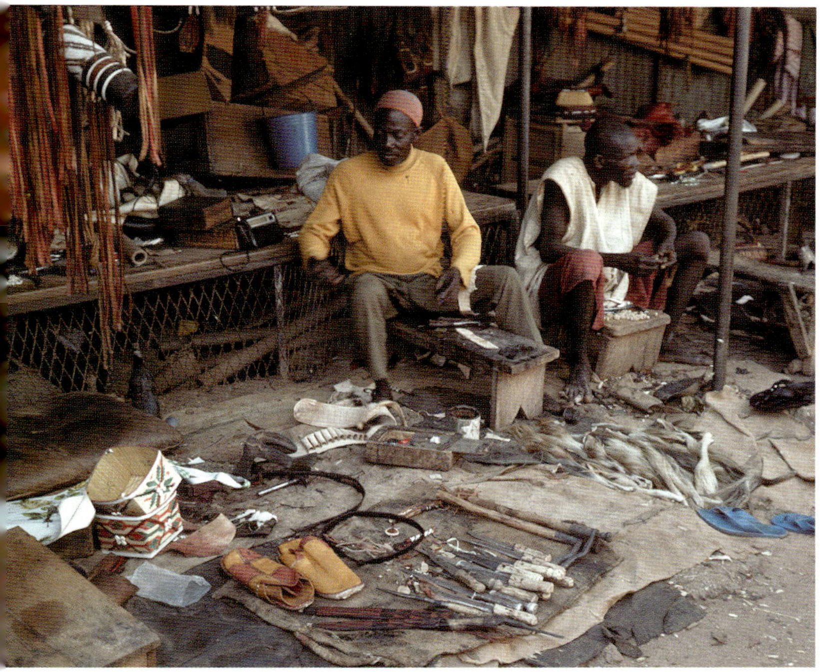

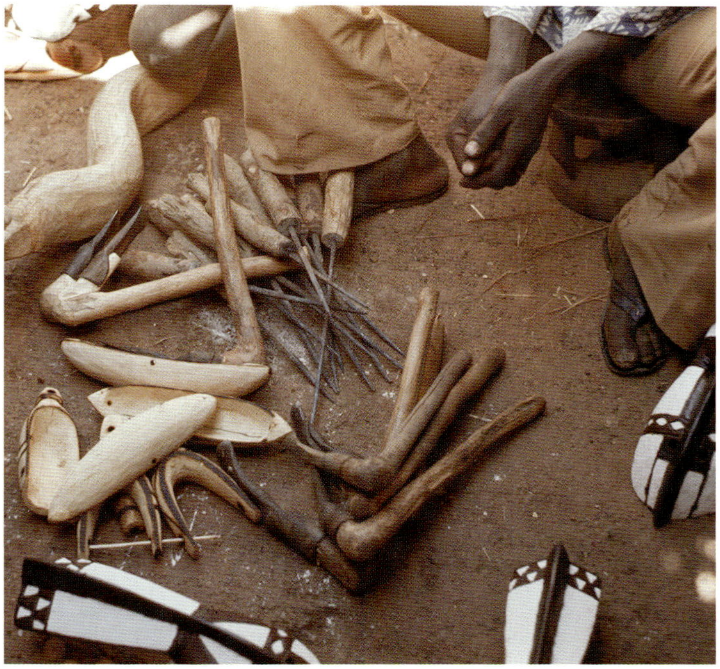

6.12 West African urban marketplaces often feature workshops like this one, in the iron and leather section of the old market of Ouagadougou (today replaced). Leatherworking can be commissioned from craftsmen awaiting clients, and the men's bladed tools are displayed as models to suggest that all is ready to produce attractive wares for customers.
PHOTOGRAPH BY CHRISTOPHER D. ROY, BURKINA FASO, 1971

6.13A–B Above, a kit of iron-bladed adzes and awls is laid out at the ready by the Mossi artist Yili Wango of Bouboulou Village, northwest of Ouagadougou, the capital of Burkina Faso. Mr. Wango creates masks and ritual objects as seen here, as well as practical tools such as wooden heddles and pulleys for local strip-cloth weaving.

Below, the Mossi artist Raogo Sawadogo ("Male Raincloud"), known for the masks he masterfully chip-carved, roughs out an antelope head. He was also renowned for donning heritage clothing that was a conscious throwback to earlier times. His remarkable cap is of a sort once worn by Mossi, Dogon, and other peoples of the greater region.
PHOTOGRAPHS BY CHRISTOPHER D. ROY, KIRSI VILLAGE, BURKINA FASO, 1977.

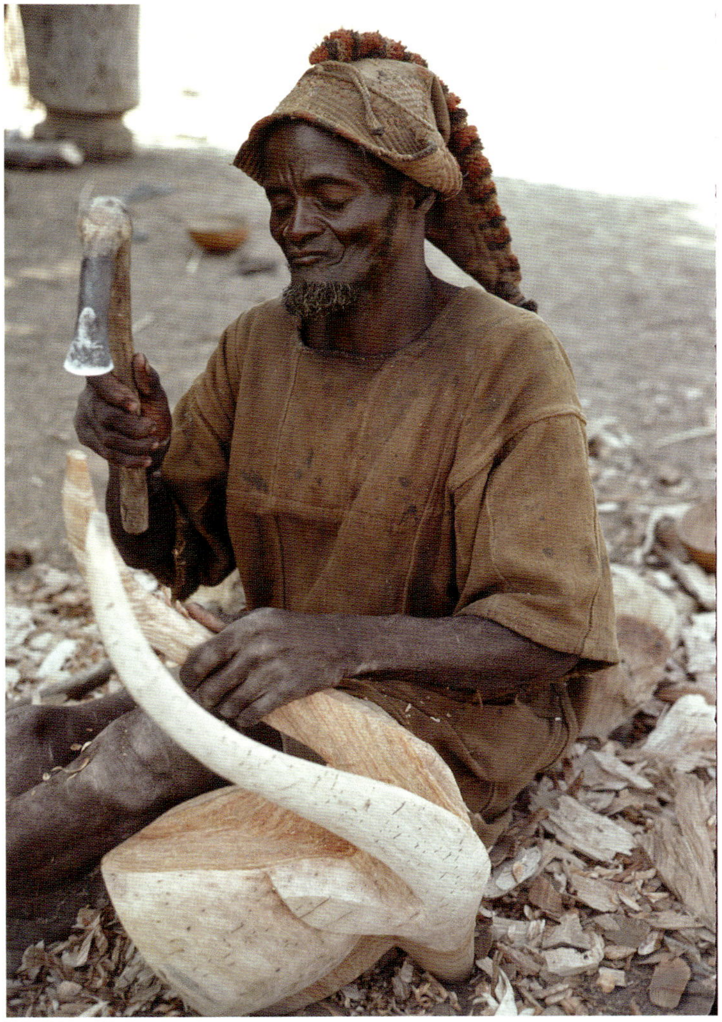

Like handy men everywhere, African craftsmen often assemble kits of bladed tools to realize complex projects (figs. 6.12, 6.13).[9] Yet in an important sense, axes and adzes are *all-purpose* tools. As Richard Sennett holds, such implements "admit of all manner of unfathomed possibilities" and can "expand our skills if only our imagination rises to the occasion." Furthermore, they "can be described as sublime.... In craftwork, that sentiment focuses especially on objects very simple in form that seemingly can do *anything*."[10]

6.14
Artist unknown (Chokwe, Lunda peoples, Democratic Republic of the Congo)
Ceremonial axe, early 20th century
Iron, wood, copper
H: 38 cm
MUSÉE DU QUAI BRANLY-JACQUES CHIRAC, PARIS, 71.1948.15.29
PHOTOGRAPH BY THIERRY OLIVIER. © MUSÉE DU QUAI BRANLY-JACQUES CHIRAC, DIST. RMN-GRAND PALAIS / ART RESOURCE, NY

This tour-de-force ceremonial axe was meant to be a showstopper, as its artistry of design and execution—including the inset copper devices of the blade and the remarkable handle clad in copper—more than exceeds the accomplishments of ordinary blacksmiths. Refined chiseling, punching, inlaying, chasing, engraving, scoring, raising, wrapping, stamping, embossing, and many more proficiences could only be expressed to this degree in a single work after many years of experience. Such a blade added to a blacksmith's honor and influence in the community, with both owner and artist distinguished every time it was displayed in public.

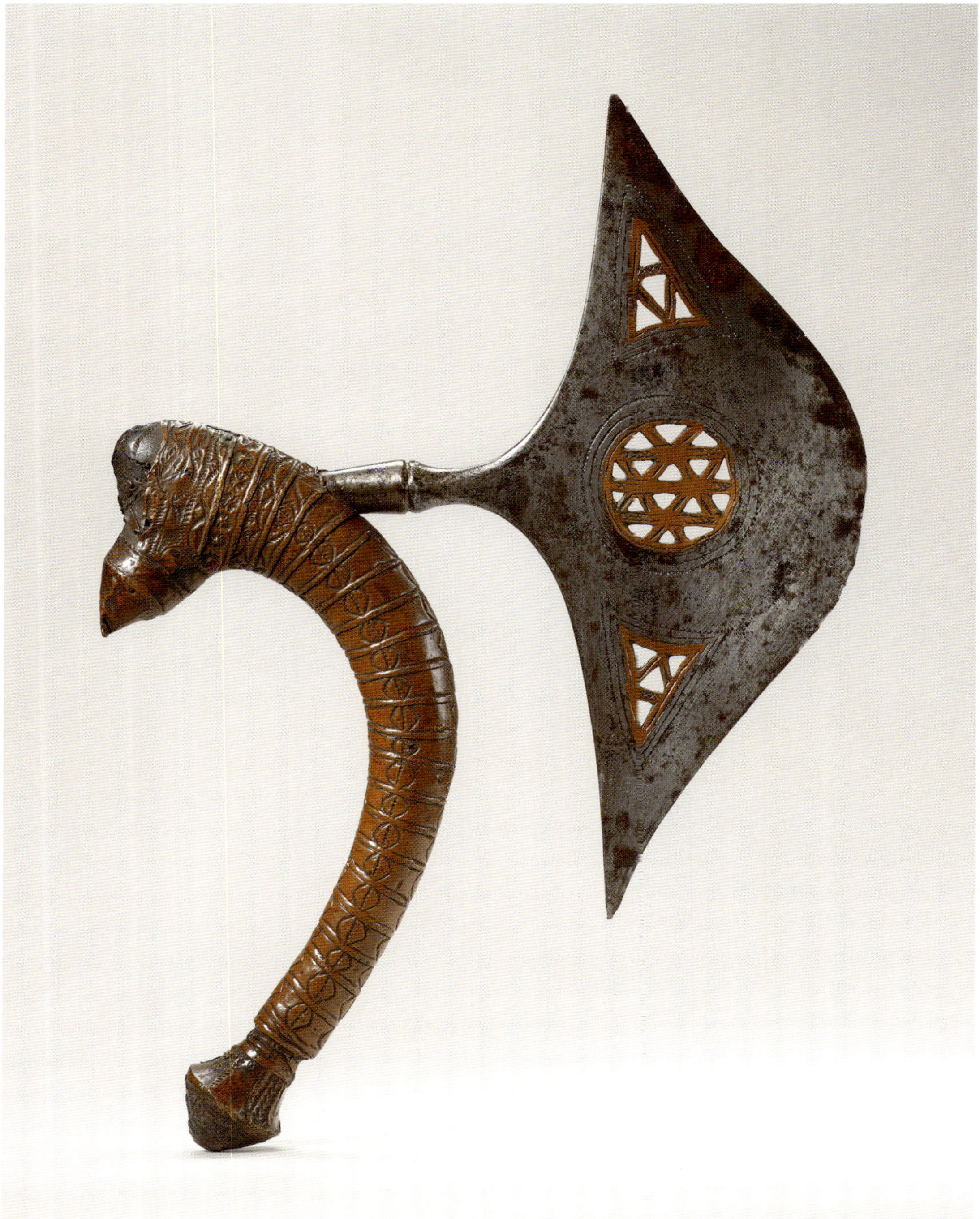

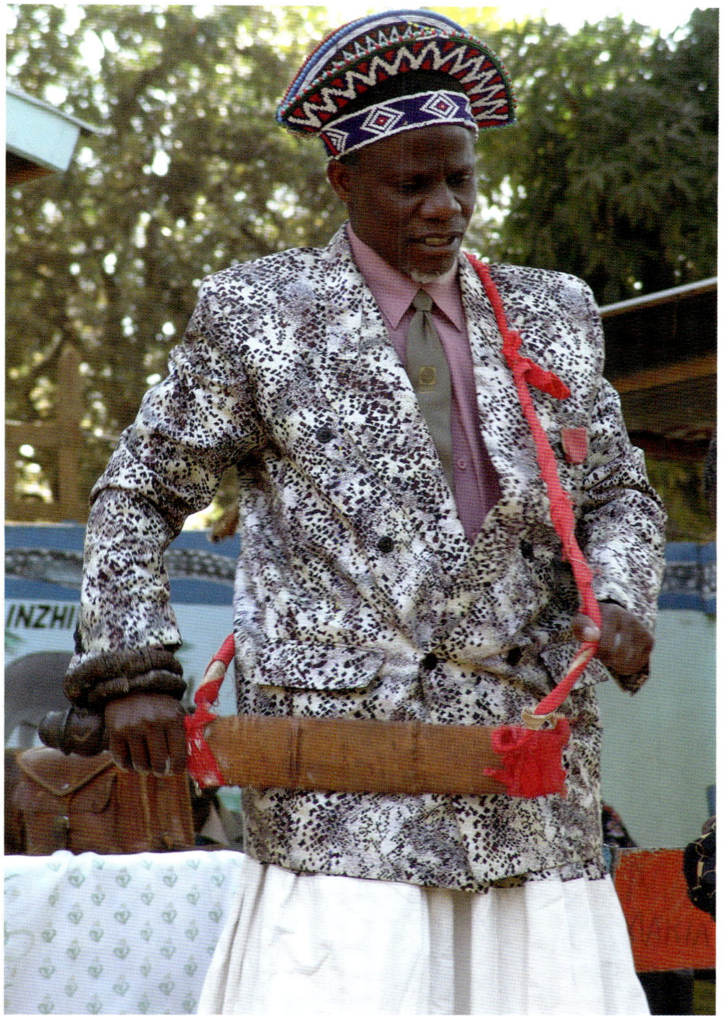

BLADES OF POWER AND PRESTIGE

Because of the "wisdom" of iron as a material and the transformative genius of smiths, and the essential tasks one can accomplish with blades—what Sennett might find "sublime" in their all-purpose potentiality—common tools have often been stylized to produce potent symbols of political power and prestige. Indeed, some of the most striking works of central African blacksmiths' known repertoires are elaborated axes, adzes, knives, swords, and spears. One example of astounding artistry is a showy Chokwe or Lunda axe (fig. 6.14); its complex iconography refers to the hunter and culture-hero Chibinda Ilunga, in whose honor performers still dance at the courts of kings and chiefs, axes in hand (fig. 6.15). Openwork triangles in the blade refer to life (upward) and death (downward), the circle is the all-powerful sun, and the "red" of the copper-wrapped handle contrasts with the "whiteness" of the iron blade, as do violent transformations with ancestral blessings.[11] Essential forces were brought to bear through performance, and such a tour-de-force achievement must have affirmed the prestige of the blacksmith who forged the blade and designed, carved, and embellished the haft, as well as the person fortunate enough to possess and display such a masterpiece.

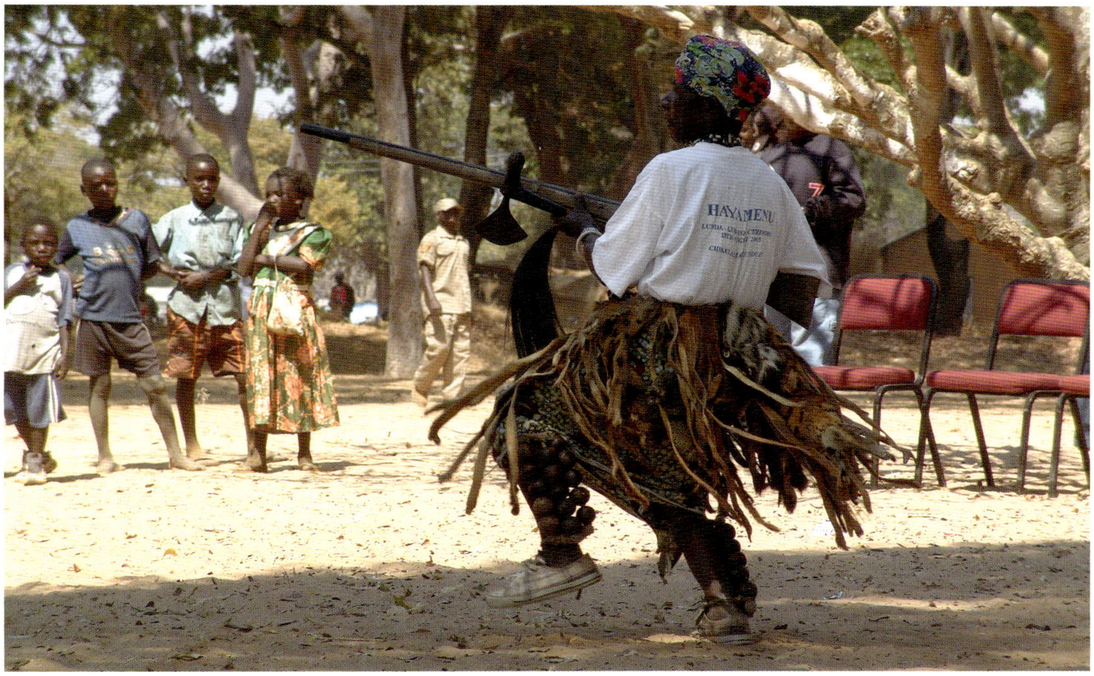

6.15A–B Above, Southern Lunda Paramount Chief Ishindi prepares to receive dignitaries from Angola whom he will invest with new royal titles as the emissary of the Mwata Yamvo, the great Lunda king residing in the southwestern corner of the Democratic Republic of the Congo. Paramount Ishindi wears a royal beaded hat as well as a *lukano* bracelet, both powerful emblems, while he grips the figurative handle of an heirloom sword that was once seized from Salampasu adversaries living north of the Mwata Yamvo.

Below, a royal diviner for Paramount Chief Ishindi dances with a musket and an heirloom ceremonial axe with a wide, flaring blade (see, for example, fig. 6.14). Such designs are rarely seen today in northwestern Zambia. The gentleman carries additional devices bearing efficacies: he holds a fly whisk in his left hand; wears animal pelts around his waist; and the gourd rattles that festoon his lower legs set the rhythm of his animated choreography.

PHOTOGRAPHS BY MANUEL JORDÁN, PARAMOUNT CHIEF ISHINDI'S PALACE, ZAMBIA, 2006.

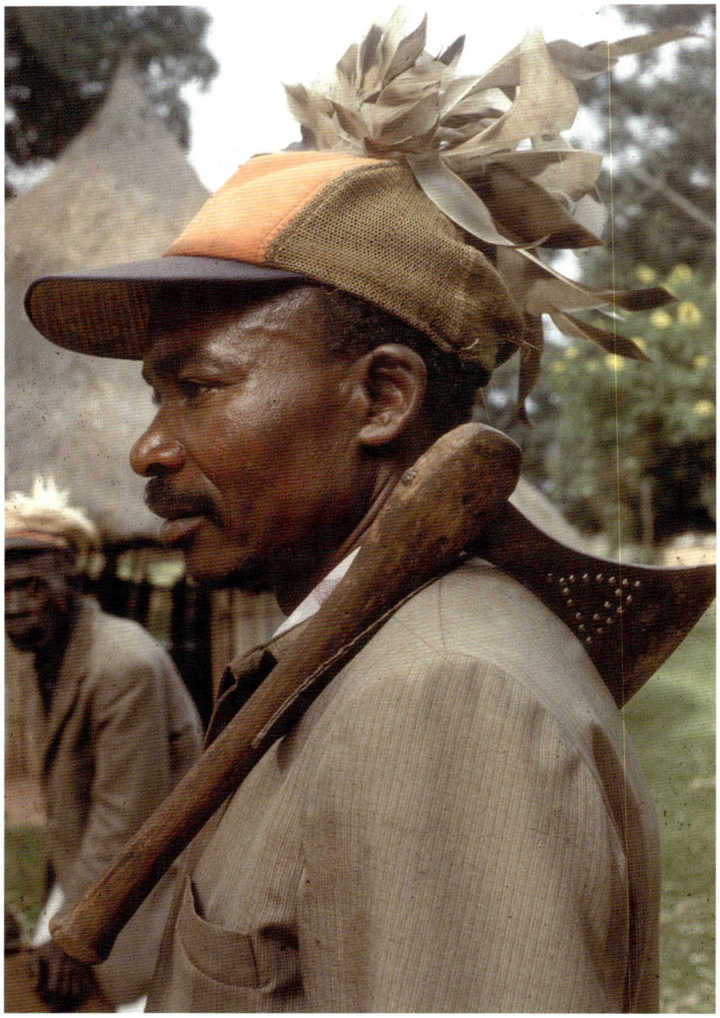

6.16 A high-ranking titleholder, called the Twite of King Kabongo's court in the Luba Heartland, carries a ceremonial axe over his shoulder as a mark of his status. Here he anticipates dancing with the axe to honor the king during performances of Kutomboka, a ritual through which the Luba Epic is recounted through complex choreography.

PHOTOGRAPH BY MARY NOOTER (ROBERTS), KABONGO, DEMOCRATIC REPUBLIC OF THE CONGO, 1989.

Among Luba, ceremonial axes with incised blades and sculpted handles that depict female tutelary spirits—making them present—were possessed by kings, chiefs, high-ranking titleholders, spirit mediums, diviners, and members of certain secret associations. Dignitaries still wear such axes over their shoulders as emblems of title and rank in dances and events at court (fig. 6.16).[12] Such gracious implements permit noble persons to commemorate the origins of Luba society, and specifically Mbidi Kiluwe and Kalala Ilunga, the culture heroes who introduced ironworking and big-game hunting as well as the sophistication of royal bearing. Kalala's emblems are his axe, spear, knife, bow-stand, and bells—all used today when tradition is performed to meet intellectual and political needs.

Allusions made to spirits and heroes are not merely visual and narrative devices, however, for they also signal efficacies of axes and other bladed implements deployed in such ways. Among Bantu-speaking peoples like Luba, such objects are *nkisi* that possess creative energies to affect change. Furthermore, evocative metaphors derived from the work of axes in cutting wood and blazing trails also inform rituals of Mbudye, a society of Luba historians, as they do problem-solving by spirit mediums and diviners. In such challenging intellectual endeavors, meaning is discerned and histories proposed by recognizing a "path" linking one significant moment or circumstance with another, "blazing" selected "trees" in a "forest" of known but mostly irrelevant data.[13]

Scrutiny of Luba axes has led metallurgist S. Terry Childs and art historian William J. Dewey to suggest an analogy between ironworking and human birth. "Evidence suggests that the transformative actions of repeated hammering, annealing, folding, and welding at the forge were metaphorically linked to the growth and maturation of individual members of society and to the reproduction of society in general." Luba vocabulary reinforces the assertion, for the verb *kusunga* means "to fabricate a weapon and to carry out a ritual" such as that "announcing a woman's pregnancy, the death of a spouse, the installation of a new chief, the end of circumcision ceremonies, and the purification of a new diviner." Childs and Dewey go on to note that the "manufacture of weapons involves analogous transformations, refinements, and adornments . . . hence use of the same verb [*kusunga*]."[14]

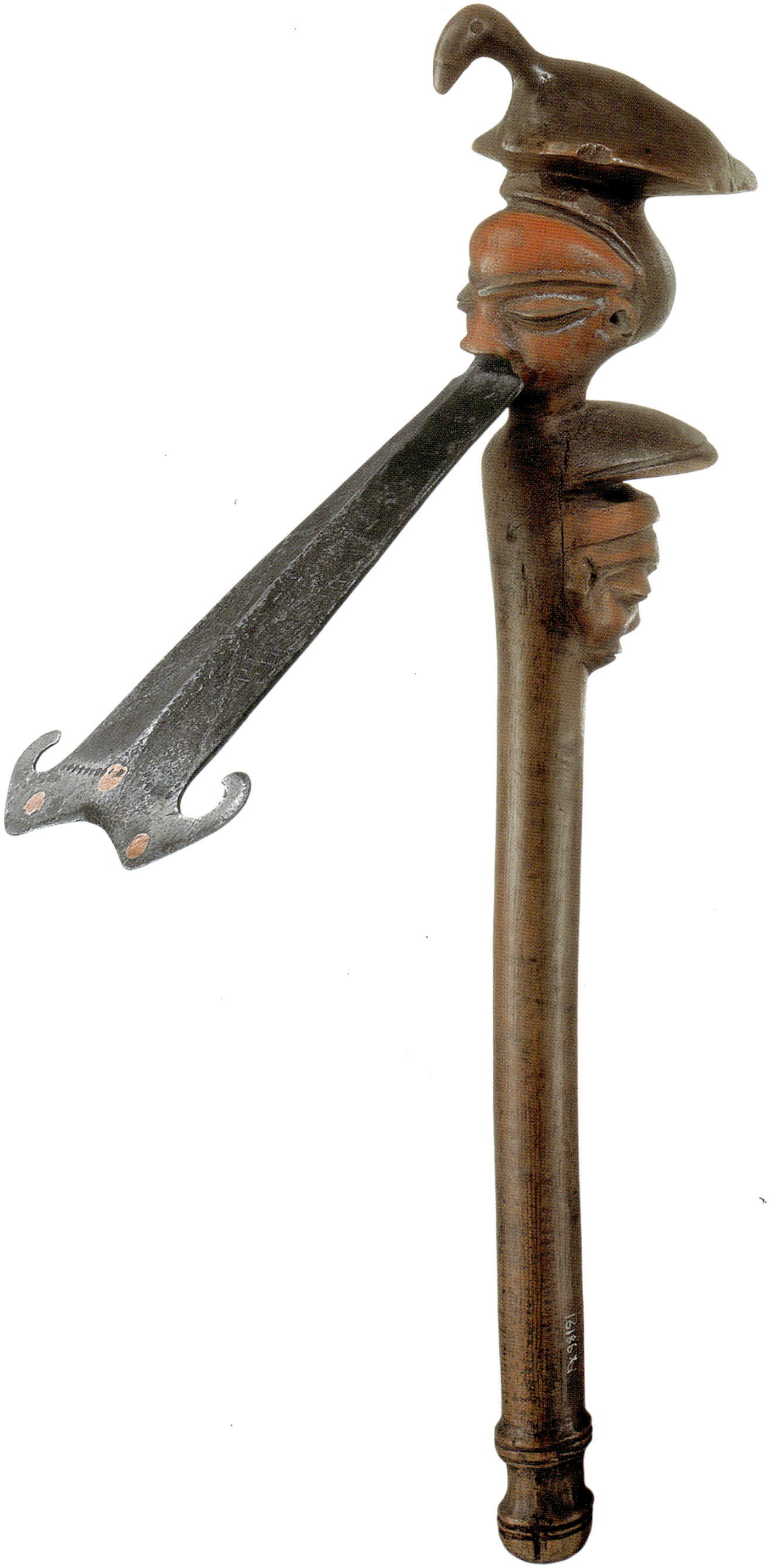

6.17
Artist unknown (Central Pende peoples, Democratic Republic of the Congo)
Adze, early 20th century
Iron, wood, copper, pigment
H: 44 cm
FELIX COLLECTION
PHOTOGRAPH BY DICK BEAULIEUX, 2005.

Ceremonial adzes like this were once carried over the shoulders of Pende chiefs as emblems of office, and the heads looking forward and back "allude to spirit workers who watch over the chief and the chieftaincy, including his people, land, and animals" (Z. S. Strother pers. comm., 2018). The blade of this adze—a remarkable feat of smithing in its own right—reflects the form of implements used during investiture. The bird at the finial refers to the ritual house of the bearer, but also bespeaks a chief's ability to shape-shift into a bird to travel long distances to surveil his community.

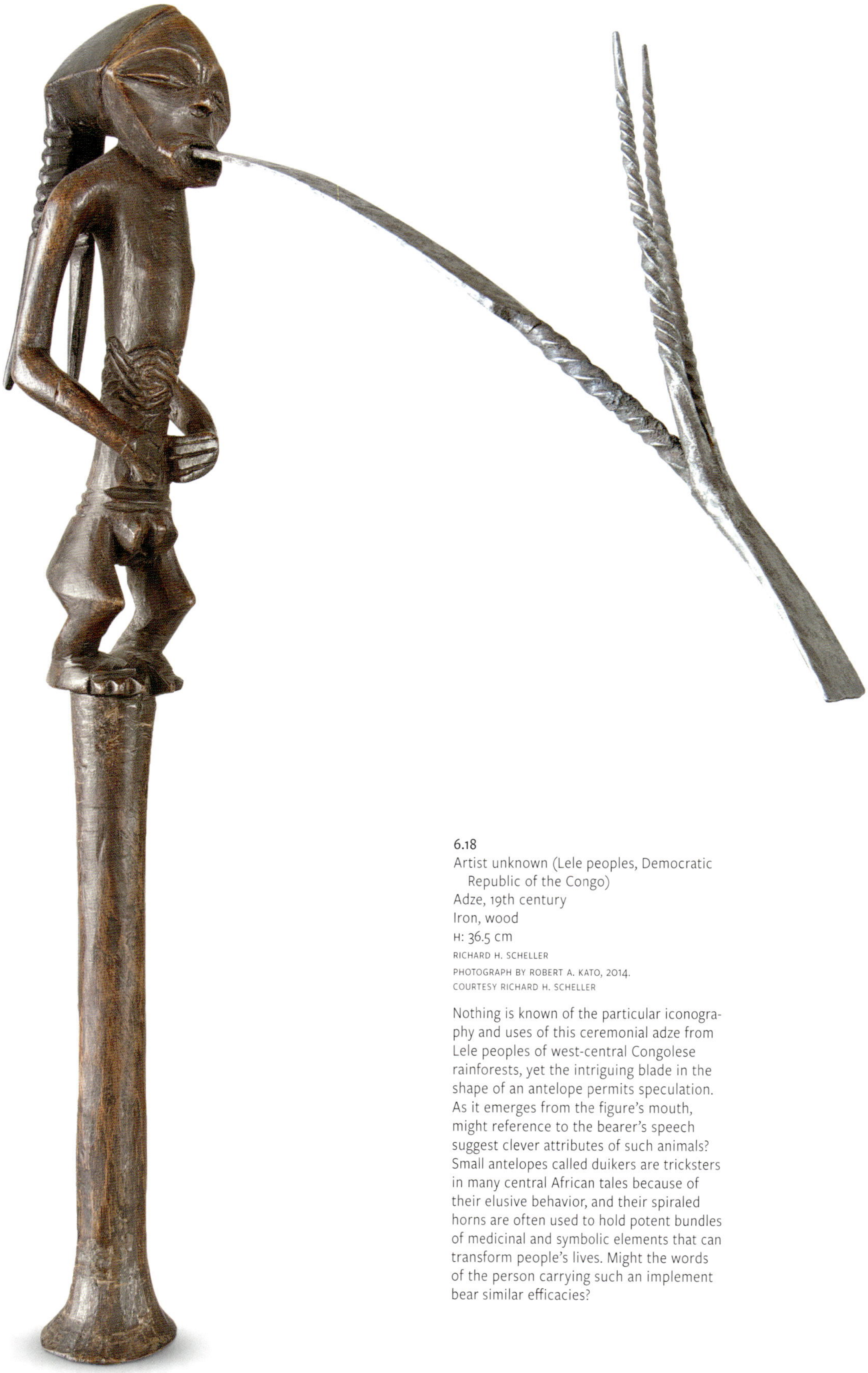

6.18
Artist unknown (Lele peoples, Democratic Republic of the Congo)
Adze, 19th century
Iron, wood
H: 36.5 cm
RICHARD H. SCHELLER
PHOTOGRAPH BY ROBERT A. KATO, 2014.
COURTESY RICHARD H. SCHELLER

Nothing is known of the particular iconography and uses of this ceremonial adze from Lele peoples of west-central Congolese rainforests, yet the intriguing blade in the shape of an antelope permits speculation. As it emerges from the figure's mouth, might reference to the bearer's speech suggest clever attributes of such animals? Small antelopes called duikers are tricksters in many central African tales because of their elusive behavior, and their spiraled horns are often used to hold potent bundles of medicinal and symbolic elements that can transform people's lives. Might the words of the person carrying such an implement bear similar efficacies?

Certain central African ceremonial adzes add a nuance to such materialization of thought and the senses, for their blades emerge from the mouths of anthropomorphic heads sculpted at the ends of their wooden handles (figs. 6.17–6.21). Particular ethnographic data on how such scepters were used is scant, but we can imagine that they portrayed the "cutting-edge" speech of spiritual agencies, and so informed and empowered the narrative arts of those privileged to bear them. When such adzes were worn over shoulders or displayed as extensions of performers' hands, arguments were broached, histories proposed, praises expressed, notorieties affirmed. Yet, as Herbert Cole suggests, "the words of leaders, like weapons and the metal of which they are made, are authoritative. Wise speech is a sharp tool" (1989, 111)—which, we would add, bears the wisdom of iron and the genius of smiths. Whether sage or not, "the fiery power of the word of the ruler, as suggested by the spitting of the edge of a ceremonial blade from the mouth" of the figure of such an adze, "mimed... the release of heat-forged weapons and... blades of war and death." Hard-edged messages could be implied, that is, and when sent as his envoy, such an instrument "spoke" to a ruler's keen resolve in addressing matters of local-level and regional politics.[15]

Central African knives and swords present similar intellectual complexities by no means limited to functionality. Most such blades were designed with practical needs in mind, of course, including how "sublime" they might be in their wide range of applications, and as made "right for the hand"; yet even when forging humble implements, blacksmiths were guided by the aesthetics of their cultures and times. Swords, for instance, might do adversaries great harm, yet some were created to be held in hands as scepters, with magnificence in mind. Similarly, the forms of knives were sometimes greatly elaborated, and as suggested by Mangbetu examples (figs. 7.13, 7.14), an easily lethal weapon could become the pointed extension of a king's grasp (fig. 7.12). Such designs shared deliberate asymmetries central to Mangbetu senses of beauty and purpose, and even head-binding shared their style (Schildkrout and Keim 1990). In other words, this was an instance of "a distinctive sensibility, the whole of life participated in forming," in order to "materialize a way of experiencing; [and] bring a particular cast of mind out into the world of objects, where men [and women] could look at it" (Geertz 1976, 1478, slightly paraphrased). The synaesthetic arts of African blacksmiths have few more striking exemplars (although see Thompson 1973b on the complexities of Yorùba artistry).

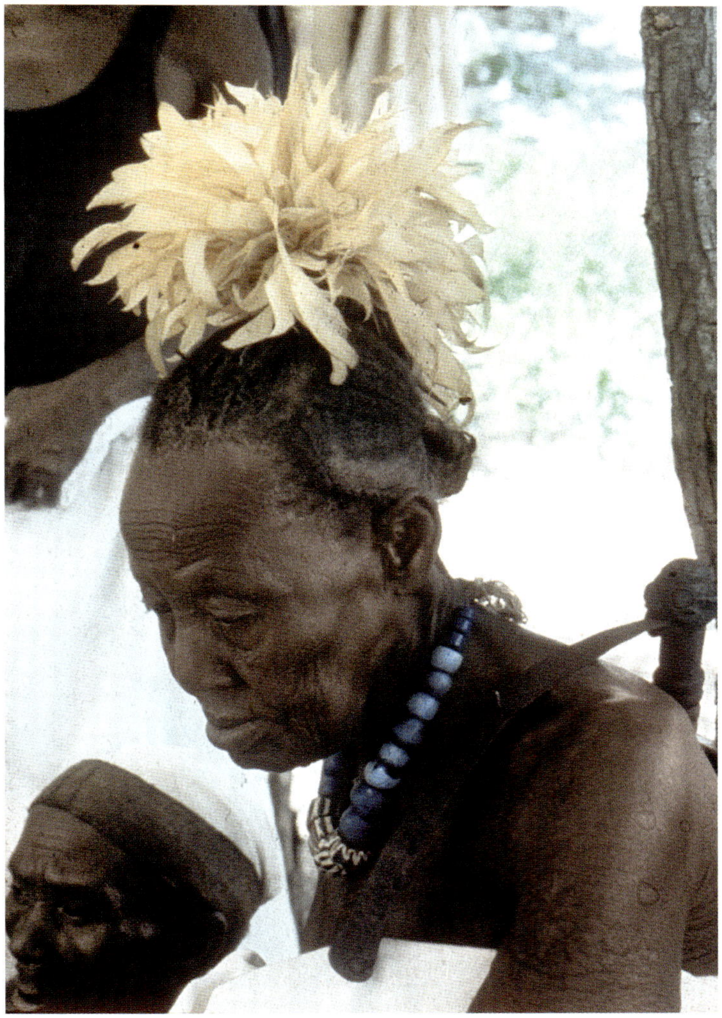

6.19 This is the last known Mwadi, who, until her death in the 1970s, incarnated the spirit of King Kasongo Niembo (d. 1931). The blade of the ceremonial adze she bears over her shoulder emerges from the sculpted head of a spirit, suggesting "cutting-edge" decisive speech, associated with the much regretted king.

PHOTOGRAPH PROVIDED MARY NOOTER (ROBERTS). COURTESY OF THOMAS Q. REEFE, CA. 1973.

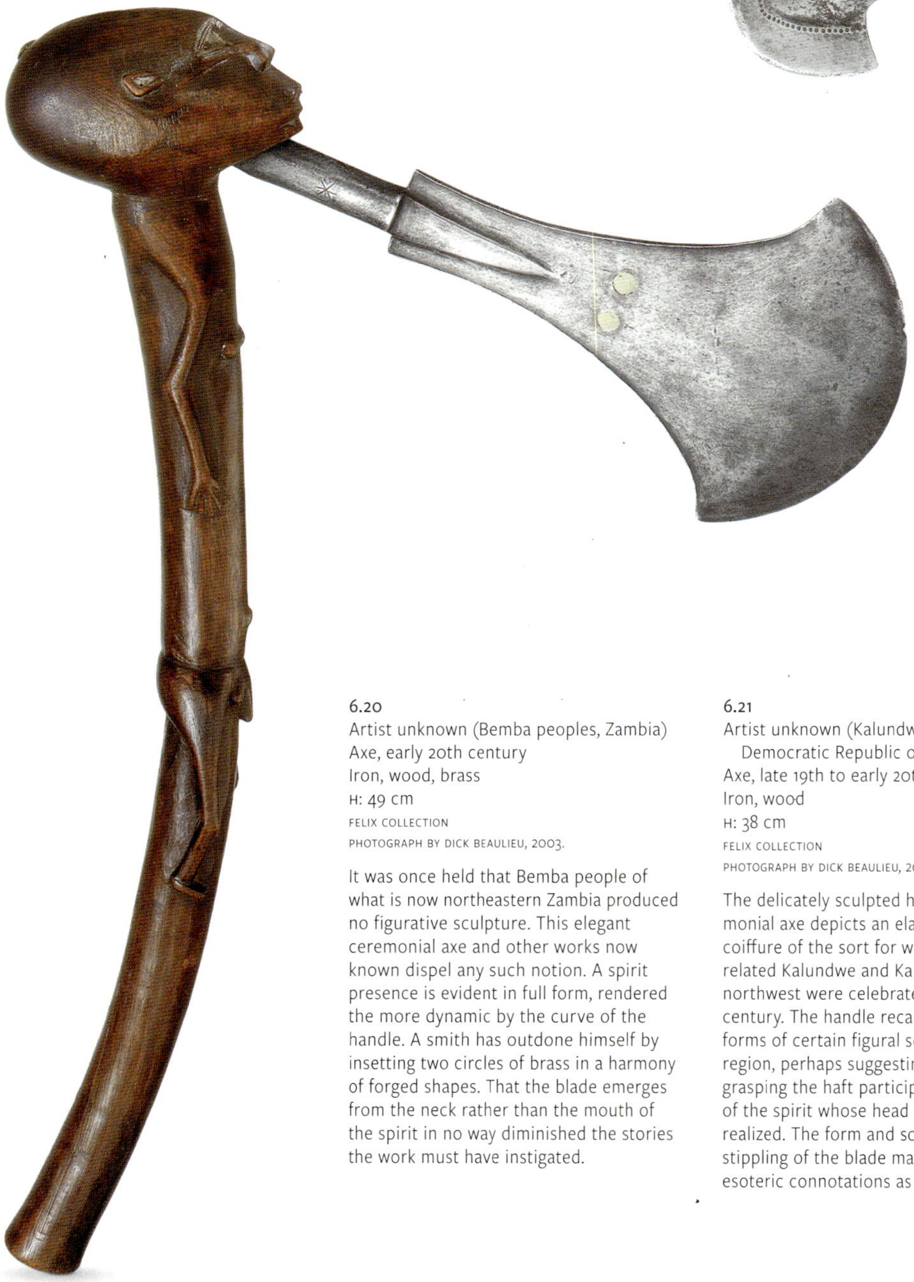

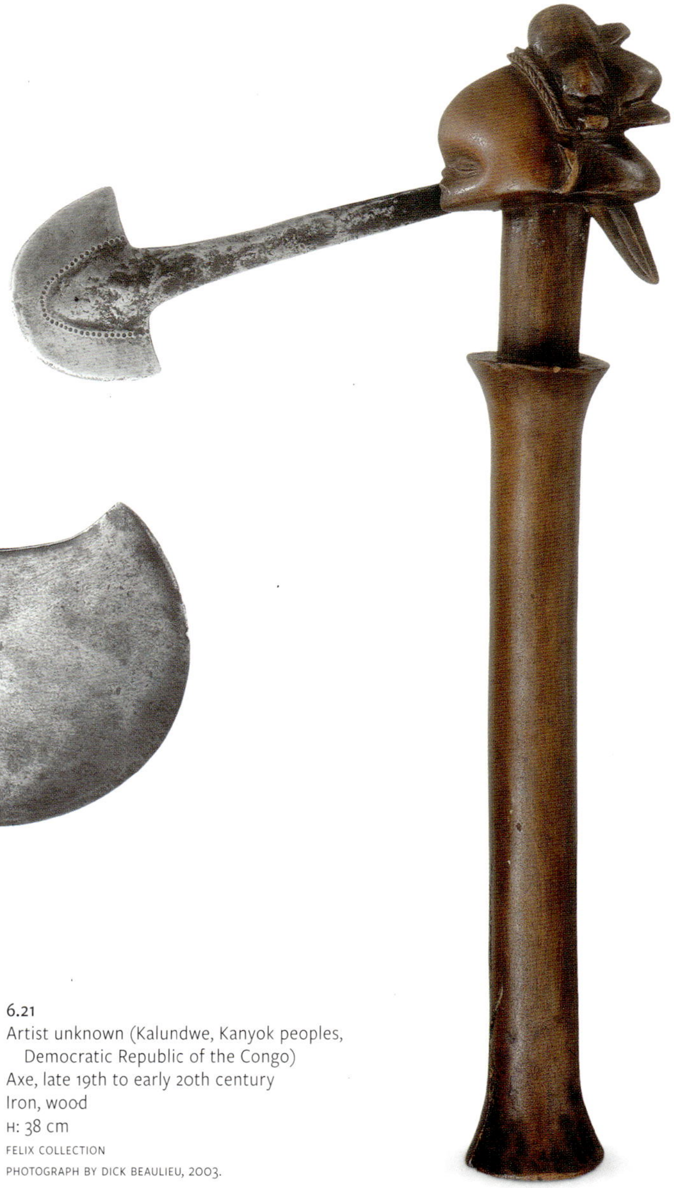

6.20
Artist unknown (Bemba peoples, Zambia)
Axe, early 20th century
Iron, wood, brass
H: 49 cm
FELIX COLLECTION
PHOTOGRAPH BY DICK BEAULIEU, 2003.

It was once held that Bemba people of what is now northeastern Zambia produced no figurative sculpture. This elegant ceremonial axe and other works now known dispel any such notion. A spirit presence is evident in full form, rendered the more dynamic by the curve of the handle. A smith has outdone himself by insetting two circles of brass in a harmony of forged shapes. That the blade emerges from the neck rather than the mouth of the spirit in no way diminished the stories the work must have instigated.

6.21
Artist unknown (Kalundwe, Kanyok peoples, Democratic Republic of the Congo)
Axe, late 19th to early 20th century
Iron, wood
H: 38 cm
FELIX COLLECTION
PHOTOGRAPH BY DICK BEAULIEU, 2003.

The delicately sculpted head of this ceremonial axe depicts an elaborate cascading coiffure of the sort for which Luba and related Kalundwe and Kanyok to their northwest were celebrated in the 19th century. The handle recalls the attenuated forms of certain figural sculptures of the region, perhaps suggesting that a person grasping the haft participated in attributes of the spirit whose head is so superbly realized. The form and scarification-like stippling of the blade may have borne esoteric connotations as well.

OF BLADES PROJECTED

Iron arrowpoints are another important genre of African blades. Arrowheads of different designs are often carried by rural hunters, as striking birds requires a technology different from that of pursuing mammals or fish (figs. 6.22, 6.23). Mangbetu hunters in the heavy forests of northeastern Democratic Republic of the Congo sometimes incise the ends of arrow shafts to permit a bowman to identify a particular arrow so that its specialized point can be quickly recognized among those in his quiver and then swiftly dispatched (Schildkrout and Keim 1990, 107). Some arrows are fletched for longer, more accurate flight, while others are not, because tiny iron points can carry organic poisons as they often do among hunter-gatherer peoples like Mbote of southeastern DRC. Arrow poison produced by Mbote specialists was once widely traded throughout east-central Africa (Terashima 1980; Angenot 1978).

Among San hunter-gatherers of Botswana, "socially, politically, and economically,... arrows have greater import than any other... artifact" because they permit provision of meat to be shared communally. Arrows also feature in San myth and are deployed in sacred eland hunting (Wiessner 1983, 261). In more centralized societies, hunting guilds sometimes marshalled esoteric technologies necessary to capture the largest and most dangerous of beasts. Among eastern Luba and their Tabwa neighbors, such a guild was called Buluwe in recognition of Mbidi Kiluwe as the great hunter and culture hero. Carefully camouflaged pit traps fitted with upward-pointing spears were enhanced by potent magical medicines, and with proper propitiation, spirits—perhaps Mbidi Kiluwe himself, riding lead elephant or buffalo—could be persuaded to direct a herd to its demise in the pits, with coups de grace administered by closely aimed arrows or spear thrusts.[16]

In the past, warriors sometimes used iron-pointed arrows, and "archers formed an important infantry element of most Sudanic armies" that could be very effective against cavalry.[17] Shooting an arrow was a two-handed task, leaving bowmen vulnerable to quick counterstrike, however, even though shields were commonly used—occasionally forged of iron. Agility and wiliness were vaunted aspects of such tactical combat, with individual valor long recalled. Yet it seems that archery was deployed more often in defense or ambush than in open combat. This was the case in the nineteenth century, when loosely organized Zaose peoples of what is now eastern Burkina Faso were harassed by marauders from powerful Mossi and Gurma polities to their either side. Taking refuge in a rocky outcropping, men rained flaming arrows upon equestrian aggressors, and their legendary successes against all odds were still celebrated in the 1980s.[18]

Arrows extend the eye as well as the hands as projectile points are aimed and fired. The prowess so implied, and especially the need to hit a moving target eager for counterattack, lends itself to metaphor. By the mid-1970s, few if any Tabwa hunted with bows and arrows anymore, for instance, but when illness or other misfortune struck and need for etiology presented itself, an "arrowhead" (*mshale*) was sent to a diviner-healer (*ng'anga*) to initiate consultation. Such an *mshale* was in fact a coin, but its currency still carried the sense of targeting the source of affliction, and the *ng'anga* would place it under his head to instigate spirit-inspired dreams to determine cause and begin redress. That arrowheads were miniature yet powerful instruments of "inbetweenness" was and is significant as well, for as John Mack might suggest, such "smallness was a visualization... as close to the efficacious world of the unseen as could be attained."[19]

Somewhat similarly, among Luba the term *miketo* for iron points is based on a verb meaning "to cut short or settle." Exchange of arrowheads figured importantly in initiating marriage, when a fiancé's father would send one to his counterpart to initiate bridewealth exchange.[20] As with spears and other iron implements, Luba understand that the "wisdom" of iron arrowheads includes reference to origin stories, for the great hunter and culture hero Mbidi Kiluwe left an especially fine, distinctive point he himself had forged as a message to his newborn son Kalala Ilunga, that with patience and cunning Kalala could initiate the sophisticated political order now known as Luba (Womersley 1975, 40).

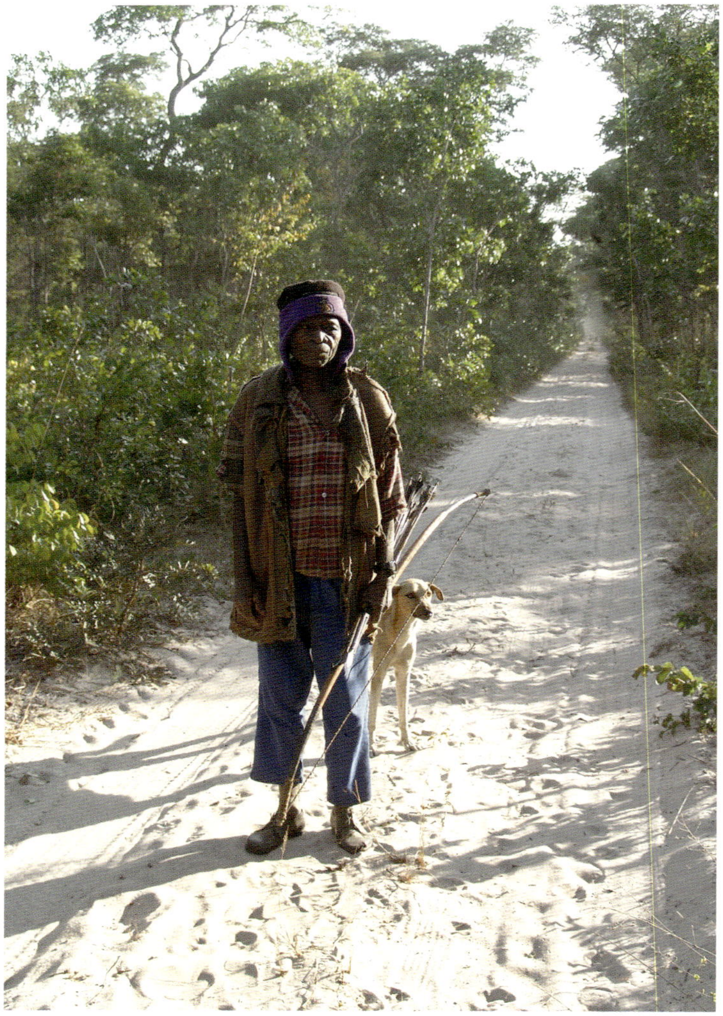

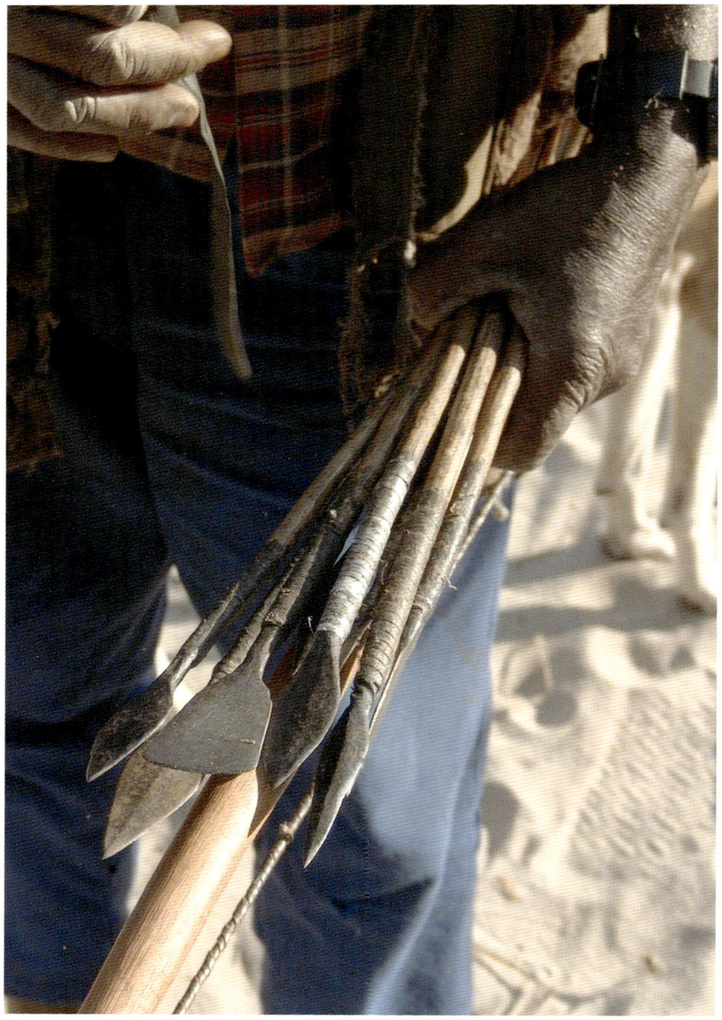

6.22A–B Hunters in many parts of rural sub-Saharan Africa employ bows and arrows fitted out with iron points of various sizes and shapes. This Luchazi man, on a very rural road west of Kayombo in northwestern Zambia, holds carefully fletched arrows meant to bring down small game. A flat-tipped arrowhead is used for impact, perhaps for birds or very small animals that the hunter is not certain of striking effectively with a first shot. Besides being his loyal companion, the man's dog assists in tasks such as flushing birds and animals from brush. Although this one does not, the dogs of many central African hunters wear forged iron bells around their necks to permit their human counterparts to know their whereabouts in dense underbrush. PHOTOGRAPHS BY MANUEL JORDÁN, ZAMBIA, 2004.

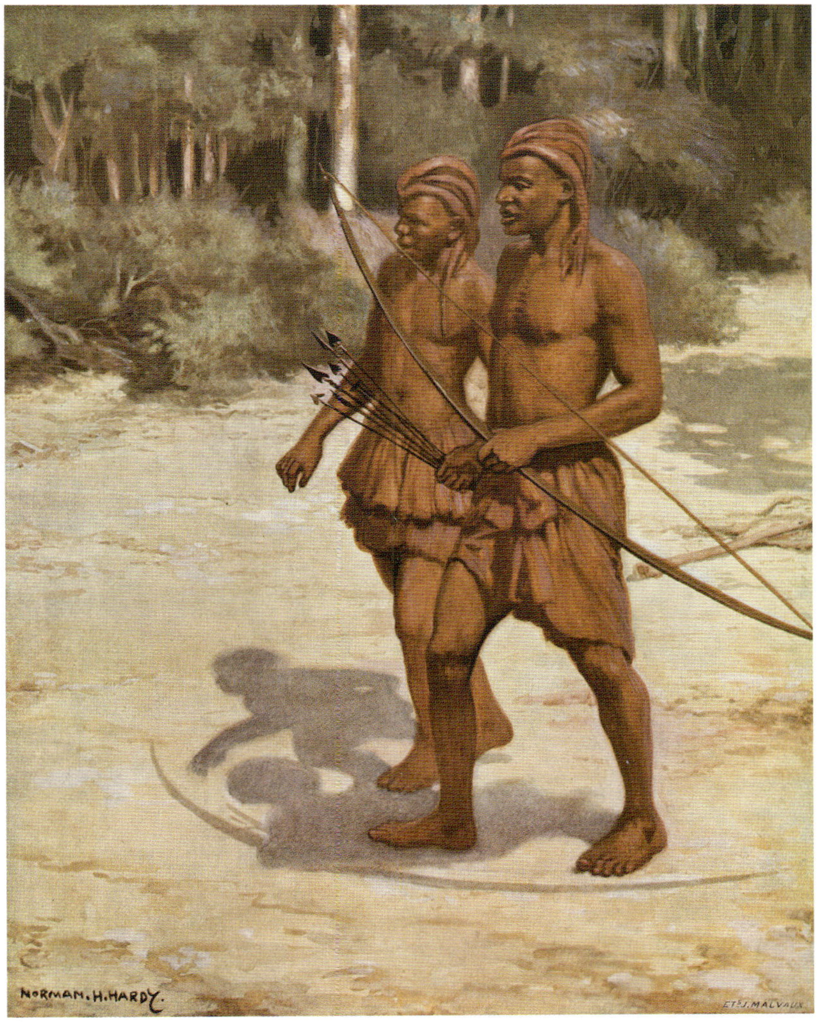

6.23 Mbala of present-day west-central Democratic Republic of the Congo—their name meaning "the red people" because of the clay with which they adorned their bodies and elaborate coiffures—migrated to remote forested lands due to persecution by brigands (Torday and Joyce 1905). The two men portrayed here may be "warriors," but defensively so, as they used the same iron-pointed arrows to hunt in what were often important communitywide activities led by Mbala chiefs (Ntole 1996, 137).

NORMAN H. HARDY, GUERRIERS BAMBALA (*DU SUD*), CA. 1911, WATERCOLOR. PUBLISHED IN EMIL TORDAY AND T. A. JOYCE, *NOTES ETHNOGRAPHIQUES SUR DES POPULATIONS HABITANT LES BASSINS DU KASAÏ ET DU KWANGO ORIENTAL, DOCUMENTS ETHNOGRAPHIQUES CONCERNANT LES POPULATIONS DU CONGO BELGE*, VOL. 2 (BRUSSELS: PUBLIÉES PAR LE MINISTÈRE DES COLONIES, 1922).

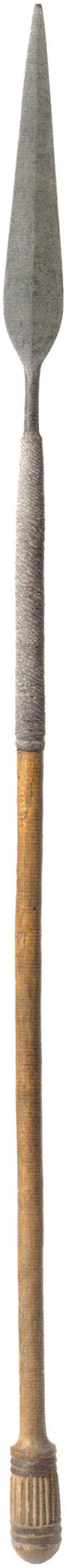

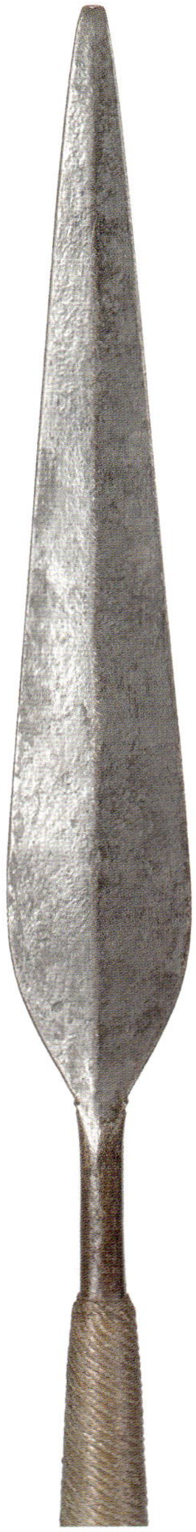

6.24A–B
Artist unknown (Zulu peoples, South Africa)
Spear (*iklwa*), early 20th century
Iron, wood
H: 135 cm
FOWLER MUSEUM AT UCLA, X98.1.2231; MUSEUM ACQUISITION

No African strategist of close combat is more legendary than the early 19th-century ruler Sigidi kaSenzangakhona (or Shaka Zulu). He organized his men into lethally efficient forces armed with *iklwa*, short-hafted stabbing spears with slender leaf-shaped iron blades. A man maneuvered his oxhide shield to push aside his enemy's weapon, followed by a deadly thrust of his *iklwa*. So effective was the tactic that Shaka prohibited his warriors from throwing their weapons, and his men conquered vast territories.

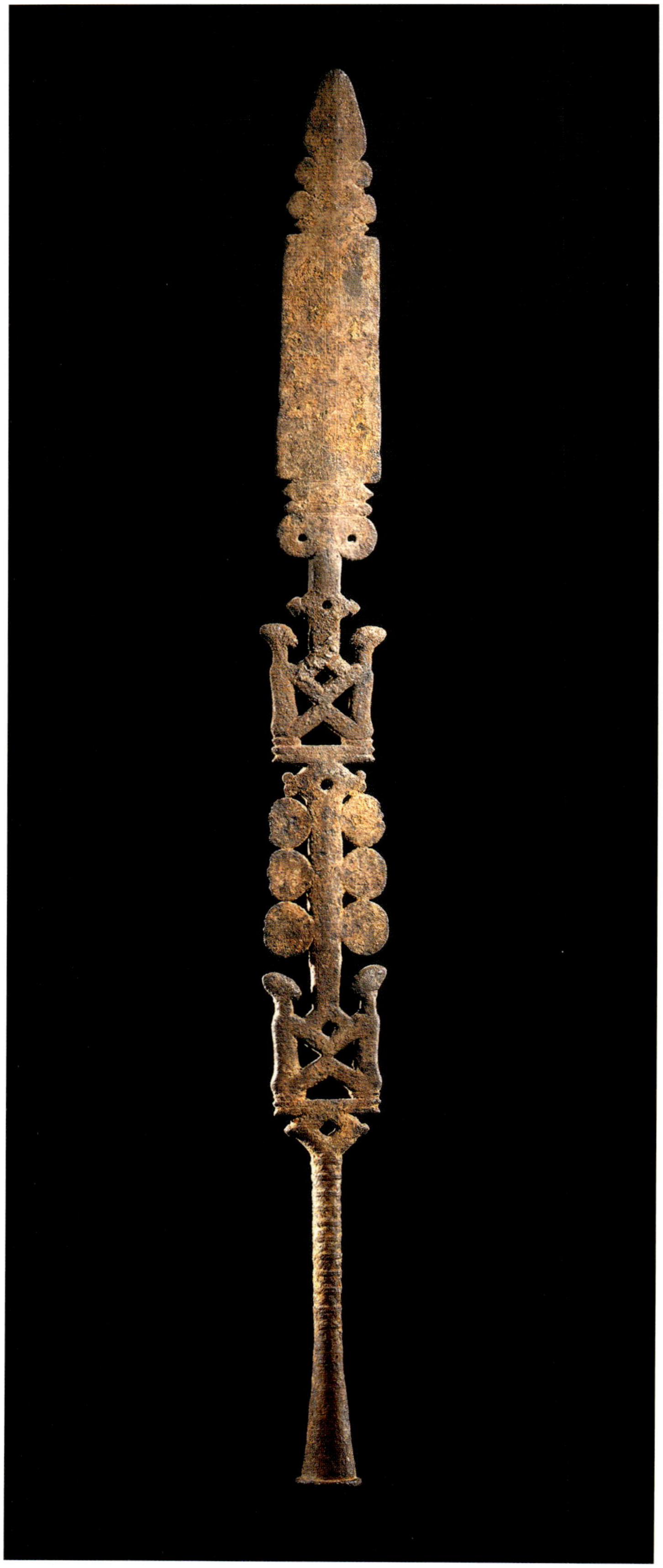

Spear blades of astounding variety were also forged in many precolonial African societies, enabling some forms of hunting but more widely used in battle (figs. 6.24–6.26). Hausa and Kanem-Bornu of present-day Nigeria, Chad, and Cameroon mounted predatory cavalries armed with spears (Reyna 1990). As the military historian John Ringquist notes of Kongo kings, "in times of war the ability to supply large numbers of warriors with [iron] weapons of war was a true and tangible symbol of a ruler's power" (2008, 12). Armor of thickly padded cloth or iron mail made Sahelian warriors largely invulnerable to archers as they attacked less organized communities lacking horses. Mail of riveted links was produced in the Sudan into the twentieth century and "provided an effective defense against slashing blows by edged weapons and penetration by thrusting and piercing weapons."[21]

Among African strategists of close combat, none is more legendary than the early nineteenth-century warlord Sigidi kaSenzangakhona (or Shaka Zulu), who organized his men into lethally efficient forces armed with short-hafted stabbing spears called *iklwa* (see fig. 6.24). Held in the right hand, a buckler in the left pushed aside an adversary's shield, permitting an *iklwa*'s deadly thrust. So effective was

6.25
Artist unknown (Numu, Abron/Akan peoples[?], Côte d'Ivoire and Ghana)
Staff finial, 18th century
Iron, copper
H: 75.5 cm
COLLECTION MINA AND SAMIR BORRO

Purchased in Côte d'Ivoire in the 1970s, this intricate spear blade may be from the Bondoukou region at the border with west-central Ghana. As early as the 16th century, Numu peoples, renowned for superb ironworking, settled around Bondoukou as a vibrant trading center. Paired figures are a common motif in Akan arts, suggesting that the work may have been produced for an Akan client by a Numu smith.

the tactic that Shaka prohibited his warriors from throwing their weapons, and the men conquered vast territory.[22] Zulu fighting technology, developed in what is now South Africa, was adopted by aggressive bands pillaging as far northward as the shores of Lake Tanganyika, where in the mid-1970s a few Tabwa remembered that the marauders had been called Batuta Kyalo—"those who break the country" (Roberts 2013a, 67).

As attested in the material record, spears were of such social significance that their forms were adapted to signify political power and prestige as well as potential through bridewealth exchange (see chapter 5, this volume). Among Samburu pastoralists of northwestern Kenya, men mediate social relations through "warrior theater" that is "sharply focused on body arts and weapons as highly visible and volatile metaphors for virility and bravery." Samburu spears have long been most significant "extensions into physical, symbolic, and social space," for they possess "rhetorical value and convey panache." Even as colonial and postcolonial regimes have altered forms and purposes of Samburu spears, they remain potent "symbols of cultural memory" bespeaking, once more, honor, pride, and valor (Kasfir 2007, 27, 103, 105, slightly paraphrased, 111).

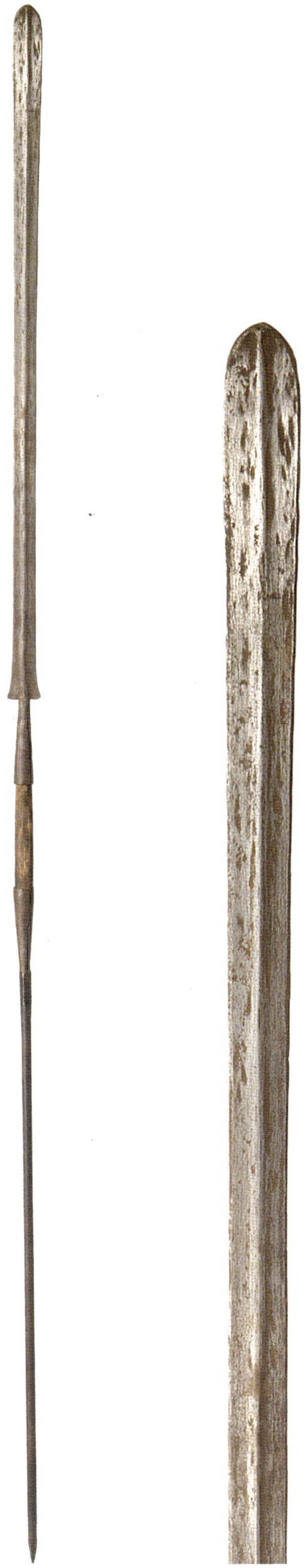

6.26A–B
Artist unknown (Maasai peoples, Kenya and Tanzania)
Warrior spear, early to mid-20th century
Iron, wood
H: 204 cm
FOWLER MUSEUM AT UCLA, 398.17; MUSEUM PURCHASE

Young men undergoing age-grade initiations among Maasai and related peoples engage in what Sidney Kasfir calls "warrior theater... sharply focused on body arts and weapons as highly visible and volatile metaphors for virility and bravery" (2007, 27). Spears are carried in defense of cattle herds threatened by rustlers, and they can be hurled or quickly planted blade-point forward to impale a pouncing lion.

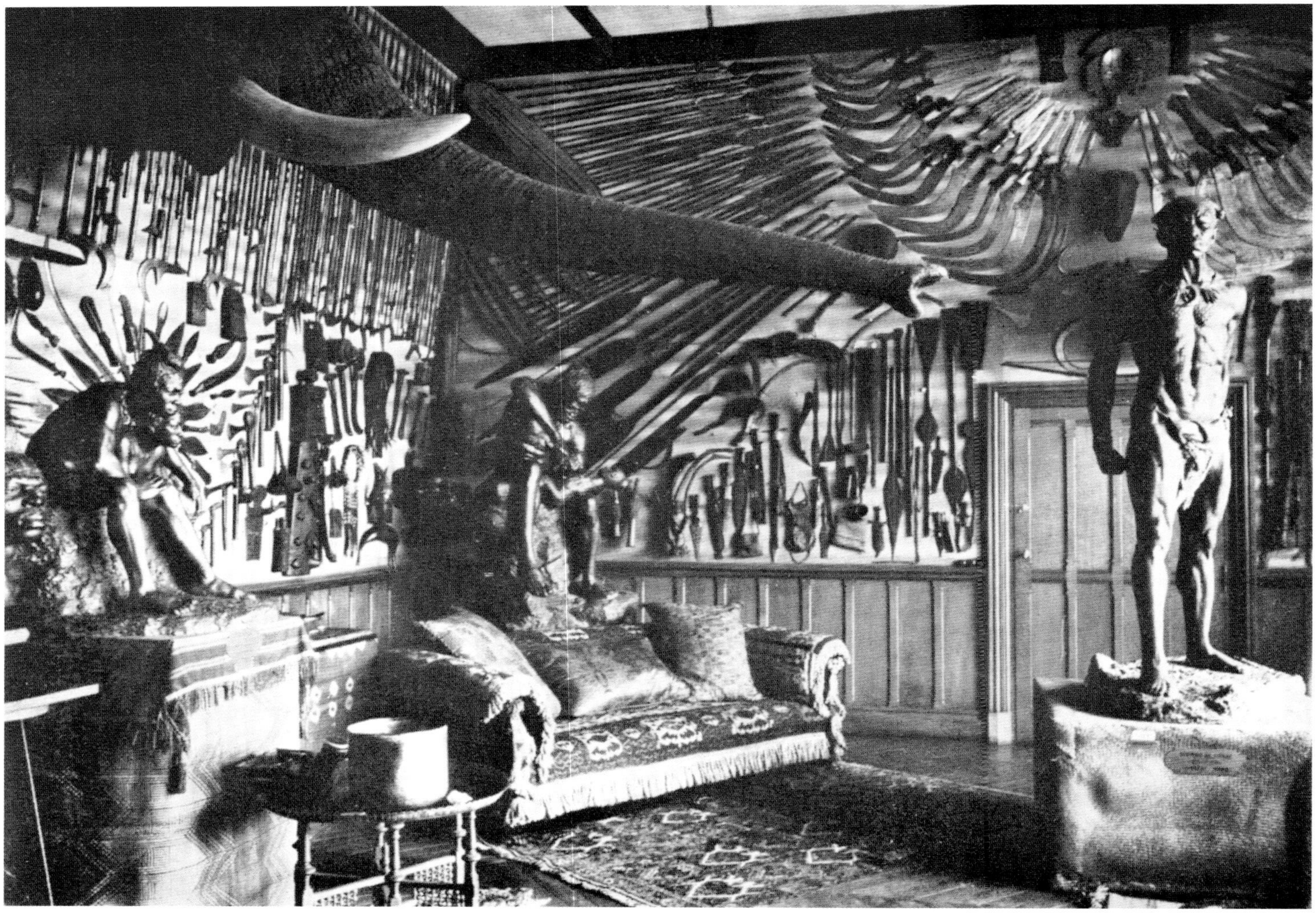

6.27 In the 1880s, Herbert Ward accompanied Henry Morton Stanley on a bellicose expedition in what would become the Congo Free State. He turned to sculpture when he returned to his native England, and his massive depictions of Congolese "natives"—with the colonial connotations of that word explicit—were long on display in the U.S. at the Smithsonian Institution and in Belgium at the Royal Museum for Central Africa. This photograph shows the "African" phantasmagoria that Ward created in his studio, with African blades mounted on the walls in precise geometrical displays suggesting imposition of colonial order on unruly people deemed in need of European "civilization."
PHOTOGRAPH BY HARRY C. ELLIS, EARLY 20TH CENTURY. PUBLISHED IN HERBERT WARD, *A VOICE FROM THE CONGO: COMPRISING STORIES, ANECDOTES, AND DESCRIPTIVE NOTES* (LONDON: WILLIAM HEINMANN, 1910).

POSTSCRIPT ON INTERPRETING AFRICAN USE OF BLADES

As significant as iron blades have been to work, defense, dignity, and senses of self as understood through African systems of thought, these same implements have been interpreted in derogatory ways by those who would invent Africa for less charitable purposes (see Mudimbe 1988). Rather than signifiers of nobility and self-reliance, spears, in particular, have been deemed gauges of primitive violence through distortions of Social Darwinism. As Hal Foster suggests of such circumstances, from a Eurocentric perspective, the spear-wielding "fantasmatic figure of the savage elicits an oscillation between esteem and disgust, with murderous envy somewhere in between" (2004, 4). In violently reductive, sundering processes, Africans have been subjected to what Susan Stewart calls "synecdochal diremption," whereby some small aspect of person or circumstance—a spear, say—is wrested from everyday life and made to stand for vast social complexities.[23] In the early days of triumphalist colonization, for example, spears and other iron weapons were displayed in World's Fairs, museums, and other vaunted venues. Torn from original contexts, blades were arranged in fanciful harmonies imposing order on the presumed chaos of "Darkest Africa," informing the "need" to colonize and so "civilize" (fig. 6.27).[24]

The case studies and illustrations of the present book challenge such reprehensible paradigms by demonstrating the richly symbolic purposes and complex cultural contexts in which these objects have long figured prominently. Let us appreciate the superbly forged iron blades and other works discussed in these pages as reflecting honor through hard work, wisdom, valor, and artistry.

INTERLEAF C

SHONA BLADES FOR THE ANCESTORS

William J. Dewey

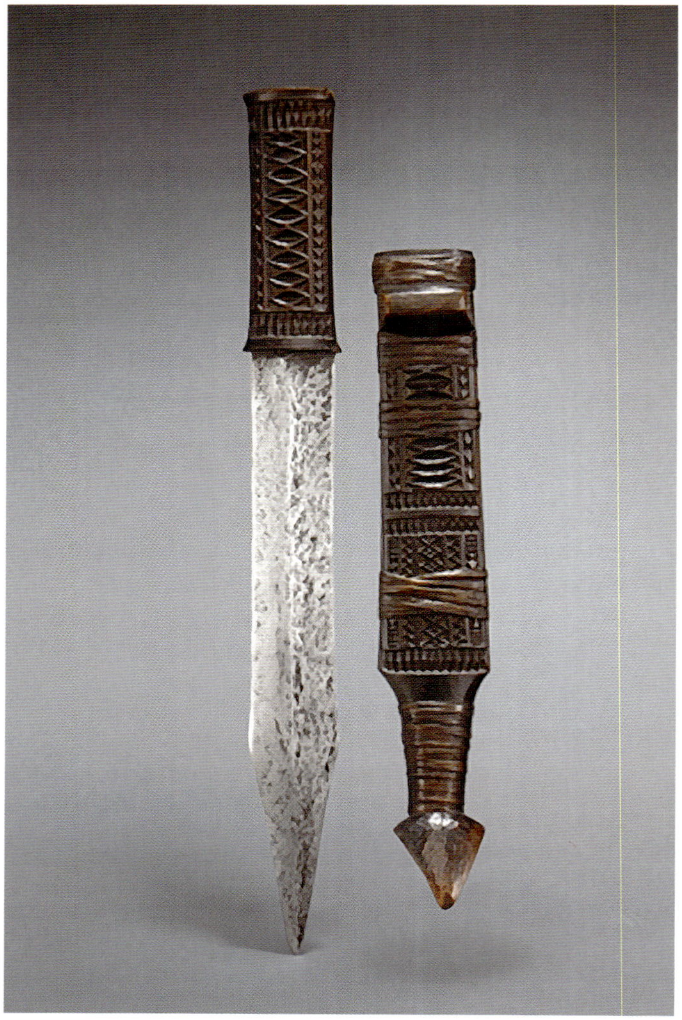

On the day when the spirit to forge posesses my husband,
The bellows do everything but speak.
Just look at what he has forged!

FROM A SHONA PRAISE-POEM[1]

Shona-speaking peoples of Zimbabwe and Mozambique have forged iron implements for centuries. Here we shall consider knives called *mapakatwa* (s. *bakatwa*;[2] fig. C.1) and tanged, crescent-shaped axes known as *makano* (s. *gano*; fig. C.2); both have histories of some six hundred years of production and use.

Mapakatwa range in size from a few inches long to full-scale swords, and their double-edged blades and the decoration of their handles and sheaths distinguish them from more ordinary single-edged knives. As can be seen in modern and archaeological examples, the forged blades of *mapakatwa* possess stepped midribs (S-shaped in cross-section) produced by alternate hammering of the blade on either side. *Mapakatwa* are held in wooden sheaths bound together with elaborately woven wirework or sinew and are often ornamented with chip-carving. Rare examples with ivory sheathes and handles also exist (Dewey 2017).

Shona smiths forge several types of axe. Ordinary ones for chopping wood have triangular blades inserted through their wooden hafting. *Makano* differ from these practical tools in their crescent blades and rectangular tangs, and their purposes have long extended to rituals associated with ancestral and other spiritual forces. Another type of ceremonial axe known as *tsotsa* (fig. C.3) has a bent-back shaft, and is most popular among northern Shona and their Ndebele neighbors.

C.1
Artist unknown (Shona peoples, Zimbabwe and Mozambique)
Knife (*bakatwa*) and sheath, 20th century
Iron, wood, fiber
H: 52 cm
COLLECTION OF SHAUN MIRAMS
PHOTOGRAPH BY JAMES WORRELL. © COURTESY SHAUN MIRAMS

Shona *bakatwa* invariably have a two-piece sheath bound together with either wirework or fiber, and a carved loop on one side for belt attachment. Chip-carved geometric ornamentation often consists of diamond and X patterns. Such decorative motifs are common on many Shona items, including snuff containers, headrests, and axes.

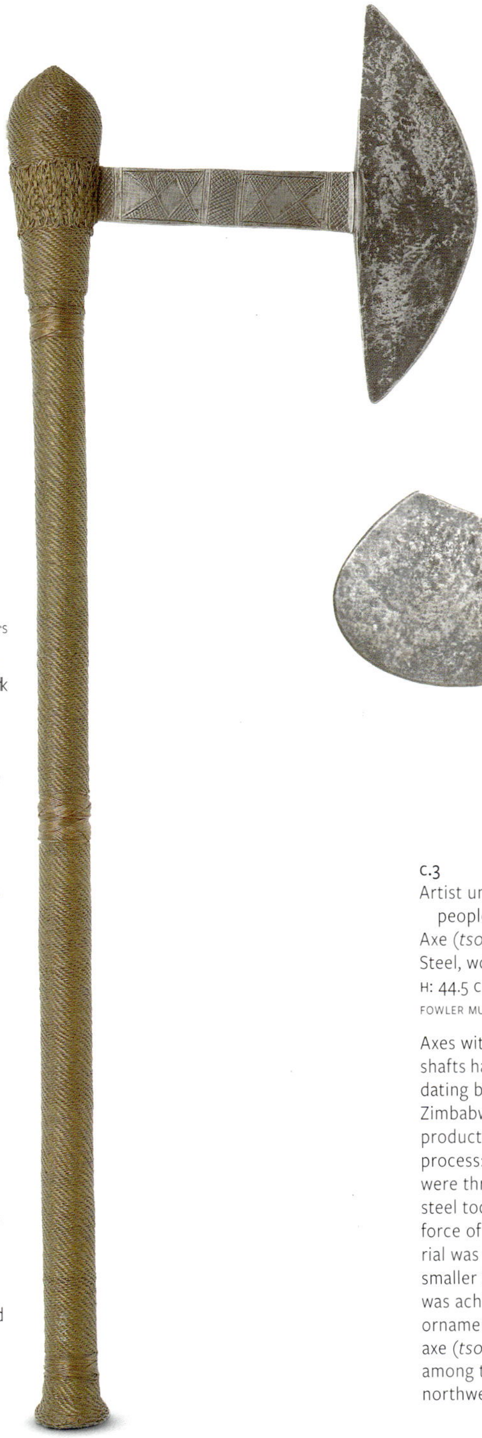

C.2
Artist unknown (Shona peoples, Zimbabwe and Mozambique)
Axe (*gano*), 19th century
Wood, iron, copper wire
L: 60 cm
COLLECTION WERELDMUSEUM ROTTERDAM
PROVENANCE: JANSSEN SCHOLLMAN, 1892
(NOT IN EXHIBITION)
PHOTOGRAPH BY THEO VAN PINXTEREN, 2017.
© WERELDMUSEUM ROTTERDAM

This *gano* axe displays typical wirework ornamentation on its haft. One plaiting technique (over three wires, then under three wires) results in series of parallel spiraling lines. Applied beside the tang are superimposed circles of wire over simple wire wrapping, a technique that dates back as far as the 15th century on Kame spear shafts. The opposing triangle-shaped incisions on the tang of the blade are also characteristic of *gano*-type axes.

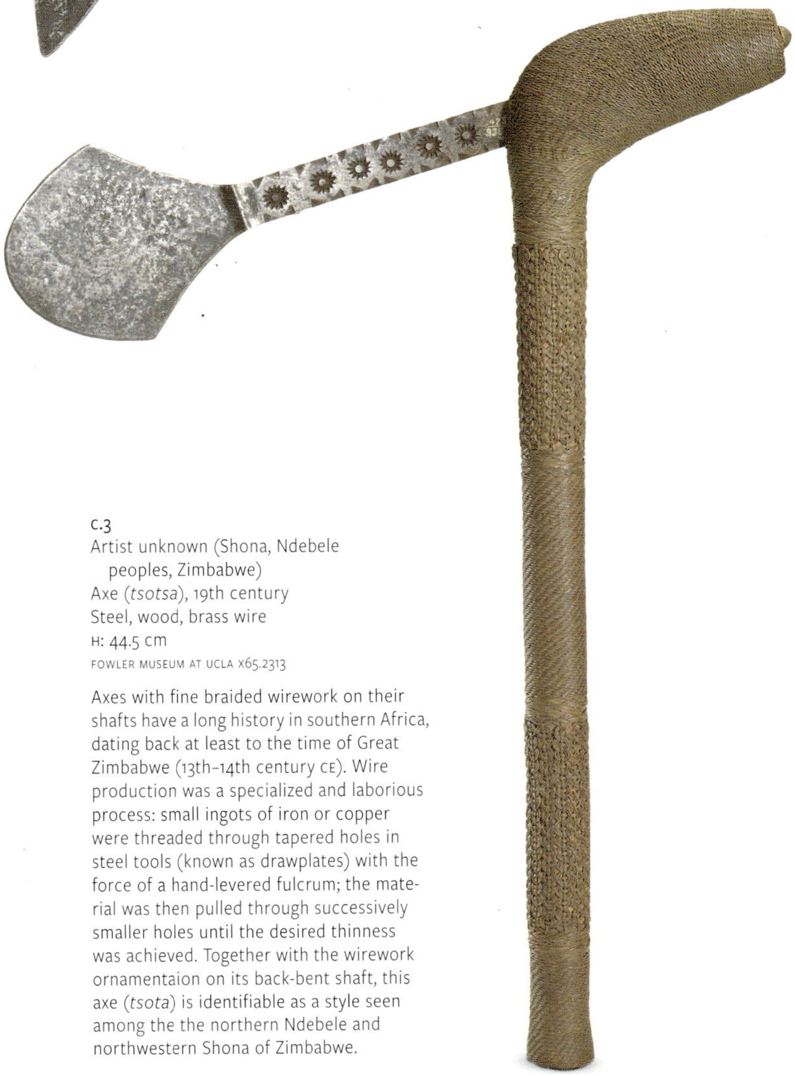

C.3
Artist unknown (Shona, Ndebele peoples, Zimbabwe)
Axe (*tsotsa*), 19th century
Steel, wood, brass wire
H: 44.5 cm
FOWLER MUSEUM AT UCLA X65.2313

Axes with fine braided wirework on their shafts have a long history in southern Africa, dating back at least to the time of Great Zimbabwe (13th–14th century CE). Wire production was a specialized and laborious process: small ingots of iron or copper were threaded through tapered holes in steel tools (known as drawplates) with the force of a hand-levered fulcrum; the material was then pulled through successively smaller holes until the desired thinness was achieved. Together with the wirework ornamentaion on its back-bent shaft, this axe (*tsota*) is identifiable as a style seen among the the northern Ndebele and northwestern Shona of Zimbabwe.

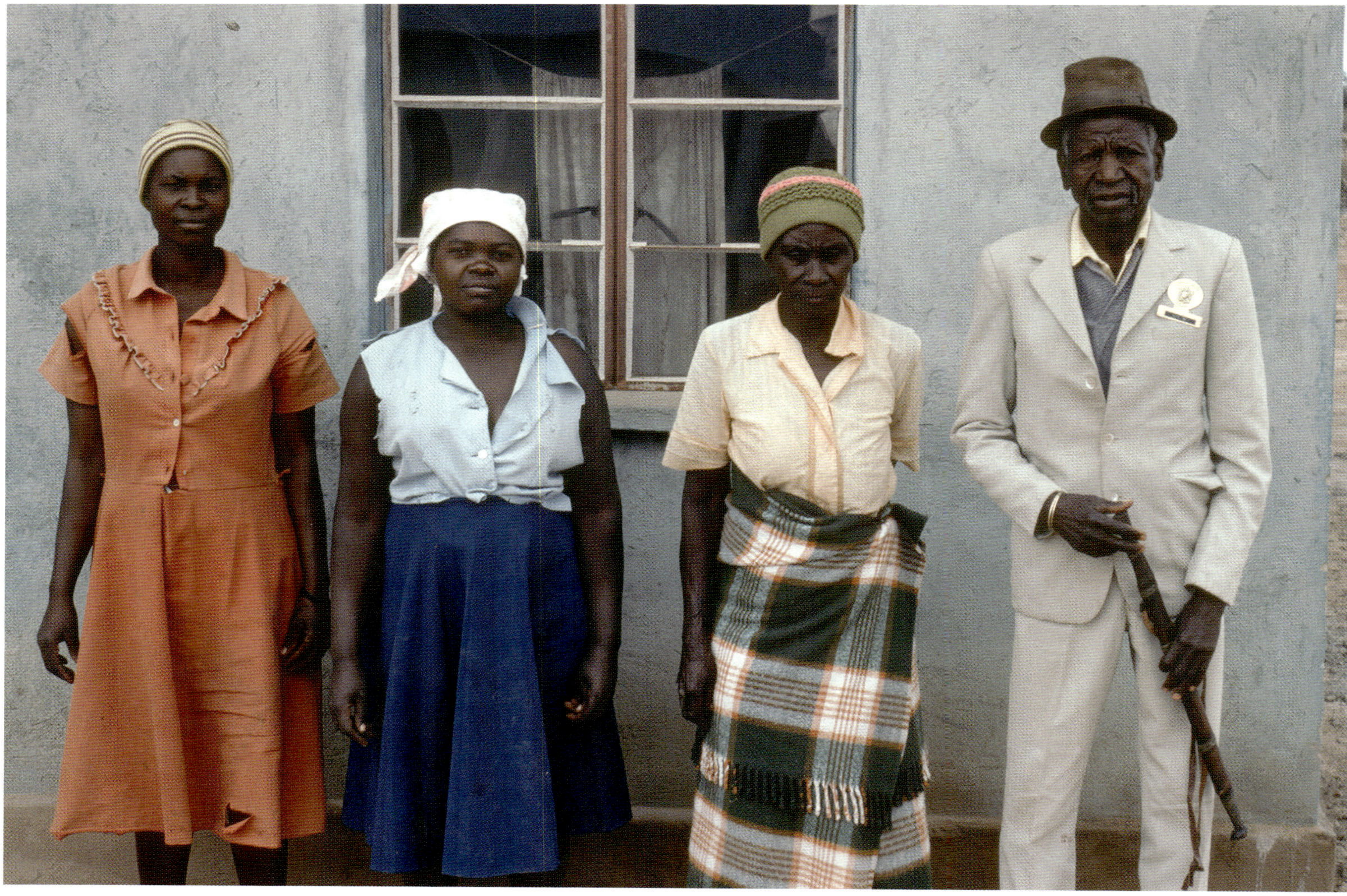

C.4 Holding a ceremonial *bakatwa* knife that came down to him from his uncle, Chief Chibi stands with wife and daughters. In the past, all Shona men carried elegant knives, but now they are kept mostly as heirlooms passed down to male descendants and used ritually to address the specific ancestors who once owned them.
PHOTOGRAPH BY WILLIAM J. DEWEY, CHIBI COMMUNAL AREA, ZIMBABWE, 1984.

Knife blades similar in form and forging to *mapakatwa* have been discovered through archaeological study of Zimbabwean Iron Age sites from at least the thirteenth century.[3] *Gano*-type axes have also been produced in Zimbabwe for many years, especially in thirteenth- and fourteenth-century Great Zimbabwe.[4] Such blades have been found at the later sites of Khami and Danangombe (formerly Dhlodhlo) as well, dating from 1450 to 1700 (see Childs and Dewey 1996; Pwiti 1997).

Sixteenth-century Portuguese visitors to the area wrote about the use of *makano* for display and warfare (Dewey 1991,131) as well as pursuit of big game. Hunters would "come up to each [elephant] in turn and strike with their axes at both feet if possible, or if not, at one, and at the least wound, the weight of their bodies breaks their feet and they fall."[5] Additional historical usage of knives and axes includes practical, political, prestige, and religious emphases. Colonial and Christian missionary activities as well as importation of Western-made knives and axes in the early twentieth century decreased the political and religious importance of these items. However, the upsurge of Zimbabwean nationalist politics in the 1960s, the war of liberation in the 1970s, and Independence in 1980 saw the revival of traditional medicine and religion, and manufacture and symbolic use of knives and axes has also become more common (Dewey 1991, 1994).

Mapakatwa and *makano*, if not buried with a person (Magava 1973), are passed down through generations and deployed when ancestors are addressed. Such personal items are distributed to descendants at inheritance rituals after their deaths (fig. C.4). In the 1980s, a blacksmith-carver named Albert Choto explained to me that "they are now kept in houses to reflect the presence of spirits or the dead ancestor who used to use them."[6]

Many blacksmiths say that they gained their abilities from their ancestors (Dewey 1986) and that the most common ritual implements they possess are *makano* and *mapakatwa*, as required by the spirits. A blacksmith-carver named Munyikwa Titiya explained that when he was young he became very ill, whereupon his deceased grandfather came to him in a dream to instruct him to take up the profession. His family held a ritual for him, he acquired the ritual weapons, and he has worn the red, white, and black cloth (*retso*) associated with the ancestors ever since (fig. C.5). He attributed his being cured of his troubling illness to such significant processes.[7]

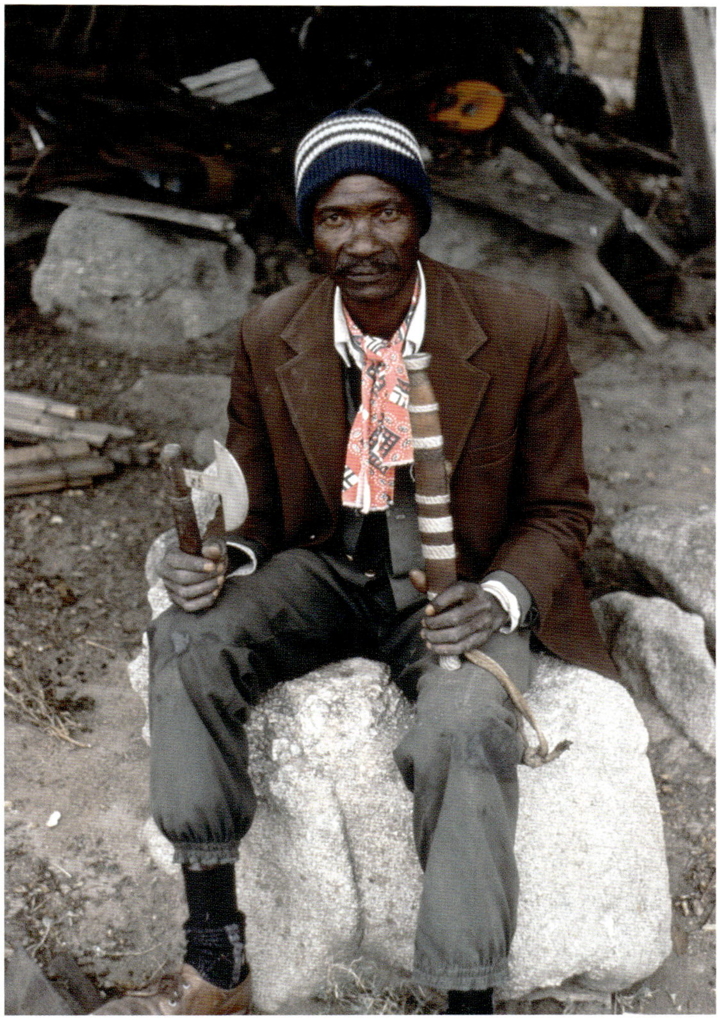

C.5 Blacksmith Munyikwa Titiya with ceremonial knife and axe in his hands and *retso* cloth—also associated with the ancestors—around his neck.

PHOTOGRAPH BY WILLIAM J. DEWEY, MAKONI RESETTLEMENT AREA, ZIMBABWE, 1984.

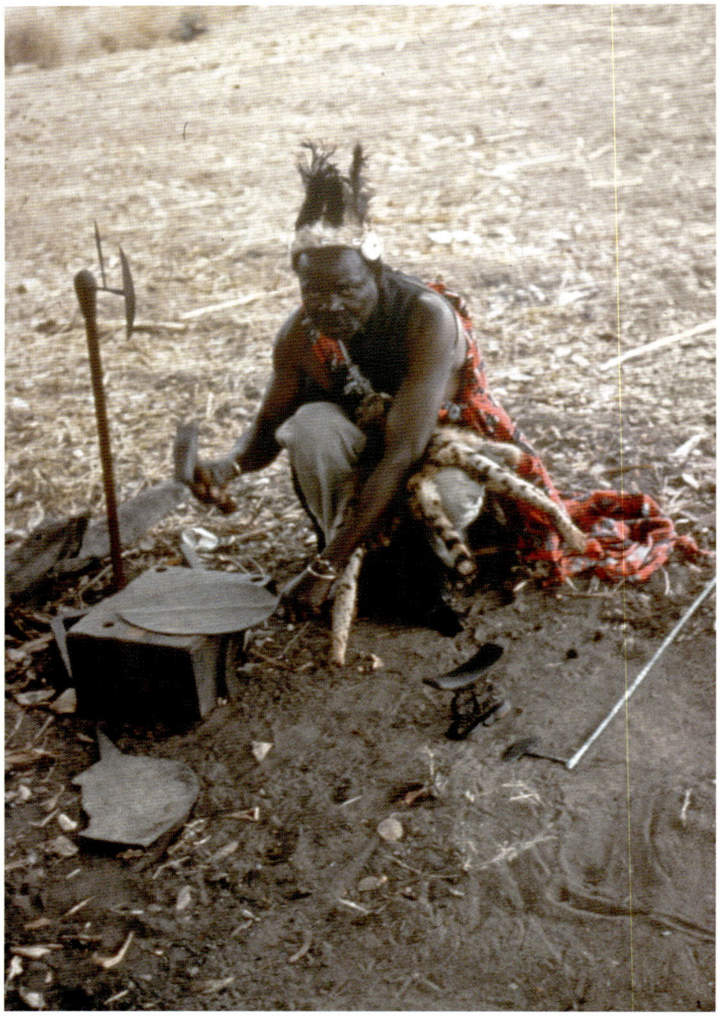

c.6 Spirit medium Sekuru Bwanya with ritual paraphernalia, axes, hoe blades, and headrest found in the cave-burial of the possessing spirit/former owner.
PHOTOGRAPH BY WILLIAM J. DEWEY, SHAMVA, 1984.

N'anga[8] traditional healers and diviners and *svikiro* spirit mediums invariably possess sacred axes as symbols of their trades. Spirit-possession rituals are the most important contexts in which axes appear. Those who become possessed are often supplied with *makano* with which to dance. Spirit medium Sekuru Bwanya explained that the ancestral spirit that possessed him had actually used, when it was alive, the headrest, axes, and hoes Sekuru Bwanya displayed during seances. When the possessing spirit first manifested itself, the spirit told Sekuru Bwanya to go and collect its things in the cave where it had been buried, thus authenticating Sekuru Bwanya's mediumship (fig. c.6).

Associations between *makano* and ancestral wisdom, morality, and valor were made clear during the independence struggles of the 1960s and 70s. In 1962, a veteran from late nineteenth-century battles against European colonizers presented a *gano* heirloom axe to Joshua Nkomo and Robert Mugabe as the new leaders of the liberation struggle (fig. c.7). The choice of the axe as the symbolic vehicle with which to pass on the mantle was not accidental—clearly the participants were aware of the many levels of meaning in operation. The axe as weapon of war was the most obvious, and a strong expression of their aspirations to regain political power.[9] Always, however, the axe is most powerful as a link to the ancestors, implying their moral sanction and active interest in political events.

A blacksmith-carver named Johwani Nyamukapa indirectly explained the potency of *makano* when asked about his own renown. He had once been commissioned by a spirit medium to forge *makano*, and when the man came to collect them, he fell into trance at the mere sight of the axes. The possessing spirit told those accompanying the medium to give Mr. Nyamukapa more money, for over and above the aesthetic quality of the axes he had forged and hafted, the objects possessed their own efficacies that had summoned the spirit. Such are the powers of *makano*![10]

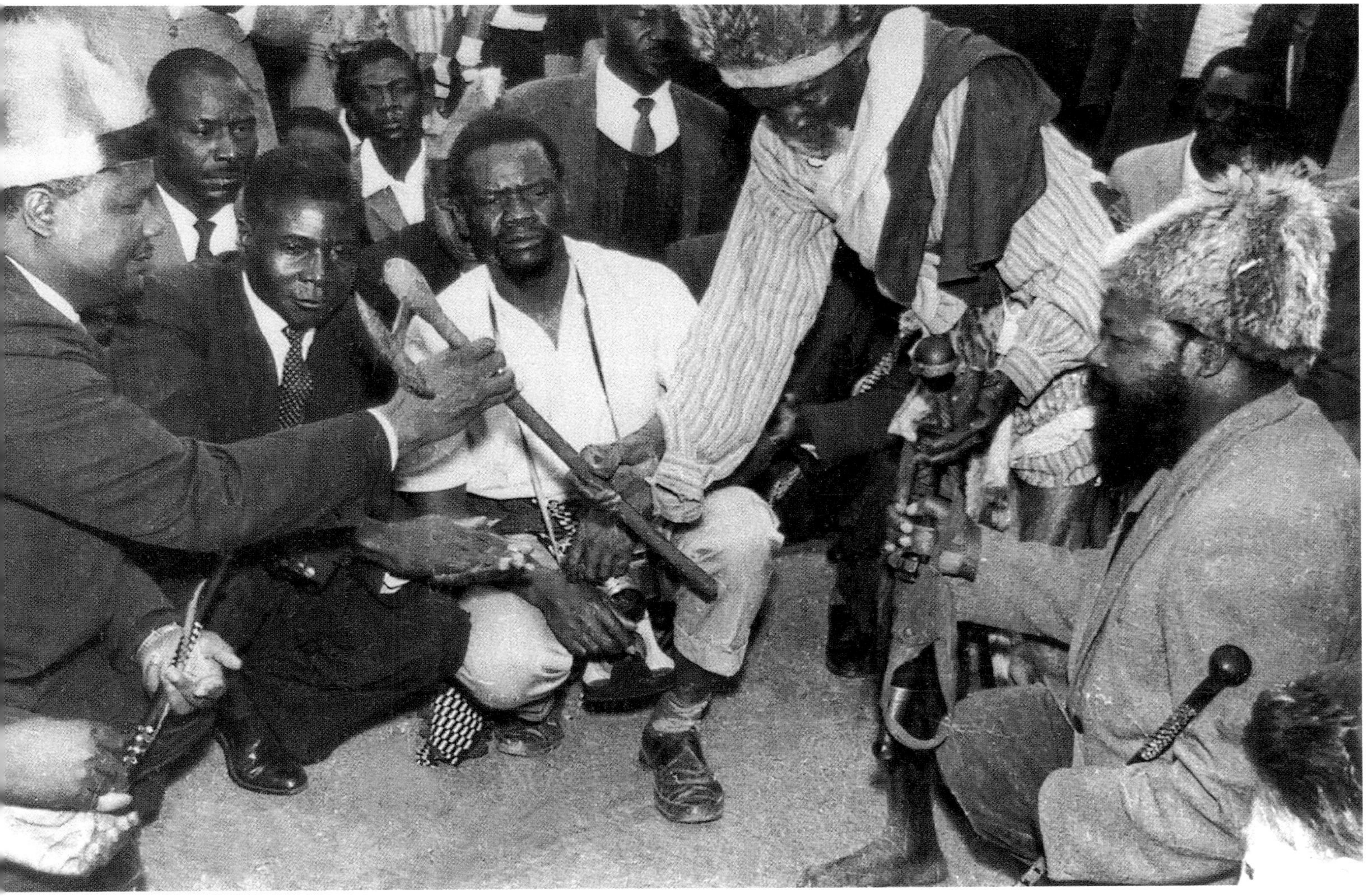

C.7 A veteran of the 1896–97 war in Zimbabwe (then Rhodesia) handing a *gano*-type ceromonial axe to the leaders of the nationalist movement in 1962.
COURTESY ZIMBABWE NATIONAL ARCHIVES, HARARE

CHAPTER 7

BEAUTY THAT BITES: CREATIVE VIRTUOSITY IN AFRICAN THROWING KNIVES

Patrick McNaughton

> The hand- or throwing-iron appeared as the almost indispensable weapon, of which many carried several specimens (up to five) in scabbards of hide, the hair of which was turned to the outside.
>
> GUSTAV NACHTIGAL, *SAHARA AND SUDAN* (1834–85), DESCRIBING THE MILITARY FORCE OF KING ABU SEKKIN OF BAGIRMI[1]

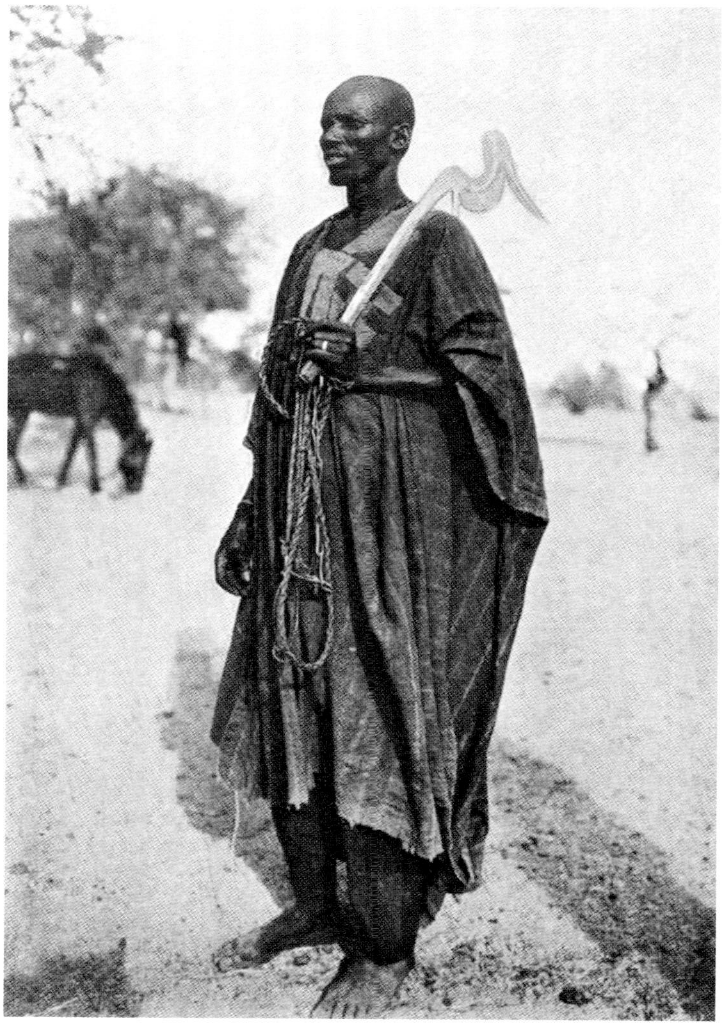

Throwing knives comprise a stunning category of African blacksmithing. They are a potent argument against old assertions that creativity easily divides into art and craft. Huge numbers of these implements are disarmingly elegant creations made by master smiths across a swath of central Africa two-thirds the size of the contiguous United States. Though used less frequently today, until the early decades of the twenty-first century they occurred in profusion over much of central and equatorial Africa and were employed across a range of activities from war to pomp and circumstance, personal adornment to sacred ritual (fig. 7.1).

While we know little about these creations' rich histories, we know more about their encounters with the West. Much of our information about throwing knives comes from foreigners visiting or working in Africa: their writings, illustrations, and the actual objects they acquired. The information is often tainted with Western prejudices and incomplete knowledge, but in aggregate helps us form a partial picture of distribution and use. By the late decades of the nineteenth century, throwing knives were visual magnets for overseas visitors. The British adventurer-artist Herbert Ward wrote that "in the manufacture of these weapons, the Upper Congo tribes

7.1
Man from the area of the Wadai Sultanate carrying a throwing knife on his left shoulder. The Wadai Sultanate (1635–1912; located in what is now Chad and the Central African Republic) was one of the large political and commercial powers of central Africa, situated between the Darfur Sultanate to the east and the vast Kanem-Bornu Empire to the west.

PHOTOGRAPH BY CLICHÉ LACOMBE. PUBLISHED IN MAURICE DELAFOSSE, *ENQUÊTE COLONIALE DANS L'AFRIQUE FRANÇAISE OCCIDENTALE ET ÉQUATORIALE* (PARIS: SOCIÉTÉ D'ÉDITIONS GÉOGRAPHIQUES, MARITIMES ET COLONIALES, 1930).

7.2
Artist unknown (Teda, Daza, Zaghawa peoples, Chad, Niger, and Sudan)
Long-handled throwing knife, late 19th to early 20th century
Iron, plant-fiber cordage
H: 62.5 cm

BARBIER-MUELLER COLLECTION, 1027-155
IMAGE COURTESY BARBIER-MUELLER MUSEUM, GENEVA.
© STUDIO FERRAZZINI-BOUCHET PHOTOGRAPHIES, 2009

Note to the reader on attributions:

Throwing knives in museums and private collections are difficult to attribute accurately. Some attributions were made by travelers, others by scholars examining the literature. Acquisition information can be vague or misleading, and very frequently similar or identical knives have been made, used, or traded among multiple peoples, sometimes over large areas. In addition, some attributions identify a particular language-speaking people, while others identify a cluster of languages or dialects shared by several peoples. Such confusion is common in colonial-era sources due to different understandings of what constitutes "ethnicity" and other aspects of social identity. Distribution maps possess the same difficulties. Attributions for this essay have been compiled from collection data and several scholarly sources, in particular Duvosquel and Deraeve (1992), Felix (1991), Schmidt and Westerdijk (2006), Spring (1993), and Westerdijk (1988). The clearest maps can be found in Duvosquel and Deraeve (1992) and Felix (1991).

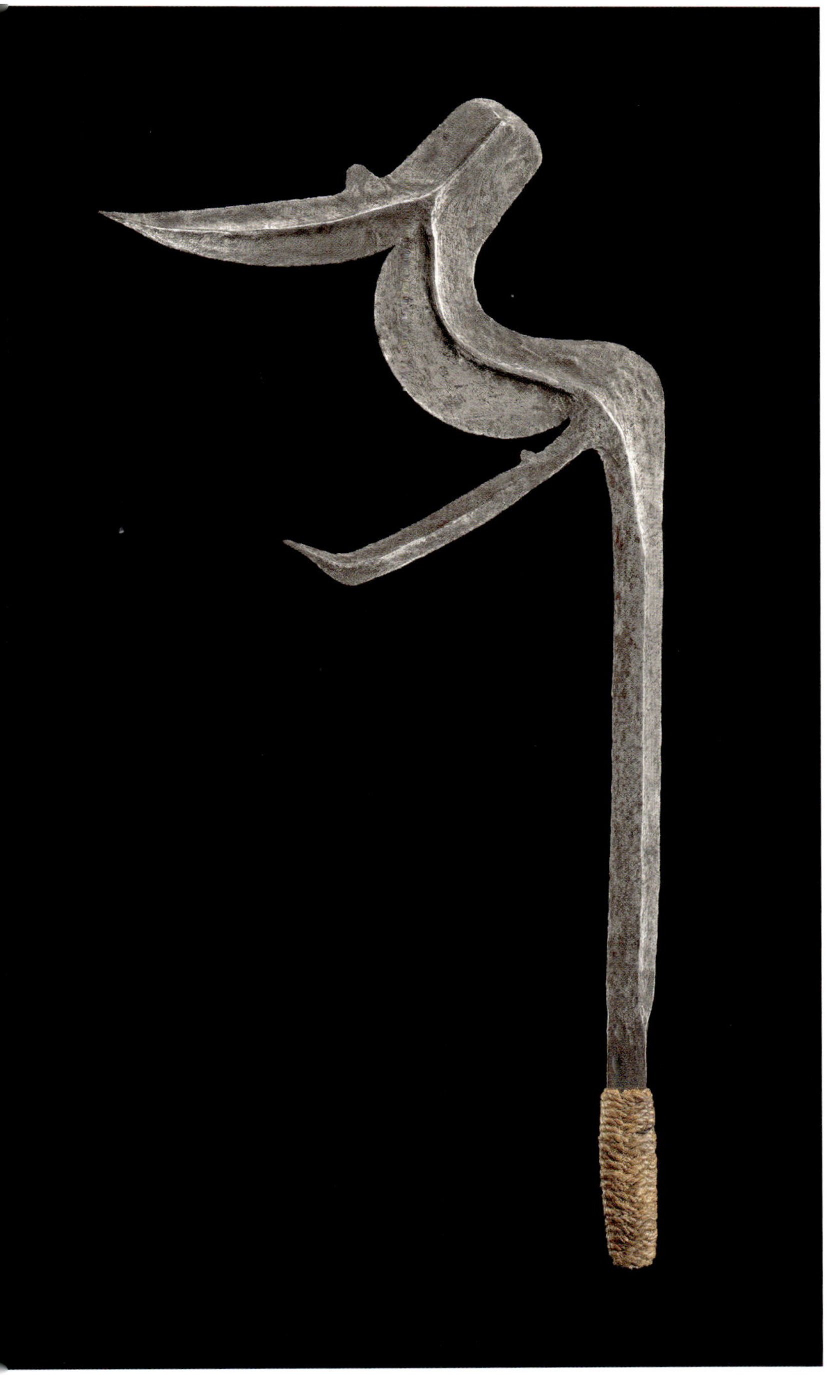

display a remarkable artistic taste and mechanical ingenuity. Most of the fighting knives manufactured possess an infinite grace of form, and display a high sense of decorative art" (1910, 264). In 1869, the German doctor and explorer Gustav Nachtigal (Fisher and Humphrey [1879–89] 1971; 1974; 1980; 1987) undertook a five-year journey through the Sahara to the Sudan, and he mentions throwing knives frequently in his very readable accounts. Around the Tibesti Mountains, men carried them daily as a matter of course. Boys received wooden replicas for practice until they were old enough to handle sharp-bladed iron. As Nachtigal traveled across central Africa from Lake Chad to the Nile, through the empires of Kanem, Bornu, Borku, Ennedi, Bagirmi, Wadai, and Darfur, he found such knives nearly ubiquitous. He witnessed their use in two horrendous battles, and called the weapons throwing irons or hand irons because they were used as missiles and in hand-to-hand combat.

Across the enormous area Nachtigal traversed, domains had been linked through travel, trade, and aggressive state-building for centuries, which explains the diffusion of a weapon type that proved so useful. Northern and southern styles may be differentiated, the former being longer and often forged into F shapes with a pointed spur projecting horizontally above the midpoint (figs. 7.2–7.5; see Spring 1993, 69). Sharp edges may start at the curved top and run to the spur's upper edge and down the back to the grip (see fig. 7.2). The spur conveniently permits the blade to be carried on the left shoulder of a right-handed man so that he can easily seize it. F-shaped throwing knives range in size from sixty centimeters to nearly a meter, and some were intended for slashing rather than being thrown.

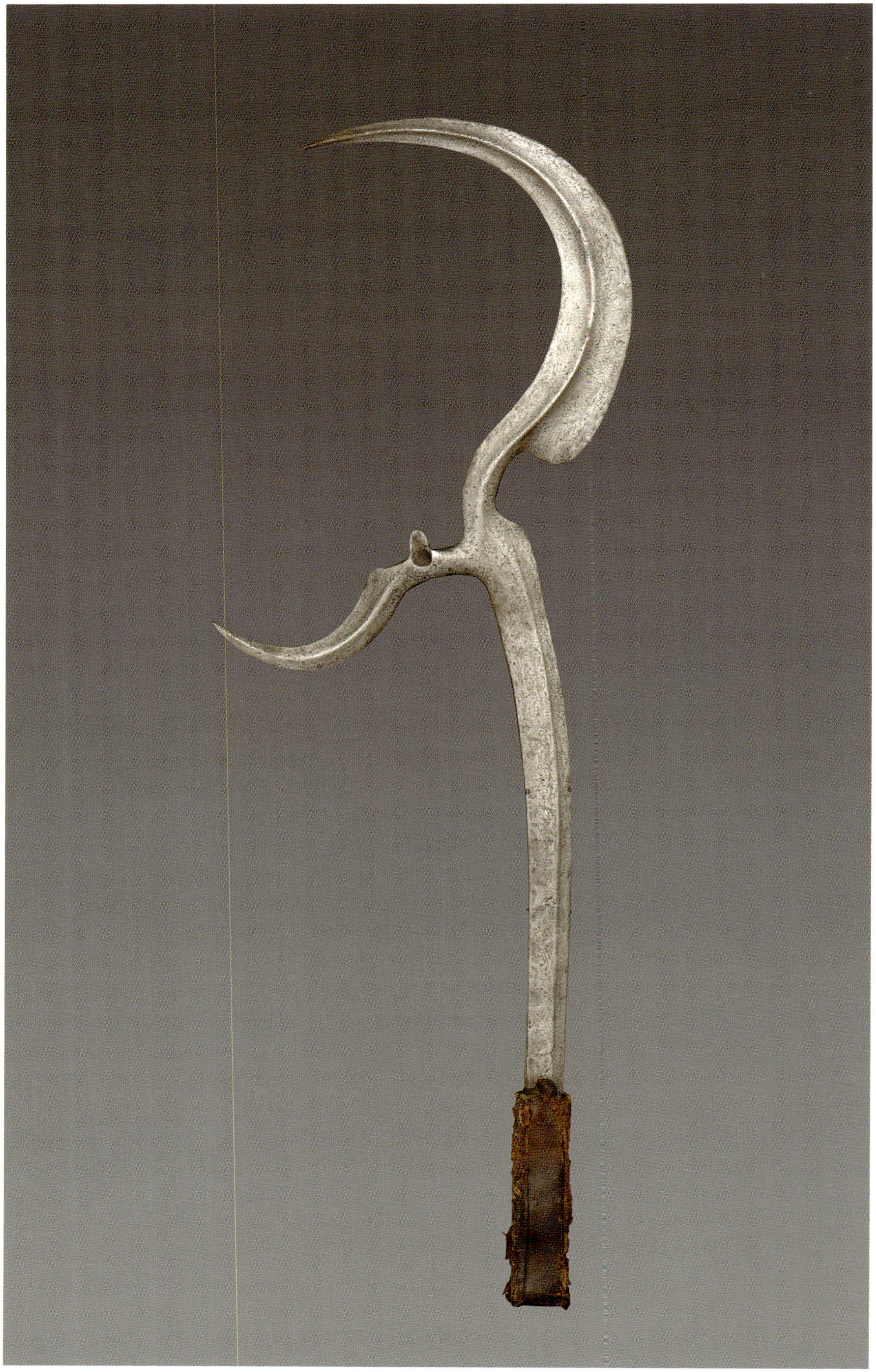

7.3
Artist unknown (Teda, Daza, Zaghawa peoples, Chad, Niger, and Sudan)
Long-handle throwing knife, 19th century
Iron, leather
H: 70 cm
PRIVATE COLLECTION
PHOTOGRAPH BY NATHALIE AND ALAIN SPELTDOORN, 2017.
COURTESY OF PRIVATE COLLECTION

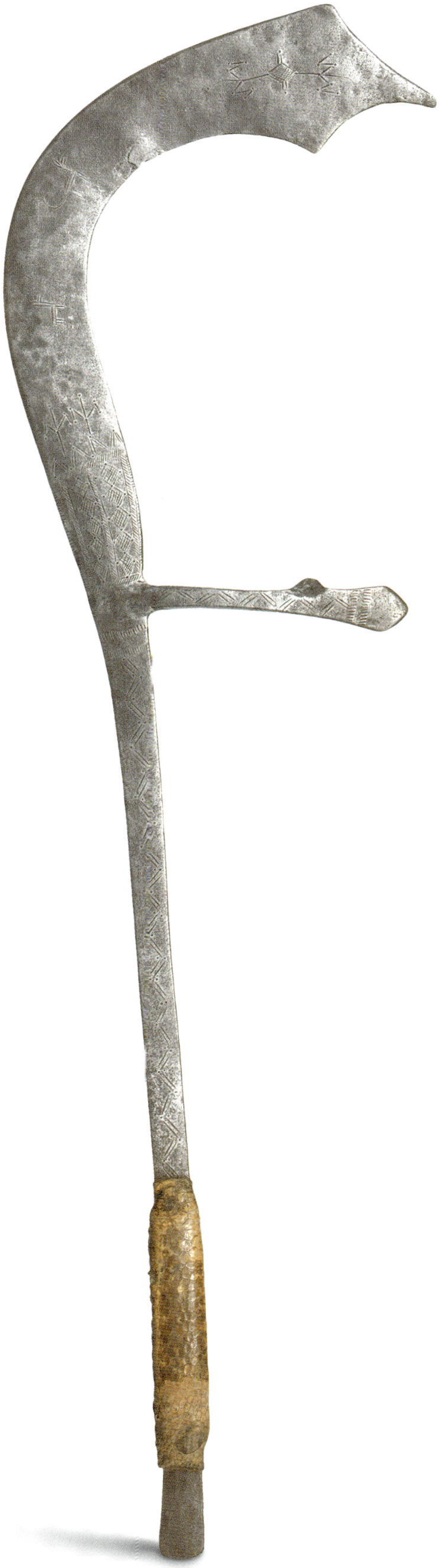

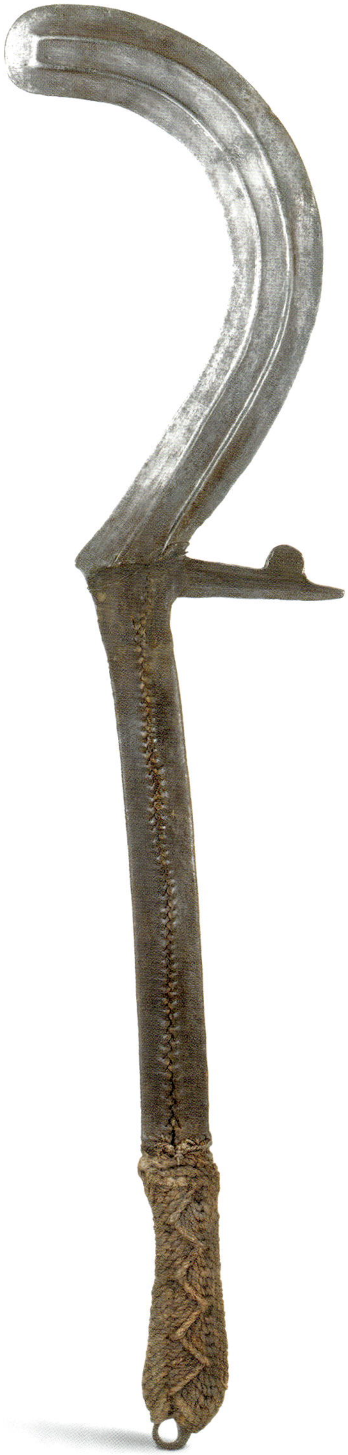

7.4
Artist unknown (Ingessana peoples, Sudan)
Throwing knife, early 20th century
Iron, lizard skin
H: 82.5 cm
FOWLER MUSEUM AT UCLA, X65.3581;
GIFT OF THE WELLCOME TRUST
(NOT IN EXHIBITION)

7.5
Artist unknown (Gbaya-speaking peoples, Cameroon and Central African Republic)
Throwing knife, early 20th century
Iron, hide
H: 66.7 cm
FOWLER MUSEUM AT UCLA, X65.3580;
GIFT OF THE WELLCOME TRUST

7.6
Artist unknown (Southern Gbaya peoples, Central African Republic)
Throwing knife, late 19th to early 20th century
Iron, leather
Dimensions: 40.5 cm
PRIVATE COLLECTION
PHOTOGRAPH BY NATHALIE AND ALAIN SPELTDOORN, 2017.
COURTESY OF PRIVATE COLLECTION

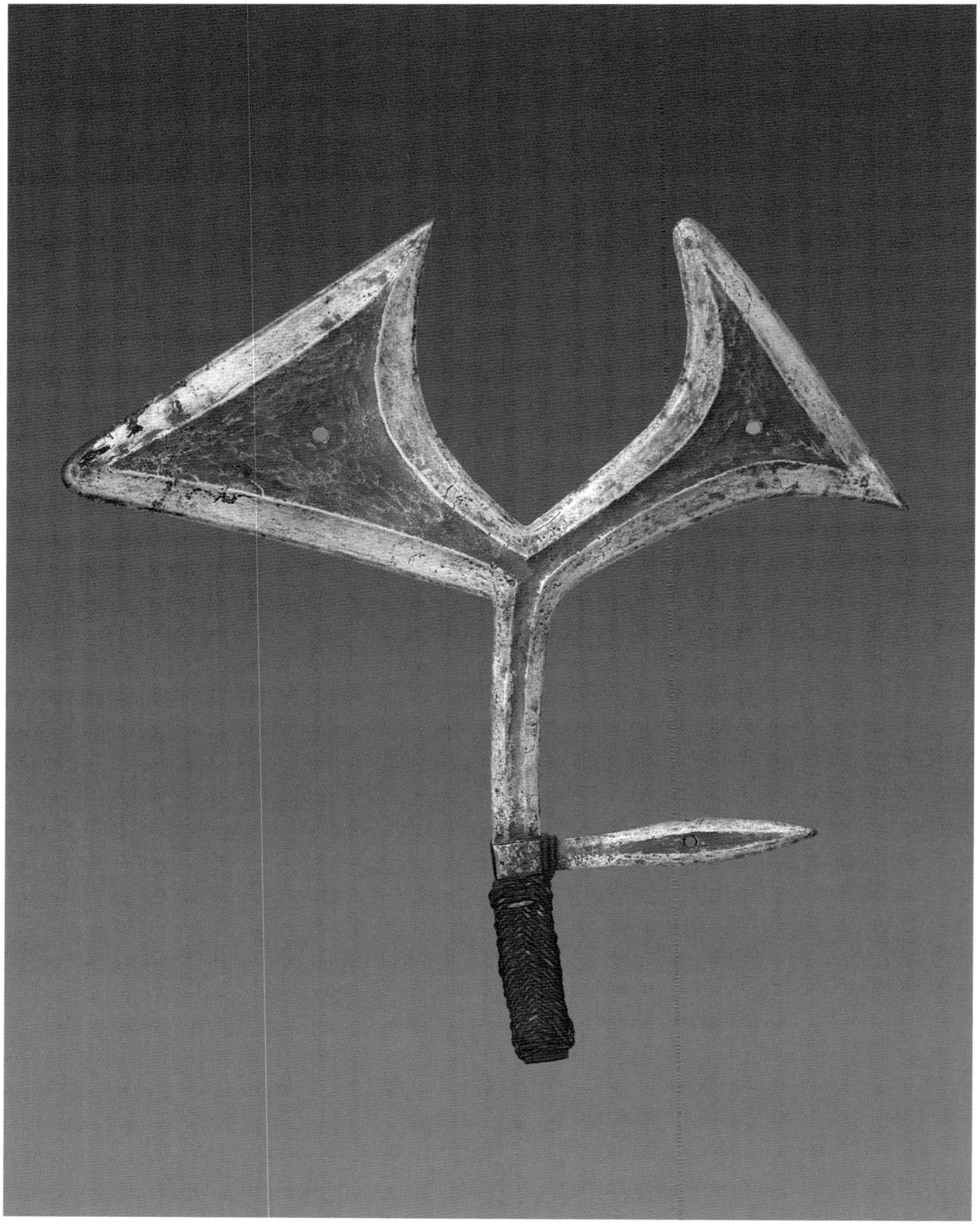

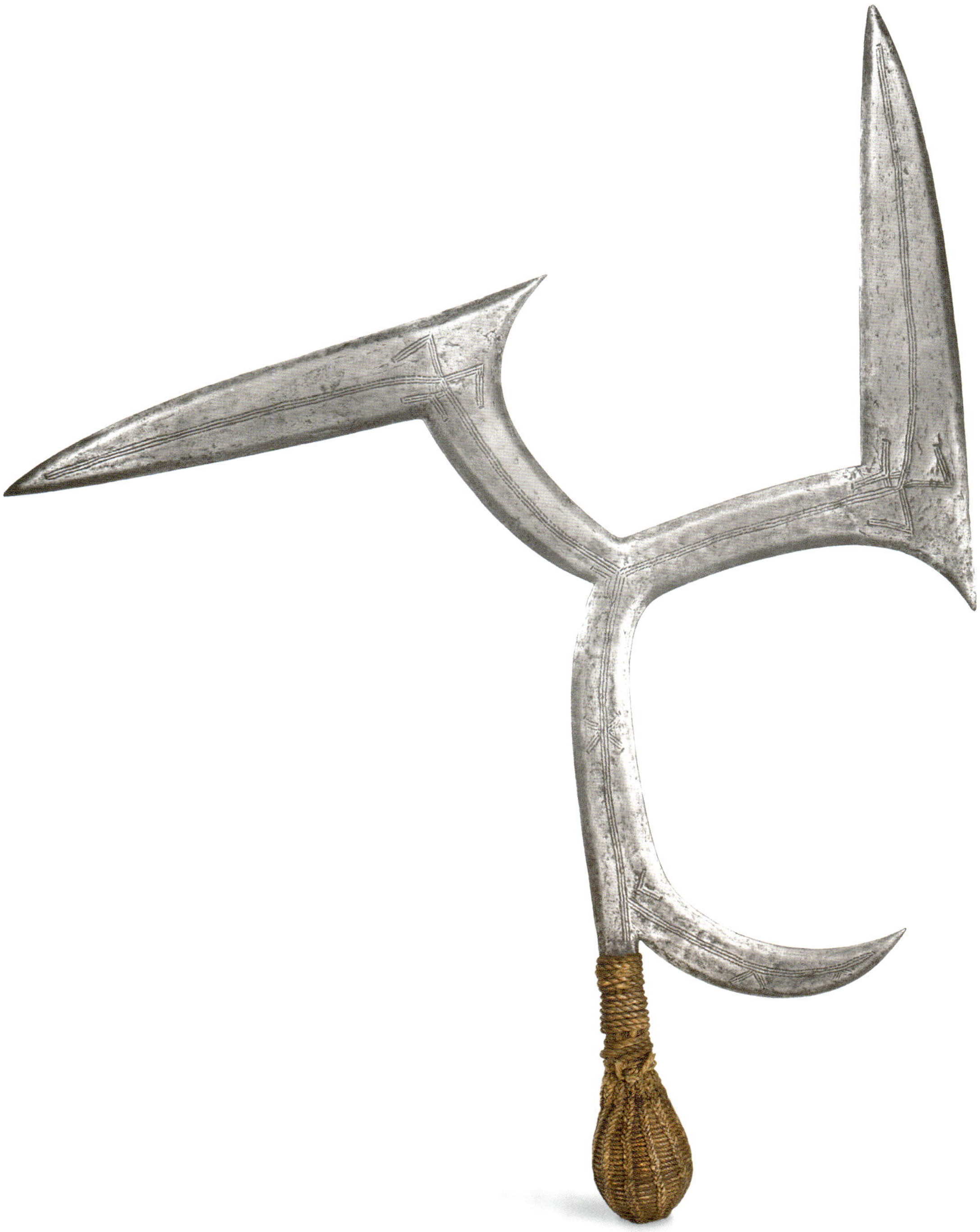

7-7
Artist unknown (Mabo, Mbanja, Ngbaka, Nzombo peoples, Democratic Republic of the Congo)
Throwing knife, early 20th century
Iron, plant fiber
H: 47 cm
FOWLER MUSEUM AT UCLA, X65.2989; GIFT OF THE WELLCOME TRUST
(NOT IN EXHIBITION)

Southern "winged-type" knives (see Thomas 1925) are generally smaller (between thirty and sixty centimeters), and while most have a central axis, it is shorter and less visually prominent, serving as starting point for an outward flaring of multiple impressive blades (figs. 7.6–7.10). Such weapons were (and still may be) distributed from what is now South Sudan to parts of Cameroon and Gabon. In 1869, the German botanist and ethnographer Georg Schweinfurth spent two and a half years in the eastern part of this region, and the illustrations reproduced alongside his writings (figs. 7.11, 7.12) feature Zande throwing knives (see fig. 7.10) and Mangbetu sickle-shaped scepters (figs. 7.13, 7.14) to reveal high levels of technological and aesthetic expertise.

Throwing knives certainly captured Europeans' attention for their exotic appearance, but the fact remains, they are spectacular technical creations. To begin with, tremendous skill is required to shape hot iron with a hammer. It is as if little waves of solid matter are being guided by very measured, careful blows into the desired shape. Beyond that, virtually invisible forge-welds, iron folded over on itself, lengths of iron split into diverging wings, and graceful swells and curves are just some of the techniques and effects that blacksmiths used to produce these knives. Especially in southern examples, the metal can be as thin as fine cutlery, with blades that sweep and soar in balanced compositions. The carefully conceived shapes of both F and winged types confirm that technical virtuosity and pronounced aesthetic acumen were (and are) prominent components of central African smithing.[2]

Aesthetic acumen is in evidence in the patterns that often grace throwing knives, painstakingly achieved in a variety of ways including notched edges; punched dashes and dots (see figs. 7.4, 7.7, 7.9); chiseled lines (see fig. 7.5), some parallel to create a central raised effect; and motifs produced by leaving the black iron scale from hammer forging on the blades' surfaces (see fig. 7.6). Additional complexity often comes from the ways knife cross sections are shaped: rectangular, in various thicknesses (sometimes very thin); diamond, triangular, or lozenge; flat with a narrow rectangle in the middle, slightly cupped, flat on the bottom side, and gently rounded on top; and several more intricate articulations into ridges and associated valleys (see Westerdijk 1988). While surface embellishments generally occur on only one side of the blade, the anthropologist Marian Charles Jedrej described a symbolic system for Ingessana peoples in Sudan, who embellish both knife sides (1975, cited in Spring 1993, 76–77); one can assume that such ornamentation served more than aesthetic satisfaction (see fig. 7.5).

Weight was also an important aspect, particularly for throwing knives designed for fighting. The contemporary scholar and collector Peter Westerdijk has noted that smiths had to discover and then adhere to "a subtle equilibrium between optimal shape and adequate weight" (1988, 19). Versions used by Sara peoples of southern Chad are very light, even delicate in appearance, whereas Ingessana knives from the Blue Nile area are quite heavy (Spring 1993, 71–73, 76–77; Westerdijk 1988, 118–33, 170–83). Handles can simply be the metal itself, sometimes with a delicately articulated little tab, or a haft wrapped with fiber (see fig. 7.2, 7.7) or copper or copper alloy wire (figs. 7.8, 7.9), or covered with hide (figs. 7.3–7.6). All of these elements, from shape to weight to all manner of embellishments, add up to a sophisticated mix of technical and artistic components that play out differently in each object. Clearly much skillful experimentation transpired, a fact that characterizes histories of ironworking across sub-Saharan Africa.

In spite of great variation in size, shape, and weight, many of these knives were remarkably designed with regard to aerodynamics. The experimentation and processes of observation and discovery that blacksmiths and throwing knife users no doubt carried out are not part of the literature. Westerdijk describes experiments by Europeans in throwing these knives, and gives his assessment of their effectiveness in battles. He is skeptical of their findings, including the best known, in 1922, by famous Belgium museum ethnographer Joseph Maes, who reported that centrally balanced examples could pierce a half-inch pine board from fifty feet (15 meters). Further, they could be thrown with accuracy up to 160 feet (49 meters) or without accuracy to over 300 feet (91 meters; Maes 1922; Thomas 1925, 130). According to Westerdijk's own experiments, F-shape examples could be thrown with effectiveness some sixty-five feet (20 meters) while winged examples were effective to nearly one hundred feet (31 meters) and could fly twice that with a good bit of fluttering near the end (Westerdijk 1988, 22–27; Schmidt and Westerdijk 2006, 35–36).

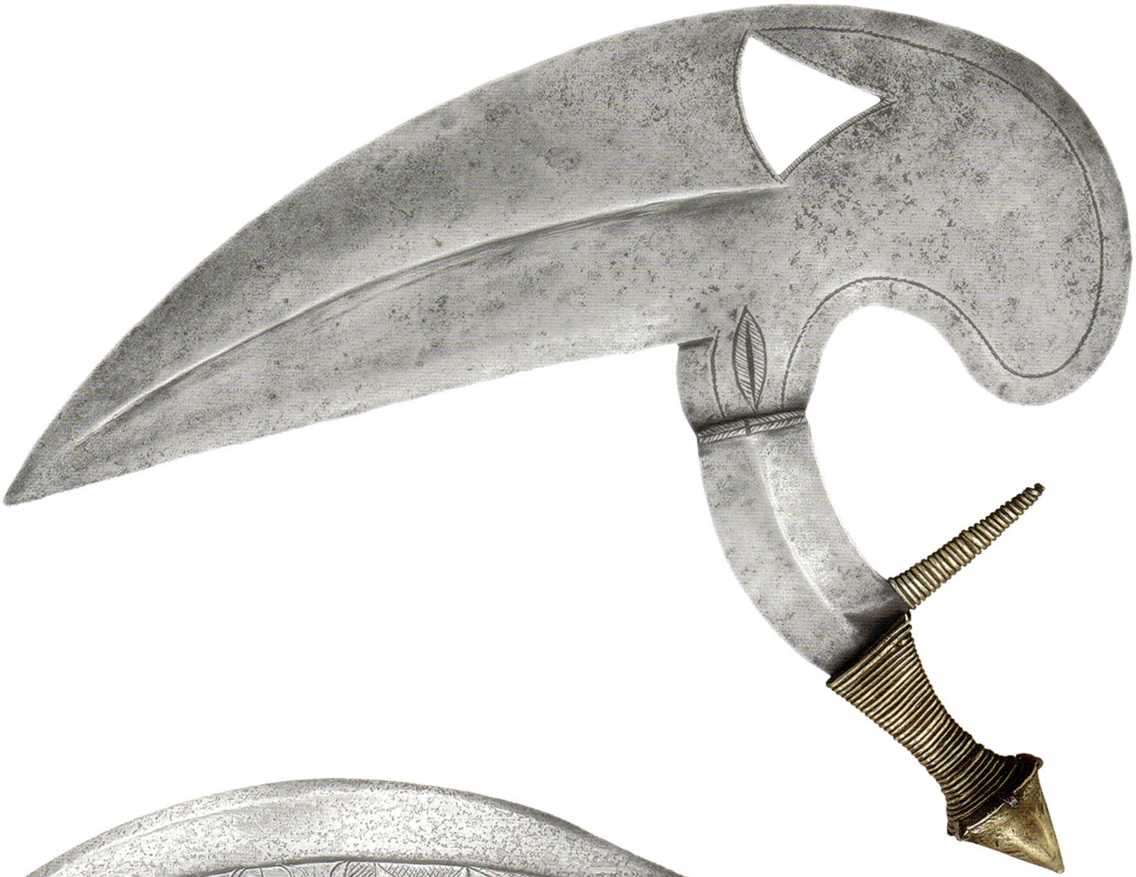

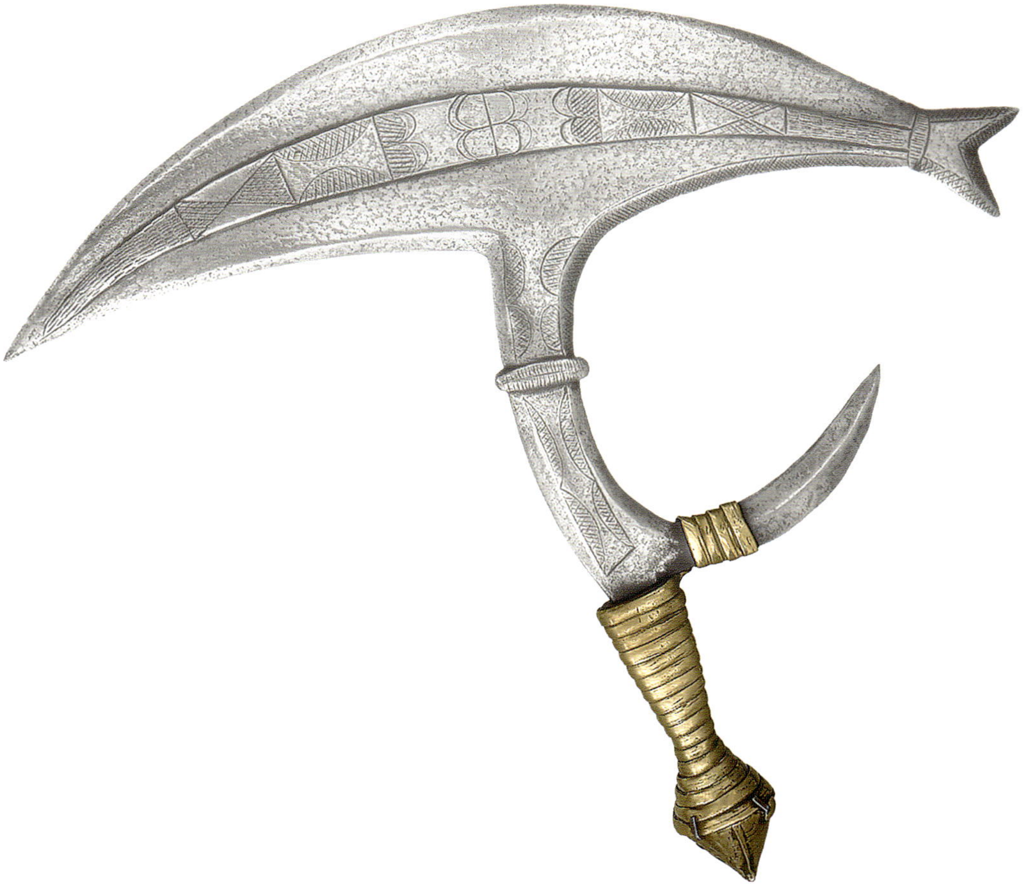

7.8
Artist unknown (Kota, Ndzabi, Ndassa, Mbede, Wumbu peoples, Gabon)
Bird-headed axe (*musede*, *osele*), early 20th century
Iron, brass, wood
W: 44 cm
ETHNIC ART & CULTURE LTD.
PHOTOGRAPH BY DICK BEAULIEU, 2009.
COURTESY OF ETHNIC ART & CULTURE LTD.

7.9
Artist unknown (Kota peoples, Gabon)
Bird-headed axe (*musede*, *osele*), early 20th century
Iron, brass, wood
H: 34 cm
JAN ELSEN COLLECTION
PHOTOGRAPH BY DICK BEAULIEU, 2009.
COURTESY OF JAN ELSEN

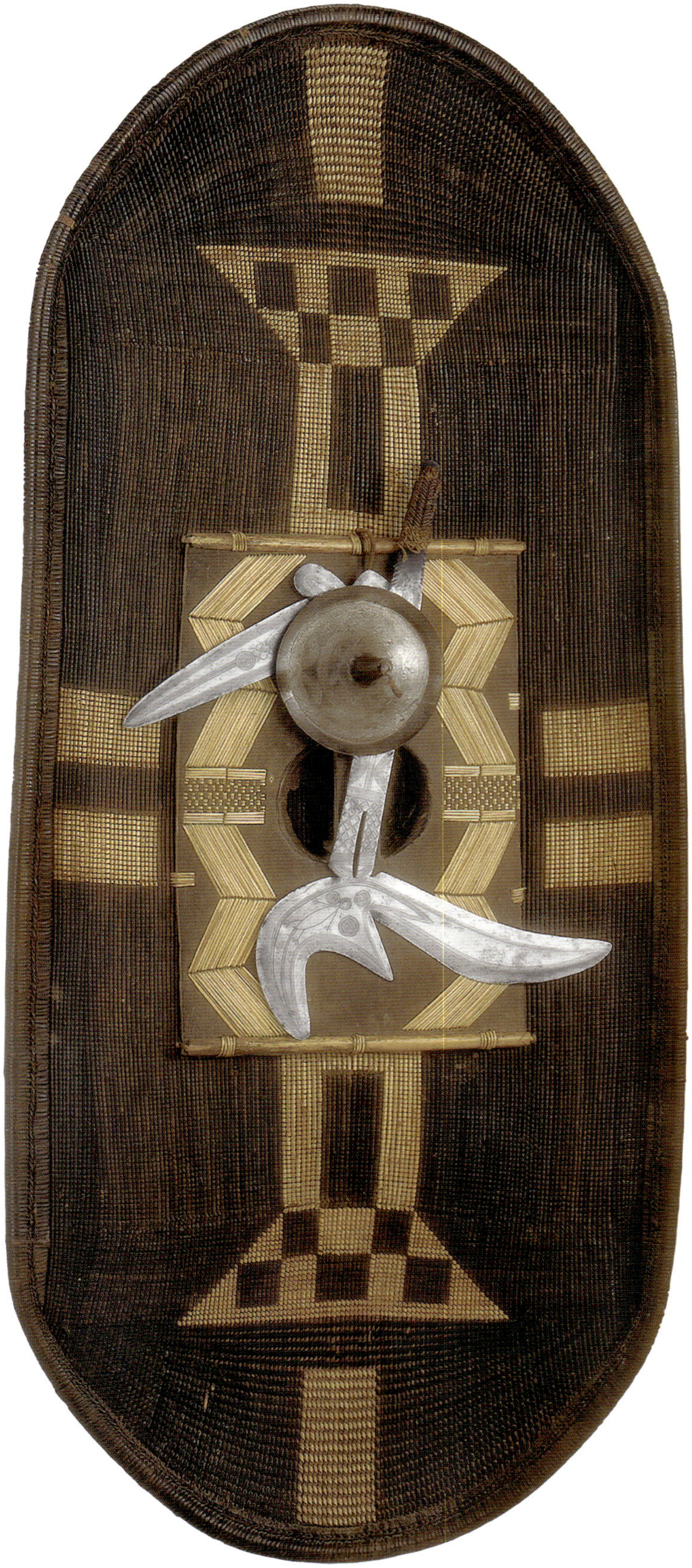

7.10
Artist unknown (Zande-Avungara, Barambo peoples, Democratic Republic of the Congo)
Shield with throwing knife (*vura*), late 19th century
Iron, leather, wood, plant fiber
H: 115 cm
JAN ELSEN COLLECTION
PHOTOGRAPH BY DICK BEAULIEU, 2009. COURTESY OF JAN ELSEN.

7.11
This array of elegant throwing knives punctuates the skill and aesthetic imagination that central African blacksmiths applied to their work. The image's composition shares a sensibility with the wall displays of African weaponry that were popular in Europe during and after the period when Western nations were carving out colonies on the African continent (see fig. 6.27).
ENGRAVING BY J. D. COOPER. PUBLISHED IN GEORG SCHWEINFURTH, *ARTES AFRICANAE: ABBILDUNGEN UND BESCHREIBUNGEN VON ERZEUGNISSEN DES KUNSTFLEISSES CENTRALAFRIKANISCHER VÖLKER* (LEIPZIG: F. A. BROCKHAUS, 1874).

Early European visitors wrote about throwing knives in tales of horrific deaths, some personally witnessed, others imagined. Reporting to the Royal Society of Edinburgh at the end of the nineteenth century, Robert Felkin placed throwing knives in an array of local weaponry and noted that they are "curiously curved, and are thrown with unerring aim" (1896, 244). Many authors made similar statements about their accuracy, but it is not always clear if these remarks were based on observation, hearsay, assumption, or fancy. For example, Felkin stated that his report came from personal observations and information from local peoples. In contrast, the mid-nineteenth century zoologist and ethnographer Paul Du Chaillu offered no basis for his imaginative and erroneous assertions that Fang throwing knives, which resemble bird heads (see figs. 7.8 and 7.9 for similarly shaped blades made by nearby groups), were aimed at a victim's head so that the point would penetrate the brain and kill instantly, after which the knife would be used to remove the head as a trophy.[3]

Many early authors, often based on information from local people, held that throwing knives were not thrown to kill, but instead were aimed at the legs to immobilize game animals or battle adversaries and their horses.[4] They still painted grisly pictures (e.g., Lander 1907, 2: 96). Ernest Thomas, who assembled early sources on throwing knives as an assistant curator at the Pitt Rivers Museum in Oxford, stated they could cut off a leg at twenty yards (18.3 meters; 1925, 130). Several authors described the fighting techniques of Zande speakers, renowned state-builders in the upper east of the Congo River system, who possessed many beautiful throwing knife types and were also adept at dodging them.[5]

In the kingdoms of Bornu and Bagirmi, whole units of foot soldiers specialized in using throwing knives as weapons (see Reyna 1990). Southern knives were carried in multiples, sometimes attached inside a shield (fig. 7.10). In what is now east-central Democratic Republic of the Congo, the early twentieth-century ethnographer Emil Torday asserted that a leading Kuba clan bears the name Bushoong, "the people of the lightning/throwing knife," and Kuba peoples hold that throwing knives were once prominent in their arsenal and facilitated migration to their present location (1925, 165). Likened to lightning, throwing knives were considered so dangerous that Kuba King Shamba Bolongongo outlawed them in the late sixteenth century (Torday 1925, 141).

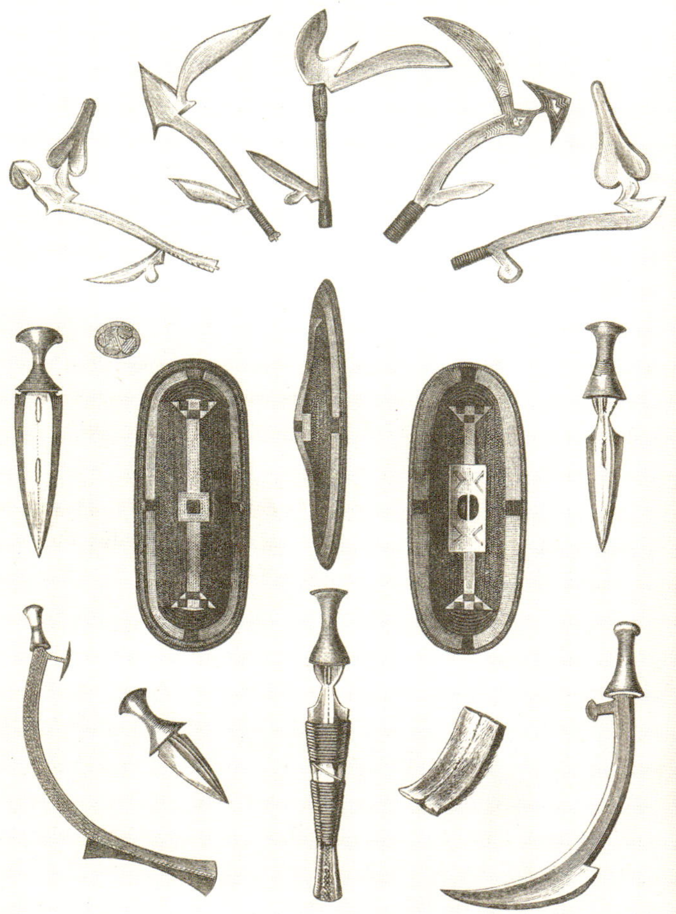

In the early twentieth century, Ward was among several authors stationed in African colonies who published notes with an eye towards making meaning of the proliferation of throwing-knife styles.[6] Writing about the upper Congo, he asserted that peoples could be "classified not alone by the distinguishing cicatrisation designs upon their faces and bodies, but also by their local designs and forms of fashioning the iron knives and spear heads, which constitute their weapons" (1910, 264). Complexities qualify such classification, however. Blacksmiths often worked for several different ethnic groups and could be highly mobile. Smiths in certain areas, for example in the Ennedi Mountains of Chad (see figs. 7.2, 7.3) and among Mangbetu peoples of the DRC (see figs. 7.13, 7.14), had reputations for making very good knives that were traded great distances. People traveling for work, war, or on pilgrimages to Mecca also helped spread knife types, and throwing knives were frequently used in commodity exchange for food and other goods. (Nachtigal, for example, had difficulty acquiring enough throwing knives to obtain the grain he needed.) Westerdijk notes a general state of shape sharing among ironsmiths (1988, 139). It is easy to imagine that a conversation of form among smiths, or between smiths and clients, would have resembled the kind of inspired improvisation of jazz musicians, making the forge a site of creativity significant to knives' histories of aesthetic development and use.

By the middle of the nineteenth century, European visitors were reporting the use of throwing knives to be in decline. Nachtigal frequently observed that Arab weaponry (guns and swords) and Islamic conceptions of fighting were supplanting such weapons (for example, 1971, 174). The British archaeologist and colonial officer Anthony John Arkell found that swords, spears, and rifles were considered more efficient than throwing knives (1939, 257). Others assumed that colonialism reduced the frequency of fighting, and so the need for the knives, ignoring Europe's war-waging and general aggression that brought on colonialism in Africa. Captain Guy Burrows, a colonial officer, could therefore report that throwing knives and other weapons were gathered up by Europeans after battles as trophies of conflict with "Congo natives," as though they would no longer be of any local use (1898, 79).

Nevertheless, other writers saw ample numbers of these weapons in many areas. In Bornu and Bagirmi, as many as five knives were carried in hide sheaths with a strap for the shoulder, or several were balanced with ease directly on the shoulder, even on a galloping horse (Nachtigal 1889, 275; Spring 1993, 71, 73). Arkell thought the knives may have been invented to fight cavalry. Both Arkell (1939, 252) and the famous nineteenth-century explorer Heinrich Barth (1857 III, 198) stated that throwing knives were aimed at cavalry horses' fetlocks, to bring both horse and rider down. Arkell thought the knives may in fact have been invented to fight cavalry, and also that heavy and often highly colorful cloth-quilted armor, famously part of royal heraldry across north-central Africa, originated as protection for horse and rider from throwing knives, though it also protected against poisoned arrows (Nachtigal 1980, 119; see also Spring 1993, chapter 2).

The impression that the central African literature repeatedly gives is that throwing knives were ubiquitous. Well south of Lake Chad near the border of the Bagirmi Kingdom, Nachtigal spent the night in a town where he found the men lounging "with the inseparable throwing-iron on their shoulders" (1987 III, 310). He mentions throwing knives in the hands of men or on their shoulders so frequently that they might be thought of as extensions of the person, as if men simply could not imagine being without them. Nachtigal's examples are copious. In some desert towns where iron weapons were forbidden in public, men carried the wooden training replicas they had received as young boys (Nachtigal 1978, 413). Given the constant presence of these knives, their usefulness and value, it is not surprising they were used as media of exchange. In fact, in many areas currency was actually iron forged into throwing knife shapes (see fig. 5.9) as imaginative as the actual knives (Spring 1992, 82–83).

7.12
The Mangbetu King Munza at court, 1870, holding a scepter suggestive of throwing knife shapes.
ENGRAVING BY J. D. COOPER. PUBLISHED IN GEORG SCHWEINFURTH, *ARTES AFRICANAE: ABBILDUNGEN UND BESCHREIBUNGEN VON ERZEUGNISSEN DES KUNSTFLEISSES CENTRALAFRIKANISCHER VÖLKER* (LEIPZIG: F. A. BROCKHAUS, 1874).

7.13
Artist unknown (Mangbetu peoples, Democratic Republic of the Congo)
Sickle-shaped knife (*mambele*, *trumbash*, *nambele*), late 19th to early 20th century
Copper alloy, wood
H: 44 cm
FELIX COLLECTION
PHOTOGRAPH BY DICK BEAULIEU, 2001.
COURTESY OF THE FELIX COLLECTION

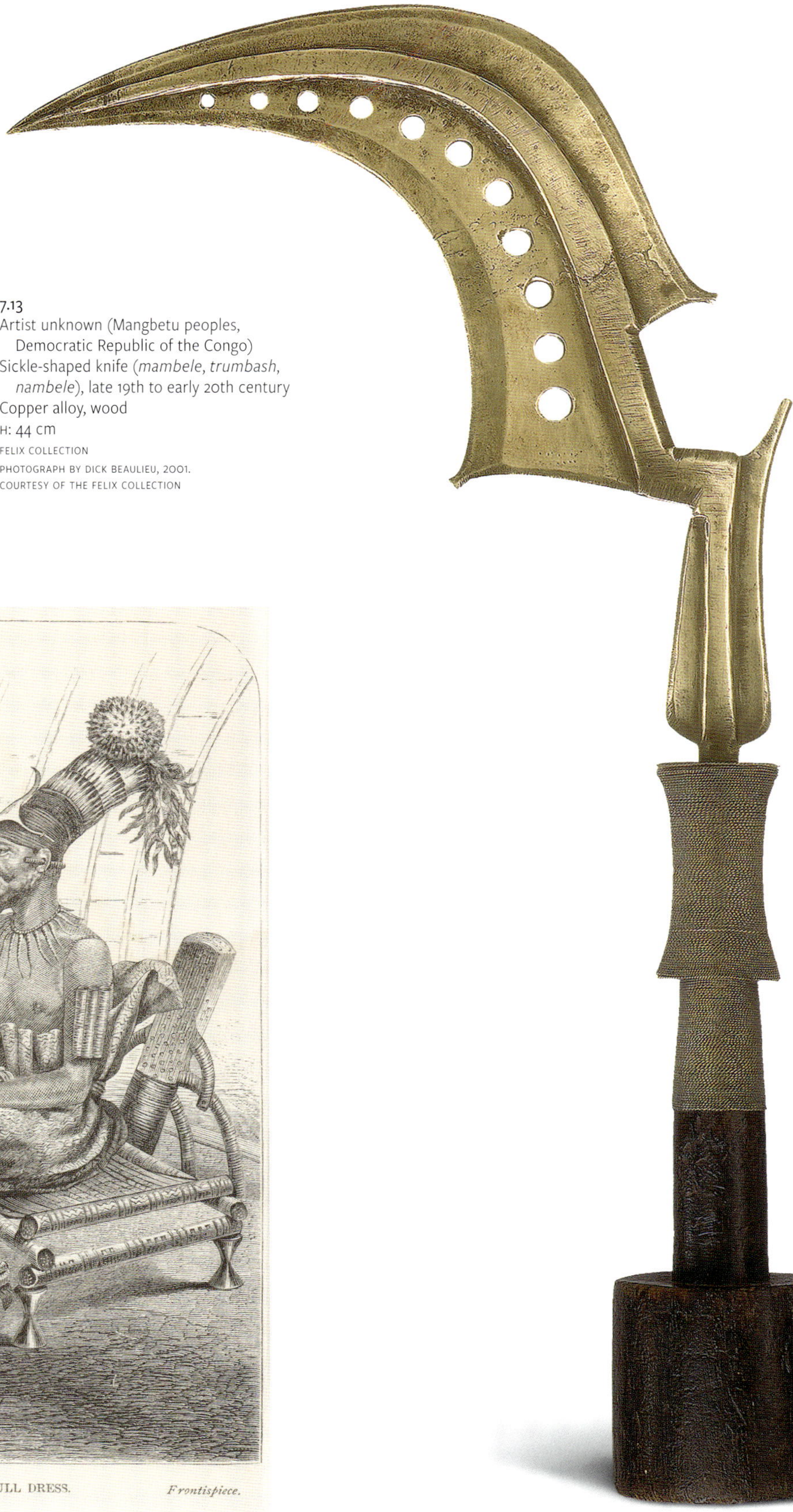

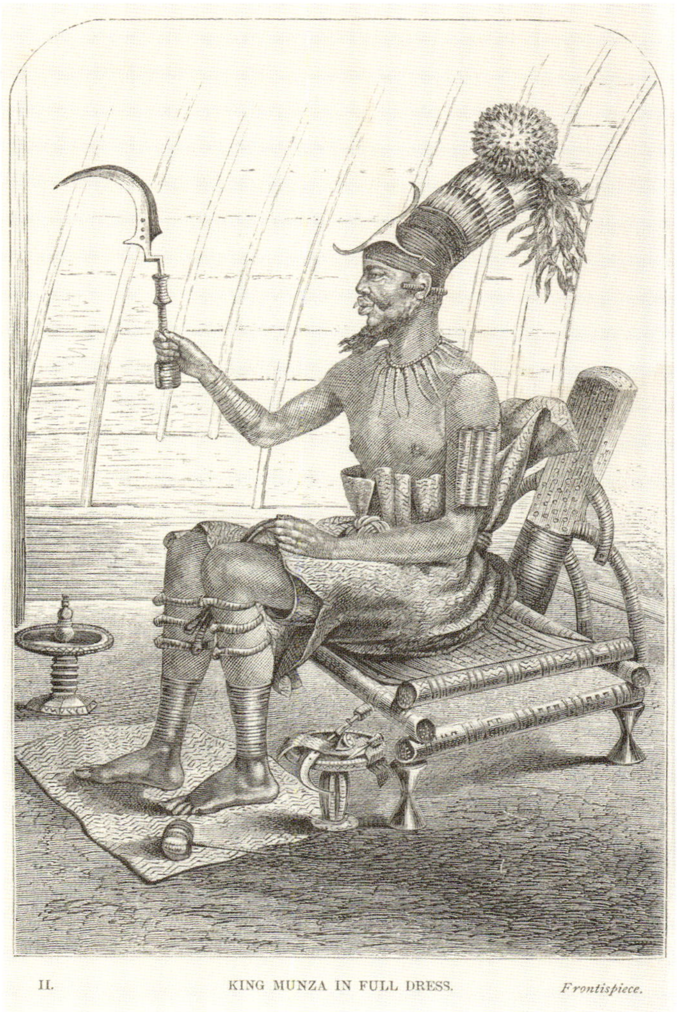

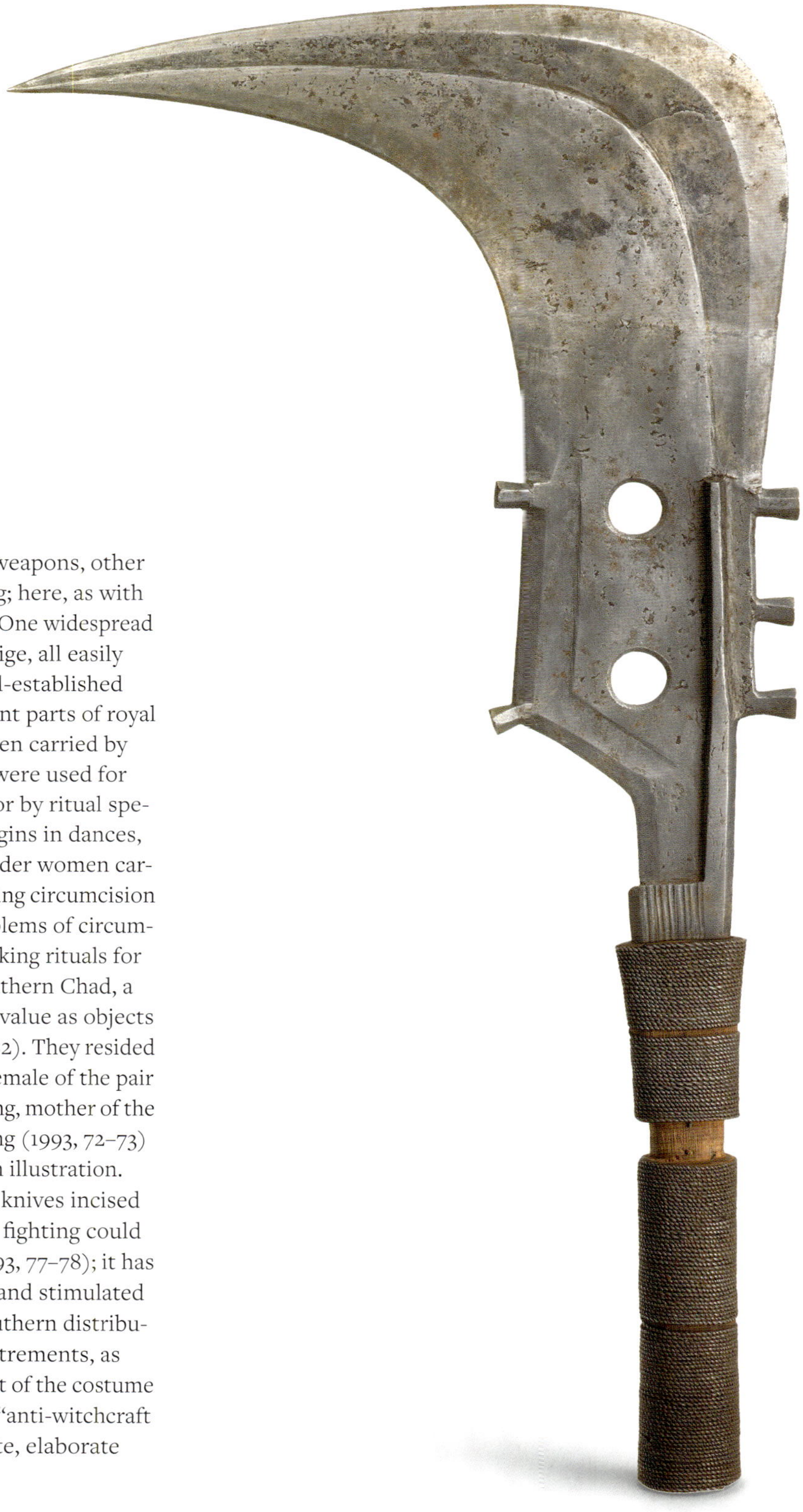

7.14
Artist unknown (Mangbetu peoples, Democratic Republic of the Congo)
Sickle-shaped knife (*mambele, trumbash, nambele*), late 19th to early 20th century
Iron, wood
H: 41.9 cm
FOWLER MUSEUM AT UCLA, X65.2953; GIFT OF THE WELLCOME TRUST

While Nachtigal emphasized throwing knives as weapons, other writers reported numerous functions beyond fighting; here, as with throwing knife shapes, there is tremendous variation.[7] One widespread use was as insignia of leadership, authority, and prestige, all easily imagined as a consequence of these instruments' well-established value as expertly crafted weapons. They were important parts of royal processions and displays of wealth and power, and even carried by special throwing knife officers. In various areas they were used for "magical purposes," carried by Islamic learned men, or by ritual specialists who controlled locusts. They were held by virgins in dances, and by young men trying to impress young women. Elder women carried them ceremonially and clashed them together during circumcision feasts for their young relatives. Men used them as emblems of circumcision and high social rank. They were used in rain-making rituals for the community or fecundity rituals for women. In southern Chad, a male and female throwing knife pair held great ritual value as objects of sacrifice sanctioning the king's authority (Fortier 1982). They resided in their own room and were understood as alive. The female of the pair was conceptualized as a beautiful woman, wife of the king, mother of the people. The British Museum curator Christopher Spring (1993, 72–73) provides a detailed description of these knives, with an illustration.

At the east end of the northern distribution, large knives incised with designs of snakes and scorpions but not used in fighting could have been a kind of male fashion accessory (Spring 1993, 77–78); it has also been suggested that they served as dance wands and stimulated cultural contemplation (Jedrej 1975, 42–48). In the southern distribution, knives were used as bridewealth, as dance accoutrements, as symbols of rank and status, and even as an essential part of the costume worn by the principal ritual specialist of an important "anti-witchcraft cult" (Spring 1993, 78). Far north, in the Mahdist state, elaborate

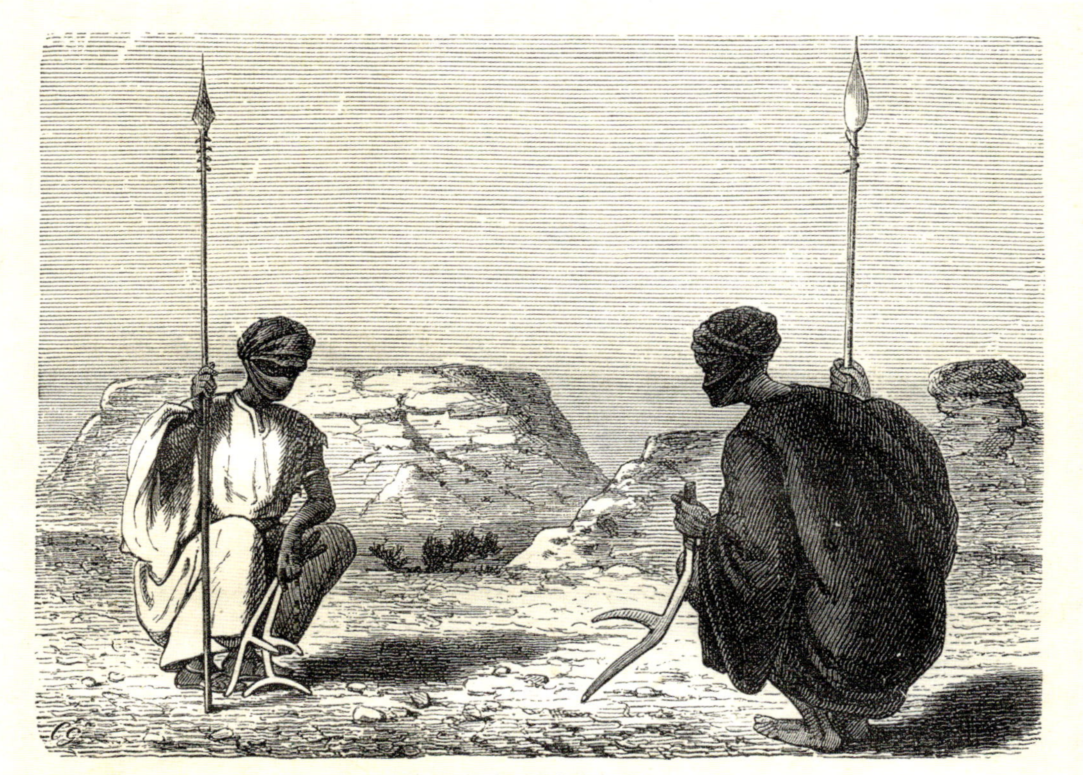

7.15
This illustration from Gustav Nachtigal's multivolume account of his travels between 1869–73 shows two Tubu-speaking people greeting each other in the Tibesti Mountains area, and suggests how ubiquitous throwing knives were in this region in the mid-nineteenth century. The men Nachtigal encountered in central Africa were generally well armed; in this case, one is in possession of two throwing knives.
PUBLISHED IN GUSTAV NACHTIGAL, *REISEN IN DER SAHARA UND IM SUDAN* (LEIPZIG: F. A. BROCKHAUS, 1887).

throwing knife–shaped scepters may have been used to call jinn to war (see interleaf D, this volume).

Given all this information, limited and spotty as it is, we can say with certainty that throwing knife history is extremely complex and multifaceted. What we cannot say is how old these forms are or where they originated. It has been argued that throwing sticks, known in ancient Egypt and found very widely in sub-Saharan Africa, were the prototype from which throwing knives emerged (Arkell 1939). Elsewhere the assertion has been made that throwing knives used around the Tibesti Mountains of Chad were derived from a copper weapon in ancient Mesopotamia that passed through Egypt and diffused across Africa (Robinson 1936, 190). Some have suggested that simpler northern types were invented first and then developed into more complex forms (see McNaughton 1970, 1971; Spring 1993). In our own century, using throwing knife vocabulary and an analysis of distribution and technology, Christopher Ehret has suggested that throwing knives were invented sometime between 800 and 1200 CE, somewhere between Lake Chad and the Marrah Mountains in the Darfur region of western Sudan (2002, 340–41). Then, in the first half of the second millennium, they spread south to the Ubangi River region.

The earliest of these scenarios suffer from the defunct assumption that everything noteworthy in Africa was either Egyptian or Hamitic, while the rest assume that technological inventions develop by small increments. All are really just conjecture without adequate historical evidence. People are far too experimental and entrepreneurial and the knives themselves far too varied and steeped in creativity for us to be able to talk about origins with any degree of confidence.

What we *can* say is that throwing knives fall among those rare products of human imagination that seem wholly original. As creations in iron, they reflect the virtuosity and supple bent of mind that permeate blacksmith's activities throughout Africa. Nachtigal was not the first to write about throwing knives, but his descriptions of encounters with these instruments—their easy presence in people's lives, the various purposes they served, and the horrendous battles in which they played a role—provide the context for further research. Archaeology and other sources of scholarship may someday shed more light on how these metal masterpieces were developed and spread through so large and varied a landscape.

INTERLEAF
D

FOR WHOSE EYES WERE THEY? CALLIGRAPHIC BLADES OF MAHDIST SUDAN

Allen F. Roberts

There are ways to experience art that require a more active collaboration between artist and observer than will be familiar to many Westerners.

SUSAN VOGEL (1997, 72)

Even the sagacious and educated may be illiterate with respect to divine signs.

SHABBIR AKHTAR (2008, 219, DISCUSSING QUR'AN 29:38)

Six acid-etched calligraphic blades from late nineteenth-century Sudan (figs. D.1–6) are captivating in design and execution, as is a headpiece of forged iron mail graced by prayer packets (fig. D.7). All are in the Wellcome Collection of the Fowler Museum at UCLA, and all have been described as bearing "pseudo-Arabic" or "arabesque" inscriptions that appear to be largely illegible.[1] To whom, though? This question will be broached here, as will ways that these marvelous objects "give shape to forms of thought" and practice (Ingold 2007, 14–15). Indeed, the blades and headpiece attest to entangled histories of the Sudan as local factions vied for advantage and the scramble of European colonizers drew to a close in 1899, when the blades were likely collected.

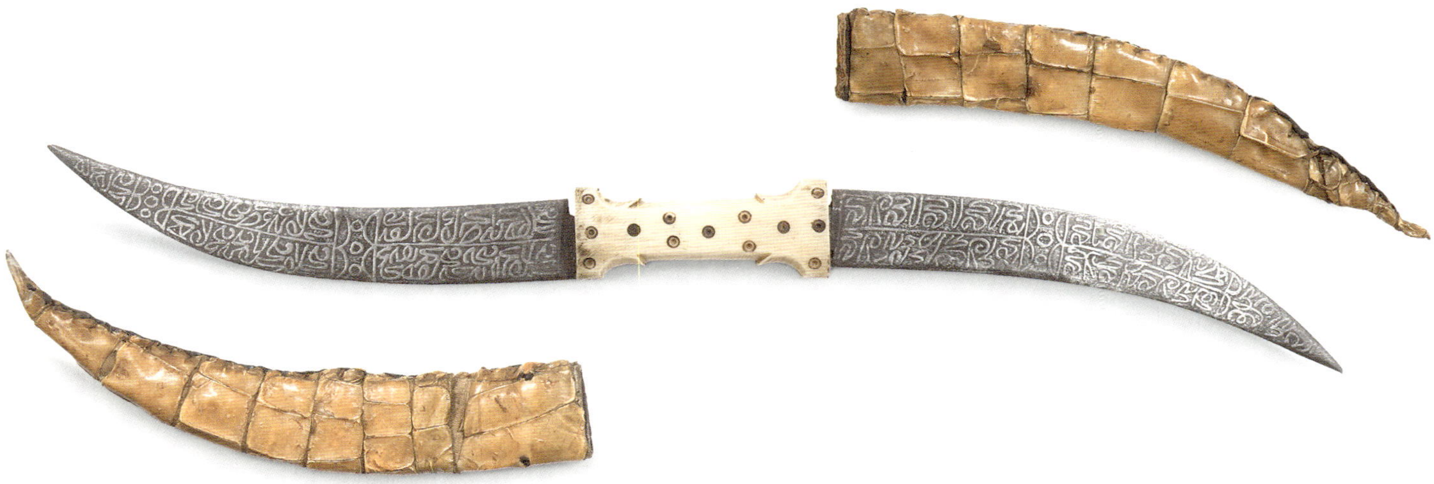

D.1
Artist unknown (Sudan)
Acid-etched double-bladed dagger (*haladie*), late 19th century
Iron, bone, crocodile skin
L (sheathed): 58.5 cm
FOWLER MUSEUM AT UCLA, X65.8674 A-C; GIFT OF THE WELLCOME TRUST

This double-bladed, keenly edged weapon is called a *haladie* after its South Asian prototype, perhaps introduced from the Red Sea port of Suakin or via Syria to the Sudan, the only region of sub-Saharan Africa where its use is documented (J. McDougall pers. comm., 2017). Such movement of technologies speaks to the Sudan as a great crossroads, but what of the sheaths? Their leather is from a crocodile's belly; does such a choice suggest local symbolism as well as practicality?

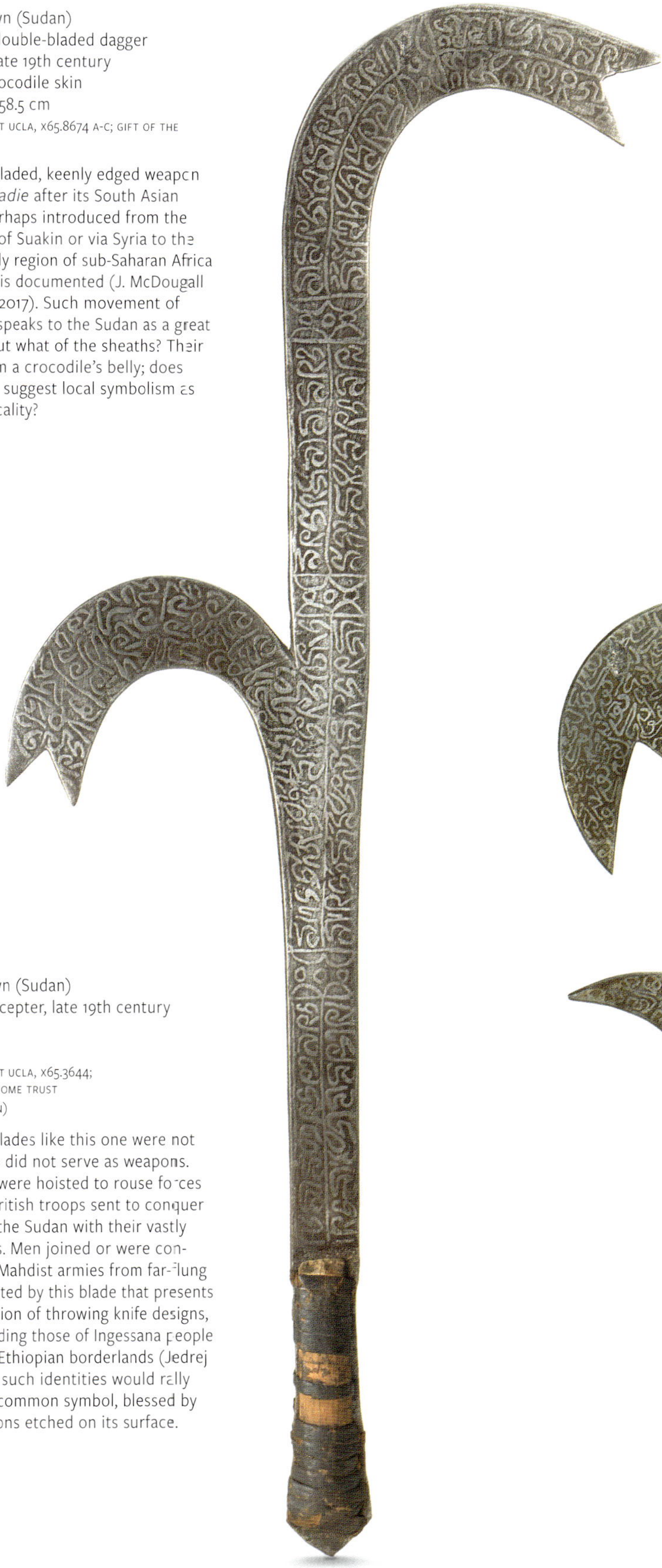

D.2
Artist unknown (Sudan)
Acid-etched scepter, late 19th century
Iron, leather
H: 63.5 cm
FOWLER MUSEUM AT UCLA, X65.3644; GIFT OF THE WELLCOME TRUST
(NOT IN EXHIBITION)

Calligraphic blades like this one were not edged, and so did not serve as weapons. Instead, they were hoisted to rouse forces to confront British troops sent to conquer and colonize the Sudan with their vastly superior arms. Men joined or were conscripted into Mahdist armies from far-flung lands, as attested by this blade that presents an amalgamation of throwing knife designs, perhaps including those of Ingessana people of the Sudan-Ethiopian borderlands (Jedrej 1995). Men of such identities would rally around their common symbol, blessed by the exhortations etched on its surface.

D.3
Artist unknown (Sudan)
Acid-etched throwing knife, late 19th century
Iron, lizard skin
H: 34.3 cm
FOWLER MUSEUM AT UCLA, X67.561; GIFT OF THE WELLCOME TRUST

The remarkable form of this calligraphic blade is based upon throwing knives called *ndumo*, once made by Ngbaka-Mabo people living along the triple border of present-day Central African Republic, the Democratic Republic of the Congo, and Congo-Brazzaville. Crafted in Omdurman, Sudan, from sheet-iron scrap (see Spring 1993, 78–79), the blade bears repetitive inscriptions in what linguistic anthropologist Fallou Ngom calls "under-specification" (pers. comm., 2018)—that is, they convey protective messages not readily legible to human eyes but which may have summoned jinn to assist Mahdist forces in dire need of their intervention as they faced British cannons and machine guns.

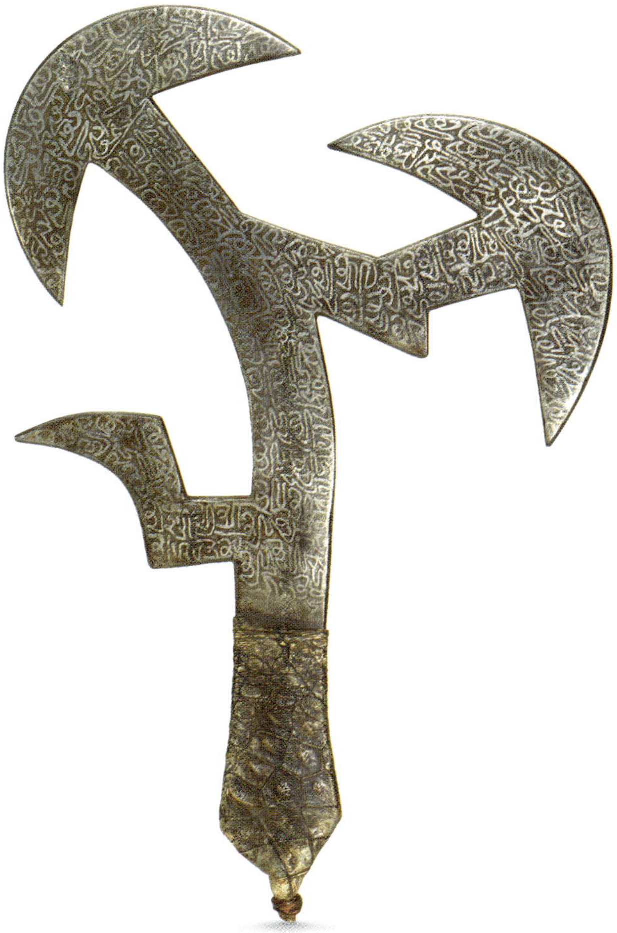

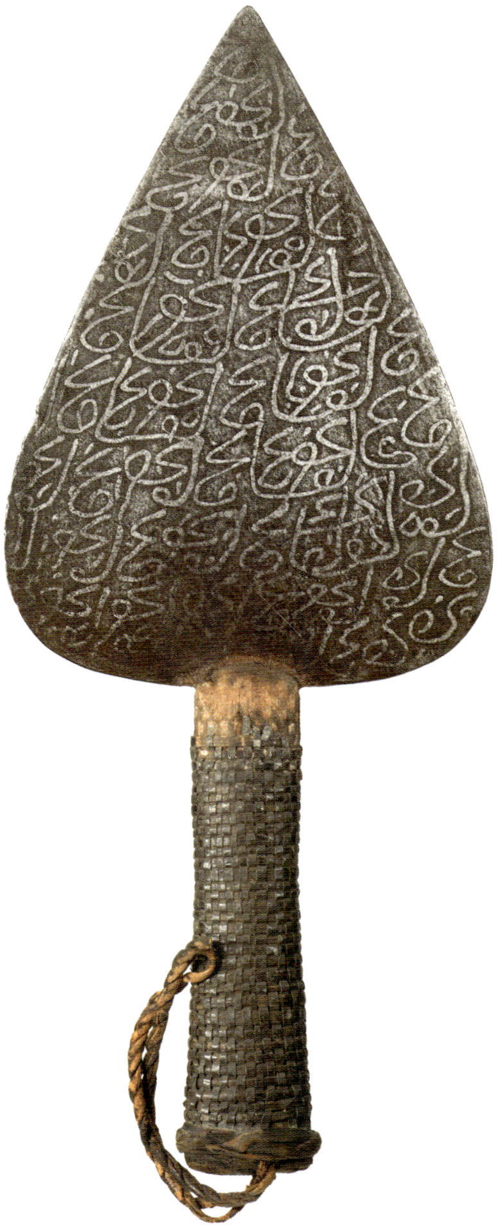

D.4
Artist unknown (Sudan)
Acid-etched knife, late 19th century
Iron, wood, leather, tin, plant-fiber cordage
H: 39.5 cm
FOWLER MUSEUM AT UCLA, X65.1998;
GIFT OF THE WELLCOME TRUST
(NOT IN EXHIBITION)

A shape derived from a spear or knife blade perhaps from Mongo peoples of west-central Democratic Republic of the Congo (Elsen 2003, 78–79) has been adapted to serve as a scepter to rally Mahdist forces hailing from lands well south of the Sudan. The cadenced inscriptions in flowing Thuluth script were meant to elicit recitations of dhikr prayers reciting the holy names of God, with a further possibility of appealing to jinn for assistance on the "horrorscapes" of anticolonial battle.

The Sudan has long been a pivot for two-way trade extending from the Red Sea to kingdoms of Sahelian west Africa, and along the Nile from the Mediterranean to central and eastern Africa. The Ottoman Empire held Egypt for nearly four centuries, and in the 1820s their forces pushed up the Nile to occupy the Sudan. Ottoman authorities employed ruthless international mercenaries locally loathed for their cruelty, devastating taxation, and lascivious behavior. Famine and epidemics also wreaked havoc. In the midst of such apocalyptic tumult, a man of Nubian heritage and Sufi devotion named Muhammad Ahmad ibn Abdullah (ca. 1848–1885) was heralded as Mahdi in 1881, come to defend Islam and local sovereignty. The advent of a Mahdi is not mentioned in the Qur'an, but hadiths predict that a reformer will appear "at the beginning of each Muslim century…[to] rectify deviations that have slipped into Islamic beliefs and practices" (Mbacké 2005, 62; cf. Boahen 1989, 47–48). Fervent contemporaries found Muhammad Ahmad "a saint in all his conduct" who was humble, generous, and pious; and they knew that his summoning by God at the turn of the fourteenth Muslim century (1883 CE) was presaged by a mighty comet.[2]

The Mahdi attracted an immense following to his noble cause. His legions resoundingly drove Ottoman forces from the Sudan, but with shocking loss of life on all sides. Soon after razing the colonial capital of Khartoum in 1885, Muhammad Ahmad perished of illness and was succeeded by Khalifa Abdullahi, who set about routinizing a Mahdist state. In the same year, sub-Saharan Africa was partitioned and European colonial authorities strove to define, extend, and defend new borders, with the Sudan among the very last prizes seized (see Lewis 1987; Holt 1970).

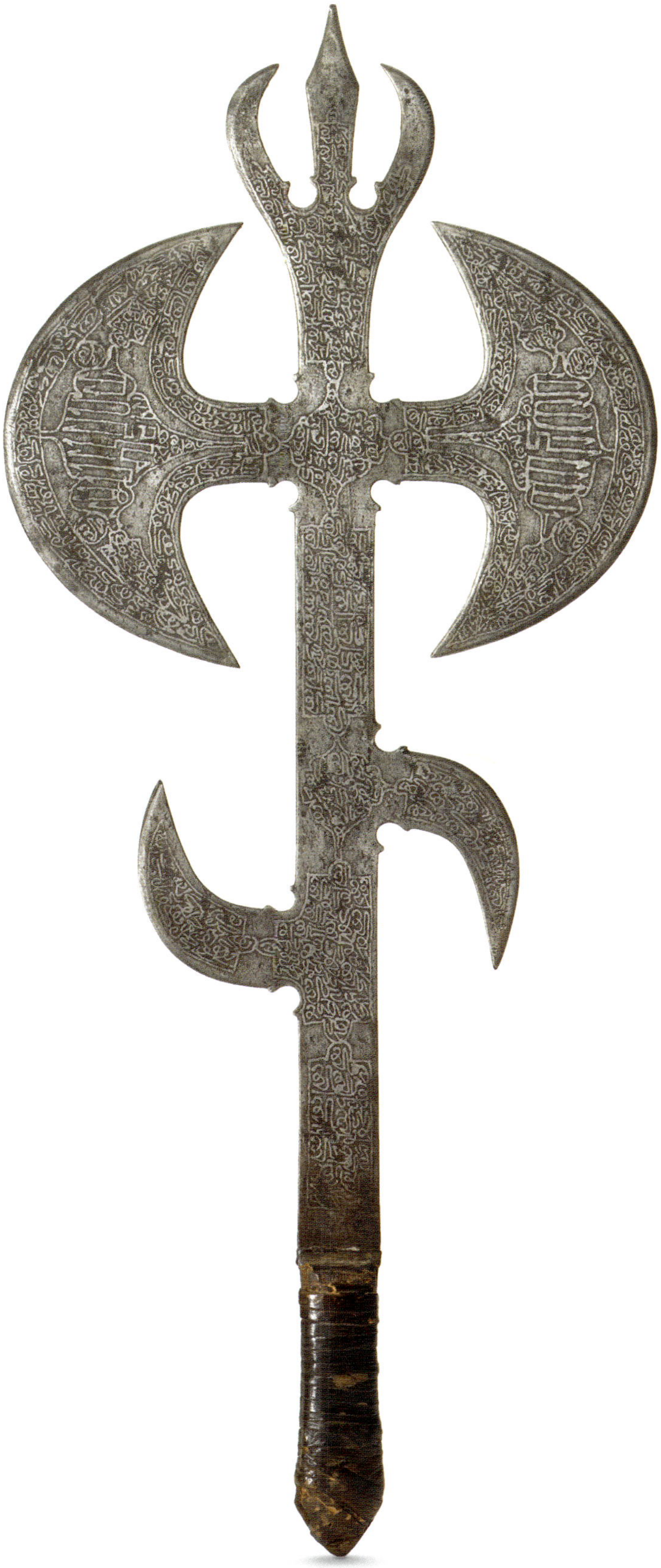

D.5
Artist unknown (Sudan)
Acid-etched scepter, 19th century
Iron, leather, wood
H: 70 cm
FOWLER MUSEUM AT UCLA, X65.3646;
GIFT OF THE WELLCOME TRUST

This elegant scepter recalls Middle Eastern designs associated with the Umma, or community of all Muslims. Its rich calligraphy includes the stylized exhortations "No God but God" (S. Hassan pers. comm., 2017) and the Qur'anic verse Ayat al-fath, "Victory is from God and triumph is near." As Christiane Gruber suggests, "typically such inscriptions are...both apotropaic and predictive, channeling and propelling divine intervention" (pers. comm., 2017). Less readily legible writing suggests *batin*, "the hidden dimension" of existence that may have addressed jinn who, as witnesses to unimaginably distant events, transcend human memory (Taneja 2013).

Mercenaries deployed to the Sudan by the Ottoman Khedive of Egypt included European officers, and the English soldier-of-fortune Charles "Chinese" Gordon was appointed Governor General of the Sudan. When Gordon was killed by Madhist forces, the "humiliation and anger" and "implacable hostility aroused" left many British, including Queen Victoria herself, it was said, "rancorous over a decade later" (paraphrasing Holt 1970, 223–24). Revenge was expected by the Khalifa, and British General Horatio Kitchener's "re-conquest" of the Sudan annihilated tens of thousands of Mahdist fighters with machine guns and other weapons of mass destruction. In late 1899, the Khalifa fell as did his state. An eyewitness reported that Mahdist "arms of all sorts and broken banner-staves were scattered over the field," including "spears in hundreds, some of enormous length, javelins, knobkerries, hatchets, swords and knives." The Fowler's calligraphic blades and mail coif may have been among trophies taken from this "horrorscape."[3]

Only one of the Fowler's Mahdist blades is edged: a double-dagger "*haladie*" based upon an Indian weapon of hand-to-hand combat (see fig. D.1). The others were cut from sheet steel, perhaps repurposed from railroad and river-shipping scrap (Spring 1993, 78), and hafted as standards and scepters to rouse and rally. The blades reflect the crossroads of Mahdist ethnic identities: two suggest Middle Eastern iconographies (see figs. D.2, D.5); one is shaped like a spear point (see fig. D.4); and two mime throwing knives, one Congolese (see fig. D.6), the other perhaps Ethiopian (see fig. D.3). They may have been carried as emblems of rank by central African men pressed into the Mahdi's service, and similar blades were probably sent southward as prestige gifts.[4]

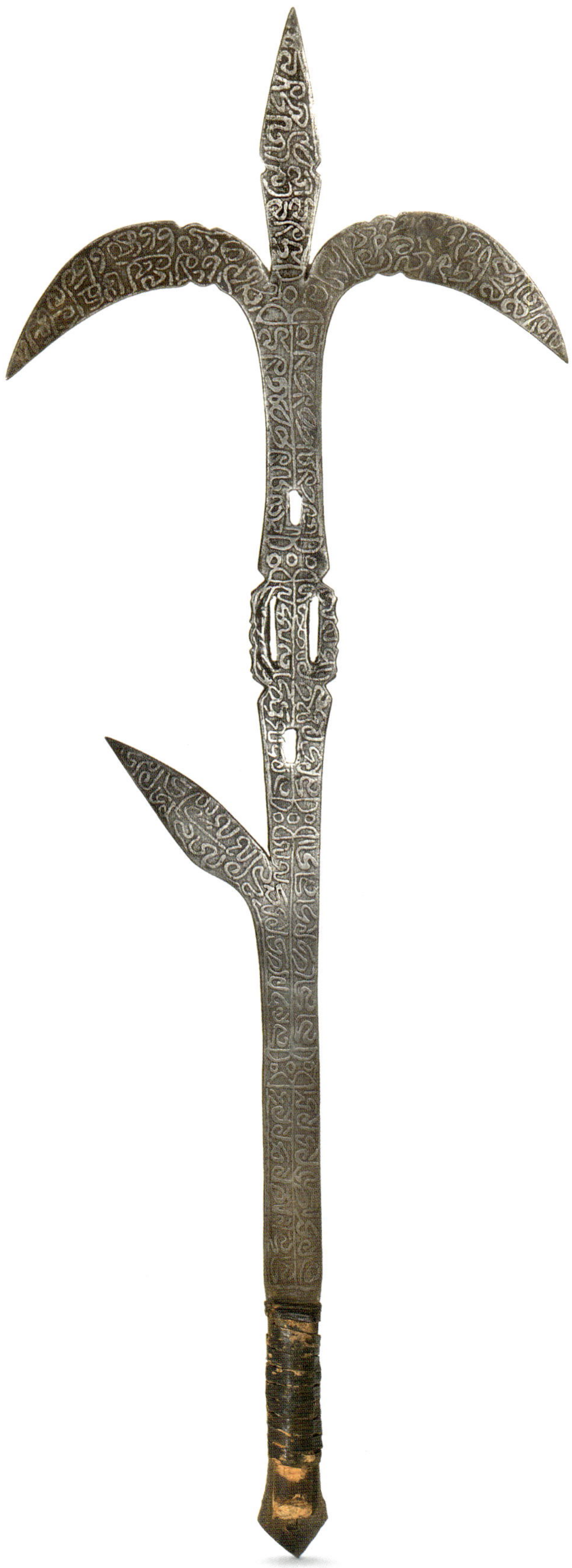

D.6
Artist unknown (Hausa peoples, Nigeria)
Acid-etched presentation blade, late 19th century
Iron, leather
H: 73.6 cm
FOWLER MUSEUM AT UCLA, X65.3645; GIFT OF THE WELLCOME TRUST (NOT IN EXHIBITION)

While documentation does not explain why this scepter is attributed to Hausa, this possibility suggests intriguing histories of trade, politics, and devotion. The Sword of Bayajidda is famed among Hausa of the Daura Emirate of northern Nigeria (Bivar 1964), yet its design and acid-etched calligraphy suggest Sudanese manufacture and long-distance trade. Such inscriptions would have been deemed miraculous by Hausa, bringing blessings to any privileged to see, hold, and, especially, possess what may have been understood as a divine Sword of Truth.

Acid-etched calligraphy in Thuluth script was achieved using a resist. The blades were highly heated to "fire-blue" their surfaces and so set off their inscriptions. As they glinted in the fierce sun of battlefields, the blades implied enlightened supplications as "an inspired attempt to impress the sounds of God upon…article[s] of war," through what might be understood as a *quranan 'ajaban*, or "marvelous recital."[5]

Most inscriptions on the blades and the prayer packets of the mail coif are not readily legible. As René Bravmann has presciently asserted regarding the Fowler's Mahdist artifacts, "it would be easy enough to dismiss these signs as the unsuccessful attempt of an illiterate to emulate Arabic script, but this seems too simple an explanation, blinding one to the possibility of deeper intent. One feels, somehow, that these etched shapes have definite meaning…[and] communicate the desire of the craftsman to capture the grace and flow of God's language and His inherent presence" (1983, 54). Again though, for whose eyes might they have *been* legible?

Striking in the calligraphy are rhythmic repetitions that evoke *dhikr* chanting of Allah's holy names as well as the cadenced iterations of sura 55 of the Qur'an, which Allah addresses to men and jinn. Inscriptions like these bear "unusual long successions of identical consonants often interrupted by mysterious underspecified (covert) letters," incantatory diacritics, "strategic omissions,"

D.7
Artist unknown (Sudan)
Warrior's helmet, late 19th century
Iron, leather, cotton textile, string
H: 61 cm
FOWLER MUSEUM AT UCLA, X65.8684;
GIFT OF THE WELLCOME TRUST
(NOT IN EXHIBITION)

This remarkable helmet from the Mahdist period of the Sudan complements the calligraphic blades of the Wellcome Collection. Three sorts of imported and locally forged iron mail are combined (Arkell 1956; Bivar 1964), assembling histories of extensive trade and pious purpose. Prayer packets bring active efficacies to the headpiece, as do devices tooled into the leather holding them. The thirteen downstrokes of one packet recall an *alif*, the first letter of Allah's name, and are "a living reminder of divine creativity and its manner of manifesting the unseen" (Sharify-Funk n.d.). So prepared, courage was without limit, sacrifice assumed, valor assured.

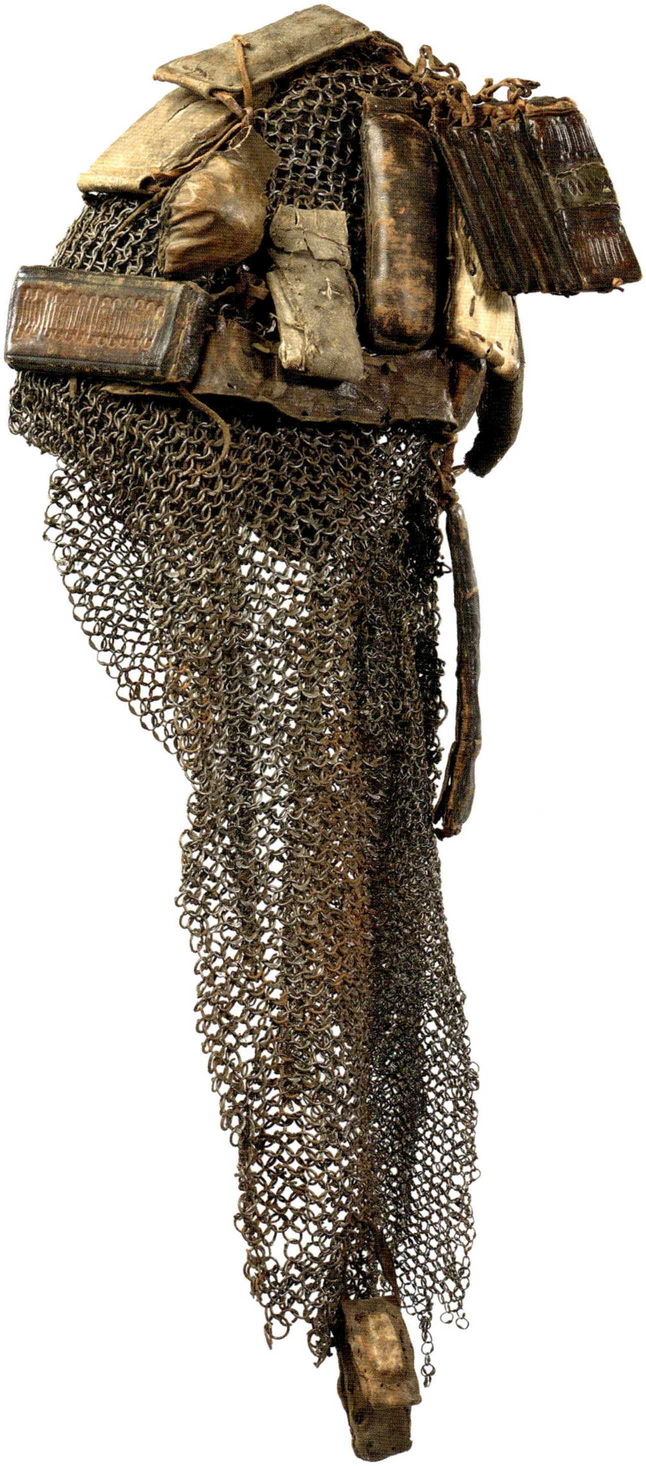

mirrored words and letters, and enigmatic designs that are all "important features of supernatural languages in Muslim Africa" used in talismans and prayers.[6] Might the hidden side (*batin*) of the Mahdist calligraphy have summoned jinn?

Jinn are mentioned frequently in the Qur'an and are "a part of life for all God-fearing Muslims" (Shah 2010, xii). Through visions, the Mahdi knew that a host of angels would guide and protect his followers, as they had the Prophet's (Farwell 1989, 45); but like Muhammad and the Prophet Solomon before him (Qur'an 72, 34:12), one suspects that the Mahdi was also master of jinn. Allah created jinn of "smokeless fire" (Qur'an 15:27), and they are far stronger than humans, live much longer lives, travel vast distances instantaneously, and infiltrate the interstices of our world (Taneja 2013, 140; El-Zein 2017, 10). Like humans and unlike angels, jinn exercise free will, and some are helpful, some baleful, and all are unpredictable. In the Khalifa's final battles when slaughter was inevitable, might a company of jinn respond to ardent exhortations and join Mahdist forces as they had the armies of King Solomon (Qur'an 27:17)? Might they direct their immeasurable strengths against infidel enemies and permit the righteous to prevail in some small measure? There could be no guarantee that jinn would answer any such entreaties; yet who would not hope that they might?[7]

PORTFOLIO

I

CURRENCIES

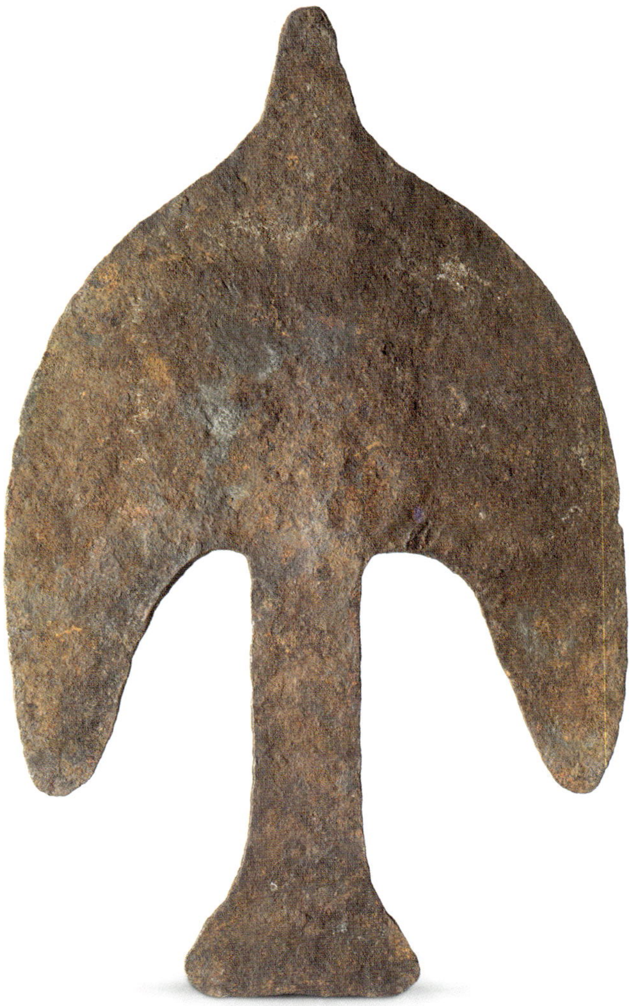

The shapes of precolonial iron currency tokens were typically derived from blade forms characteristically similar to tools and weapons; the work accomplished by forged iron objects was so essential to local economies that representations of them were synonymous with value itself. A wide variety of currency denominations were produced throughout the African continent and often configured as blades or bars in a range of suitable sizes for exchange in measured bundles of several units or more. Some were elaborate in design and dramatic in scale—as delicate as a tendril, as tall as a person. They drew worth from the quality and weight of the iron used and gained efficacy from blacksmiths' forging achievements in producing them. Though most common types were used in producing goods and compensating for services, the most highly valued tokens stood as payment in the exchanges that mattered most in life: marriage; litigation; ransom of battle captives; and the purchase of horses, enslaved persons, and other prized commodities.

I.1
Artist unknown (Karamajong peoples, Uganda)
Spear-blade-shaped currency, mid-19th century
Iron
H: 38 cm
PRIVATE COLLECTION

1.2
Artist unknown (Nkutshu peoples, Democratic Republic of the Congo)
Axe-blade-shaped currency (*iwenga*), 20th century
Iron
H: 55.9 cm
COLLECTION OF MR. AND MRS. HANSEN, BELGIUM

1.3
Artist unknown (Ngbaka, Ngala peoples, Democratic Republic of the Congo)
Spear-blade-shaped currency (*mbili*), late 19th to early 20th century
Iron
L: 50.5 cm
FOWLER MUSEUM AT UCLA, X65.9164; GIFT OF THE WELLCOME TRUST

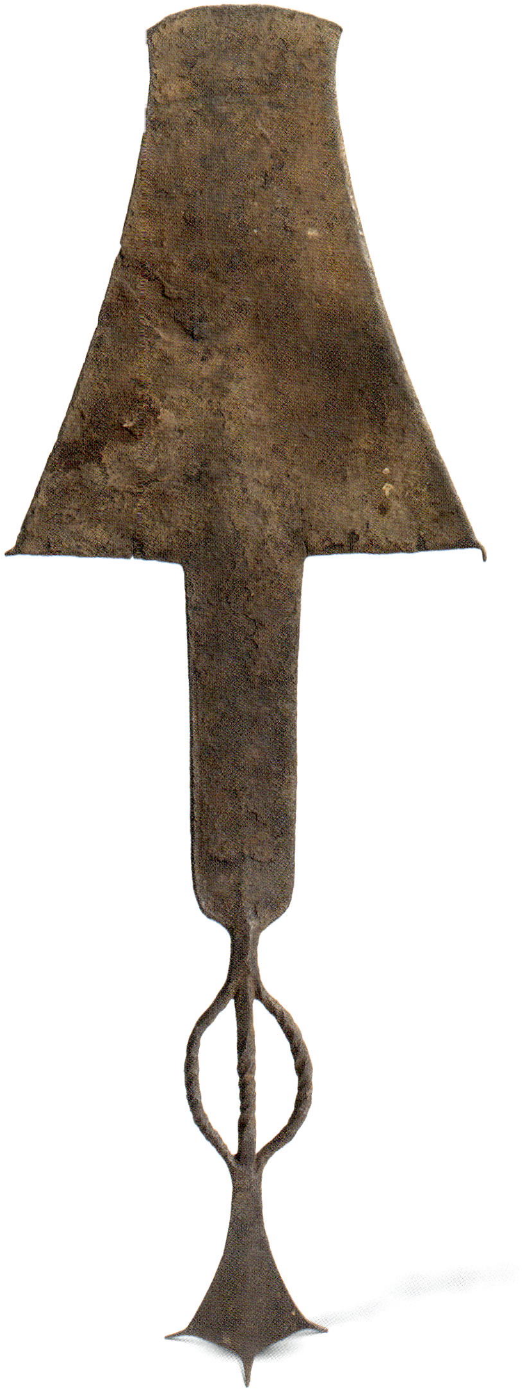

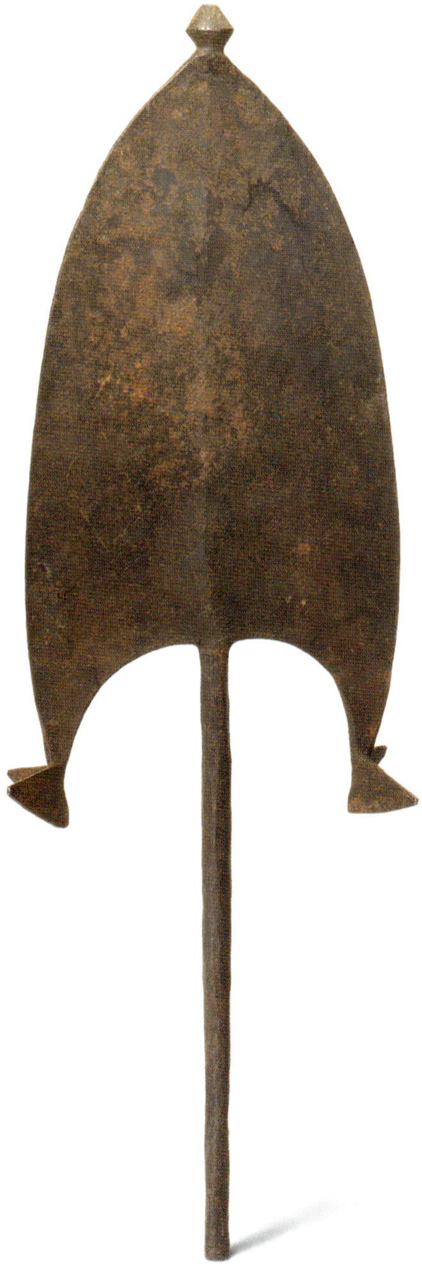

1.4
Artist unknown (Bamileke peoples, Cameroon)
Hoe-blade-shaped currency, late 19th to early 20th century
Iron
H: 47 cm
PRIVATE COLLECTION

1.5
Artist unknown (Niyamwezi peoples, Tanzania)
Hoe-blade-shaped currency, early 20th century
Iron
H: 74 cm
ITZIKOVITZ COLLECTION
PHOTOGRAPH BY BRIGITTE CAVANAGH, 2017.

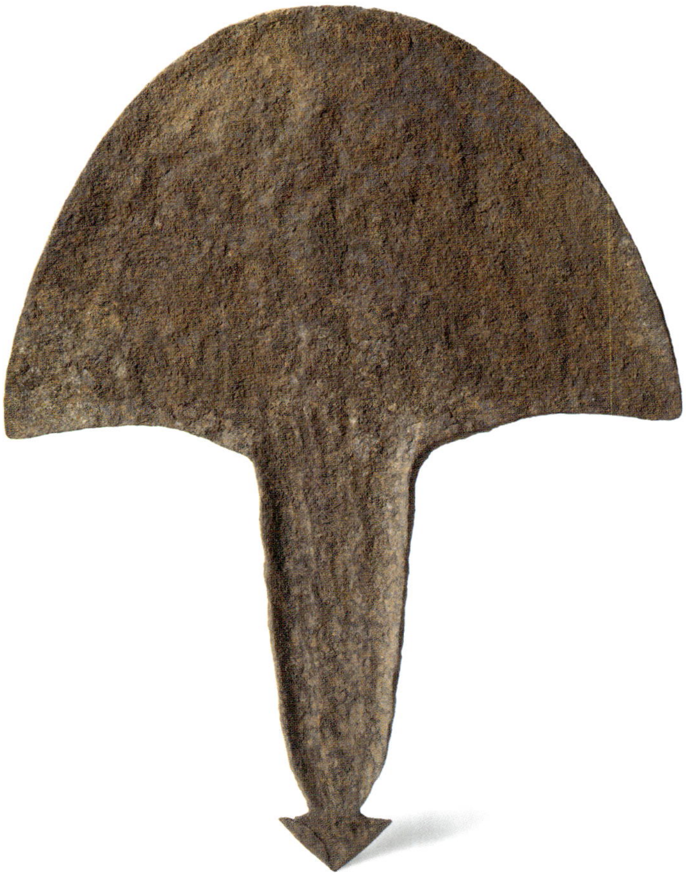

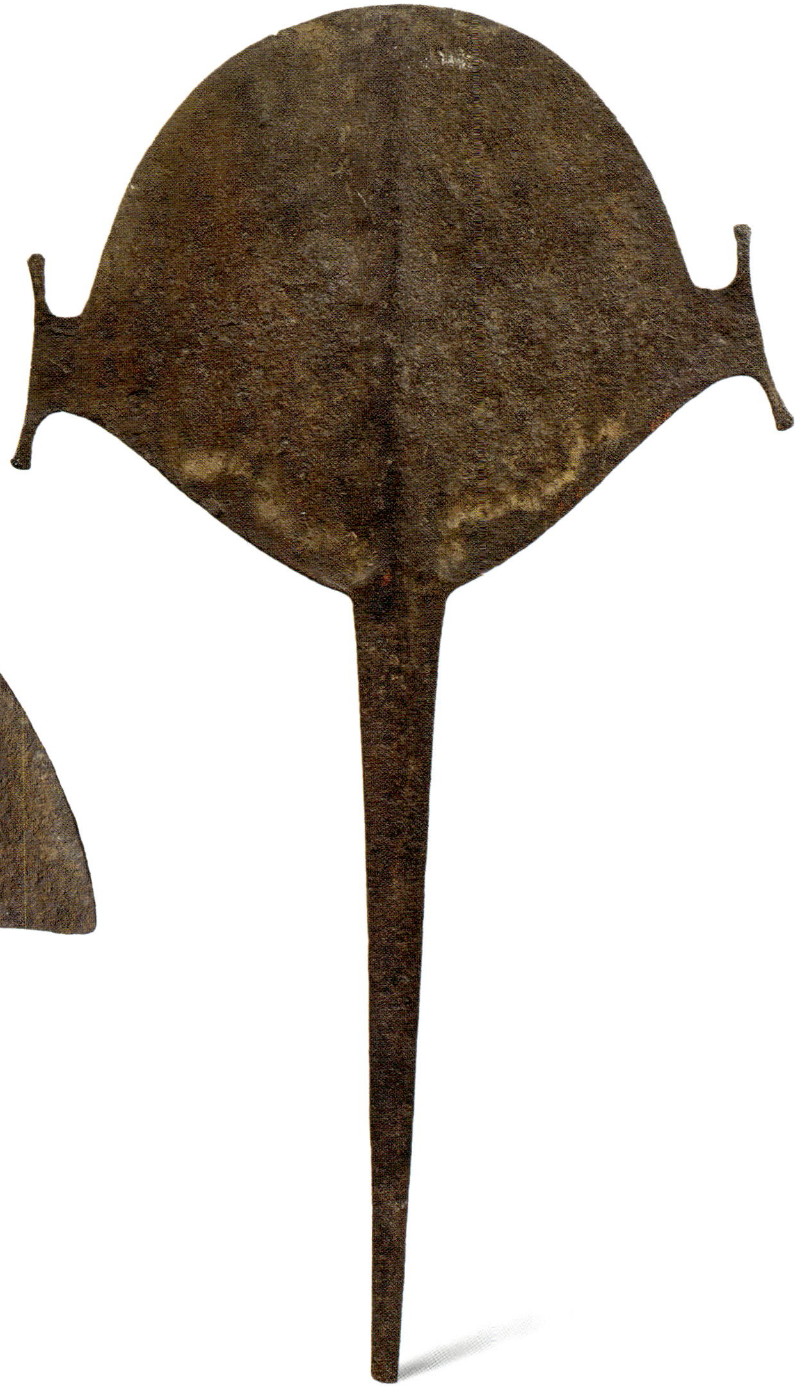

1.6
Artist unknown (Mbugbu peoples, Central African Republic)
Throwing knife-shaped currency (*ngindza*), late 19th century
Iron
H: 27.5 cm
BARBIER-MUELLER COLLECTION, 1026-195

1.7
Artist unknown (Ngongo-lyembe/Ngelima peoples, Democratic Republic of the Congo)
Bridewealth currency, late 19th century
Iron
H: 118 cm
PRIVATE COLLECTION

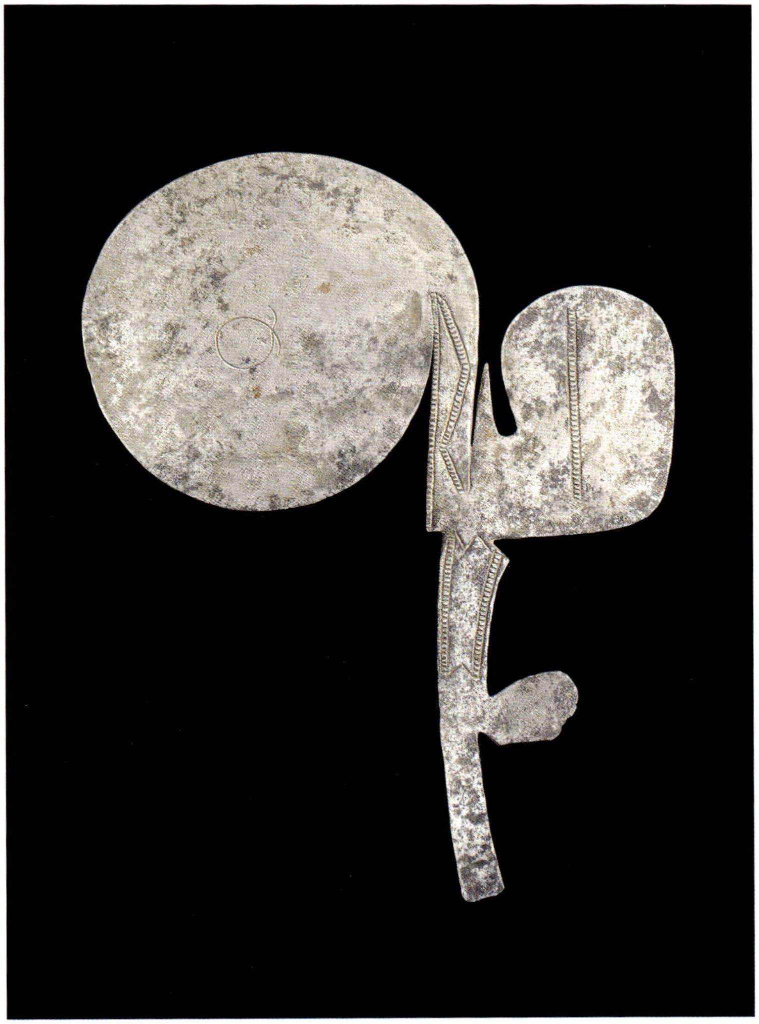

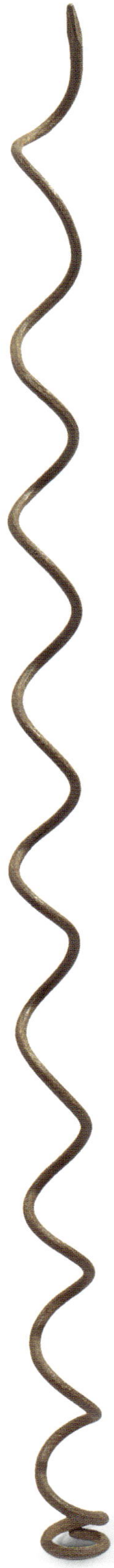

PORTFOLIO

II

BLADES

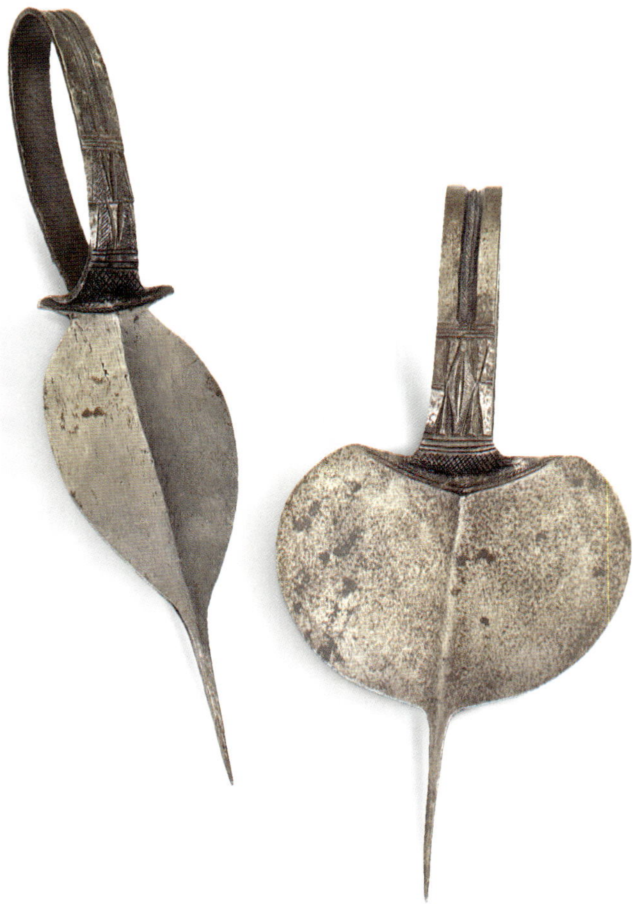

The efficacies of African blades include their capacity to transmit potentialities—social, political, and spiritual. Blades produced as weapons have served defensive purposes and achieved warlike ends. Blacksmiths have also transformed such weapons into insignia capable of performing political power. Blades have been deliberately removed from practical use by having their edges dulled; their aerodynamic balance decreased; their forms made larger, thinner, and more ornate than could serve ordinary purposes. These formidable and often magnificent forgings offered sensual pleasures of handling and also aestheticized power to communicate key ideas about honor, prosperity, prestige, and sophistication.

II.1
Artist unknown (Tiv peoples, Nigeria)
Armband dagger, late 19th to early 20th century
Iron
H: 21 cm; 23.5 cm
FOWLER MUSEUM AT UCLA, X65.3606, X65.3607; GIFT OF THE WELLCOME TRUST

II.2A–B
Artist unknown (Kota-Obamba peoples, Gabon)
Ceremonial blade, 19th century
Iron, wood, copper
H: 64 cm
Private collection
PHOTOGRAPH BY ANNE DE KNOCK, 2008. COURTESY OF PRIVATE COLLECTION

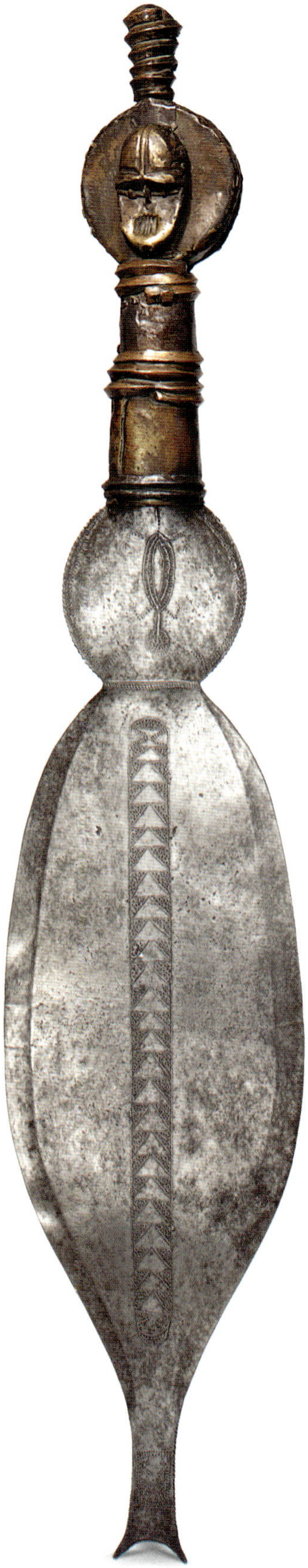

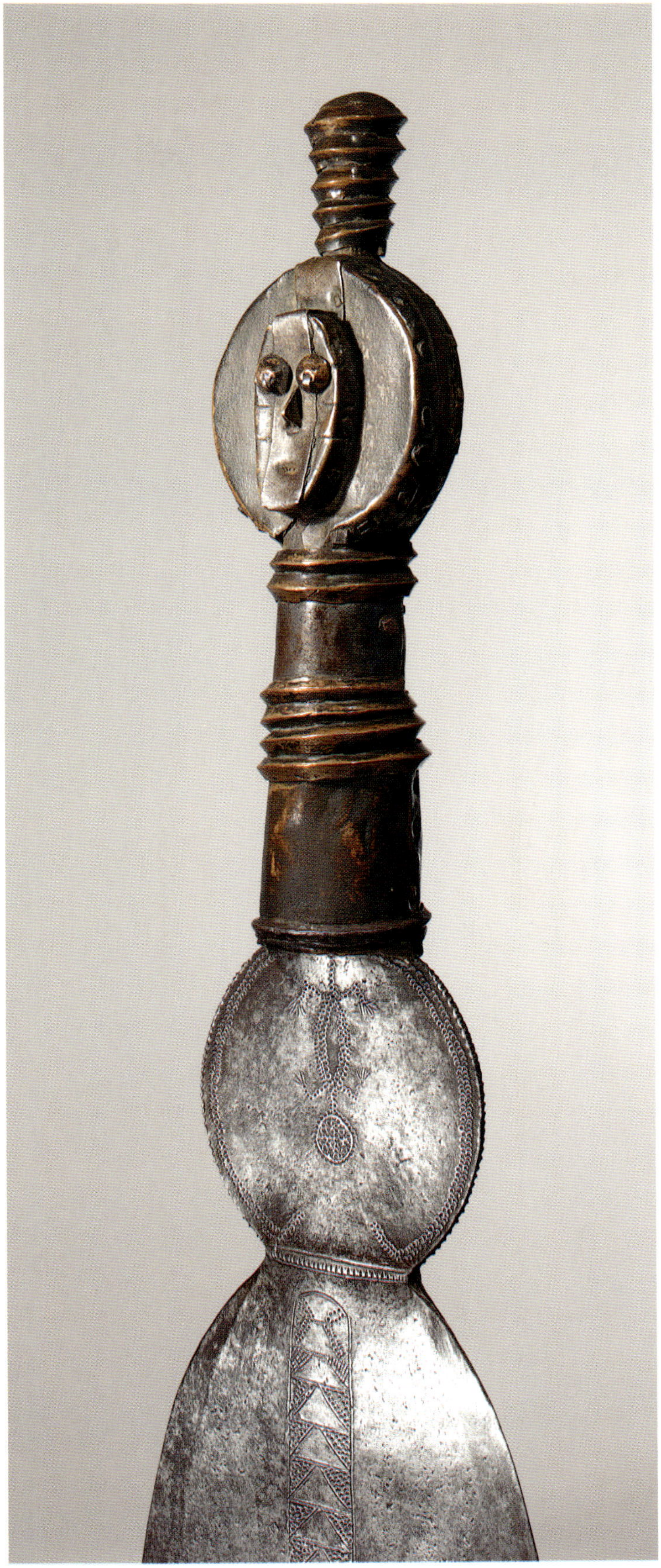

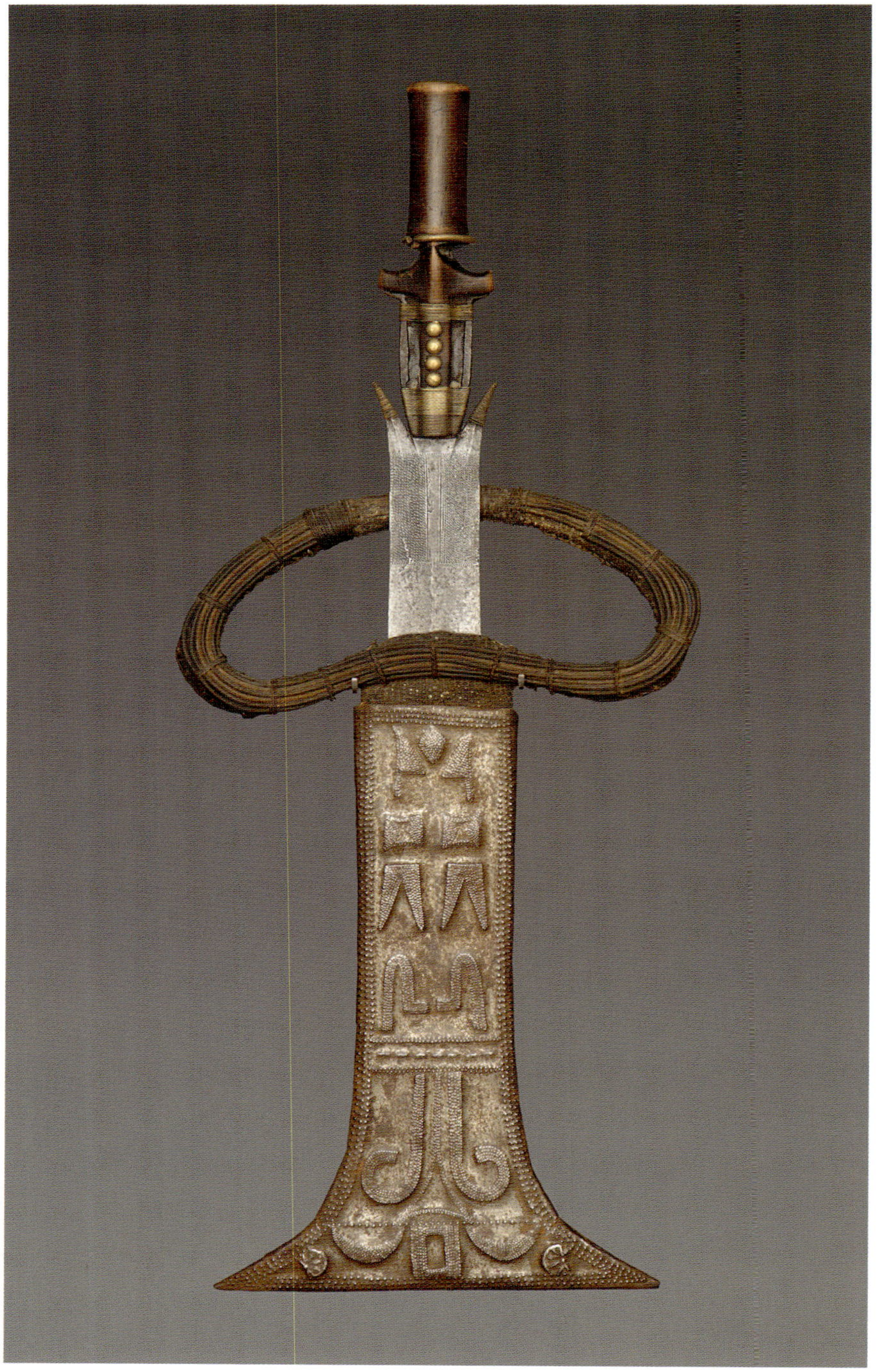

II.3
Artist unknown (Fang peoples, Gabon)
Sword with sheath, late 19th to early 20th century
Iron, brass, wood, copper, leather, plant fiber
H: 45 cm
PRIVATE COLLECTION
PHOTOGRAPH BY NATHALIE AND ALAIN SPELTDOORN, 2017.
COURTESY OF PRIVATE COLLECTION

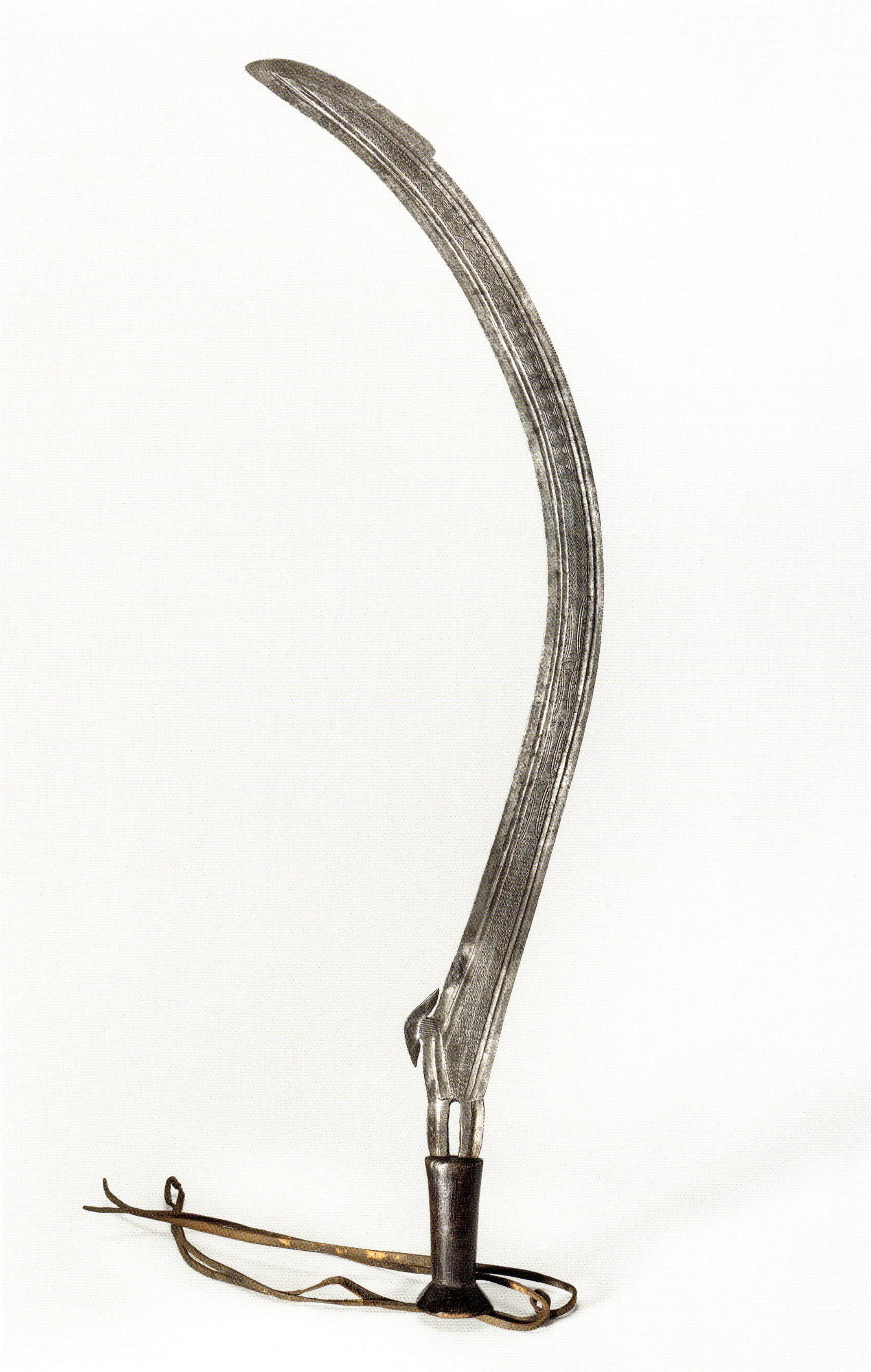

II.4
Artist unknown (Boa, Binja, Bandia, Nsakara peoples, Democratic Republic of the Congo and Central African Republic)
Sickle sword (*mbonga*), early 20th century
Iron, wood, leather
H (without fringe): 85 cm
PRIVATE COLLECTION

II.5
Artist unknown (Boa, Binja, Bandia, Nsakara peoples, Democratic Republic of the Congo and Central African Republic)
Sickle sword (*asumbi*, *eneno*), early 20th century
Iron, wood
H: 89 cm
PRIVATE COLLECTION

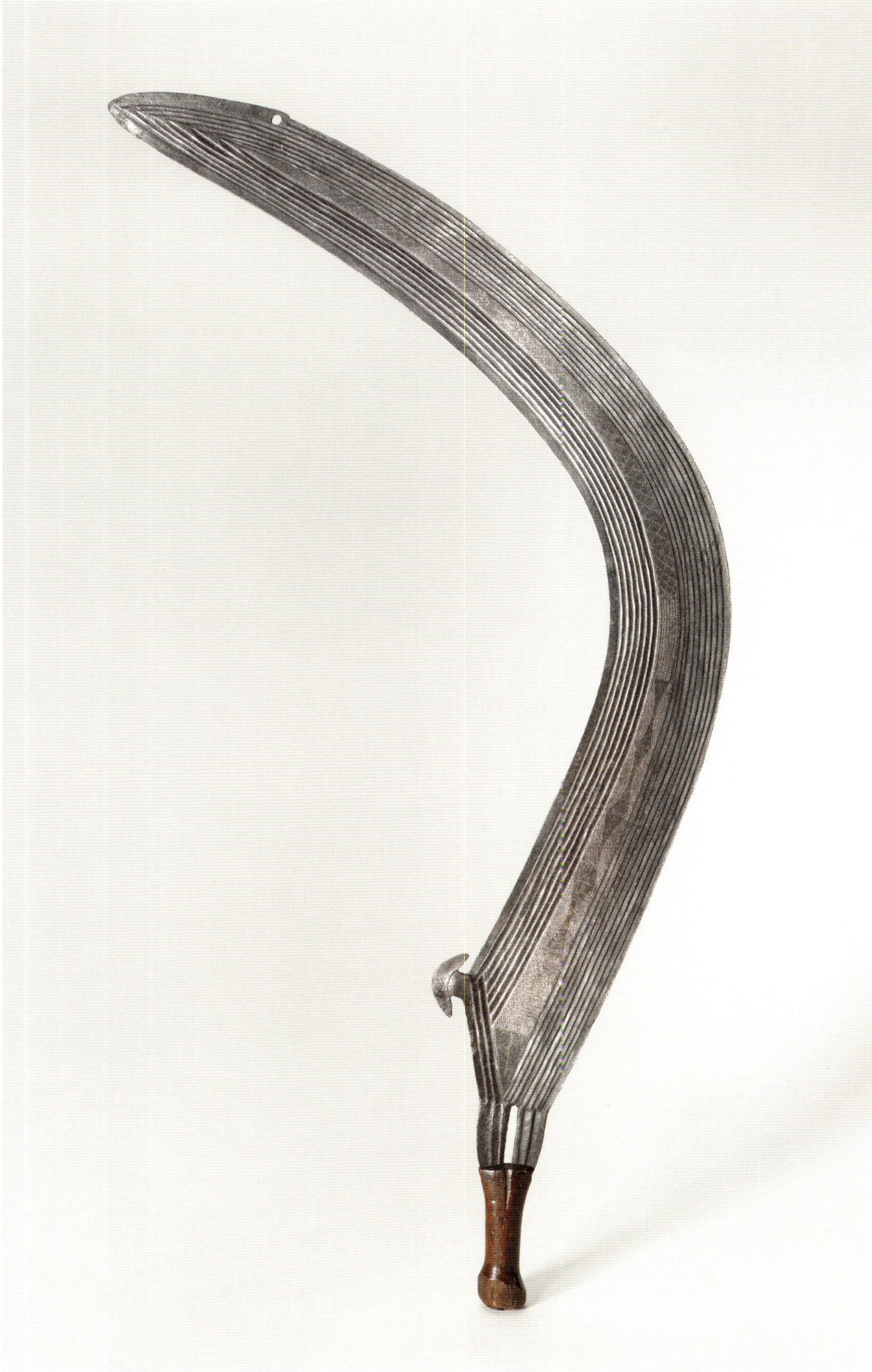

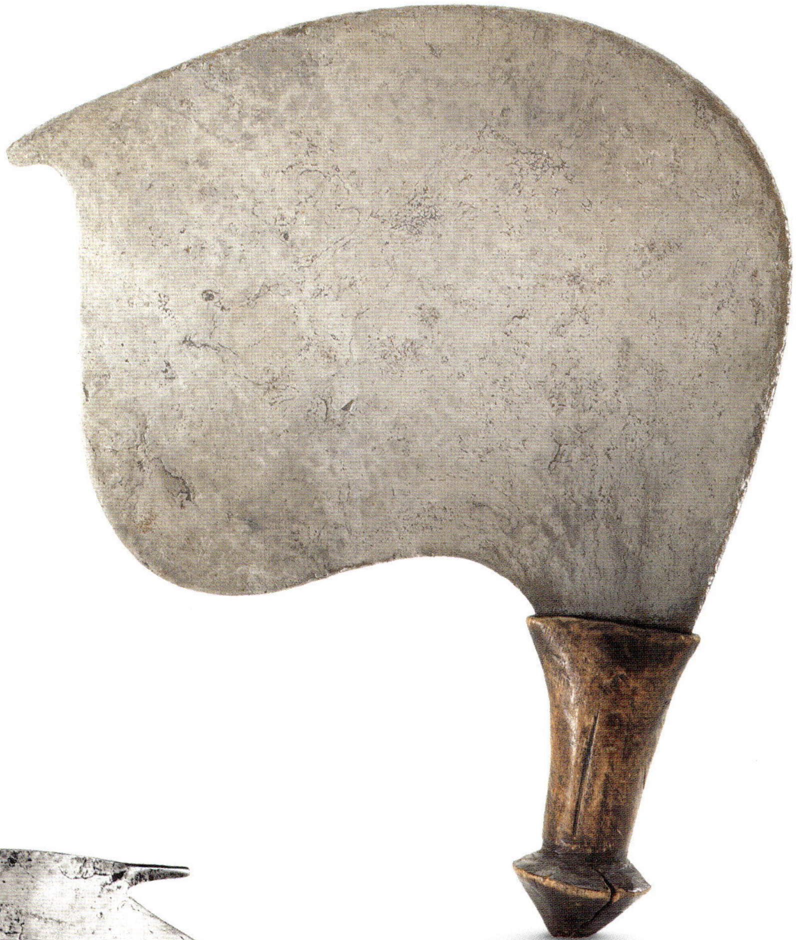

11.6
Artist unknown (Ngul, Yanzi peoples, Democratic Republic of the Congo)
Knife (*epalang*, *ukanglate*), 19th century
Iron, wood
H: 28.5 cm
FOWLER MUSEUM AT UCLA, X65.2335; GIFT OF THE WELLCOME TRUST

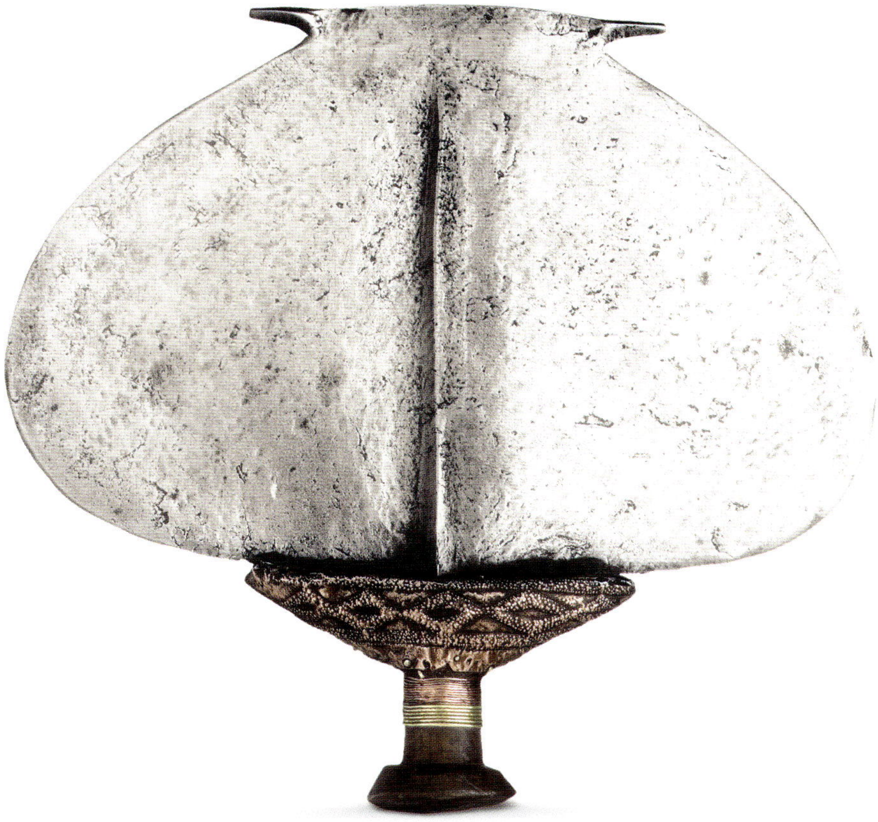

11.7
Artist unknown (Mbuun, Yanzi, Ngul, Lwer, Dzing peoples, Democratic Republic of the Congo)
Knife (*mossal*), 19th century
Iron, wood, copper, brass
H: 23 cm
JAN ELSEN COLLECTION
PHOTOGRAPH BY DICK BEAULIEU, 2009.

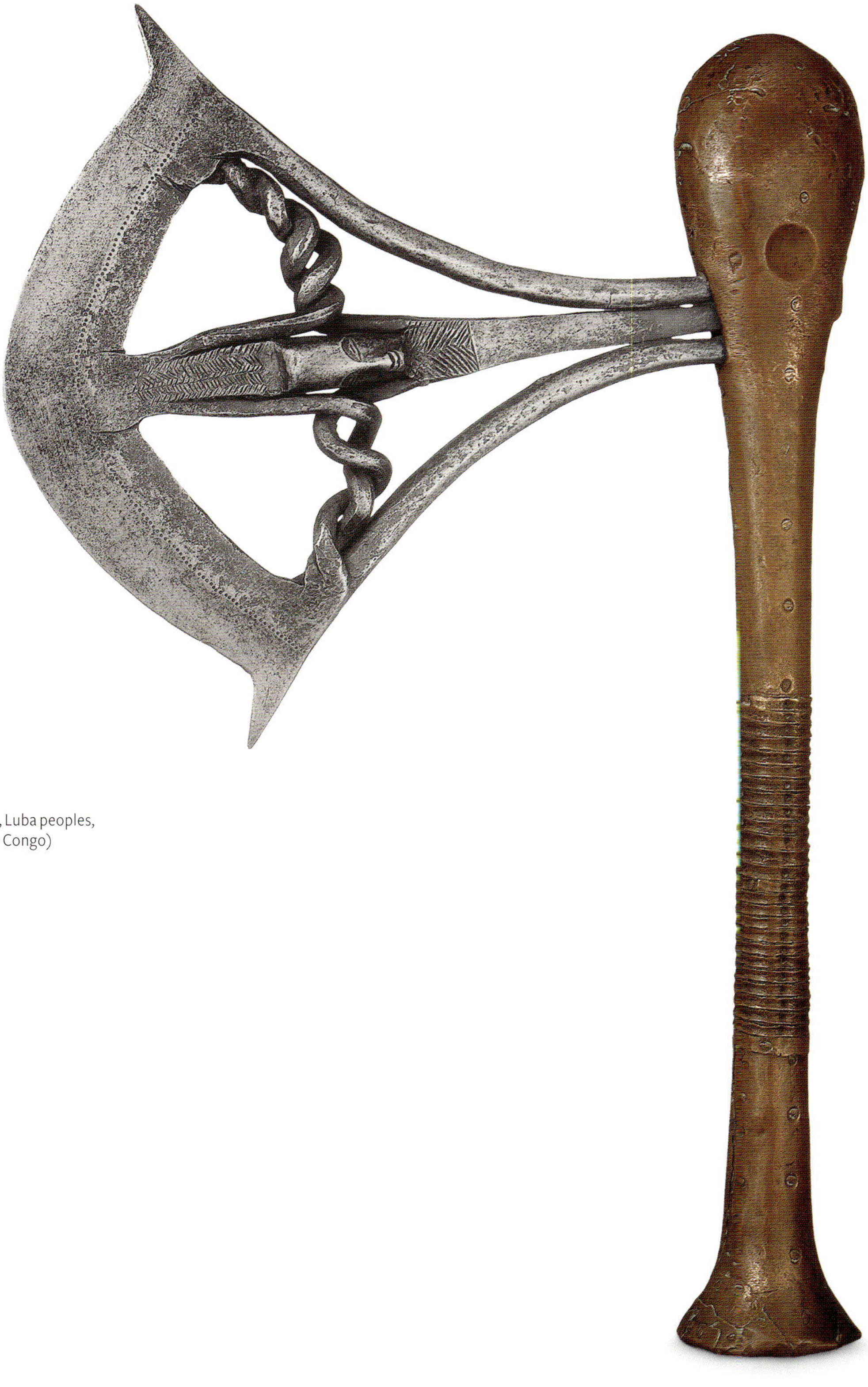

II.8
Artist unknown (Nsapo/Songye, Luba peoples, Democratic Republic of the Congo)
Axe (*kilonda*), 19th century
Iron, copper, wood
H: 43 cm
JAN ELSEN COLLECTION
PHOTOGRAPH BY DICK BEAULIEU, 2009.

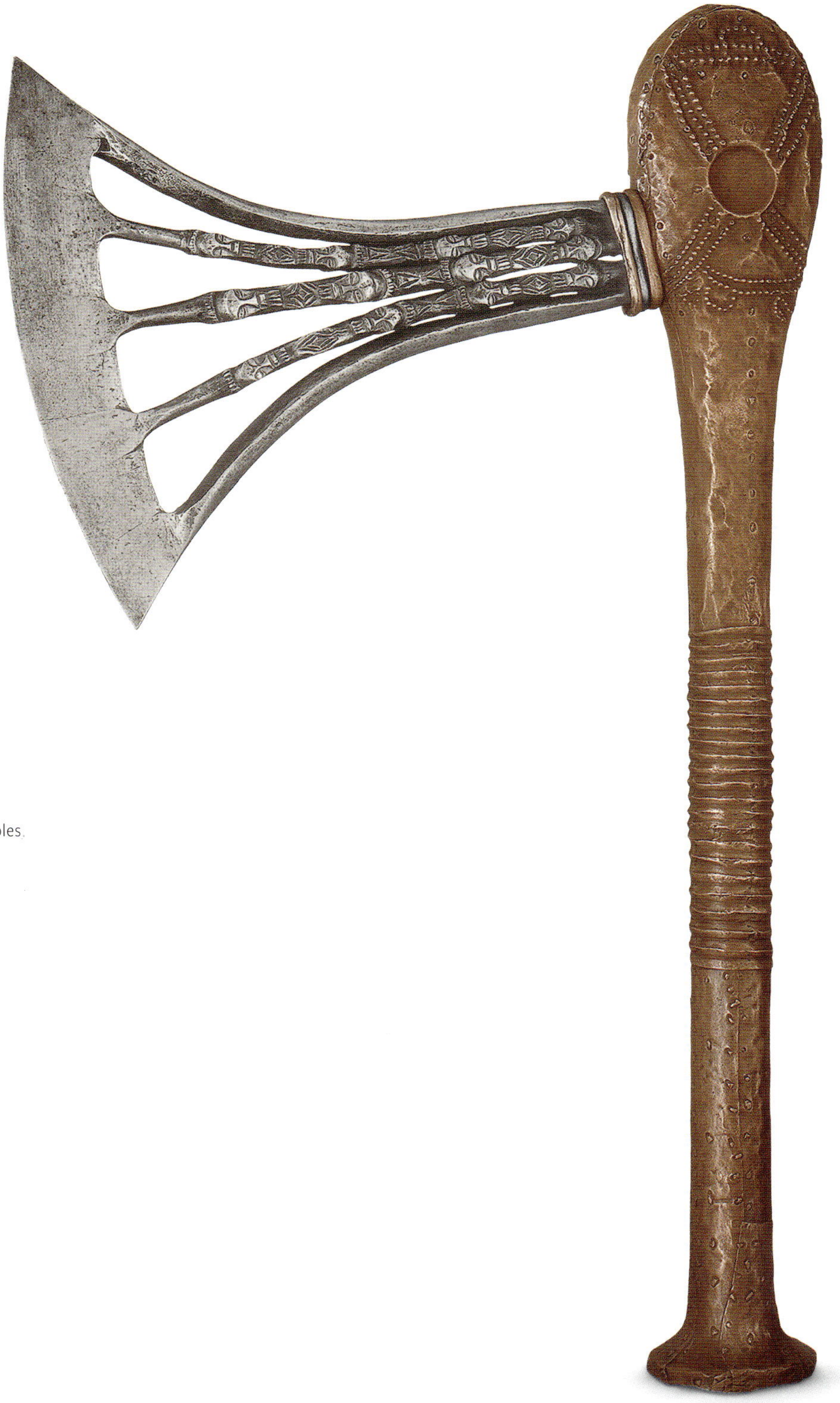

11.9
Artist unknown (Nsapo/Songye, Luba peoples,
Democratic Republic of the Congo)
Axe (*kilonda*), 19th century
Iron, copper, wood
H: 43 cm
ETHNIC ART & CULTURE LTD.
PHOTOGRAPH BY DICK BEAULIEU, 2009.

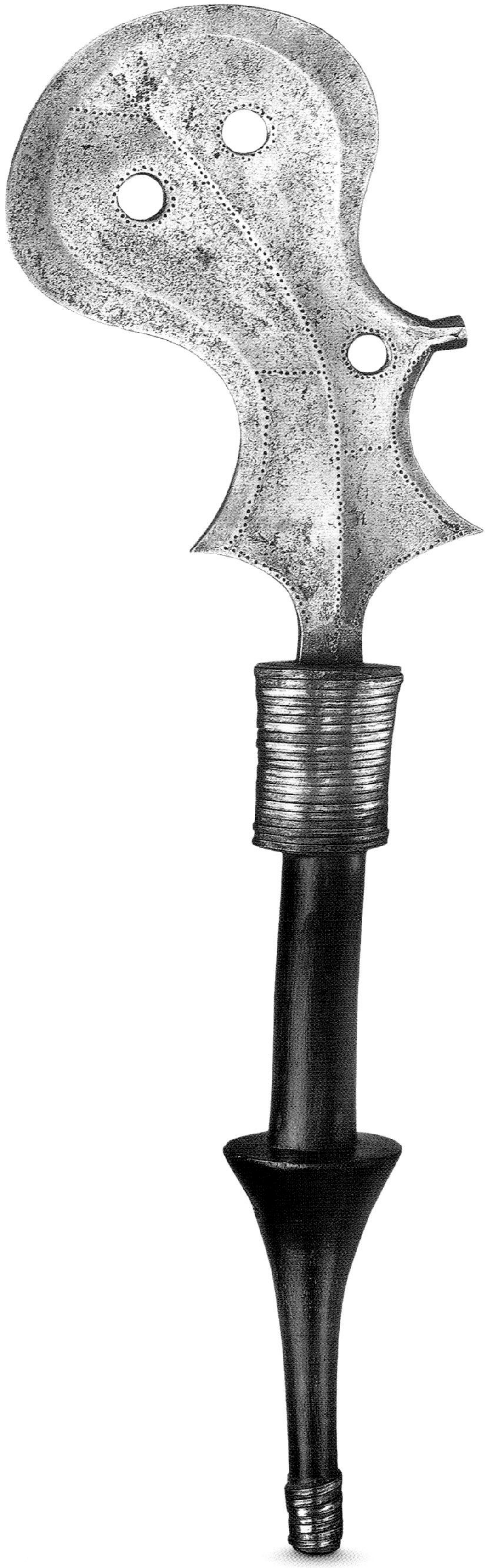

II.10
Artist unknown (Momvu, Makere peoples, Democratic Republic of the Congo)
Sickle knife (*uggo*, *nenole*), early 19th century
Iron, wood
H: 49 cm
FELIX COLLECTION
PHOTOGRAPH BY CHARLES MEUR, 2010.

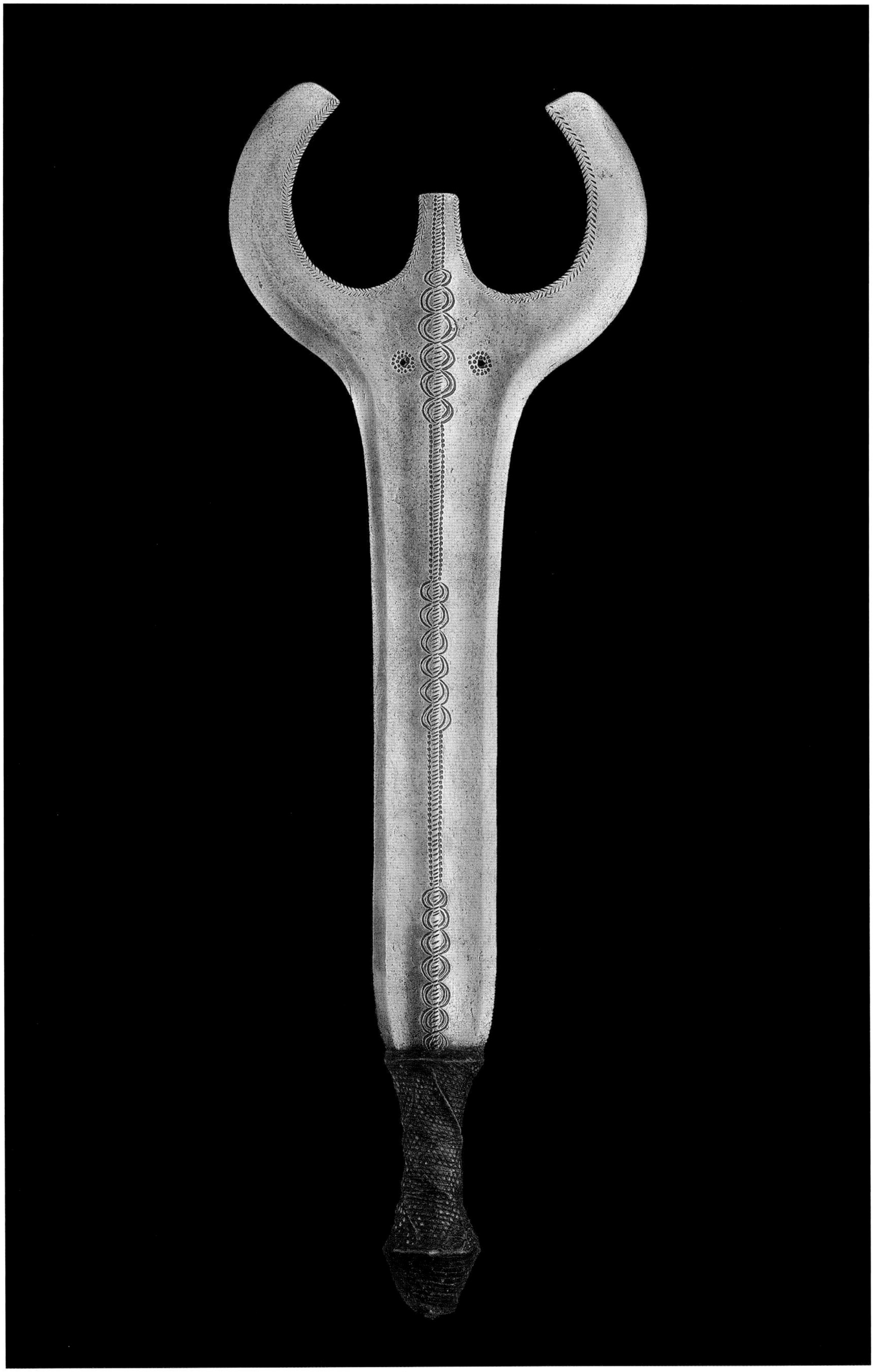

II.11
Artist unknown (Mondzombo, Lobala, Ngbaka peoples, Democratic Republic of the Congo)
Short sword (*bango*), late 19th to early 20th century
Iron, wood, Nile monitor skin (*Varanus niloticus*)
H: 59 cm
JAN ELSEN COLLECTION
PHOTOGRAPH BY DICK BEAULIEU, 2009.

II.12
Artist unknown (Kusu, Songye peoples, Democratic Republic of the Congo)
Knife, 19th century
Iron, copper, wood
H: 49.3 cm
PRIVATE COLLECTION
PHOTOGRAPH BY NATHALIE AND ALAIN SPELTDOORN, 2017.
COURTESY OF PRIVATE COLLECTION.

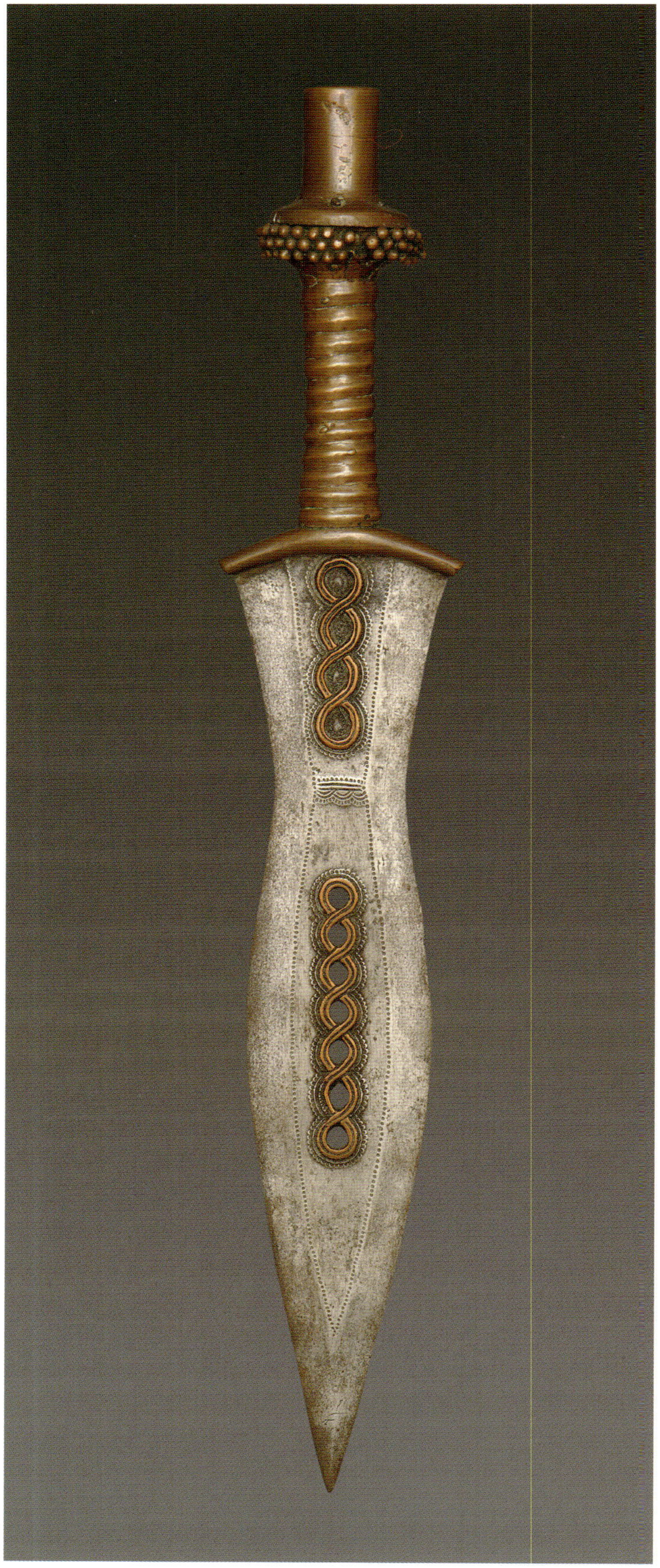

II.13
Artist unknown (So, Ngombe peoples, Democratic Republic of the Congo)
Sword, mid- to late 19th century
Iron, copper, wood
Private collection
H: 80 cm
PRIVATE COLLECTION
PHOTOGRAPH BY NATHALIE AND ALAIN SPELTDOORN, 2017.
COURTESY OF PRIVATE COLLECTION.

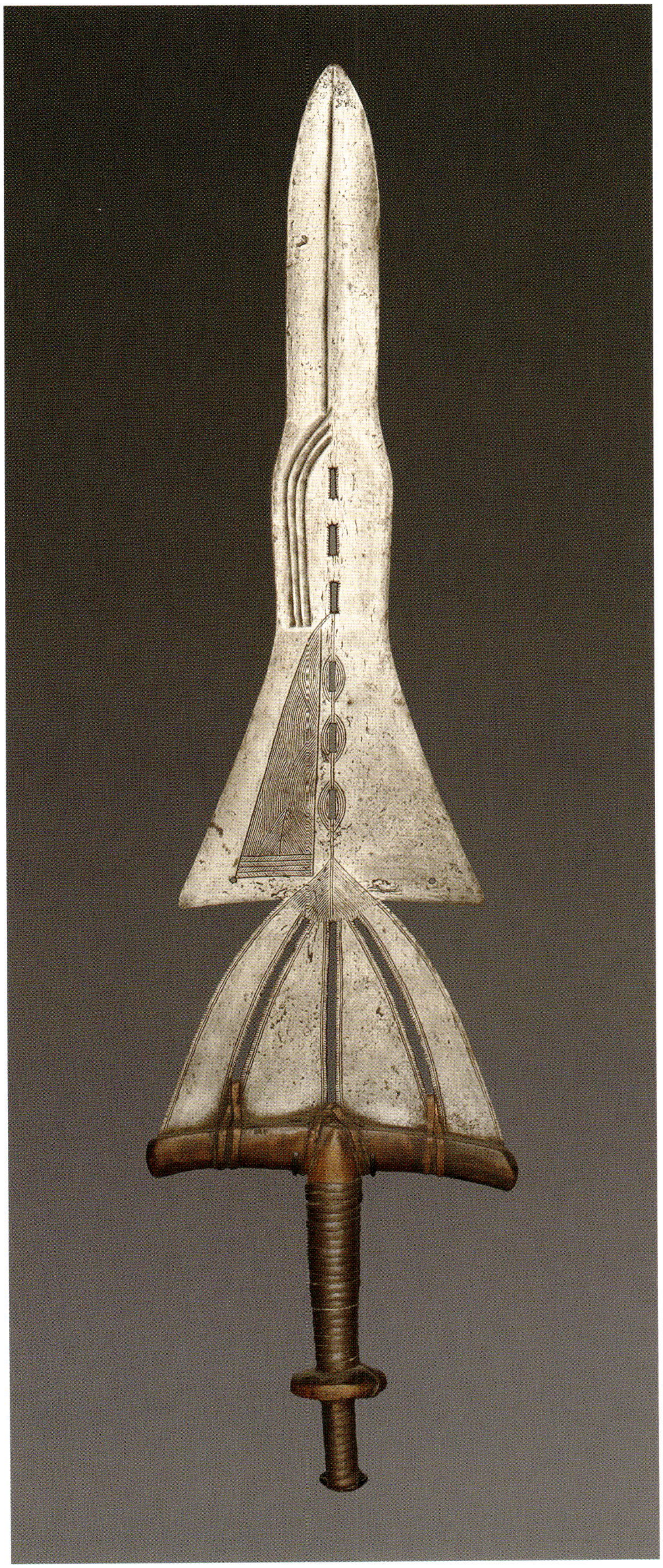

II.14
Artist unknown (Ngombe, Doko, Iboko, Mongo peoples, Democratic Republic of the Congo)
Sword (*ngulu*), late 19th century
Iron, copper, brass, leather, wood, plant fiber
H: 61 cm
FOWLER MUSEUM AT UCLA, X65.2993; GIFT OF THE WELLCOME TRUST

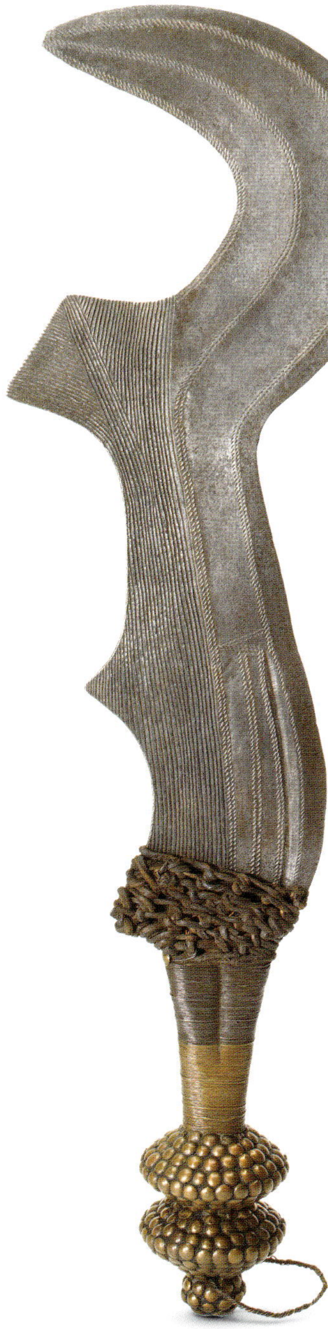

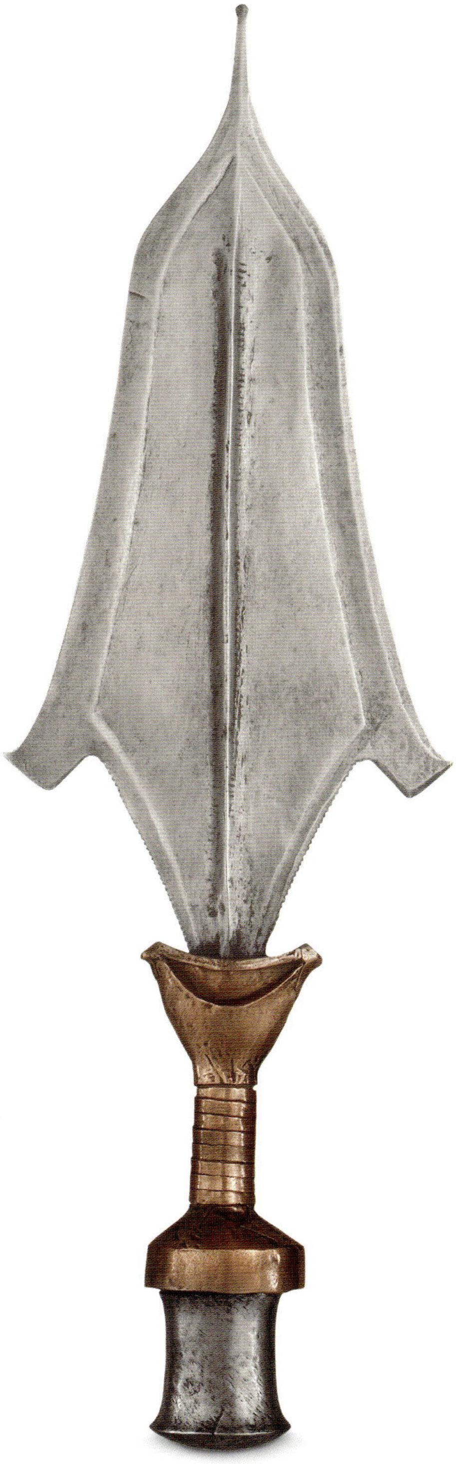

II.15
Artist unknown (Tetela, Nkutshu people, Democratic Republic of the Congo)
Knife (*balingbwa*), early 20th century
Iron, copper, wood
H: 37 cm
JAN ELSEN COLLECTION
PHOTOGRAPH BY DICK BEAULIEU, 2009.

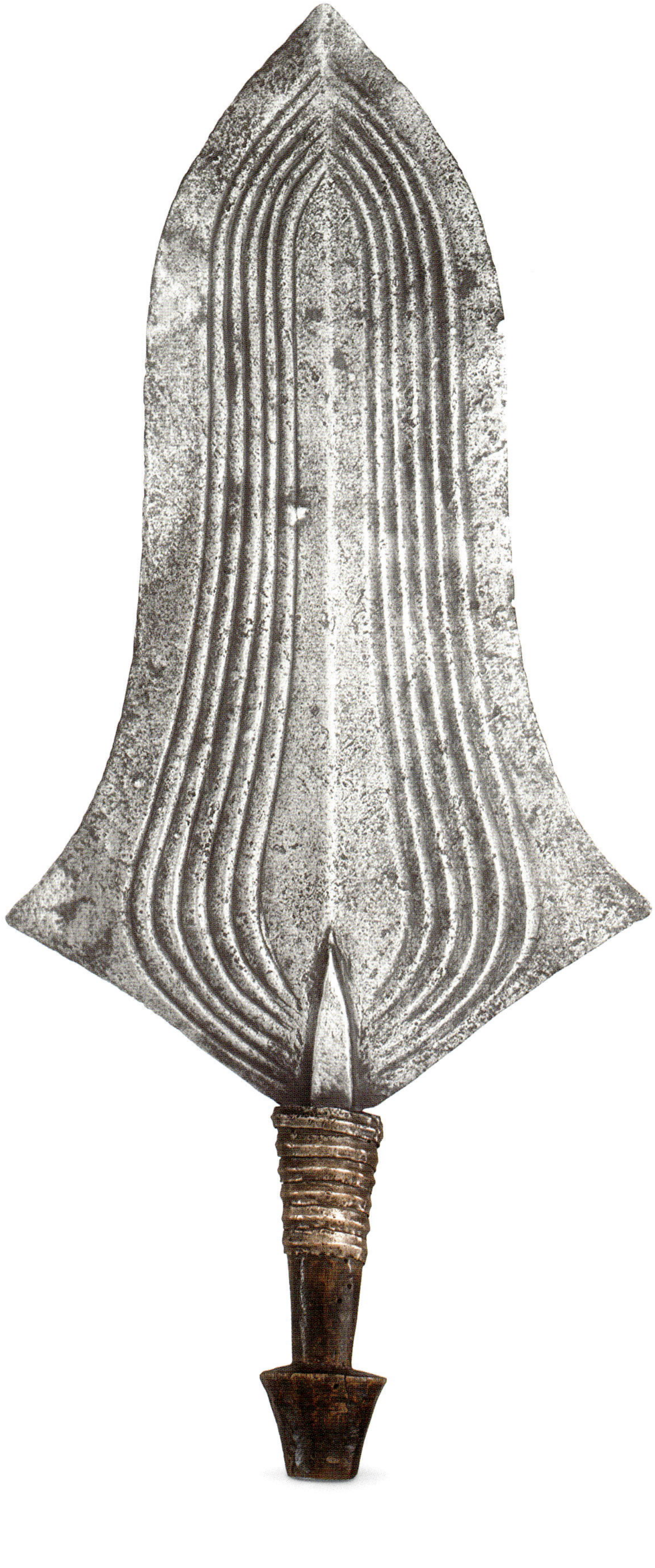

II.16
Artist unknown (Nkutshu, Basogo-Meno peoples, Democratic Republic of the Congo)
Short sword, late 19th to early 20th century
Iron, wood, copper
H: 47 cm
JAN ELSEN COLLECTION
PHOTOGRAPH BY DICK BEAULIEU, 2009.

II.17
Artist unknown (Lokele, So, Mba, Topoke, Genya peoples, Democratic Republic of the Congo)
Knife (*empute*), late 19th to early 20th century
Iron, wood, leather, copper, brass
H: 59.6 cm
BARBIER-MUELLER COLLECTION, GENEVA
PHOTOGRAPH BY STUDIO FERRAZZINI-BOUCHET. COURTESY OF THE BARBIER-MUELLER MUSEUM

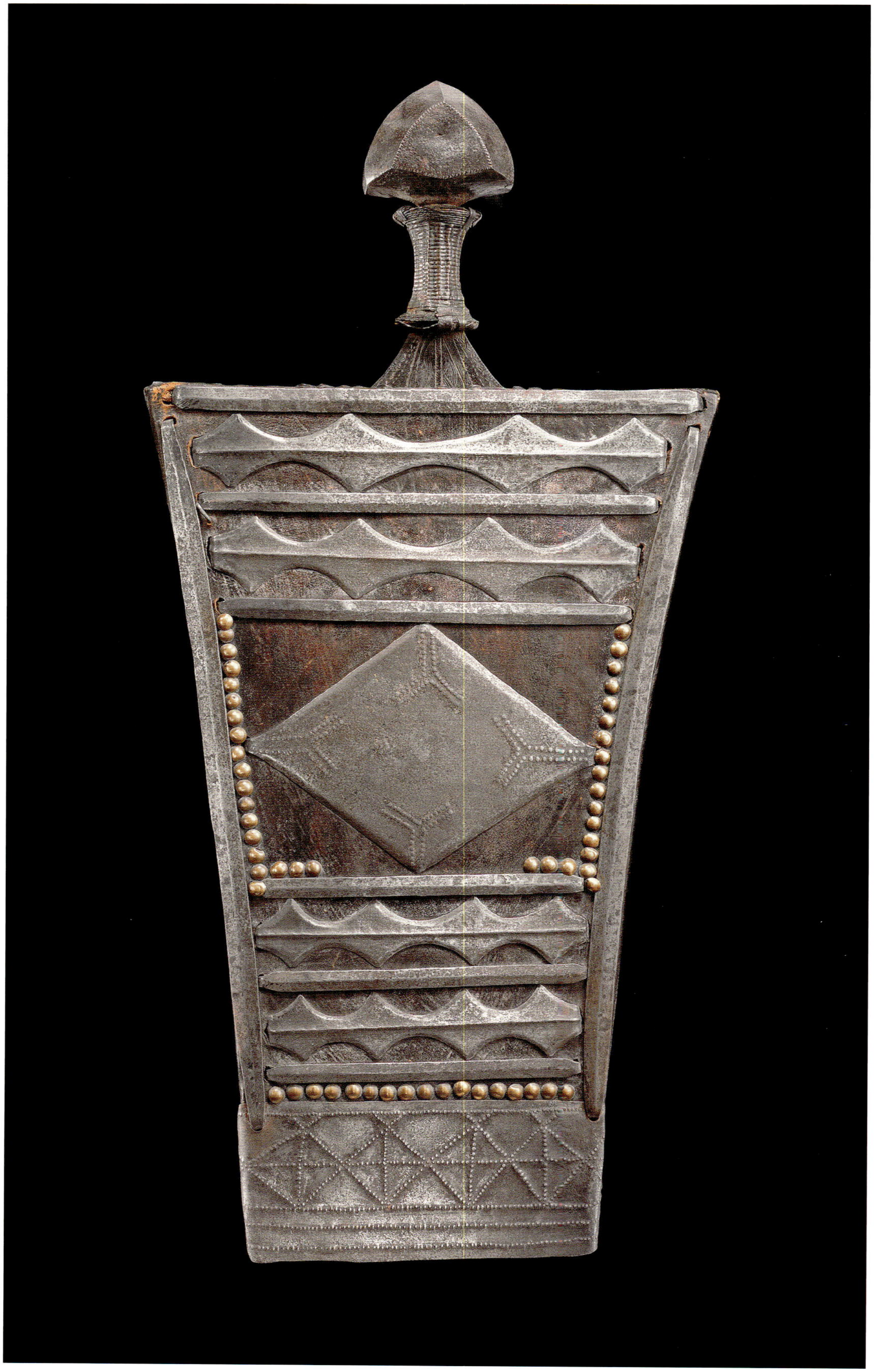

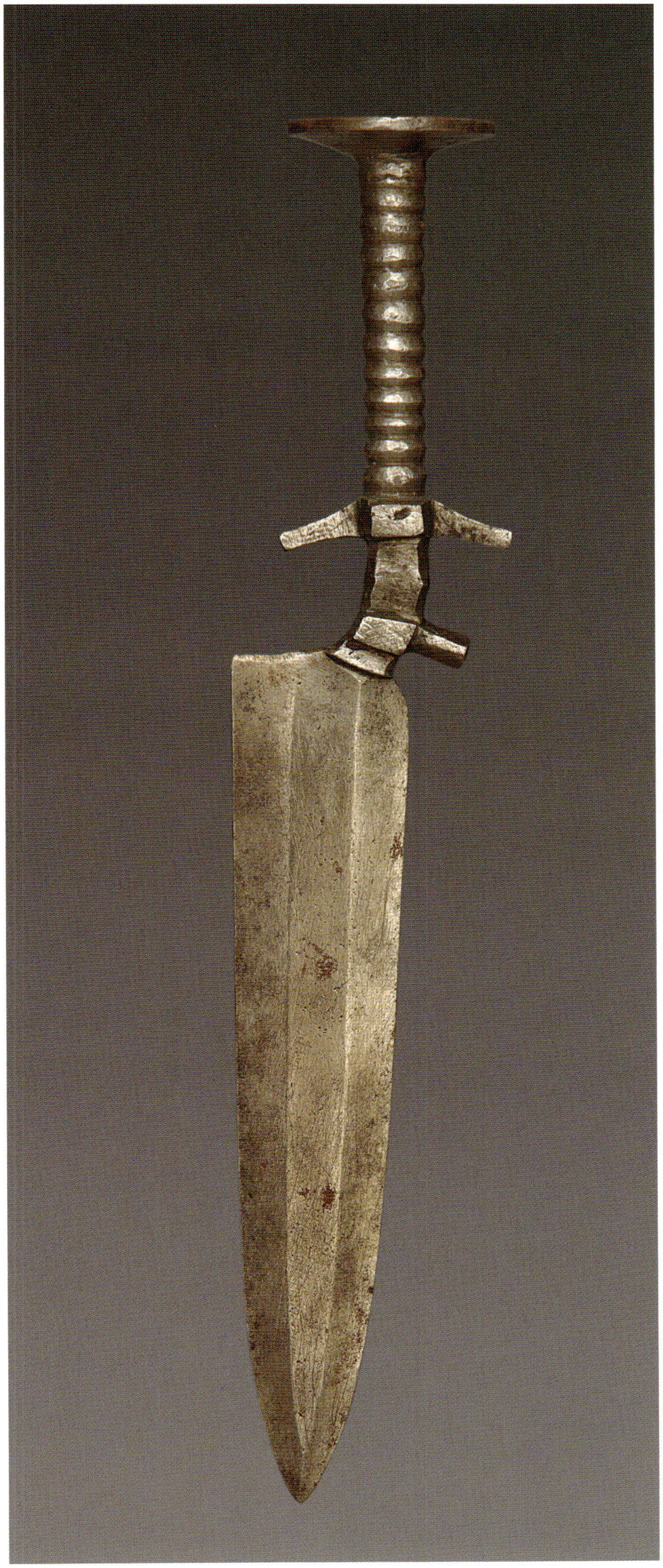

II.18
Artist unknown (Boa peoples, Democratic Republic of the Congo)
Dagger, early 20th century
Iron
H: 38 cm
PRIVATE COLLECTION
PHOTOGRAPH BY NATHALIE AND ALAIN SPELTDOORN, 2017.
COURTESY OF PRIVATE COLLECTION

II.19
Artist unknown (Ekonda peoples,
Democratic Republic of the Congo)
Ritual staff, early 20th century
Iron, wood
H (with base): 75 cm
COLLECTION OF JOEL AND ZACHARY COONER
IMAGE COURTESY OF JOEL AND ZACHARY COONER, 2005

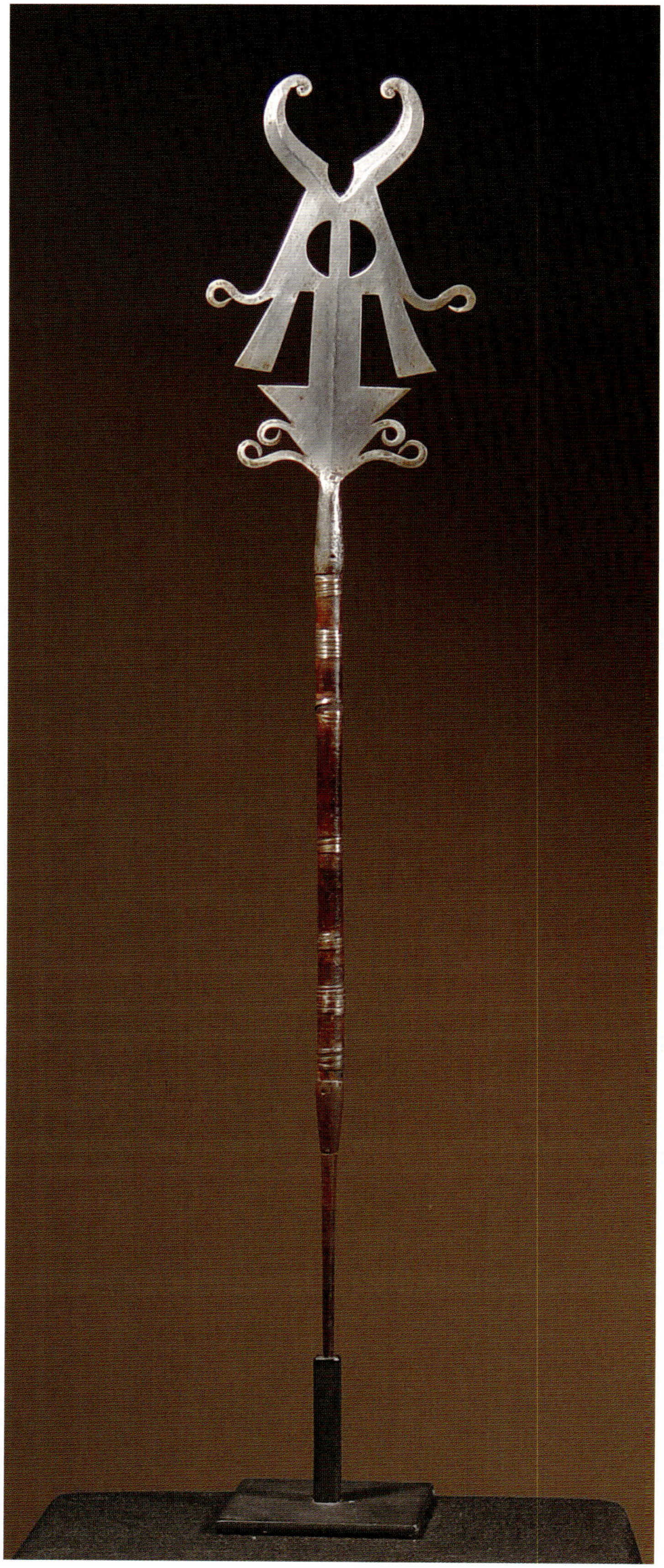

II.20A–B
Artist unknown (Ekonda peoples,
Democratic Republic of the Congo)
Ritual staff, early 20th century
Iron, wood
H: 70 cm
COLLECTION OF JOEL AND ZACHARY COONER
IMAGE COURTESY OF JOEL AND ZACHARY COONER, 2005

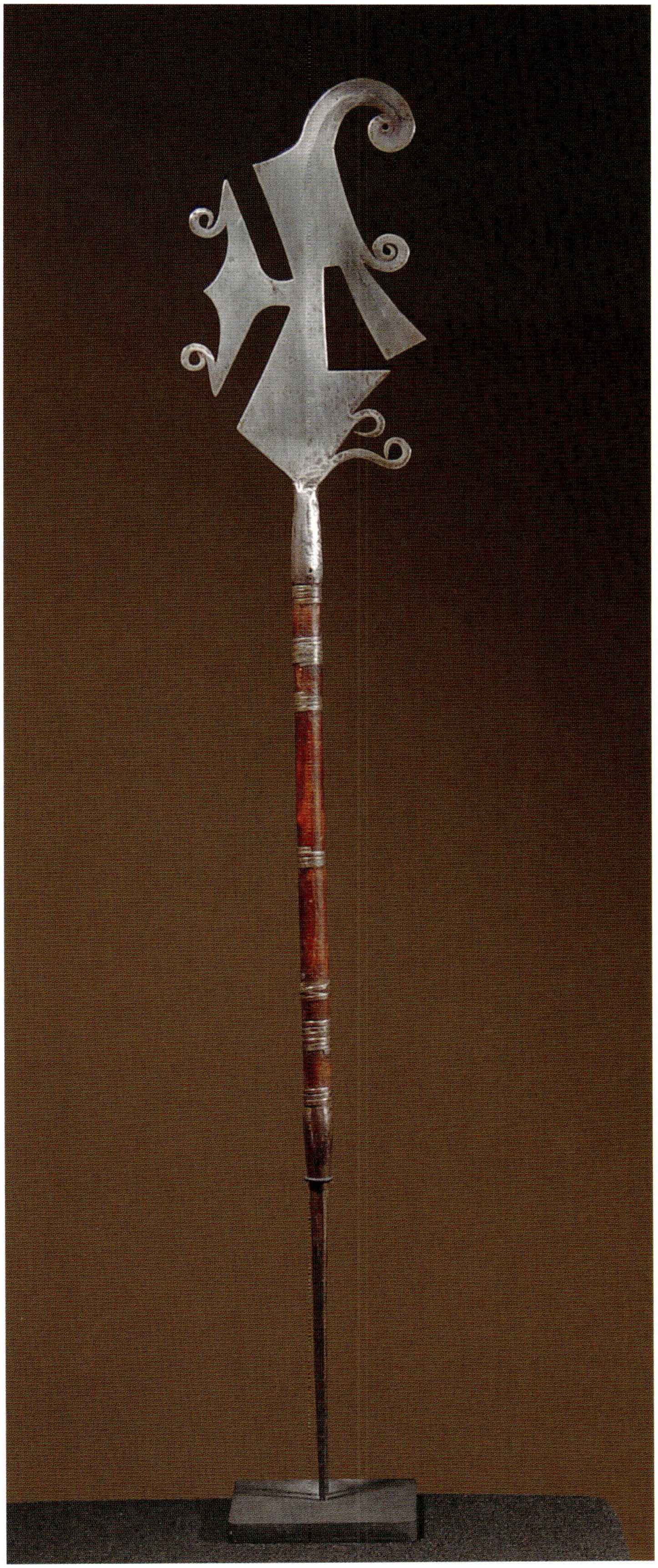

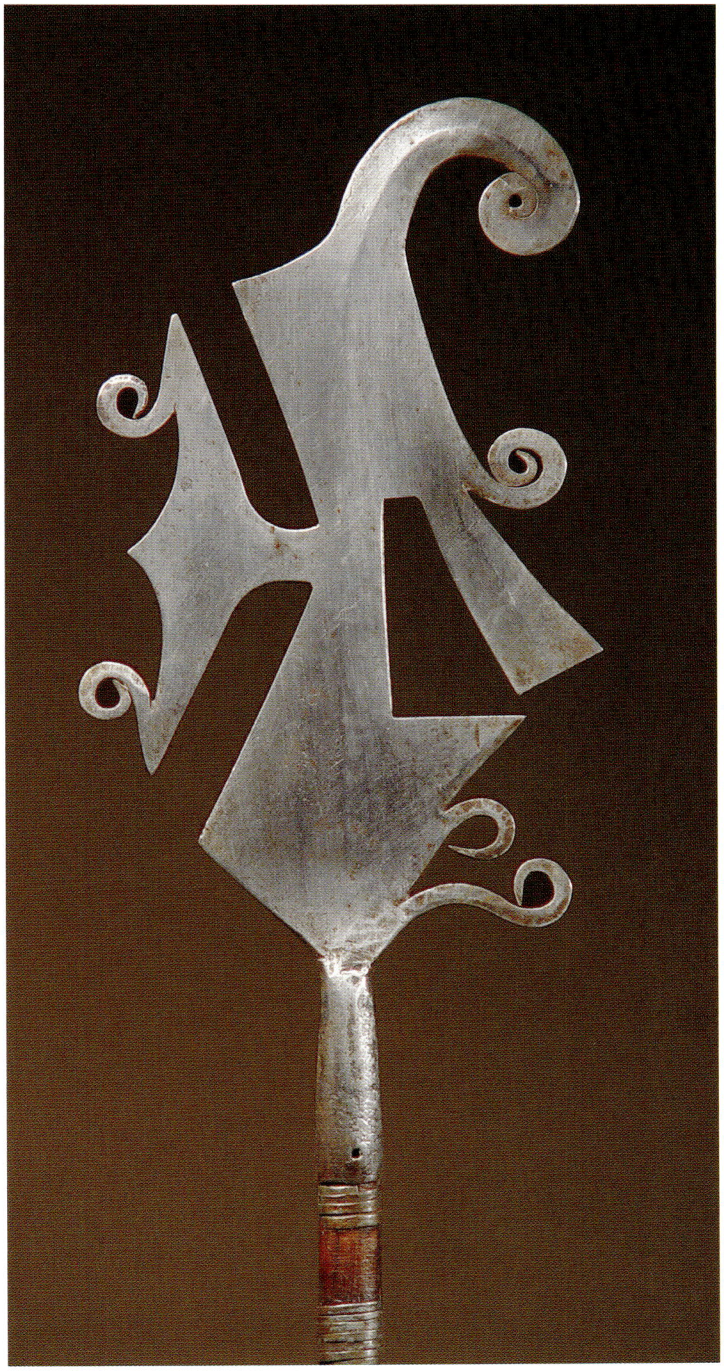

II.21
Artist unknown (Makaraka, Adio peoples, Sudan)
Knife, mid- to late 19th century
Iron, wood
H: 44.5 cm
PRIVATE COLLECTION
PHOTOGRAPH BY NATHALIE AND ALAIN SPELTDOORN, 2017. COURTESY OF PRIVATE COLLECTION

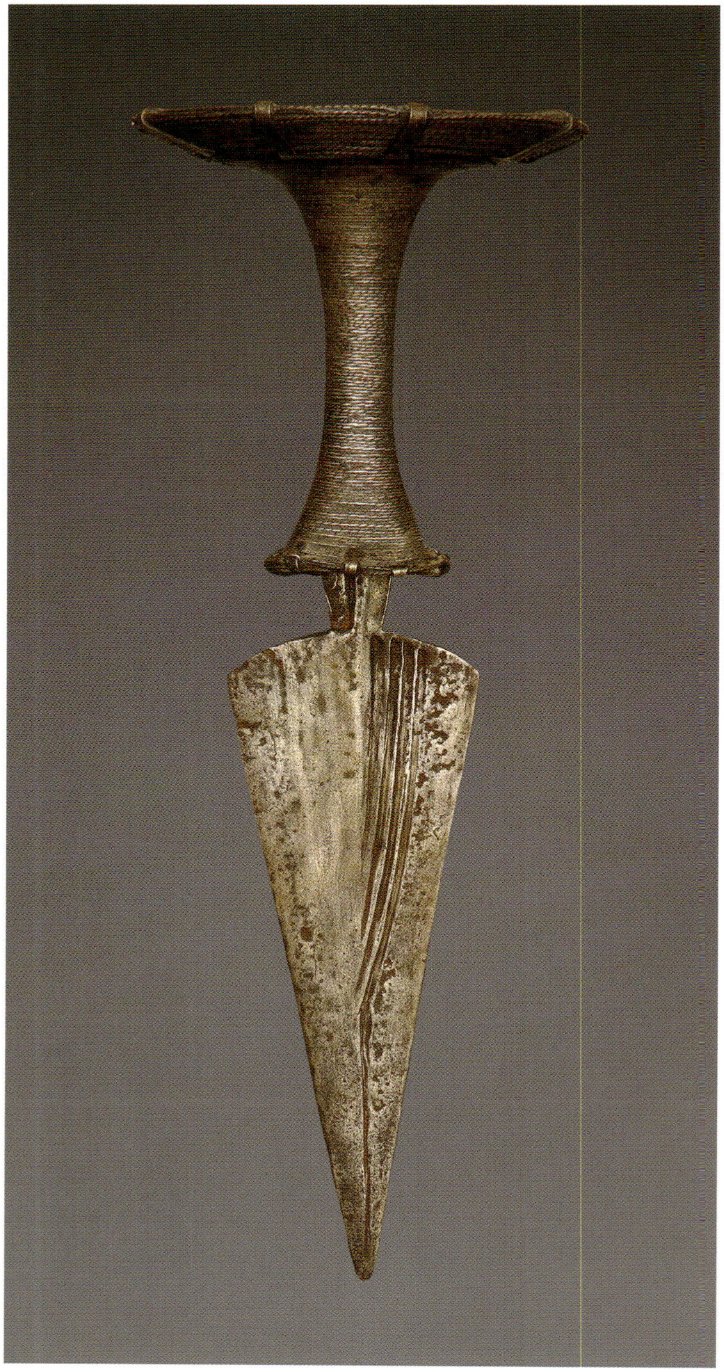

II.22
Artist unknown (Somali peoples, Republic of Somalia)
Short sword (*belewa*), late 19th to early 20th century
Iron, ivory, zinc alloy, leather
H: 62.9 cm
FOWLER MUSEUM AT UCLA, X65.3639A; GIFT OF THE WELLCOME TRUST

II.23
Artist unknown (Somali peoples, Republic of Somalia)
Knife (*belewa*) and sheath, late 19th to early 20th century
Iron, horn, zinc alloy, leather, wood
H (knife): 42 cm
FOWLER MUSEUM AT UCLA, X65.1717A–B; GIFT OF THE WELLCOME TRUST

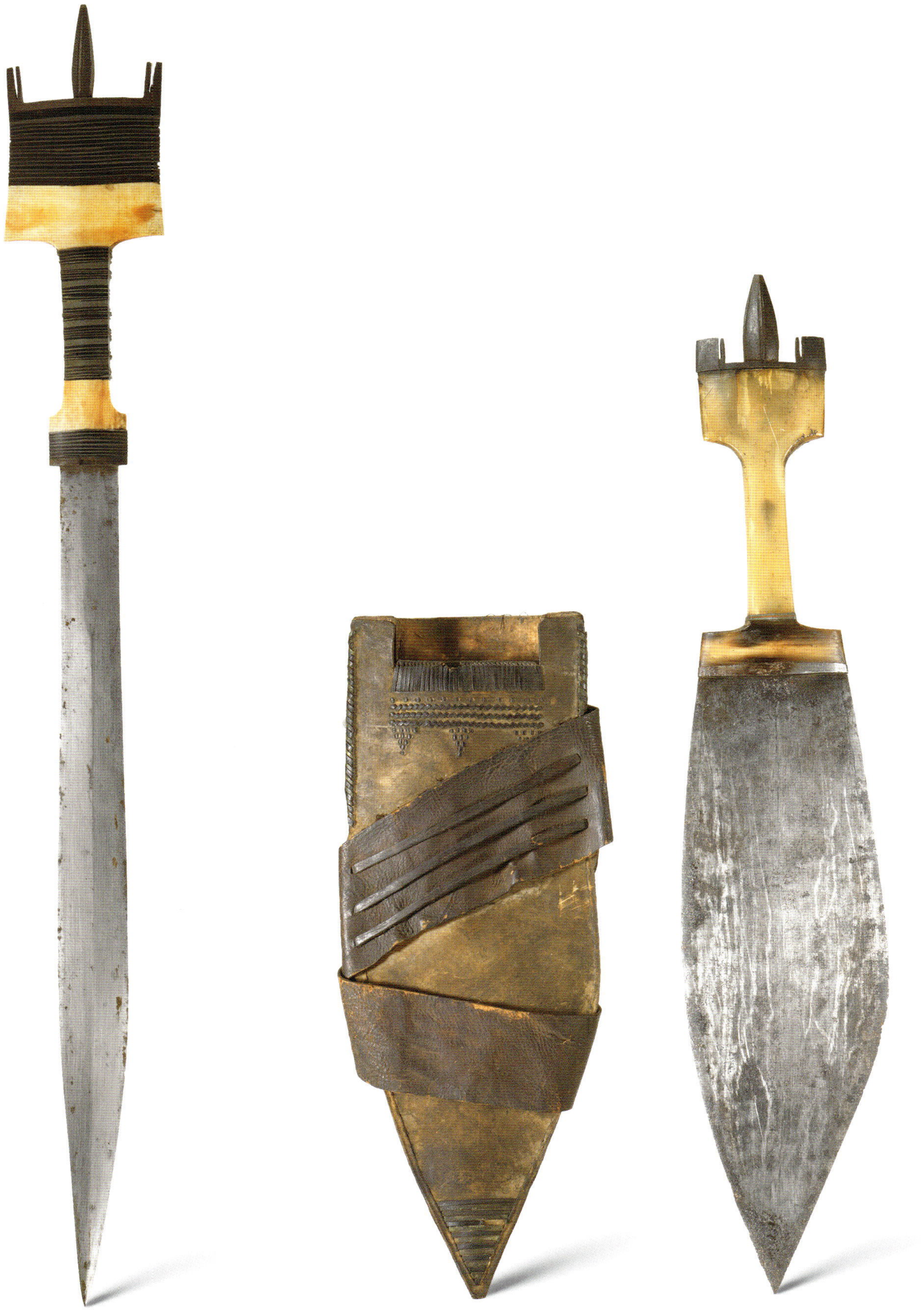

PART II

EARLY IRONWORKING AND ARCHAEOLOGY

CHAPTER

8

AGES OF IRON: ORIGINS OF THE BLACKSMITH'S ART IN AFRICA SOUTH OF THE SAHARA

Shadreck Chirikure William J. Dewey Scott MacEachern

When and how did iron smelting begin in Africa south of the Sahara? World history is often categorized as a progression from the Stone to the Bronze to the Iron Age, but in sub-Saharan Africa the world's earliest use of stone technology progressed directly into the African Iron Age.[1] A commonly held view is that Anatolia was the first place in the world where, around 1800 BCE, iron ore was purposefully smelted in atmosphere-controlled furnaces, and that by 1200 BCE iron production was common in the Levant (Pringle 2009). Sub-Saharan Africa does not share with other parts of the world (e.g., the Levant and Egypt) a long development of metalworking, in copper or bronze particularly; as a consequence, researchers disagree on whether ironworking was developed indigenously in the region, or whether it was introduced from elsewhere (fig. 8.1). Such questions have been the subject of considerable scholarly debate and are nowhere near resolution (see Pringle 2009; Killick 2009; Zangato and Holl 2010).

Copper was being mined and worked at sites like Akjoujt, in Mauritania, in the first millennium BCE (Vernet 2000) and in Niger probably some centuries earlier (Grébénart 1983, 117). The origins of these technologies are unclear, but they may have been inspired by trans-Saharan contact with the Iberian Peninsula. Alioune Deme and Susan McIntosh (2006) have suggested a possible connection between west African metallurgy and that in Mauritania.

One area of interest has been the Nile Valley, because many scholars, prior to the 1960s, assumed that knowledge of iron metallurgy was brought from Anatolia to Egypt and on to Meroë, in what is now the Sudan, whence it was transmitted throughout the African continent. Several empirical impediments have led to the rejection of these early assumptions; smelting technologies from Meroë and Egypt are different from sub-Saharan examples, but do date back to at least the sixth century BCE (Humphris and Schreibner 2017).[2]

David Phillipson (2005) offers another perspective, asserting that ironworking was established in northern Africa before being transmitted across the Sahara to west Africa from the first millennium BCE as part of an active commerce of ideas and goods. Reviewing published material on early metallurgy in Africa, Stanley Alpern (2005) holds that Phoenician settlements such as that at Carthage, now in Tunisia, most likely had iron metallurgy by 800 BCE—before ironworking began in ancient Egypt, but not long before Meroë or the earliest sites south of the Sahara. While trade networks certainly extended across the Sahara by the early first millennium CE (Fenn et al. 2010; Magnavita 2010; MacEachern 2016), the question is whether iron might have been moved across such networks a millennium earlier.

The history of metallurgy in the Sahara itself is no less complicated. Researchers have found evidence of ironworking in the Termit and Egaro areas of east-central Niger dating from the second millennium BCE (Paris et al. 1992), well before iron was being used in Carthage. However, there are questions about the associations of the iron artifacts with radiocarbon-dated pottery. In the Sahara, wind constantly moves and removes soil, and as a result it is difficult to prove that materials excavated close to one another coexisted in the past (Killick 2004, 103–04). The only ironworking furnaces known from the region, at Do Dimmi 15 in the Termit massif of south-central Niger, 1400 kilometers south of Carthage, seem to date to the mid- to late *first* millennium BCE (Killick 2004, 104). As with other regions, disagreement remains about the kinds of evidence that can substantiate claims of early ironworking in Sahelian lands.

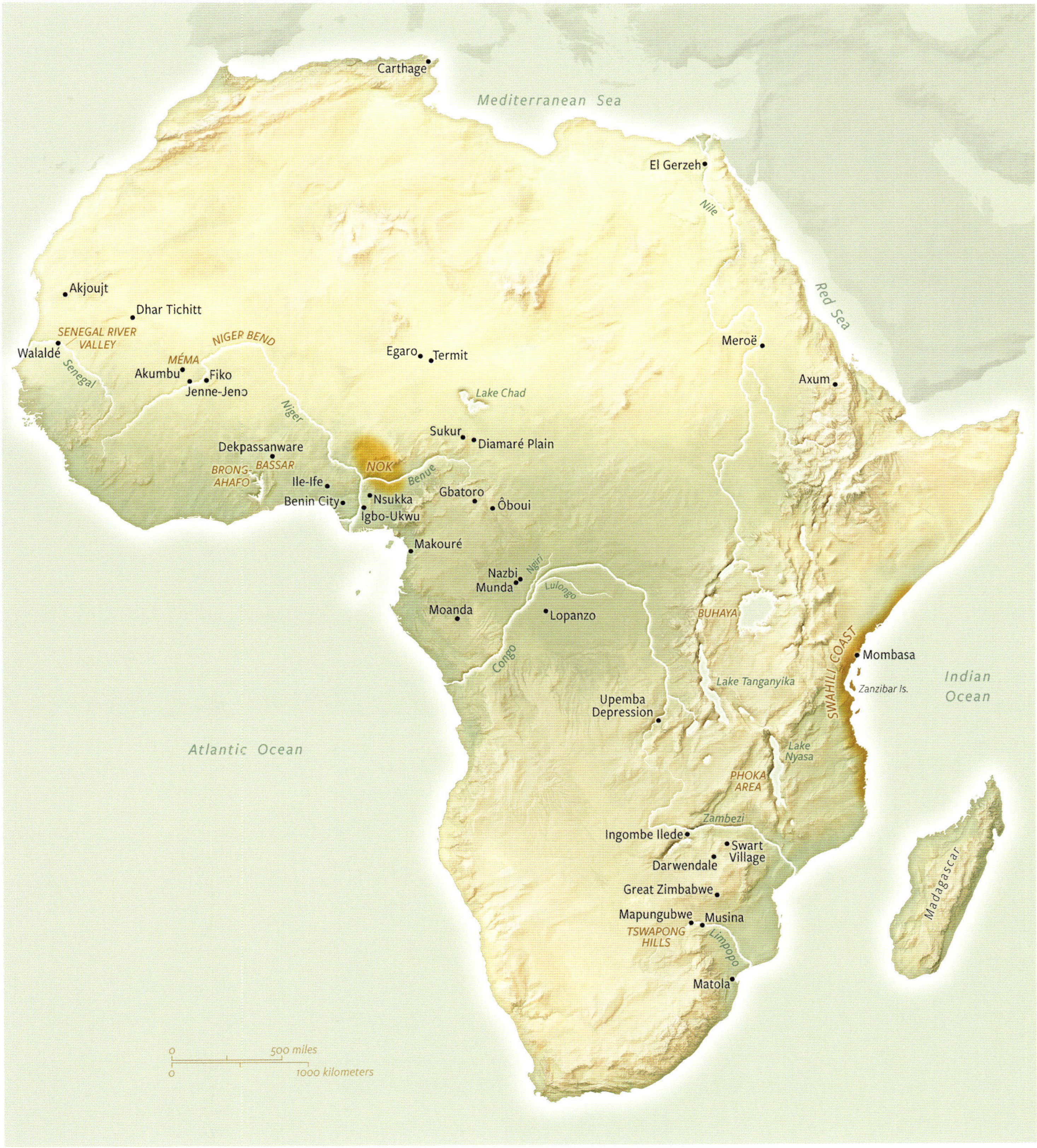

8.1 Significant sites and regions for the study of ancient African ironworking.

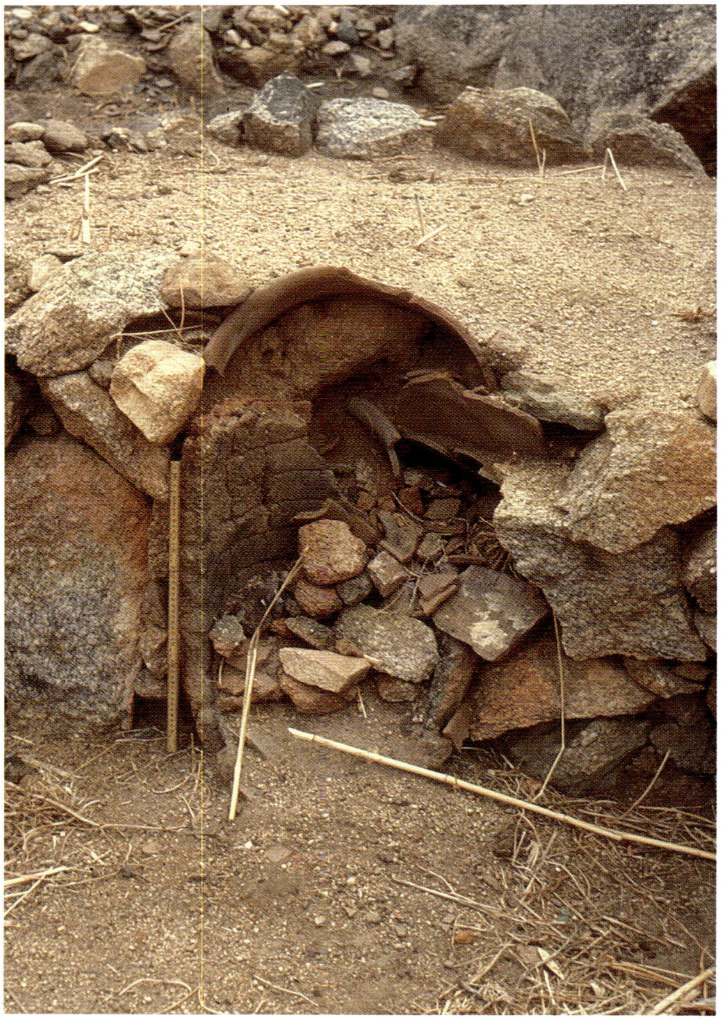

8.2 Remains of an iron smelting furnace probably dating to the early 20th century in an agricultural terrace wall. Studying local smelting furnaces and the tools produced in them helps us understand the capabilities of traditional blacksmiths and changes in ironworking technologies over time and regionally.
ALL PHOTOGRAPHS IN THIS CHAPTER BY SCOTT MACEACHERN, MAYO PLATA, NORTHERN CAMEROON, 1989.

Another unanswered question is why no metal technologies other than iron production were introduced to sub-Saharan Africa at this time?[3] According to Augustin Holl (2009), there is no hard evidence demonstrating that a technological transfer and exchange took place between the supposed donor regions to the north and the receiving areas to the south. On the contrary, Holl asserts that iron metallurgy in Africa south of the Sahara is autochthonous. Again, researchers debate what kinds of evidence are needed to establish the existence of iron metallurgy at a particular place and time. Can we indeed infer the existence of iron technologies from other kinds of material culture—ceramics, for example?

There are also significant problems in obtaining precise age estimates for archaeological sites from the period between about 2700–2400 years ago (Wright 2017). This seems to be when ironworking was becoming important in different regions of the continent, but the characteristics of the radiocarbon calibration curve for those years make it difficult to convert radiocarbon dates to well-demarcated calendar dates.

Such challenges play out at a number of archaeological sites in Africa south of the Sahara, where researchers claim to have recovered evidence for very early mastery of iron technology, from the second millennium BCE or even earlier. Perhaps the most interesting are two sites about a hundred kilometers apart: Oboui, in the Central African Republic, and Gbatoro, in Cameroon (Zangato and Holl 2010). These were certainly ironworking facilities, and each has yielded multiple radiocarbon dates from the third millennium BCE. Such dates would make ironworking at these sites contemporary with or earlier than iron production in other parts of the world. A third site, Gbabiri, located just eight kilometers from Oboui, has yielded dates indicating that ironworking also took place there somewhat later, in the early first millennium BCE. Scholars worry, however, that these radiocarbon dates might have come from disturbed sediments, and thus reflect earlier phases of human occupation or natural processes on the site rather than episodes of ironworking. They also note that the iron artifacts recovered are extraordinarily well-preserved for objects that would have spent more than four millennia buried in a wet, tropical environment.

Oboui and Gbatoro are excellent examples of the relationship between extraordinary claims and definitive evidence. Were they identified as two thousand years old instead of four thousand, there would probably be little argument about their status. But given their age and the relative lack of other archaeological data from the region, these sites continue to generate debate. Other ironworking sites in west and central Africa similarly have yielded radiocarbon datings from the beginning of the first millennium BCE and earlier—either just one or two dates, very imprecise dates, or dates from potentially disturbed archaeological contexts (see, for example, Essomba 1992; Eze-Uzomaka 2009; Zangato and Holl 2010, 20). They have not provided as substantial evidence for early ironworking as Oboui and Gbatoro. The dates from Niger and the Central African Republic have generated further questions, such as whether it is iron smelting and/or forging that is being assessed, and whether older charcoal that has been found and

8.3A–B Reconstructed iron smelting furnace in use by Plata Kapa peoples in the Mandara Mountains of northern Cameroon. Plata traditionally smelted iron and made iron tools for themselves and their neighbors. By the 1950s, they had largely switched to making tools out of Western iron, but knowledge of smelting persisted until the 1980s.

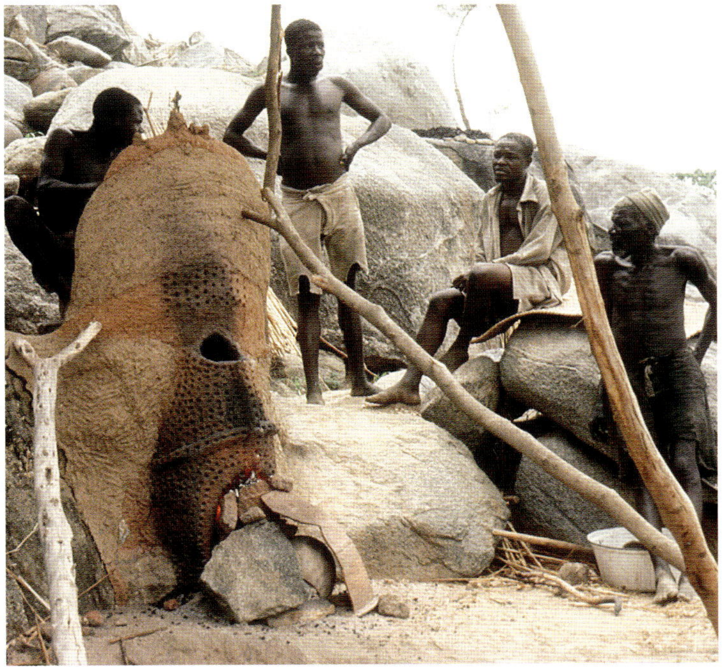

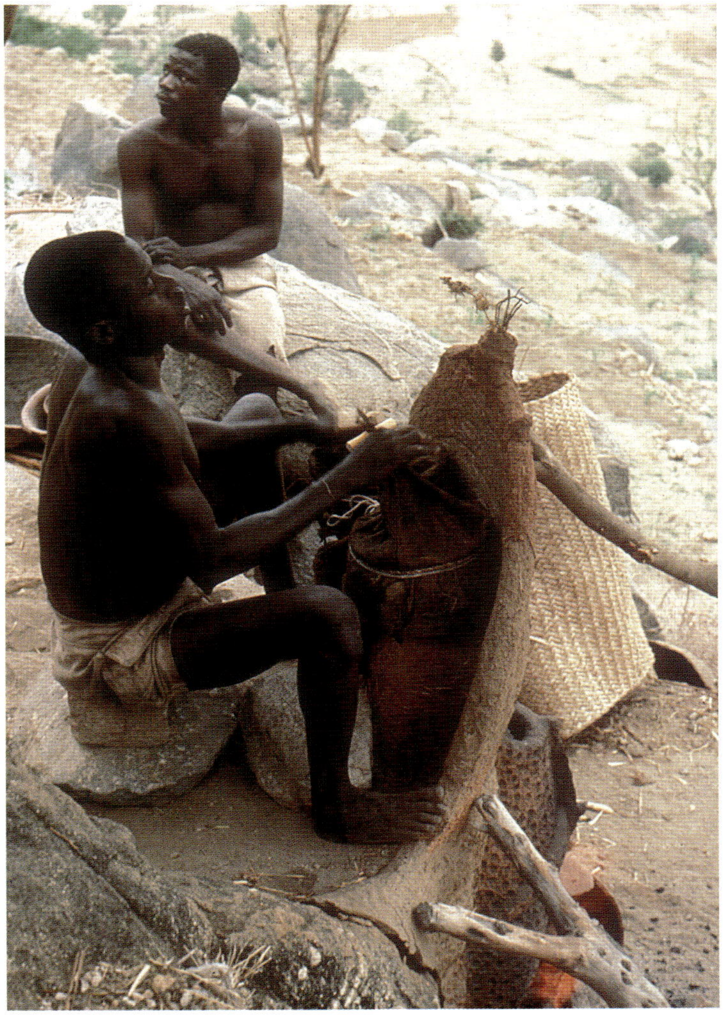

was used by the ancient smelters has falsely skewed the radiocarbon measurements to suggest the sites are older than they actually were.[4]

Whatever the points and counterpoints for external and local origins, a more cautious position places the origins of ironworking in west Africa between 800–400 BCE.[5] East African beginnings date to the same period, while southern Africa was later.[6] Without further fieldwork and additional data, the debate as to whether ironworking in Africa south of the Sahara is indigenous may have reached a standstill, with no side prepared to concede or accept basic limitations inherent in its position. Origins aside, that iron (once established) became a major metal that significantly shaped the continent's communities of different time periods is not a matter for debate. The story of iron production and use in Africa is punctuated with a great diversity of brilliant inventions and innovations, attesting to local experimentation and adaptation (figs. 8.2–8.4).

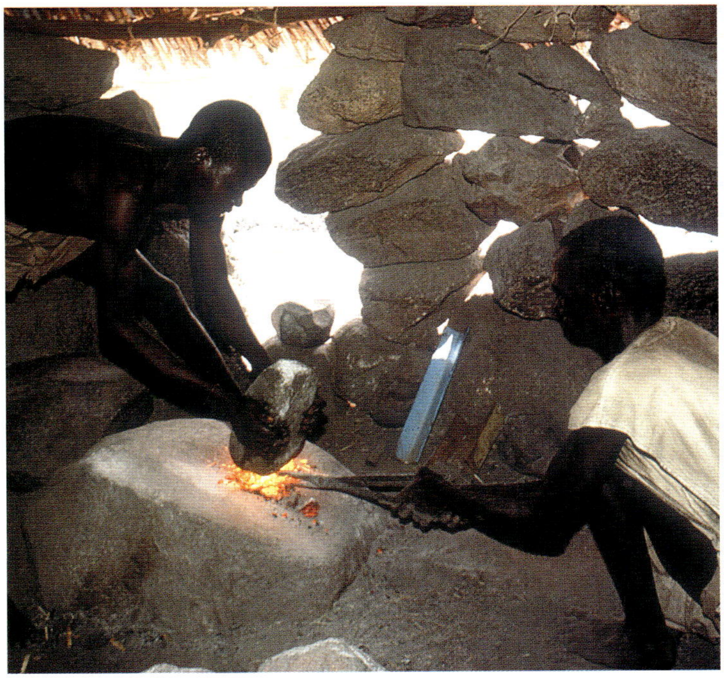

8.4 Blacksmiths using a stone hammer (*amtsal*) to amalgamate fragments of smelted iron into a single mass for toolmaking.

CHAPTER
9

TRANSFORMING CULTURE: WEST AND CENTRAL AFRICAN IRON ARCHAEOLOGY

Scott MacEachern

The goal of this chapter is to examine our present knowledge of precolonial histories of ironworking and iron use in west and central Africa, and their relationships with practices in other parts of the continent (see fig. 8.1). Most of our knowledge of the creative work of smelting iron from ore and forging iron tools—distinct activities not necessarily carried out by the same persons or in the same communities—is derived from archaeological data. Written sources (by outsiders) on west and central Africa are limited to the last millennium and primarily concerned with elite activities, almost never mentioning technical processes. Archaeological research on iron technologies involves analysis of the metallurgical characteristics of ancient iron objects and slag, which provides information on production processes, as well as ethnoarchaeological reconstructions of ironworking practices that have survived until recent times, and comparisons of these two sources of data.

DEVELOPMENTS IN THE FIRST MILLENNIUM BCE

It is widely accepted that iron came to be smelted, forged, and used across different areas of west and central Africa by the middle of the first millennium BCE at the latest. Unfortunately, major challenges present themselves when calibrating radiocarbon dates for the period between about 2700 and 2400 years ago, making detailed chronologies hard to establish. The earliest widely accepted example of ironworking in the region is the Walaldé mound in north-central Senegal, which was occupied by iron-using cattle pastoralists during 800–500 BCE, with a second occupational episode between roughly 500 and 200 BCE. Iron artifacts and some evidence for smelting have been found in close association with datable materials, in deposits that do not seem badly disturbed. Pottery and copper artifacts from Walaldé show similarities to sites further to the north, with the copper coming from mines around Akjoujt in what is now western Mauritania.[1]

Nearly contemporary with Walaldé are Nok sites in central Nigeria. Nok culture is chiefly known for its artistic accomplishments, primarily elaborate terra-cotta figures depicting humans and animals (Breunig 2014). Early Nok works were discovered during tin-mining activities (Fagg 1990). In the 1970s, however, archaeological excavations of Taruga and Samun Dukiya yielded iron artifacts and furnace remains dating to the mid- to late first millennium BCE (Connah 1969; Rustad 1980). A much more extensive survey and excavation effort on Nok sites since 2005 has added a great deal of knowledge, including more than 160 new sites, and 174 radiocarbon and 27 thermoluminescence dates (Breunig and Rupp 2016; Franke 2016).[2] Nok pottery appeared in the region by perhaps 1500 BCE, but the most distinctive elements of the society—terra-cotta figures and ironworking—are only found on sites dating after 900 BCE (figs. 9.1, 9.2; see also figs. 3.2, 3.3).

The florescence of Nok seems to have been from 900 to 400 BCE, although a few sites and artifacts are dated to the beginning of the Common Era. This far more detailed chronology provides a richer appreciation of the dynamics of cultural change, and it has implications for how we think of ironworking in other parts of sub-Saharan Africa. At Nok, ironworking appeared after 900 BCE within a cultural tradition that had already existed for centuries—not through the wholesale replacement of a Neolithic culture by an Iron Age culture. Such recognition undermines the assumption that other categories of material culture such as pottery necessarily heralded the coming of iron.

Iron also appears during the first millennium BCE at other sites in the region, for example around Nsukka, further to the south in Nigeria (Okafor and Phillips 1992), and at Dekpassanware, in the

9.1 Remains of three iron ore reducing furnaces at the site of Janjala, Nigeria, photographed in 2013. The infills of the two furnaces in the foreground contained one Nok terra-cotta each.
IMAGE © Goethe-UNIVERSITÄT FRANKFURT, INSTITUTE FÜR ARCHÄOLOGISCHE WISSENSCHAFTEN, ARCHÄOLOGIE UND ARCHÄOBOTANIK AFRIKAS, GERMANY

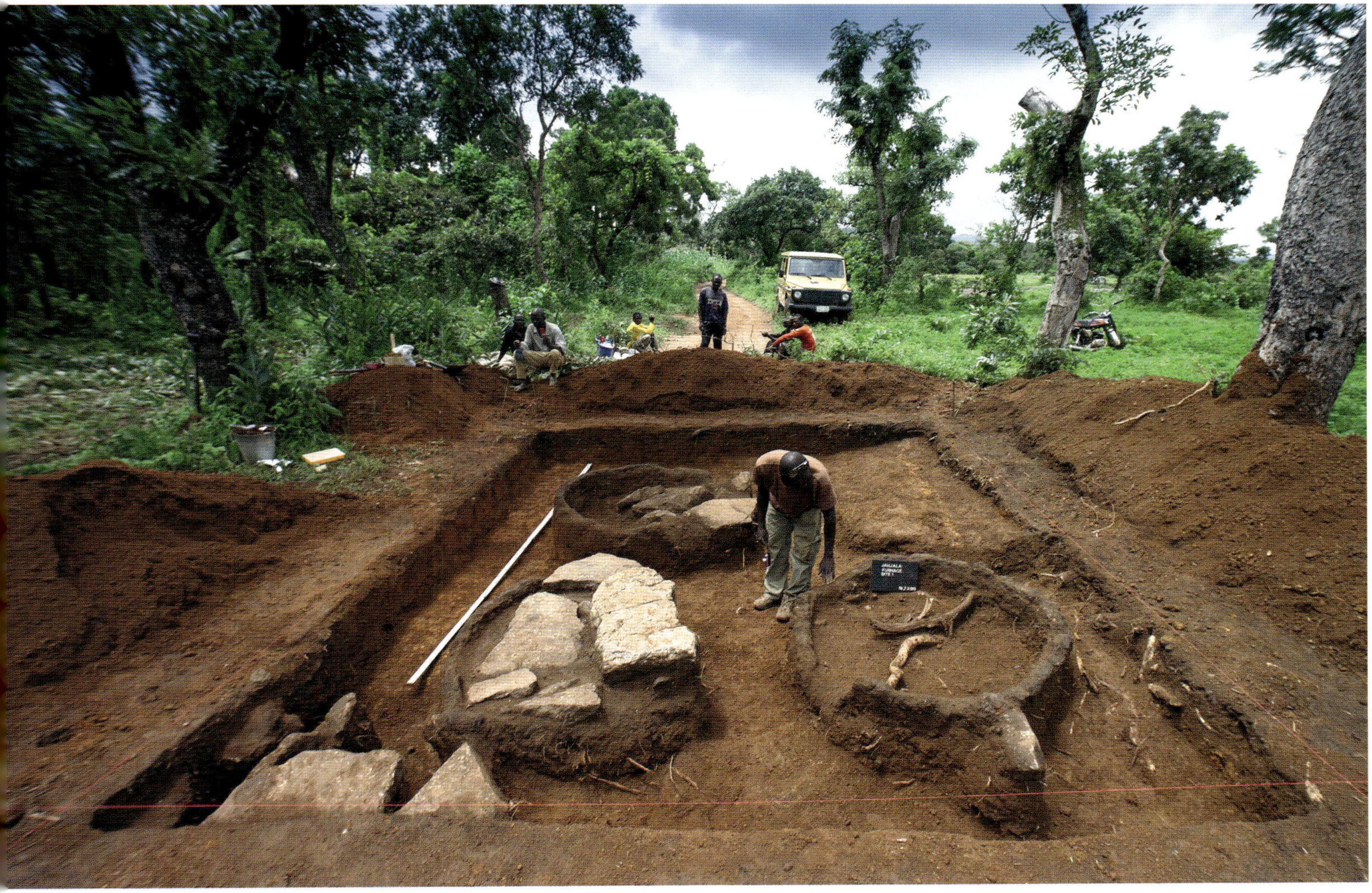

Bassar area of northern Togo (de Barros 2003), where ironworking seems to have been generally contemporary with Nok sites. In some parts of the Lake Chad Basin of northeastern Nigeria and northern Cameroon, the first evidence of iron smelting and iron artifacts dates to the mid-first millennium BCE (MacEachern 1996), and thus is generally comparable with the Nok sites six hundred kilometers to the southwest. At the same time, iron began to be used only in the initial half of the first millennium CE just a hundred kilometers to the south, on the Diamaré Plain of Cameroon (Langlois 1995, 515, 603–09; Marliac et al. 2000, 72; see also chapter 13, this volume). It is notable that, in the southern Lake Chad Basin, early ironworking does not seem to be associated with major social or economic transformations; those would come some centuries later, as formerly seminomadic populations began to settle in permanent communities and adopt earthen architecture in the early first millennium CE. The same kinds of differences in the chronology of iron's adoption in neighboring regions are seen in Mauritania, also in the early first millennium CE (Vernet 1993, 353–63).

9.2
Artist unknown (Nok culture, Nigeria)
Male figure, 500 BCE–500 CE
Terra-cotta
H: 73 cm
(NOT IN EXHIBITION)

Male figure kneeling on an upturned vessel, excavated in 2011 at the Nok site Daji Gwana 1, Nigeria.

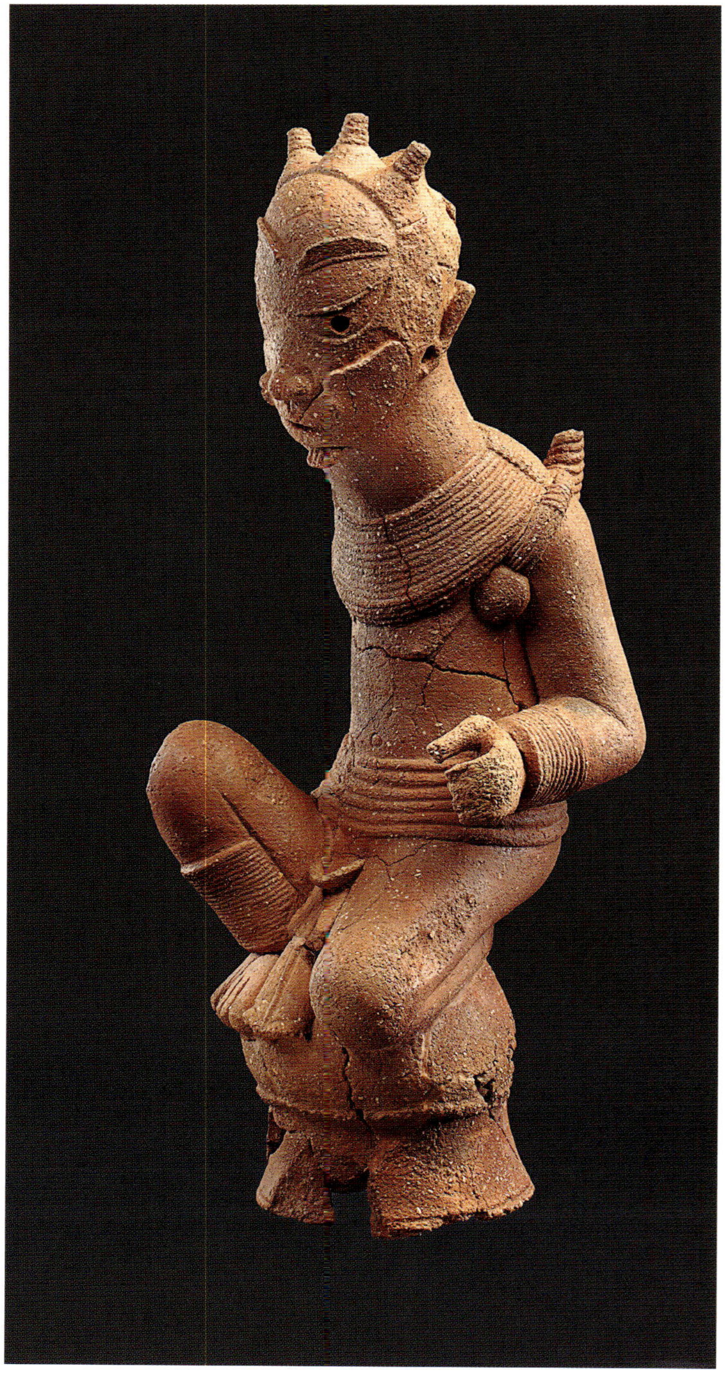

Salvage archaeology undertaken in the early 2000s to examine sites threatened by the construction of the Chad-Cameroon Oil Pipeline illustrates the challenges of interpreting evidence for early ironworking (Lavachery et al. 2010). A furnace on the Mban,/ETA-69 site in Chad initially yielded a radiocarbon date of the mid-second millennium BCE, but two additional charcoal samples taken later from the same furnace provided overlapping dates from much later—around 1000 CE. This implies that the first sample dated was of older charcoal that had been introduced into a later furnace through natural processes. Two other sites, Bissiang/ECA-138 and Meyang/ECA-335, just across the border in Cameroon, yielded early first millennium BCE dates and pottery similar to that found in later ironworking communities—but no iron. This situation might be analogous to that of the pre-iron Nok sites. The site of Zili/ECA-323, in southern Cameroon, yielded a single date of the early first millennium BCE from a pit containing iron slag. This is again comparable to dates from early Nok ironworking sites. But pits are complex environments where materials like charcoal may be moved around by the actions of gravity, water, or other natural processes. In such a case, one should probably not draw grand conclusions from a single radiocarbon date. The best evidence for early ironworking is probably found in a smelting furnace from the site of Makouré/ECA-124, again in southern Cameroon, which dates to the last half of the first millennium BCE, and there are numerous dated sites from after that time. Where on this continuum of evidence through time should we accept that societies in this part of central Africa were making use of iron technologies, though? In other parts of west and central Africa, unequivocal evidence for early ironworking most frequently dates to the last few centuries BCE. That is the case for the Makouré/ECA-124 site, as noted above, and at sites including Moanda, in Gabon (Clist 2012).

Such questions point to a major challenge to our understanding of the adoption of ironworking throughout west and central Africa. The uneven distribution of archaeological research across this vast area means that we do not really know whether these early sites were part of a larger regional pattern of iron use, or whether they were isolated communities within a matrix of societies that still used stone tools. The proximity of Nok to the Lake Chad Basin sites, Nsukka, and southern Cameroonian locations may suggest some cultural contacts, especially since other common economic features can be discerned at a number of these sites, particularly the use of millet as a staple crop. Millet was originally a southern Sahara domesticate (Manning et al. 2011), and its occurrence in southern

Cameroon in very different environments is therefore striking. In general, however, we should not assume that ironworking spread as a single wave rolling outward from particular centers of technological innovation to transform adjacent societies inexorably. Iron's adoption was probably a far more complex and piecemeal process, with some communities seeing no particular advantage to its use for centuries after their neighbors had embraced the new metal.

THE SPREAD OF IRON IN THE FIRST MILLENNIUM CE

By the early first millennium CE, iron was being produced and used far more widely through the lands around the Gulf of Guinea and at sites such as Munda and Nazbi in the Republic of the Congo (Eggert et al. 2007). Extraordinarily, recent archaeological work has shown that iron was already playing an important role in social and political hierarchies in Cameroon and Equatorial Guinea in the third or fourth centuries CE. In this region, researchers have found elaborate and beautifully made iron objects—axes, spears, sickles, spoons, bracelets, and other artifacts—included among rich grave goods in burials, which surely signaled high status, at sites including Campo (Meister and Eggert 2008; Meister 2010; González-Ruibal et al. 2012). We do not know at this point if the technological or economic possibilities of this new material sparked social revolutions in the societies that adopted iron, or whether the metal represented and supported preexisting hierarchies. These burials have, in any case, revolutionized our understanding of the social complexity of societies that existed around the Gulf of Guinea two thousand years ago. By the end of the tenth century CE, similarly elaborate iron artifacts would be found in rich graves on the other side of the Congo forest, more than two thousand kilometers away in the Upemba region of the southeast Democratic Republic of the Congo (de Maret 1977; 1999). We do not know whether there is any historical relationship between early sites in Cameroon and Equatorial Guinea and the later classic Kisalian sites of the Upemba, but they probably share a common cultural heritage in Central Bantu-speaking communities (see chapter 15, this volume).

In the succeeding centuries, the use of iron spread among societies in west and central Africa. The assumption has usually been that knowledge of ironworking was carried through the tropical forests of central Africa, along rivers that provided axes of communication and interaction in the region. This hypothesis is largely based upon intensive use of these same rivers by more recent populations and upon much-debated models of the productivity and habitability of the Congo forests (Hart and Hart 1986; Vansina 1986; Klieman 2003), and it makes a good deal of sense. However, direct archaeological evidence for the spread of ironworking is very sparse indeed. Interpretations depend primarily upon the recovery of ancient pottery seen as stylistically similar to that used by later ironworking groups (or simply as generalized "Bantu" pottery, with the misleading assumption that all Bantu-speaking peoples practiced ironworking), and upon the progressive disappearance of evidence for the making and use of stone tools (Eggert 1993, 1994; Clist 2005, 2012). We have already seen with Nok that there is not necessarily a close association between the adoption of particular kinds of pottery and the beginnings of ironworking; the history of pottery use by west African populations who also made stone tools is complex, dating to the tenth millennium BCE at Ounjougou in Mali (Huysecom et al. 2009). The spread of ironworking likely *was* associated to a significant degree with the spread of Bantu-speaking populations and the wider use of pottery in central Africa (Ehret 1995; Vansina 2006), but this was probably a very complex and variable process, involving a diversity of technological and social relations among communities. Recent discoveries that ancient climate changes probably influenced the extent of tropical forests complicate our interpretations, because people may have traveled along different routes through the forest than those existing now (Bostoen et al. 2015).

If central African archaeology sometimes seems dominated by the issue of Bantu migrations, west Africa provides a very different situation. There is significant evidence across the region of iron use through the first millennium CE, particularly in Nigeria, Ghana, Mali, and Senegal—although some communities appear to have continued to use stone tools until the end of the first millennium CE (DeCorse 2005). In both Ghana and Nigeria, dates for the beginnings of ironworking are slightly later in tropical forest zones along the Atlantic coast than they are in the more open central and northern environments. In central Ghana, for example, iron seems to have been in use by two thousand years ago (Shinnie and Kense 1989), but at the coast the earliest dates are five hundred years later (DeCorse 2005). This disparity is puzzling, since, as we have seen, rainforest environments were not an obstacle to early adoption of ironworking in central Africa. The difference may be related to the difficulty archaeologists have in discovering sites in tropical environments, resulting in a paucity of fieldwork in such places. In Ghana, recent research indicates that through the first millennium CE, southern forests were more densely settled than previously thought (Chouin and Decorse 2010; Chouin 2012). It is likely that further work will produce earlier radiocarbon dates for iron.

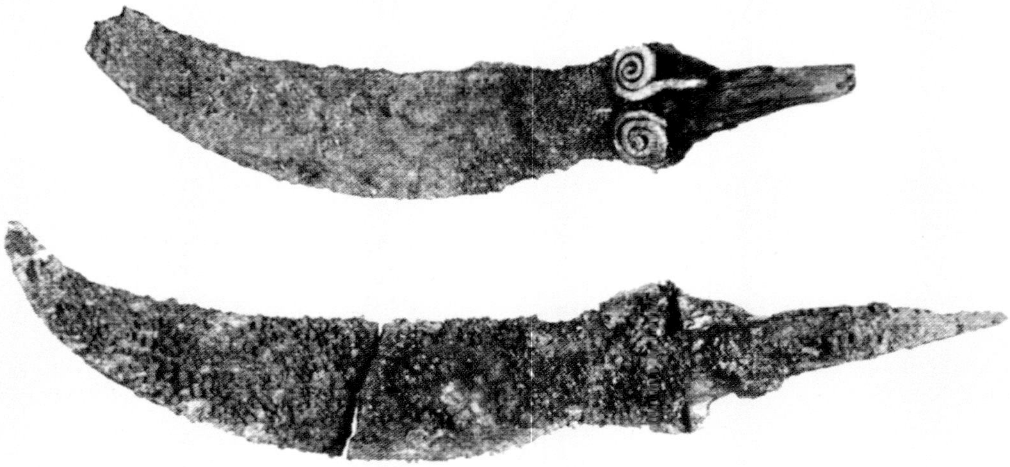

9.3 Iron blades, possibly razors or scarification knives, with tangs and copper wire ornamentation on their handles, excavated in southeastern Nigeria.
PHOTOGRAPH BY THURSTON SHAW, IGBO-UKWU, NIGERIA, CA. 1970.

In southern Nigeria the same situation obtains. Recent research on complex earthworks at Ile-Ife and Benin City have yielded radiocarbon dates of the twelfth and thirteenth centuries CE (Chouin 2013). In later times, these would demarcate urban areas; we do not know if that was the case almost a thousand years ago, but it seems very likely that such architecture would have been associated with substantial settlements with long histories in the region. The elaborate ritual complex at Igbo-Ukwu in southeastern Nigeria, for instance, is dated to the beginning of the second millennium CE (Shaw 1970, 1993; Craddock et al. 1997). The extremely rich artifacts found in different parts of this ritual complex, including iron blades (fig. 9.3), suggest the region had been occupied by iron-using communities for much of the first millennium CE. This may have been the case across rainforest zones of west Africa, where large iron-using populations of the first and early second millennium AD may have been decimated by diseases, among them bubonic plague, introduced from northern Africa via trans-Saharan trade routes (Chouin 2013).

SOCIAL AND CULTURAL CONSEQUENCES OF IRON USE

Archaeological investigations of second millennium CE sites in west and central Africa tend to focus upon a limited number of issues: trade routes and the emporia they generated; processes of state formation and urbanism; and European contact, trade, and colonialism (MacEachern 2005). For this period, when it is assumed that all communities were using iron to some degree, ironworking recedes from archaeological consciousness, becoming a background technology rather than a topic of research in its own right. Ironworking does sometimes remain important to archaeologists, although usually for socioeconomic rather than technological reasons. This is true of large-scale iron production, when iron is exported to neighboring communities. As noted above, iron smelting, smithing, and tool use are distinct activities, and not necessarily undertaken in the same ways across societies. From ethnohistorical data, for instance, we know that the community of Sukur, at the edge of the Mandara Mountains in northeastern Nigeria, specialized in large-scale production and regional export of iron, and aspects of Sukur's sociopolitical system were adapted to the labor requirements of such large-scale work (David 1996; David and Sterner 1997; see also chapter 13, this volume). The origins of such production are unknown, but may date to the seventeenth century CE. Ethnoarchaeology in the 1980s and during excavations in 2008 to 2011 allowed comparison of recent tools forged of smelted iron with five-hundred-year-old iron artifacts found only about twenty kilometers away (MacEachern et al. 2013).

The Babungo chiefdom of southern Cameroon was producing and exporting iron in similar magnitudes in the nineteenth century (Fowler 1990), and such large-scale regional trade may date to the fourteenth century CE in the Bassar region of the Republic of Togo (de Barros 1986; 1988). Elsewhere, vast amounts of iron production debris at west African sites argue for extensive production and export. As examples, well over forty thousand smelting furnaces have been counted in one eighty-kilometer section of the Senegal River Valley in Mauritania (Robert-Chaleix 1994); a fifty-by-six-hundred-meter slag mound was found at one site in the Méma region of south-central Mali (Haaland 1980), probably dating to the late first to early second millennium CE; and more than three hundred thousand square meters of slag at the Fiko site in Mali date to the early second millennium CE (Robion-Brunner 2010). There is less evidence for large-scale ironworking in central Africa. However, researchers found a very dense cluster of ironworking sites along twenty kilometers of the Chad-Cameroon Oil Pipeline in southwestern Chad, dating to approximately 800 to 1100 CE (Lavachery et al. 2005). That, and the iron mines and slag heaps recently discovered in southern Central African Republic (Lupo et al. 2015), may be traces of specialized production and export. These would be comparable to sites found in southeastern DRC and to production practices among related peoples living in southwestern Tanzania. Large-scale iron production implies the existence of complex exchange among communities in a region, as well as local efficiencies of scale and some degree of craft specialization in the communities where iron was produced. Unfortunately, archaeologists can only rarely determine the areas that such exchange networks covered.

Iron production and use in west and central Africa have never been purely technological phenomena. Iron-making is an intrinsically impressive process, involving high temperatures and the transformation of different substances—charcoal and different kinds of ore—into a radically new material—iron. This, of course, accounts for some of the debate about whether iron technologies could have been invented de novo south of the Sahara. It is not particularly surprising, then, that iron-making is often associated with diverse and complex symbolic systems that connect this particular activity and its participants to other cultural concerns.

The identity of blacksmiths as powerful and potentially dangerous transformers is central to the complex social systems[3] that extend through Sahelian and Sudanic zones of west Africa (Sterner and David 1991; Tamari 1997, 2012). This aspect of iron production is also related to the close association between ironworking and political power that appears early in histories of African ironworking. We do not know the exact statuses of the individuals buried with iron grave goods in southern Cameroon and Equatorial Guinea in the early first millennium CE, but they probably occupied the top ranks of local social hierarchies. This is almost certain for the later classic Kisalian graves in southeastern DRC and the Igbo-Ukwu site of southern Nigeria, where iron was accompanied by other exotic materials in elaborate burials (see figs. E.1–E.3). Over all of west and central Africa, blacksmiths are associated in complex and paradoxical ways with chiefs and kings—paradoxical because the technological and ritual power of blacksmiths could in some cases become a threat to the ruler (de Maret 1985; Herbert 1993).

CONCLUSION

The question of the historical origins and development of ironworking has been a preoccupation in Africanist archaeology for more than the last half century. It is likely that more research and new dating technologies will clarify the matter. Much more interesting, though, is the broader question of how ironworking technologies were adopted by ancient communities, and the extent to which iron smelting, smithing, and use transformed them. It is certain that iron was associated with far-reaching changes in African societies as its usage spread, but we should be careful of any assumptions that ironworking *caused* particular changes in any straightforward sense. We must be especially critical of assumptions that iron was adopted purely for reasons of economic efficiency or utility, given the complex symbolic systems that encompass modern African ironworking and the fact that so many of the early iron artifacts that archaeologists recover are nonutilitarian in nature. It may well be that some communities adopted iron in large part because of the impressive pyrotechnologies involved, or because they admired the way that iron was transforming gender relations among neighboring groups (see chapter 13, this volume). Such questions will be much more challenging—but not impossible—to address, and they will provide us with more stimulating perspectives on the African past.

INTERLEAF

E

ANCIENT PRESTIGE WORKS FROM THE UPEMBA REGION OF CENTRAL AFRICA

William J. Dewey

In a burial in what is now the heartland of Luba peoples along the Lualaba River of southeastern Democratic Republic of the Congo, an iron axe blade and pins that likely once decorated its wooden shaft lie next to the remains of a man with a forged iron anvil beside his skull (figs. E.1–E.3). This burial, which dates to the eighth to tenth centuries CE, was excavated at the Kamilamba site on the northern shore of Lake Kabamba, where people settled in the seventh century CE. While hundreds of graves from these early times have been excavated and researched, no occupation sites have yet been discovered. Grave goods at Kamilamba include pottery, iron tools, and iron and copper jewelry, the latter probably traded from the Katanga Copperbelt to the south. A few individuals were buried with considerably more objects, and in some cases with rare items such as the richly engraved ceremonial axe and anvil under consideration here. Archaeologists believe such graves are indicative of the development of sociopolitical differentiation in consolidating polities (Maret 1992; Childs and Maret 1996). There seem to be cultural continuities between the Kamilamba graves, graves from the second millennium CE found by archaeologists, and historical practices among Luba peoples living in the region today.

While we will never know who the individual in this Kamilamba grave was, the anvil beside his head suggests that he may have been a blacksmith. However, no additional forging tools have been identified. This burial may be an early example of a widespread cultural idiom in central Africa through which kings and blacksmiths share the transformative powers of iron as a material, ironworking as an essential technology, and forged objects as important to local and regional trade. Among Luba living in the region today, anvils are royal regalia. Historically, during the investiture of at least certain Luba kings, a ritual called "the beating of the anvils" saw nobles striking the king-to-be's knees, to "forge" a mortal into a sacred ruler (Womersley 1985, 71; see also see chapters 1, 6, and 15, this volume).

Ceremonial axes are emblems of Luba spirit mediums, court historians, and other officials, with the most ornate ones reserved for kings. Among them was a special axe to be held by the king during investiture, understood to have been passed down from the Luba culture hero and first blacksmith, Kalala Ilunga. Decorative incisions on such blades (figs. E.4, E.5) are called *ntapo*, the same term used to refer to the scarifications through which women perfected their bodies. The analogy is explicit, for Luba hold that only a woman's body is capable of holding the spirits of kings (M. N. Roberts 2013). Among Luba, iron pins like those found in the Kamilamba grave are called *vinyundo* (s. *kinyundo*), or "little anvils" (Nooter 1991; Dewey and Childs 1996). They adorn memory boards (*lukasa*; see fig. 15.27), staffs, ritual objects, and tools, assuring community prosperity through the transformative efficacies of iron.

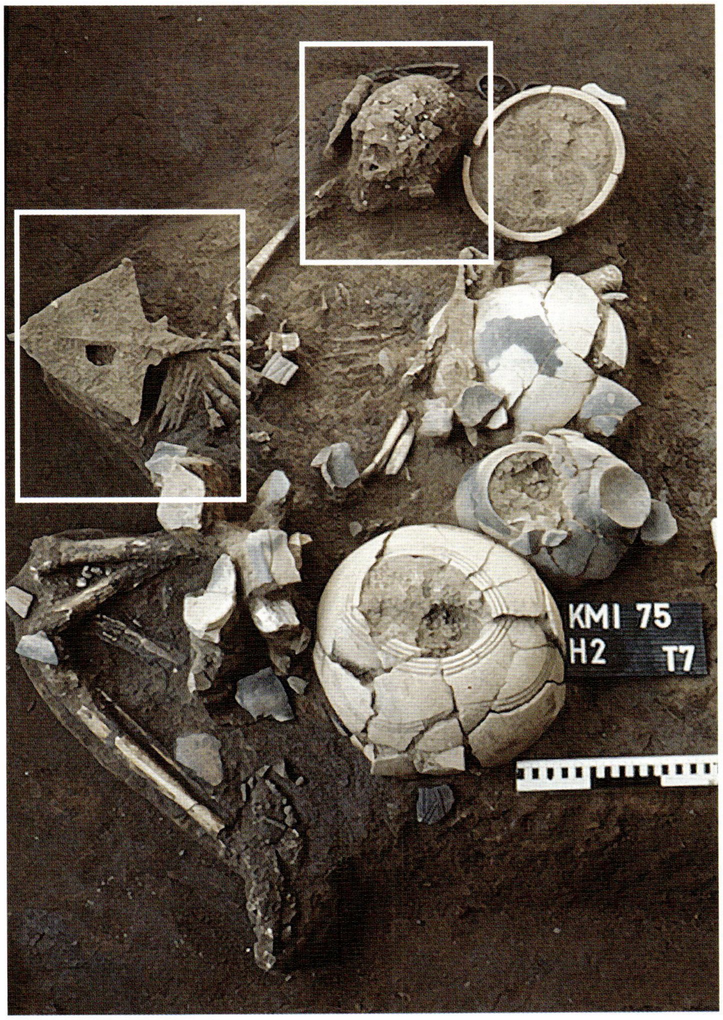

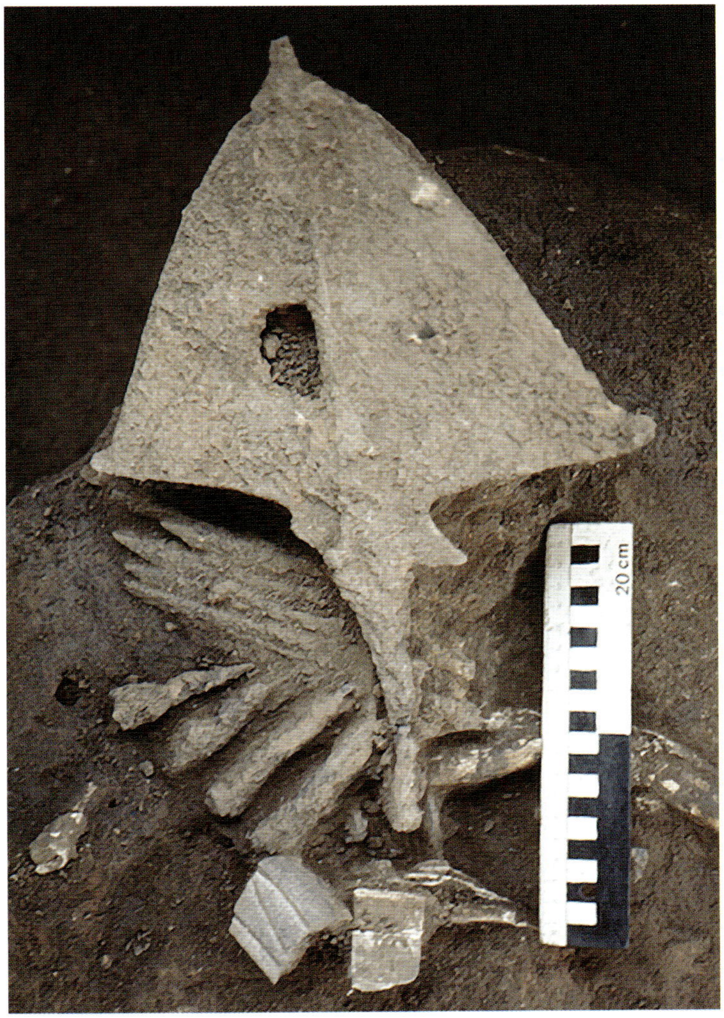

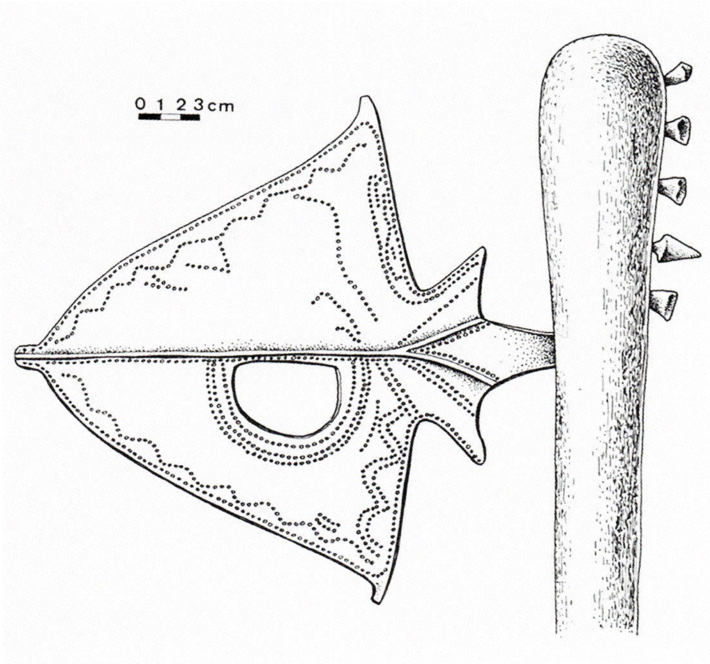

E.1 At Tomb 7, Kamilamba, archeologists have identified numerous objects buried alongside a human body. Strikingly, they include iron artifacts: an axe blade, pins, and an anvil. In this documentary photograph, the blade and pins are located above the figure's bent knees; the anvil is to the left of his skull.

PHOTOGRAPH BY PIERRE DE MARET, KAMILAMBA, UPEMBA REGION, DEMOCRATIC REPUBLIC OF THE CONGO, 1992. © PIERRE DE MARET AND ROYAL MUSEUM FOR CENTRAL AFRICA, TERVUREN

E.3 Rendering of axe blade with wooden handle ornamented by pins (based on artifacts at Tomb 7).

DRAWING BY YVETTE PACQUAY, 1992. © ROYAL MUSEUM FOR CENTRAL AFRICA, TERVUREN

E.2 Detail of iron axe blade with pins at Tomb 7.

PHOTOGRAPH BY PIERRE DE MARET, 1992. © PIERRE DE MARET AND ROYAL MUSEUM FOR CENTRAL AFRICA, TERVUREN

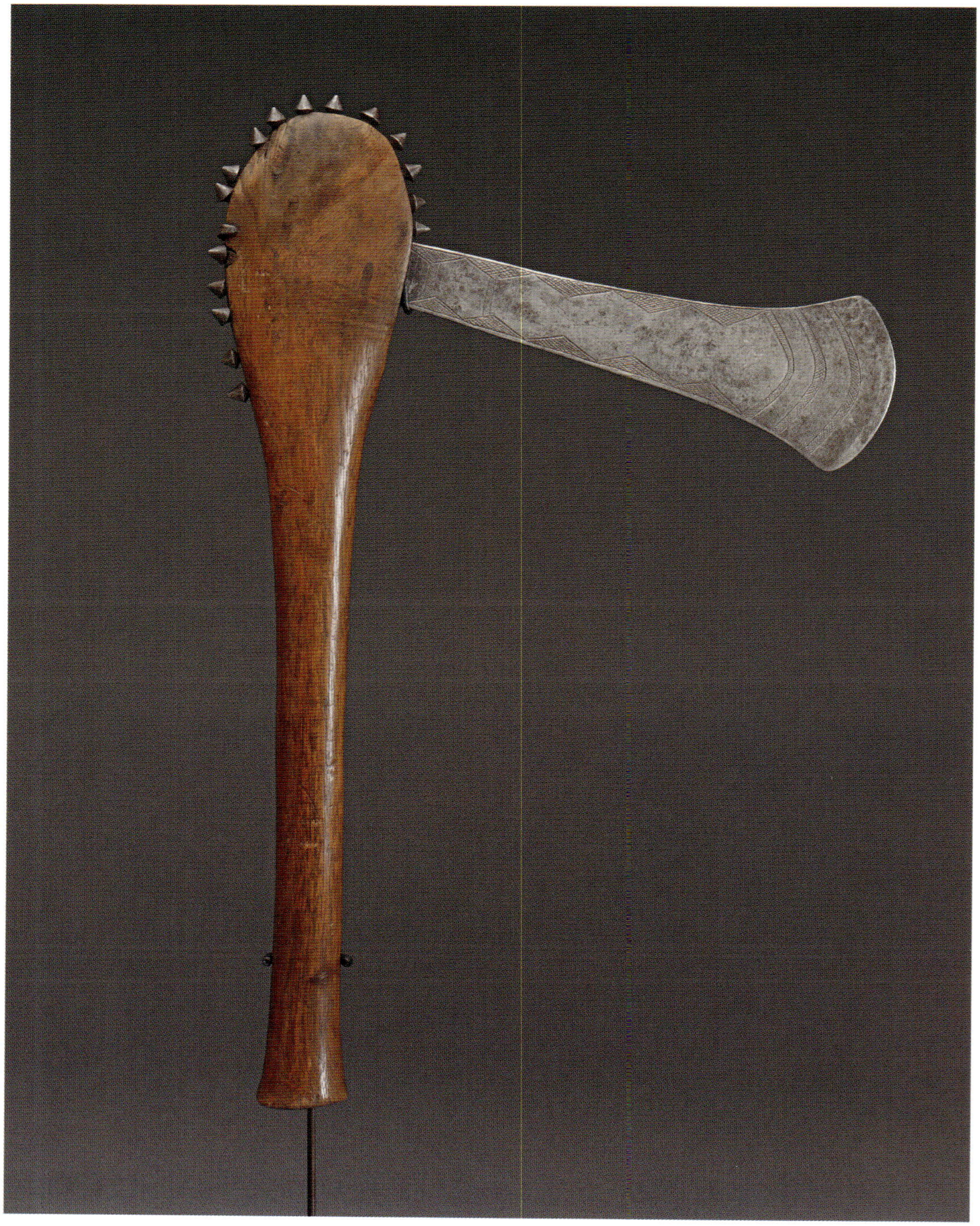

E.4
Artist unknown (Luba peoples, Democratic Republic of the Congo)
Ceremonial axe, early 20th century
Wood, copper, iron
H: 38 cm
MUSEUM FÜNF KONTINENTE, MUNICH, 13-57-126
PROVENANCE: HUGO DEININGER, MUNICH, 1913
(NOT IN EXHIBITION)
PHOTOGRAPH BY NICOLAI KAESTNER, 2017.

This Luba ceremonial axe is decorated with pins called *vinyundo* ("little anvils") in a similar manner to the Kamilamba ancient example (see figs. E.2, E.3).

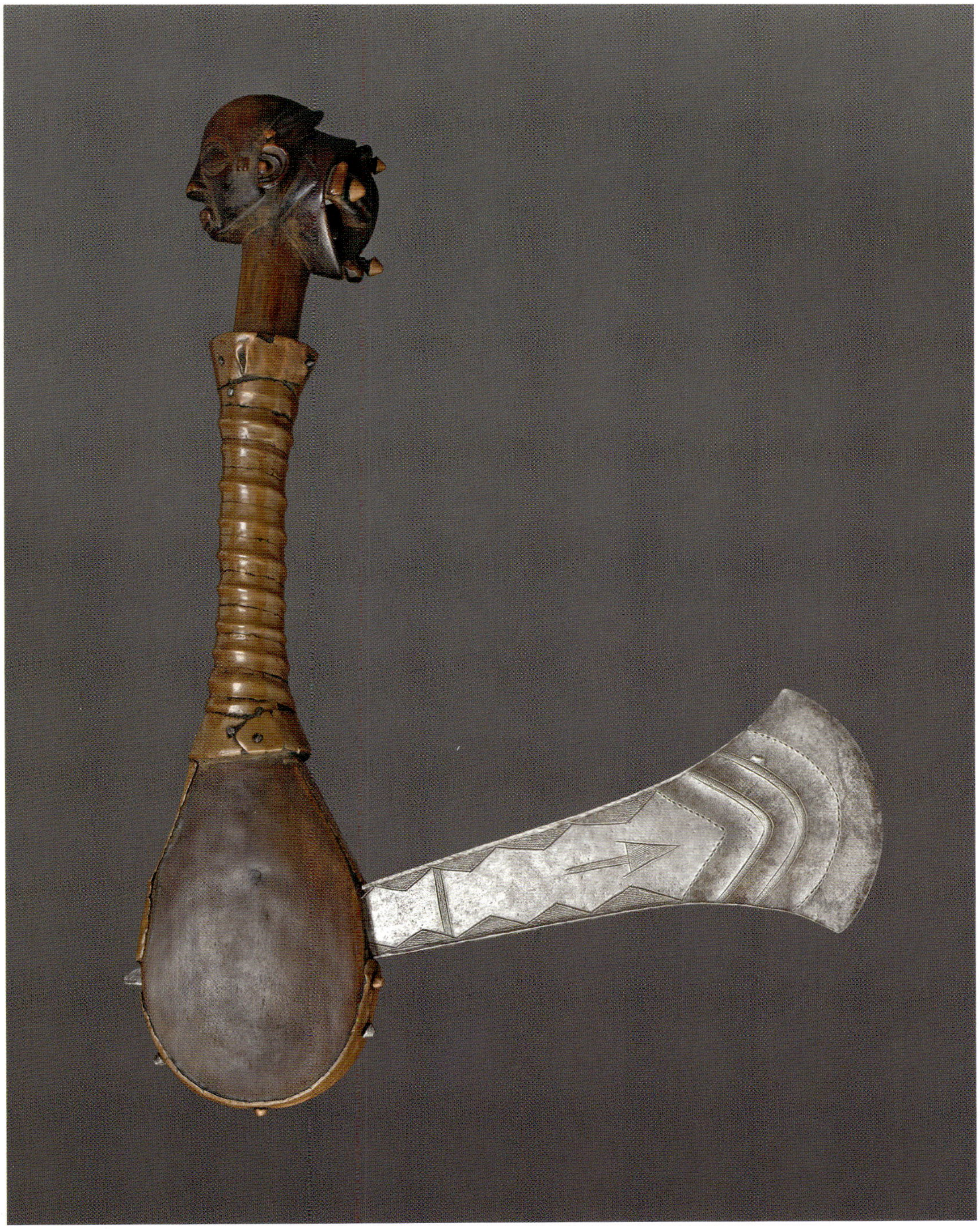

E.5
Artist unknown (Luba peoples, Democratic Republic of the Congo)
Ceremonial axe, early 20th century
Wood, copper, iron
H: 38.5 cm
MUSEUM FÜNF KONTINENTE, MUNICH, 13-57-127
PROVENANCE: HUGO DEININGER, MUNICH, 1913
(NOT IN EXHIBITION)
PHOTOGRAPH BY NICOLAI KAESTNER, 2017.

A relatively recent Luba ceremonial axe shows how the incisions referred to as *ntapo*, or "scarification," have been applied as blade decoration. Decorating the blade's sculpted handle are iron pins, including in the coiffure of the female head at its end.

CHAPTER
10

CULTURE AS TECHNOLOGY: EAST AND SOUTHERN AFRICAN IRON ARCHAEOLOGY

Shadreck Chirikure   William J. Dewey

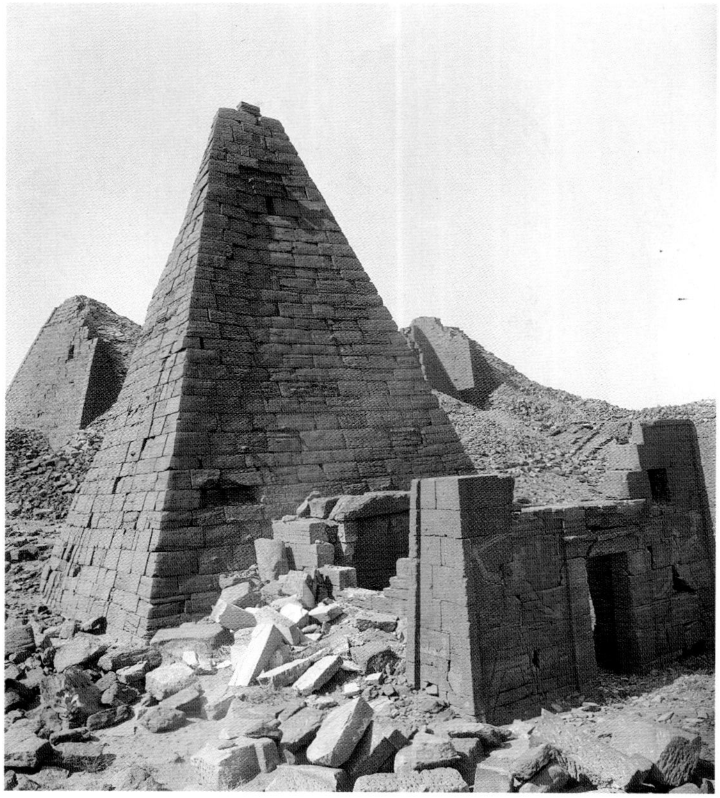

For the past two and a half millennia, terminating with the introduction of colonialism, there have been multiple, temporally shifting ironworking practices in the vast region of eastern and southern Africa, involving a mix of natural-draft-powered and bellows-driven furnaces operated in environments conditioned by local topographic and cultural factors. While the technological principles remained the same (i.e., bloomery processes) across space and time, the sizes of furnaces differed, from the more than two-meter-high natural-draft furnaces of Phoka peoples in Malawi to the shallow bowl furnaces of Zulu in South Africa. And yet, despite the diversity of types and operational parameters, these furnaces could all produce grades of soft iron and steel for conversion into regionally and culturally specific finished objects. Pointing to sustained and layered histories of interaction, technological cross-borrowing, and innovation, the practices of preindustrial iron production in this wide region, where communities were neither technologically nor culturally isolated, are of special interest in the global study of early iron production.

THE SUDAN, ETHIOPIA, AND THE HORN OF AFRICA

From the late 1800s to the 1960s, it was assumed that iron technologies had diffused southward into sub-Saharan Africa from the kingdom of Kush and its capital, Meroë (fig. 10.1; see also fig. 8.1), on the upper Nile in what is now the Sudan (ca. 300 BCE–350 CE). Huge mounds of slag were to be found there, representing, by a

10.1 Nubian royal pyramids at Meroë in the Kingdom of Kush, Sudan (first millennium BCE). Considerable amounts of slag found nearby were once thought to be evidence of the most ancient iron smelting in Africa. Meroë was proposed as the source of iron-working technologies that spread to the south. Smelting technologies across the continent, however, are different, and there is no evidence of diffusion from Meroë to the rest of Africa.

THE EGYPTIAN EXPEDITION, PHOTOGRAPHED IN 1906. COURTESY OF THE ORIENTAL INSTITUTE OF THE UNIVERSITY OF CHICAGO

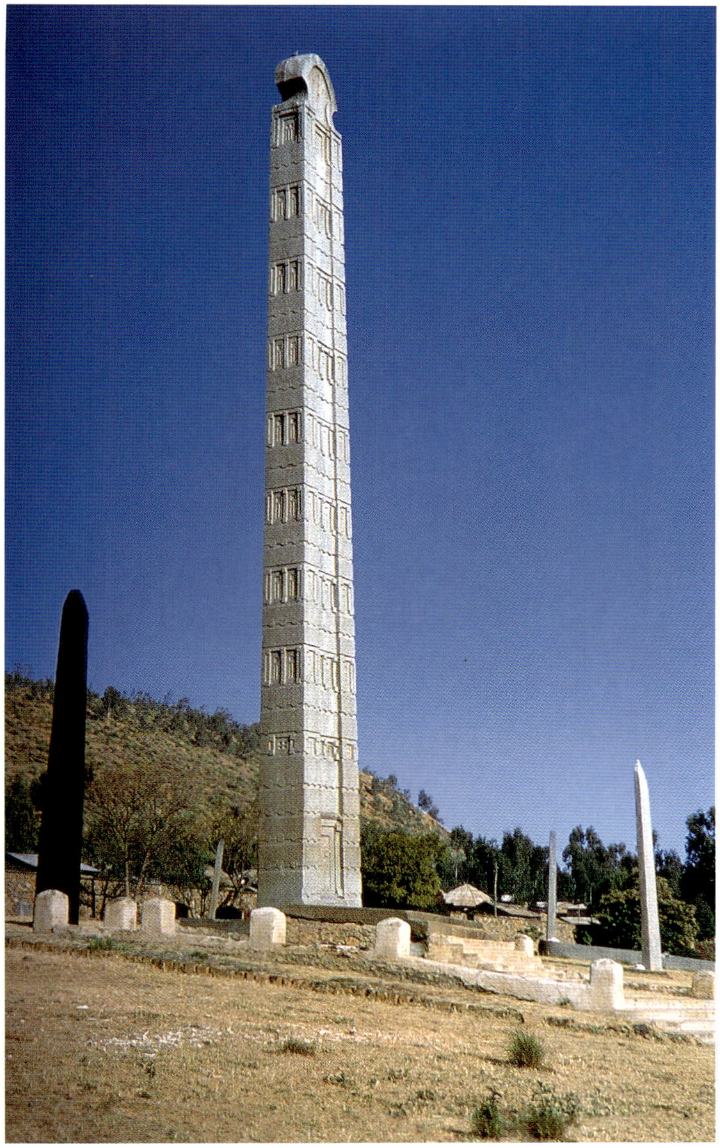

10.2 Stela at Axum (third to fifth century CE), marking where kings were buried in underground chambers. The first extensive use of iron in Ethiopia occurred at Axum.

PHOTOGRAPH BY DAVID W. PHILLIPSON, 1992–93. COURTESY OF DAVID W. PHILLIPSON

rough calculation, the production of five thousand tons of iron bloom (Rehren 2001). With the advent of radiocarbon dating it was realized that the furnaces discovered at Meroë dated to the first half of the first millennium CE (Shinnie 1985). These furnaces' technology of draining slag from furnaces during smelting (versus permitting slag to collect in a pit under the furnace) was probably derived from Roman prototypes (Trigger 1969). If so, Meroë technology may not have been comparable to iron smelting south of the Sahara.

Current analysis of sites in and around Meroë suggests that the slag was produced over a long period, probably through several different smelting technologies, some different from Roman prototypes (Humphris and Carey 2016). Much is still to be discovered about Meroitic ironworking, but we now know that iron smelting dates back to at least the sixth century BCE (Humphris and Scheibner 2017). So much iron was produced compared to the number of iron objects found through archaeological excavations of Meroitic sites to date that most iron must have been sent away in trade, but destinations are unknown (Phillipson 2005, 225).

Iron may have been introduced into Ethiopia from either Meroë or from across the Red Sea in the first millennium BCE. Though some has been found archaeologically at Yeha (Mapunda 1997), it was likely imported. The first extensive local use of iron in Ethiopia occurred at Axum (fig. 10.2) in the first centuries of the first millennium CE (Phillipson 2002), but where the production took place still remains unclear.[1]

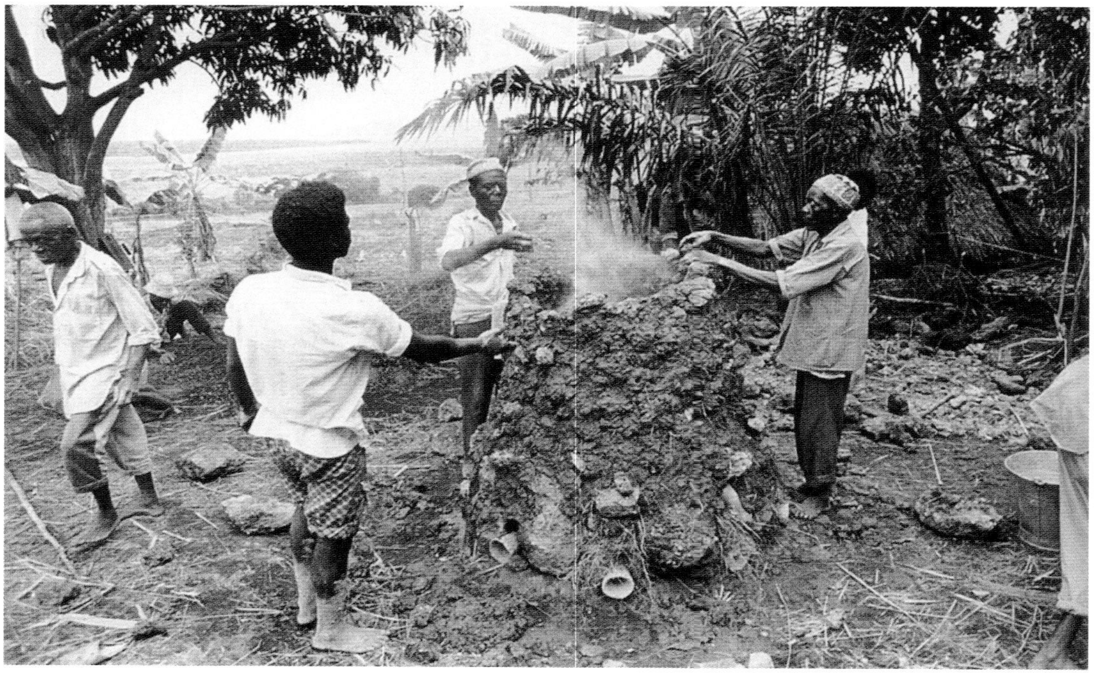

10.3 In 1976, elderly smelters at a furnace reconstruction in Buhaya, Tanazania (organized by archaeologist Peter R. Schmidt) remembered technologies that closely matched those used in the area more than two thousand years before.
PHOTOGRAPH BY PETER R. SCHMIDT, 1976.

THE GREAT LAKES REGION TO THE SWAHILI COAST

The most extensive record of iron production for this area is associated with Bantu-speaking peoples (see chapter 15, this volume). It was once thought that these groups, which dominate most of eastern, southern, and central Africa, spread from northeastern Nigeria/Cameroon to central and southern Africa two to three thousand years ago, primarily because their knowledge of ironworking gave them distinct agricultural, economic, and military advantages over peoples already living there (see, for example, Oliver and Fagan 1975). However, such narratives are now widely considered to oversimplify complex processes that did not involve unidirectional mass movement. Indeed, early Bantu speakers likely acquired knowledge of ironworking and agriculture from their non-Bantu-speaking neighbors (Childs and Herbert 2005; Eggert 2005; Maret 2013). The earliest dates for iron production and usage in east Africa are in the middle centuries of the first millennium BCE; in southern Africa, both are estimated to have begun by 200 CE (Chirikure 2013).[2]

Movement of Bantu speakers in eastern and southern Africa and the associated societal characteristics of early farming communities (settled occupation, cultivation, herding, pottery, metallurgy) are often labeled the Chifumbaze Complex (Phillipson 2005, 249–61). They are also assigned to the generic categories Early Farming Communities (EFC) and the Early Iron Age (EIA). Between the eighth and sixth centuries BCE, pioneer metallurgists and agriculturalists settled to the west of the Great Lake Nyanza (also known as Lake Victoria) in lands that are now in Rwanda, Burundi, southwestern Uganda, western Kenya, and northwestern Tanzania. These peoples made the pottery commonly known as Urewe (Lane 2015, 488).

Peter Schmidt and S. Terry Childs (1985) have undertaken archaeological studies in the Buhaya area of northwestern Tanzania, focusing on Urewe-associated iron smelting sites (fig. 10.3). All were slag-pit furnaces of about eighty to one hundred and thirty centimeters in diameter with multiple tuyères inserted into them. The pits were lined with grass before being charged with iron ore and charcoal, a practice clearly aiding the efficiency of production for the smelt. Archeometallurgical studies (e.g., Avery and Schmidt 1979; Childs 1996) have confirmed the ancient ability to have the furnaces reach at least 1250° Celsius and to efficiently drain fluid slag from the iron to produce steel blooms.[3] Very similar early furnaces have been found in Rwanda and Burundi (van Grunderbeek et al. 1982; 2001). These sites' dating has been rather contentious over the years; they are now believed to range from approximately 800 to 400 BCE (Killick 2009, 406).[4]

In the interlacustrine areas to the west of Lake Nyanza, iron production continued to be important after the early Urewe period, and was critical to the emergence of later states (Reid 2013). Several important studies now show that iron smelting techniques varied significantly even when in proximity (Humphris et al. 2009), and that iron smelting responded to and stimulated political consolidation and state-development activities (Humphris and Iles 2013). Detailed ethnographic collection of smiths' memories of smelting activities (Childs 2000; Wright 2002) and synthesis of ethnographic and archaeological work conducted in Uganda (Iles 2013) are helping us gain understanding of the diversity of techniques in the region.[5] The interlacustrine region offers prime examples of strong interdependence, both symbolic and real, between kings and ironsmiths, which is typical of Bantu-speaking peoples. For example, in one of the most famous excavations of a royal tomb in Rwanda, two iron anvils were found at the head of King Rugira (Van Noten 1972; see also chapter 15, this volume).

The Swahili Coast provides further information on early ironworking in Africa south of the Sahara. The mercantile character of this part of the continent, with trade and cultural ties and interactions with the Middle East and India, make it different from sub-Saharan Africa inland from the Indian Ocean (fig. 10.4).[6] While for many years coastal east Africa was incorrectly perceived as being more Arabic than African, Adria LaViolette points out that "what has been established is the important knowledge that Swahili ironworking grew out of broader eastern African trends, based on bowl-furnace smelting and smithing of the resulting bloom (Mapunda 2018). This key finding underscores the indigenous origins of Swahili society (Kusimba et al. 1994)" (LaViolette 2018, 321). Felix Chami (1992; 1994) has investigated a first-centuries-CE pre-Swahili site called Limbo, located in Tanzania just south of Dar es Salaam, where at least three tons of slag have been found. This level of iron production is only comparable to a few east African sites (e.g., Meroë, Axum), but unfortunately we do not know how broadly iron was distributed from the Limbo site nor why production ceased.

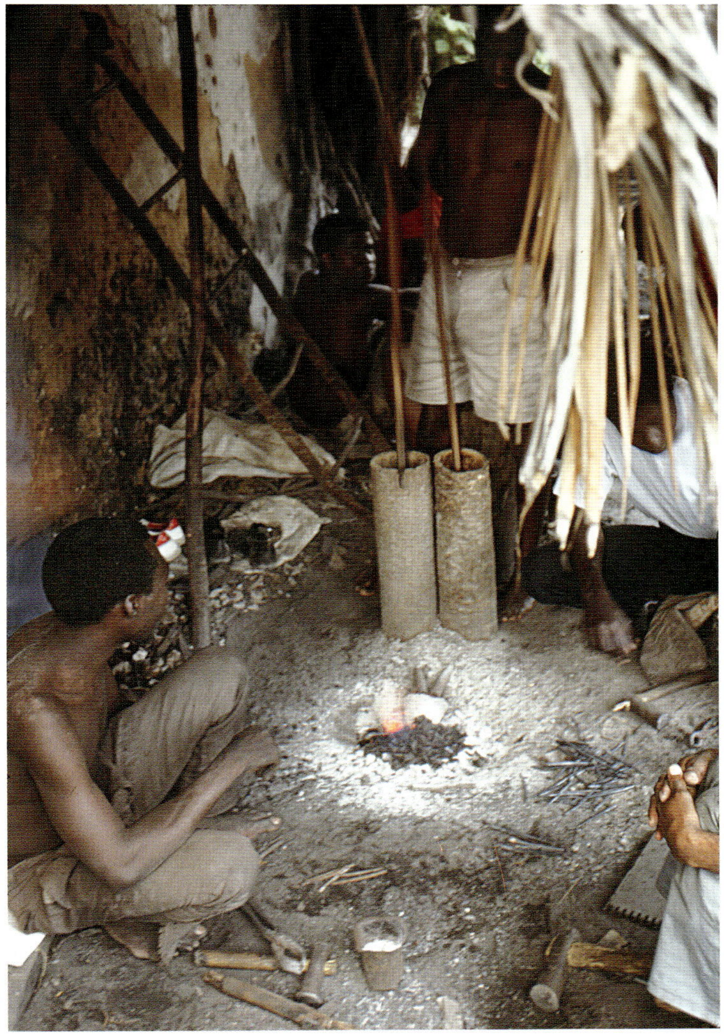

10.4 Swahili blacksmiths using piston-type bellows to make nails for dhow boat construction. The variety of Swahili cultural ties, trade, and interactions is mirrored in their arts and technologies. Piston bellows are also seen in Madagascar and Indonesia. PHOTOGRAPH BY WILLIAM J. DEWEY, NEAR STONE TOWN, ZANZIBAR, 1993.

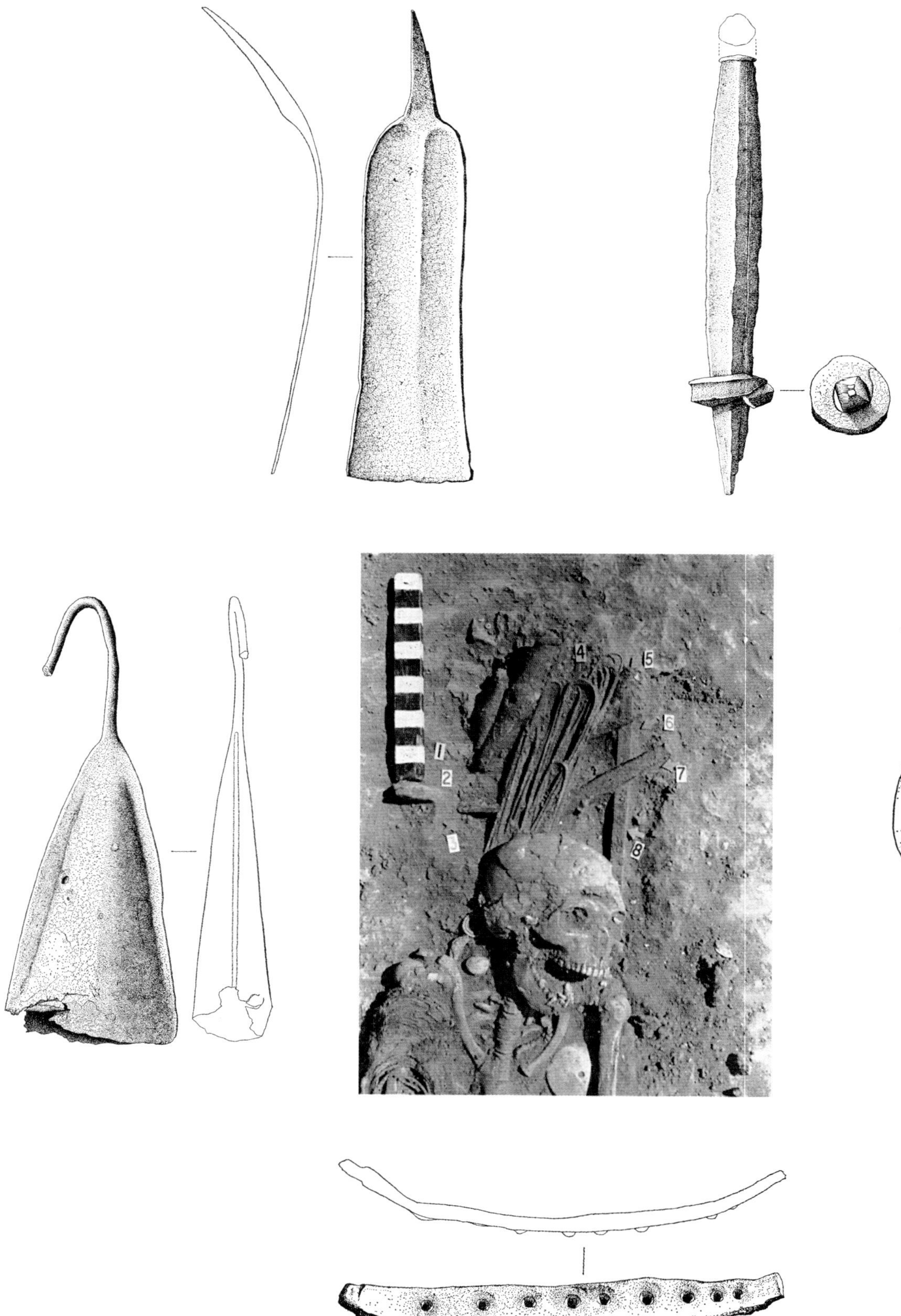

10.5A–F The 16th century witnessed a period of increased interaction among the Mwene Mutapa confederacy of northern Zimbabwe, Swahili peoples, and Portuguese traders at Ingombe Ilede, along the middle Zambezi River (in present-day Zambia). A series of burials there contain rich accumulations of grave goods, including metal items. For example, at Grave 11/3 (center), a pair of tongs, a hoe blade, a socketed hammer-head, a draw plate for making wire, and a bell have been documented (clockwise from top left). IMAGES REPRODUCED IN BRIAN M. FAGAN, "EXCAVATIONS AT INGOMBE ILEDE 1960–2." IN BRIAN M. FAGAN, D. W. PHILLIPSON, AND S. G. H. DANIELS, *IRON AGE CULTURES IN ZAMBIA*, VOL. 2 (LONDON: CHATTO & WINDUS, 1969).

An important study of one hundred fifty iron objects from five Kenyan Swahili coastal sites dates them from the eighth to the sixteenth century CE (Kusimba et al. 1994). While most were forged of locally smelted iron, a few pieces of crucible steel were also found. These findings have split archaeological opinion, with David Killick (2009a) arguing that they were imported from India and Chapurukha Kusimba (1997) maintaining that they were locally made. Nevertheless, evidence indicates that Swahili technologies of ironworking were remarkably stable for much of the two thousand years before nineteenth-century Omani and European colonization of eastern Africa (Killick 2009b).

Adding to the archaeological record, some historical sources comment on iron's importance to political economies of the Swahili Coast. One early Arabic source, "commenting on the technical quality of iron produced in East Africa... noted that it was an important trade item for southern Indian merchants, who confessed of its superior malleability to Indian iron" (Kusimba 1997, 507, citing Al-Idrisi [ca. 1100–1166 CE]). Even as late as the nineteenth century, "Indian and Arab Merchants rated iron locally produced near Mombasa [Kenya] more highly than they could import from Sweden" (Allen 1993, 57), a sentiment shared with an early European observer of Tabwa smelting along the southwestern shores of Lake Tanganyika (see chapter 1, this volume).

In southwestern Tanzania, extensive attention has been given to the tall natural-draft furnaces of Fipa peoples. There, the rather late persistence of iron smelting during the colonial era has meant that master smelters such as Mzee Stefano could still be interviewed (Wright 2002) and that both archaeological investigations (Mapunda 2010) and ethnoarchaeological studies and reconstructions could be undertaken (Barndon 2010). Fipa also used smaller forced-air furnaces to consolidate blooms, making theirs a rather unusual two-furnace process. Such achievements were similarly realized by Tabwa peoples living directly across Lake Tanganyika in the southeast of present-day Democratic Republic of the Congo. Given cultural and historical relations between Tabwa and Fipa, it is not clear who may have influenced whom.

In Malawi, Killick (1990) undertook a combined archaeometric, ethnographic, and historical study of bloomery iron smelting practiced in the central region of the country between 1860 and 1940. His most significant finding is that Kasungu smiths were able to convert very low-grade ore into steel in their furnaces, thus confirming that African smiths have been highly innovative in adapting technologies to meet their resources.

A variety of iron smelting and ironworking traditions in Zambia have been surveyed by David Phillipson (1968; 1969) and Brian Fagan (1961; 1962). Their most interesting discoveries concern graves found at Ingombe Ilede, once a thriving town at the convergence of the Zambezi and Lusitu rivers where trade relations for the larger region were once maintained (Fagan 1969; Phillipson and Fagan 1969). Burials of eleven individuals dating from the sixteenth century CE have become famous for their rich accumulations of grave goods (McIntosh and Fagan 2017). The deposited iron objects include hoe blades, axes, clapperless bells, hammer heads, tongs, spikes, and draw plates for producing copper wire; also included were conus shell ornaments imported from the Indian Ocean world and wire and cruciform ingots of copper from the Hurungwe district of northwestern Zimbabwe (fig. 10.5). The wealth of individuals so interred certainly derived from long-distance trade of locally produced copper, iron, and salt, suggesting the importance of metal and related commodities to complex economies of the region.

Iron tools found as grave goods were used for drawing wire (Steel 1975). Bundles of copper wire were a form of currency found at several ancient sites, and wire was also important for jewelry and for ornamenting ceremonial objects throughout central and southern Africa (see Matenga and Dewey forthcoming; see also interleaf C, this volume). The inclusion of metalworking tools in graves, however, has raised questions as to whether the buried were actual metalworkers. Or, as among numerous central Bantu groups, was this an example of blacksmith-king metaphorical associations? In other words, were these elite individuals leaders rather than smiths?[7]

10.6A–B Iron artifacts have been found throughout the site of Great Zimbabwe, among them hoe blades (above) and a central-tanged axe blade (below). A particularly large hoard of items (especially hoe blades) found just outside of the Great Enclosure perhaps represented tribute or trade.

ABOVE: PUBLISHED IN J.T. BENT, *THE RUINED CITIES OF MASHONALAND, BEING A RECORD OF EXCAVATION AND EXPLORATION IN 1891* (LONDON: LONGMANS, GREEN, AND CO., 1895); BELOW: PUBLISHED IN P.S. GARLAKE, *GREAT ZIMBABWE* (NEW YORK: STEIN AND DAY, 1973).

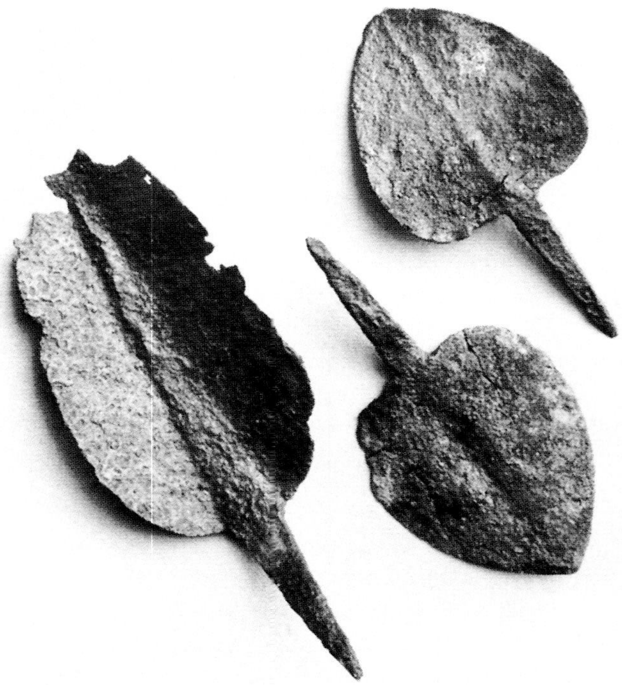

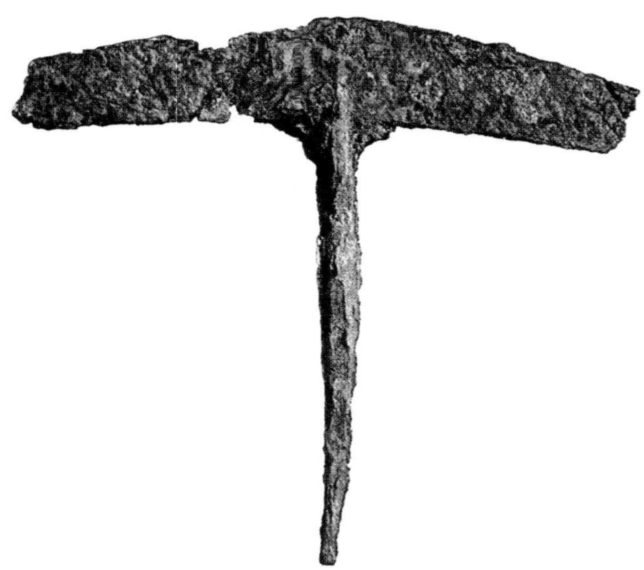

SOUTH OF THE ZAMBEZI RIVER: ZIMBABWE, BOTSWANA, SOUTHERN MOZAMBIQUE, AND SOUTH AFRICA

The earliest evidence of iron production and use south of the Zambezi is associated with sites belonging to the Early Iron Age and dating to the early centuries of the first millennium CE (Phillipson 2005). The archaeological site of Matola (ca. 200–400 CE) in Mozambique has yielded evidence of what is currently accepted to be the earliest iron production in southeastern Africa. More remains of iron production and early millet cultivation have been recovered at Silver Leaves, in modern-day South Africa (Mitchell 2002). It is possible that hoes and axes played important roles in subsistence practices, but most available data suggest that what little iron was available was used to make small ornamental items. From Zimbabwe through southern parts of Mozambique and on to South Africa and Botswana, most sites are associated with slag, remnants of ore, broken tuyères, and in some cases collapsed furnaces, suggesting that iron production became more abundant after the first half of the first millennium CE (Chirikure 2013).

It is not clear whether this increase in iron production necessarily represents growing populations and political consolidation (Chirikure 2015). Possibly, agriculturalists had now mastered smelting technologies and could produce more iron while retaining lineage-based societies. Indeed, by the end of the first millennium CE, iron production was deeply entrenched in social, economic, and political systems south of the Zambezi. During this period, Arabic writers such as Al-Masudi (ca. 893–956 CE) commented that communities of southern Africa exchanged iron with merchants from the Indian Ocean world in return for desirable trade goods including cloth, glass beads, and sea shells (cited in Summers 1969).

According to Gilbert Pwiti (1996), available data suggest that towards the end of the first millennium CE, the size of farming community sites increased remarkably, pointing towards social differentiation. Indeed, long-distance trade goods such as conus shells, glass beads, and, in some cases, elephant ivory, were recovered from sites such as Kadzi, KwaGandaganda, and Schroda. Other late first-millennium sites including Swart Village in northern Zimbabwe and the Tswapong Hills of Botswana offer plentiful evidence of iron production. Southern Africa's early states developed towards the interface of the late first and early second millennia CE, and there is no doubt that iron sustained communities through food production, hunting, trade, and exchange relationships as well as protective and/or aggressive activities with adversaries. This trend continued into the second millennium, when relatively large populations—concentrated at centers such as Mapela, K2, and Mapungubwe (in present-day South Africa) and at Great Zimbabwe (fig. 10.6)—suggest increasing demand for food and, consequently, for iron to produce useful tools and weapons.

As far as organization of production is concerned, to date no large-scale concentrations of slag remains have been found near specialist iron-producing landscapes such as the Botswana's Tswapong Hills or Musina in South Africa (Chirikure 2015). Neither are centers of power such as Great Zimbabwe associated with huge dumps of metal production waste. As a comparison, some historical numbers powerfully drive this point home. Historical data collected by Beach (1974) among the Njanja of Zimbabwe indicates that the production of iron sufficient for a one-kilogram iron hoe produced one kilogram of slag. The average duration of use for such a hoe was between two to five years. Assuming that an individual would have required one hoe blade for cultivation, an estimated population of ten thousand adults at Great Zimbabwe would, every five years, have consumed one metric ton of iron and created one ton of slag (Beach 1974; Chirikure 2007). Because no large slag mounds have been found at Great Zimbabwe or the surrounding area, it appears that production was more spread out than concentrated, resulting in numerous small-scale iron production sites scattered across the landscape (Mtetwa 2011). A tributary system is also suggested, through which taxation of dispersed homestead production units would have met the demand for iron at and beyond Great Zimbabwe (see interleaf F, this volume).

To put the dispersed mode of production and its scale in southern Africa into perspective, it is essential to consider other historical landscapes associated with specialized iron production in the region. For example, the Zulu king Sigidi kaSenzangakhona, or Shaka Zulu, is believed to have raised an army that numbered as many as fifty thousand soldiers. While Shaka had an unknown number of iron producers based at Nkandla (fig. 10.7), no large mounds of slag have been recorded from Zulu production sites (Maggs 1992). Therefore, different organization systems—dispersed (as at Great Zimbabwe) or concentrated (as at Meroë)—could achieve specialization and surplus for trade and exchange in different contexts.

Shifting to technology, surviving evidence suggests a mix of furnaces was deployed south of the Zambezi: low-shaft furnaces that were bellows-driven, tall-shaft furnaces that were natural-draft powered, and forced-air-driven bowl furnaces. While some low-shaft furnaces required slag pits, others had a frontal opening. Known as the rake channel, the aperature was closed during smelting and opened upon completion so that slag and bloom could be removed. While historical evidence of natural-draft furnaces is plentiful north of the Zambezi (e.g., among Phoka of Malawi and Fipa of Tanzania), so far the only known cases of natural-draft furnaces to its south are archaeological. For example, Martin Prendergast (1975) excavated a natural-draft furnace near Darwendale in northern Zimbabwe, while Weber Ndoro (1994) excavated a possible natural-draft furnace at Chigaramboni, near Great Zimbabwe. The Tswapong Hills of Botswana have yielded tuyères fused in multiples, suggesting that natural-draft furnaces were used during the Zhizo phase of the Early Iron Age (Chirikure 2013), while bowl furnaces were deployed in the fifteenth century near Rhenosterkloof in northeastern South Africa. The evolution of these furnace types is hardly chronological; instead, it is a mix that speaks to context-specific selection based on cultural and topographic features (Childs 1991; Chirikure et al. 2009).

The evidence for iron objects is sparser when compared to that of debris such as slag, tuyères, and collapsed furnaces, largely because the soils in this region are very acidic, leading to decomposition of iron artifacts. The earliest iron objects from southern Africa appear to be mostly decorative beads and bracelets, as discovered at sites such as Broederstroom, South Africa (Miller and van der Merwe 1994). More such objects were found at Nqoma, in the Tsodilo Hills of Botswana, dating to after 500 CE. Heavy utilitarian tools mostly appear in abundance towards the end of the first millennium, and are known from Divuyu, also in the Tsodilo Hills, and other places. By the second millennium, evidence for finished iron objects becomes more plentiful in the form of arrowpoints, spearheads, and hoe blades, as well as bracelets, bangles, and beads, which all no doubt supported the subsistence economy. Across southern Africa, hoes, in particular, were widely used but appear in distinct regional designs (see chapter 6, this volume).

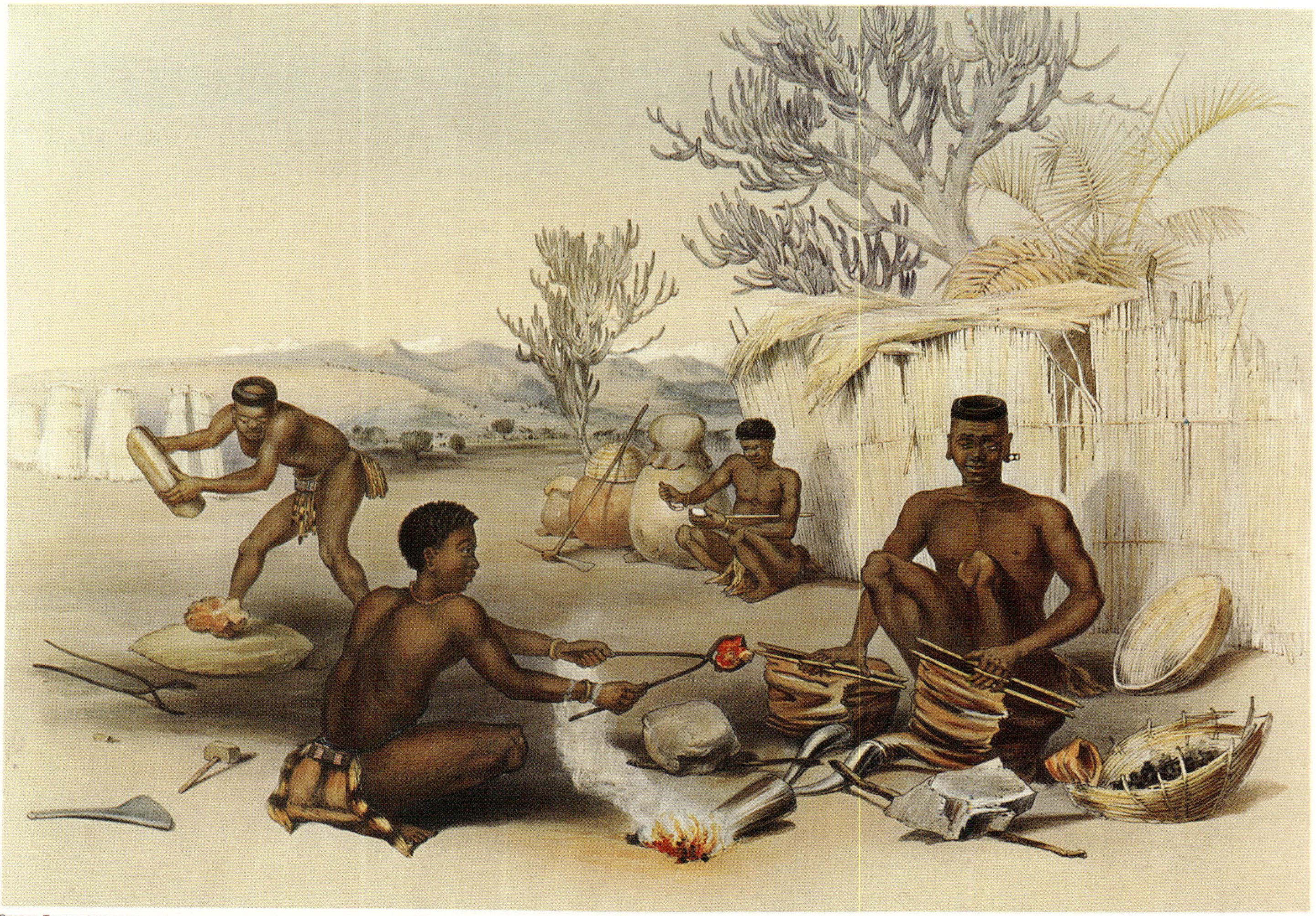

10.7 The military prominence of the 19th-century Zulu Kingdom was fueled by huge armies of warriors. George Angas, an English explorer, naturalist, and painter who visited South Africa in the late 1840s, recorded the activities of a group of smiths refining bloom and manufacturing weapons in 1849, during the reign of King Mpande.
PUBLISHED IN GEORGE FRENCH ANGAS, *THE KAFIRS ILLUSTRATED: A FACSIMILE REPRINT OF THE ORIGINAL 1849 EDITION OF HAND-COLOURED LITHOGRAPHS* (CAPE TOWN: A. A. BALKEMA, 1974).

In terms of distribution, well-networked trade and exchange relationships involving forged objects such as hoes implied social transactions including bridewealth. More interestingly, iron clapperless bells (fig. 10.8) similar to those known north of the Zambezi in Zambia and the DRC were recovered at Great Zimbabwe and two other sites in present-day Zimbabwe, Chiumnungwa (historically recorded as Umnukwana), and Danamombe (formerly known as Dhlodhlo). Because the instruments from these sites are the only ones known from archaeological and historical sites in regions south of the Zambezi (southern Zambezia), they may have been presented to the rulers as gifts in larger networks of exchange and cultural contact extending far to their northwest.

CONCLUSION

Once established, iron production and use played critical roles in the development of communities in eastern and southern Africa. The whole production sequence—from raw-material collection to transformation of iron ore in furnaces and forges, to iron's use, reuse, and dispersal—suggests that ironworking was deeply embedded in early societies, which adapted to accommodate and exploit such complexities. Basic implements including hoes, arrowheads, axes, knives and spears must have served ceremonial and ritual functions as well as practical needs. Forged objects were also exchanged and situationally important to interactions between adjacent and widely separated communities. Amidst all these interactions were countless cases of local experimentation and innovation, with performance of "enchanted technologies" deemed as necessary as smelting per se. Such creativity makes histories of iron production, use, and trade in southern and eastern Africa intellectually stimulating as we learn more about the lives of early peoples.

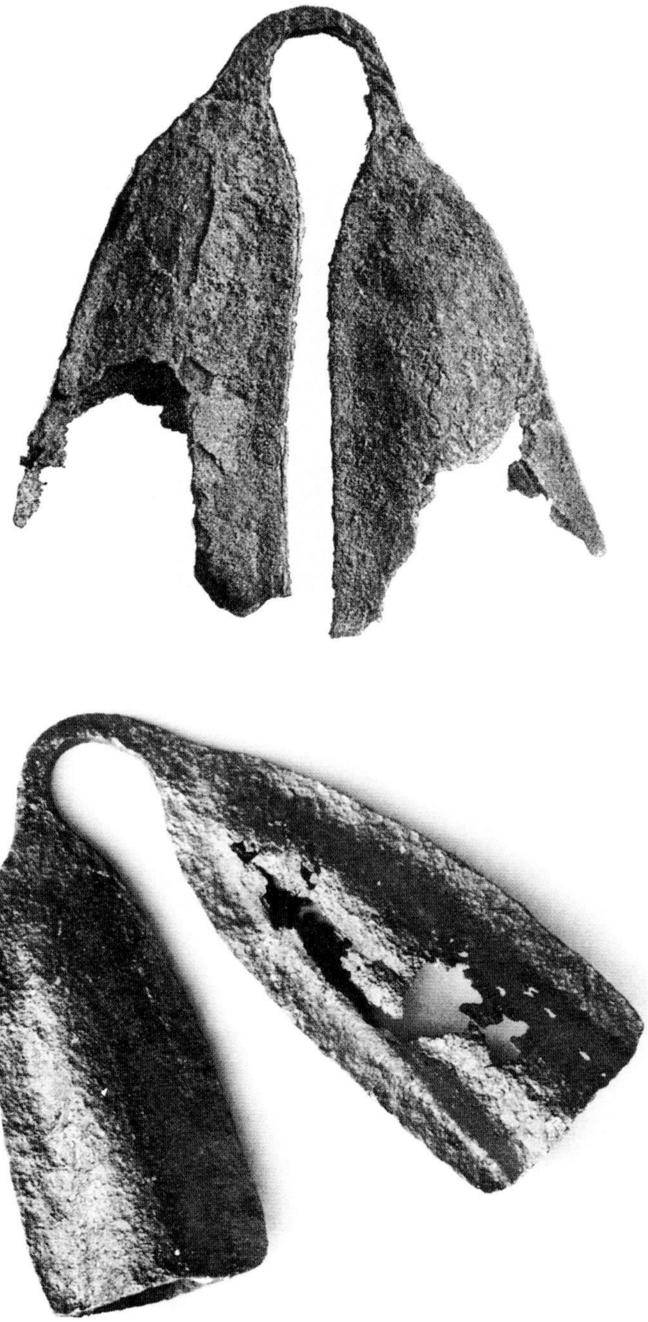

10.8A–B Double bells found at Great Zimbabwe were probably trade items or gifts from central Africa, where they signified status.

ABOVE: PUBLISHED IN J. T. BENT, *THE RUINED CITIES OF MASHONALAND, BEING A RECORD OF EXCAVATION AND EXPLORATION IN 1891* (LONDON: LONGMANS, GREEN, AND CO., 1895); BELOW: PUBLISHED IN P. S. GARLAKE, *GREAT ZIMBABWE* (NEW YORK: STEIN AND DAY, 1973).

INTERLEAF
F

FORGED VALUE: HOES IN THE ARCHAEOLOGY AND HISTORY OF SOUTHERN AFRICA

Shadreck Chirikure

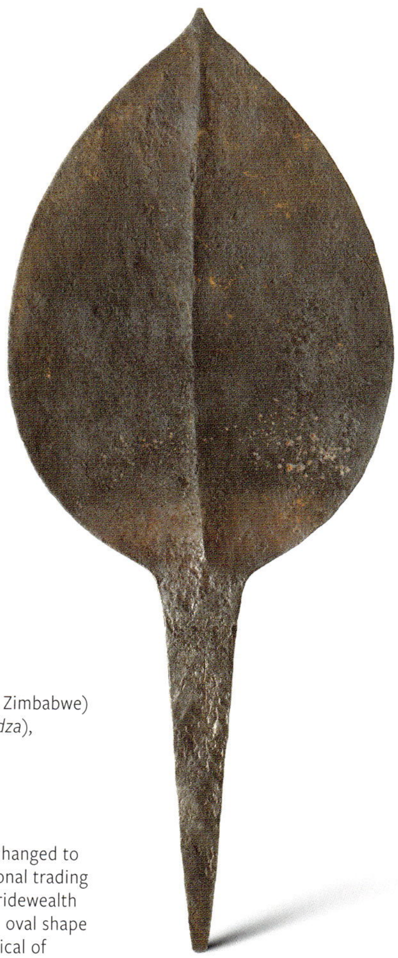

F.1
Artist unknown (Shona peoples, Zimbabwe)
Hoe-blade-shaped currency (*badza*), late 19th century
Iron
H: 62 cm
PRIVATE COLLECTION

Historically, hoe blades were exchanged to suit the needs of particular regional trading contexts and also accepted as bridewealth throughout southern Africa. The oval shape of the blade pictured here is typical of Shona peoples.

> My husband is a craftsman in iron,
> Truly a wizard at forging hoes.
>
> FROM A SHONA PRAISE-POEM[1]

Hoes are among the most significant objects of value forged of iron in southern Africa, where their production is to this day embedded in utilitarian, subsistence, economic, and symbolic categories of use (fig. F.1). Hoe blades have long figured into different trade and exchange transactions as dictated by particular contexts in the region extending south of the Zambezi River to the Cape of Good Hope. These blades, and the currency tokens derived from them, are variously shaped, for example oval ones for Shona peoples, diamond-shaped for Phalaborwa. All such blades taper to narrow tangs set into wooden handles to create composite pieces of equipment for digging, weeding, and ground-clearing. Archaeologically, iron blades have a far higher survival rate than their perishable wooden hafting.

Iron hoes were widely valued for subsistence and economic activities in the roughly two-thousand-year-long Iron Age (ca. 200–1900 CE) of southern Africa. However, archaeologists have not found much evidence of iron hoes from the first millennium CE. Presumably, after reaching the end of their use-lives, hoes were recycled into smaller objects such as knives and beads, examples of which have been excavated at sites such as Broederstroom (ca. 550–700 CE) in South Africa. From 1000 CE onwards, iron hoes of various shapes appeared in increasing abundance at Great Zimbabwe (ca. 1000–1700 CE), Mapungubwe (ca. 1220–1290 CE), Khami (ca. 1400–1650 CE), and other prominent sites (Miller 2002). The relatively high populations associated with these communities raise the importance of hoes in achieving food security, and, directly and indirectly, hoes contributed to and sustained political centralisation.

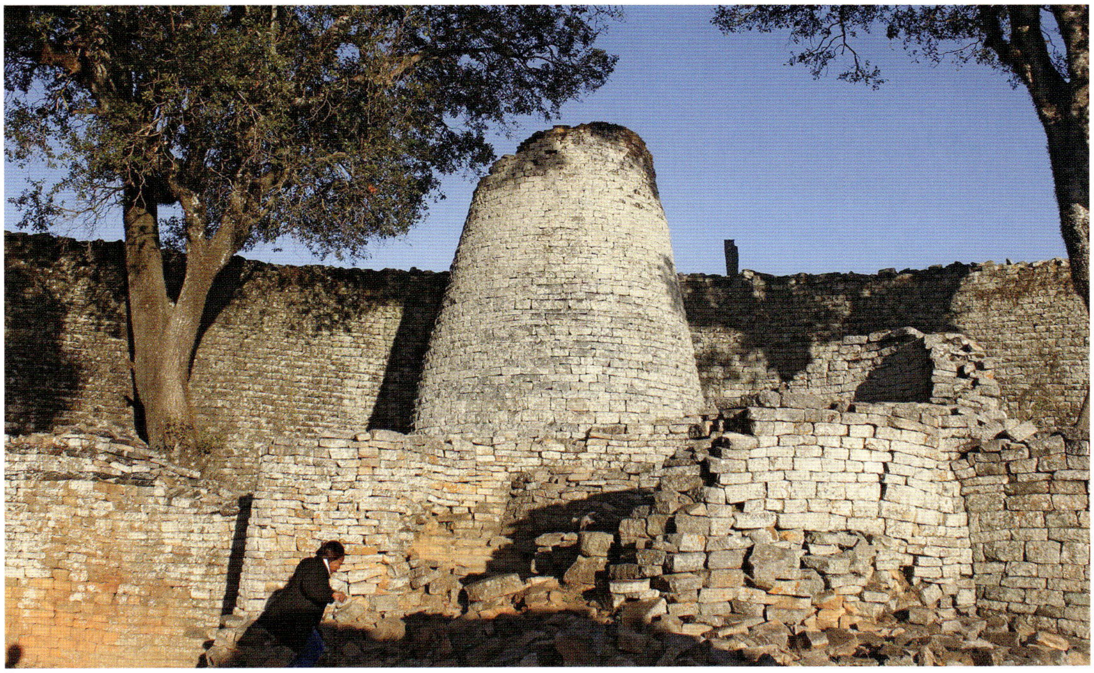

F.2 From the 14th to the 15th century CE, Great Zimbabwe's Great Enclosure was probably the residence of the elite. The conical tower (shaped like Shona graneries) perhaps symbolized tribute or the king's ability to care for his people.
PHOTOGRAPH BY SHADRECK CHIRIKURE, 2014.

At Great Zimbabwe, a sizeable number of hoes have been found in archaeological contexts associated with households inside and outside the famed stone walls of the ancient community (fig. F.2). An unknown number were recovered from the Valley Enclosures; they have been described as a "hoard" insofar as it is assumed that the blades were assembled for a purpose, likely trade and exchange (Garlake 1973). In later political formations including the Mutapa state, based in northern Zimbabwe and adjacent regions of Mozambique (ca. 1450–1900 CE), kings levied taxes (tribute) in the form of hoes blades and other commodities while their subjects traded and exchanged hoes among themselves (Mudenge 1988). The proceeds from tribute were used to cultivate the kings' fields as part of a labor organisation and cultural practice known as *zunde*, under which subjects worked to produce food kept in reserve for the community. Kings also exchanged hoes to obtain various commodities both local and imported. Hoes were, therefore, valued by everyone—kings and commoners, rulers and ruled alike.

Combined ethnohistorical and oral historical evidence gathered from various parts of southern Africa indicates that from the seventeenth century onwards the significance of iron hoes started at the forge, where they created value for smiths and then consumers acquiring them for various purposes (Beach 1977). Although wooden or bone hoes were sometimes used by Xhosa and Zulu peoples of South Africa, they were not as prized, effective, or durable as their iron counterparts (Moffett, Maggs, and van Schalkwyk 2017). Beyond subsistence, the utility of hoes extended to shallow-pit mining: they helped dig surface earth, making it easy to retrieve ores. Depending on frequency of use, a good-quality iron hoe blade lasted between two and five years (Beach 1974), after which it might be recycled to produce an axe or adze. Assuming a life expectancy of roughly sixty years, individuals would have consumed between ten and twenty hoes during their lifetimes.

As elsewhere in Africa, the use-value of hoes often made them central to trade and exchange systems, in which they assumed currency status (see chapter 5, this volume). Between the eighteenth and late nineteenth century, Njanja, Venda, and Phalaborwa peoples developed extensive and complex trade networks with hoes as principal media of exchange. Unlike other groups, Njanja, who are Shona peoples, developed a distribution system for which they travelled within a two-hundred-kilometer radius from their home area to Bocha, Shurugwi, and Gutu to exchange hoes for cattle, goats, salt, and elephant ivory, producing trading adventures still recalled (Beach 1977; Mackenzie 1975). The farther a trader ventured from the Njanja heartland, the more expensive the hoes became; this in turn encouraged people to travel to Njanja country to trade for hoes.

The wife of the blacksmith in the poem that opens this interleaf praises her husband as a "wizard" in making iron, because his occupation brings fortune to the family. Njanja converted wealth acquired from trade into political power locally, founding a centralized political confederacy made up of various lineages (Mackenzie 1975). As tokens of value, hoes figured into bridewealth negotiations (*roora*, for Shona; *lobola*, for Zulu and Xhosa). They were symbols exchanged for the reproductive power and productivity of the bride, and understood as "material praise" by her new in-laws.

With the advent of industrially manufactured iron in the late nineteenth and early twentieth centuries, hoes made of bloomery iron became obsolete, but traditional forms and symbolic associations have continued to inspire modern industrial hoe designs. Some blacksmiths continue to forge hoes using scrap iron. In summary, then, iron hoes have created and oiled social relations, and have fulfilled subsistence needs of southern African communities, varying, to be sure, from community to community, and generation to generation.

PART III

REGIONAL STUDIES

CHAPTER
11

BLACKSMITH SOCIETIES OF A MANDE WORLD

Patrick McNaughton

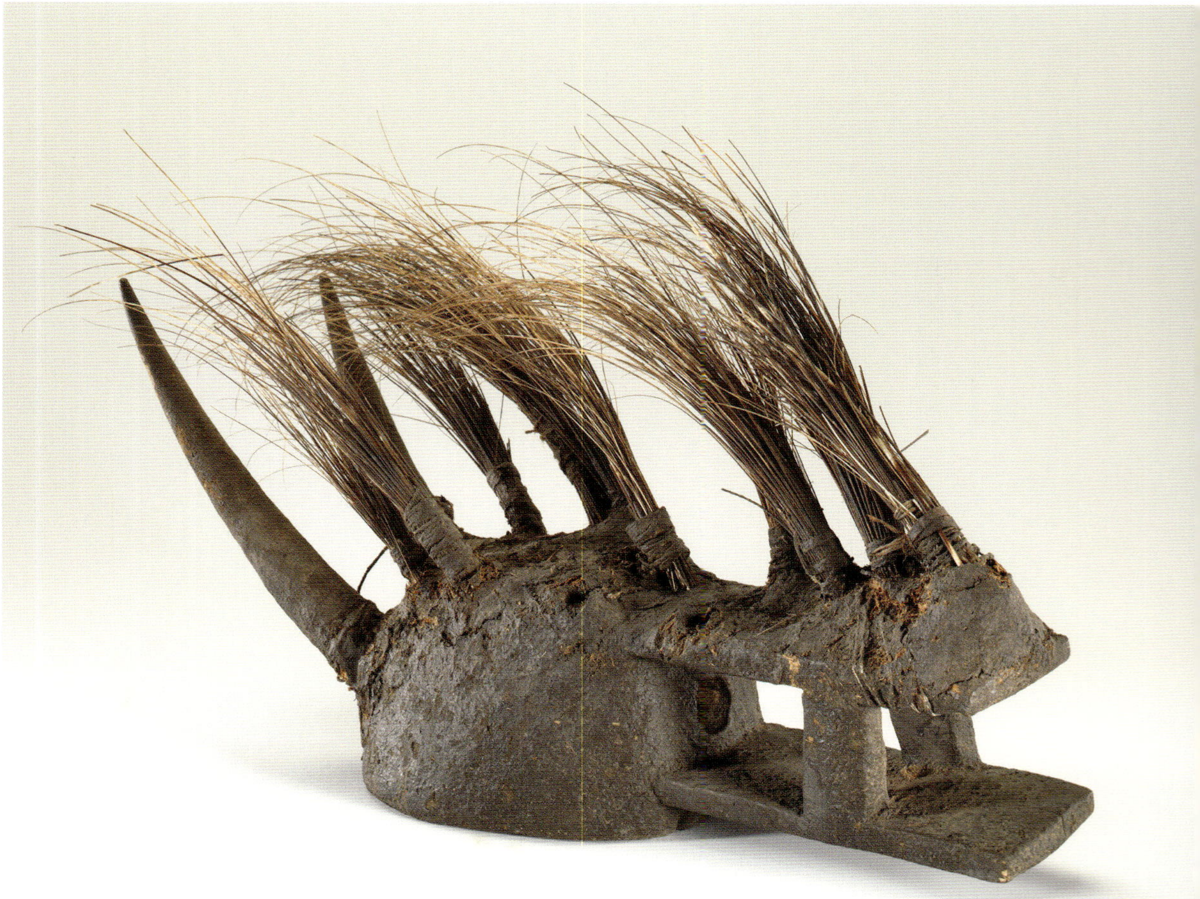

11.1
Artist unknown (Bamana peoples, Mali)
Kòmò mask, early to mid-20th century
Wood, horn, quills, string, assorted organic and sacrificial materials
L: 38.1 cm
FOWLER MUSEUM AT UCLA, X77.392; GIFT OF DR. AND MRS. JOEL BREMAN

Masks worn in dramatic performances are focal points of the Kòmò association's members-only meetings. This wooden mask presents a composite animal with a long open maw, helmet-shaped head, and horns. Its surface is studded with porcupine quills enhanced with medicinal materials, and beneath these organic additions lies exacting adze and knife work. Such masks are used in conjunction with other objects such as wood or iron flutes (see fig. 11.4) and *boliw* (see fig. 11.5). One can imagine the startling impact of such a mask being danced by an association member wearing a large masquerade costume covered in feathers and loaded with amulets.

11.2
Artist unknown (Bamana peoples, Mali)
Staff with equestrian figure, early 20th century
Iron
H: 52 cm
NICOLE AND JOHN DINTENFASS
PHOTOGRAPH BY VINCENT GIRIER DUFOURNIER, 2017.

Mande smiths consider figural staffs their most skill-demanding sculptures, and this spear-shaped one was assembled from over a dozen individually fashioned parts forge-welded together at near-molten temperatures. It features a rider wearing a hunter's hat and a horse with a bird's head to invoke potent levels of power derived from esoteric knowledge. The horse additionally suggests wealth and military might.

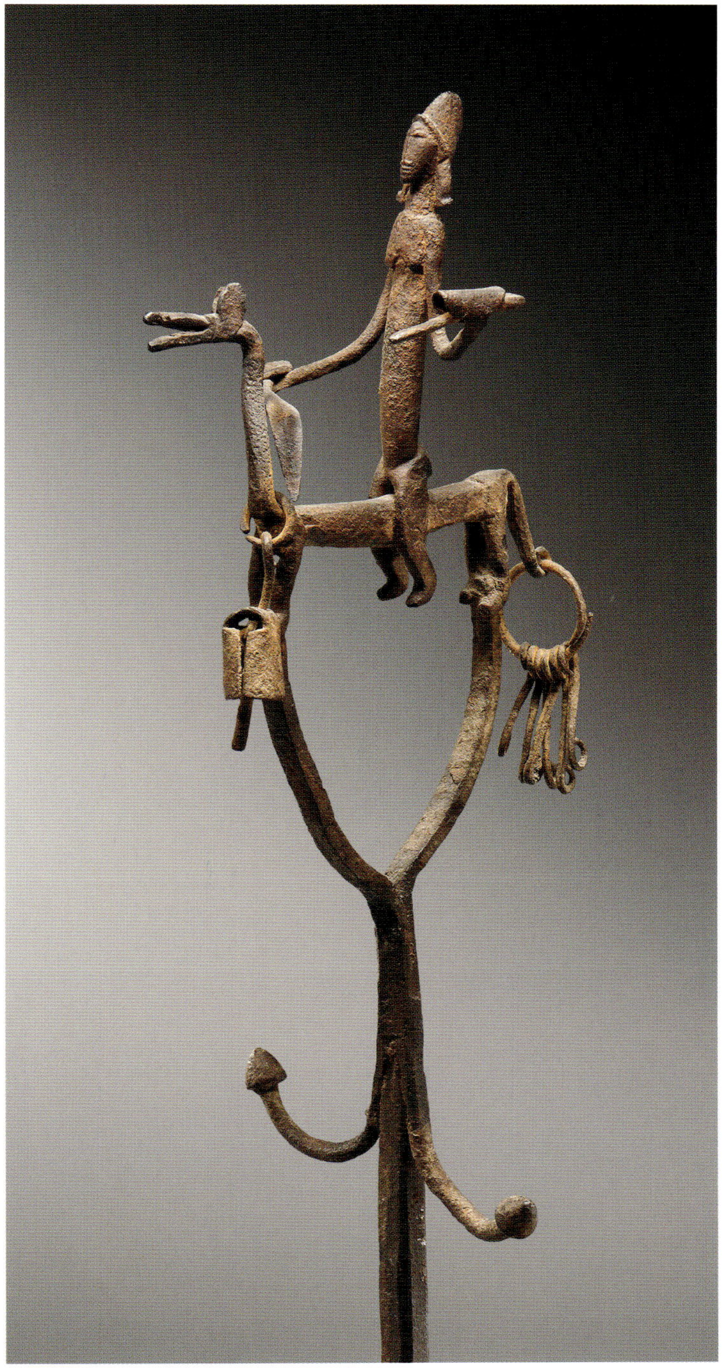

Being a blacksmith is no simple affair.

Some spend the night hungry.
Some leave the remnants of their meals.
Ah, the smiths are full!

The first son of the world is a blacksmith.

SUNG AT PUBLIC PERFORMANCES TO HONOR BLACKSMITHS IN THE WASULU AREA OF SOUTHERN MALI (KALILOU TERA PERS. COMM., 1977)

In the arid Sahel and grassy Savanna lands across the western half of west Africa, from the middle Senegal River to beyond Timbuktu, Mande-speaking peoples (such as Bamana, Maninka, Wasuluka, Marka, and many others) have long engaged with their neighbors in commerce and statecraft to create a dynamic and distinctive history. Mande blacksmiths (*numu*, pl. *numuw*) have been major players in that history, and for good reason, because they possess impressive résumés. They are an assemblage of endogamous clans with multiple professional activities. Women of these same clans are potters with imposing expertise and a dynamic history that may include the famous ancient Middle Niger terra-cotta sculptures.[1] Because the present volume addresses blacksmithing and the practices that support or spring from it, my focus is on men who specialize in wood carving and ironworking.[2] They are also often expert herbal doctors, practitioners of arcane sciences, and diviners, and they lead the powerful Kòmò association. In addition, they advise political leaders and mediate important social situations. Much literature calls them a "caste," but their social situation is far too complex to validate such quick categorization.[3]

MATERIAL ABILITIES

Mande smiths are woodworkers par excellence. Using adzes and smaller finishing tools, they carve household items such as stools, doors, figural door locks, mortars, and pestles. For farmers and hunters, they shape handles for knives, axes, sickles, harvest blades, and numerous types of women's and men's hoes. For drummers, they make an assortment of drums, including the world-famous *jembe*. And they sculpt a large array of masks and figures used across contexts ranging from popular, public, and secular ones, to those that are secret, sacred, and charged with extraordinary potency (fig. 11.1). Such works can be exquisite, but even hoe handles and stools can be sleek and elegantly shaped.

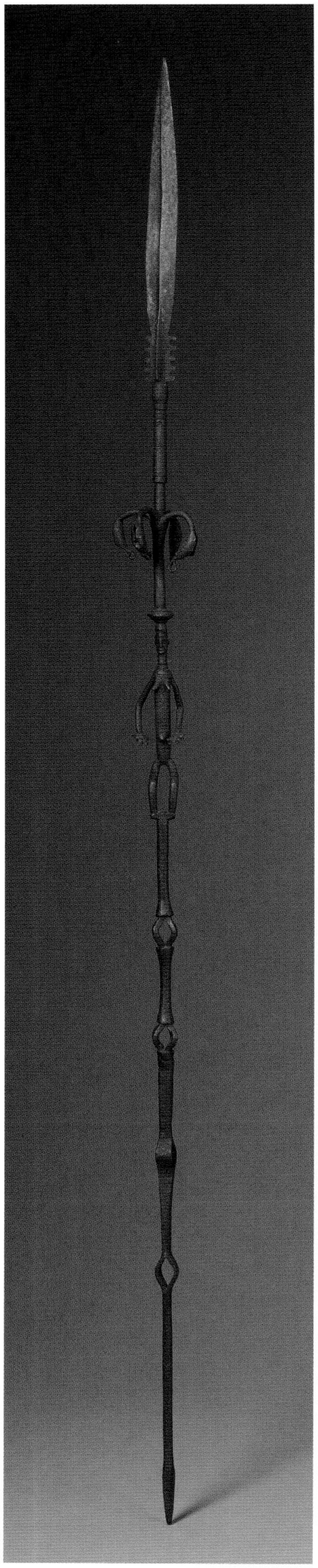

11.3
Artist unknown (Bamana peoples, Mali)
Anthropomorphic lance (*nege muso*), late 19th to early 20th century
Iron
H: 161.4 cm
MUSÉE DU QUAI BRANLY-JACQUES CHIRAC, PARIS 70.2000.17
PHOTOGRAPH BY PATRICK GRIES. © MUSÉE DU QUAI BRANLY-JACQUES CHIRAC, DIST. RMN-GRAND PALAIS/ART RESOURCE, NY

In addition to gracing Mande initiation associations' altars, iron staffs honor ancestors and herald chiefs. They can be owned by individuals and are often kept with other power objects. Virtuosic blacksmithing is evident in the details of the spear blade, figure, and shaft.

Ironwork by Mande smiths is equally accomplished. They forge all sorts of blades, for hoes, knives, adzes, and axes. Some make tools for boatbuilders of the great Niger River and its tributaries. Some make guns for hunters. In the past, they made complex iron lamps with cups to hold oil and cotton wicks, mounting them on forged iron frameworks, as well as elaborate staffs with figural finials (fig. 11.2) or spearheads at the top or bottom (fig. 11.3), to be planted on altars to enhance power with their beauty. Their iron flutes (fig. 11.4) were used by Kòmò leaders in esoteric power structures that included *boli* sculptures (fig. 11.5) and masks.

It is inspiring to watch Mande blacksmiths work, using adzes and knives for wood, or hammers, anvils, and tongs for iron. Every movement is carefully considered and controlled no matter how swiftly executed. Adze strokes make wood shavings fly with such precision that smiths could shave a person with their razor-sharp blades and never scratch the skin. With knives, they go back over their adze work to smooth off little ridges and create crisp, clean surfaces that bring form to exacting life. Ironworking is as impressive. Iron is heated in a charcoal basin until it is white hot, and then shaped and refined on an anvil with hammer blows and taps (see chapter 4, this volume).

Adzes and hammers are beautifully balanced. Through training and then experience, talented smiths make them extensions of their hands. This takes time, effectively facilitated by many years of apprenticeship, usually with a father or other close relative. First, bellows are mastered over hours of pumping air through tubes, which builds strength and endurance, all wrapped up in patience. To make it enjoyable, smiths develop rhythms on the bellows by timing their thrusts on goatskin bags or skin-covered pots so that air rushes through like syncopated drumbeats (see chapter 17, this volume). Bellows rhythms can be so distinctive that blacksmiths of renown are remembered by them.

After the bellows, apprentice-smiths learn the principles and precise skills of carving wood with an ability to think quickly in "taskscapes" (Ingold 2010, 97) keeping both details and end goals in mind. Smiths learn to block out major shapes and refine them by increments of reduction to their final form. Last comes working iron, when metal is guided with great finesse into useful forms. For large objects such as massive axe heads, two smiths, even three, will work together, using hammers so heavy that they require two hands to raise and control.

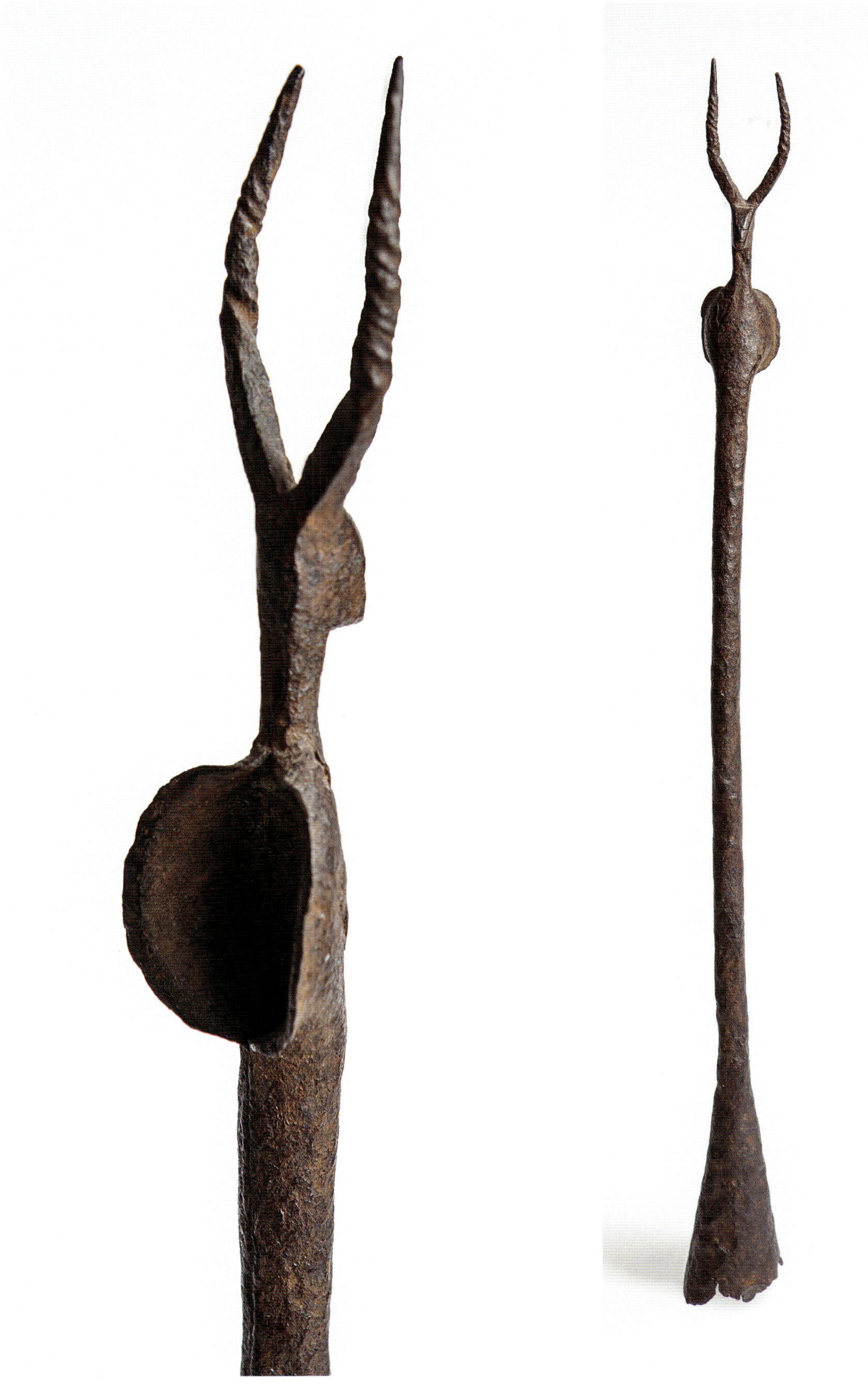

11.4A–B
Artist unknown (Bamana peoples, Mali)
Flute, late 19th to early 20th century
Iron
H: 58 cm
COLLECTION OF SUSAN AND RICHARD ULEVITCH
PHOTOGRAPH BY JOE BELCOVSON, 2017.

Often used in divination, forged iron and carved wood flutes were used by Kòmò leaders and members in conjunction with such power objects as masks (see fig. 11.1) and *boliw* (see fig. 11.5). Flutes stored with *boliw* occurred in sets of three. Individual Kòmò members could also own small versions, which were considered part of a secret communication network.

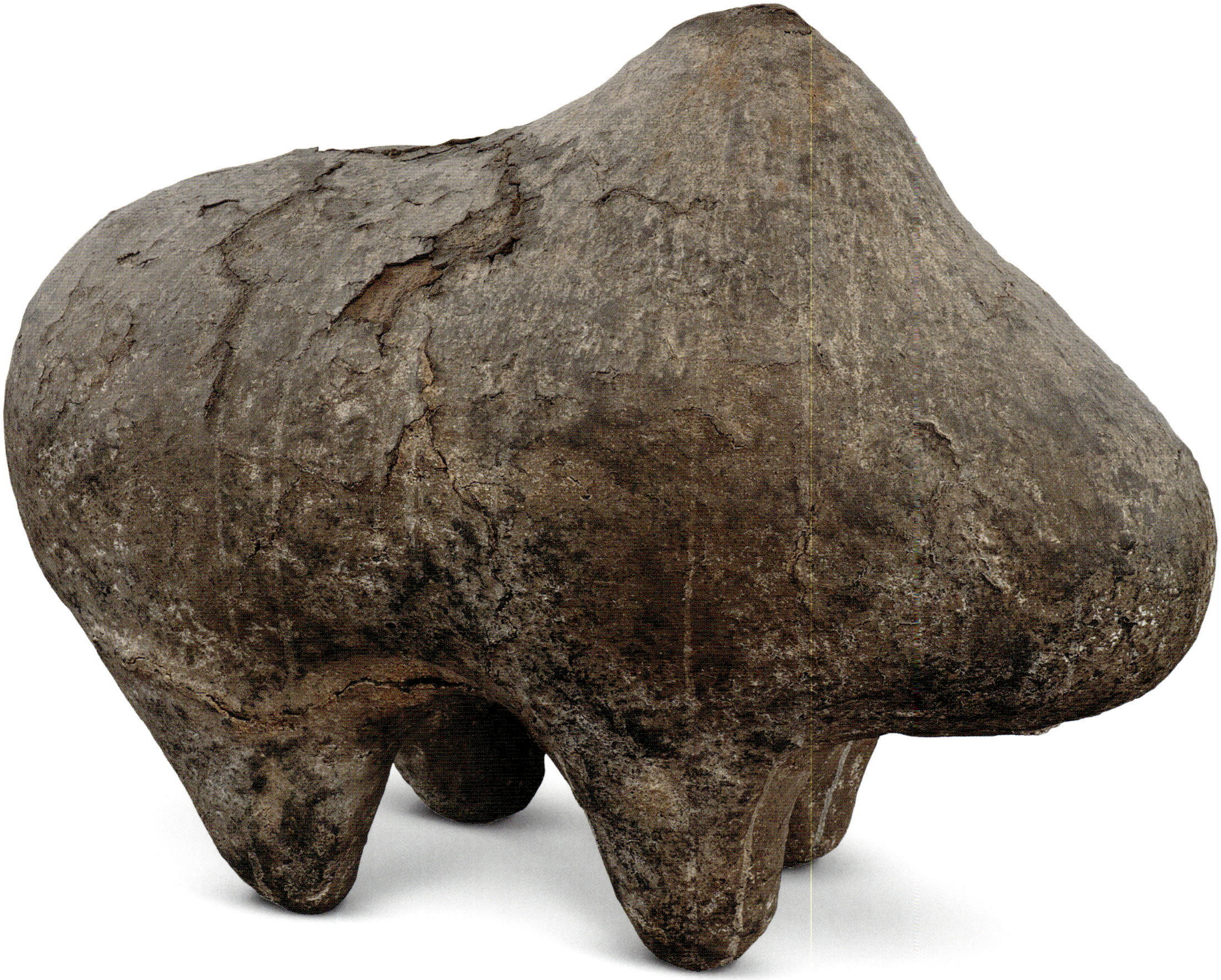

11.5
Artist unknown (Bamana peoples, Mali)
Altar (*boli*), early to mid-20th century
Wood, clay, sacrificial materials
L: 56 cm
COLLECTION OF MRS. JEROME (ELLEN) STERN
PHOTOGRAPH BY STEVEN TUCKER, 2017.
COURTESY OF MRS. JEROME (ELLEN) STERN

A *boli* altar can be characterized as a potent zoomorphic sculpture to be handled only by Kòmò association members with the highest levels of knowledge. Created using secret procedures and materials prescribed by "the science of the trees" (*jiridon*), it plays an essential role in Bamana spiritual practices. This *boli* has been built up over a wooden armature with layer upon layer of sacrificial materials. Iron embedded within or added to a *boli* is often invisible to the naked eye yet present to activate and control the life force *nyama*, thereby enhancing community well-being.

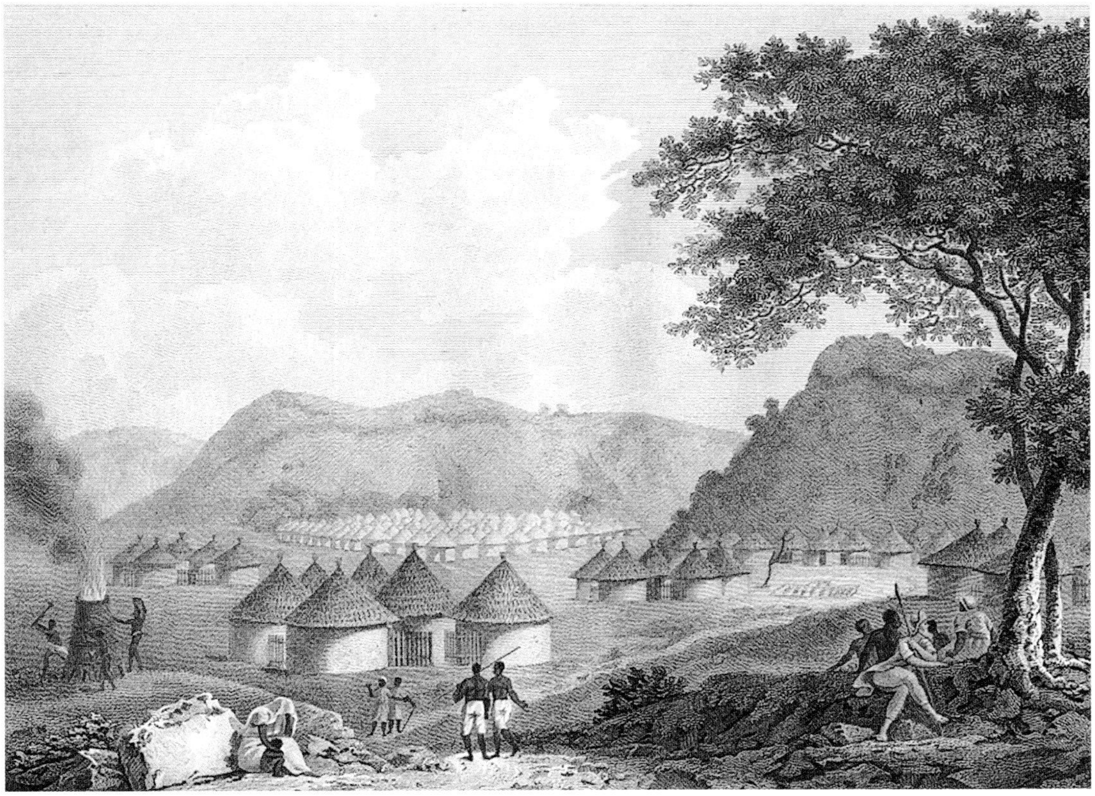

11.6 In his book *Travels in the Interior of Africa* (1799), the Scottish explorer Mungo Park included the illustration "A View of Kamalia" (in southern Mali, west of present-day Bamako), which shows a smelting furnace in operation to the far left. According to Park's description, the furnace was three meters in height and produced iron for local use and commercial exchange. He observed that most blacksmiths smelted and worked gold as well.
ENGRAVING BY W. C. WILSON FROM A DRAWING BY J. C. BARROW, BASED ON A SKETCH BY MUNGO PARK.

Before (and to a sharply reduced extent after) colonial contact, blacksmiths also smelted iron from ore. These skills included building furnaces and air-induction tubes called tuyères, making charcoal, mining and refining the ore, charging the furnace with careful mixtures of ore and fuel, and then monitoring and refueling the furnace through the reduction process (see chapter 2, this volume). In many towns, individual smith families smelted iron once a year, and the turn-of-the-nineteenth-century Scottish explorer Mungo Park published an illustration of one such event (fig. 11.6). In other areas, the operation was on a far larger scale, with iron produced in numerous furnaces (Binger 1892, 1, 261) to enter west Africa's complex commerce systems. Local residents say that in the thirteenth-century war that created the Mali Empire, the protagonists Sunjata and Sumanguru had enormous fields of smelting furnaces dedicated to producing weapons.

Part of the popular aura surrounding smiths still focuses on their mastery of fire, furnaces, and technologies of transformation that produced useful objects from ore. Festival songs praise the heritage of blacksmithing through the smiths of old who smelted iron, likening them to the objects they created, asserting that both smiths and the fruits of their labor are indispensible.

ESOTERIC ABILITIES

Apprenticeship frequently extends beyond working wood and iron into arcane realms. Herbal medicine and divination are not the exclusive domain of blacksmiths, and not every smith chooses to learn such sciences, but young smiths are often given an opportunity to study *jiridon*, "the science of the trees" (fig. 11.7). This is a massive and constantly expanding body of knowledge that teaches how to extract ingredients from the natural world, process them, and recombine them to produce effective medicines (Imperato 1977; Traore 1965). Such knowledge is organized into units called *daliluw* that function at two overlapping planes. In the tangible, visible world (in which you are reading this book), *daliluw* help diagnose and cure ailments like sore throats, menstrual cramps, or infections from animal bites. In the less tangible, less visible world, *daliluw* assemble ingredients to harness energies that help shape life: they assure successful marriages, business ventures, or harvests; they fight antisocial behavior; they prevent animals from biting people (McNaughton 1982a; 1988, 42–46, 56–64). Over many centuries, blacksmiths have become so expert at using such knowledge that most members of broader society have the indelible conviction that Mande smiths are almost by definition doctors, practitioners of arcana, and soothsayers.

There is more. Mande consider the world to run on an energy called *nyama* that is rather like electricity and can be translated as "the energy of action, the energy that animates the universe" (Bird and Kendall 1980; McNaughton 1988, 15–18).[4] *Nyama* abounds in the visible and less visible worlds. Every action requires this energy, and releases it in direct proportion to degree of difficulty and danger. Hunters need large amounts of *nyama* to safely bring down game. Soldiers need large amounts to survive the horrors of war. Blacksmiths possess huge amounts by virtue of their occupation and inheritance. Their endogamous clans harbor *nyama* in great quantity, just as they harbor a monopoly on the rights to their profession, including *gundow*, meaning secret expertise (McNaughton 1982a; 1987; 1988; 2008). Smithwork requires and emits large volumes of this energy. As smiths hammer, they plow *nyama* back into their products. This makes smiths extraordinary and immensely powerful. Blacksmiths are full participants in Mande society, but they see themselves and are seen by others as a nation apart, in large measure because of this power.

In addition, many smiths have special relationships with wilderness spirits often associated with particular geographic features and sometimes conceived of as having communities and social organizations resembling those of people. Mande commonly identify them with Arabic loan words such as *jinn* (pl. *jinnw*; also *jenè*, *wokolo*).[5] These beings populate less visible spaces where energies abound, but materialize at will in the visible world, especially in wilderness or at night. People must be careful because *jinnw* can be delightful and benign but also treacherous, fearsome, and dangerous. Blacksmiths spend time in wild places gathering wood or herbal ingredients; in earlier times, they spent even more time there when they mined or gathered ore, made charcoal, and smelted iron. They retain privileged relationships with wilderness spirits as masters of natural resources who can facilitate smiths' work.[6]

As if all this were not enough, smiths are the leaders, and indeed the inventors, of the exceedingly powerful Kòmò initiation association.[7] Kòmò is mandated to prevent crime in the visible world and to control antisocial human and malevolent spirit behavior in the less visible world. Kòmò members also manipulate *nyama* to benefit people and communities. When blacksmith leaders, as masters of divination, don their *nyama*-laden masks (see fig. 11.1) and costumes to perform at meetings, a featured agenda item is high-powered soothsaying. In addition, Kòmò teaches its initiates essential knowledge about how the world works and how they can be successful in it.

Kòmò encompasses a triangle of power: blacksmith leaders, the masks and many other power objects, and wilderness spirits. Its blacksmith-leaders are a potent force. The society's masks and costumes are ripe with the energy generated by the science of the trees, and regular sacrifices add even more *nyama*. Kòmò leaders also maintain beneficial relationships with wilderness spirits of astounding power, for they are never under anyone's control but are often willing to work in league with smiths. This formidable ensemble focuses on wilderness, where masks are made and *jinnw* reside. Kòmò is a daunting thing of the bush, brought into civilized spaces by ironsmiths.

THOUGHTS ON HISTORY

The histories of west Africa can give broader perspective—especially through archaeology and oral traditions—to Mande smiths and their metal, wood, and arcane creations. Metal and Kòmò histories are insufficiently understood, but there are clues, beginning with copper mining and production near Akjoujt, Mauritania, between 800 and 200 BCE (Deme and McIntosh 2006, 343; Lambert 1972 and 1983). Copper products were widely diffused and reached the ancient Dhar Tichitt culture, where iron smelting began between 800 and 300 BCE.[8] Evidence suggests that climate change and neighbors' aggression pushed Dhar Tichitt peoples southward to become the Soninke founders of Wagadu, the great gold-trading polity often called the Ghana Empire (MacDonald 1996; Munson 1968; 1976; 1980; 1989).

In the Middle Niger, iron production became a major enterprise and iron an important trade commodity as of the middle of the first millennium BCE (Bedaux et al. 2001, 843). Smithing appears in the earliest phases of society at Jenne-jeno, circa 250 BCE (McIntosh 2005, 147–48; LaViolette 1987, 161, 173). Nearby, in the Méma region, iron was smelted on a near-industrial scale at numerous production sites. Archaeology has shown that Akumbu, for one, featured extensive smelting from 548 to 1401 CE, and imported goods testify to long-distance and trans-Saharan trade.[9]

Iron helped fuel the Méma's consequential place in history. Akumbu appears linked to the rise of Wagadu, while Soninke and Maninka traditions emphasize Méma's importance to the political development of both Wagadu and the Mali Empire.[10] When Wagadu declined, Méma was the most structured polity in the Mande world (Togola 2008, 82–83). Traditions say that the hero Sunjata lived there for many years honing his military skills. Local chronicles and Arab texts, including Ibn Battuta's from the mid-fourteenth century, refer to numerous long-distance trade routes through the region (Togola 1996, 92, 95–96).

During the middle and later half of the first millennium CE, a significant shift in the physical location of smithing sites occurred across the Jenne-jeno hinterland, the Méma, the Lakes Region to the north, and the Niger Bend region further east. Across this enormous area, smithies moved to the edges of settlements or to satellites away from larger habitation centers. The archaeologists Susan McIntosh (1995, 278) and Roderick McIntosh (1998, 177–81; 2005, 155–56) think such shifts may represent the consolidation of the professional group we know as *numuw*. Moving may have helped maintain blacksmiths' monopoly on secret metalworking and herbal expertise. It might also have reflected their "separate nation" status, or protected communities from potentially dangerous *nyama*.

We see corroboration in oral traditions, which vary greatly in detail but can be very suggestive, and in the case of Mande smiths and Kòmò, such histories project a common theme of power, authority, and inventiveness. Wagadu origin stories describe Dinga, the ancestor of all Soninke and the father of Wagadu's founder, Djabe (or Diabé Sissé). A hunter, warrior, and powerful practitioner of esoterica, Dinga conquered much territory and controlled rain and spirits, and his blacksmith, Domba, introduced ironworking. Dinga had special relationships with animals, especially hyenas and vultures, which are important symbols and bearers of potent and dangerous knowledge in Kòmò.

Some oral traditions suggest that Dinga sired a monstrous snake named Bida, with whom Djabe negotiated a pact that gave gold and rain to the Soninke in exchange for sacrificing one maiden annually.[11] In other traditions, the first smiths discovered copper by watching a pebble melt in the hearth. They quickly invented tools and weapons, and then discovered iron and gold. They kept secret their knowledge, which from the start included herbal sciences and sorcery. For all these reasons, smiths were the first to dominate the world (Doumbia 1936, 334–43; N'Diayé 1970, 66).

Many narratives hold that smiths were among the world's first people and not originally endogamous. One suggests that the first smith was Domajiri, sired by two deities. He ended drought by petitioning the Creator to bring rain and create the Niger River, shared much potent and dangerous knowledge with people, invented initiation associations, and mediated between forces of chaos and stultifying equilibrium.[12] Another tradition asserts that the first smith brought tools, millet, and rain to earth (Pâques 1964, 151–52), thus inventing agriculture.

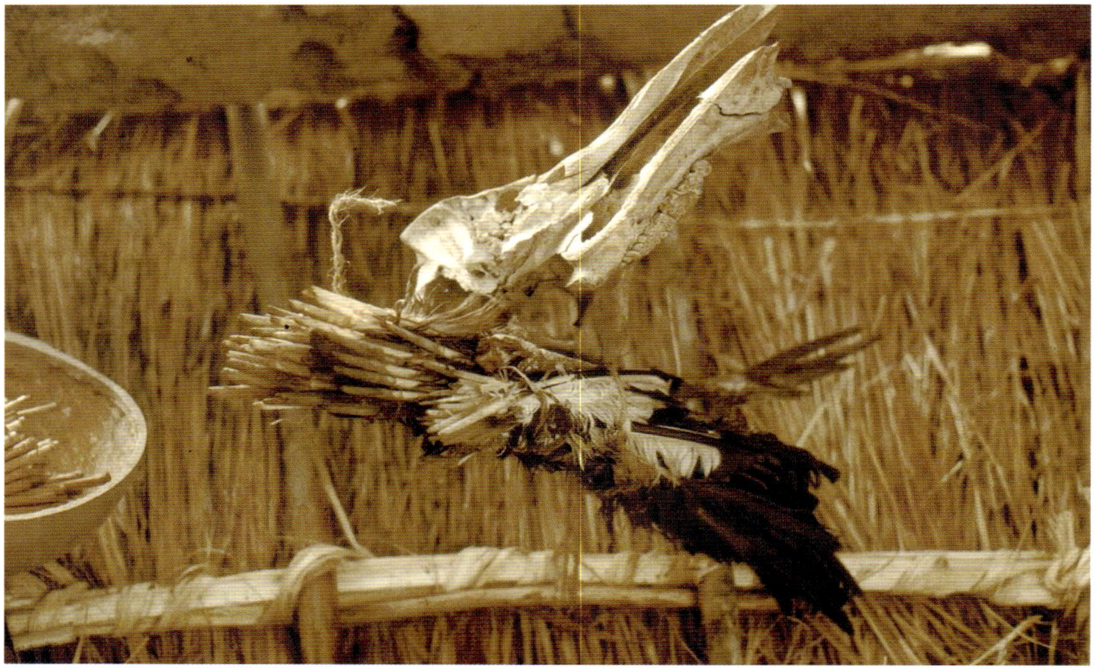

11.7 Inside the entrance to his forge in southern Mali, blacksmith Sedu Traore hung organic materials processed according to the science of the trees (*jiridon*) and intended to prevent theft.
PHOTOGRAPH BY PATRICK MCNAUGHTON, SENU, MALI, 1971.

The third-generation direct descendent of that first smith, Domajiri, was N'Fajigi, who gave people Kòmò, the women's counterpart society called NyaGwan, and the very concept of *basiw*, that is, power devices based on the science of trees (*jiridon*). Famous and powerful Kòmò masks have names, and N'Fajigi's is named for a tree whose flower is rare and always hidden, but whose fruit benefits everyone and is forever available (Kassim Kone pers. comm., 2015; Kalilou Tera pers. comm., 1978).[13]

Many traditions describe Kòmò's invention and spread. Some say its original form was a mystical creature named Bintu that was half antelope. A blacksmith captured and transformed it into the association, after much adventure and ritual activity (Pâques 1964, 143–48). Other accounts have it that Soninke invented Kòmò, and spread it when Wagadu fell. Because eleventh-century Arab travelers reported that the probable capital of Wagadu, Kumbi Saleh, held sacred groves in which priests managed powerful objects (Levtzion 1973, 25), it is tempting to wonder if an association akin to Kòmò existed there.

Most blacksmiths say Kòmò existed before Sunjata fought Sumanguru Kante to found the Mali Empire circa 1230 CE. Fakoli, one of Sunjata's great generals and military strategists and a powerful Kòmò leader, spread the organization everywhere he fought. Once the empire was established, three clans of smiths expanded Kòmò's distribution under the impetus of Fakoli (Conrad 1992; 1999; McNaughton 1979b, 18–19). The historian George Brooks writes that Mande smiths spread power associations patterned after Kòmò to many regions of west Africa (1993, 73–79), while the art historian René Bravmann presents a specific example in the borderlands between Côte d'Ivoire and Ghana, where Mande smiths and merchants participated in the great gold trade from as early as the fifteenth century (1974, 59–73, 119–46). Another historian, Marie Perinbam, argues that gold was controlled by smiths through extensive and elaborate ritual restrictions that focused on Kòmò, and she emphasizes that the *soma* smiths of the Kamara clan were masters of extraordinary knowledge and power (1996). Not surprisingly, many Kamara clan smiths are depicted as political leaders in the Sunjata Epic (McNaughton 2011; 2012).

That epic depicts the rise of the Mali Empire led by Sunjata Keita and the fall of the Soso state led by blacksmith-king Sumanguru Kante, known as a virtuoso master of arcane sciences.[14] In the aftermath of these events, smiths were barred from political office and became part of a larger grouping of specialized professions, along with bards (*jeliw*) and leather workers (*garankew*), whose name, *nyamakala*, highlights the capacity to manage and articulate the energy of action (Niane 1984, 133–35). While they lost the right to rule, their professional ken and Kòmò left smiths with plenty of power.

ACCOUNTING FOR POWER

The coalescence of technology, arcane potency, and leadership around smiths is widespread across Africa. While the challenging question of metallurgy's African origins stimulates much scholarly debate, the equally vexing issue of blacksmiths' links to power and leadership are not often investigated.[15] In Tanzania, the archaeologist Bertram Mapunda (2011) suggests that healing in Fipa smiths' repertoire strengthened their professional monopolies and added social-political clout and economic diversity. Such sensible professional practice would apply to the powers of Mande smiths, too, but what conceptual and experiential foundations allowed it? We have hints.

First is people's experience of smiths. Watching Mande blacksmiths work iron or wood is arrestingly impressive, because shapes are generated and objects created with an economy of action that seems almost magical. Smelting iron was exponentially more impressive, because cause and effect were not observable. Producing iron from ore through technical and ritual procedures would easily generate awe in popular imagination and oral traditions. The same is true for deploying the science of the trees, experienced by the general public as complex and exacting activity steeped in ritual and secrecy, and producing medicines and *nyama*. Thus iron ken and herbal ken, though dramatically different, bear apparent similarities.

There are also conceptual relationships. Extracting elements from nature and recombining them into fruitful medicines finds parallels in iron crafts, especially smelting. Both are parts of histories of observation and experimentation. They demand intensive training, time, and personal sacrifice. They are steeped in secrecy that protects the public but also protects proprietary expertise while enhancing the drama of such activities.

Hard work, earned expertise, much knowledge, and secrecy are all ideas that resonate in Mande imaginations and are archetypically possessed by blacksmiths. They are summed up in another Mande concept, *nya*, or "the means to accomplish difficult things" (Bird and Kendall 1980). It is instructive that the Sunjata Epic focuses on how things get done and not on actual battles. Having or creating "means" is critical in Mande social action, and comes via many avenues, such as destiny and inheritance, knowledge, allegiances, and numerous types of power and authority (Lukes 1978).

Ironsmiths overflow with means. Through their clans they inherit heavy levels of *nyama*. Endogamy protects an abundance of professional secrets. Lengthy apprenticeships provide knowledge, skill, and often virtuosity. And, being adventuresome and entrepreneurial, smiths are proactive and productive. We cannot really account for such an amassing of means. We know too little about west African histories. We can, however, say with certainty that Mande smiths have played significant, proactive, enterprising roles in those histories. No wonder the punch line in a praise-song recorded by linguist Kalilou Tera is "The world belongs to blacksmiths" (pers. comm., 1978).

CHAPTER 12

ÒGÚN/GU'S RESONANCE IN YORÙBÁ, EDO, AND FON WORLDS

Rowland Abiodun Henry J. Drewal

> Ògún, possessor of a massive stock of iron, suffices me as a divinity to give worship to.
>
> FROM A YORÙBÁ PRAISE-POEM (BABALỌLA 1989, 159)

Yorùbá, Edo, and Fon—neighboring peoples in west Africa—share multiple historical and cultural elements evident in systems of knowledge, practices, and material objects linked with divination, medicines that cure and kill, warfare, and the honoring of ancestors. They recognize iron as a medium possessing a performative power called *àṣẹ* by Yorùbá and Edo and *se* by Fon.[1] Ògún or Gu, as the deification of iron, contributes *àṣẹ/se* to all things requiring iron components—from objects of art, to keychains and farm implements, military weapons, computerized automobiles, bullet trains, and space shuttles. The creative energies of Ògún/Gu orient a person's life, and as a *cosmic connector*, a guarantor of certain cultural necessities, iron insures the efficacy of sacred acts and actors.

YORÙBÁ

According to Yorùbá myth, striking iron began in the otherworld. Ògún, patron deity of blacksmiths, was well-established there. Equally so were his priests, among them Ṣokòtí Alágbẹ̀dẹ ọ̀run, whose praise-poem (*oríkì*) may be translated as "the veritable blacksmith in the otherworld" (Fagunwa 2005, 3). An Ìjálá chant—a genre of spoken art practiced mainly by the Ọ̀yọ́ Yorùbá of western Nigeria and ritually associated with the worship of Ògún—reports that it was with an iron chain cast by Ògún in the otherworld that all 401 deified ancestors (*òrìṣà*) were able to descend from the otherworld above (Orun-Òkè) to the Mountain of Wonders (Òkè-Àrà) in Ilé-Ifẹ̀, the ancestral home of Yorùbá peoples and where the world began.[2]

Ògún is one of four most-influential *òrìṣà* after whom the four-day Yorùbá week is named. The others are Ifá, Ṣàngó, and Òrìṣàńlá. Regardless of a person's religion or affiliation with any *òrìṣà*, it is impossible to escape the influence of Ògún because of his presence in, and authority over, everything in life. As Ìjálá verse attests:

> To Ògún belong calabashes and plates
> To Ògún belong hoes and knives
> Without Ògún, the King of Ìrè, there would have been no farms
> Without Ògún, the King of Ìrè, there would have been no roads
> Unfortunately, without the implements of Ògún, King of Ìrè,
> No one would ever have been enslaved.[3]

Ògún is ageless and ever present. Appropriately, the longevity of ancient Ifẹ̀'s legendary culture hero Ọbalùfọ̀n II is often likened to that of iron. His *oríkì* describes him as "lasting and enduring as iron," alluding also to his state of mental and physical soundness (Abiodun 2014, 219).

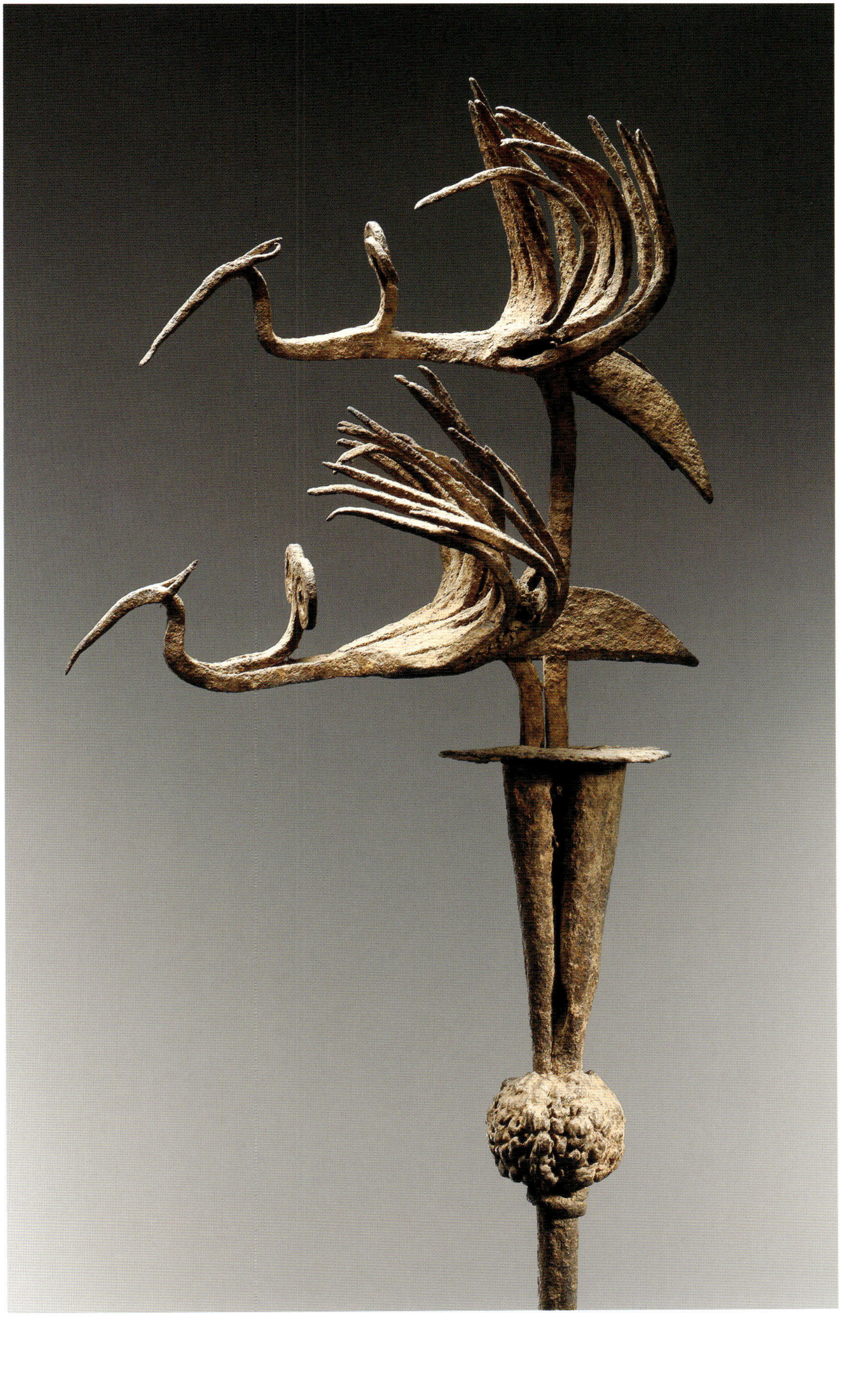

12.1A–B
Artist unknown (Yorùbá peoples, Nigeria)
Diviner's staff (Osùn *babaláwo*), early 20th century
Iron
H: 109.2 cm
NICOLE AND JOHN DINTENFASS
PHOTOGRAPH BY VINCENT GIRIER DUFOURNIER, 2017.

As a ritual instrument, the Osùn *babaláwo* is used to implement Ọ̀rúnmìlà's directives in Ifá divination processes.

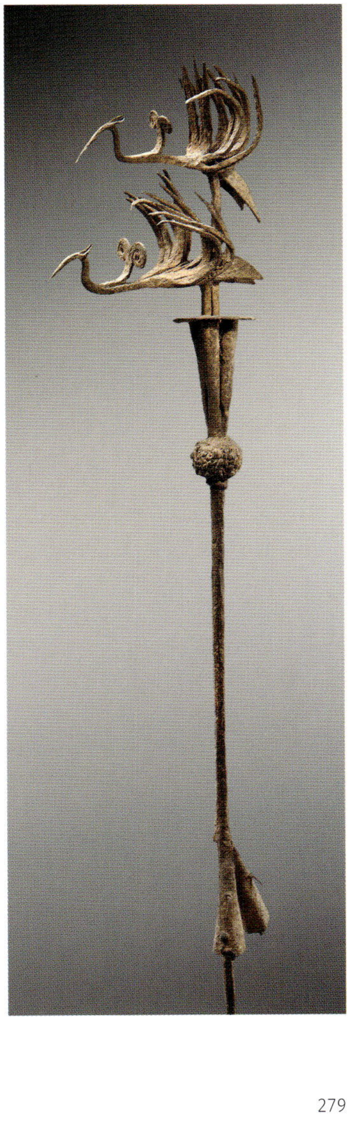

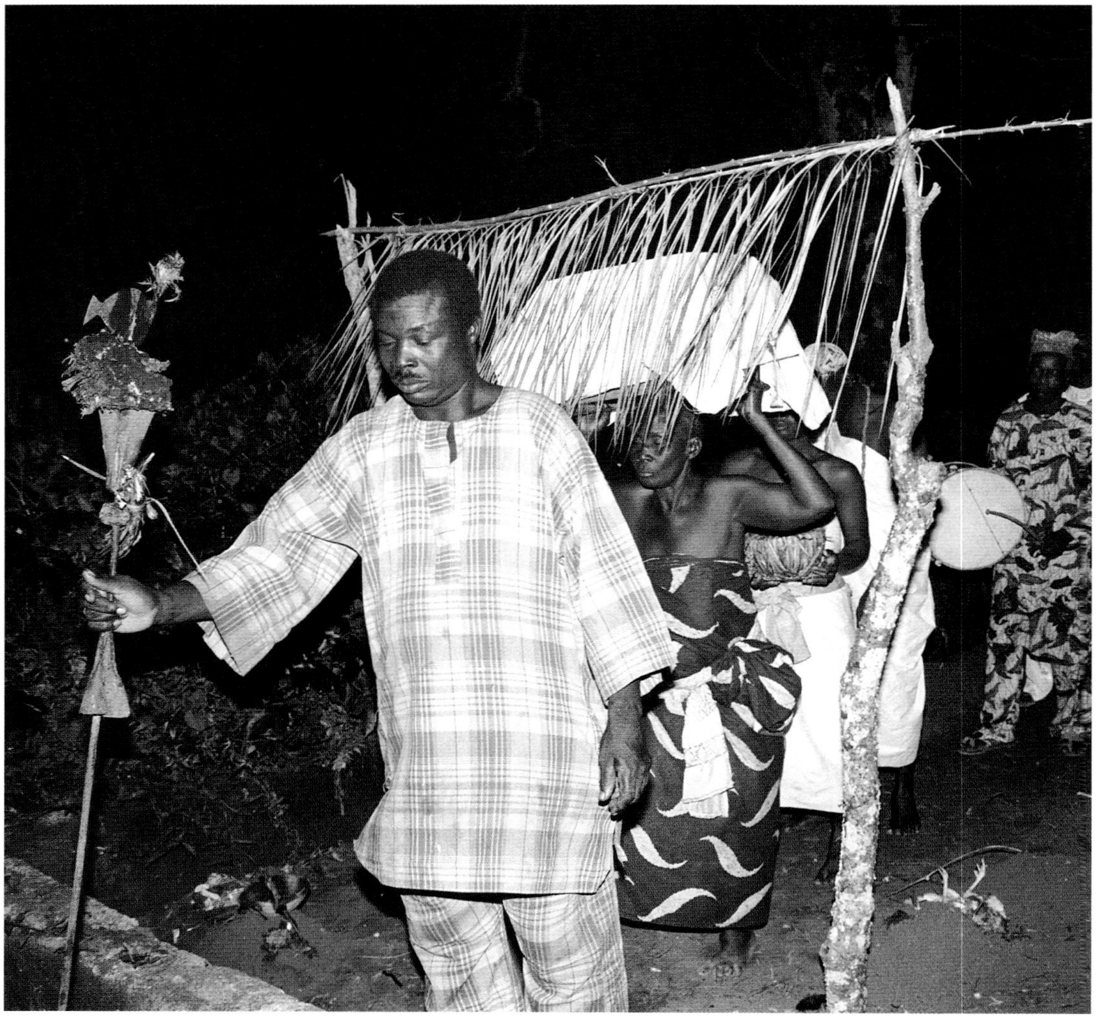

12.2
When undertaking an "Ifá journey"—an important divinatory ritual that involves a procession—the staff (Osùn *babaláwo*) leads the way and serves as a weapon to drive away destructive forces.
PHOTOGRAPH BY MARGARET THOMPSON DREWAL, ÌJẸ̀BÚ-YORÙBÁ, NIGERIA, 1988.

12.3A–B
When not in ceremonial use, the Osùn *babaláwo* stands erect in front of a diviner's house next to a tree, called the *akóko* (*Newboldia laevis*), whose leaves are used in bestowing titles and other ritual matters.
PHOTOGRAPHS BY HENRY J. DREWAL, ÌJẸ̀BÚ-YORÙBÁ, NIGERIA, 1988.

In ritual contexts, Ògún's *àṣẹ* is crucial to successful performance of the duties of all other *òrìṣà*. For example, a diviner's iron staff (fig. 12.1) known as the Osùn *babaláwo* (also called *ọ̀pá òṣoorò* or *ọ̀pá ọ̀rẹ̀rẹ̀*) is required by the Ifá deity Ọruúnmìlà to carry out any task. Typically measuring between 80 and 150 centimeters in length, it is carried vertically in the right hand by a senior Ifá priest (*babaláwo*) during processions (fig. 12.2) and may be planted in the ground in an upright position during important gatherings.[4] In front of a diviner's house, the Osùn *babaláwo* staff appears standing erect next to a tree called the *akóko* (*Newboldia laevis*; fig. 12.3). Among their various functions, *akóko* leaves are used in bestowing titles.[5] Besides marking the residence of a diviner, the staff also refers to Òrúnmìlà as the source of the unfathomable knowledge Ọsanyìn needs to identify herbs for healing or harming. Ọ̀rúnmìlà's friendship with the goddess of the sea, Olókun, is equally important, since water is an indispensable component of all medicines. This relationship is manifest in Ọ̀rúnmìlà's epithet: "*Akóko* tree by the sea." When a diviner departs from this world, a staff is laid down on its side and the diviner's journey to the watery realm of Olókun commences. The staff embodies the diviner himself, for an Ifá saying goes "The Osùn staff is always found erect and vertical as long as its owner lives"; there is likewise this prayer for a diviner's long life: "Osùn stand up, don't fall down."

The Osùn *babaláwo* has multiple identities. It represents the power of the diviner to conquer. When used with incantations, the iron staff acts as a weapon against death and destructive forces. The distinctive performative power of its medium reinforces its aspect as a weapon, recalling a story of Ọ̀rúnmìlà's victory over three warriors noted for their mastery of iron. The substances that give the staff much of its potency are buried at its base, or "tail" (*ìrù*), which is stuck into the ground to absorb power through its end. Because the staff's efficacy is in its *ìrù*, it is often likened to the power of a pangolin, a scaly anteater that curls its tail around its belly to make itself impervious to attack.

When undertaking an "Ifá journey"—that is, when performing an important divinatory ritual that involves a procession—the staff leads the way (see fig. 12.2). Encountering a crossroads during such a "journey," an Ifá entourage stops and chants incantations as the diviner in the lead swings the end of the staff outward to drive away destructive forces. As the diviner Kolawole Oshitola once explained, an "enemy that stands in the way of Osùn is ready to die." He also noted, "history told us that if we want something to act quickly, we should do it in the name of the hawk" (pers. comm.,

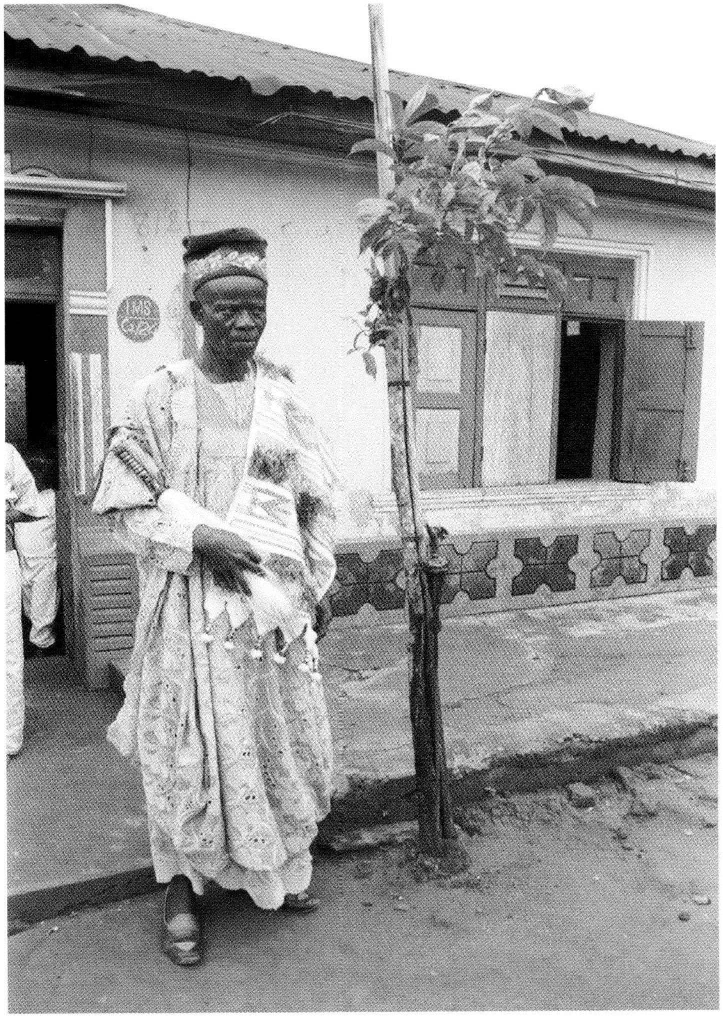

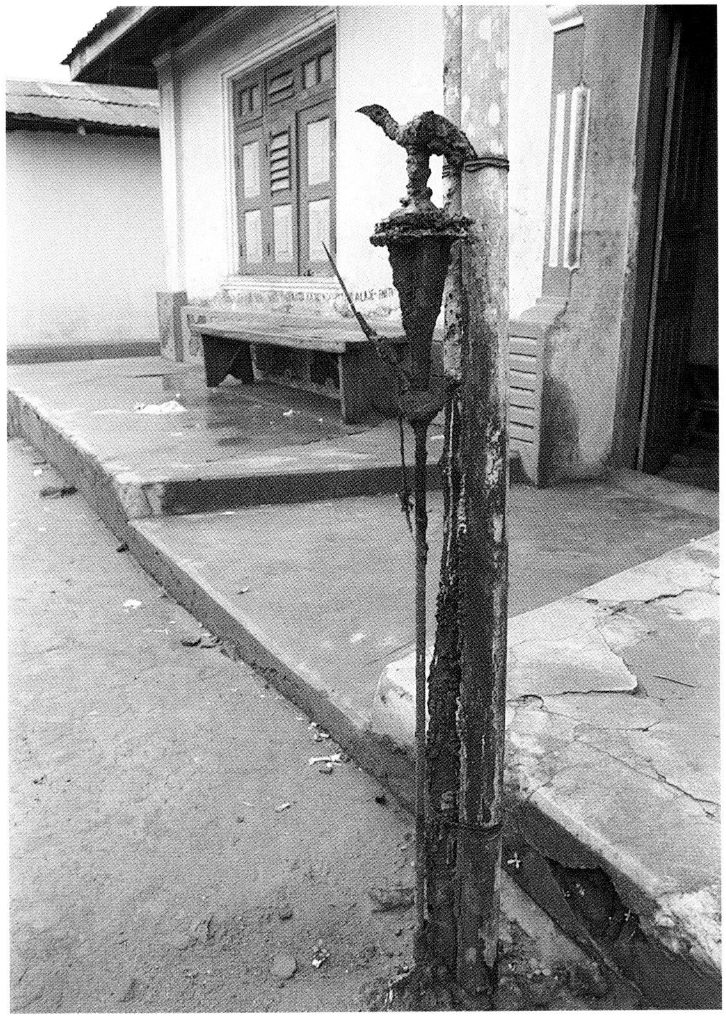

1982). Accordingly, the bird motif on a staff indicates that it is quick acting and efficient like a hawk (*àṣá*).[6] If a diviner wants to take the staff along in procession, he chants: "Stand up, let's go because the hawk is alert. He acts quickly." This chant refers to how the sharp pointed end of the Osùn staff acts when it strikes the ground, transforming whatever is beneath it.

The bird atop an ọ̀pá Osún staff refers to the conquering diviners and their transformational powers as visualized in the *ẹyẹkàn* (the lone or single bird) that became two. The number of birds on a staff has numerological significance, for when the *ẹyẹkàn* becomes two—*ẹyẹlé*—that is *àṣẹ* in action. An exemplary iron staff (see fig. 12.1) has at its finial *ẹyẹlé* standing on a flat disc resting on an inverted cone/bell. The pair descend from the *ẹyẹkàn* that is the source of Ọ̀rúnmìlà's *àṣẹ*. The story of the *ẹyẹkàn* from Èjì-Ogbè, a chapter in the Ifá divination corpus, relates how this "wild pigeon" (*ẹyẹ-oko*) was once hermaphroditic, lived in the forest, and remained childless for a long time. When the *ẹyẹ-oko* consulted Ifá and performed the prescribed ritual sacrifices, it had two offspring. Henceforth, the *ẹyẹ-oko* became *ẹyẹ-oko* (pronounced *ẹyẹlé*), or "domesticated pigeons" (Abiodun 2014, 77–80). The birds atop the Osùn staff refer to transformation and elevation of status by alluding to Ifá priests' attainment of high socioeconomic positions in Yorùbá society. The slim metal cones/bells forged onto the staff at different levels along its height refer to coiled leaves as well as the musical instruments that sound greetings to Ògún and his followers (figs. 12.4–12.6).

Ògún/iron is the instrument that allows culture to master nature, from cutting paths to healing. Yorùbá medicine-workers (*ọlọ́sanyìn*) provide knowledge and wisdom to guide people, like Ògún. Their distinctive emblem, the *ọ̀pá* Ọ̀sanyìn (figs. 12.7, 12.8), identifies them with powerful prescriptions—leaves, herbs, and other substances from deep in the fertile forests of Yorùbáland—and with Ọ̀sanyìn, the deity of medicine. Such staffs are typically surmounted by a circle of sixteen birds surrounding a larger central one (Abiodun 2014, 78). The *ọ̀pa* Ọ̀sanyìn is not to be confused with the Osùn *babaláwo* of diviners, for the birds that surround Ọ̀sanyìn staffs present potentially aggressive forces with which human beings must cope. Generally short and most often placed on altars with objects for other divinities, *ọ̀pa* Ọ̀sanyìn recall trees filled with birds, evoking the forest as a source of sacred medicines as well as a site of potentially dangerous forces. A staff's encircling birds remind us that destructive as well as auspicious forces can come from any direction—one must be prepared to defend or attack with weapons of iron such as these empowered staffs.

12.4
Artist unknown (Yorùbá peoples, Nigeria)
Bell (*ọ̀mọ̀*), 20th century
Iron
H: 37 cm
PRIVATE COLLECTION

Iron instruments occupy the first place in music and, by the tempos they create, tell those listening what to do and how to move. *Ọ̀mọ̀* bells feature prominently in Yorùbá community meetings, when they are rung by official overseers at tempos to enforce obedience and speak to the final order of legitimate action. The *ọ̀mọ̀* pictured here has smaller bells beneath its anvil-shaped finial; these funnel and amplify its resonant sounding, thus recalling the ringing of hammers put to use at a blacksmith's forge.

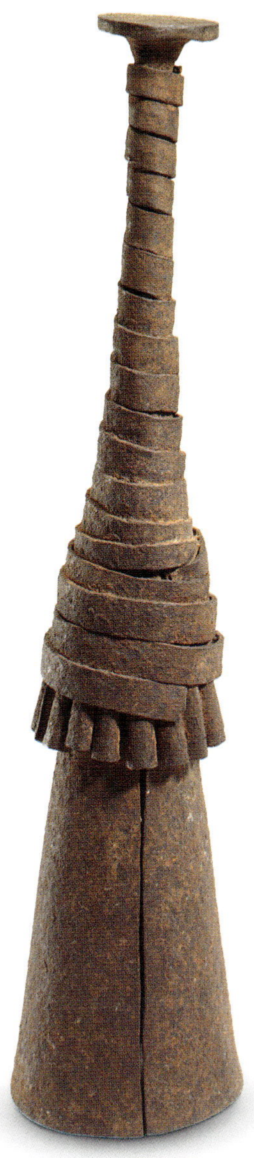

12.5
Artist unknown (Yorùbá peoples, Nigeria)
Bell (*ọ̀mọ̀*), 20th century
Iron
H: 38.1 cm
PRIVATE COLLECTION

This *ọ̀mọ̀* may have been wielded by a blacksmith, for the handle terminates in a smith's tongs, a tool used to safely hold hot iron and extend the smith's reach toward the work.

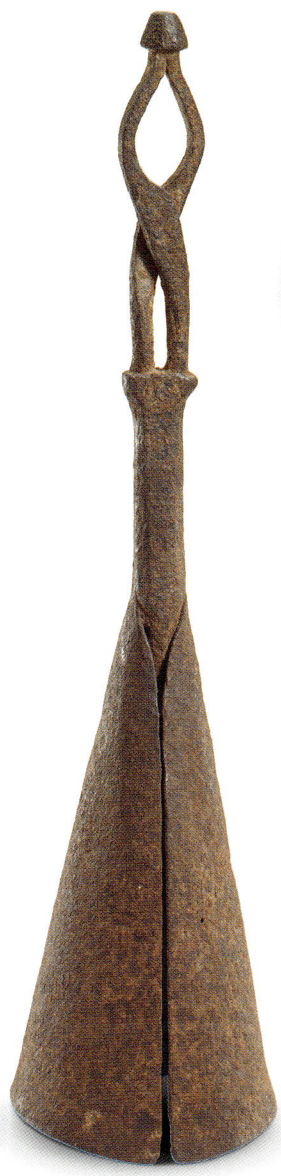

12.6
Artist unknown (Yorùbá peoples, Nigeria)
Rattles (*ikù*), mid-20th century
Iron
H: 29 cm (each)
PRIVATE COLLECTION

These rattles (*ikù*) are elaborations of ankle rattles that are an essential element of Gẹlẹdẹ masquerades honoring the mystical powers of women and celebrating positive social contributions (including those of blacksmiths). They were likely held in the hand and shaken. The sounds made by the pebbles loose inside the two opposing pods and six dangling bells at each end would have transmitted the powers of Ògún, Yorùbá deity of iron and war, indispensable to combatting disease, sickness, and other calamities.

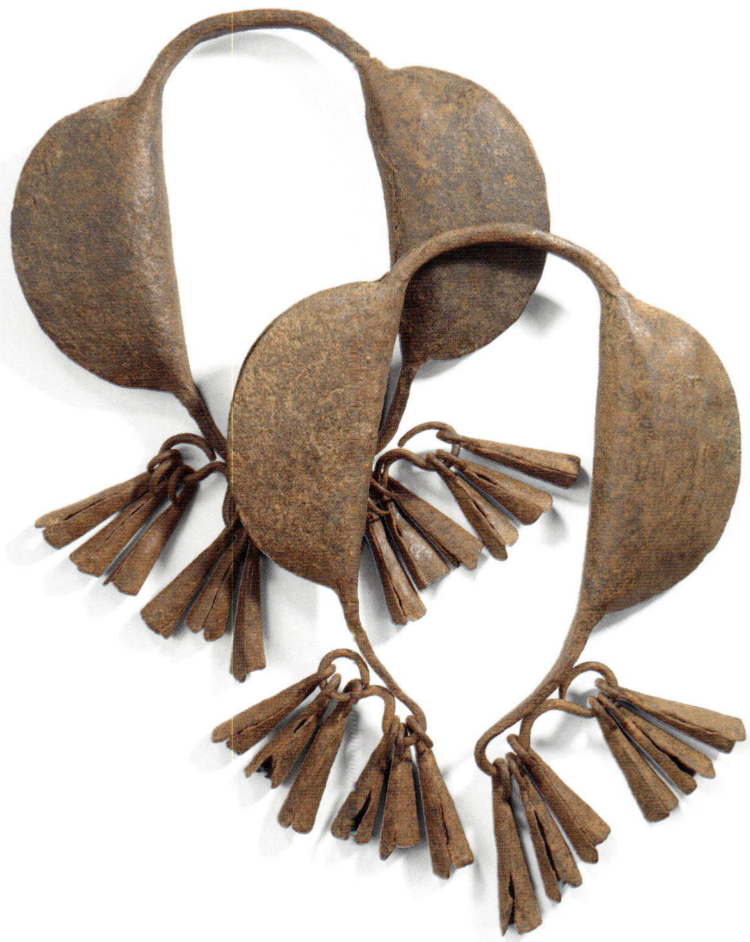

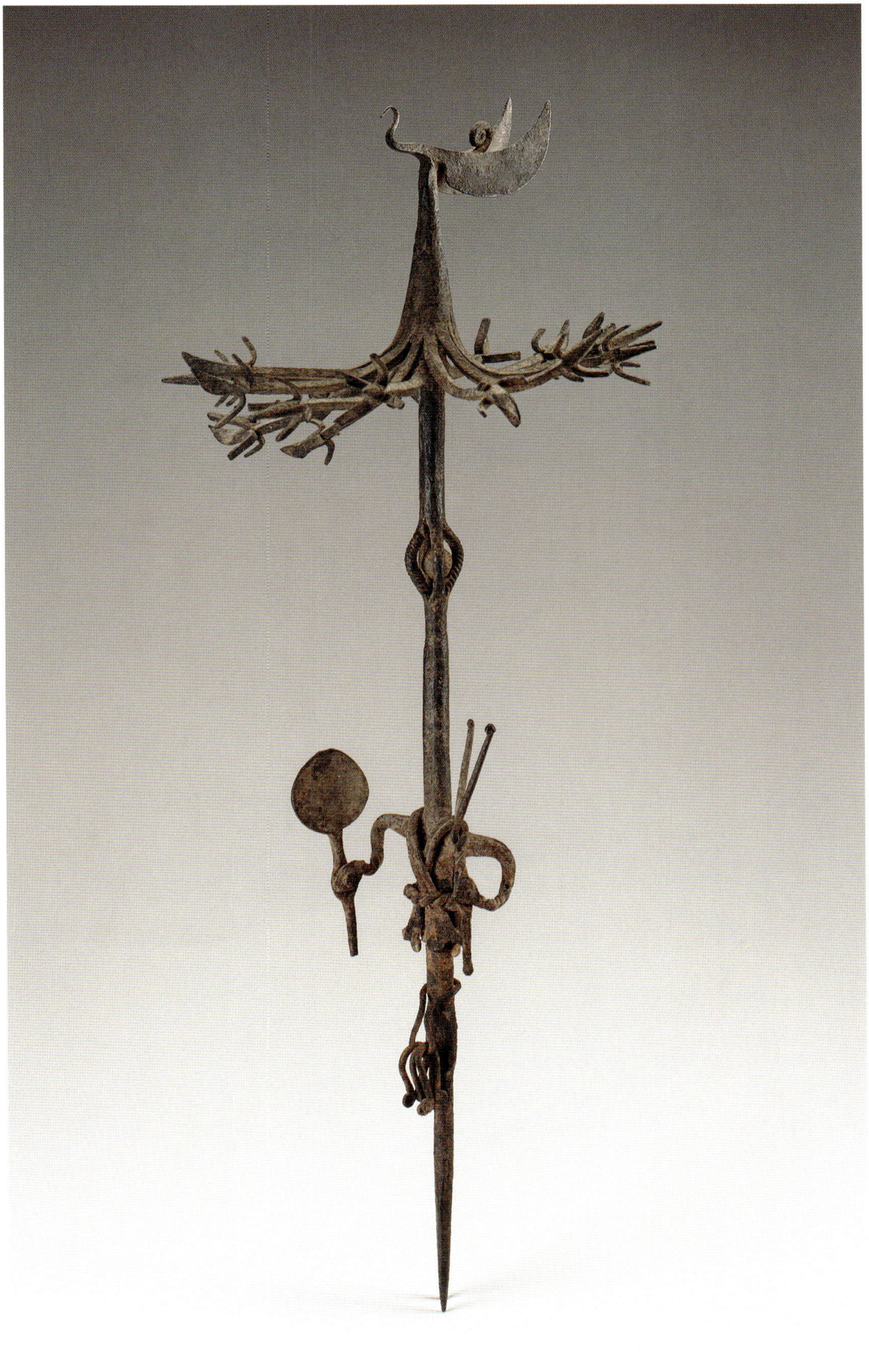

12.7
Artist unknown (Yorùbá peoples, Nigeria)
Herbalist's staff (*ọ̀pá* Ọ̀sanyìn), late 19th to mid-20th century
Iron, stone
H: 71 cm

NATIONAL MUSEUM OF AFRICAN ART, SMITHSONIAN INSTITUTION, GIFT OF AMBASSADOR AND MRS. BENJAMIN HILL BROWN JR., 76-3-4

The (*ọ̀pá* Ọ̀sanyìn of medicine-workers displays birds that represent various potentially aggressive forces that humans must confront. Generally short and often placed on altars for other divinities, the staff recalls trees filled with birds, evoking the forest as a source of sacred medicines. In the example pictured here, below the birds is a stone encased in the iron shaft—a remarkable feat of forging. It is as if the round stone becomes the head of a figure: the shaft can be seen to transform into a medicine-worker who holds a fan in his right hand and a *Y*-shaped staff with twisted handle (or tongs?) in his left. He wears crisscrossing sashes on his chest and a belt around his waist (see fig. 12.13 for similar attire).

12.8
Artist unknown (Yorùbá peoples, Nigeria)
Herbalist's staff (*ọ̀pá* Ọ̀sanyìn), 19th century
Iron
H: 61.5 cm
COLLECTION MINA AND SAMIR BORRO

Birds symbolizing aggressive forces cluster around the tree-like form of a medicine-worker's *ọ̀pá* Ọ̀sanyìn. This staff is exceptional for having birds perched along the upper length of its shaft, in a dramatic presentation. More typically, an *ọ̀pá* Ọ̀sanyìn has at its very top sixteen birds encircling a central one (see fig. 12.7).

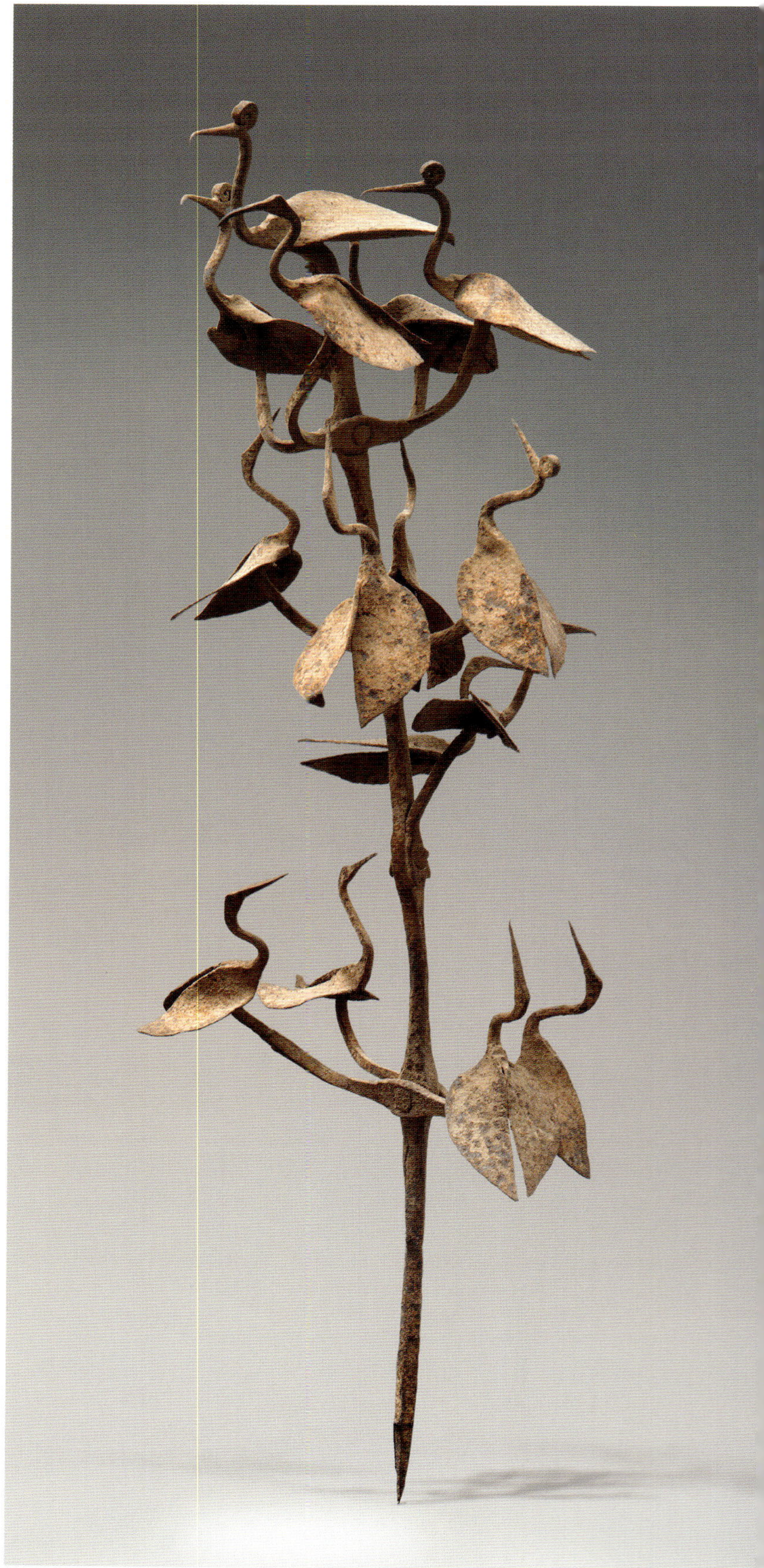

Ògún's association with weaponry is evoked by a ceremonial sword called the *udàmalóre* and its beautifully beaded sheath (*ẹ̀wù*), adorned with imagery celebrating iron weapons, equestrian warriors, and clever animals such as birds and monkeys (fig. 12.9).[7] It is worn by the sovereign warrior-chief (Ọjọmọ) during the Igogo festival in Ìjẹ̀bú-Ọwọ to honor chiefs of military prowess such as the Ọjọmọ of Ìjẹ̀bú-Ọ̀wọ̀, whose *oríkì* includes the line "the sharp sword that draws blood." The Ọjọmọ is an Ògún devotee often praised as "possessor of the great sword" (Abiodun 1989).

An elaborate garment likewise covers the *ọ̀pá* Òrìṣà Oko, an iron staff and spiritual weapon against such issues as antisocial behavior, barrenness, disease, and poverty (figs. 12.10–12.12). The *ọ̀pá* Òrìṣà Oko embodies the deity associated with agricultural fertility and sustenance, and is exquisitely chiseled with designs and mounted onto a studded wooden shaft. Forged from many repurposed hoes, the staff possesses the amassed *àṣẹ* of many farmers—their blood, sweat, and tears—as well as the *àṣẹ* of the master blacksmith who made it. Such extraordinary power gives this object a distinctive aura. Placed on altars in brass bowls—since the staff possesses, like the feet of royalty, extraordinary *àṣẹ* —the *ọ̀pá* Òrìṣà Oko must remain elevated to show appropriate respect to both Òrìṣà Oko and the earth, the sustaining abode of the ancestors and the resting place of the dead. The phallus-like finial, abstract faces, and seed-like studs on the shaft evoke the themes of fecundity. An eternal cycle of life, departure, and return is visualized in the interlacing patterns on the elaborate garments, sheaths and caps or crowns, that often dress the staffs. Such lavish displays of beads and cowries—emblems of wealth—evoke prosperity as well as the Yorùbá tradition of distributing new yams to the poor (Drewal, Pemberton, and Abiodun 1989, figs. 187, 189, 247; Drewal and Mason 1998, 258–62).

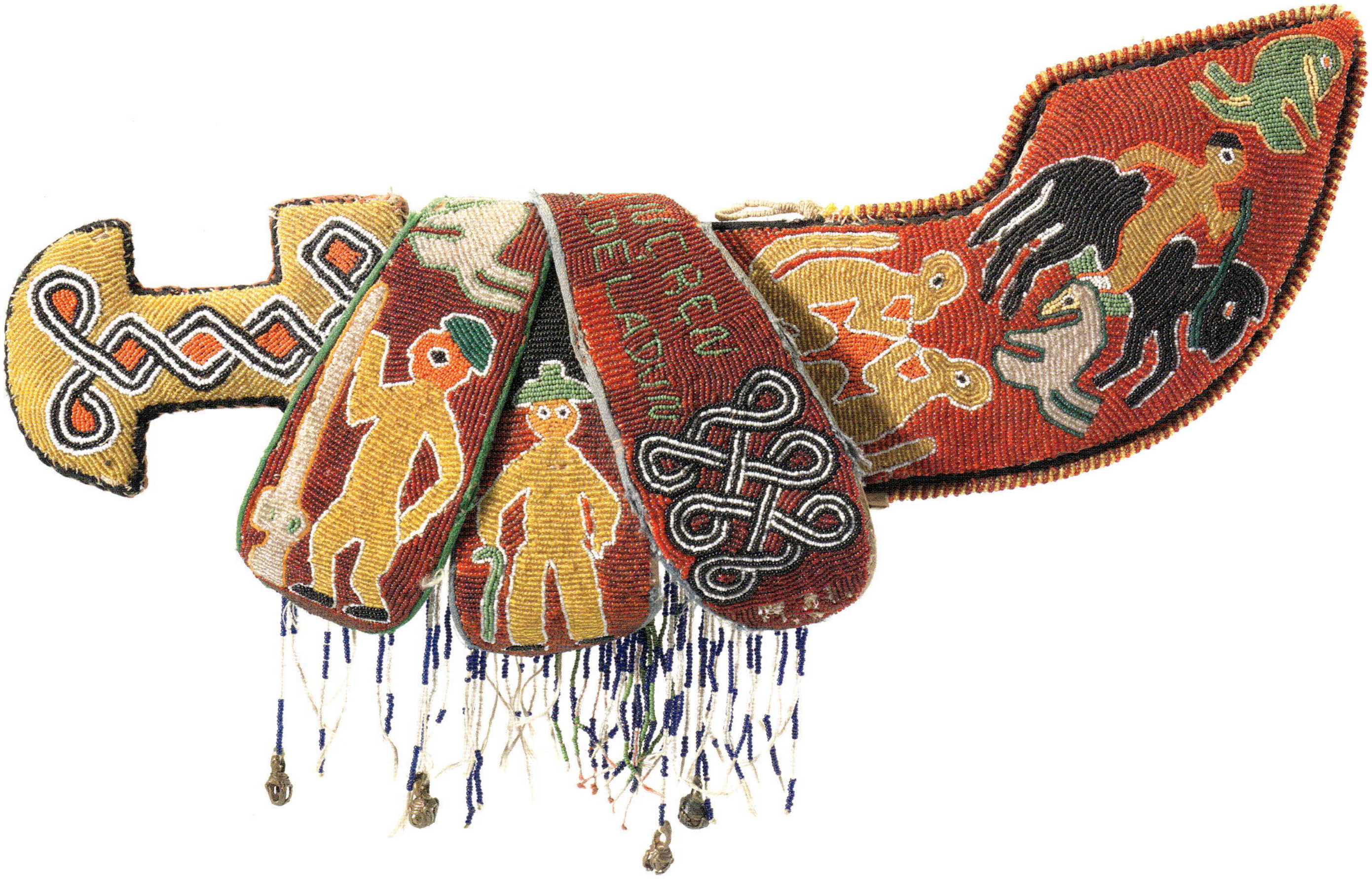

Just as a blacksmith's poker protectively extends the hand and stoke and regulate the fire for creative forging, Ògún mediates between humans and the potential dangers of iron implements (fig. 12.13). In judicial matters brought before members of the Ògbóni Society (referred to in some parts of Yorùbáland as the Ọ̀ṣùgbó Society), defendants swear to truthfulness on iron. As those wise elders closest to ancestors—those eternally present spiritual beings continually engaged in human affairs—the Ògbóni Society is a judicial body responsible for deciding disputes and criminal cases in communities as well as for supervising the installation and/or dethronement of an *ọba*, or sacred ruler (Drewal 1989). It deals with the duality, complementarity, and essential unity of men and women in their roles as guardians of the law and ancestral presence in the earth, and members own, use, and wear *ẹdan* Ògbóni/Ọ̀ṣùgbó, which are lost-wax castings of men and women made of copper alloy and joined at the heads by a metal chain attached to iron shafts (figs. 12.14, 12.15). The iron epitomizes maleness, Ògbóni, and long-lasting durability, while the copper alloy evokes femaleness and the goddess Ọ̀ṣun. Together, they symbolically encompass the entire community and work to judge and solve social problems.

12.9
Artist unknown (Yorùbá peoples, Nigeria)
Ceremonial sword (*udàmalóre*) and sheath (*ẹ̀wù*), early 20th century
Wood, cotton, leather, rawhide, glass beads, thread, copper alloy
L: 51.5 cm
FOWLER MUSEUM AT UCLA, X91.1636;
GIFT OF DR. AND MRS. JEFFREY HAMMER

The beaded sheath on this *udàmalóre* includes depictions of animals, which in Yorùbá art often serve as metaphors for the attributes and actions of humans and as allusions to proverbial and divinatory wisdom. We might, therefore, imagine that the owner who brandished this sword and from its sheath to the delight of his audience intended to suggest his own wit, cunning, courage, and audacity as a celebrated follower of Ògún. On the hilt of the *udàmalóre*, the interlace motif signifies the eternal cycle of life, death, and return.

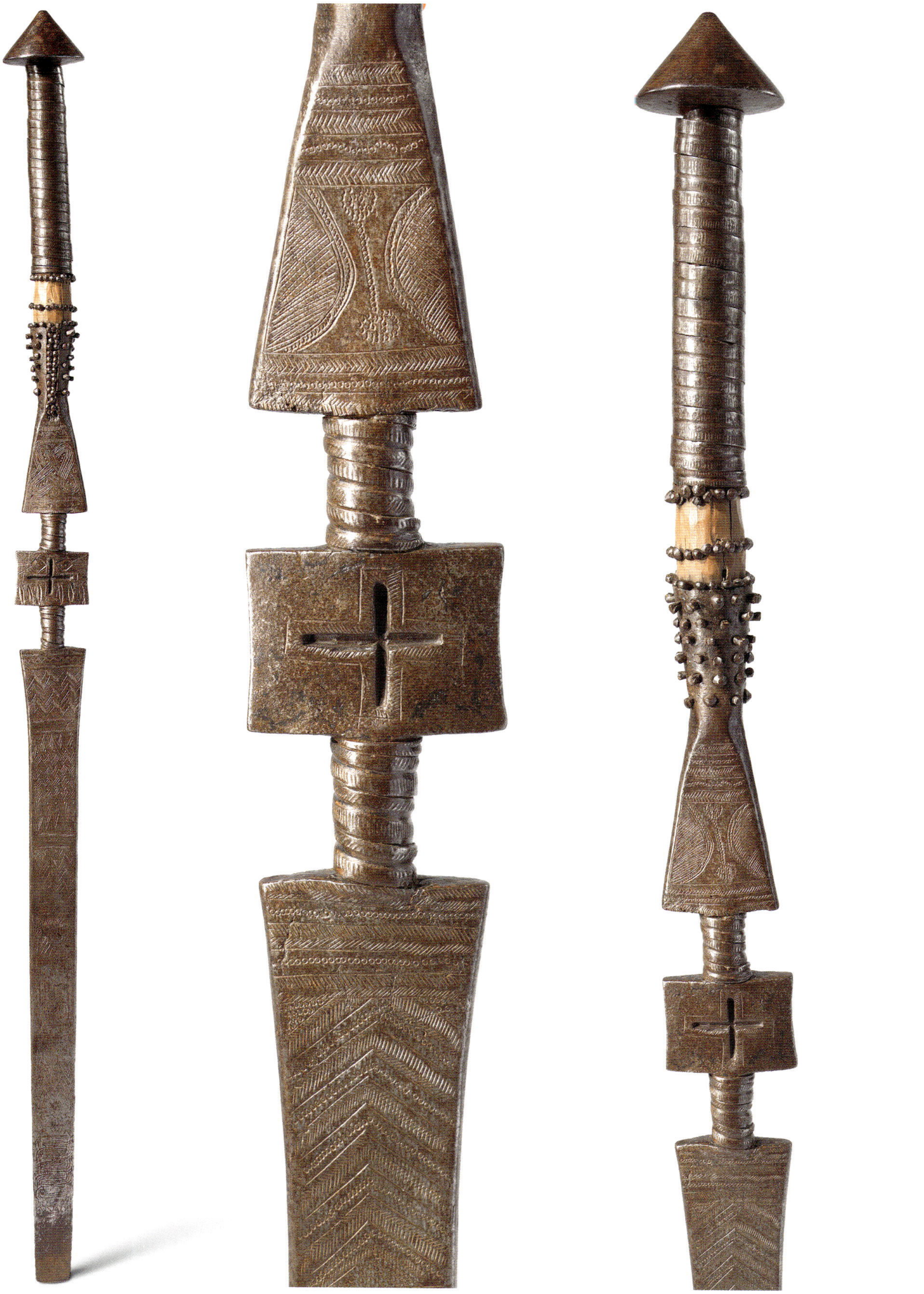

12.10A–C
Artist unknown (Yorùbá peoples, Nigeria)
Staff (*ọ̀pa* Òrìṣà Oko), 19th to early 20th century
Iron
H: 146.5 cm
PRIVATE COLLECTION

The elaborate imagery on this staff and sheath refers to the eternal cycle of life and death. Among Yorùbá peoples, the staff called *ọ̀pá* Òrìṣà Oko honors and embodies Oko, a deity associated with agricultural fertility and sustenance. The efficacy of an *ọ̀pá* Òrìṣà Oko is enhanced by the extraordinary amount of labor a blacksmith devotes to shaping and decorating its every surface. This particular staff's ability to confer blessings of community prosperity is strengthened by an investment of worn-down iron hoe blades, forged together to create its figurative "torso." The staff's embellishments (as well as those on its sheath, see fig. 12.11) present elaborate imagery referring to the eternal cycle of life and death. On its front side, a small rectangular section presents its "face" (*ojú*), with two diamond-shaped eyes and three vertical marks on each "cheek"; an indented cross signifies the *oríta méta*, or crossroads of the world and otherworld. The eternal knot—an interlacing form—is incised on the section just above, which extends into a conical shape reminiscent of a Yorùbá ruler's crown and relating to *ìbọrí*, an individual's spirit essence.

12.11
Artist unknown (Yorùbá peoples, Nigeria)
Crown and sheath for the staff of Òrìṣà Oko (*adé ati èwu* Òrìṣà Oko), early 20th century
Cotton, beads, cardboard, leather, wood, thread, metal
H: 152 cm
FOWLER MUSEUM AT UCLA, X88.289 A-D; GIFT OF MR. PETER J. KUHN

The beaded sheaths and crowns that dress *ọ̀pà* Òrìṣà Oko emphasize themes of royalty and prosperity derived from the essential work of farmers. This sheath's two female figures wear white baldricks and hold a bowl (above) or bird (below) to signal generosity and "our mothers." The triangular panels along the sheath's sides recall the ubiquitous amulets (*tiraa*) worn by Yorùbá. As objects of spiritual protection and empowerment, they seal in and secure the *àṣẹ* of the iron staff. The colors of Òrìṣà Oko balance hotness and coolness: red is a hot, advancing color, evocative of energy and danger, while white is cool and calm, evoking patience and wisdom. Moyo Okediji (2003) writes that for Yorùbá all color begins at the smith's forge, when heated black charcoal sparkles as varied hues begin to appear.

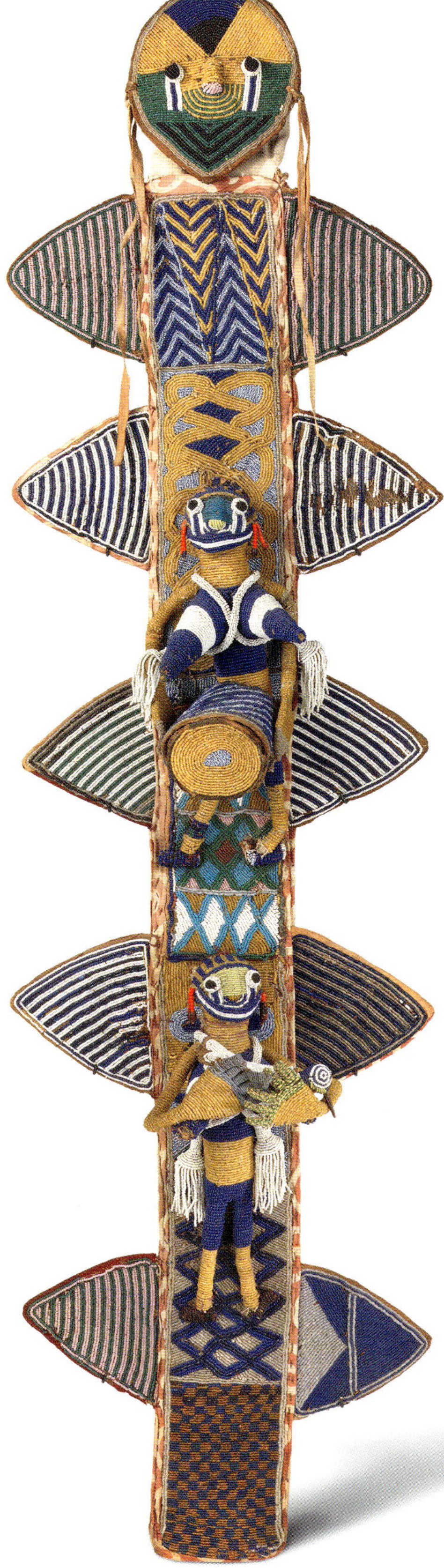

On a striking set of *ẹdan* staffs (see fig. 12.15), one iron shaft ends in the form of a small spoon, a reference to the sustenance of community and the Ògbóni/Òṣùgbó female titled elder Ìlá Erelú; or it might be a blacksmith's paddle-rake, as the other shaft ends with a blacksmith's poker, suggesting a male titled elder and a reminder of Ògún's role in matters of ancestral law and the imperative of truthfulness. The iron shafts of most *ẹdan* terminate in simple points that recall the Yorùbá saying that describes Ògún as "a needle that is sharp at both ends." The iron chain that joins female and male heads (enlarged to emphasize their knowledge and wisdom) ensures that both perspectives join, adding, as Yorùbá say, "salt" (or taste) to a world of challenges, problems, and questions demanding collective action. The chain allows Ògbóni members to drape the *ẹdan* around their necks (Drewal in Drewal, Pemberton, Abiodun 1989, 136–43, fig. 136), and its iron parts serve as a reference to a mystical/invisible third presence—the earth as the abode of the ancestral spirits who witness all acts of the living and judge them according to moral precepts.[8]

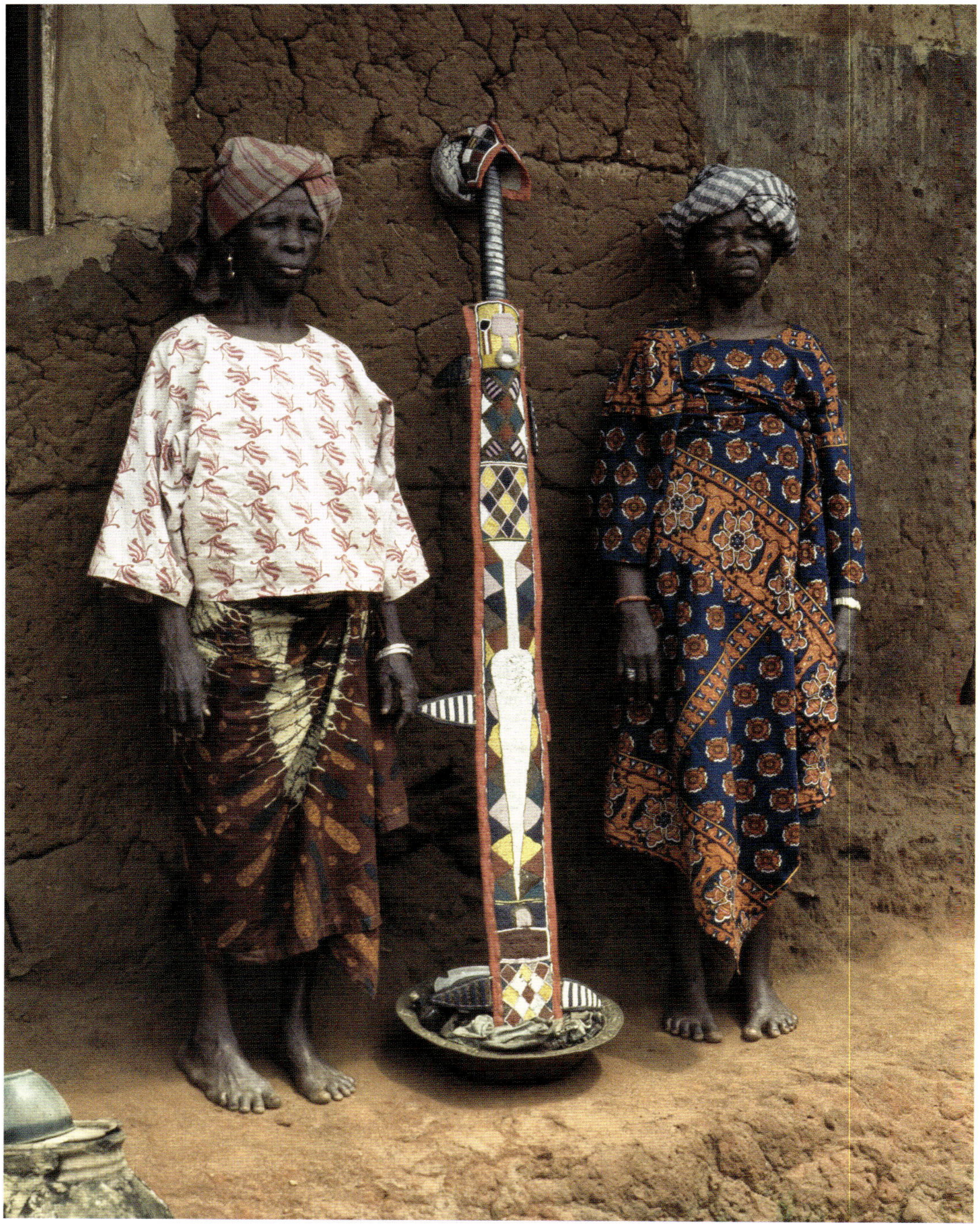

12.12 Òrìṣà Oko's devotees are primarily female, and some scholars identify the deity as female. For the "goddess of agriculture," according to R. C. Abraham, "[t]he priestesses form a secret society and nobody dares to offend them. She is worshipped at the time of the new moon. Some new yams are presented to this deity, others being eaten by the worshippers, while still others are given to the poor. Òrìṣà Oko was, according to tradition, a Chief's wife who introduced cultivation to a previously hunting community" (1958, 466). Abraham's commentary recalls Yorùbá attitudes about the mystical powers of women, including their ability to transform into birds, and to conceive as well as destroy, linking them to the judicial role of the Òrìṣà Oko in matters of antisocial behavior.

PHOTOGRAPH BY HENRY J. DREWAL, ÈGBÁDÒ-YORÙBÁ, NIGERIA, 1975.

12.13
Artist unknown (Yorùbá peoples, Nigeria)
Title staff of head of blacksmiths (*iwana* Ògún), late 19th century
Iron, copper alloy
H: 31.8 cm
FOWLER MUSEUM AT UCLA, X70.690; GIFT OF THE RALPH B. LLOYD FOUNDATION

This figurative ceremonial copper alloy poker refers to a key blacksmithing tool that manages fire and keeps danger contained and creative/destructive force controlled, evoking Ògún as a "double-edged sword." The figure, seated on a blacksmith's poker or rake, has an array of accoutrements that proclaim his privileged status while suggesting battle: gun on one shoulder, fly whisk on the other, and daggers/knives at the hips. His crown-like headdress, beard, and baldric belts all commemorate military rule, warfare, and the unrest that marked much of the nineteenth century for Yorùbá and Fon peoples (and others in the region as well).

EDO

The second Benin dynasty, of Edo peoples, was founded by a Yorùbá prince, Ọ̀rànmíyàn of Ilé-Ifè, in the fourteenth century, and the two groups overlap in aspects of their deities and rituals. Edo celebrate Osun as the god of sacred medicines in iron staffs that proclaim fearsome powers that can heal or harm. A locus of an Osun priest's power resides in the staff known as the Osun *ematon*, or "iron Osun," understood to be a weapon to protect followers and destroy enemies (fig. 12.16). At its summit is a commanding bird, sometimes astride a horse, surrounded by a descending array of encircling birds and other animals associated with powers of transformation such as snakes and chameleons. Miniature iron tools of blacksmiths and various iron weapons (swords, lances, knives) are also featured. Hoes and blades refer to tools for gathering forest medicines as weapons to attack negative forces. Thus, the iron staff links an Osun specialist with blacksmiths, hunters, and warriors—all Ògún devotees. With branches shooting outward at a raking angle like flames from a fire, its praise-name is Osun *nigiogio*, meaning "the Osun is burning up with heat," perhaps associating the transformation of adversaries into conquered peoples with the activities of blacksmiths' fiery forges (Girshick Ben-Amos 1995, 74).

Such ferocity is also dramatically displayed in the faces of copper alloy Osun bell heads (see fig. 1.4, this volume; see also Peek and Picton 2016). Snakes, other reptiles, and frogs represent *ihen*, famous warriors and magicians (Girshick Ben-Amos 1999, 108) who are "magically powerful" avatars and warriors of Osun. They are often shown issuing from nostrils or mouths and they "vomit out snakes when setting out to destroy their enemies."[9]

FON

Dahomey, the eighteenth- and nineteenth-century kingdom of Fon peoples, had an intimate if fraught relationship with Yorùbá to the east. For several centuries they vied for territory and access to African riches and to European merchants engaged in the transatlantic slave trade. As the power of the Ọ̀yọ́-Yorùbá empire waned and Dahomey grew, so too did royal arts associated with the king at Abomey.

One object epitomizes this era of death, destruction, and slavery: the Gubasa, a massive royal war sword of the Fon deity of iron and war, Gu (fig. 12.17). Its long surface bristles with lethal blades thrusting out in all directions—death by weapons of war is its fearful message. Now missing, this royal sword may be recognized as capturing the essence of the Dahomey Kingdom's rise to prominence through military prowess, and its ultimate demise by French colonial forces armed with superior firearms.[10]

Remarkable as well are the iron ancestral altars called *asen* or *assanyin* (figs. 12.18–12.20): iron-and-copper-alloy staffs topped with circular platforms richly decorated with cast and forged human figures, animals, plants, and cultural objects, all of deep and resonant significance and often related to proverbs and praise-poetry (Bay 2008). Created to honor and celebrate ancestors and the vast pantheon of gods of Vodun, as Fon religion is known, *asen* serve as conduits of sacred power and the accumulated life forces of ancestors. Like so many African objects forged of iron, they are cosmic connectors. For such connections to remain vibrant, humans must offer prayers and food to ensure life, prosperity, and good fortune (figs. 12.21, 12.22).

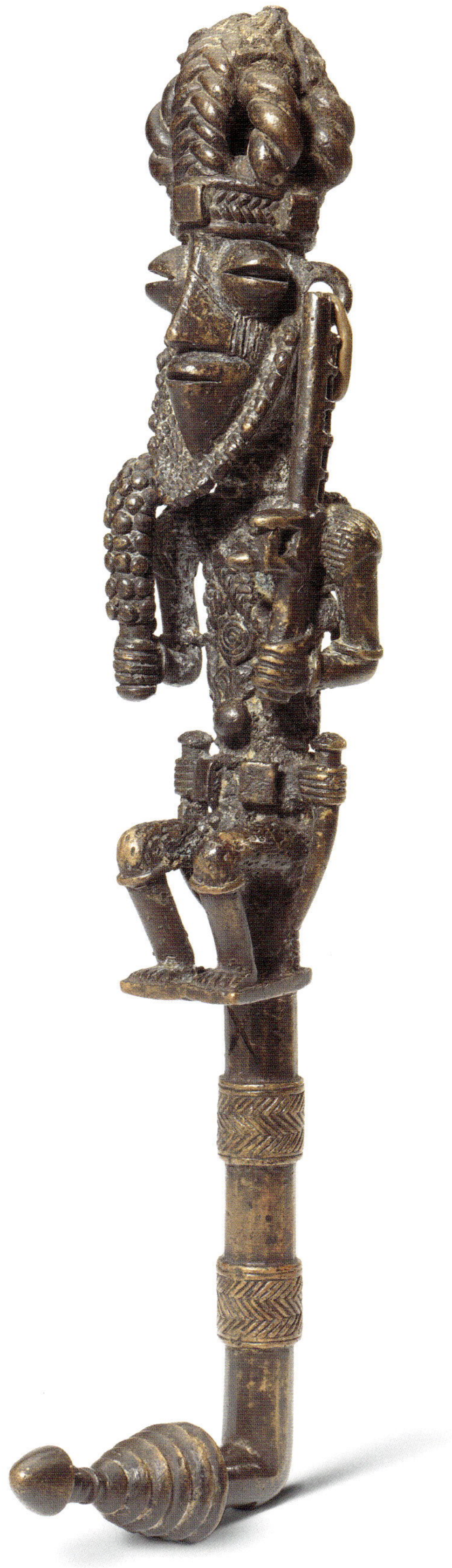

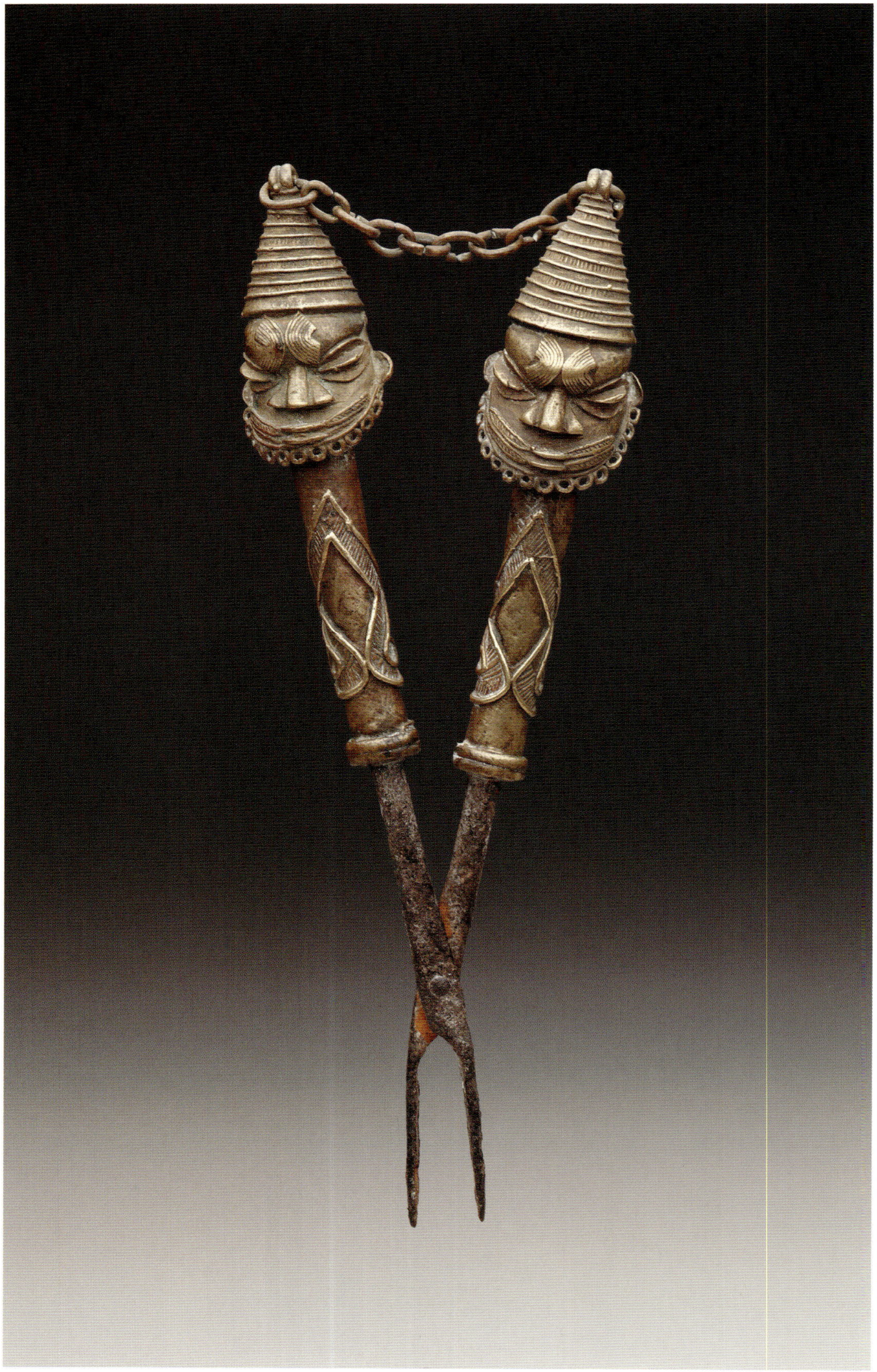

12.14
Artist unknown (Yorùbá peoples, Nigeria)
Figurated staffs for Ògbóni Society rituals (*ẹdan* Ògbóni), early to mid-20th century
Copper alloy, iron
H: 30.5 cm
DALLAS MUSEUM OF ART, 2005.92, GIFT OF GEORGE AND SIDNEY PERUTZ IN HONOR OF ROSLYN A. WALKER
PHOTOGRAPH BY BRAD FLOWERS, 2013. COURTESY OF DALLAS MUSEUM OF ART

The Ògbóni/Oṣùgbó Society is the Yorùbá judicial body responsible for deciding disputes and criminal cases as well as for supervising the installation or dethronement of sacred rulers. Elder members possess, wear, and direct the efficacies of *ẹdan*—intricate brass castings on iron shafts depicting the heads or full bodies of female and male figures joined by an iron chain. The two shafts of one pair exhibited here terminate as tongs (fig. 12.14) while another's become a spoon or paddle-rake and poker (fig. 12.15). Such blacksmith's tools would have lent additional capacity to Ògún's presence in Ògbóni/Òṣùgbó activities, and to the ability to judge wisely and solve community problems.

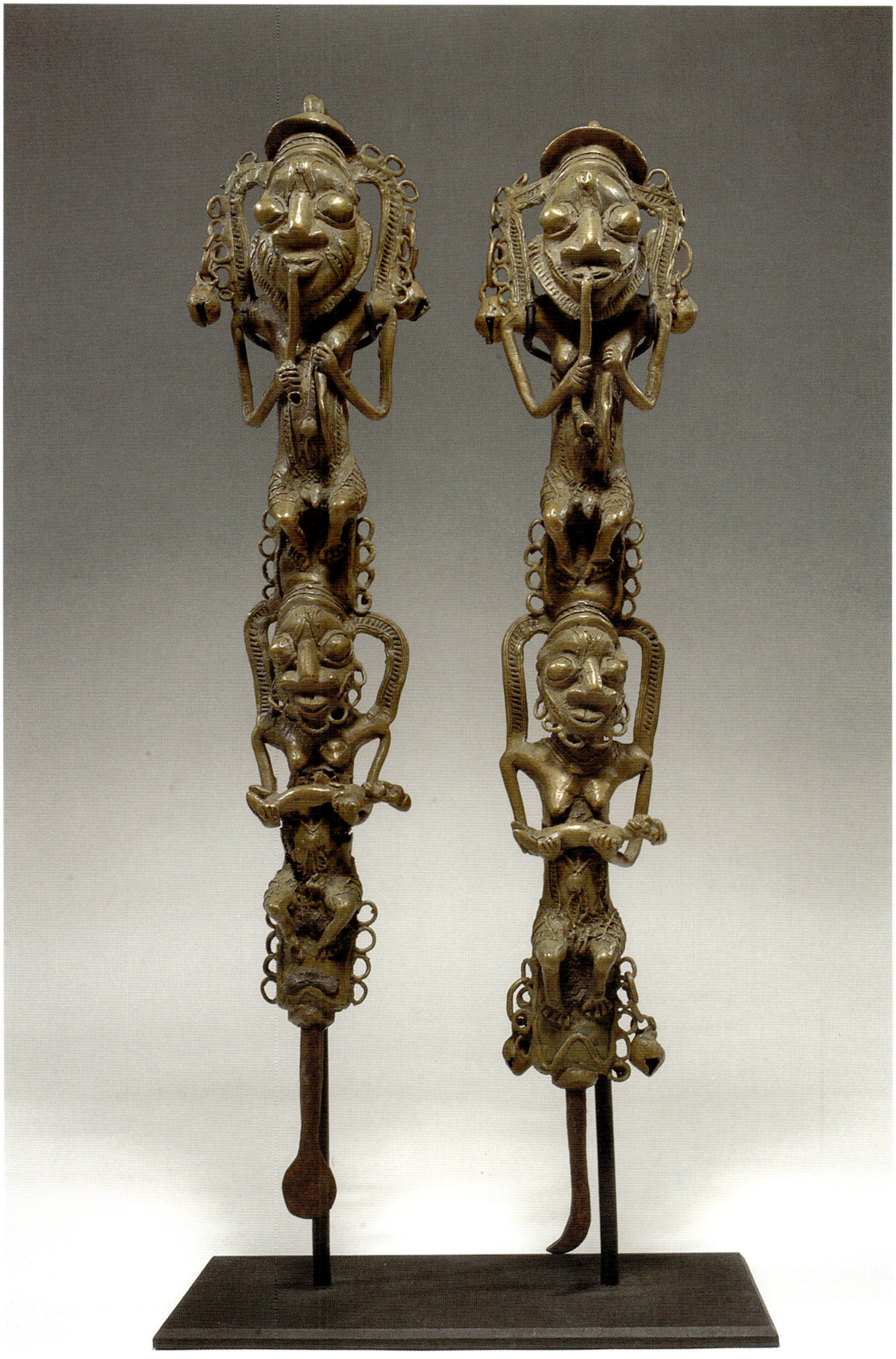

12.15
Artist unknown (Yorùbá peoples, Nigeria)
Figurated staffs for Ògbóni Society (*ẹdan* Ògbóni), 19th century
Iron, copper alloy
H: 42 cm
CHARLES AND KENT DAVIS
PHOTOGRAPH BY CHARLES DAVIS, 2010.
COURTESY OF DAVIS GALLERY

Dahomean arts are renowned for their eclectic, incorporative forms, imagery, and media. The iconography of an *asen* is often derived from local, transcultural, and international sources. European items of trade (guns, machetes, liquor bottles, sewing machines, cars, bicycles) figure prominently, attesting to long histories of exchange. An *asen*'s tree-like form suggests the awesome presence of the sacred *ìrókò* tree (*Milicia excelsa*). As its branches flare outward to support the platform, an *asen* also recalls a massive umbrella, a primary accoutrement of powerful kings and prominent citizens (see fig. 3.15), as inspired by dignitaries from overseas and neighboring kingdoms.

The form also resonates with the staffs of Yorùbá diviners and medicine-workers, just as the term *assanyin* suggests a relationship to the *ọ̀pá* Ọ̀sanyìn (see figs. 12.7, 12.8). Indeed, many blacksmiths forging *asen* in the city of Abomey hail from Yorùbá families who were war captives. Among the most famous was Djegun Akankossi, said to have been from the royal family at Ọ̀yọ̀, and his descendant, Akpele Kendo Akati, is credited with having created a famous forged iron figure of Gu captured by the French on the battlefield beach at Ouidah, in the present-day Republic of Bénin, in 1894 (see interleaf H, this volume). This masterful figure, created by an enslaved Yorùbá blacksmith, was carried into battle to inspire and protect Dahomian soldiers resisting French colonial conquest. Art's power to inspire bravery and fortitude, like the iron Osun staffs of Edo priests, was essential in battle, yet in the end Gu's enlivened figure became a spoil of war. Another sculpted Fon image reiterates the Gu warrior theme (fig. 12.23) as a soldier with sword and gun mounted on a staff and resembling the famous Gu figure captured at Ouidah as well as the tableau figures on *asen*.

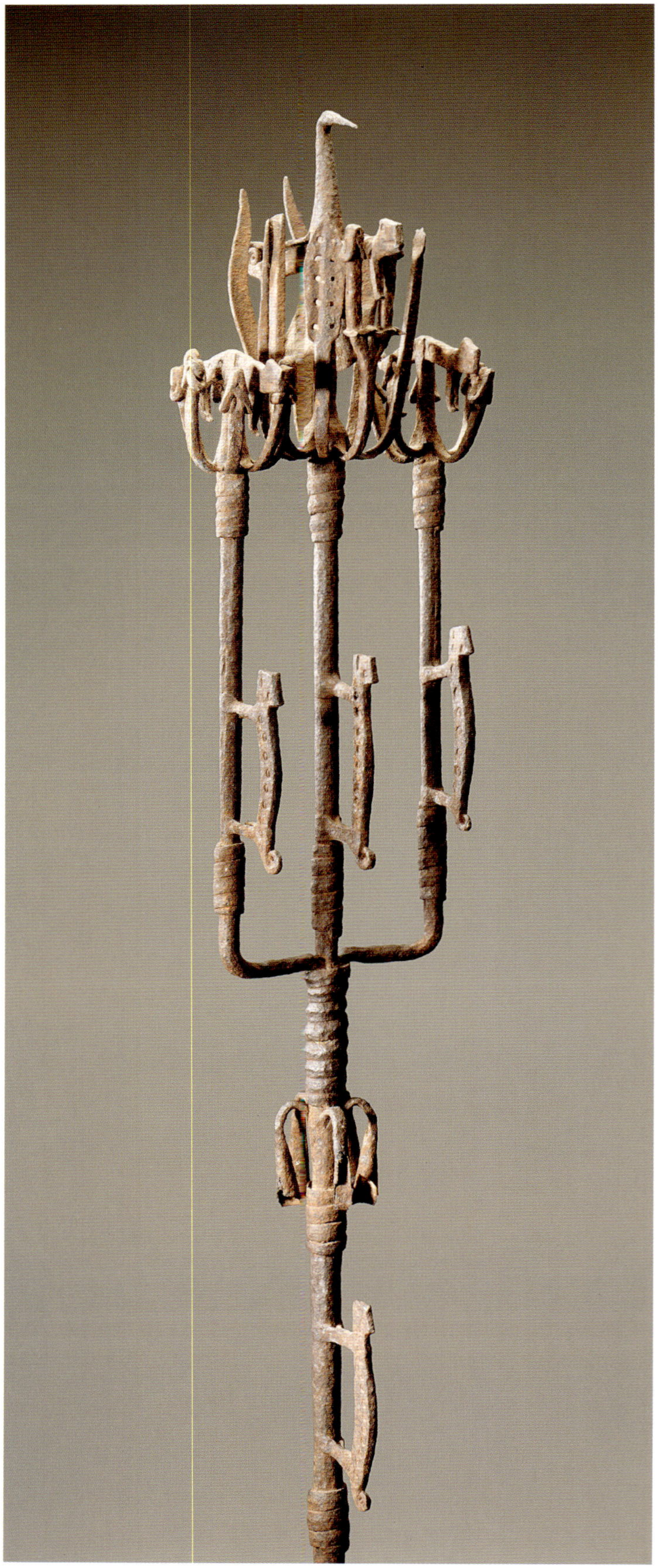

12.16
Artist unknown (Edo peoples, Benin City, Nigeria)
Medicine staff (Osun *ematon*), 18th to 19th century
Iron
H: 168.8 cm
NATIONAL MUSEUM OF AFRICAN ART, SMITHSONIAN INSTITUTION, MUSEUM PURCHASE AND GIFT OF DR. WERNER MUENSTERBERGER, 96-29-1
PHOTOGRAPH BY FRANKO KHOURY. © NATIONAL MUSEUM OF AFRICAN ART, SMITHSONIAN INSTITUTION

The power of a priest of Osun, Edo god of sacred medicines, resides in his staff (*ematon*) as a weapon to protect followers and destroy enemies. This Osun *ematon*, which evokes a tree, displays powerful creatures that dwell in the dense rainforests of Edoland, the source of awesome substances and beings. The commanding bird at its summit is surrounded by a descending array of more birds (associated with antisocial persons termed *azen*) and snakes and chameleons, animals possessing extraordinary powers of sight, sound, and transformation. Also represented are cone-shaped leaves, on the shaft-trunk, and miniature smithing tools along with various weapons and implements.

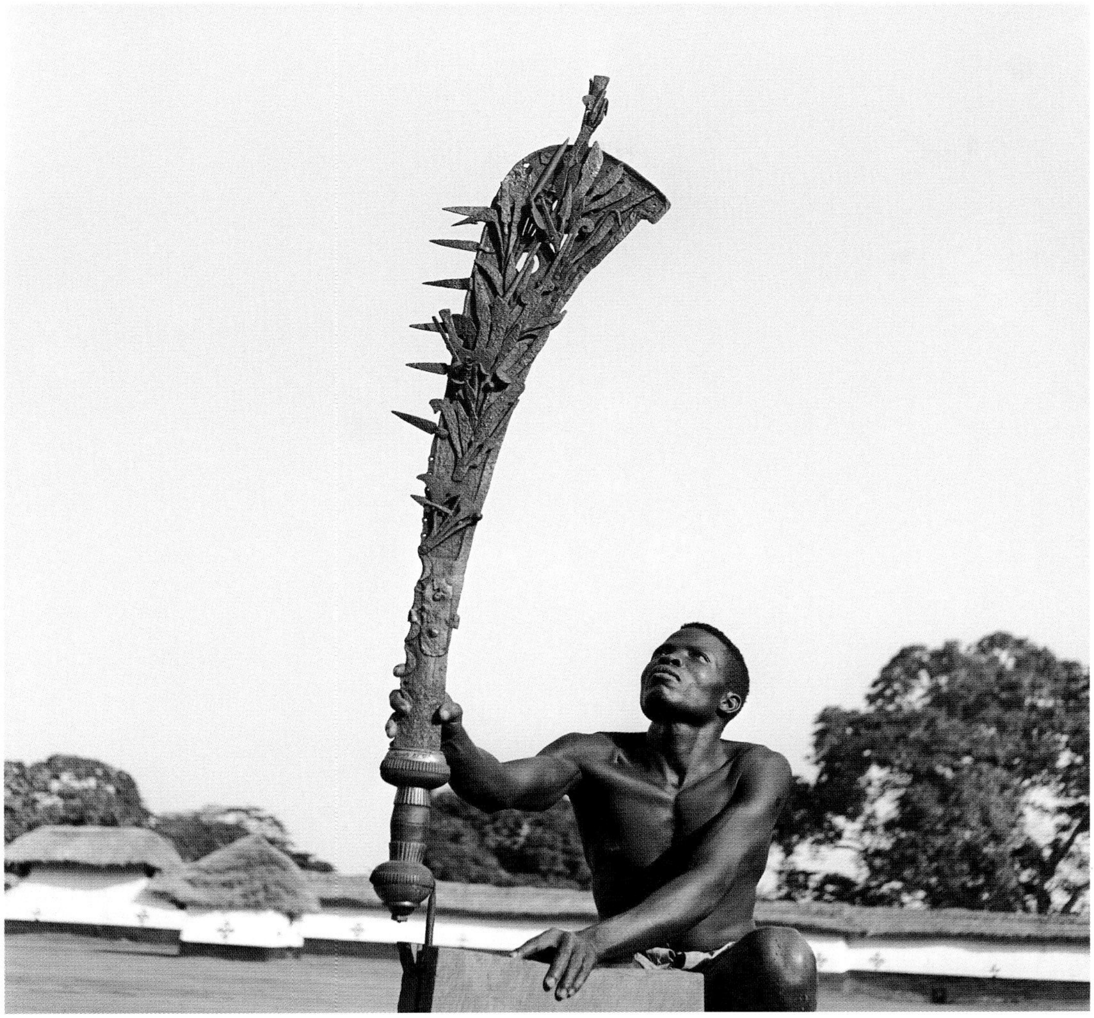

12.17 The Gubasa attests to the military glory of Fon peoples' Kingdom of Dahomey in the 17th to 19th centuries. A massive royal sword celebrating Gu as the god of iron and war, the Gubasa is said to have been forged by Akpele Kendo Akati, Master of the Long Horned Ram, for King Glele (1858–1889) during an era of colonial aggression, death, destruction, and slavery. The Gubasa was once displayed in the Musée Historique, Abomey, Republic of Bénin, but is now missing.
PHOTOGRAPH BY PIERRE VERGER, ABOMEY, DAHOMEY, 1952.

Ògún/Gu's iron finds its warrior role again in *bociọ* (figs. 12.24–12.26), "empowerment objects" of "fury and force" meant to protect oneself or to attack enemies worldly and otherworldly (Blier 1995, 4).[11] Materializations of action and used to enhance their makers' or owners' lives, they are assemblages of many materials, both seen and unseen, such as iron, wood, bones, straw, leather, pottery, cloth, fur, feathers, blood, beads, as well as libations, prayers, curses, songs, and activating gestures. *Bociọ* abound "with textures, tastes, smells, colors, and references to sounds" that provoke deep, visceral emotions often beyond words (ibid., 1; see also Drewal forthcoming). These forms are beyond explication. They are meant to be secrets known only to their makers, for if knowledge of a *bociọ* 's makeup were to spread, its potency could be compromised or even negated (see Blier 1993; Nooter 1993). Among Ewe of southwestern Republic of Bénin and coastal Togo, as peoples closely related to Fon and Yorùbá, those who are empowered to activate *bociọ* are called "masters of fire" (*dzotọ*), for they deal with the world of heat, anger, danger, and war (Blier 1995, 78–79). This is the domain of Ògún/Gu and blacksmiths, who are associated with transformation and sacred activation through material objects.

CONCLUSION

Yorùbá, Edo, and Fon (and Ewe) peoples have interacted, shared, and exchanged many ideas and material forms through trade and conflict—as Ògún/Gu's principal domains. As the deity of iron, Ògún/Gu cuts to final forms and truths. Blacksmiths' hammers, anvils, tongs, and pokers permit the creation of culture's tools and weapons. Sculptors' axes, adzes, and knives permit hidden realities to be imagined as they embellish and focus worship at altars for devotion (*ojúbọ*) (Thompson 1993). Farmers' hoe blades dig earth and build conical yam mounds to feed families, as celebrated in the *ọ̀pá* Òrìṣà Oko. Iron-and-brass emblems of Ògbóni/Ọṣùgbo test truthfulness to settle social ills. The iron staffs of diviners and medicine-workers are spiritual weapons, markers of wisdom possessing sacred powers to wage war against enemies by penetrating forests for medicines with awe-inspiring *àṣẹ* to cure ailments and destroy obstacles. Fon *asen* staffs venerate and celebrate ancestors' deeds in a medium that endures. *Bociọ* assemblages attest to the agency of individuals and communities acting to shape their lives using the performative power of iron and other empowered substances. To all these aspects of culture-making and sustaining, Ògún/Gu's iron is essential.

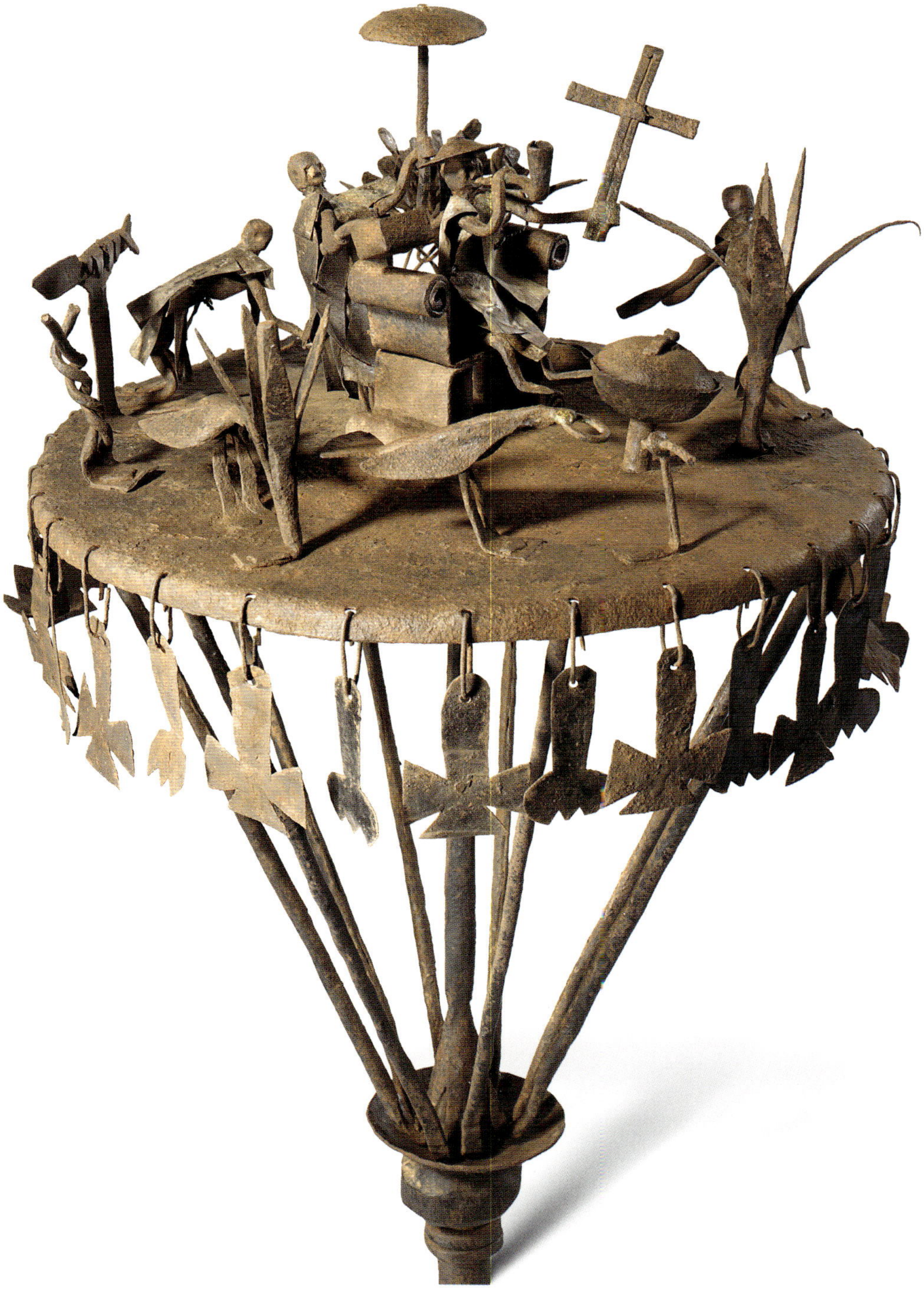

12.18
Artist unknown (Fon peoples, Republic of Bénin)
Ancestral altar (*asen*), 20th century
Iron
H: 151 cm
PRIVATE COLLECTION

Fon commission blacksmiths to produce umbrella-shaped *asen* staffs, kept in shrines to celebrate ancestors and the Fon pantheon. *Asen* staffs are surmounted by circular platforms and richly decorated with cast and forged human figures, animals, plants, and cultural objects, signifying proverbs and praise-poetry to honor ancestors and the vast pantheon of Vodun gods. *Asen* may feature hanging rattles that make gentle sounds in the breeze, attracting ancestors' positive intervention.

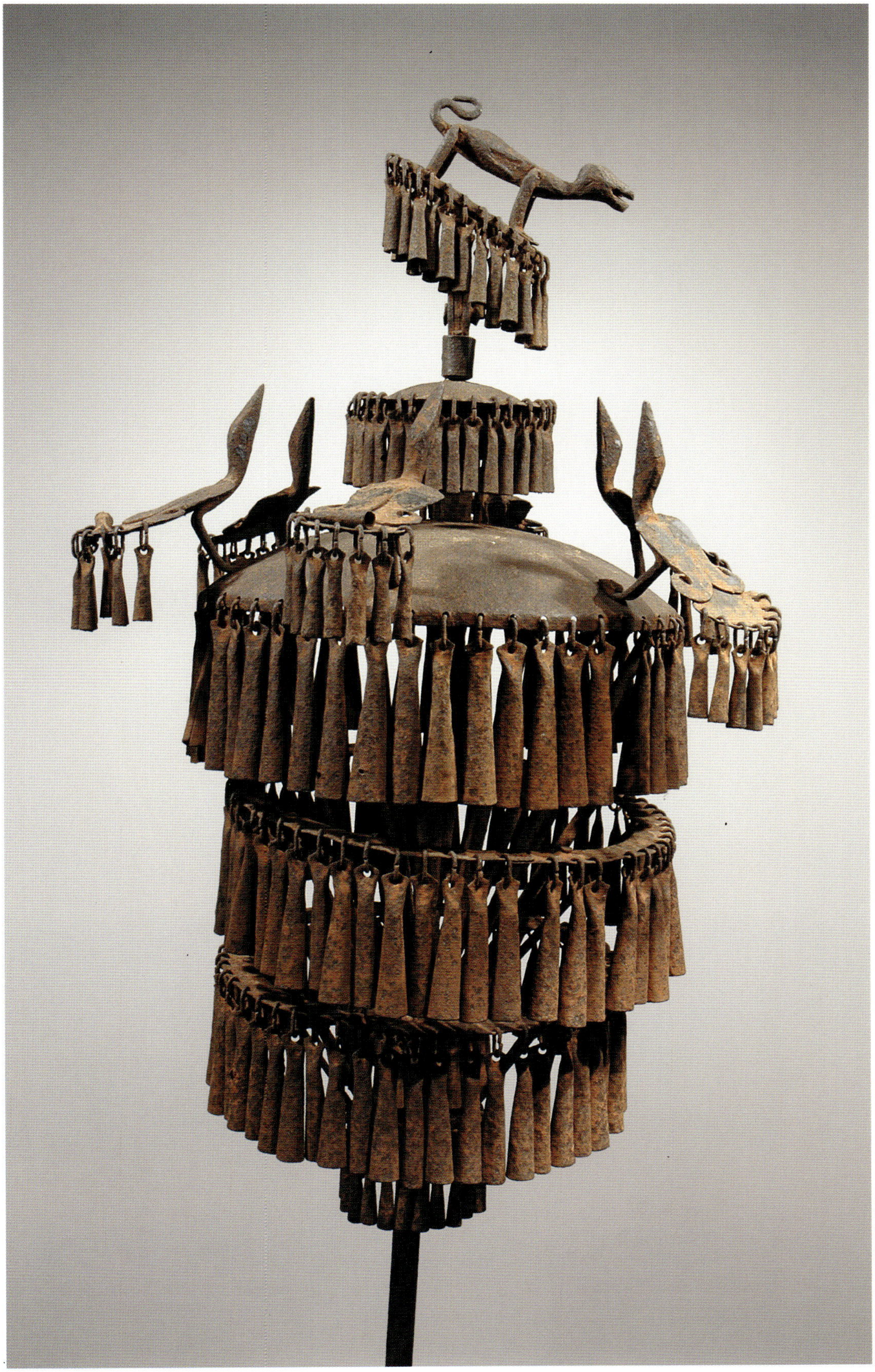

12.19
Artist unknown (Fon peoples, Republic of Bénin)
Ancestral altar (*asen*), early 20th century
Iron
H: 145 cm, W: 41 cm, D: 41 cm
CHARLES AND KENT DAVIS
PHOTOGRAPH BY CHARLES DAVIS, 2010.
COURTESY OF DAVIS GALLERY

A commanding animal at the summit, a circle of birds, and three tiers of clapperless bells would have sounded praise for an ancestor before this *asen* joined others set in a family shrine (*dejo*).

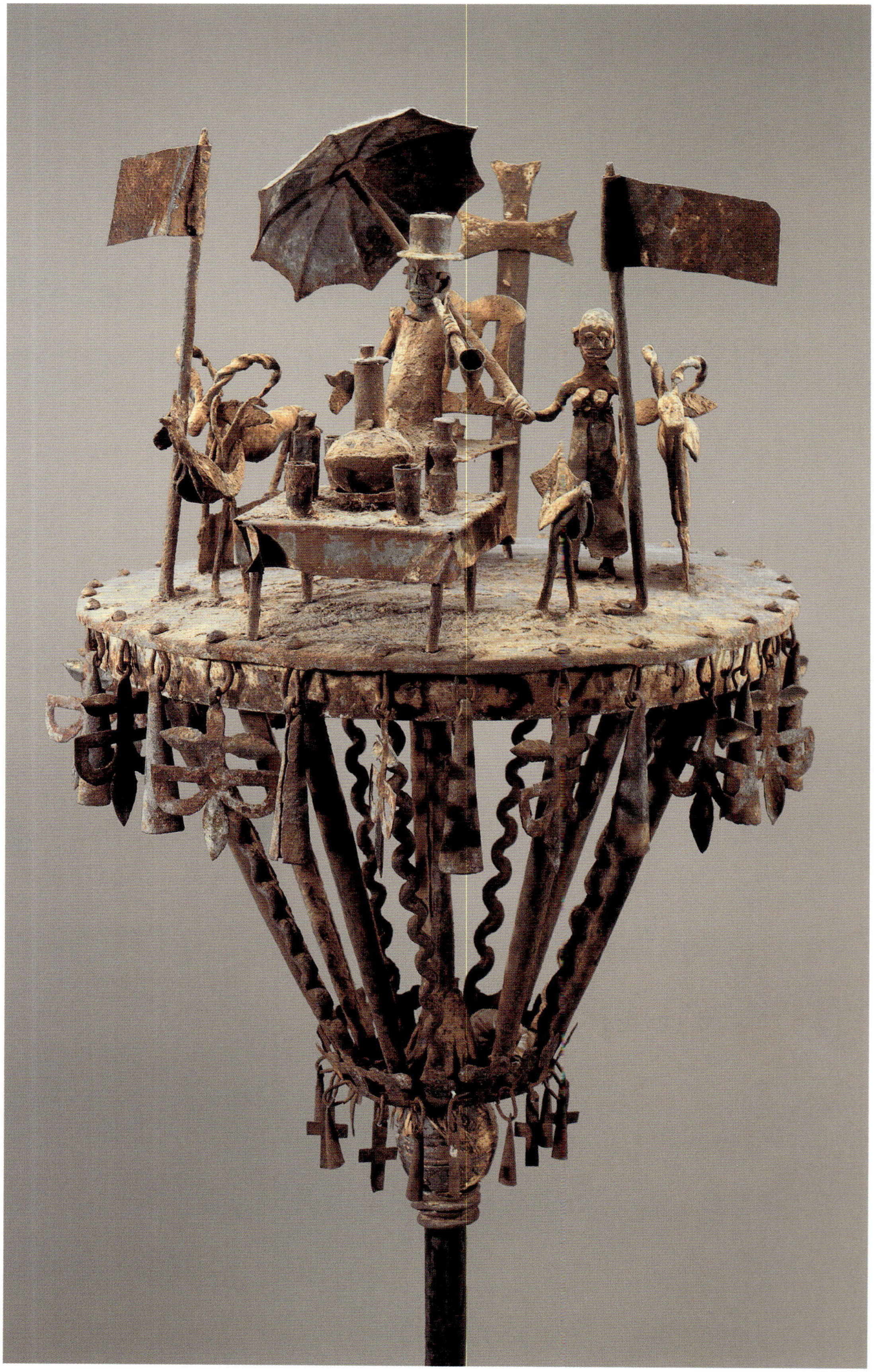

12.20
Attributed to Akpele Kendo Akati (active 19th century; Fon peoples, Dahomey)
Ancestral altar (*asen*), mid-19th century
Iron, raffia, wood, organic materials
H: 142.2 cm
NEW ORLEANS MUSEUM OF ART, GIFT OF FRANÇOISE BILLION RICHARDSON, 89.257
IMAGE COURTESY OF THE NEW ORLEANS MUSEUM OF ART

Asen serve as conduits of sacred power and cosmic connection, linking descendants to their forebears via exquisite iron altars activated with songs, prayers, food, and drink to ensure life, prosperity, and good fortune. On this *asen*'s platform, a covered gourd sits on a table as a food offering to the departed, who is shown seated on his chair and wearing a top hat symbolic of his elevated status. He is flanked by a wife and flags of identity, and the cross behind him suggests his faith not in Christianity but rather "Mawu, the female portion of the [Fon] creator couple" (Bay 1986, 17).

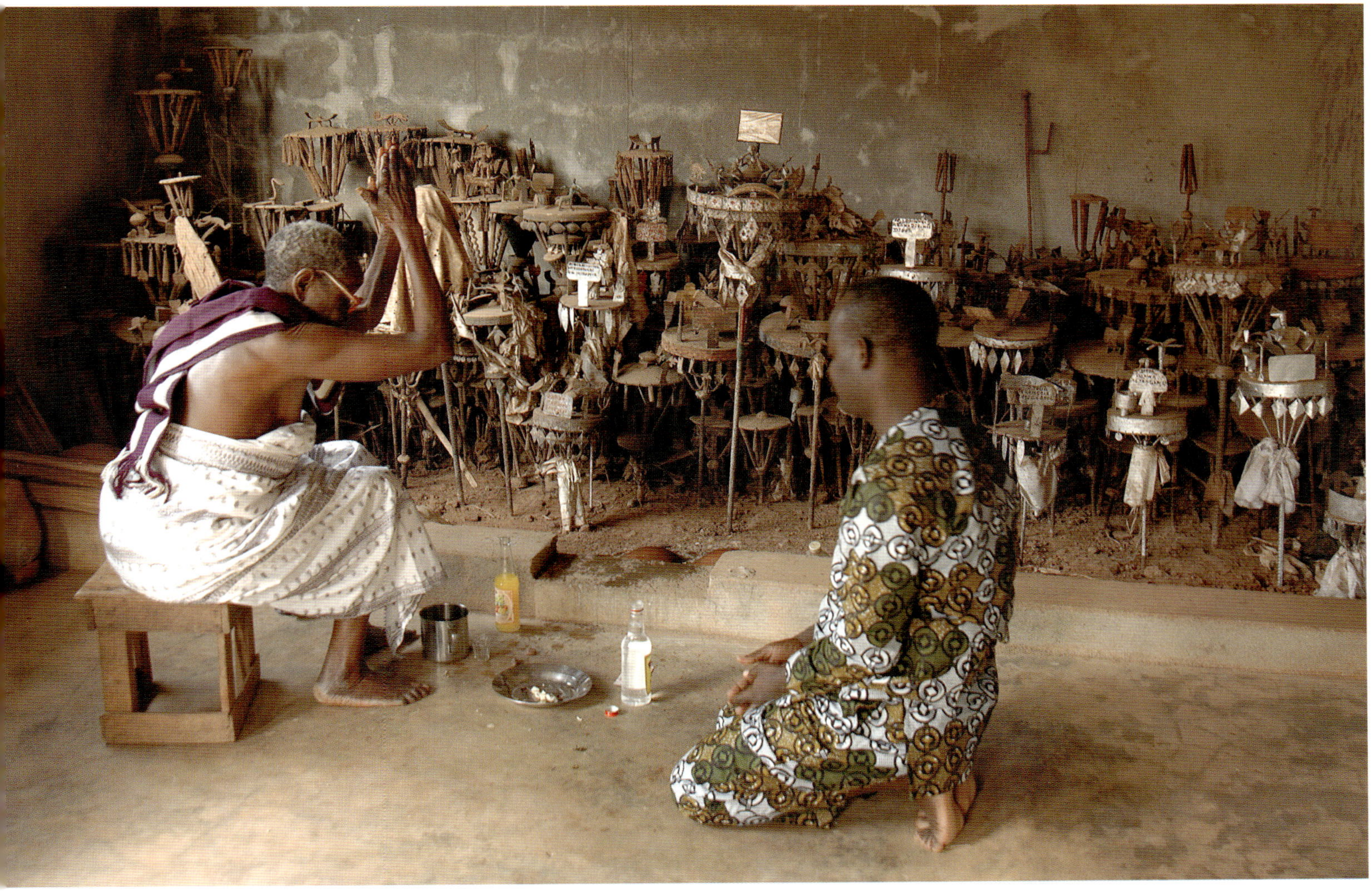

12.21 Djoua Atchawenon and her nephew Pierre Dagba venerate *asen* in the Yovogan Hounnon Dagba family compound. Family shrines where offerings are made on a regular basis to offer praise and nourish the deceased sometimes have large accumulations of *asen*.
PHOTOGRAPH BY TOM JOYCE, OUIDAH, REPUBLIC OF BÉNIN, 2010.

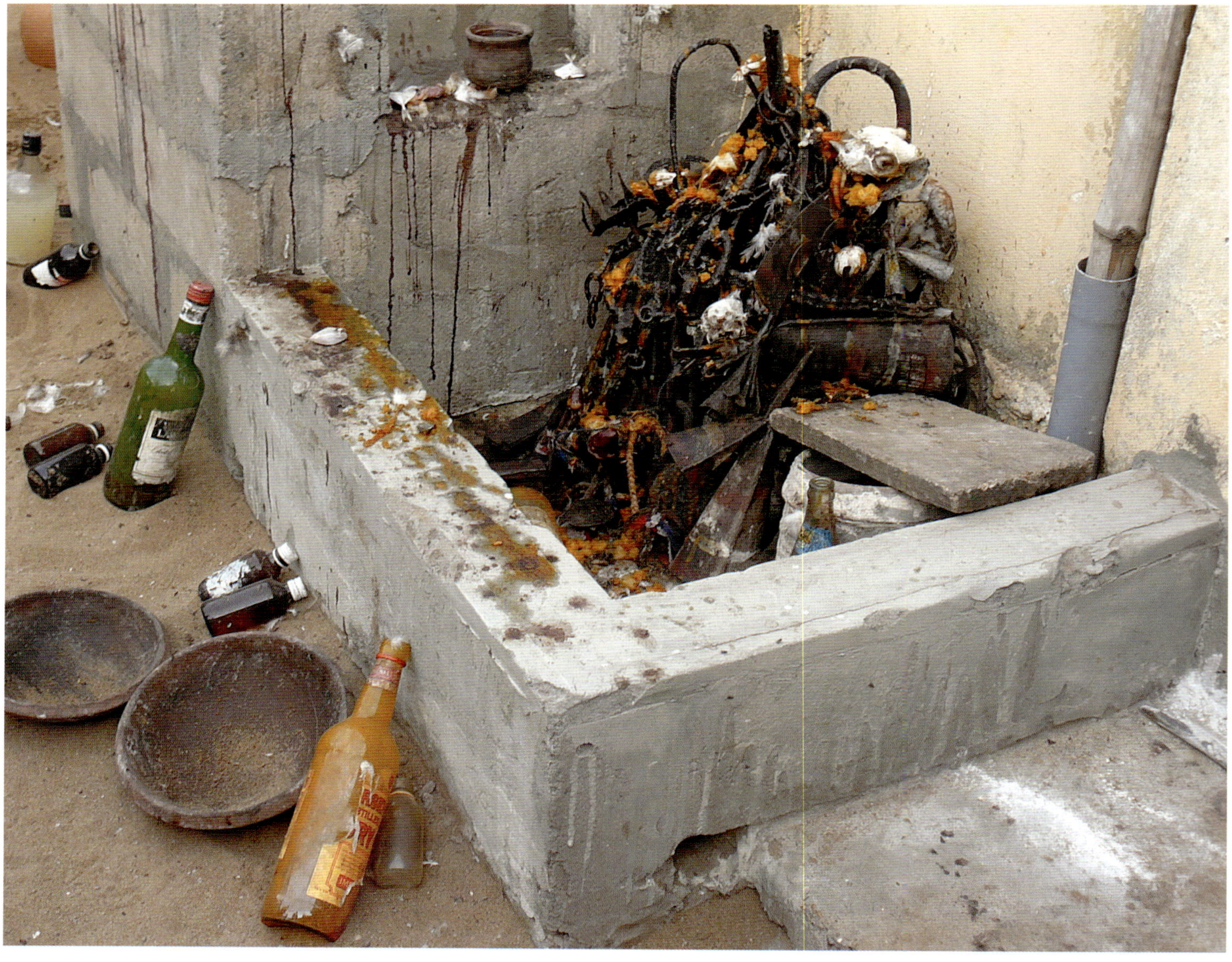

12.22 A shrine to Gu, Fon god of iron and transformation, in a Mina temple compound dedicated to the deities Mami Wata and Mama Tchamba. Maintained by a priestess known as Hounbonon, the shrine's objects of ritual practice and active devotion, including several bells, are enlivened by offerings of yellow palm oil and animal sacrifice, made to calm Gu's dangerous ardor and so assure his protection and promotion of Hounbonon. The bells are painted in stripes of alternating red and white, perhaps so that the red of Gu is "cooled."
PHOTOGRAPH BY ANDOCHE ADJOVI, ANÉHO, TOGO, 2017.

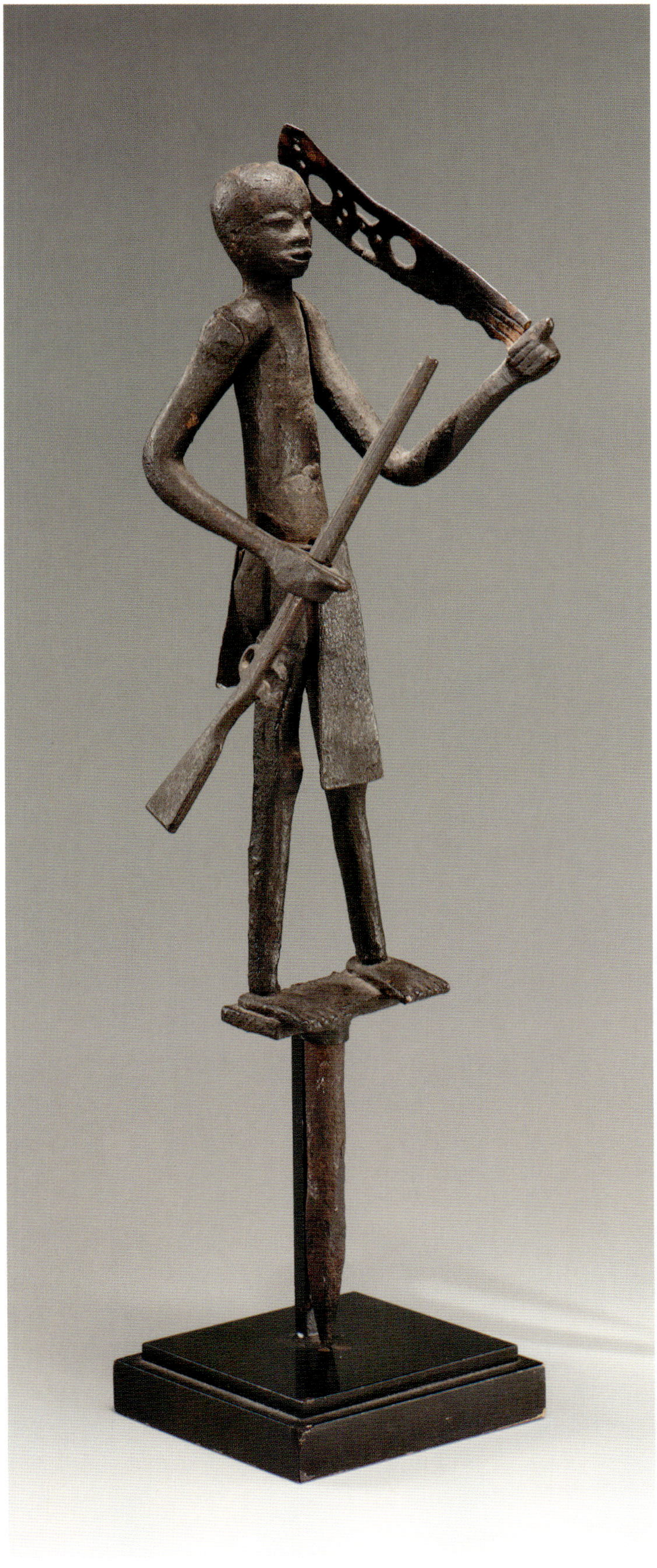

12.23
Artist unknown (Fon peoples, Republic of Bénin)
Gu, late 19th century
Iron
H: 47 cm
THE COLLECTION OF AMY AND ELLIOT LAWRENCE
PHOTOGRAPH BY STEVEN TUCKER, 2016.

This figure of Gu was brought to France around 1900, after the French defeated and exiled the King of Dahomey, Béhanzin, in 1894. It celebrates the warrior Gu, with long-barreled rifle and sword raised aloft, ready for battle. Gu can be creative and destructive, securing order and fostering chaos (see also interleaf H, this volume).

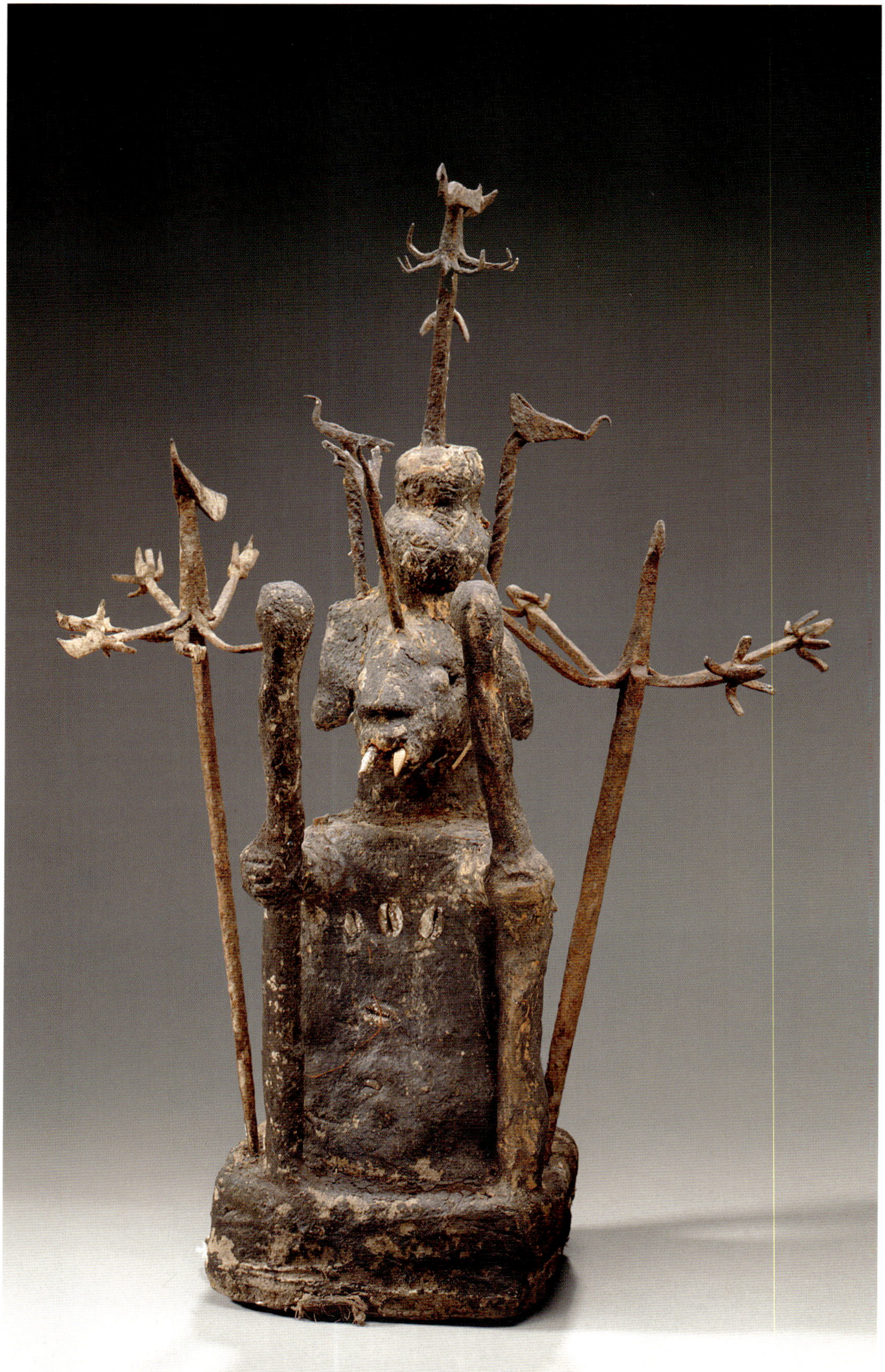

12.24
Artist unknown (Gun/Yorùbá peoples, Republic of Bénin)
Ọ̀sanyìn's altar for Ṣoponna, early to mid-20th century
Iron, clay, cloth, cowrie shell (*Cypraea moneta*), mixed materials
H: 62.3 cm
COLLECTION NATIONAAL MUSEUM VAN WERELDCULTUREN, AM-382-1
IMAGE © NATIONAAL MUSEUM VAN WERELDCULTUREN

Surrounding this *bociọ*'s somewhat amorphous figure-grasping-two-staffs are five miniature Ọ̀sanyìn medicine-worker staffs with bird finials. They are like barbed wire, intended to caution intruders. The lumpy body and bulbous forms of the central figure's "head" may connote a body diseased by pestilence, and the identification of this altar with Soponna links it to the Fon deity of smallpox, called upon by healers for a cure.

12.25
Artist unknown (Ewe peoples, Togo)
Power figure (*bociọ*), mid-20th century
Wood, metal, padlocks, keys, beads, fabric, feathers, hair or fur(?), soil
H: 55.9 cm (including mount)
MINNEAPOLIS INSTITUTE OF ART, THE MARY RUTH WEISEL ENDOWMENT FOR AFRICA, OCEANIA, AND THE AMERICAS 2014.91
(NOT IN EXHIBITION)
PHOTOGRAPH BY CHARLES WALBRIDGE, 2015. COURTESY OF THE MINNEAPOLIS INSTITUTE OF ART

This *bociọ* is a roughly carved human figure carrying an amorphous mound of matter in a large container on its head. In it are two iron staffs with curving, bifurcated blades for the thunder god Hevioso; an *asen*-like form supported by iron; and a red cloth tied and draped onto this summit-ensemble. The wooden figure is encased in layers of metal padlocks and multiple keys strung together; the locks signify "irreversibility," whether as curse or blessing. Below this is a "skirt" of iron cones and bells—sounding forms that call to Gu—and empowered leaves that cure or kill.

12.26
Artist unknown (Fon peoples, Republic of Bénin)
Power figure (*bociọ*), early to mid-20th century
Wood, iron, fabric, plant fiber, snail shell
H: 67 cm
COLLECTION NATIONAAL MUSEUM VAN WERELDCULTUREN, COLLECTIE CONGREGATIE VAN DE HEILIGE GEEST, AM-490-2
IMAGE © NATIONAAL MUSEUM VAN WERELDCULTUREN

A *bociọ* with an *asen*-like disk on top and an iron chain joining the figure with a cone/bell of Ògún/Gu sealed with the double-axe of Hevioso. The sum of all these spiritually charged elements (ancestors and deities) is more than their parts, for a *bociọ* is an ensemble created and empowered to take fearsome action on behalf of its owner.

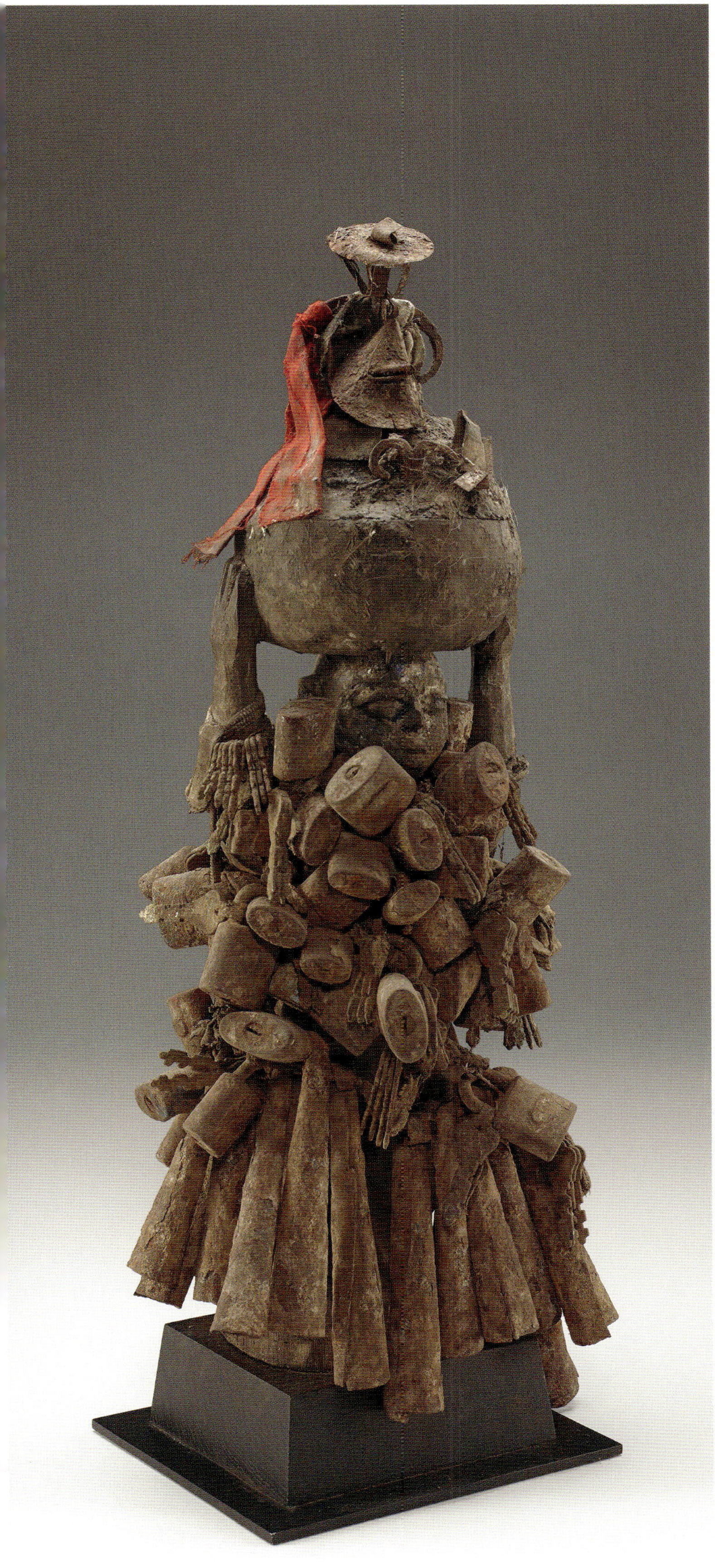

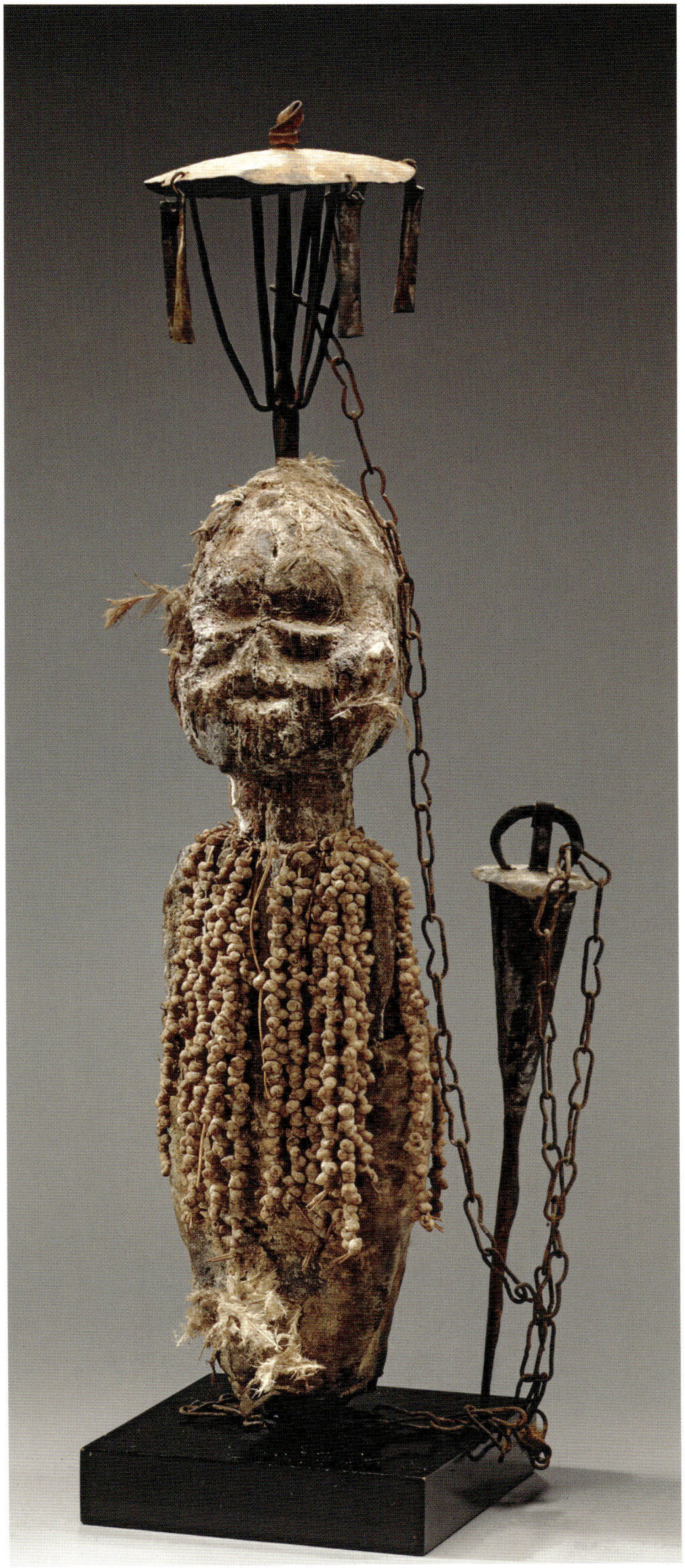

INTERLEAF G

DANCING FOR ÒGÚN OǸDÓ

Henry J. Drewal

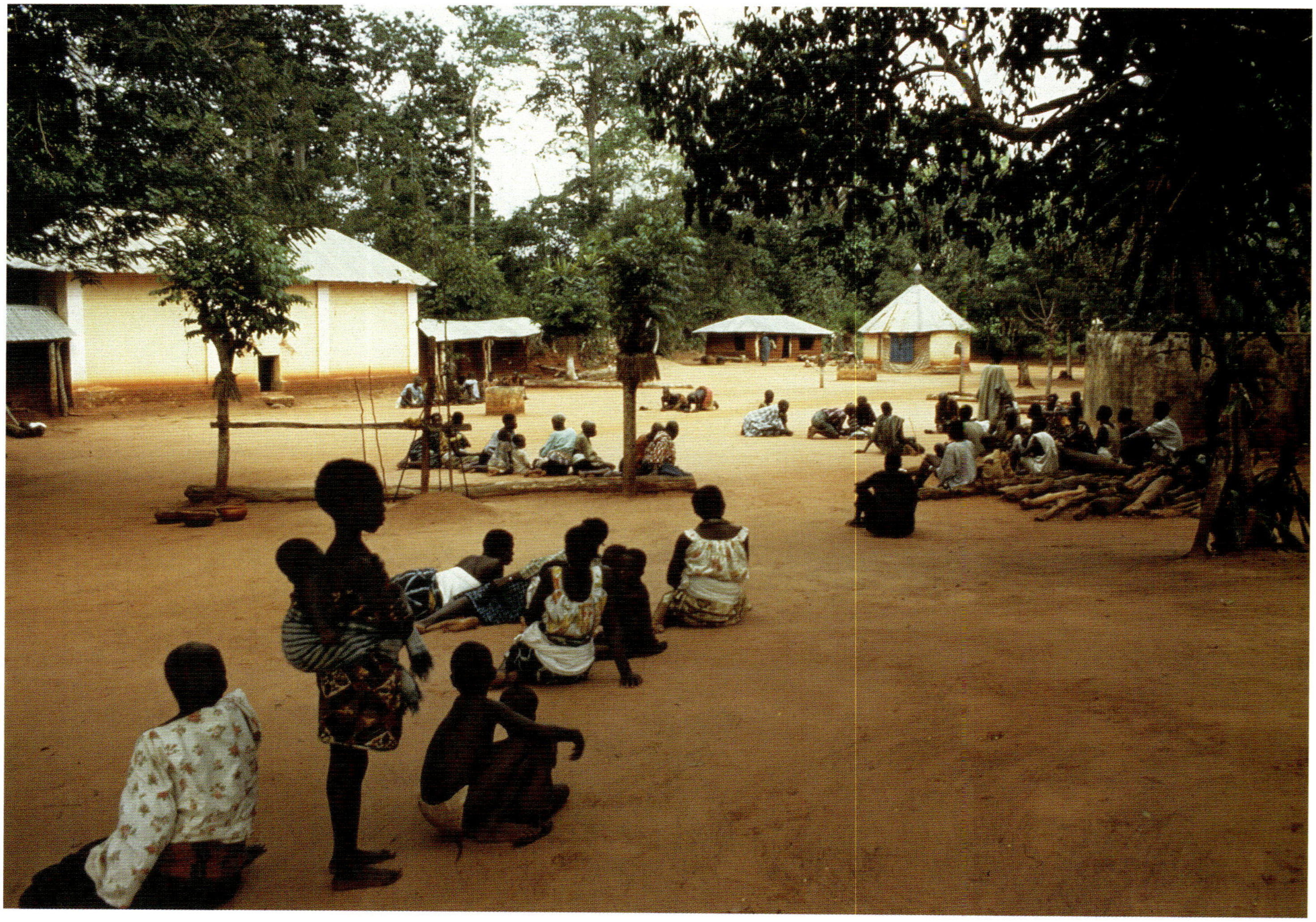

On Ògún's day (the final day of the four-day Yorùbá week), sacred works in iron—Osùn staffs, impressive swords, bells small and large—appear at ceremonies honoring Ògún Oǹdó among Yorùbá in the Republic of Bénin. Near Pobè, in a spacious forest grove (fig. G.1), devotees enact an elaborate series of processions (fig. G.2), dances, offerings, and communal feasts for Ògún Oǹdó; his divine wife and partner, Arè; guardian of the cosmic crossroads, Èṣù-Ẹlẹ́gba; and a founding deity, Dùduà. Among the gathered are priestly leaders, junior priests (*ọkẹrẹ*), the priest and priestesses (*ìyàwórìṣà*) responsible for, Arè's and Dùduà's shrines, "soldiers" of Ògún (*ẹgbẹ́ńlá*) bearing wooden staffs (fig. G.3), and a chorus of women who sing the praise-poems (*oríkì*) of Ògún. Directly in front of Ògún Oǹdó's large and rectangular temple is the *ìdómosùn*, an open altar. (Arè's, Èṣù-Ẹlégba's, and Dùduà's are smaller, circular buildings with conical roofs.) On the *ìdómosùn*, Osùn staffs belonging to deceased priests are placed and fed with palm oil (*èrọ̀*; literally, "softener") to sooth, cool, and quiet Ògún's hot temper.

The ceremony begins with offerings of yams and beans in palm oil. The priests divine with kola nuts to determine if the gods are satisfied. After confirming a positive response, they return to their respective seats and sit motionless. Drums (*aposi*, *ògìdán*, and *kélé*) and bells begin to invoke the gods with sharp, loud, and rapid rhythms that intensify the atmosphere sonorously (fig. G.4). Suddenly the percussion stops. After a few minutes, it starts again—with even more force. On the seventh iteration, with tension and expectation at its highest point, Olúpọ̀nà, the priest of Èṣù-Ẹlẹ́gba, gives a loud shout and plunges into trance as his divinity descends (fig. G.5). He takes his bell and club (*ògọ*), touches the ground three times with both, and leaps into the air to begin his rapid march toward Ògún's temple (fig. G.6). The *okere* and priests of Ògún and Arè enter trance in a similar way, joining Olúpọ̀nà to move together as they salute the four cardinal points of the cosmos.

The entranced priests (*elégùn*) proceed to greet devotees in order of seniority and in a very particular way. Holding a large iron bell (*ààjà*), wooden club (*ògọ*), and sword (*adá*) in the right hand (fig. G.7), they shake these objects three times in salutation to create "the veritable music of blacksmiths that recalls the sound of a hammer striking an anvil" (Verger 1981, 93). They then shake left hands—the sacred side of the body, used only for such spiritual moments of *cosmic connection*—and voice prayers and blessings (fig. G.8).

As the music increases in pace, the priests move in a kind of march-dance, greeting all assembled worshippers before taking their seats again as the drumming ceases. Slowly, with diminishing body tremors, they return to their worldly, human identities as the gods depart. Ògún has come and gone—honored with tastes, songs, dances, and the ringing sounds of iron.

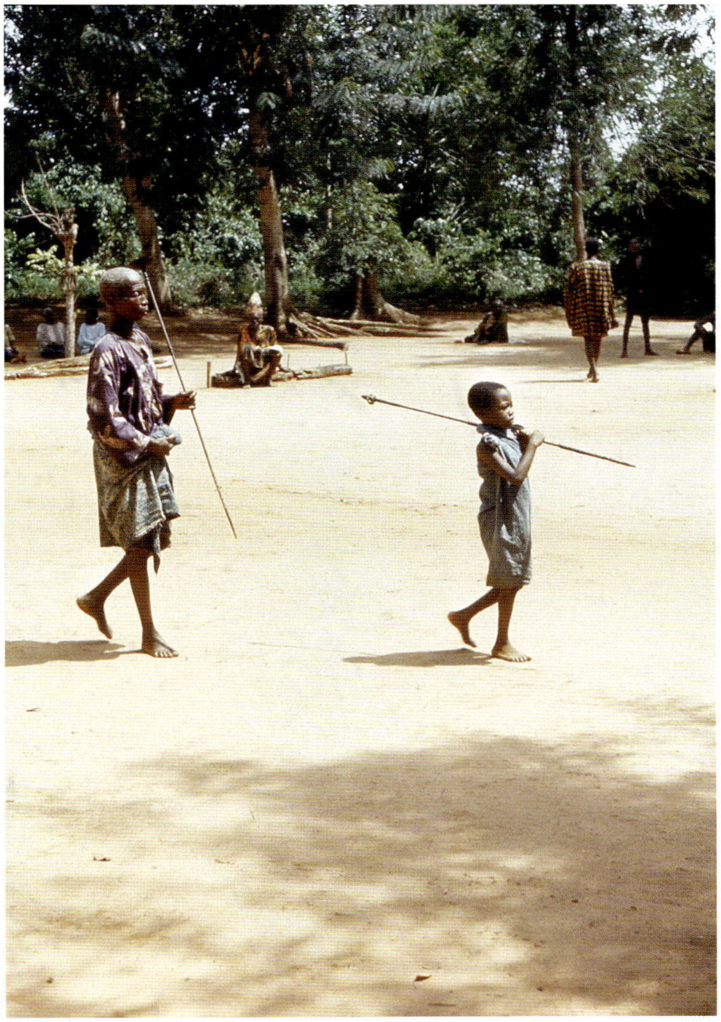

G.1 In a spacious forest grove near Pobè, Republic of Bénin, rites take place for Ògún Oǹdó among other deities on the fourth day of the Yorùbá week.
PHOTOGRAPH BY MARGARET THOMPSON DREWAL, 1973.

G.2 *Ọ̀pà* Osùn staffs are carried to the *ìdómosùn*, the altar for departed Ògún priests.
PHOTOGRAPH BY MARGARET THOMPSON DREWAL, POBÈ, 1973.

G.3 The "soldiers" of Ògún (*ẹgbẹ́nlá*) with their sharpened wooden staffs.
PHOTOGRAPH BY MARGARET THOMPSON DREWAL, POBÈ, 1973.

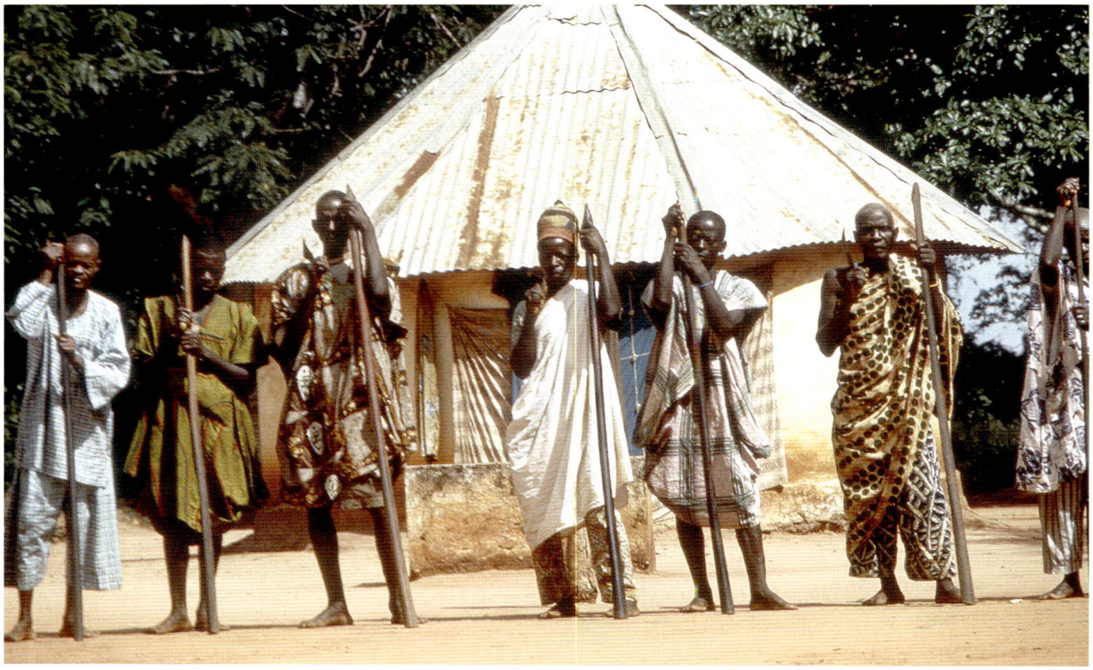

G.4 During the rights, the lead drummer plays a long *ògìdán* drum, its phallic form clearly rendered. Other musicians sound bells of different sizes.
PHOTOGRAPH BY MARGARET THOMPSON DREWAL, POBÈ, 1973.

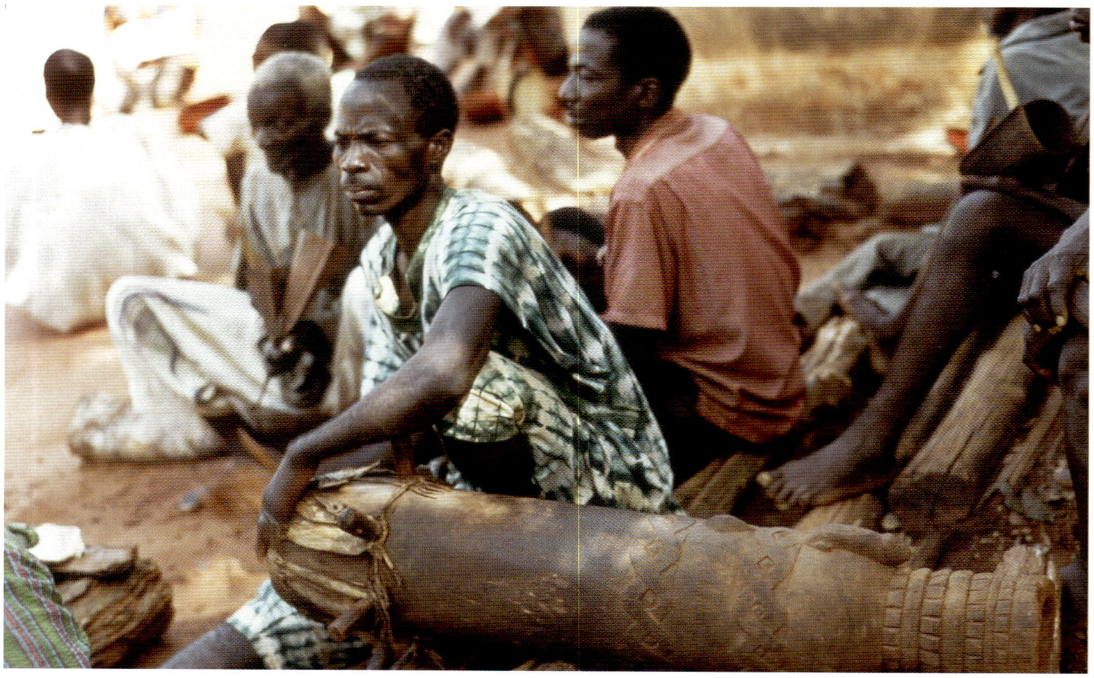

G.5 The French-born photographer Pierre Verger, who documented the Ògún Oǹdó ceremony in the mid-twentieth century, captured the trance shouts of two junior priests (*ọkẹrẹ*).

PHOTOGRAPH BY PIERRE VERGER, POBÈ, DAHOMEY, 1952.

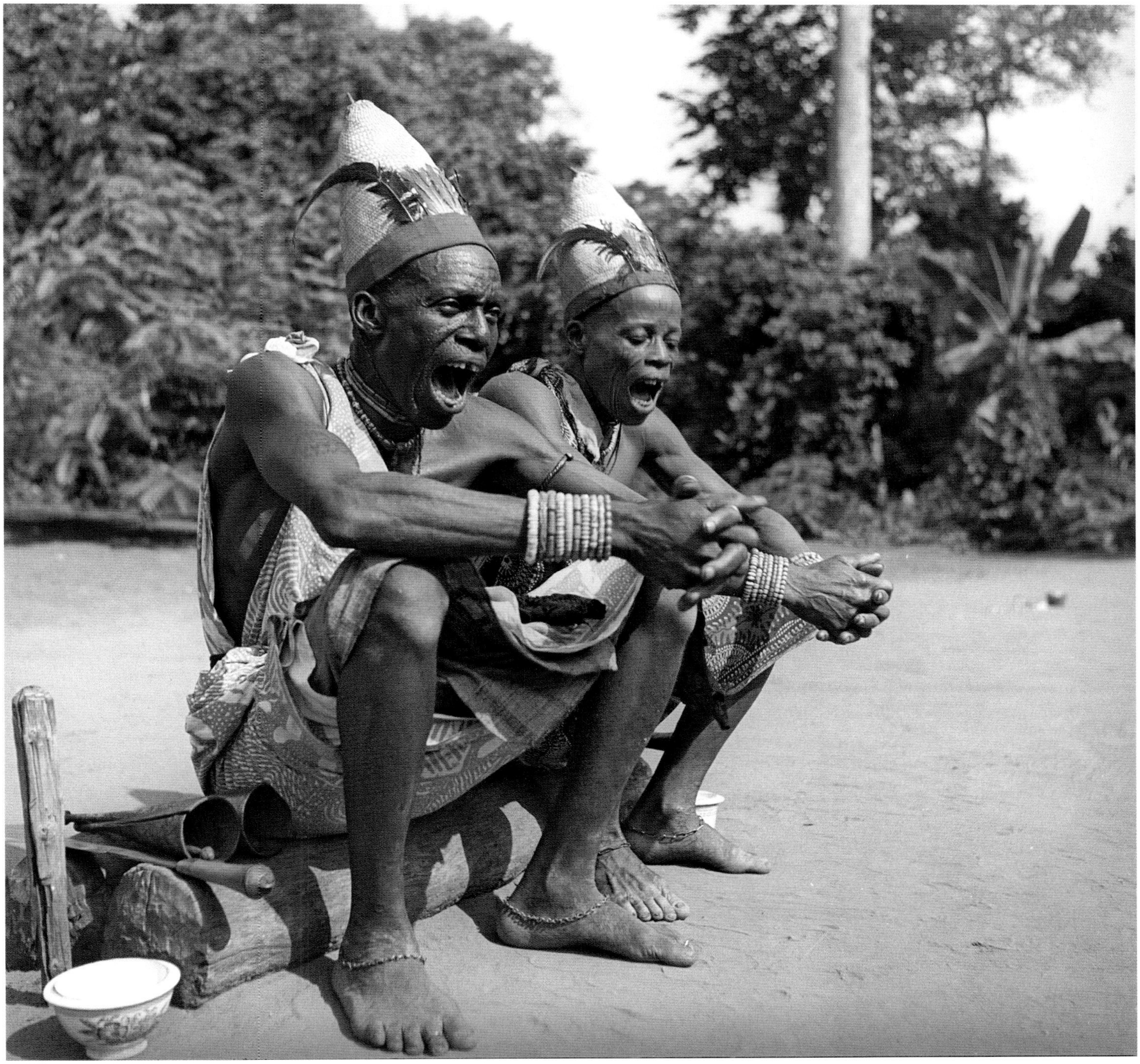

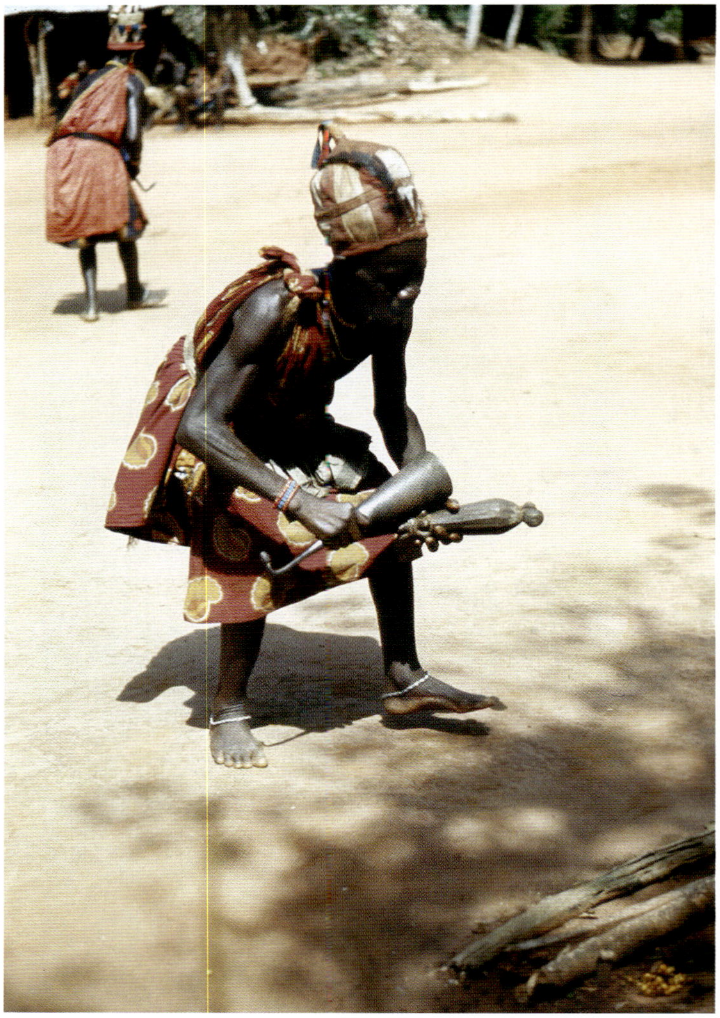

G.6 Entranced, the priest of Ẹṣù-Elẹ́gba (Olúpọ̀nà) holds an iron bell and wooden club as he moves rapidly toward Ògún's temple.
PHOTOGRAPH BY MARGARET THOMPSON DREWAL, POBÈ, 1973.

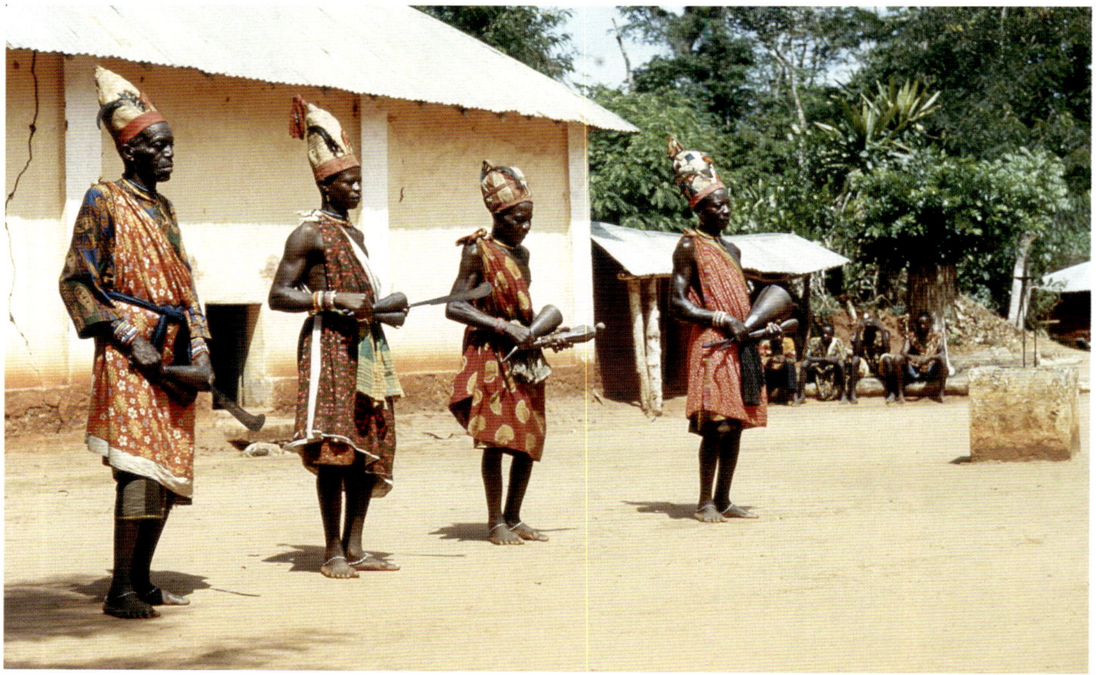

G.7 Entranced priests (*elégùn*) hold iron bells (*ààjà*) that terminate in blacksmith's fire-tending rakes as handles, wooden clubs (*ọ̀gọ*), and swords (*adá*) and embody Ẹṣù-Elẹ́gba, Ògún, Arè, or Dùduà.
PHOTOGRAPH BY MARGARET THOMPSON DREWAL, POBÈ, 1973.

G.8 Olúpọ̀nà greets the most senior priest (*alaàṣẹ*) voicing prayers and blessings as they shake hands using their left—the sacred side of the body—with bells held in the right.
PHOTOGRAPH BY PIERRE VERGER, POBÈ, 1952.

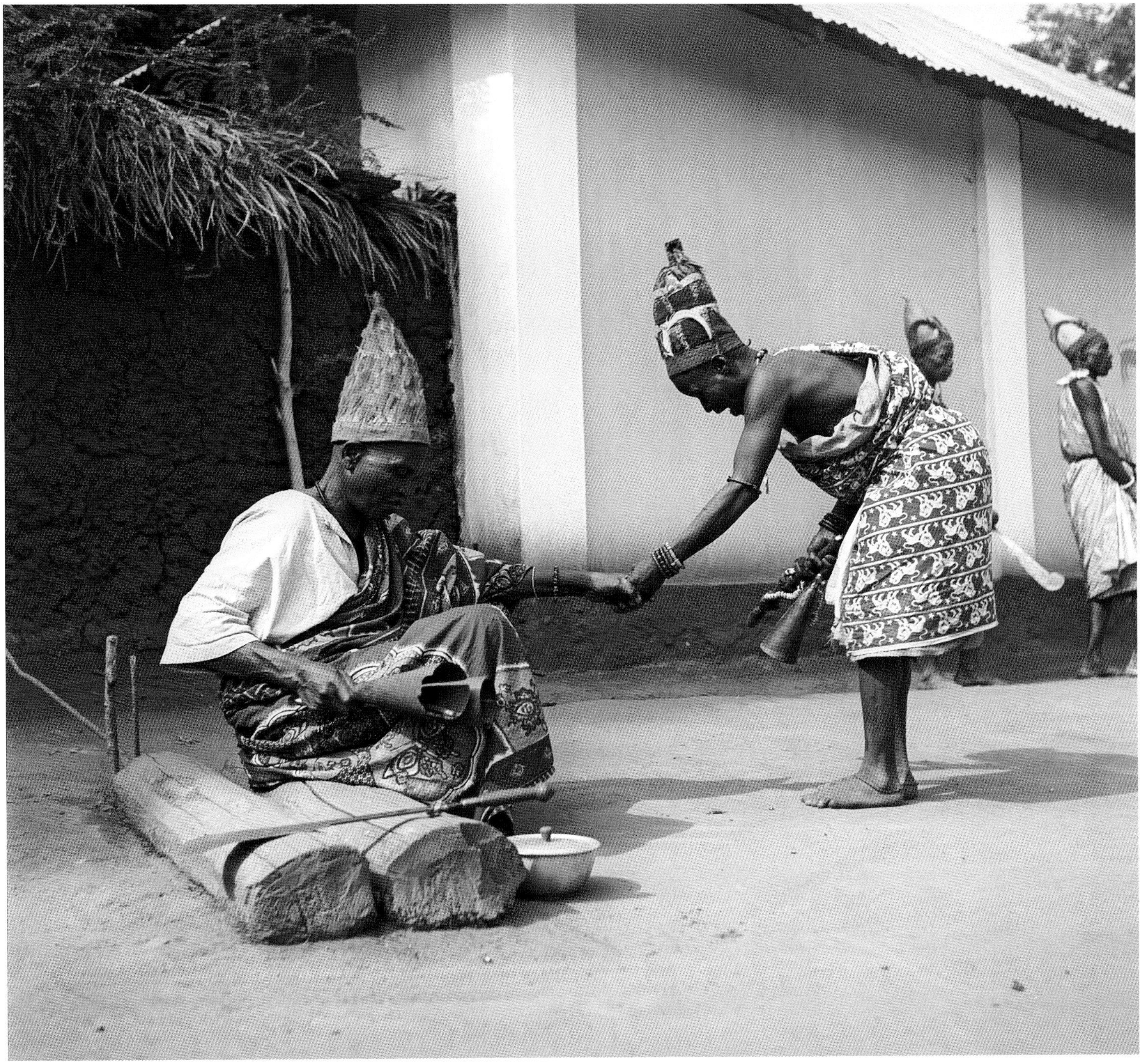

INTERLEAF

H

GU'S JOURNEY TO THE LOUVRE

Colleen E. Kriger

Standing midstride and measuring 165 centimeters in height, this commanding male figural sculpture is known as Gu (also Gou; figs. H.1, H.2), god of iron and war among speakers of Fongbe and Yorùbá languages in the kingdom of Dahomey and neighboring Yorùbá kingdoms (in present-day Republic of Bénin and western Nigeria). It is made of specially selected mild steel and wrought iron elements, all of them identifiable as recycled sheet and bar iron originally made in Europe and exported to Africa's western coast. Remarkable for its large-scale and wide-ranging smithing techniques—forging, cutting, splitting, fire welding, chiseling, bending, repoussé, riveting, punching, piercing, incising, collaring—Gu is justly recognized as an extraordinary work of African art, now in the collection of the Musée du quai Branly-Jacques Chirac in Paris and exhibited at the Louvre. In form, the sculpture is additive, with iron elements assembled around an armature of heavy iron bars, the feet firmly secured to a flat iron base (Tom Joyce pers. comm., 2017). With neatly folded iron loincloth securely held in place, appropriately kept from public view beneath a tunic, Gu exudes the stature, strength, and bravery of a god of war.

The advancing figure appears ready for battle, dressed in a royal war tunic (*kansawu*), its characteristically wide skirt flared with gussets fashioned in carefully hammered sheet iron. Atop Gu's head rests a circular platform supporting an array of iron objects in miniature, turning the sculpture into the embodiment of an ancestral *asen*, that is, a vehicle for honoring the dead. Speakers of Fongbe might have recognized this platform and its placement on Gu's head as a visual pun referring to the specific type of ancestral *asen* known as *asen gbadota* (see figs. 12.18–12.20), meaning "*asen*-like a hat" (Bay 2008, 2–3). Among the small items on the platform are a hoe blade, a spear blade, several types of sword blades, a dagger, fish hooks, a hatchet blade, and a symbol of Dan, the rainbow serpent deity in Vodun religious belief. Traces of blood on some of these items, noted at the time the sculpture arrived in Paris in 1894, suggest that it had received at least some sacrificial offerings in the past (Biton 2000, 110–12). Gu's equipage—the imposing Gubasa sword (see fig. 12.17) in his right hand and iron bell in his left—round out the figure's multiple references to iron and their richly interlocking meanings, both sacred and secular.

Edna Bay's invaluable study (2008) considering the wide variety of metal forms *asen* may take reveals how ancestral *asen*, dedicated to individual human forebears, are conceptually and historically related to *asen* associated with worship of Vodun deities. Religious practice has generated a seemingly endless range in the scale, form, and placement of *asen*, whether worn or held on the body of a devotee or as part of an assemblage of offerings made by many individuals over time on altars dedicated to one or another deity. One vivid example Bay describes is a shrine located in the compound of a lineage of blacksmiths in Abomey, Dahomey's capital, where they made offerings honoring the Vodun Gu. Among the many iron items placed there were automobile parts, a Gubasa sword (sword of Gu), a dozen conical ancestral *asen*, and small replicas of iron tools and weapons including hoe blades, fish hooks, files, swords, knives, spears, and other blades (Bay 2008, 25–30). Viewed in light of this rich cultural and religious context, the Gu statue in Paris can be admired as an unprecedented work of art that nevertheless is firmly rooted in long-standing communal beliefs and values.

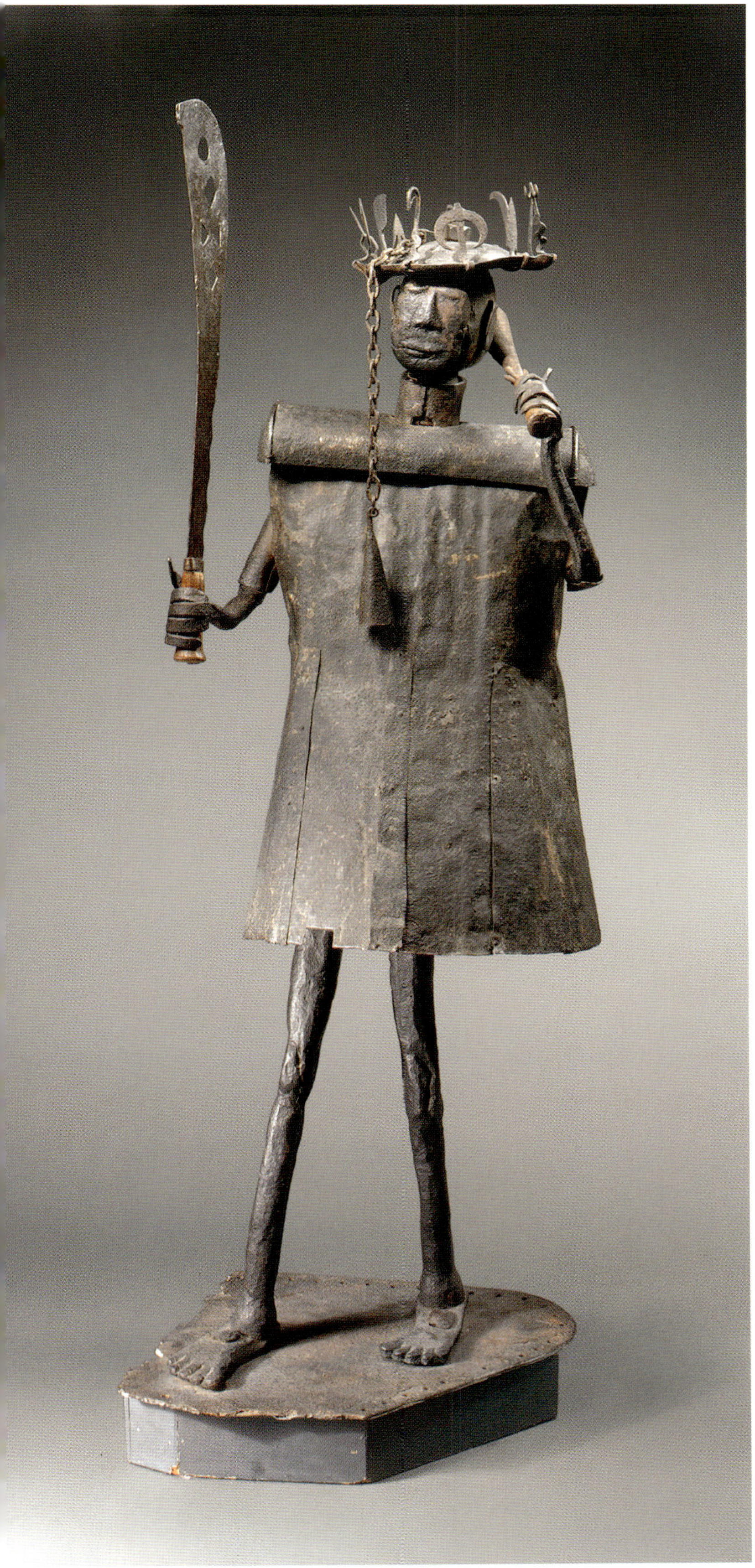

H.1
Ekplekendo Zomabodo Glenegbe Akati (active 19th century; Fon peoples, Republic of Bénin)
Sculpture dedicated to Gu, god of iron and warfare, before 1858
Iron, wood
H: 165 cm
MUSEE DU QUAI BRANLY-JACQUES CHIRAC, 71.1894.32.1 (NOT IN EXHIBITION)

A sculpture embodying the Fon god Gu was commissioned by King Glele of Dahomey to honor his father, King Gezo, whose final funeral took place in 1860, when Glele and his warriors were aggressively asserting his right to succession.

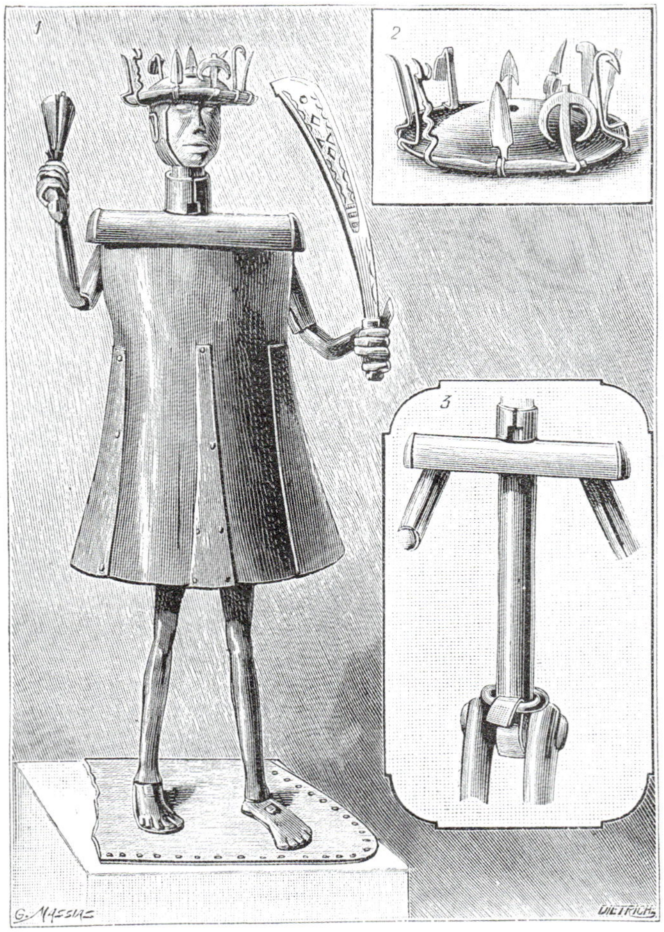

Statue dahoméenne rapportée au Musée du Trocadéro par M. le capitaine Fonssagrives.
1. Vue d'ensemble. — 2. Détail du casque. — 3. Armature intérieure.

H.2 Near to the fall of the Dahomey Kingdom, the Gu sculpture was removed from Africa by a French army captain. It entered the collections of the Musée d'Ethnographie du Trocadéro in Paris in 1894, and sketches of its armature and other details were published in the French journal *La Nature* that same year. (Note that in printing directly from an etched plate the image is reversed.)
IMAGE BY G. MASSIAS AND DIETRICH. PUBLISHED IN MAURICE DELAFOSSE, "UNE STATUE DAHOMÉENNE EN FONTE," *LA NATURE*, 1894.

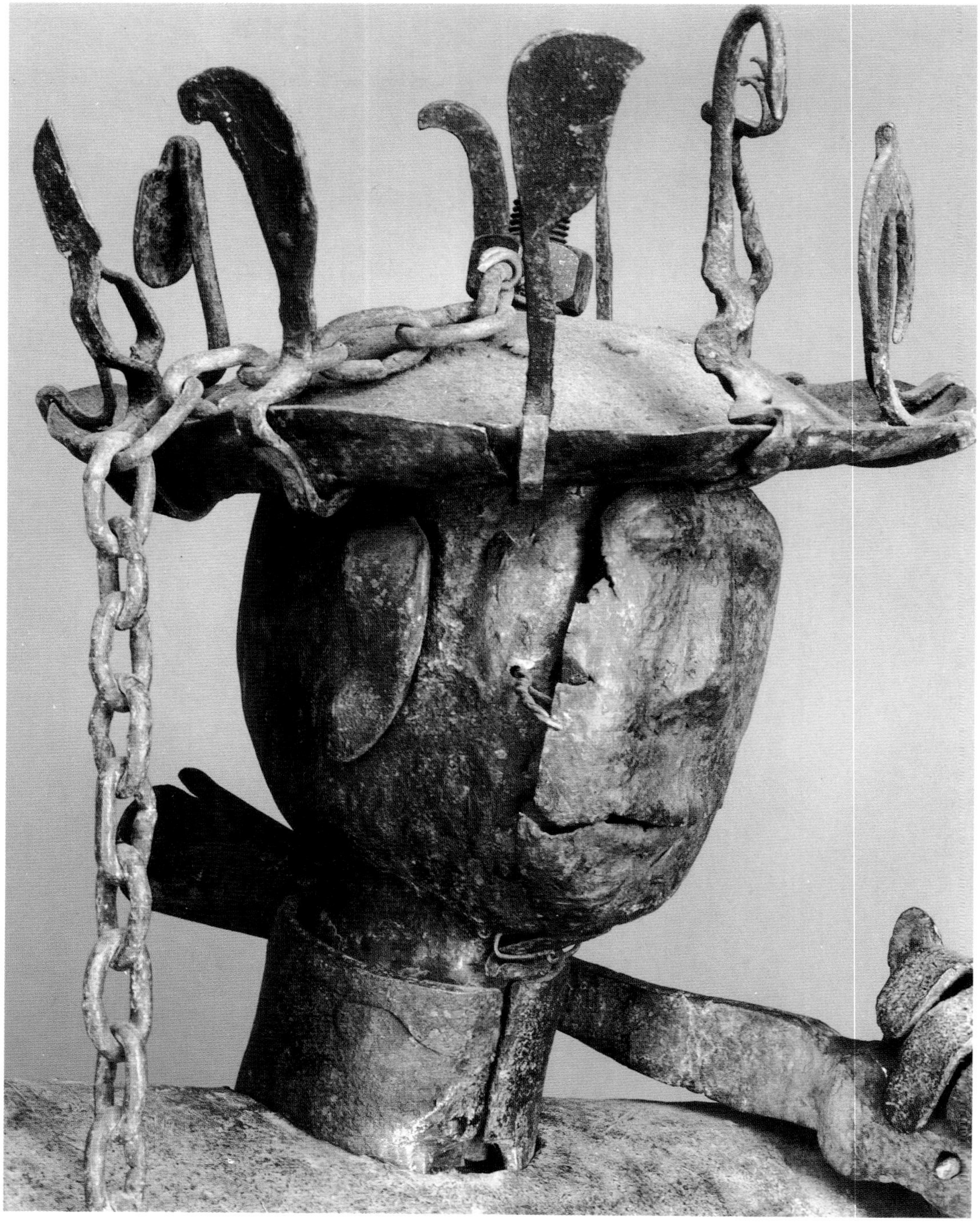

H.3 When the Gu sculpture was exhibited at the Museum of Modern Art in New York in 1935, it was photographed along with other African works by the American photographer Walker Evans.
IMAGE © WALKER EVANS ARCHIVE, THE METROPOLITAN MUSEUM OF ART, COURTESY THE MICHAEL C. ROCKEFELLER MEMORIAL COLLECTION, GIFT OF ROBERT GOLDWATER, 1961 AND 1962 (1978.412.2005)

H.4
A photograph of Gu taken in 1965 and published in 1985 by the Musée de l'Homme in Paris shows Gu with his sword and bell removed. Both have since been returned to the figure.
IMAGE © MUSÉE DU QUAI BRANLY–JACQUES CHIRAC, DIST. RMN-GRAND PALAIS / ART RESOURCE, NY

Times of conflict and turbulence marked the genesis of this iron sculpture and its later travels. It was commissioned by King Glele of Dahomey to be an *asen* for the final funeral of his father, King Gezo, in 1860. The death of Gezo in 1858 ushered in an uncertain interregnum during which Glele had to prove himself capable of succession. Prowess in war being one such proof, Glele led a campaign against the Yorùbá town of Anagodume, just east of Abomey. Oral traditions of this event note that the Dahomean army arrived during Vodun ceremonies, and so among the war captives were chiefs, priests, devotees, and skilled artisans including blacksmiths. Master blacksmith Ekplekendo Zomabodo Glenegbe Akati agreed to create an *asen* that would commemorate Glele's victorious war campaign and honor the achievements of his father (Bay 2008, 27–30). Thusly the figure Gu came to be. It is a multicultural sculpture born of complex histories of migration, territorial expansion, intermarriage, assimilation of captives, and cultural exchange, with additional nineteenth-century upheavals spawned by the end of the Atlantic slave trade and the beginnings of European colonial intrusions.

How and where the Gu figure was captured by the French army and how it came into the hands of a captain, Eugène Fonssagrives, is unclear. During the 1880s, Dahomey was able to maintain its sovereignty despite increasing threat from a newly established French military presence on the coast. But France then launched a determined campaign, from March 1892 to January 1894, bringing an end to Dahomey as an independent kingdom (Manning 1982, 163–65). In Paris, on April 30, 1894, the Gu sculpture was registered into the collections of the Musée d'Ethnographie du Trocadéro as a gift from Capitaine Fonssagrives, and erroneously described as Ebo, the tutelary deity of Dahomey's "port" city of Ouidah (Biton 1994, 25).

Over the many years since that gift, Gu has been seen, photographed, and shown in publications and exhibitions, but not always accurately. Late nineteenth- and early twentieth-century photographs feature him with his equipage intact. However, a 1931 postcard printed by the Musée Trocadéro shows him disarmed, his Gubasa sword missing. It was still missing when Gu traveled across the Atlantic for the landmark 1935 exhibition *African Negro Art* at the Museum of Modern Art in New York (fig. H.3). A photograph taken in 1965 (fig. H.4) and published in 1985 by the Musée de l'Homme in Paris (which inherited the collection of the Musée du Trocadéro) shows Gu with both sword and bell missing (Webb 2000, 70; Vogel and N'Diaye 1985, 80–81). Thankfully, as documented in photographs since 2000, the world is once again able to see and appreciate Gu with his full power and dignity restored.

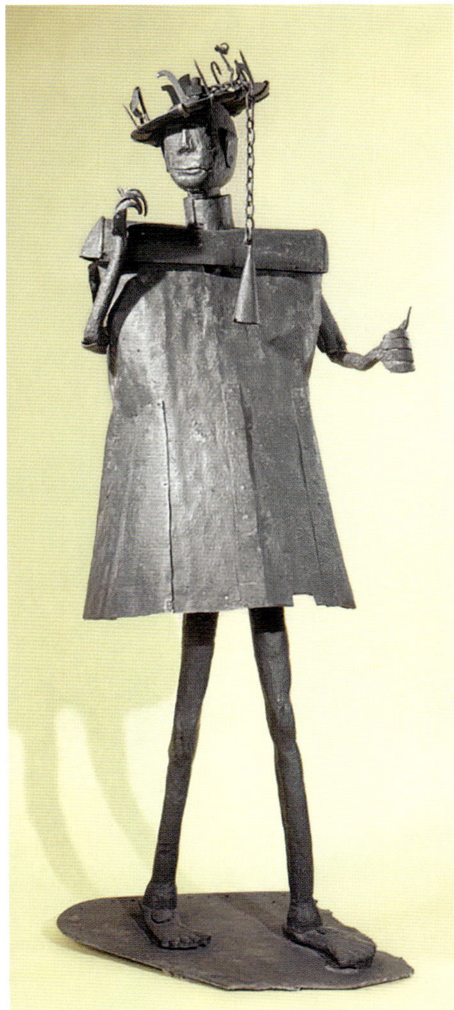

INTERLEAF

I

SYMBOLS OF AUTHORITY: IGBO *ỌFỌ* IN WOOD AND IRON

Philip M. Peek

Ọfọ are the primary representations of Igbo individuals' ritual authority and personal integrity. Mostly found among northern and central Igbo communities in southern Nigeria, these sacred staffs are held while their owners make prayers or swear oaths. While *ọfọ* is the usual term for this object in its various forms, some Igbo groups use *ọvọ*, as do neighboring Isoko groups. The basic wooden type is a thick columnar form topped with a human head surmounted by an iron hoe-blade shape; the face and armless body are decorated by iron bands or wire in various patterns (figs. I.1–I.3). This human-body-and-head form may derive from the copper alloy cast *ọfọ* forms found among other Igbo groups, which in turn replicate the original *ọfọ* twig shape.

Although Europeans have noted *ọfọ* since the mid-nineteenth century (Espenel and Houdart 2001, 17), most outside attention to Igbo shrine images has focused on *ikenga* and other wooden carvings. Nevertheless, virtually all Igbo religious and political activities involve the ritual use of *ọfọ* staffs. All adult free men have their own and each social unit, from the family to the whole lineage, has an *ọfọ* staff as part of its official regalia. Whoever holds the *ọfọ* is that unit's leader. Richard Henderson discusses *ọfọ* in his study of kingship in Onitsha, the king's *ọfọ* being called the "anvil *ọfọ*" (1972, 268). Christopher I. Ejizu (1987) and Eli Bentor (1988) examine *ọfọ* among other Igbo groups.

The tree at the heart of the *ọfọ* complex is found throughout western Africa. Originally recorded in Senegal—as its botanical name, *Detarium senegalense*, indicates—it is utilized in various ways, from its popular fruit and wood to its medicinal and cosmetic properties. How it came to be a sacred tree among Igbo and Isoko is unknown, although its red sap and unique manner of dropping its branches may have caused such attention. The tree's remarkable twigs are four to six inches in length and shaped like the knotty joints of human bones; with a slight depression at one end, the twig is reminiscent of a human forearm and cupped hand. These twigs can be collected only after they have fallen to the ground. They may be employed individually but are usually gathered in bunches of seven or nine, bound (often with wire), and placed on altars (fig. I.4). Such configurations, usually covered with offerings, are the most common *ọfọ* among all Igbo and Isoko and are even found represented on Lower Niger bronze crown bell heads (fig. I.5).

I.1
Artist unknown (Igbo peoples, Nigeria)
Staff (*ọfọ*), early 20th century
Iron, wood
H: 39.8 cm
COLLECTION OF MARK CLAYTON
(NOT IN EXHIBITION)

Adorned with a variety of fanciful patterns of iron-strip decorations, *ọfọ* may display prominent facial features atop a columnar armless body. Virtually all Igbo religious and political activities involve ritual use of *ọfọ*, and men enshrine them to draw on the collective power of ancestors. *Ọfọ* are made of iron and the wood of a tree (*Detarium senegalense*) considered sacred by Igbo peoples.

I.2
Artist unknown (Igbo peoples, Nigeria)
Staff (*ọfọ*), late 19th to 20th century
Iron, wood
H: 50 cm
COLLECTION OF CLAUDIE AND ALAIN LEBAS

I.3
Artist unknown (Igbo peoples, Nigeria)
Staff (*ọfọ*), early 20th century
Iron, wood
H: 52.7 cm
FOWLER MUSEUM AT UCLA; GIFT OF THE WELLCOME TRUST
PROVENANCE: COLLECTED BY M.D.W. JEFFREYS, AGULERI, NIGERIA, 1930
(NOT IN EXHIBITION)

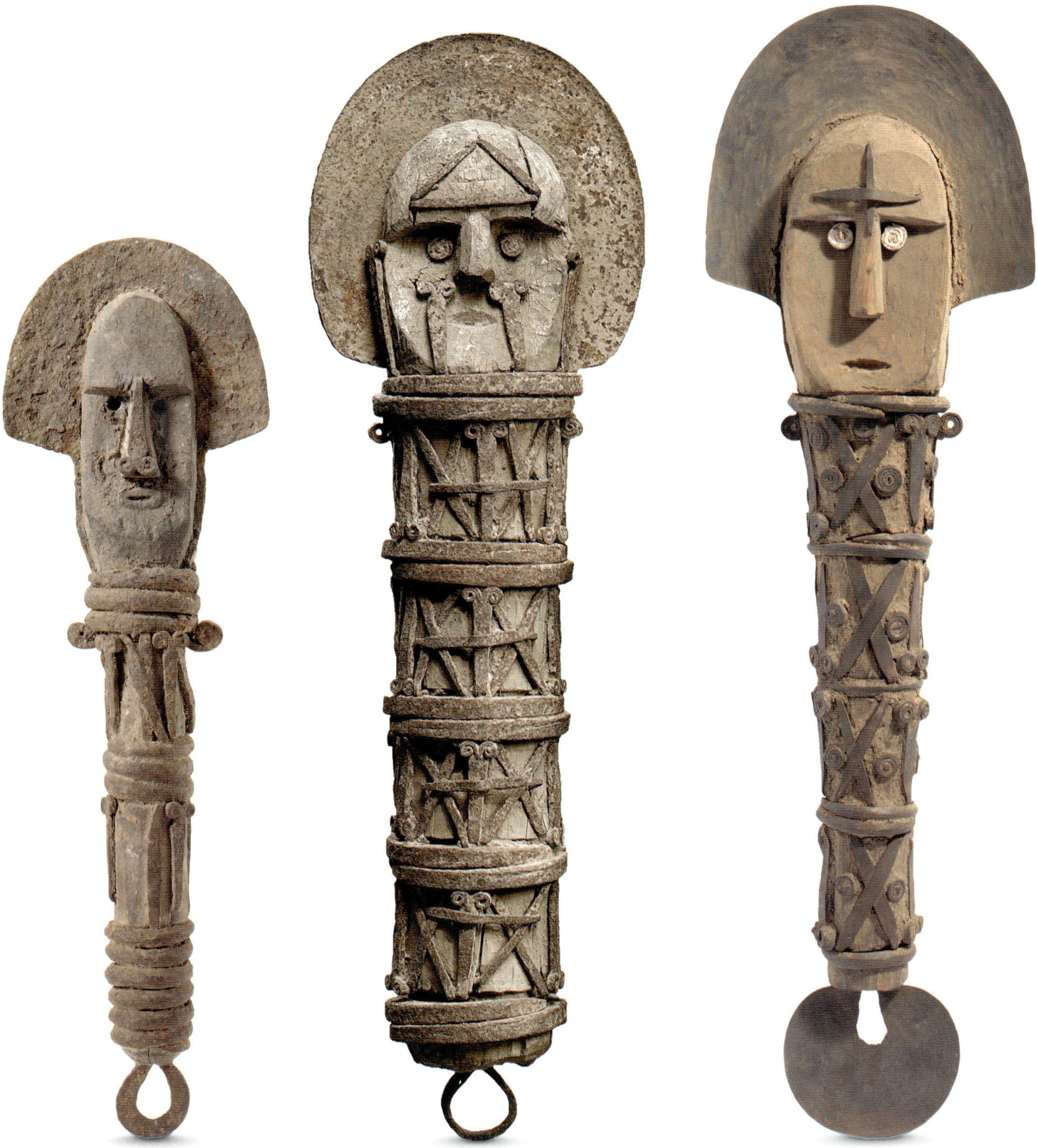

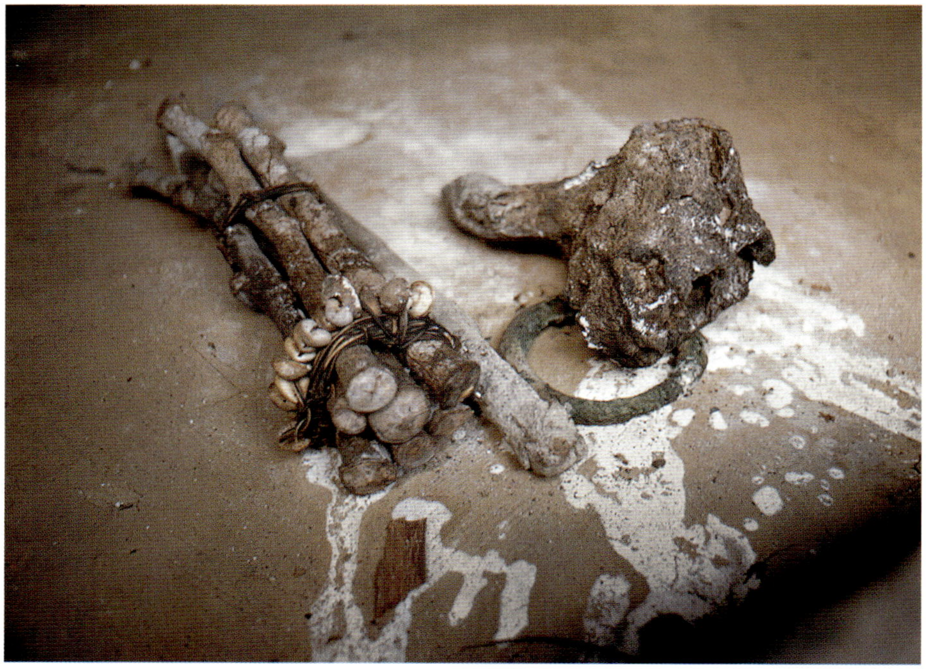

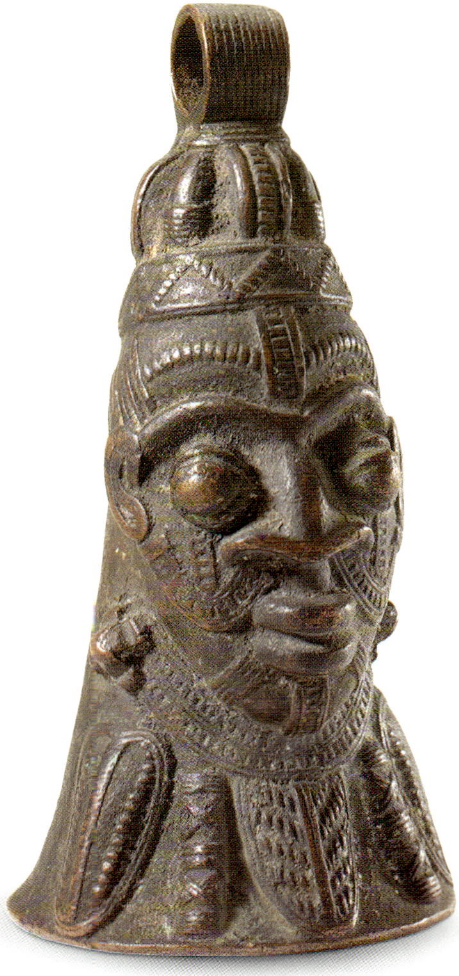

This tree is the namesake of the wood-and-iron forms, but *ọfọ* staffs can take a number of different shapes. Communities in which bunches of *ọfọ* twigs are wrapped in metal wire might be where "iron-bound" *ọfọ* originated (see Bentor 1988, 66–67). In the Nsukka Igbo area, horse tails were attached to staffs' ends. Some staffs are basically lumps of wood decorated with wire configurations and red ribbons. Casters created the copper alloy forms while blacksmiths produced the elaborate iron decorations as well as the hoe blades that form dramatic aureolas around the heads of these sacred objects.

G. T. Basden, writing in the 1930s, provided some background on these important objects:

> The manufacture of hoes is the main business of the majority of the smiths, followed next in quantity by axe-heads. There are examples of more artistic merit such as the staves which form part of the insignia of titled men. For the minor degrees they are little more than plain shafts, spear pointed. The higher degrees entitle the holders to use staves of more elaborate pattern either of iron or brass, or in combination. Some of the twisted iron effects on these are neatly executed.[1]

Igbo blacksmiths, creating essential items from the earth, are highly regarded. But beyond its critical utilitarian role in farmer's hoes and other common tools and weapons, iron does not have the rich symbolic associations it holds among Yorùbá and Edo, who share the important deity Ògún (see Aremu 2004, 160–64). Nevertheless, it is clear that iron is a fundamental element in the unique Igbo tradition of the Mbari houses, where iron bars play key symbolic as well as structural roles. Iron bars are placed within the walls and iron rods provide armatures for the clay figures decorating these elaborate buildings (Cole 1982).

This practice of embedded iron is found in another Igbo context. Collector Mark Clayton discovered that running down the interior length of many copper alloy *ọfọ* is a thin iron wire. While these disparate examples of embedded iron are not replicated by the wrought ironwork decorating the exterior of the wood-and-iron staff *ọfọ* forms, they do underscore the importance of iron in Igbo culture. The practice of attaching bronze figures to iron rods is well-known from the Yorùbá Ògbóni/Ọ̀ṣùgbo Society's *edan* (see figs. 12.14, 12.15) as well as that judicial body's earthen sculptures, and has been more recently recorded in southern Nigeria with Lower Niger "Aro Knot" finials (Peek forthcoming).

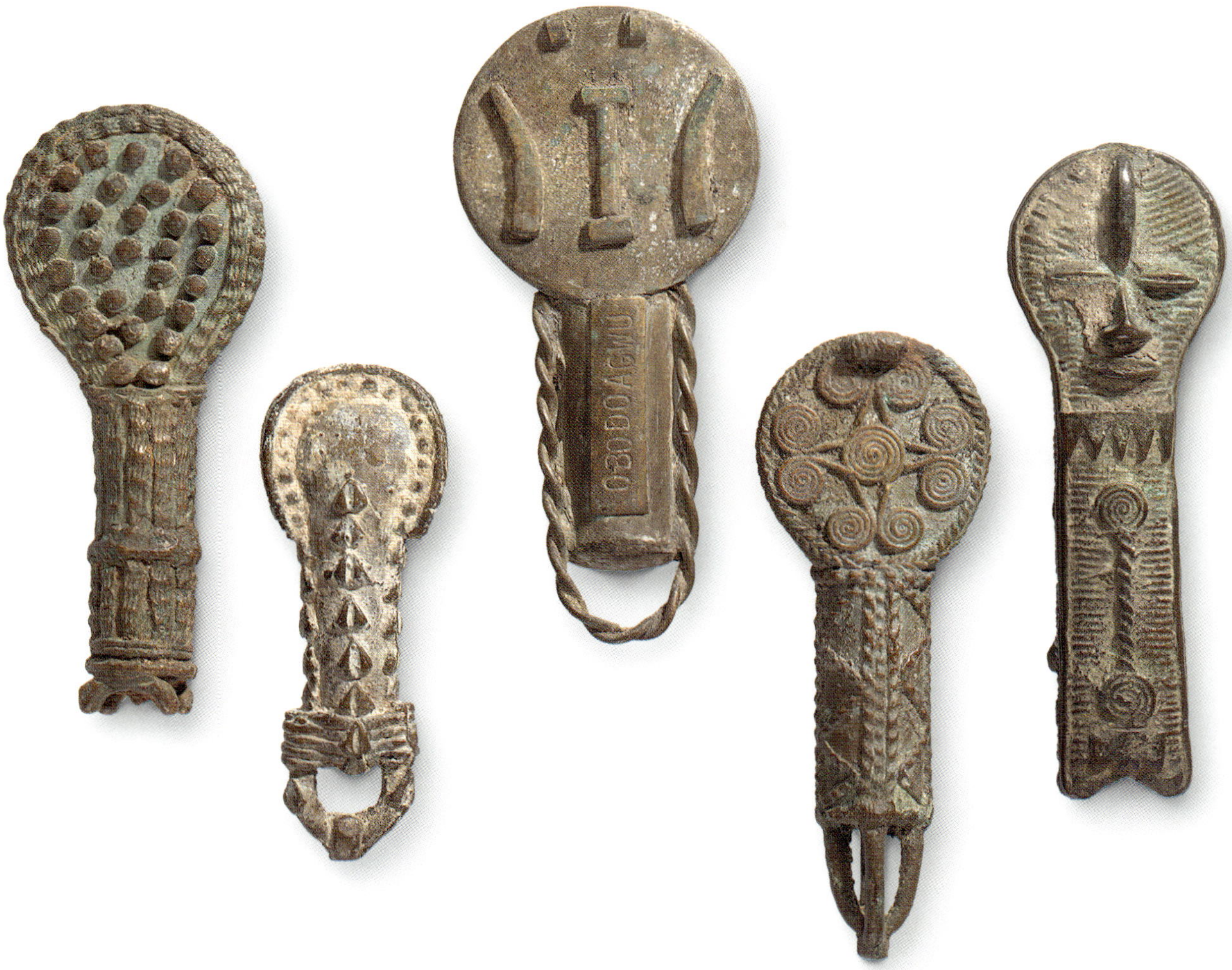

1.4 A bundle of *ọvọ* (as Isoko term them) twigs next to an animal skull fragment and a small bronze bracelet—a typical grouping for an Isoko shrine.
PHOTOGRAPH BY PHILIP M. PEEK, UMẸ, ISOKO, NIGERIA, 1971.

1.5
Maker unknown (England)
Bell, early to mid-20th century
Copper alloy
H: 16 cm
COLLECTION OF MARK CLAYTON
(NOT IN EXHIBITION)

Several *ọvọ* bundles depicted on the base of this bell head's attest to its use in ancestral veneration. This type of bell head was so popular in southern Nigeria in the 19th century that traders had copies mass-produced in Birmingham, England. Such mold-cast copies have telltale side seams and production numbers on their bases.

1.6
Artists unknown (Igbo peoples, Nigeria)
Shrine objects for offerings (*ọfọ*), early to mid-20th century
Copper alloy, iron
H (from left to right): 17 cm; 14.5 cm; 18 cm; 18.5 cm
COLLECTION OF MARK CLAYTON
(NOT IN EXHIBITION)

The basic tubular form of these *ọfọ* emulates the shape of the *ọfọ* tree twig, and each has a concave end for the placement of chalk offerings (the top end as oriented in this image). Their embellishment, however, is greatly varied.

In a widespread *ọfọ*/*ọvọ* type found among Riverain Igbo, Isoko, and Ijo (here it is termed *ọvuọ*) in the Niger Delta (fig. 1.6), native chalk is placed in the depression as part of attendant rituals. More elaborate *ọfọ*/*ọvọ* forms have slightly raised designs and even facial features depicted in the concave end. Another common feature, perhaps marking a regional sub-style, is the three or four "legs" placed underneath to support the image.

The *ọfọ*/*ọvọ* complex is at once an ancient tradition and a contemporary marker of Igbo cultural identity. Today, there are online title disputes concerning the correct holder of a town's *ọfọ* as well as lively Igbo-centric debates about the nuances of *ọfọ*. A recently founded Nigerian studies journal is called *ỌFỌ: Journal of Transatlantic Studies*, reminding all of the value of ancestral authority. The history of objects seldom provides such striking evidence of the commonality of ethnic traditions over such large distances of time and space. *Ọfọ* are not simply wooden staves wrapped in decorative iron bands; they are evidence of still vibrant and essential cultural practices across western Africa.

CHAPTER
13

WORN AND HELD: FORGING LIFE'S PASSAGES IN NORTHERN CAMEROON AND NORTHEASTERN NIGERIA

Scott MacEachern   Marla C. Berns

This two-part chapter considers ironworking histories in a region south of Lake Chad that includes the Mandara Mountains of northern Cameroon and northeastern Nigeria and the Gongola River Valley, a part of the larger Benue River Valley. In part I, archaeologist Scott MacEachern assesses findings in the region since the late twentieth century, which have informed his understandings of the use of iron in social transitions around the Mandara Mountains. Art historian Marla C. Berns, in part II, recounts her observation of ritual practices and sacred objects among related peoples in the Gongola Valley, where virtually no archaeological studies have yet been undertaken. The authors' two approaches provide a comprehensive view of the transformational role and efficacy of iron among neighboring peoples in west-central Africa.

PART I

Scott MacEachern

ORIGINS OF THE FORGE: DEEP HISTORIES OF IRON SOUTH OF LAKE CHAD

The region of the southern Lake Chad Basin (fig. 13.1) is one of the natural crossroads of Africa where east-west communications between the Atlantic and the Nile are constricted into two corridors north and south of the Mandara Mountains, through the plains south of Lake Chad and the Benue River Valley respectively. Around Lake Chad itself, trans-Saharan trade routes emerge into the Sudanic zone south of the desert. The region is complicated culturally, with a multitude of different ethnic groups speaking languages from three of the four great African language families, and with precolonial political structures ranging from centralized and predatory plains states to relatively egalitarian farming communities. In this cultural mosaic, the production, exchange, and use of iron has been significant to many domains; indeed, the metal is not simply a technological product, but also functions in social relations, ideology, and politics.

Ironworking's long histories in the lands south of Lake Chad exemplify the challenges of understanding how historically known systems of use have developed over time. Archaeologists do not know whether ironworking was introduced to the Lake Chad Basin, nor by what route it may have arrived, although recent work (Humphris 2017) around Meroë (in modern Sudan) suggests that initial knowledge of the material may have been diffused westward across grasslands stretching from the Nile Valley to Lake Chad. This hypothesis remains untested, as little archaeology has been undertaken in the intervening Kordofan, Darfur, and Wadai regions.

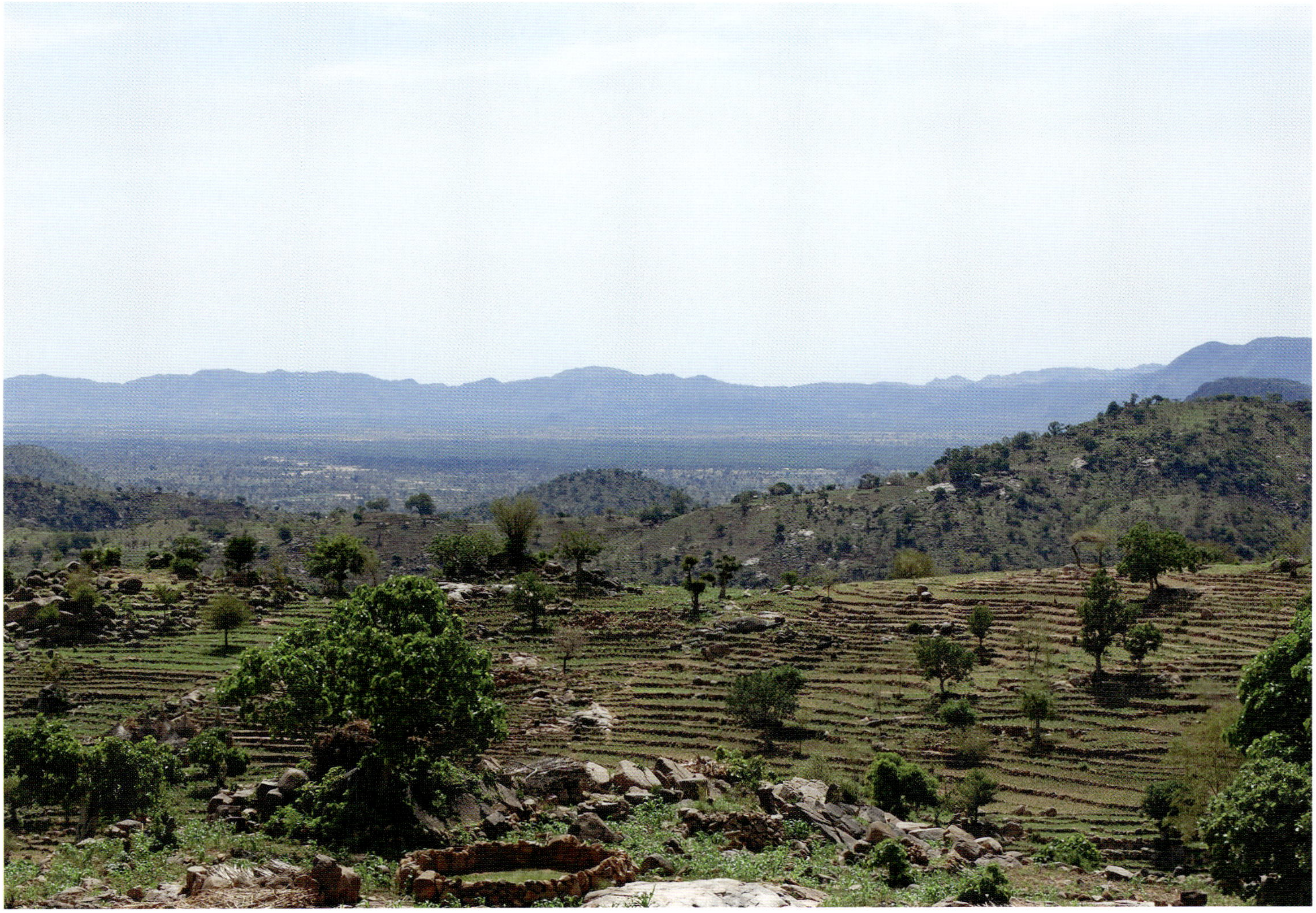

13.1 Dispersed settlements and agricultural terraces at Kuva, Cameroon, in the Mandara Mountains.
PHOTOGRAPH BY SCOTT MACEACHERN, CAMEROON, 2008.

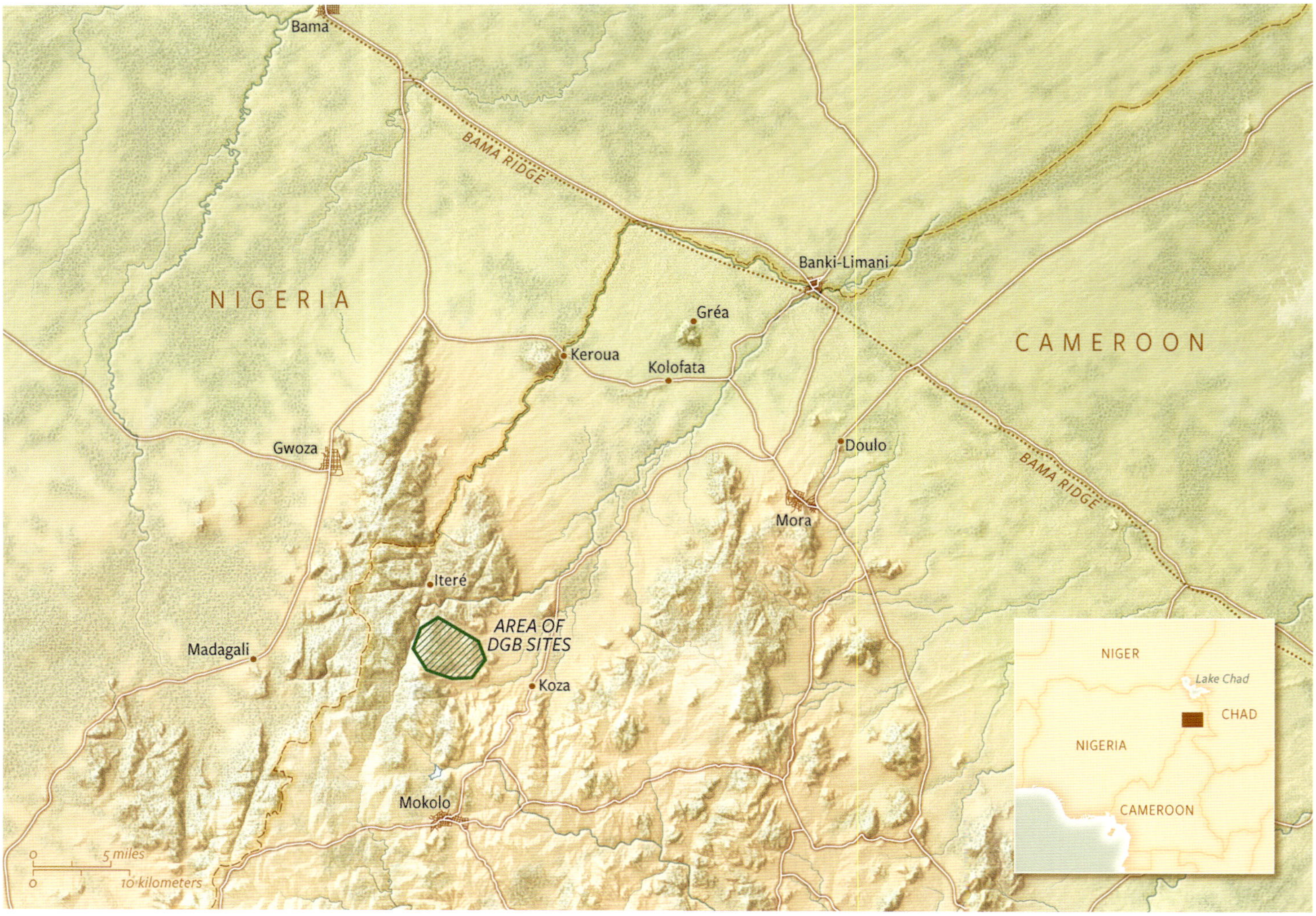

13.2 Map of the area around the northern Mandara Mountains, where traces of iron from the first millennium BCE have been recorded at archaeological sites.

By whatever route, iron seems to have spread rapidly throughout central Africa through the middle of the first millennium BCE. The frustrating uncertainty of the radiocarbon calibration curve during this period makes delineation of the exact chronology of iron's introduction difficult (see chapters 8–10, this volume). South of Lake Chad, the earliest evidence for ironworking comes from sites in Nigeria and Cameroon close to the Mandara Mountains (fig. 13.2). The deepest levels at these sites contain iron and slag from the first half of the first millennium BCE (MacEachern 2012)—dating similar to the earliest for ironworking around Nok in Nigeria (Junius 2016).[1] Such

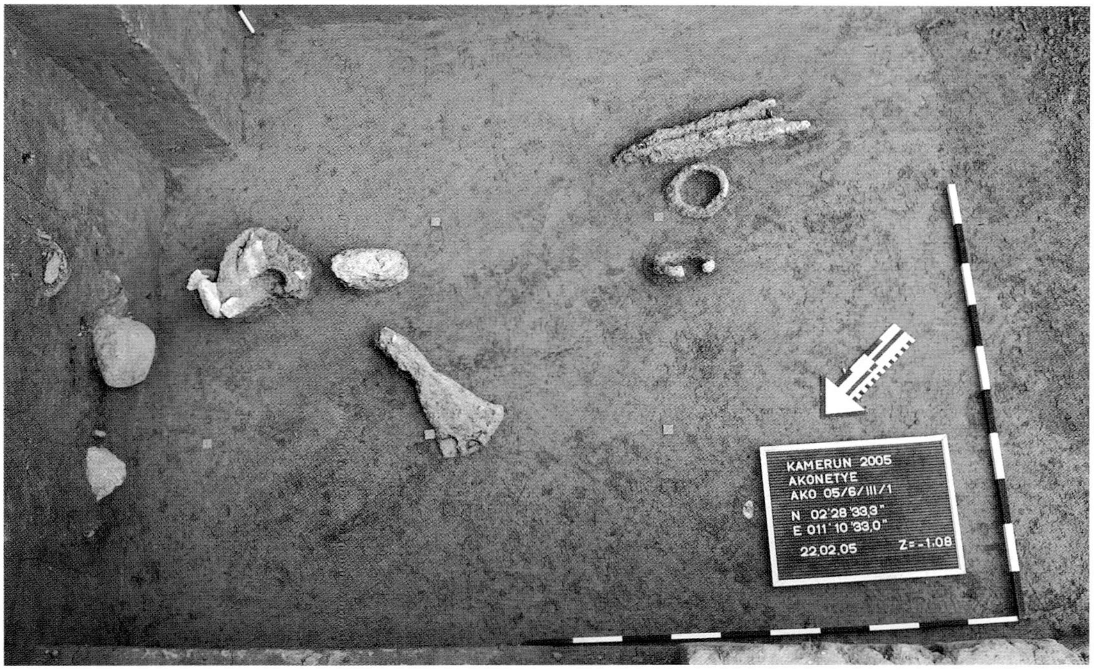

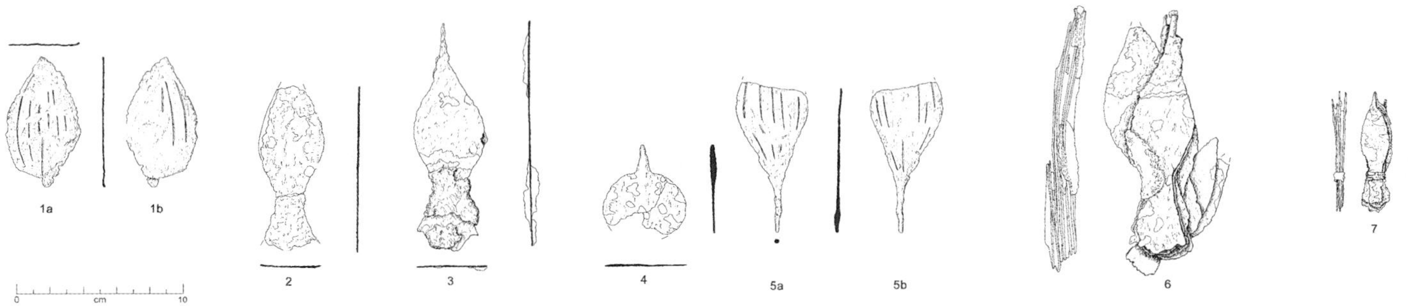

13.3A–B Bracelets, an axe, and other iron objects recovered from a grave at the site of Akonétye, Cameroon, dating to the first century CE.

IMAGES REPRODUCED IN C. MEISTER AND M. K. H. EGGERT, "ON THE EARLY IRON AGE IN SOUTHERN CAMEROON: THE SITES OF AKONÉTYE," *JOURNAL OF AFRICAN ARCHAEOLOGY* 6 (2), 2008.

concurrence between sites seven hundred kilometers apart is particularly notable given that iron-bearing sites just a hundred kilometers south of the Mandara sites are more than five hundred years more recent (Langlois 1995); there, ironworking seems to have coexisted with the use of stone tools during the early first millennium CE.

Striking variation in adoption of ironworking in adjoining regions that do not seem to have been isolated from one another otherwise should make us cautious of any assumption that iron spread simply because of its technological and economic advantages. Indeed, we cannot be entirely sure about the domains of social and cultural life that first experienced iron's revolutionary impact. Archaeologists have often presumed that iron was first adopted for utilitarian purposes, but without proof. Indeed, such a theory somewhat contradicts much human behavior: materials that are new, rare, and exotic are often incorporated into cultural systems through social status and display (Helms 1988). In other words, iron may well have originally spread for nonutilitarian reasons. Rich early Iron Age burials in southern Cameroon and surrounding regions have yielded iron (and other) objects probably associated with wealth and status, not workaday tools (figs. 13.3, 13.4; Meister 2010; González-Ruibal 2013).[1]

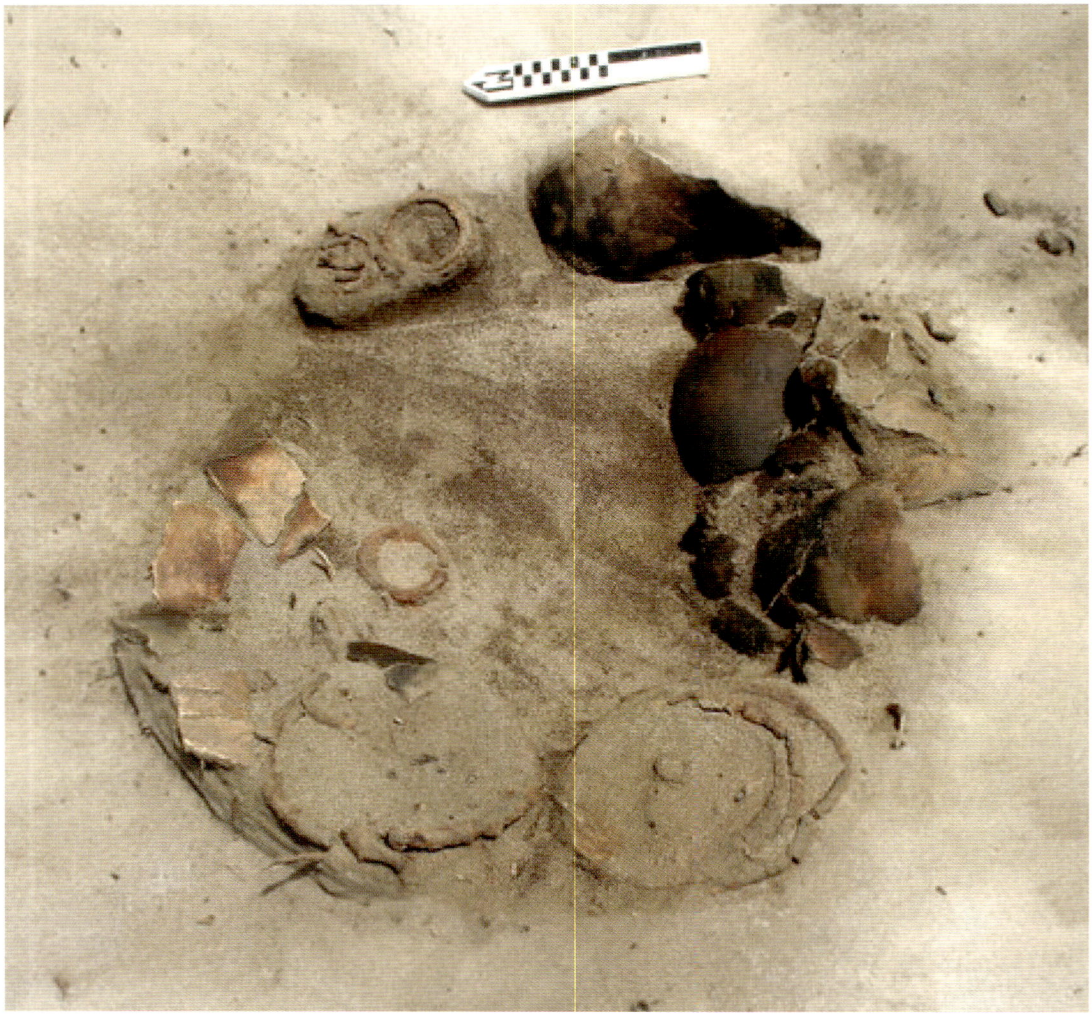

13.4a–b Above: Iron axe, bracelet, and ceramics as grave goods at the site of Nandá, Corisco, Equatorial Guinea, dating to the early first millennium ce. Below: Renderings of objects at the same site.
images reproduced in gonzález-ruibal, a., m. sánchez-elipe, and c. otero-vilariño, "an ancient and common tradition: funerary rituals and society in equatorial guinea (first–twelfth centuries ad)," *african archaeological review* 30 (2), 2013.

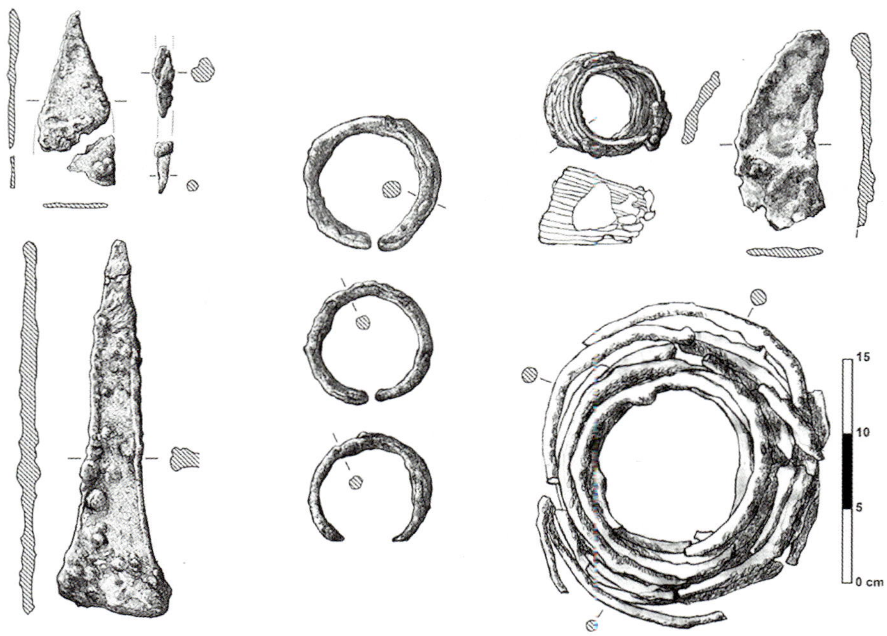

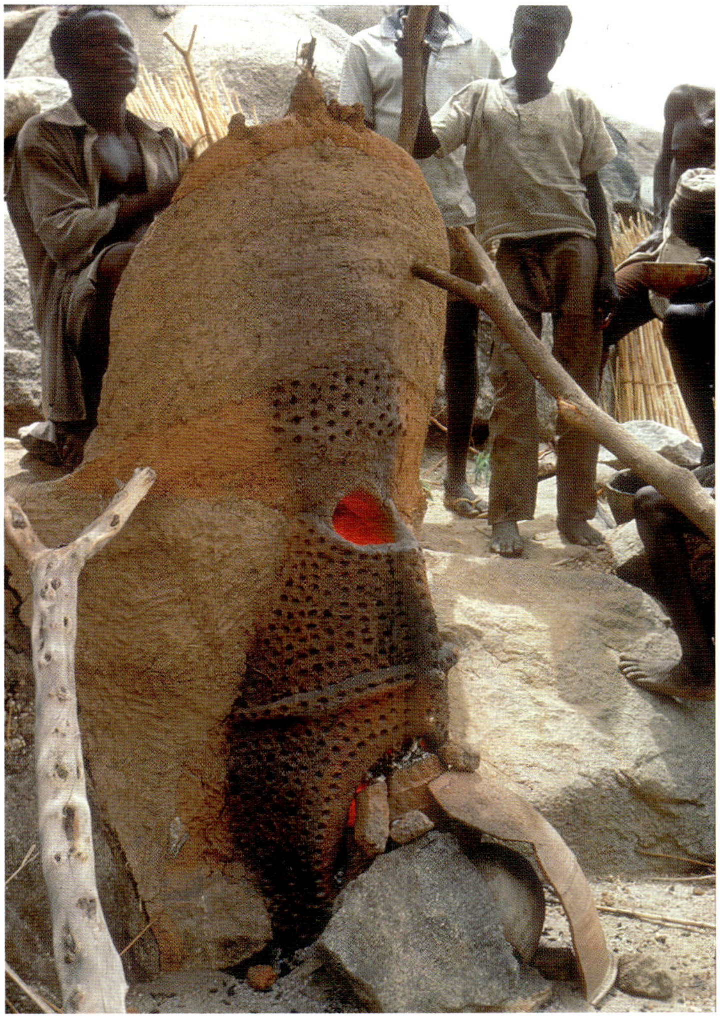

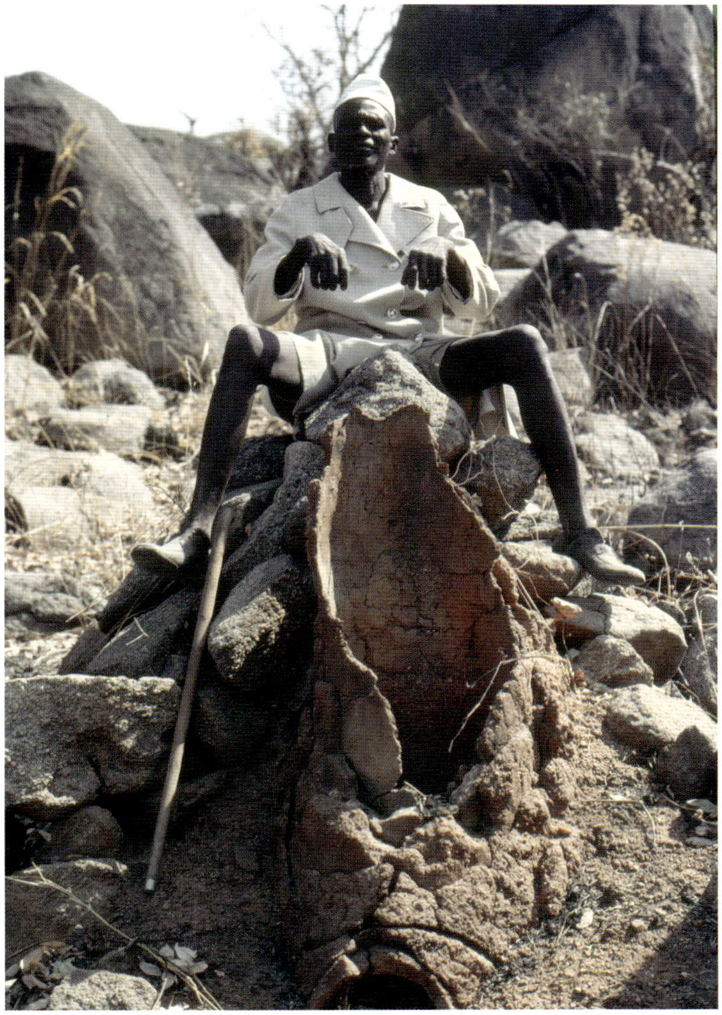

Archaeologists observe that from the early first millennium CE onward, iron fragments and especially slag become increasingly common on sites around the Mandara Mountains and Lake Chad Basin. That no ancient furnaces have been found on archaeological sites in the region probably indicates that iron production took place away from dwelling areas, as was the case more recently, and indeed more widely, in sub-Saharan Africa. Furnace remains located in terrace walls around the Mandara Mountains (see fig. 8.2) are probably quite recent—perhaps between one and two hundred years old—but they cannot be easily dated given how readily such sites are disturbed by farming. Nonetheless, Mandara smelting furnaces seem generally similar to those of the Gongola River Valley, with a bellows worker sitting on a platform above the shaft, pumping air down through a tuyère to increase interior temperatures (figs. 13.5–13.7). This furnace type is found nowhere else in the world; its origin is a mystery. In the Mandara Mountains, it has at least three distinct subtypes differing somewhat in size and configuration and in the length and frequency of smelts produced.

13.5 Plata Kapa smelting furnace, reconstructed for use in 1989 (see also fig. 8.3).
PHOTOGRAPH BY SCOTT MACEACHERN, MAYO PLATA, CAMEROON.

13.6 In the Biki Hills—just east of the Gongola River and north of its confluence with the Benue, populated today by Bəna peoples—are the remains of a mud shaft furnace (*jibo*) at a former smelting site. An elder demonstrates how a bellows worker would have sat atop the mound of rocks supporting the double-walled furnace.
PHOTOGRAPH BY MARLA C. BERNS, NIGERIA, 1981.

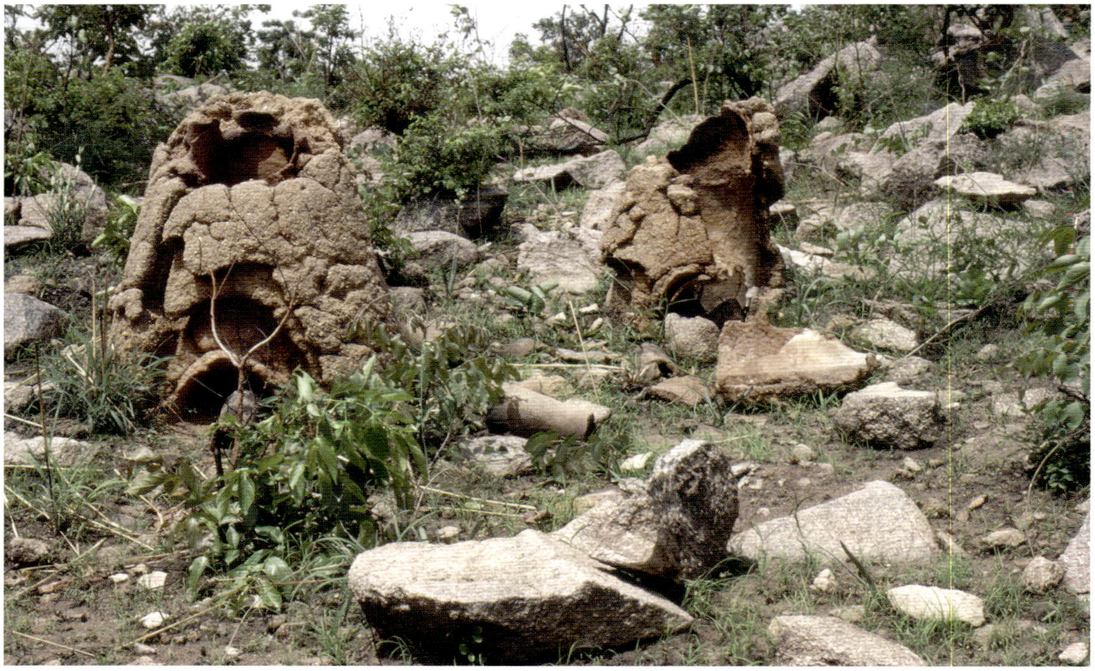

13.7 Two partially intact Yungur mud shaft furnaces (*zibo*) located at an upland site called Kaira, near the westernmost extent of Yungur occupation. Visible at the base of the furnace to the left (approximately 135 centimeters high, with double walls about 17 centimeters thick) is a ceramic tuyère through which air was originally forced into the top of the furnace.

PHOTOGRAPH BY MARLA C. BERNS, NIGERIA, 1981.

13.8A–D

Clockwise from upper left: Iron artifacts found in the Mandara Mountains in a cache on the DGB-1 site dating to roughly the 16th century CE; outline of iron artifacts recovered from DGB-1 cache; socketed and tanged iron artifacts recovered from DGB-1 cache after conservation.

UPPER LEFT: PHOTOGRAPH BY SCOTT MACEACHERN, KUVA, CAMEROON, 2008; UPPER RIGHT AND BOTTOM: RENDERINGS AND PHOTOGRAPHS BY MOLLY O'GUINESS CARLSON, 2010–13.

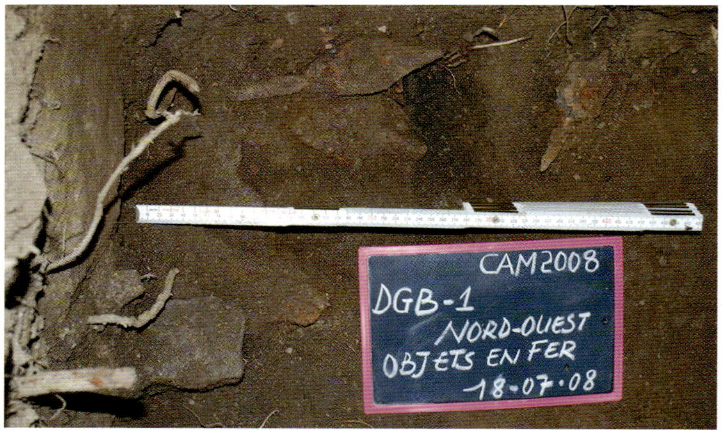

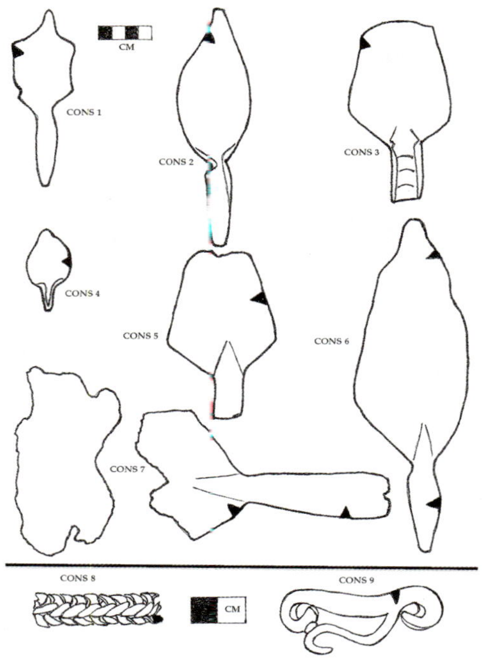

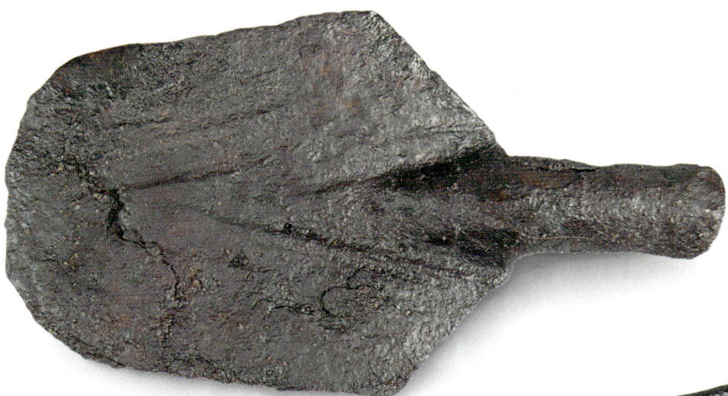

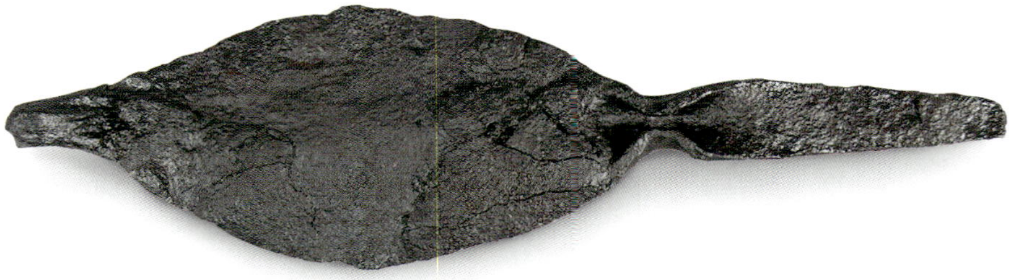

Complete objects from the last two millennia are not often found in the region, for iron was always relatively valuable, and iron tools were used until they were reduced to nubs and fragments, then reworked into new objects including symbolic ones. Close to Lake Chad, iron artifacts are sometimes found in burials from after about 800 CE, but exotic imports from the Sahara and north Africa—carnelian and glass beads, items of bronze and brass—seem to have had greater prestige as grave goods by that time. Discovered buried carefully under the floor of a sunken courtyard at the DGB-1 archaeological site—just under occupation levels—a cache of objects provides rare examples of prehistoric iron from the Mandara Mountains region. The objects date to approximately the sixteenth century CE and show evidence of ritual, not domestic, activity (fig. 13.8); four among them seem to be a tanged spear and arrow points, albeit not entirely completed, while two are curious socketed artifacts (MacEachern 2013). The exact function of these objects, and the reason for their burial, remains obscure at this point.

Researchers have been able to document direct technological parallels between the five-hundred-year-old objects discovered at DGB-1 and modern tools made by Mafa blacksmiths. Such long-term analyses, unparalleled in research elsewhere in Africa, testify to technological continuity in iron production systems in the Mandara Mountains. Cultural and ideological meanings ascribed to iron and use of iron tools in different circumstances presents another question entirely.

TRANSFORMATION, POWER, AND LIFE'S COURSE: THE SOCIAL WORK OF IRON IN THE MANDARA MOUNTAINS

In the 1980s, my colleagues and I worked with Mafa blacksmiths to recreate traditional iron smelting processes in northern Cameroon, and observed that making iron was far from a simple technical process. It was a complex cultural endeavor, with a diversity of magical substances as important to success as the calculations of ore, heat, and structure, and with metaphors of reproduction, gender, and power applied to the iron smelting furnace itself. The day of the smelt was a musical performance and a "time for war," involving the brute production of hot metal. It culminated in the evening with a yard-long tongue of flame pulsing from the furnace into the dusk (see fig. 13.5), in time with the music of zither and bellows (David 1988)—iron-making as performance, as well as technology.

Today, blacksmiths are important members of rural communities throughout the Mandara Mountains, making tools and ceremonial objects out of scraps of industrially produced metal and repairing broken and worn-out tools. In the precolonial period, and indeed into the 1950s, locally produced iron played many roles. Notably, in societies with some degree of political hierarchy, the status of a chief or ruler was often traditionally associated with ironworking. Thus, sovereigns in the Wandala state, north of the Mandara Mountains, were inaugurated accompanied by the sound of smiths beating on their anvils (Forkl 1989), while across the mountains and in the surrounding plains deposits of charcoal and iron ore were frequently placed in tombs, surrounding chiefs' bodies in "tomb-furnaces" (Langlois 2010). The central economic and military roles that hoes, axes, and weapons played in Mandara societies indicate the close association of iron and power, but the connection was also more intimate and mundane.

The status of blacksmiths can be identified with broader processes of social and ideological transformation, as throughout most of the mountain region blacksmithing families formed an endogamous caste, strictly differentiated from the far more numerous farmers of Mandara societies. Blacksmiths could not marry non-smiths upon pain of death, and prohibitions often prevented social contact. Historically, blacksmiths did not merely work iron—they acted as morticians, musicians, barbers, makers of specialized clothing, diviners, curers, and, in general, ritual experts. Their wives made all of the pottery for their communities and acted as midwives, ritual experts, and healers (Sterner 1993; David, 2012). Among many Mandara groups, blacksmiths broadly defined were producing virtually all of the durable material record.

A number of the primary technical functions of these people involved changes of status: iron-rich sand into iron, clay into pottery, the unborn into living humans, children into adults, and the living into the ancestral dead. (The ritual impurity ascribed to blacksmiths is often justified by pointing out that they handled dead bodies.) Some scholars use the term "transformers" to describe such people, to better encompass all of their roles, and because "blacksmith" is not inclusive of equally important activities of female members of the caste (David 1988; 2012; see also chapter 2, this volume). Indeed, many Mandara societies could be divided into two castes: "transformers" (the 2–5 percent of the population who handled the specialized tasks listed above) and "farmers" (everyone else).

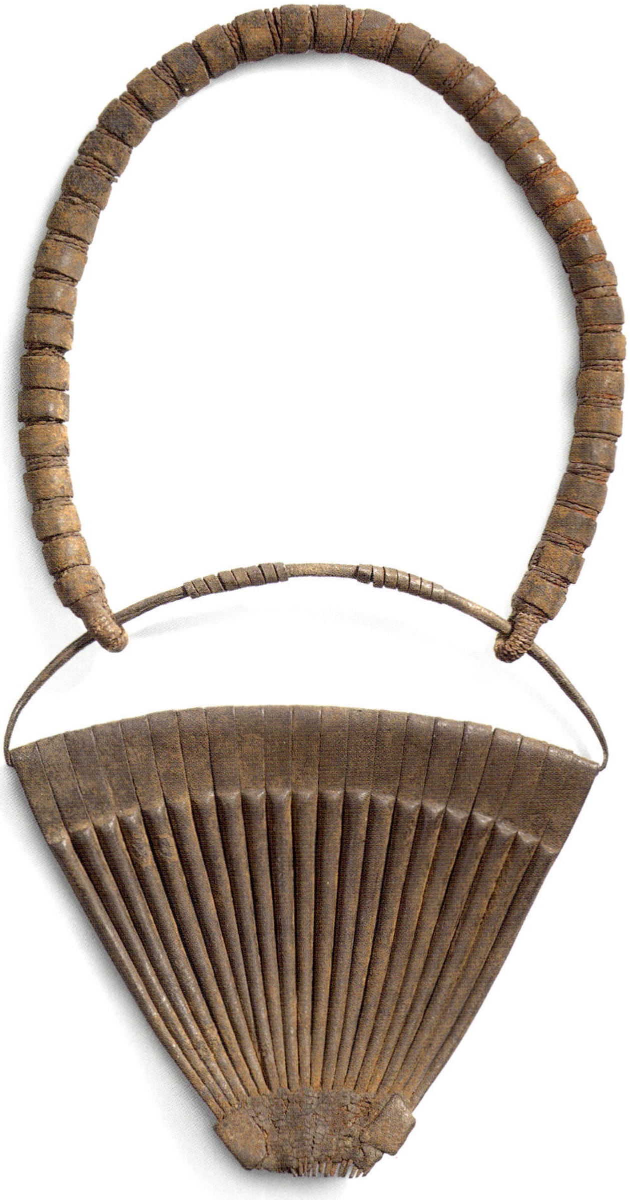

13.9
Artist unknown (Mafa peoples, Cameroon)
Woman's pubic apron (*motsoɗ yam*), mid-20th century
Iron, leather, cotton cordage
H: 36 cm
PRIVATE COLLECTION

Comprised of a cluster of finely forged iron strips held together in a tight triangle, a Mafa woman's *motsoɗ yam* is worn point down and facing away from the body. In these societies, iron is closely associated with adult women's fertility.

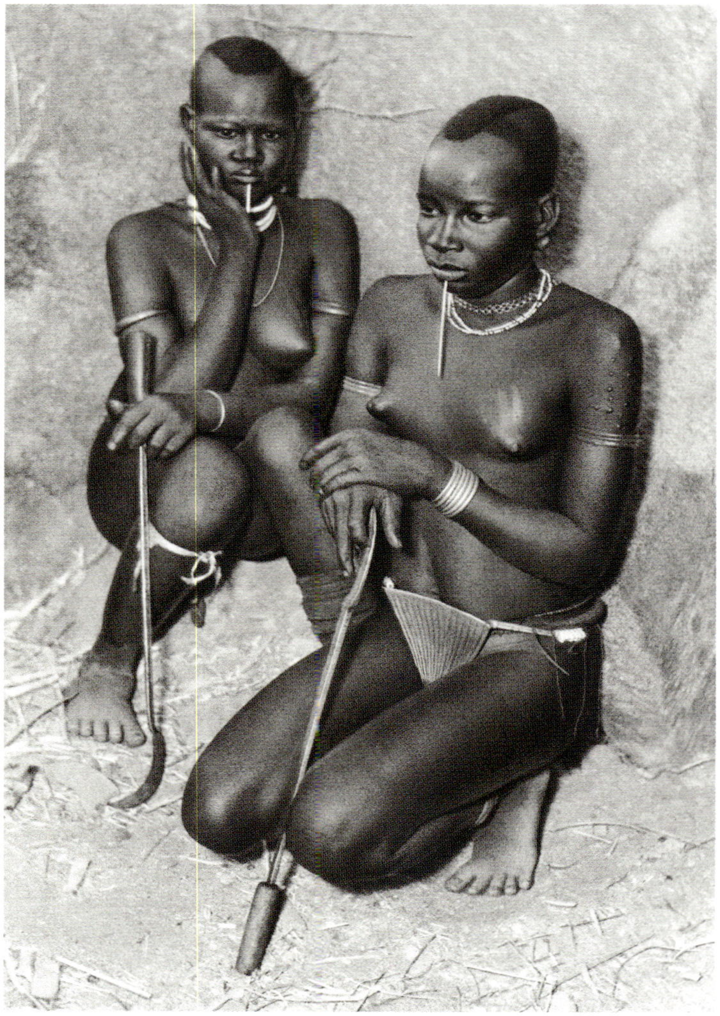

13.10 A pubic apron similar to that shown in fig. 13.9 can be observed on the young Mafa woman in the foreground. The women wear additional metal adornments and carry ceremonial dance sickles that would have been used on special occasions.
PHOTOGRAPH BY RENÉ GARDI, MOKOLO, CAMEROON, 1953. © BERNHARD GARDI

Given these circumstances, it is hardly surprising that iron objects, ironworking, and blacksmiths figure prominently in early ethnographic writings about the region—especially since indigenous smelting traditions were still thriving into the mid-twentieth century. Jean Mouchet, a French colonial officer and ethnographer, wrote some of the first systematic cultural comparisons of different Mandara populations, including descriptions of smelting and blacksmithing activities (see, for example, 1947a; 1947b; 1948). Bertrand Lembezat discussed the role of iron in Podokwo society (1952).

The great diversity of Mandara populations has been addressed by various ethnographic approaches, but with less attention paid to the role of iron in more recent periods. Exceptionally, the long-term research of Nicholas David, Judith Sterner, and Gerhard Müller-Kosack has frequently focused on the statuses and actions of blacksmiths as well as on the material culture that they produce in the central Mandara Mountains, primarily among Sirak and Mafa populations in Cameroon, and in Sukur in Nigeria, on the western edge of the massif. Also, Walter E. A. van Beek's four decades of ethnographic fieldwork has documented Kapsiki/Higi communities

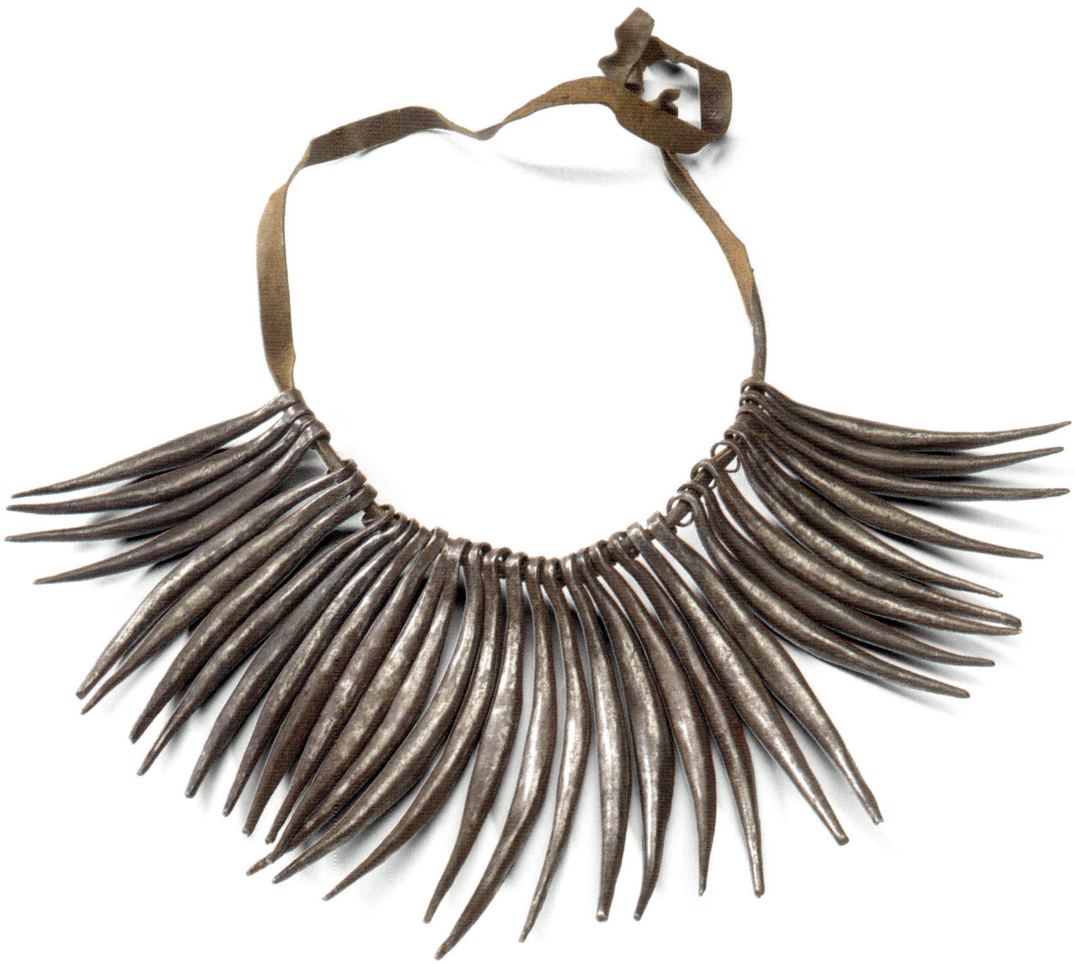

13.11
Artist unknown (Sukur peoples, Nigeria)
Pubic apron (*mbambazai*), 20th century
Iron, leather
L (each iron part): approx. 12 cm
PRIVATE COLLECTION

In the Mandara Mountains region, women wear delicately forged pubic aprons when marrying and throughout the rest of their lives. Sukur women thus adorned promenade with long-handled ritual iron sickles in evening dances held outside the chief's residence during biennial male initiation ceremonies (*ber*). This Sukur apron consists of thirty-four hollow iron pods strung on a leather belt to be tied around the waist. Each tapering pod curves in two directions simultaneously, demanding meticulous skill from the smith to create its distinctive convex curvature.

in the western Mandara Mountains, on both sides of the border between Cameroon (Kapsiki) and Nigeria (Higi). He has specifically considered the central role that iron objects play in social transitions from childhood to adulthood, adulthood to death.

In Mandara societies, iron is closely associated with women and, in different ways, with the status of being an adult. A Mafa woman's pubic apron (*motsod yam*) is made up of a triangular cluster of forged iron strips and worn upon marriage and on ceremonial occasions throughout the rest of the woman's life (figs. 13.9, 13.10; see Müller-Kosack 2003); among Sukur peoples similar pubic aprons are found (*mbambazai*; figs. 13.11, 13.12). In Kapsiki/Higi communities, the object most evocative of girls' transition to mature status is an iron "skirt" (*livu*) worn around the waist and constructed from a row of linked iron chains with a second skirt on top made of hammered iron strips (figs. 13.13, 13.14; see Beek 2012, 285–93). At particular points during the weeks-long ceremonial cycle through which young Kapsiki/Higi women (*makwa*) become wives, they wear nothing else. At other times during the cycle, they don additional iron ornamentation. Throughout a Kapsiki/Higi woman's married life, on ritual occasions her *livu* will be brought out and worn, symbolizing her success in childbearing, as the *motsod yam* does for Mafa women.

The importance of iron for *makwa* contrasts with its absence among boys (*gwela*) in Kapsiki/Higi communities undergoing a parallel transition to manhood; they instead wear brass ornaments. Thus, iron symbolizes female fecundity, domesticity, and membership in a community, while brass is associated with the wild lands beyond the village, where men will eventually hunt, carve out new fields, and go to war. Iron sickles, both utilitarian and ceremonial, are carried by adult women during festivals throughout the region (figs. 13.15, 13.16). This may reflect the ways in which men and women of the blacksmith caste are associated with femininity in Kapsiki/Higi society. It is likely that similar associations hold widely throughout the Mandara region. Iron bracelets and staffs are presented to men when they become chiefs (Beek 2012, 294–95), and it is notable that accession to that status is supposed to involve a change in behavior, as chiefs are expected to be soft-spoken and self-controlled—characteristics usually attributed to women and blacksmiths.

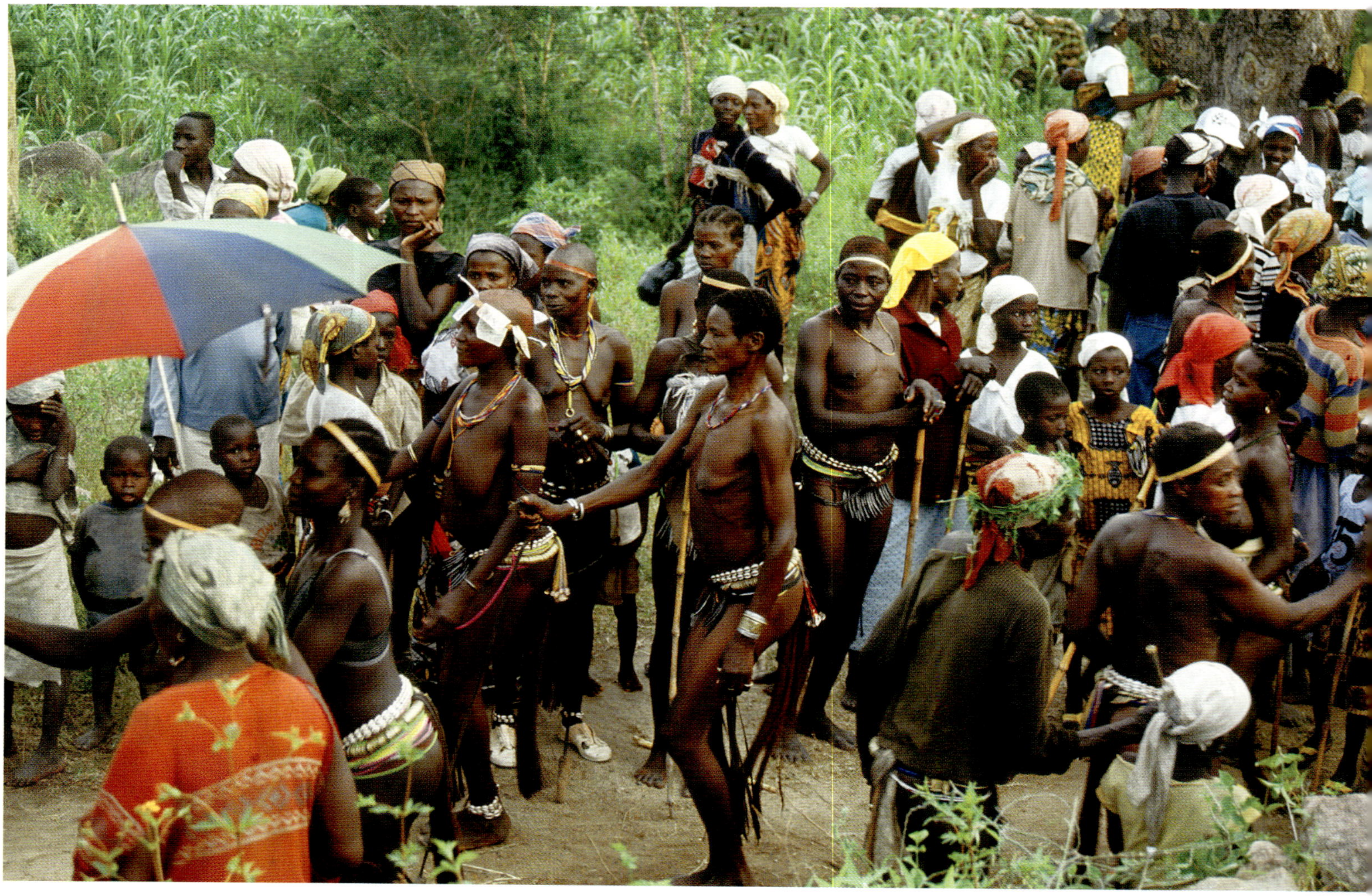

13.12A–B As they perform during a male initiation ritual (*ber*), Sukur women wear iron pubic aprons (see fig. 13.11) along with various accouterments that speak to both tradition and modernity.

PHOTOGRAPHS BY BRANWEN J. HENNIG, SUKUR, NIGERIA, 1992.

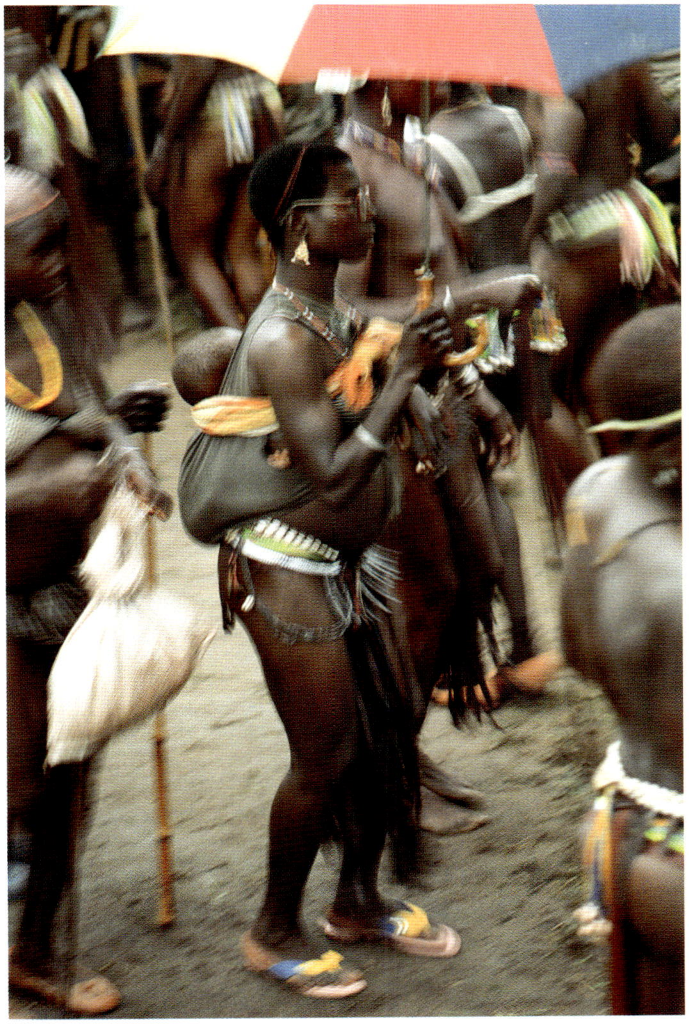

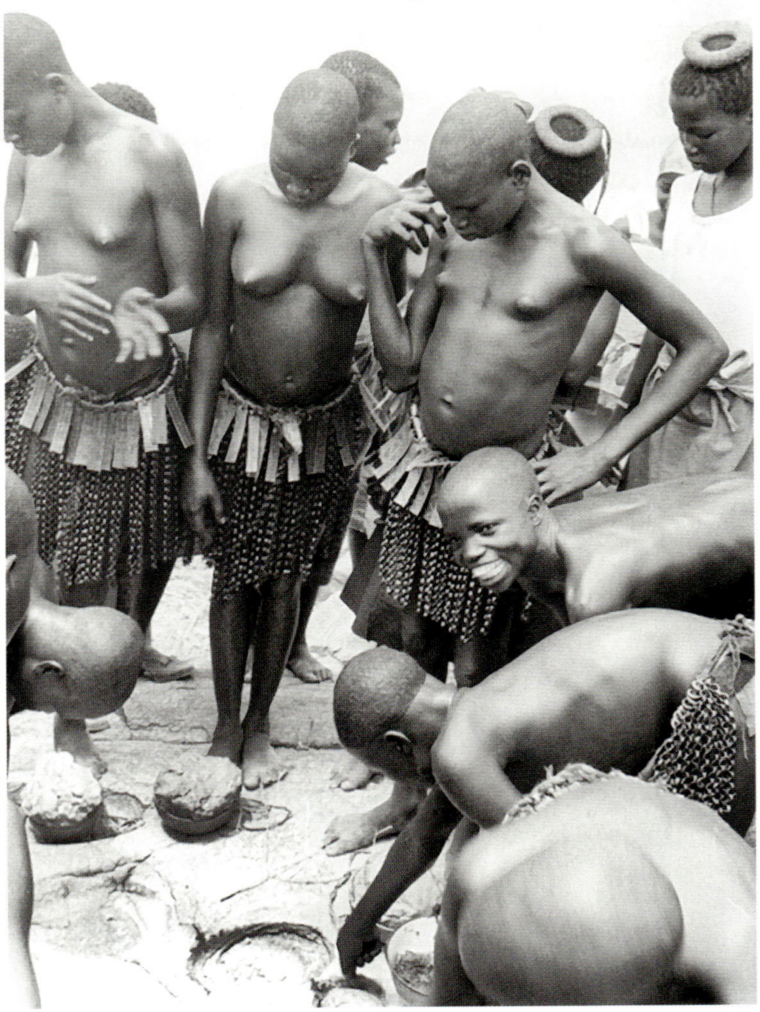

13.13
A group of Kapsiki girls wearing their bridal *livu*, made up of a row of linked iron chains under a second skirt of hammered iron strips.
PHOTOGRAPH BY WALTER E. A. VAN BEEK, MOGODE, CAMEROON, 1973. COURTESY OF ARCHIVES WALTER E. A. VAN BEEK

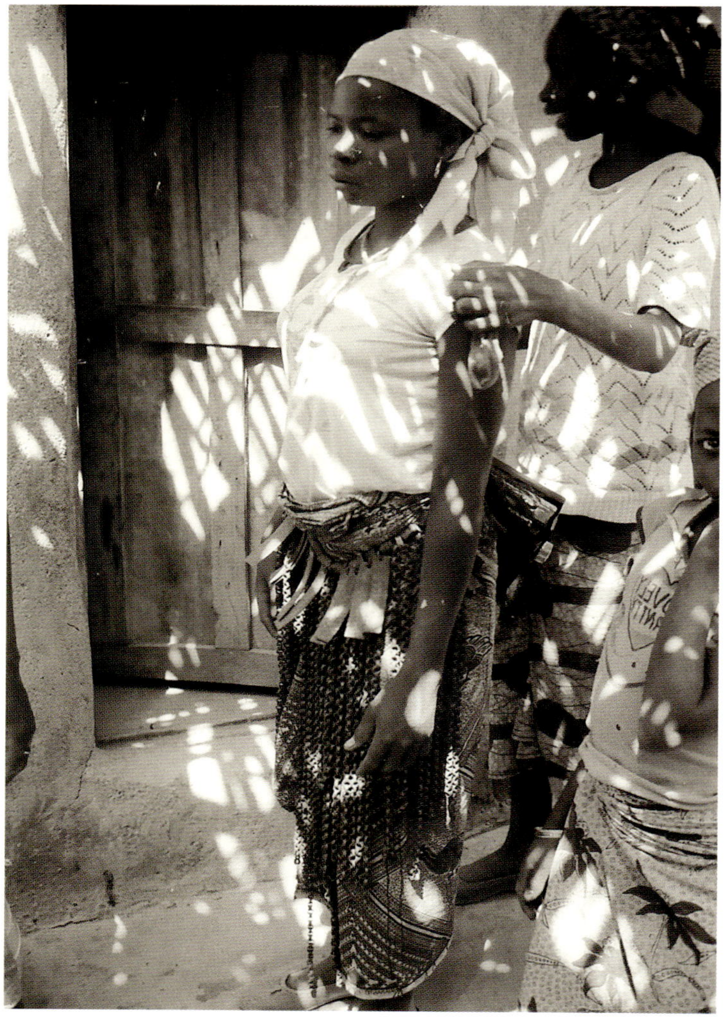

13.14
Kapsiki pubic apron (*livu*) worn with Western clothing.
PHOTOGRAPH BY WALTER E. A. VAN BEEK, MOGODE, CAMEROON, 1973. COURTESY OF ARCHIVES WALTER E. A. VAN BEEK

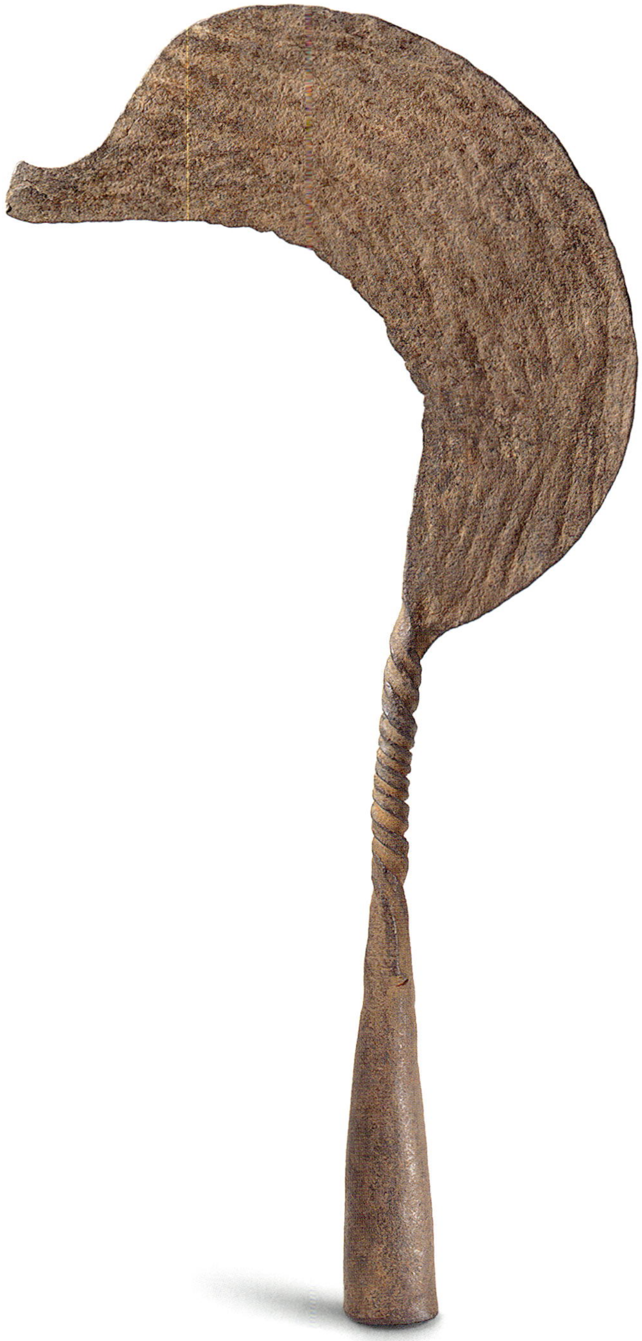

13.15
Artist unknown (Mafa, Podokwo peoples, Cameroon)
Ceremonial sickle, mid-20th century
Iron
H: 39.5 cm
PRIVATE COLLECTION

Women danced holding these iron sickles aloft during ceremonial occasions including agricultural festivals and possibly funerary rites.

13.16 Intense dancing with ceremonial sickles became, by the 1950s, a spectacle for tourists in Oudjila, a Podokwo community in northern Cameroon near the town of Mora.
PHOTOGRAPH BY MICHEL HUET, OUDJILA, CAMEROON, CA. 1950S.

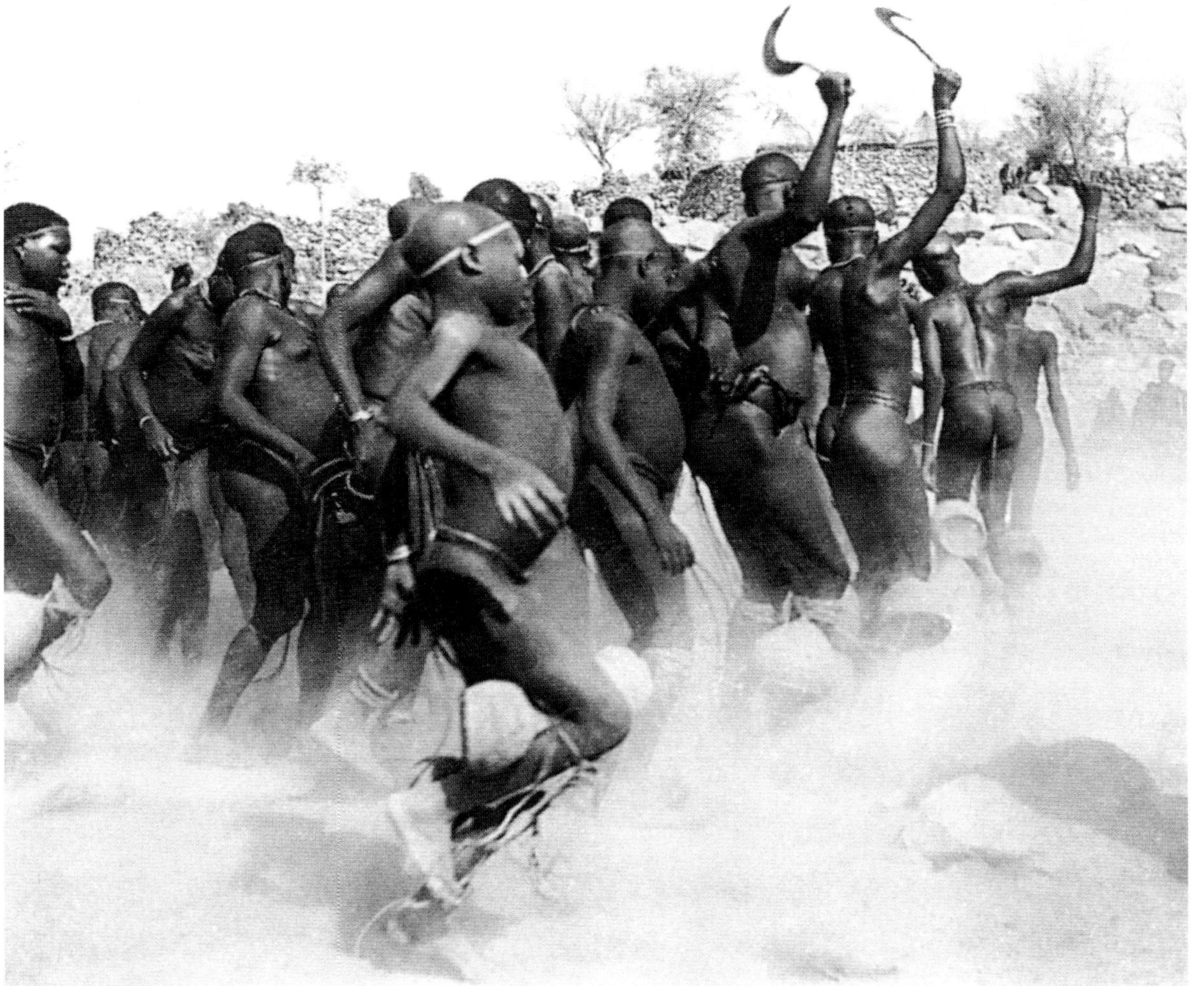

Iron plays very different roles during Kapsiki/Higi burial, when blacksmiths act as undertakers and directors of complex ceremonies that transform living humans into departed (and still active) ancestors. In these rituals, iron is notable in its near absence (ibid.). Although graves are dug with iron-tipped tools, only their wooden handles are left on the graves. Corpses are adorned with different materials—particularly intricate clothing—but not with iron, and no iron objects are placed in graves. Women of the community wear *livu* as they dance to honor the dead, but it seems that iron's associations with wealth and fecundity limit its role in the transition through death, that is, the end of fertility and of efforts to become wealthy and powerful.

Iron's importance in social transitions does not exhaust its cultural and ideological functions in Mandara communities. Iron containers (*mblaza*) and bracelets (*takase*) are necessary in traditional Kapsiki/Higi medicines because iron stabilizes and contains such substances (Beek 2012, 292–93). Also, because iron was traditionally associated with wealth, not merely for Kapsiki/Higi but more widely in the region, standard-sized bars functioned as a currency, and it is possible that some of the iron objects found in a cache on the DGB-1 archaeological site (see fig. 13.8) were formed out of such bars only partially finished. These bars are called "hoes" and sized to provide enough material to make one of these foundational agriculture tools. As representations of the fertility of fields tilled by hoes, they are still an essential part of payments made to brides' families during marriage negotiations (see chapter 5, this volume).

On the western flanks of the Mandara mountains, Sukur peoples specialized in precolonial iron production, with blacksmiths twice as numerous as in neighboring societies. Sukur produced tens of thousands of iron bars annually and exported them throughout the southern Lake Chad Basin, thus playing a central role in the region's agriculture (David 1996; 1997). Specialization in iron production for utilitarian purposes seems also to have involved the loss of many ritual and ideological aspects central to the process in neighboring societies: Sukur blacksmiths were much less involved in other "transformer" statuses, whether as undertakers or ritual experts; women were often involved in iron smelts; and there was far less use of other empowering materials in the smelts themselves. In the case of Sukur, wealth production superseded many of iron's cultural functions among other Mandara populations.

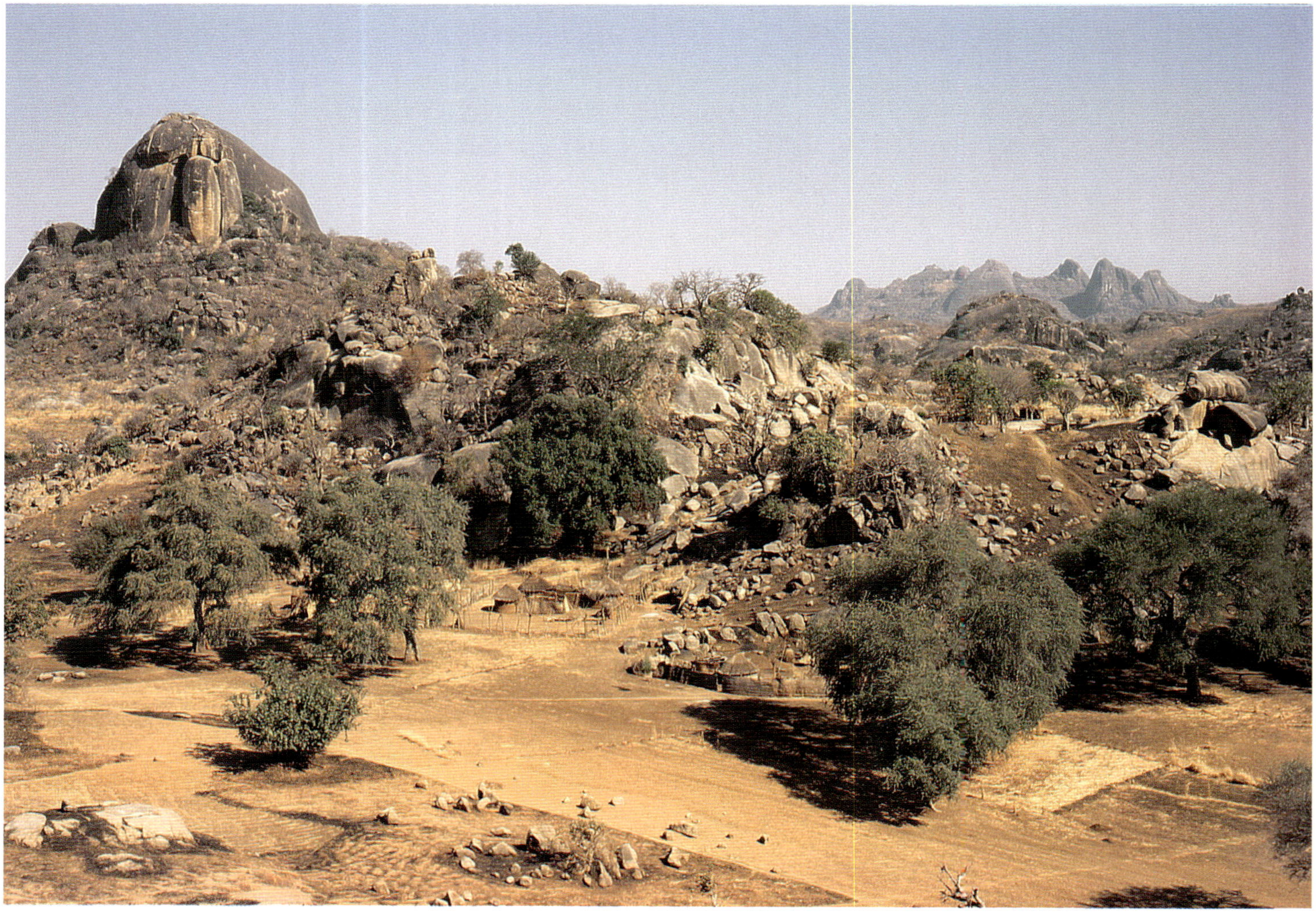

PART II

Marla C. Berns

IDENTITY AND TRANSFORMATION POWERED BY IRON: THE GONGOLA RIVER VALLEY

The region west of the Nigeria-Cameroon border and north of the Benue-Gongola River confluence is home to peoples who have long lived in small hamlets sheltered by hills and dramatic granite outcroppings (fig. 13.17). Many speak Biu-Mandara languages closely related to those of their eastern neighbors, and despite the absence of archaeological investigation, historical and cultural concurrences across the region, including significant use of forged iron tools, regalia, and adornment, support the connections that their shared languages imply. Moreover, Biu-Mandara speakers moved into an area already dominated by people probably autochthonous to the region. They, too, share widespread practices of forging iron and may also have served as regional producers of smelted bloomery iron.

Aside from reports by colonial anthropologists and district officers in the early twentieth century, little sustained field research has been carried out in this region (see, for example, Meek 1931). An overarching goal of my art historical research in the early 1980s was to survey extant genres of artistic activity and cultural practices among the Gongola River Valley's diverse peoples and to analyze their correspondences as indicators of historical relationships.[2] Most if not all these peoples used forged iron objects in domestic, ceremonial, and sacred spheres of activity, and the importance of iron tools in agriculture, hunting, and, formerly, warfare was likely the basis for the transformation of hoes, sickles, adzes, and spears into ritual implements capable of announcing and activating various social and spiritual transitions and processes (figs. 13.18, 13.19). Through their aestheticization, forged iron adornments in turn gained symbolic weight through use in culturally significant contexts, especially initiations, funerals, and investitures. Drawing on efficacies of iron, Gongola River Valley smelters and smiths transformed the material into objects of life-sustaining utility and often stunning originality.

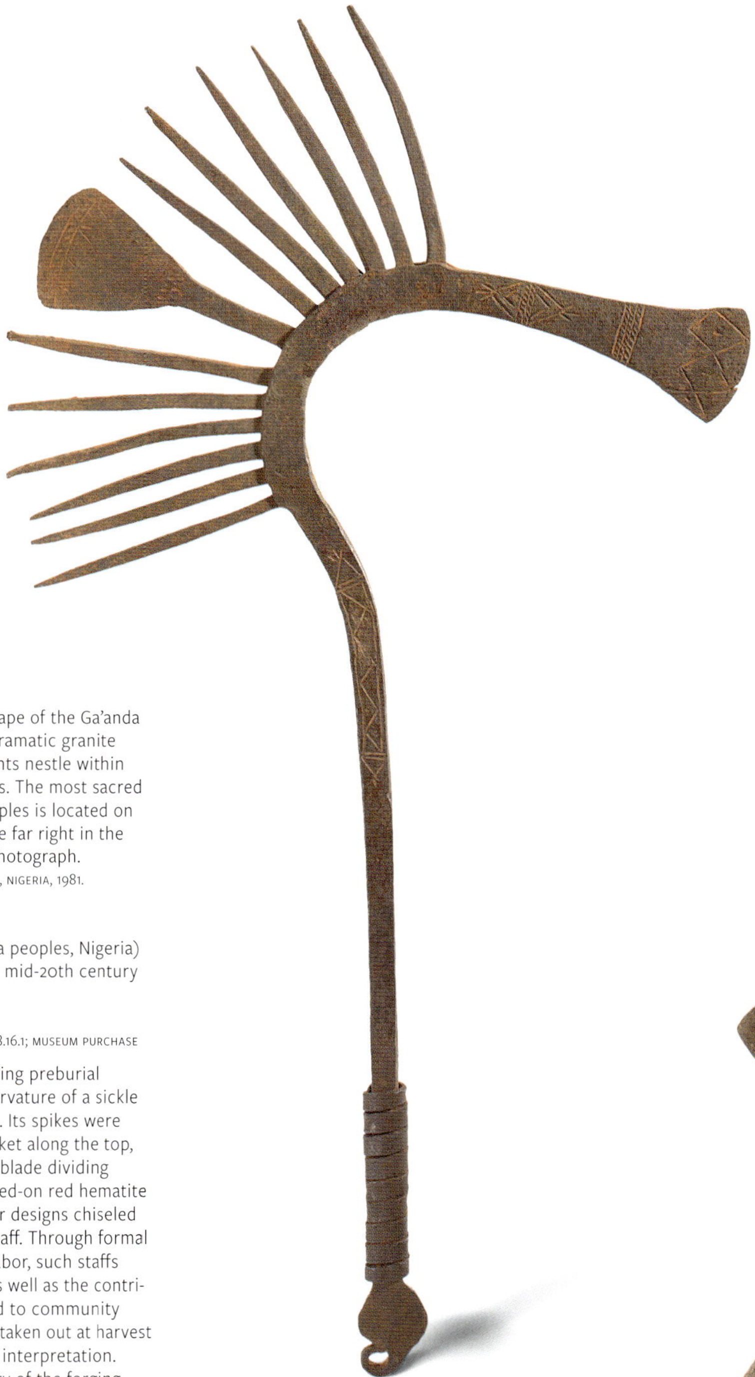

13.17 The rugged landscape of the Ga'anda Hills is punctuated by dramatic granite outcroppings. Settlements nestle within their protective shadows. The most sacred precinct of Ga'anda peoples is located on Makwar Mountain, to the far right in the middle ground of this photograph.
PHOTOGRAPH BY MARLA C. BERNS, NIGERIA, 1981.

13.18
Artist unknown (Ga'anda peoples, Nigeria)
Ritual staff (*wanshipta*), mid-20th century
Iron
H: 54.5 cm
FOWLER MUSEUM AT UCLA, X2008.16.1; MUSEUM PURCHASE

This staff, held aloft during preburial rituals, combines the curvature of a sickle with the blade of an axe. Its spikes were forge-welded into a pocket along the top, with an additional adze blade dividing them. A residue of rubbed-on red hematite remains in the triangular designs chiseled into both sides of the staff. Through formal references to tools of labor, such staffs symbolized hard work as well as the contributions of the deceased to community survival. That they were taken out at harvest festivals reinforces this interpretation. Moreover, the complexity of the forging surely proclaimed the skill of the laboring blacksmith. And, what of the symbolism of the "spiky mane"? It remains unclear whether it represented anything. The sharp projections may have assured the capacity to ward off danger, securing the divine intervention that protected the deceased as he or she transitioned into the world of the ancestors (see also fig. 13.25).

13.19
Artist unknown (Verre peoples, Nigeria)
Ceremonial double hoe, early 20th century
Iron
H: 30.5 cm
PRIVATE COLLECTION

Verre peoples live south of the Gongola River Valley and are known for their metal-smithing skills (iron forging and copper alloy casting). This "hoe" with gently curved handle and scrolled terminus is forked at one end to receive two socketed blades almost identical to single varieties sold at regional markets for practical use in weeding. Its decorative flourishes and thinly forged blades, too light for use in agriculture, suggest its use in ceremonial dance contexts, where it was likely associated with the fertility of the land.

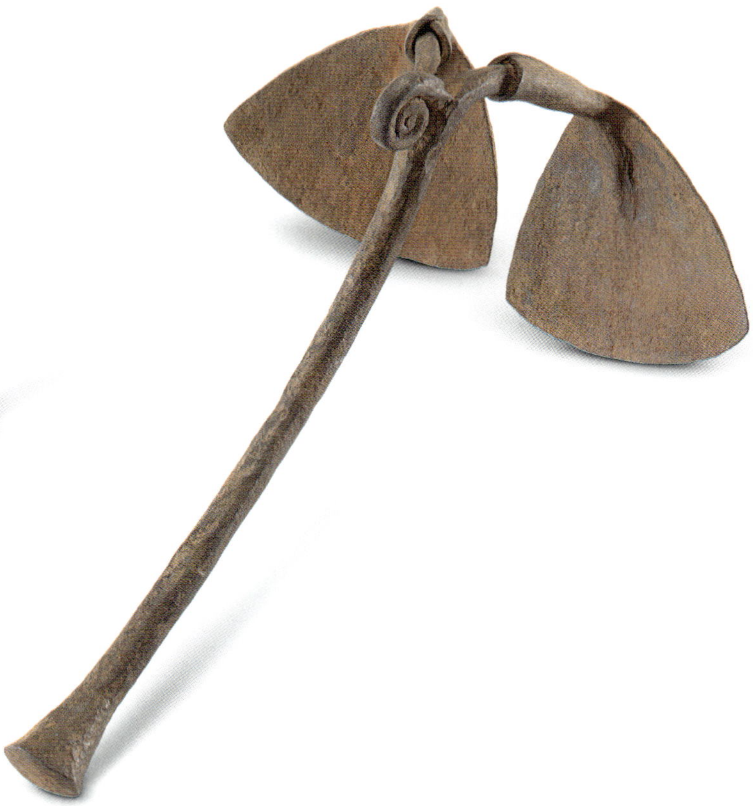

13.20 Two Ga'anda brides, presented to the community during the annual harvest festival Ho'mbata, with Thleta scarifications just visible down their torsos and on their foreheads. They did not complete the final stage, however, because Thleta was prohibited in 1978 by the local government authority. Both brides wear a profusion of metal adornments, including iron ankle rattles, bundles of iron pods strung on strips of cotton, and wrist- and armbands, mostly of cast aluminum (formerly cast in copper alloy).
PHOTOGRAPH BY MARLA C. BERNS, GA'ANDA TOWN, NIGERIA, 1980.

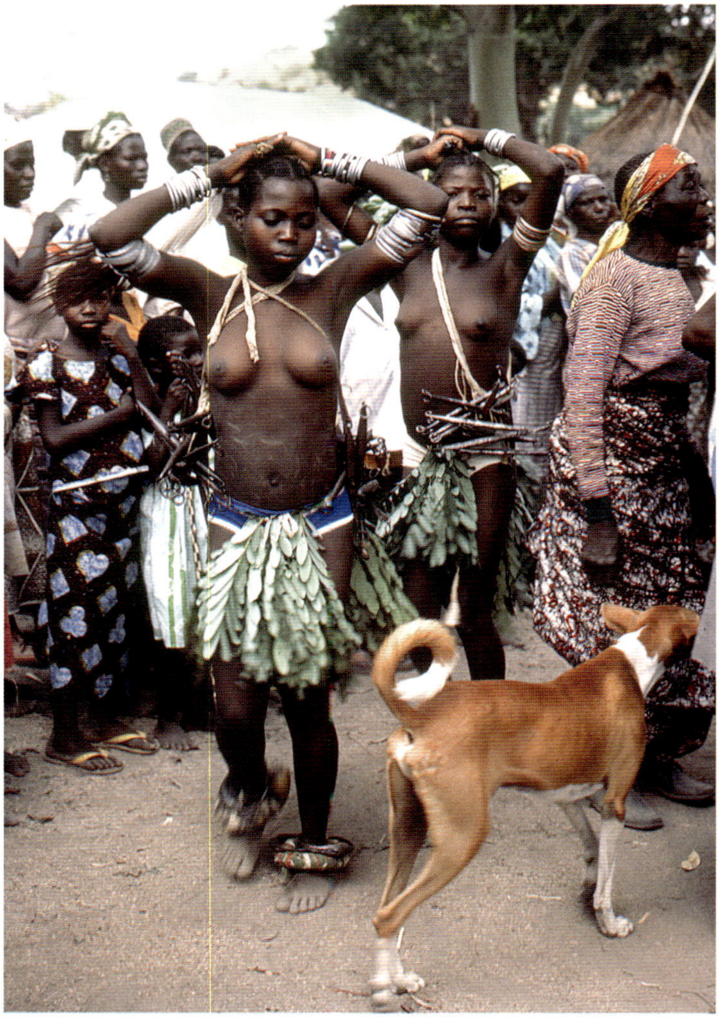

The remains of earthen furnaces in scattered rocky outposts, especially in the area occupied today by Bəna and Yungur, suggest that iron smelting was practiced using furnace forms similar to those of the Mandara Mountains, with a bellows worker sitting behind the furnace's vertical shaft (see figs. 13.5–13.7).[3] Though smelting had disappeared by the early twentieth century, Gongola Valley blacksmiths were still actively forging iron in the early 1980s, using imported scrap metal to create basic utilitarian tools and ceremonial objects in demand and described here.

Among Ga'anda, the empowering presence of iron was especially important to the transition of young men and women into adulthood and marriageability. Negotiations for betrothals began during infancy, when families of prospective brides and grooms initiated the exchange of household items, for example iron hoe blades, given in stages by the groom's family to the bride's. As elsewhere, the use of hoes as bridewealth, which stood for agricultural work and the promise of sustenance they ensured (see chapter 5, this volume), fell out of favor by the mid-twentieth century.

As a prerequisite to marriage, all Ga'anda females underwent a multistage program of scarification called Thleta. Elderly women effected the scarification using a small iron hook to lift the skin, and then a small iron razor to make rows of short, shallow, regular cuts. In the process of healing, the cuts scarred, forming subtle patterns of tiny raised bumps, slightly lighter in color than the unmarked skin around them.[4] The final stage of Thleta was the most extensive, and at the end of the process, which took years to complete, a woman's body was marked from head to foot with prescribed designs.

At the event called Yowo, seven days of feasting and celebration to mark Thleta's completion, the young women appeared, their bodies rubbed with the ritual cosmetic *mesaktariya*, made from powdered red iron oxide (hematite) mixed with oil.[5] Applied over the course of Thleta as a kind of insulating barrier, *mesaktariya* protected against infection as the scars took "proper" form (while failure to heal fully was viewed as the work of disapproving or vindictive spirits or other possible adversaries). Initiates wore iron ankle rattles (*chakita*) and, diagonally across their chests, bundles of long, folded iron pods (*sanglicha*) roughly twenty to twenty-five centimeters in length and strung on strips of cloth (fig. 13.20). As the young women danced, the rattles announced their adult status. This Ga'anda rite resembled a practice of related Kapsiki peoples, for whom "iron seems to be associated with women and with femininity, especially in situations where a transition has to be effected" (Beek 2012, 290). By wearing multiple forgings on their bodies, girls were transformed into marriageable women, and "non-fertility was rendered into fertility" (ibid.).

In the initiation called Sabta, held at seven-year intervals, Ga'anda boys were subjected to various physical trials—mirroring the pain of girls' scarification—to test their strength and endurance (see Berns 1986, 110–21). At community festivals called Yoxiiwa, held to celebrate Sabta graduates, boys' skin was rubbed with *mesaktariya*, and they wore an array of prescribed adornments: cast bronze and aluminum bracelets that stretched from their wrists to their upper arms, and iron *chakita* and *sanglicha*. They also carried iron staffs (*chichawa*; fig. 13.21) of a form resembling an F-shaped throwing knife (see fig. 7.4), and rhythmically pounded the ground with them to contribute to Yoxiiwa's sacred soundscape. When initiates performed, the iron adornments' specific tones and resonances aroused the attention of spirits whose blessings were sought and that signified the responsibilities and aspirations of social belonging.

Women who completed Thleta in the same year that Yoxiiwa was performed also participated in the latter, holding ornamental tools and implements associated with successfully running a domestic

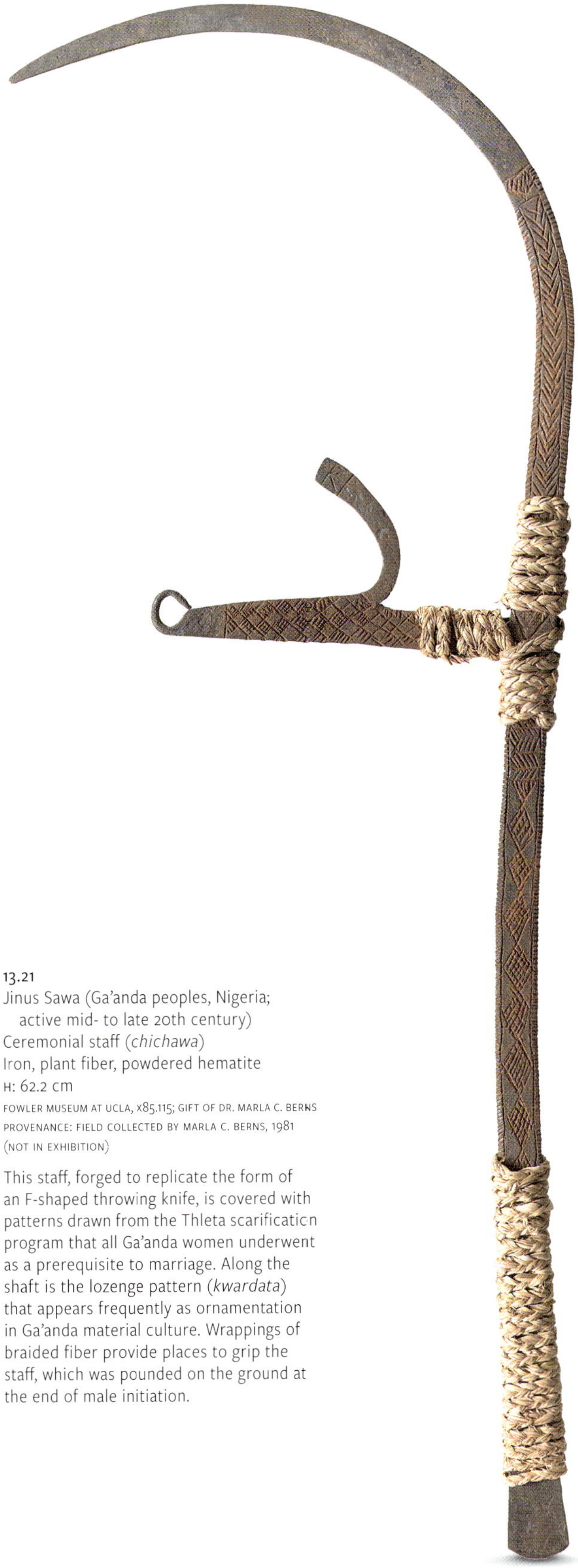

13.21
Jinus Sawa (Ga'anda peoples, Nigeria;
active mid- to late 20th century)
Ceremonial staff (*chichawa*)
Iron, plant fiber, powdered hematite
H: 62.2 cm
FOWLER MUSEUM AT UCLA, X85.115; GIFT OF DR. MARLA C. BERNS
PROVENANCE: FIELD COLLECTED BY MARLA C. BERNS, 1981
(NOT IN EXHIBITION)

This staff, forged to replicate the form of an F-shaped throwing knife, is covered with patterns drawn from the Thleta scarification program that all Ga'anda women underwent as a prerequisite to marriage. Along the shaft is the lozenge pattern (*kwardata*) that appears frequently as ornamentation in Ga'anda material culture. Wrappings of braided fiber provide places to grip the staff, which was pounded on the ground at the end of male initiation.

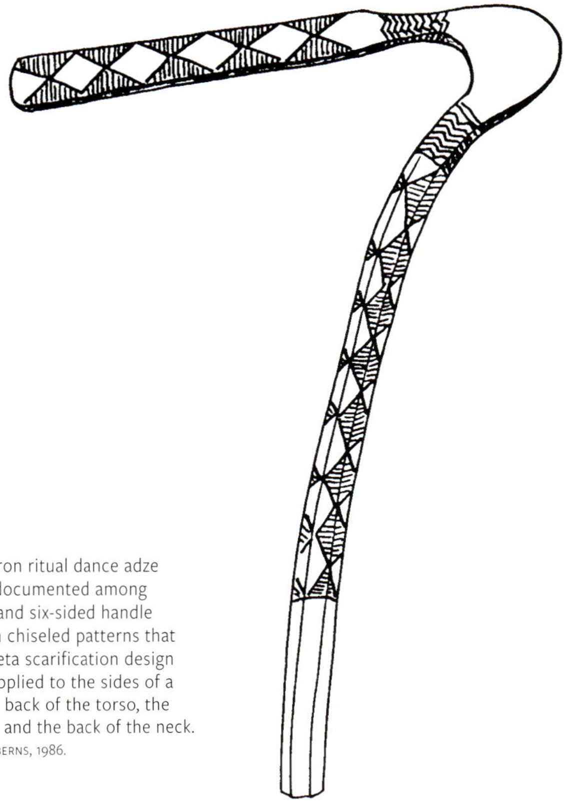

13.22 Drawing of iron ritual dance adze (*wurta*; H: 36 cm) documented among Ga'anda. Its blade and six-sided handle are decorated with chiseled patterns that reproduce the Thleta scarification design called *kwardata*, applied to the sides of a woman's body, the back of the torso, the front of the thighs, and the back of the neck.
RENDERING BY MARLA C. BERNS, 1986.

household: an iron hoe (*wanketa njoxwa*) hooked over the left shoulder, an iron adze (*wurta*) in the right hand, and a decorated gourd-bowl (*njoxtiteba*) in the left. These objects were transformed for ceremonial use through the application of designs otherwise inscribed on the female body, and especially chained lozenges and closely spaced lines (*kwardata*).[6] A *wurta's* long, rectangular blade and six-sided handle are covered in lengths of chiseled *kwardata* (fig. 13.22), and the surface of a *chichawa* is chiseled with the same patterns. Adorned with powdered hematite, the *chichawa's* blade recalls the reddened bodies of Sabta graduates walking in bent position, echoing the curvature of the throwing knife's outline. Metamorphosed through aestheticized shapes and incised with scarification patterns, these ceremonial tools show that the smith's transformative processes invoked "the human model as a measure of all things," to borrow the thesis of anthropologist Eugenia Herbert (1993, 5). Basic to human existence everywhere (with formulations specific to each culture), gender intersects with key moments in the life cycle to define social identity, while creative enterprises, Herbert argues, "revolve around complex and inseparable notions of power and fertility" (ibid.).

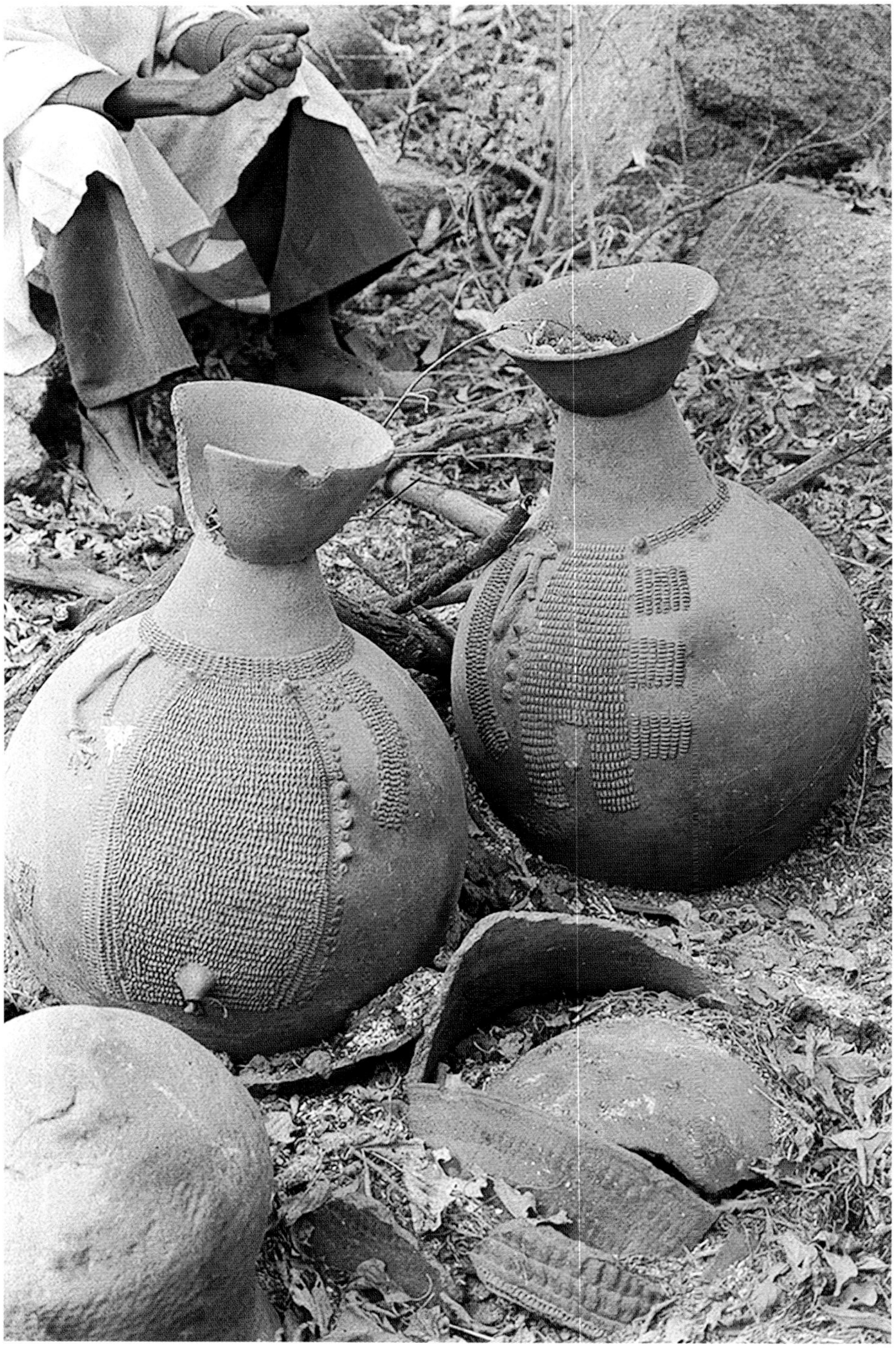

13.23 These two ceramic Ngum-Ngumi vessels were kept by the Wetu family of the Ga'anda subdistrict Boga, located in a hilly location far from their current villages. In 1981, when this photo was taken, the thatched enclosure that once would have covered them had not been refurbished by its custodians, and various forces of change (including the spread of world religions) had reduced the urgency of regularly supplicating Ngum-Ngumi spirits. The decoration shows the deity carrying an axe or adze in its right hand, and its torso covered with elaborate "corrugations" emulating the textured cicatrices on Ga'anda women's bodies.
PHOTOGRAPH BY MARLA C. BERNS, NIGERIA.

For Ga'anda, social transitions also required spiritual confirmation, and before a young bride *(perra)* was able to enter her husband's household and consummate marriage, she attended an annual post-harvest festival called Hom'bata, held in October or November. Hom'bata was the only time in a woman's life when she could enter the sacred groves where male ritual leaders maintained Ga'anda shrines as sites of supplication to the most powerful spirit forces. After returning, the brides danced once more to celebrate the harvest with the rest of the community (see fig. 13.20). Their fully scarified, now-healed bodies—adorned as for the Yowo and Yoxiiwa coming-of-age festivals, but with the notable absence of *mesaktariya*—signaled their permanently changed status. Wearing iron rattles again transferred to them the smith's procreative capacity.

A sacred precinct at Makwar Mountain (see fig. 13.17) features shrines to two foremost spirit benefactors, Ngum-Ngumi and Mbir'thleng'nda.[7] The former is the Ga'anda culture bearer, conceived as a pot and credited with having moved of its own volition to lead Ga'anda from the "east" to the impressive granite inselberg of Makwar. The tutelary spirit Mbir'thleng'nda also figures in early migration stories, and is contained within its own ceramic vessel type (see fig. 13.23).[8] It was at such sacred sites that Ga'anda maintained connections to ancestral roots; brides were taken to them before entering their husbands' households in completing their marriages.

The decorative programs for Ngum-Ngumi's and Mbir'thleng'nda's ceramic representations include iconic elements signifying the transitional moment of maturation, both male and female. Each spirit pot is anthropomorphized with clear human features (arms, nipples, umbilicus, mouth, etc.) and carries an axe or adze over its right shoulder in the same manner as young initiates of either sex (fig. 13.24). The torsos of some vessels were modeled with motifs of small knives, gourd-bowls, and bows and quivers, aligning the tools necessary to fulfill household responsibilities with the potential of these spirits, equally armed, to secure the positive conditions for communities to survive.[9] Most dramatically, both Ngum-Ngumi and Mbir'thleng'nda have "scarified" clay skins, suggesting their ancestral roles as templates for physical and social transformation and belonging (see Berns 2011). That these motifs are repeated on ritual accoutrements forged in iron (as well as on other categories of Ga'anda artistic production) extends the efficacies of such designs in inscribing meaning (see M. N. Roberts, 2008).

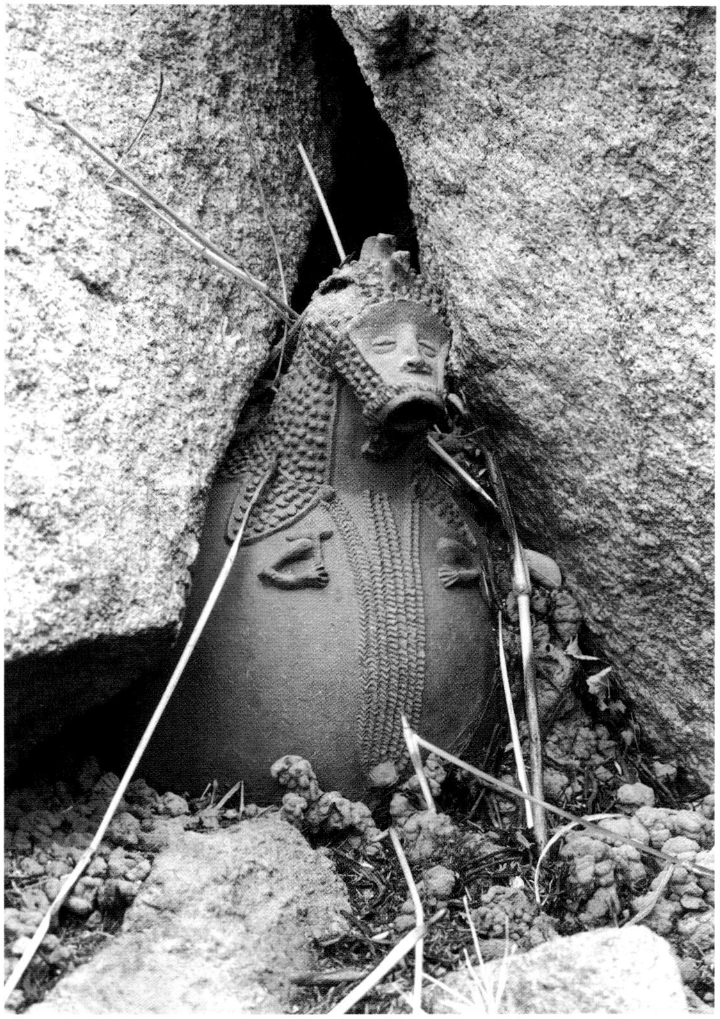

13.24 The spirit vessel Mbir'thleng'nda was wedged into a rock cleft above the hamlet of Hurwire to benefit its Dingai-Ga'anda inhabitants. The iron axe the deity holds in its right hand is visible. Is it holding it to be at the ready to fend off enemies, whether actual or spiritual? Is this an indicator that the labor of the deity mirrors that of the humans it protects? The vessel was photographed again (by Marla C. Berns in 1980) before it was removed and exported from Nigeria, entering the collection of the Musée Barbier-Mueller in Geneva in 2002. PHOTOGRAPH BY ARNOLD RUBIN, 1970.

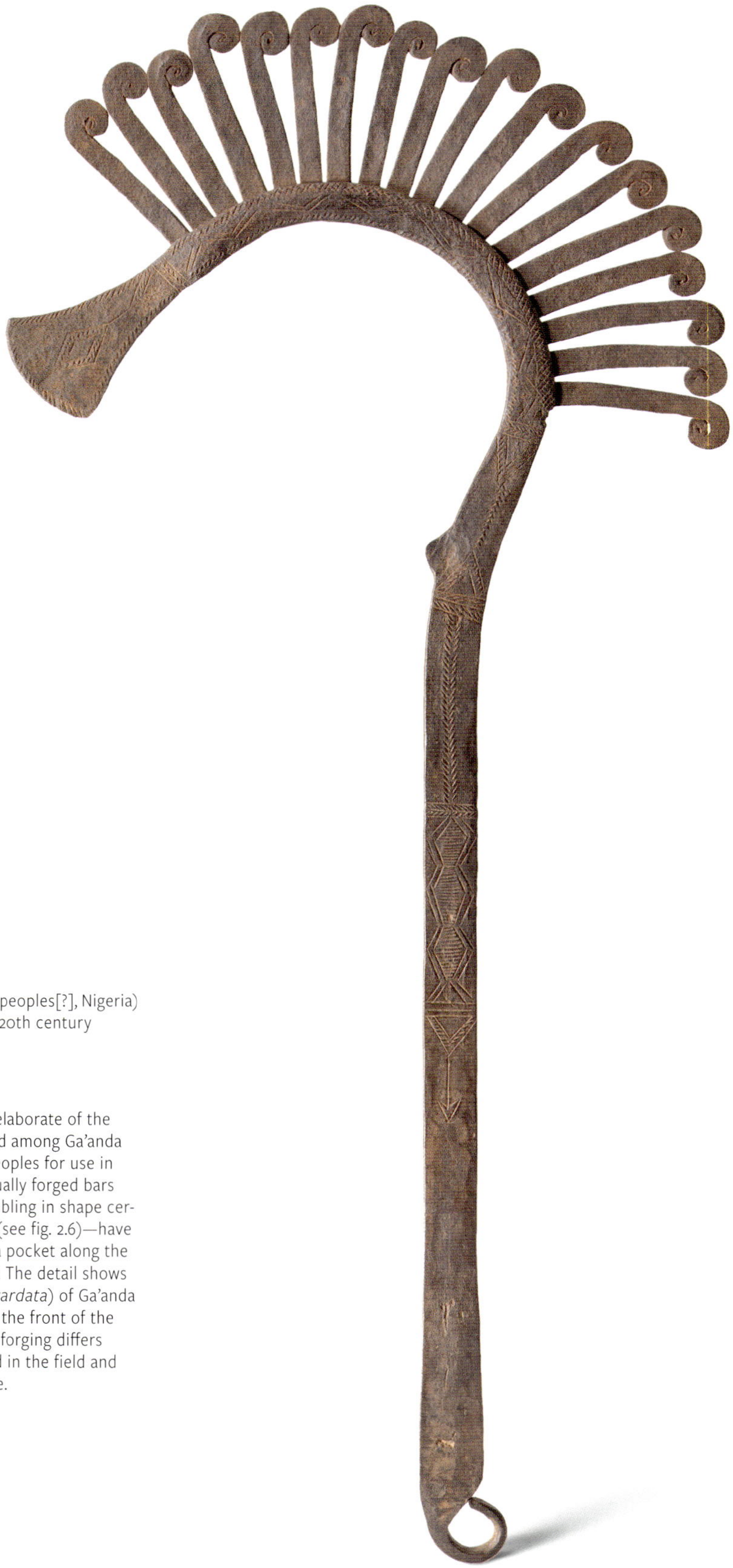

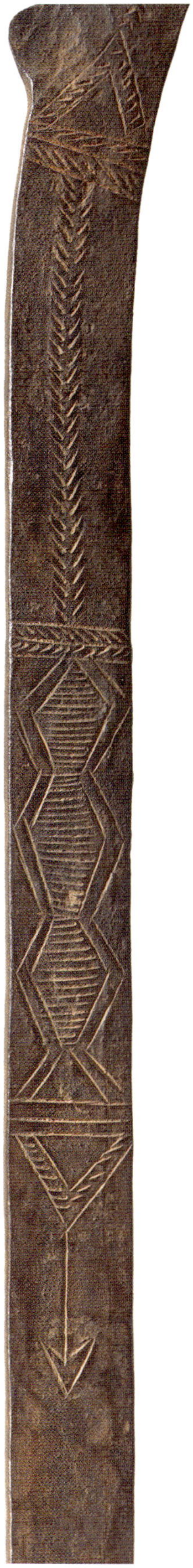

13.25A–B
Artist unknown (Ga'anda peoples[?], Nigeria)
Ritual staff (*wanshipta*), 20th century
Iron
H: 71.2 cm
PRIVATE COLLECTION

This is among the most elaborate of the spiked staffs documented among Ga'anda and neighboring Bəna peoples for use in preburial rituals. Individually forged bars with curled ends—resembling in shape certain iron bar currencies (see fig. 2.6)—have been forge-welded into a pocket along the upper curve of the shaft. The detail shows the lozenge patterns (*kwardata*) of Ga'anda scarification chiseled on the front of the shaft. This tour-de-force forging differs from others documented in the field and may be a unique example.

13.26
Artist unknown (Dadiya peoples, Nigeria)
Ritual sickle (*jən'nyi*), mid-20th century (before 1970)
Iron
H: 38 cm
FOWLER MUSEUM AT UCLA, X86.2615; GIFT OF GIFT OF JIM AND JEANNE PIEPER
PROVENANCE: FIELD COLLECTED BY ARNOLD RUBIN, 1970

This heavily decorated sickle was carried by young men during Kal, the festival held every five years to mark the culmination of male initiation. Such iron tools, so transformed via decoration, speak to the value of work and the importance of belonging. Distinctive to the forgings of Dadiya smiths is the iron strip "wrapped" around the handle, which provided an easy-to-grip surface when held in performance.

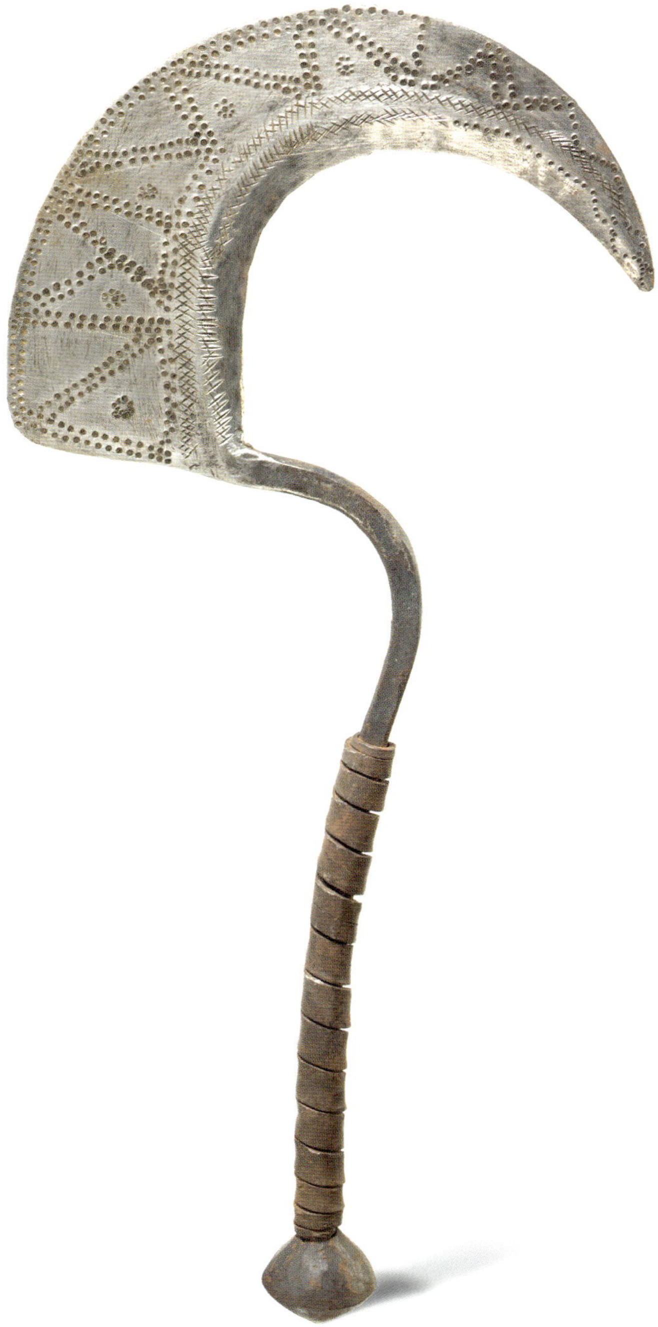

Iron objects also marked the final passage to death. Ga'anda smiths forged dance staffs for funerals that took highly distinctive forms. The *wanshipta* (fig. 13.25), which combines the curvature of a long-handled sickle with the gentle flair of an adze blade, is distinguished by a crown of spikes forge-welded along its top to appear like a horse's mane, and may be covered with the finely chiseled scarification patterns seen on other Ga'anda dance blades.[10] The *parwiita*, a second funerary staff type, has a wide, flat, sickle-shaped blade with an elongated handle. Both blade types were carried by male relatives of the deceased during three days of preburial rites as families paid their last respects. Kept within local shrines dedicated to Mbir'thleng'nda, the blades, as the embodiment of the collective life force of the ancestors, were taken out of the shrines annually at Hom'bata, when their presence lent power to securing the next year's agricultural bounty.

During a funeral, the corpse was wrapped in cotton cloth and displayed on a low wooden bier, with the head resting on a hoe blade as the measure of a person's social and economic contributions to family and community life. A colonial officer reported that before burial, the body of a male elder was laid out with implements associated with the transaction of a proper life: a quiver hung over the left shoulder, a bow in the left hand, an axe hooked over the right shoulder, and a fly whisk in the right hand (Boyle 1913, 37).[11] Thus, in death the body was adorned in a manner that resembled the positioning of tools on Ngum-Ngumi, the ancestor from whom all Ga'anda have descended.

Across the Gongola Valley, other dance staffs based on agricultural tool shapes are used in performances associated with coming-of-age rituals, harvest festivals, and funerals. For example, among Dadiya people to the west, events called Kal were held every five years to celebrate the initiation of young men, who performed with forged, single-piece dance sickles. Called *jən'nyi*, such a sickle joins a wide, flat blade covered in chiseled and punched designs with a handle "wrapped" in an iron strip and finished with a conical cap (fig. 13.26; see Berns 1986b, 60–61). During Kal, an adult male also carried a dance rattle (*kichibyok*) made up of a stirrup-shaped iron handle attached to a heavily pyro-engraved gourd with strips of leather and surmounted by an elaborately forged grip and crest (fig. 13.27). This complex forging transformed the gourd—a food bowl decorated by women for their husbands' use—into an attribute of male power (figs. 13.28). As the *kichibyok* was swung through the air, its jangling rings called forth spirits and announced the initiates.

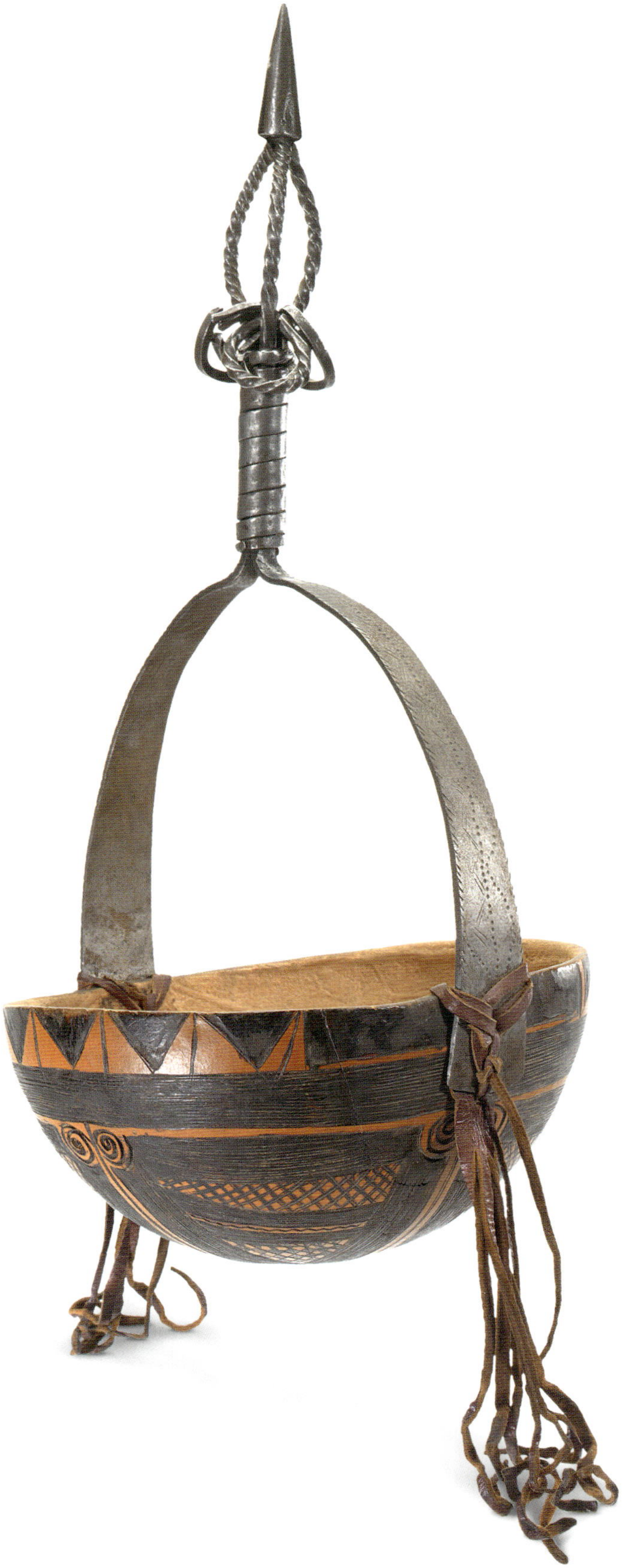

13.27
Artist unknown (Dadiya peoples, Nigeria)
Ritual rattle (*kichibyok*), mid-20th century (before 1970)
Iron, gourd, leather, thread
H: 40 cm
FOWLER MUSEUM AT UCLA, X83.796; GIFT OF BARBARA RUBIN-HUDSON
PROVENANCE: FIELD COLLECTED BY BARBARA RUBIN-HUDSON, 1970

This rattle—a union of hemispherical gourd bowl and iron handle—was used during Kal to signify the initiates' future households, which would depend on the work of women as well as men. The two sides of the stirrup, punched with designs echoing those on the sickle used in Kal (see fig. 13.26), are joined by a handle with iron wrapping and surmounted by a small open basket of twisted iron strands threaded with rings and capped with a pointed tip. In this instance, fine pyro-engraved designs, applied by women, and masterful forging, a male smith's task, are combined in an innovative sculptural and symbolic form.

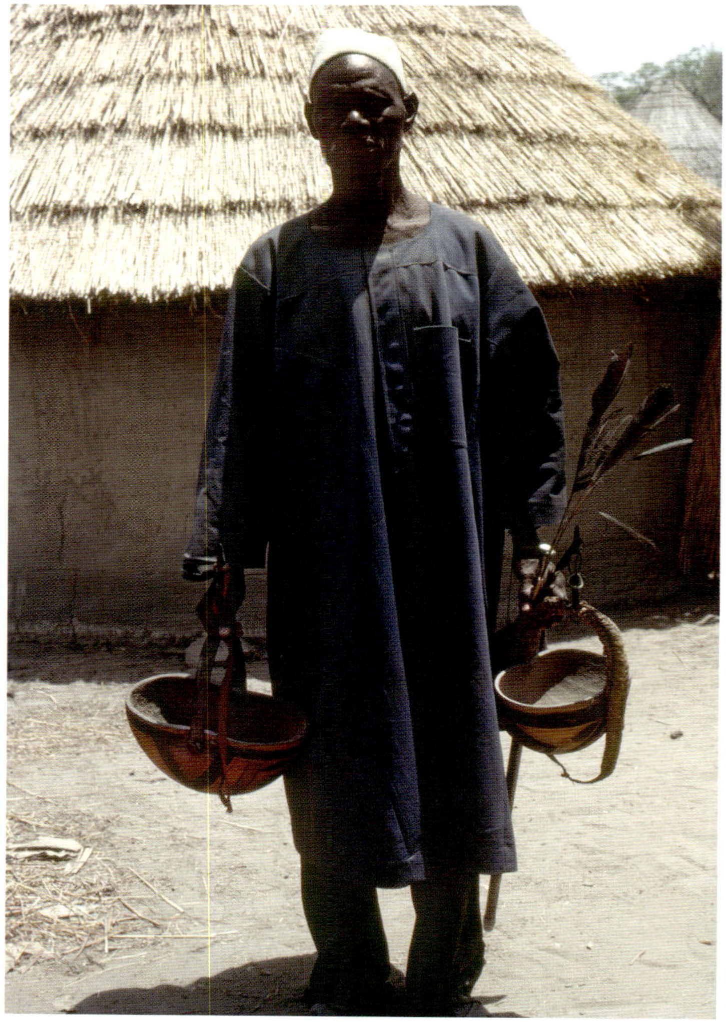

13.28 Among Kwa peoples, who live just south of Dadiya, men called "war chiefs" historically carried gourd-rattles (*kishibiya*) to exhibit bravery in war and, in the 1980s, to drink locally made beer when a lion was killed. Here, Bagude, the chief of Gyakan Village, carries two. Their iron elements likely refer to weapons; the complexity of the forging may symbolize the prowess of those who wielded such weapons successfully.
PHOTOGRAPH BY MARLA C. BERNS, NIGERIA, 1982.

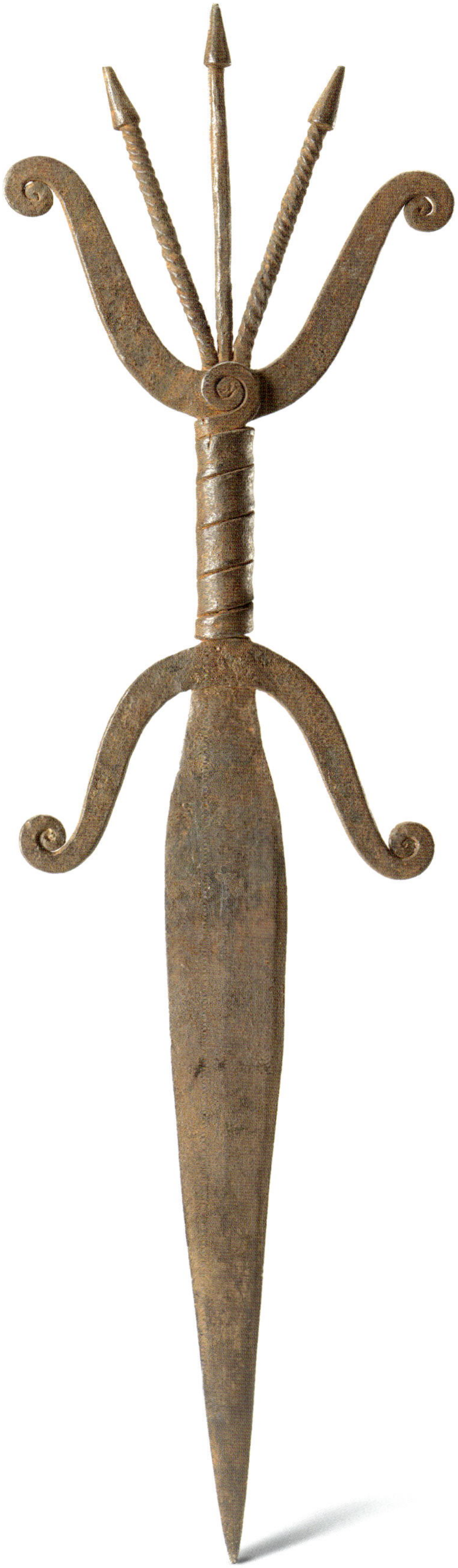

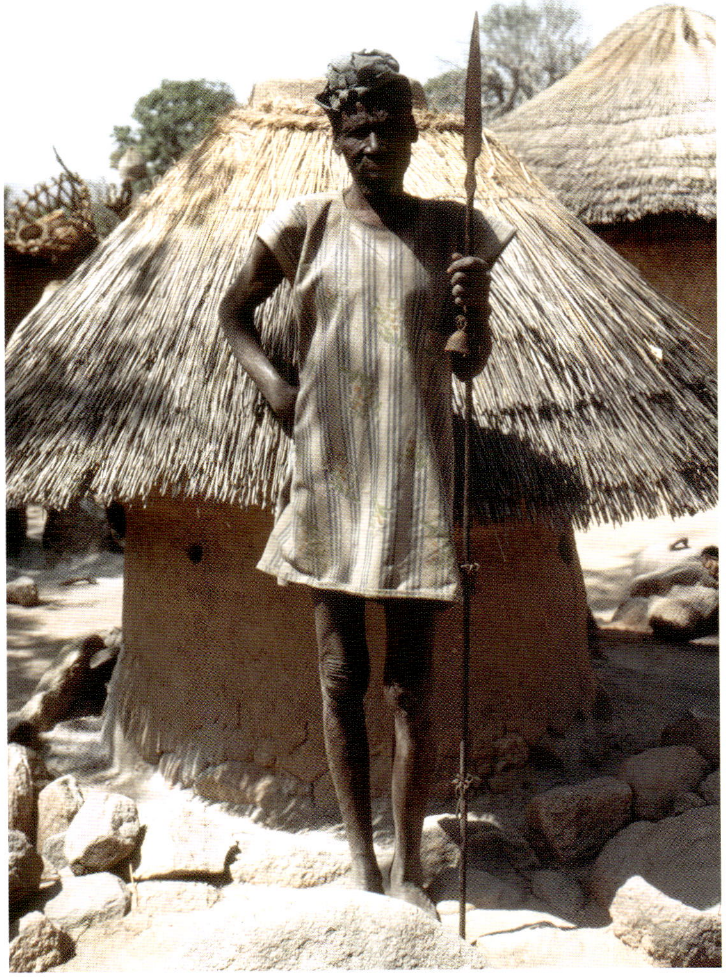

13.29
Artist unknown (Dadiya, Tula peoples, Nigeria)
Ritual dagger (Dadiya: *nyansanye*;
Tula: *tochile*), mid-20th century
iron
H: 46.5 cm
PRIVATE COLLECTION

Forged iron ceremonial daggers were carried by young Dadiya and neighboring Tula men in performance contexts that acknowledged their skill and bravery as hunters or warriors.

13.30 The chief of Kilake, a remotely located Bəna hamlet in the Ga'anda Hills, holds his iron spear of office (*kəda*). This tall spear is provided with loops at three places along the shaft for the suspension of short forged chains and a clappered bell. With such attachments, traditional chiefs who were responsible for positive interventions could announce their presence and call spirits to action.

PHOTOGRAPH BY MARLA C. BERNS, NIGERIA, 1981.

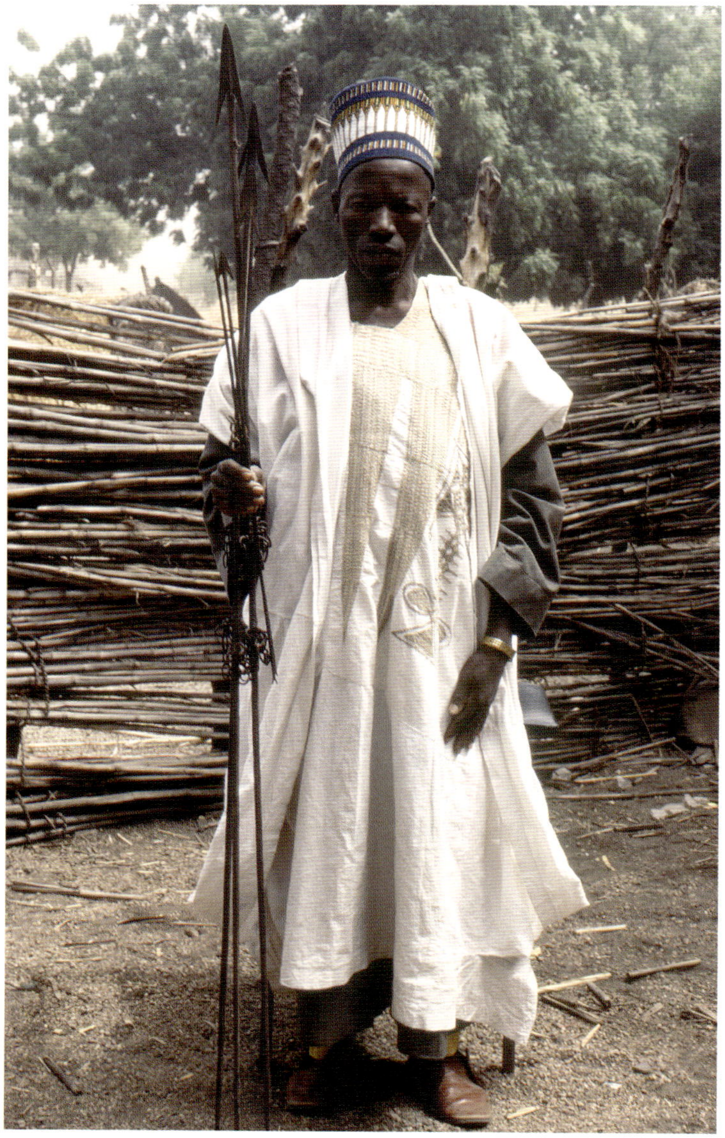

13.31 The Wihliwihli chief of Walama, an Islamized Pidlimndi town north of the Gongola-Hawal confluence, stands holding three spears (*rũm*) and two arrows (*hafa*) of office. Wihliwihli is the name of one of three lineages of Pidlimndi chieftaincy; each has its own set of regalia. Despite his adoption of Islam and distinctive style of dress associated with Hausa who dominate Nigeria's north (the elaborately embroidered robe and tall, fez-shaped cloth cap), this chief still used these iron blades in the early 1980s to signify his rule.
PHOTOGRAPH BY MARLA C. BERNS, NIGERIA, 1980.

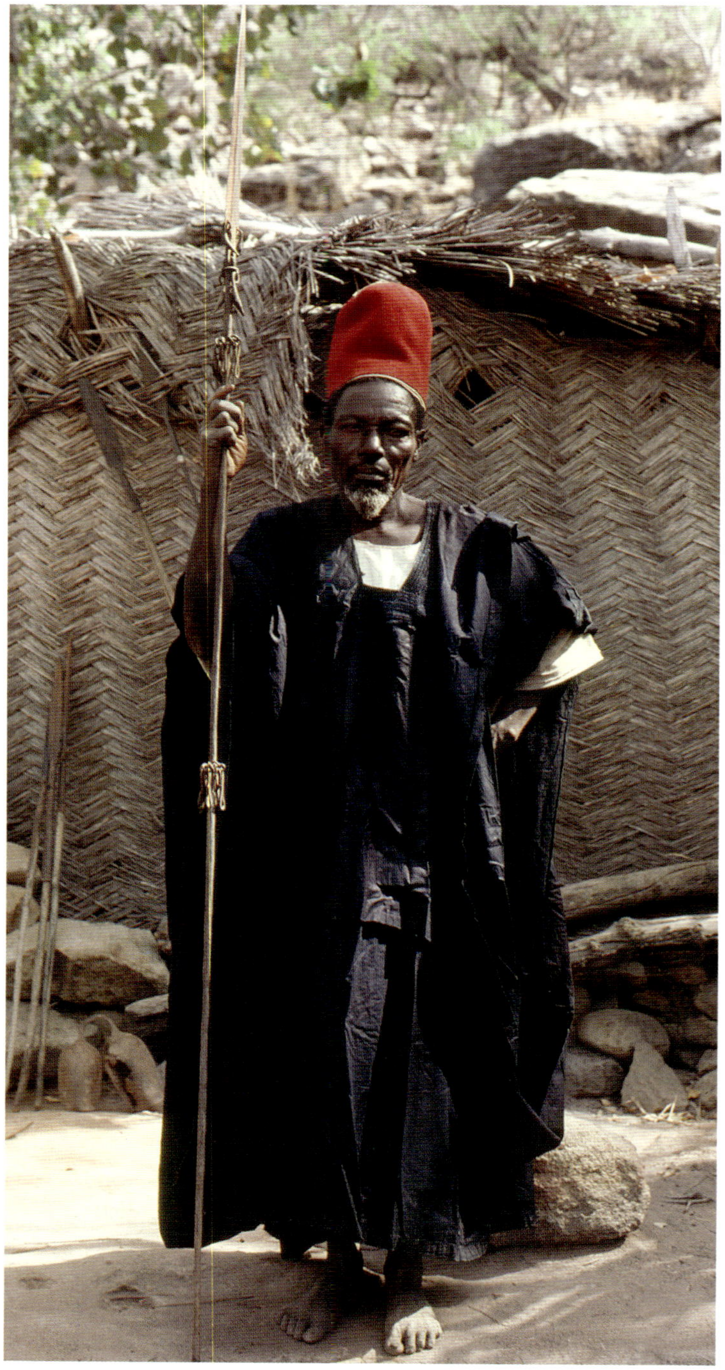

13.32 Chief Ndena of Gwamba, an Mbula town just north of the Benue River and east of its confluence with the Gongola, holds his spear of office (*jar darah*) and wears a tall red fez (*jeko bang'a*) of a type restricted to chiefs. He would have carried his spear to the place where Mbula spirits were offered supplication; the sounds of chains and pods suspended from loops along the shaft would have helped summon them. To the left behind Chief Ndena are spears used in rituals to bring rain and secure good harvests. Next to them is a small double bell, struck during funerals of chiefs.
PHOTOGRAPH BY MARLA C. BERNS, NIGERIA, 1982.

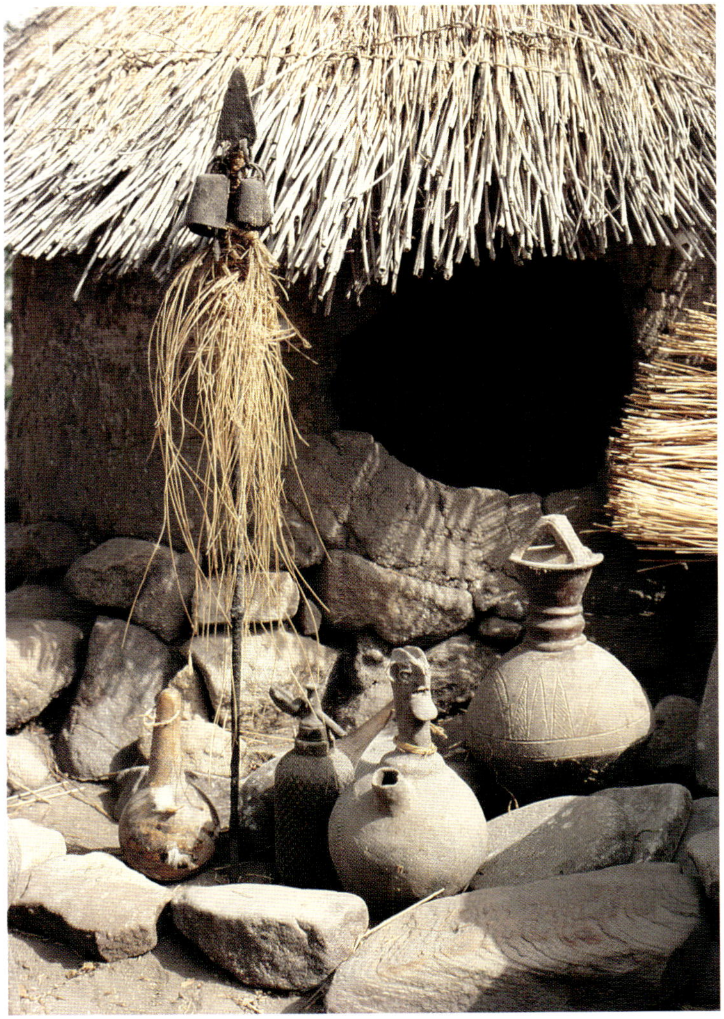

13.33 An iron spear (*swe mam*) is a dominant element of the contents of the Kan Mam (House of Mam spirits) maintained by the Dadiya chief of the Lokwilla family. The figurative vessel (*magula*) in front was used in healing rites enacted by chiefs at the site, providing a point of contact between them and the spirits they called forth. The iron spear, however, was the most important healing medium. It was circled over a person's head in what one healer described as the symbolic act of "killing" disease-causing spirits. Sounds made by the cluster of large clappered bells at the top added to the weapon's efficacy. PHOTOGRAPH BY MARLA C. BERNS, NIGERIA, 1982.

Throughout the region, iron weapons of war, once used to defend communities and inflict harm, were deployed in investiture to confer positive agency to chiefs and ritual workers (figs. 13.29–13.32). Tall spears were festooned with iron rings, bells, chains, and folded amulets. Producing sound was central to their efficacy, whether mobilizing spirits or calling people to action.

During festivals held at the end of the dry season to pray for ample rain and again to offer thanks for a successful harvest, Dadiya ritual priests wielded iron spears (*swe mam*) adorned with large clappered bells of bent iron; on these, grasses were tied to activate the spirit protector Mam (fig. 13.33). Yungur similarly considered annual renewal events—Mam Nsara, a post-harvest festival in October and November, and Mama, a pre-planting festival in April—as opportunities to appeal to ancestral spirits for sufficient rainfall and agricultural abundance. During Mama, rain chiefs held crook-shaped iron staffs to facilitate invocations to spirits inhabiting anthropomorphized ceramic portrait vessels called *wiiso* (fig. 13.34; see Berns 1990; 2011, 536–47).

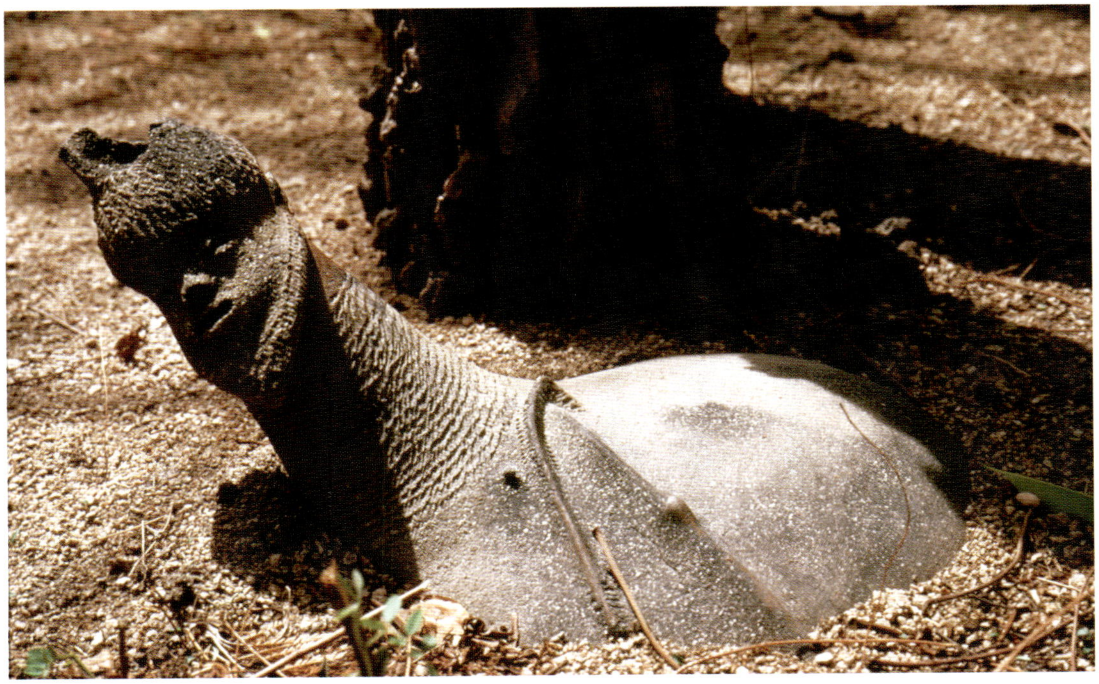

13.34 Yungur *wiiso* vessels contained ancestral spirits. The largest among them held the spirit of a former chief (*gubo*). This one, named Tentire, was buried in the ground to serve as sentry at the entrance to the most important sacred precinct at the Yungur district of Diterra. The staff visible over the spirit's left shoulder represents the bent iron staff (*salawa*) that a *gubo* carried during rainmaking rites. PHOTOGRAPH BY MARLA C. BERNS, NIGERIA, 1981.

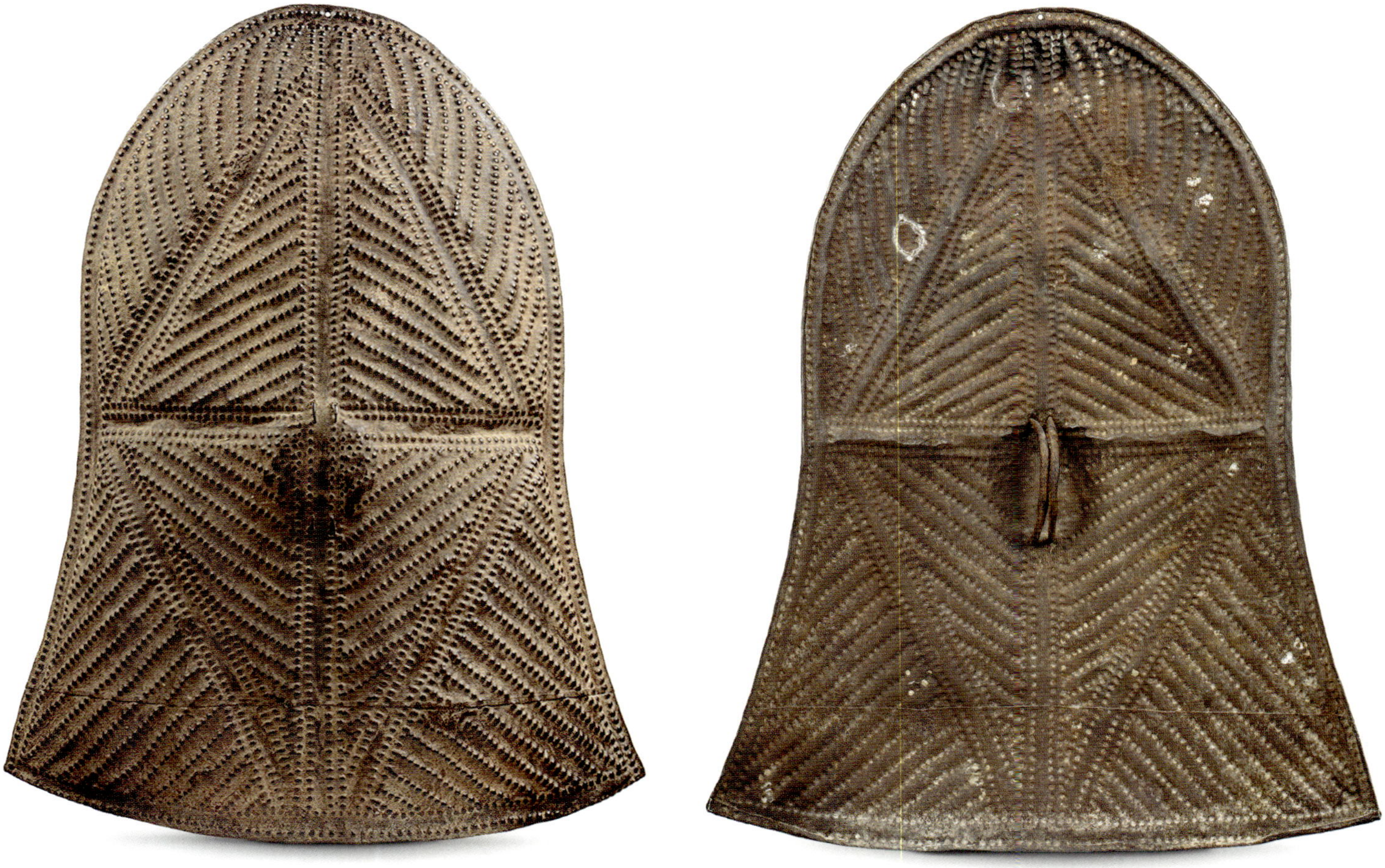

13.35A–B
Artist unknown (Fali peoples, Nigeria)
Shield (*kərmin*)
Iron
H: 90 cm
AGNÈS WOLINER, PARIS
PHOTOGRAPH BY BRIGITTE CAVANAGH, 2017. COURTESY OF AGNÈS WOLINER

This ogival Fali iron shield is covered with lines of lavish, cold-punched bumps in a design formation consistent with the rare examples found in collections and others documented in the field. The central protrusion on the front (right) allows room from behind to attach a carrying handle (left); it can also be read as a "navel" surrounded by a "torso" with body scarification.

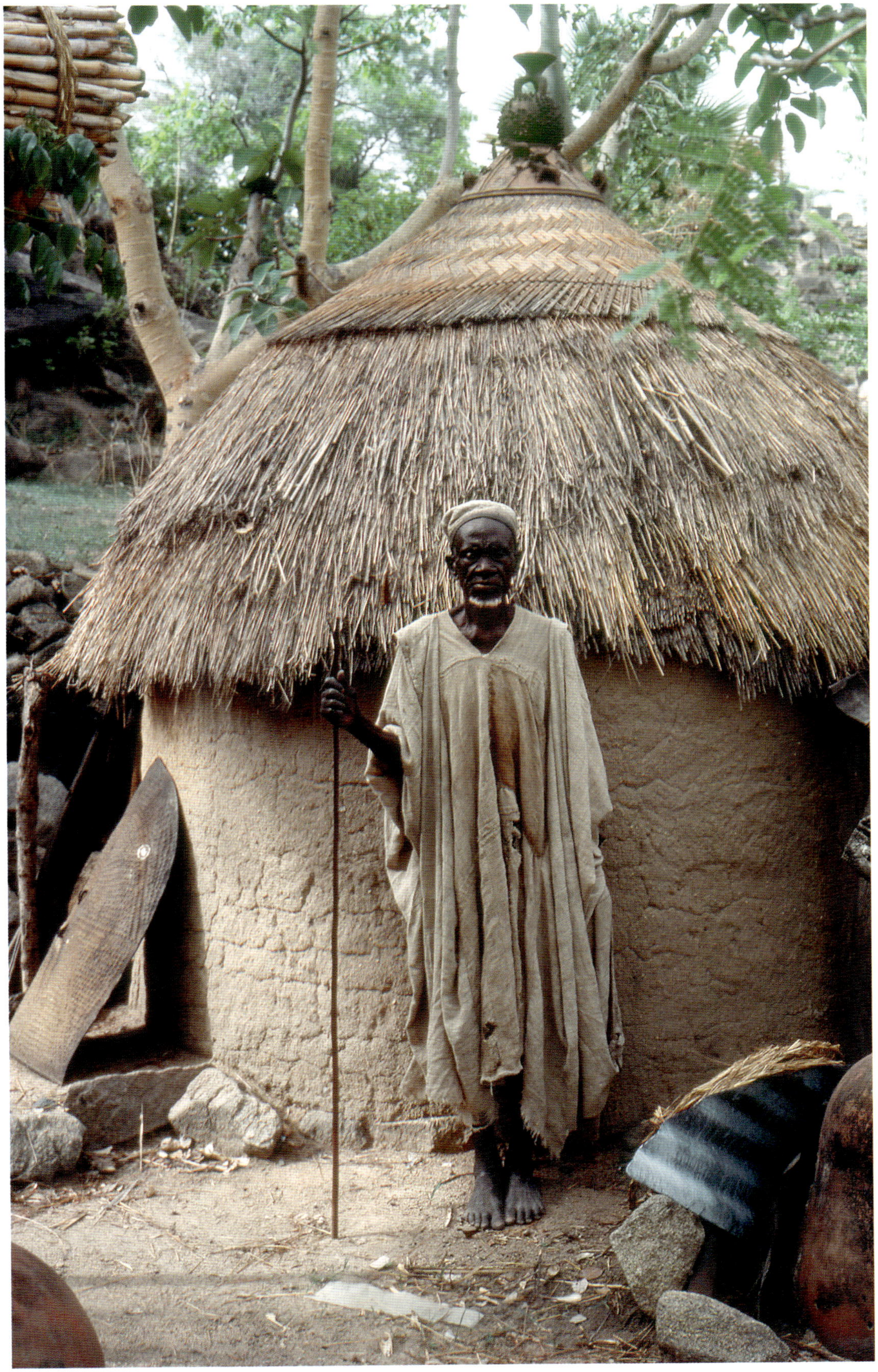

13.36 The Fali chief priest of Mwəniv, a hamlet in Vimtim, stands with a wooden staff of office in front of his residence. On its roof, a ceramic finial elaborately modeled with small figures and animals denotes his status. Leaning against the house is his iron shield, a prized cultural heirloom because of its use in activities associated with key social transitions, for example male initiation, and its former role in battle.
PHOTOGRAPH BY MARLA C. BERNS, NIGERIA, 1982.

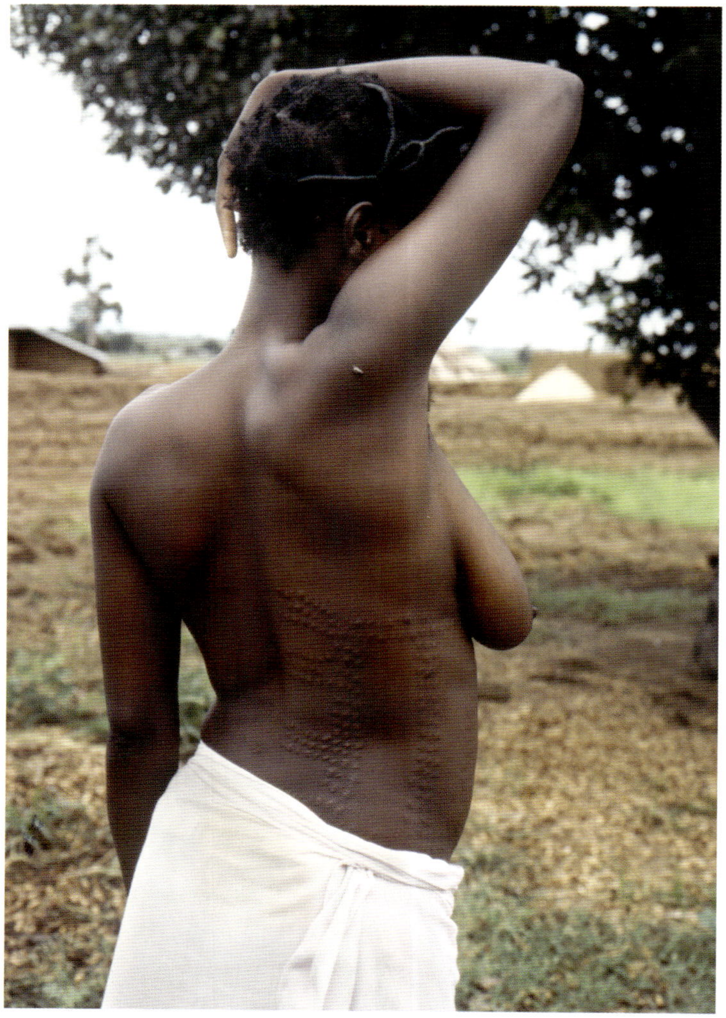

13.37 Young Fali females and males completed a two-stage sequence of scarifications called Jom'mon before marriage. The wives of blacksmiths who belonged to the Mixin caste did the cutting using small, triangular iron razors. This photograph shows a grid pattern incised in the abdomen and back in the first stage. The cuts healed to form raised bumps somewhat larger and more widely spaced than those on Ga'anda women's bodies (see fig. 13.20). The texture of scars bore strong similarity to the surfaces of *kərmin* shields (see fig. 13.35), carried during young men's transition into adulthood.
PHOTOGRAPH BY MARLA C. BERNS, NIGERIA, 1982.

Among Vimtim and Muchella Fali subgroups of Fali peoples in the western valley and foothills of the Mandara Mountains, blacksmiths were called *mixin* and were responsible for burying the dead and seeing to related ritual activities. Their wives were potters. Among the objects these smiths produced, ogival shields covered in embossed decoration were easily the largest and most impressive works in iron, as documented among other Mandara peoples, including Sukur (fig. 13.35). In earlier times, shields (*kərmin*) were shaped from animal hide, but an influx of salvaged sheet metal from automobiles and oil barrels seems to have led to their production in iron.[12] *Kərmin* that may have once offered warriors cover in battle were repurposed to offer their protective capacity to Fali youth during an initiation activity called Tagirala (Vimtim) or Mandrijin (Mucella). According to a ritual chief of the Mwəniv hamlet of Vimtim (fig. 13.36), young males were taken to a high rock outcropping twice a day for thirty days and covered from head to foot in oil and red iron oxide (*umshidin*). The initiates wandered in the hills during the day, carrying the shields as they learned skills associated with hunting (formerly warfare). They then proceeded into the second of two stages of body and facial scarification, called Jom'mon. Young Fali females were scarified similarly as part of their transition to adulthood (figs. 13.37, 13.38). It is easy to draw a close comparison between the embossed surface of the shields and the raised cicatrices on the faces and bodies of Fali. In communities throughout the Gongola River Valley, the act of inscribing meaning and memory in skin imbricates these shields and other iron objects chiseled with scarification designs with significance, potency, and enduring shared histories.[13]

This chapter has taken the transformative capacity of blacksmiths as a central metaphor for understanding the critical social shifts young men and women make as they move biologically into maturity. That all rituals marking such progressions (as well as those involving death and ancestor status) involve the use of iron implements and adornments shows that it is not the material itself that possesses potency, but the fact of its making as a complex cultural endeavor. Peoples across the Mandara Mountains and the Gongola River Valley closely associate iron with fertility, domesticity, and community belonging. The iron things that symbolically affect such identity formations are often forgings of extravagant creativity and technical innovation, supporting the arguments made throughout this volume that the virtuosity of smiths enables and enhances iron's potentialities.

13.38 The second stage of Fali Jom'mon involved inscribing three parallel lines across the forehead and down the cheeks in a zigzag pattern. Both men and women wore these same facial scarifications.
PHOTOGRAPH BY MARLA C. BERNS, NIGERIA, 1982.

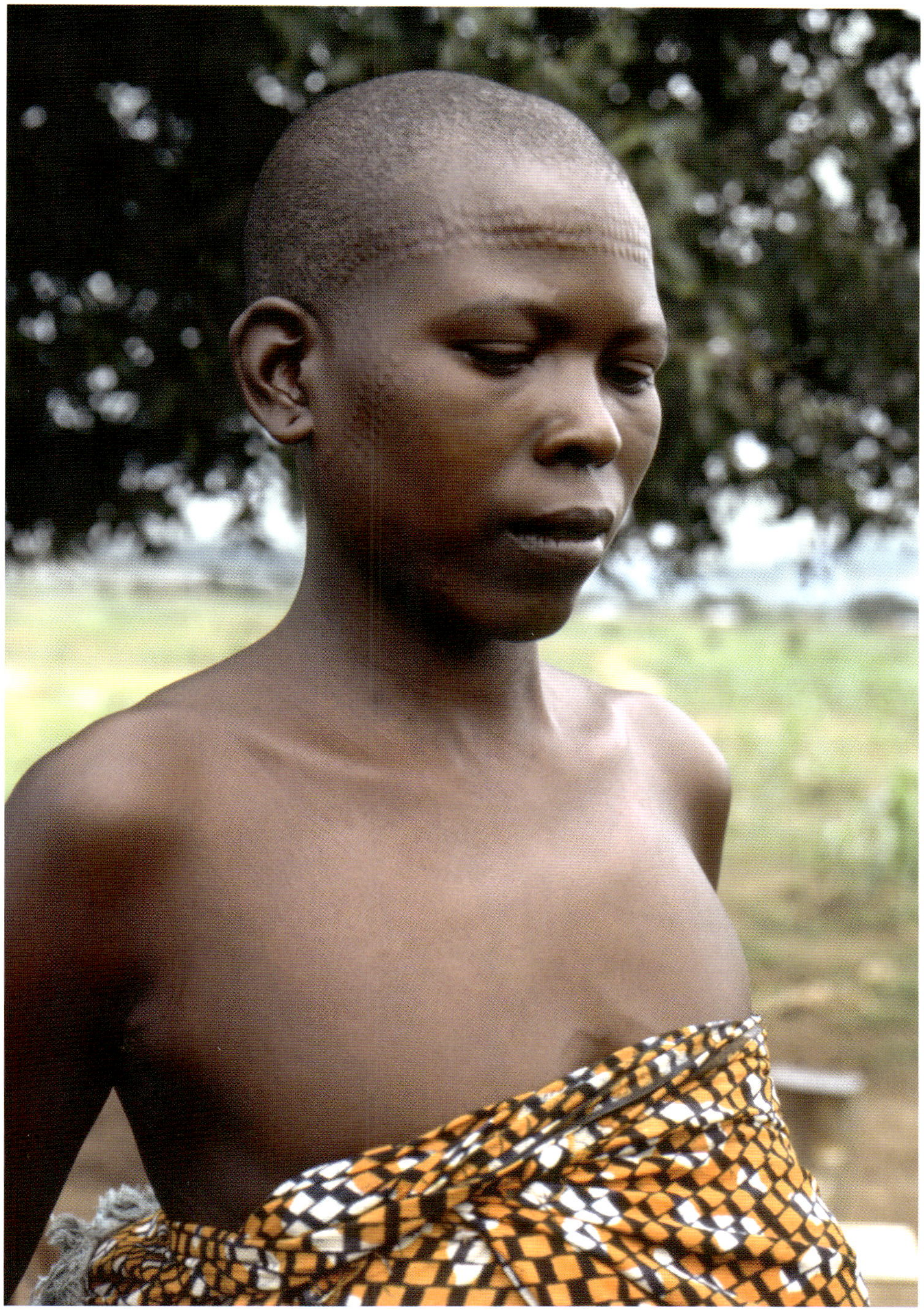

INTERLEAF
J

PERSONAL ADORNMENT OF HIMBA WOMEN

John Watson

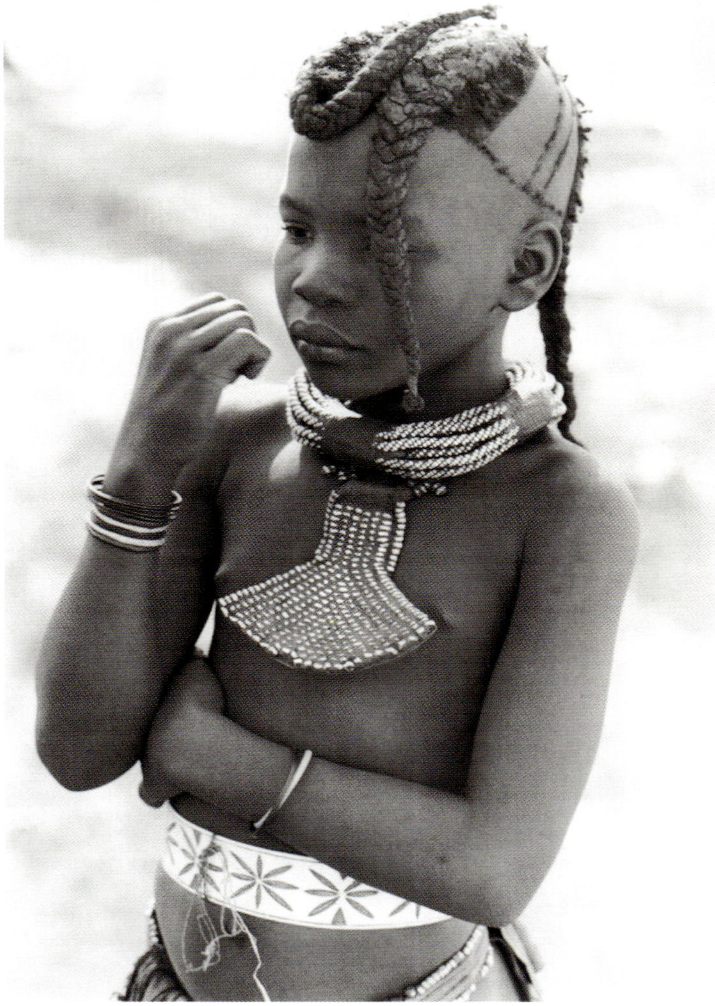

At childhood, Himba girls of northern Namibia and southern Angola begin wearing leather-backed, stranded iron bead necklaces that signify their age and changing social status (fig. J.1). At puberty, they add long pendant necklaces (*ohumba*; figs. J.2, J.3) from which they hang the large sea shell of the *Conus imperialis* and place the prepubescent necklace so that it hangs on the back of the necklace. Married women wear calf-leather headdresses (fig. J.4); after the birth of a first child, a more ornate belt (*epanda*; fig. J.5) is added. All of the adornments illustrated here were rubbed with *otjize*, a cosmetic mixture of butterfat and red ochre that colors the leather and keeps it supple; the same cream can be applied to the body as protection from the harsh desert climate. Further enhancing the beauty of the wearer, *otjize* is often perfumed with aromatic resin. Himba are pastoral peoples, and leather used as women's adornment is a reminder of the exchange of cattle as bridewealth in marriage, since wealth and well-being are measured in livestock.

J.1 A preadolescent Himba girl, identified by her double-plaited hairstyle and the necklace with flared pendant (*eha*) that she will reconfigure to wear down her back upon puberty. She also wears a belt of plastic engraved with designs and rubbed with charcoal and ochre paste.
PHOTOGRAPH BY JOHN WATSON, KAOKOLAND, NAMIBIA, 1996–97.

J.2
Artist unknown (Himba peoples, Namibia and Angola)
Necklace with pendant (*ohumba*), ca. 1920
Iron, leather, aluminum, shell (*Conus imperialis*), ochre pigment (*otjize*), plant-fiber cordage
L: 67 cm
GHYSELS COLLECTION

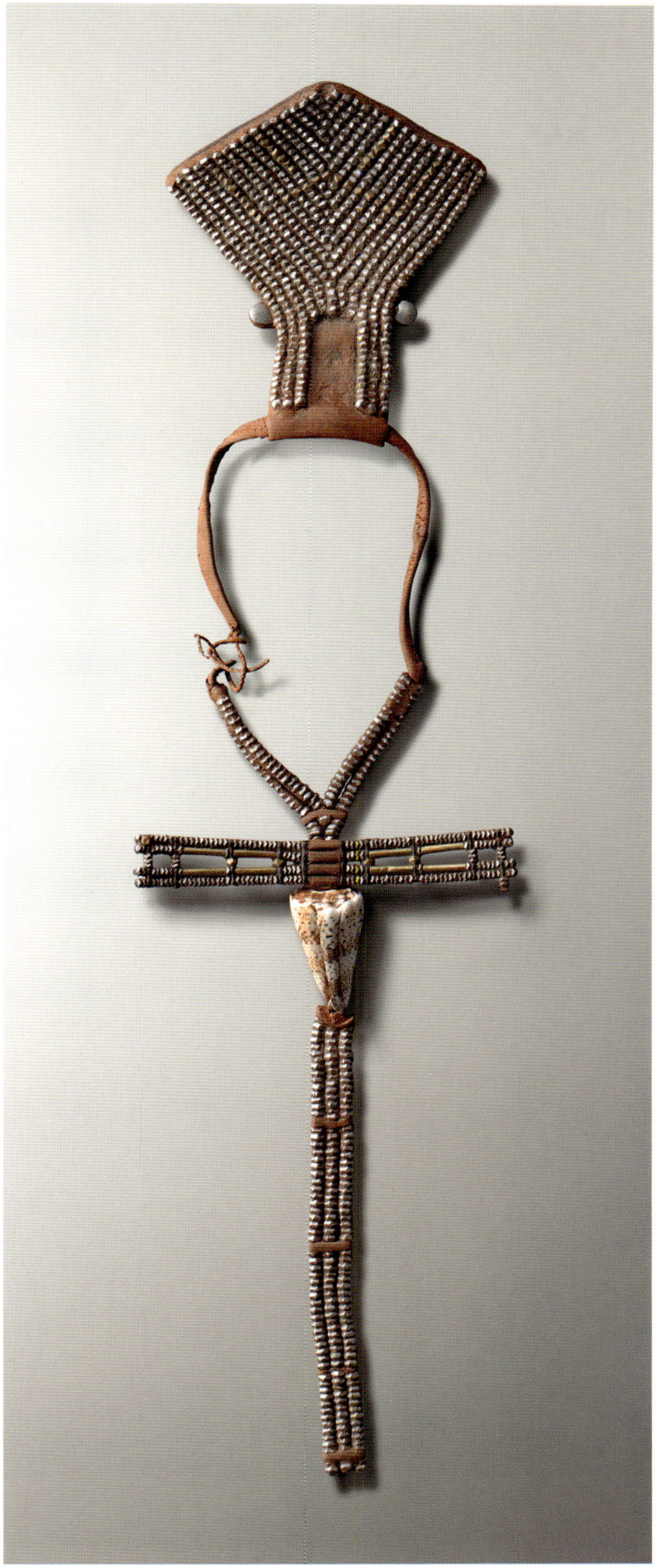

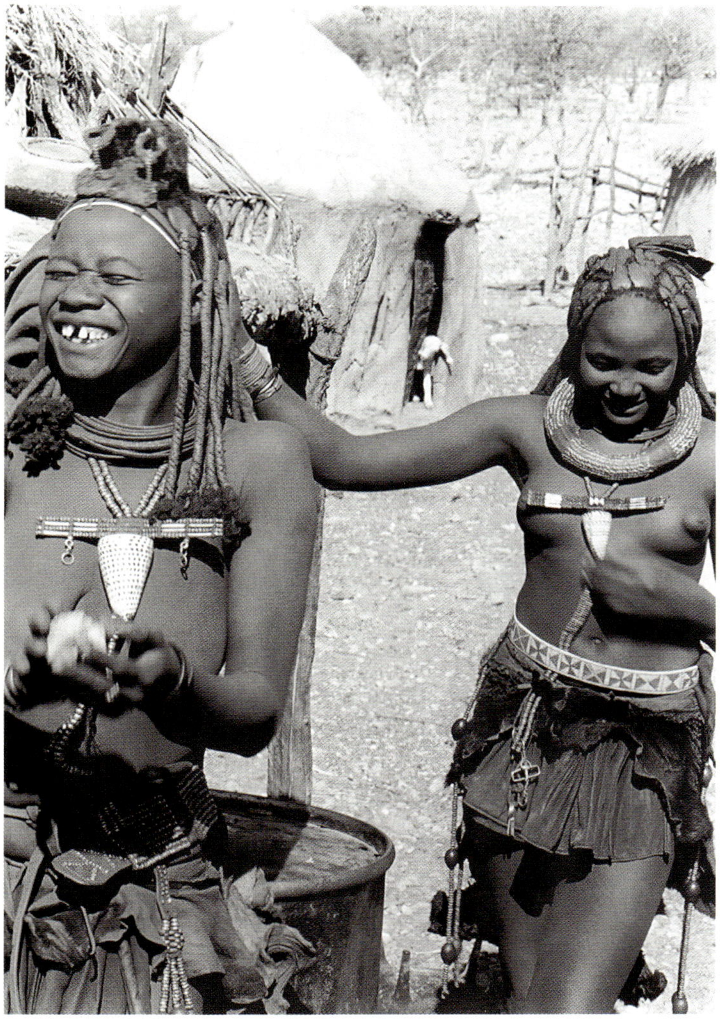

J.3 Two Himba women wearing leather and iron adornments as well as coiffures identifying their age and social standing. The thick, rolled necklace on the younger woman to the right consolidates the iron-beaded leather strands she wore as a child. The beaded iron belt worn by the woman to the left indicates she has children.
PHOTOGRAPH BY JOHN WATSON, KAOKOLAND, NAMIBIA, 1996–97.

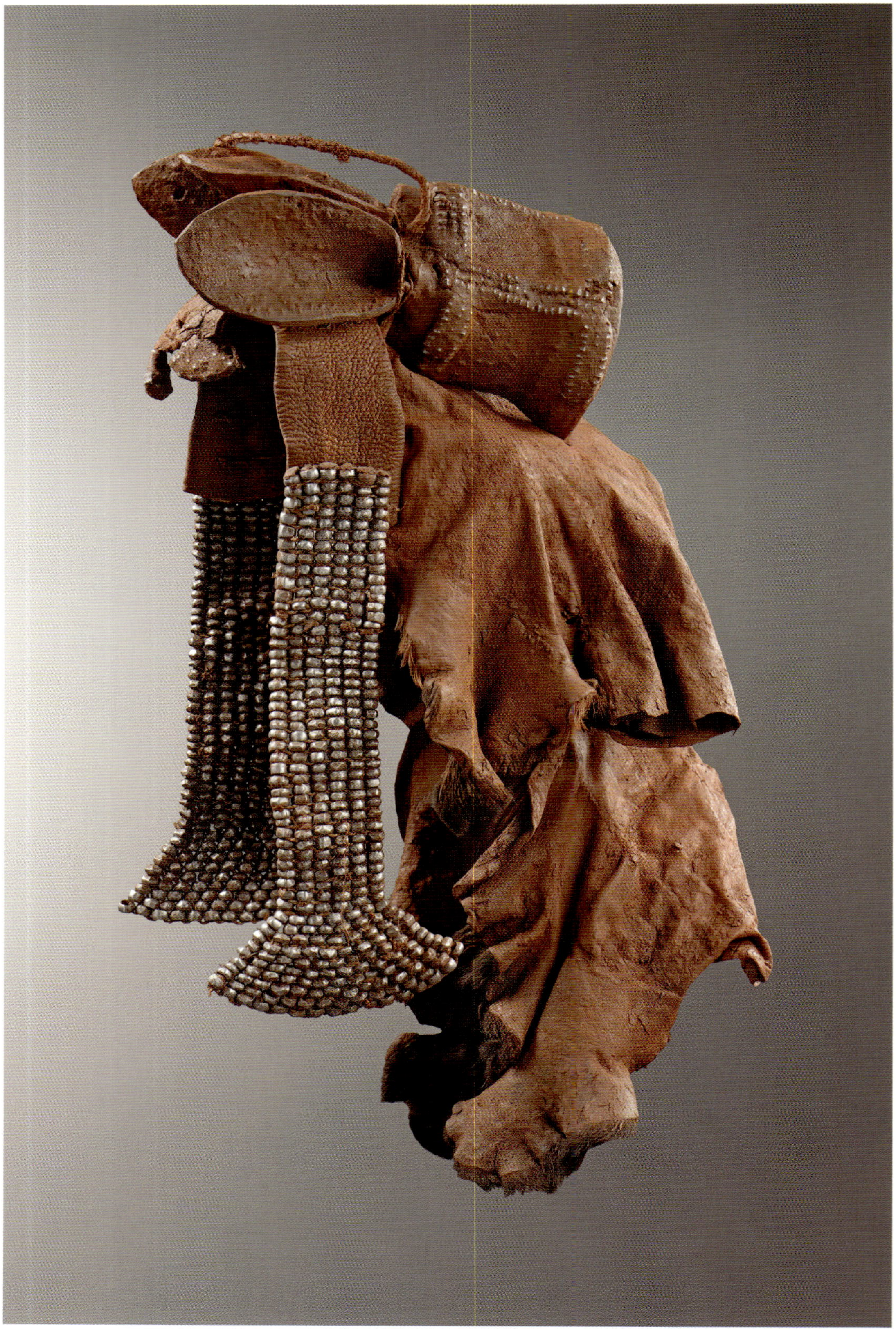

J.4
Artist unknown (Himba peoples, Namibia and Angola)
Headdress (*ekori*), ca. 1920
Iron, leather, ochre pigment (*otjize*), plant-fiber cordage
H: 55 cm
GHYSELS COLLECTION
IMAGE © STUDIO R. ASSELBERGHS-FRÉDÉRIC DEHAEN, BRUSSELS, 2018

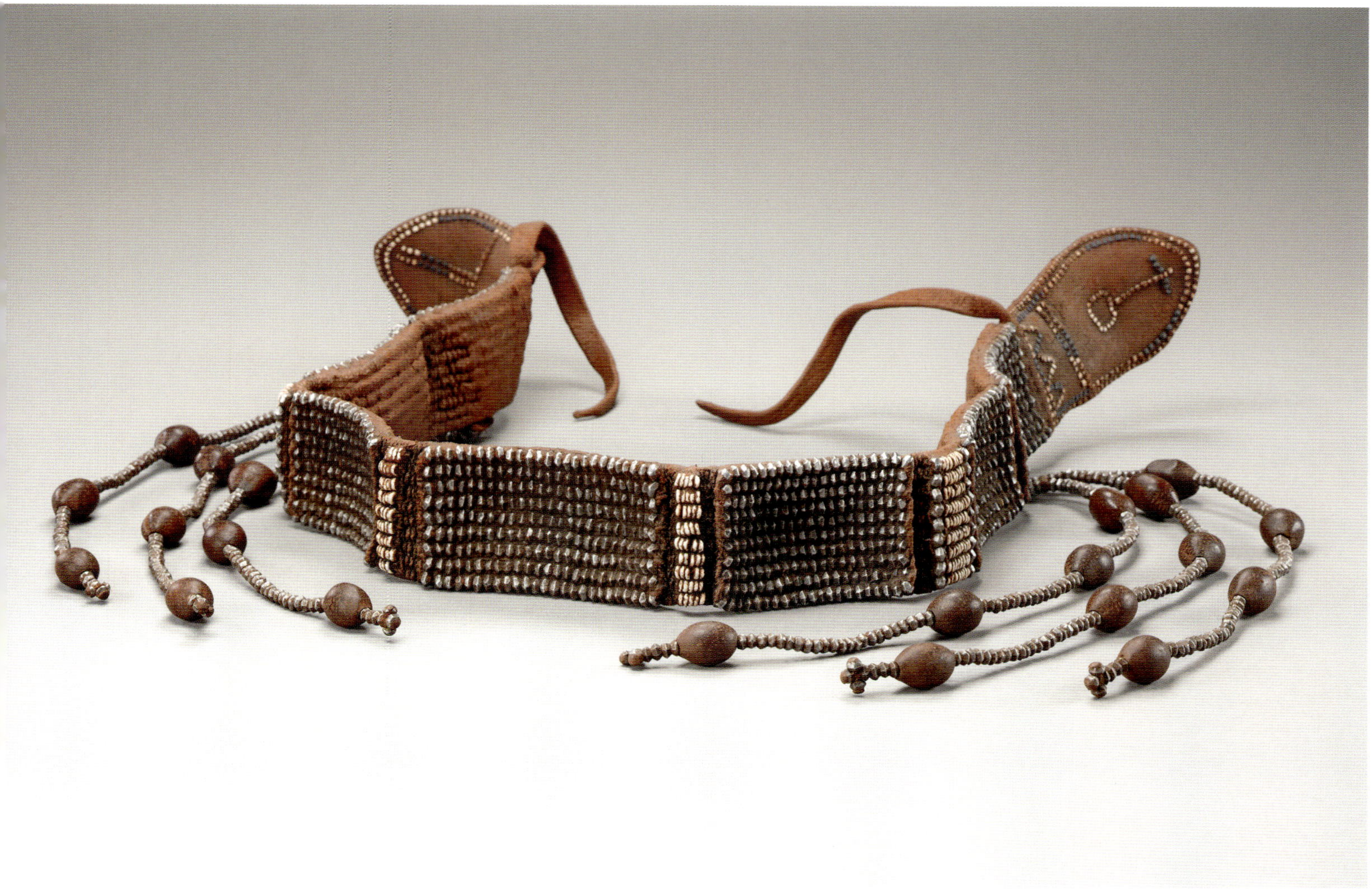

J.5
Artist unknown (Himba peoples, Namibia and Angola)
Belt (*epanda, ozonhumba*), ca. 1920
Iron, leather, lead, ostrich eggshell, seeds, ochre pigment (*otjize*), plant-fiber cordage
L: 87 cm
GHYSELS COLLECTION

CHAPTER

14

FORGING POWER AND MEMORY IN WEST-CENTRAL AFRICA

Colleen E. Kriger

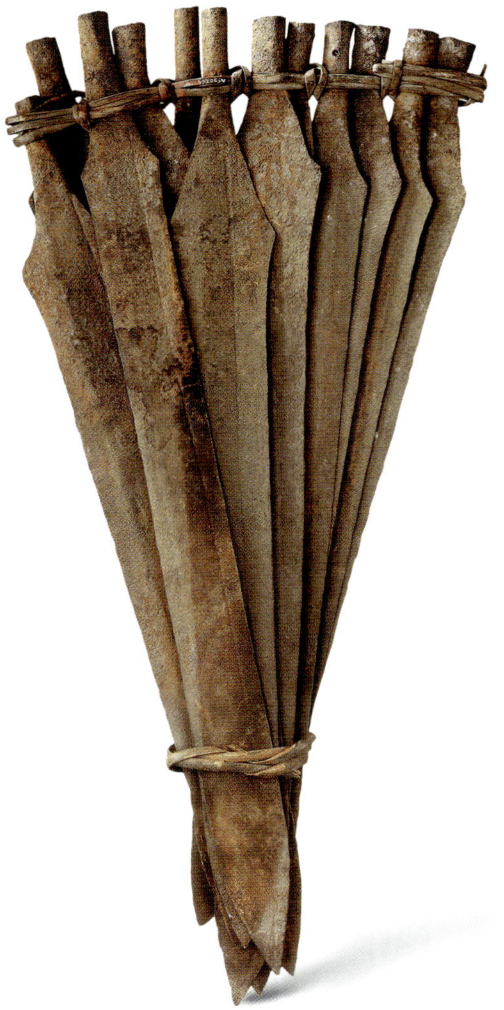

> The variety of historical evidence is nearly infinite. Everything that one says or writes, everything that one makes, everything one touches can and ought to teach us about humankind.[1]

Smelting and smithing iron in west-central Africa's Congo River basin goes back to the last centuries BCE. Over time, as skills in working iron became more widely practiced and as blacksmiths' products became more specialized, iron traveled among and between individuals at all levels of society. Iron was at the center of tropical agriculture as the handheld hoe fueled seasonal cycles of planting and cultivating crops. Exhausted fields left fallow were opened up again by iron as farmers wielded heavy axes and sharp machetes to clear overgrowth and prepare new ground. Iron was important also to skilled hunters, whose specially designed arrowheads or spear blades proved effective in felling wild game to supplement domesticated sources of protein—the goats and fowl villagers harbored within their households. Thus, iron was long an economic necessity over much of west-central Africa, especially in precolonial times, providing tools for supporting and sustaining peoples' communities. That basic contribution to food production was, from the beginning, a major hallmark of iron and the blacksmith.

Widespread value and demand for iron could be seen in circulation of many forms of iron currency units. Interconnected and overlapping currency zones featured three-dimensional forms of bar iron as well as semifinished blades in a range of shapes and sizes. People used them as money—that is, as standards of value

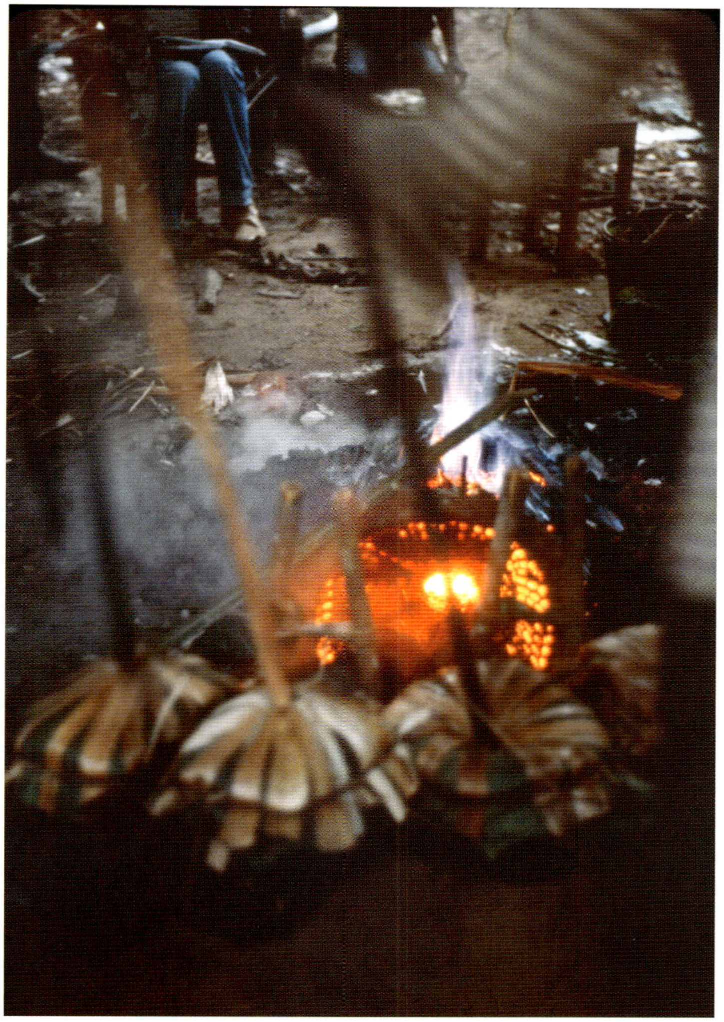

14.1
Artist unknown (Budja peoples, Democratic Republic of the Congo)
Bridewealth bundle of spear-blade-shaped currencies (*adjenge*), late 19th century
Iron, plant-fiber cordage
H (each currency): approx. 58 cm
FELIX COLLECTION

Currency forms were produced often as bars or blades at sizes suitable for exchange in bundles of several or more. They were formed into standardized units and drew value in part from the amount of iron required.

14.2
Bowl furnace of the kind used by Libinza blacksmiths, with a wide tuyère opening so that the bellows operator can gauge furnace temperature by the fire's color.
PHOTOGRAPH BY COLLEEN E. KRIGER, LOPANZO, DEMOCRATIC REPUBLIC OF THE CONGO, 1989.

in pricing and also as means of exchange in payments for goods or services (fig. 14.1). In the late nineteenth and early twentieth centuries, ethnographers, explorers, and missionaries traveling or living in the region took note of market prices, especially for foodstuffs, quoted in iron currencies. In 1894, the price of a good-sized chicken was one unit of the hoe currency *temo* in the central highlands of Angola, for example, while in the upper Lukenie watershed the price in 1907 was one unit of the blade currency known as *ikonga*. Integration of these local iron currencies into region-wide copper currency zones by this time was evidenced by the establishment of currency conversion rates. Four upper Lukenie *ikonga* blades were considered the equivalent value of one cast copper cross from the Katanga Copperbelt, located roughly four hundred miles to the southeast (sources cited in Kriger 1999, 89, 103–04).

The use of iron currencies as payment ranged from simple arrangements such as chiefs charging traders a toll fee in hoe currency or spear blades paid to settle a dispute, to much larger and more complex exchanges combining goods and currencies in the sale of exports such as ivory and slaves.[2] Iron currencies were also some of the most preferred items included in bridewealth payments, when goods passed from the family of a man to that of his prospective wife. Such transfers took place formally and in the presence of witnesses to guarantee legitimation of the marriage contract (Kriger 1992, 198–204; Kriger 1999, chapter 4; see also chapter 5, this volume).

Thus, blacksmiths became known as makers and changers of money; the smithy was both a workplace and a form of bank where, on occasion, people could borrow iron, exchange worn-out iron tools for new ones, or exchange copper currency for iron, or vice versa. From what we know about west-central Africa in more recent times—from at least the nineteenth century—smithing was a highly respectable male occupation and considered the "pride of men." Blacksmiths' work, especially with iron and iron money, gave them a revered position in society, and their useful skills could be parlayed into success and wealth. Well-connected in and beyond their communities, many became prominent and powerful as heads of large households and lineage groups (Kriger 1999, chapter 4).

It is therefore not surprising that blacksmiths would be among the significant people remembered in west-central African oral traditions and in lineage and family histories. Some of these histories provide us with descriptions of blacksmiths' workshops, their careers over time, and aspects of their social and geographical mobility. A particularly vivid example is the history of Bobaka and Bongenge, two villages of blacksmiths that monopolized iron-working up the Ngiri River, a tributary of the Congo, during the nineteenth century. Both villages were founded by Libinza, or river peoples, whose men had changed their primary occupation from fishing to smithing. In the late eighteenth century, as export trade in slaves and ivory was developing on the middle Congo River, groups of Libinza smiths moved up the Ngiri River to found new settlements away from the violence and predation of trading and transporting enslaved people. Not content to limit themselves to everyday repair work, these smiths made contact with smelters some distance away and arranged to learn from them the specialized skills of how to win iron from ore (fig. 14.2).

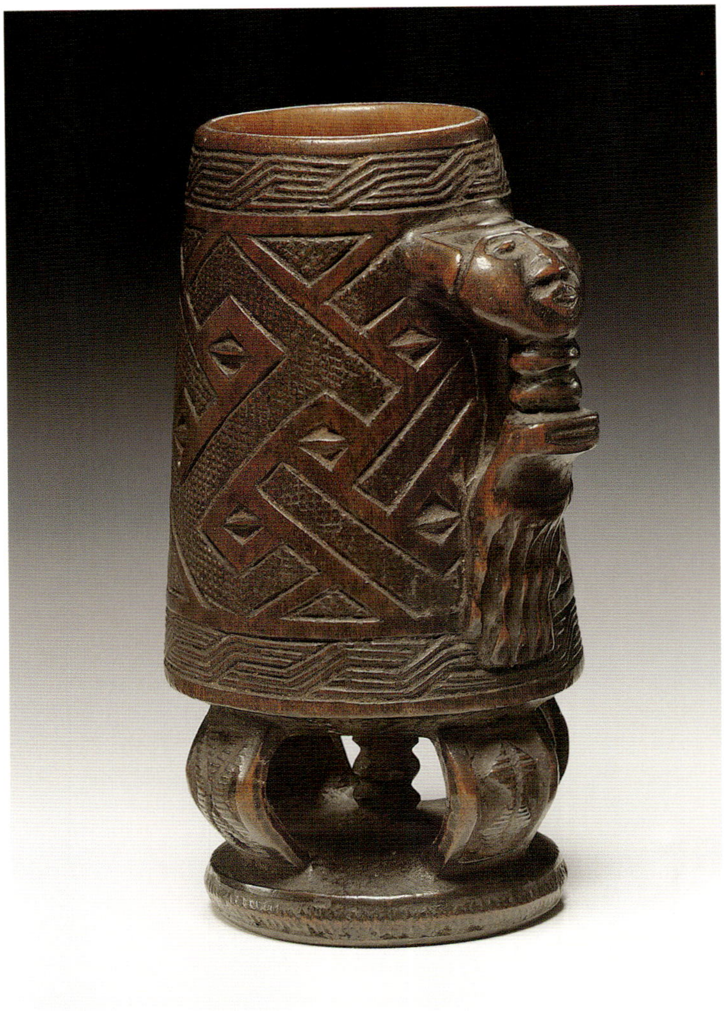

14.3
Unknown artist (Kuba peoples, Democratic Republic of the Congo)
Cup, 19th century
Wood
H: 19 cm
ETHNOLOGISCHES MUSEUM, STAATLICHE MUSEEN, BERLIN, GERMANY, INV. NO. III C 19764
(NOT IN EXHIBITION)
IMAGE © BPK BILDAGENTUR, ETHNOLOGISCHES MUSEUM, STAATLICHE MUSEEN, BERLIN/ART RESOURCE, NY

An exemplar of Kuba artistry, this ceremonial palm wine drinking vessel was carved to replicate the shape of a specific drum. Such fine shaping and modeling and the intricacy of the interlacing patterns are testament to the woodcarver's mastery and the high quality of his specialized iron tools.

At first such smiths smelted iron from their own local ore deposits, but as their settlements grew and prospered they contracted river traders to bring them additional supplies of iron ore from a major source some two hundred miles up the Lulonga River, another tributary of the Congo. As smiths in these villages became wealthy, they made investments that compounded their wealth. Enterprising individuals broadened their skills and toolkits by learning to make new products after traveling to work for a time with well-known master smiths in other communities. Among the wide range of resulting wares were the local bar iron currency *ebuni*, hatchets, chisels, razors, bells, five different types of fishing harpoons, three types of lances, two types of adzes, and copper ornaments. The most important Libinza iron form was the knife, of which four sorts were designed for utilitarian purposes, two for ceremonial display, and two as currency blades. Smiths also invested in family. Because their productivity in ironworking allowed them to offer more iron-rich bridewealth payments, they could acquire more wives than ordinary men (Mumbanza mwa Bawele 1989, 114–32). Thus, some blacksmiths became founders of lineages, with ever-growing numbers of descendants ensuring proud remembrance of names and careers.

BLACKSMITHS REMEMBERED IN KINGDOMS

Blacksmiths also rose to positions of political power, and in the west-central African past, individual smiths could be found among titleholders, governing officials, and kings in the region. One of the most prominent examples is the Kuba Kingdom, whose oral traditions called attention to the importance of ironworkers. Kingship and the Bushong ruling dynasty are estimated to go back to the early seventeenth century; by the time of the kingdom's apogee in the 1800s, iron and iron products were key elements in its economy, social structure, and governance. Kuba rulers built their wealth on unusually high agricultural productivity, setting them apart from neighboring kingdoms. Widespread in the region were new crops introduced from the Americas such as maize and cassava, which offered greater yields than sorghum and yams as earlier staples. Unusual to Kuba, however, was a system of double-cropping, which intensified their food production even further and brought them much larger annual harvests (Vansina 1978, 175–78). Supported by this strong subsistence base, artisans in turn became increasingly specialized and extremely skilled, creating over time the dazzling corpus of royal and official iron artworks; carved wooden images commemorating kings; masks and masking paraphernalia embodying religious and cultural beliefs; and richly patterned material culture, from carved objects (fig. 14.3) to textiles and beadwork.[3]

14.4
King Kot a-Mbweeky III of the Kuba Kingdom, in state dress with royal sword and drum.
PHOTOGRAPH BY ELIOT ELISOFON, MUSHENGE, DEMOCRATIC REPUBLIC OF THE CONGO, 1971. ELIOT ELISOFON PHOTOGRAPHIC ARCHIVES, EEPA EECL 02139, NATIONAL MUSEUM OF AFRICAN ART, SMITHSONIAN INSTITUTION

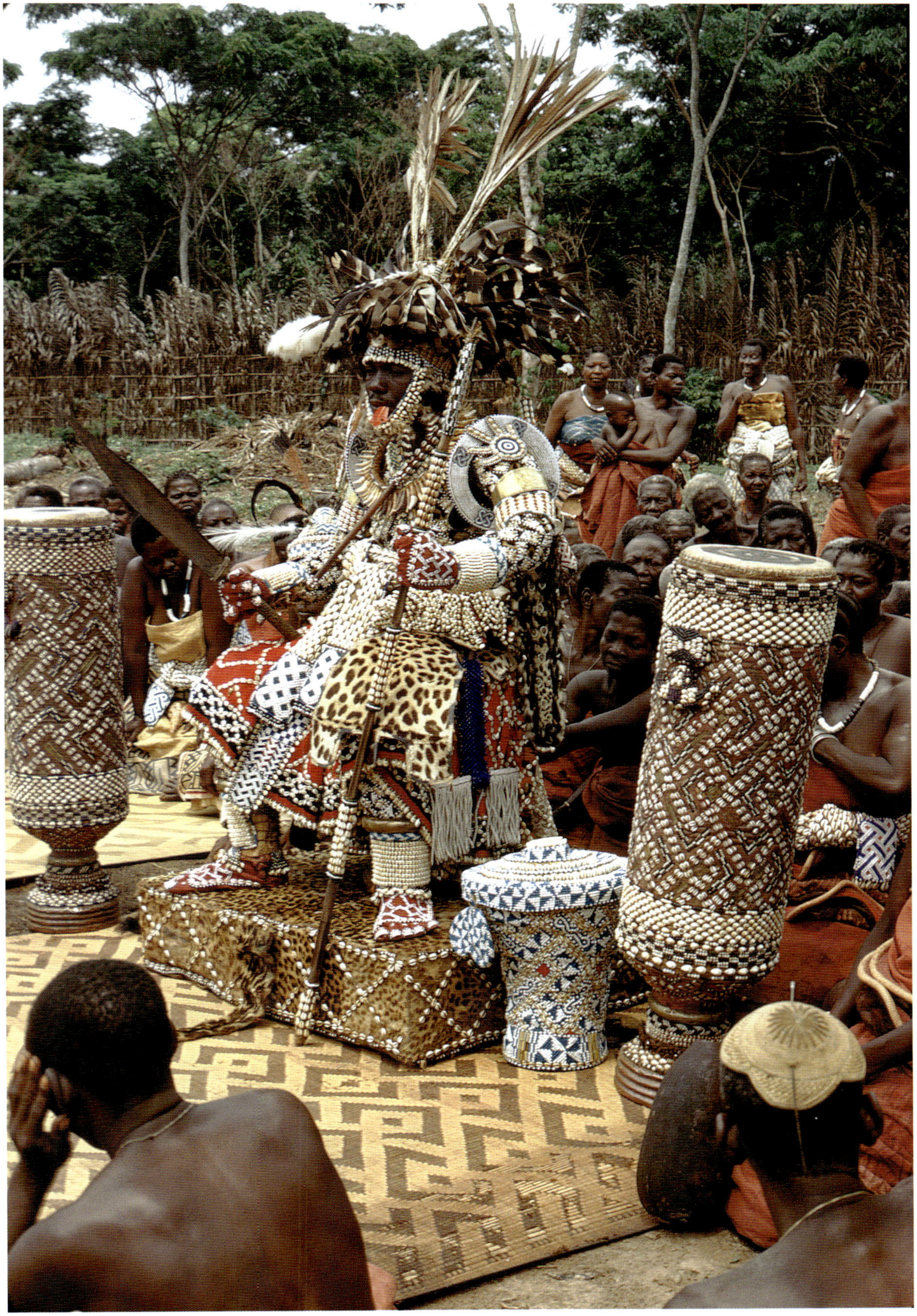

14.5
Unknown artist (Kuba peoples, Democratic Republic of the Congo)
Knife (*ikul*), 19th century
Iron, additional metals, wood
L (blade): 26 cm
ROYAL ONTARIO MUSEUM, CANADA, 939.2.188
(NOT IN EXHIBITION)
PHOTOGRAPH BY BRIAN BOYLE. © ROYAL ONTARIO MUSEUM

This fine example of the classic *ikul* knife was probably collected by the African American missionary William Sheppard between 1890 and 1910. It came into the possession of the Canadian missionary T. Hope Morgan, and changed hands again before entering the collections of the Royal Ontario Museum in Toronto in 1939 (Kriger 1992).

14.6
Unknown artist (Kuba peoples, Democratic Republic of the Congo)
Knife (*ikul*), late 19th to early 20th century
Iron, brass, wood
H: 35.5 cm
PRIVATE COLLECTION
PHOTOGRAPH BY NATHALIE AND ALAIN SPELTDOORN, 2017.
COURTESY OF PRIVATE COLLECTION

14.7
Unknown artist (Kuba peoples, Democratic Republic of the Congo)
Knife (*ikul*), late 19th to early 20th century
Iron, copper, brass, wood
H: 38.5 cm
PRIVATE COLLECTION
PHOTOGRAPH BY NATHALIE AND ALAIN SPELTDOORN, 2017.
COURTESY OF PRIVATE COLLECTION

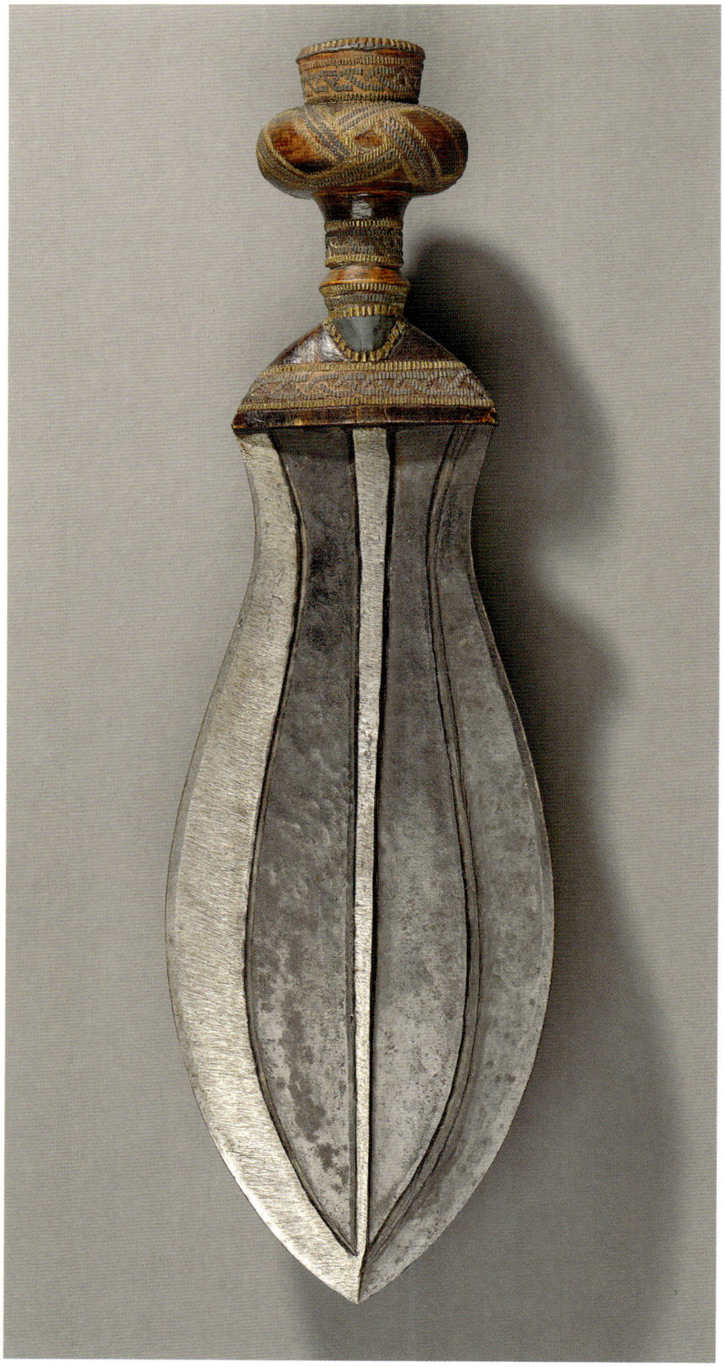

Kuba's Bushong ruling elite considered the art of smithing to be their forte, and oral traditions claim that all men of royal lineage had been trained in it to some degree. At least one king, Mbop Pelyeeng, is said to have been a blacksmith, and other individual smiths such as Myeel and Pyeekol reportedly lived and worked at the court in the seventeenth century. Indeed, some oral traditions describe Myeel as a member of the royal house. What can be said for certain is that production of bar iron and semifinished iron goods was carefully monitored by Kuba officials, who made sure that the capital received tribute from iron-producing areas in the provinces.[4] The enduring social and cultural significance of iron and smithing played out dramatically in the domains of politics and diplomacy as well, through displays of official iron insignia and the deployment of iron masterworks by royals in and around and even beyond the court (fig. 14.4; see Kriger 1999, chapter 6).

One prominent Kuba knife is an innovation known as *ikul* (figs. 14.5–14.8), a special item of costume designating an esteemed class of men. The Reverend William Sheppard, an African American missionary who lived near the Kuba Kingdom from 1890 to 1910, learned the Bushong language and created the earliest collection of Kuba art. He recounted how important it was for a free man not to be seen in public without his *ikul* displayed as a personal ornament at his waist (fig. 14.9), and above all how crucial it was for him not to lose it.[5] *Ikul* are modest in size, the blade and handle together usually measuring about thirty-three centimeters in length. The most frequently made version, and probably the original prototype, has a blade whose silhouette is shaped like a broad, gently curving leaf, with each side blackened and selectively polished to highlight the midrib and one of the two beveled edges (see fig. 14.5). Despite its apparent simplicity, close examination reveals the blade's exceptional structure. Its precise cross-sectional geometry—that is, the elegant design and careful modeling of the midrib and bevels—offers compelling evidence of Kuba blacksmiths' singular artistry (Kriger 1999, 172–76). Novel variations of this blade were invented over time, some with special surface patterns, others with slightly different silhouettes (see figs. 14.6–14.8). That there would be such richly developed variations within the *ikul* genre indicates the degree to which elite Kuba men were connoisseurs of special iron products.

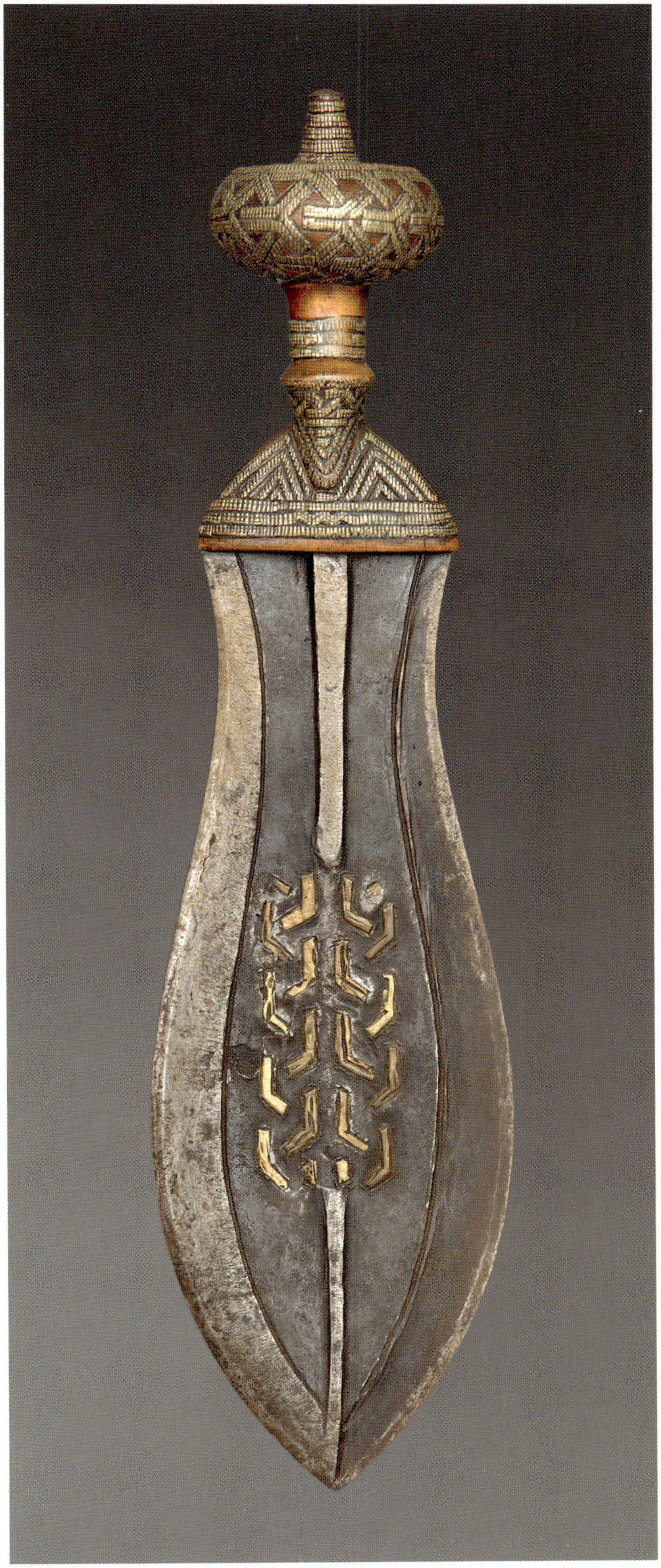

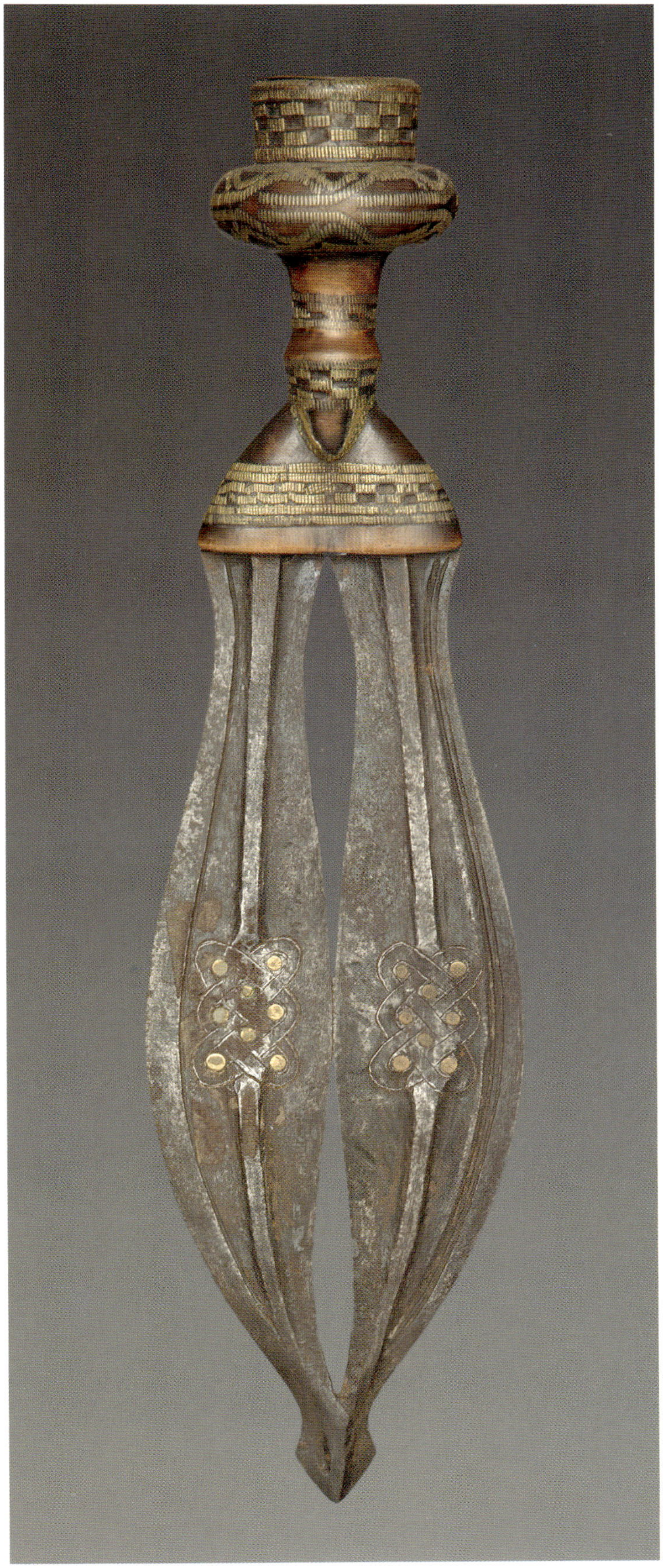

14.8
Unknown artist (Kuba peoples, Democratic Republic of the Congo)
Knife (*ikul, ikulbapyaang*), early 20th century
Copper, pewter, wood
H: 45.5 cm
PRIVATE COLLECTION
PHOTOGRAPH BY NATHALIE AND ALAIN SPELTDOORN, 2017. COURTESY OF PRIVATE COLLECTION

In the Kuba Kingdom a great variety of iron blades and related implements were forged by royal court members. One prominent Kuba knife, the *ikul*—worn by men in public as insignia (see fig. 14.9)—has a curved blade shaped like a broad leaf. Both its sides are blackened and selectively polished to highlight the midrib and one of its two beveled edges (see fig. 14.5). Seemingly simple in design but stunning in technical precision, *ikul* knives demonstrate the talent of Kuba blacksmiths and the connoisseurship of elite Kuba men.

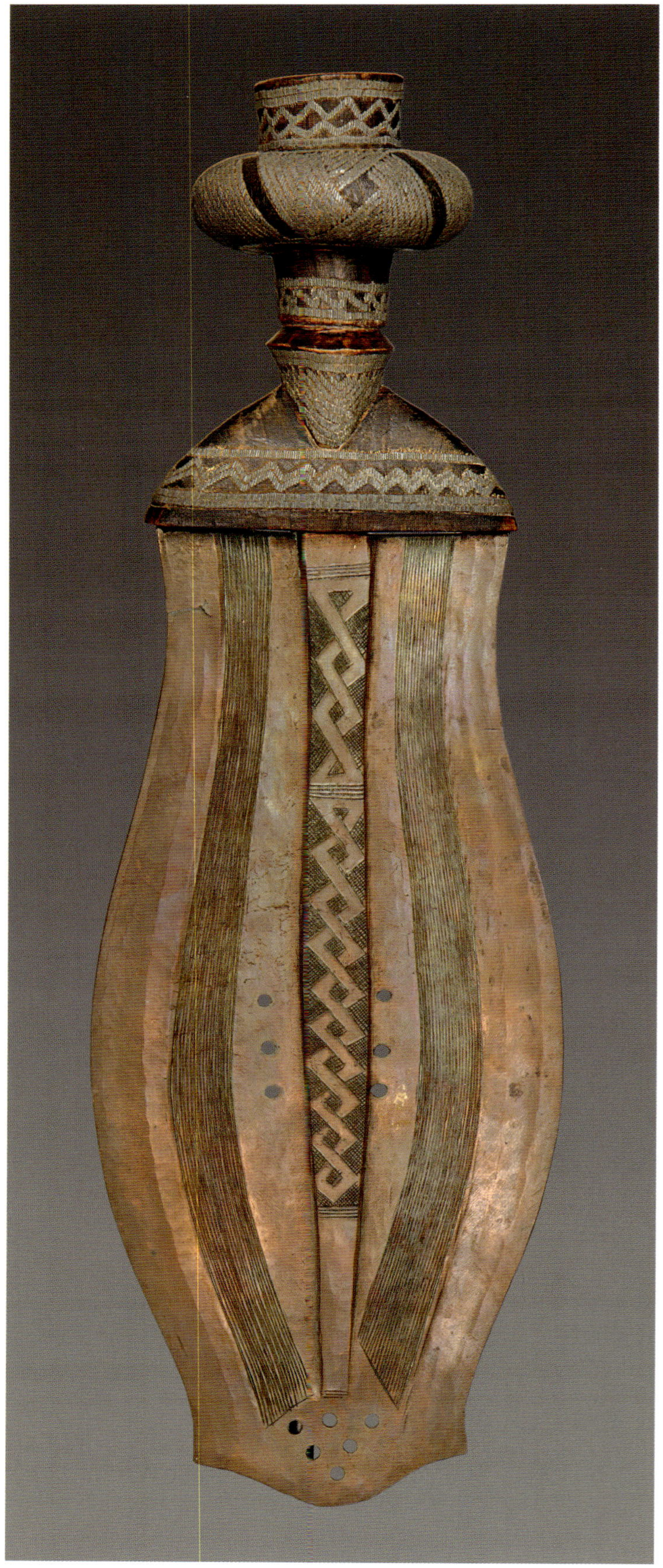

Selected from smithies in the capital and outside of it, singular and spectacular works of art in iron were kept in the royal treasury, where, along with the finest products by various kinds of artisans, they represented the kingdom's wealth; some of the glories of its history; and the taste, wit, and acumen of its foremost blacksmiths. Amassed gradually over time, revered accomplishments of many individual smiths were held as Kuba treasure and telescoped by memory and oral tradition into the products of one great forge and one great master, the legendary seventeenth-century virtuoso Myeel. As clear and effective glorifications of the king, the kingdom, and the skill of Kuba blacksmiths, such masterpieces served more practical purposes as well. Well-known objects belonging to the treasury were shown and used in all sorts of ways on major ceremonial occasions and on official business. Some were carried, for example, as proof of identification by royal messengers as they relayed confidential information too sensitive to be sent by drum or other public means. Others verified the legitimacy of royal officials such as government tribute collectors embarking on periodic travels around the kingdom (Kriger 1999, 179–81). As visual confirmation—in the form of technically complex objects that could not be easily replicated—such iron objects were effective administrative tools used to govern this complex oral society.

14.9
Titled Kuba soldiers with *ikul* knives at waist-level.
PHOTOGRAPH BY ELIOT ELISOFON, BUNGAMBA VILLAGE, DEMOCRATIC REPUBLIC OF THE CONGO, 1970. EEPA EECL 05268, ELIOT ELISOFON PHOTOGRAPHIC ARCHIVES, NATIONAL MUSEUM OF AFRICAN ART, SMITHSONIAN INSTITUTION

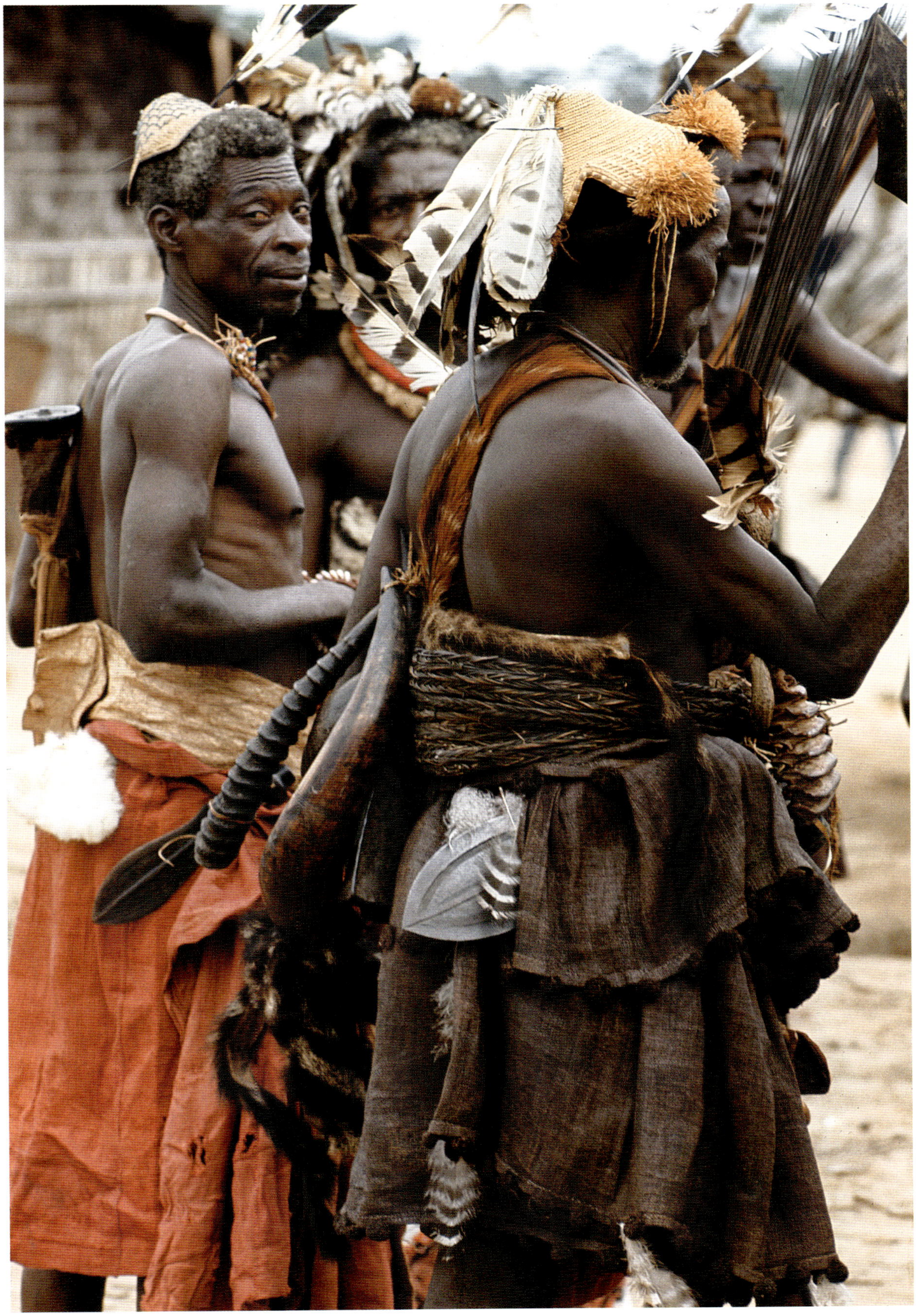

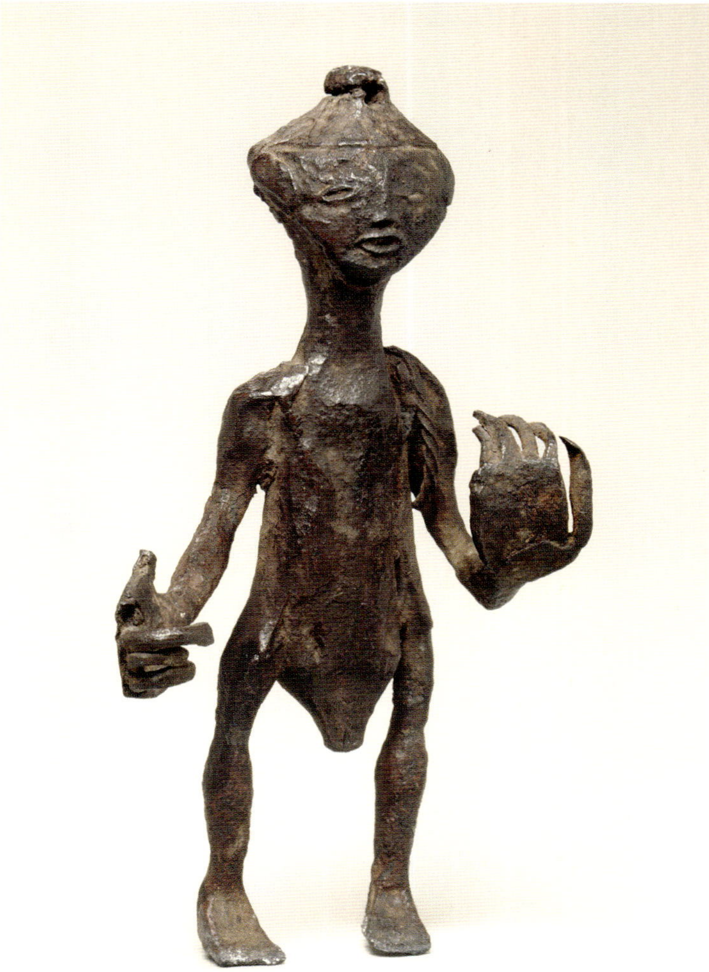

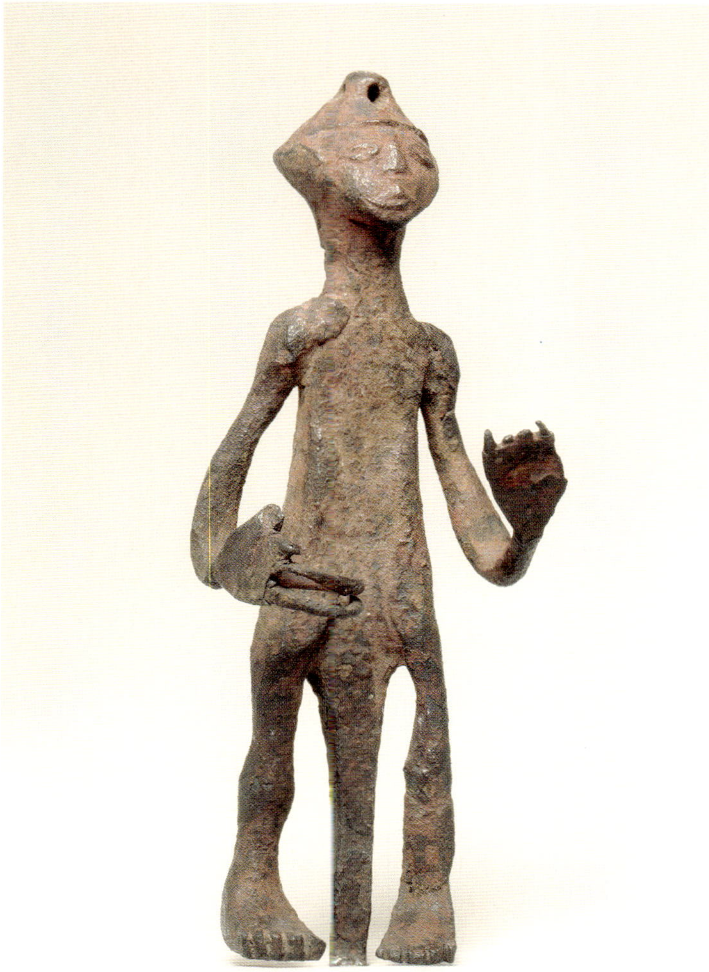

14.10
Unknown artist (Kuba peoples, Democratic Republic of the Congo)
Figurine, 17th century(?)
Iron, red pigment
H: 19.5 cm
COLLECTION OF THE MAS, ANTWERP, BELGIUM (AE. 0773)
PHOTOGRAPH BY MICHEL WUYTS, 2009. © MAS | MUSEUM AAN DE STROOM, ANTWERP

A masterpiece once kept in the Kuba treasury, this male figure is among the artworks attributed by oral traditions to the renowned blacksmith named Myeel. Remembered for his uncommon skill in forging sculptures and combining iron with copper and other metals, he is believed to have resided at the Kuba royal court in the 17th century. Remembered also for being too strict to govern, Myeel might have been a royal contender who never succeeded in becoming king (Vansina 1978, 182). The figure's most distinctive and prominent features are its hands, each with separate fingers and thumbs meticulously forge-welded onto sturdy palms.

14.11
Unknown artist (Kuba peoples, Democratic Republic of the Congo)
Figurine, 17th century(?)
Iron, pigment
H: 18.7 cm
COLLECTION OF THE MAS, ANTWERP, BELGIUM (AE. 0774) (NOT IN EXHIBITION)
PHOTOGRAPH BY MICHEL WUYTS, 2009. © MAS | MUSEUM AAN DE STROOM, ANTWERP

This figure and its homologue (see fig. 14.10) are credited to the Kuba blacksmith Myeel, but there is not sufficient evidence to confirm him as the artist nor to explain the original purpose of the figures. Damage to the extension from the lower torso in fig. 14.10 may have occasioned the creation of fig. 14.11 as a replacement.

Among a group of remarkable iron objects that went missing from the Kuba treasury by the early years of the twentieth century are two forged human figurines, each just under twenty centimeters high (figs. 14.10, 14.11). They are additive sculptural forms, with the parts forged separately and then joined together by high-heat welding, a technique associated with the making of iron hammers and anvils (Claerhout 1976, 60–64, 92). Who or what the figures represent is not known. What seem to be precise and significant gestures of the arms and hands could very well represent an iconographic message or cue. It has been suggested by Claerhout that the extensions from the figures' torsos were for the purpose of mounting them on a support for display (ibid.). The pierced opening at the top of each figure's head might have allowed the object to be carried suspended at the waist or from the neck. Indeed, it is probable that these special figures were intended for public viewing, perhaps atop a staff or to be worn on the king's person or by certain of his emissaries (Kriger 1999, 180).

A centerpiece of Reverend Sheppard's collection of Kuba art is the extraordinary sword (fig. 14.12) he received from the Kuba treasury, given to him as a gift by one of the king's sons. The prince had taken it with him as he fled the capital after his father's death, at the onset of an intensely competitive and potentially violent interregnum period. Having been offered sanctuary at the Presbyterian mission station at Ibaanc, he presented the sword to Sheppard in gratitude. Resembling in scale and silhouette the special sword reserved for the king, the blade is a startling visual and structural innovation within the corpus of Kuba art. Choosing to eschew the basic conventions of blacksmithing, this artist has created the *semblance* of an iron blade rather than one that can be used as such. It is made instead to be publicly seen and admired, and indeed it makes a strong visual statement with wit and remarkable skill. The Kuba love of pattern—units of repeating elements arranged into larger patterns—is presented here unexpectedly as additive sculpture, with the repeating elements meticulously forged in iron filigree. Sheppard was told the sword had been deployed in an annual public ceremony at court, where each chief in turn would pick up the sword, salute the king, dance with it before the assembled, and then present his report on events of the preceding year (Kriger 1999, 181–83).

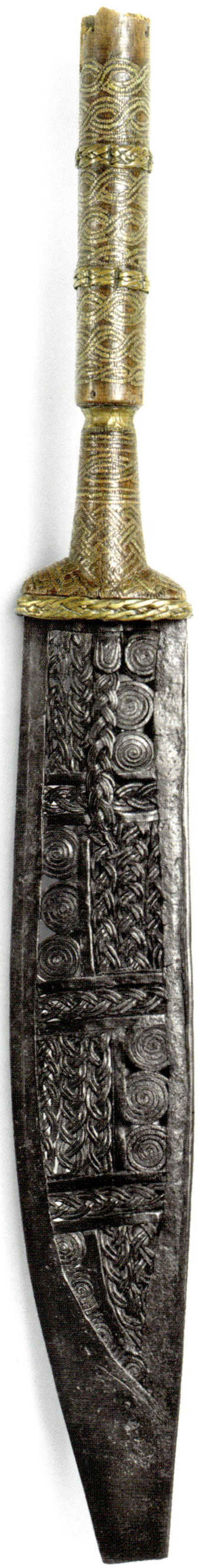

14.12
Unknown artist (Kuba peoples, Democratic Republic of the Congo)
Sword, 19th century
Iron, copper, brass, wood
L (blade): 74.5 cm
WILLIAM H. SHEPPARD COLLECTION, 15.226, HAMPTON UNIVERSITY MUSEUM, HAMPTON, VIRGINIA
(NOT IN EXHIBITION)

This ceremonial sword from the Kuba treasury is an assemblage of individual pieces of bravura smithing. Tightly wound coils and delicate interlaces feature techniques known as filigree, commonly associated with more malleable gold or silver wire, and sometimes copper, but not normally associated with iron. Here, the master smith applied filigree to iron strips and wire, integrating the blade into the kingdom's signature aesthetic, one based on repeat patterning and intricate graphic motifs. This sword must have made an eloquent visual statement when displayed in the Kuba capital, for it would have appeared both familiar and dramatically original at the same time.

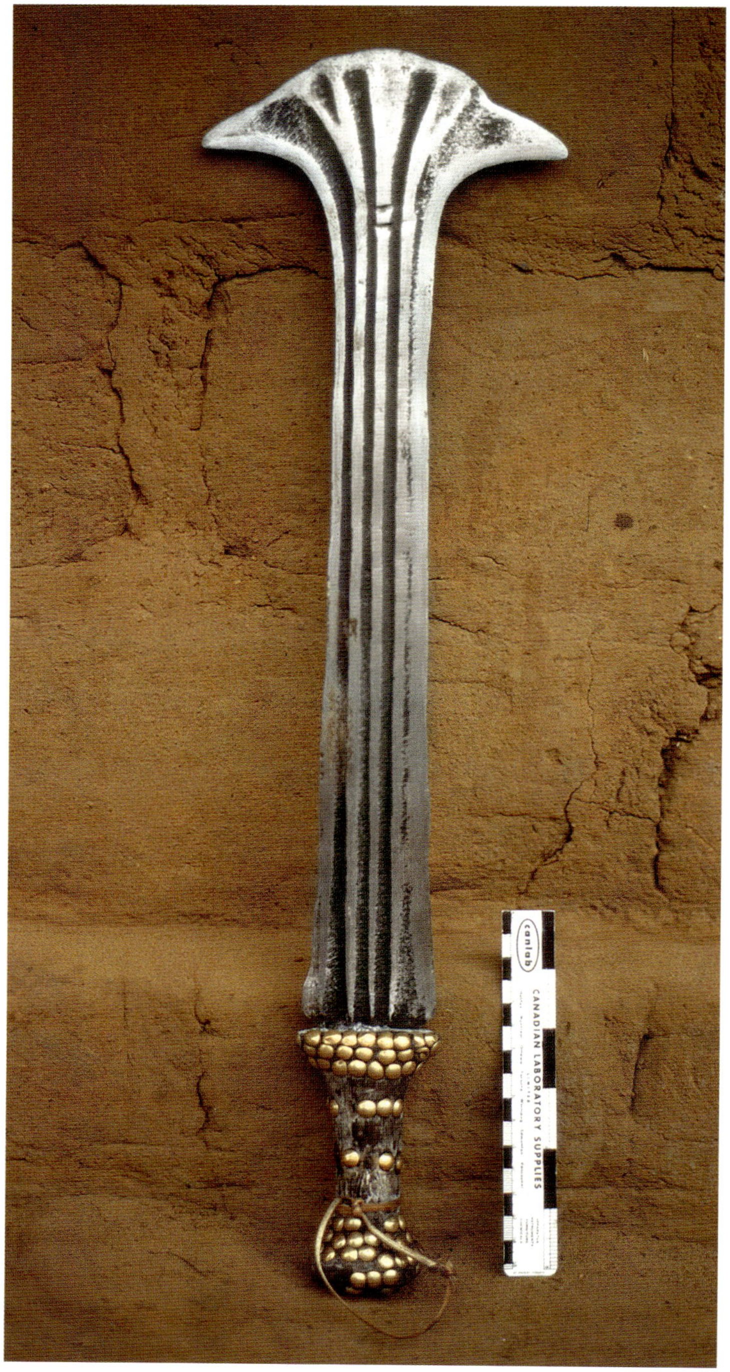

14.13 *Ebale* type of *ingonda* knife for elite men, made by Nyekuli Bana type of *ingonda* (ca. 1900–1980).
PHOTOGRAPH BY COLLEEN E. KRIGER, LOPANZO, DEMOCRATIC REPUBLIC OF THE CONGO, 1989.

BLACKSMITHS REMEMBERED IN TOWNS

Masterworks of past blacksmiths were preserved as treasures in west-central Africa's decentralized, lineage-based societies as well, though on a smaller and more intimate scale. Examples could be seen in the Ekonda town of Lopanzo, near Lake Tumba, during a research visit that I made with four other scholars in July and August of 1989 to what was then the Republic of Zaïre (now the DRC).[7] On a number of occasions, elder members of each of the town's three founding lineages brought out special treasures for show, including iron objects. Oral traditions of Lopanzo's founding lineages mention several sites where their ancestors resided before establishing the town in the nineteenth century. A local ore deposit, showing signs of having been mined in the past, and slag heaps from past smelts at the town's former site nearby confirm claims that their forebears engaged in both smelting and smithing. One of the most important prerogatives of Lopanzo's senior lineage was the right to grant access to the town's shallow pit mine, which they exercised until the last iron smelt in the 1930s. In 1989, elders in Lopanzo still associated their local deposit of iron ore with the origin of their mothers, that is, with bridewealth payments sealing their families' marriages.

One of Lopanzo's most famous smelter-smiths, Nsabonzo Engala of the senior Nzelika lineage, whose peak productive period was circa 1880 to 1910, was remembered as a wealthy, powerful, and wise man, a "founder of a lineage" who had twelve wives, nine of whom bore him children. A special ceremonial knife made by him was brought out and displayed during our visit; despite heavy mineralization on its surface, it was possible to recognize Nsabonzo Engala's accomplished skill and control. It was the most challenging version of an *ingonda*, the most highly significant knife form made by past Lopanzo smiths. In silhouette, an *ingonda* blade is characterized by a splayed, fan-like termination at the tip. Its central midrib branches out into two or more ribs at that end and its two cutting edges have bevels on both sides. Lopanzo's three main lineages kept *ingonda* knives in their treasuries, each one associated with a renowned master blacksmith. Whoever was the current eldest male in the lineage served as the privileged keeper of the treasury, and in naming past keepers of the knives, he recounted his predecessors. Thus, in public displays *ingonda* honored the achievements of their blacksmith makers and also served as mnemonic devices for remembering important aspects of lineage history.

14.14 Wielding an *ebale* ("the best") *ingonda* knife as a weapon, this *nkumu* titleholder assumes a dramatic pose for the camera. Recognizable by his tall, multitiered leadership cap and cast copper medallions, his individually achieved position combined the attributes of judge, diviner, and political leader.
PHOTOGRAPH BY GUSTAAF WAUTERS, ÉQUATEUR PROVINCE (IN PRESENT-DAY DEMOCRATIC REPUBLIC OF THE CONGO), CA. 1930. IMAGE COURTESY OF THE ARCHIVE OF THE MISSIONARIES OF THE SACRED HEART, BORGERHOUT, BELGIUM

14.15 This elite man displays an *ingonda* knife with two midribs—not as exceptional as *ebale* but more esteemed than the most common *ingonda* with one midrib. Most likely he is traveling, perhaps to a social event where he will show off his *ingonda* in dance.
PHOTOGRAPH BY GUSTAAF WAUTERS, ÉQUATEUR PROVINCE (IN PRESENT-DAY DEMOCRATIC REPUBLIC OF THE CONGO), CA. 1930. IMAGE COURTESY OF THE ARCHIVE OF THE MISSIONARIES OF THE SACRED HEART, BORGERHOUT, BELGIUM

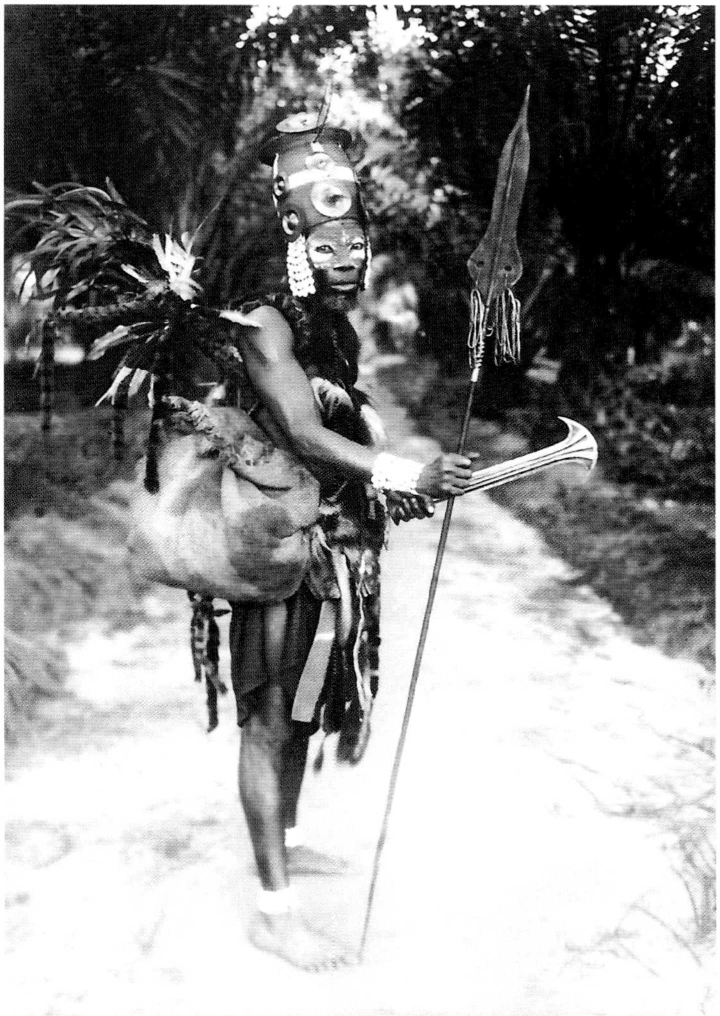

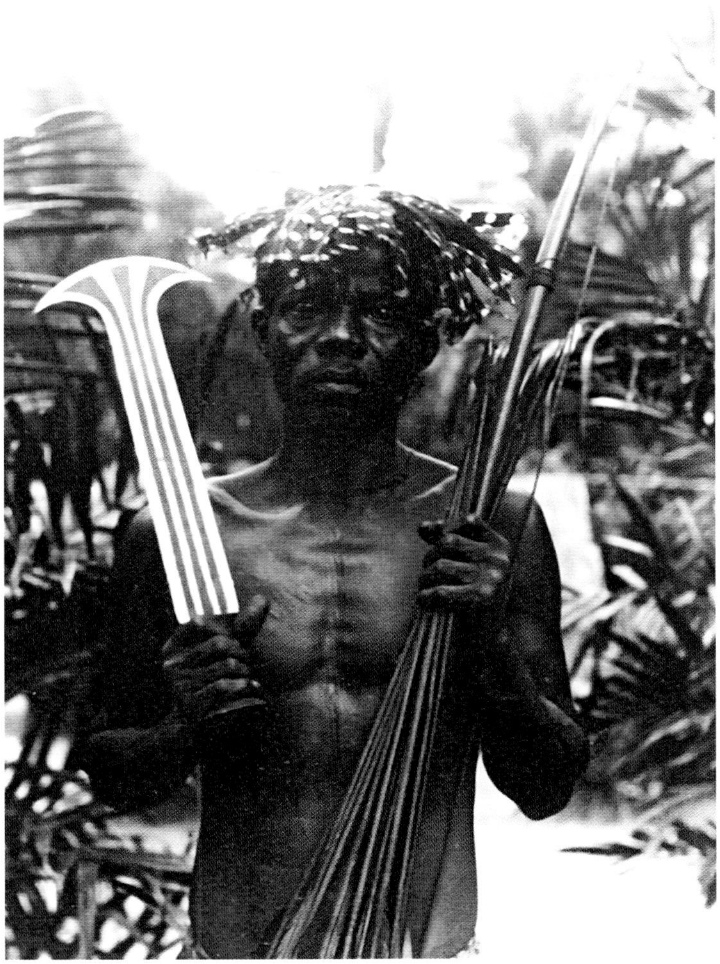

An *ingonda* (fig. 14.13) made by Nsabonzo Engala's son Nyekuli Bana exemplifies the most prized variant of this knife type, distinguished by the number of midribs in the blade. While most *ingonda* knives have one midrib, others two, more rare is this version that has three midribs and a specific name: *ebale*. The *ebale* made by Nyekuli Bana was not kept in the Nzelika treasury; it was instead in the possession of his son Nsabonzo Bilanga, who in 1989 was the current capita (head) of Lopanzo. Bilanga stated that he displayed the knife on important public occasions as the "Knife of Lopanzo," which explains why it was kept in such carefully polished and oiled condition. He claimed he also used it for traveling, and carried it as a sign of his high position and status when attending a *bonkunda* (funeral of a lineage chief). Archival photographs taken in this same general locale during the early to mid-twentieth century show formal ceremonial groupings of male notables, chiefs, diviners, and territorial judges, oftentimes holding an *ingonda* for display (figs. 14.14, 14.15; see Kriger 1999, 197–210;). *Ingonda* thus represents another exceptional knife form that, like *ikul*, was a special entitlement of west-central Africa's powerful elite men.

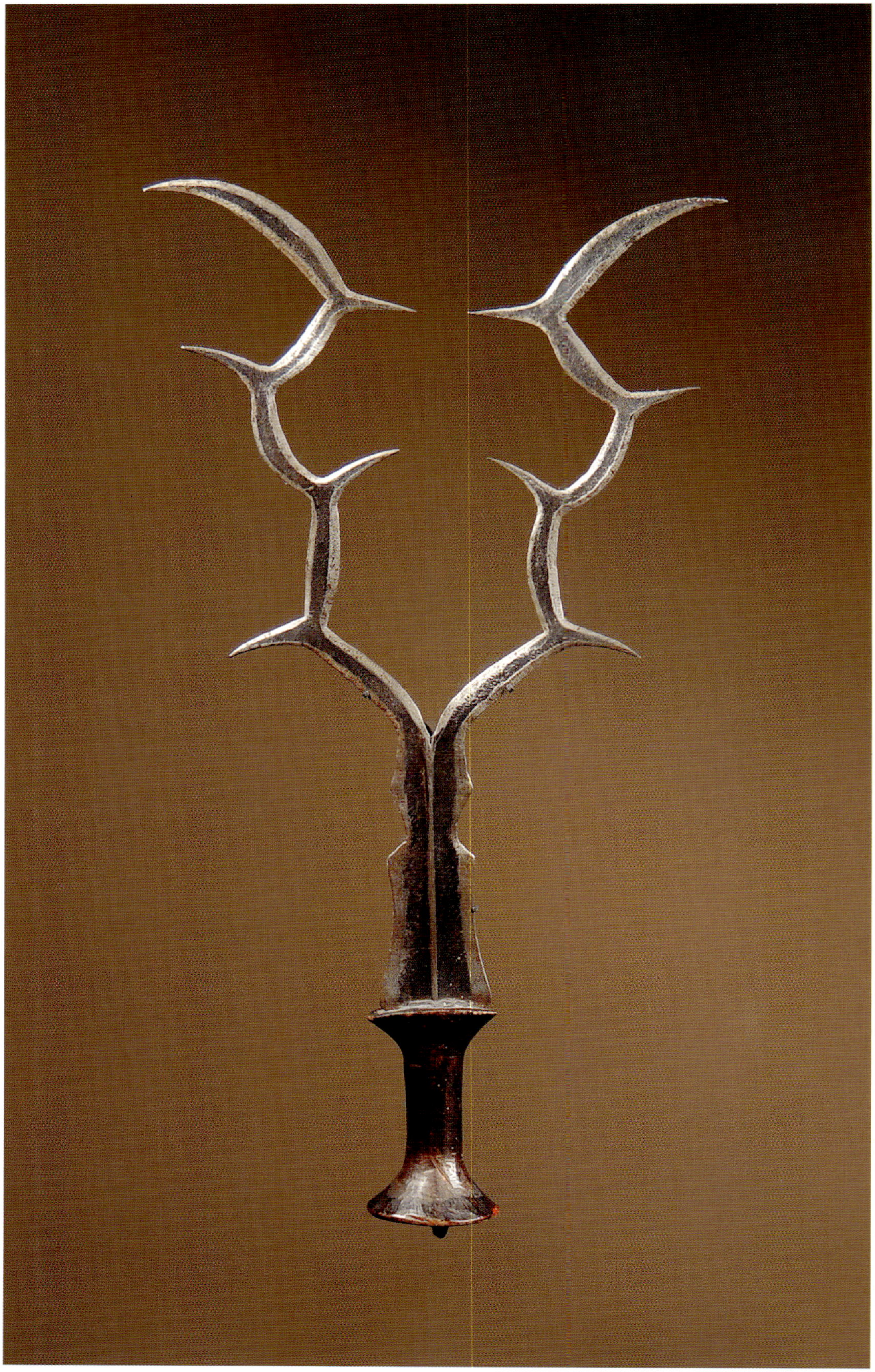

14.16
Artist unknown (Ekonda peoples,
Democratic Republic of the Congo)
Prestige knife (*ikakalaka*), early 20th century
Iron, wood
H: 53 cm
COLLECTION OF JOEL AND ZACHARY COONER
IMAGE COURTESY OF JOEL AND ZACHARY COONER, 2005

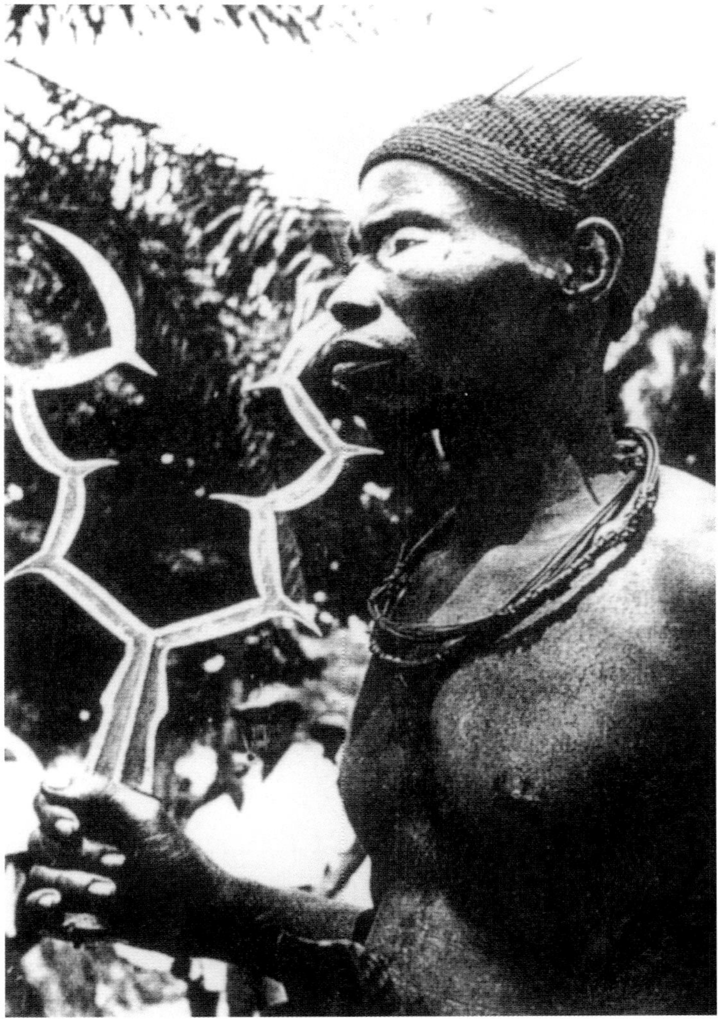

1989 was the twenty-fourth year of the military dictatorship of Mobutu Sese Seko, which would last another eight years. At the time of our visit, much of Zaïre's former system of paved roads had reverted to footpaths, leaving many local towns and villages relatively isolated and dependent on hunting, fishing, and farming for subsistence. Lopanzo was fortunate to have an active forge in operation daily, with four blacksmiths repairing hoes and producing harpoons, arrowheads, and lance blades. Members of our research group who were staff from Zaïre's national museum hoped to help the local economy by making purchases for the museum's ethnographic collections. Among the items they acquired were examples of women's pottery and a handsome three-tiered fiber cap customarily worn in the past by territorial judges. But not a single *ingonda* knife was available for purchase—they were considered inalienable.

Knives that were available were of another kind, called *ikakalaka*, with very elaborate blades but no set form. They were made instead by improvisation, with smiths using a variety of ornamental cutting, splitting, and sometimes punching techniques to demonstrate their prowess in controlling and balancing design elements (figs. 14.16–14.18). Access to these "show knives" or "dance knives" was not restricted to men. Prominent women were known to carry, perform with them at funerals, or display them in public ceremonies honoring the birth of a first child (Kriger 1999, 210–11). Although exceptional and much-admired iron blades, *ikakalaka* were not inalienable.

14.17 A high-ranking Ekonda official holding double-bladed *ikakalaka* knife.
PHOTOGRAPH BY ANDRÉ SCOHY, INONGA (IN PRESENT-DAY DEMOCRATIC REPUBLIC OF THE CONGO), CA. 1950.

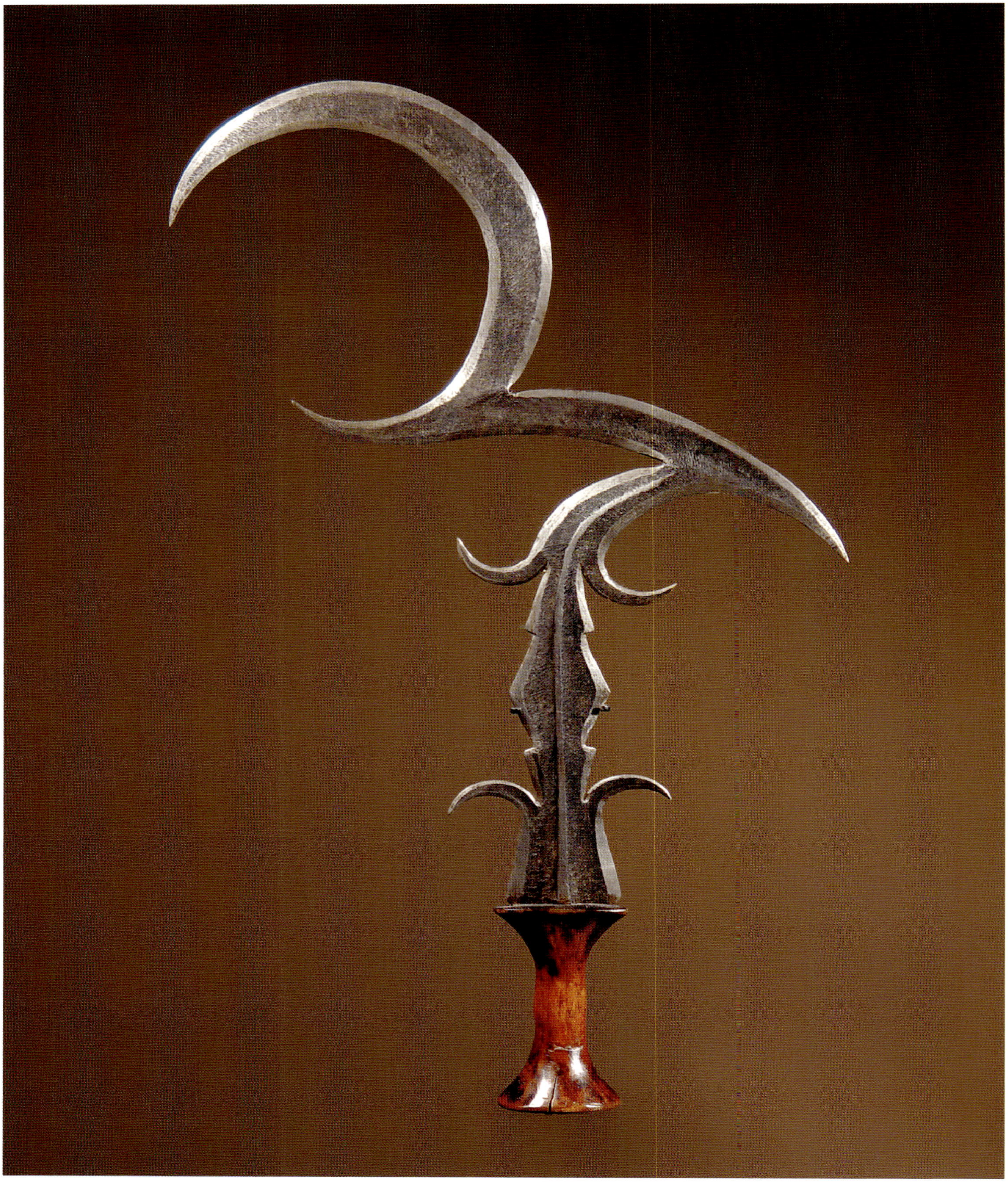

14.18
Artist unknown (Ekonda peoples, Democratic Republic of the Congo)
Prestige knife (*ikakalaka*), early 20th century
Iron, wood
H: 48 cm
COLLECTION OF JOEL AND ZACHARY COONER
IMAGE COURTESY OF JOEL AND ZACHARY COONER, 2005

Among the creations of Ekonda blacksmiths—revered as founders of lineages—are *ikakalaka* knives prized for the ornamentation and improvised forms of their blades. Both men and women carried these "show knives" for display on public occasions.

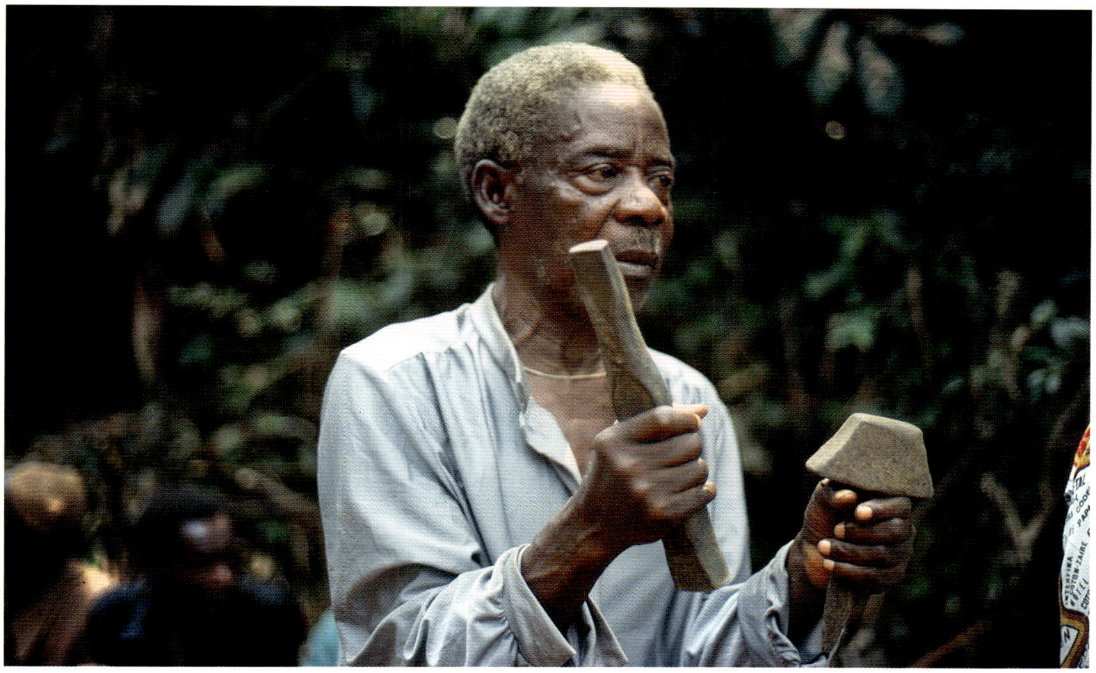

14.19
Keeper of the Nzelika lineage treasury, holding a hammer and anvil made by his forebear, master blacksmith Nsabonzo Engala (active 19th century) and passed down to Nsabonzo's son, master blacksmith Nyekuli Bana (ca. 1900–1980). By proudly and publicly recounting their names and ironworking prowess he honors their memory and Nzelika lineage history.
PHOTOGRAPH BY COLLEEN E. KRIGER, LOPANZO, DEMOCRATIC REPUBLIC OF THE CONGO, 1989.

In Lopanzo, we saw that iron tools were capital goods establishing smithies and marking highly skilled blacksmiths. Especially treasured were hammers and anvils of specifically known and remembered master blacksmiths. The master smith of Lopanzo in 1989, Lotoy Lobanga, claimed that in 1961 his own master, Ndzou, made the two anvils and at least four hammers still used in his workshop. Lotoy allowed his three apprentices, all members of his own lineage, to each keep a hammer to bring to the smithy daily; the anvils, however, remained under his care.

Early colonial reports convey how much blacksmiths valued their tools. One missionary on the middle Congo River lamented in 1892 that a local blacksmith had steadfastly refused to take any price for the purchase of his tools and inventory. An ethnographer made the same complaint in 1908, stating that it was impossible to buy a hammer or anvil (sources cited in Kriger 1999, 121–28, 151, 200).

And so it was in Lopanzo in 1989, when an elder of the Nzelika lineage brought out precious blacksmithing tools—a hammer and anvil, masterworks from their treasury (fig. 14.19). It was not to sell them or even to simply display them visually in public. It was above all to voice praises and honor by name the master who made them, Nsabonzo Engala, and the name and mastery of his son and heir Nyekuli Bana (Kriger 1992, 281–87). This treasured hammer and anvil were still very much in use as vehicles for keeping alive the memory of a lineage's powerful and revered master blacksmiths.

CHAPTER 15

OF BLACKSMITHS AND KINGS: A CENTRAL BANTU WORLD

William J. Dewey *Allen F. Roberts*

> [Among Luba people of the Democratic Republic of the Congo, the Supreme Being is] The One Who Forges Suns, Kafula Moba. Certain chiefs and diviners give themselves this . . . name, Lufula, Excellent Blacksmith.
>
> VAN AVERMAET AND MBUYA (1954, 164)

Some of the continent's most virtuosic precolonial works in iron were forged by central African blacksmiths. Aside from such brilliance that is well worth considering in its own right, an unusual characteristic of ironworking is an "obviously old and intricate relationship . . . between leadership and metallurgy" in what can be understood as a "world" of closely interrelated groups sharing language and logic.[1] Indeed, cosmologies explain the centrality of ironworking to political histories, and a number of societies have proclaimed that their first chiefs and kings (and sometimes prominent ones to follow) were themselves blacksmiths (fig. 15.1). In other words, the forging of iron implements has legitimized leadership and rule, and vice versa.[2]

Control of iron ore, smelting, and trade in forged implements was important to the rise of certain central African states, but so was regulation of resources such as salt production, hunting and fishing, and regional trade routes, contributing to the rise of the Luba state (in present-day Democratic Republic of the Congo), for example (see Reefe 1981). Yet as Pierre de Maret has suggested, such factors alone cannot account for the signal importance of the blacksmith-king relationship among Central Bantu peoples (1985, 76–77). If smelting and forging were two distinct technologies, why is the latter so intimately associated with politics despite the significance of smelting to local and regional economies?

An educated guess is that as a highly esoteric "enchanted technology" (after Gell 1992), smelting constituted carefully guarded knowledge and its processes were undertaken at a remove from daily life. Analogies between the "birth" of malleable iron from rocky ores and human procreation brought further prohibitions and practices to bear that were in need of careful oversight by a master smelter (fig. 15.2). Forging, on the other hand, was and still is central to community life, as when, in earlier days, Tabwa blacksmiths (in present-day southeastern DRC) built open-sided workshops that secondarily served as places for men to gossip and boys to sleep. Furthermore, iron tools and weapons have been of enormous practical value, but in some ways forged currencies and other largely symbolic objects were even more significant to social life when used in bridewealth or rituals that assured community cohesion.[3] Little wonder, then, that analogies should be drawn between the transformative achievements of blacksmiths and those of rulers.

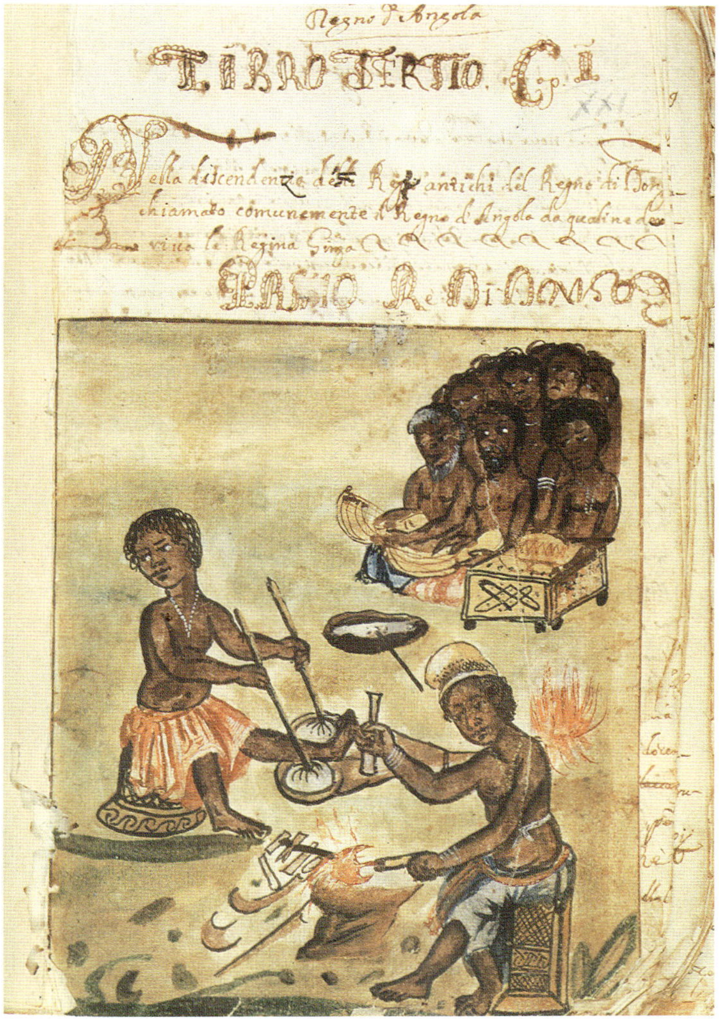

15.1 In the 1660s, the Capuchin missionary Giovanni Antonio Cavazzi da Montecuccolo (1621–1678) painted this watercolor, included in his account "Missione evangelica nel Regno de Congo," completed in 1668. In the foreground, the blacksmith-king of Ndongo, a polity of Mbundu ethnicity that was an uneasy tributary of the Kongo Kingdom, strikes a red-hot ceremonial axe blade with a hammer-anvil; other blades lie at his feet while he dons a royal cap and sits on upon an elevated prestige stool. To his right, a man works bellows to heat a forge fire. He too uses a richly ornamented stool suggesting his status, but his seat is lower than the king's. Behind, two men play lutes, a young woman sits behind a crown displayed upon a relinquary inscribed with cosmograms, and a throng joins the scene.

IMAGE COURTESY OF VINCENZO NEGRO

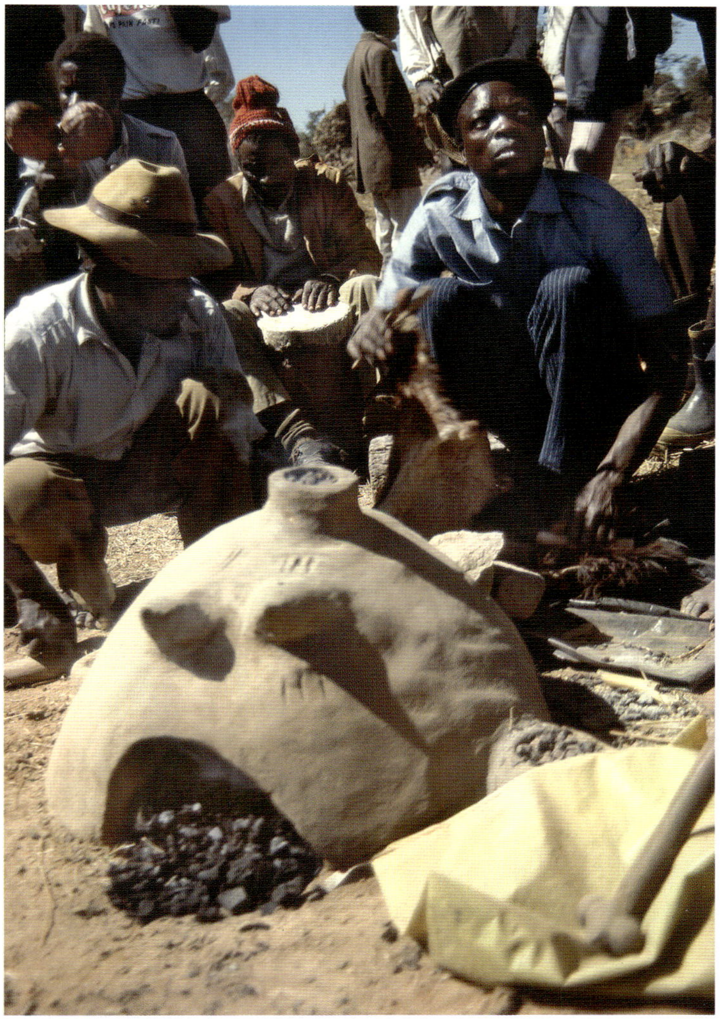

15.2 Iron smelting reconstruction organized by William J. Dewey in 1988 among the Njanja subgroup (Chief Ranga's village) of Shona of Zimbabwe. Note that the smelting furnace is made in the abstracted form of a woman's body with breasts and decorative scarification lines, making evident that, like a woman, a furnace gives birth.

PHOTOGRAPH BY WILLIAM J. DEWEY, CHIKOMBA DISTRICT, ZIMBABWE, 1988.

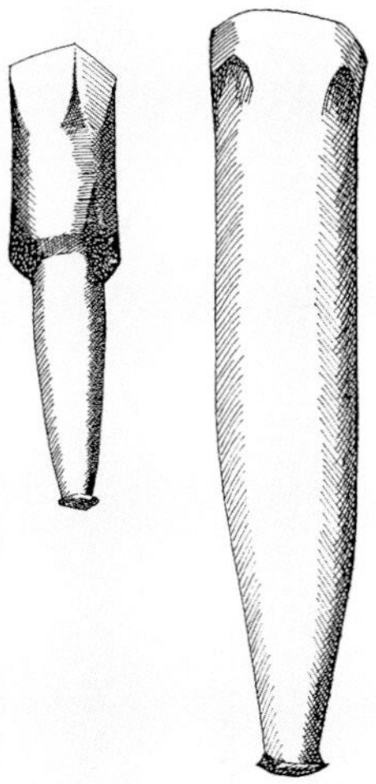

15.3 In central Africa, a *nyundo,* or "hammer-anvil" (pl. *manyundo*), could be thrust into the ground to provide a platform on which to strike hot iron, or held by the shaft for use as a hammer. The illustrated tool to the left is associated with Luba peoples; the tool to the right, with Tabwa. In many societies *manyundo* were also understood to possess their own transformative powers, and objects based upon their forms were treasured possessions of kings and chiefs.
DRAWINGS BY L. PRAET, PUBLISHED IN EMMA MAQUET, *OUTILS DE FORGE DU CONGO, DU RWANDA ET DU BURUNDI* (TERVUREN: ROYAL MUSEUM FOR CENTRAL AFRICA, 1965).

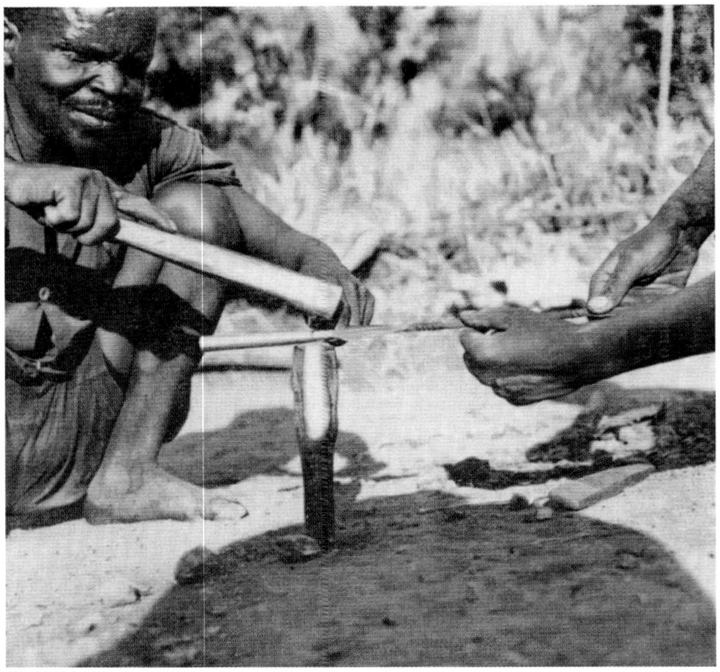

15.4 Using two *manyundo*—one held in the hand and the other stuck in the ground—Hutu blacksmiths in Rwanda put finishing artistic touches on a spear blade. Evidence of such use is often apparent as strike marks on the tops and sides of these multifunctional tools.
PHOTOGRAPH BY R. DE WILDE D'ESTMAEL, GISHAMVU, RWANDA, 1955. COLLECTION ROYAL MUSEUM FOR CENTRAL AFRICA, TERVUREN

A complementary hypothesis may be advanced that reinforces the blacksmith-king relationship by reflecting the ambiguous nature of authority as understood by societies throughout the region (see, for example, Mudimbe 1996, 245–47). Central African blacksmiths designed deliberately multipurpose tools called *manyundo* (s. *nyundo*; fig. 15.3), from the proto-Bantu word **-(j)o#ndo 9/10*; as the eminent central Africanist historian Jan Vansina asserted, such an implement was (and sometimes still is) "used during the forging process both as a hammer and as an anvil wherever this tool was used" (2006, 333–34, after Guthrie, 1967–71). In other words, a *nyundo* might be grasped as a hammer to strike red-hot iron, or it might be planted in the ground to serve as the anvil on which to strike iron (fig. 15.4). Evidence of such double purpose can often be seen in strike marks on the sides and tops of *manyundo* observed while still in use or in expatriate collections.

We here translate *nyundo* as "hammer-anvil," following archaeologists who have worked in central Africa (e.g., Childs and Maret 1996, 57). In many other parts of sub-Saharan Africa and elsewhere in the world, hammers and anvils are different implements (for example, see Pole 2010; see also chapter 4, this volume), and indeed it is quite likely that the same blacksmiths working with *manyundo* possess other sorts of hammers and anvils. A Luba smith, for instance, might use a *vitôtwâ* iron hammer and a *dibwe* stone anvil (Van Avermaet and Mbuya 1954, 722, 452) to complement the effects achieved with his *nyundo* hammer-anvil.

Vansina further suggests that a *nyundo* was often "of considerable value and/or an emblem of blacksmiths" (2006, 333). We suggest that such symbolic efficacy was due, at least in part, to how the twofold-purpose of the "hammer-anvil" resonated with ways that rulers in many Central Bantu societies were elevated beyond ordinary dualities when they were "forged" through investiture rituals. Among Luba, *manyundo* were sometimes used to gently pound the legs of would-be kings, or officiants' fists replaced them in explicit mimicry, creating a semidivine king who transcended ordinary human binaries and related ontological limitations (Dewey and Childs 1996). In logical extension, some Luba diviners summon spirits using *nyundo*-shaped rattles (now crafted from recycled tin cans); one must assume that the practitioner's task of "forging" etiological histories for those suffering misfortune is assisted by such a powerfully efficacious analogy.[4]

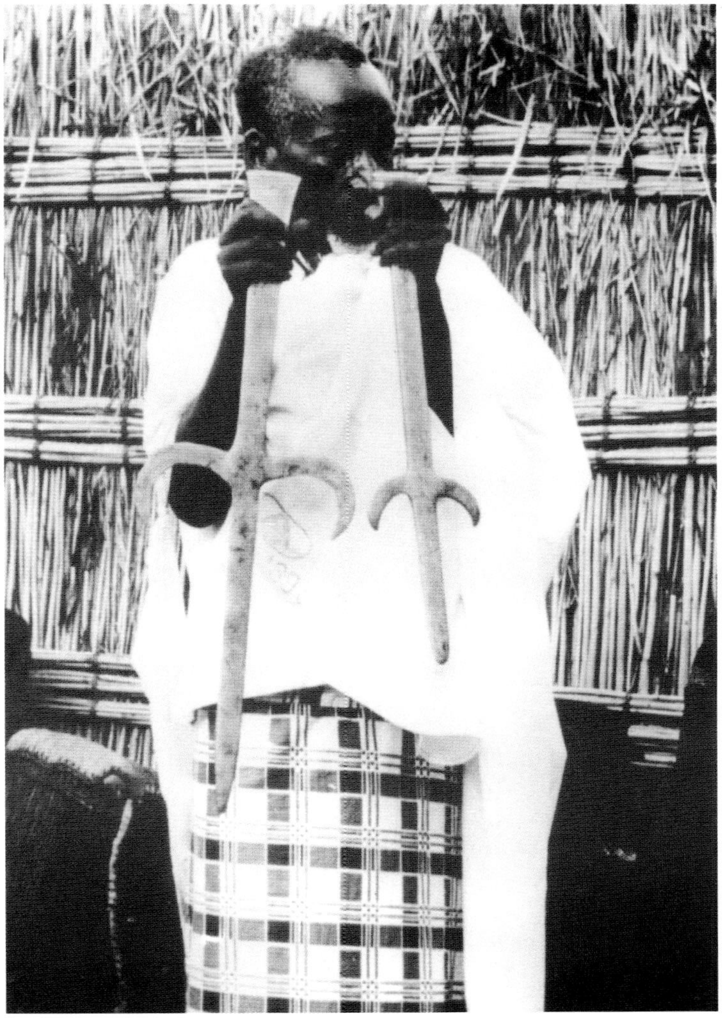

15.5 Earlier kings of Rwanda were understood as "Supreme Blacksmiths" and possessed heirloom devices based upon the form and purposes of hammer-anvils. These animated instruments were named Nyarushara and bore "horns" with reference to the cattle closely associated with the kingdom's wealth and welfare. Nyarushara efficacies contributed to the fertility of the king and his people.

PUBLISHED IN HAMO SASSOON, "KINGS, CATTLE AND BLACKSMITHS: ROYAL INSIGNIA AND RELIGIOUS SYMBOLISM IN THE INTERLACUSTRINE STATES," *AZANIA: JOURNAL OF THE BRITISH INSTITUTE IN EASTERN AFRICA* 18, NO. 1 (1983).

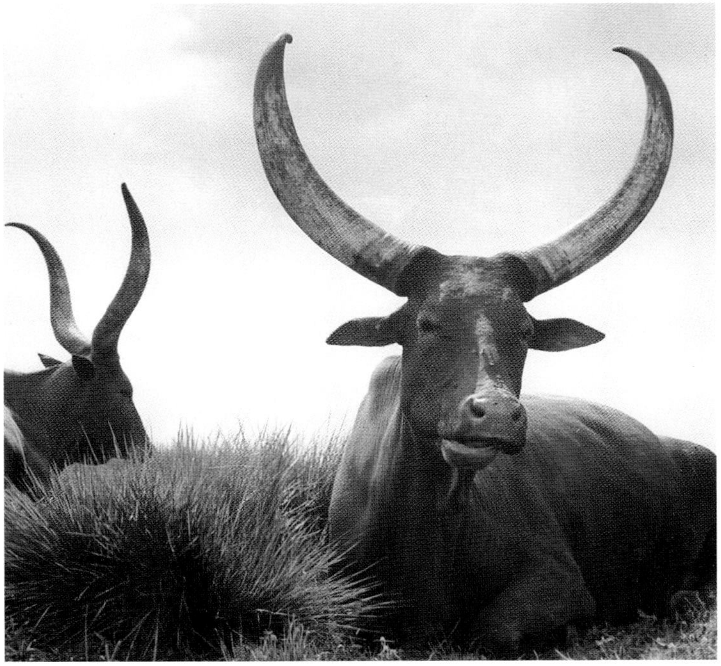

15.6 Men in the interlacustrine region of central Africa are so closely identified with their majestic cattle that a "relational ontology" exists between them. In a way, they "are" each other. The Rwandan king's Nyarushara corniform hammer-anvils refer to the centrality of this relationship via the careful shaping of the exceedingly large horns of cherished beasts like the two seen here.

PHOTOGRAPH BY JEAN-YVES CLAVREUL, GITARAMA, RWANDA, BETWEEN 1965–70. MUSÉE DU QUAI BRANLY-JACQUES CHIRAC, PARIS, FRANCE.

Evocative historical accounts evince similar relationships between blacksmiths and kings. Kuba oral traditions link metallurgy with the origins of rule, for instance, and in the early days of the Kuba Kingdom (in present-day east-central DRC) "all men of the royal lineage had to know the art of smithing."[5] Among Kwese, members of the chiefly clan often wore wooden pendants in the form of bellows (Mack 1990, 41). For Mbundu of southeastern Angola, a hammer-anvil was personified and revered as a chief; when sent to an outlying village in his stead, women did not pursue ordinary work, as though the chief himself were visiting and required their ardent hospitality (Maret 1985, 82, citing Read 1902; cf. Pole 2010, 65–66).

Among Luba, *manyundo* were "among the most potent of all the regalia of chieftaincy," and their efficacies were so secret that even high-ranking officeholders of the Luba court might not know their location (Nooter 1991, 143). Sacred hammer-anvils were paired and guarded by a spirit medium in the Luba kingdom of Kinkondja, and they were so powerful that they would blind any unauthorized person laying eyes upon them. Following complex symbolic practices, they were struck together while the king ate in utter seclusion, and iron *mazele* rattles were shaken when he drank; yet in other contexts, should royal *manyundo* touch each other inadvertently, calamity would strike, and perhaps even result in the death of the king or his child (ibid., 144).

In earlier days, the King of Rwanda was understood to be the "Supreme Blacksmith" of the land, and he possessed five hammer-anvils that were "dynastic... insignia of his high dignity" (Maret 1985, 74, citing Kagame 1956).[6] Called Nyarushara, such an implement was "the principal iron regalium of the kingship and [the] source of exceedingly strong ritual power" that assured "fertility and power of reproduction—of the land, of the people, of the kingdom: ecology, society, culture" (Newbury 2009, 242). So closely identified were Tutsi kings and hammer-anvils that when a monarch was enthroned, new ones were "ritually reforged, signifying renewal of the dynasty," and a seventeenth-century king was buried with his head resting on two such tools (Herbert 1993, 134, citing Hertefelt and Coupez 1964, 52). More recently, a nationalist newspaper reporter asserted that Nyarushara was itself considered King of Rwanda, while human monarchs were the animated hammer-anvil's "General Chiefs of Staff." In days of yore, "every Rwandan was serving God, represented by Nyarushara" (Athar 2007).

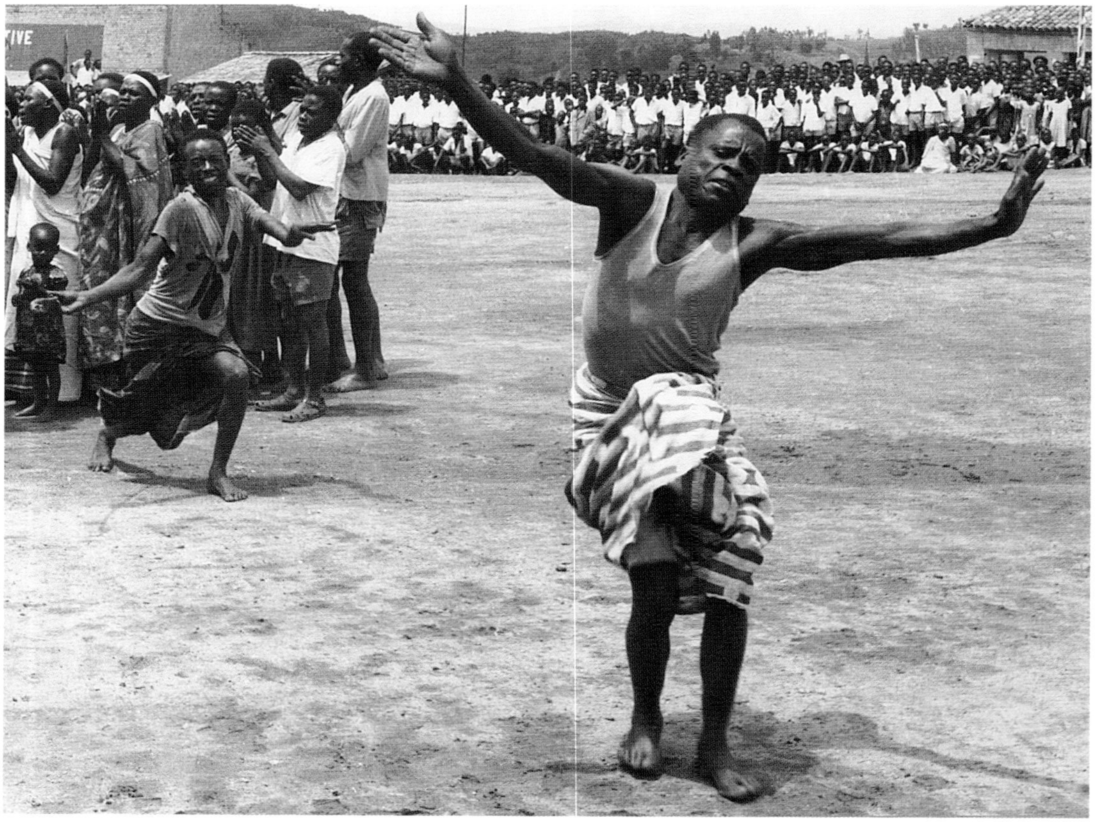

15.7 Cattle are so important to Rwandan social life that graceful choreographies permit dancers to emulate the lyrically beautiful horns of beloved animals
PHOTOGRAPH BY JEAN-YVES CLAVREUL, GITARAMA, RWANDA, BETWEEN 1965–70. MUSÉE DU QUAI BRANLY-JACQUES CHIRAC, PARIS, FRANCE. © MUSÉE DU QUAI BRANLY-JACQUES CHIRAC, DIST. RMN-GRAND PALAIS / ART RESOURCE, NY

Nyarushara (fig. 15.5) bore corniform appendages curving downward from the upper platform, and the name Nyarushara is derived from an Ankole name for prestige oxen with carefully shaped horns (fig. 15.6).[7] Staff of the Ethnographic Museum of Butare, Rwanda, suggest that "the elegant extension of dancers' arms in Rwandan traditional dance emulates the long, smooth, crescent shape of the cow horns" (fig. 15.7).[8] One can imagine that such choreography evokes sacred relationships among royalty, fertility, and wealth in cattle as mediated by iron. In turn, such associations have proven important to neighboring peoples emulating the glories of Rwandan royalty, and notably by Karagwe kings, at least one of whom is understood to have been a most masterful blacksmith.

OF COWS AND BLACKSMITH-KINGS

A remarkable royal treasury of forged objects was created by smiths of the Karagwe Kingdom of northwestern Tanzania (figs. 15.8, 15.9). The collection consisted of well over one hundred intricate standards, spears, hammer-anvils (some horned), and iron sculptures of cattle that exemplify "deliberate incorporation of iron and ironworking within a wider framework of symbolism and social power" (Reid and MacLean 1995, 146). Karagwe was the easternmost of interrelated interlacustrine states with mixed economies of farming and cattle herding. In the mid-nineteenth century, a convergence of trade brought prominence to the polity.[9] Some of the earliest known iron smelting sites in eastern Africa are found in the area (Schmidt 1997), and ironworking was a major occupation, with smelting continuing until around 1940 (Katoke 1975, 14). King Ndagara, who reigned from the 1820s until 1853, is thought to have himself forged objects in the Karagwe treasury (Ford and Hall 1947, 8). Whether this was a singular, collective, or symbolic accomplishment cannot be determined, for ironworking clans assisted Ndagara in such endeavors (Katoke 2001, 46). Members of one such body oversaw installation of a new king to the beat of a drum replicating the rhythm of a blacksmith's hammer (Katoke 1975, 159), thus "forging" an ordinary person into a king through ritual processes similar to those among Luba, Kongo, and many other Central Bantu peoples.

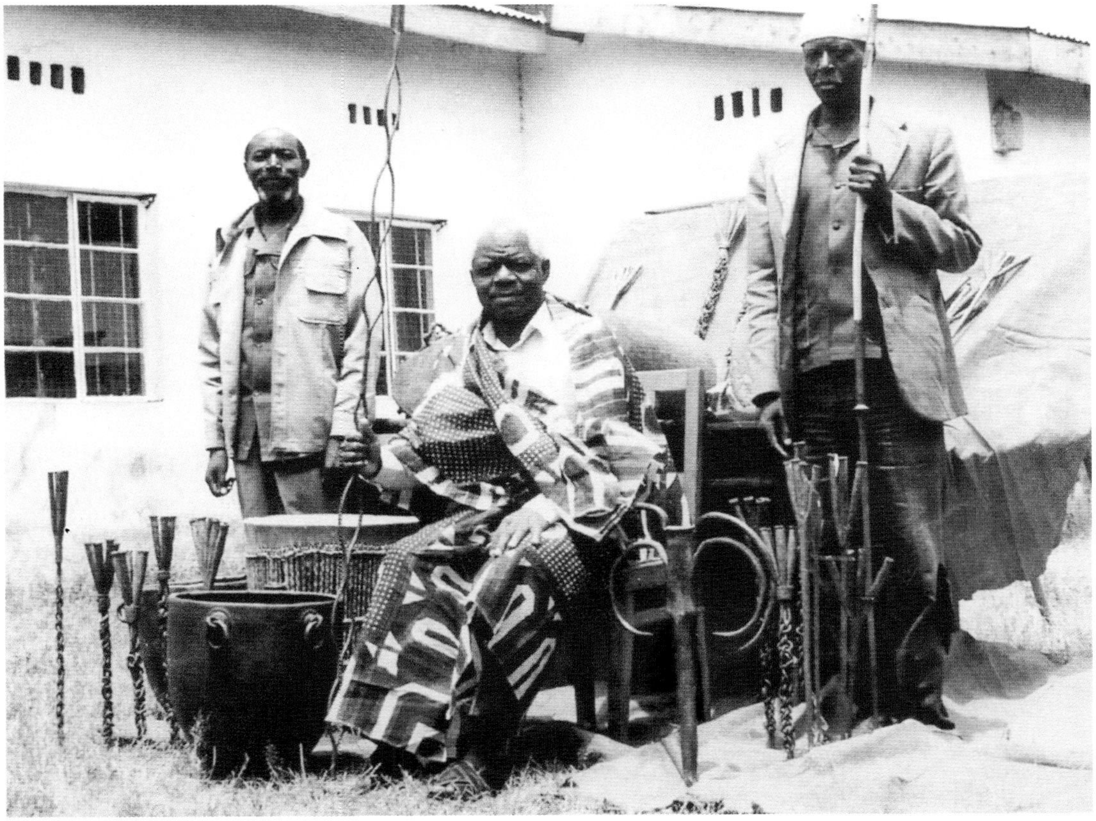

15.8 Community representatives display objects of the Karagwe Kingdom, a polity disenfranchised soon after Tanzanian independence in 1961. A horned ritual hammer-anvil is planted in the ground to the seated gentleman's left, while to his either side are intricate iron standards that once held leafy branches to screen the king's sacred possessions and activities from profane gaze. The man bears a ceremonial spear of forged iron in his right hand.
PHOTOGRAPH BY ISRAEL K. KATOKE, KARAGWE DISTRICT, KAGERA REGION, TANZANIA, MID-20TH CENTURY. IMAGE COURTESY OF THE KATOKE ESTATE, DAR ES SALAM, TANZANIA, 2018

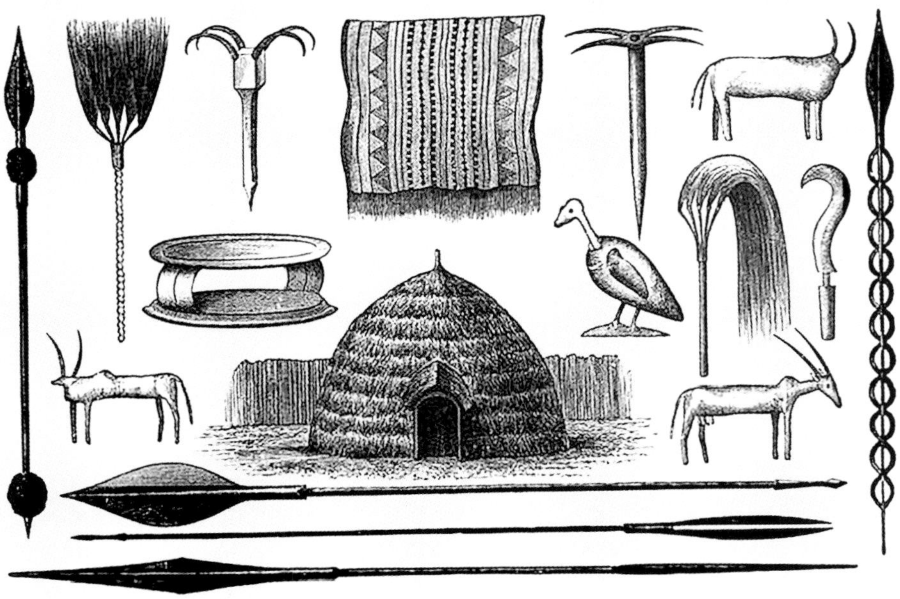

15.9 The American adventurer Henry Morton Stanley (1841–1904) was taken with the forged iron "treasures" of the Karagwe king whom he visited in the mid-1870s. As illustrated in this drawing from a hugely popular travel book by Stanley, King Rumanika's treasury included two horned hammer-anvils, three bovine sculptures, and a remarkable spear with eleven "lanterns" in its shaft (see fig. 15.11).
PUBLISHED IN HENRY MORTON STANLEY, *THROUGH THE DARK CONTINENT* (NEW YORK: HARPER AND BROTHERS, 1878).

15.10
Artist unknown (Karagwe Kingdom, Tanzania)
Royal standard, late 19th to early 20th century
Iron
H: 73 cm
IRIS & B. GERALD CANTOR CENTER FOR VISUAL ARTS AT STANFORD UNIVERSITY, 2010.35; GIFT OF ROBERT M. AND RUTH L. HALPERIN
IMAGE © IRIS & B. GERALD CANTOR CENTER FOR VISUAL ARTS AT STANFORD UNIVERSITY

Forged standards like this are unique to the Karagwe Kingdom. Karagwe kings are said to have been blacksmiths or at least closely associated with them. Leafy branches of a symbolically and medicinally significant shrub were placed in the sockets at the upper end of such standards to define a sacred perimeter screening a king's ritual performances.

15.11
Artist unknown (Karagwe Kingdom, Tanzania)
Staff/spear, mid-19th century
Iron
H: 159 cm
FELIX COLLECTION
PHOTOGRAPH BY MAUREEN VINCKE, 2005.

Iron spears are another virtuosic forging achievement of blacksmiths of the old Karagwe Kingdom. As weapons, they referred to valorous pursuits of battle and protection, while the "lanterns," or openings, of their shafts, must have impressed viewers through the beauty of their design.

15.12
Artist unknown (Karagwe Kingdom, Tanzania)
Bovine figure, mid-19th century
Iron
H: 30 cm
FELIX COLLECTION
PHOTOGRAPH BY DICK BEAULIEUX, 2005.

The treasury of the old Karagwe Kingdom once included bovine sculptures forged of iron and understood to possess their own efficacies to protect and promote the welfare of the polity. The sculptures' horns turning upward or down reflect careful shaping of those of cherished cattle. Feats of such imaginative forging bring the powers of ironworking to bear on royal aesthetics, and vice-versa.

15.13
Artist unknown (Karagwe Kingdom, Tanzania)
Bovine figure, mid-19th century
Iron
H: 40 cm
THE DONALD AND ADELE HALL COLLECTION
PHOTOGRAPH BY FRANCO KHOURY. COURTESY DONALD AND ADELE HALL

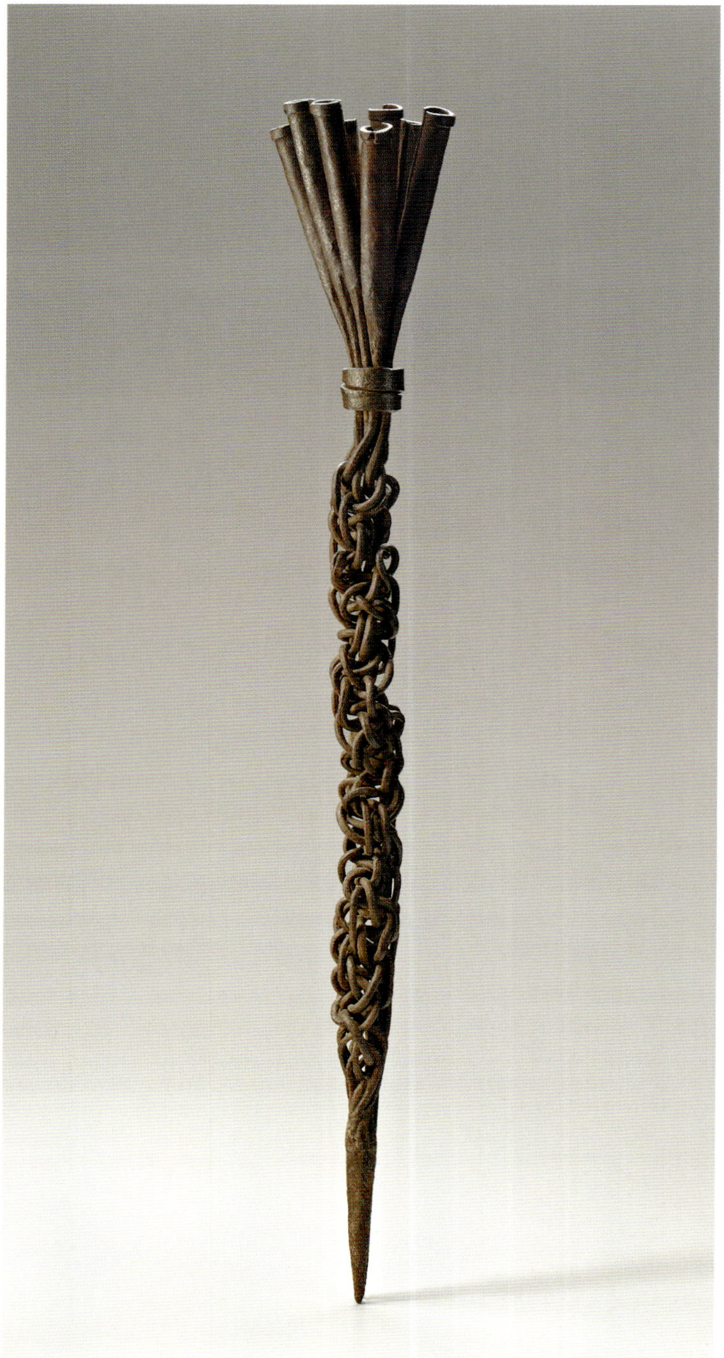

Some fifty iron standards once in the Karagwe treasury display stunning smithing skill, for they are forged of rods plaited together as though they were supple green vines. (fig. 15.10). These highly unusual objects bear sockets at their upper ends to hold branches of *omusinga* (*Hibiscus fuscus*), a tall, flowering shrub of eastern Africa with medicinal and practical applications. The king's leafy standards screened him, his great drum, and his hammer-anvils from profane gaze during community-renewal New Moon celebrations.[10] Such rites are common among Central Bantu peoples, as ancestral spirits are summoned and hope and courage renewed when the moon reappears after two or three nights of utter darkness (see Roberts and Maurer 1985).

Remarkably achieved spears entirely forged of iron also featured in the Karagwe king's treasury. A most elaborate example possesses eleven openwork lanterns in the shaft (fig. 15.11), and was described by the American adventurer Henry Morton Stanley as among the "Treasures of [King] Rumanika" (1878, 373). Such virtuosic weapons clearly departed from utility, and were emblematic of a king's military might and authority as realized through artistic mastery of iron.

As remarkable as the king's standards and spears are, however, the Karagwe treasury's masterpieces must have been a dozen forged sculptures of cattle (figs. 15.12, 15.13).[11] The works were described to early European visitors as *dawa*, or "medicine" (Silvia Dolz, in Van Wyk 2013, 217), suggesting that they surpassed an observer's dismissal as "mere ornaments" (Speke 1863, 206). Indeed, one of the last Karagwe kings told the archaeologist Hamo Sassoon that the sculptures were placed in front of the *omusinga* standards during New Moon rituals, undoubtedly because, as *dawa*, they possessed efficacies of their own.[12]

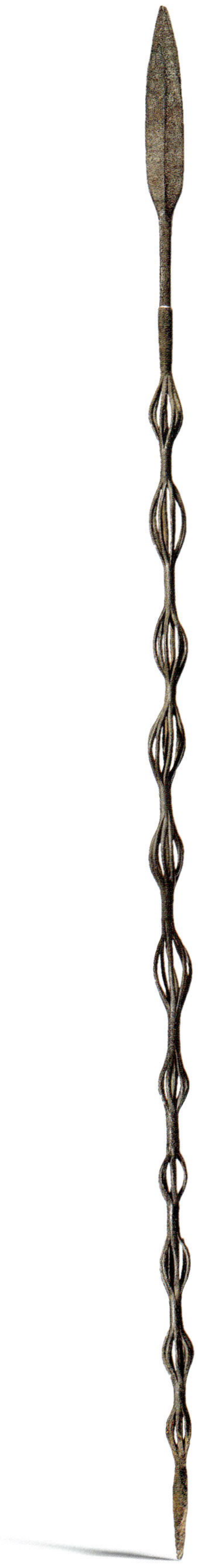

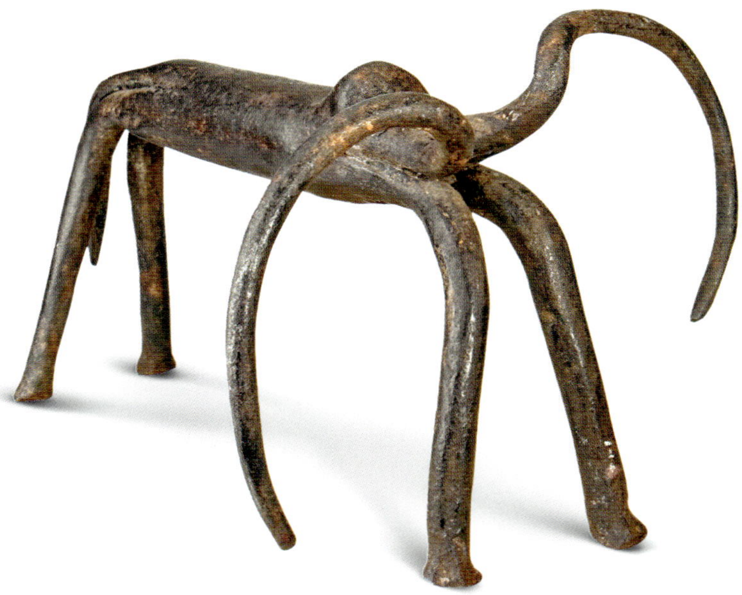

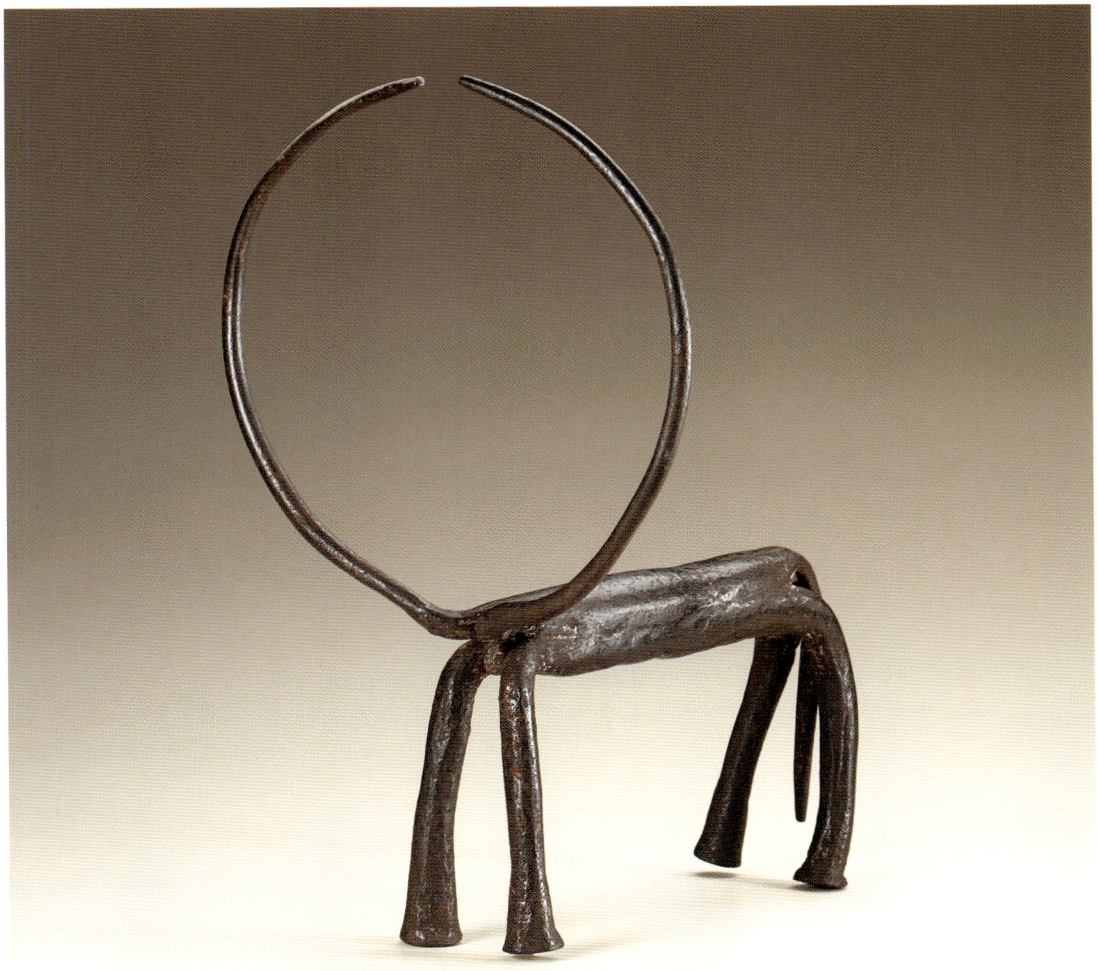

15.14
Artist unknown (Karagwe Kingdom, Tanzania)
Royal anvil, mid-19th century
Copper, brass
H: 49 cm
LINDEN MUSEUM STUTTGART
(NOT IN EXHIBITION)
PHOTOGRAPH BY DIRK KITTELBERGER AND VOLKER NAUMANN, 2017. © LINDEN MUSEUM, STUTTGART

The Karagwe king's treasury included objects that referred to the transformative potential of blacksmiths' forged iron hammer-anvils and the carefully shaped horns of prized cattle, considered the wealth of the kingdom. Because the Karagwe Kingdom was at some remove from the Copperbelt (in present-day southern Democratic Republic of the Congo and northern Zambia), copper would have been valued for the long-distance trade it represented.

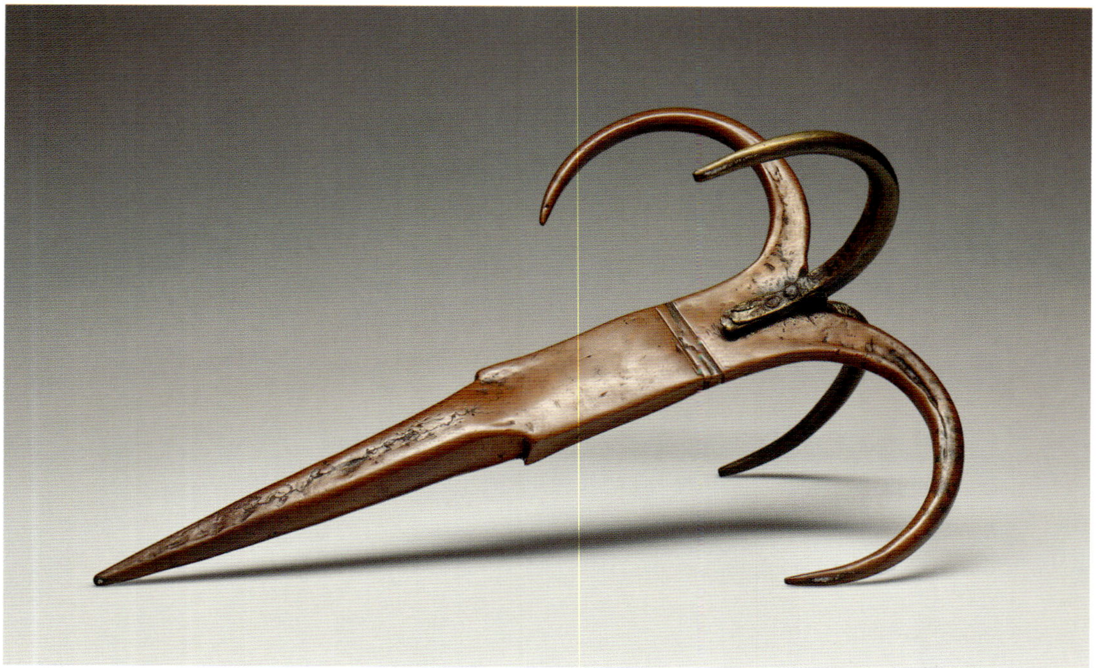

The most salient feature of Karagwe bovine sculptures are vast horns turned upward or down.[13] Horn-shaping has long been practiced by people of the "cattle complex" of agro-pastoralists of South Sudan and eastern Africa that includes Nilotic and Bantu-speaking peoples, among whom men can be so attached to their favorite beasts that a "relational ontology" exists between them. The Karagwe figures, then, "harmonize the fiery masculine principles of metalworking with the powerful fertility of women associated with cows. By employing these symbols, the king emphasized his role as guarantor of the peace and prosperity of his people and their crops and herds."[14]

In this, the figures resonate with several horned hammer-anvils of the Karagwe treasury (fig. 15.14). Sassoon X-rayed the bovine sculptures during their documentation at the National Museum of Tanzania in Dar es Salaam and confirmed that they were welded from six separate iron rods flared at one end and pointed at the other (1971). The resulting objects—hammer-anvil-like shapes transformed into figures; ritual hammer-anvils horned like cattle—combine essential economic and cultural elements of the kingdom. The powers of the one inform the other, and vice versa. Furthermore, as Brad Weiss notes of Haya communities closely related to people of the old Karagwe Kingdom, "inherited goods, or heirlooms and especially the sociocultural practices that animate them... have the capacity... to establish and concretize memory of social beings and social relations, and, therefore [to] transmit knowledge and substance across generations" (1997, 164). Would that we knew more about what iron treasures meant to Karagwe kings, then, who forged, empowered, and performed processual histories through them.

OF IRON STOOLS AND BOW STANDS

An unusual hoard of forged iron works was found among a Luba-related people of south-central Zambia known as Soli (Fagan 1961). The assemblage was placed upon the graves of two mid-nineteenth-century dignitaries, a chief and his brother, a master blacksmith. Nineteenth-century Soli were renowned for their smelting and smithing, and contemporary Soli recount histories of "a symbolic connection between metal, fire, and the fertility that the rains bring." Such relationships remain important, despite the "mutability of symbolism" to meet twenty-first-century circumstances (Simbao 2014, 47). As large accumulations of otherwise useful and valued objects, such hoards demonstrate that "they were powerful when taken out of circulation, perhaps as offerings of some kind" (Gosden 2013 , 45)—in this case, perhaps to appease the spirits of the bellicose chief and his blacksmith brother (Fagan 1961, 229).

The Soli collection, removed to Zambia's Livingstone Museum in the 1940s, consists of dozens of knives, spears, hoes, adzes, ceremonial and utilitarian axes, and bells. Among the most unusual and skillfully made objects are four sets of ceremonial tongs with handles suggesting human figures with arms raised (fig. 15.15). A rare, "ingeniously made" iron stool, also in the cache (fig. 15.16), has a supporting structure of circular forms bound together with thick wire and strips of iron (Fagan 1961, 237); it differs from the several other iron stools documented in Zambia and Malawi that have iron platforms supported by pieces welded together as divided-diamond shapes, in a motif of symbolic significance throughout central Africa (fig. 15.17).[15]

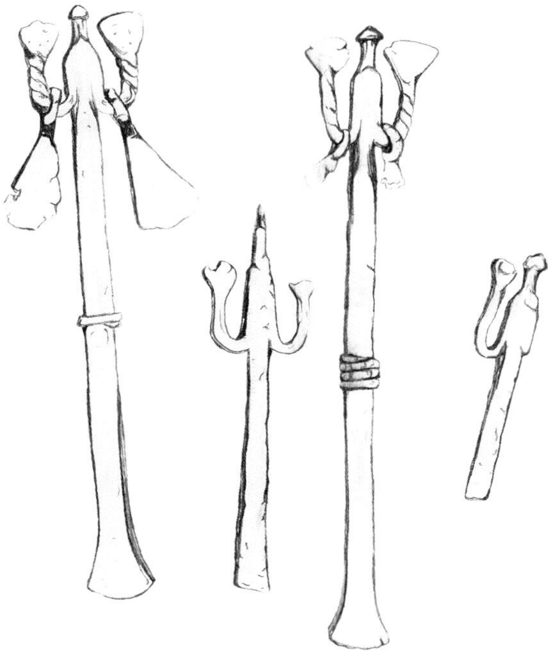

15.15 Documented here, four iron objects found among the Soli hoard, with handles suggesting raised arms, may have been used as blacksmiths' tongs. They may also have facilitated the manipulation of coals in lighting the tobacco of smoking pipes (Fagan 1961, 235).
DRAWING BY KEILY V. MALDONADO, BASED ON A PHOTOGRAPH BY WILLIAM J. DEWEY OF OBJECTS AT THE LIVINGSTONE MUSEUM, ZAMBIA, 2008.

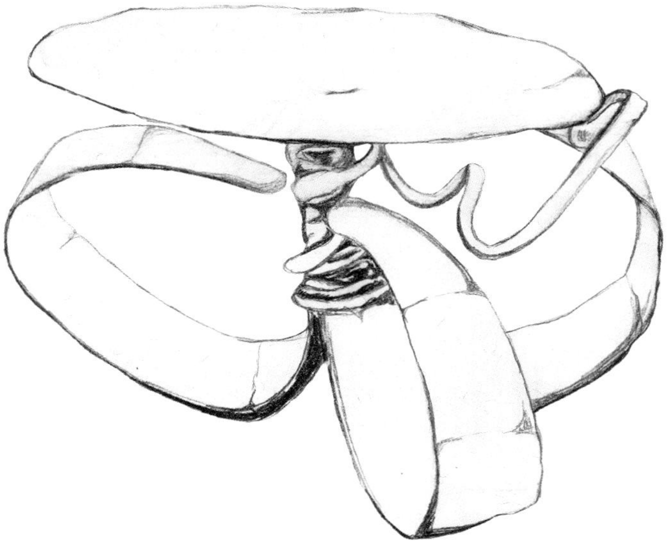

15.16 Stools of forged iron are unusual to the polities of south-central Africa. As a seat of power, an iron stool must have informed senses of political authority through its own ambiguities, for iron tools and weapons promote society yet can be terribly destructive.
DRAWING BY KEILY V. MALDONADO, BASED ON A PHOTOGRAPH BY WILLIAM J. DEWEY OF OBJECT AT THE LIVINGSTONE MUSEUM, ZAMBIA, 2008.

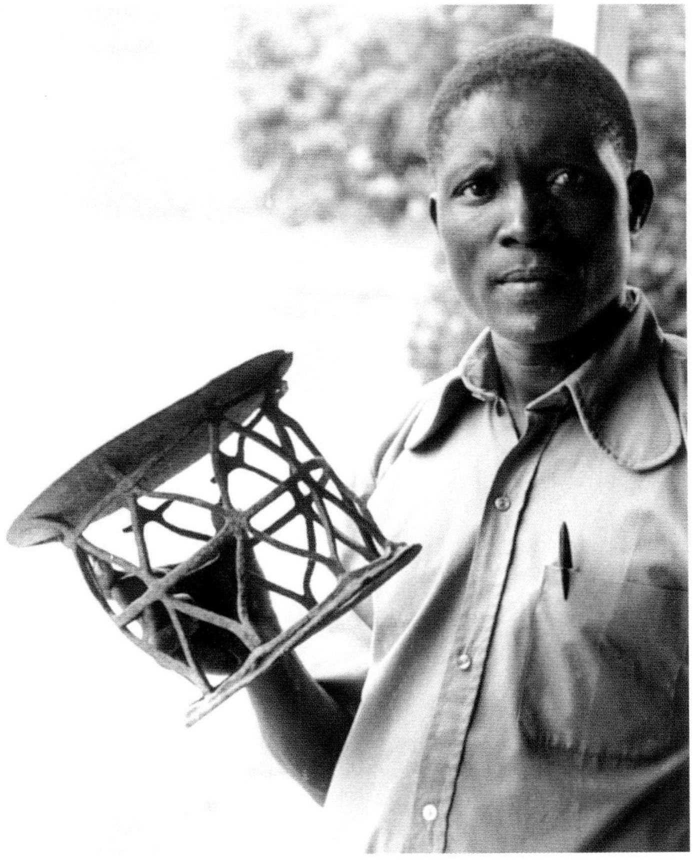

15.17 In the 1970s, the archaeologist David Killick was shown this iron stool, a precious heirloom of Chief Chulu. His community of Chewa people live in northern Malawi, yet oral traditions link them to Luba peoples of southeastern Democratic Republic of the Congo, and especially to Luba political aesthetics, and such stories suggest that their first leaders gave iron stools to affiliates as symbols of office.
PHOTOGRAPH BY DAVID KILLICK, KASUNGU, MALAWI, 1983. COURTESY OF DAVID KILLICK

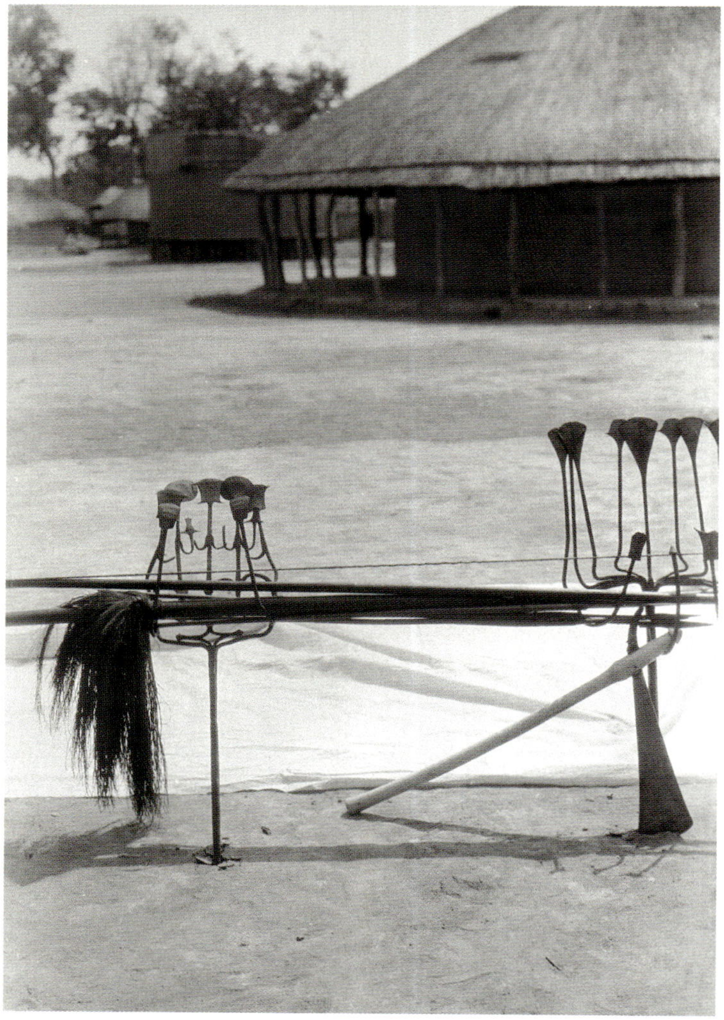

15.18 Two iron bow stands from eastern Zambia. In the 1930s, Bemba told the anthropologist Audrey I. Richards that the objects "came from Lubaland." Richards studied fourteen such iron bow stands, and others are reported from Zambia, northern Malawi, and southwestern Tanzania. These carefully kept sacred objects might hold weapons, but their own efficacies blessed phases of farming and provided links to ancestral and other spiritual powers.

PHOTOGRAPH BY AUDREY I. RICHARDS, CA. 1930S.

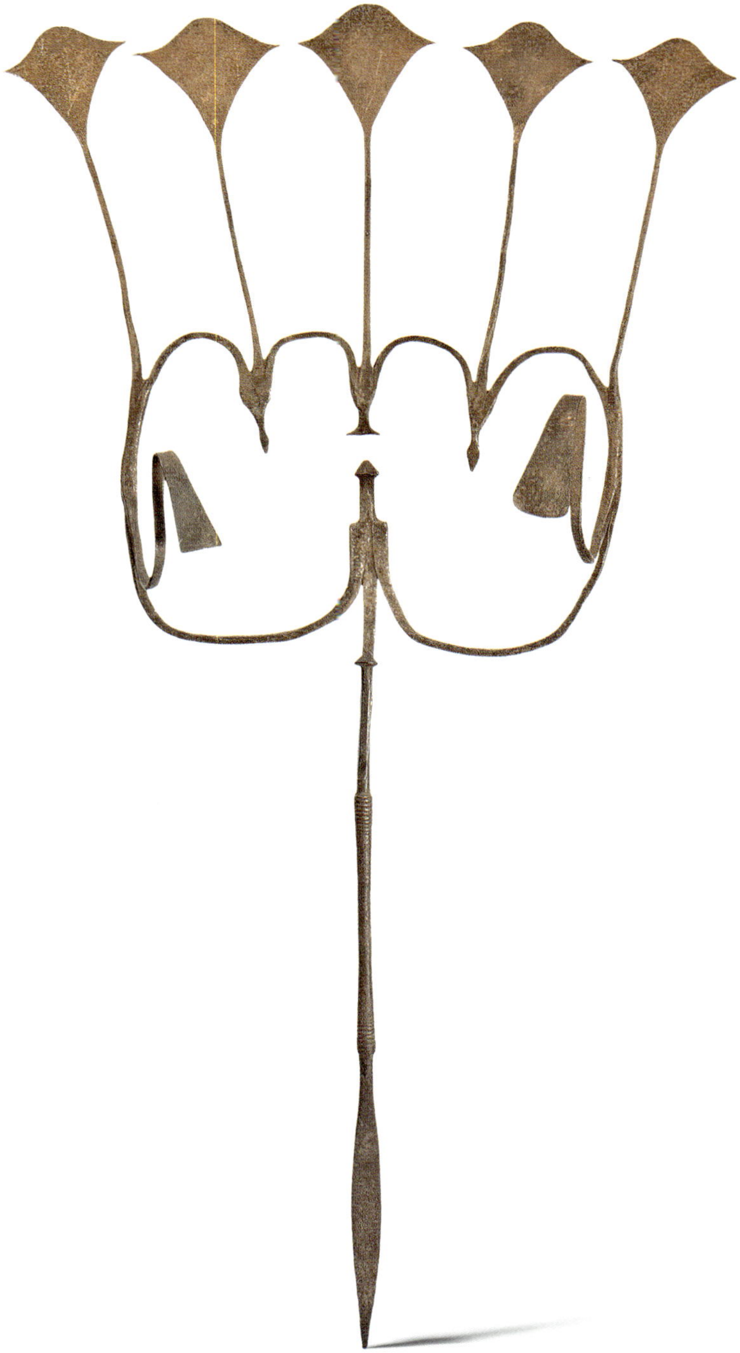

15.19
Artist unknown (Luba peoples, Democratic Republic of the Congo)
Bow stand, mid-19th century
Iron
H: 61 cm
Felix Collection
PHOTOGRAPH BY MAUREEN VINCKE, 2010.

This iron bow stand may have been forged by a Luba smith; if so, it is very rare compared to wooden ones sculpted by Luba artists (see fig. 15.20). The object is a virtuosic achievement, combining references to blades (via the upper forms that echo early razors) and the hammer-anvil (through the conical central pin) as among the transformative implements in a blacksmith's repertoire.

15.20

Artist unknown (Luba peoples, Katanga, Democratic Republic of the Congo)
Bow stand, mid- to late 19th century
Wood
H: 73.50 cm
FOWLER MUSEUM AT UCLA, X65.7489; GIFT OF THE GIFT OF THE WELLCOME TRUST
(NOT IN EXHIBITION)

Like Luba stools, staffs, and *lukasa* (see fig. 15.27), wooden bow stands are mnemonic and recall relationships between the noble person possessing such a sacred object and the origins of human culture via the heroes Mbidi Kiluwe and his son Kalala Ilunga, who were great hunters, blacksmiths, and kings. That the hands of the spirit figure gracing this bow stand end in hooves is unusual, and may refer to the great mother of Luba society, who was said to be a buffalo or to have mated with one. The pointed ends of the three prongs may refer to *vinyundo* (miniature hammer-anvils) placed in such positions on other bow stands.

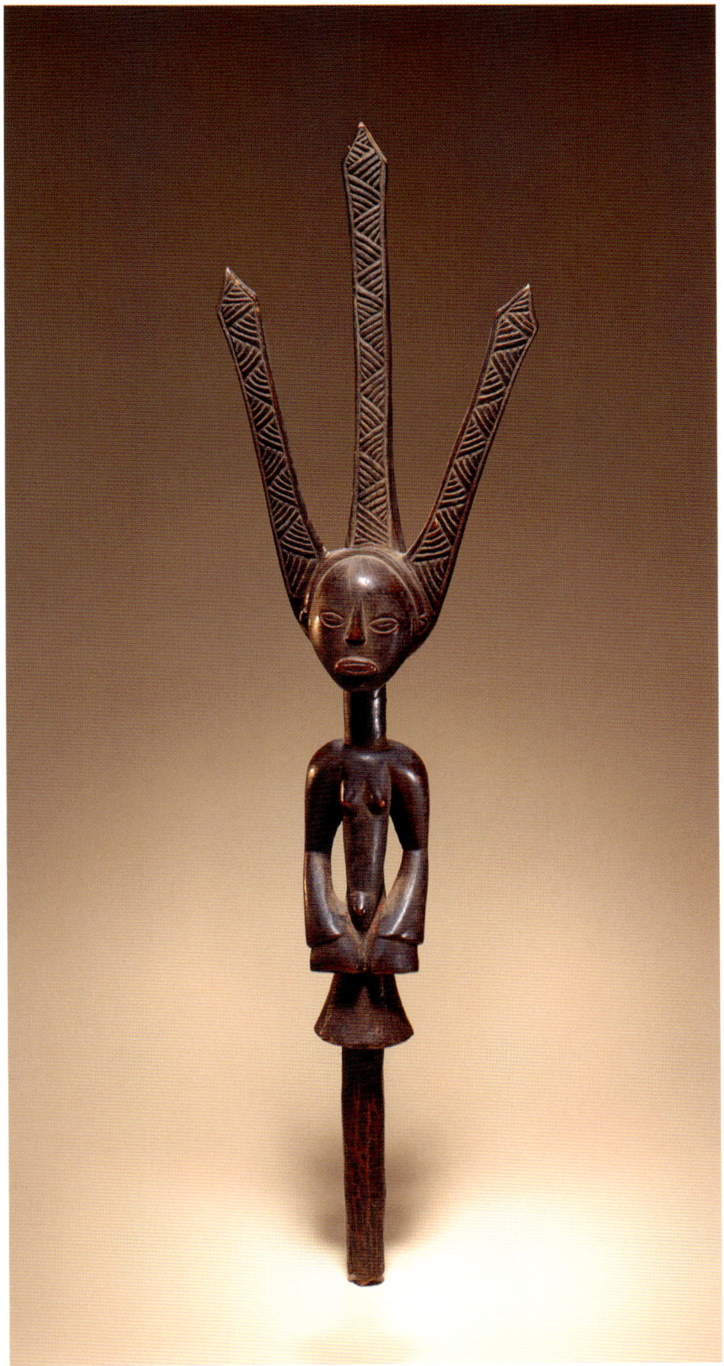

Two iron bow stands in the Soli hoard are stylistically quite different from but undoubtedly served similar purposes as ones photographed by anthropologist Audrey Richards among Bemba in what is now eastern Zambia (fig. 15.18) and others collected more recently (fig. 15.19). Called *nsakakabemba*, Bemba bow stands were sacred, secret, and kept in chiefs' "relic houses" by "specially appointed hereditary guardians" (Richards 1935, 31). In the Bemba language, *nsaka* refers to the place of assembly before a chief's residence, suggesting close identification between the forged bow stand and the leader himself as he represented his community. Richards observed Bemba Chief Mwamba publicly exhibiting two iron bow stands holding his hereditary bow, ceremonial axe, and fly whisk, but she assumed this might have been a latter-day practice. In the 1930s, Bemba iron bow stands nonetheless possessed the agency to bless seeds and first fruits, and to summon ancestral spirits.[16]

The Bemba, Bisa, and Unga interlocutors of Richards insisted that their iron bow stands "came from Lubaland" (1935, 31), and Brelsford was told that Luba Chief Mukulu forged such works (1940, 39). Mukulu is a praise name meaning "exalted," and refers to the Luba culture hero Mbidi Kiluwe, whom Luba understand to have introduced ironworking to humanity (see chapter 1, this volume]. Bemba chiefs prized their bow stands, but they were sometimes lost in battle or given or traded as gifts to dependent people living to their south in what are now northern Malawi and Zimbabwe. Although particular bow stands had their own histories and so were mnemonic, more generally such works were coveted by those who would emulate Luba political and spiritual powers. Indeed, the Bemba Paramount Chitimukulu traces his heritage to Luba origins, and Bemba kings have long cherished Luba wooden sculptures and royal accoutrements to demonstrate their participation in Luba political aesthetics.[17]

Looking more closely at an iron bow stand (see fig. 15.19) that may have been Luba or was certainly inspired by Luba royal designs, the object combines the forms of a variety of blades and hammer-anvils, as though a smith had assembled objects of his repertoire in a tour-de-force accomplishment.[18] Like their more common counterparts sculpted of wood (fig. 15.20)—which portray, and so make present, tutelary spirits—iron bow stands might hold special weapons. Far more important was their manifestation of sacred relationships between hunting and forging, two great gifts from the culture hero Mbidi Kiluwe, as reinforced by many other earlier art forms.

15.21
Artist unknown (Luba peoples, Democratic Republic of the Congo)
Female figure, 18th to 19th century
Iron, wood, cloth, plant-fiber cordage, pigment
H: 31 cm
FOWLER MUSEUM AT UCLA, X65.7488; GIFT OF THE WELLCOME TRUST

This figure's serene insight of spiritual presence is suggested by its inward gaze and gesture to the breasts, where Luba women preserve secrets and prerogatives of dignified rule bequeathed by the culture heroes Mbidi Kiluwe and Kalala Ilunga. A *kinyundo*-shaped hairpin (see figs. 15.24, 15.25) inserted into the back of the head suggests the transformative potential of interaction with the spirit that inheres in the sculpture. The lower portion of the figure was lost long ago.

15.22
Artist unknown (Kusu peoples[?], Democratic Republic of the Congo)
Half-figure (*kakudji*), late 19th to early 20th century
Iron, wood
H: 20 cm
THE CLEVELAND MUSEUM OF ART, RENÉ AND ODETTE DELENNE COLLECTION, LEONARD C. HANNA, JR. FUND, 2010.457
IMAGE COURTESY OF THE CLEVELAND MUSEUM OF ART

Truncated figures found among Luba and neighboring peoples including Kusu were probably attached to the top of a gourd or placed in a small basket holding potent medicines of healing, protection, and promotion. Hammer-anvil-shaped pins, and especially the one in the top of the head, direct the efficacies of this crucial tool to realizing creative transformations through ritual use of medicine-bearing devices like this one. They also close in and so guard the figure's secrets empowered by medicines beneath the miniature hammer-anvil.

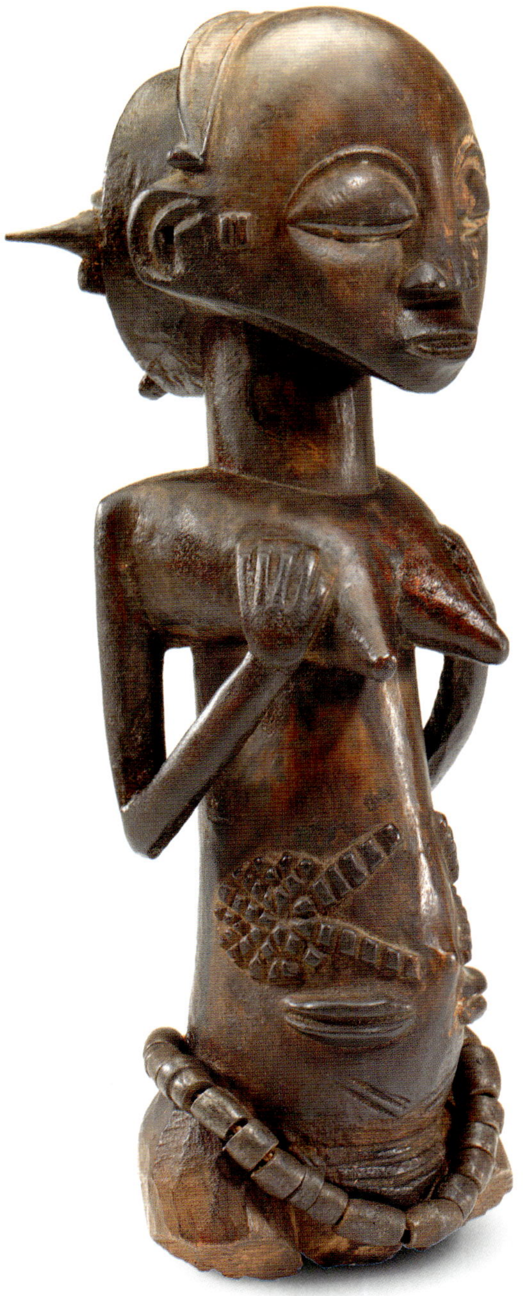

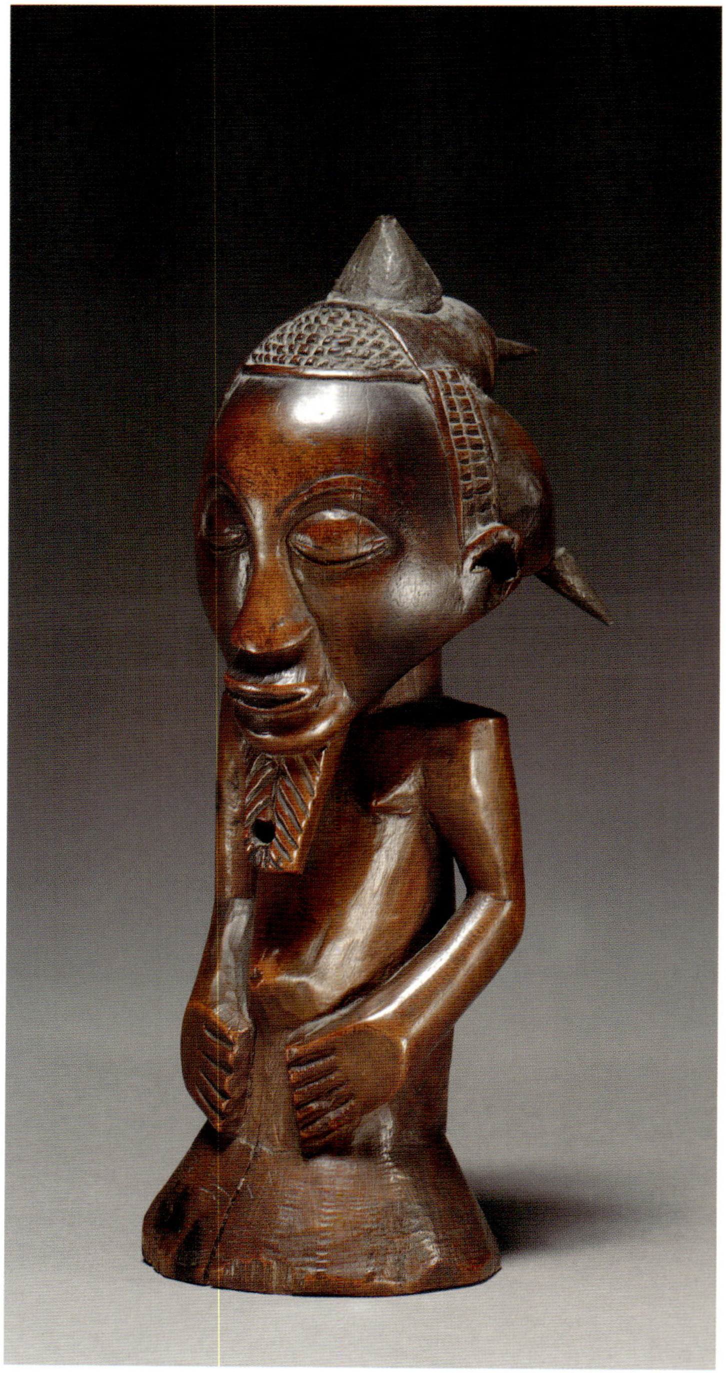

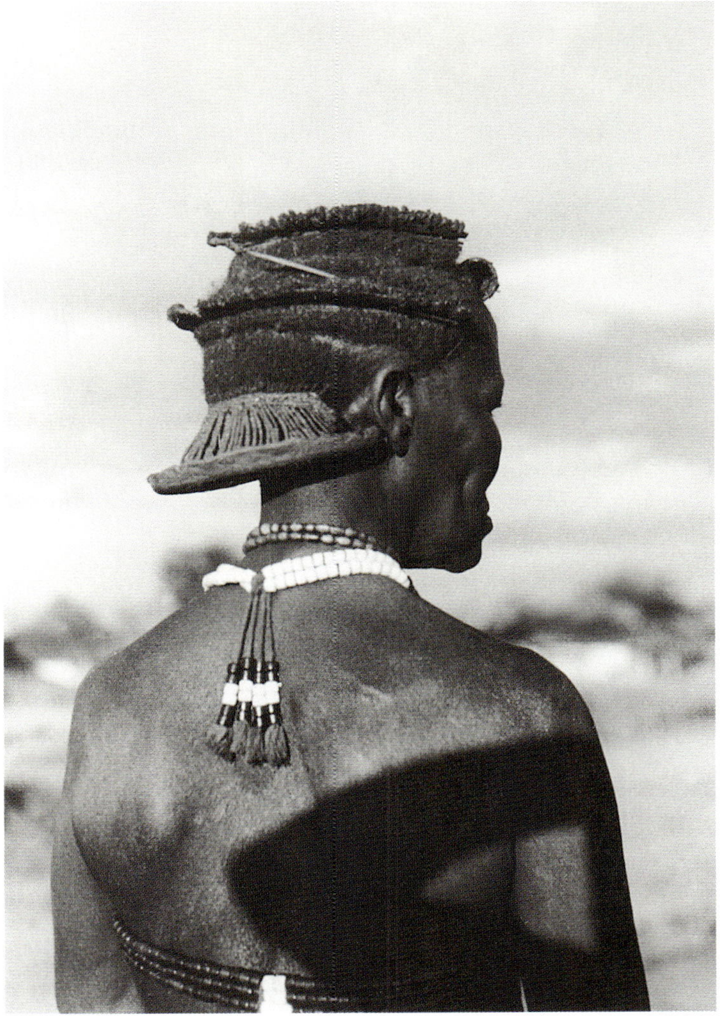

15.23 Luba coiffures were considered so chic that people came from many miles to admire and emulate them. So significant were forged iron hairpins that they have been found, in graves excavated by archaeologists, still in constellations suggesting early hairdos.

PHOTOGRAPH BY M. MORTELMANS. COLLECTION ROYAL MUSEUM FOR CENTRAL AFRICA, TERVUREN.

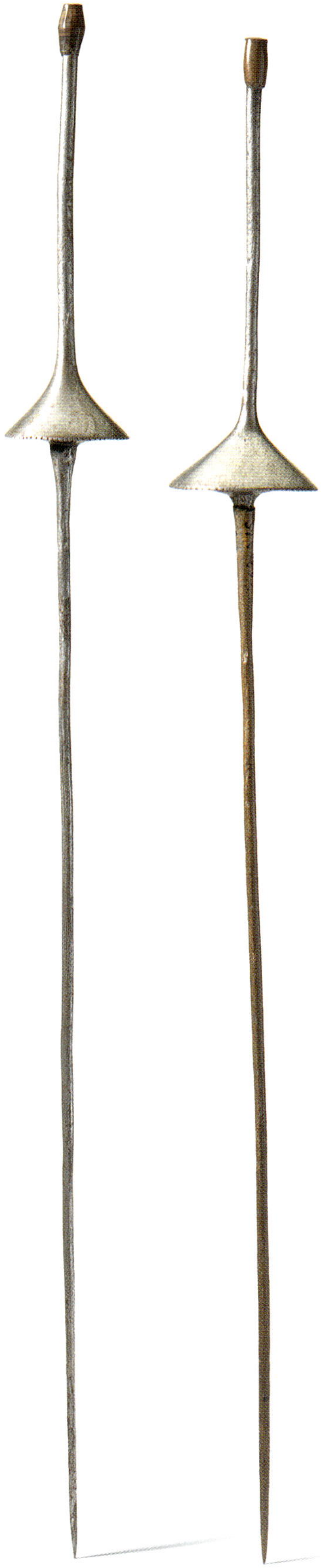

15.24
Artist unknown (Luba peoples, Democratic Republic of the Congo)
Hairpins (*vinyundo*), late 19th century
Iron
H: 34 cm, 34.5 cm
FOWLER MUSEUM AT UCLA, X67.9191, X67.715; GIFT OF THE WELLCOME TRUST

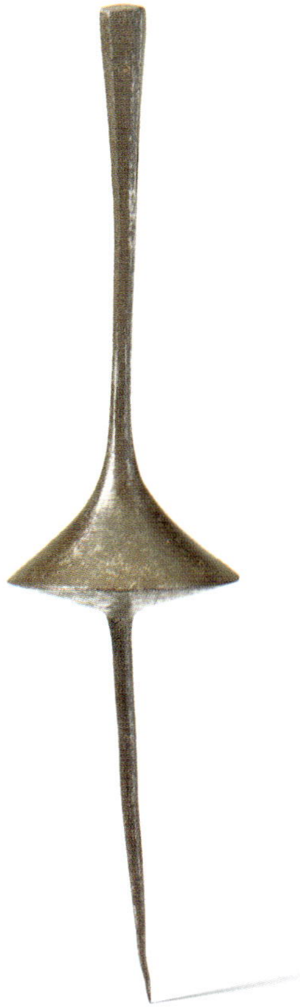

15.25
Artist unknown (Luba peoples, Democratic Republic of the Congo)
Hairpin (*kinyundo*), late 19th century
Iron
H: 14.6 cm
FOWLER MUSEUM AT UCLA, X65.10060; GIFT OF THE WELLCOME TRUST

Nineteenth-century Luba men and women were famous for their coiffures that took days to perfect and required headrests to keep them up off the ground during sleep. Iron hairpins were among the devices used to secure these elaborate hairdos. Aside from this practical function, such pins provided their own efficacies, as did the Luba hammer-anvils they miniaturized. Little wonder, then, that hairpins of this sort have been found in important proto- and early Luba graves.

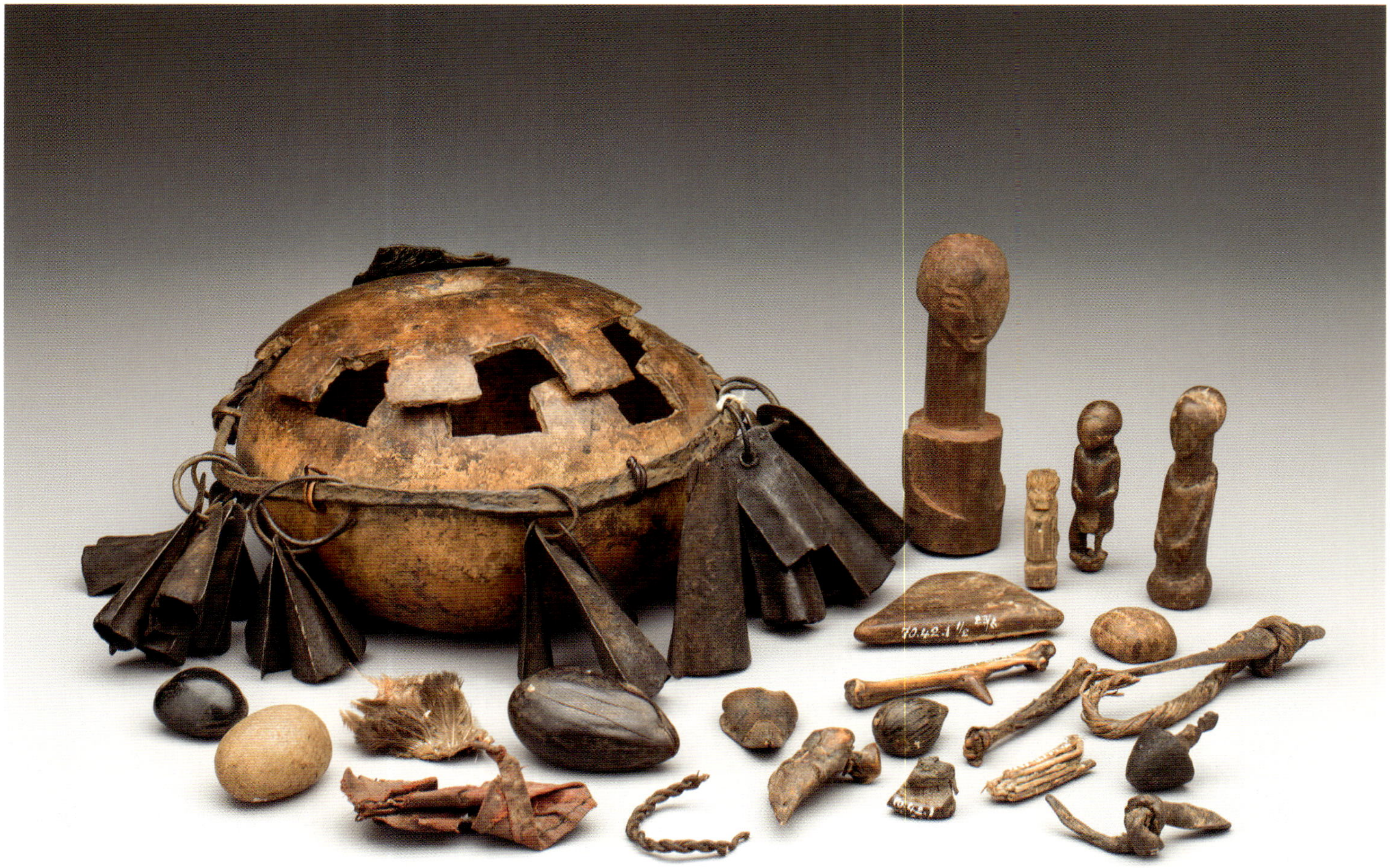

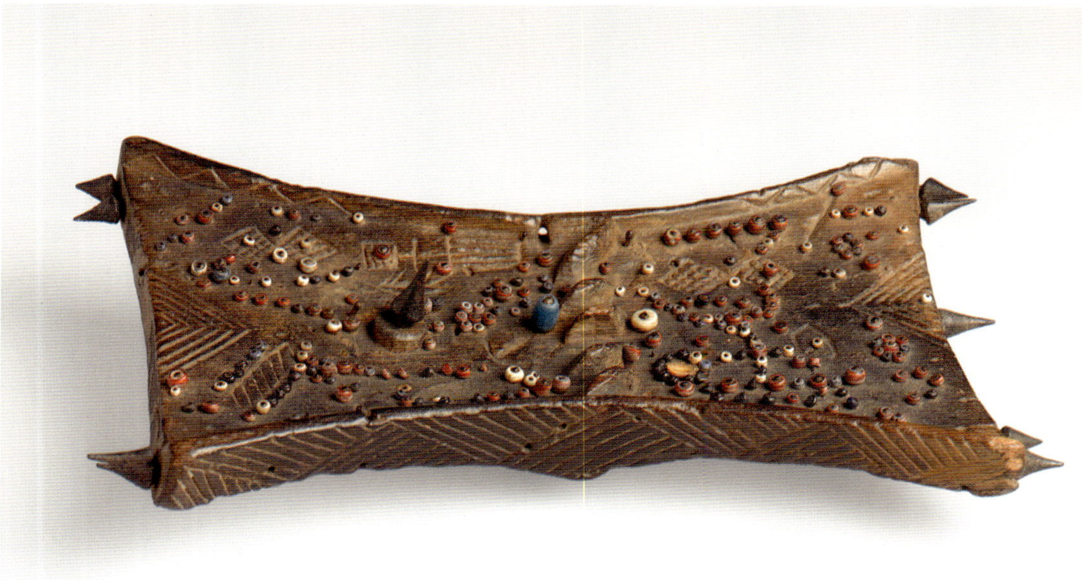

15.26 A Luba divination gourd and its contents, including iron bells, a small hammer-anvil, and a piece of bloomery iron.
COLLECTION ROYAL MUSEUM FOR CENTRAL AFRICA, TERVUREN
PHOTOGRAPH BY J. VAN DE VYER.

15.27
Artist unknown (Luba peoples, Democratic Republic of the Congo)
Memory board (*lukasa*), late 19th to early 20th century
Wood, iron, glass beads
L: 25.4 cm
BROOKLYN MUSEUM, 76.20.4, GIFT OF MARCIA AND JOHN FRIEDE
IMAGE COURTESY OF THE BROOKLYN MUSEUM

This hourglass-shaped wooden board, called a *lukasa*, is a mnemonic device adorned with incised and raised designs. Pegged to its surface with acacia thorns are glass beads and miniature hammer-anvils (*vinyundo*) that empower it as complex data sets are ordered for recall. Touching these with a forefinger, members of the Mbudye Society recount histories of grandeur, including epic tales of the culture hero Mbidi Kiluwe and his son Kalala Ilunga, who established Luba royal bearing and behavior and introduced ironworking to Luba.

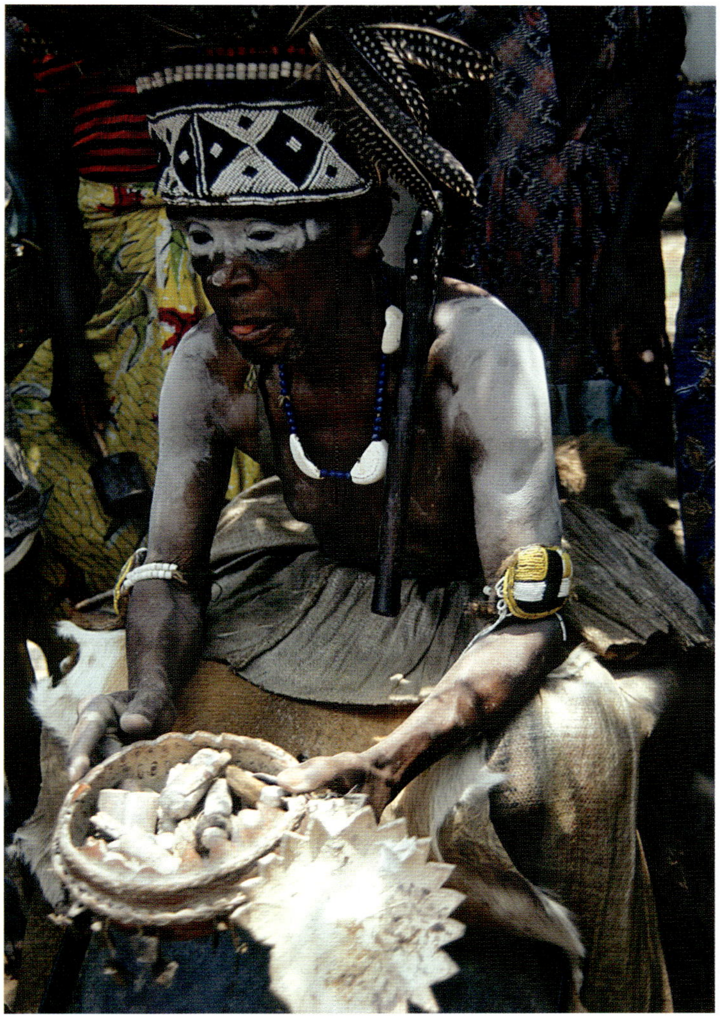

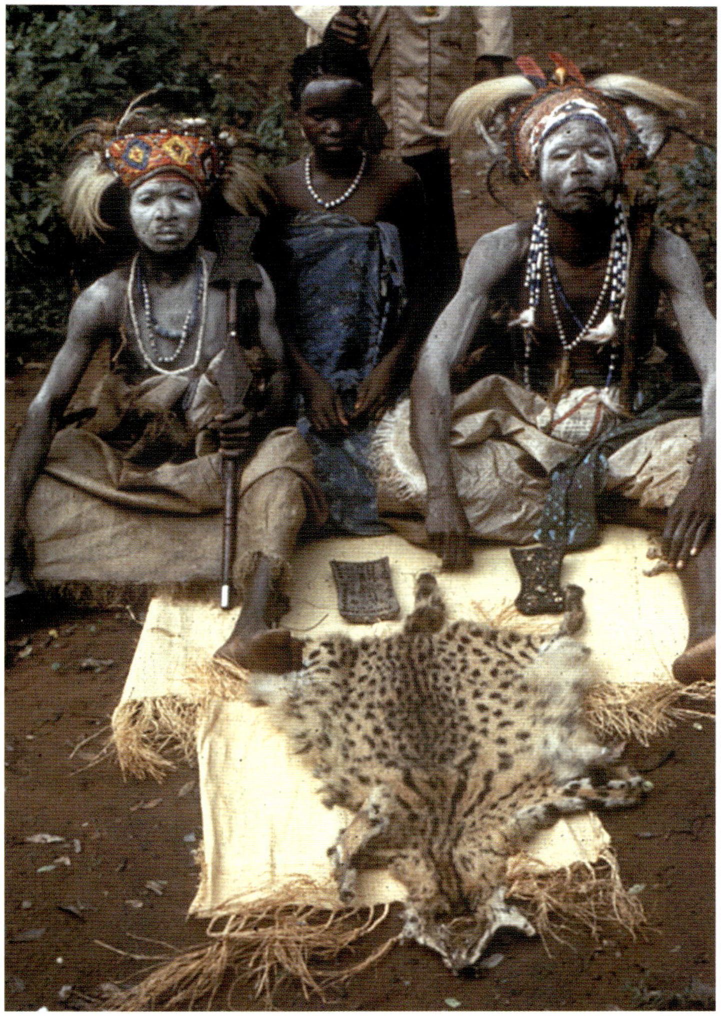

In particular, Luba and neighboring peoples' figural sculptures were often endowed with conical pins in the tops or backs of the heads of the spirits they made present (figs. 15.21, 15.22). Luba called such pins *vinyundo*, "little hammer-anvils" (s. *kinyundo*). Hairpins extended the idiom to the elaborate coiffures for which nineteenth-century Luba were celebrated (figs. 15.23–15.25). *Vinyundo* have been found so associated in pre-Luba graves (Childs and Dewey 1996, 158) and in some Luba communities, a grave-digger touched the forehead of the person being buried with a *kinyundo* pin to mark the transition to the beyond (Petit 1993, 345). A miniature hammer-anvil is sometimes among the diverse objects in an *mboko* divination gourd, assuring that histories will be forged to explain misfortune (figs. 15.26, 15.28);[19] and in a most esoteric application of the *device* (for active efficacies are so implied), *vinyundo* pins sometimes adorn and empower *lukasa* memory boards (figs. 15.27, 15.29).

15.28 Ilunga Mande Kombe performing gourd divination. Luba peoples have perfected several forms of divination to determine the cause of afflictions and so assist in problem-solving. The gourd seen here holds many artifacts of natural and human life, which appear in different juxtapositions as the gourd is shaken. Such associations are then narrated to explain what may be wrong so that rectifying actions can be taken.
PHOTOGRAPH BY WILLIAM J. DEWEY, MASANGU VILLAGE, KINKONDJA, SOUTHEASTERN DEMOCRATIC REPUBLIC OF THE CONGO, 1988.

15.29 The esoteric knowledge and know-how needed to use a *lukasa* (see fig. 15.27) is learned through initiations within the Mbudye Society of historians associated with Luba royalty. The face of the Mbudye member seen here is chalked to enlighten him while possessed by spirits who guide him as he reads the constellations of the board's surface. He carries an iron-bladed axe over his shoulder as further proof of his authority.
PHOTOGRAPH BY MARY NOOTER (ROBERTS), SOUTHEASTERN DEMOCRATIC REPUBLIC OF THE CONGO, 1989.

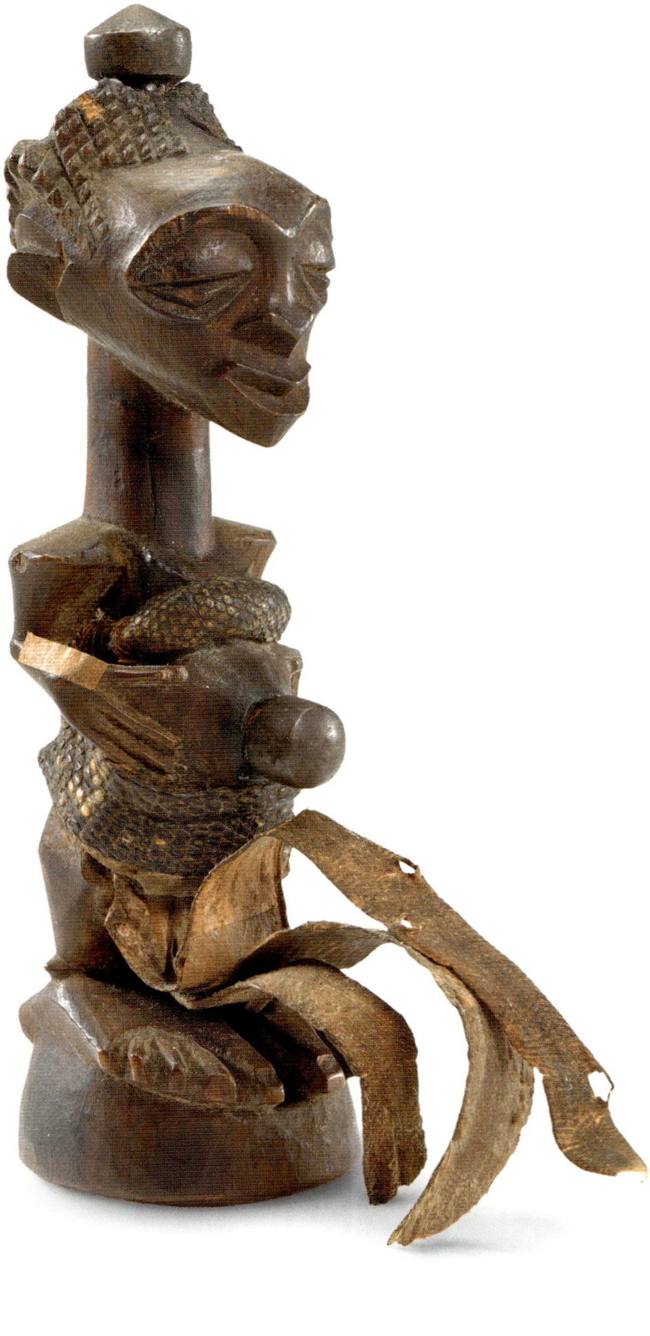

Members of the Mbudye Society recall and recount histories of grandeur using a *lukasa*, including epic tales of Mbidi Kiluwe and his son Kalala Ilunga, who established royal bearing and behavior as known and documented in the late 1980s by Mary Nooter (1991). *Vinyundo* associated with a *lukasa* are understood as *bitenta* (s. *kitenta*), or "spirit capitals," where legendary kings are buried and spirit mediums preside. A hammer-anvil pin embedded in a *lukasa* denotes the *kitenta* of a king or the culture hero he embodies, and medicines are inserted beneath the *kinyundo* to empower the memories invoked as a *lukasa* is narrated (ibid., 87). The same *kinyundo* cum *kitenta* is understood as a Luba king's throne, so that logically, a hammer-anvil is his seat of power (ibid., 157). In all these cases and more, *vinyundo* pins "closed," and so held and protected, secret precepts while activating the transformative capacity of such potent knowledge and practice.[20]

Songye are northern neighbors of Luba and share a great deal of language, culture, and history with them (Hersak 1986; Reefe 1981). The paramount of an important Songye polity called Kalebwe was transformed into a "blacksmith-lion" during investiture rituals. Hammer-anvils were used in this process; the king-to-be was instructed to sleep in shelters around smelting furnaces or blacksmiths' ateliers as his appropriate domain when he traveled; and he was said to use a large stone anvil as his throne, for arrowheads essential for hunting and defense had been forged thereupon (Anthoine 1996, 37–38).

Following aesthetic logic shared with Luba, *vinyundo* were driven into and so empowered some Songye *mankishi* figural sculptures (s. *nkishi*; figs. 15.30–15.32) and nearly cover a few (see Anderson and Kreamer 1989, 120–21; Koloss 2002, 162–63). As Dunja Hersak asserts, these pins "signal the dual capacity of the *nkishi* to counteract and redirect mystically construed aggression," thus making use of the express ambiguity of the hammer-anvil in practical circumstances of local-level politics (2002, 225). Indeed, proliferation of *vinyundo* on some *mankishi* may suggest that the figures' "powers were summoned repeatedly" in troubled times (Anderson and Kreamer 1989, 120).

Miniature blacksmiths' tools and forged accouterments also embellished some *mankishi*. What may have been diminutive rakes modeled on a blacksmith's implement are attached to some (see fig. 15.32), permitting a person to manipulate an *nkishi* deemed "too hot to handle"—that is, too perilously and unpredictably agentive.[21] Once again, the ambiguities of iron as a material and the tools and processes of ironworking came to the fore in equipping this figure with implements to facilitate its interventions in everyday lives.

15.30
Artist unknown (Songye peoples, Democratic Republic of the Congo)
Power figure (*nkishi*), 19th century
Iron, wood, reptile skin, leather
H: 16.5 cm
FOWLER MUSEUM AT UCLA, X65.7493; GIFT OF THE WELLCOME TRUST

Songye peoples living northeast of the Luba heartland have long shared elements of political aesthetics and ritual practice with Luba, as the hammer-anvil-shaped insertions of this powerful *nkishi* figure suggest. The top of a person's head and his/her navel (as framed by the hands) are understood as points of transformation but also of vulnerability to malevolent intrusion. In this sculpture, these anatomical locations were activated through *vinyundo*, but they were also protected by the implement's distinctly multipurpose efficacies. Potent medicinal bundles were embedded in both locations and then closed by the hammer-anvils, at the ready to assist in striking decisions and in vigilance against wrathful adversaries.

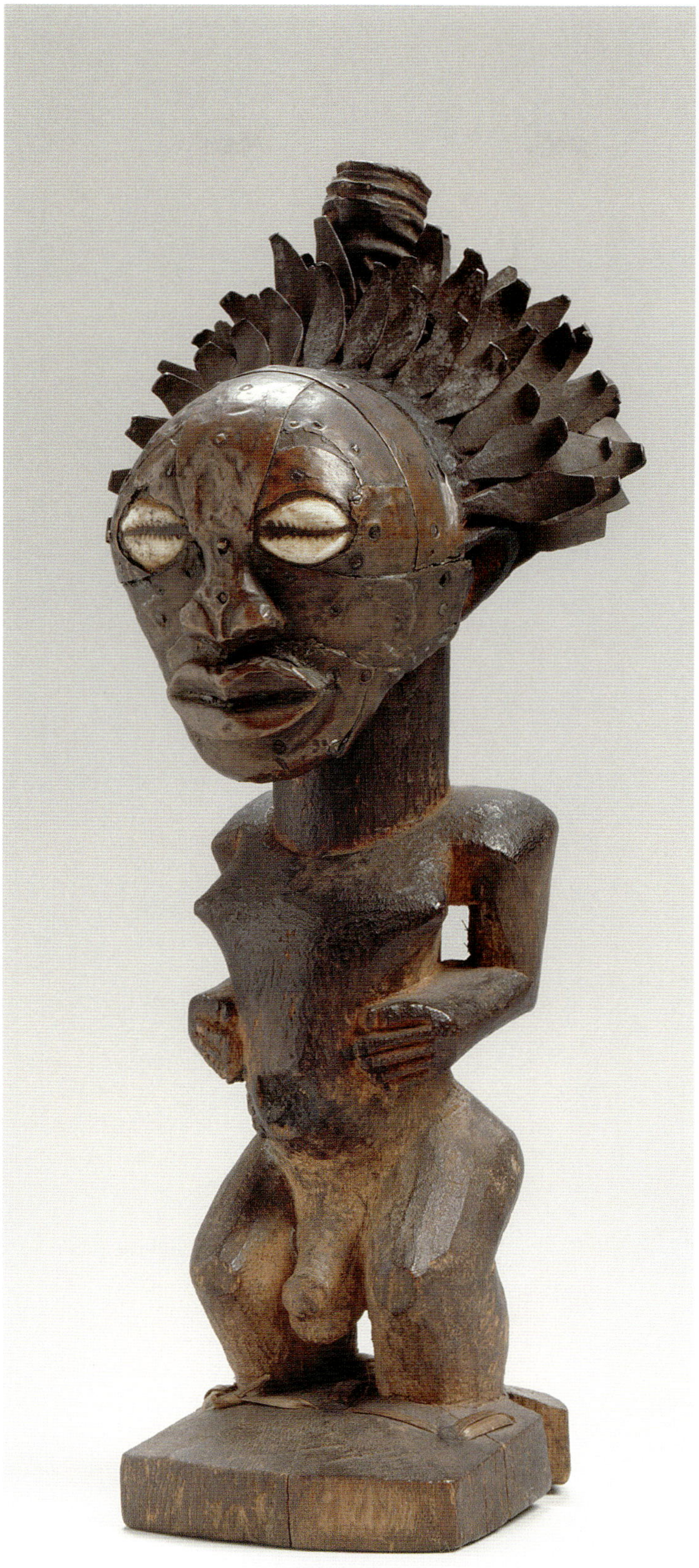

15.31
Artist unknown (Songye peoples, Democratic Republic of the Congo)
Power figure (*nkishi*), mid- to late 19th century
Wood, iron, copper alloy, cowrie shell (*Cypraea moneta*), horn
H: 22.9 cm
COLLECTION OF THE MAS, ANTWERP, BELGIUM (AE.0720), GIFT OF LOUIS FRANCK, ANTWERP, 1920
PROVENANCE: FIELD COLLECTED IN 1920 BY LOUIS FRANCK (1868–1937)
SELECTED FOR THE BELGIAN PAVILION AT THE 1939 NEW YORK WORLD'S FAIR (NICKNAMED "THE MAN WITH IRON HAIR")
PHOTOGRAPH BY MICHEL WUYTS, 2016. © MAS | MUSEUM AAN DE STROOM, ANTWERP

Sadly, no particular information exists about this or most other early central African sculptures, so who sculpted, commissioned, or used them remains unknown, as do purposes and contexts of their use. Generally, such works were points of power enlivened by spiritual energies, but how can one not wonder about how such needs were conceived and met in actual circumstances? Smelters and smiths must have looked to such sources of spiritual power—why wouldn't they? This figure made an assertive statement through iron: blades protected its head and projected its potentialities, as did the potent medicines placed in the inverted duiker (small antelope) horn that the blades surround; might such a device have once been cherished by an ironworker?

15.32
Artist unknown (Songye peoples, Democratic Republic of the Congo)
Power figure (*nkishi*), mid- to late 19th century
Iron, wood, cowries, leather, plant fiber, sacrificial material
H: 22.9 cm
MUSÉE AFRICAIN DE NAMUR
PHOTOGRAPH BY PIERRE WACHHOLDER, 2017. COURTESY MUSÉE AFRICAN DE NAMUR

This remarkable figure bristles with an impenetrable coiffure of miniature iron blades. It is additionally endowed with medicinal bundles around its neck, chest, and waist. As a power figure (*nkishi*) deemed too "hot" to touch due to its supernatural capacities, it required iron intermediaries for safe handling and was carried and manipulated by two articulated iron rods forged like the ember rakes used to tend a blacksmith's fire. The accumulation of iron accoutrements strengthened the *nkishi*'s protective potential and activated divine forces residing within it. The figure was found by General Henry de la Lindi in 1897, who reported it abandoned on the battlefield after a violent skirmish between Belgian forces and local anticolonial rebels including Songye warriors.

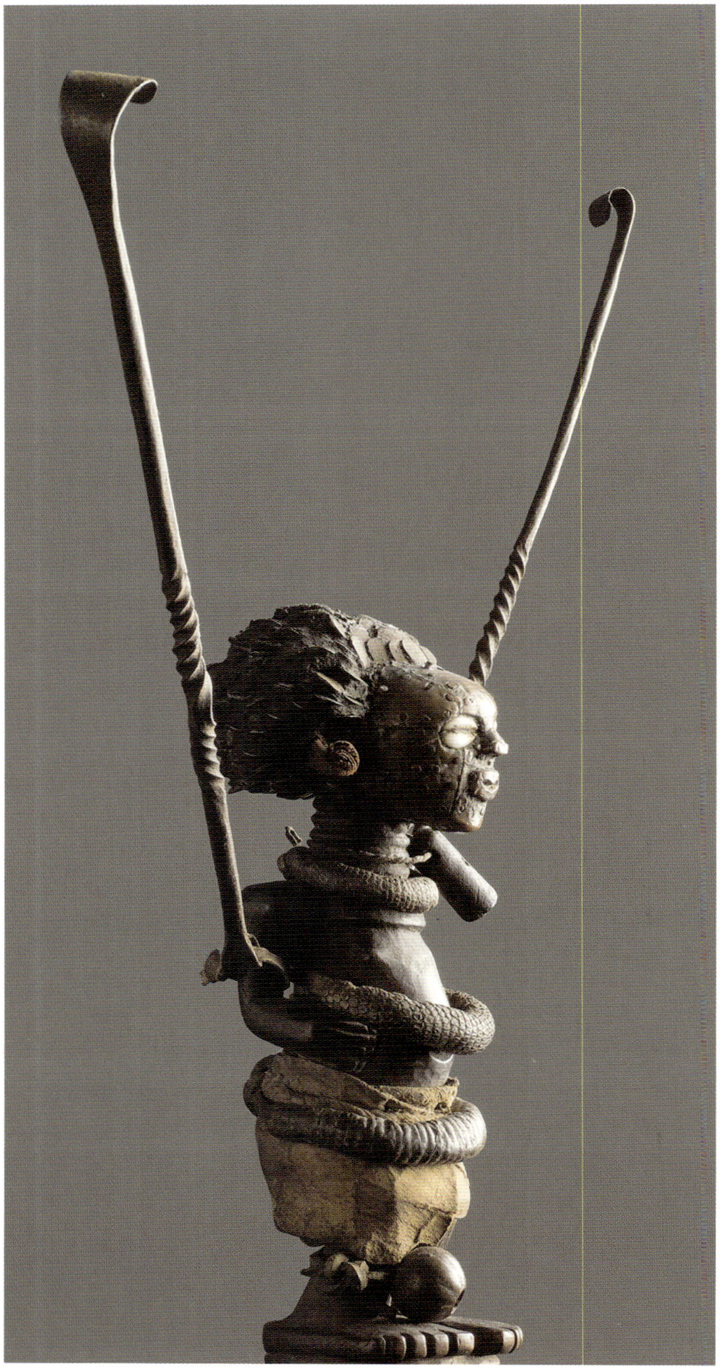

15.33
Artist unknown (Kongo peoples, Democratic Republic of the Congo)
Sword/cross (*mbele a lulendo*), 18th century
Iron
H (on base): 89 cm
COLLECTION OF JOEL AND ZACHARY COONER
IMAGE COURTESY OF JOEL AND ZACHARY COONER

Kongo "swords of honor" were explicitly anthropomorphic, reflecting their own agencies to transform the circumstances of those wielding them while referring to weapons introduced to the Kongo by Portuguese visitors beginning in the late 15th century. The pommels of such swords present wide-eyed faces below caps of the sort once donned by Kongo kings, while the gesture of this sword's "arms" is both a dance gesture and a sign of God's sheltering sky.

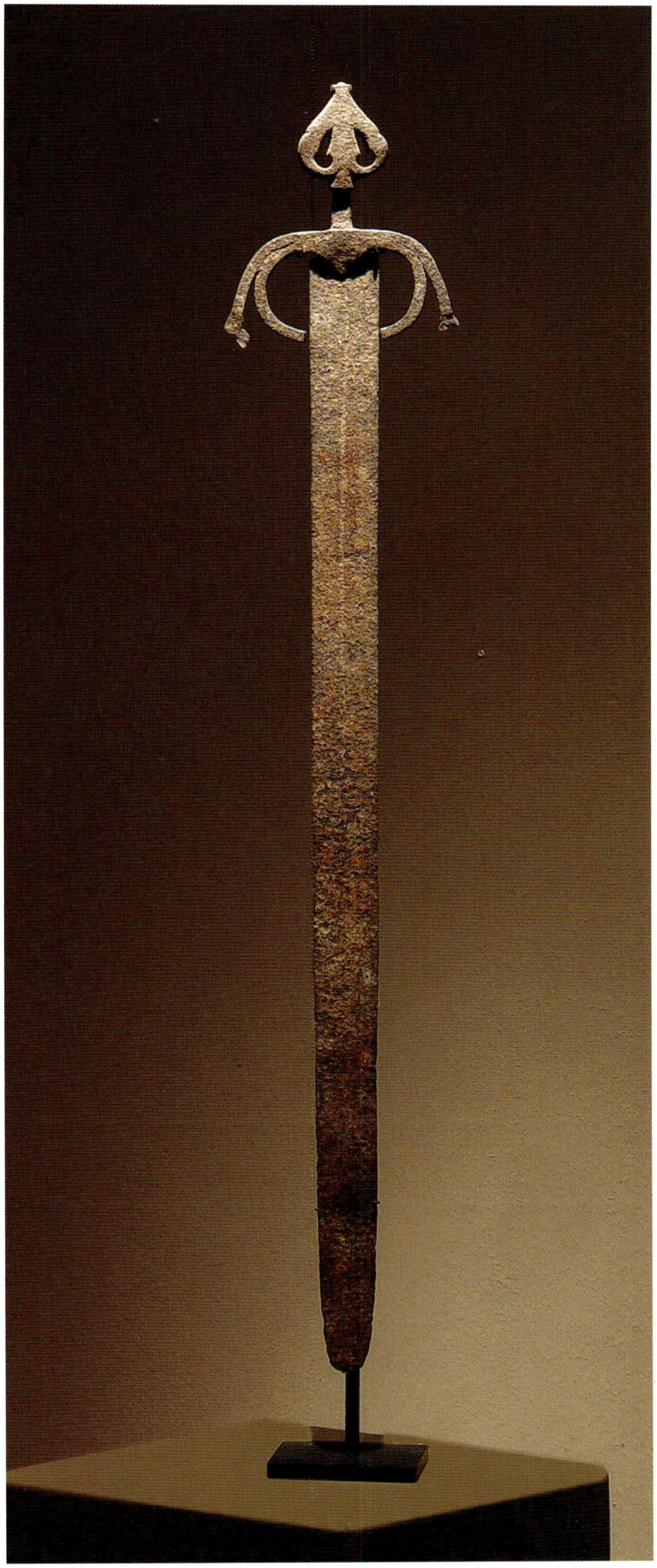

KONGO DIA NTADI—"KONGO OF IRON"

Kongo peoples of northwestern Angola and adjacent lands provide a most evocative last study, for "the tie between politics and ironworking is particularly strong in the Congo Basin" (Maret 1985, 75). Oral traditions have it that "Lukeni, a young foreign prince, left his homeland north of the Congo River, marched south with an army of followers, and conquered his future [Kongo] kingdom in a great military feat made possible by his command of iron technology, until then unknown to people of the southern shore. In addition to military skill and iron smelting, Lukeni brought new wisdom that forms the basis of the kingdom's organization"—hence the epithet "Kongo dia ntadi" (Kongo of Iron).[22]

Ironworking processes and products also informed earlier installation rituals of Kongo kings and rulers of dependent states. Hammer-anvils were struck together as "'royal and very old insignia,'" bringing to bear transformative rhythms of the forge; "blacksmiths' whistles" were played as instruments of royal privilege (Fromont 2014, 40); iron bells were sounded as in battle and pomp, to call on "the ubiquitousness of the ancestors," of whom the king was "the protector and interpreter" (Balandier 1968, 111); and *malunga* bracelets were presented to the new monarch to "endow the king with attributes derived from a rich web of semantic and visual associations linked to its name, material, and form."[23] As hammer-anvils clanged, a blacksmith opened the reliquary holding remains of former kings and their *malunga*, and he chose one of the iron bracelets for the new king to wear and so sustain the continuity of rule. Once "born" through ritual, the new monarch danced exuberantly, sword in hand (Fromont 2013, 30). That a blacksmith was the "midwife" of such a process recalls the "enchanted technologies" and "danced causality" of smelting, with ironworking and rule informing each other once again.[24]

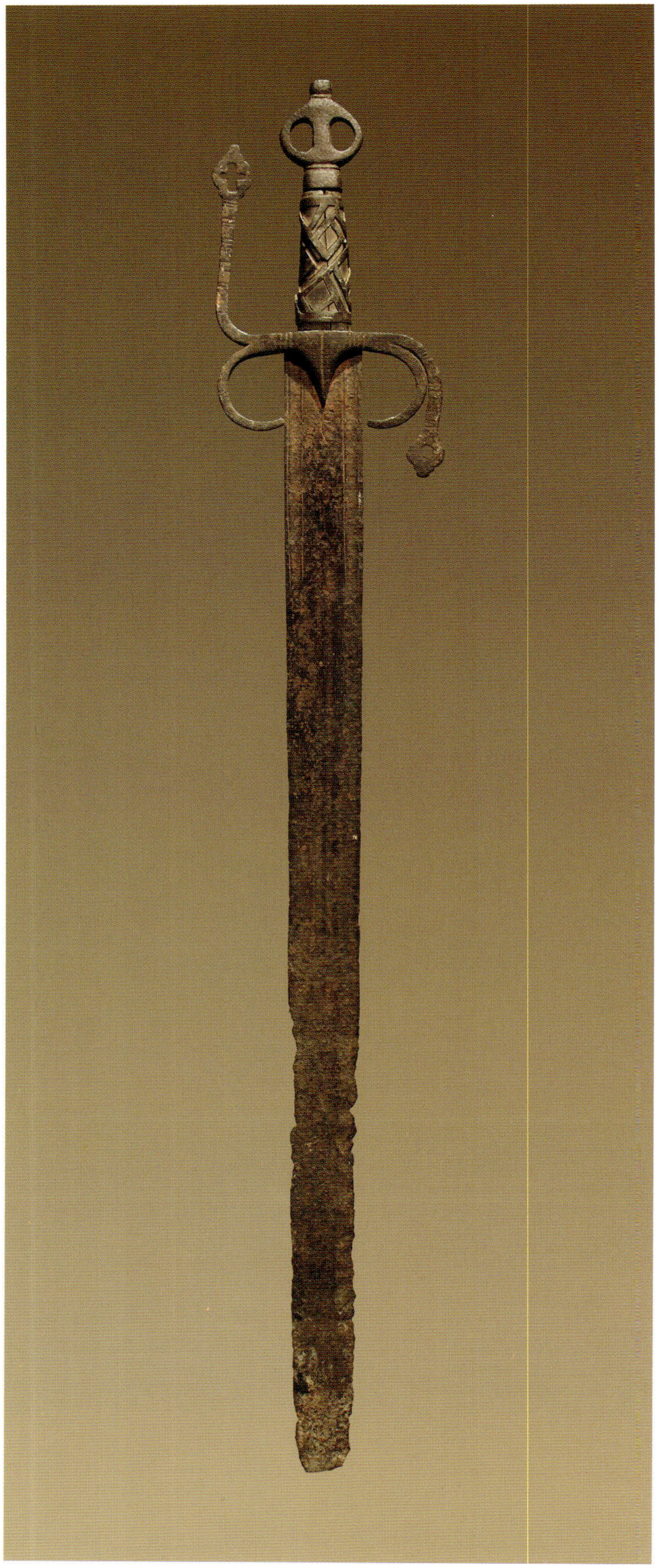

15.34A–B
Artist unknown (Kongo peoples, Democratic Republic of the Congo)
Sword/cross (*mbele a lulendo*), 18th century
Iron
H (on base): 85 cm
COLLECTION OF JOEL AND ZACHARY COONER
IMAGE COURTESY OF JOEL AND ZACHARY COONER

The hand guards of "swords of honor" materialize gestures of Kongo being-in-the-world. The upward "arm," as art historian Robert Farris Thompson (1981) explains, "hails the law" of the living, the downward arm "cools the community" through deference to the ancestors beneath our feet, and the horizontal structure is the threshold between the two realms. Dancing with such a potent implement placed the performer at the center of human existence, as mediated by iron.

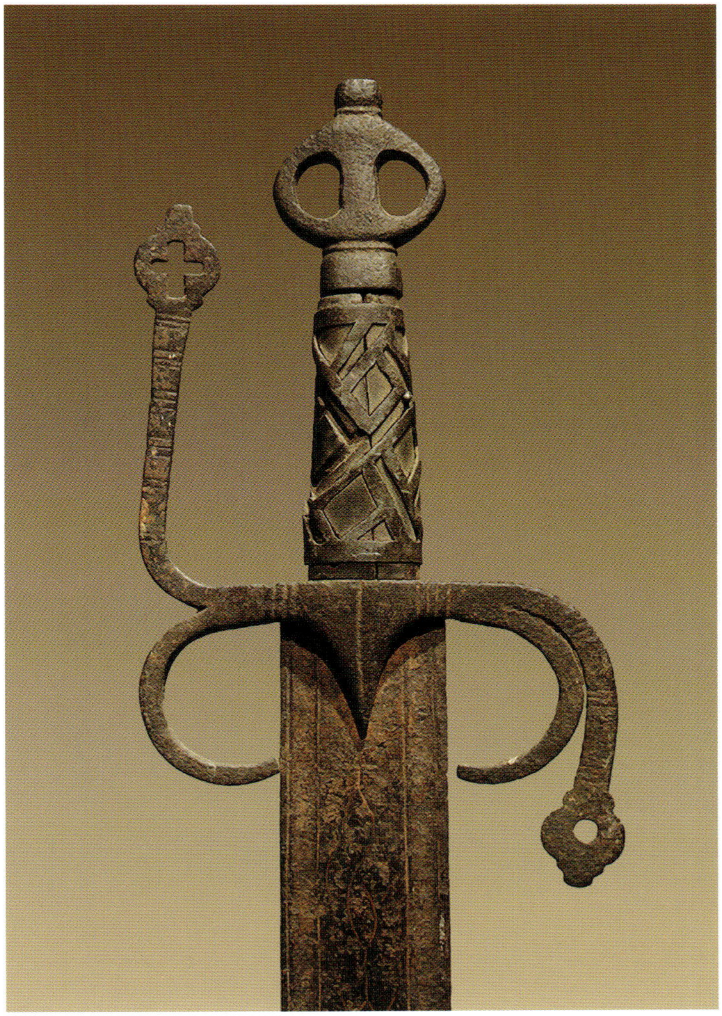

Portuguese adventurers first visited Kongo kingdoms in the late fifteenth century, and soon thereafter a synthesis of Kongo and European royal aesthetics was realized in locally forged "swords of honor" (*mbeele a lulèndo*). Based upon arms borne by Portuguese, these magnificent blades (figs. 15.33, 15.34) "memorialized the conversion of the kingdom [to Christianity] that soon followed this first encounter" (Fromont 2013, 28). As oral narratives have it,

> iron swords providentially appeared in the hands of Saint James and an army of knights under a resplendent celestial cross, to help the great Christian king of Kongo, Afonso, secure the throne and impose Christianity as the realm's official religion. The miraculous swords…gave legitimacy to the incoming king in both local and Christian terms; they associated his accession…with deep-rooted central African metaphors of iron kingship as well as with the European-derived symbolism of the weapons as emblems of nobility and Christian knighthood. These stories proved very successful, and the swords became paramount symbols of Kongo Christian might and prestige for centuries.[25]

Mbeele a lulèndo were of signal importance to politics but were not used for actual fighting. If, for instance, in the course of litigation brought before an important turn-of-the-twentieth-century Kongo chief named Makonde Mambutu, a defendant was deemed guilty but refused to submit to punishment, Mambutu's sword of honor would be thrust into the ground. Seeing this, "the recalcitrant . . . would immediately agree to pay his fine because" the gesture "signaled that the chef had 'arisen'" and would unleash his fearsome powers. Anyone who continued to resist would be held for ransom to be paid by his people (MacGaffey 2000, 68).

Robert Farris Thompson has explained how the hilts of Kongo swords of honor bear complex iconography (1981, 28). Their pommels suggest heads wearing royal caps like that of the blacksmith-king of Ndongo in Cavazzi's watercolor (see fig. 15.1), and they feature piercing, vigilant eyes.[26] The "arms" of most swords' quillons strike "crossroads poses," playing upon the Kalunga line demarcating the "up" of the living and the "down" of the dead. The position "hails the law" with its upward-pointing hand and "cools the community" as it directs attention to the ancestors with its other hand (ibid., 62). Indeed, "the arms indicate power (*lèndo*) and signal a flow of energy (*ngolo*) between the two worlds" (Martínez-Ruiz 2009). The downward crescent below the "arms" also bears cosmogonic reference, for it is a sign of the sky (Martínez-Ruiz 2012, 200). More generally, as Bárbaro Martínez-Ruiz would add, materialized gestures like these are "scripture in space,…a form of spatial writing complemented by sounds, phrases, or secret rhythms which, when used within ritual performance, invoke powerful spiritual forces from nature and the cosmos."[27]

Gripping such a hilt empowered the bearer and extended his reach. Thinking of King Afonso's visitation by Saint James, one can imagine that whenever such a weapon was struck on a battlefield, "the fire imprisoned within it leapt forth and the sword shone, not because it reflected light, not because of some violent blow, but because of its own valiant metal,…[for] the sword that is forged and tempered with all the smithy's dreams is heroism made material. It is legendary in its very substance, before it ever belongs to the hero." Yet when it was wielded, one witnessed "the interplay of the material and the moral, the cosmic and the human."[28] And in sum, such inspiring relationships of honor and courage characterize the "obviously old and intricate relationship…between leadership and metallurgy" in a Central Bantu World.

INTERLEAF

K

INTERFORMANCES OF A KONGO NAIL FIGURE

Allen F. Roberts

> Putting in a nail is like this: If a man has had something stolen and he asks who took it but nobody owns up, he goes to the *nganga* [practitioner] and asks permission to nail his *nkisi*. The *nganga* says "Very well." Ko! Ko! Ko! Lord bite his chest, bite his ribs. If he stole from me, go for him!
>
> KONGO ELDER BABUTIDI TIMITIO, CA. 1916[1]

More than a century ago, Marcel Mauss urged ethnographers to eschew the term "fetish" because of cruel misunderstandings it bears (1907, 244–45, cited in Surgy 1987, 1). The word retains currency nonetheless, largely because "fetishes" so readily foster invention of "Africas" to suit many purposes, and especially hegemonic, non-African ones.[2] No so-called "fetishes" are more iconic than figural sculptures known as *minkisi nkondi* (s. *nkisi nkondi*) that were once created by Kongo peoples of west-central Africa, bristling as they are with nails and fragments of iron.[3] What fantasies they have conjured! Yet for Kongo, they served as "monuments to cultural confidence," visibly and supernaturally asserting that "there is no crisis that cannot be weighed and solved by chiefs and ritual experts, secure in their knowledge of ancient antidotes and ancestral insights kept in everlasting trust" (paraphrasing Thompson 1978, 219–20).

In all likelihood, an imposing nail figure (fig. K.1) now in the Wellcome Collection of the Fowler Museum at UCLA was created in, near, or brought to the Congolese town of Boma in the mid- to late nineteenth century to meet personal and community needs, as was a remarkably similar and far-better-documented *nkisi nkondi* (fig. K.2) in the holdings of Belgium's Royal Museum for Central Africa, Tervuren.[4] Boma lies sixty miles up the Congo River from the Atlantic, and nine small Kongo kingdoms grew in wealth and prominence as European traders established themselves there. Early colonial occupation proved horrific, as overt slavery was replaced by deadly corvée labor and "the word 'ivory' rang in the air, was whispered, was sighed" while "a taint of imbecile rapacity blew through it all, like a whiff from some corpse."[5] We can surmise that the *minkisi nkondi* now at the Fowler and the RMCA were Kongolese devices meant to push back against harsh expatriate machinations while kings aggressively vied with local competitors for opportunities to profit handsomely.[6]

"Interformance" is a neologism that helps to understand the iconography and votive practices of these sculptures that, in all their "spectacular complexity," "advertise[d] their inner secrets by... outer flamboyance" (MacGaffey 2000, 89, 86). The prefix *inter* is Latin for "between" or "among," while *formance* "furnishes" experience (Morris 1967, 683, 974). An *interformance*, then, permits one to begin the betweens of dynamic transaction.[7] Indeed, an interformative *nkisi nkondi* served as "the visible knot which tied together an invisible skein of relations fanning out in social space and social time" (Gell 1998, 62, writing thus of such figures), and in a sense, these sculptures were and remain archives of intentionality.

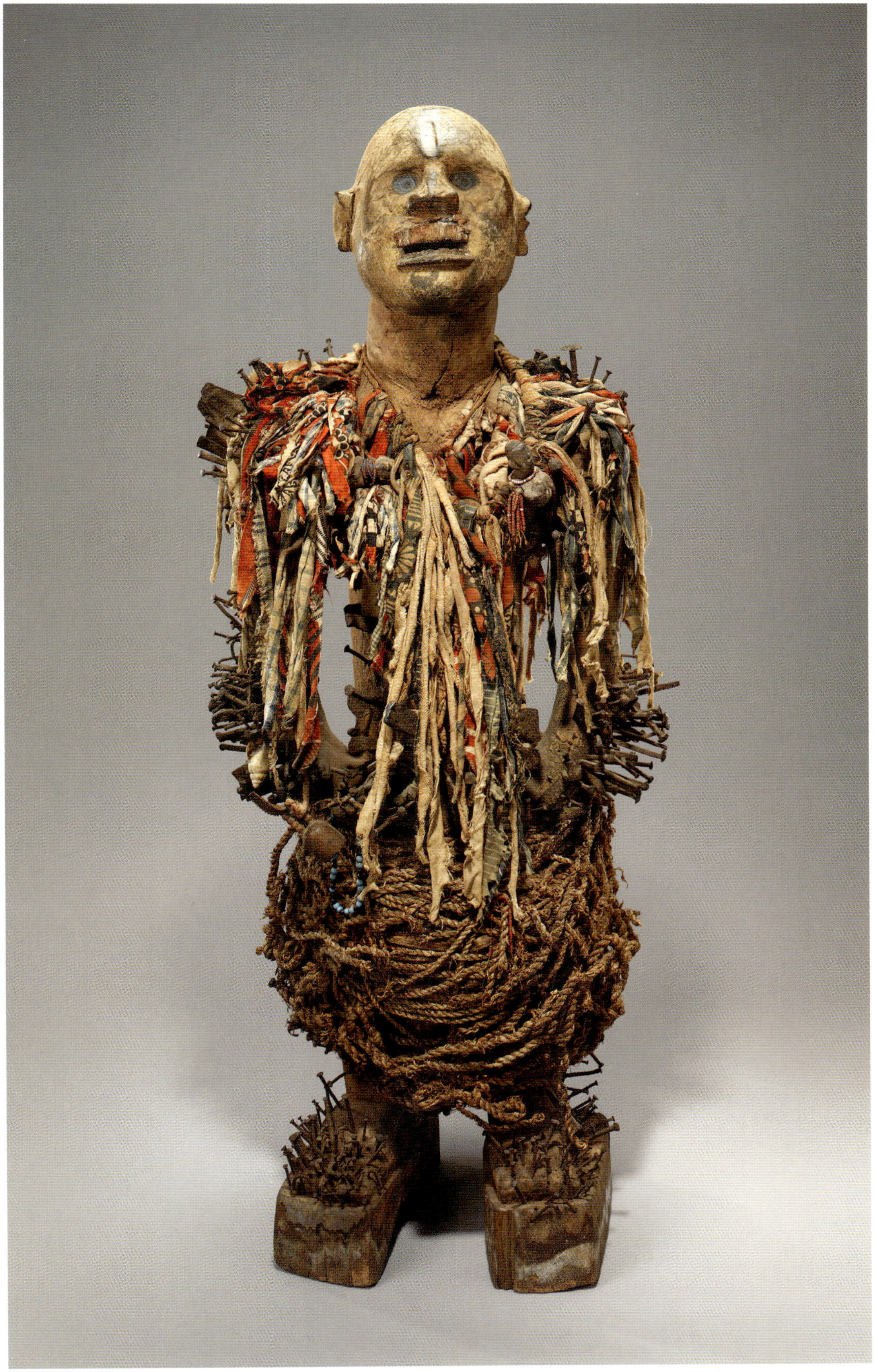

K.1
Artist unknown (Yombe peoples, Democratic Republic of the Congo)
Power figure (*nkisi nkondi*), 18th to 19th century
Wood, iron, copper alloy, mirrors, cloth, cordage, glass beads, cowrie shells (*Cypraea moneta*)
H: 114.3 cm
FOWLER MUSEUM AT UCLA, X65.5837; GIFT OF THE WELLCOME TRUST

Nothing is known of the provenance of this Kongo nail figure other than that it was acquired at the turn of the 20th century by Sir Henry Wellcome. Its purposefully spectacular assemblages nonetheless reveal the work to be an archive of intentionality.

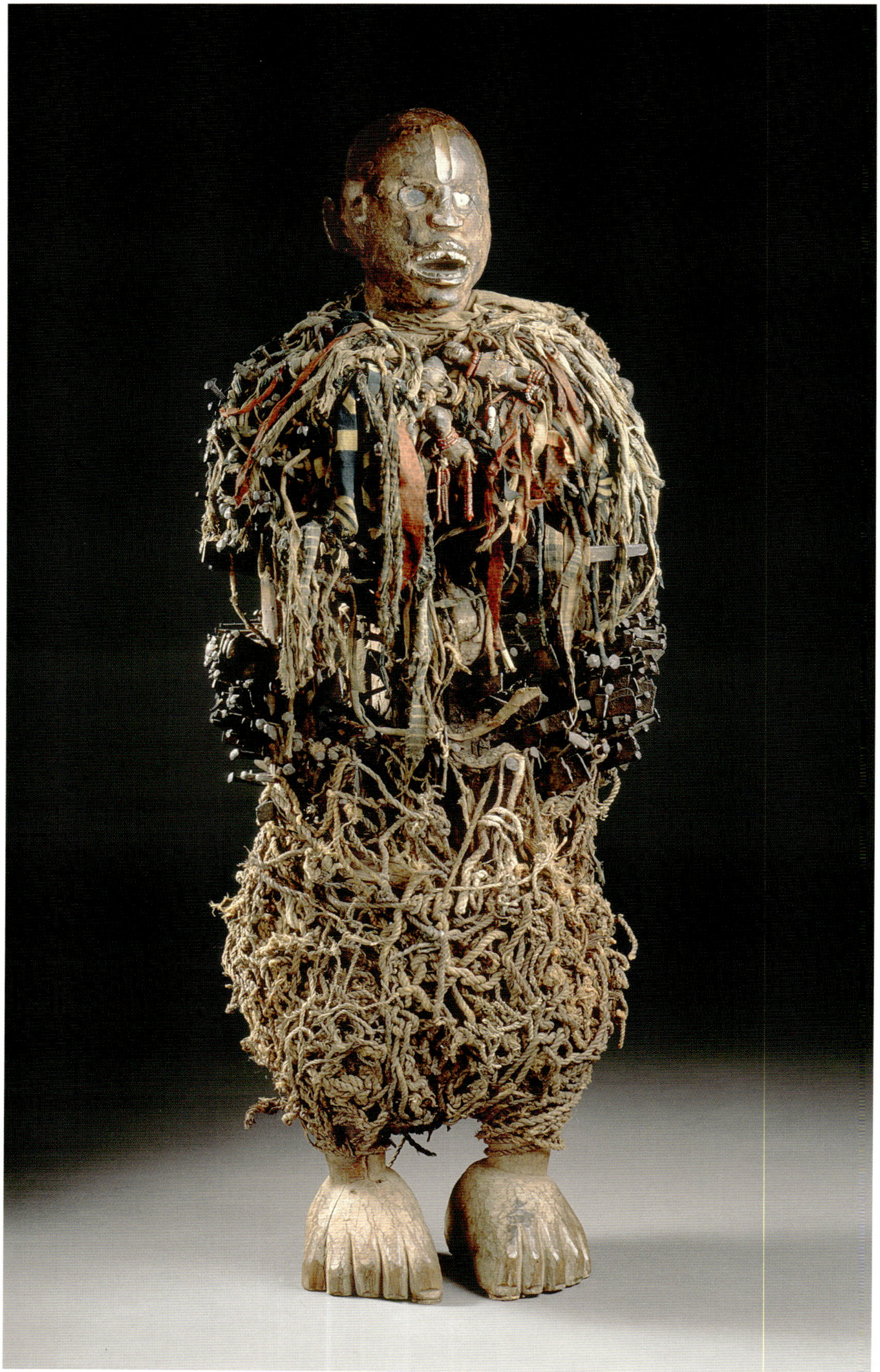

K.2
Artist unknown (Kongo Peoples, Democratic Republic of the Congo)
Power figure (*nkisi nkondi*), 19th century
Iron and additional metals, wood (*Canarium schweinfurthii*), pigment, glass, cloth, plant fiber
H: 115 cm
COLLECTION ROYAL MUSEUM OF CENTRAL AFRICA, TERVUREN, EO.0.0.7943
(NOT IN EXHIBITION)
PHOTOGRAPH BY PLUSJ. © RMCA, TERVUREN

An *nkisi nkondi* seized at the Congolese town of Boma in 1878 by Alexandre Delcommune, a Belgian businessman.

Such transactional potentialities start with the "heart" (*ntima*) of the figure, as activated by potent medicines held in a bundle now hidden by festooning fabric (fig. K.3).[8] Such a device was a "little world" composed of myriad herbal and symbolic substances—the missionary K. E. Laman listed ninety medicines in one.[9] Unidentifiable shavings or crumbs of elements would be included, and because only the person confecting the device would know its ingredients, what it *might* contain was as important as what it did. Such is the nature of secrecy to conceal and reveal (Nooter 1993). Most significantly, human relics and earth taken from graves and incorporated into the "heart" assured that one or more spirits would be summoned as the *nkisi* was activated (Mulinda 1987, 211).

Nkondi refers to hunting, and the forces of an *nkisi nkondi* "lurk[ed] in ambush for liars, thieves, adulterers, for all sorts of persons who might undermine societal structure" (Thompson 1978, 211). Kongo hunting dogs mediate between village and forest, the living and the dead (MacGaffey 1990, 55), and they have inspired Janus-headed *nkisi kozo* figures to extend vigilance beyond human ken (fig. K.4). The net around the *nkisi nkondi*'s waist emphasized hunting, for ne'er-do-wells would be trapped and dispatched by its spiritual and medicinal agencies, while the net's knotting added conceptual potency through cadenced binding and closure.

The Kikongo verb *kukoma* means "to nail," but also "to arouse, to activate" (MacGaffey 2014/2015, 152). In piercing a figure with iron, a supplicant "'signed' his vow" and "inflamed" the agencies of an *nkisi loko*—that is, a figure conceived to receive imprecations directed toward an adversary perceived as having committed an injustice.[10] A nail might be licked to make purpose intimate, for in a sense, his or her personhood was extended to the blade as an oath was taken or a promise made (Thompson 1978, 215). An "aesthetics of nailing" was at play, and differently shaped iron blades may have connoted severity of perceived harm and intended retribution.[11] Furthermore, the act was undertaken with sufficient public notice that the accused, knowing the potency of a particular *nkisi loko*'s success in rendering justice (just look at *all* those blades!) might offer redress so that the nail could be removed and harmony restored (Mulinda 1987, 215). As a totality, then, the figure provided "moral medicine," "moral revelation" (Thompson 1978, 211, 212).

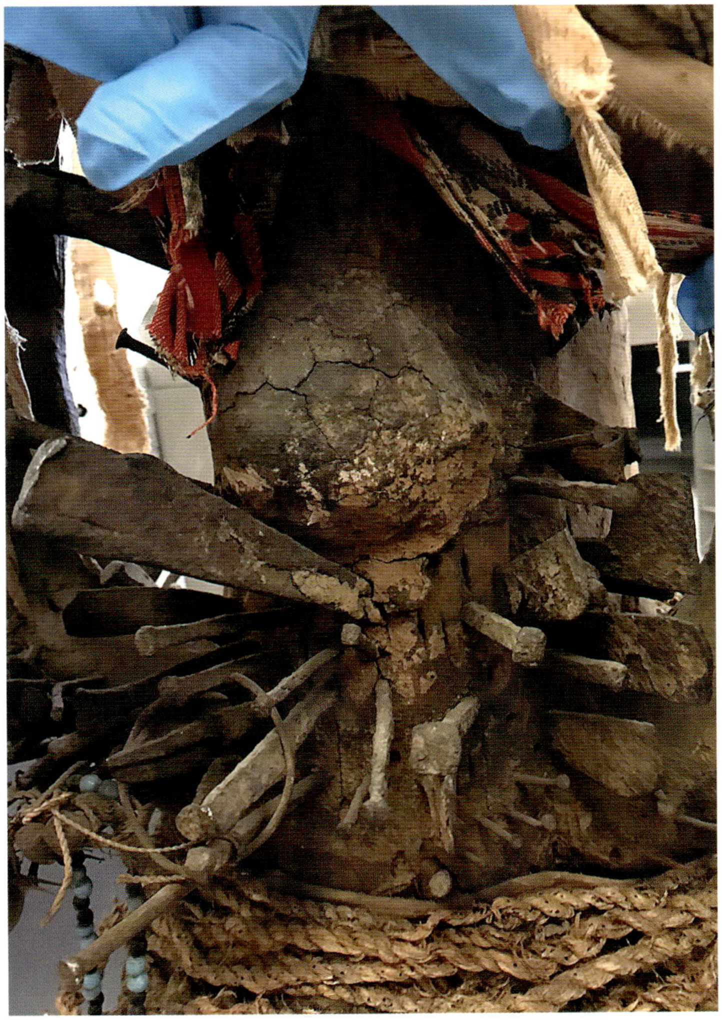

K.3 Gentle lifting of festooning fabric by blue-gloved fingers reveals the "heart" (*ntima*) of the *nkisi nkondi* at the Fowler Museum (see fig K.1) as a bundle of empowering substances held in organic mastic.

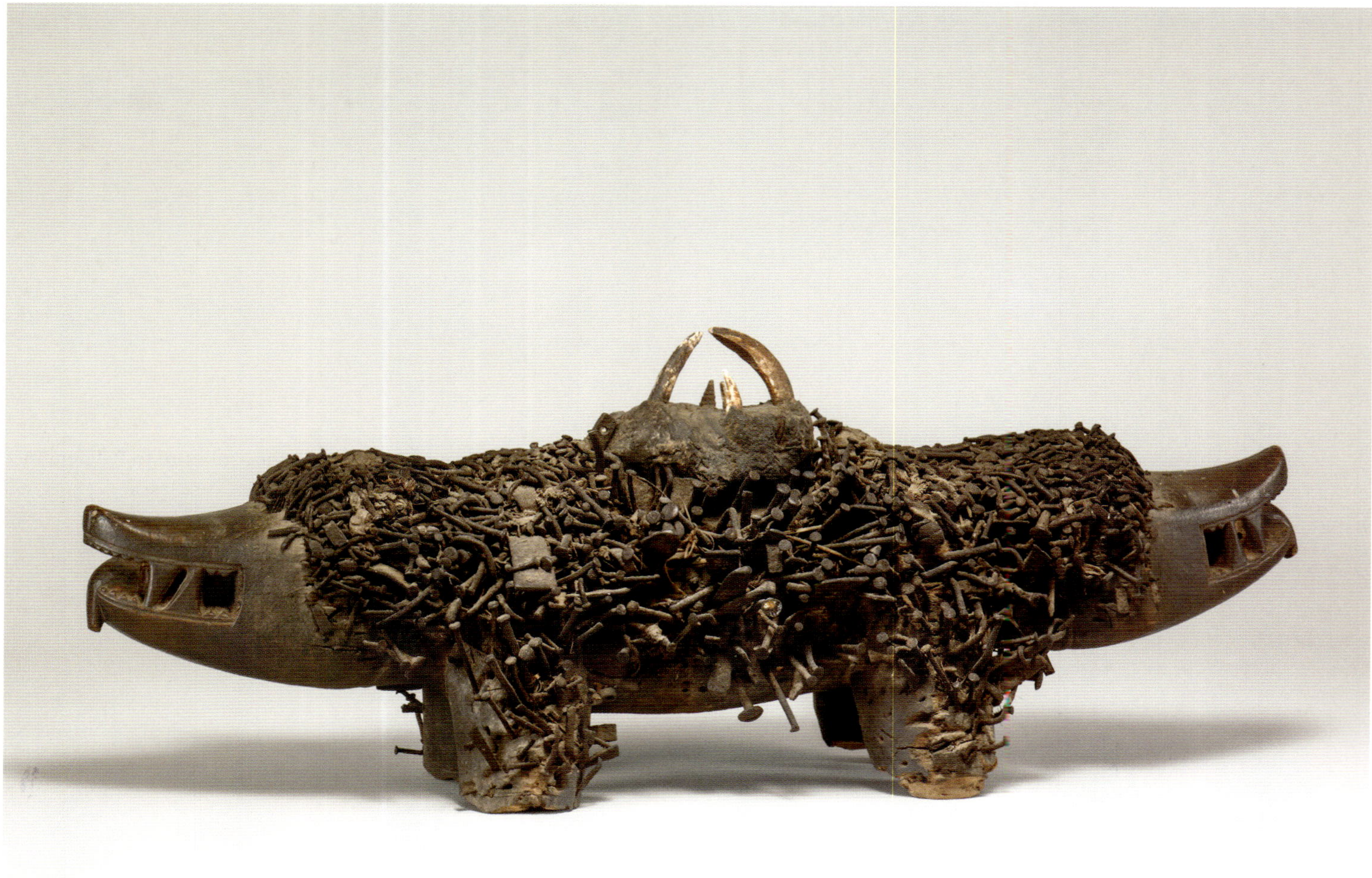

K.4
Artist unknown (Vili, Woyo, Yombe peoples, Angola and Democratic Republic of the Congo)
Janus power figure of a dog (*nkisi nkondi mbwa*, *nkisi kozo*), 19th century
Iron, wood, cloth, plant fiber, resin, warthog tusks, shell
L: 72.4 cm
THE MENIL COLLECTION, HOUSTON, 1977-03 DJ
PHOTOGRAPH BY PAUL HESTER. COURTESY OF THE MENIL COLLECTION

Hunters' dogs perceive game and avoid danger (see fig. 6.22), and Janus-headed canine nail figures called *nkisi kozo* use their uncanny senses to protect and pursue malefactors approaching from any direction. Such figures were sometimes used in conjunction with *minksi nkondi* like those of figs. K.1 and K.2.

Historical analysis of the iron insertions of the *nkisi nkondi* at the Fowler has yet to be completed.[12] Smiths' iron scraps join a plethora of nails, some forged, some manufactured. Histories of trade may be discerned in such numbers and variety (Gurstelle n.d.). The figure's *assemblage* of iron, cloth, beads, small *minkisi*, and other activated objects is also intriguing; how might these elements have interacted with each other as "living, throbbing confederations" whose properties were "emergent in their ability to make something happen" in ways "distinct from the sum of the vital force of each materiality" (Bennett 2010, 24)? And how might such a concatenation be "chimeric" through "a way of organizing differences that does not recognize any pre-established center of epistemological priority" (Severi 2015, 12)?

Features of the *nkisi nkondi* adding to its imposing presence include piercing eyes that "saw through" any mere mortal, yet reflected evil back on itself. Ocular surfaces provided a threshold between mundane matters and the figure's interior agencies, and medicines may have been packed behind them. A groove along the midline of the forehead is finger-patinated as testament to an *nganga* practitioner's repeated stroking during invocations to activate the seat of inspiration that the forehead midpoint is understood to be.[13] The teeth are bared in aggression, but also to reveal the perfection of their filing (cf. Roberts 2016). But what of the upper lip? Its injury may not have been accidental, and instead revealed the sort of grievous injury a miscreant might expect if humble pardon were not sought and accorded (MacGaffey 2000, 101). The head and neck are covered with a now-chalky, yellowish substance, perhaps to articulate transformation as more overtly red symbols did (Jacobson-Widding 1979). The taut, hands-on-hips stance suggests preparedness (cf. Thompson 1973), while the feet, raised from the ground as a sign of nobility, are articulated to permit stealthy ambulation in pursuit.

How and why these nail figures at the Fowler and the RMCA are so uncannily similar remains to ponder. To date, the provenance of the first is unknown, but we can extrapolate from the second following recent research by RMCA curator Maarten Couttenier (2018). The figure at the RMCA was "captured" in 1878 by Alexandre Delcommune, an agent dealing in wild rubber and ivory. Drought had decreased revenues for the kings of Boma; when they increased transport taxes to recoup losses, European traders took issue and altercations ensued. Delcommune and his fellows considered such "insolence" a "'declaration of war,'" and they mustered mercenaries who razed the kings' villages. In the course of melee, "one of the most reputed idols of the whole region" was seized. Delcommune already knew the "'God,'" as he had "rented" it to drive a nail and apprehend six thieves who gave themselves up upon learning that their doom had been so sealed. Once in his possession, Delcommune used the figure to protect his warehouse so it could be left unlocked. King Ne Kuko demanded his *nkisi*'s restoration, but "ransom" was denied and Delcommune removed the sculpture to Belgium in 1883. There it served "colonial propaganda," and despite repeated requests for its repatriation, the *nkisi* has never returned to the Congo (Couttenier 2018, 84–85, 87, based upon Delcommune 1922).

Does the nail figure at UCLA have a similarly interformative history? Was it produced in the same workshop and/or dressed and manipulated by the same practitioner? Was it a replacement for the figure seized by Delcommune, or might both sculptures be simulacra of some other "original"? An hypothesis is offered by Wyatt MacGaffey, who notes that while possessing a spectacular figure protected a king's interests, it also attracted clients from adjacent communities whose fees greatly benefited the proprietor's kin. Neighbors noticing such business success might seek "children" of the *nkisi*, "franchised" as fully prefabricated yet utterly spectacular displays of nails, textile strips, and all (MacGaffey 2014/2015, 152–53). "The whole procedure, after all, depended on drama" (MacGaffey pers. comm., 2018). Was the *nkisi nkondi* at the Fowler so commissioned for local enterprise?

PORTFOLIO

III

PERSONAL ADORNMENT

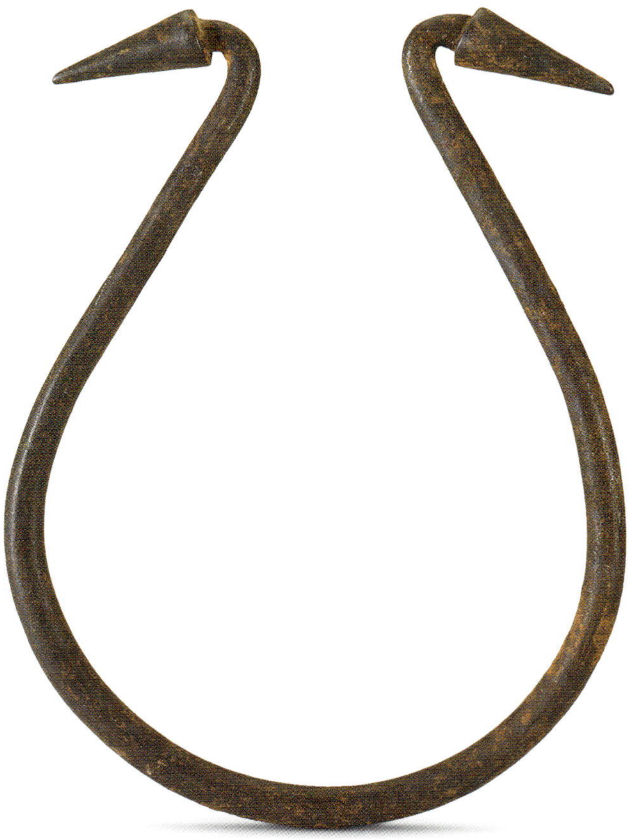

In sub-Saharan Africa, decoration of the body is part of highly localized, culturally specific systems of beautification. A wide array of forged iron adornments—necklaces, bracelets, elaborate headdresses, belts, and hair ornaments—have marked social transition, status, and identity among various peoples. Congolese men and women historically wore their hair shaped, teased, and/or braided into elaborate coiffures held in place and embellished by the insertion of delicately forged iron hairpins (see chapter 15, this volume). Personal decoration of the body has involved ephemeral applications as well as more permanent alterations (see chapter 13, this volume). Exemplifying the latter, irreversible transformation of human skin through scarification has been a potent medium for expressing social and biological change as well as belonging in certain societies. Often exquisitely forged by blacksmiths, knives are used to make precise incisions that heal into patterns of raised bumps or lines, and iron thus lends social and cultural weight to such practices. Thin-edged razors have also been used by men for head and facial shaving, sometimes in patterns prescribed for ceremonial occasions.

III.1
Artist unknown (Shi peoples, Democratic Republic of the Congo)
Neck torque, mid-19th century
Iron
H: 19 cm
FOWLER MUSEUM AT UCLA, X80.154; MUSEUM PURCHASE

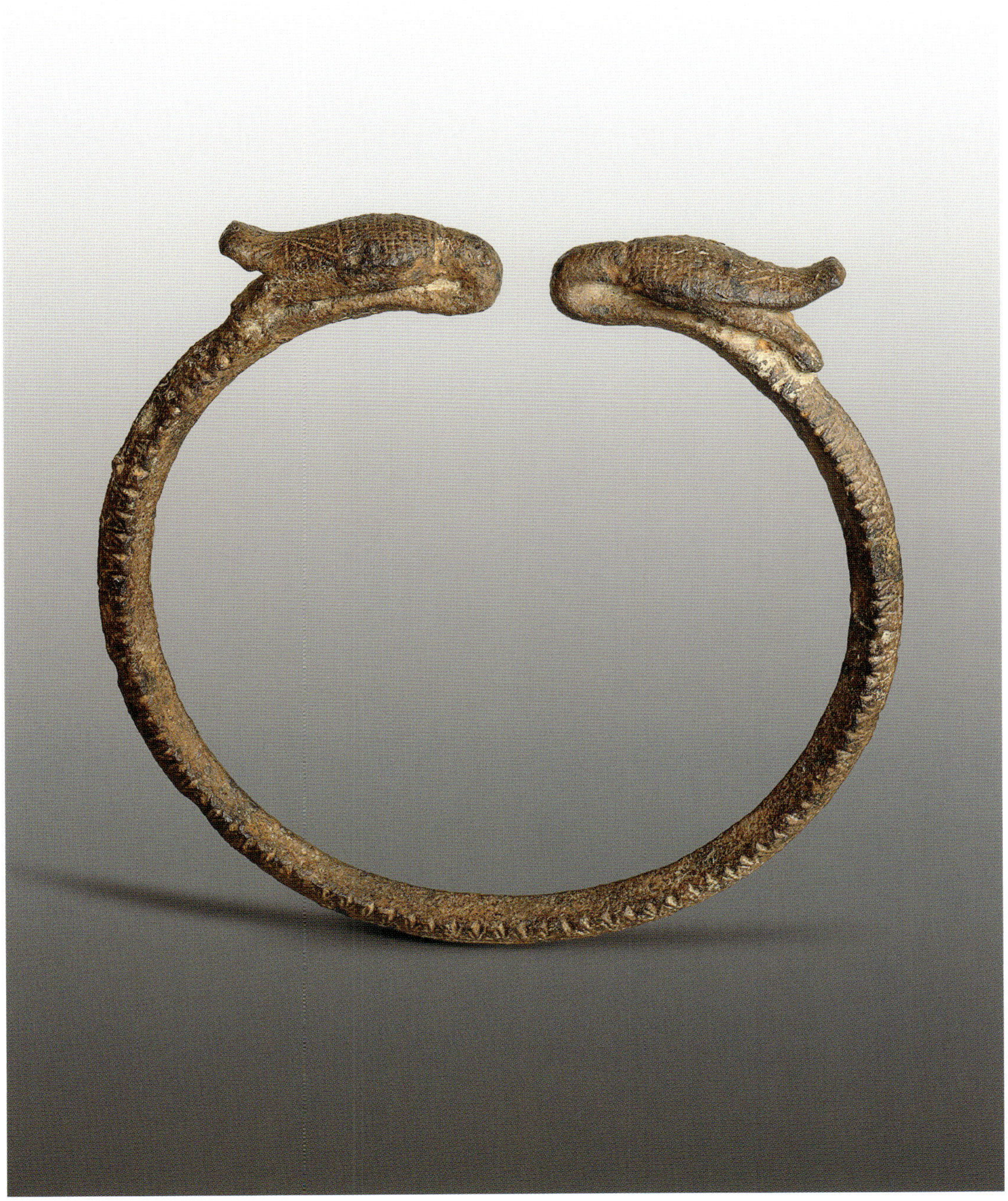

III.2
Artist unknown (Gan peoples, Burkina Faso)
Neck torque, early 19th century
Iron
H: 18 cm
ITZIKOVITZ COLLECTION
PHOTOGRAPH BY BRIGITTE CAVANAGH, 2017.

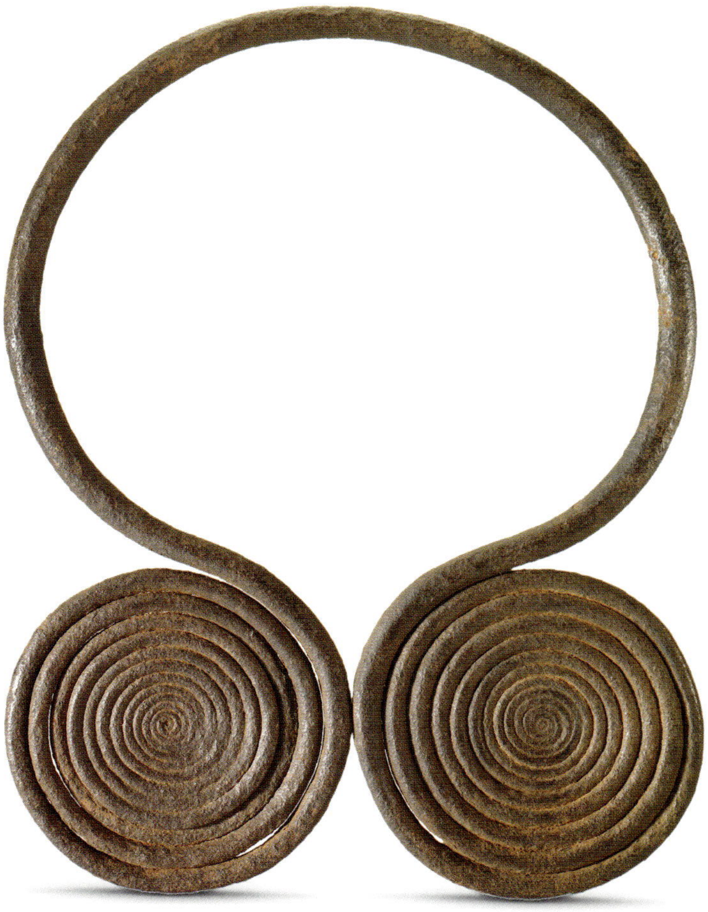

III.3
Artist unknown (Mongo[?] peoples,
Democratic Republic of the Congo)
Neck torque, mid-19th century
Iron
H: 26.7 cm
PRIVATE COLLECTION

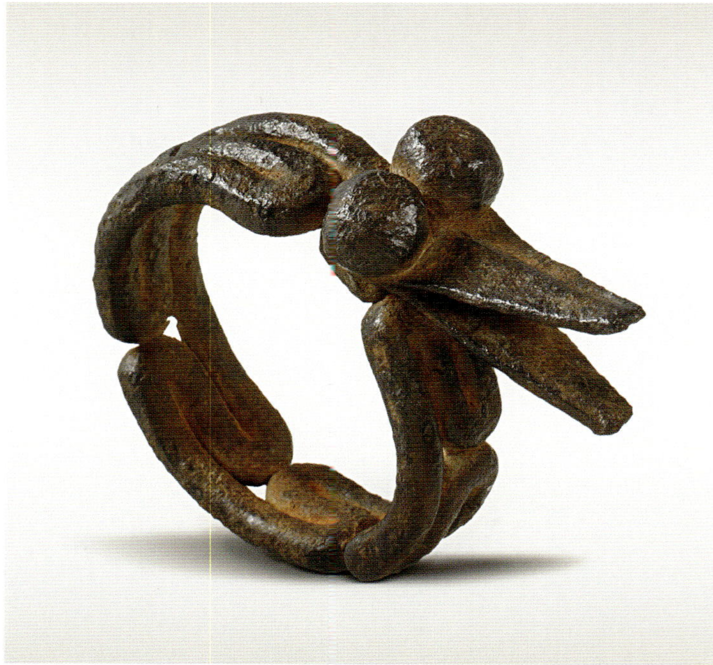

III.4
Artist unknown (Gan peoples, Burkina Faso)
Bracelet, mid-19th century
Iron
D: 14.5 cm
ITZIKOVITZ COLLECTION
PHOTOGRAPH BY BRIGITTE CAVANAGH, 2017.

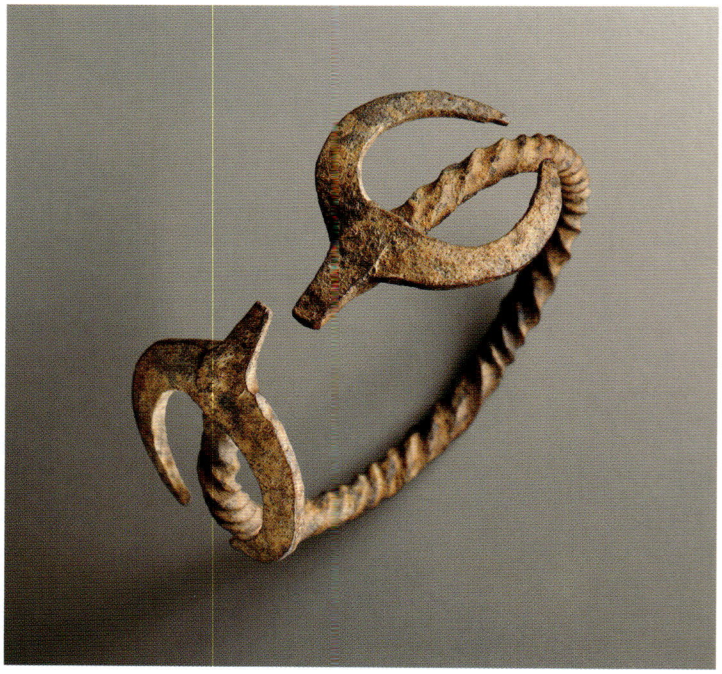

III.5
Artist unknown (Senufo peoples, Mali)
Bracelet, early 20th century
Iron
D: 12.4 cm
GHYSELS COLLECTION
IMAGE © STUDIO R. ASSELBERGHS–FRÉDÉRIC DEHAEN, BRUSSELS, 2018.

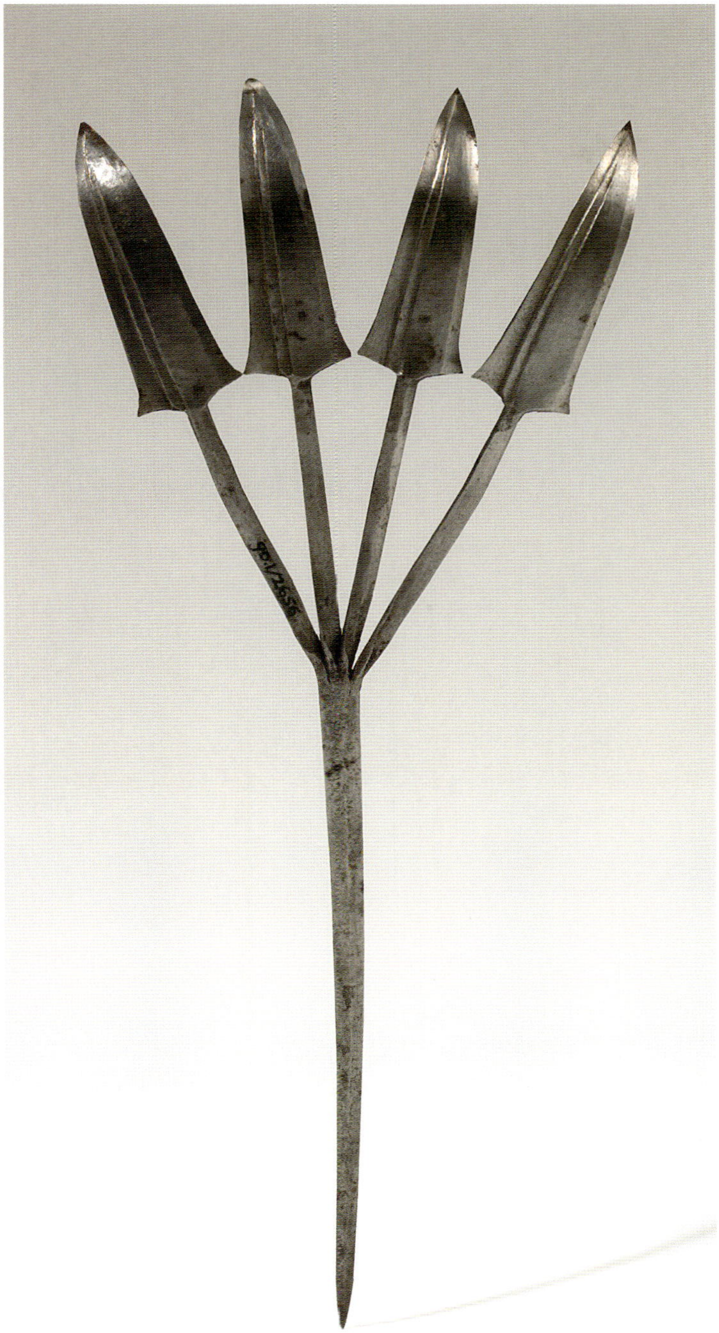

III.6
Artist unknown (Zande[?] peoples, Democratic Republic of the Congo)
Hair pin (*mangilli*), late 19th to early 20th century
Iron
H: 28 cm
AMERICAN MUSEUM OF NATURAL HISTORY, 90.1/2656; GIFT OF HERBERT LANG
IMAGE COURTESY OF THE DIVISION OF ANTHROPOLOGY, AMERICAN MUSEUM OF NATURAL HISTORY

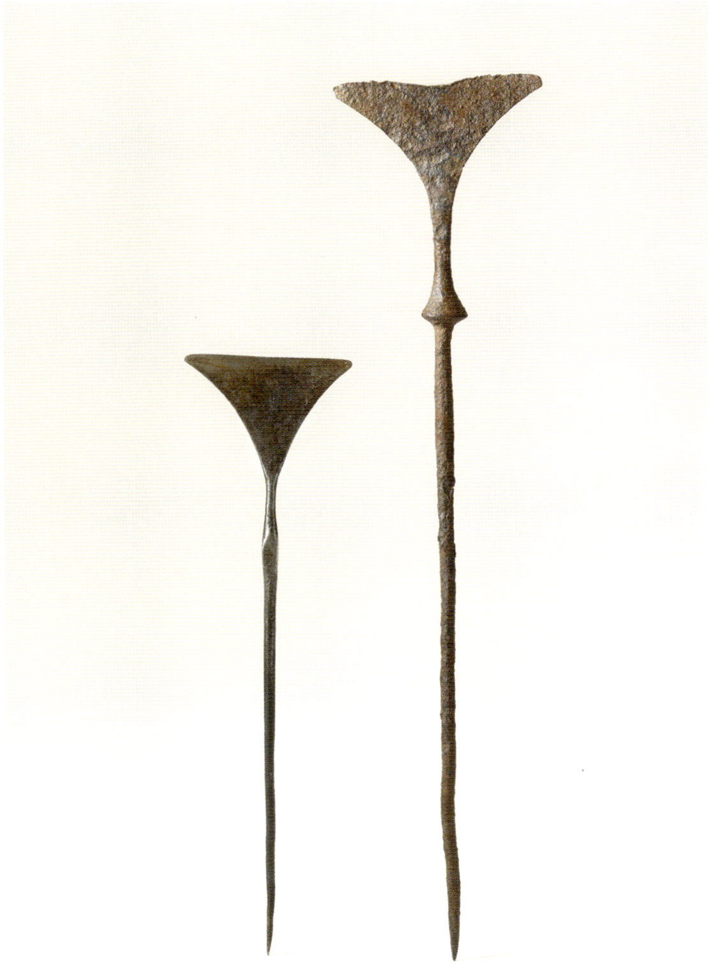

III.7
Artist unknown (Luba peoples, Democratic Republic of the Congo)
Hair pins (*vinyundo*; s. *kinyundo*), late 19th century
Iron
H: 14.5 cm; 20 cm
FOWLER MUSEUM AT UCLA, 382.40, 382.41; MUSEUM PURCHASE

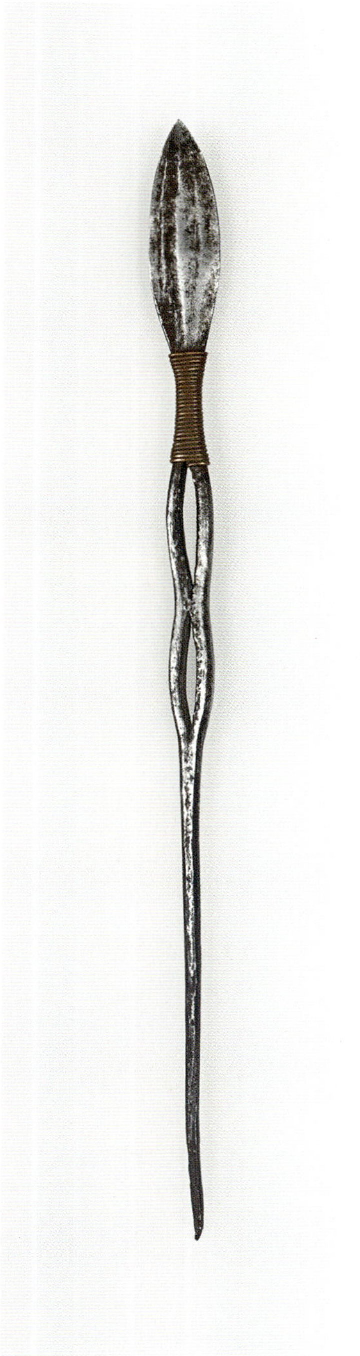

III.8
Artist unknown (Ngbandi, Yakoma peoples, Democratic Republic of the Congo and Central African Republic)
Scarification knife, late 19th century to early 20th century
Iron, copper
H: 16 cm
ITZIKOVITZ COLLECTION
PHOTOGRAPH BY BRIGITTE CAVANAGH, 2017.

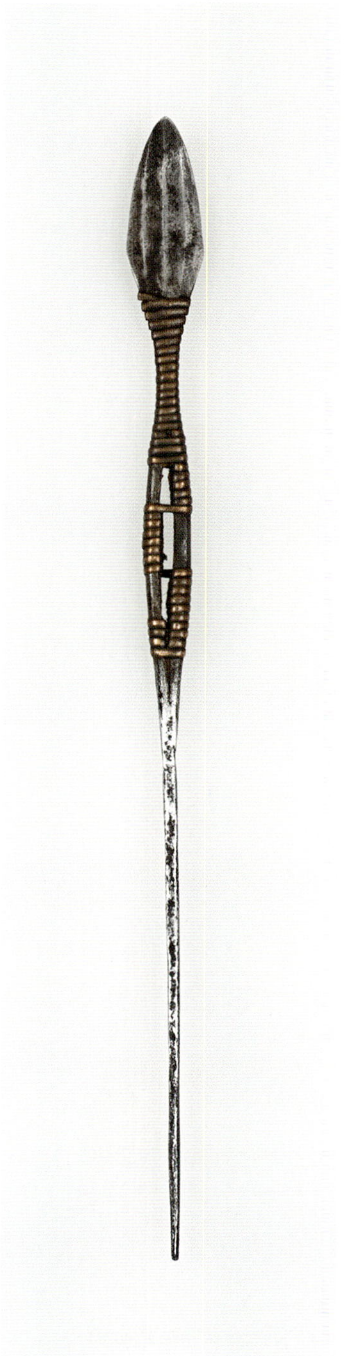

III.9
Artist unknown (Ngbandi, Yakoma peoples, Democratic Republic of the Congo and Central African Republic)
Scarification knife, late 19th century to early 20th century
Iron, copper
H: 18.5 cm
ITZIKOVITZ COLLECTION
PHOTOGRAPH BY BRIGITTE CAVANAGH, 2017.

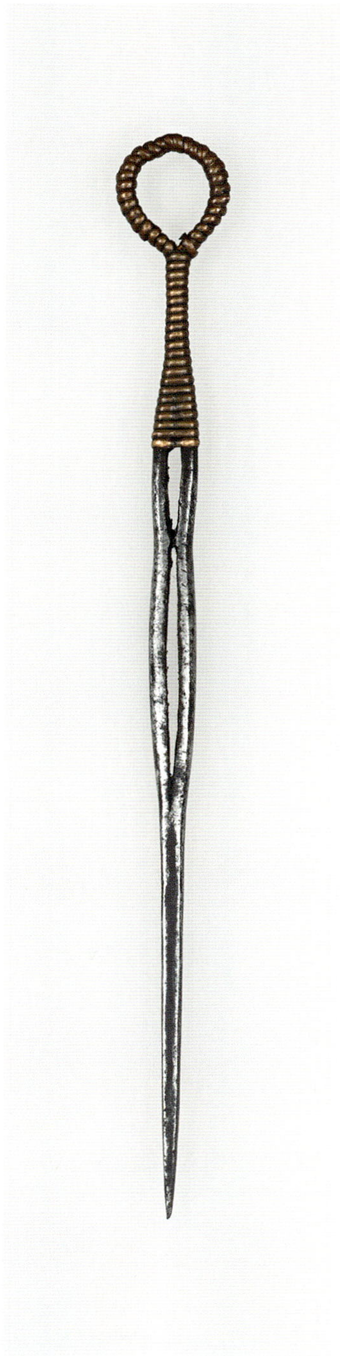

III.10
Artist unknown (Ngbandi, Yakoma peoples, Democratic Republic of the Congo and Central African Republic)
Scarification knife, late 19th century to early 20th century
Iron, copper
H: 15 cm
ITZIKOVITZ COLLECTION
PHOTOGRAPH BY BRIGITTE CAVANAGH, 2017

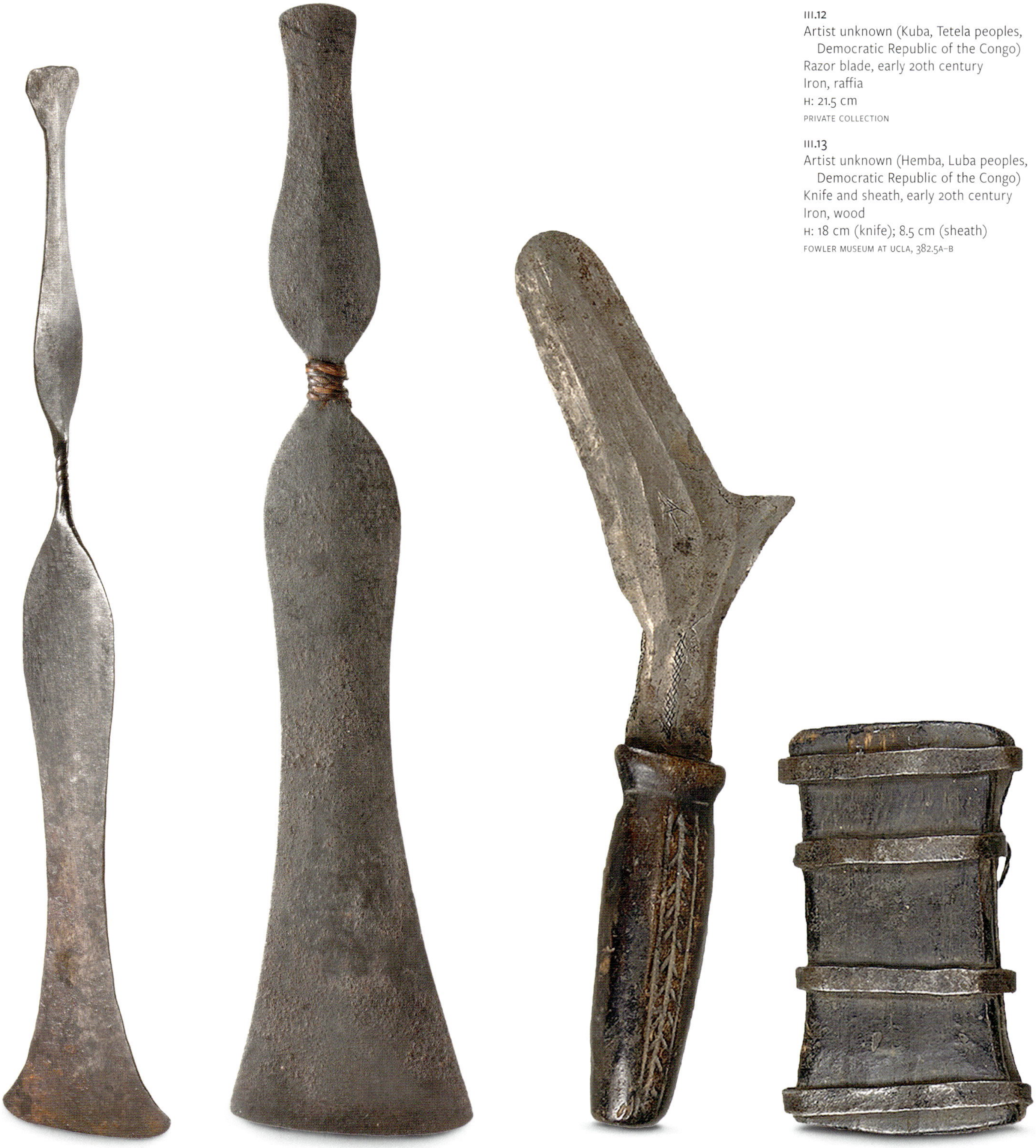

III.11
Artist unknown (Kuba, Tetela peoples, Democratic Republic of the Congo)
Razor blade, early 20th century
Iron
H: 19.7 cm
PRIVATE COLLECTION

III.12
Artist unknown (Kuba, Tetela peoples, Democratic Republic of the Congo)
Razor blade, early 20th century
Iron, raffia
H: 21.5 cm
PRIVATE COLLECTION

III.13
Artist unknown (Hemba, Luba peoples, Democratic Republic of the Congo)
Knife and sheath, early 20th century
Iron, wood
H: 18 cm (knife); 8.5 cm (sheath)
FOWLER MUSEUM AT UCLA, 382.5A–B

PART IV

AFRICAN IRONWORKING AND CHANGE

CHAPTER 16

IRON ON THE ANVILS OF CHANGE

Candice Goucher

> Quantities of raw iron are made... and thereby a great Trade made of [iron] and when manufactured [it is] sent into most parts of the world.
>
> ANDREW YARRANTON (B. ASTLEY, ENGLAND; 1616–1684), CIVIL ENGINEER AND IRONWORKS OWNER (1681, 57–58)[1]

The *Henrietta Marie*, an English merchant-slaver ship, was built at the end of the seventeenth century for the express purpose of carrying precious cargoes of metals and men. It traversed the seas in the so-called Triangular Trade that connected English ports to African, the Atlantic world to the Americas. Within a year of setting sail for Africa in 1699—its captain and crew hoping to reap vast profits—the vessel sank in Florida waters. By far the largest and most valuable of the goods carried from London were thirty-three tons of iron, in the form of more than eighteen thousand bars to be used in the African trade.

By the time the *Henrietta Marie* disappeared on the other side of the Atlantic Ocean, many of its goods had been traded for enslaved Africans. Only twenty-eight bars were recovered by archaeologists (Malcom 1998; see interleaf L, this volume). The underwater excavation of the *Henrietta Marie* shipwreck now gives voice to iron's deadly route, linking three continents in the history of iron technology and the trade in metals and enslaved Africans. Long before and for more than a century after the voyage of the *Henrietta Marie*, African iron objects and those who practiced the art of making iron goods had become valued commodities on the move. The continent's ironworkers, their technologies, and the so-called "traditions" of African ironworking were neither stationary nor unchanging in their decidedly global contexts.

By exploring the history of African ironworking from a wider perspective of change over time, it is possible to place the achievements of African industries in the context of other global production, to explore their impact and meaning, and to examine how the impact of global forces shaped local ironworking. The African continent is one of the places in the world where preindustrial practices survived into the twentieth century; recent case studies of African technology have been used to conceptualize ancient Greek and modern Japanese metallurgies (Blakely 2006).[2] Ancient trade in African metals, including iron, copper, and gold, reached much of Eurasia, not just the transatlantic African diaspora of the early modern world. The long history of African iron production provides evidence of key strategies used by smelters and smiths to adapt to their changing economic and environmental circumstances.

CHANGING TECHNOLOGIES, CHANGING ENVIRONMENTS

African ironworkers designed a bewildering variety of furnace shapes, from lined pits dug into the ground to two-person-tall natural-draft furnaces. Forges could range from ardent embers held in pottery shards to female shapes elaborately sculpted in clay. Besides furnaces and anvils, processes utilized refining crucibles, tuyères, clay plugs, quenching vessels, and bellows of all shapes and sizes. Although the chemical and pyro-technical parameters for smelting and smithing were theoretically constant, ores and fuel were dynamic factors that might alter such processes. Technologies for smelting and forging iron also changed over time (see Killick 2004, 97–112; see also chapters 8–10, this volume). Archaeologists and historians have identified some specific contexts, suggesting that establishment of ironworking technology was subsequently opened to innovations forged by opportunities of trade, talent, and environment.

Between 750 and 1750 CE, Botatwe speakers of what are now eastern Zambia and adjacent republics participated in social networks

16.1 This tall, natural-draft smelting furnace was reconstructed from parts of an ancient one at Banjeli, in the Bassar region of Togo (see fig. 2.1). Clay plugs could be inserted in the furnace holes at the base of the furnace.
PHOTOGRAPH BY CANDICE GOUCHER, 1984.

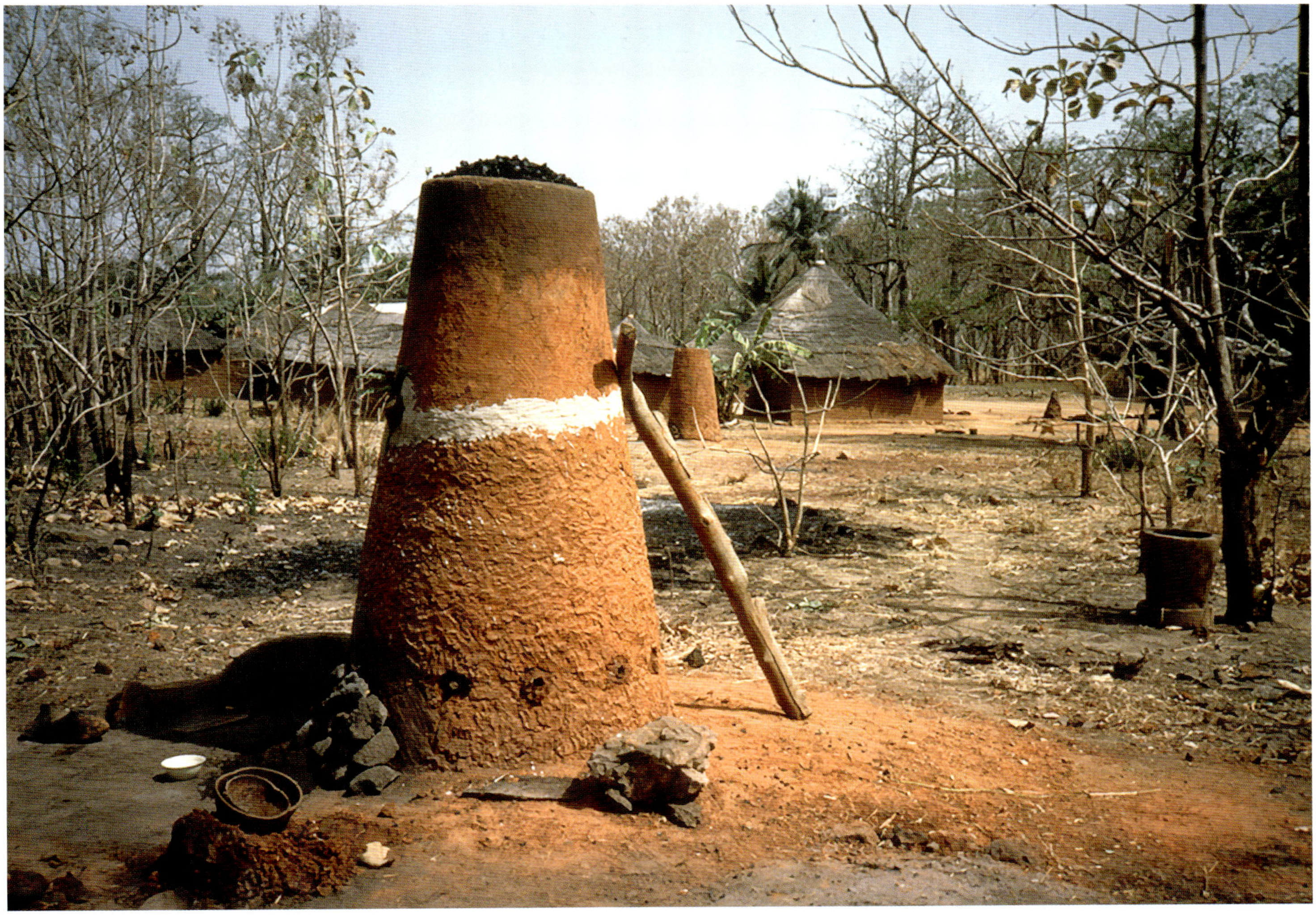

of prestige and trade that made it possible to borrow new forms and techniques for making iron. Just as iron axes had been critical to successful implementation of slash-and-burn farming in the early Iron Age, iron hoes enabled efficient cultivation of cereal crops. Changes in landscapes were wrought by the activities of farmers, smiths, and smelters. Historian Kathryn de Luna has identified two distinct metallurgical traditions that met in the frontier of the Congo and Zambezi rivers of south central Africa in the late first millennium CE: utilizing bellows, and the manipulation of furnace holes to tap slag and subsequently produce ingots for trade from the furnace blooms. Analysis of shifts in language has suggested that early smelting was by small-scale movement of itinerant specialists (de Luna, 83). Later innovations by regional industrial complexes at some distance from smelting villages created more abundant and possibly larger-scale products to meet local demands and the requirements of faraway markets.

16.2A–B Making charcoal near Hani Village in the Brong Ahafo region of Ghana.
PHOTOGRAPHS BY CANDICE GOUCHER, CA. 1979.

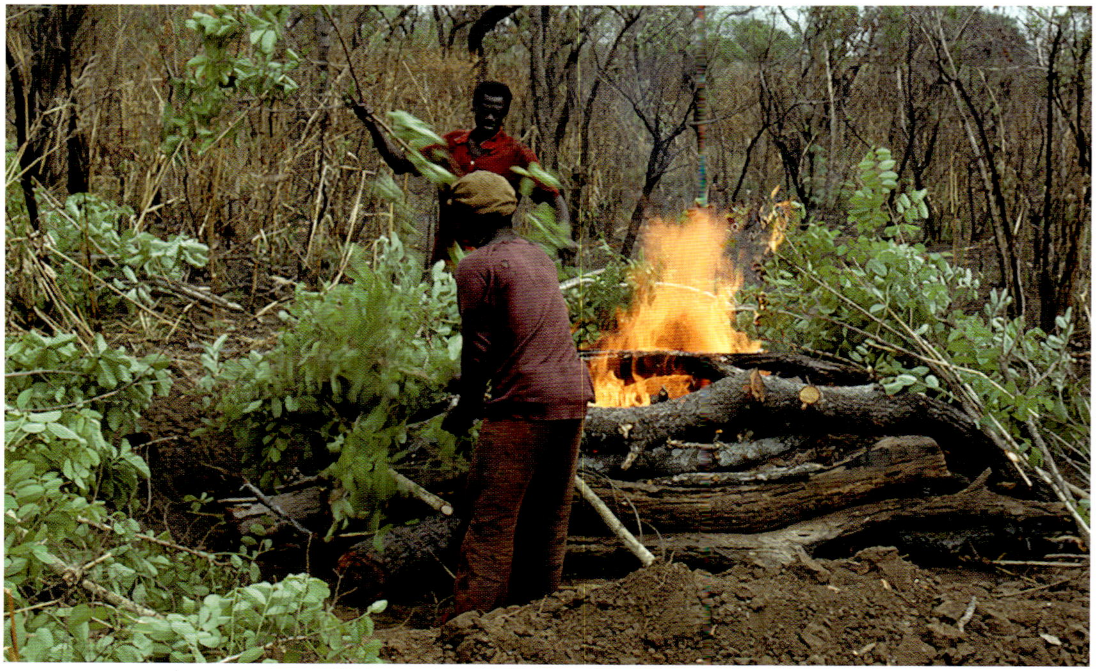

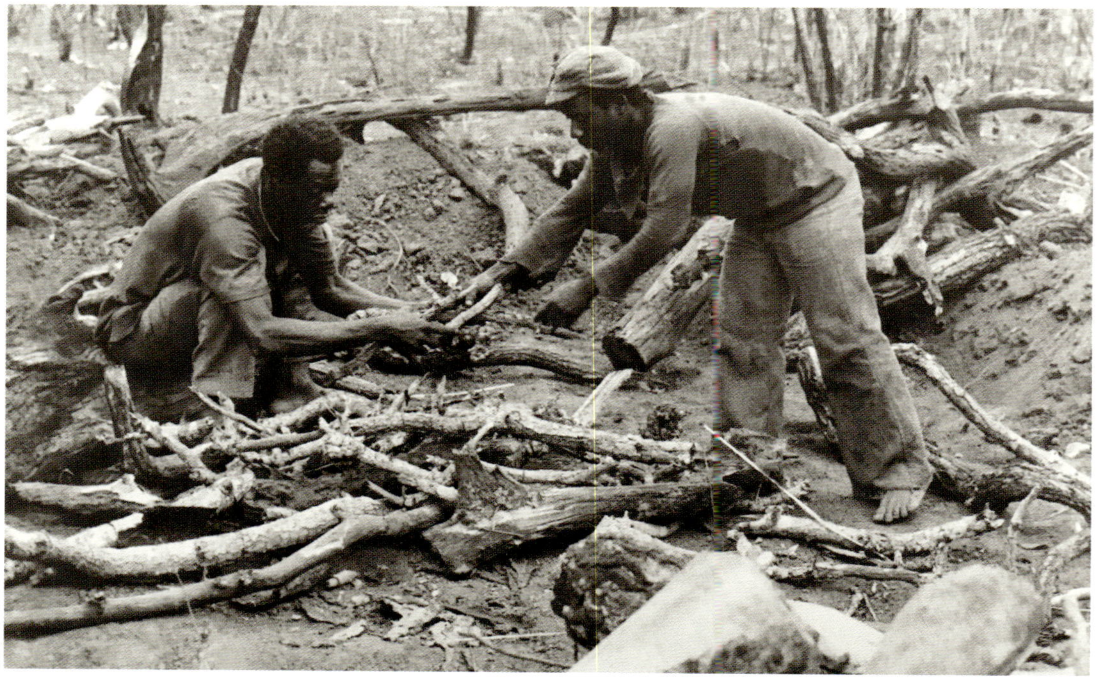

16.3 Loading charcoal into the furnace during a reconstruction of iron smelting at Banjeli, Togo.
PHOTOGRAPH BY CANDICE GOUCHER, 1984.

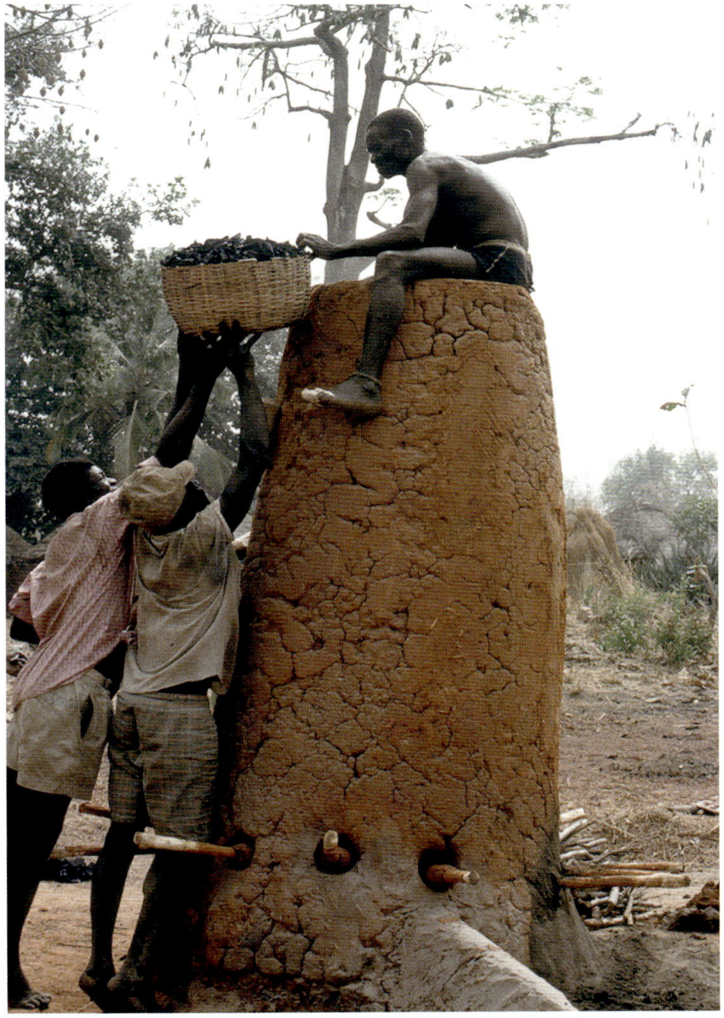

A similar transfer of technology may have enhanced regional centers of iron production in west Africa. One of the best-studied regions is that of Bassar, in northern Togo (Goucher 1988, 56–69; Saltman, Goucher, Herbert 1986; de Barros 1986, 148–74). Oral traditions and archaeological investigations confirm continuous smelting of iron between the twelfth and mid-twentieth centuries CE. Large-scale production began in the fourteenth century with the introduction of tall shaft furnaces and the specialization of aspects of the industry that served the region's large states, including Dagomba, Gonja, and Asante, in present-day Ghana (fig. 16.1). Archaeologist Philip de Barros has estimated that the four centers of the Bassar region may have produced as many as thirty-six million large hoes during these centuries of operation, which led eventually to elaborate political hierarchies such as that of Kabu and Bassar itself. No doubt west African industries struggled with the inevitable depletion of slow-growing hardwoods needed for charcoal fuel, as did their counterparts in China and England.

Observations of charcoal-making in the region have revealed its complex science. Among the most desirable species for charcoal production were *Burkea africana* and *Daniella thurifera* trees, both possessing dense wood that, when burned, permitted ironworkers to achieve the high temperatures needed for smelting. Near the village of Hani in the now-deforested Brong Ahafo region of west-central Ghana, descendants of metalworkers remember how to make the specialized charcoal for metallurgical operations by piling wet logs in a pit and covering them with branches and earth (fig. 16.2).[3] Slowly burning off the moisture was itself a labor-intensive expenditure of fuel, but the charcoal produced was longer lasting and more efficient for burning than fresh wood. Although women were excluded from smelting sites, they often played key roles in making charcoal for their husband-smelters, as was the case in the Bassar region of neighboring Togo (see chapter 2, this volume). Eventually the destruction of forests for fuel and the pollution spewing from tall standing furnaces led to denuded landscapes (see Goucher 1988). Over centuries, African industries adjusted by shifting locations and adopting technical innovations and fuel-saving measures such as taller, more efficient, and slower-burning types of natural-draft furnaces. In the face of limited European competition, Bassari smelters continued to produce high-quality iron from their natural-draft furnaces (fig. 16.3).[4] Smiths persisted in working the locally smelted blooms and carried headloads of iron hoes south to the urban complex of Lomé on the coast, where, during the early colonial period, they were exchanged for salt. While output at some west African production sites increased, even during the competitive period of European trade in iron, other formerly key sites ceased smelting altogether, and the industries eventually were banned by colonial policy edicts meant to reduce self-sufficiency and engage local people in market economies controlled by authorities.

MARITIME CONNECTIONS

Iron produced in sub-Saharan Africa was traded widely across the continent through a complex interplay of regional networks and long-distance trade routes on land and sea. Iron trade also reached beyond the continent via Indian Ocean, Red Sea, trans-Saharan, and Mediterranean maritime trade networks in place since before the first century CE. A mariner's guide known as *The Periplus of the Erythraean Sea* (Red Sea) describes a lively trade in iron with other metals exported and imported through northeastern African ports during the early Christian era. The Horn of Africa was a particularly significant crossroads for technological exchanges. Via the Red Sea and Indian Ocean, it is possible that elements of iron smelting and crucible technologies were shared between South Asia and eastern Africa. At the time of the introduction of Islam, after the seventh century, iron was still a major commodity aboard merchant ships plying the Indian Ocean. The Arab geographer Muhammad al-Idrisi, who lived in the court of the Norman King Roger II of Sicily, recorded iron being carried from eastern Africa to what is now India and adjacent islands during the twelfth century CE.

North African iron technology reached across the Mediterranean from antiquity. A common steel technology existed by the time of al-Andalus, as evidenced in thirteenth-century knife blades and an iron thimble excavated at Castillo de la Torre Grossa, in Alicante. The thimble suggests the use of iron needles, whose technology reportedly made its way to England from Spain in the mid-sixteenth century, during the reign of Henry VIII.[5] The wire-drawing process of making thin wire from a bar or rod was claimed to have been introduced by a "Spanish Negro," or Moor, who began producing wire locally in 1563, and whose image long appeared at the center of English needle-makers' crest of arms. Wire-drawing of soft metals had been known since antiquity, but the African method was one in which hard, or "heavy," iron or steel was hammered repeatedly. African copper wire-drawing also used ironworking techniques and is best known from early archaeological contexts in southern and eastern Africa, such as the site of Ingombe Ilede in Zambia.

When they first navigated along the coastline of west Africa in the middle of the fifteenth century, European travelers described their appreciation of local iron production, and noted, with considerable awe, the iron weapons and tools they encountered. The region was thought to have "much iron of good quality" already being traded at an enviable profit (Fernandes 1938, 134; Pereira 1937, 98). European merchants came to rely on African blacksmiths at local slave castles and forts, and purchased inland iron made from native iron. Above all, they sought skilled blacksmiths for the slave trade, transporting them to far-flung posts of the Atlantic world, where they demanded higher prices than unskilled Africans. By the dawn of the sixteenth century, Portuguese explorers had circumnavigated the African continent. Along the Swahili Coast they discovered that iron was exchanged for other goods. Merchants thought so highly of the quality of the iron produced by Shona smelters and smiths in what are now the republics of Zimbabwe and Mozambique that they carried it to India for use in the manufacture of weapons, as had Muslim and other traders centuries earlier.[6] Even Atlantic-era wooden sailing ships required a remarkable number of iron working parts and equipment, making one or more blacksmiths necessary onboard, with African specialists among such numbers.

IRON CURRENCIES

As European merchants observed the value of iron to African communities, they sometimes mimicked the aesthetic forms in which it might be traded, matching hoe and spear shapes and bar size and content to appeal to specific African markets (figs. 16.4; see also figs. 5.11–5.13). Export companies labeled their wares to match the intended destination, distinguishing among specific hoe shapes desired in west African ports and those of Barbados, Jamaica, and the Carolinas. Examining the shipment of Swedish iron to England and then African destinations, Chris Evans and Goran Ryden have found that African consumers of this Baltic "voyage iron" in faraway ports such as Calabar, in present-day Nigeria, dictated the composition and size of the metal they preferred (2007, 166). The early eighteenth-century merchant Jean Barbot, a French Huguenot who settled in England, recorded the concerns of careful African consumers:

> The Blacks of the Gold Coast . . . examine and search very narrowly all our merchandize, piece by piece, to see [that] each be of the quality and measure contracted for by samples. . . . They measure iron bars with the sole of the foot . . . from the iron bars the Smiths forge out all their weapons, and country and household tools, and utensils.[7]

Global commodity chains of the sort relied on systems of coerced labor, including serfs from the Urals and enslaved Africans in the Atlantic world.

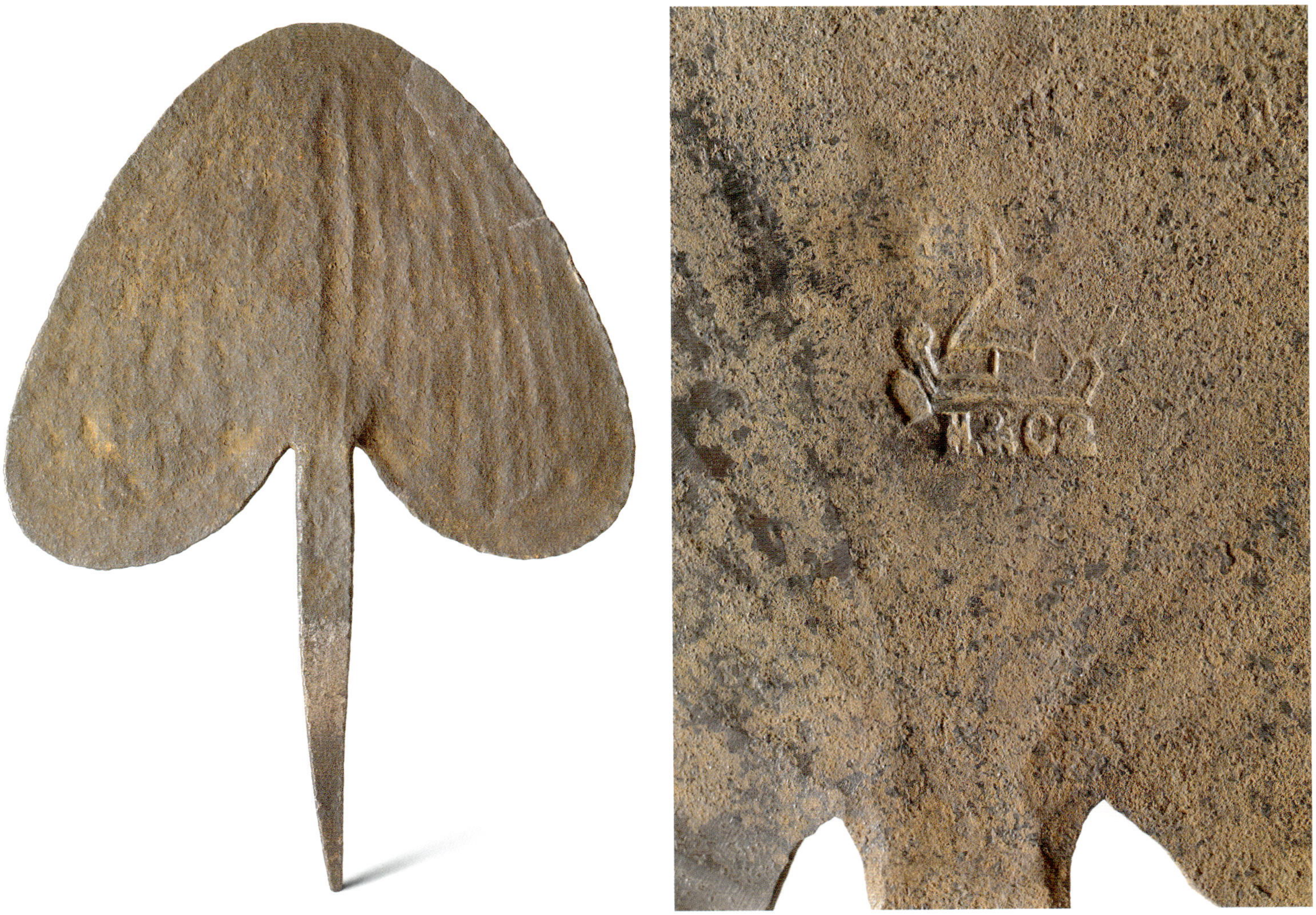

16.4A-B
Maker unknown (Europe)
"Iramba-like" hoe-blade-shaped currency, early 20th century
Iron
H: 40 cm
PRIVATE COLLECTION

After observing the value of iron to African communities, European merchants began mimicking currency forms exchanged locally in west and central Africa. One such currency, based on a hoe-blade form, was first forged by Iramba smiths in Tanzania and then mimicked by European smiths for African trade. European manufacturers labeled their wares to indicate their intended destination in Africa; a detail of the front of the "Iramba-like" currency shows its identifying mark.

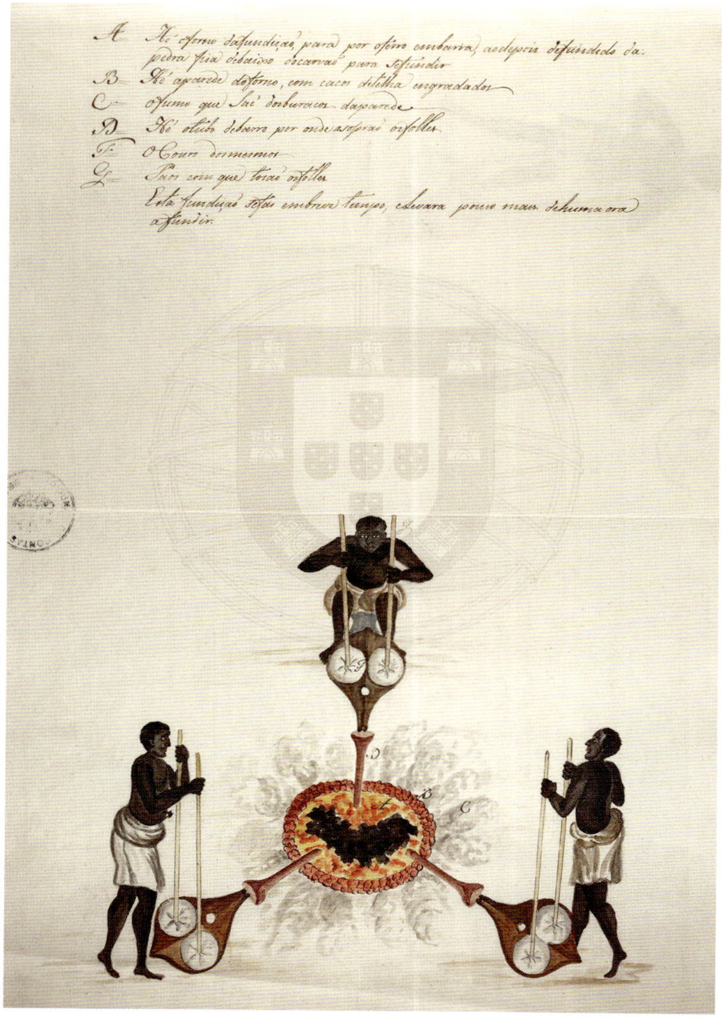

16.5
A Portuguese report depicts workers using bellows to smelt iron in small pits in late 18th-century Angola. After the construction of the Royal Iron Factory there, circa 1766, the Portuguese took great interest in local technologies, hoping to recruit and employ Africans in their operations.
JOSÉ ÁLVARES MACIEL, "NOTÍCIA DA FÁBRICA DE FERRO DA NOVA OEIRAS DO REINO DE ANGOLA, SÃO PAULO DE ASSUNÇÃO DE LUANDA," DECEMBER 15, 1797. ARQUIVO HISTÓRICO DO TRIBUNAL DE CONTAS (AHTC), ERÁRIO RÉGIO (ER), 4196.

Within its global context, the iron bar was equal to human cargo as understood through an equivalency of multiple meanings. Indeed, the very consumption of African iron could be a measure of a man or a woman, representing status, wealth, or an equivalency of shared value. At slave ports along the Guinea coast, an enslaved African could be exchanged for iron bars or a combination of goods equal to the value of those bars. However, such "currencies" also served as a convenient way to trade metal destined for the blacksmith's forge to be manufactured into weapons, tools, and other implements. Without further working, forged iron products might be given as bridewealth, compensating a family for the loss of a daughter in marriage, or as emblems of office, carried in processions or danced in rituals.

Africans used iron currencies to facilitate the movement of other goods. At times, iron slag and iron bars could be used as ballast for ships or payment. Certainly, annual importation of European firearms (an estimated 180,000 in 1730 alone) no doubt enabled the "production" of enslaved Africans on the continent. While guns fueled warfare and entrapped captives, imported hoes helped maintain levels of agricultural productivity that fed people during their tragic transport across the Middle Passage. Bar iron consumed in African workshops was used to fashion shackles and chains as much to shape cutlasses used in resistance and uprisings. There is little doubt that availability of imported iron did much to forge the transatlantic slave trade. As the Scottish traveler Mungo Park observed in the eighteenth century, "All these fetters and bolts [for securing enslaved captives] are made from native iron; in the present case they were put on by the blacksmith, as soon as the slaves arrived" ([1831] 1931, 635). As the demand for enslaved persons grew, so did the demand for bar iron and local blacksmiths skilled in forging technologies.

As mentioned, African metallurgists were highly valued for their knowledge and skills during the Atlantic slave trade. Early European trading forts and castles built along the west African coast utilized blacksmiths' services for the trafficking of the slave trade and the industries of harbors required by maritime trade. Around 1561, the English merchant William Towerson praised the "handsome darts and various instruments of iron" in Guinea (quoted in Williams 1974, 72). Observing the ports of the seventeenth-century Gold Coast, Wilhelm Johan Muller noted that "the blacksmiths in the Fetu country work in a different way from those in Europe" (quoted in DeCorse 2001, 125). Visiting merchant ships relied on the services of blacksmiths of coastal communities. According to Crislayne Alfagali, Portuguese explorers beheld local iron industries and trade and eventually tried to construct, between 1766 and 1772, an iron factory at Nova Oeiras, in the hinterland of present-day Angola.[8] The royal factory would have used the skills and labor of Portuguese and local smelters, but the project was never fully realized. After the four immigrant Portuguese artisans died of scurvy and fevers, their African counterparts returned to their own enterprises (fig. 16.5). Some other metallurgical collaborations in the Atlantic world were more successful.

IRON IN THE AFRICAN DIASPORA

In the Americas, because early iron and steel industries relied heavily on the expertise of enslaved Africans, as chattel blacksmiths fetched higher prices in auctions. In early American industries, enslaved African ironworkers were employed in foundries in Maryland, Virginia, Louisiana, Pennsylvania, the Carolinas, and other colonies. These enterprises ranged from large-scale ironworking to service plantations to small, mobile blacksmith jobbers. Skilled African ironworkers were particularly valued during and following the Revolutionary War, when iron pricing went sky-high. For example, as of 1773, iron was smelted in Rocky Mount, Virginia, at John Donelson's bloomery; the number of enslaved blacksmiths there increased from thirty-eight in 1809 to eighty-three in 1837, when they "did everything except manage the works" (Gordon 1996, 82).

In Morant Bay, Jamaica, an eighteenth-century iron foundry successfully employed 276 Africans, some enslaved and some free laborers. According to the owner, Englishman John Reeder, the foundry's key metallurgists were Africans "principally employed in carrying on the Operations of the Foundry," whose skills were "perfect in every branch of iron manufacture."[9] Archaeological excavations at the site suggest a significant African presence and a legacy of local iron forging (fig. 16.6). A few years after the foundry was in full swing, Reeder received permission from the island Assembly to erect a smelting furnace near Bath, but it is unclear whether his plans were carried out. The Jamaican Governor General Archibald Campbell ordered the foundry closed in 1782. He dismantled the industry out of fear that an invasion by Spanish and French forces could place the strategically valuable operations in enemy hands. Reeder spent much of the rest of his life attempting to obtain compensation for the loss of his business and equipment, and in futile attempts to cancel his public debt and interest owed. Ironically, because of the foundry's destruction, historians have learned more about its operations than any comparable site in the Atlantic world. The slaves suffered considerably, according to Reeder:

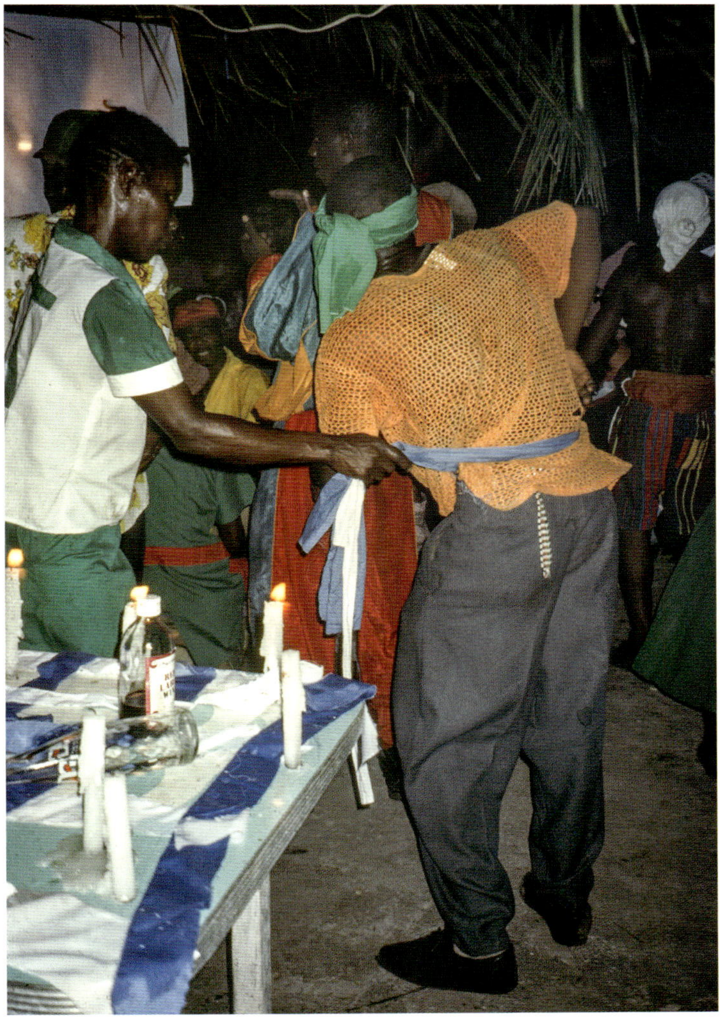

16.6 Near Morant Bay, Jamaica, in a Kumina ritual, followers acknowledge the role of the ancestors in the author's archaeological excavation of an 18th-century African cutlass, called locally by its Akan name, *afana*.
PHOTOGRAPH BY CANDICE GOUCHER, 1995.

> In consequence of the dismantling [of] my foundry, I have lost capital, the most valuable of my Negroes are dead. . . . [F]rom my being under the necessity to employ them in Husbandry, despondency ensued. Trades Negroes consider themselves of rank superior to others and cannot brook the disgrace of being employed in the cultivation of Lands.[10]

The release of such a large number of skilled African artisans—some enslaved, some Maroons hired for wages, and some freed Africans—would have had a regional impact. While the Reeder foundry operations never recovered, some evidence of the movement of its African labor force may be seen in the growth of iron industries in North America at the end of the eighteenth century.

Sacred knowledge and rituals as "enchanted technologies" (after Gell 1992) and related pathways to empowerment accompanied the transport of African metallurgical skills and techniques across the Atlantic world. Followers of Ògún, Yorùbá god of iron, spread to Trinidad, Brazil, Haiti, Cuba, and other lands of the Black Atlantic, where the performance of rituals provided a framework for remembering complex processes. According to the late Haitian blacksmith Georges Liautaud (1899–1991), the dancing of ironworkers at the forge extends power to others:

> When a person becomes possessed by the spirit of Papa [Ogou Feray in Haiti] an iron bar standing in the earth near the altar dedicated to his name is heated in a fire. The possessed person will then take this staff or some other equally heated bar and dance with it in his or her bare hands to prove that the possession is genuine. Deep mastery of self is the point of the play with fire and heat. [Quoted in Morris 1995, 385–86]

In Winkel Village of Berbice, British Guiana, enslaved African ironworkers were so empowered that their solidarity during a 1796 uprising helped usher in their freedom in 1821, long before the colony's formal emancipation of enslaved and bonded Africans.

Colonies attempted to control the movement of iron goods and skilled artisans. The ironmasters of the American colonies were adamant that they required African labor in their operations. In Pennsylvania, many or even all of the workers at a number of forges, foundries, and furnaces were enslaved Africans:

> Negroes were used in the ironworks from the early establish ment of the Pennsylvania industry. In 1727 the shortage of labor was so acute that the ironmasters in the colony petitioned the Assembly for permission to import Negroes free of duty to labor at their works. . . . A bill permitting Negroes imported into the colony for the express purpose of laboring at ironworks to enter duty free failed by the deciding vote of the Speaker. Two years later, however, the duty of £5 on each Negro brought into the Province was reduced to £2. . . . While Negro slaves and freed Negroes usually worked at menial tasks, at many ironworks they were skilled workmen. [Bining 1938, 100–01]

The shortage of smiths and smelters continued to plague North American industry for the next century. The foundry of Richard Smith in Colebrook, Connecticut, built in 1771, produced iron not only for American colonists during the Revolutionary War, but also for shipment to the Caribbean, where local demands were consistently high.[11] Particularly at charcoal-fired forges and blast furnaces, Africans' techniques for producing high-quality, dense, and efficient charcoal fuel; their knowledge of preheating the air blast; and their ability to handle crucibles proved essential to the development of successful enterprises. African ironworkers converted pig iron to bar iron and helped feed the iron hunger across the modern Atlantic world.[12]

ADAPTING TO CHANGE

Changing landscapes on either side of the Atlantic ushered in an age of global industrialization that relied on African markets and skills. In sub-Saharan Africa, smelters and smiths were periodically forced to move in response to climatic change and depletion of resources. Ore beds could give out and regions could be deforested, requiring that industries shift. Hills of slag surrounding the ancient site of Meroë in the Sudan give silent testimony to the rise and decline of early centers of iron production. Many blacksmiths traveled regularly to regional villages and towns, and sometimes they accompanied armies on long-distance marches to repair weapons in the field. While African smiths and smelters played significant roles in technological changes across the Atlantic world, many local African industries were forced to contend with the competition from importation of bar and scrap iron by European merchants.

Such interactions led to systems of iron production spanning from the slaving ports of Africa and the Americas to the manufacturing districts of Birmingham, England, and on to forges in rural Sweden, where production of charcoal iron persisted. Environmental conditions, shortage of charcoal fuel due to African and European deforestation, and cheap imported bar iron ultimately reduced the large-scale iron producing regions' capacity to compete. Exceptions to decline included markets supplying larger African states such as Dahomey and Asante, in present-day Republic of Bénin and Ghana respectively, and as well as more remote communities such as those of Fipa and Tabwa peoples living along the shores of Lake Tanganyika.

Africa's "first Birmingham" was perhaps ancient Meroë, where ironworking, among other industries, supported state structures of the Nile.[13] The origins of the Mali state also relied upon the expertise of smelters and smiths, inspiring legends of state power, from the Sunjata Epic (or Sundiata), after the legendary mind- and iron-bending ruler of the fourteenth century CE, to the cinematic griot Souleymane Cissé's film *Yeelen* (1987), a modern political tale told through the saga of a Komo blacksmith.[14] Material and oral evidence from Benin City, in present-day Nigeria, suggests a close link between visual displays of the power of the state and controls over ironworkers who trace their arrival in the state's capital to having been captured in warfare.[15] Were ironworkers more valuable to building local African states or as prisoners of war to be sold into the transatlantic slave trade? Central African states, ruled by the aptly named "blacksmith-kings," have clung to symbols of forging power for more than fifteen hundred years, which suggests that reliance on technologies of smelting and smithing and associated rituals have validated local political power and enabled state formation (Vansina 1955 and 1969; see also chapter 15, this volume).

Collecting African ironwork completed the process of European colonization of the continent. The effective South African *assegais* (short stabbing spears) of Shaka kaSenzangakhona (or Shaka Zulu) became so famous that the term, which possibly derives from the Arabic *az-zaghavah* encountered by Europeans in North Africa, embraced any sort of iron superweapon.[16] Observations of African blacksmiths inspired Humphrey Bogart and Katharine Hepburn, in the Hollywood film *The African Queen* (1951), to reproduce a charcoal forge, which their characters use to repair a ship's propeller. Indeed, such literary and cinematic references differ little from ancient African expressions of iron's most powerful symbolism, that of immutability and endurance. Balancing survival skills against the utility of his trade, an ironworker was equally the master of his own dynamic past. The multiple sources of local and imported iron available to African blacksmiths suggest that metallurgical expertise was highly flexible, constantly innovative, and adaptable to the challenging and ever-changing conditions and trade routes that brought a variety of materials to foundry and forge, there to be shaped by distinct destinies on multiple continents.

CONCLUSION: RECYCLING THE PAST

Though the lineages of African iron smelters are close to forgotten for what were once such important associations, blacksmiths remain engaged in the production of metals from rural Ethiopia to the urban capital of Mali at Bamako and Port-au-Prince in Haiti. Today, African blacksmiths are as active as ever, whether repairing hoes or operating forges as "fitters," or motor mechanics, adapting to contemporary markets and claiming their places as key strategic links in the business of recycling metals. Iron is the most recycled material in the world (see Roberts 1996), and such repurposing accounts for higher levels of iron and steel production than the smelting of ores worldwide. Automobiles, machinery, and all sorts of other metal goods are subject to assemblage as scrap metal to be reworked into new items of material culture by modern and enterprising African blacksmiths, heirs to fiery forges of the past.

INTERLEAF

L

"VOYAGE IRON," FROM EUROPE TO AFRICA

Colleen E. Kriger

Before 1500 CE, many of Africa's kingdoms, communities, and societies traded beyond the continent by way of seaports situated along her Mediterranean, Red Sea, and Indian Ocean coastlines, ports offering entry into the commercial networks and markets of southern Europe and Asia. A totally different trading pattern, one centered on Africa's western coast, opened up for the first time in the 1440s and 50s, and it ushered forth a dramatically new era in world history, the era of Atlantic trade between Afro-Eurasia and the Americas. What motivated the Portuguese mariners who achieved this breakthrough was the prospect of direct access to western Africa's gold exports and a sea route to the Indian Ocean trade in Asian silks and spices. As they began to explore the Guinea Coast, they did indeed find gold suppliers there and also commodities such as ivory, pepper, and dyewood, and then finally, all the way down the coast and around the Cape of Good Hope, an entry into the Indian Ocean basin. Over the next two centuries came the establishment and development of European colonies and plantations in the American tropics, especially to grow and process cane sugar, with African captives and their labor becoming an ever-growing export to the Americas.

African merchant-suppliers of exports on the Guinea Coast drove hard bargains with their European counterparts by demanding very particular types of overseas goods and commodities in exchange. From early on and continuing through the nineteenth century, the two most important goods Europeans needed to offer were textiles and metalwares, and figuring very prominently among the latter was a special product called "bar iron" (fig. L.1; see also chapter 16, this volume). Documentation of European bar iron exports to the Guinea Coast is sparse for the fifteenth and sixteenth centuries, when the Portuguese crown sought to prohibit their merchants from trading in iron. However, from the seventeenth century onward, northern European merchants—Dutch, English, and French (and some Flemish and German)—traded bar iron to Africa regularly and in large quantities, hence the commercial term "voyage iron." England's Royal African Company (RAC) alone shipped five thousand tons of Swedish voyage iron (approximately four hundred thousand bars) to Africa between 1673 and 1704 (Kriger 2017, chapter 2).

Generally speaking, voyage iron was consistently in strong demand all along the Upper Guinea Coast (from today's Republic of Senegal south to the Republic of Liberia).[1] Some of it was cut down into what were known as "country bars"—locally made African forms of bar iron about twenty to twenty-five centimeters in length (Jobson 1999, 160–61). Voyage iron also found markets farther south, on the Ivory Coast and Gold Coast and seasonally on the "Slave Coast" (today Togo, the Republic of Bénin, and southwest Nigeria). Cowry shells (used as currency) were the overseas commodity in highest demand by far, but RAC agents at Ouidah in Bénin recommended the company send iron bars in January and February, an iron-hungry time of year, when fields were cleared and planted using handheld axes, machetes, and hoes (Kriger 2017, chapter 1).[2]

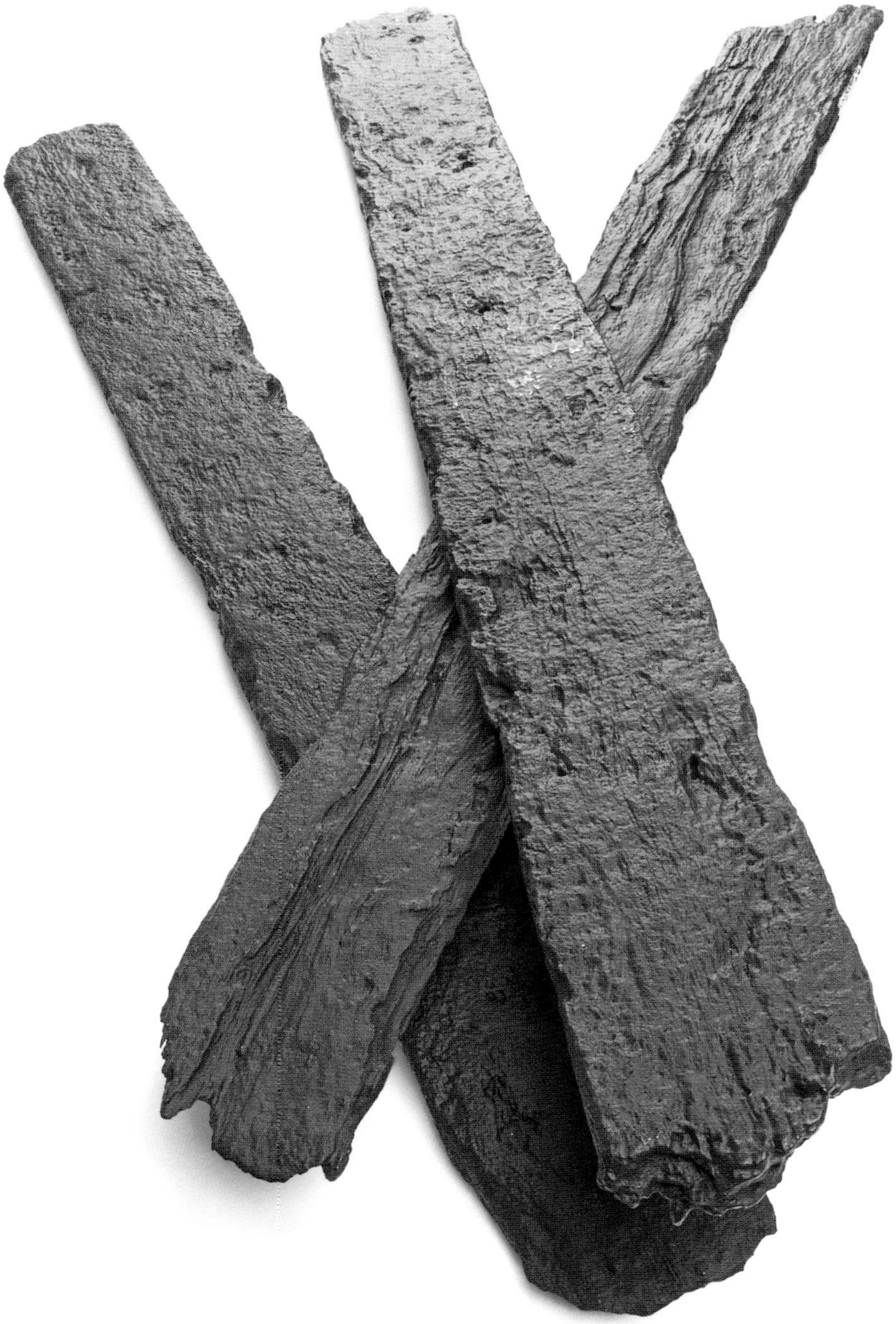

L.1
Maker unknown (England)
Iron bar stock, ca. 1700
Iron
H: 40.4 cm; H: 38.8 cm; H: 37.3 cm
FROM THE COLLECTION OF THE MEL FISHER MARITIME MUSEUM, 1986.008.1139A, 1986.008.1393A, 1986.008.1393B
PHOTOGRAPH BY DYLAN KIBLER, 2018.

Recovered from the wreckage of the slave-trading vessel *Henrietta Marie* (port of registry, London), which sank in Florida waters circa 1700, these iron trade bars were manufactured in Europe often to the specifications of African blacksmiths.

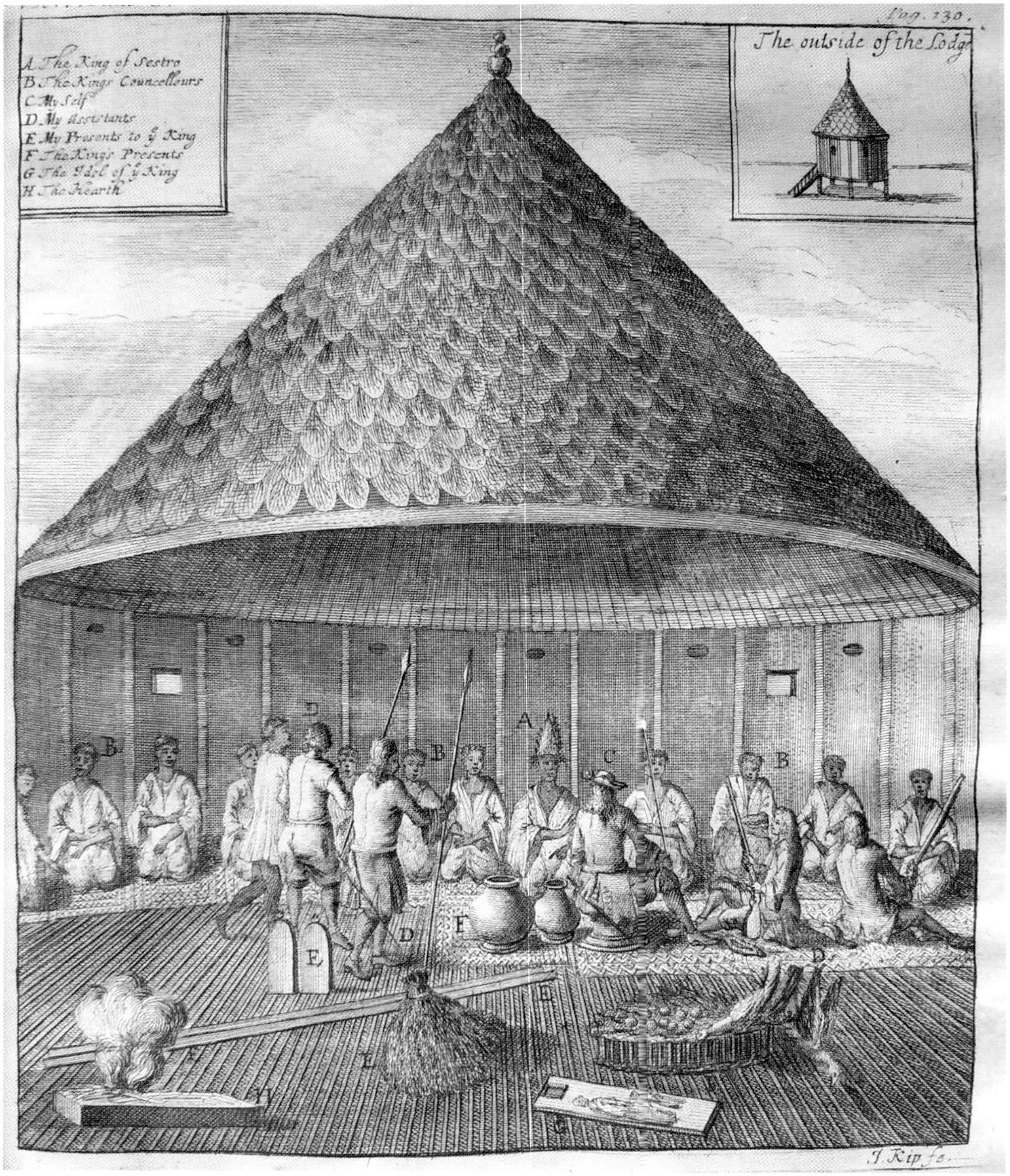

L.2 In 1682, on the Upper Guinea Coast, the King of Sestos receives Jean Barbot and his gifts of brandy, voyage iron, and trade beads (each labeled *E*). Note the lengths of voyage iron arrayed across the floor of the king's official reception lodge.

IMAGE BY JOHN BARBOT, "A DESCRIPTION OF THE COASTS OF NORTH AND SOUTH-GUINEA," IN A. AND J. CHURCHILL, *A COLLECTION OF VOYAGES AND TRAVELS* (LONDON, 1732) COLLECTION OF THE MASSACHUSETTS HISTORICAL SOCIETY.

Standard forms of European-made "bar iron" were very long (at most three to four meters), flat, semifinished pieces that any African blacksmith could turn into a wide range of items, from tools to currencies to works of art (fig. L.2). Precise dimensions varied from company to company, and over time and among Europeans the bars were sold by weight. In the seventeenth century, for example, Dutch and Swedish bars weighed about thirteen to fifteen kilograms each (thirty to thirty-two pounds), or about seventy to seventy-five bars per ton (Davies 1975, 171; Evans and Rydén 2007, 165). On Africa's Guinea Coast, bars were purchased as units by count, and in areas including the Upper Guinea Coast they were used as currency to reckon values and prices for goods and labor. Their cumbersome size meant transport of large quantities was best done in shipments along the coast by ocean vessels; smaller numbers of bars cut down to shorter lengths could make their way up rivers by boat to markets inland, and from there overland to other destinations by donkey (Kriger 2017, chapter 3). Suppliers of voyage iron learned that dealing in it required careful management, for producers had to be kept apprised of precise specifications requested by African merchants and consumers as to the quality and weight of bar iron units. In some locales along the coast, potential buyers made thorough inspections of each shipment, looking out for the trademark of their preferred bar iron manufacturers (Evans and Rydén 2007, 165).

It has sometimes been assumed or suggested that imported bar iron from overseas must have had a dramatically negative impact on iron smelting in Africa. However, on the Upper Guinea Coast, which saw steady demand for voyage iron from the seventeenth century onward, African smelters in interior regions continued to produce their own iron even in the face of competition from imports, probably because the market for iron was so extensive and varied. Iron held high value as an intermediate good for manufacturing a wide range of objects. Also important was its role as currency, or store of value; this kept iron moving in circulation, hoarded as savings, or buried ceremonially with honored personages. Yet another factor was blacksmiths' keen preferences for particular types of iron. Smiths were highly selective and would only work iron they deemed to be of the right quality. For these and perhaps other reasons it is notable that, despite two hundred years of voyage iron imports to the Upper Guinea Coast, iron smelting continued to be practiced in the Futa Jallon highlands of western Africa into the nineteenth and twentieth centuries (Mollien 1820, 1:281–89; Appia-Dabit 1965). In the Congo basin of central Africa as well, some smelters continued their skilled work into the twentieth century.[3]

Smiths remain extremely choosey about their sources of iron. Lotoy Lobanga, master blacksmith of Lopanzo, recounted to me in 1989, when I visited the Republic of Zaire, the different types of European iron imports coming in, ranking them in terms of price alongside whether they were good to work with and to what degree (Kriger 1992, 291–92). He chose, whenever possible, the iron he deemed to be of the highest and most appropriate quality.

INTERLEAF

M

THRESHOLDS: SUB-SAHARAN BLACKSMITHS IN MOROCCO

Henry J. Drewal

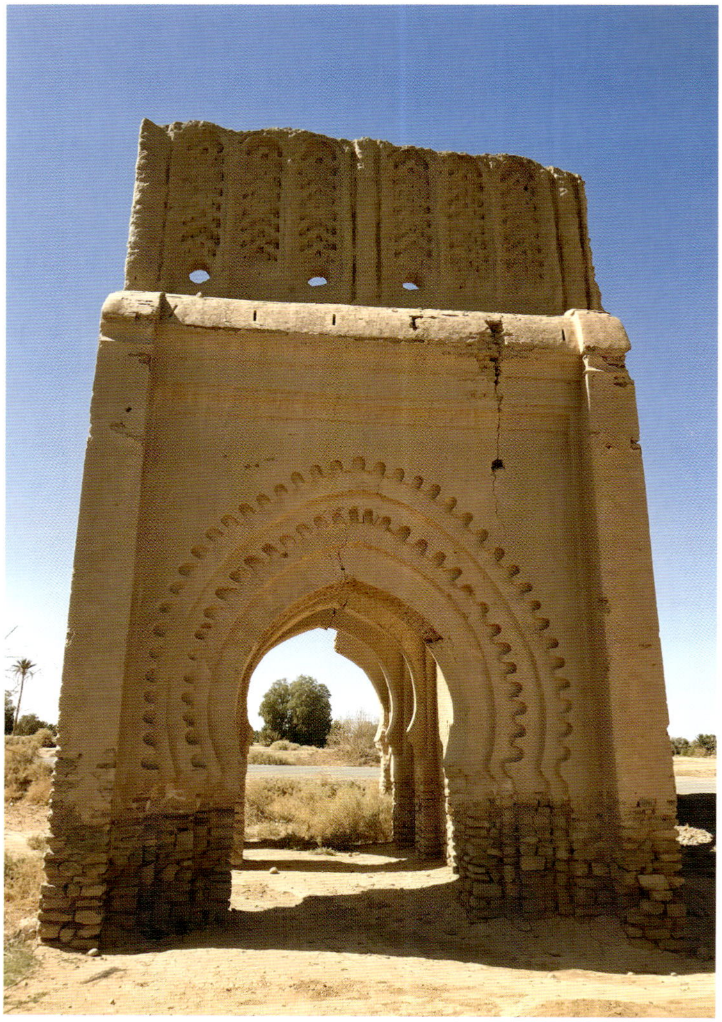

The Sahara has undergone several dramatic climatic changes over the millennia, from periods of heavy monsoon rains (10,500 and 7,300 years ago) to a sea of sand (after 3500 BCE) (Zhongshi Zhang et al., 2014). Despite this last change, the desert has continued to be a connector of peoples, arts, ideas, and goods from earliest times, especially before and during the "Golden Age" of medieval Islam in the Maghreb, specifically in Morocco and Algeria (ca. 400–1500 CE). At the center of this world of exchange was Sijilmasa, the ancient northern terminus (now in southeastern Morocco) of a major trans-Saharan caravan route for trading goods—gold, salt, and enslaved sub-Saharan people. From Sijilmasa they were dispersed to various parts of northern Africa, the Mediterranean basin, Egypt, the Levant, southern and central Europe, and as far as southeast Asia and China via long-distance silk and spice routes.

Various Muslim dynastic rulers controlled the walled and fortified city of Sijilmasa beginning in the eighth century CE, vying for control of such lucrative commerce. Legend claims an Amazigh (Berber) blacksmith founded the city in the fourth century CE, and in the ninth century one of Sijilmasa's rulers was named Aswad, meaning "black," possibly in reference to his sub-Saharan origins (Messier and Miller 2017, 70–73, 200–04; James A. Miller pers. comm., 2017). The Muslim traveler al-Bakri reported, in the tenth century, that Sijilmasa had twelve great gates, with eight forged of iron (fig. M.1). All such references suggest the prominent presence of blacksmiths and links with sub-Saharan African peoples as Sijilmasa connected sub- and supra-Saharan worlds.

Extensive histories of trans-Saharan trade and human trafficking involve African soldiers and, most likely, blacksmiths engaged as weapons makers, brought to serve the armies of various rulers based at Sijilmasa and other imperial centers. A well-documented example

M.1 Bab Fez, also known as the "Gate of the Wind," served as the northern gate of the fabled fortified city of Sijilmasa on the road to the city of Fez. The remnant seen here was built during the Merinid dynasty that ruled Morocco circa 1216–1471. Its gate may have been made of iron, like one of the eight gates of iron at Sijilmasa described by the Muslim traveler al-Bakri in the tenth century.
PHOTOGRAPH BY HENRY J. DREWAL, SIJILMASA (RISSANI), MOROCCO, 2017.

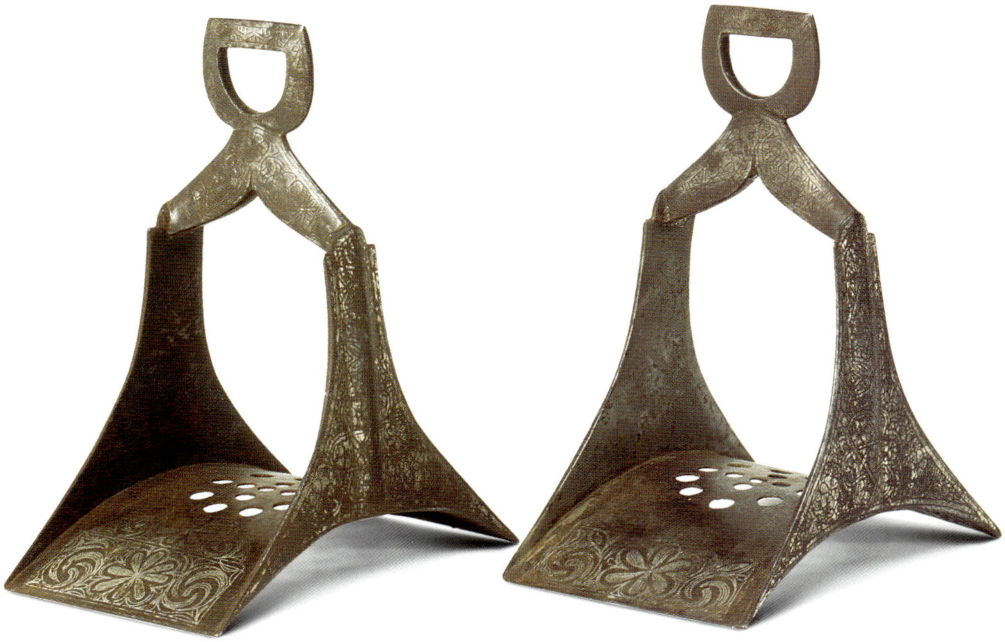

M.2
Unknown artist (Morocco)
Pair of stirrups, 20th century
Iron
H: 24.5 cm
COLLECTION OF THE ROYAL MUSEUM FOR CENTRAL AFRICA, 84.38.40
(NOT IN EXHIBITION)
PHOTOGRAPH BY J.-M. VANDYCK. © ROYAL MUSEUM FOR CENTRAL AFRICA, TERVUREN

These stirrups recall Moroccan equestrian warriors, many of whom may have come from sub-Saharan Africa along with the blacksmiths who made them.

occurred at Meknes, in northern Morocco, where Sultan Moulay Ismail (1672–1727) built a professional military by creating "an army of black slaves from the Bambara kingdom of the Sudan, new converts to Islam. They were called *abid-al-Bukhari* (slaves of al-Bukhari) because they swore an oath of loyalty to the sultan upon the book of hadith compiled by al-Bukhari" (Messier and Miller 2015, 154). The sultan's army maintained control of Sijilmasa, taking in wealth from trans-Saharan trade. Ismail built a series of *qsur* (fortified villages) throughout the country, and his son Mamun al-Kabir ruled over Sijilmasa from a *qasba* just outside the walled city, manned by five hundred "black cavalrymen" (ibid.; fig. M.2). Thirty years after the sultan's death, instability led to disbanding this cavalry and their dispersal throughout the realm. Those who were or became blacksmiths may have developed a series of grille designs called *bukricha* (from Bukhari) that display prominently spear/arrow and bent-bow motifs (fig. M.3).

These histories help explain why the blacksmiths I interviewed and worked with in Morocco in 2017, as well as those interviewed by Abdelhamid Ghandi for his recent research on Moroccan blacksmiths, are known as "people of the desert" (Ghandi 2014, 33). They emphasize their roots in Africa, south of the Sahara. Many are Sufi followers of mystical branches of Islam. Their roots in the Sahara conjure sensory memories of burning heat and mirages understood as mystical visions. Associations of dirt, smoke, and ashes, together with their "magical" gifts as transformers, combine to create deep-seated attitudes of fear, awe, and reverence for blacksmiths, not unlike those among Mande-speaking peoples of Mali (see chapter 11, this volume). Another connection between iron and mysticism is the fact that Gnawa, descendants of enslaved sub-Saharan Africans in Morocco, use iron cymbals (*qraqeb*) in their healing ceremonies to evoke spirits into possession (Becker 2011).

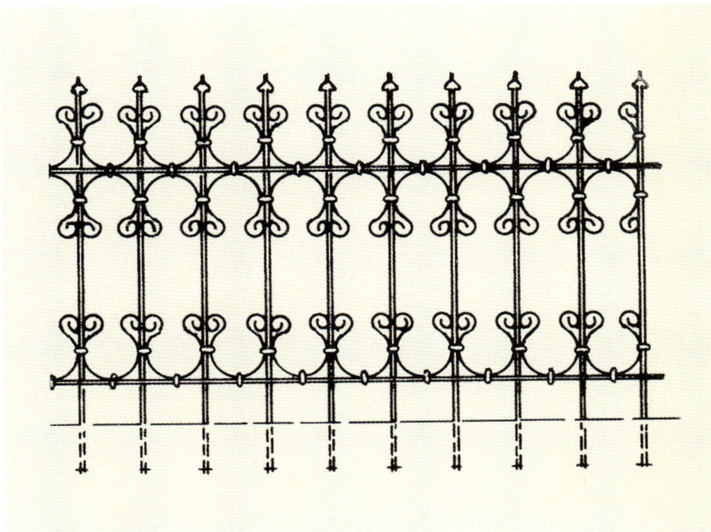

M.3 Above, an arrangement of traditional grille designs known as *bukricha*; below, its architectural realization in Morocco. Often a combination of spear/arrow and bent-bow motifs, such traditional designs may have been first forged in the 18th century by enslaved African soldiers from Mali who were or later became blacksmiths in Morocco.
RENDERING AND PHOTOGRAPH BY ABDELHAMID GHANDI, FEZ, MOROCCO, 2014.

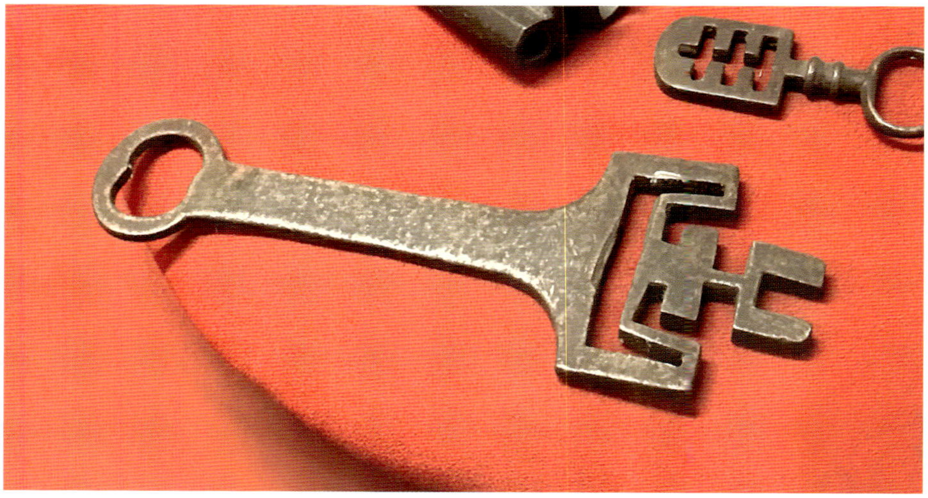

M.4 Stories about jinn describe their special relationship to iron and their reputation as *khelemkan*, or "the real/true owners of a place." Perhaps blacksmiths imagined and imaged their interactions with illusive jinn in certain 18th- and 19th-century keys such as this, forged to abstract the human figure.
PHOTOGRAPH BY HENRY J. DREWAL, DAR JAMAI MUSEUM, MEKNES, MOROCCO, 2017.

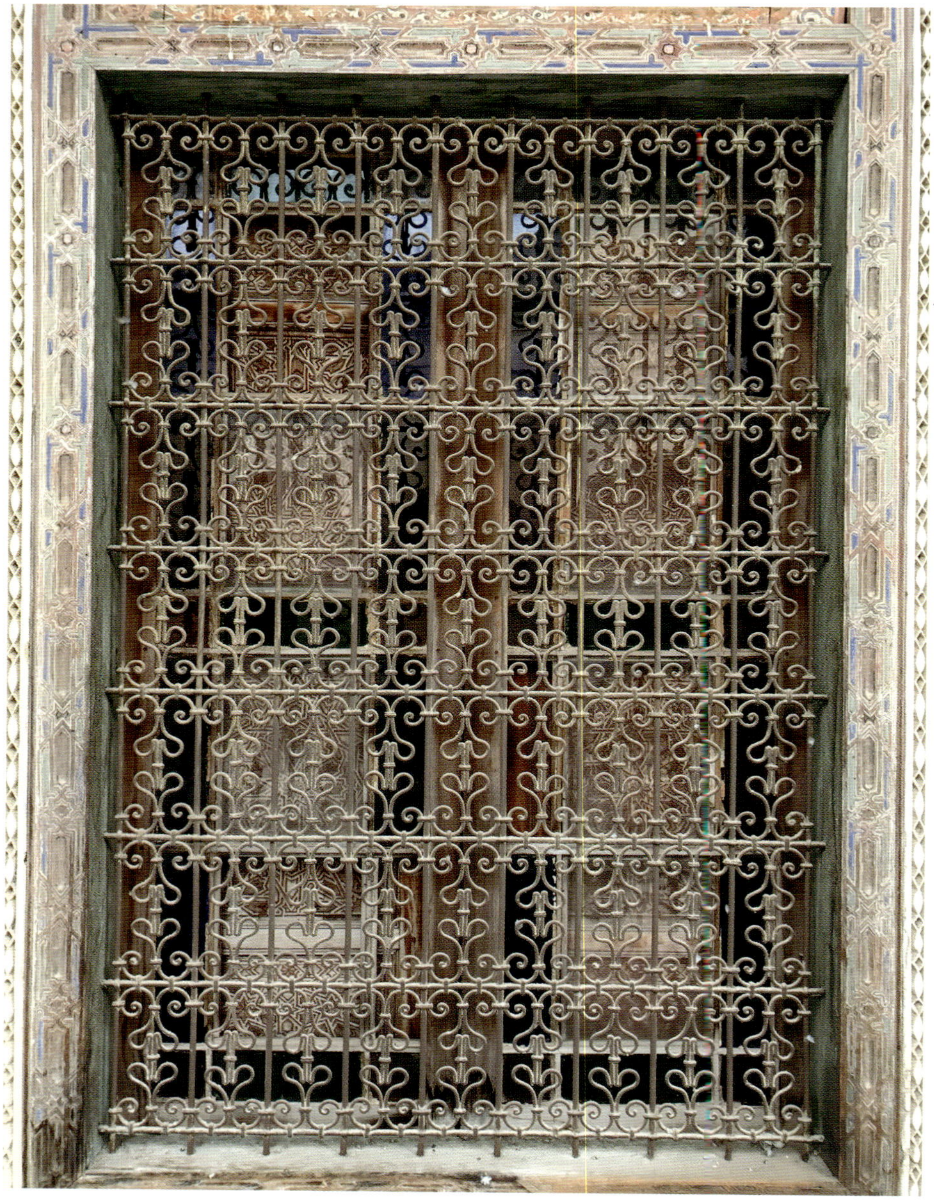

M.5 The iron grilles called *sh-bbak* may be understood as architectural "veils." Such works provide privacy and protection, and, as is typical of blacksmiths' creations in Morocco, they establish thresholds.
PHOTOGRAPH BY HENRY J. DREWAL, MEDINA OF FEZ, MOROCCO, 2017.

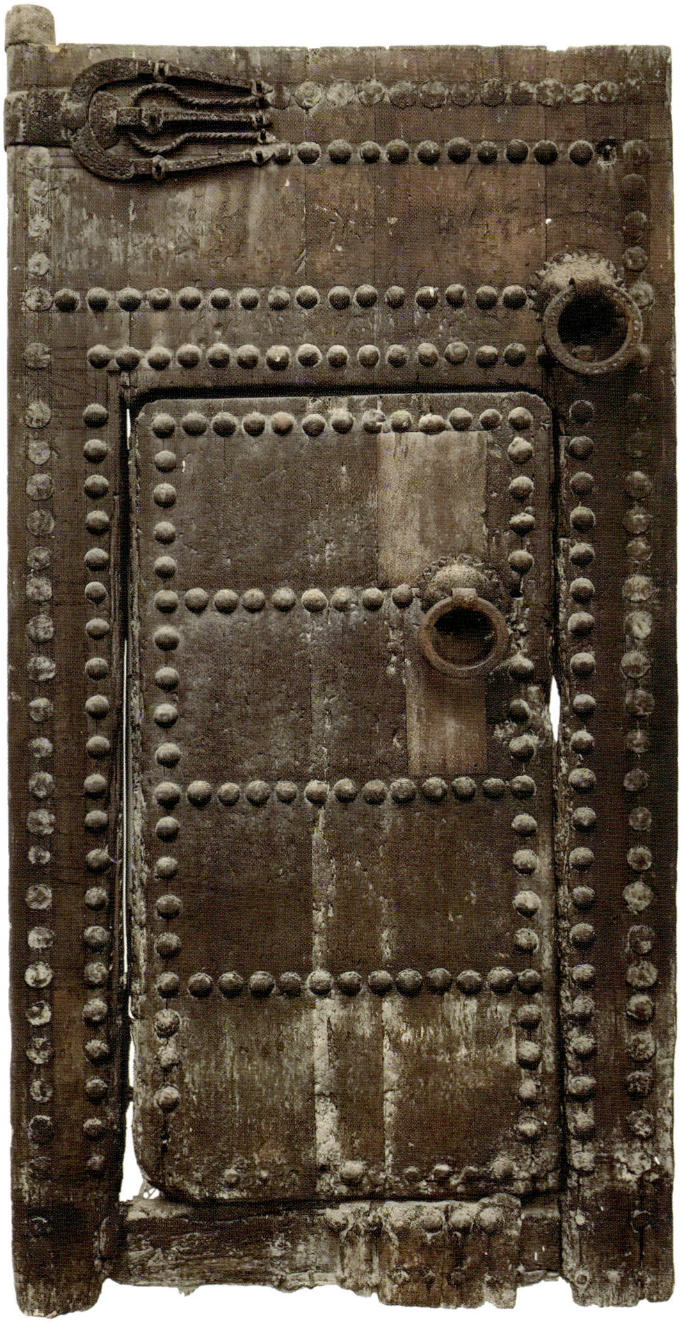

M.6 Doors to homes may be studded with iron bosses and elaborate forged hinges.
PHOTOGRAPH BY HENRY J. DREWAL, MEDINA OF FEZ, MOROCCO, 2017.

Similar feelings are associated with invisible beings known as jinn, created by God from "smokeless fire" according to sura 15:27 of the Qur'an (see chapter 5 and interleaf D, this volume). Jinn can transform themselves into humans, animals, or plants, and they influence or possess humans for positive or negative purposes. Other Qur'anic verses tell of God's gift of ironworking to King David (forged chain mail), while to Solomon God bestowed control of winds (bellows) as well as jinn, suggesting links to mystical aspects of blacksmiths' work with heat, flame, water, and bellows' "breath." As one author wrote in the early twentieth century, "there is a popular belief in Morocco that water used by a blacksmith to quench hot iron, when used to bathe, can protect one from harmful spirits (*jnoun*) [jinn]" (Westermarck 2003, 1:115).

Among numerous contemporary stories that circulate about jinn are those that tell of how a person possessed by a jinn goes into trance and must be given a key or iron object to hold. This action releases the person from the jinn's grip. Keys are for owners of places, and one jinn story describes them as *khelemkan*, "the real/true owners of a place." A friend related to me how, when his family was preparing to move from one home to another in the medina of Fez, their departure kept being delayed and their neighbors would say to them, "The *khelemkan* don't want you to leave!" (Abdelouahed Hamadou pers. comm., 2017). Are elaborate eighteenth- and nineteenth-century keys, some forged as abstracted human figures (fig. M.4), how blacksmiths may have imagined and imaged their interactions with illusive jinn?

Not unlike jinn, blacksmiths are often seen as liminal beings, living and working in the interstices of society. They deal with extraordinary, awesome, and fearful forces as they transform iron into useful objects. Reinforcing this "apartness" or "inbetweenness" (Basu 2017), Moroccan blacksmiths, like Tuareg silversmiths (see chapter 3, this volume), have their own distinctive language to talk about the things they make and the processes they embody. One must learn a new "tongue" when speaking about iron implements such as a simple yet elegant, chisel-like iron tool made for coppersmiths (Hamza El Fasiki pers. comm., 2017; Abdelhamid Ghandi pers. comm., 2017). Such linguistic matters are derived from ancient oral traditions not expressed in Moroccan Arabic (Darija), Hebrew, or a sub-Saharan tongue, but may reflect all these sources, reminding us of the region's extensive histories of intersecting cultures.

Blacksmiths' forged work often establishes thresholds—gates, door knockers, hinges, bars, locks, keys, balcony railings, ceiling grilles over small courtyards, and window grilles (fig. M.5). Such threshold works provide privacy and protection, especially the older window screens made of wood (*mashrabiya*), more recently replaced by iron grilles (*sh-bbak*) that may be understood as architectural "veils" (David Amster pers. comm., 2017). Such iron architectural works express, embody, and enable particular lifestyles and social and religious practices that stress privacy in urban settings—narrow lanes for streets, thick walls pierced by small windows, and large wooden or iron-clad doors (fig. M.6), behind which family life occurs in the privacy of interior rooms and courtyards. Blacksmiths are *inbetweeners*, cosmic connectors (see chapter 3, this volume), like the threshold iron they work.

INTERLEAF

N

FORGING FAITH: ETHIOPIAN IRON CROSSES

Raymond Silverman

As the quintessential expression of visual piety among Orthodox Christians, the cross (*mesqel*) is ubiquitous in Orthodox communities throughout Ethiopia. Crosses are worn as small pendants suspended from the neck, they are carried in liturgical processions, they are held in the hand by priests who use them to bless the faithful. The basic hand cross is composed of a finial in the form of a cross, attached to a vertical handle (or shaft) and a base.

In 2012, I had an opportunity to work briefly with Luel WoldeSelassie (1963–2014), a blacksmith who lived in the historic city of Aksum, located in the Tigray Region of northern Ethiopia. Though he was primarily engaged in making and repairing agricultural tools such as plow blades, Luel also made hand crosses. Crosses of iron have been produced in Aksum and other parts of northern Ethiopia for a very long time. Today's city rests on the ruins of the capital of the ancient state of Aksum, where Christianity was introduced in the early fourth century. The archaeological record offers evidence that iron was smelted and shaped into a variety of implements at Aksum for at least two thousand years.[1] Iron smelting in the region ceased in the early twentieth century, when blacksmiths began using imported iron as well as recycled metals from vehicles and military ordnance.

Iron crosses are difficult to date. The distinctive forms of the two crosses illustrated here (figs. N.1, N.2), and specifically their long, attenuated shafts and proportionately small finials, suggest that they are from the thirteenth to fifteenth century. In most cases, the chronology employed for dating crosses is based on their formal affinities with crosses depicted in church paintings (Chojnacki 2006). Among the small number of iron hand crosses documented in monasteries and churches is an elegant, delicately incised one (fig. N.3) said to have been owned by Abba Pantalewon, one of the Nine Saints associated with the spread of Christianity in Abyssinia in the fifth and sixth centuries; it is maintained in the monastery he founded near Aksum.

Today, there is little demand from priests for hand crosses made of iron or any other material—brass, silver, wood—historically used to fabricate such objects. Instead, most priests purchase factory-produced plastic crosses. The metal and wood crosses that are being made are primarily for souvenir merchants, who sell them to tourists visiting Ethiopia. In 2012, Luel WoldeSelassie indicated he had not made an iron cross in thirteen years. When I asked if he would make one so that I might document the process, he gladly accepted the commission.

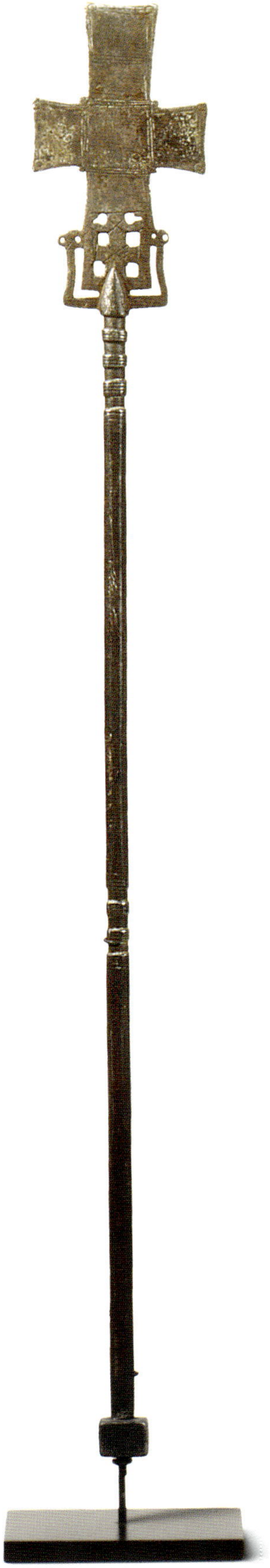

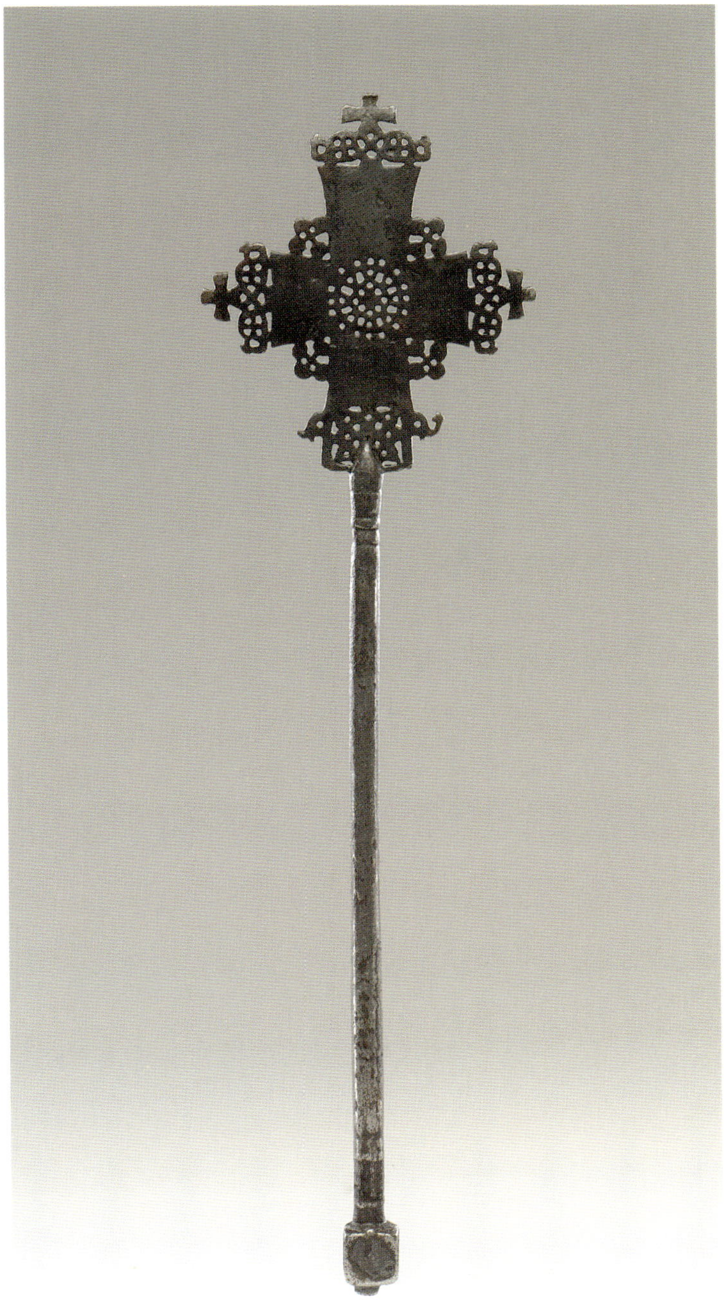

N.1
Artist unknown (Ethiopia)
Hand cross, 13th to 15th century
Iron
H: 55 cm
Amyas Naegele
PHOTOGRAPH BY AMYAS NAEGELE, 2016.

N.2
Artist unknown (Ethiopia)
Hand cross, 13th to 15th century
Iron
H: 41.9 cm
VIRGINIA MUSEUM OF FINE ARTS
GIFT OF ROBERT AND NANCY NOOTER IN CELEBRATION OF VMFA'S 75TH ANNIVERSARY, 2011.216
PHOTOGRAPH COURTESY OF VIRGINIA MUSEUM OF FINE ARTS

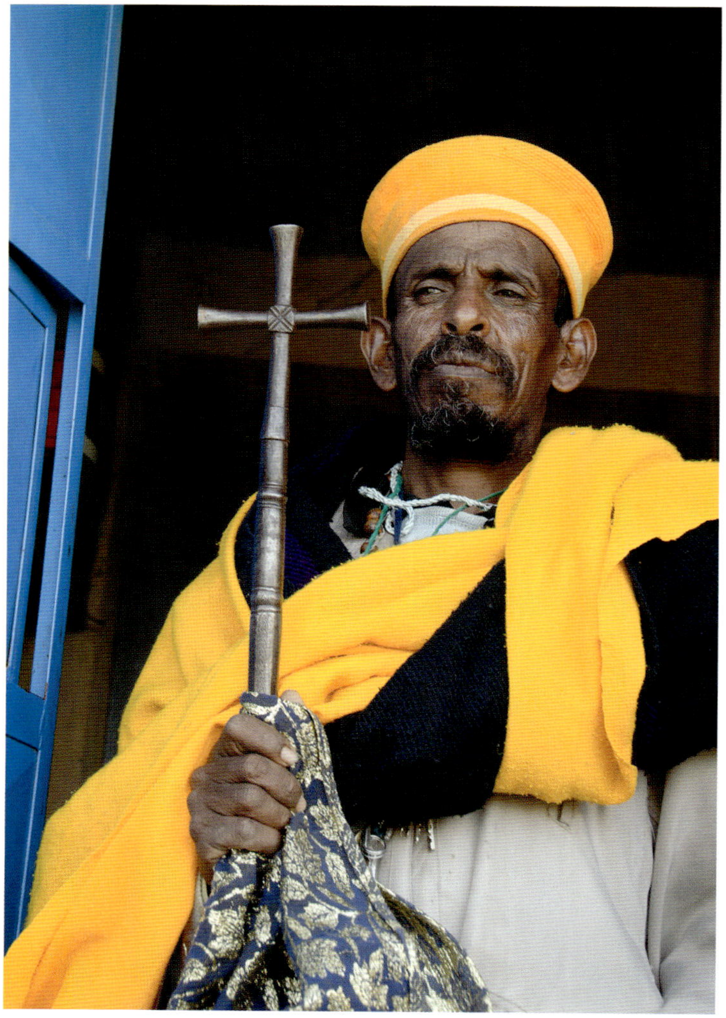

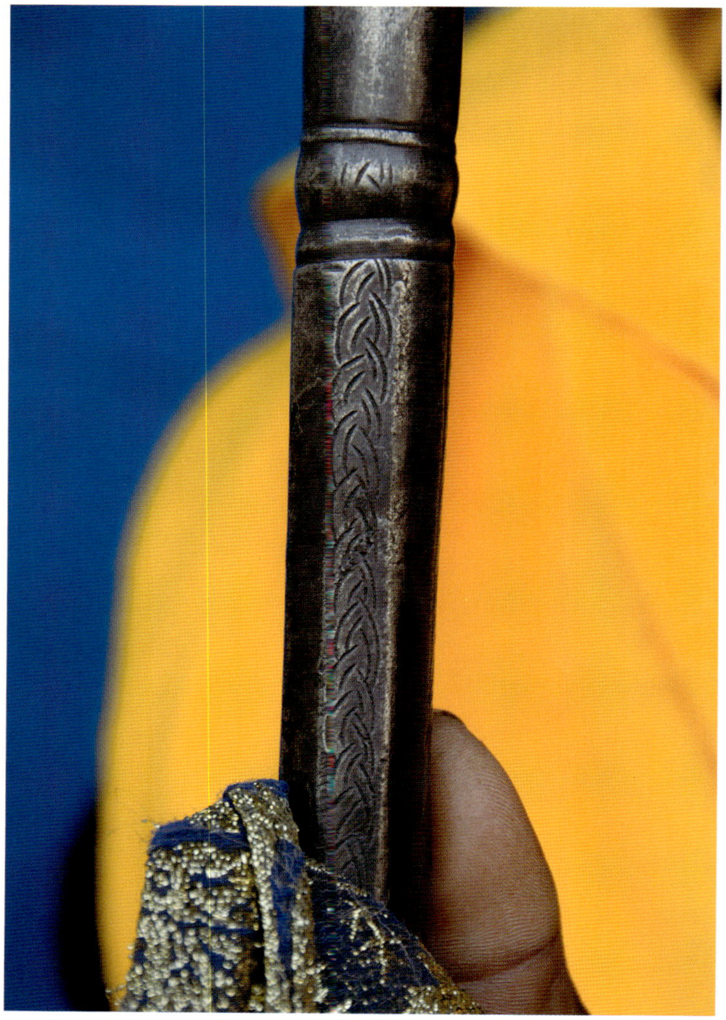

The forge at which Luel worked consisted of a shallow pit he had dug in the ground and filled with charcoal. Once lit, an assistant continuously stoked the fire using a goatskin bellows. Luel began shaping the cross from a piece of sheet iron obtained as scrap from the undercarriage of an automobile. He set it in the fire and covered it with hot coals. After several minutes of pumping the bellows to increase the temperature, the iron became red-hot, and Luel removed it with a large pair of steel tongs and placed it on an anvil. Using a steel chisel and hammer, he cut the rough form of the cross from the sheet of metal. Then, over the next hour, he moved back and forth between anvil and forge fire, striking the cross with his hammer until the metal cooled, placing it back in the charcoal fire until it again turned red-hot and malleable, and then returning it to the anvil. At the end of each cycle he closely examined the cross before returning it to the fire, and he continued this routine until he was satisfied with the results. Picking up the chisel again, Luel cut away small bits of metal from the cross, further refining its profile. He used the chisel as well as several punches to create designs on the cross's surface. The final stage involved fixing the cross in a vice and using a flat file to smooth its edges. It took Luel roughly four hours to complete the cross (fig. N.4).

Metal crosses and the contexts of their manufacture present an interesting paradox. These quintessential symbols of Orthodox piety were produced by individuals who historically have been viewed as pariahs by the farmers and clergy who make up most of northern Ethiopian society. Perceived as *buda*—possessing the evil eye—for centuries metalworkers have endured segregation, living in endogamous communities and denied rights enjoyed by other members of society such as owning land. While the origins of these social values are unknown, they have been associated with people who *create* things, including potters, woodworkers, weavers, and tanners. But during the last fifty years, the stigma associated with working iron, copper, silver, and gold has diminished considerably, largely due to government intervention. Blacksmiths have been given land on which to farm. Interestingly, a growing number of nonhereditary blacksmiths and jewelers—men coming primarily from farming families—take up these trades because they are more lucrative than farming. Significantly, unlike hereditary metalworkers, they are not *buda*. Though belief in the evil eye persists, especially among older individuals, metalworkers in Ethiopia today have greater social mobility and better lives than in the past.

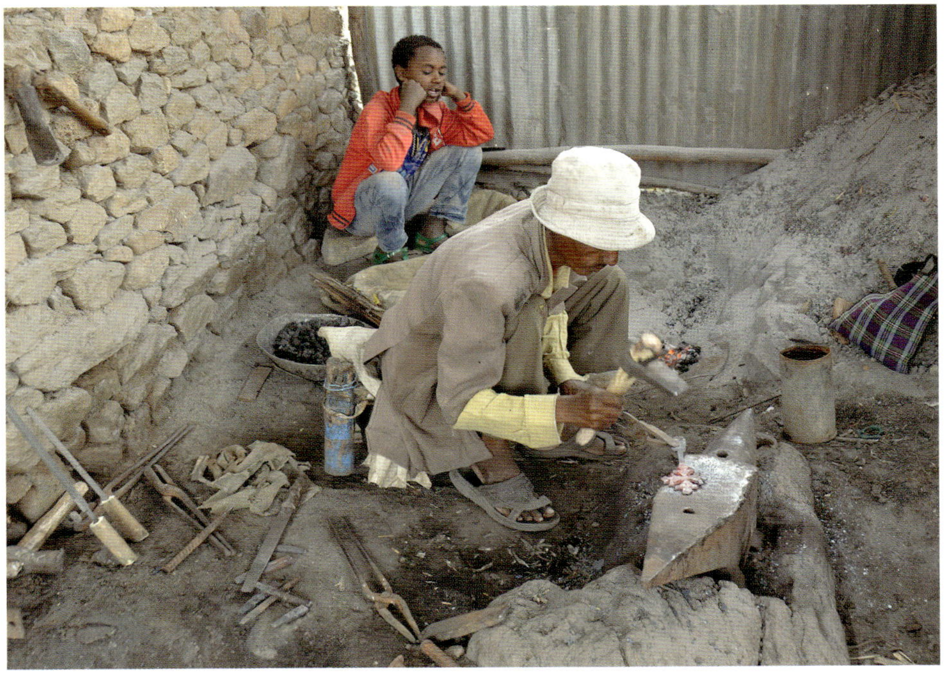

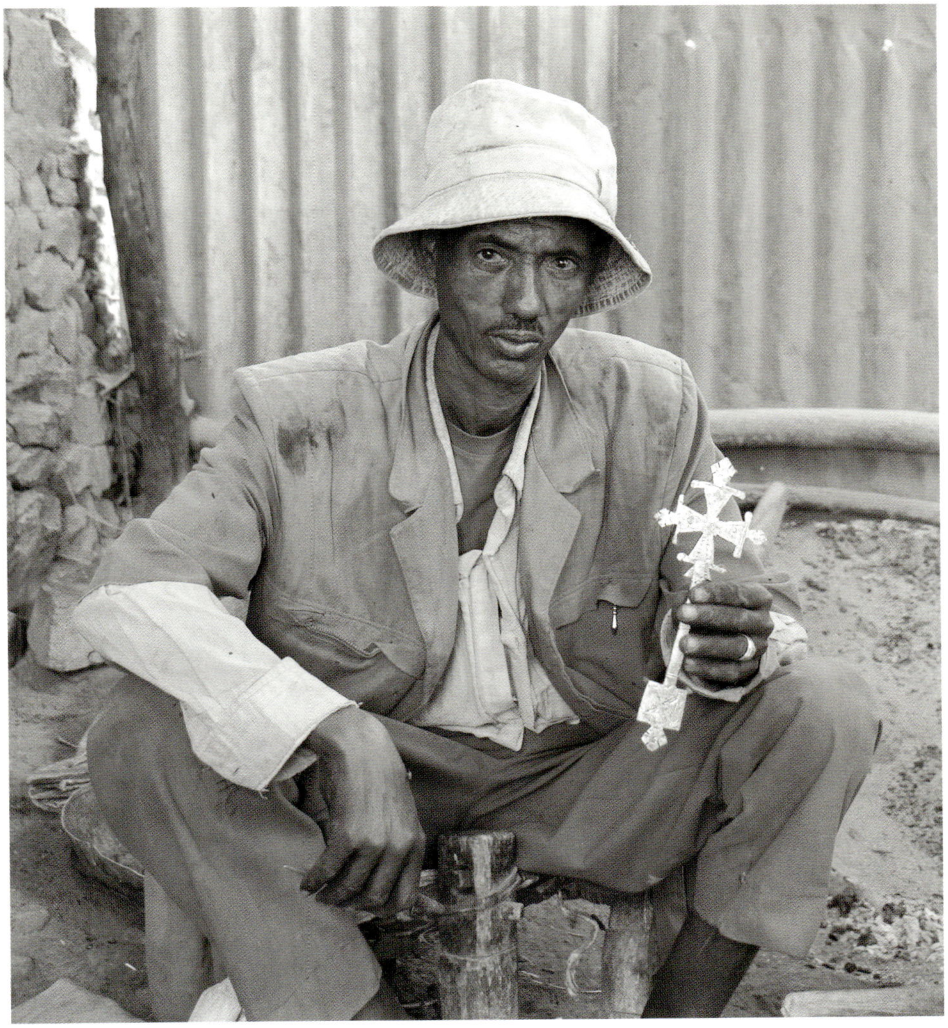

N.3A–B Abba GebreTsion Haile, a monk at the monastery of Abba Pantalewon near Aksum, holding an iron cross attributed to the Abba Pantalewon, one of the Nine Saints, associated with the spread of Orthodox Christianity in Abyssinia during the fifth and sixth centuries.
PHOTOGRAPHS BY RAYMOND SILVERMAN AND NEAL SOBANIA, 2012.

N.4A–B
At his workshop in Aksum, blacksmith Luel WoldeSelassie forges an iron hand cross.
PHOTOGRAPHS BY RAYMOND SILVERMAN AND NEAL SOBANIA, 2012.

CHAPTER 17

SOUNDING FORMS

Manuel Jordán

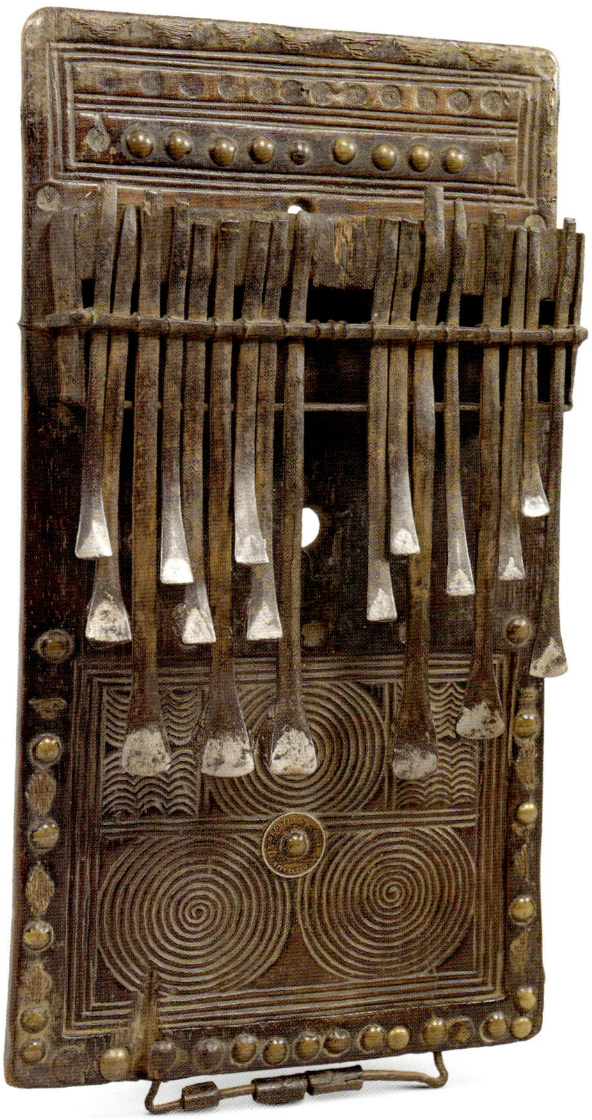

> Human life takes place . . . in the company of non-human objects, many assemblages of [which are] animate and inanimate, organic and mechanical. In part we are all . . . defined by our interactions with technologies . . . and this has tremendous implications for knowing in and through sound.
>
> STEVEN FELD (2017, 94)

In a photograph taken in the 1950s among Chokwe people of eastern Angola (fig. 17.2), a musician sits on the ground playing a lamellophone, or "thumb piano," known to Chokwe as a *chisanji* (fig.17.1, 17.3).[1] Another man pumps bellows to maintain air flow over hot coals, and a master blacksmith hammers a piece of iron into shape on his anvil. The smith's blows must have created a basic beat, a rhythmically repetitive *ding-ding-ding*, to sustain the pace of work and provide a sound-based continuum comparable to striking a bell. The blown-air "breathing" of the bellows and the "thumping" of the flexed skin membranes on its hollow structure would have added complementary, interlacing sounds. In the photographed scene, the presence of the *chisanji* player underscores the integral musicality of forging. As an instrument featuring multiple iron keys attached to a wooden soundboard, in this case placed over a large gourd resonator, the *chisanji* would have added melody and perhaps even meaningful tone poems (Berliner 1981), rounding out the forge's percussive ensemble.

17.1
Artist unknown (Chokwe peoples, Angola)
Lamellophone (*chisanji*), late 19th century
Wood, iron
H: 36.2 cm
MUSICAL INSTRUMENT MUSEUM, PHOENIX, 2013.56.1
PROVENANCE: COLLECTION OF MR. AND MRS. F. AND F. BOULANGER-BOUHIÈRE
PHOTOGRAPH BY TROY SHARP, 2016.

This *chisanji* possesses three-tiered keys and "sound-enhancing" elements including iron beads on a piece of bent wire attached to the bottom of the wooden soundboard. Under the keys are tiny balls of hornet's wax applied for tuning purposes. The rich iconography of the board presents motifs often found in Chokwe material culture, adding to the beauty of the "tone poems" played by a *chisanji* musician.

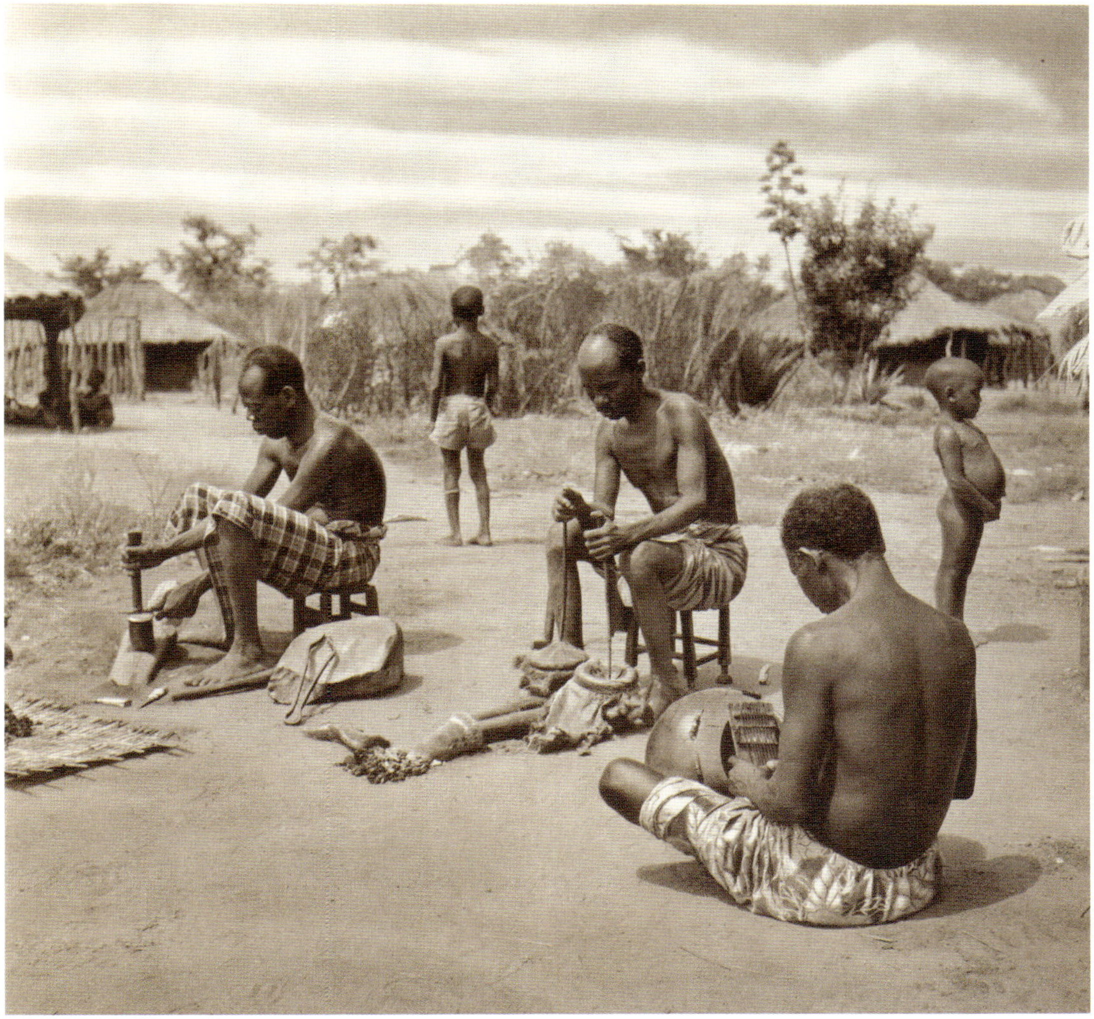

17.2 Three rhythms produce a soundscape as a Chokwe blacksmith's hammer is complemented by a bellows-man's "breathy" sounds and a *chisanji* player's tune (resonated by the instrument's open-ended gourd), perhaps accompanied by song.
PHOTOGRAPH TAKEN IN EASTERN ANGOLA, CA. 1950S. PUBLISHED IN JOSÉ OSÓRIO DE OLIVEIRA, *FLAGRANTES DA VIDA NA LUNDA* (LISBON: DIAMANG, DUNDO MUSEUM, 1958).

Although neither hammer-anvil nor bellows are musical instruments strictly speaking, they can serve as such. During my years of research in northwestern Zambia, I worked with smiths who equated their techniques of forging with drumming and maintaining certain beats. On various occasions among Chokwe, Luvale, and Luchazi peoples, I observed the blending of sounds and actions as part of blacksmiths' processes, whether working alone or with partners or assistants.[2] Taskscapes like these and the one published by Osório de Oliveira—or, indeed, that depicted in Giovanni Cavazzi's seventeenth-century painting of the king of Ndongo forging a ceremonial axe blade accompanied by two lute-players (see fig. 15.1, this volume)—imply what musician and ethnomusicologist Steven Feld terms "an experiential nexus of sonic sensation." Such a synaesthetic *event* is an "acoustemology," that is, a conjoining of acoustics with epistemology that produces "a knowing-in-action: a knowing-with and knowing-through the audible" (Feld 2017, 85).

An acoustemology may also be thought of as a relational ontology, "a connectedness of being... built on the between-ness of experience."[3] In ironworking as a process of *making*, a fugue resounds when a blacksmith strikes iron, an assistant pumps bellows, and, at least sometimes at a Chokwe forge, a musician plays a *chisanji*. Creativity is cadenced as tasks are accomplished. Interrelated motor skills and muscle memories are implied as movements are repeated so rhythmically and efficiently that they are realized without conscious effort, leaving the blacksmith free to contemplate and affect artistic nuances.

How might particular sequences of notes as "fields of corporeal potential" cue or provide counter-rhythms to specific manual phrases by a blacksmith and his bellows men? Might what Marcel Mauss called "prestigious imitation" come into play as apprentice blacksmiths (often working bellows) emulate the muscle memories of their masters so that their bodies "become inhabited... by conventional movements and gestures peculiar to that sound space, borrowed from one's teacher"?[4] The complexity of such "acoustic communities" (Feld 2007, 101) increases dramatically when multiple forges tightly occupy an area, whether in villages dedicated to smithing or in cities such as Bamako, Mali, where iron markets' cacophonous clatter resolves into mesmerizing rhythms of hyper-intense labor and purpose (see chapter 18, this volume).

Such an ontological approach to musicality-in-process serves several purposes. On the practical side, playful musical engagement relaxes what adds up to endless days of repetitious and taxing work. Furthermore, among Chokwe and related groups, manipulation of iron was highly ritualized through the smelting of ore in an anthropomorphic, female-shaped furnace (see figs. 2.8, 2.9). "Sounding forms" emphasize cosmological concepts of fertility for Chokwe, and music was important to iron production from beginning to end, as it still is when prayers and songs to the forge are preludes to a blacksmith's tasks.

Throughout sub-Saharan Africa, the playing of musical instruments made of iron, by virtue of the spiritual and supernatural qualities and potency attributed to the material, may create authoritative and transformational sounds and contexts. Equated with voices from spiritual realms or serving to "tap" into such transcendent forms of interaction, iron instruments are often kept in the treasuries of chiefs, held in the hands of ritual experts such as diviners, and used to mark important human transitions such as initiation. The acoustemology of the forge, then, as well as the essence of iron, the wisdom of form and process, and a great deal more may contribute to a relational ontology extending from forge to community to divinity itself.

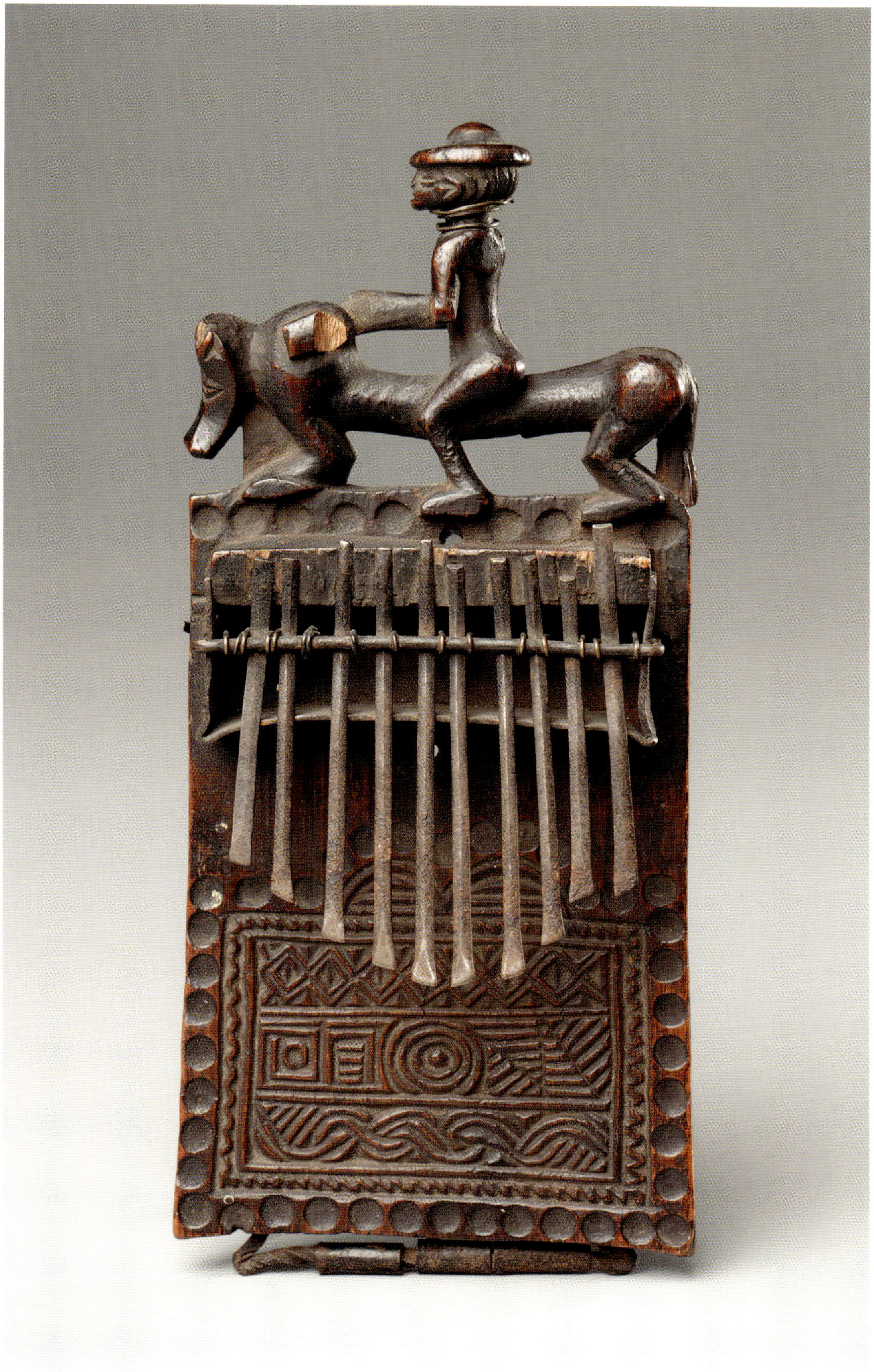

17.3
Artist unknown (Chokwe peoples,
Democratic Republic of the Congo)
Lamellophone (*chisanji*), late 19th century
Wood, iron
H: 19.1 cm
THE CLEVELAND MUSEUM OF ART, 1915.495, THE HAROLD T. CLARK EDUCATIONAL EXTENSION FUND, 1915.495

The trader (*pombeiro* in Portuguese) atop this *chisanji* recalls histories of long-distance commerce across a region divided among colonial powers: Portuguese Angola, British Northern Rhodesia, and the Belgian Congo (today Angola, Zambia, and the Democratic Republic of the Congo). During such travels, *pombeiros* brought news as well as goods, and their narratives accompanied by the tones of a *chisanji* made them the more welcome in remote communities.

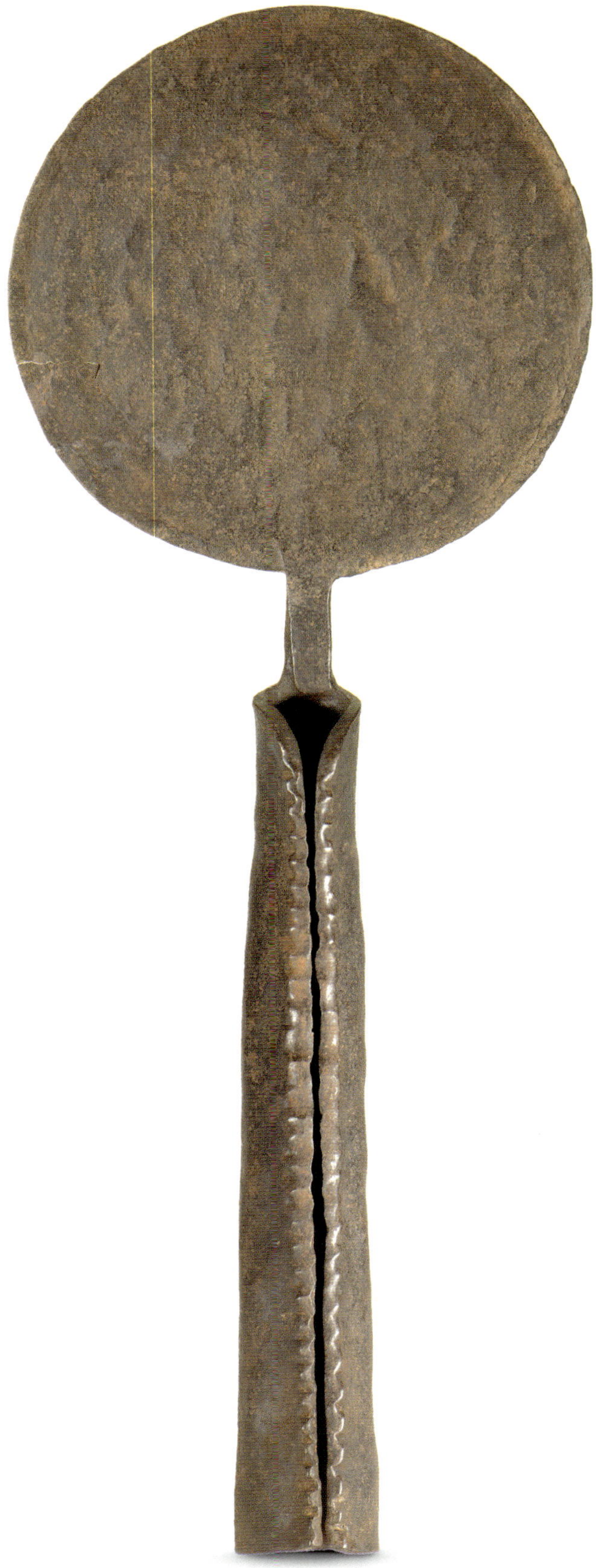

17.4
Artist unknown (Senufo peoples, Côte d'Ivoire and Mali)
Percussion rasp (*keneke*), early 20th century
Iron
H: 58 cm
PRIVATE COLLECTION

When played, this rasp rests horizontally on top of the musician's forearm while the narrow neck of the instrument lies on the palm of the hand with the round percussion plate extending beyond. A thin iron rod is run across the handle's corrugated seam to produce a ratcheting sound that is metronomically punctuated when the rod slides off the cylinder to hit the percussion plate. The smith cold-hammered the disc after shaping it to work-harden the surface and "temper" the sound produced.

MUSICAL INSTRUMENTS OF IRON OR WITH IRON PARTS

The preponderance of African musical instruments forged from iron are idiophones, or instruments that create sound through the vibrating core of their principal material, whether struck, plucked, or rubbed, and without the aid of material parts such as strings or membranes. These include a large variety of bells, rattles, keyed instruments, and rasps (fig. 17.4). Within these categories there is endless variation, which leads to an engaging question: for whose ears is the acoustemology of the forge meant, as well as the instruments made at it—those of community members, those of ancestors and other spiritual presences, or both?

LAMELLOPHONES

Lamellophones ("thumb pianos") are often called *mbira*, a generic term borrowed from a name for the instrument among Shona and related peoples of Zimbabwe. *Mbiras* may incorporate a few to more than two-dozen iron tines, or keys, set on anterior and posterior bridges, held by a pressure bar, and attached to a wooden soundboard. The board is usually held in both hands so that the keys may be depressed with the thumbs (hence, "thumb pianos"); the keys' thickness and length determine the pitch of the notes produced. These days, keys are commonly forged from conveniently available metal scrap such as bicycle-wheel spokes. Found throughout central, southern, and eastern Africa, thumb pianos may also be found farther north, from Uganda to Nigeria and beyond; throughout, they may be characterized as the quintessential instruments of storytellers, setting the pace of narration, emphasizing dramatic moments, and adding to the pleasures of listening.

The *mbira dzavadzimu* (*mbira* of the ancestors) of Shona peoples is perhaps the most recognizable form of thumb piano (see Berliner 1981). It is used in rituals to maintain a continuum of engagement with the ancestral world via spirit mediums. Somewhat similarly, lamellophones called *kasanji* by Luba peoples in the Democratic Republic of the Congo have functional and formal relationships with *lukasa* memory devices performed by royal titleholders and court historians (see figs. 15.27, 15.29). The crisp notes *kasanji* produced were (and may still be) tone poems standing as musical *lieux de mémoire*, reminding those possessing sufficiently esoteric knowledge of how the beads and other features of a *lukasa* might be "read" while recounting complex narratives (Roberts and Roberts 1996, 140, 144; cf. Berliner 1981, 2–7, 56–58).

Among neighboring Tabwa peoples, four types of thumb pianos have been documented; the most common is called a *likembe*, played to pass the time on long mountain hikes, to accompany drums in dances, or in praise-singing. The central hole in the wooden board of a *likembe* is fitted with thick webbing from the nest of a terrestrial spider; when a *likembe's* keys are played, the webbing buzzes like a kazoo, which is amplified by an open-ended small gourd held beneath the board. Another Tabwa lamellophone, called a *kankobele* and no longer used by the 1970s, has one or two spirit heads sculpted at its top end (see fig. 17.5); it was played in rituals to assure ancestral presence. In the old days, the mystical efficacies of a *kankobele* could also keep lions at bay (see Roberts and Maurer 1986, 194–95).

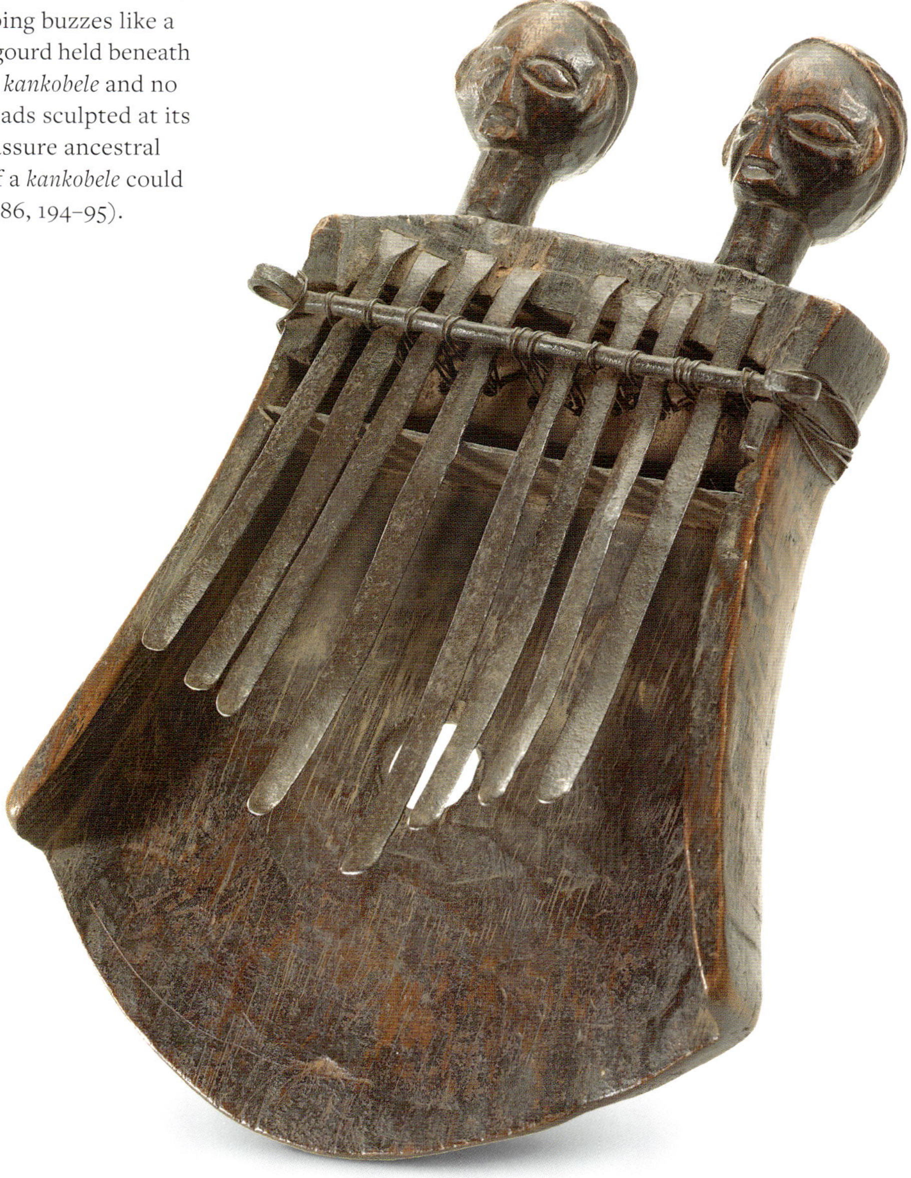

17.5
Artist unknown (Tabwa peoples, Democratic Republic of the Congo)
Lamellophone (*kankobele*), 20th century
Wood, iron
H: 20.4 cm
FOWLER MUSEUM AT UCLA, X86.1890;
GIFT OF HELEN AND DR. ROBERT KUHN

Tabwa play instruments like this on joyful social occasions or to pass the time when hiking on rugged mountain trails or traveling along the shores of Lake Tanganyika in multi-person dug-out canoes. This *kankobele*'s spirit heads contribute to the purposes of ritual music. The hole in the soundboard was once covered with a thick spiderweb so that the instrument buzzed like a kazoo when played on top of a gourd.

A tour-de-force Chokwe *chisanji* (see fig. 17.3) includes a carved full figure of an ox-riding trader (*pombeiro*) from Angola on its sound board. During long travels from the Atlantic coast to the central African interior, the instrument surely accumulated a history, transmitting information through its playing and accompanying songs. Such a melodic instrument in the hands of a skilled performer would have been well received and rewarded in villages along the way, by people welcoming entertainment and new information. Chokwe chiefs seem to have picked up on this principle and celebrated the concept by commissioning figures of chiefs playing lamellophones (fig. 17.6). Although Chokwe chiefs do not themselves play such instruments, the figures reaffirm concepts related to royalty and their control of information shared among people.

Another Chokwe lamellophone (see fig. 17.1) is a large and well-made example of the tiered instrument, designed to be true to the owner's interest in a particular range of tones in relation to a favored finger-playing technique. As is often the case, additional vibrating elements have been applied to create varied qualities of sound. Popular in central African musical aesthetics is the "buzzing" created by bottle caps or loose metal rings placed around keys, for example. Somewhat similar are the metal shakers or extensions added to the necks of west African *djembe* drums and the metal extensions with rattling rings found on instruments such as plucked lutes of Nigerian Hausa peoples and harp lutes of Malian hunters. The resonant buzzing of iron parts on African instruments may be traced to a range of music genres throughout the Black Atlantic, suggesting aesthetic and spiritual carryovers.

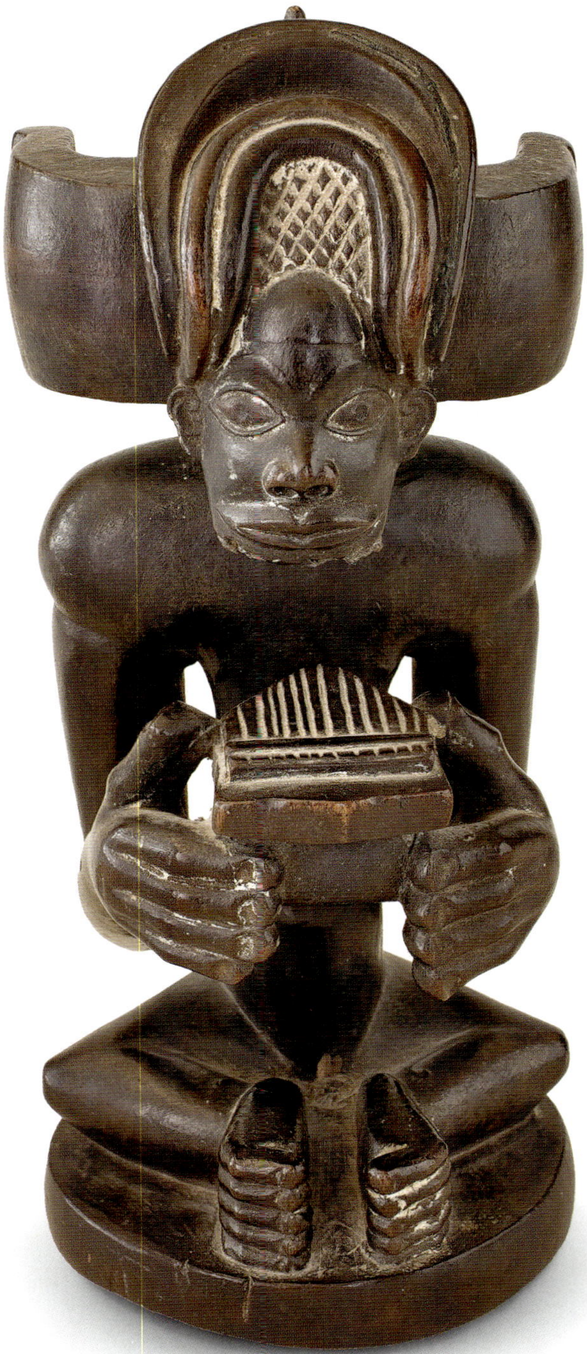

17.6
Unknown artist (Chokwe peoples, Angola)
Figure, 19th century
Wood, pigment
H: 24.1 cm
RICHARD SCHELLER COLLECTION, PALO ALTO, CALIFORNIA
PROVENANCE: ALAIN DE MONBRISON, PARIS
(NOT IN EXHIBITION)
PHOTOGRAPH BY ROBERT A. KATO

A Chokwe chief or other noble person is here portrayed playing a *chisanji*, or "thumb piano." Chiefs do not in fact play such instruments, but their praises are intoned and sung by court musicians. Indeed, so vital to political status is the relationship between music and prerogative that it was given material expression by the artist creating this figure.

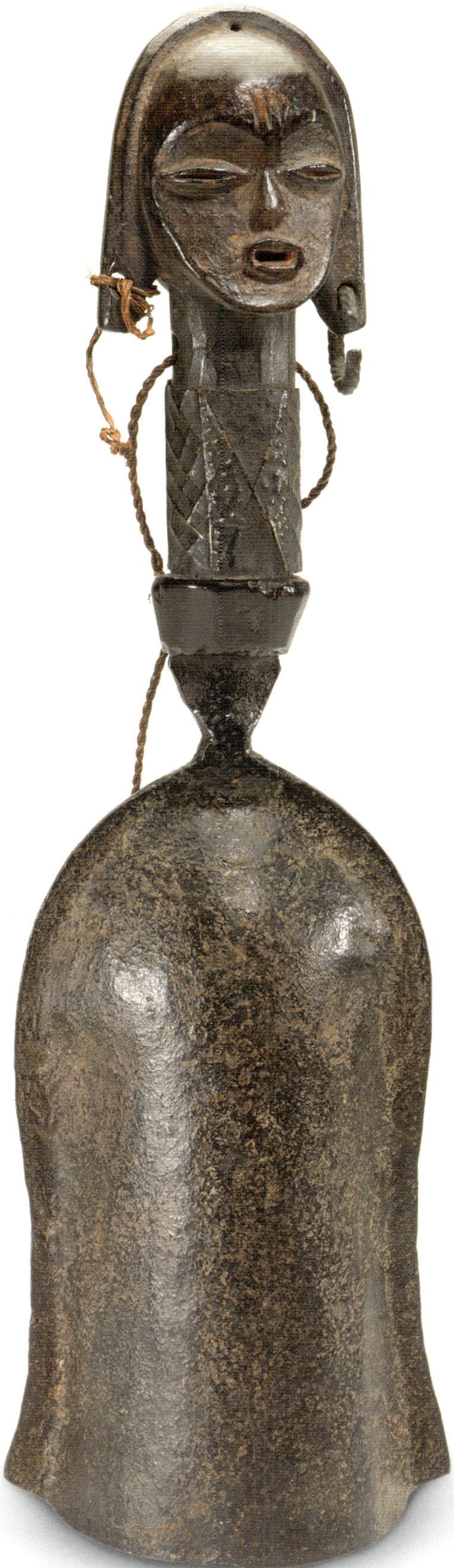

RATTLES AND BELLS

Iron rattles and bells open up worlds of inventiveness. Rattles may be devised as small parts attached to other objects, woven into arm or leg bands, or shaken while held in hand (see figs. 12.6, 0.1, 0.2). Many involve creative ways of securing loose materials—pebbles, iron pellets, chains—within single or multiple iron encasings. Bells of forged iron may be single, double, or triple (figs. 17.7–17.17). They may be struck with beaters or, in the case of smaller ones such as those secured to hunting dogs' collars in central Africa, rung by inner clappers. Bells range in size, with the largest several feet in height, and they attest to skillful smithing through which iron was beaten into sheets, forged with midribs, and then flange-welded (see fig. 4.13). As Jan Vansina noted, these same techniques were essential to forging serviceable spears and other bladed weapons and tools; he posited that the wide distribution of bells found in archaeological excavations as well as through later observation and collection accords with the formation of central African conquest states (1969, 188).[5]

While beating time, the tones of such bells could communicate praise names or send musical messages significant distances, often in circumstances of conflict. They might be displayed as prestige objects associated with chiefs (fig. 17.12), or sometimes "bells could only be owned by warriors who had shown extraordinary bravery." Among Tetela, "the bell would act as a standard which should never be allowed to fall into the enemy's hands" (Vansina 1969, 190). A 1970 photograph taken by Eliot Elisofon in west-central DRC shows a group of Mangbetu court musicians playing instruments including different types of wooden slit drums, drums, rattles, and a double bell of the type featured here (fig. 17.14). The context was probably a chief's ceremony, and the double bell musician used a beater to maintain a tempo for the rest of the orchestra.

17.7
Artist unknown (Tsogo peoples, Gabon)
Janus bell (*mokengue*), early 20th century
Iron, wood
H: 45.5 cm
FOWLER MUSEUM AT UCLA, X86.1893; GIFT OF HELEN AND DR. ROBERT KUHN

Created for the Tsogo men's initiation association called Evovi, "the judges," this forged bell would have been struck at a tempo to symbolize a man's heartbeat. The head that surmounts the handle may represent the mythical entity Kombe—the sun, the source of life, and supreme judge of humans. The finely sculpted face is typical of Tsogo figurative carving.

17.8
Artist unknown (undetermined peoples, Cameroon)
Bell, 19th century
Iron
H: 97 cm
PRIVATE COLLECTION

This large and broad clapperless bell represents the single-body type produced in many earlier sub-Saharan communities. The surface grain of the inside and outside indicates that it was pieced together from over fifty narrow strips of iron bloom. The tour-de-force result of several blacksmiths' collaborative work, the bell was among the largest in a family of community-owned bells. It was rung with cushioned mallets and held tang end down and resting in a wooden socket, so sound reverberated upward through its open mouth. With this "call"—intended to be heard beyond the village center—performers summoned ancestors whose spirit presences were essential at significant ceremonial events.

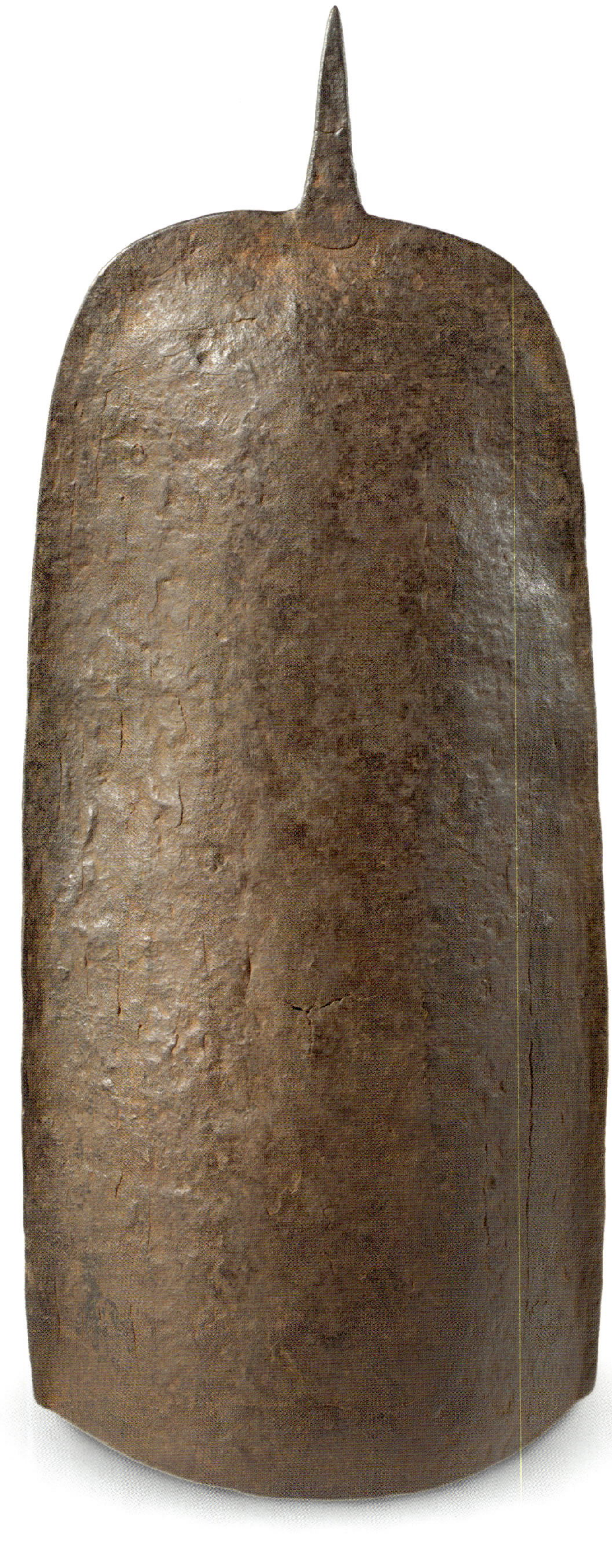

17.9
Kwaku Bempah (active early 20th century, Asante peoples, Ghana) and unknown artist(?)
Double bell (*dawuro*, *nnawuta*), ca. 1920
Iron, wood
H: 57.5 cm
FOWLER MUSEUM AT UCLA, X87.1312;
GIFT OF ELIZABETH LLOYD DAVIS

This Asante clapperless bell was often performed with drumming associations of warrior groups. It may also have been used to send messages. The specific timbres of such iron instruments and the distinctive rhythms made by striking them have culturally specific references, meanings, and resonances. The renowned sculptor Kwaku Bempah carved the bell's wooden handle, enhancing its prestige and beauty. Such attributes would, in turn, have proclaimed the holder's political and religious status.

17.10
Artist unknown (Bacham peoples, Chad and Cameroon)
Double bell, early to mid-20th century
Iron, cloth, bamboo
H: 90.2 cm
FOWLER MUSEUM AT UCLA, X86.1912;
GIFT OF HELEN AND DR. ROBERT KUHN

Clapperless bells are sometimes called gongs because they are struck on the side with drumming sticks or mallets. They are recognized as important military and political emblems throughout Africa, appearing, for instance, in masquerades directed at social control and law enforcement. This robustly proportioned double clapperless bell joined by an arched handle exemplifies prestige instruments associated with chiefs and their entourages. Each bell's range of pitch permits different areas to be struck (singly or together) to vary the musical notes produced.

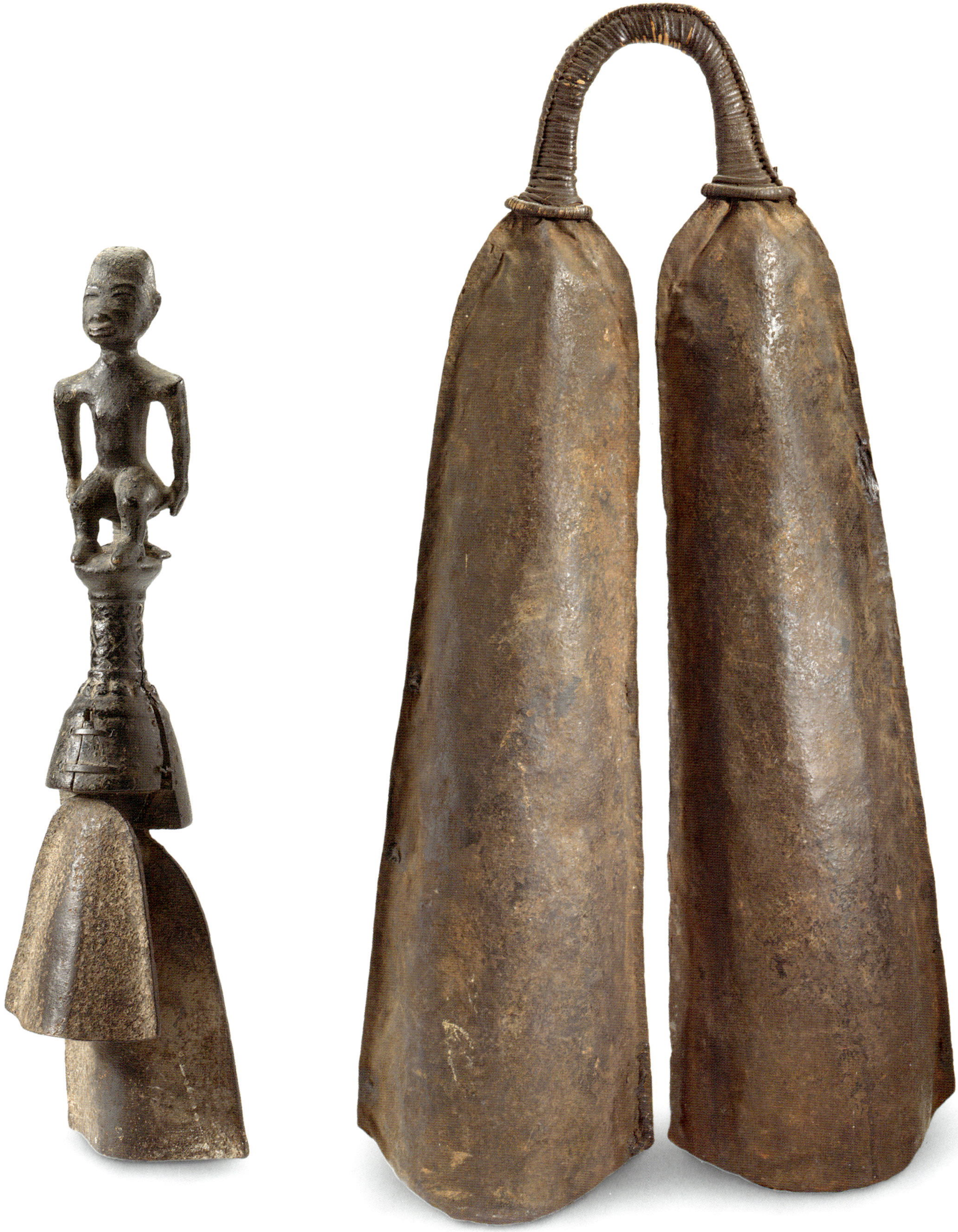

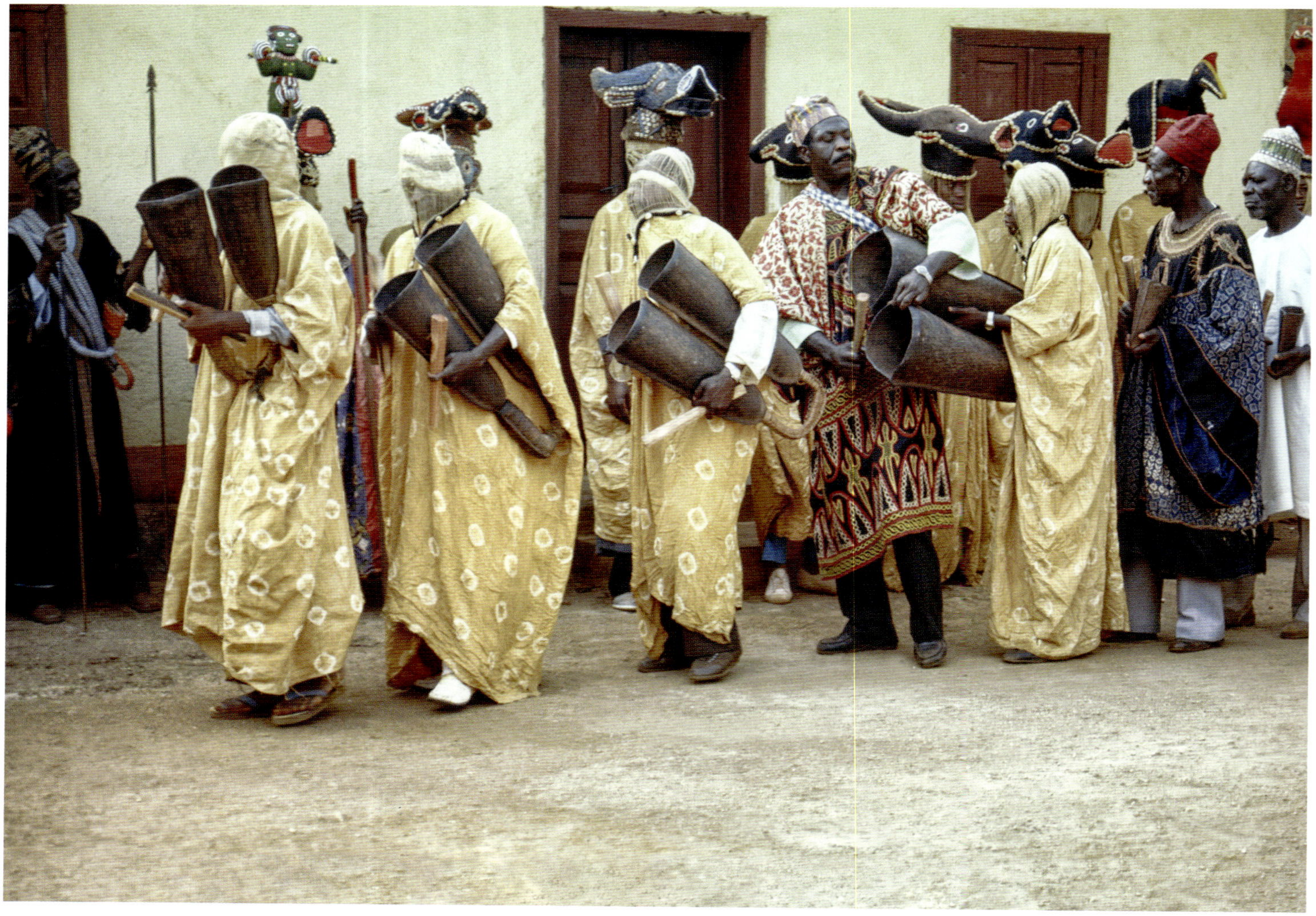

17.11 In a courtyard of the king's palace at Foumban, Cameroon, Bamum men strike large double bells to create a soundscape for the performance of a masquerade during the annual festival of Nja in December. Such masquerades are often associated with regulatory societies whose purpose is to ensure social control and law enforcement.
PHOTOGRAPH BY CHRISTRAUD M. GEARY, 1984. ELIOT ELISOFON PHOTOGRAPHIC ARCHIVES, EEPA 1993-001-1792, NATIONAL MUSEUM OF AFRICAN ART, SMITHSONIAN INSTITUTION

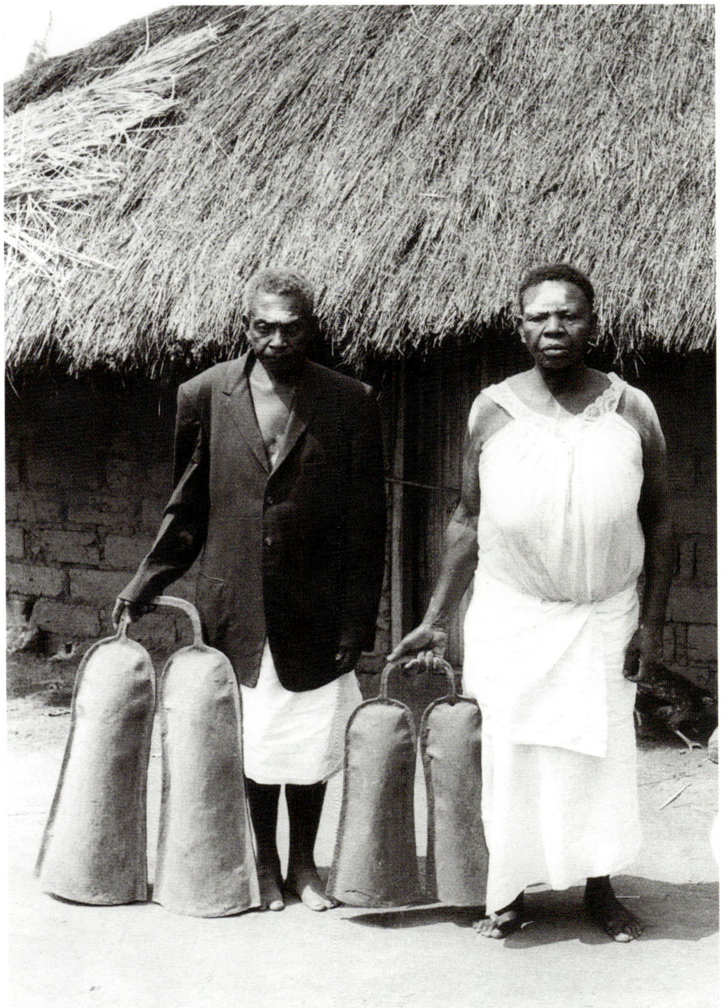

17.12 Iron double clapperless bells held by Lunda members of the royal court.

PHOTOGRAPH BY JOS GANSEMANS, KATANGA PROVINCE, DEMOCRATIC REPUBLIC OF THE CONGO, CA. 1970. COLLECTION ROYAL MUSEUM FOR CENTRAL AFRICA, MP.O.O.1566.

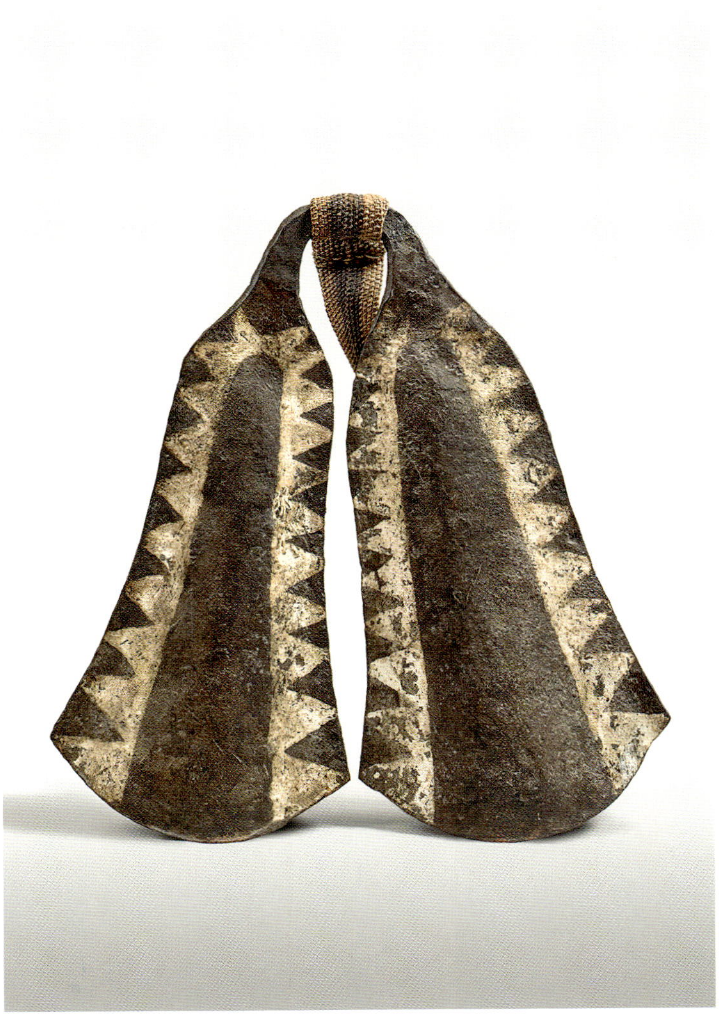

17.13
Artist unknown (Mangbetu peoples, Democratic Republic of the Congo)
Double bell (*nengbongbo*), early 20th century
Iron, plant fiber, pigment
H: 42 cm
ITZIKOVITZ COLLECTION
PHOTOGRAPH BY BRIGITTE CAVANAGH, 2017.

Painted double bells have been found among the instruments of Mangbetu royal musical ensembles (see fig. 17.14). Such a bell was played solo to accompany praise-singing or produce cadences for choreographies, or it was rung in combination with other bells to contribute to the tempo of an orchestral ensemble. In a processional performance, it would have been held with its mouths facing upward; when played sitting down, it was rocked side to side to alternate its tonal density when struck with a mallet.

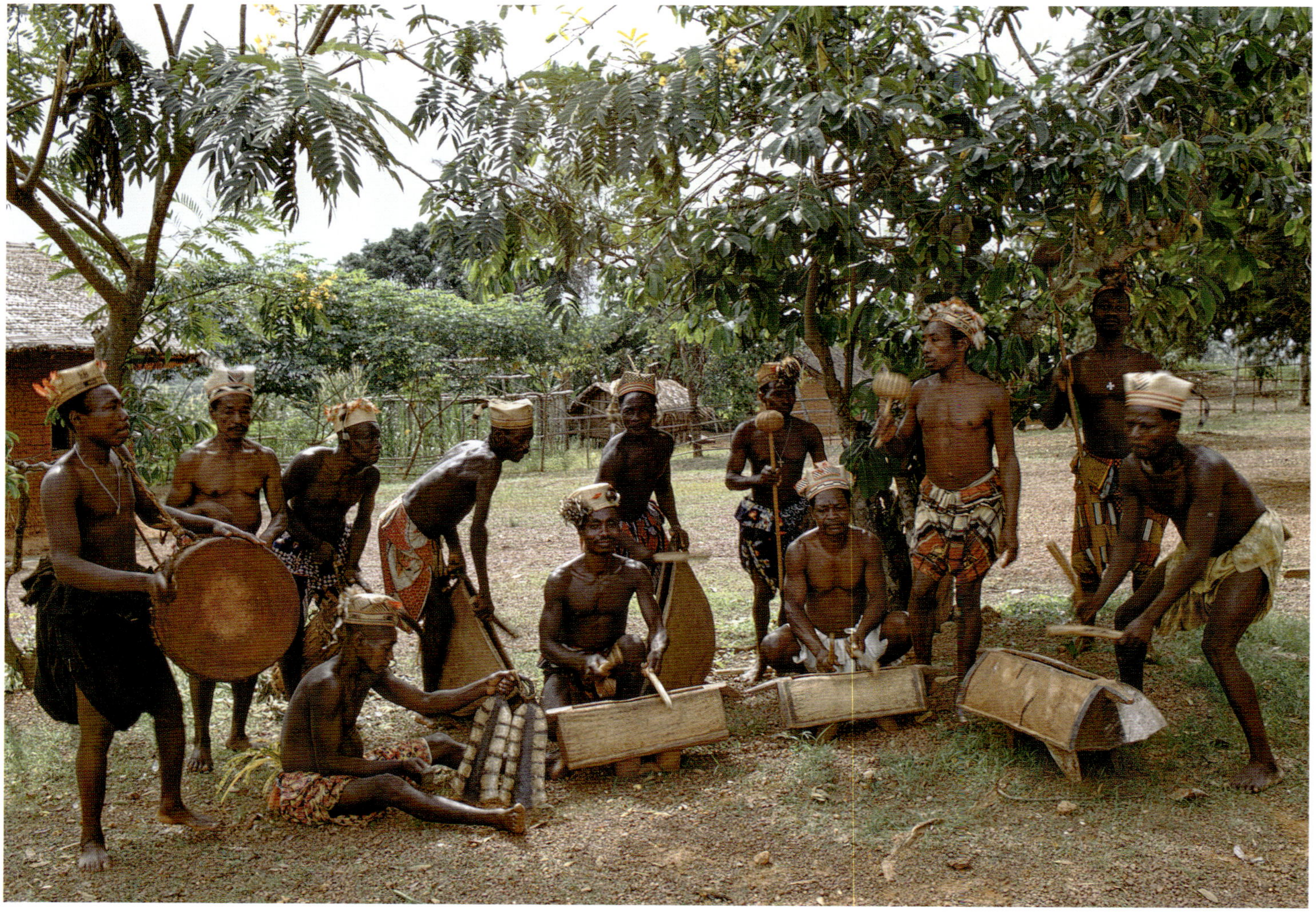

Handheld double bells of Ewe peoples of Togo (fig. 17.15) are commonly found in musical ensembles that include multiple drummers. Throughout west Africa, this style of instrument—with a large bell sharing a handle with an angled smaller one—is familiar in ceremonial ensembles of Akan people of Ghana that include royal drums. With minor variations, similar instruments are played in world music ensembles to keep time. Industrial steel versions may be attached to percussion drum kits or played held in hand, and similar bells are also manufactured as Latin American percussion instruments, again suggesting aesthetic links of the Black Atlantic.

It is feasible to characterize the sound of a bell as having a readily recognizable timbre, or "sound color." Striking an iron bell with a stick or with an iron beater may result in a crisp sound (depending on the bell's qualities) that "cuts through air" with audible sharpness. It is partly because of such qualities that bells often serve as rhythmic instruments. In the case of a triple bell from Nigeria (fig. 17.16), the overall elegance of the object—its elongated principal bell and formally balanced smaller bells at the crown's end—suggest a sound quality for an instrument designed primarily as a three-tone idiophone that is a prestige object in its own right.[6]

An emphasis on bell form in correspondence with other aesthetic attributes leads to instruments with remarkably specific ritual functions. Two Yorùbá bells (see figs. 12.4, 12.5) are extraordinary in that their visual appearance indicates that their sound range is limited to the lower portion of the instruments' length. They are "capped" with handles elaborating metalsmithing tools, suggesting that these ritual bells symbolically addressed the important role of blacksmiths. It is pertinent to note that among Yorùbá, certain bells and other art forms were and still are maintained in the context of altars. These bells may have belonged to a Yorùbá ruler or a metalsmiths' guild, and part of their symbolic discourse may have been to honor Ògún—warrior, king, and deity of iron and smithing (see Drewal and Pemberton 1989, 224–25, 227).

17.14 This staged photograph shows Mangbetu court musicians playing with wooden slit drums and leather-headed percussion instruments as well as a painted double bell (see fig. 17.13).
PHOTOGRAPH BY ELIOT ELISOFON, MEDJE VILLAGE, DEMOCRATIC REPUBLIC OF THE CONGO, 1970. ELIOT ELISOFON PHOTOGRAPHIC ARCHIVES, EEPA EECL 5648, NATIONAL MUSEUM FOR AFRICAN ART, SMITHSONIAN INSTITUTION

17.15
Artist unknown (Ewe peoples, Togo)
Double bells (*gangokui*), mid-20th century
Iron
H: 13.5 cm; 37 cm
PRIVATE COLLECTION

Gangokui are double bells forged in various sizes and with considerable tonal ranges. Early most mornings in Ewe communities, the bells can be heard, struck with wooden sticks by elder women as they clear malevolent spirits from pathways leading to market centers, water sources, and agricultural fields.

17.16
Artist unknown (Yorùbá peoples, Nigeria)
Bell, early 20th century
Iron
H: 85 cm
PRIVATE COLLECTION

At the heart of this complex instrument, comprised of seven idiophones, a resonant call and response is at play. A large central bell is flanked by two smaller bells that, when struck with a mallet, activate tiny clappers dangling from the mouths of the slender conical rattles emerging from four corners at the center. The blacksmith designed every bell and rattle to produce variable tones, perhaps to suggest, in performance, the tonality of spoken language.

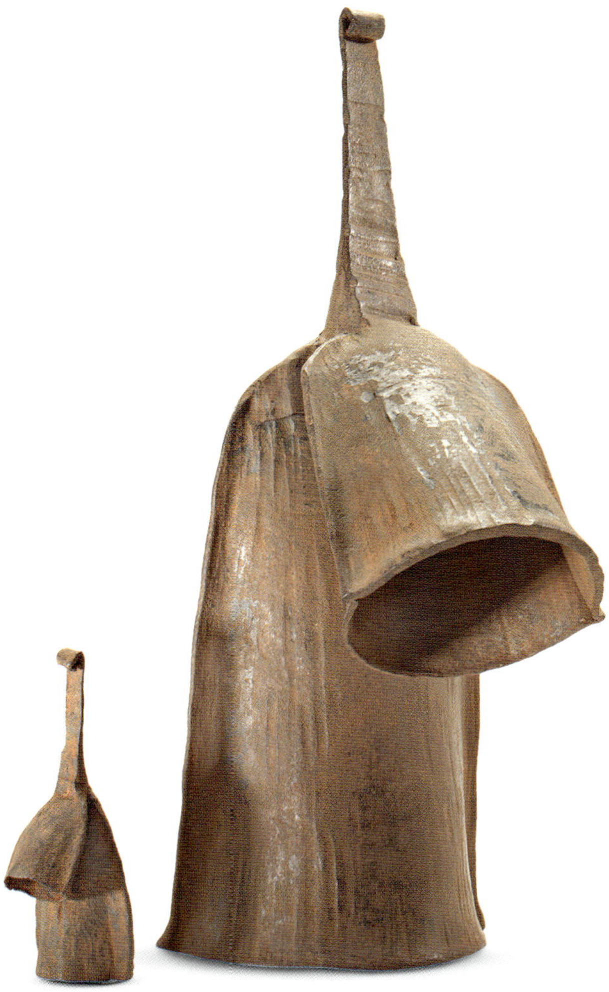

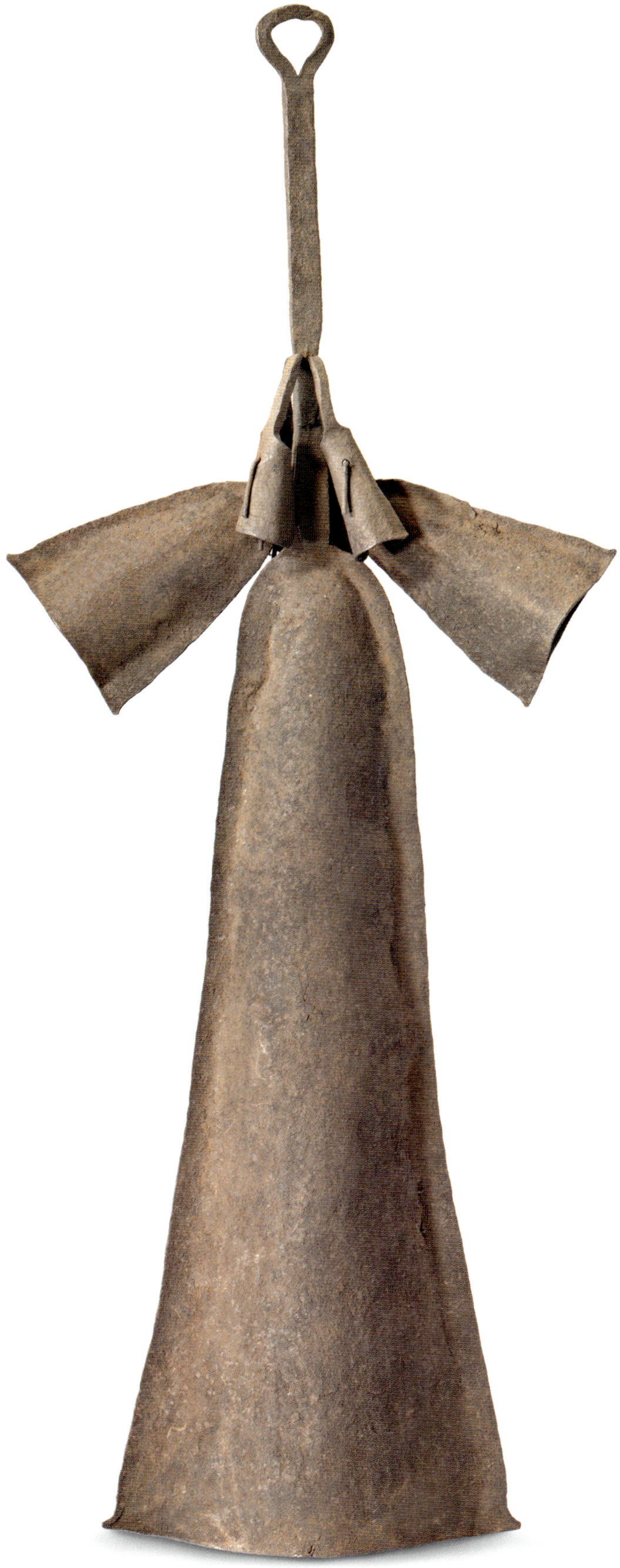

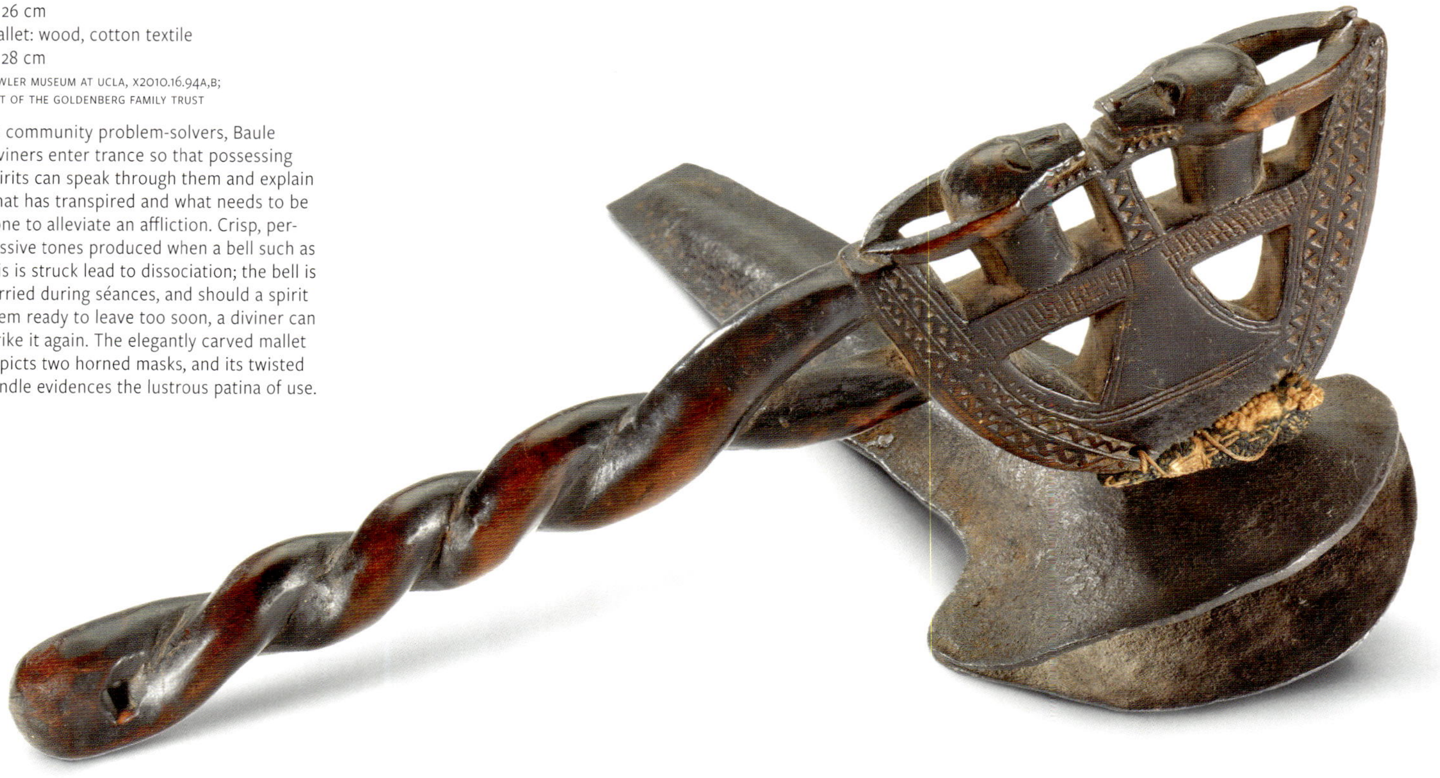

17.17
Artist unknown (Baule peoples, Cote d'Ivoire)
Divination bell (*lawre*) and mallet (*lawre waka*), late 19th to early 20th century
Bell: iron
H: 26 cm
Mallet: wood, cotton textile
H: 28 cm
FOWLER MUSEUM AT UCLA, X2010.16.94A,B; GIFT OF THE GOLDENBERG FAMILY TRUST

As community problem-solvers, Baule diviners enter trance so that possessing spirits can speak through them and explain what has transpired and what needs to be done to alleviate an affliction. Crisp, percussive tones produced when a bell such as this is struck lead to dissociation; the bell is carried during séances, and should a spirit seem ready to leave too soon, a diviner can strike it again. The elegantly carved mallet depicts two horned masks, and its twisted handle evidences the lustrous patina of use.

Returning to the sound quality of iron idiophones, their musicality can easily be "reversed," as repetitive metallic sounds can be particularly aggravating. Bells and particularly rattles may be struck or sounded to invite spiritual intervention, and "annoying through sound" is a strategy sometimes used by diviners for problem-solving as they seek to gain and sustain the attention of tutelary spirits, as explained to me by the Southern Lunda diviner and village chief Mr. Chitofu Sampoko.[7] Spirits are often "difficult," and while a bell's sounds may be initially pleasing, after a few tries a diviner may abruptly rattle or beat on an instrument to coerce the spirit's attention through irritation. Once received, the sounds are maintained by the diviner or an assistant to make sure the spirit is not distracted (see Roberts and Roberts 1996, 194). For this reason, some divination and other ritual implements, for example *mboko* gourds of Luba and Songye peoples of the DRC, feature attached rattling iron elements (see fig. 15.26).[8]

Among Baule peoples of Côte d'Ivoire, when diviners hear the regular striking of a bell (fig. 17.17), they "go into a trance state… in private but they carry the gong [bell] and mallet in performance to use if they start to lose trance during the session, or wish to deepen the trance" (Vogel 1997, 229). Writing on cosmology and ritual in western Cameroon, Viviane Baeke points to the use of different bell types through a photograph of an initiation leader holding a bell in front of an initiate's forehead; its sound "boosts the efficacy of the officiating member's pronouncement" (1995, 57–91). Use of percussive instruments (fig. 17.18) by initiates themselves is illustrated in photographs taken by Allen F. Roberts in 1965 among Kabre people in Tchitchao, Togo (fig. 17.19), documenting initiates "climbing up the side of the steep hill in a duck-walk, a ritual axe in one hand and a clapperless bell [*ekpande*] in the other. Every once in a while they would stop, and they would flip the gongs up in the air to strike them against iron rings on their thumbs" (pers. comm., 2016). Research conducted among the Kabre in 2008 and 2010 by Tom Joyce and Steven Feld shows that they continue to forge and use *ekpande* in initiation contexts (fig. 17.20; Joyce pers. comm., 2018). Baeke's, Roberts's, and Joyce and Feld's accounts are helpful in illustrating the centrality of sounding forms during critical ritual instances when engagement with spiritual realms is necessary. In the Cameroon example, bells offer a form of the "supernatural bringing of clarity" to meaningful speech; in Togo, they are held in the hands of initiates transitioning between stages of ritual, moving towards new lives of achieved social status.

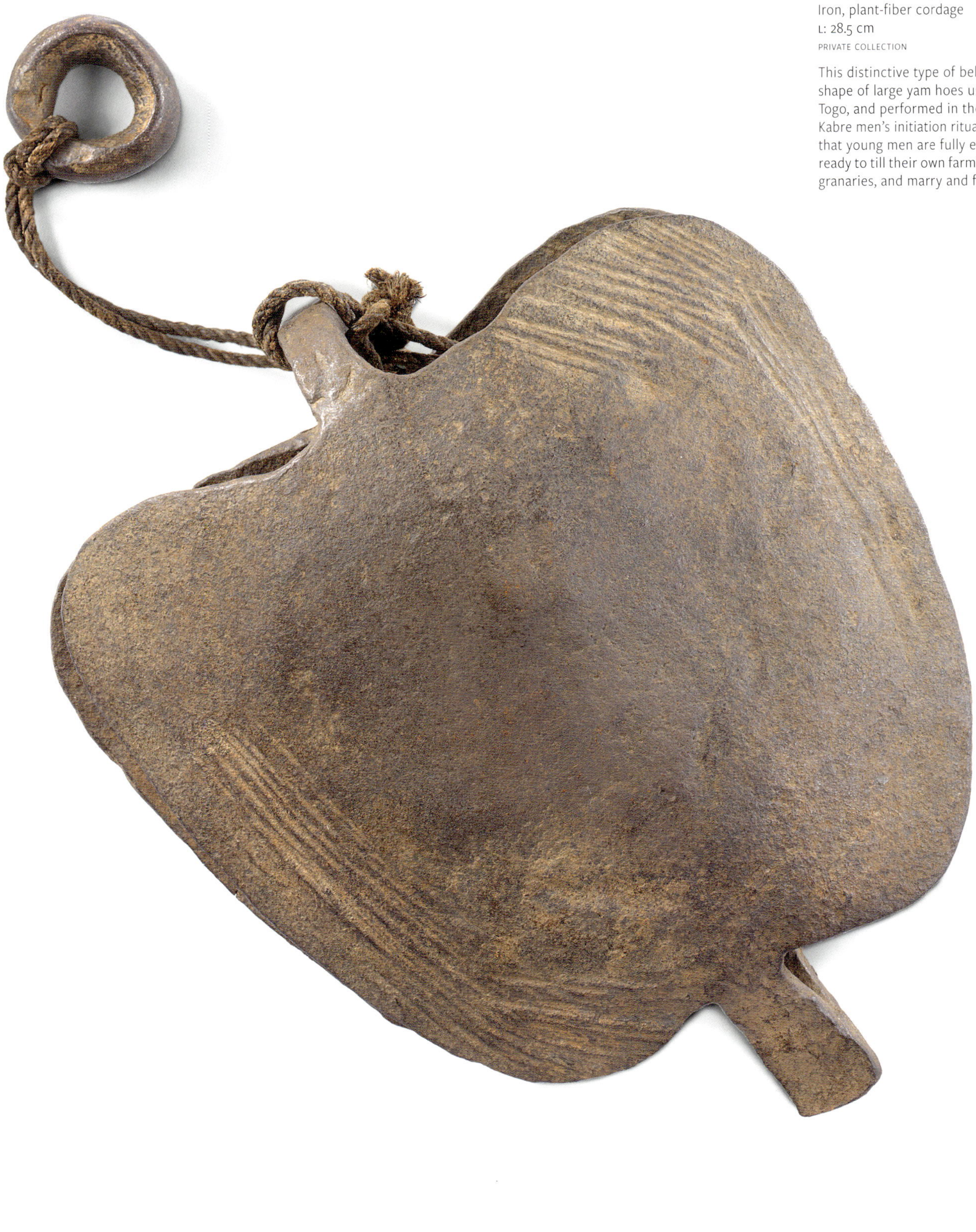

17.18
Artist unknown (Kabre peoples, Togo)
Bell and percussion ring (*ekpənde*),
mid-20th century
Iron, plant-fiber cordage
L: 28.5 cm
PRIVATE COLLECTION

This distinctive type of bell is based on the shape of large yam hoes used in northern Togo, and performed in the last phase of Kabre men's initiation rituals to demonstrate that young men are fully educated adults ready to till their own farmland, fill their own granaries, and marry and found families.

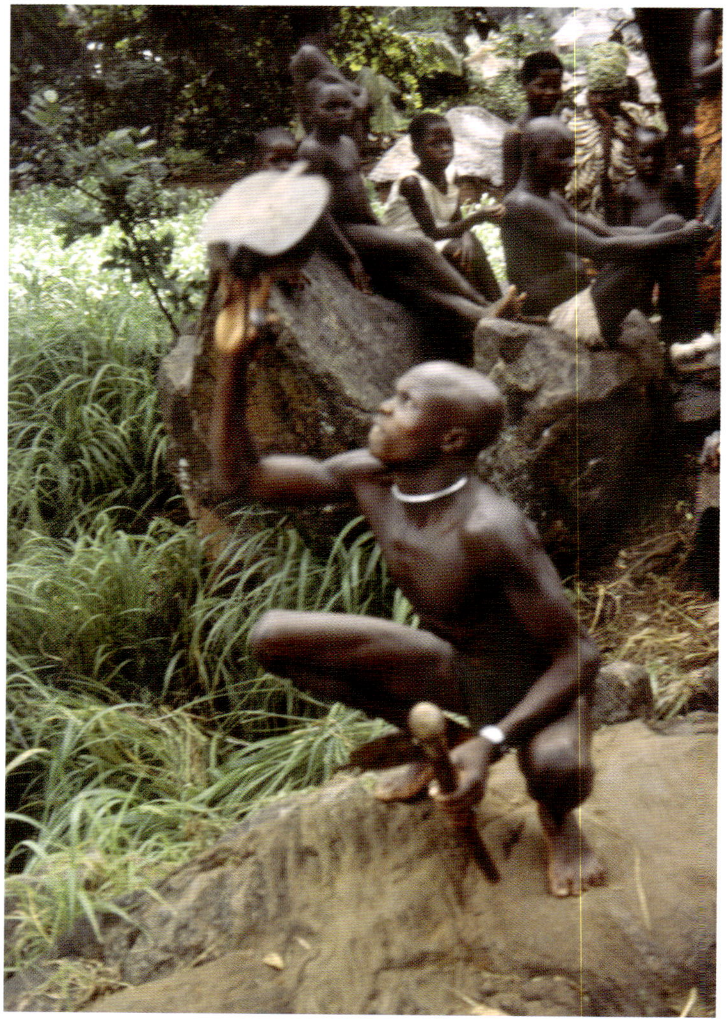

17.19A–B Kabre boys undergo a ten-year passage to manhood through phased initiation rituals that juxtapose spectacles emphasizing individual abilities, as in wrestling tournaments and dance, with carefully controlled deference to elders and authority more generally. Here, young men in the last phases of an arduous initiation climb a sacred mountain in a pose of submission that conveys confidence to the community that as adults, they will be wise and trustful masters of ancestral knowledge. Pausing as they climb a sacred mountain in single file, each young Kabre man holds a ritual axe in his left hand while with the right he tosses an *ekpande* (see fig. 17.18). A spectacle of prowess and arcane power results as the man proves possession of the resolve and humility necessary to marry and establish a family.

PHOTOGRAPHS BY ALLEN F. ROBERTS, TCHITCHAO VILLAGE, TOGO, 1965.

17.20
When a young Kabre initiate performs using an *ekpande*, he tethers it loosely to his right wrist with a thong. The bell is then energetically released in a downward spiral that recoils the iron form with a snap (like a yoyo) back into his open palm, where it strikes an iron ring on the thumb with a resounding tone. Although initiates' dress has changed over time, the tradition is ongoing.

PHOTOGRAPH BY TOM JOYCE, KUWDÉ, TOGO, 2010.

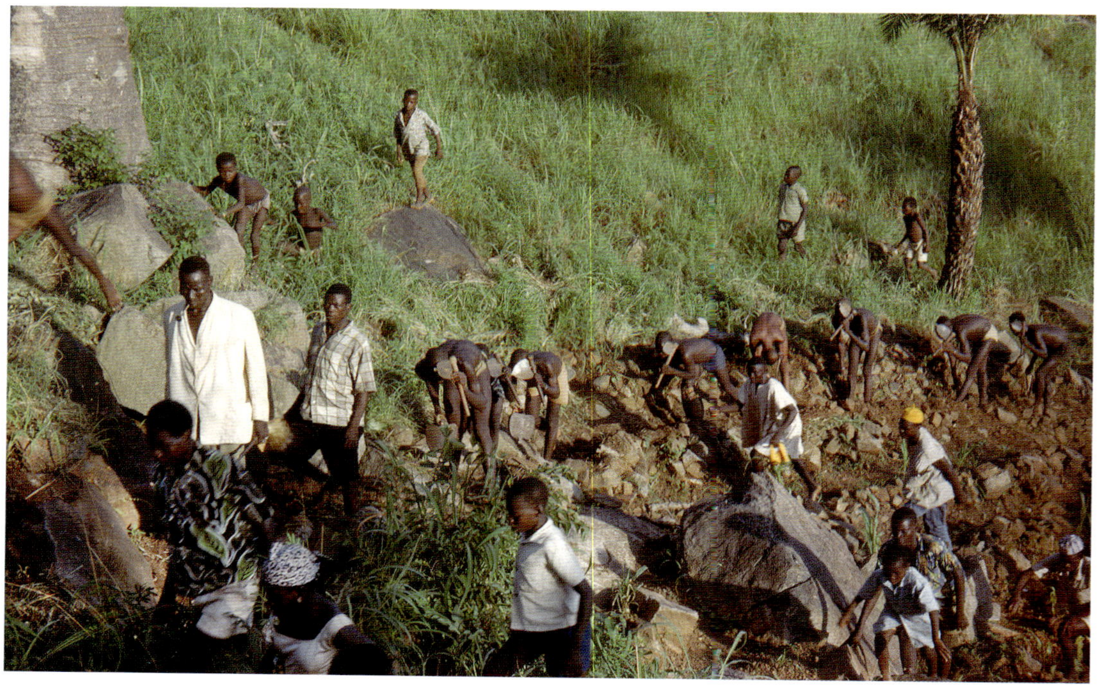

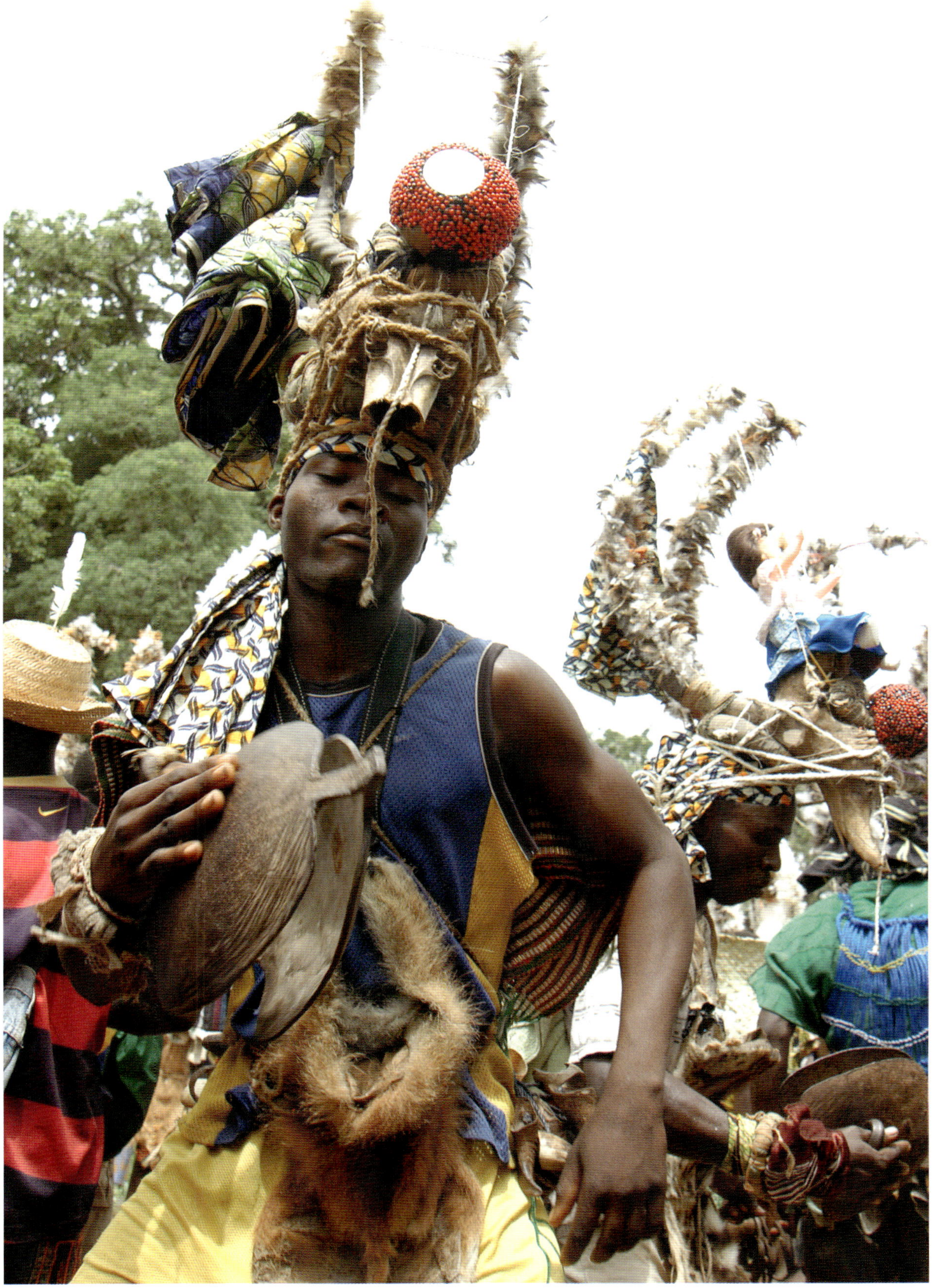

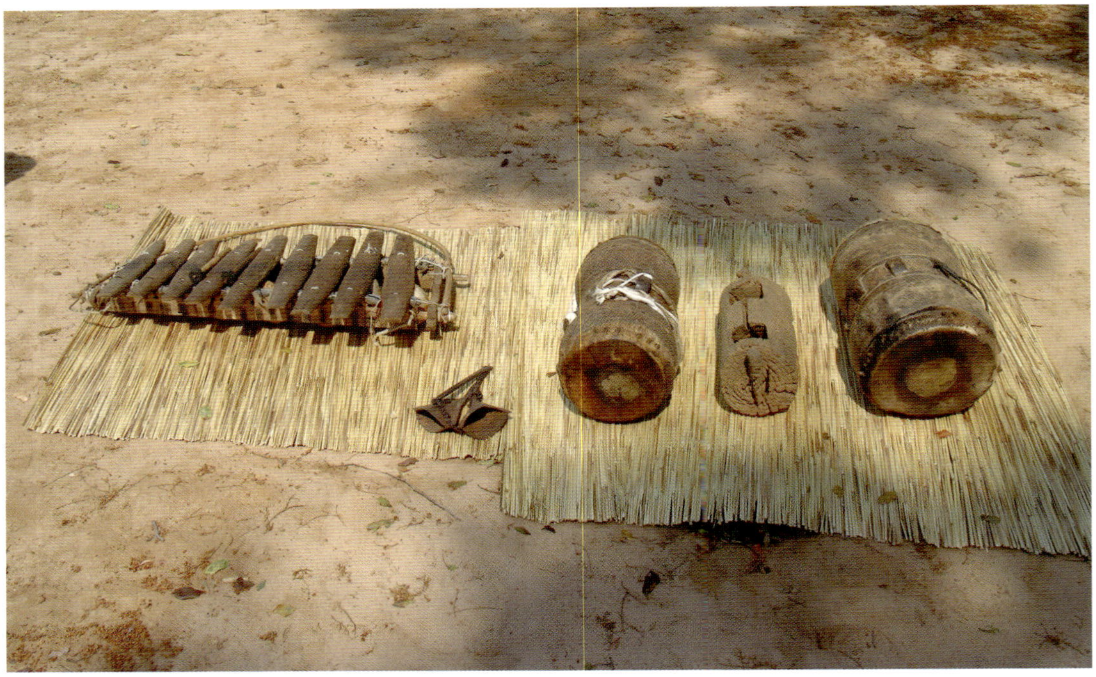

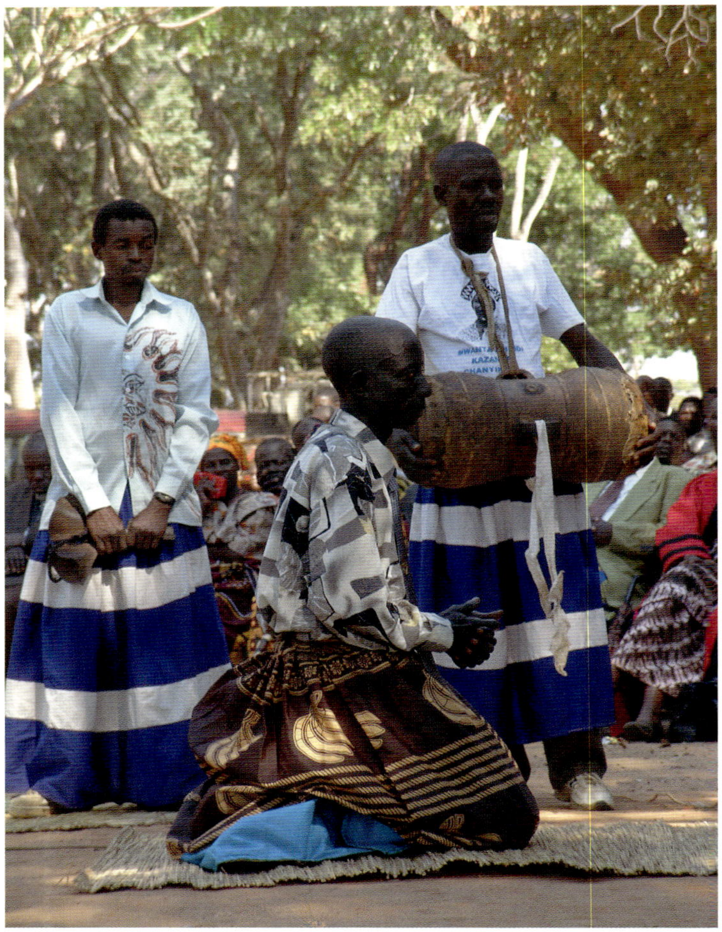

17.21A–C Instruments kept within the royal treasury of Senior Southern Lunda Chief Ishindi are maintained by musicians who are also titled members of his court. Pictured in fig. 17.21A, from left to right: a xylophone, an iron double bell, a double membrane drum, a slit drum, and a double membrane drum.

In fig. 17.21B, Lunda musicians hold a double bell and drum during the investiture of a new chief (kneeling) with Senior Chief Ishindi presiding (not pictured). The iron double bell is struck at the ritual moment when the candidate is anointed with white clay to signify his change of status. Bells are amongst the most sacred instruments of the Lunda.

Senior Chief Ishindi, in fig. 17.21C, is seated on his palanquin during an annual confirmatory ceremony with musicians taking part in the procession. Drummers provide tempo accented by a struck bell to establish a pace for the procession of hundreds of people.

PHOTOGRAPHS BY MANUEL JORDÁN, NORTHWESTERN ZAMBIA, 2006.

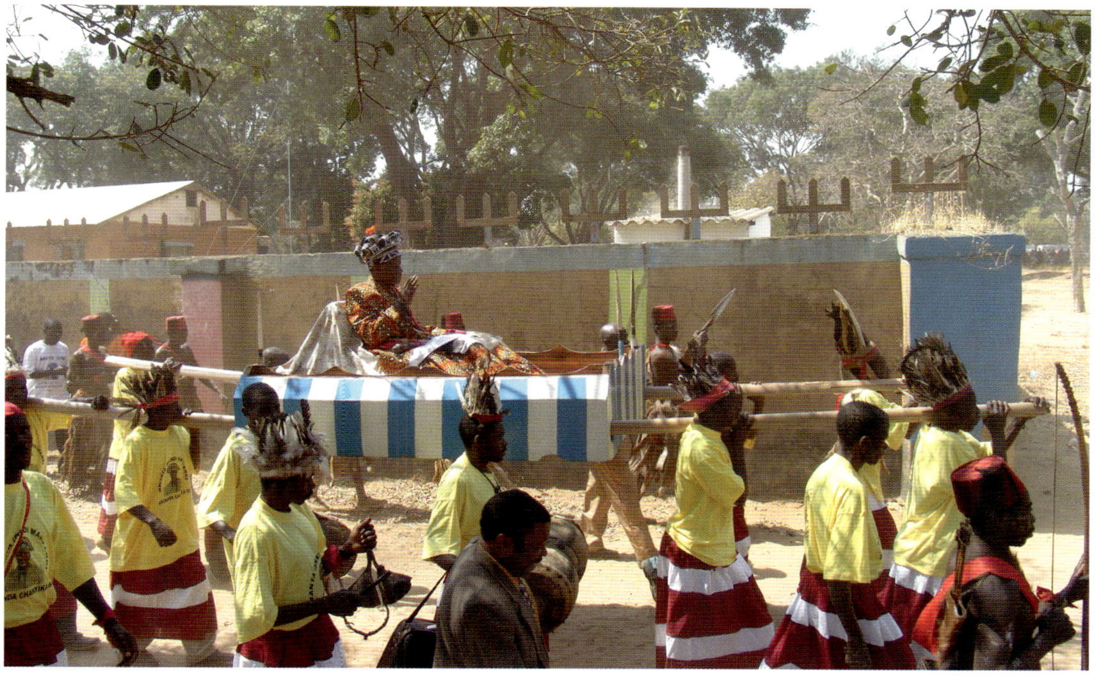

SOUNDING FORMS IN NORTHWESTERN ZAMBIA

In northwestern Zambia in 2006, Senior Southern Lunda Chief Ishindi graciously facilitated my interviewing court musicians at his palace, where they showed me a set of royal instruments kept within an enclosure inside the grounds (fig. 17.21; see also fig. 6.15). The musicians brought out a xylophone, two double membrane drums, a small slit drum, and an iron double bell. The slit drum and double bell were pointed at as I was told, "These two are the most important ones." Translated from Lunda to English, the words that were shared in explanation were "oldest" and "most sacred." The bell was discussed in practical ritual terms, whereas the slit drum was said to be "rarely used." Knowing the extraordinary material and symbolic attributes of iron objects and the complex ritual processes involved in creating them for Lunda-related peoples, it came as no surprise to me that the double bell was especially valued.

My trip coincided with the chief's annual confirmatory ceremony. For his royal procession, he was carried on a palanquin around the village grounds accompanied by members of the court, warriors, and musicians. Most significantly, the performance happened at a time when an Angolan Lunda delegation was visiting Chief Ishindi's court in the company of their own Lunda senior chief, for the purpose of holding investiture rituals and "crowning" nine new Southern Lunda chiefs from Angola.

A relatively private setting for the investitures was achieved by setting a table in front of the entrance to the palace, surrounded by chairs for Angolan and Zambian chiefs. Court performers and dignitaries encircled the main event area, which was "open" towards the back to the general public. Two double membrane drums and the bell were used to bring Chief Ishinde to the grounds and to escort each of the new chiefs to individually face him for their crowning rituals. The drummers and the bell musician stood to the right of the senior chief and only played at select instances during Ishindi's speech, including, significantly, when the new chiefs were anointed with white clay in the most sacred and transitional part of the ritual. It is pertinent to note that the musician's double bell featured a handle in the shape of an iron axe blade. Playing into the symbolism of the chief's royal sword, this was direct affirmation through the power of the weapons that the chief's voice or mandate is law.[9]

On the day after the investitures, Chief Ishindi was carried once more through the grounds adjacent to the palace on his royal palanquin, his wife on another behind him, and the procession accompanied by court officials and hundreds of the chief's people. Directly to his right, again, walked the two drummers and double bell musician, this time influencing the pace of the event, with the bell informing all that the chief was engaged in the annual ceremony.

The primary role of early nineteenth- to twentieth-century iron musical instruments such as Chief Ishindi's royal bell remains vital, as the instrument's vivid sounds recall those made generations ago by earlier chiefs who fulfilled similar courtly functions with the exact same heirloom bell. Yet for all peoples exposed to new ideas and technologies, the instruments, materials, and concepts of musicality continue to be responsive to change. Nowadays in northwestern Zambia, celebratory occasions may be accompanied by recycled tire rims played like bells, bicycle chainrings as rasps, and metal pots and pans used as drums or struck as idiophones. In all, iron instruments and their "voices" retain their crucial context-specific identities and character while they contribute to a broader and continually retextured sense of musicality that reflects daily and ceremonial life.

INTERLEAF

O

RITUAL ACTIVATION WITH SOUND: CHAMBA OF THE MIDDLE BENUE, NIGERIA*

Marla C. Berns

Among Chamba, forged iron objects were costly, labor-intensive to reproduce, and essential to many ritual activities, together with medicines and sacrifices. Chamba and Verre smiths produced sonorous instruments including large double bells that were a prerogative of chiefs (see fig. 17.10), rattles held in the hand or worn on the leg when dancing (fig. 0.1), and double hand bells to accompany masquerades, as well as less public paraphernalia of men's and women's associations.

Among the most impressive ritual forgings are iron rattles that vary greatly in dimension and consist of similarly shaped clappers suspended from an iron ring, sometimes arranged symmetrically on either side of a central bell (fig. 0.2). The size and sort of rattle varies according to the ritual association (*jup* in Chamba Daka; *vom* in Chamba Leko) to which it belongs. In performance, rattles, as the "bones of the *jup*," are "shaken rhythmically, in different combinations of two motions—up and down from the elbow, and twisting from the wrist" (fig. 0.3). Their "uncannily hollow, loud clanging" warned away nonmembers from performances. Such persons averted their gaze when rattles were brought to funerals (Fardon 2011, 386–87). A fully initiated Mapeo Chamba man might be buried with a clapper removed from the ring of his *jup*'s rattle, and when his grave was later reopened to remove his skull, the clapper too would be taken, and returned to the ring. Each *jup* had distinctive practices and instruments, and generalization is difficult; however, as elsewhere in northern Nigeria, "iron was pervasively linked to the powers of death" (ibid.; see also chapter 13, this volume).

Women's initiation associations (*jeem* in Chamba Daka) possessed iron paraphernalia too, including, sometimes, miniature versions of rattles seen only by the eldest women. Pairs of small hoe blades (*jeem ban*, *ban* meaning "hoe" in Chamba Daka) were struck together to accompany songs during a woman's wake and at harvest festivals (figs. 0.4, 0.5). Leaders of *jeem* carried upright all-iron, long-handled staffs quite distinct from functional tools but said to be elaborated versions of ancient hoes (fig. 0.6). Their form culminates in an asymmetrically shaped spade blade that served as insignia of office (*baan daka*). Augmenting the hoe blade, handle, and coil-wrapped collar is a lightning arrow that resembles the same device on the tip of the ceremonial sickle (*lama*) of some male ritual masters in Mapeo. The incorporation of hoe blades into activities of *jeem* linked women, with iron as an intermediary, to agricultural fertility and reproduction generally.

* These notes on Chamba iron forgings are taken from Fardon (1990, 69–71; 2011, 384–89). The editors thank Richard Fardon for his generosity in sharing his research for this volume.

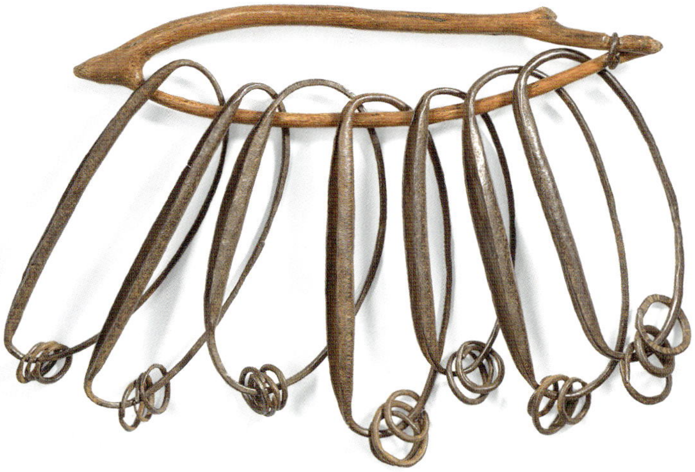

0.1
Artist unknown (Middle Benue region[?], Nigeria)
Rattle, 20th century
Iron, wood
L: 46.9 cm
PRIVATE COLLECTION

We do not have specific data on this particular type of rattle, but based on others in the region it was likely shaken in ritual contexts, when its rhythms may have announced the presence of nonhuman powers and provided accompaniment to ceremonial dancing. Its sound carried distances, potentially to enable communication.

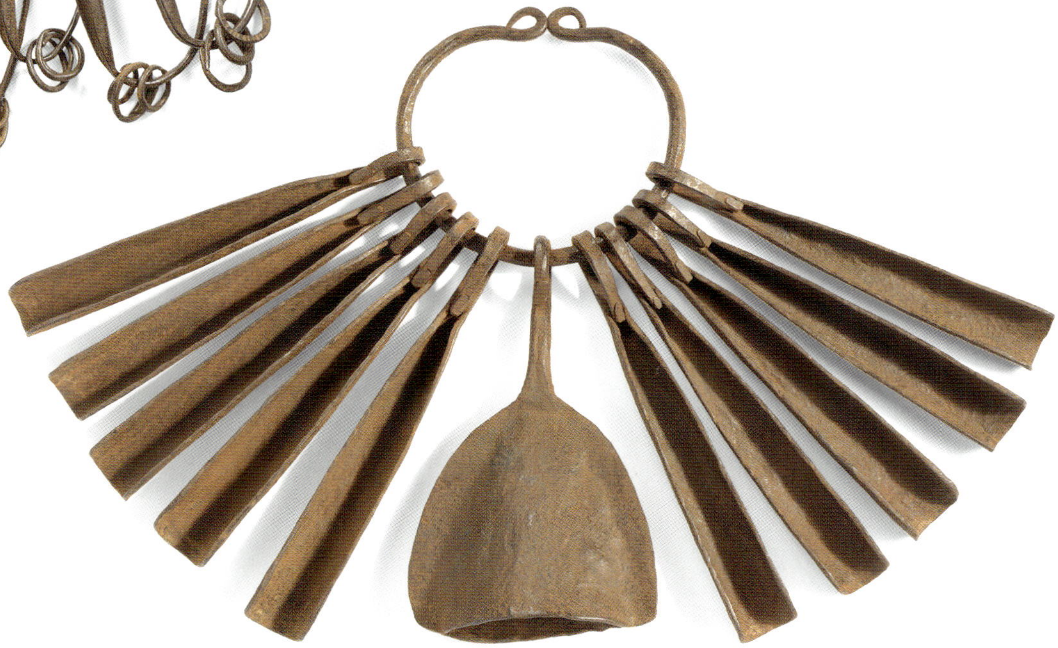

0.2
Artist unknown (Chamba, Verre peoples, Nigeria)
Rattle, 19th to 20th century
Iron
W: 55 cm
PRIVATE COLLECTION

Twelve trough-shaped rattles are hung on either side of a central bell on the ring of this large instrument, used to produce loud rhythmic sounds when held by its ring and thrust up and down during Chamba and also neighboring Verre rituals. Such instruments were kept by members of male ritual associations and used to announce their rituals and warn off noninitiates.

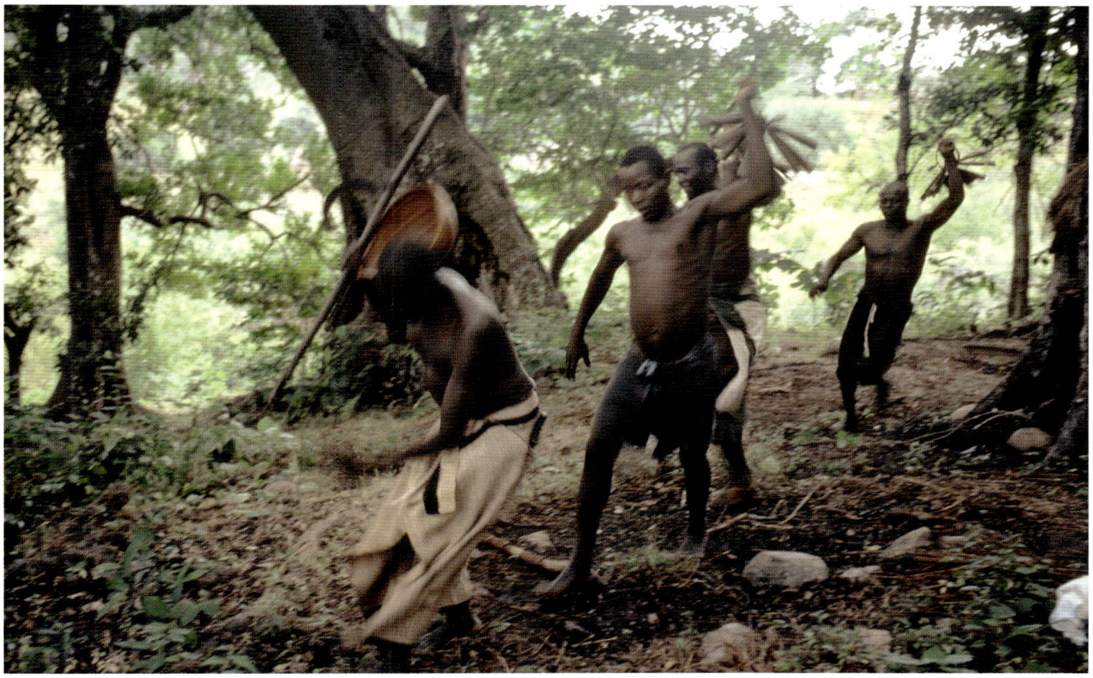

0.3 In advance of the Jup Kupsa festival, members of a male Chamba ritual association (*jup*) shake large rattles (see fig. 0.2) as they are led by a ritual specialist (*nwaan*) holding a lightning-pointed sickle (*lama*).
PHOTOGRAPH BY RICHARD FARDON, MAPEO, NIGERIA, 1976.

0.4
Artist unknown (Chamba, Verre peoples, Nigeria)
Ritual hoe blades (*jeem baanbu*; s. *jeem baan*), 20th century
Iron
H: 10.5 cm; 22.5 cm
PRIVATE COLLECTION

Pairs of these small hoe blades were struck and rubbed together as rhythm instruments during performances of Chamba women's secret associations (*jeem*). Each was forged with a short handle finished with a spiral finial to facilitate handling.

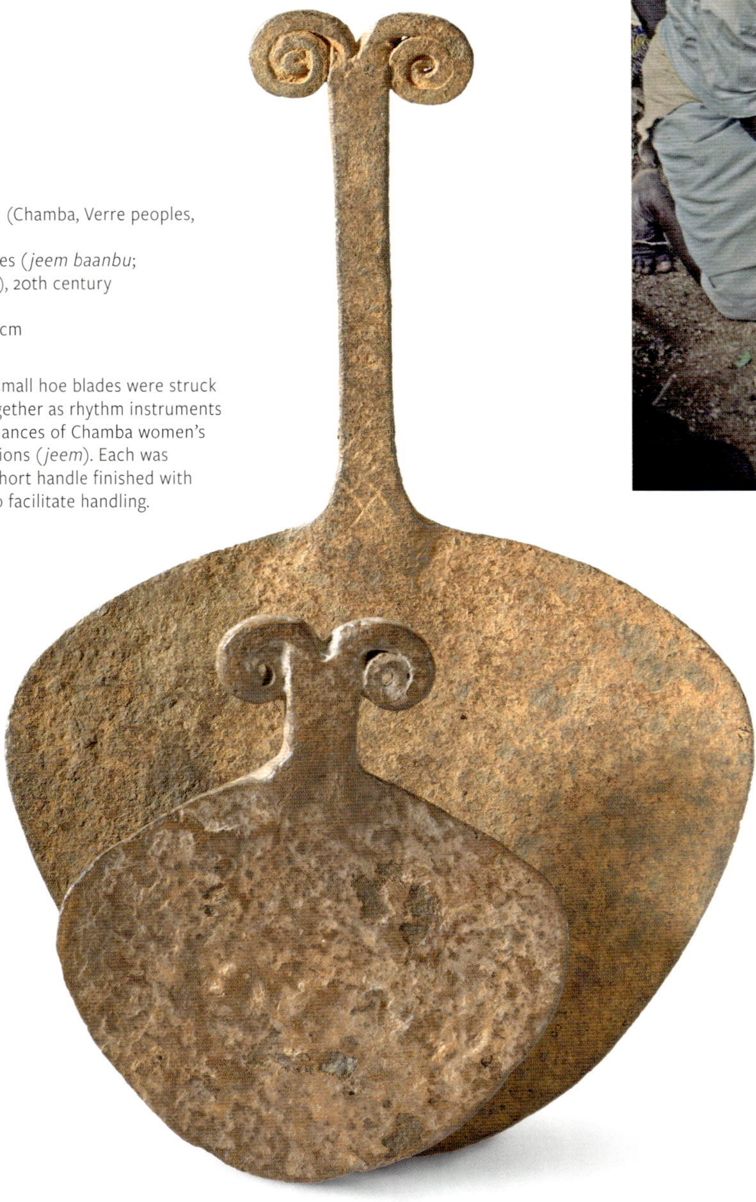

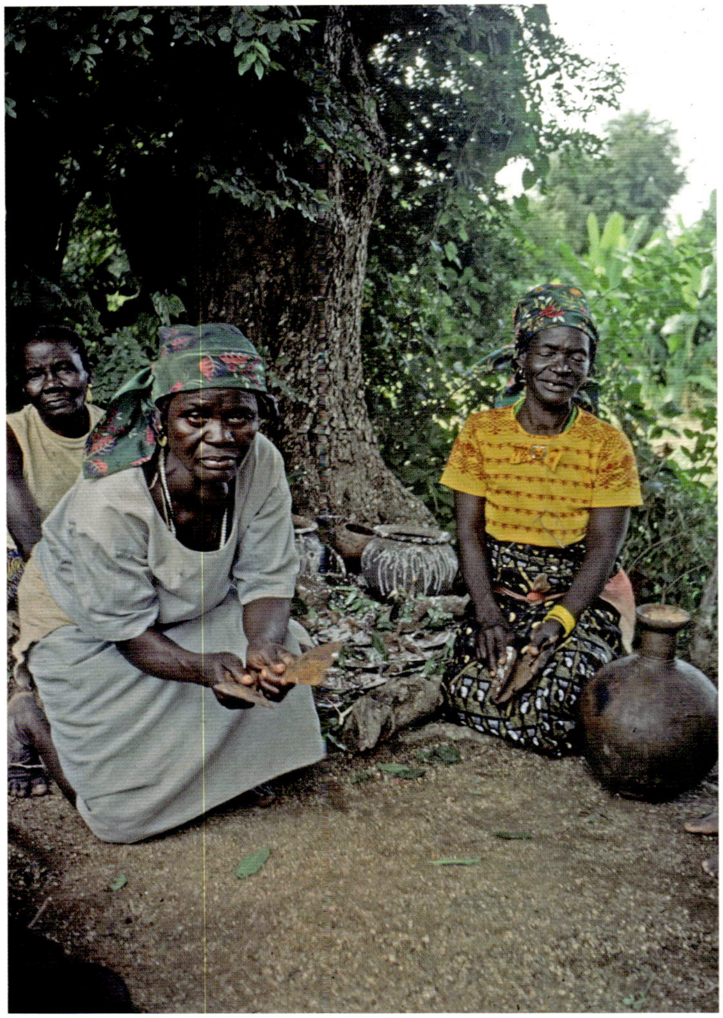

0.5 At a women's initiation society (*jeem*) meeting during a harvest festival, the head of the association holds *jeem banbu* (see fig. 0.4).
PHOTOGRAPH BY RICHARD FARDON, NIGERIA, 1976.

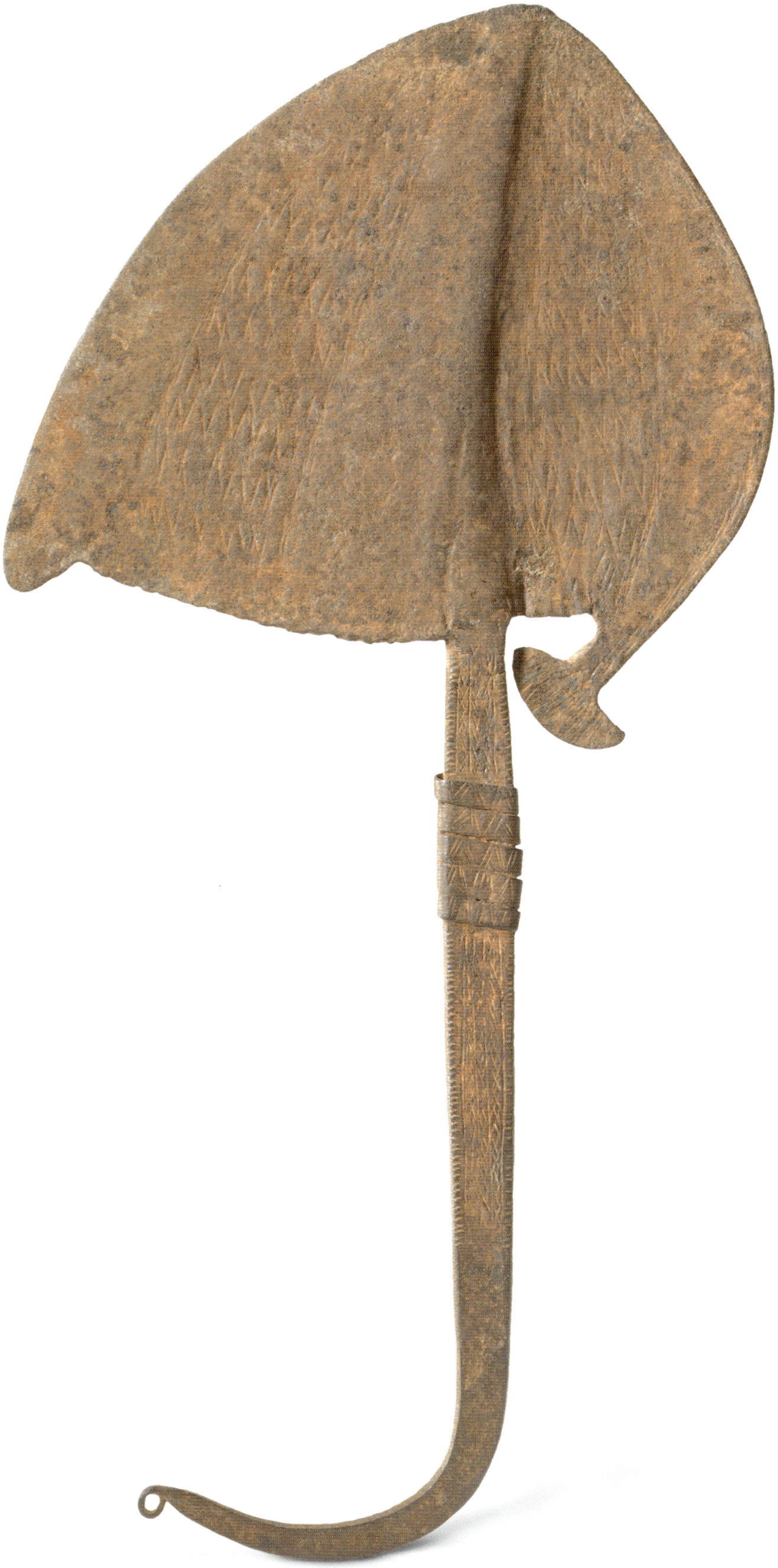

0.6
Artist unknown (Chamba peoples, Nigeria)
Insignia of office (*baan daga*), 20th century
Iron
H: 47 cm
COLLECTION OF MARK CLAYTON

This ceremonial hoe blade with long handle was forged in one piece with delicate flourishes at the edges of the blade's base and finely chiseled designs on its handle. These elaborations are evidence of the blacksmith's skill and of the transformative processes that turn a basic tool into an insignia of office.

CHAPTER 18

FORGED IN THE MARKETPLACE: MIGRANT BLACKSMITHS IDENTITIES IN BAMAKO

Isaie Dougnon (*translated from the French by Allen F. Roberts*)

If migration to cities in Mali by young farmers from Dogon and Bamana communities has been the subject of many social science studies, that of blacksmiths as a specific professional group remains overlooked.[1] Historically, like other artisans, many blacksmiths have been sedentary while others have been mobile, seeking sites to obtain iron ore or villages and hamlets where their services are needed. In the 1950s and 60s, the growth of cities profoundly changed their patterns of movement: first, because farmers and hunters began to settle in urban areas and their peripheries, and second, because these places offered blacksmiths new sources of scrap iron and constant clientele (cf. Osborn 2009). Furthermore, postcolonial governments intensified the modernization of agriculture by creating workshops to create modern agricultural tools (Jonckers 1979). To avoid marginalization, many blacksmiths have adapted to these new contexts by moving their workshops to urban marketplaces (fig. 18.1). Bamako, as the political and economic capital of the Republic of Mali, is the primary destination for Dogon and Bamana smiths, whose original forges are often hundreds of miles away. Urban relocation offers access to tools, inexpensive recycled materials, and a more diverse clientele than can be found in rural settings (fig. 18.2).[2]

My field research among blacksmiths who moved to Bamako as early as the 1970s and as recently as the present day pursued multiple questions.[3] How have rural smiths transferred their knowledge and ritual practices associated with ironworking to urban settings? How have they adapted to marketplaces where others may work with iron but are not blacksmiths as understood through Dogon and Bamana cultural terms? And how do the social identities of blacksmiths change when they enter into urban commercial production? Six key points emerged that define relationships between market and forge:

(1) There are two groups of migrant blacksmiths—those who have now settled in Bamako for decades, and young people who come to the city to work as apprentices on a seasonal basis.
(2) Despite now dwelling in Bamako permanently, certain blacksmiths maintain more traditional workshops in their home villages. In such cases, the two forges nourish each other in technological and symbolic ways that revalorize blacksmiths' cultural traditions in contexts of migration.
(3) Migration changes social relations among blacksmiths themselves and between blacksmiths and non-blacksmiths regarding the disappearance of matrimonial prohibitions, options for other professions, and like factors associated with Dogon and Bamana cultural practices.
(4) Transfer of forges to Bamako marketplaces reinforces individualism, entrepreneurial spirit, and competition among immigrant blacksmiths, as well as between them and non-blacksmiths.
(5) Exercising iron arts in marketplace circumstances transforms technical knowledge and beliefs.
(6) Immigrant blacksmiths conserve practices tied to mystical knowledge that is not contradictory to earlier Islam as practiced in Mali, such as certain therapeutic functions and social mediation.

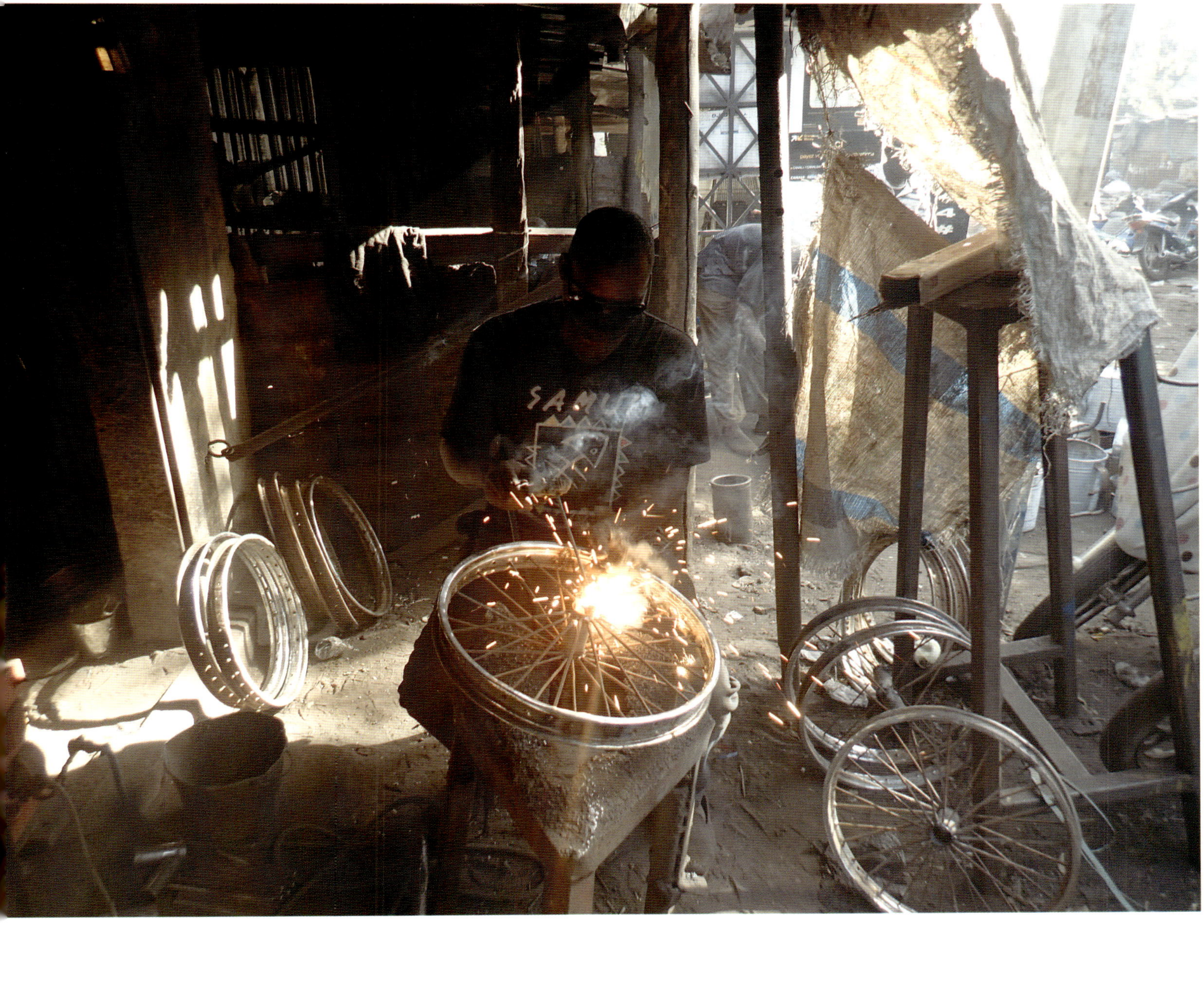

18.1 At his forge in the Valley of Blacksmiths, a smith welds cold-shaped parts to manufacture wheel rims for rickshaws, an inexpensive mode of transportation mainly relied on by local businesswomen to transport merchandise. Metalworkers of Bamako are able to adapt earlier technologies to meet contemporary needs of the sort.
PHOTOGRAPH BY ISAIE DOUGNON, SUGUNIKURA MARKET, RACETRACK NEIGHBORHOOD, COMMUNE II, BAMAKO, MALI, 2017.

18.2 A young migrant apprentice in front of the forge of his master. Piles of recycled iron bars (foreground) will serve to create spare parts for carts, ploughs, and other equipment.
PHOTOGRAPH BY ISAIE DOUGNON, SUGUNIKURA MARKET, RACETRACK NEIGHBORHOOD, COMMUNE II, BAMAKO, MALI, 2017.

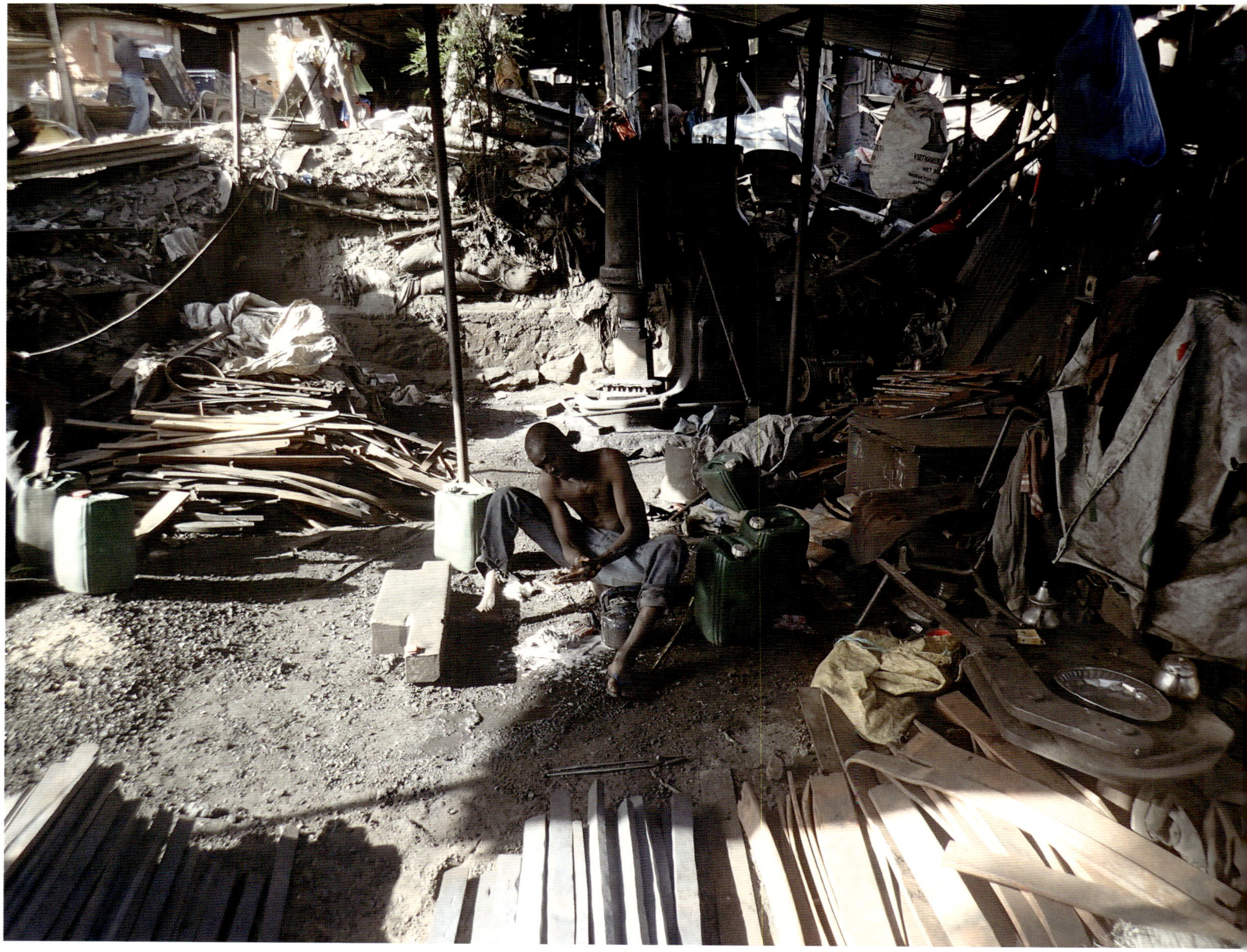

18.3 At his small-scale forge, Mr. Sekou Samassékou, originally of Dogon country, employs a Dogon apprentice originally from his same rural village.

PHOTOGRAPH BY ISAIE DOUGNON, SUGUNIKURA MARKET, RACETRACK NEIGHBORHOOD, COMMUNE II, BAMAKO, MALI, 2017.

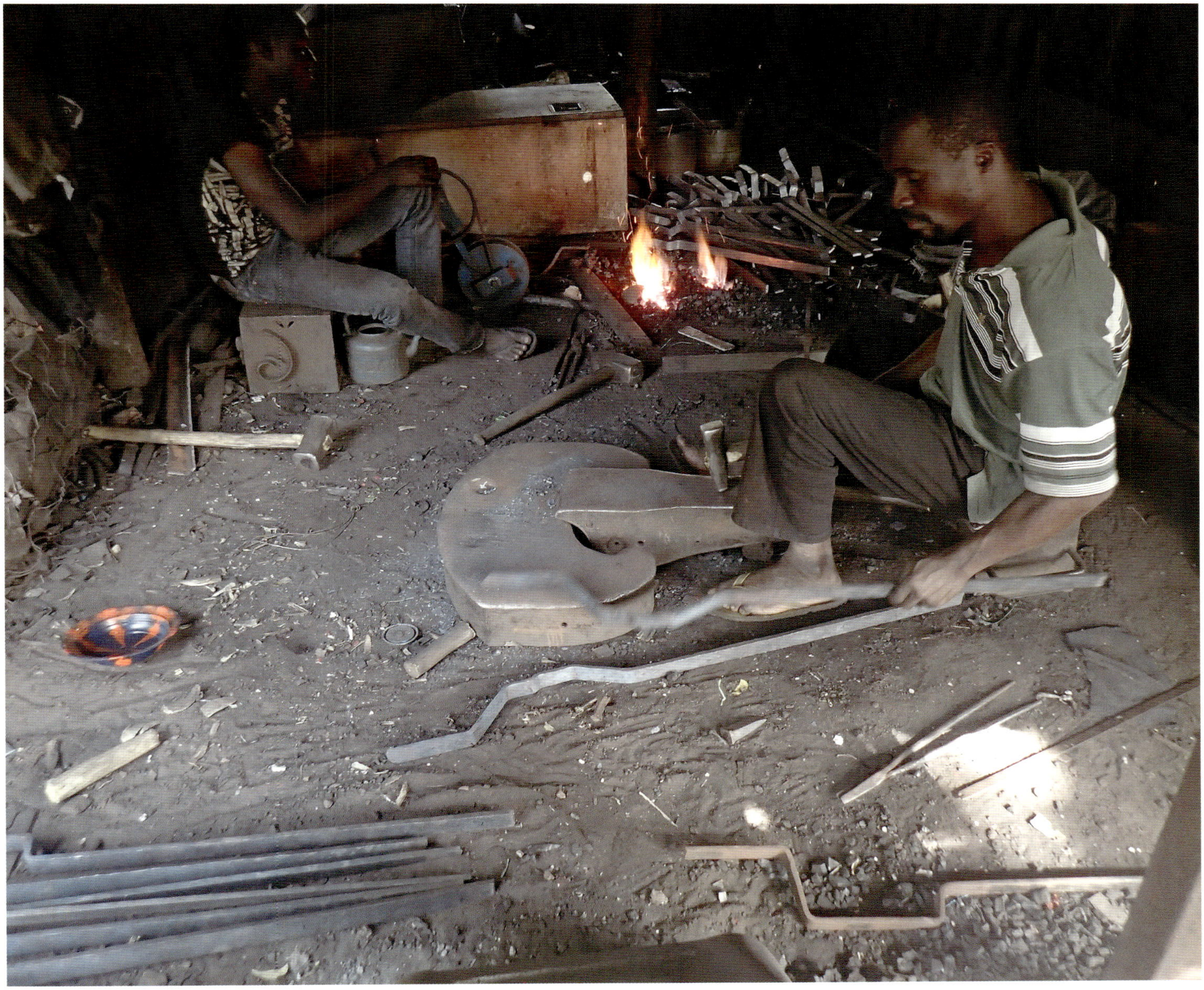

While the intensification of agriculture through mechanization and plowing brought on a period of disinterest in traditional ironworking in some areas of Mali, and especially in the "Inner Delta" of the Niger River north of Bamako, has produced some social marginalization of blacksmiths, urban migration and the relocation of forges to city marketplaces has not only permitted the profession to be sustained, it has made it dynamic in new ways.[4] The present-day migration of people from rural blacksmith families includes young people coming to the city seasonally to work as apprentices at the forges of Bamako's marketplaces (fig. 18.3). These young people want to learn and have a desire to innovate. A blacksmith brings to the city his centuries-old arts, and transfers new means and methods to the forge in his natal village. Migration revolutionizes ancestral models of ironsmithing, in other words. Younger generations of migrants adapt very well to circumstances of forging practices new to them, and such innovations bring renewed élan to iron crafts in both urban and rural settings.

Immigrant smiths understand that thanks to the combination of innovation and ancestral knowledge, they can attract an ever-increasing clientele of urbanites as well as European and African tourists. In effect, many people living in the city want blacksmiths' products, from those who are well-off to small-scale gardeners to those engaged in agriculture, building trades, and hunting. But who are these immigrant blacksmiths, whence do they come from Dogon and Bamana territories, and where have they established their workshops in Bamako?

My research addresses two groups of immigrant blacksmiths, one from the Dogon Plateau, the other from Bamana country. To compare forges according to articles they produce for sale in city marketplaces, I chose three different sites: workshops in Sugunikura, also known as "the new little market"; a workshop in a family compound; and others in zones peripheral to Bamako proper.

SUGUNIKURA MARKET, OR THE VALLEY OF BLACKSMITHS

When entering Sugunikura, one is struck by a veritable valley of several hundred metalworkers engaged in frenetically fervent work (*en ébullition*)—indeed, more than fifteen hundred toil in some four hundred ateliers. A most unusual ambiance reigns in this place, marked by every sort of noise produced by striking iron. Clanging, banging, and the sparks, flames, and ardent embers of so many forges overcome any casual visitor.[5] One wanders among mountains of recuperated metal scrap and sees tools of all sorts at work. Truly, the smiths' valley of Sugunikura presents "chaos on a thin leash" (Roberts 1996, 91), but where better can there be to find the forge in all its contemporary splendor and diversity? The "valley" is a place of tradition, modernity, and the evolution of iron workmanship. A number of the forges function as complex, multipurpose businesses: everything is for sale, from raw materials to finished products. Clients teem; pathways through working places are difficult to navigate. Visitors cannot but feel that they have plunged into some sort of initiatic place charged with cosmogonic significance, where one returns to the sacred—as historians and anthropologists have long described West African ironworking (see Échard 1983). Yet, in important ways the symbolism of metal and its creative powers are here replaced by hopes of gain and commercial success.

In Sugunikura, blacksmiths are organized in a grand cooperative called Fantogo (the word for "forge" in the Bamana language). Fifty workshops participate, and the principal objective of Fantogo, according to its president, is to reinforce solidarity among immigrants and so preserve ancestral forms of work and identity among blacksmiths. "Seventy-five of one hundred smiths working in Sugunikura are 'blacksmiths of blood,'" the president of Association Fantogo, Sekou Samassékou, holds—that is, they have been born into lineages of smiths (figs. 18.4, 18.5). Fantogo aims to promote the most significant sacred and social functions that smiths provide in village milieus, and above all, the safeguarding of their spiritual and magical knowledge and powers. The association hopes to preserve the cosmogonic foundation of blacksmiths' work among Dogon and Bamana peoples. Members also want public authorities to offer them particular attention by supporting them, for example in their attempts to obtain financial credit permitting them to equip their forges with modern machinery.[6]

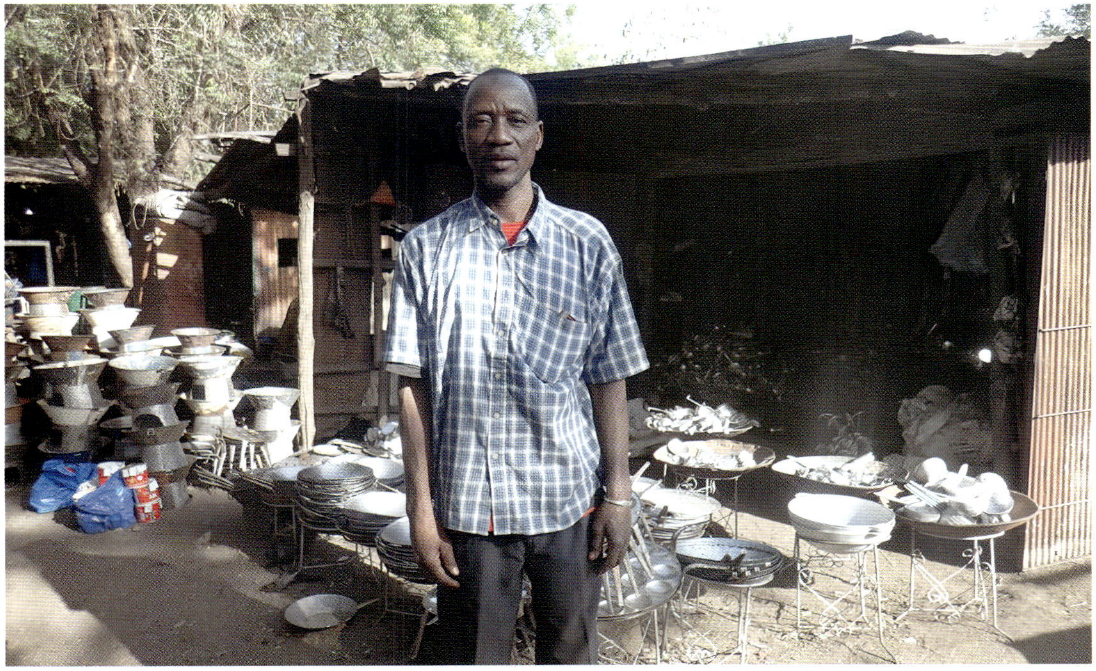

18.4 Mr. B. Berthé, a blacksmith originally from the region of Sikasso in the south of Mali, stands in front of his forge. He is a member of Fantogo, a cooperative association of blacksmiths by birth whose goal is the preservation of the highest social duties sacred to blacksmiths. On view are cold-worked and soldered cooking implements largely made from sheet metal. PHOTOGRAPH BY ISAIE DOUGNON, SUGUNIKURA MARKET, RACETRACK NEIGHBORHOOD, COMMUNE II, BAMAKO, MALI, 2017.

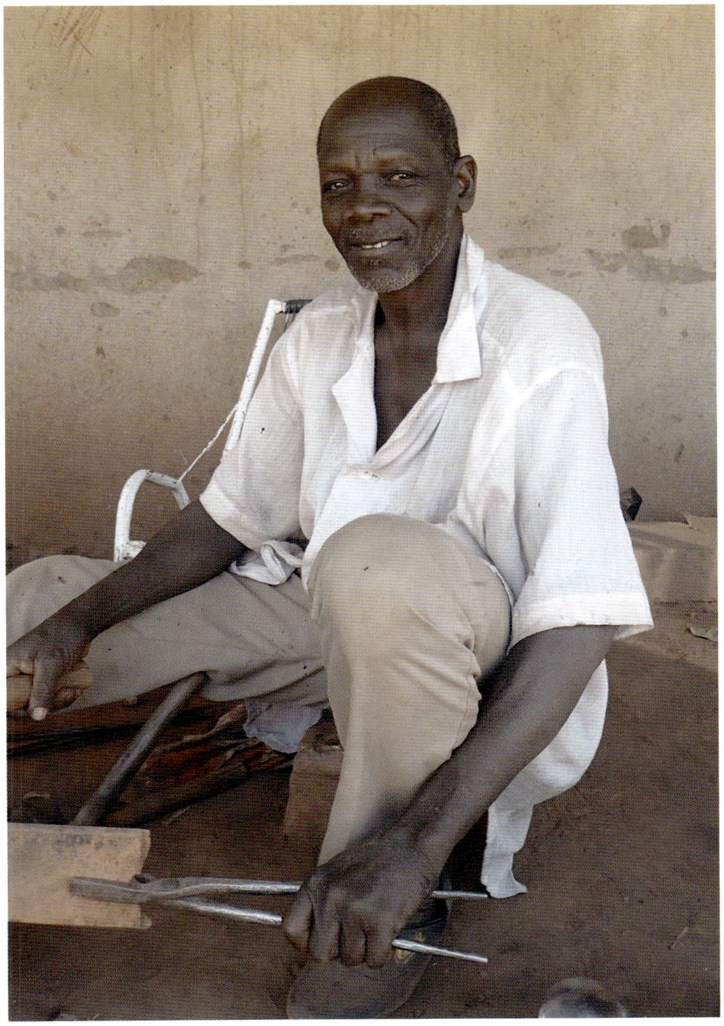

18.5 At the forge of his elder brother (see fig. 18.4), Mr. Issoufou Berthé, a migrant blacksmith, has set up to work on the front porch. He allocates his time in different forges according to the wishes of his clients. During winter, he works in his village; during the dry season, he travels between Sugunikuna, Bamako, and Burkina Faso. Mr. Berthé buys iron in Sugunikura that he transports to Burkina Faso to make hoes, carts, plows, and other equipment. PHOTOGRAPH BY ISAIE DOUGNON, SABALIBOUGOU NEIGHBORHOOD, COMMUNE V, BAMAKO, MALI, 2017.

THE FAMILY FORGE AND THE MARKET FOR ARMS

The market for arms is one in which blacksmiths excel. Such was the case with the late Youssouf Yanogo, a Dogon smith who migrated to Bamako in 1985. Agents of the state, hunters, and others sought his weapons. Monsieur Yanogo was born in 1938 in the village of Nigary, and died in Bamako in 2015. Three of his children told me of the past and present status of his forge, which he left to his youngest grandson. During my visits with the family, it was clear that his sons have maintained their father's flame. They told me that their ancestors came from Mande country, bringing with them the expertise necessary to fabricate firearms, as an old family métier maintained by each generation from Mande to Dogon lands.[7]

M. Yanogo came to Bamako not to settle, but to obtain an official permit to fabricate firearms. In rural Mali at the time, security forces were arresting and punishing those thus employed who lacked such papers. In 1985, after an unfortunate incident with the authorities, M. Yanogo decided to travel to the capital to seek the necessary license. Once there, he learned that the market for firearms in Bamako was much more lucrative than in his rural town. He was told that Bamako was surrounded by Malinke and Bamana communities where many hunters had need of inexpensive guns of the sort he could make. M. Yanogo quickly seized the opportunity, and his permit was signed by President Moussa Traoré himself, and renewed in 1992 by Traoré's successor, President Alpha Oumar Konaré. The elderly M. Yanogo called the model of musket he produced the "Dana," which means "hunter" in the Dogon language. Even today, despite ready access to European arms, the Dana remains popular.

FORGES ON URBAN PERIPHERIES

Not all smiths who have migrated to the capital have settled in the city center, nor are they solely engaged in forging modern tools. Those settling on the peripheries may not feel able to fabricate the cooking vessels, wheelbarrows, carts, tools, household utensils, and vehicle parts needed by urban dwellers. On the urban outskirts, they instead provide agricultural implements to other rural migrants, while also selling goods to workers in building trades and to those needing particular tools, for example for well digging. They often situate their workshops along roads leading into and out of the city, where peasant farmers and others heading toward fields and construction sites can easily see and obtain smiths' ironwares. Indeed, roadside work is not far removed from what these same smiths undertook in their own natal villages in the countryside.

At workshops set up near roads leading to fields, farms, and construction zones, the forge is often a meeting spot for people working in the area or new to it. A place called Noumouké carré (Street of Blacksmiths) constitutes a hub for the neighborhood under construction; once building is complete, smiths hoping to attract similar clientele will move to subsequent peripheries as the city inexorably spreads outward from its center.

MIGRATION AND IDENTITIES IN TRANSITION

Blacksmiths interviewed for this essay were unanimous in saying that nowadays, the forge is undergoing rapid change. They affirmed that smithing is a trade that is very difficult and demanding, but that in the city center, substantial profits are to be gained if one works hard. Nonetheless, some smiths who immigrated to Bamako have abandoned their forges for more lucrative activities, and some do not want others to know they were once so occupied.[8]

Emphasis must be given to such rapid change in the forge and the identities of blacksmiths in migration. Smiths admit, though, that even given this crisis, some do fight to revalorize their identities. The recently deceased Dogon man introduced above as author of the Dana musket prototype was a champion of such struggles. Until his death, he excelled not only in the fabrication of firearms, but also in therapeutic practices. He healed people, especially those suffering from burns or hemorrhoids. M. Yanogo was also a grand master of social mediation. The tool from his forge that he used to impose peace between disputants was his *sènè*, or traditional hammer. This function defies more ordinary Dogon associations between iron and danger, death and blood. Among blacksmiths, a *sène* can compel respect during conflict, social strife, or dissimulation. It is the instrument of truth and peace. In new urban traditions, when one purchases a means of transportation like a bicycle, motorcycle, or automobile, one prays "that God protect you [the vehicle] from the misfortunes of iron." In contrast, when rural people call upon a blacksmith to mediate cases of conflict or illness, the smith uses his iron hammer to pacify and heal.

In litigation, should someone be accused of theft or other malfeasance, a blacksmith will dip his *sènè* in water that he then asks the disputants to drink. Anyone who has been lying will die within twenty hours. Even in Bamako, M. Yanogo maintained the hammer's role as social mediator until the day he died. "One is only half a blacksmith if one no longer holds the mystical knowledge associated with the *sènè*," the old man's children told me. The hammer holds greater powers, they added, than any medicines possessed by farmers or others who come to purchase tools at the forge.

An implement that has been at the core of blacksmith identity, the *sènè* is no longer accorded such importance by young blacksmiths. Immigrant blacksmiths who have left more traditional ways of understanding the forge, and who are perhaps influenced by today's Islamic radicalism and looking to new agricultural technologies, risk losing ironworking as a particular profession as known in the past.

CONCLUSION

Reflecting today on relationships between blacksmiths and marketplaces refers us to concepts of innovation, dynamic social change, social development, and a politics directed toward finding employment for young people. All these are associated with imported or recycled metals from Europe and tools that are traditional, modern, or hybrids of the two. Blacksmiths are aware they must follow such changes in order to adapt to contemporary markets. Some even wish to learn how to fabricate machines as do "European blacksmiths." In Sugunikura Market, the wares on display offer a sense of how much blacksmiths look to such progress. They hope to have modern workshops with machines powered by electricity, diesel engines, or solar energy.[9] Yet some wish to become entrepreneurs without neglecting social and cultural values associated with the forge in rural settings. Two essential elements have been added to their indispensable savoir faire to attract clientele in Bamako: innovation and affordable pricing.

It should be underscored that many immigrant blacksmiths also find success in work far removed from metallurgy. In Bamako, some combine ironworking with carpentry, well-digging, grave- or ditch-digging, healing and providing minor surgeries, religious offices, and social mediation. The question is, what does a blacksmith become if he abandons his cultural traditions? In other words, can one be "modern" without losing a part of oneself? This is an age-old problem of modernity and Africanity, of course, but blacksmiths as a specific professional group experience a dilemma greater than their farming counterparts in this regard.

What are the real impacts, then, of migration upon changes of social identity? We have seen that one group of migrant blacksmiths that settled on the outskirts of the city continues to produce traditional farming tools and other implements necessary for rural life, while another group has converted itself to produce more modern firearms, plows, and parts for trucks and other vehicles. Among the latter, some retain ancient instruments, even though their fabrication has been forgotten for several generations now. Keeping such heirlooms is a means to demonstrate their own development; they hold regard for these things and the mode of life to which they were once essential. In Sugunikura Market, several hundred blacksmiths work in such proximity to one another that they seem almost anonymous as individuals. But looking more closely, one sees that each owner of a forge attaches capital importance to his means of making a living while valorizing his family, ancestors, ethnic heritage, and region of origin.

Among Dogon blacksmiths in Bamako, metalsmithing remains a noble profession, but one that is above all a combination of local and global references that are the basis for the new smith as entrepreneur. Young rural blacksmiths follow such footsteps. Certain ones among them are very young, and they move to the city in large numbers despite strong governmental campaigning against child labor and migration. In the sheds and hangars of forges in Sugunikura Market, they learn modern techniques slowly but surely, combining simplicity with harmony. In general, they do not rush to independence as do apprentice automobile mechanics, for their métier demands patience to be viable.

Evidence from Dogon and Bamana blacksmiths evokes the very particular work of immigrant smiths with far more ready access to modern tools than most economically disadvantaged rural people can hope to have. What would become of the vast garden plots and other agricultural sites around Bamako without parts for plows, harrows, and the like? What would become of hunters without the Dana-style muskets produced by Bamako smiths? Without the improved woodstoves fabricated by the blacksmith Samessékou, what would become of poor families? In the capital, the work of blacksmiths is visible downtown and in peripheral neighborhoods. The forge is an important part of the technological patrimony of Bamako, and of Mali.

PORTFOLIO

IV

BAMAKO IRON MARKET

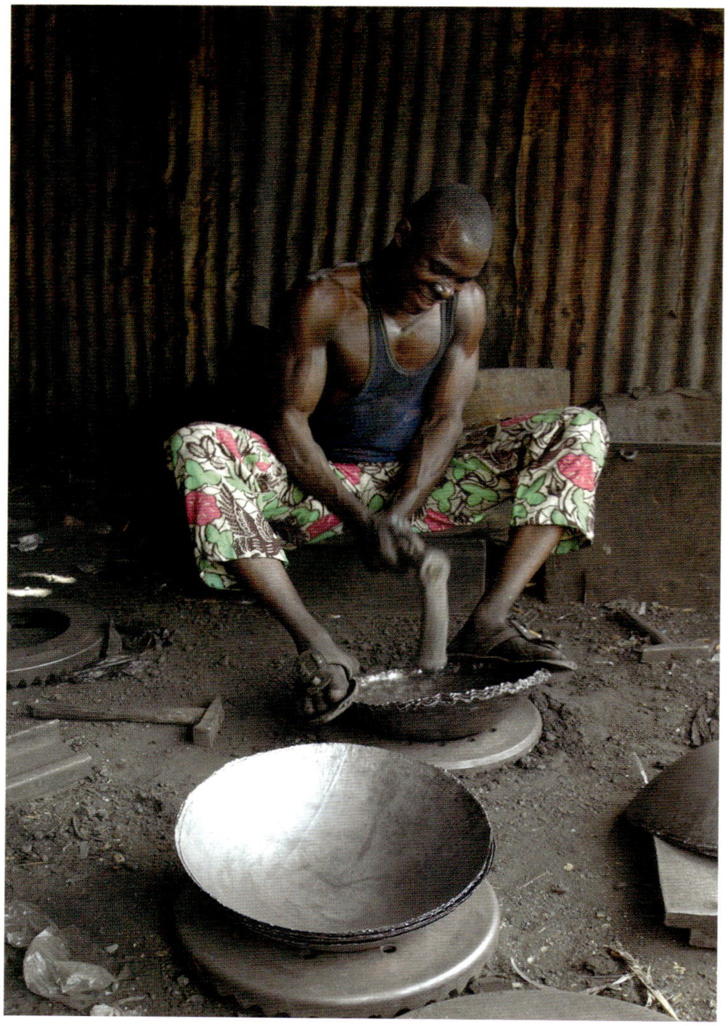

In 2001, a united consortium of metalsmiths in Bamako, Mali, were awarded legal status by the government and—after six evictions from previous locations near the city center—artisans were enabled to conduct business as a professional cooperative founded on municipal property. Over a thousand practitioners set up shop adjacent to one another, forming the Coopérative des Forgerons et Ferblantiers Diama Djigui, Marché Médine, to "promote and protect members' interests" (Sekou Samassékou, association president, pers. comm. to Tom Joyce, 2012). By 2012, when the photographs of this portfolio were taken, the cooperative had swelled to 1,527 members: blacksmiths, tinsmiths, coppersmiths, sheet metal workers, wire weavers, cutlers, joiners, welders, solderers, gunsmiths, machinists, mechanics, boiler makers, scrap cutters, stock preparers, heat treaters, aluminum casters, brass founders, fabricators, and more. These specialists work in solidarity, encouraging high standards of craftsmanship within their respective fields as together they produce impressive quantities of functional forms.

In Bamako, the iron market is a cacophonous, sprawling landscape of vibrant intensity, where the ringing of hammers fills the air as ironworkers young and old—many members of esteemed lineages—shape mostly recycled metals within an encyclopedic concentration of skilled metal arts (see also chapter 18, this volume).

Text by Tom Joyce and Henry J. Drewal
Photographs by Tom Joyce

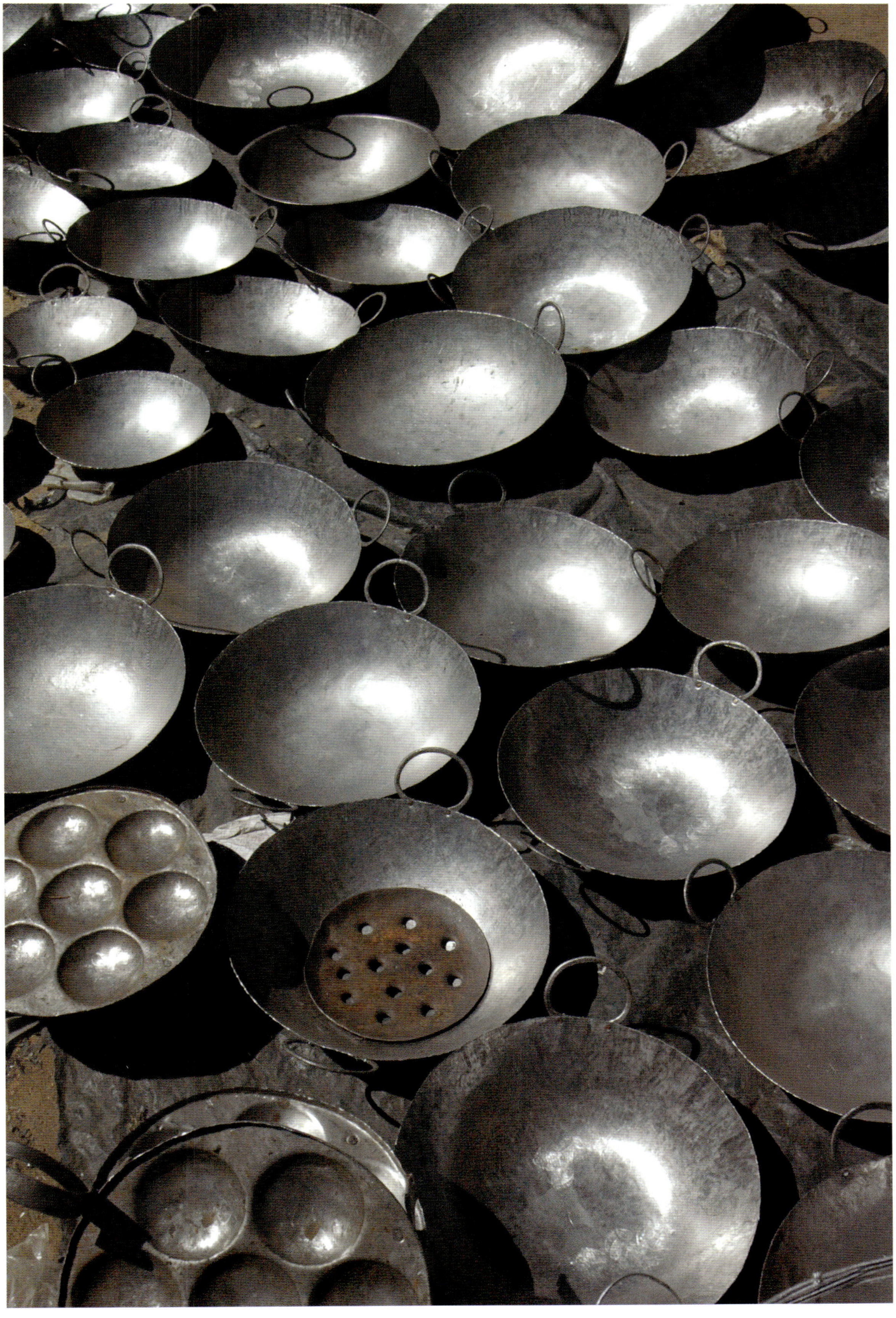

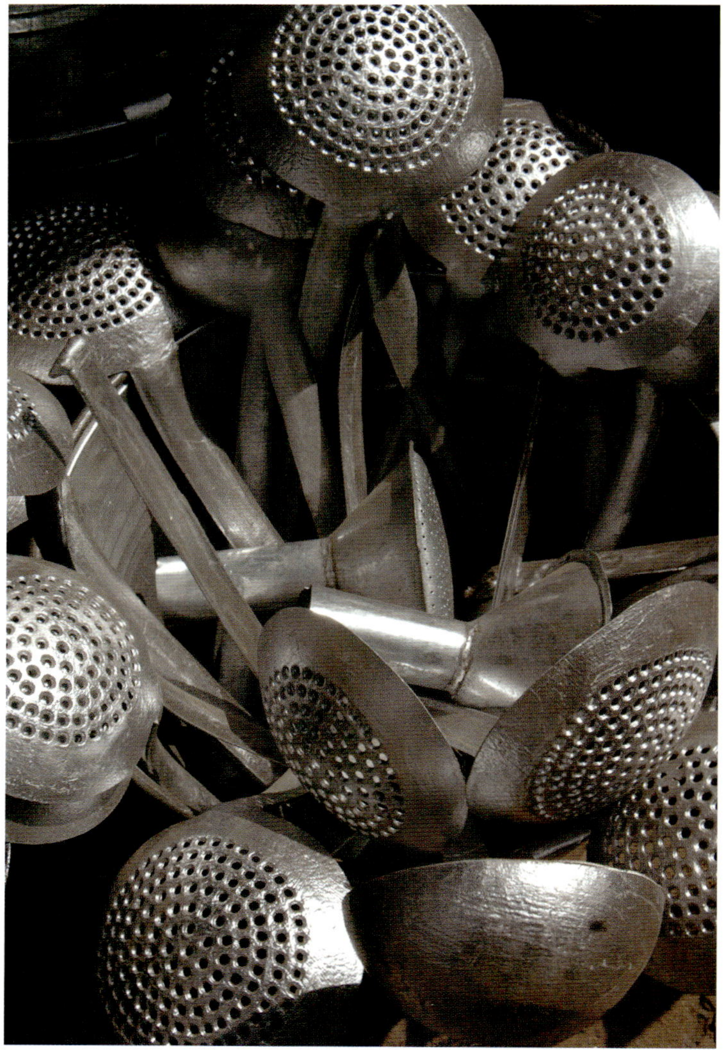

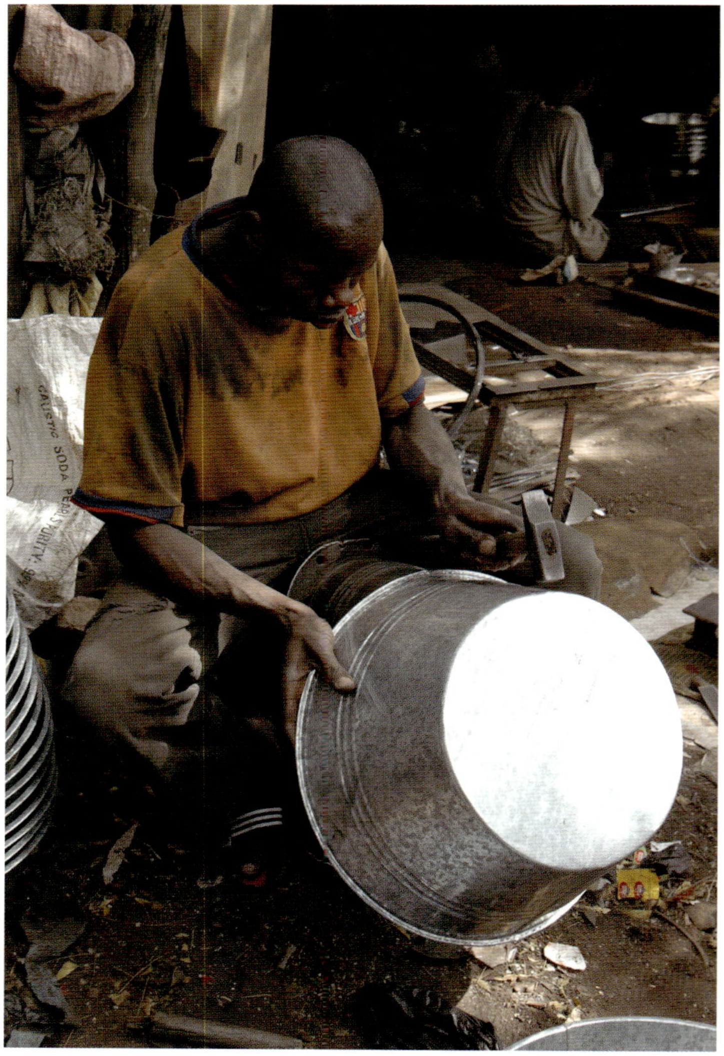

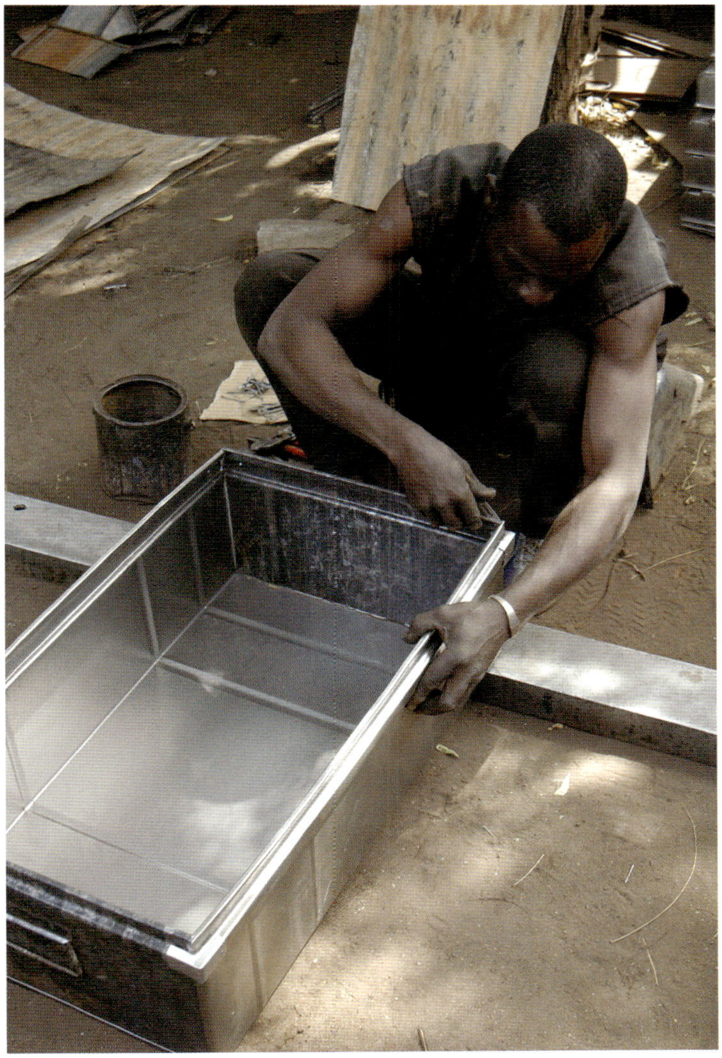

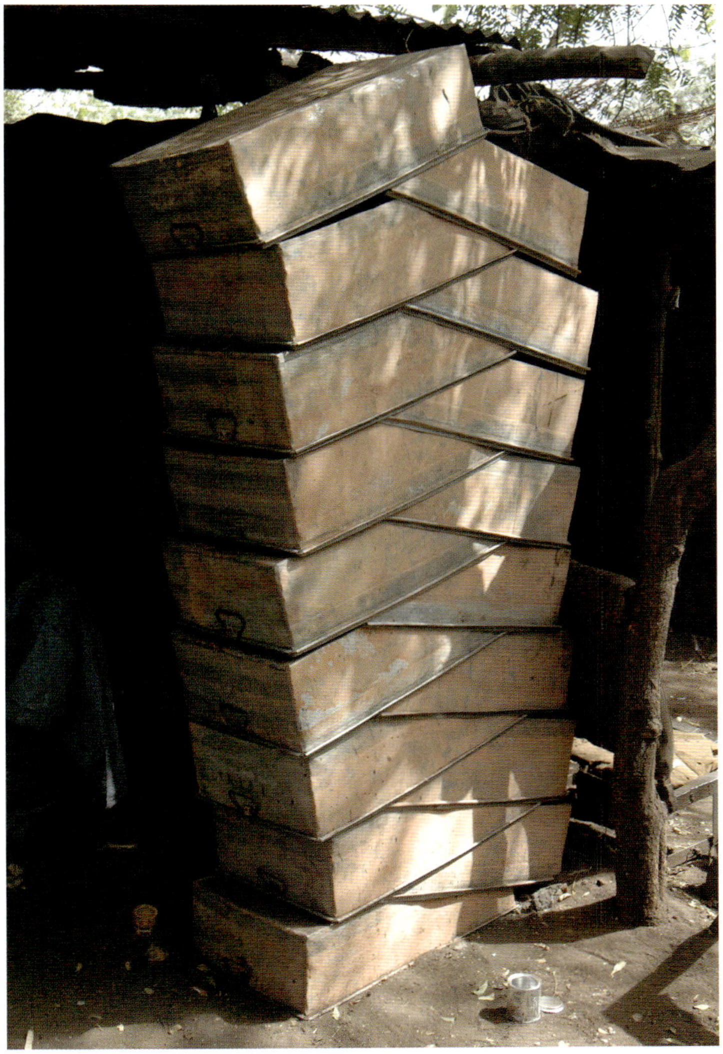

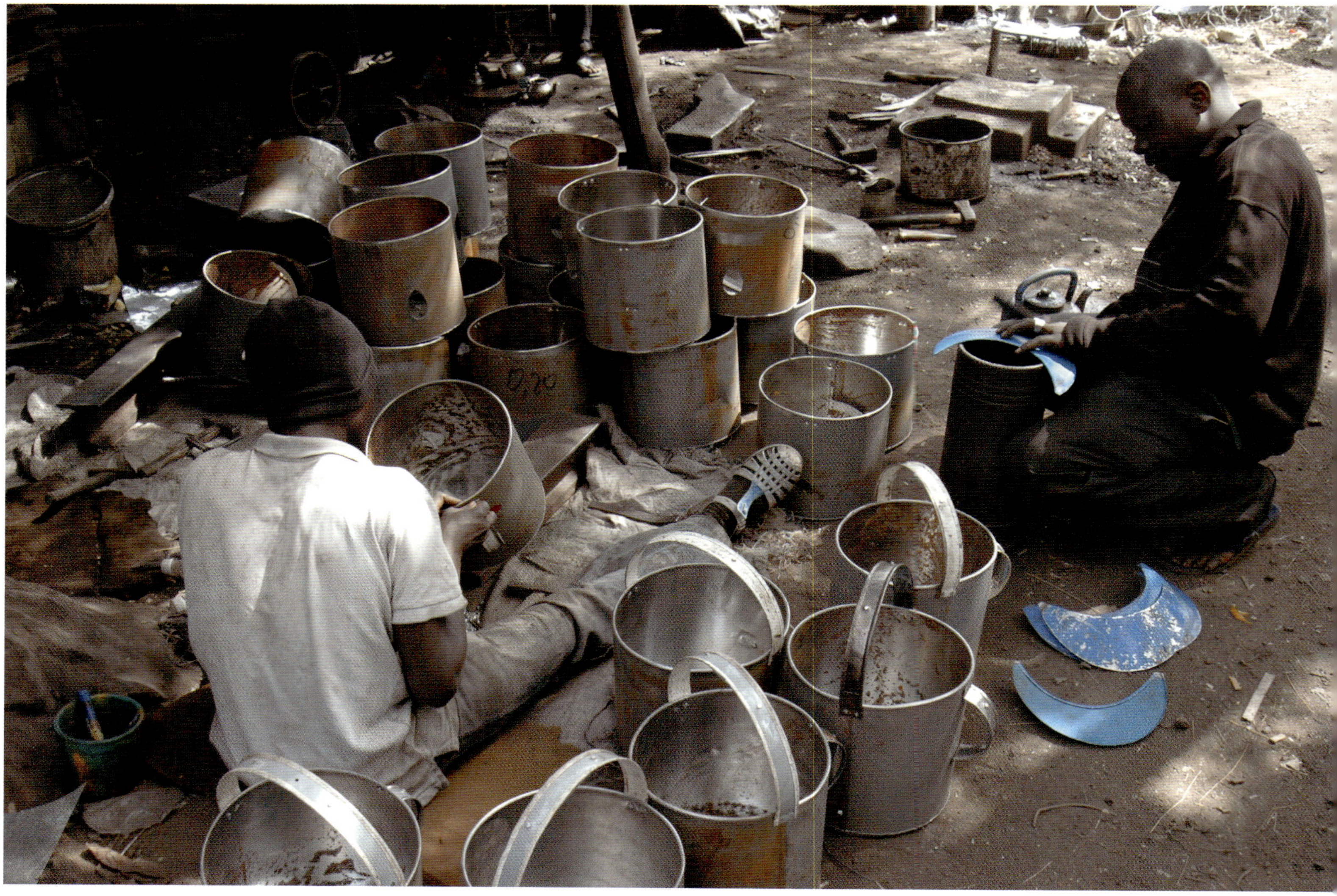

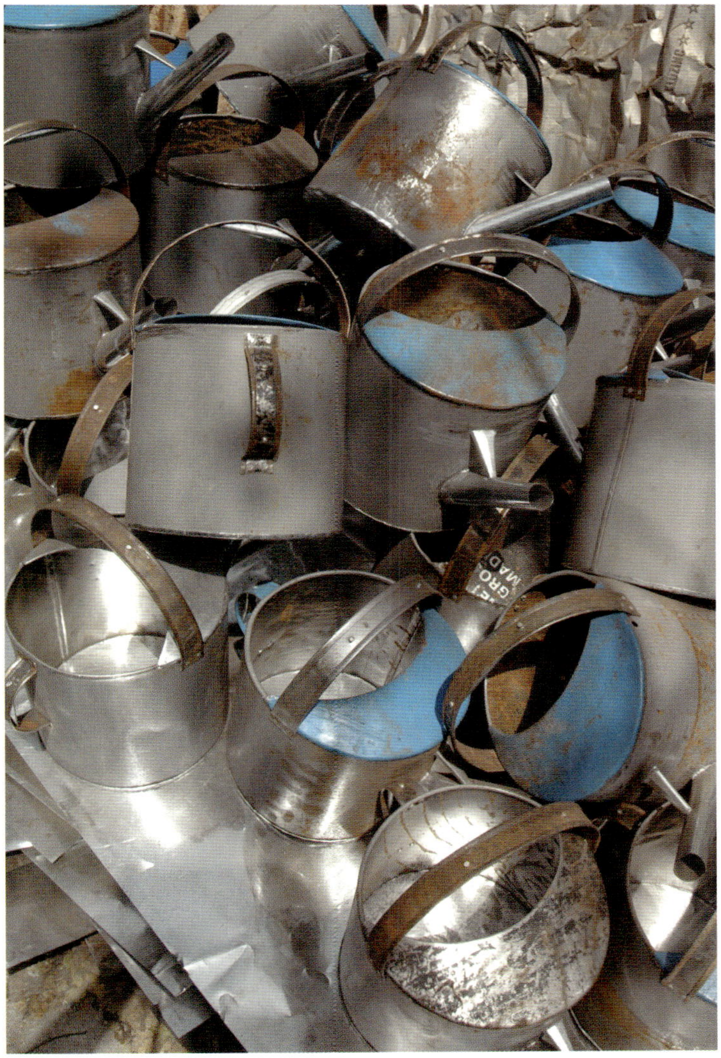

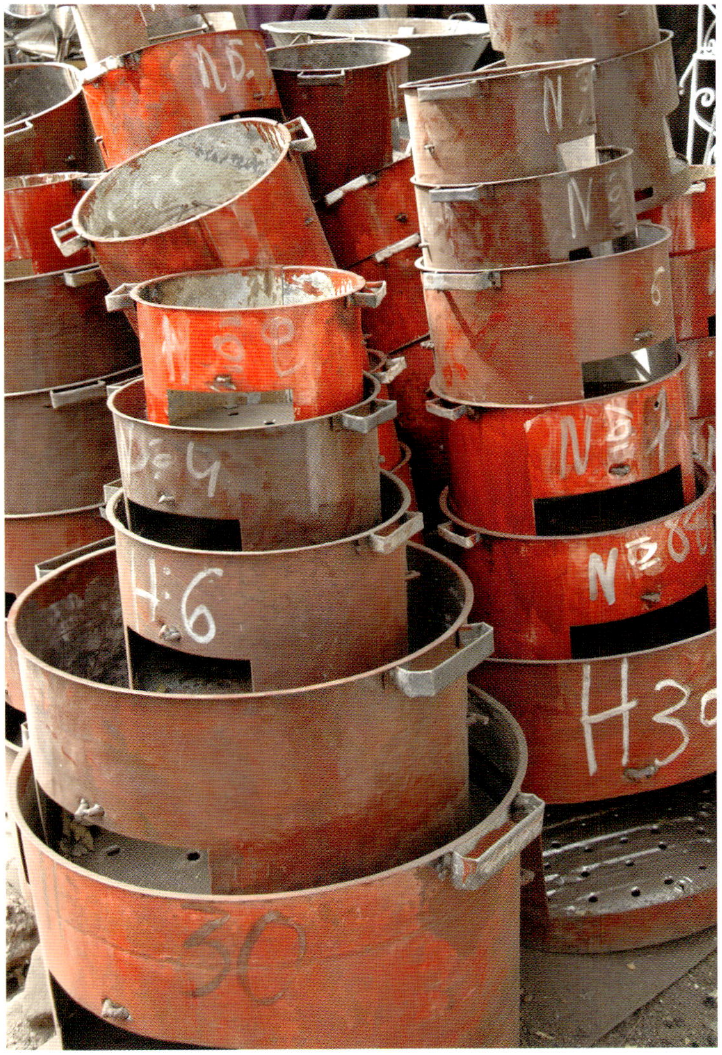
H:6
30

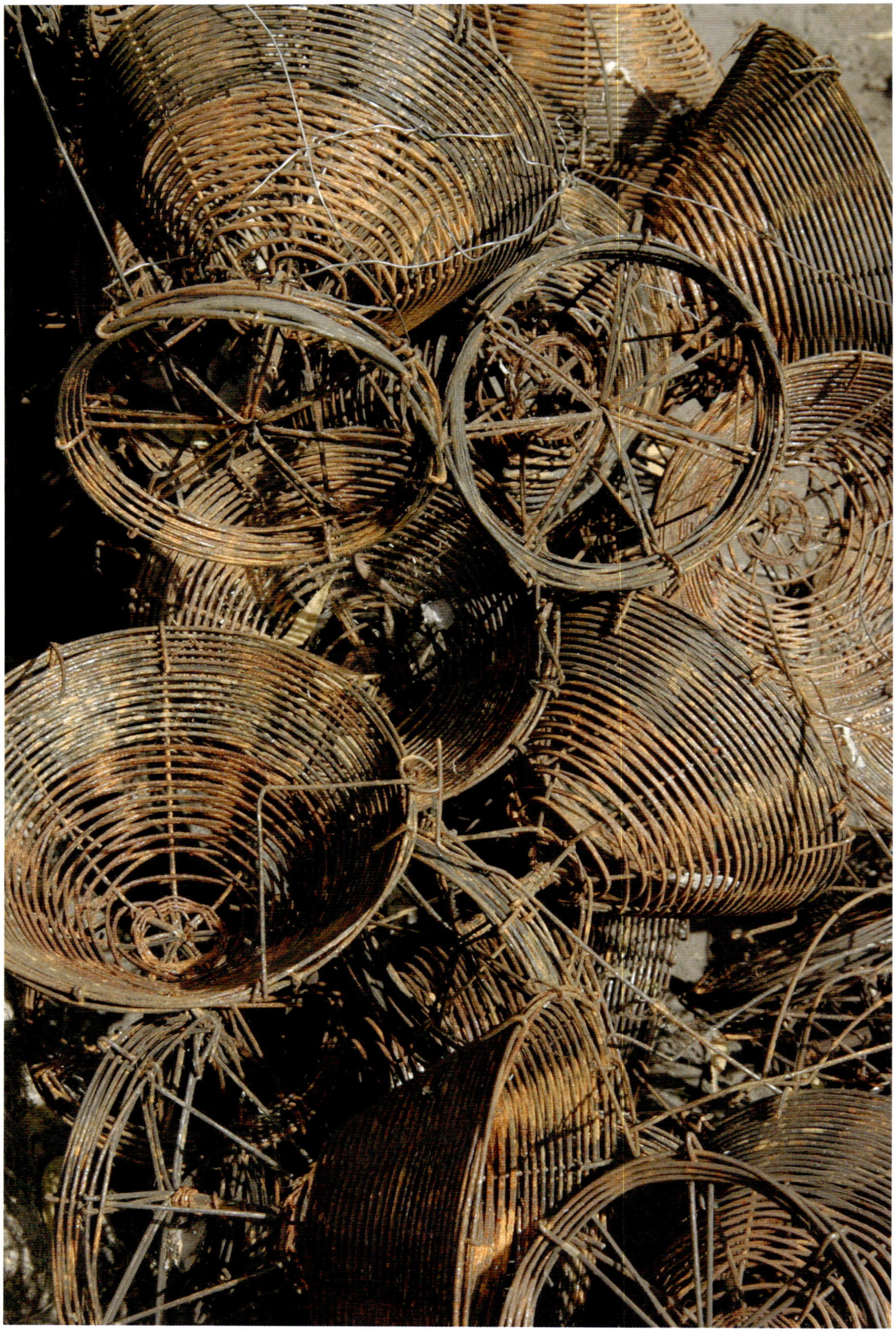

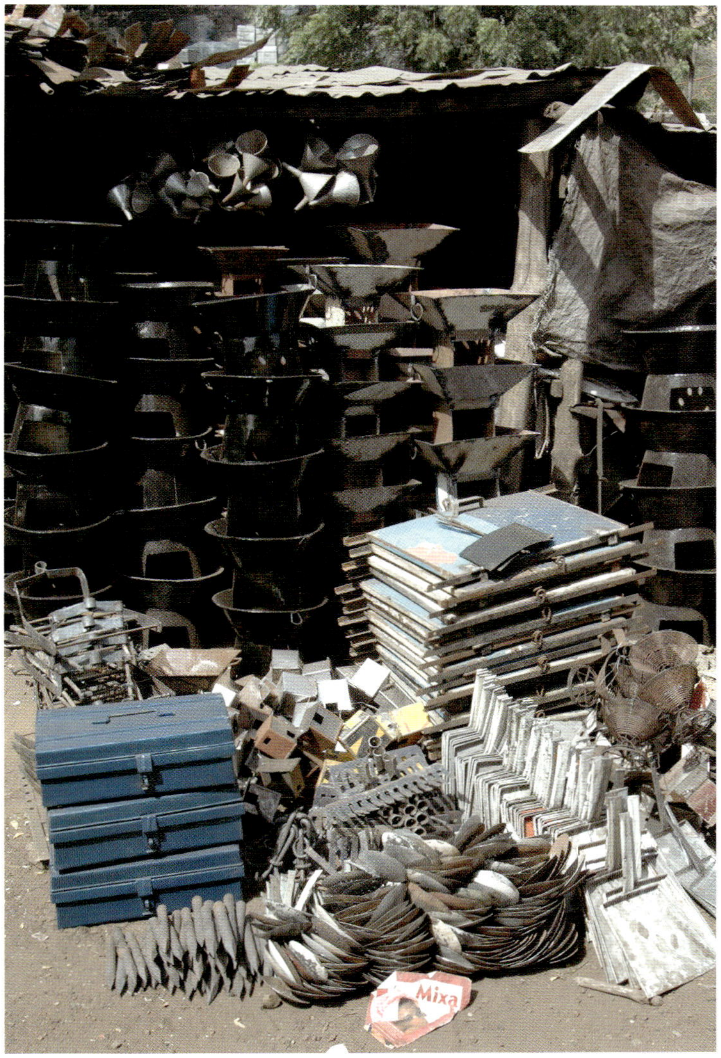
Mixa

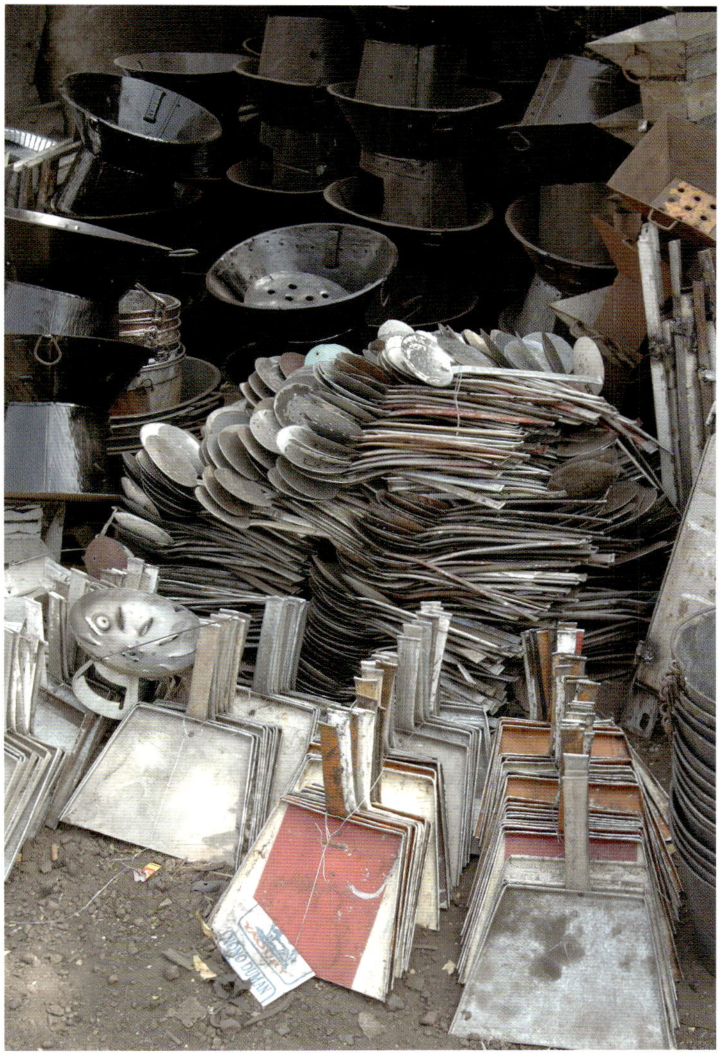

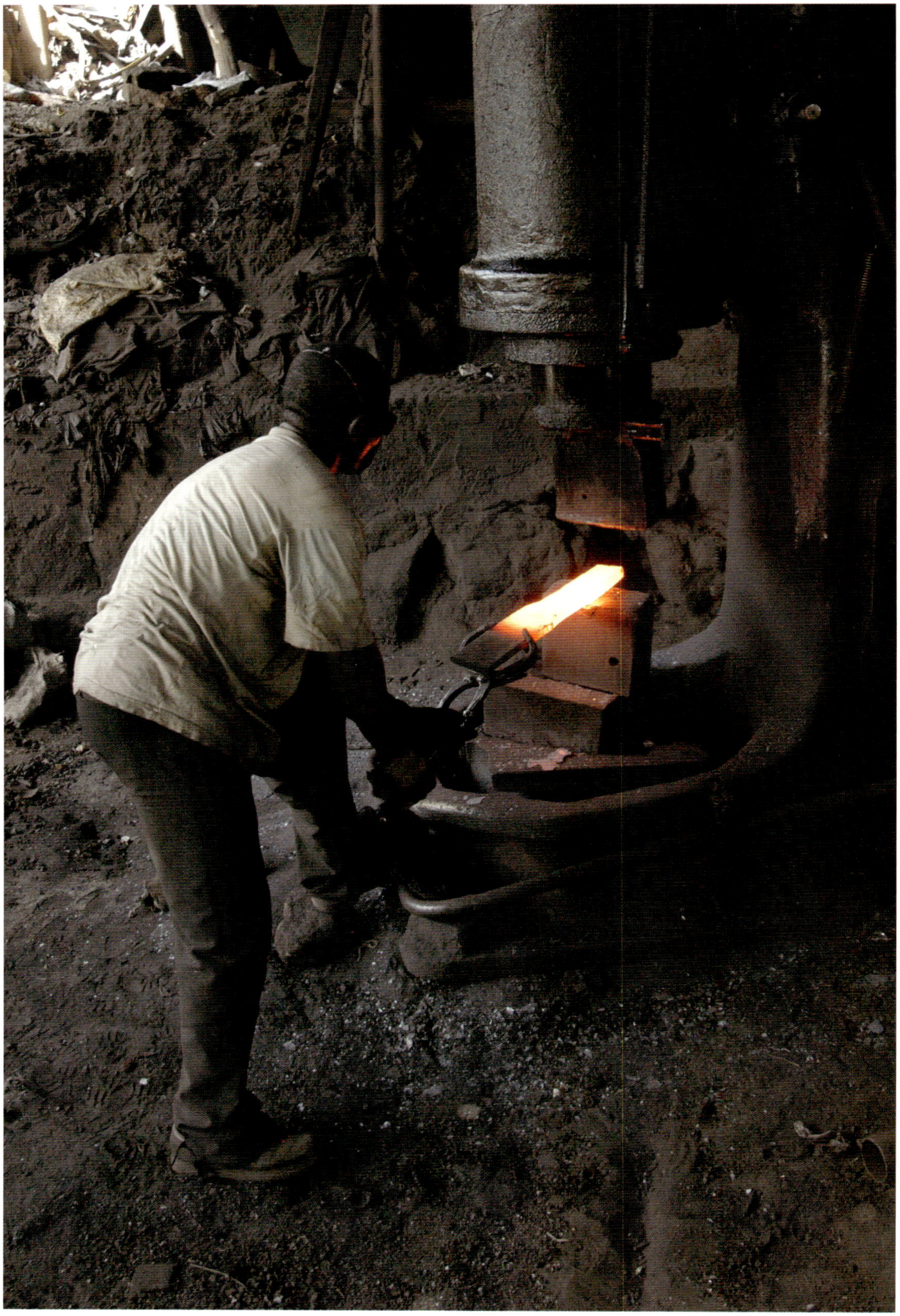

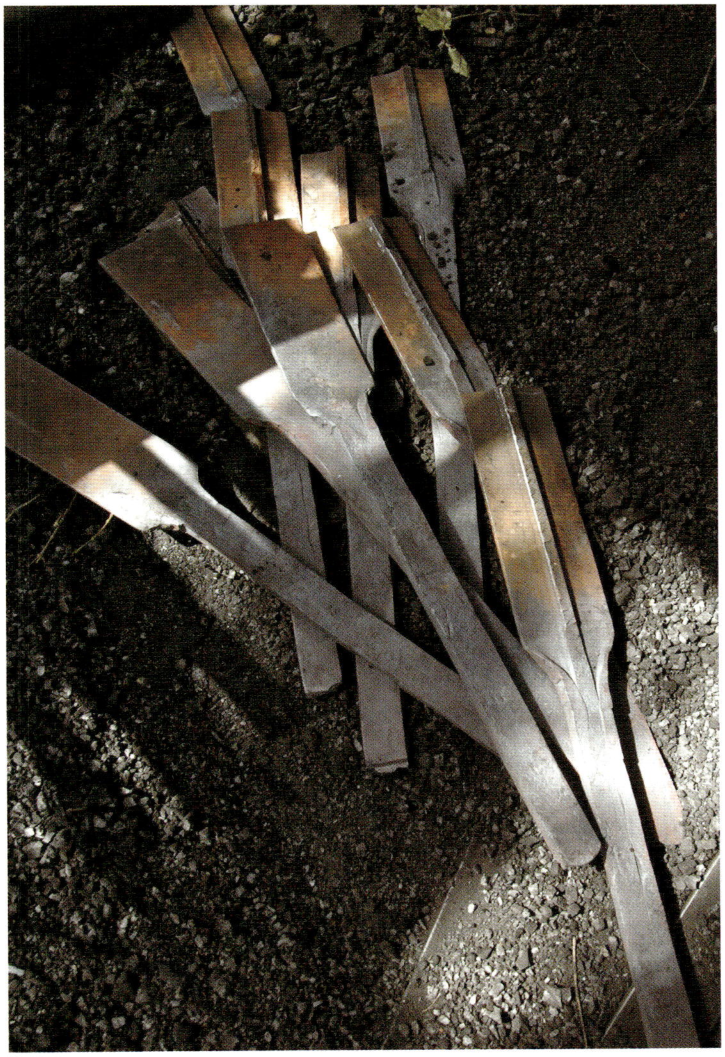

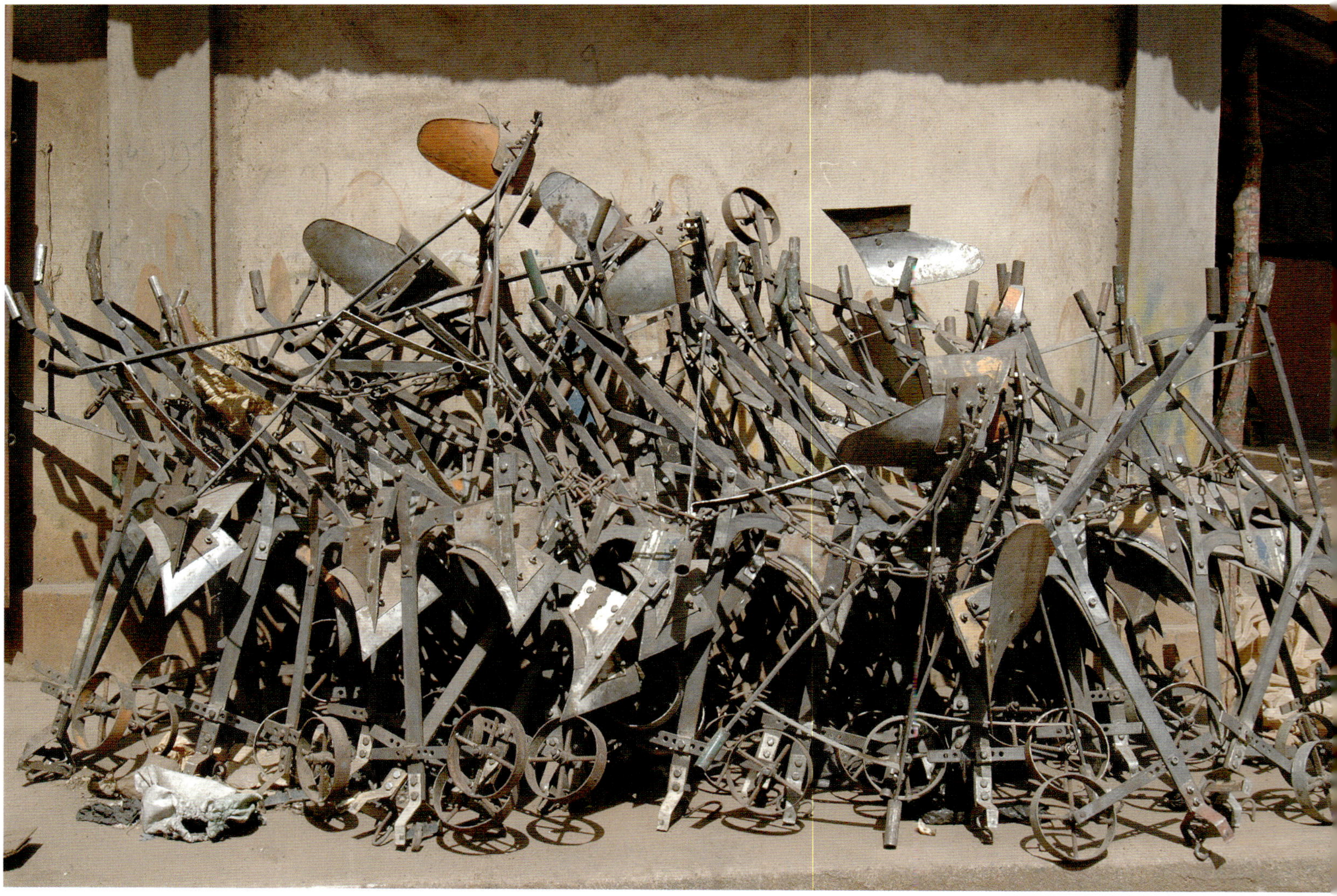

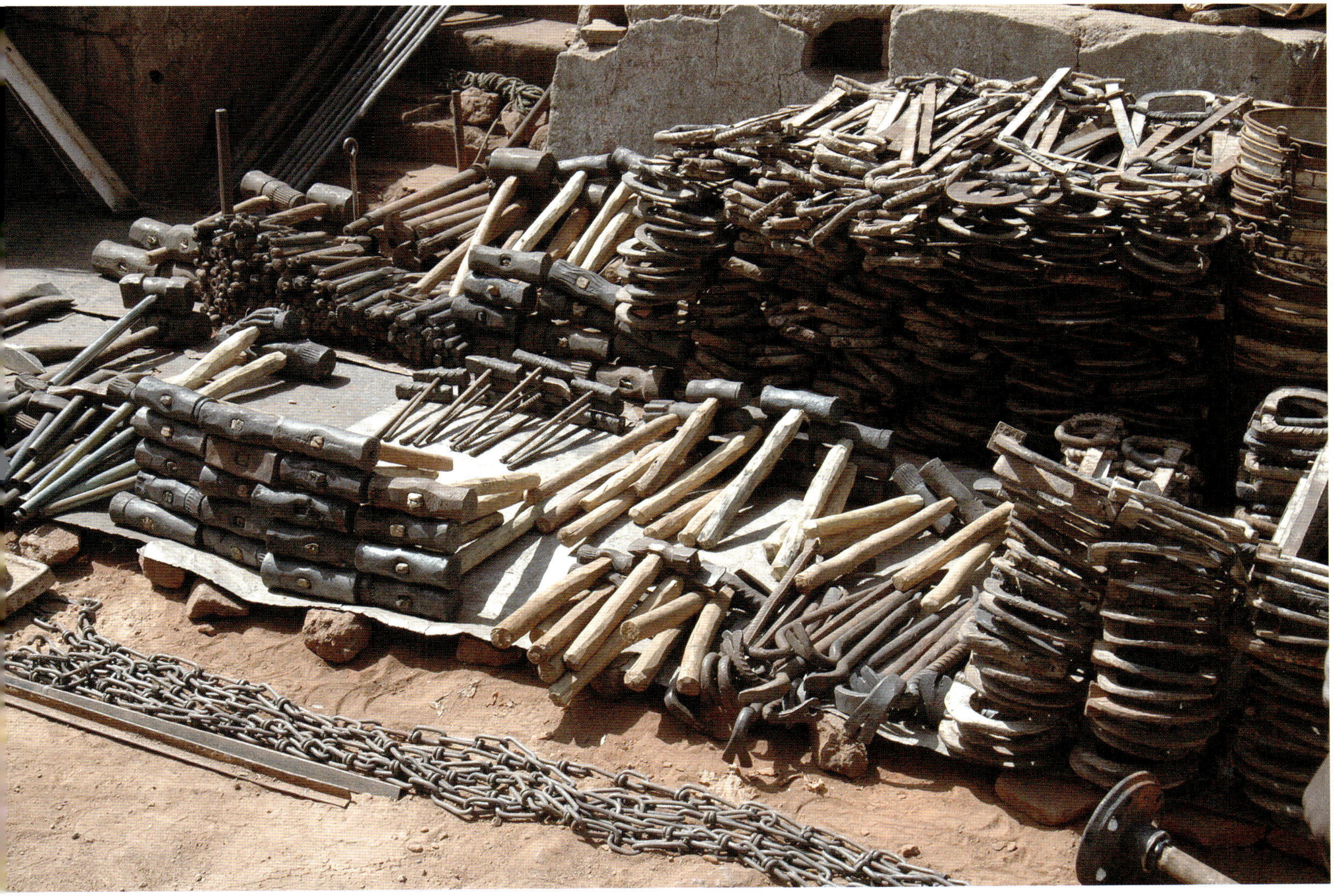

NOTES TO THE TEXT

PREFACE 1 (JOYCE)

1. Johansson (1967, 36–42). Three sources of information (Eyo 1979; Johansson 1967; Quiggin 1949), compiled from library archives rather than field research, propose several cultural attributions, offer a variety of local names, and slightly different interpretations of this currency bar's use and value. More recent attribution from local sources in Nigeria consider it to have been forged and used by Mumuye peoples.

PREFACE 2 (KRUPP)

1. Thanks to David Killick as well as to S. Terry Childs, Duncan Miller, John Wasson (founder of the UCLA Collection of Meteorites), and project colleagues for discussion of these points in 2017, and especially to Helke Mocke of the National Earth Science Museum in Windhoek, Namibia, for her thoughts concerning the lack of evidence of Nama peoples' use of iron from Gibeon meteorites. I am certainly grateful to Allen F. Roberts for broad direction and for collecting essential information on meteoritic iron and early iron technology in sub-Saharan Africa.

PREFACE 3 (ABIODUN)

1. Eugenia W. Herbert addresses this very important phenomenon in her book *Iron, Gender, and Power* (1993).

2. Ìrókò is often called "African teak," and its wood is very hard, durable, and termite-resistant. Only experienced and accomplished artists can use such wood to carve palace doors, veranda posts, thrones, and the like.

3. All subsequent translations of Yorùbá names are from Babalola and Alabi (2003).

INTRODUCTION (ROBERTS)

1. Sennett (2008, 8–9). The *art* of African blacksmiths is implicitly addressed in all essays of this volume, yet the controversial question of what constitutes "art" in the first place is not within our purview, let alone how such a term might be defined and applied in any given African community. As contributor Rowland Abiodun (2014) asserts, we must seek the African in "African art" while avoiding any spurious sense that art defined in one community extends to other peoples elsewhere on the continent or in the world. In this regard, we take the Congolese philosopher and social critic V. Y. Mudimbe (1986) to heart and leave "African art as a question mark," except insofar as whatever it may constitute in the eyes of particular African beholders responds to that "enduring, basic human impulse…to do a job well for its own sake," as Richard Sennett aptly states.

2. Although many sub-Saharan peoples and regions are considered in these pages, the continent is obviously too vast for the present volume to be comprehensive. Madagascar is not included, for example, despite the strength of research on ironworking there (see Radimilahy 1988), nor are smelting and smithing, earlier or contemporary, in entire present-day republics. A purpose of *Striking Iron* is to encourage more research, writing, teaching, and exhibition activities by African and Africanist students.

3. Rabaté (2018, 2, 12, 40). Iron enthusiasts will enjoy Jean-Michel Rabaté's little book, especially for its chapter "Rust Aesthetics," in which he explores the homonymy of the Japanese terms *sabi*, "rust," and *sabi*, "beautiful imperfection." As he suggests, "Zen existential loneliness, or sabi," readily instigates "a sweet melancholia associated with the contemplation of objects that evoke impermanence" (ibid., 94, 96–97). John Ruskin's 1858 lecture "The Work of Iron, in Nature, Art, and Policy" is also well worth pondering, for as he asserted, "it is not a fault of iron, but a virtue, to be so fond of getting rusted, for in that condition it fulfils its most important functions in the universe, and most kindly duties to mankind. Nay, in a certain sense, and almost a literal one, we may say that rusted is Living; but when pure or polished, Dead" (Ruskin 1905, 376–77). Ruskin's praise of rust is thought to have influenced certain poems of William Carlos Williams and Marianne Moore, such as Moore's antiwar poem "In Distrust of Merits," published in *The Nation* in 1943: "Hate-hardened heart, O heart of iron / iron is iron till it rust." See http://www.ruskinartclub.com/calendar/sunday-january-22-2017-lecture-reading-by-robert-pinsky; cf. Goucher (1981). One can imagine that at least a few African savants and other intellects around the world have reflected on similarly poetic analogies.

4. Tim Ingold (2010, 97–98) is quoted but also paraphrased in this paragraph, for I have taken the liberty to change Ingold's focus on the work of a carpenter sawing a plank to that of a blacksmith forging an implement. Ingold's logic and wording fit our explicit purposes here, as well as his own, without betraying the particularities of African ironsmithing (see chapter 4, this volume).

5. Gẹ̀lẹ̀dẹ́ masks—with this term referring to the headpiece and costuming as well as the choreography, drumming, and songs composed for particular performances—are created in response to a community's current events and felt needs. In many instances, a given mask is only danced once; as a consequence, we cannot know particular purposes this sculpture may have fulfilled for its original audience. On Gẹ̀lẹ̀dẹ́ performance more generally, see Drewal and Drewal (1990); Lawal (1997).

6. That *sympoiesis* is Greek for "making or creating together" and in turn is from an Indo-European root that leads to "poem" in Greek and "body" in Sanskrit adds depth to the word's application to African ironworking (Morris 1967, 1012, 1304, 1524). Pertinent comparison is provided by Alexander Langlands (2018) through his compelling studies of medieval European *craefts*, as the Old English spelling had it, that present the sympoiesis of people working closely with each other as well as with their environments through acute observational sciences brought to bear on rural life.

CHAPTER 1 (ROBERTS)

1. See "African Sage Philosophy," February 2[illegible], 2016, *Stanford Encyclopedia of Philosophy*, https://plato.stanford.edu/entries/african-sage/.

2. Griaule and Dieterlen (1975). See also Griaule and Dieterlen (2015). It must be stressed that no "correct" or "universal" cosmology or sequence of particular details is to be expected, and that Dogon culture—like any other—constantly changes to meet new circumstances. On Dogon cultural dynamism, see Richards (2005). Ogotemmeli was among the remarkable Dogon intellectuals "manipulating exegeses" that Dieterlen had in mind (1989, 43), and the "speculation" of which Dieterlen wrote suggests dynamic problem-solving as stories are recounted for particular reasons. Regarding debates about inventions of Dogon peoples by non-Dogon writers, see Clifford (1988); Ezra (1988); Apter (2005); Jolly (2007; 2013); Kreamer (2012, 115–19). For recent ethno-archaeological research on Dogon ironworking, see Robion-Brunner (2010).

3. Griaule and Dieterlen (1975, 41–42). See also Duchâteau (2013, 50–55). It seems unlikely that Nommo possessed gender, and the pronoun "he" used here might be more appropriately "it," at least until the Seventh Nommo reached the ground and became humanity's first blacksmith and so male.

4. Duchâteau (2013, 67) reports forged iron *yo domolo* crooks but does not illustrate any.

5. Layton (1991, 10–11), after Horton (1965). See also Griaule (1994, 332 and passim); Roberts (1988a). On "reading" the iconographies of mnemonic devices such as *yo domolo* to compose pertinent histories among a different sub-Saharan group, see Roberts and Roberts (1996).

6. Laude (1994, 189). On Dogon figures with upraised arms, see Ezra (1989, 59–60). Not all scholars agree that this gesture necessarily suggests pleading for rain. See Paudrat (1994, 145); Jolly (2013, 84). Dogon "cloud hooks" may take other sculptural forms. See Blandin (1992, 78); Ezra (1988, 78). Footage of a Dogon blacksmith forging a spirit figure (very similar to that of fig. 1.13) is presented in "Behind the Mask," an episode of the BBC program *The Tribal Eye* first aired in 1975. Filmed during a Dama masquerade to celebrate the life of a notable person who has passed away, it shows performers pausing to dance before the same iron figure stuck into the ground at a juncture in their rocky path.

7. On Luba political histories, see Reefe (1981). On Luba religion, see Burton (1961); Petit (1993). On the structure of Luba myths compared to those of neighboring peoples, see Heusch (1982). On the complexities of ethnicity and other situationally determined social identities, including how some Tabwa may also be Luba and vice versa see Roberts (1980; 1989; 2013a). On how personified precepts have been performed in Tabwa and Luba rituals and related arts, see Roberts (1985); Nooter (1991); Roberts and Roberts (1996; 2007); M. N. Roberts (2013); and Roberts (2016).

8. See Roberts (1980; 1989). Ethnography presented here is from my forty-five months of doctoral research in socio-cultural anthropology in the mid-1970s among Tabwa living along the southwestern shores of Lake Tanganyika and archival work in Lubumbashi, Kalemie (both DRC), Brussels, and Rome.

9. On *ngulu* earth spirits recognized by Tabwa and as a cultural phenomenon once important in this part of central Africa, see Roberts (1984). In writing of the "ecological cults" once offered to such deities in northeastern Zambia, Wim van Binsbergen has suggested that such practices offer "an actional, symbolic and spatial counterpoint" to the ways in which people exploit their environments (1981, 97–98). Lodestones are composed of magnetite and, unlike most examples of that mineral, they are naturally magnetic, perhaps because of strong magnetic fields created by repeated lightning strikes of the sort one can expect on a summit like that of Mount Kalolo. See Livingston (1996, 14–20).

10. See Roberts (1993a). Christopher Ehret (1995) traces the roots of particular central African ironworking terms such as *mutapo* from proto-Bantu roots, suggests their distribution over time and space, and draws inferences regarding attendant technological histories over the centuries. Complementary historical views and detailed distribution lists of ironworking terms are presented in Vansina (2006). On color symbolism among central African peoples, see Jacobson-Widding (1979).

11. Examples of many of these forged objects are now in the collections of the Royal Museum for Central Africa in Tervuren, Belgium, and a small selection is illustrated in Roberts and Maurer (1985, 172–74, 194–95, 203, 205, 257–59, 72).

12. Weghsteen (1962, 214–16). On Tabwa origin stories, see Roberts (1980). Tabwa consider the aardvark (*Orycteropus afer*) a culture hero because the anomalies they perceive in the animal's anatomy and behavior help define the order of nature to which they provide such ready—and amusing—contrast. An aardvark's snout, perceived as comically phallic, suggests ironic thoughts about masculine prerogative among peoples observing matrilineal descent. See Roberts (1995, 81, 83); Maret (2005). The covered basket of this story is of a sort used by Tabwa spirit mediums of the 1970s and understood as a model of the universe not so different from the Dogon granary/ark in this essay. See Roberts (2012).

13. Kizumina Kabulo, born in the late 1870s or early 80s, was very elderly when I knew him in the mid-1970s. He was very much the critically important sage of the sort whom Odera Oruka (1990) would have recognized and "a scholar in his own [Tabwa] fashion," as Germaine Dieterlen (1989, 43) described Dogon savants like Ogotemmeli. Kizumina's wisdom greatly informs Roberts (2013a). Other Tabwa friends recounted stories of Kyomba to me, and versions have been recorded by missionary authors including Joseph Vleugels (19[illegible]3).

14. Roberts (1980, 344, 414). There were several models of Tabwa smelting furnaces, and terminology for their features as well as tools and processes of smelting are discussed in detail in Roberts (1993a).

15. Womersley (1975, 83). On the "forging" of Luba kings, see also Dewey and Childs 1996. The "ethnographic present" of these paragraphs is based on early-twentieth-century ethnography and

more recent research by Mary Nooter (1991) and William J. Dewey (1996) in the late 1980s. Nooter (1991, 150–52, 181 n. 22) was offered exegeses similar to Womersley's account of "forging the king," and was told that while in the Luba kingdom of Kabongo, where Womersley worked, dignitaries beat the king-to-be's knees with their fists, in other Luba polities hammer-anvils (*manyundo*) were used—lightly! Ritual and practical uses of this unusually multipurpose tool are discussed in chapter 15, this volume. Acute civil strife in the DRC has disrupted and altered many cultural practices these last years, and I am not sure if ritual processes continue as described here.

16. Roberts (1980, 206). Nfwimina, like its Fipa counterpart Ngufwila (Robert 1948, 235–40, 245), was wounded through a most elaborate deception so that its blood and flesh could be used to activate several sorts of exceedingly potent arcane sciences, including iron smelting (Roberts 1980, 233–57). Because the great healers who could achieve such miracles are long gone, in the 1970s my Tabwa interlocutors acknowledged using several structural equivalents to affect Nfwimina's crucial transformations. See Roberts (1980, 258–86).

17. On Tabwa health and healing, see Davis (2000). Of comparative ritual processes of smelting, those of Fipa peoples directly across Lake Tanganyika are especially pertinent. Fipa hold that they are "descended from Tabwa" (Willis 1964, 340–42) and political relations were still maintained across Lake Tanganyika in the 1970s. On Fipa ironworking, see Wyckaert (1914); Robert (1948); Wise (1958a; 1958b); Wembah-Rachid (1969); Wright (1985); Barndon (1996; 2004; 2012); Blakely (2006); Mapunda (2010); Ngonadi and Eze-Uzomaka (2015). Brelsford's (1949) study among Tabwa-related Chishinga of eastern Zambia and that of Merwe and Avery (1987) among Tumbuka of northern Malawi offer valuable details from which one may speculate about now-lost Tabwa ritual technologies of ironworking. See also Schmidt and Childs (1985); Davison and Mosley (1988); Lemelle (1992); Schmidt (2009).

18. How apprenticeship of Tabwa ironworkers was undertaken could not be determined during my research in the 1970s. Was there a "guild" or society of arcane knowledge, as there was for masters of big-game hunting? Tabwa-related peoples including Chishinga seem to have possessed ironworking guilds of mutual support. See Brelsford (1949, 28); Musambachime (2016, 75–81, 257). Also, how ironworking rituals were integrated into other religious practices remains unclear. A single account (Stuer 1903) hints that Tabwa ironworkers may have created or commissioned *nkisi* figural sculptures through which they interacted with tutelary spirits of their industry.

19. After Gell (1992). Several elderly men, including an *ngezia* whose praise name, Kapalamoto, means "the one blessed by fire," explained medicines of smelting to me in the 1970s, which they recalled from their time as apprentices many decades earlier. Such knowledge, so assiduously guarded in earlier times, was remembered then by very few and differed from one man to another. Indeed, various smelters undoubtedly would have possessed and deployed different medicinal elements and practices, as Tabwa diviner-healers do today. In this regard, Marcia Wright (2002) is correct to point out that, when possible, histories and exegeses of particular ironworkers should be studied. Doing so permits one to avoid flattening particular data to produce the generic accounts so common to, and in some ways inevitable in, historical ethnographic discourse.

20. Warburg was writing about rain dances among early twentieth-century Pueblo Indians (1923, 221), from Michaud (2007, 293–331), as cited in Kosky (2013, 32). It must be understood that smelting was not always successful due to its complexity. When it failed, explanations were sought, blame directed, and measures taken to assure a fruitful bloom the next time.

CHAPTER 2 (GOUCHER)

1. In many parts of sub-Saharan Africa, ores were surface-collected. In the Bassari region, women mined ores from shallow pits and dangerously worked the deeper shafts dug into the earth with tools.

2. Slags are by-products of both smelting and smithing activities. See Hans-Gert Bachmann (1982) for a fuller discussion of the importance of the identification of slags from archaeological sites and for examples of archaeometallurgical slags that have been useful indicators of process, ore source, and smelting conditions.

3. The production of charcoal took place through a process of dry distillation, in which the absence of oxygen facilitated the removal of water from wood. The relatively slow growth of certain hardwoods created denser and more complex structures (with higher carbon content) that burned more slowly and reached higher temperatures once made into charcoal. Softer woods typically used for fuel were also needed, and these were employed to fire the charcoal inside the furnace.

4. I am grateful to Louis Iles for sharing her unpublished paper (2017). This mixing of male and female ores as opposite forces has also been reported by Randi Barndon among Pangwa smelters of Tanzania, where the bodily fluids of sexual intercourse are viewed much the same, and the ores are washed together to become "as one, like a man and woman" (2012, 40); and by Allen F. Roberts among Tabwa of the DRC, as discussed in chapter 1, this volume.

5. According to Haaland et al. (2000), in southwestern Ethiopia these powers might also lead to the social and ritual isolation of the smelter and his wife. See also Barndon (2012); Dewey (1990); Mapunda (1995).

CHAPTER 3 (DREWAL)

1. Rupp et al. (2005); Atwood (2011); Abraham (2013); Tylecote (1975a; 1975b); Willett (1971); Junius (2016); Franke (2016). Gabriele Franke has recent radiocarbon dates of 750–550 BCE for ironworking at Nok (pers. comm., 2018).

2. Although men sometimes sculpt in clay, it is a medium ordinarily worked by women. This raises an intriguing question: Who were the sculptors of Nok? I believe the evidence strongly suggests women were the artists, pottery sculptors, of Nok.

3. Tom Fenn cited in Breunig and Rupp (2016). Although the mineral is found in astonishing profusion in the Copper Belt of today's south-central Democratic Republic of the Congo and north-central Zambia, what would be considered Copper Age tools, weapons, or regalia elsewhere in the world was not developed in central Africa (Allen F. Roberts pers. comm., 2017).

4. Whether performed or not, such masks are spiritually charged. When a forest spirit requests its mask-owner to act, to dance or engage in other actions, culture is shaped in specific ways. For example, the forest spirit Deangle enters the village to ask women for food during initiation camps. Tankagle and Bagle entertain and enliven the community with dances, skits, and mimes, whereas Bugle, a fighting spirit embodying the courage of warriors, historically led men into battle.

5. See Fischer (1978). Over time, any respected Dan mask can be elevated to the category *gunagle*, as a mask that represents a village quarter, or *gle wa*, a judicial mask. Dan people swear oaths on spirit-masks and the presence of iron may reinforce and seal such oaths, as among Yorùbá peoples.

6. This synthesis is based in the excellent field-based research of Glaze (1981), Gagliardi (2010a; 2010b; 2010c; 2015), and Gagliardi and Petridis (2015).

7. Based on linguistic evidence, Tal Tamari suggests that Tuareg endogamous "castes" of blacksmiths and bards were developed with Sudanic peoples. Tuareg terms for blacksmith and bard are of non-Berber origin and "the designation for the endogamous blacksmiths among the southern Tuareg is *gargassa* (a cognate of the Songhay *garaasa* and Fulani *garkasaae*), whereas it is *enaden* among the northern Tuareg (meaning 'the other')" (Tamari 1991, 244).

8. See Abiodun (1994); Ben-Amos (1995, 25); Drewal, Pemberton, and Abiodun (1989); Ezra (1992, 192).

9. Ezra (1992, 179, citing Roth 1968, 61) notes that in the nineteenth century *eben* were sent to Birmingham, England, for silver-plating and then returned to Benin, presumably to make their flash even greater when the blades twirled in the air.

10. This same balancing of *ada* and *eben* images (blades for different purposes essential to maintaining the Benin Kingdom and social hierarchy) occurs on a chief's carved door in Benin City (Curnow n.d.).

CHAPTER 4 (JOYCE)

1. So critical is the study of iron in the human body that many medical facilities and universities around the world have created divisions or separate institutions dedicated to advancing knowledge about the biological and medical roles iron, and the processes in which it participates, plays (for example, UCLA's recently founded Center for Iron Disorders). Annual congresses organized by the International BioIron Society (IBIS) provide international forums for discussions among peers and collaborators to promote research, public education, and biological studies of iron chemistry in humans.

2. Water remained a gas until earth cooled below 100 degrees Celsius (the boiling point of water) approximately 3.8 billion years ago, when water vapor was able to gather into droplets and release as rain to fill primordial oceans. See the website of the National Oceanic and Atmospheric Administration, a division of the U.S. Department of Commerce (www.noaa.gov).

3. This geologic stratum is 1.8–2.5 billion years old and accounts for 60 percent of the available iron reserves worldwide. In May 2018, the World Steel Association reported the world's crude steel production at 154.9 million tons, a 6.6 percent increase compared to May 2017. See https://www.worldsteel.org/media-centre/press-releases/2018/may-2018-crude-steel-production.html. See also Logan (1995, 10–13).

4. Anderson (1989, 147–55). Due to the density of earth's iron core, the relative percentage of iron diminishes closer to the mantle, where the mass abundance of the nine most prevalent elements in the earth's crust is approximately: 5.6 percent iron, 46 percent oxygen, 28 percent silicon, 8.2 percent aluminum, 4.2 percent calcium, 2.5 percent sodium, 2.4 percent magnesium, 2 percent potassium, and 0.61 percent titanium. Other elements occur at less than 0.15 percent.

5. In 1908, *The American Blacksmith*, a professional trade journal, published a three-part story on "The Blacksmith of Central Africa and His Work." Editor Walter O. Bernhardt reviewed tooling, forging, and smelting methods, illustrating the article with blades, currencies, farming implements, musical instruments, blacksmithing tools, and furnace designs. On smelting, the author states: "In regard to its homogeneousness and its malleability, the iron produced…is quite equal to the best forged iron of our own country. The dexterity of the smiths…is wonderful…and the short space of time in which they convert the raw material into spades and lances is unrivaled. The masterpieces… in refinement of form and neatness of finish…bear comparison with the productions of American and European smiths" (146–50, 183–85, 200–03). See also Celis (1991); Killick and Gordon (1993, 243–70); chapter 16, this volume.

6. Blacksmiths are, it should be further stressed, indispensable in their communities. Because the objects and services they provide can have decisive effects on individuals, blacksmiths tend to hold implicit power over people in need, at times eliciting fear. When sustenance is at stake, where lives are on the line, and wherever an ancestor's intervention or support is requested, the smith is at the forefront. The positive and negative potential embodied in many objects a blacksmith makes is understood by those who use his tools, weapons, ritual paraphernalia, etc., and staying on a blacksmith's favorable side engenders respectful appreciation and reverence. See Drewal (1997, 235–60).

7. In Yohonou, Togo, the blacksmith-chief Togbui Raphael Yovo Agosso relayed that young women in his community have increasingly picked up the hammer and been welcomed into the tutoring cycle to learn the art along with young men—not only to assist family members at established forges, but also to organize sole proprietorships and cooperative ventures (pers. comm., 2010).

8. When iron is heated in a forge, its surface naturally oxidizes from contact with blasts of air pumped by bellows. Each time iron is heated, approximately 3 percent of its thickness is lost to this flaking by-product called scale. It is jolted loose by hammer blows and accumulates in a scattering of thin flakes around an anvil's base as the blacksmith works. Occasionally airborne flakes land on the smith's clothing or skin.

9. Palm oil pressed from palm fruit at artisanal and commercial production sites throughout west and central Africa leave behind thick-shelled kernels high in thermal energy. Many workshops now fuel their forges with locally produced charcoal made in small batches from these palm fruit hulls, providing a high-heat, regionally grown, renewable alternative to conventional charcoal fuels. Adewumi and Ogedengbe (2005, 1082–87).

10. This process very likely lent gravitas toward iron becoming one of the most important bridewealth offerings in marriage contracts, providing a stable metaphor to launch and bless a marriage. This, and the fact that iron is "pregnant" with potential to become anything the bride's family chooses to have made from it, also stands as a wish and a promise for the prosperous continuation of the newlyweds' bloodlines. See chapter 5, this volume.

11. In Tchare, Togo, in 2010, village historians and retired blacksmiths Yom Pyim and Amourou Kondo, octogenarian uncles of Kossi Kao, explained to me, through interpreter Pakétam Kourakoma, that knowledge of how to forge, among Kabre blacksmiths, is transferred by the spirit of those who founded the village, and therefore cannot be passed to someone outside the familial lineage. Libations are offered to secure the work site so that ancestral relatives know their knowledge is respected and their voices are heard.

12. Iron and all the various steel alloys made from it are bound by distinct forging temperature ranges based on each material's chemical composition. No matter what the alloy, the material will begin to melt when upper-end temperatures are exceeded; oxygen will attack the surface as carbon, which makes iron strong, will sizzle away. What remains of the ferrous compound, will be irreparably damaged and rendered too fragile to make anything of use.

13. The dynamic physics occurring at impact zones in forging is overtly understood and applied as an additive presence on wooden sculptures that incorporate iron accoutrements—worn or secured—

onto various surfaces of a figure. Examples include, to name but a few: small iron staffs, replicas offull-sized versions used in ritual contexts, emerging from the crown of the head on some Gun/Yorùbá figures (see fig. 12.24); miniature anvils with tangs imbedded into potent points of the body on Luba and Songye carvings (see figs. 15.21, 15.30); and shards of iron, blades, tools, and nails driven into animal and human figures of the Songye, Vili, Woyo, Yombe, and Kongo peoples in central Africa (see figs. K.1–K.4). What these carved sculptures all have in common is the activating principle of iron, which works as a channeling conduit to amplify and aim with precision the power source embodied in the object and emanating from the forge. The strength of iron, the decisive action of the smith, and the heat of the process are mediated by the sculpture's ritual handler to join forces for a common goal, delivered with fearsome purpose. See also Griaule (1965, 87); Arens and Karp (1989, xi–xxix); LaGamma (2015, 220–65).

14. A hammer's blow "collides" with a stationary piece of iron on the anvil and the effects of this collision are governed by Newton's laws of motion. The momentum of a moving object is the product of its mass plus velocity. A two- or three-fold increase of momentum benefits forging's effectiveness as it moves material, while at the same time creating a rebound in the opposite direction of the force exerted.

15. A forging may look homogenous on the surface, but inside, under the skin, its strength may be compromised by internal cracks that can cause breakage, soft spots that weaken and bend, or incorrect temper that can cause an object to snap in two. All of these failures, and many more, result from improper forging techniques, and they are quite difficult to detect. However, the smith intimately knows when something has gone awry, and it is up to him to rectify the breach of security these weaknesses promote.

16. Sensory perception transmitted through hands applies to the iron rods sometimes used to "handle" power figures known as *nkisi* or *nkishi*, such as a Songye example from the Musée Africain de Namur (fig. 15.32) that has iron "reins" attached to the figure; like a blacksmith's tongs, they provide safe mediation when handling potentially dangerous forces within and beyond one's control.

17. Though archaeological excavations of early Iron Age smelting sites indicate the likelihood that delivery systems for preheated air had been widespread in Africa for centuries, it was not until 1828, in Scotland, that a similar discovery was made in the West. Called "hot blast" by the inventor, James Beaumont Neilson, his patented version for introducing preheated air into blast furnaces was considered the most important technology developed for iron smelting during the European Industrial Revolution. Belford (2012, 32–44).

18. See Oliver and Fagan (1975, 70–105, 203–14); de Barros (2000); chapter 2, this volume.

19. Surplus oxygen will heavily scale iron, imparting a crusty layer of oxidation that contaminates if hammered into the working surface. Many anvils have a slightly crowned or sloping face that combats this by-product by encouraging oxidized flakes to slough off the working surface as soon as scale falls from the piece, jolted loose from impact.

20. During a research trip to Togo in 2008, while visiting blacksmiths in the rural forging community of Yohonou, I noticed on the first day of fieldwork that the distinctive overlay of sound from bellows pumping and hammers striking was uncannily similar to African drumming rhythms. The changing tempo and tonal fluctuation communicated to the smith when a piece of iron was ready in the fire. Throughout the day, the bellows' hiss and the hammer's blow never lost their melodious and relational intonation. Driven by the timbre of the work, it seemed no surprise that percussive rhythms would have originated in workshops such as this. Curious whether a musician would recognize formal musical foundations in the sonic textures produced at the forge, I reached out to friend, musician, and ethnomusicologist Steven Feld, who was then working with jazz musicians in Accra, Ghana. Accepting the invitation, he joined me for a few days to listen, assess, and record the sounds of making, while Ewe blacksmiths Galbert Atakpa and Hodenou Noglo produced a classic double bell called a *gangokui* (see figs 4.13, 4.31, 17.5). Thus began a rewarding two-year collaborative research project among Ewe smiths in southern Togo and with equally gifted Kabre blacksmiths working in the north. See also McNaughton (1988, 24–25, 30).

21. Strikers are especially necessary when forging large or cumbersome work, and I have seen as many as five additional strikers conducting a rhythmic wave of steady blows to move material quickly.

22. Proprioceptors are sensory receptors in muscles, tendons, joints, and other body tissues that detect motion, signaling to the nervous system to maintain balance and orient one's body in space.

23. In 2010, during the second of my two interviews with blacksmith-chief Togbui Raphael Yovo Agosso of Yohonou, Togo, I was invited to visit his family's shrine. Inside the circular, thatched, adobe structure was an altar built up of bellows, hammers, tongs, chisels, punches, and other blacksmithing tools used by his predecessors. Glistening with fresh libations and food offerings, it was clear the venerated ancestors were respectfully remembered, named, and regularly fed by the family.

24. Since I was handed a hammer at age thirteen by my own mentor, iron and the tools used to shape it have, as if they were living entities, metamorphosed in my mind into emblematic symbols gathered together into a personal *tenet*. Beyond a principle or a belief, I identify with that term's earlier Latin root in the verb *tenere*, meaning to grasp, to possess or maintain, to hold in the mind, to take in, to understand and to stretch—these are the attributes the hammer and its curriculum teach me. The functioning ideology the tools of the forge manifest when put to use provides a particularly discrete resource for understanding materials and processes that transform an element in the universe. African blacksmiths have implemented innumerable variations along this line of reasoning with brilliant results in the objects illustrated throughout this volume.

25. Michel Foucault reflects on this notion, where things come sufficiently close to one another to be in juxtaposition: "their edges touch, their fringes merge, the extremity of the one also denotes the beginning of the other. So that in this hinge between two things a resemblance appears. A resemblance that becomes double as soon as one attempts to unravel it: a resemblance of place, the site on which nature has placed the two things, and thus…in this natural container: the world, adjacency is not an exterior relation between things but the sign of a relationship, obscure though it may be" (1994, 18).

INTERLEAF B (JOYCE)

1. See Ndiaye (2012, 34). During a symposium organized by the International Network of Ethnography Museums and World Cultures (RIME), with representatives from ten European institutions holding significant ethnographic collections (including many field photographs) amassed during the colonial era, Belgian anthropologist Ken Kelountang Ndiaye, born into the Manjaco population of Senegal, had this to report: "The role of subjectivities is essential in the museum chain. From the producer of the artifact, to the researcher, to the conservators, and curators, and guides, to the visitor—the exposed truth changes constantly. The attitude is arrogant but also tiresome, always waving away, with the back of one's hand legitimate demands of the source populations, co-narrating histories on the grounds of expertise that are, in fact, deficient or fatuous."

CHAPTER 5 (ROBERTS)

1. Smith continues: "[W]e tend to become most conscious of our own evaluative behavior when the need to select among an array of alternative 'goods' and/or resolve an internal 'contest of sentiments' moves us to specifically verbal or other symbolic forms of accounting.…Most of these 'calculations,' however, are performed intuitively and inarticulately, and many of them are so recurrent that the habitual arithmetic…comprises the very style of our being and behavior, forming what we may call our principles or tastes."

2. Wiredu as cited in Santos (2016, 203–04) without particular reference, but see Wiredu (1997).

3. African currency tokens have been the subject of a number of engaging museum exhibitions, including *Nigeria and the Evolution of Money* (1979), curated by Ekpo Eyo; *The Artistry of African Currency* (2000), at the Smithsonian's National Museum of African Art; Karel Nel's *Bridewealth, Currency and Other Assets* (2000), at the Standard Bank of Johannesburg; and *The Art of Money* (2007), by Eric Robertson. For a children's introduction to "what is money" with reference to African currency tokens, see Avery and Allen Roberts (1988), in a special issue of *FACES: The Magazine About People*.

4. Today's crypto- and other digital currencies suggest how fragile such associations may be. The quotations of this paragraph are from Hart (2005, unpaginated), reflecting on Georg Simmel's classic treatise of 1900, *The Philosophy of Money*. Hart reminds us that the word "money" comes to us from Moneta, Roman goddess of memory and mother of the Muses. "Her name was derived from the Latin verb *moneo*, whose first meaning is to remind, bring to one's recollection. For the Romans, money, like the arts, was an instrument of collective memory that needed divine protection. As such, it was both a memento of the past and a sign of the future." See also see Mack (2003, 26).

5. The first lines of this paragraph are from Ingold (2007, 14–15) with partial sentences combined to make this a paraphrase. The proposition of a "radical ontology" is from Ingold (2010, 91, 95).

6. "Cultural entailments" is my extension of the now-classic propositions of Lakoff and Johnson (1980, 9, 193) that "metaphorical entailments…characterize a coherent system…of concepts and a corresponding system of…expressions for those concepts" that implicate both rationality and poetic imagination. Thanks to Rowland Abiodun for discussion of this point.

7. On Alison Saar's sense of the "wisdom" of materials, see M. N. Roberts and Saar (2000, 41); cf. M. N. Roberts (1994). A most poignant example of the wisdom of iron is a work by *Striking Iron* lead curator Tom Joyce for the National September 11 Memorial and Museum in lower Manhattan: he created letters by reforging steel recovered from the Twin Towers to spell out "no day shall erase you from the memory of time," a line from Virgil's *Aeneid*. See Nathanson (2012).

8. Paraphrasing Bennett (2010, vi–viii), who borrows the term "actant" from Latour (1999). Ògún as "the Lord of the Cutting Edge" is from Thompson (1984, 53), while *àṣẹ* as "performative power" is from Abiodun and Drewal in chapter 12, this volume.

9. For overviews of such matters, see Douglas (2004) and Dalton (1969). For a detailed account diverging from some of the conclusions drawn here concerning currencies and commodities, see Kriger (1999, 86–117) and chapter 5, this volume.

10. "Wealth in people" is a phrase developed by Miers and Kopytoff (1977, 55), as qualified and developed here by Guyer (1993, 243–44), also citing Joseph Miller (1988, 43, 46, 50).

11. The essays in Radcliffe-Brown and Ford (2015) remain valuable resources, although such customs have been adapted to cope with postcolonial politics, radical urbanization, long civil unrest, and other contemporary factors. Bridewealth in African societies has often been deliberately misunderstood by outside observers bent upon denying humanity—that is, love, intellect, and generous purpose—to people deemed in need of "saving" and "civilization" via Eurocentric paradigms. See Bunting et al. (2014) for discussion of "contestation over consent and coercion" in contemporary African marital circumstances.

12. Decades of civil strife and recent petroleum exploitation have brought radical change to some segments of Chadian society, and practices described here may no longer be observed or even readily remembered.

13. Both *kokio* and *kuur* currency tokens were displayed in an exhibition case labeled "Numismatique" at the Chadian National Museum in the southern city of Sarh, as viewed in 2016 on a website no longer available.

14. On the Bagirmi kingdom of the early sixteenth to late nineteenth century, see Reyna (1990). When I spent three years teaching middle school in southern Chad in the 1960s, men of several ethnic groups routinely carried throwing knives over their shoulders, especially when they traveled on foot far from their homesteads. In 1966, a homicide in the town of Pala occurred when a Sara-Ngambaye man, upon returning home at night, discovered another man leaving his wife's sleeping house. Seizing the F-shaped throwing knife over his shoulder, he struck the man in the chest with its medial blade, killing him instantly. In 1969, Alan Etzel, then a Peace Corps Volunteer stationed in the Chadian town of Laï, was invited to go hunting with local friends who managed to break the legs and dispatch an antelope by skillfully slinging their throwing knives sidearm, low and parallel to the ground (pers. comm., 2017).

15. Fortier (1982, 52, 122–24, 127, 142, 148, 163). Joseph Fortier, a Catholic missionary and ardent ethnographer in Chad for many years, also published a 1967 monograph about myths of the trickster-hero Suu. Suu brought gifts to humanity including an iron balafon—a xylophone ordinarily made with wooden keys and resonators made from long bottle gourds—and he became the first blacksmith among Sara. A palladium is "a sacred object having the power to preserve a city or state possessing it," and is a term derived from "the fabled statue of Pallas Athena that assured the safety of Troy as long as it remained in the city" (Morris 1967, 944–45).

16. Dupré (1995, 83–85). Georges Dupré, who conducted his ethnographic research in the last decades of the twentieth century, discusses other aspects of Kwele use of iron currencies, including careful counting of tokens bound in packets of ten, with tallies kept by using sticks of two different lengths (ibid., 85, 94). Arithmetic as an aspect of exchange that could become complex when large numbers of tokens and other objects were promised through bridewealth (see Guyer 1986, 1995) is a significant feature of such objects and practices that cannot be given proper attention here, but see Iroko (1988).

17. Basu (2017, 14), discussing Thomas (1991, 27–30). Such "promiscuity" continues as objects are removed from their original contexts, translated and so commodified as "art," and brought to international art markets. Such was the case in 1988, when Michael Ward, a prestigious dealer in ancient and medieval arts, mounted an exhibition at his gallery in Manhattan called *Symbols of Wealth: Abstractions in African Metalwork*. It featured a *ndjong* token he sold for many thousands of dollars (Ward 1988, fig. 8; pers. comm., 1988). In the early 1980s, I purchased four *mandjong* from a street-side African *antiquaire* in Libreville, Gabon, for fifteen U.S. dollars for the lot, which was, at the time, a sufficient sum locally.

18. Dupré (1995, 87–88) offers a version of this hypothesis, acknowledging that his thoughts were influenced by a conference paper he heard me present called "The Uncommon Currency of Masks" (Roberts 1992).

19. The iconography of Ekuk masks sometimes suggests the curved horns of rams and a few other animals. It is not clear to me to what degree Ekuk masked performance continues among Kwele. See Perrois (2001); LaGamma (2007, 288–95).

20. In 2004, the late master blacksmith Russell Jaqua forged an iron gate of greatly oversized Kissi penny shapes that now serves as the threshold to my family's home in Los Angeles. Russell was a good friend and remarkable artist, and this chapter is dedicated to his memory.

CHAPTER 6 (ROBERTS)

1. Fraser and Cole (1972, 308) have considered how swords and other implements may serve as "symbolic (or actual) extensions of the leader's physical power." "Tool-being" is a concept developed by Graham Harman in his reflections on Martin Heidegger's approach to "tool-analysis" as well as the philosopher's further sense of *Zuhandenheit*, or "readiness to hand," as explicated in *Being and Time*. Harman suggests that "philosophy must reawaken its lost talent for unleashing the enfolded forces trapped in the things themselves" (2002, 1, 98). One is also reminded of Alfred Gell's sense of "person-things" (1998, 232), and Jane Bennett's consideration of how "human being and thinghood overlap...and...slip-slide into each other" (2010, 4).

2. Cf. Derrida (1993), on a blind person's stick as an "optical prosthesis."

3. M. N. Roberts (1996, 179). Luba spears are mnemonic devices, and proleptic understanding that all royal spears and other emblems "are" those of the culture hero is presented in Roberts and Roberts (1996; 2007), as based upon Nooter (1991, 165–67 and passim).

4. Cole (2017b, 143). On associations between Gwan figures, smithing, and smelting furnaces, see Colleyn (2009, 33), as discussed in Cole (2017, 142–45); see also Ezra (1986). Needless to say, "manliness" and "womanliness" are culturally determined. Luba, for example, hold that "the king is a woman." Noblemen once wore women's coiffures, and figural sculptures depicting women refer to men at a second register, for Luba women hold the precepts and secrets of royal men's bearing and prerogatives in their breasts. See M. N. Roberts (2013).

5. The Tohoessou case study is presented in detail in Roberts (1996, 94–100). The ethnographic present of these sentences is 1994, when Dr. C. Joseph Adande (now Emeritus Professor of Art History, Université Nationale du Bénin/Abomey-Calavi) and I conducted a brief research project with the Tohoessou family of blacksmiths.

6. The phrase from the *oríkì* (praise-poem) is Ògbógbódirin (Abiodun 2014, 219). On *ara* inventiveness as an element of Yorùbá aesthetics, see Adepegba (1983). The Tohoessou family is of a western Yorùbá community called Aja. *Assanyin* (following Dr. Adande's orthography) are similar to *asen* commemorative works forged by Fon blacksmiths, but the term also links such important material culture to Osanyin healers' staffs among Yorùbá, thus exemplifying movement and adaptation of ideas and practices among regional peoples. See also chapter 12, this volume.

7. For compelling examples of tools varying across cultures and ecologies, see Rubin (1989); Janzen (2009); Kasfir (2007). On how blade use is learned, see Dira and Hewlett (2016).

8. See Schildkrout and Keim (1990, 107); Simoons (1958, 398). Tim Ingold offers an evocative reflection on how the ancient Greek word "for the skill of the practitioner, *techne*, is derived from the Sanskrit words for axe, *tasha*, and the carpenter, *taksan*," that also lead to "technique" and "technology" (2010, 92).

9. For an example among Songye peoples of the DRC, see Hersak (1986, 57). On blade use in the production of figural sculpture, see Ottenberg (1975); Drewal (1980); Rubin (1985); Burssens (2003). The essays in *The Traditional Artist in African Societies* (1975), edited by Warren L. d'Azevedo, still have few peers; chapters by William Bascom on the Yorùbá master carver Duga of Meko, John Messenger on how wood carving is learned among Anang, and James Fernandez on four Fang carvers remain especially stimulating.

10. Sennett (2008, 195; emphasis added). Though he is writing of screwdrivers, Sennett's thoughts can easily be extended to at least some African bladed tools. On "polyvalent" African farming tools, see M.-C. Dupré (2000) and Blench (2006).

11. Thanks to Manuel Jordán for this information (pers. comm., 2017). He notes that very few such remarkable Chokwe blades are known (but see Elsen and Felix 2009, 343, 348) and their perfection is such that some have assumed they were commissioned and even designed by Europeans—as though Africans could not create and possess the most magnificent crafts of their own cultures. Chibinda Ilunga is conceptually related to the Luba hero Kalala Ilunga, and in some accounts he is Kalala's junior kinsman. On arts of Chokwe and related peoples, see Jordán (1998). On commonalities of Chokwe, Luba, and other central African origin myths, see Heusch (1982).

12. M. N. Roberts (1996, 179–80), as derived from Nooter (1991). Archaeologists have discovered ceremonial axes as first millennium BCE grave goods in lands where Luba now live, suggesting long cultural continuities and histories of "tool-being." See Childs and Maret (1996, 51, 56). See also interleaf E, this volume.

13. See M. N. Roberts (2000) on how Luba diviners "set meaning before the eyes" in this manner; cf. Victor Turner's classic *A Forest of Symbols* (1970). Luba senses of bladed implements possessing efficacies as *nkisi* complement other examples of "iron's empowering presence" discussed in chapter 3, this volume.

14. Slightly paraphrasing Childs and Dewey (1996, 154–55), who cite Van Avermaet and Mbuya (1954, 649–50) regarding the Luba verb *kusunga*. Childs and Dewey also discuss incising, copper cladding, and other embellishments of Luba axes, drawing upon their own research and that of Mary Nooter (1991).

15. Thompson (1973, 142, slightly paraphrased), in reference to Jan Vansina's discussion of Tio (Teke) kings of the Congo, whose "messengers would go when ordered, bearing the king's adze to bring his orders to squires and village leaders" (1973, 335). Cole has written of these same "tongue-as-blade adzes" and how they manifest a "link between...powers of leadership and those of commanding speech" (1989, 110–11). Would that we knew of particular "speech acts" performed as singular narrative events instigated by such evocative instruments, following the reasoning of J. L. Austin (2005).

16. The significance of such hunting for practical and symbolic ends is suggested by an episode of the Luba Epic, when the "drunken king" Nkongolo Mwamba instructs his sister's son, Kalala Ilunga, to dance before him on a mat that hides just such a pit trap fitted out with spears. Kalala becomes aware of and eludes the trap as a triumph still recalled and performed. See M. N. Roberts (1996b). For examples of Luba sculpture depicting a heroic spirit standing upon or riding astride an elephant or buffalo, see Roberts and Roberts (1996, 216–19).

17. Spring (1993, 45). Christopher Spring's *African Arms and Armor* (1993) remains a useful survey with pertinent bibliography. See also the studies of R. S. Smith (1989); Thornton (2000); R. J. Reid (2012). A number of catalogues from exhibitions of African weapons offer sumptuous photography and provocative ideas; notable are Fischer and Zirngibl (1978); Elsen (2003); Elsen and Felix (2009).

18. From my research of the early 1980s in Tangaye, Burkina Faso. See Roberts (1987).

19. Mack (2007, 144). On miniaturization, see also Stewart (1993). On Tabwa divination, see Davis (2000) and Roberts (2000); on Luba divination, M. N. Roberts (2000).

20. Van Avermaet and Mbuya (1954, 248–50), where vocabulary for types and parts of arrows is offered, as well as explanation of how iron arrow points were (are?) used symbolically in purposeful, cut-to-the-chase decision-making in marriage and litigation.

21. Spring (1993, 33–46). On mail produced in the Sudan, see Arkell (1956) and interleaf D, this volume. In the mid-1960s, the present author saw mail worn by equestrian guards of the Lamido of Rey Bouba, a Fulani kingdom of northern Cameroon, now a UNESCO World Heritage Site.

22. See Hamilton (1998) on the veritable industry of histories, polemics, novels, plays, theme parks, tourist souvenirs, and more, all based upon Shaka and his legendary exploits. Because of such romanticism, even seemingly straightforward assertions like those presented here, as based upon Spring (1993, 126–38), must be accepted with caution. See Maggs (1991). As always, there is no single way to understand historic persons or events, and Zulu approaches to Shaka differ from positivist accounts. See the late Zulu poet Masizi Kunene's recently reissued *Emperor Shaka the Great* (2017), written during the apartheid period to give courage to the oppressed.

23. Stewart (1994, 204), as developed with regard to African examples in Roberts (2013a, 108). The defiant display and use of spears and other blades in the wildly popular superhero film *Black Panther* (2017) is a response; for an overview of the movie, see Johnson (2017).

24. See Coombes (1994). An immense scholarship has addressed such fraught matters, with the Congolese philosopher V. Y. Mudimbe's groundbreaking reflections in *The Invention of Africa* (1988) and *The Idea of Africa* (1994) still excellent places to start one's reading.

INTERLEAF C (DEWEY)

1. Paraphrased from a praise-poem sung by a blacksmith's spouse as recorded in 1965 at Kambasha in Bindura District, Zimbabwe. See Hodza and Fortune (1979, 351–56).

2. See Hannan (1981) for such terminology.

3. See Caton-Thompson (1931, 110, 214); Randall-MacIver (1906, 45, 79–80, pls. 20 31); Garlake (1973, pl. 66).

4. See Bent (1892, 181–82); Hall (1905, pl. 2); Caton-Thompson (1931, 110, 214, pl. 21 [1:3]).

5. Fernandes (1560, 74). Arthur Pendered found a hunter in 1930 who recalled how axes were still used in this manner to hunt elephants by "cutting their hamstrings" (1930, 47).

6. Phrasing throughout is from my interviews in Zimbabwe in the mid-1980s.

7. Illness or distress is a common theme in stories related by blacksmiths and other artists about the origin and nature of their talents, as is the case with spirit mediums and diviners. See Fry (1976); Bourdillon (1982).

8. *N'anga* is the same as *nganga* in other Bantu languages. Standard Shona language (Hannan 1981) uses *n'* to represent the nasal sound, such as in the English word *singing*.

9. David Lan gives the following interpretation for the same event: "The ceremonial axe with which they are being presented is identical to those used by spirit mediums and symbolizes the legitimate ownership of the land" (1985, pl. 5).

10. I am here thinking of contributions to understanding the efficacies of objects by such provocative authors as Gell (1998), Miller (2005), and Houtman and Birgit Meyer (2012).

CHAPTER 7 (MCNAUGHTON)

1. Nachtigal (1987, 3: 323). Bagirmi was southeast of Lake Chad in what is now the nation of Chad, at the border with northern Cameroon.

2. Peter Westerdijk (1988, 18–21; 2006, 32–33) discusses throwing knife creation, as does S. Terry Childs (2016), who carried out a careful laboratory analysis on a zigzagging three-bladed knife collected by Herbert Ward and now in the collection of the Smithsonian Institution, Washington, D.C. Using microscope and X-ray, she discovered impressive expertise, including several welds invisible even under the microscope and a central shaft split to create an elegant curve just above the handle.

3. Du Chaillu (1861, 79). Du Chaillu was a fanciful writer, as Westerdijk notes, citing James Fernandez (1982, 33–35).

4. For example, see Schweinfurth (1874, II: 9); Thomas (1925, 130); Nachtigal (1991, I, 453). An eyewitness account from southern Chad in the 1960s is mentioned in chapter 5, note 14, present volume.

5. Westerdijk (1988, 22–42); Schmidt and Westerdijk (2006, 33–39). See also Spring (1993, 79–82); Schweinfurth (1874, II, 9–11); Evans-Pritchard (1957).

6. For examples, see Arkell (1939); Nalder (1935); Olderogge (1934); Philipps (1926); Ringdom (1936); Thompson (1925). More recently, Westerdijk (1988) offers an exacting stylistic analysis, a great deal of technical information, and an encyclopedic bibliography. See also McNaughton (1970); Spring (1993, 68–83). It should be noted that earlier sources often reflect imperialist and racist perspectives of their times.

7. For excellent summaries of these expanded realms of use, see Arkell (1939); Westerdijk (1988, 42–44); Schmidt and Westerdijk (2006, 39–41); Spring (1993).

INTERLEAF D (ROBERTS)

1. On the Wellcome Collection of the UCLA Fowler Museum, see Berns (2014). One of the Fowler Mahdist blades is presented in McNaughton (1970, 58) and in Berns (2014, 96); two others are discussed by Bravmann, as is the Mahdist headpiece (1983, 53–54, 57). The latter is also presented by Arnoldi and Kreamer (1995, 32–33). Threads about Mahdist blades and their calligraphy on www.vikingsword.com are informative, and I am grateful to the colleagues who have contributed to this piece, and especially to Jim McDougall and James Moore.

2. Reid (1936, 72); cf. Grant (ca.1895, 66–67, 79). On Mahdism in sub-Saharan Africa, see Monteil (1965, 73). On the Great Comet of 1882 auguring the advent of the Sudanese Madhi, see Wingate (1893, 44). On comets and Islamic messianism, see Cook (2009, 1, 9–10); cf. Roberts (1982; 2013c). The Qur'an (72:8–9) warns that anyone doubting God's revelations "will find a comet in wait for him" (Fakhry 1997, 382), and the shapes of comets have sometimes been associated with divine "swords of truth" (Schechner 1997, 35–37).

3. Count Gleichen, cited at http://www.victorianmilitarysociety.org.uk/reserach/2012-09-09-11-20-22/archive/63-sudanese-weapons-of-the-late-19th-century. This website offers discussion of the hand-held weapons strewn upon Kitchener's killing field. Charles Glass (2017) coined "horrorscape" to describe recent obliteration of Aleppo, Syria.

4. See Elsen (2003, 78–81); Spring (1993, 78–79).

5. Bravmann (1983, 54), of one of the Fowler's blades (see fig. D.6). The Arabic phrase is from Akhtar (2008, 374, n. 3), with reference to sura 72:1, which tells of the Prophet's revelation of the Holy Qur'an to a company of jinn. On "fire-bluing" and other technical characteristics of similar acid-etched Mahdist blades, see Grady (2007).

6. Quotations are from a paper in progress by Fallou Ngom and Mustapha Kurfi. They analyze a Mandinka manuscript dated 1789 CE, now in the Harvard University Library, in which a specific jinni (singular of jinn) is invoked through repetition and inversion of the Arabic letter *kaf*. In scrutinizing photographs of the Fowler's Mahdist blades, Christiane Gruber finds that "in a number of instances" the Arabic "word for victory (*al-fath*) is repeated in a stylized and almost square-shaped format. If this reading is correct, then the blades are 'encrusted' with patterns proclaiming triumph.... Typically such inscriptions are believed to be both apotropaic and predictive, channeling and propelling divine intervention in situations of warfare" (pers. comm., 2017).

7. The inscriptions may have been meant to repel or otherwise keep at bay evil jinn who were conniving with invading enemies, and it is understood by some that ill-intentioned jinn fear iron like that of these calligraphic blades (Lebling 2010, 109–10). In west African oral narratives, however, jinn figure prominently and positively in key historic battles (Fallou Ngom pers. comm., 2017; see Camara 2010, xxvi). Jinn are also associated with wars of our times. In Tunisia, some feel that jinn helped people resist everyday French colonial injustices (William Gallois pers. comm., 2017), and in Afghanistan, they helped thwart Soviet occupation (Anon. 2006).

CHAPTER 8 (CHIRIKURE ET AL.)

1. See Alpern (2005); Chirikure (2015). These broad terms are used routinely by many disciplines; they do not imply that stone tools ceased to be used, or were necessarily less functional or aesthetically less desirable than iron ones. Even now, African blacksmiths use stone anvils and hammers when they are more efficient and can be wielded with greater accuracy than their iron counterparts (see chapter 4, this volume).

2. While Meroitic iron production could be earlier than that of Egypt, and Meroitic smelting techniques could have come from other places such as India (see Haaland 2014), well-known cultural connections among Egypt, Nubia, and Kush make such possibilities unlikely.

3. See Chirikure (2015). Gold and bronze (metal and alloy) technologies were introduced to Africa south of the Sahara at later dates, yet were known for millennia in regions of northern Africa after demonstrable contact with the Roman and Islamic worlds (Fenn et al. 2010) and later with Islamic societies via Indian Ocean trade to eastern and southern Africa. Why would Africans to the south only adopt iron and not gold and bronze technologies? That these questions are unanswered exposes the limitations of simplistic evolutionary models for technological development and change.

4. See Alpern (2005); Chirikure et al. (2010); Holl (2009); Killick (2004); Pringle (2009); Zangato and Holl (2010). Henry Louis Gates, Jr., in the first episode of his PBS television series *Africa's Great Civilizations* (2017), opines that iron smelting began in the CAR sometime between 1800–1500 BCE, making the discovery of iron in Africa either earlier than or at the same time as in Anatolia. Gates makes no mention of the problematic nature of these CAR sites.

5. In Nigeria, there are sites in the Nok and Nsukka areas that date to the 800–400 BCE range (Breunig 2014; Fagg 1969; Jemkur 2006; Okafor 1993; Eze-Uzomaka 2013). In Cameroon, there are sites with dates ranging between 550–400 BCE (Essomba 2004; MacEachern 1996; Woodhouse 1998; Pole 2010). From Senegal, there are dates from the eighth to the third century BCE (Deme 2003). From Niger, there are sites that have been dated to 800 BCE and to between 500 BCE and the cusp of the Common Era (Grébénart 1988; Quenchon 2004).

6. For Tanzania, Rwanda, and Burundi, see Schmidt (1997); van Grunderbeek (2001); Killick (2009a). Iron appeared in southern Africa at a later date, around CE 200, and this is associated with the southward spread of Bantu-speaking peoples.

CHAPTER 9 (MACEACHERN)

1. This tends to substantiate the idea that the southern Sahara saw important developments in different kinds of metallurgy in the early first millennium BCE.

2. Radiocarbon dating measures the decay of the radioactive carbon 14 isotope of carbon, which is found in any organic material—charcoal, bone, seeds. Thermoluminescence dating measures the gradual accumulation of trapped electrons in materials including ceramics and sand by measuring photons emitted when such materials are heated. The two dating techniques are complementary, since they use different processes and materials; comparisons between them are thus very useful to archaeologists.

3. These complex social systems involve social hierarchy, endogamy, ideas about ritual purity and impurity, and occupational specialization. For example, blacksmiths might be considered "polluted," of a lower status, and only marriageable to offspring of other blacksmiths (see chapter 11, this volume).

CHAPTER 10 (CHIRIKURE AND DEWEY)

1. Humphris (2018), however, is exploring whether the nearby sites of Beati Neqwal and Mariam Kadih (with evidence of large-scale iron production and Axumite dates) could have been the sources of their iron.

2. See Iles and Lyaya (2015) for an excellent overview of research on metals in East Africa from the 1960s onward.

3. Steel is an alloy of iron and carbon, in which carbon ranges from just under 0.1 percent for low-carbon steel to 2.1 percent for high-carbon steel. Wrought iron, with a very low-carbon content, usually less than 0.08 percent, is not considered to be steel. It was, however, the more common product in Africa. Wrought iron is more malleable; steel holds a sharper edge on weapons and tools (Chirikure 2015).

4. See also Schmidt (1996; 1997; 2013). The most accessible source for this archaeological research is the film *The Tree of Iron* (Muhly et al. 1988), which documents octogenarian Haya smelters reconstructing traditional iron smelting furnaces and demonstrating technological principles, from charcoal preparation to furnace construction, that the ancients also used to obtain high furnace temperatures and produce a high-carbon steel.

5. Marcia Wright's work among Fipa with the former master smelter Mzee Stefano and Terry Childs's interviews with Toro smelter and smith Adyeri are among the most prominent examples.

6. See the essays in Meier and Purpura (2018).

7. See Maret (1985, 86).

INTERLEAF F (CHIRIKURE)

1. Published in Hodza and Fortune (1979), the full praise-poem is available at http://www.africanpoems.net/praise/in-praise-of-the-ironsmith/.

CHAPTER 11 (MCNAUGHTON)

1. See Berns (1993); Frank (1993; 1994; 1998; 2002; 2007a; 2007b); Mayor (2010a; 2010b). These sources offer a cross-sectional glimpse of the technical sophistication and aesthetic acumen of Africa's ceramic traditions. They also suggest that women played major roles in the history of African ceramic sculpture

2. See Doumbia (1936); Kanté (1993); McNaughton (1979a; 1987; 1988; 1995); N'Diayé (1970); Sidibé (1959).

3. As examples, Conrad (2001) and most of the authors in Conrad and Frank (1995, 1–17) consider in detail the limitations and colonialist trappings of the caste designation; Johnson (1986, 22, 68, n. 1) points out critical differences with castes in India; McNaughton (1988, 7–21) argues against the term and its implications; and Tamari (1991) puts the word in quotes.

4. In important ways, *nyama* is similar to *àṣẹ* as understood by Yorùbá peoples and the energies of *nkisi* as defined by Bantu-speaking peoples of central and eastern Africa. See chapters 12 and 15 as well as interleaf K, this volume.

5. For smiths and other Mande citizens, wilderness spirits are extremely potent beings capable of invisibility and shape changing. They possess super-human capabilities for understanding and acting in the world. Their behavior towards humans can be playful, helpful, capricious, vengeful, or malevolent and dangerous. Many are associated with particular geographic features, and they are sometimes conceived of as having communities and social organization resembling that of people. Mande commonly identify them with Arabic loan words such as *jinn*. See also chapter 5 and interleafs D and M, this volume.

6. See Appia (1965, 319–20); Kanté (1993, 44–45); McNaughton (1988, 18, 134–40).

7. See Dieterlen and Cissé (1972); Diamitani (2008); McNaughton (1979b; 1988, 129–45); O'Hern et al. (2016); Travélé (1929).

8. See Deme and McIntosh (2006, 341); MacDonald et al. (2003; 2009); Vernet (1993, 331, 356; 2000).

9. See Haaland (1980); MacDonald (1996); Téréba Togola (1996, 92–107; 2008, 79–83).

10. Wagadu and the Mali Empire were great polities grounded in complex trade networks and effective statecraft. Both were long-lived and are celebrated in oral traditions, most notably the Sunjata Epic.

11. Association SCOA (1980, 70–73; 1984, 113, 117); Bathily (1975); Dieterlen and Sylla (1992, 11–16, 48–56); Levtzion (1973, 16–17); McIntosh (1998, 259); McNaughton (1979b, 32–33); Monteil (1953, 369).

12. Dieterlen (1951, 1–33; 1955; 1957, 127, 130; 1959); Imperato (1983, 29–31, 34, 42–45, 47–49); Johnson (1986, 21); Kalilou Tera (pers. comm. 1977, 1973); Zahan (1960, 54–57, 74, 78, 306). Kalilou Tera is a linguist who worked for many years as a Malian government researcher gathering folklore and information about social practices across the country.

13. Kassim Kone is an anthropologist with extensive knowledge of Mande oral tradition, folklore, and history. See also McNaughton (1979b).

14. Conrad (1984; 1999; 2004); Johnson (1979; 1986); Kone (2001); Levtzion (1973); Niane (2006).

15. But see Alpern (2005); Bocoum (2004); Childs and Herbert (2005); Childs and Killick (1993); Chirikure (2010b, 2015); David (2001); Herbert (1984, 1993); Mapunda (2011); Rehren et al. (2007); Vogel (2000).

CHAPTER 12 (ABIODUN AND DREWAL)

1. On *àṣẹ*, see Abiodun (1994); Drewal, Pemberton, and Abiodun (1989). On *se*, see Maupoil (1943, 381); Blier (1995, 186–87), citing Montilus (1972).

2. See Abímbọ́lá (2005). Iron and stone are of central importance in Yorùbá creation myths and in the distinctive art of Ilé-Ifẹ̀ that developed in the first millennium CE (see Drewal and Schildkrout 2010). Both media possess *àṣẹ* associated with named entities: the warrior deity of iron Ògún; and Ọ̀ràmfẹ̀ (or Jàkúta), the Ifẹ̀ thundergod and "thrower of stones." The presence of iron and stone in myths, and combined in several remarkable ancient objects, may recall certain ages, technologies, and historical personalities who later became *òrìṣà*. One of the most dramatic pieces of early stone/iron work is the Ọ̀pa Ọ̀rànmíyàn, the "staff of Ọ̀rànmíyàn" (ibid., fig. 17). Oranmiyan was the son of Ògún and Odùduwà, and founder of dynasties in Benin and Ọyọ. Carved of granite gneiss, Ọ̀pa Ọ̀rànmíyàn is almost six meters in height and studded over much of its length with spiral-headed iron nails. Another sculpted stone, Adá Elédisi (Sword of Elédisi), is a monolith the top end of which curves to resemble an ancient sword form shared by Yorùbá, Edo, and Fon peoples; swords, like all iron implements, embody Ògún (ibid., fig. 9).

3 Abímbọ́lá (2005, 121). Ìrè is the name of an Èkìtì town (Ìrè-Èkìtì).

4. From an Ifá text, we learn how an Ifá client, Baba Àwúsí, used the iron Osun staff ritually to solve his problem. He struck the earth with it and found a large quantity of precious beads (Abímbọ́la 1969, 127–28). Similarly, a back was made straight when an Ifá Osun staff was used to touch it (Lijadu 1972, 71).

5. The tree's association with diviners and staffs is also found among Edo, who know it as the *ikhinmwin* tree. As an Osun priest explained, "The tree is very important to the Edo people. According to one Edo proverb 'there is no tree older than Ikhinmwin.'... No palace, village, or native doctor can be without one. It is fundamental to most of our religious and magical rites. The Ikhinmwin tree is traditionally planted when a new village, settlement, is made" (Daryl Peavy pers. comm., 2016).

6. For both Yorùbá and Edo, birds symbolize the mystical powers of women. Among Edo, such women are termed *azen*. Osun ritual specialists (*ebo*) find and concoct medicinal preparations to cure, divine, administer ordeals to suspected evildoers, and deal with the fearsomely destructive *azen*. Osun priests announce to *azen* that a ceremony is about to begin using ivory horns (*oko*) shaped like long beaks (see Girshick Ben-Amos 1995, 46, 73, fig. 55). In other contexts, birds are associated with prophecy and omens (see Nevadomsky 1988). In all such cases, they possess awesome, sacred powers, whether beneficent or deadly.

7. Animals in Yorùbá art often perform the attributes and actions of humans, in allusions to proverbial and divinatory wisdom. The image of a seated monkey eating is a widespread motif in Yorùbá sculpture, yet its connotations are poorly understood. Abímbọ́lá (1976, 200–04) gives a story from the Ifá corpus that reveals popular ideas associated with monkeys, including their quick wit: a thieving monkey, too slow to run away, pleads innocent by reasoning that he sat and ate the corn in the farmer's presence. See also Beier (1970, 83–84).

8. David Doris (pers. comm., 2018) has suggested that Ogboni iconography, particularly the "male" and "female" figures of edan's bronze superstructures, diverts viewers from perceiving "secret" truths about Ogboni power and that those truths, plainly shown in the *ẹdan*'s plain iron shafts, go unseen and unknown. Iron is present not only in *ẹdan* Ògbóni, but is also embedded in clay sculptures that decorate some Ògbóni lodges (*ilédì*) (see Drewal, Pemberton, and Abiodun 1989, figs. 147, 148).

9. Girshick Ben-Amos (1995, 46). See also Nevadomsky (1988, 82); Blackmun (1984b, 17). The killing powers of Osun priests were closely allied with those of warriors on the battlefield. Staffs were carried into battle to ensure success (Nevadomsky 1986; Peavy pers. comm., 2016), and blacksmiths would have been there too, as is recorded in accounts of their traveling with Edo armies to repair armor and weapons (Bradbury 1961, 133). An Osun priest explains: "Osun is our god of medicine and magic—the two terms are interchangeable. He owns all of the leaves in the forest—a reference to powerful magico-medicines that can heal or kill. He is a hot and fiery deity and closely associated with the elders of the Night. Osun is Osanyin on steroids, with vast magical powers and with a really bad temper.... The imagery of Birds on the staff correlates with Edo Osun's close association with the Night world (Witches and Wizards)" (Peavy pers. comm., 2016).

10. This remarkable symbol of war, now "missing in action" (stolen), is represented by a copy at the Abomey Historical Museum, Republic of Bénin.

11. See a *bociọ* said to be empowered by Heviosc, Fon god of thunder, in Blier (1995, 17).

INTERLEAF I (PEEK)

1. Basden (1966, 321). For M. W. D. Jeffreys, a colonial District Officer in Igbo country, wood-and-iron *ọfọ* were proof of the "devolved" Egyptian traditions of the Igbo. He was convinced that the iron-banded form depicted wrapped mummies (1956a) and that these were related to the worship of the sun as practiced in ancient Egypt (1956b)—intriguing but incorrect attributions. The Fowler Museum owns three *ọfọ* originally collected by Jeffreys.

CHAPTER 13 (MACEACHERN AND BERNS)

MACEACHERN:

1. Ironworking in the Middle Nok period appears at the same time as the elaborate pottery sculpture figurines for which Nok culture is famous (Franke 2016; see also chapters 3 and 9, this volume, and figs. 3.1, 9.1, 9.2). In the period when iron was first found on archaeological sites in the Lake Chad Basin, it seems the identities of particular (male) individuals were first being memorialized in rare anthropomorphic statues as well (MacEachern 2013)—although these do not compare in sophistication of execution to the Nok terra-cottas. It may not be a coincidence that the arrival of iron into the southern Lake Chad Basin coincided with the abandonment of the beautiful pottery traditions found there during the earlier Neolithic period: the spectacular pyrotechnologies of ironworking may have supplanted quieter and more domestic ceramic production systems as a locus of cultural value, with all of the implications for cultural changes and gender relations that may involve.

BERNS:

2. The results of my research are detailed in my doctoral dissertation, Berns (1986), and summarized in Berns et al. (2011). With the exception of my more intensive work among the Ga'anda, Bəna, and Yungur, most of my survey research documented things people made and identified under the general rubric of "traditional" (*kayan gargajiya*, Hausa for things of "olden times"). My objectives and time frame did not allow for the kind of ethnographic research that might have embedded art production across the broader Gongola Valley region more deeply within social structures and systems of meaning. Thus, despite the many material indicators I observed of likely cultural continuities, without more rigorous historical or archaeological data connections across the Mandara-Gongola area remain speculative if still highly intriguing.

3. A colonial report filed anonymously in 1926–28 named the location occupied today by Chadic-speaking Gudu (and where Adamawa speakers formerly lived) as a major smelting center (NAK 2710-H, no. 225/1926/8). The area occupied today by northern and western Bena hamlets was also a site of intensive smelting, as evidenced by remains of furnaces and colonial reports (NAK G2.Q, no. P.147/13). See also David (2012, 11–12).

4. These raised cicatrices eventually fade and become nearly imperceptible as a woman ages. For a full discussion of Ga'anda scarification, see Berns (1988; forthcoming).

5. *Məsaktariya* also activated spiritual intervention in a number of Ga'anda propitiatory contexts and functions as an interface between people and their spirit benefactors. For example, smeared on a spirit vessel, it could heal, when touched, via a transfer of the iron oxide's power. Hunters (formerly warriors) dipped their bows and arrows in *məsaktariya* and were likewise protected. The potency of this substance may be the literal application of powdered red iron, but it signified blood and its fundamental life force. The loss of blood in both male and female initiation is proof of the willingness to bear pain to gain social acceptability. *Məsaktariya* was also linked to menstrual blood and its association with women's reproductive fertility. Mafa women also formerly anointed themselves with this same substance (calling it *mber mbesak*) to protect against the possible ill will of humans or spirits (David et al. 1988, 370). Young Ga'anda males applied a mixture of white ground-sprouted sorghum and water (*chikta*) as a protective ingredient during initiation, which referred to semen and male potency. *Chikta* was also used in spirit activations. At Yowo and Yoxiiwa, young females wore necklaces of bundled strands of red and white beads (*musurta*) to underscore the importance of this symbolic color combination. For more on the color symbolism of ritual adornments in Ga'anda male and female initiation, see Berns (1988, 63–67).

6. *Kwardata* are part of the final stage of scarification (*thlee'ngup*, or allover scarification). Women I spoke with considered *kwardata* to be the "finest" part of the design program, and indeed, these lozenges appear on other Ga'anda public arts, such as the woven patterns of architectural walls, basketry, mud granaries, and pottery. See Berns (1988).

7. When I was in communication with Ga'anda men in 2011 via email during the production of the exhibition *Central Nigeria Unmasked*, elders were still attending to the shrines at Makwar. See Berns et al. (2011, 31–32).

8. In the Ga'anda historical imaginary, a narrative trope identifies ceramic spirit vessels as the "first ancestors," with authority to decide where people should live and how they should thrive. See Berns (forthcoming).

9. See Berns et al. (2011) for discussion of these representations. Ngum-Ngumi is covered with the greatest variety of motifs, including an enigmatic circle filled with pellets that seems to represent gourd-bowls perhaps filled with food.

10. I documented at least five *wanshipta* in the field, in various states of disrepair: two in the Ga'anda town of Kwanda, two in the Ga'anda-Dingai hamlet of Tontomra (where they were called *kateh-kateh*), and another across the Gongola River in the Longuda hamlet of Jiu (*kankaka*), where the form was somewhat different and included the iron "wrapping" seen on other Longuda and Dadiya forgings (see figs. 13.26, 13.27, 13.29). A number of others have appeared on the art market, including the two shown here, which were purchased within the past ten years.

11. Removed before burial, these items were kept in the deceased's room along with a ceramic vessel to temporarily hold his spirit (*thlef'nda*). Ga'anda *thlef'nda* ancestor vessels are covered with densely impressed decoration emulating the designs worked over a female torso during Thleta. The final separation ceremony occurred a year later, after Hom'bata harvest festivals, when the vessel could be safely shattered to release the spirit and sever its connection to the living. See Berns (2011c, 519–21).

12. According to the information I collected among Fali elders, iron had been smelted locally in the past. Blacksmiths had forged the blooms into long iron bars with curved and flattened ends (*xwatin*) that were used as a local exchange currency.

13. For additional discussion of scarification's ability to "inscribe meaning," see Roberts and Roberts (1996) and M. N. Roberts (2008).

CHAPTER 14 (KRIGER)

1. Paraphrased from Bloch (1953, 66).

2. For an overview of the precolonial social and economic history of the Congo basin and analysis of how people used currencies and carried out trade transactions, see Harms (1981, 85–108, 163–68). See also Jerven (2016, 32) for a succinct summary of current research by economic historians about Africa's precolonial economies in world history. He concludes: "While [capitalism] was not the dominant system of production as per the orthodox definition, as has been documented here, this does not mean that there were no markets or economic growth according to formal definitions.... Africa was integrated with the rest of the world economy through the Indian Ocean, the Mediterranean, the Atlantic trade and the trans-Saharan trade—a flow of ideas, goods and people *for centuries*" (emphasis added).

3. For a splendid introduction to Kuba art, see Binkley and Darish (2009).

4. On the question of Myeel and his status in Kuba court politics, see Vansina (1978, 67, 187, 218).

5. Invaluable information about Sheppard, photographs of him, and a selection of African art objects he brought back to the U.S., now in the collections of Hampton University, can be found in Hultgren and Zeidler (1993).

6. A smelting project, directed by Eugenia Herbert and Kanimba Misago, focused on the town's history and documenting an iron smelt, assisted by Boilo Mbula and Carlyn Saltman. My complementary project centered on smithing and documenting iron products and artworks, as described in this chapter. For more on the research I carried out in Lopanzo, see Kriger (1992, chapter 8; 1999, chapter 7).

CHAPTER 15 (DEWEY AND ROBERTS)

1. Maret (1985, 86). Herbert (1993, 132–63) provides a masterful overview of the relationship throughout sub-Saharan Africa. "Bantu" in this chapter's title is the word for "people" in a vast family of closely related languages extending from the borderlands of Nigeria and Cameroon across the continent to the Swahili Coast of Kenya and Tanzania, and southward to the Cape of Good Hope (Hinnebusch 1989). Bantu languages share grammar, a great deal of vocabulary, and an intellectual platform—that is, logic—differently developed over the centuries to meet particular opportunities and needs of what have become situationally identifiable ethnic groups.

2. While the Central Bantu relationship between ironworking and political power is unquestionable, whether that between king and blacksmith was a literal one, with the king himself a blacksmith, has been appropriately challenged by several scholars. See Vansina (1973, 124–29); Herbert (1993, 132, 149); Kasfir (2007, 135 and passim). In a related vein, John Thornton asserts that "histories of the origins of kingdoms in central Africa are essentially political documents, and like politics they change over time, subject to the hard whim of pragmatic realities" (2004, 32). As an example pertinent to our concerns, Thornton discusses how, in 1710, when the Kongo Kingdom was in turmoil, unification was realized by "'a wise and skillful blacksmith' who managed to win election to the position of king" because he "harkened to an origin story of a blacksmith king" as based upon "widespread central African imagery of blacksmiths as mediating figures." In other words, the man cleverly referred to associations between blacksmiths and kings—whether or (probably) not true of every Kongo monarch—to establish his political bona fides.

3. Maret (1985, 76–77). See also Maret (1980; 1993); Herbert (1994, 116); Childs and Maret (1996); Heusch (1956).

4. In the late 1980s, Mary Nooter Roberts filmed Bilumbu spirit mediums at work in a number of Luba communities of southeastern Democratic Republic of the Congo. The footage includes a diviner's wife playing a rattle with a distinctly conical head based upon the form of *manyundo* hammer-anvils and *vinyundo* miniature hammer-anvils. Other instances of such resonance include the conical covers of *dikumbo* baskets, in which the relics and certain precious heirlooms of Luba kings were (and are?) preserved; one is reminded of the Kongo reliquary seen in the watercolor by Cavazzi (see fig 15.1).

5. Vansina (1978, 49–51, 182), as cited in Maret (1985, 74). For detailed discussion of ironworking in east-central DRC, see Kriger (1999) and chapter 14, this volume.

6. Bantu languages are spoken by Hutu and Tutsi peoples of Rwanda, notwithstanding contentious conjecture about non-Bantu origins of Tutsi and the horrifically fraught ethnic politics that have led to genocide.

7. Herbert (1984, 252). Ankole was a kingdom just north of Rwanda in present-day Uganda, disenfranchised by the government in the 1960s. Cattle called Ankole-Watusi are an ancient breed understood by many to be "the most striking members of the bovine kingdom" because of their exceedingly large horns that can measure three meters from tip to tip. See https://livestockconservancy.org/index.php/heritage/internal/ankole-watusi.

8. Thanks to Talia Lieber for sharing her museum research in Rwanda (pers. comm., 2017). The same museum colleagues told her that the king kept Nyarushara with him at all times, even when sleeping, perhaps to "forge" his dreams.

9. "Interlucustrine" refers to lands among the Great Lakes of central Africa formed along the western extension of the Rift Valley. Historically important interlacustrine kingdoms are Buganda, Ankole, Bunyoro, Rwanda, and Burundi, with Karagwe located farthest to the east. The past tense is used here because soon after Tanzanian independence, kingdoms and other polities were disenfranchised and brought into a socialist democracy.

10. Sassoon (1995). Straight *omusinga* sticks were also used as spear shafts, and less affluent people gave them as tribute to the king (ibid.). *Hibiscus fuscus* has been used by healers of eastern Africa to treat a great many ailments, to fortify expectant mothers, and to minister to newborns. The plant's medicinal efficacies informed symbolic uses against aggressive spirits and persons of evil intentions. See http://www.africamuseum.be/collections/external/prelude/view_plant?pi=06840. Might the choice of *omusinga* branches as a screen for the Karagwe king have had more than overtly practical purposes, then? Similarly, might the layout of the Karagwe king's screen have borne mnemonic and/or performative dimensions?

11. In the early 1900s, R. Hall (1938, 1) saw eight such sculptures at the Bweranyange Tribal Museum (an institution that still exists in the old capital of the Karagwe Kingdom). Three more were accessioned by the Linden Museum, Stuttgart, by 1906. Another was collected in 1899 by the chief physician in the German colonial service of Tanganyika (Van Wyk 2013, 216–17). Whether more once existed is unknown. Cows are celebrated in the interlacustrine area for their milk and fertility, but although several of the iron figures display udders, the gender of others is indeterminate.

12. Sassoon (1971, 114). Incorporated into Sassoon (1983) is some but not all of the data from this unpublished paper, gathered during his distinguished career as a conservator of antiquities for British East and West African governments (Stamp 2008, 171). *Dawa*, as a loan word from Arabic into Swahili, bears important mystical as well as medicinal connotations that most likely were lost on early European visitors whose mastery of Swahili was rudimentary (at best).

13. Sassoon (1983, 105) speculated that down-turned horns may have referred to animal sacrifice or some form of deference. As Ittai Weinryb (2017, 102–03) has suggested, "animal sacrifice was one facet of ancient votive giving" in Europe and the Middle East; might Karagwe iron sculptures have been ex-votos "deposited as a substitute for actual animal sacrifice already made" or avoided? Or if sacrifice was not an iconographic reference, might the figures have been forged "on behalf of the livestock, the object…[as] material agent, acting for the safekeeping of farm animals"?

14. Slightly paraphrasing Van Wyk (2013, 221), who follows Sassoon (1983, 105) and Reid and MacLean (1995)—again presuming that the bovine sculptures are all gendered as "cows." On "relational ontologies" between men and their cattle, see Insoll (2015, 146–48) and Coote (1994).

15. In the 1880s, Hermann Von Wissman collected an iron stool among Bemba that is now at the Ethnological Museum of Berlin; Audrey Richards (1935) discovered an iron stool among Bisa or Bemba peoples on Chilubi Island in Lake Bangweulu; Cullen Young (1935) published a 1932 photograph of Chief Kanyenda of Maravi peoples of Malawi seated on his "hereditary stool" of iron; and W. V. Brelsford (1942) found the remains of two old iron stools at Bisa chief Mungulube's in Zambia that seem to have had the more looped supports of the Soli stool. In the 1970s, David Killick examined and photographed an iron stool (see fig. 15.15) in the possession of Chief Chulu of the Southern Chewa people of Kasungu, Malawi, and another, which once belonged to Chief Kanyenda, is in the collection of the National Museum of Malawi, Blantyre (pers. comm., 2017). Oral traditions suggest that these chiefdoms were founded by Luba people or those who would emulate their political aesthetics, and that the first such leader gave iron stools to his affiliates as symbols of office (Killick 1990, 49–50).

16. Richards (1935, 31). In the 1930s, Richards studied fourteen iron bow stands in situ and asserted that they were found "all over" Bemba territories. Iron bow stands were sometimes placed on chiefs' graves among neighboring Bisa and Unga peoples. Others were recorded by Brelsford (1940) in eastern Zambia and northern Malawi, and he illustrated several, reporting that Lala and Mambwe blacksmiths were producing and selling new ones in the 1930s.

17. A. D. Roberts (1973, 49–51, 344–45). Fagan writes of an iron bow stand once "treated like a god" among "Baka Chimobe" people in what is now northern Zimbabwe (1961, 236, citing Nicolle 1936). On "peripheral visions" of Luba political aesthetics among neighboring peoples, see Roberts and Roberts (1996, 211–43), where mnemonic capacities of bow stands and other material arts are also discussed. While sculpted wooden bow stands seem a Luba artistic invention with a few examples known among closely related peoples such as Tabwa, William J. Dewey saw "spear rests" of wood with iron hooks covered in beads at the Ugandan Museum in Kampala in 1993. They had been collected when President Milton Obote abolished kingdoms in 1967 and are illustrated in Sassoon (1971), but were returned to their communities on the restoration of kingships under President Yoweri Museveni in 1993.

18. A magnificent iron bow stand at the Ethnological Museum of Berlin makes the point even more dramatically. See Roberts and Roberts (1996, 81–82). Luba bow stands were secluded in special houses maintained by women dignitaries who provided prayers and sacrifices to these most sacred of objects (Nooter 1990, 65, citing Alfred Maesen).

19. *Mboko* is a form of basket divination found among neighboring Central Bantu peoples. On this and other devices of Luba etiology, see M. N. Roberts (2000). On related practices among Chokwe and related peoples, see Jordán (2003), Silva (2014).

20. For detailed discussion of Luba ironworking past and present, use of *manyundo-* and *vinyundo*-shaped pins, see Dewey and Childs (1996), Childs and Maret (1996), and the other essays of Roberts and Roberts (1996), with this latter based upon research first presented in Nooter (1991).

21. Hersak (1986, 131; idem 2002). As Dunja Hersak (2010, 42) notes, other Songye figures were carefully manipulated with wooden poles because of their "great potency" and "ambivalent potential" to instigate both good and evil, which made them dangerous to touch, and, indeed, "elevated them to a type of sacred ruler." Symbols are always multireferential, as Victor Turner (1970) elucidated, and the miniature blacksmiths' "rakes" associated with this and other small, probably personal *mankishi* (Hersak 2002) may have instigated other potent allusions, as Viviane Baeke (2010, 338–39) suggests. Looking to Songye cosmology, she posits a relationship between the iron rake-like iron accompaniments of this very figure with the personified rainbow that Luba would understand to be the anti-hero Nkongolo Mwamba (see chapter 1, this volume). Such a proposition is consistent with ideas about heat and transformation shared with other regional peoples. See Heusch (1982). Of equal interest is Baeke's presentation of the fraught context in which this figure was collected, after a battle of colonial resistance when many Songye—presumably including the person who had brought this sculpture to the fray—fell to the overwhelming firepower of Congo Free State forces. Might the Songye man have hoped the spirit(s) inhering in the powerful *nkishi* would protect him in the heat of lethal battle?

22. Paraphrasing Fromont (2014, 24–25, 40). On stories of the Kongo culture hero Lukeni, also see MacGaffey (1986), while Herbert (1993, 137–42) offers an overview of histories of Kongo blacksmith-kings.

23. Slightly paraphrasing Fromont (2014, 40). How widespread these ritual practices may have been is unclear. Pierre de Maret reported that "two hammers/anvils were struck against each other during enthronement rituals. The Mpangu [a Kongo subgroup] have, until recently, retained this aspect of the ritual as it was performed in 1624 at the Kongo capital for the enthronement of King Garcia I" (1985, 80). That at least some Kongo blacksmiths wore *malunga* bracelets suggests strong identification between kings and ironworkers, and vice versa (Heusch 2000, 108, referring to MacGaffey 1986, 196).

24. Heusch (2000, 103–05); MacGaffey (1986, 196). "Enchanted technologies" (after Gell 1992) and "danced causality" (after Warburg 1923, 221, as discussed in Kosky 2013, 32) are presented in chapter 1, this volume. Such a Kongo reliquary must have been a most potent *nkisi* with its own efficacies that further informed such rituals as the enthronement of a new king. See MacGaffey (1991; 1993).

25. Fromont (2013, 29). In the course of the important "Kongo Kingdom" project (Sengelov et al., 2018), archaeologists have recently excavated four "swords of honor" in the Kongo Christian/elite cemetery of Kindoki. Fromont (2014) offers detailed discussion of Christian visual arts in early Kongo kingdoms, while Fromont (2011) emphasizes Kongo performance arts, as when swords of honor were danced in choreographies. On Kongo religion and society, see MacGaffey (1986).

26. See Thompson (1981, 30, 47), where the Cavazzi watercolor is discussed and an analogy drawn between cosmograms on the reliquary and similar inscription among Chokwe, Lunda, and other Angolan peoples southeast of Ndongo/Kongo lands. On histories of Ndongo and related Kongo kingdoms, see Heywood and Thornton (2007). Thornton (2008) is a blog providing an English translation of the Araldi Manuscript of Cavazzi's writings. Bassani (1987) reproduces Cavazzi's watercolors; the same images are available at http://slaveryimages.org. More recent Kongo iron-working is discussed in Janzen (2009) and Ringquist (2008). On Kongo lutes of the sort depicted by Cavazzi, see Jadinon (2013).

27. One is reminded of how Julio González, the Spanish sculptor of the 1930s, sought to "communicate the full dialectical strength of iron sculpture as *dessiner dans l'espace* (drawing in space)" (Giménez 1993, 14). The "hands" of the cross guards (also called quillons) of many Kongo swords are graced with crosses that at once refer to the *dikenga* cosmogram that "represents the conception of all living beings in the universe" (Martínez-Ruiz 2007, 187) and the Christian cross (Fromont 2013, 30–31). See Heimlich (2016) for archaeological insights and bibliography. Wooden and cast-brass crucifixes are also recalled by the pierced handguards and sometimes the incising on sword blades, as a most significant achievement of Kongo Christian arts that Kongo understood as *nkisi* power objects. On this latter congruence, see MacGaffey (1993); on Kongo crucifixes more generally, see Poyner (2013) and Volper (2016).

28. Bachelard (1991, 132) paraphrased, followed by Mary McAllester Jones's comment (ibid., 113). Bachelard was reflecting on Ilmarinen, the blacksmith of Finnish legend, a reminder that insights concerning ideas and practices of a Central Bantu World are widely resonant (cf. Blakely 2006).

INTERLEAF K (ROBERTS)

1. Paraphrasing Babutidi Timitio, writing circa 1916 for the Swedish missionary Karl Edward Laman, as cited in MacGaffey (2000, 104).

2. On the nature of "the fetish," see Pietz (1985, 1987); on the invention of Africas, Mudimbe (1988, 1994). Insightful discussion of whether the term "fetish" can still be useful is provided by Bouttiaux (2011) and Heusch (2000).

3. See McClusky (2002), Mulinda (1987), and Thompson (1978), among others; and for a particularly cogent overview, Petridis (2008, 20–29). *Nkisi nkondi* (or *nkonde*) is the singular of *minkisi* (*mi*) *nkondi*. MacGaffey (1991, 6) warns that "it is a mistake to try to classify *minkisi* exactly, since no mechanism existed to produce or enforce taxonomic orthopraxy," and one can further assume that what an *nkisi* "was" depended upon who was deploying or supplicating it, as well as particular goals of a determined interformative event. Mulinda (1987, 209) notes that Woyo people, as a Kongo group living from the mouth of the Congo River to around the town of Boma, do not recognize the word *nkondi* as a descriptor of *minkisi*. Granted "there are many Kongo worlds" (Hersak 2001), but I shall use *nkisi nkondi* for practical reasons, while acknowledging a desire to celebrate such cultural, individual, and circumstantial differences.

4. See Couttenier (2018). Wording of the present essay reflects an effort to decolonize object attributions and so avoid suggesting that a work's identity is determined by naming the person by or for whom it was collected or the institution in which it is now held.

5. Conrad (1967, 232). *Heart of Darkness*, from which these famously evocative lines are taken, was based on Joseph Conrad's firsthand observations of Boma and other European stations in the 1890s. For context, see Hochschild (1999). Political and religious terminology has been inconsistent in colonial and academic discourse concerning sub-Saharan Africa, and it is not always clear how a "king" may be differentiated from a "chief," or a "god" from a "spirit."

6. Similar reasoning is developed by Alissa LaGamma (2015, 221–22 and passim). Care must be taken to avoid anachronism, for art histories should not suggest that local people held broad understandings of colonial impact before on-the-ground organization in the second decade of the twentieth century made such politico-economic realities evident. Instead, Boma kings engaged in local contests with expatriates and peers.

7. "Interformance" was coined in a brief essay (Roberts 2004, 179) addressing the perplexities of an intermedia work by the late Hans Breder. *Blond Shadow* created a stroboscopic splintering of sequential experience when digital cameras captured a woman dancing improv in response to images of herself projected onto a screen displaying her performance an instant earlier. A disorienting, ever-deepening mise en abyme resulted. Breder's work was always more about processes than outcomes.

8. The bundle is situated over the navel rather than the physical heart of the figure, and may have articulated symbolism of matrilineal descent, even as it centered presences and powers as "the heart of the matter." Textiles cascading over the *nkisi nkondi* may have been "identifying relics" (MacGaffey 2000, 107) from the clothing of persons seeking protection or being warned of retribution by the figure's agencies. "Cloth was a basic resource" for Kongo peoples throughout their histories (Martin 1986, 1, 3–4), and through a "politics of costume" (Picton and Mack 1979, 175), elaborate fashions articulated social status and self-presentation. Histories of textiles imported to other parts of sub-Saharan Africa suggest how much may be gained through close consideration of this *nkisi*'s fabrics (James Green pers. comm., 2018). See Kriger (2009; 2017, especially chapter 2). Study of the figure's beads may be similarly revelatory of early trade and function.

9. MacGaffey (2000, 84); cf. Mulinda (1987, 206–07). "Little world" (*kadunia* in Swahili, as adapted from Arabic) is how Tabwa people describe bundles of the sort. Despite living hundreds of miles away in southeastern DRC, Tabwa conceive their medicine bundles through shared central-Bantu logic deployed in composing the "heart" of a Kongo *nkisi nkondi*. A "little world" anticipates all eventualities while protecting the person(s) possessing it, whose own relics (a few hairs, a drop of blood, earth from a footprint) have been included in the ingredients.

10. Mulinda (1987, 209), with reference to what he calls *minkisi d'imprécation* in a mix of KiKongo and French. "Imprecation" bears only negative senses in English and French usage, as an invocation of evil or a curse; yet the word's Latin root, *precari* (to pray, entreat, ask), suggests ambivalent potential to assist the supplicant, even as the entreaty may be harmful to the person so targeted (Morris 1967, 661; NPL 1972, 530). See Batulukisi (1999, 77) for pertinent discussion of such ambiguities. "'Signed' his vow" and "inflaming" an *nkisi nkondi* with "metaphoric insult (wedges, pegs, thorns, and nails)" are from Thompson (1978, 207, 214).

11. MacGaffey writes of the "aesthetics of nailing" (2014–15, 149) but is skeptical of Thompson's assertion (1978, 216–17) that given shapes were chosen for particular genres of iniquity, feeling instead that the insertions were simply "handy scrap." Although "there's a scholarly premium on classifying and decoding…perhaps we should just admire the energy of creative bricolage" (pers. comm., 2018).

12. In a preliminary study of the *nkisi nkondi* now at the Fowler, Tom Joyce has identified small parts of flintlock long guns, as he also discovered inserted into an *nkisi* Mangaaka now in the collections of the Metropolitan Museum of Art in New York (LaGamma 2015, 264, 282). Parts include "a sear screw, a rifle barrel pin, a trigger guard screw, a broken trigger, a single-ear pin for securing the barrel to the stock, a barrel tang, and a cock jaw screw" (Joyce pers. comm., 2018). Black-powder muskets have long been used for hunting across much of sub-Saharan Africa, and are still fabricated and fixed by local blacksmiths in many places; a first hypothesis would be that these artifacts further empowered the hunting capacities of the *nkisi nkondi* in the collection of the Fowler Museum.

13. Wyatt MacGaffey (pers. comm., 2018). Among Bisi Kongo, the forehead is regarded as "the seat of intelligence, longevity, judgment, and good fortune" (ibid.). Kongo nail figures from the Atlantic coast sometimes bear similarly patinated lines bisecting the forehead, as does a striking *nkisi* Manyangu collected in 1915 at Mayombe, DRC (Vanhee 2013, 196–97), and now at the Royal Museum for Central Africa in Tervuren, Belgium (EO.0.0.22485). Might the figure at the Fowler have been created by a Yombe artist, outfitted by a Yombe *nganga*, and then brought to Boma at the behest of politically and economically powerful people there? In these regards, René Bravmann's thesis (1973) about the mobility of African arts and artists across social and geographical distances should be recalled, as wood carvers, blacksmiths, practitioners of arcane sciences, and other artists strove to sell their wares and services to make their livings.

CHAPTER 16 (GOUCHER)

1. See Yarranton (1681, 57–58).

2. See the commentary of sword-maker Akira Kihara: "The furnace is like a human being. We can think of the iron and coal as food we feed her. She digests it so that she will bear good *tamahagane* [iron]." NOVA, "Secrets of the Samurai Sword," aired October 9, 2007, on PBS, http://www.pbs.org/wgbh/nova/ancient/secrets-samurai-sword.html.

3. In addition to the hardwoods required to achieve sufficiently high temperatures, other (softer) woods were needed to start the furnace fires and for ordinary domestic use, for example cooking.

4. The area encompassing the Bassar region became a German protectorate between 1884 and 1914. It was formally divided between France and Great Britain in the Treaty of Versailles (1922). German exports in all metals and metalware remained around twenty-five tons in 1918, a fraction of local production.

5. See Morrall (1862, 3–4 and 44), on the Moor's likeness in profile on the crest of needle-makers.

6. See Blench and Spriggs (1999, 271–72).

7. Barbot (1732, 273–74). Barbot claimed the Dutch and French also traded in the same size and weight of iron bars along the Guinea coasts.

8. Crislayne Gloss Marao Alfagali, "*Ngangula*, *Pulungu* and bellows workers: the ironworkers according to Nova Oeiras' Royal Iron Factory data," https://www.academia.edu/16160524/Ngangula_Pulungu_and_bellows_workers1_the_ironworkers_according_to_Nova_Oeiras_Royal_Iron_Factory_data.

9. John Reeder, quoted in letter from Stephen Fuller to Lord Sydney, 28 June 1788, John Reeder Correspondence, Jamaica 1160M/C/J/1–30, Devon Archives, Heritage Trust, Exeter, UK. See also Goucher (1993; 1990).

10. Letter from John Reeder to William Pitt, 24 June 1789, Reeder Papers, Devon. See also the petitions of Reeder for compensation in Great Britain. Parliament. House of Commons. *Journals*, vols. 35–55 (London, 1774–1800), 257, 596, 738.

11. See Landgraf and Banks (2008, 6). Iron was certainly included in the cargoes of the nine ships Smith purchased in 1775 and filled "with New England goods, which were to be traded in the Caribbean for rum, sugar, and molasses and then sailed to England."

12. Pig iron was produced by large blast furnaces to make cast-iron products. The conversion of pig iron into more malleable wrought or bar iron that could be hammered by blacksmiths was sometimes profitable. This was accomplished in a smaller refinery furnace using charcoal and bellows. For a discussion of converting pig iron into wrought iron, see Gordon (1996, 125–28).

13. This comparison is widely attributed to A. H. Sayce (1911), and has since been repeated by many scholars.

14. For more on the historical Mande, see chapter 11, this volume. The power of the African cinematic blacksmith is also explored in MacRae (1995).

15. For ideas about the role of technology and ritual in west African empire-building across the Guinea coast, see Barnes (1997, chapter 3).

16. "Assegai spears," mentioned in Joseph Conrad's novel *Heart of Darkness* (1899), are manufactured today by the U.S. company Cold Steel. See http://www.coldsteel.com/assegai-with-long-shaft.html.

INTERLEAF L (KRIGER)

1. The best introduction to the history of the Upper Guinea Coast is Wright (2015).

2. For a richly rewarding precolonial history of Ouidah, coastal port of the Kingdom of Dahomey (today's Republic of Bénin), see Law (2004).

3. See the website for the Society of Africanist Archaeologists (http://www.safa.rice.edu/) for reports regularly written up in *Nyame Akuma Bulletin* on iron smelting and other archaeological research from 1972 to the present.

INTERLEAF N (SILVERMAN)

1. Severin et al. (2011). There is also documentary evidence from the first century CE that iron from South Asia served as an important trade commodity in the ancient state (Phillipson 1998, 64).

CHAPTER 17 (JORDÁN)

1. The photograph published by José Osório de Oliveira in 1958 was staged and does not depict a working forge, yet it shows the relationships among blacksmith, bellowsman, and musician of importance here. For lamellophone terminology and definitions as well as matters of use, instrument classification, and musicality in different regions of Africa, see Jordán (2012); Kubik (1999); Berliner (1981). Paul Berliner (1980) provides a close study of just such a polymath, a Shona man named John Kunaka, famed in the 1970s for all three skills. Berliner also documents and illustrates all phases of lamellophone construction, tuning, and playing.

2. In northwestern Zambia in 1993, I filmed a Chokwe metalsmith who sang initiation, hunting, and drumming songs while "playing" his bellows.

3. Feld (2017, 86). A different sort of relational ontology is discussed in chapter 15, this volume, regarding how blacksmith-kings among Intralucustrine peoples such as Haya of the old Karagwe Kingdom of Tanzania were so closely associated with cattle-keeping that their pillar anvils were sometimes adorned with iron "horns" like those of cherished oxen, and the king is said to have forged bovine sculptures.

4. The following sentences slightly paraphrase ideas about capoeira from Greg Downey (2002, 494, 499, 502, citing Mauss 1973, 73). Thanks to Johanna Kirk for bringing this work into focus for this chapter. "Muscle," "implicit," "procedural," or "performative memory," as well as the related "somaesthetics," have engendered an immense interdisciplinary literature, including neurological studies; Shusterman (2011) is a useful review.

5. Vansina held that it is erroneous to call such clapperless bells "gongs," though that term is still often used in nonmusicological literature (1969, 189).

6. In describing comparable royal bells among Lunda of the DRC, Henrique de Carvalho (1890) suggested that the bells were suspended from a wooden frame and used for important audiences. He added that in playing the bells with an iron beater, a court master musician was able to produce the tonal range one may find in instruments such as lamellophones or xylophones by striking up and down the instrument's walls. This was in relation to the different parts of the bell and the wider-to-narrower diameters of the bell from crown (top) to mouth (open end).

7. As explained to me by Mr. Chitofu Sampoko while I was conducting field research in Zambia during 1993–96.

8. The multiple found and natural parts and miniature carvings inside Luba and Songye *mboko* gourds as well as Chokwe divination baskets are meant to be shaken within their containers as symbolic elements to be "read" by diviners in conflict resolution. "Tossing" the contents of such baskets extends the rattling sounds, serving to maintain a spirit's attention throughout the divination session (see Jordán 1996).

9. During a field trip I made in 2004, the late Chief Chibwika of Mwinilunga, Zambia, explained to me the relationship between weapons (axes and swords) and flywhisks in "wielding power" and the implementation of "law" among his people.

CHAPTER 18 (DOUGNON)

1. Translator's note: Bamana are a congeries of peoples living in the "Inner Delta" of the Niger River in central Mali, who are also known as Bambara in French literature to this day. Dogon live along the Bandiagara Escarpment to the east, where some communities are of mixed Bamana and Dogon heritage, with northern Senufo and other groups sometimes represented as well. On the place of iron in Dogon cosmology, see chapter 1, this volume. On blacksmith societies and related cultural constructions concerning their marginalization and endogamy among Mande-speaking peoples including Bamana, see chapter 11.

2. On recycling in Africa and elsewhere in the world, see the essays and bibliographies of Cerny and Seriff (1996) and Ferera (1994).

3. The present essay is based upon the author's ethnographic research undertaken in winter 2016.

4. As Tom Joyce notes from his own research in Mali, in more rural regions where mechanized farming is not prevalent, if it exists at all, "blacksmiths seem to be thriving." They "sell their wares locally and send, through intermediaries, surplus finished goods to markets" in towns and the city of Bamako, where they are competitively marketed (pers. comm., 2018).

5. The term "forge" is used here following its Latin root of *fabrica*, and may refer to a place where blacksmiths forge iron or "an artisan's workshop" more generally (Morris 1966, 515). The extended use is important to note, for much of the work being undertaken by metalsmiths in the Bamako market is through "cold-working processes that require no hot forged components," while metalworkers "include welders, sheet metal workers, machinists, mechanics, cooking implement makers, cutlers, architectural grill and gate makers, scrap metal processors," and men seeing to still other tasks as well (Joyce pers. comm., 2018).

6. Also at a more practical level, Fantogo "provides on-site medical care and standardized pricing to guarantee viable livelihood," as a union might in certain other societies (Joyce pers. comm., 2018).

7. The firearms in question are black-powder muskets of the sort still fabricated and used by small-game hunters in many parts of rural Africa. See Spring (1993, 63, 102).

8. TN: Such reticence is based upon marginalization of blacksmiths following cultural practices among Mande-speaking peoples. See chapter 11, this volume.

9. TN: As an indication of how Malian metalworking is adaptive, in 1964, soon after independence from colonial control, President Modibo Keïta inaugurated the very innovative National Solar Energy Laboratory in Bamako, through which local, self-sufficient solutions to energy needs have been sought using forging and other traditional technologies in combination with imported methods. See Sékou Oumar Traoré, interview by *L'Indépendant*, July 22, 2011, http://www.maliweb.net/interview/skou-oumar-traor-directeur-du-cnesoler-le-mali-a-le-plus-grand-potentiel-solaire-en-afrique-dici-2015-10-de-llectricit-produite-au-mali-proviendront-des-nergies-renouvelables-26385.html.

REFERENCES CITED

Abímbọ́lá, Kọ́lá

2005 *Yorùbá Culture: A Philosophical Account*. Birmingham: Ìrókò Academic Publishers.

Abímbọ́lá, Wándé

1969 *Ijinle Ohun Enu Ifa, Apa Keji*. Glasgow: Collins.

1976 *Sixteen Great Poems of Ifa Divination*. Paris: UNESCO.

1994 "Làgbàyí: The Intinerant Wood Carver of Ọ̀jọwọ̀n." In Abiodun et al. 1994, 137–42.

Abiodun, Rowland

1989 "The Kingdom of Owo." In Drewal et al. 1989, 91–115.

1994 "Understanding Yorùbá Art and Aesthetics: The Concept of *Ase*." *African Arts* 27, no. 3: 68–78, 102–03.

2014 *Yoruba Art and Language: Seeking the African in African Art*. New York: Cambridge University Press.

Abiodun, Rowland, Henry Drewal, and John Pemberton III, eds.

1994 *The Yoruba Artist: New Theoretical Perspectives on African Arts*. Washington, D.C.: Smithsonian Institution Press.

Abraham, Curtis

2013 "Nigeria: Nok has more wonders yet for the world!" *New African*, December 3. https://newafricanmagazine.com/news-analysis/arts-culture/nigeria-nok-has-more-wonders-yet-for-the-world/.

Abraham, Roy Clive

1958 *Dictionary of Modern Yoruba*. London: University of London Press.

Abram, David

1996 *The Spell of the Sensuous in a More-Than-Human World*. New York: Pantheon Books.

Ackerman, J. Kyle, David Killick, Eugenia Herbert, and Colleen Kriger

1999 "A Study of Iron Smelting at Lopanzo, Equateur Province, Zaïre." *Journal of Archaeological Science* 26, no. 8: 1135–43. DOI: 10.1006/jasc.1999.0409.

Adande, Codjovi Joseph

2000 "Le Dieu Gou de Musée de l'Homme: ses origins, sa significance historique comme oeuvre d'art." In Adande et al 2000, 1–7.

Adande, J., Gabin Djimasa, Nicolas Garnier, Claude Savary, and Serge Tornay, eds.

2000 *Une figure emblématique de l'ancien Musée de l'Homme: Le Dieu Gou de Danhomè*. Paris: Musée de l'Homme.

Adepegba, Cornelius

1983 "*Ara*: The Factor of Creativity in Yoruba Art." *The Nigerian Field* 48, nos. 1–4: 53–66.

Adewumi, I. K. and M. O. Ogedengb

2005 "Optimizing Conditions for Activated Charcoal Production from Palm Kernel Shells." *Journal of Applied Sciences* 5: 1082–87.

Akhtar, Shabbir

2008 *The Quran and the Secular Mind: A Philosophy of Islam*. New York: Routledge.

Alde, Annick

n.d. "A Voice From the Congo." *Museum Menagerie* (blog). http://museummenagerie.blogspot.com/2016/05/herbert-ward.html.

Aldersey-Williams, Hugh

2011 *Periodic Tales: A Cutural History of the Elements, from Arsenic to Zinc*. New York: Harper Collins Publishers.

Alfagali, Crislayne Gloss Marao

n.d. "Ngangula, Pulungu and Bellows Workers: The Ironworkers According to Nova Oeiras' Royal Iron Factory Data." Unpublished paper. https://www.academia.edu/16160524/Ngangula_Pulungu_and_bellows_workers1_the_ironworkers_according_to_Nova_Oeiras_Royal_Iron_Factory_data.

Allen, James de Vere

1993 *Swahili Origins*. London: James Curry.

Alpern, Stanley B.

2005 "Did They or Didn't They Invent It? Iron in Sub-Saharan Africa." *History in Africa* 32: 41–94.

Anderson, Don L.

1989 *Theory of the Earth*. Boston: Blackwell.

1999 "Planet Earth." In Beatty et al. 1999, 111–24.

Angell, Norman

1929 *The Story of Money*. Garden City, NY: Garden City Publishing Company.

Angenot, L.

1978 "Les poisons de flèches africains." *Africa-Tervuren* 24, no. 2: 36–51.

Anonymous correspondent

2006 "Born of Fire." *The Economist*, 19 December. http://www.economist.com/node/8401289.

Anthoine, Christophe

1996 "Le forgeron qui devint roi: Idéologique politique de a chefferie songye des Kalebwe (Zaïre)." *Civilisations* 43, no. 2: 19–44.

Anthon, Charles

1847 *A Classical Dictionary: Containing an Account of the Principal Proper Names Mentioned in Ancient Authors*. New York: Harper & Brothers.

Appadurai, Arjun

1986 "Introduction: Commodities and the Politics of Value." In *The Social Life of Things: Commodities in Cultural Perspective*, edited by Arjun Appadurai, 3–63. New York: Cambridge University Press.

Appia-Dabit, Béatrice

1965 "Les Forgerons de Fouta Jallon." *Journal de la Société des africanistes* 35: 317–52.

Apter, Andrew

2005 "Griaule's Legacy: Rethinking 'la parole claire' in Dogon Studies." *Cahiers d'Etudes Africaines* 45, no. 177: 95–130.

Aremu, David A.

2004 "Iron Roads in Africa: A Contribution from Nigeria." In Bocoum 2004, 149–64.

Arens, W., and Ivan Karp, eds.

1989 *Creativity of Power: Cosmology and Action in African Societies*. Washington, D.C.: Smithsonian Institution Press.

Arkell, A. J.

1939 "Throwing-Sticks and Throwing-Knives in Darfur." *Sudan Notes and Records* 22, no. 2: 250–67.

1956 "The Making of Mail at Omdurman." *Kush: Journal of the Sudan Antiquities Service* 4: 83–84.

Arnoldi, Mary Jo

1998 "Where Art and Ethnology Met: The Ward African Collection at the Smithsonian." In Schildkrout and Keim 1998, 193–216.

2016 "A Throwing Knife in Motion: The Journey from the Congo to the Smithsonian." In *Engaging Smithsonian Objects Through Science, History, and the Arts*, edited by Mary Jo Arnoldi, 175–91. Washington, D.C.: Smithsonian Institution Scholarly Press.

Arnoldi, Mary Jo, Christraud M. Geary, and Kris L. Hardin, eds.

1996 *African Material Culture*. Bloomington: Indiana University Press.

Arnoldi, Mary Jo, and Christine Mullen Kreamer

1995 "Focus on Twelve African Hats." In *Crowning Achievement: African Arts of Dressing the Head*, edited by M. J. Arnoldi and C. M. Kreamer, 26–51. Los Angeles: UCLA Fowler Museum of Cultural History.

Ashley, C. Z.

2010 "Towards a socialized archaeology of ceramics in Great Lakes Africa." *African Archaeological Review* 27: 135–63.

Association SCOA

1980 *Actes du colloque: histoire et tradition orale: projet Boucle du Niger, 3e année: L'Empire du Mali, L'Empire du Ghana, L'Empire du Songhay: troisième Colloque International de l'Association SCOA*. Niamey, Niger 30 novembre–6 décembre 1977. Paris: Association SCOA.

1984 *Actes du séminaire: rapport du moyen Niger avec le Ghana ancien. Niamey, 16–21 janvier 1981*. Paris: Association SCOA.

Athar, Amani

2007 "Last days of Gihanga and his legacy, descendants, economic & social policies." *The New Times* (Rwanda), September 15. http://www.newtimes.co.rw/section/article/2007-09-15/78231/.

Atwood, Roger

2011 "The Nok of Nigeria." *Archaeology* 64, no. 4. https://archive.archaeology.org/1107/features/nok_nigeria_africa_terracotta.html.

Austen, J. L.

(1962) 2005 *How to Do Things with Words*. Cambridge, MA: Harvard University Press.

Avery, Donald, and Peter Schmidt
1979 "A Metallurgical Study of Iron Bloomery, Particularly as Practised in Buhaya." *Journal of Metals* 31:14–20.

Babalọla, Adeboye
1989 "A Portrait of Ògún as Reflected in Ìjálá Chants." In Barnes 1997, 146–72.

Babal la, Adeboye, and Olugboyega Alabi
2003 *A Dictionary of Yoruba Names*. Lagos: West African Book Publishers.

Bachelard, Gaston
1991 "Earth, Fire, and Water: Images of the Smithy." In *Gaston Bachelard, Subversive Humanist: Texts and Readings*, by Mary McAllester Jones, 129–34. Madison: University of Wisconsin Press. Excerpted from Bachelard. 1948. *La terre et les rêveries de la volonté: essai sur l'imagination des forces*. Paris: Corti.

Bachmann, Hans-Gert
1982 *The Identification of Slags from Archaeological Sites*. Occasional Publication 6. London: Institute of Archaeology.

Baeke, Viviane
1995 "Water Spirits and Witchcraft: Rituals, Myths and Objects Mfumte-Wuli, Western Cameroon." In *Objects: Signs of Africa*, edited by Luc de Heusch, 57–91. Ghent: Snoeck-Ducaju.
2004 *Le temps des rites: Ordre du monde et destin individuel en pays wuli (Cameroun)*. Nanterre, France: Société d'Ethnologie.

Baker, Huston
1994 "Beyond Artifacts: Cultural Studies and the New Hybridity of Rap." In *Cultural Artifacts and the Production of Meaning: The Page, The Image, and the Body*, edited by Margaret J. M. Ezell and Katherine O'Brien O'Keeffe, 183–97. Ann Arbor: University of Michigan Press.

Balandier, Georges
1968 *Daily Life in the Kingdom of the Kongo, from the Sixteenth to the Eighteenth Century*. New York: Pantheon.

Barbot, Jean
1732 *A description of the coasts of North and South-Guinea, and of the Ethiopia inferior* [. . .]. London.

Barndon, Randi
1996 "Fipa Ironworking and Its Technological Style." In Schmidt 1996b, 58–73.
2004 *An Ethnoarchaeological Study of Iron-Smelting Practices among the Pangwa and Fipa in Tanzania*. Oxford: Archaeopress.
2012 "Technology and Morality: Rituals in Iron Working among the Fipa and Pangwa Peoples of Southwestern Tanzania." *Archeological Papers of the American Anthropological Association* 21, no. 1: 37–48.

Barnes, Sandra T., ed.
1997 *Africa's Ogun: Old World and New*. 2nd ed. Bloomington: Indiana University Press.

Baroin, Catherine, and Yves Moñino, eds.
1991 *Forges et forgerons*. Paris: Éditions de l'ORSTOM.

Barth, Heinrich
1857 *Travels and Discoveries in North and Central Africa: Being a Journal of an Expedition Undertaken Under the Auspices of H.B.M.'s Government, in the Years 1849–1855*. Vol. 2. New York: Harper and Brothers.

Barros, Philip de
1986 "Bassar: A Quantified, Chronologically Controlled, Regional Approach to a Traditional Iron Production Centre in West Africa." *Africa* 56, no. 2: 152–74.
1988 "Societal Repercussions of the Rise of Large-scale Traditional Iron Production: A West African Example." *African Archaeological Review* 6: 91–114.
2000 "Iron Metallurgy: Sociocultural Context." In *Ancient Africa Metallurgy: The Sociocultural Context*, edited by Michael S. Bisson and Joseph O. Vogel, 147–98. Walnut Creek, CA: Alta Mira Press.
2003 "Recent Early Iron Age Research in Bassar, Togo." *Nyame Akuma* 59: 76–78.

Basden, G. T.
(1938) 1966 *Niger Ibos*. London: Cass.

Bassani, Ezio
1987 "Un Cappuccino nell'Africa near del seicento: I disegni dei Monoscritti Araldi del Padri Giovanni Antonia Cavazzi da Montecuccolo." *Quarderni Poro*, 4.

Basu, Paul
2017 *The Inbetweenness of Things: Materializing Mediation and Movement between Worlds*. London: Bloomsbury.

Bathily, Abdoulaye
1975 "A Discussion of the Traditions of Wagadu with Some Reference to Ancient Ghana." *Bulletin de l'Institut Fondamental d'Afrique Noire, Serie B: Sciences humaines* 37: 1–94.

Baudrillard, Jean
(1968) 1996 *The System of Objects*. New York: Verso.

Bay, Edna
1986 *Asen: Iron Altars of the Fon People of Benin*. Atlanta: Emory University of Art and Archeology.
2008 *Asen, Ancestors, and Vodun: Tracing Change in African Art*. Champaign: University of Illinois Press.

Beach, David N.
1974 "The Shona economy: branches of production." Paper presented for Henderson Research Seminar Series, History Department, University of Rhodesia (Zimbabwe).

Beatty, J. Kelly, Carolyn Collins Petersen, and Andrew Chaikin, eds.
1999 *The New Solar System*. 4th ed. Cambridge, MA: Sky Publishing.

Becker, Cynthia
2011 "Hunters, Sufis, Soldiers, and Minstrels: Trans-Saharan Derivations of the Moroccan Gnawa." *Res: Anthropology and Aesthetics* 59–60: 124–44.

Beek, Walter van
2012a *The Dancing Dead: Ritual and Religion Among the Kapsiki/Higi of North Cameroon and Northeastern Nigeria*. New York: Oxford University Press.
2012b "The Iron Bride: Blacksmith, Iron, and Femininity Among the Kapsiki/Higi." In David 2012, 285–302.

Beer, Ferdinand Pierre, Elwood Russell Johnston, Jr., and John T. DeWolf
(1981) 1992 *Mechanics of Materials*. London: McGraw-Hill.

Beier, Ulli, ed.
1970 *Yoruba Poetry: An Anthology of Traditional Poems*. Cambridge: Cambridge University Press.
2002 *Yoruba Poetry*. Bayreuth African Studies Series 62. Bayreuth, Germany: Bayreuth University.

Belford, Paul
2012 "Hot Blast Iron Smelting in the Early 19th Century: Reappraisal." *Historical Metallurgy* 46, no. 1: 32–44.

Ben-Amos, Paula Gershick
1995 *The Art of Benin: Revised Edition*. Washington, D.C.: Smithsonian Institution.

Bennet, Jane
2010 *Vibrant Matter: A Political Ecology of Things*. Durham, NC: Duke University Press.

Bent, James Theodore
1892 *The Ruined Cities of Mashonaland*. London: Longmans, Green.

Bentor, Eli
1988 "Life as an Artistic Process: Igbo Ikenga and Ofo." *African Arts* 21, no. 2: 66–71, 94.

Berliner, Paul
1981 *The Soul of Mbira: Music and Traditions of the Shona People of Zimbabwe*. Berkeley: University of California Press.
1980 "John Kubaka, Mbira Maker." *African Arts* 14, no. 1: 61–67, 88.

Bernhardt, Walter O.
1908 "The Blacksmith of Central Africa and His Work." *The American Blacksmith* 7, no. 7: 147–50.

Berns, Marla C.
1986 "Art and History in the Lower Gongola Valley, Northeastern Nigeria." PhD diss., University of California, Los Angeles.
1988 "Ga'anda Scarification: A Model for Art and Identity." In Rubin 1988, 57–76.
1990 "Pots as People: Yungur Ancestral Portraits." *African Arts* 23, no. 3: 50–60, 102.
1993 "Art, History, and Gender: Women and Clay in West Africa." *The African Archaeological Review* 11: 129–48.
2011a "Introduction to Upper Benue Arts." In Berns et al. 2011, 465–75.
2011b "Containing Power: Identities in Clay in the Eastern Gongola Valley, the Bəna, Yunger, and Dera." In Berns et al. 2011, 529–49.
2011c "Containing Power: Identities in Clay in the Eastern Gongola Valley, the Ga'anda." In Berns et al. 2011, 529–49.
2014 *World Arts Local Lives: The Collections of the Fowler Museum at UCLA*. Los Angeles: Fowler Museum at UCLA.
Forthcoming "History in Clay and Skin: Strategies of Remembrance among Ga'anda of Northeastern Nigeria" (working title). In *The Oxford Handbook of History and Material Culture*, edited by Ivan Gaskell and Sarah Carter. New York: Oxford University Press.

Berns, Marla C., Richard Fardon, and Sidney Littlefield Kasfir, eds.
2011 *Central Nigeria Unmasked: Arts of the Benue River Valley*. Los Angeles: Fowler Museum at UCLA.

Berns, Marla C., Doran H. Ross, and Mary Nooter Roberts
2011 "African Art in the Fowler Museum at UCLA: Collections and Research." In *Representing Africa in American Art Museums: A Century of Collecting and Display*, edited by K. B. Bickford and, C. Clarke, 184–204. Seattle: University of Washington Press.

Berns, Marla C. and Barbara Rubin Hudson
1986 *The Essential Gourd: Art and History in Northern Nigeria*. Los Angeles: Museum of Cultural History, UCLA.

Bevan, Alex, and John de Laeter
2002 *Meteorites: A Journey through Space and Time*. Washington, D.C.: Smithsonian Institution Press.

Binger, Louis-Gustave
1892 *Du Niger au Golfe de Guinée par le pays de Kong et le Mossi*. Vol. 1. Paris: Hachette.

Bining, Arthur Cecil
1938 *Pennsylvania Iron Manufacture in the Eighteenth Century*. Philadelphia: Pennsylvania Historical Commission.

Binkley, David A., and Patricia Darish
2009 *Kuba*. Visions of Africa Series. Milan: 5 Continents.

Bird, Charles S., and Martha B. Kendall
1980 "The Mande Hero." In *Explorations in African Systems of Thought*, edited by Ivan Karp and Charles S. Bird, 13–26. Bloomington: Indiana University Press.

Biton, Marlène
1994 "Sculpture dédiée à Gou, divinité du fer travaillé et de la guerre." In *Sculptures. Afrique, Asie, Océanie, Amériques*, edited by J. Kerchache, 110–13. Paris: Réunion des musées nationaux.
2000 "Question de 'Gou.'" *Arts d'Afrique Noire* 91: 25–34.

Bivar, A. D. H.
1964 *Nigerian Panoply: Arms and Armour of the Northern Region*. Lagos: Department of Antiquities, Federal Republic of Nigeria.

Bjorkman, Judith K.
1973 "Meteors and Meteorites in the Ancient Near East." *Meteoritics* 8, no. 2: 91–130.

Blackmun, Barbara
n.d. "Art and Rule in the Benin Kingdom." Art & Life in Africa, University of Iowa Stanley Museum of Art. https://africa.uima.uiowa.edu/topic-essays/show/4?start.

Blakely, Sandra
2006 *Myth, Ritual, and Metallurgy in Ancient Greece and Recent Africa*. New York: Cambridge University Press.

Blandin, André
1992 *"Fer Noir" d'Afrique de l'Ouest. Salon-de-Provence*, France: Imp. B. Amigo.

Blench, Roger
2006 "African agricultural tools: implications of synchronic ethnography for agrarian History." Paper presented at Fifth International Workshop for Agrican Archaeobotony, London, July. http://www.homepages.ucl.ac.uk/~tcrndfu/IWAA/Blench.pdf.

Blench, Roger, and Matthew Spriggs, eds.
1999 *Archaeology and Language III: Artefacts, Languages and Texts*. Vol. 1. New York: Routledge.

Blier, Suzanne
1984 "Antelopes and Anvils: Tamberma Works of Iron." *African Arts* 17, no. 3: 58–63, 91.
1990 "King Glele of Danhom Part One: Divination Portraits of a Lion King and Man of Iron." *African Arts* 23, no. 4: 42–53, 93–94.
1993 "Art and Secret Agency: Concealment and Revelation in Artistic Expression." In Nooter 1993, 180–94.
1995 *African Vodun: Art, Psychology, and Power*. Chicago: University of Chicago Press.

Boahen, A. Adu
1989 *African Perspectives on Colonialism*. Baltimore: Johns Hopkins University Press.

Bocoum, Hamady
2001 "Samori's Smithies: From Craft Production to Attempted Manufacturing, or a Draft Plan for Technological Independence." *Mande Studies* 3: 55–63.

Bocoum, Hamady, ed.
2004 *The Origins of Iron Metallurgy in Africa: New Light on its Antiquity: West and Central Africa*. Paris: UNESCO.

Bognolo, Daniela
2007 *Lobi*. Milan: 5 Continents.

Bostoen, Koen, Bernard Clist, Charles Doumenge, Rebecca Grollemund, Jean-Marie Hombert, Joseph K. Muluwa, and Jean Maley
2015 "Middle to Late Holocene Paleoclimatic Change and the Early Bantu Expansion in the Rain Forests of Western Central Africa." *Current Anthropology* 56, no. 3: 354–84.

Bouttiaux, Anne-Marie
2009 *Persona. Masks of Africa: Identities Hidden and Revealed*. Tervuren, Belgium: Royal Museum for Central Africa, 5 Continents Editions.
2010 "Grasping a Terra Still Incognita." In *David Adjaye's Geo-Graphics: A Map of Art Practices in Africa, Past and Present*, edited by Emiliano Battista, Anne-Marie Bouttiaux, et al., 22–25. Brussels: Bozar Centre for Fine Arts.
2011 "The Porosity of Objects." In Bouttiaux and Seiderer 2011, 152–57.

Bouttiaux, Anne-Marie, and Anna Seiderer, eds.
2011 *Fetish Modernity*. Tervuren, Belgium: Royal Museum for Central Africa.

Bourdillon, Michael F. C.
1982 *The Shona Peoples*. Harare: Mambo Press.

Boyle, C. V.
1913 "Lala District: Historical and Anthropological Notes." National Archives of Nigeria, Kanuda, J18 [1912–1928: 12–38].

Bradbury, Raymond E.
1961 "Ezomo's Ikegobo and the Benin Cult of the Hand." *Man* 56, no. 165: 129–37.

Bravmann, René A.
1974 *Islam and Tribal Art in West Africa*. London: Cambridge University Press.
1977 "*Gyinna-Gyinna*: Making the Djinn Manifest." *African Arts* 10, no. 3: 46–52, 87.
1983 *African Islam*. Washington, D.C.: Smithsonian Institution Press.

Brelsford, W. V.
1940 "Notes on Some Northern Rhodesian Bow Stands." *Man* 40: 35–41.
1949 "Rituals and Medicines of Chishinga Ironworkers." *Man* 49: 27–29.

Breunig, Peter
2014 *Nok: African Sculpture in Archaeological Context*. Frankfurt: Africa Magna.

Breunig, Peter and Nicole Rupp
2016 "An Outline of Recent Studies on the Nigerian Nok Culture." *Journal of African Archaeology* 14, no. 3: 237–55.

Briggs, Lloyd Cabot
1965 "European Blades in Tuareg Swords and Daggers." *Journal of the Arms and Armour Society* 5, no. 2: 37–92.

Brooks, George E.
1993 *Landlords and Strangers: Ecology, Society, and Trade in Western Africa, 1000–1630*. Boulder, CO: Westview Press.

Buchan, James
1997 *Frozen Desire: An Inquiry into the Meaning of Money*. London: Picador.

Buchwald, Vagn Fabritius
2005 *Iron and Steel in Ancient Times*. Copenhagen: Det Kongelige Danske Videnskabernes Selskab.

Bunting, Annie, Benjamin N. Lawrance, and Richard L. Roberts, eds.
2016 *Marriage by Force: Contestation over Consent and Coercion in Africa*. Athens, Ohio: Ohio University Press.

Burka, Temesgen
2008–09 "Qaalluu, Smith and Metal: Traditional Conflict Resolution Mechanisms in the Medium of Metals among the Oromo of Northeast Wollega, Ethiopia." *Ejossah* 6, nos. 1–2: 25–42.

Burrows, Captain Guy
1898 *The Land of the Pigmies*. London: C. A. Pearson.

Burssens, Herman
2003 "Sculpting in Wood, Ivory, and Stone." In Herreman 2013, 14–62.

Burton, William Frederick Padwick
1961 *Luba Religion and Magic in Custom and Belief. Annales* 35. Tervuren, Belgium: Royal Museum for Central Africa.

Cabakulu, Mwamba
2004 *Maxi Proverbes Africains*. Paris: Marabout.

Calame-Griaule, Geneviève
1965 *Ethnologie et langage: la parole chez les Dogon*. Paris: Gallimard.

Camara, Sirifo
2010 *The Epic of Kelefaa Saane*. Bloomington: Indiana University Press.

Cancel, Robert
2013 *Storytelling in Northern Zambia: Theory, Method, Practice, and Other Necessary Fictions*. Cambridge: OpenBook Publishers.

Carvalho, Henrique
1890 *Expedição Portuguesa ao Muatianvua: Ethnographia e história tradicional dos povos da Lunda*. Lisbon: Imprensa Nacional.

Caton-Thompson, Gertrude
1931 *The Zimbabwe Culture: Ruins and Reactions*. Oxford: Claredon Press.

Celis, Georges
1987 "Fondeurs et Forgerons Ekonda (Equateur Zaire)." *Anthropos* 82: 109–34.
1989 "La métallurgie traditionnelle au Burundi, au Rwanda et au Buha: Essai de synthèse." *Anthropos* 84, nos. 1/3: 25–46.
1991 *Eisnehütten in Afrika: les Fonderies Africaines du Fer*. Frankfurt: Museum für Völkerkunde.

Celis, Georges R., and E. Nzikobanyanka
1976 *La Métallurgie traditionelle au Burundi (Techniques et Croyances)*. Tervuren, Belgium: Royal Museum for Central Africa.

Cerny, Charlene, and Suzanne Seriff, eds.
1996 *Recycled, Reseen: Folk Art from the Global Scrapheap*. New York: Harry Abrams for the Museum of International Folk Art, Santa Fe.

Chaiken, Shelly
1980 "Heuristic Versus Systematic Information Processing and the Use of Source Versus Message Cues in Persuasion." *Journal of Personality & Social Psychology* 39, no. 5: 752–66.

Chami, Felix A.
1992 "Limbo: Early iron-working in south-eastern Tanzania." *Azania* 27: 45–52.
1994 *The Tanzanian Coast in the First Millennium AD: An Archaeology of the Iron-Working, Farming Communities*. Studies in the African Past 7. Uppsala: Societa Archaeological Upsaliensis.

Chen, Serena, Kimberly Duckworth, and Shelly Chaiken
1999 "Motivated Heuristic and Systematic Processing." *Psychological Inquiry* 10, no. 1: 44–49.

Childs, S. Terry
1991 "Style, technology, and iron smelting furnaces in Bantu-speaking Africa." *Journal of Anthropological Archaeology* 10, no. 4: 332–59.
1996 "Technological history and culture in western Tanzania." In Schmidt 1996b, 277–320.
2000 "Traditional Iron Working: A Narrated Ethnoarchaeological Example." In Vogel 2000, 199–253.
2016 "A Throwing Knife through Production and Use." In *Engaging Smithsonian Objects Through Science, History, and the Arts*, edited by Mary Jo Arnoldi, 159–74. Washington, D.C.: Smithsonian Institution Scholarly Press.

Childs, S. Terry, and William J. Dewey
1996 "Forging Symbolic Meaning in Zaire and Zimbabwe." In Schmidt 1996b, 145–71.

Childs, S. Terry, and Eugenia W. Herbert
2005 "Metallurgy and Its Consequences." In Stahl 2005, 276–300.

Childs, S. Terry, and David Killick
1993 "Indigenous African Metallurgy: Nature and Culture." *Annual Review of Anthropology* 22: 317–37.

Childs, S. Terry, and Pierre de Maret
1996 "Re/Constructing Luba Pasts." In Roberts and Roberts 1996, 48–59.

Chirikure, Shadreck
2010a *Indigenous Mining and Metallurgy in Africa*. Cambridge: Cambridge University Press.
2010b "On Evidence, Ideas and Fantasy: The Origins of Iron in Sub-Saharan Africa—Thoughts on É. Zangato & A. F. C. Holl's 'On the Iron Front.'" *Journal of African Archaeology* 8, no. 1: 25–8.
2013 "The Archaeology of African Metalworking." In Mitchell and Lane 2013, 131–43.
2015 *Metals in Past Societies: A Global Perspective on Indigenous African Metallurgy*. Cham, Switzerland: Springer.

Chirikure, Shadreck, Paul T. Craddock, Manfred. K. H. Eggert, and Scott MacEachern
2010 "Reactions to 'On the Iron Front: New Evidence from North-Central Africa.'" *Journal of African Archaeology* 8, no. 1: 24–41.

Chojnacki, S.
2006 *Ethiopian Crosses: A Cultural History and Chronology*. Milan: Skira.

Chouin, Gerard
2012 "The 'Big Bang' Theory Reconsidered: Framing Early Ghanaian History." *Transactions of the Historical Society of Ghana* 14: 13–40.
2013 "Fossés, enceintes et peste noire en Afrique de l'Ouest forestière (500–1500 AD)." *Afrique: Archéologie et Arts* 9: 43–66.

Chouin, Gerard, and Christoper R. Decorse
2010 "Prelude to the Atlantic Trade: New Perspectives on Southern Ghana's Pre-Atlantic History (800–1500)." *The Journal of African History* 51, no. 2: 123–45.

Claerhout, Adriaan
1976 "Two Wrought-iron Statuettes." *African Arts* 9, no. 4: 60–4: 92.

Clifford, James
1988 *Predicament of Culture: Twentieth Century Ethnography, Literature, and Art*. Cambridge, MA: Harvard University Press.

Cline, Walter Buchanan
1937 *Mining and Metallurgy in Negro Africa*. General Series in *Anthropology* 5. Menasha, WI: George Banta Publishing Company.

Clist, Bernard
2005 "Des premiers villages aux premiers européens autour de l'estuaire du Gabon: quatre millénaires d'interactions entre l'homme et son milieu." PhD diss., Université libre de Bruxelles.
2012 "Vers une réduction des préjugés et la fonte des antagonismes: un bilan de l'expansion de la métallurgie du fer en Afrique sud-Saharienne." *Journal of African Archaeology* 10: 71–84.

Cole, Herbert M.
1982 *Mbari: Art and Life Among the Owerri Igbo*. Bloomington: Indiana University Press.
1989 *Icons: Ideals and Power in the Art of Africa*. Washington, D.C.: Smithsonian Institution Press.
2013 *Igbo*. Milan: 5 Continents.
2017a "The Complexity of Maternity in Africa." *Tribal Art* 84: 110–21.
2017b *Maternity: Mothers and Children in the Arts of Africa*. Brussels: Mercatorfonds.

Colle, Pierre
1903 "La métallurgie dans l'Ourua." *Missions d'Afrique (Pères Blancs)*, 244–45.
1913 *Les Baluba*. 2 vols. Antwerp, Belgium: Albert Dewit.

Colleyn, Jean-Paul
2002 "Les Jow." In *Bamana: Un art de savoir-vivre au Mali*, edited by J.-P. Colleyn, 95–236. Zurich: Museum Rietberg (also published as *Bamana: The Art of Existence in Mali* [New York: Museum for African Art, 2001]).
2009 *Visions of Africa: Bamana*. Milan: 5 Continents.

Comelli, D., Massimi D'orazio, Luigi Folco, Mahmud El-Halwagy, Tommaso Frizzi, Roberto Alberti, Valentina Capogrosso, Abdeirazek Elnaggar, Hala Hassan, Austin Nevis, Franco Porcelli, Mohamed B. Rashed, and Gianluca Valentini
2016 "The Meteoritic Origin of Tutanhkamun's Iron Dagger Blade." *Meteoritics and Planetary Science* 51, no. 7: 1301–09.

Connah, Graham
1969 "The Coming of Iron: Nok and Daima." In *Lectures on Nigerian Prehistory and Archaeology*, edited by Thurstan Shaw, 30–36. Ibadan, Nigeria: Ibadan University Press.

Conrad, David C.
1984 "Oral Sources on Links between Great States: Sumanguru, Servile Lineage, the Jariso, and Kaniaga." *History in Africa* 11: 35–55.
1992 "Searching for History in the Sunjata Epic: The Case of Fakoli." *History in Africa* 19: 147–200.
1999 *Epic Ancestors of the Sunjata Era: Oral Tradition from the Maninka of Guinea*. Madison: African Studies Program University of Wisconsin–Madison.
2001 "Categorizing Artisans and Bards of West Africa." *Canadian Journal of African Studies* 35, no. 1: 140–46.
2004 *Sunjata: A West African Epic of the Mande Peoples*. Cambridge: Hacket Publishing Company.

Conrad, David C., and Barbara E. Frank
1995 "Nyamakala: Contradiction and Ambiguity in Mande Society." In Conrad and Frank 1995, 1–23.

Conrad, David C., and Barbara E. Frank, eds.
1995 *Status and Identity in West Africa: Nyamakalaw of Mande*. Bloomington: Indiana University Press.

Conrad, Joseph
(1899) 1967 "Heart of Darkness." In *Great Short Works of Joseph Conrad*, 211–92. New York: Harper & Row.

Cook, David
2009 "Messianism and Astronomical Events During the First Four Centuries of Islam." *Revue des Mondes Musulmans et de la Méditerranée* 91–94: 1–22.

Cooksey, Susan, Robin Poynor, and Hein Vanhee, eds.
2013 *Kongo across the Waters*. Gainesville: University Press of Florida.

Coombes, Annie E.
1994 *Reinventing Africa: Museums, Material Culture, and Popular Imagination in Late Victorian and Edwardian England*. New Haven: Yale University Press.

Coote, Jeremy
1994 "'Marvels of Everyday Vision': The Anthropology of Aesthetics and the Cattle-Keeping Nilotes." In Coote and Shelton 1994, 245–73.

Coote, Jeremy, and Anthony Shelton, eds.
1994 *Anthropology, Art, and Aesthetics*. New York: Oxford University Press.

Couttenier, Maarten
2005 *Congo tentoongesteld: Een geschiedenis van de Belgische antropologie en het Museum van Tervuren (1882–1925)*. Leuven, Belgium: Acco.
2018 "EO.0.0.7943." *BMGN – Low Countries Historical Review* 133, no. 2: 91–104. http://doi.org/10.18352/bmgn-lchr.10553.

Craddock, Paul, Janet Ambers, Duncan Hook, Ronald Farquhar, V. Emenike Chikwendu, Alphonse Umeji, and Thurstan Shaw
1997 "Metal Sources and the Bronzes from Igbo-Ukwu, Nigeria." *Journal of Field Archaeology* 24, no. 4: 405–29.

Curnow, Kathy
n.d. "Benin Kingdom Leadership Regalia." *Art & Life in Africa*, University of Iowa Stanley Museum of Art. https://africa.uima.uiowa.edu/topic-essays/show/17?start=0.
2007 "Cultural Flows and Cultural Breakwaters: Art Connecting the Benin Kingdom with the Coast." In *Benin Kings and Rituals: Court Arts from Nigeria*, edited by Barbara Plankensteiner, 170–83. London: Faber & Faber.

Dalton, George
1969 "Theoretical Issues in Economic Anthropology." *Current Anthropology* 10, no. 1: 63–80, with commentaries 80–99.

David, Nicholas
1996 "A New Political Form? The Classless Industrial Society of Sukur (Nigeria)." In Pwiti and Soper 1996, 593–600.
2001 "Lost in the Third Hermeneutic? Theory and Methodology, Objects and Representations in the Ethnoarchaeology of African Metallurgy." *Mediterranean Archaeology* 14: 49–72.

David, Nicholas, ed.
2012 *Metals in Mandara Mountains' Society and Culture*. Trenton, NJ: Africa World Press.

David, Nicholas, and Yves Le Bléis
1988 *Dokwaza, Last of the African Ironmasters*. Calgary: Department of Communications Media, University of Calgary.

David, Nicholas, and Judith Sterner
1991 "Gender and Caste in the Mandara Highlands: Northeastern Nigeria and Northern Cameroon." *Ethnology* 30, no. 4: 355–69.
1993 "Transformers Transformed: Aspects of Caste and Iron Technology in the Mandara Mountains." Paper presented at the Transformations, Technology, and Gender in African Metallurgy conference, Oxford.
1997 "Water and Iron: Phases in the History of Sukur." In *L'homme et l'eau dan le bassin du Lac Tchad*, edited by Daniel Barreteau, Hermann Jungraithmayr, and Uwe Siebert, 255–70. Paris: Editions de L'ORSTOM.
2012 "Smith and Society: Patterns of Articulation in the Northern Mandara Mountains." In David 2012, 87–113.

David, Nicholas, Judy Sterner, and Kodzo Gavua
1988 "Why Pots are Decorated." *Current Anthropology* 29, no. 3: 365–89.

Davies, K. G.
1975 *The Royal African Company*. New York: Octagon Books.

Davis, Christopher
2000 *Death in Abeyance: Illness and Therapy among the Tabwa of Central Africa*. Edinburgh: Edinburgh University Press.

Davis, Neil
1992 *The Aurora Watcher's Handbook*. Fairbanks: University of Alaska Press.

Davison, Spana, and P. N. Mosley
1988 "Iron-Smelting in the Upper North Rukuru Basin of Northern Malawi." *Azania: Archaeological Research in Africa* 23, no. 1: 57–99. https://doi.org/10.1080/00672708809511387.

d'Azevedo, Warren L., ed.
1973 *The Traditional Artist in African Societies*. Bloomington: Indiana University Press.

DeCorse, Christopher
2001 *An Archaeology of Elmina: Africans and Europeans on the Gold Coast, 1400–1900*. Washington, D.C.: Smithsonian Press.
2005 "Coastal Ghana in the First and Second Millennia AD: Change in Settlement Patterns, Subsistence and Technology." *Journal des Africanistes* 75, no. 2: 43–54.

Deatrick, Elizabeth
2016 "Pharaoh's Iron Dagger Made from a Meteorite, Study Confirms." *Eos* 97, no. 14: 4–5.

Dekmejian, Richard, and Margaret Wyszomirski
1972 "Charismatic Leadership in Islam: The Mahdi of the Sudan." *Comparative Studies in Society and History* 14, no. 2: 193–214.

Delcommune, Alexandre
1922 *Vingt années de vie africaine. Récits de voyages d'aventures et d'exploration au Congo belge, 1874–1893*. Brussels: Ferdinand Larcier.

Deleuze, Gilles, and Félix Guattari
1987 *A Thousand Plateaus: Capitalism and Schizophrenia*. Minneapolis: University of Minnesota Press.

Delfosse, Daniel, Gilles Pageau, Roger Bennett, and Anoush Poursartip
1993 "Instrumented Impact Testing at High Velocities." *Journal of Composites Technology and Research* 15, no. 1: 38–45.

Delmonte, John
1985 *Origins of Materials and Processes*. Lancaster, PA: Technomic Publishing.

Deme, Alioune, and Susan Keech McIntosh
2006 "Excavations At Walaldé: New Light On The Settlement Of The Middle Senegal Valley by Iron-Using Peoples." *Journal of African Archaeology* 4, no. 2: 317–47.

Dempster, Beth
ca. 1998 "Sympoietic and Autopoietic Systems: A New Distinction for Self-Organizing Systems." Paper for MES thesis, University of Waterloo, School of Planning. https://pdfs.semanticscholar.org/4429/9317a20afcd33b0a11d3b2bf4fc196088d45.pdf.

Derrida, Jacques
1993 *Memoirs of the Blind: The Self-Portrait and Other Ruins*. Chicago: University of Chicago Press.

Dewey, William J.
1986 "Shona male and female artistry." *African Arts* 19, no. 3: 64–67, 84.
1990 *Weapons for the Ancestors*. Iowa City: University of Iowa Video Production Unit. Video, 25 min.
1991 "Pleasing the Ancestors: The Traditional Art of the Shona People of Zimbabwe." PhD diss., Indiana University.
1994 "AK-47s for the Ancestors." *Journal of Religion in Africa* 24, no. 4: 358–74.
2009 "Blacksmiths and Kings and Their Work in Iron." In Elsen and Felix 2009, 18–28.
2013 "Iron Sculpture in Africa." In Herreman 2013, 64–77.
2017 "Knife (*Bakatwa*) with ivory handle." In *Arts of Global Africa: The Newark Museum Collection*, edited by Christa Clarke, 142–45. Newark: The Newark Museum.

Dewey, William J., and S. Terry Childs
1996 "Forging Memory." In Roberts and Roberts 1996, 60–83.

Dewey, W. J., and Els De Palmenaer, eds.
1997 *Legacies of Stone, Zimbabwe: Past and Present*. Vol 1. Tervuren, Belgium: Royal Museum for Central Africa.

Diamitani, Bureima Tiékoroni
2008 "Observing Komo among Tagwa People in Burkina Faso: A Burkinabe Art Historian's Views." *African Arts* 41, no. 3: 14–25.

Didi-Huberman, Georges
2005 *Confronting Images: Questioning the Ends of a Certain History of Art*. University Park, PA: Pennsylvania State University Press.

Dieterlen, Germaine
1938 "Le Duge signe d'alliance chez les Dogons de Sanga (Soudan Français)." *Bulletin du Comité d'Etudes Historiques et Scientifiques de l'Afrique Occidentale Française* 21, no. 1: 108–29.
1951 *Essai sur la religion Bambara*. Paris: Presses Universitaires de France.
1955 "Mythe et organisation sociale au Soudan Français." *Journal de la Société des Africanistes* 25, nos. 1–2: 39–76.
1957 "The Mande Creation Myth." *Africa* 27, no. 2: 124–39.
1959 "Mythe et organisation sociale en Afrique occidentale (suite)." *Journal de la Société des Africanistes* 29, no. 1: 119–38.
1964 "Contribution à l'étude des forgerons en Afrique Occidentale." *École pratique des hautes études, Section des sciences religieuses*. Annuaire 73: 3–28.
1965 *Textes sacrés d'Afrique Noire*. Paris: Gallimard.
1989 "Masks and Mythology among the Dogon." *African Arts* 22, no. 3: 34–43, 87–88.

Dieterlen, Germaine, and Youssouf Cissé
1972 *Les fondements de la société d'initiation du Komo*. Paris: Mouton.

Dieterlen, Germaine, and Diarra Sylla
1992 *L'empire de Ghana: le Wagadou et les traditions de Yéréré*. Paris: Karthala-ARSAN.

Dira, Samuel Jilo, and Barry S. Hewlett
2016 "Learning to Spear Hunt among Ethiopian Chabu Adolescent Hunter-Gatherers." In *Social Learning and Innovation in Contemporary Hunter-Gatherers: Evolutionary and Ethnographic Perspectives*, edited by Hideaki Terashima and Barry S. Hewlett, 71–81. Tokyo: Springer Japan.

Douglas, Mary
(1967) 2004 "Primitive Rationing: A Study in Controlled Exchange." In *Themes in Economic Anthropology*, edited by Raymond Firth, 119–48. London: Routledge.

Doumbia, Paul Emile Namoussa
1936 "Étude du clan des forgerons." *Bulletin du Comité d'études historiques et scientifiques de l'Afrique occidentale française* 19, nos. 2–3: 334–60.

Downey, Greg
2002 "Listening to Capoeira: Phenomenology, Embodiment, and the Materiality of Music." *Ethnomusicology* 46, no. 3: 487–509.

Drewal, Henry J.
1977 *Traditional Art of the Nigerian Peoples: The Ratner Collection*. Washington, D.C.: Museum of African Art.
1988 "Beauty and Being: Aesthetics and Ontology in Yoruba Body Art." In Rubin 1988, 83–96.
1989 "The Meaning of Oshugbo Art: A Reappraisal." In *Man Does Not Go Naked: Textilien und Handwerk aus Afrikanishen und Anderen Landern*, edited by Beate Engelbrecht and Bernhard Gardi, 151–74. Basler Beitrage zur Ethnologie 29. Basil: Wopf.
1994 "Introduction: Yoruba Art and Life as Journeys." In Abiodun et al. 1994, 193–99.
1997 "Art or Accident: Yorùbá Body Artists and Their Deity Ògún." In Barnes 1997, 235–60.
2008 *Mami Wata: Arts for Water Spirits in Africa and Its Diasporas*. Los Angeles: Fowler Museum at UCLA.
Forthcoming "Sensiotics: Senses in Understandings of Material Culture, History, and the Arts." In *Oxford Handbook of History and Material Culture*. Oxford: Oxford University Press.

Drewal, Henry John, ed.
1980 *African Artistry: Technique and Aesthetics in Yoruba Sculpture*. Atlanta: The High Museum of Art.
2008 *Sacred Waters: Arts for Mami Wata and other Divinities in Africa and the Diaspora*. Bloomington: Indiana University Press.

Drewal, Henry J., and John Mason
1997 "Ogun and Mind/Body Potentiality: Yoruba Scarification and Painting Traditions in Africa and the Americas." In Barnes 1997, 332–52. Bloomington: Indiana University Press.
1998 *Beads, Body and Soul: Art and Light in the Yoruba Universe*. Los Angeles: UCLA Fowler Museum of Cultural History.

Drewal, Henry J., and John Pemberton III, with Rowland Abiodun
1989 *Yoruba: Nine Centuries of African Art and Thought*. Edited by Allen Wardwell. New York: The Center for African Art/Abrams.

Drewal, Henry J., and Enid Schildkrout
2010 *Dynasty and Divinity: Ife Art in Ancient Nigeria*. New York: Museum for African Art. Published in the U.K. as *Kingdom of Ife: Sculptures from West Africa* (London: The British Museum).

Drewal, Margaret T., and Henry J. Drewal
1983 "An Ifa Diviner's Shrine in Ijebuland." *African Arts* 16, no. 2: 60–67, 99–100.
1990 *Gelede: Art and Female Power among the Yoruba*. Bloomington: Indiana University Press.

Du Chaillu, Paul Belloui
1861 *Explorations and Adventures in Equatorial Africa: With Account of the Manners and Customs of the People, and of the Chace of the Gorilla, Crocodile, Leopard, Elephant, Hippopotamus, and Other Animals*. London: John Murray.

Duchâteau, Armand
2013 *Sakrale Eusebskulpturen der Dogon/Sacred Iron Sculptures of the Dogon*. Dettelbach, Germany: J. H. Röll.

Dupré, Georges
1995 "The History and Adventures of a Monetary Object of the Kwélé of the Congo: Mezong, Mondjos, and Mandjong." In Guyer 1995, 77–96.

Dupré, Marie-Claude
2000 "Un outil agricole polyvalent: le couteau de culture dans les monts Du Chaillu (Congo, Gabon)." In *Outils aratoires en Afrique*, edited by C. Seignobos, Y. Marzouk, and F. Sigaut, 191–225. Paris: Karthala.

Duvosquel, Jean-Marie, and Jacques Deraeve
1992 *Beauté fatale: armes d'Afrique Centrale*. Brussels: Crédit communal.

Échard, Nicole
1983 "Scories et Symboles: Remarques sur la Métallurgie Hausa du Fer au Niger." In *Métallurgies Africaines: Nouvelles*, edited by Nicole Échard, 209–24. Memoires de la Société des Africanistes 9. Paris: Societé des Africanistes.

Eggert, Manfred K. H.
1993 "Central Africa and the Archaeology of the Equatorial Rainforest: Reflections on Some Major Topics." In Shaw et al. 1993, 289–329.
1994 "Pots, Farming and Analogy: Early Ceramics in the Equatorial Rainforest." In *The Growth of Farming Communities in Africa from the Equator Southwards*, edited by John Edward Giles Sutton, 332–37. Nairobi: British Institute in Eastern Africa.
2005 "The Bantu Problem and African Archaeology." In Stahl 2005, 301–26.

Eggert, Manfred, Alexa Hohn, Stephanie Kahlheber, Conny Meister, Katharina Neumann, and Asstrid Schweizer
2007 "Pits, Graves and Grains: Archaeological and Archaeobotanical Research in Southern Cameroun." *Journal of African Archaeology* 4, no. 2: 273–98.

Ehret, Christopher
1995 "The Establishment of Iron-Working in Eastern, Central, and Southern Africa: Linguistic Inferences on Technological History." *Sprache und Geschichte in Afrika* 15: 1–39.
2002 *The Civilizations of Africa: A History to 1800*. Oxford: James Currey.

Eliade, Mircea
1962 *The Forge and the Crucible: The Origins and Structure of Alchemy*. Chicago: University of Chicago Press.

Elsen, Jan
2003 *De fer et de fierté: Armes blanches d'Afrique noire du musée Barbier-Mueller*. Milan: 5 Continents.

Elsen, Jan, and Marc Leo Felix, eds.
2009 *Fatal Beauty: Traditional Weapons of Central Africa*. Brussels: Tribal Arts Brussels, for the National Museum of History, Taiwan.

El-Zein, Amira
2017 *Islam, Arabs, and the Intelligent World of the Jinn*. Syracuse, NY: Syracuse University Press.

Ejizu, Christopher I.
1987 "The Taxonomy, Provenance, and Function of Ofo—A Dominant Igbo Ritual and Political Symbol." *Anthropos* 82, nos. 4–6: 457–67.

Emsley, John
2002 *Nature's Building Blocks: An A–Z Guide to the Elements*. New York: Oxford University Press.

Espenel, Alexandre, and R. Houdart
2001 *Ofo Anam: Regards sur Statuaire Igbo*. Paris: Galerie l'Accrosonge & Editions Rage.

Essomba, Jean-M.
1992 *Civilisation du fer et société en Afrique centrale. Le cas du Cameroun méridional: Histoire ancienne et archéologie*. Paris: Editions l'Harmattan.

Evans, Chris, and Göran Rydén
2007 *Baltic Iron in the Atlantic World in the Eighteenth Century*. Leiden and Boston: Brill.

Evans-Pritchard, E. E.
1957 "Zande Warfare." *Anthropos* 52: 239–62.

Eyo, Ekpo
1979 *Nigeria and the Evolution of Money*. Lagos: The Central Bank of Nigeria and the Nigerian Federal Department of Antiquities.

Eze-Uzomaka, Pamela I.
2009 "Iron Age Archaeology in Lejja, Nigeria." In *Dimensions of African Archaeology: Studies in the African Past*, edited by Gilbert Pwiti, C. Radimilahy, and S. Macamo, 41–51. Dar es Salaam: E & D Vision Publishers.
2014 "Iron and Its Influence on the Prehistoric Site of Lejja." In Humphris and Rehren 2014, 3–9.

Ezra, Kate
1989 *Art of the Dogon: Selections from the Lester Wunderman Collection*. New York: Harry Abrams.
1992 *Royal Art of Benin: The Perls Collection*. New York: The Metropolitan Museum of Art.

Ezra, Kate
1986 *A Human Ideal in African Art: Bamana Figurative Sculpture*. Washington, D.C.: Smithsonian Institution Press.

Fabian, Johannes
1996 *Remembering the Present: Painting and Popular History in Zaïre*. Berkeley: University of California Press.

Fagan, Brian
1961a "A Collection of Nineteenth Century Soli Ironwork from the Lusaka Area of Northern Rhodesia." *Journal of the Royal Anthropological Institute of Great Britain and Ireland* 91, no. 2: 228–50.
1961b "Pre-European Ironworking in Central Africa with Special Reference to Northern Rhodesia." *The Journal of African History* 2, no. 2: 199–210.
1962 "Two Soli Smelting Furnaces from Lusaka, Northern Rhodesia." *The South African Archaeological Bulletin* 17, no. 65: 27–28.
1969 "Excavations at Ingombe Ilede, 1960–62." In Phillipson et al. 1969, 55–161.

Fagg, Bernard
1968 "The Nok Culture: Excavations at Taruga." *West African Archaeological Newsletter* 10: 27–30.
1990 *Nok Terracottas*. London: Ethnographica.

Fagunwa, David O.
2005 *Ogboju Ode Ninu Igbo Irunmole*. London: Nelson.

Faïk-Nzuji, Clémentine
1992 *Symboles graphiques en Afrique noire*. Paris: Eds. Karthala.

Fakhry, Majid
1997 *The Qur'an: A Modern English Version*. Reading, UK: Garnet.

Falgayrettes-Leveau, Christiane
1994 *Dogon*. Paris: Musée Dapper.

Fardon, Richard
1990 *Between God, the Dead and the Wild: Chamba Interpretations of Ritual and Religion*. Edinburgh: Edinburgh University Press for the International Africa Institute, London.
2011 "Heavy Metal: The Prestige of Iron in the Middle Benue." In Berns et al. 2011, 384–89.

Fariello, M. Anna
2005 "The Language of Objects." In Fariello and Owen 2005, 148–73.

Fariello, M. Anna, and Paula Owen, eds.
2005 *Objects and Meanings: New Perspectives on Art and Craft*. Lanham, MD: The Scarecrow Press, Inc.

Farwell, Byron
1989 *Prisoners of the Mahdi*. New York: W. W. Norton.

Feld, Steven
2017 "On Post-Ethnomusicology Alternatives: Acoustemology." In *Perspectives on 21st Century Contemporary Musicology: Ethnomusicology or Transcultural Musicology?*, edited by Francesco Giannattasio and Giovanni Giuriati, 82–98. Udine, Italy: NOTA – Valter Colle.

Felix, Marc L.
1991 *Kipinga: Throwing-Blades of Central Africa / Wurfklingen aus Zentralafrika*. Munich: Verlag Fred Jahn.

Felkin, Robert W.
1884 "Notes on the For Tribe of Central Africa." *Proceedings of the Royal Society of Edinburgh* 13, no. 38: 205–65.

Fenn, Thomas, David. Killick, John Chesley, Sonja Magnavita, and Joaquin Ruiz
2010 "Contacts Between West Africa and Roman North Africa: Archaeometallurgical Results from Kissi, Northeastern Burkina Faso." In Magnavita et al. 2010, 119–46. Frankfurt: Africa Magna.

Ferera, Lisette, ed.
1994 *Ingénieuse Afrique: Artisans de la récupération et du recyclage*. Québec: Éditions Fides for the Musée de la Civilisation.

Fernandes, A.
1898–1903 Letter to Brother Luiz Froes at College of Goa, June 25, 1560. In *Records of South-Eastern Africa*, edited by George McCall Theal. 9 vols. Cape Town: Government of the Cape Colony.

Fernandez, James
1966 "Principles of Opposition and Vitality in Fang Aesthetics." *Journal of Aesthetics and Art Criticism* 25, no. 1: 53–64.
1982 *Bwiti: An Ethnography of the Religious Imagination in Africa*. Princeton, NJ: Princeton University Press.
1986 *Persuasions and Performances: The Play of Tropes in Culture*. Bloomington: Indiana University Press.

Fernandes, Valentim
1938 *Description de la Cote d'Afrique de Ceuta au Senegal*. Paris: A. Thoyon-Thèze.

Fisch, Rudolf
1911 *Nord-Togo und seine wetsliche Nachbarscaaft*. Basel: Verlag der Basler Missionbuchhandlung.

Fischer, Eberhard
1978 "Dan Forest Spirits: Masks in Dan Villages." *African Arts* 11, no. 2: 16–23, 94.

Fischer, Eberhard, and Hans Himmelherber
1984 *The Arts of the Dan in West Africa*. Zurich: Museum Reitberg.

Fischer, Werner, and Manfred A. Zirngibl
1978 *African Weapons: Knives, Daggers, Swords, Axes, Throwing Knives*. Passau, Germany: Prinz-Verlag.

Flint, Richard F., and Brian J. Skinner
1977 *Physical Geology*. New York: John Wiley & Sons.

Ford, J., and R. de Z. Hall
1947 "The History of Karagwe (Bukoba District)." *Tanganyika Notes and Records* 24: 3–27.

Forkl, Hermann
1989 "The Development of Urban Structures with Reference to Political Evolution in the Wandala Kingdom (17th to 19th Century, Northern Cameroon)." In *Households and Communities: Proceedings of the 21st Annual Chacmool Conference*, edited by Scott MacEachern, Richard Garvin, and David Archer, 542–50. Calgary: Archaeological Association, University of Calgary.

Forni, Silvia, and Doran H. Ross
2017 *Art, Honor, and Ridicule: Fante Asafo Flags from Southern Ghana*. Toronto: Royal Ontario Museum.

Fortier, Joseph
1967 *Mythes et contes de Sou en pays mbaï-Moïssala*. Paris: Juillard.
1982 *Le couteau de jet sacré: Mémoire des Sar et de leurs rois au sud du Tchad*. Paris: Harmattan.

Foster, Hal
2004 *Prosthetic Gods*. Cambridge, MA: MIT Press.

Foucault, Michel
(1966) 1994 *The Order of Things: An Archaeology of the Human Sciences*. New York: Vintage Books Editions.

Fowler, Ian
1990 "Babungo: A Study of Iron Production, Trade and Power in a Nineteenth-century Ndop Plain Chiefdom (Cameroons)." PhD diss., University of London.

Frank, Barbara E.
1993 "Reconstructing the History of an African Ceramic Tradition: Technology, Slavery and Agency in the Région of Kadiolo (Mali)." *Cahiers d' Études africaines*, Année 33, no. 131: 381–401.
1994 "More than Wives and Mothers: The Artistry of Mande Potters." *African Arts* 27, no. 4: 26–37, 93–94.
1998 *Mande Potters and Leather Workers: Art and Heritage in West Africa*. Washington, D.C.: Smithsonian Institution Press.
2002 "Thoughts on Who Made the Jenne Terra-Cottas: Gender, Craft Specialization, and Mande Art History." *Mande Studies* 4: 121–32.
2007a "Marks of Identity: Potters of the Folona (Mali) and their "Mothers." *African Arts* 40, no. 1: 30–41.
2007b "Field Research and Making Objects Speak." *African Arts* 40, no. 1: 13–17.

Franke, Gabriele
2016 "A Chronology of the Central Nigerian Nok Culture—1500 BC to the Beginning of the Common Era." *Journal of African Archaeology* 14, no. 3: 257–89.

Fraser, Douglas, and Herbert M. Cole
1972 "An Overview." In *African Art and Leadership*, edited by D. Fraser and H. M. Cole, 295–327. Madison: University of Wisconsin Press.

Freeman-Grenville, Stewart Parker
1962 *The East Africa Coast: Select Documents from the First to the Early Nineteenth Century*. Oxford: Clarendon Press.

Frei, Robert, Claudio Gaucher, Simon W. Poulton, and Don E. Canfield
2009 "Fluctuations in Precambrian Atmospheric Oxygenation Recorded by Chromium Isotopes." *Nature International Journal of Science* (September 10), 250–53.

Friede, H. M., and R. H. Steel
1986 "Traditional Wooden Drum Bellows of South-Western Africa." *The South African Archaeological Bulletin* 41, no. 143: 12–16.

Fretham, Stephanie J. B., Erik S. Carlson, and Michael K. Georgieff
2011 "The Role of Iron in Learning and Memory." *Advances in Nutrition* 2, no. 2: 112–21.

Lemert, Charles
2011 "Foreword to the Routledge Classics Edition: Money? Really!" In *The Philosophy of Money*, by Georg Simmel, viii–xiii. New York: Routledge.

Fromont, Cécile
2011 "Dance, Image, Myth, and Conversion in the Kingdom of the Kongo, 1500–1800." *African Arts* 44, no. 4: 52–63.
2013 "By the Sword and the Cross." In Cooksey et al. 2013, 28–31.
2014 *The Art of Conversion: Christian Visual Culture in the Kingdom of Kongo*. Chapel Hill: University of North Carolina Press.

Fry, Peter
1976 *Spirits of Protest: Spirit-Mediums and the Articulation of Consensus Amongst the Zezuru of Southern Rhodesia*. Cambridge: Cambridge University Press.

Gagliardi, Susan Elizabeth
2010a "Arts of Power Associations in West Africa." *Heilbrunn Timeline of Art History*, The Metropolitan Museum of Art, New York. http://www.metmuseum.org/toah/hd/powe/hd_powe.htm.
2010b "Divination and Senufo Sculpture in West Africa." *Heilbrunn Timeline of Art History*, The Metropolitan Museum of Art, New York. http://www.metmuseum.org/toah/hd/divi/hd_divi.htm.
2010c "Senufo Arts and *Poro* Initiation in Northern Côte d'Ivoire." *Heilbrunn Timeline of Art History*, The Metropolitan Museum of Art, New York. http://www.metmuseum.org/toah/hd/poro/hd_poro.htm.

Gagliardi, Susan Elizabeth, and Constantine Petridis
2015 *Senufo Unbound: Dynamics of Art and Identity in West Africa*. Milan: 5 Continents Editions.

Gansemans, Jos
2009 *Collections of the RMCA: Musical Instruments*. Tervuren, Belgium: Royal Museum for Central Africa.

Garlake, Peter
1973 *Great Zimbabwe: New aspects of archaeology*. London: Thames & Hudson.

Garrard, Timothy F., and Foungnigué Idrissa Soro
2001 *La sagesse d'un peuple: 2000 proverbes senoufo*. Malvern, England: Garrard Timothy F. Garrard and Foungnigué Idrissa Soro.

Geary, Christraud M.
2003 *In and Out of Focus: Images from Central Africa, 1885–1960*. London: Philip Wilson Publishers.
2007 "Mondes virtuels: les représentations des peuples d'Afrique de l'ouest par les cartes postales, 1895–1935." *Le Temps des médias* 1, no. 8: 75–104.

Geertz, Clifford
1973 "Religion as a Cultural System." In *The Interpretation of Culture: Selected Essays*, by C. Geertz. New York: Basic Books, 87–125.
1976 "Art as a Cultural System." *MLN* (*Modern Language Notes*) 91, no. 6: 1473–99.

Gell, Alfred
1992 "The Technology of Enchantment and the Enchantment of Technology." In Coote and Shelton 1994, 40–63.
1998 *Art and Agency*. New York: Oxford University Press.

Germann, P.
1922 "Afrikanische Wurfeisen und Wurfhölzer im Völkermuseum zu Leipzig." In *Jahrbuch des städtisches Museum für Völkerkunde zu Leipzig* 8 (1918–21): 41–50.

Gettens, Rutherford J., Roy S. Clarke, Jr., and W. T. Chase
1971 *Two Early Chinese Bronze Weapons with Meteoritic Iron Blades*. Occasional Papers 4, no. 1. Washington, D.C.: Freer Gallery of Art.

Ghandi, Abdelhamid
2014 *L'art du fer forgé à travers l'architecture au Maroc*. Fez, Morocco: printed by the author.

Gilbert, Michelle
n.d. "Akan Leadership Art and Ceremony." *Art & Life in Africa*, University of Iowa Stanley Museum of Art. https://africa.uima.uiowa.edu/topic-essays/show/41?start=10.

Giménez, Carmen
1993 *Picasso and the Age of Iron*. New York: Guggenheim Museum.

Girshick Ben-Amos, Paula
(1980) 1995 *The Art of Benin*, rev. ed. Washington, D.C.: Smithsonian Institution Press.

Glass, Charles
2017 "In the Horrorscape of Aleppo." *New York Review of Books* 64, no. 9: 18–20.

Glaze, Anita J.
1981 *Art and Death in a Senufo Village*. Bloomington: Indiana University Press.

Goldsmith, Werner
1960 *Impact: The Theory and Physical Behaviour of Colliding Solids*. London: Edward Arnold.

González-Ruibal, Alfredo, Manuel Sánchez-Elipe, and Carlos Otero-Vilariño
2012 "An Ancient and Common Tradition: Funerary Rituals and Society in Equatorial Guinea (First–Twelfth Centuries AD)." *African Archaeological Review* 30(2): 115–43.

Gordon, Robert B.
1996 *American Iron, 1607–1900*. Baltimore: Johns Hopkins University Press.

Goucher, Candice
1981 "Iron is Iron 'til It is Rust: Trade and Ecology in the Decline of West African Iron-smelting." *Journal of African History* 22, no. 1: 179–89.
1984 "The Iron Industry of Bassar, Togo: An Interdisciplinary Investigation of African Technological History." PhD diss., University of California, Los Angeles.
1988 "The Impact of German Colonial Rule on the Forests of Togo." In *World Deforestation in the Twentieth Century*, edited by John F. Richards and Richard P. Tucker, 56–69. Durham, NC: Duke University Press.
1990 "John Reeder's Foundry: A Study of Eighteenth-century African-Caribbean Technology." *Jamaica Journal* 23, no. 1: 39–43.
1993 "African Metallurgy in the Atlantic World." *The African Archaeological Review* 11: 197–215.

Goucher, Candice L., and Eugenia W. Herbert
1996 "The Blooms of Banjeli: Technology and Gender in West African Iron Making." In Schmidt 1996b, 40–57.

Goucher, Candice, Eugenia Herbert, and Carlyn Saltman
1986 *The Blooms of Banjeli: Technology and Gender in West African Iron-making*. Watertown, MA: Documentary Educational Resources. Video, 28 min.

Grabar, Oleg
1992 *The Mediation of Ornament*. Princeton, NJ: Princeton University Press.

Grady, Anne R.
2007 "Technical Analysis and Treatment of a Sudanese Kaskara Sword." Paper presented at the annual conference of the Association of North American Graduate Programs in Conservation, Harvard University Art Museums, Cambridge, MA. http://resources.conservation-us.org/anagpic-student-papers/anagpic-2007-student-papers-posters/.

Grant, James
ca. 1895 *Cassell's History of the War in the Soudan*. Vol. 2. New York: Cassel.

Gray, John
1957 *The British in Mombasa 1824–1826*. London: Macmillan.

Grébénart, Danilo
1983 "Les métallurgies du cuivre et du fer autour d'Agadez (Niger), des origines au début de la période médiévale." In *Métallurgies africaines: nouvelles contributions*, edited by N. Echard, 109–25. Paris: Société des africanistes.
1988 *Les Premiers Metallurgistes en Afrique Occidentale (les origins de la metallurgie en Afrique Occidentale)*. Paris: Errance.

Griaule, Marcel
(1948) 1965 *Conversations with Ogotemmêli: An Introduction to Dogon Religious Ideas*. London: Oxford University Press, International African Institute.
(1938) 1994 *Masques dogons. Travaux et mémoires de l'Institut d'Ethnologie* 33. Paris: Institut d'Ethnologie.

Griaule, Marcel, and Germaine Dieterlen
1975 *Conversations with Ogotemmeli: An Introduction to Dogon Religious Ideas*. New York: Oxford University Press.
(1965) 2015 *Le renard pâle – religion ancienne d'Afrique de l'Ouest: Histoire de la création de l'univers pas le people dogon*. Paris: Editions Anté-Matière.

Gurney, Oliver Robert
1954 *The Hittites*. 2nd ed. Harmondsworth, England: Penguin Books.

Gurstelle, Andrew
n.d. "Archives in Iron: Re-evaluating the Historical Conxt of *Minkondi*." Unpublished paper.

Guyer, Jane I.
1985 "The Iron Currencies of Southern Cameroon." *Symbols* (December): 2–5, 15–16.
1986 "Indigenous Currencies and the History of Marriage Payments: A Case Study from Cameroon." *Cahiers d'Études africaines*, no. 104: 577–610.
1993 "Wealth in People and Self-Realization in Equatorial Africa." *Man*, New Series 28, no. 2: 243–65.

Guyer, Jane I., ed.
1995 *Money Matters: Instability, Values and Social Payments in the Modern History of West African Communities*. Social History of Africa Series. London: James Currey.

Guyer, Jane I., and Samuel M. Eno Belinga
1995 "Wealth in People as Wealth in Knowledge: Accumulation and Composition in Equatorial Africa." *Journal of African History* 36, no. 1: 91–120.

Haaland, Gunner, Randi Haaland, and Rijahl Suman
2002 "The Social Life of Iron: A Cross-Cultural Study of Technological, Symbolic, and Social Aspects of Iron." *Anthropos* 97: 35–54.

Haaland, Randi
1980 "Man's Role in the Changing Habitat of the Méma During the Period of the Old Kingdom of Ghana." *Norwegian Archeaological Review* 13, no. 1: 31–46.
1985 "Iron Production, Its Socio-Cultural Context and Ecological Implications." In Haaland and Shinnie 1985, 50–72.
2014 "Iron Working in an Indian Ocean Context." In Humphris and Rehren 2014, 146–55.

Haaland, R., and P. Shinnie, eds.
1985 *African Iron Working*. Oslo: Norwegian University Press.

Haaland, Randi, et al.
2000 "Ethnoarchaeological Research on Iron Smelting in Southwest Ethiopia." *Nyame Akuma* 54: 6–12.

Hall, R. de Z.
1938 "A Tribal Museum at Bweranyange, Bukoba District." *Tanganyika Notes and Records* 5: 1–4.

Hamilton, Caroline
1998 *Terrific Majesty: The Powers of Shaka Zulu and the Limits of Historical Invention*. Cambridge, MA: Harvard University Press.

Haraway, Donna J.
2016 *Staying with the Trouble: Making Kin in the Chthulucene*. Durham, NC: Duke University Press.

Hardin, Kris L., and Mary Jo Arnoldi
1996 "Introduction: Efficacy and Objects.' In Arnoldi et al. 1996, 1–13.

Harman, Graham
2002 *Tool-Being: Heidegger and the Metaphysics of Objects*. Peru, IL: Open Court.

Harms, Robert W.
1981 *River of Wealth, River of Sorrow: The Central Zaire Basin in the Era of the Slave and Ivory Trade, 1500–1891*. New Haven: Yale University Press.

Hart, Keith
2001 *Money in an Unequal World*. New York: Texere.
2005 "Notes towards an Anthropology of Money." *Kritikos* 2. http://intertheory.org/kritikos.

Hart, Terese B., and John A. Hart
1986 "The Ecological Basis of Hunter-gatherer Subsistence in African Rain Forests: The Mbuti of Eastern Zaire." *Human Ecology* 14, no. 1: 29–55.

Head, James W., III
1999 "Surfaces and Interiors of the Terrestrial Planets." In Beatty et al. 1999, 157–73.

Headrick, Daniel
1981 *The Tools of Empire: Technology and European Imperialism in the Nineteenth Century*. New York: Oxford University Press.

Heaney, Seamus
2007 *District and Circle: Poems*. New York: Farrar, Strauss and Giroux.

Heidegger, Martin
(1927) 2008 *Being and Time*. Translated by John Macquarrie and Edward Robinson. New York: Harper Perennial.

Heimlich, Geoffroy
2016 "The Kongo Cross Across Centuries." *African Arts* 49, no. 3: 22–31.

Helms, Mary
1988 *Ulysses' Sail: An Ethnographic Odyssesy of Power, Knowledge and Geographical Distance*. Princeton, NJ: Princeton University Press.

Henderson, Richard
1972 *The King in Every Man*. New Haven: Yale University Press.

Hentze, Matthias.W., Martina U. Muckenthaler, B. Galy, and Clara Camaschella
2010 "Two to Tango: Regulation of Mammalian Iron Metabolism." *Cell* 142, no. 1: 24–38

Herbert, Eugenia W.
1984 *Red Gold of Africa: Copper in Precolonial History and Culture*. Madison: University of Wisconsin Press.
1993 *Iron, Gender, and Power: Rituals of Transformation in African Societies*. Bloomington: Indiana University Press.

Herreman, Frank, ed.
2013 *Material Differences: Art and Identity in Africa*. New York: Museum for African Art.

Hersak, Dunja
1986 *Songye Masks and Figure Sculpture*. London: Ethnographica.
2001 "There Are Many Kongo Worlds: Particularities of Magico-Religious Beliefs among the Vili and Yombe of Congo-Brazzaville." *Africa: Journal of the International African Institute* 71, no. 4: 614–40.

Hertefelt, Marcel d', and André Coupez
1964 *La Royauté sacrée de l'ancien Rwanda. Annales du Musée royal de l'Afrique centrale, sciences humaines* 52. Tervuren, Belgium: Royal Museum for Central Africa.

Hesiod
1973 "Works and Days." In *Hesiod and Theogirus*, translated by Dorothea Wender, 59–86. Harmondsworth, U.K.: Penguin Books.

Heusch, Luc de
1956 "Le Symbolisme du Forgeron en Afrique." *Reflets du Monde* 10: 57–70.
1982 *The Drunken King, or, The Origin of the State*. Bloomington: Indiana University Press.
1991 "Le roi vient d'ailleurs." In Jacobson-Widding 1991, 109–17.
2000 *Le roi du Kongo et les monstres sacrés*. Paris: Gallimard.

Heywood, Linda, and John Thornton
2007 *Central Africans, Atlantic Creoles, and the Foundation of the Americas, 1585–1660*. New York: Cambridge University Press.

Hinnebusch, Thomas
1989 "Bantu Languages." In *The Niger-Congo Languages: A Classification and Description of Africa's Largest Language Family*, edited by John Bendo-Samuel and Rhonda L. Hartell, 450–73. Lanham, MD: University Press of America.

Hochschild, Adam
1998 *King Leopold's Ghost*. New York: Houghton Mifflin Company.

Hodder, Ian
2012 *Entangled: An Archaeology of the Relationships Between Humans and Things*. Malden, MA: John Wiley and Sons.

Holt, Peter M.
1970 *The Mahdist State in the Sudan 1881–1898: A Study of Its Origins, Development, and Overthrow*. Oxford: Clarendon.

Homann, Lisa
2011 "When Muslims Masquerade: *Lo Gue* Performance in Southwestern Burkina Faso." PhD diss., UCLA Department of Art History.

Horton, Robert
1965 *Kalabari Sculpture*. Lagos: Department of Antiquities.

Hultgren, Mary Lou, and Jeanne Zeidler
1993 *A Taste for the Beautiful: Zairian Art from the Hampton University Museum*. Hampton, VA: Hampton University Museum.

Humphris, Jane
2018 "Comparing Kushite and Aksumite iron production: chronology, technology and broader considerations." Presentation at the Society of Africanist Archaeologist's Biennial Meetings, Queens College, University of Toronto. June 20.

Humphris, Jane, and Chris Carey
2016 "New methods for investigating slag heaps: Integrating geoprospection, excavation and quantitative methods at Meroe, Sudan." *Journal of Archaeological Science* 70: 132–44.

Humphris, Jane, and Louise Iles
2013 "Pre-Colonial Iron Production in Great Lakes Africa: Recent Research at UCL Institute of Archaeology." In Humphris and Rehren 2013, 56–65.

Humphris, Jane, Marcos Martinon-Torres, Thilo Rehren, and Andrew Reid
2009 "Variability in single smelting episodes – a pilot study using iron slag from Uganda." *Journal of Archaeological Science* 36: 359–69.

Humphris, Jane, and Thilo Rehren, eds.
2014 *The World of Iron*. London: Archetype Publications.

Humphris, Jane, and Thomas Schreibner
2017 "A New Radiocarbon Chronology for Ancient Iron Production in the Meroe Region of Sudan." *African Archaeological Review* 34, no. 3: 377–413.

Holl, Augustin F. C.
2009 "Early West African Metallurgies: New Data and Old Orthodoxy." *Journal of African Archaeology* 22: 415–38.

Huysecom, Eric, Michel Rasse, L. Lespez, Katharina Neumann, Fahmy, A. Ballouche, Sylvain Ozainne, M. Maggetti, Chantal Tribolo, and S. Soriano
2009 "The Emergence of Pottery in Africa during the Tenth Millennium Cal BC: New Evidence from Ounjougou (Mali)." *Antiquity* 83, no. 322: 905–17.

IAAPB (Institut apostolique des Pères Blancs)
1892 *Près du Tanganika par les Missionnaires de Son Éminence le Cardinal Lavigerie*. Antwerp: Imp. H. Majoo.

Iles, Louise
2013 "Applying Ethnographic Presents to Archaeological Pasts: Preliminary Thoughts on Memories of Iron Production from Western Uganda." In Humphris and Rehren 2013, 281–87.
2017 "Knowledge Networks and Interactions in the Iron Production Landscape of Precolonial Western Uganda: Understanding the Spread of Past Technologies." Unpublished paper.

Iles, L., and E. Lyaya
2015 "Making metals in East Africa and beyond: Archaeometallurgy in Azania." *Azania: Archaeological Research in Africa* 50, no. 4: 481–94.

Iliffe, John
2005 *Honour in African History*. New York: Cambridge University Press.

Imperato, Pascal James
1977 *African Folk Medicine: Practices and Beliefs of the Bambara and Other Peoples*. Baltimore: York Press.
1983 *Buffoons, Queens and Wooden Horsemen: The Dyo and Gouan Societies of the Bambara of Mali*. New York: Kilima House.

Imagawa, Misaki
2015 "Welcome to Burkina FassSSsso—Snake Form Currency." *Primitive Living and Collecting* (blog), March 20. http://www.beprimitive.com/blog/welcome-to-burkina-fasssssso.

Ingold, Tim
2007 "Materials against Materiality." *Archaeological Dialogues* 14, no. 1: 1–16.
2010 "The Textility of Making." *Cambridge Journal of Economics* 34: 91–102.
2011 *Being Alive: Essays on Movement, Knowledge and Description*. Abington, UK: Routledge.
2013 *Making: Anthropology, Archaeology, Art and Architecture*. Milton Park, UK: Routledge.

Insoll, Timothy
2015 *Material Explorations in African Archaeology*. New York: Oxford University Press.

Iroko, Abiola Félix
1988 *Les cauris en Afrique Occidentale du dixième au vingtième siècle*. Lilles, France: Atelier national de reproduction de theses.

Jackson, Michael
2005 *Existential Anthropology: Events, Exigencies, and Effects*. New York: Berghahn.

Jacobson-Widding, Anita
1979 *Red-White-Black as a Mode of Thought: A Study of Triadic Classification by Colours in the Ritual Symbolism and Cognitive Thought of the Peoples of the Lower Congo*. Uppsala Studies in Cultural Anthropology 1. Stockholm: Almqvist & Wiksell International.
1990 "The Fertility of Incest." In *The Creative Communion: African Folk Models of Fertility and the Regeneration of Life*, edited by A. Jacobson-Widding and W. van Beek, 47–73. Uppsala Studies in Cultural Anthropology, 15. Stockholm: Academiae Ubsaliensis.

Jacobson-Widding, Anita, ed.
1991 *Body and Space: Symbolic Models of Unity and Division in African Cosmology and Experience*. Uppsala Studies in Cultural Anthropology 16. Stockholm: Almqvist and Wiksell.

Jacquot, Jeremy
2009 "The Earth's Great Oxidation Event may have ended in a big dip." *Ars Technica*.https://arstechnica.com/science/2009/09/dips-and-swells-may-have-shaped-oxygens-early-rise/.

Jadinon, Rémy
2013 "On Nsambi Pluriarcs." In Cooksey et al. 2013, 143–47.

Janzen, John M.
2009 "Postcolonial Collecting in Lower Congo: Categories, Perspectives, and Negotiations." Paper presented at annual African Studies Association conference, New Orleans.

Jedrej, M. C.
1975 "Ingessana Throwing Knives (Sudan)." *Anthropos* 70, nos. 1–2: 42–48.
1995 *Ingessana: The Religious Institutions of a People of the Sudan-Ethiopia Borderland*. Leiden: Brill Academic.

Jeffreys, Mervyn David Waldegrave
1955 "A Bamenda Meteorite." *Man* 55: 167–68.
1956a "The Degeneration of the *ọfọ* Anam." *Nigerian Field* 21, no. 4: 173–77.
1956b "The Anam Ofo: A Cult Object Among the Ibo." *South African Journal of Science* 52, no. 10: 227–33.

Jenniges, Em
1909 *Dictionnaire Français-Kiluba*. Brussels: Spineux & Cie., for the Belgian Ministry of Colonies.

Jerven, Morten
2016 "Capitalism in Pre-Colonial Africa: A Review." *African Economic History Working Paper Series*, no. 27. https://www.aehnetwork.org/wp-content/uploads/2016/06/AEHN-WP-27.pdf.

Jewsiewicki, Bogumil
2008 "Mami Wata/Mamba Muntu Paintings in the Democratic Republic of the Congo." In Drewal 2008, 126–33.

Jobson, Richard
(1623) 1999 *The Discovery of River Gambra*. Edited by David P. Gamble and P. E. H. Hair. London: The Hakluyt Society.

Johnson, Diane, Joyce Tyldesley, Tristan Lowe, Philllip J. Withers, and Monica Grady
2013 "Analysis of a Prehistoric Egyptian Iron Bead with Implications for the Use and Perception of Meteorite Iron in Ancient Egypt." *Meteoritics & Planetary Science* 48, no. 6: 997–1006.

Johnson, Frederick
(1939) 1988 *A Standard Swahili-English Dictionary*. New York: Oxford University Press.

Johnson, John William
1979 *The Epic of Sun-Jata According to Magan Sisòk*. 2 vols. Bloomington, IN: Folklore Publications Group.
1986 *The Epic of Son-Jara: A West African Tradition*. Bloomington: Indiana University Press.

Johnson, John William, Thomas H. Hale, and Stephen Belcher
1993 *Oral Epics from Africa: Vibrant Voices from a Vast Continent*. Bloomington: Indiana University Press.

Johnson, Tre
2017 "Black Superheroes Matter: Why a 'Black Panther' Movie is Revolutionary." *Rolling Stone*, October 17. https://www.rollingstone.com/movies/movie-news/black-superheroes-matter-why-a-black-panther-movie-is-revolutionary-198678/.

Jolly, Éric
2007 "Rêveries exotiques sur les Dogon." *L'Homme* 182: 187–214.
2013 "Exhibition Review: 'Dogon,' Musée du quai Branly, April 15–July 24, 2011." *African Arts* 46, no. 3: 82–84.

Jonckers, Danielle
1979 "Notes sur le forgeron, la forge et les métaux en pays minyanka." *Journal des Africanistes* 49: 103–24.

Jones, Kellie
2011 *EyeMinded: Living and Writing Contemporary Art*. Durham, NC: Duke University Press.

Jordán, Manuel
1996 "Tossing Life in a Basket: Art and Divination Among Chokwe, Lunda, Luvale and Related Peoples of Northwestern Zambia." PhD diss., University of Iowa.
1998 *Chokwe! Art and Initiation Among Chokwe and Related Peoples*. Birmingham, AL: Birmingham Museum of Art.
2003 "Tupele: Basket Divination Symbols of the Chokwe." *Tribal Art* 8-1, no. 30: 96–106.
2012 *Sanza: African Thumb Pianos from the Collections of F. & F. Boulanger-Bouhière, the Royal Museum for Central Africa, and MIM*. Phoenix: Musical Instrument Museum.

Joyce, Tom
1998 *Life Force at the Anvil: The Blacksmith's Art from Africa*. Asheville, NC: Biltmore Press.

Junius, Henrik
2016 "Nok Early Iron Production in Central Nigeria—New Finds and Features." *Journal of African Archaeology* 14, no. 3: 291–311.

Kagame, Alexis
1956 "Les Hamites du Rwanda et du Burundi sont-ils des Gallas?" *Bulletin de l'Académie Royale des Sciences Coloniales* 11, no. 3: 351–66.

Kanté, Nambala
1993 *Forgerons d'Afrique Noire: Transmission des Savoirs Traditionnels en Pays Malinké*. Paris: L'Harmattan.

Karp, Ivan, and Steven D. Lavine, eds.
1991 *Exhibiting Cultures: The Poetics and Politics of Museum Display*. Washington, D.C.: Smithsonian Institution Press.

Kasfir, Sidney
2007 *African Art and the Colonial Encounter: Inventing a Global Commodity*. Bloomington: Indiana University Press.

Kasfir, Sidney Littlefield, and Till Förster, eds.
2013 *African Art and Agency in the Workshop*. Bloomington: Indiana University Press.

Kassam, Aneesa, and Gemetchu Megerssa
2004 "Iron and Beads: Male and Female Symbols of Creation: A Study of Ornament among Booran Oromo." In *The Meaning of Things: Material Culture and Symbolic Expression*, edited by Ian Hodder, 23–32. New York: Routledge.

Katoke, Israel K.
1975 *The Karagwe Kingdom: A History of the Abanyambo of North-West Tanzania*. Nairobi: East African Publishing House.
2001 "The Karagwe Collection." In *From Ritual to Modern Art: Tradition and Modernity in Tanzanian Sculpture*, edited by Manfred Ewel and Anne Outwater, 42–60. Dar-es-Salaam: Mkuki na Nyoto Publications.

Kaufman, Brett Sanford
2014 "Empire Without a Voice: Phoenician Iron Metallurgy and Imperial Strategy at Carthage." PhD diss., University of California Los Angeles.

Killick, David
1990 "Technology in its Social Setting: Bloomery Iron-Smelting at Kasungu, Malawi, 1860–1940." PhD diss., Yale University.
1996 "On Claims for 'Advanced' Ironworking Technology in Precolonial Africa." In Schmidt 1996b, 247–66.
2004 "What Do We Know About African Iron Working?" *Journal of African Archaeology* 2, no. 1: 97–112.
2009a "Cairo to Cape: The Spread of Metallurgy through Eastern and Southern Africa." *Journal of World Prehistory* 22, no. 4: 399–414.
2009b "Agency, dependency, and long-distance trade: East Africa and the Islamic world." In *Polities and Power: Archaeological Perspectives on the Landscapes of Early States*, edited by S. E. Falconer and C. L. Redman, 179–207. Tucson University of Arizona Press.
2015 "Invention and Innovation in African Iron-smelting Technologies." *Cambridge Archaeological Journal* 25, no. 1: 307–19.

Killick, David J., and Robert B. Gordon
1993 "Adaptation of Technology to Culture and Environment: Bloomery Iron Smelting in America and Africa." *Technology and Culture* 34, no. 2: 243–70.

Klieman, Kairn
2003 *"The Pygmies Were Our Compass": Bantu and Batwa in the History of West Central Africa, Early Times to c. 1900 CE Social History of Africa*. Portsmouth, NH: Heinemann.

Konadu, Kwasi
Forthcoming *Our Own Way in This Part of the World: Biography of an African Community, Culture, and Nation*. Durham, NC: Duke University Press.

Kone, Kassim
2001 "Review of Search of Sunjata: The Mande Oral Epic as History, Literature, and Performance, edited by Ralph A. Austen." *African Studies Review* 44, no. 1: 156–58.

Kosky, Jeffrey
2013 *Arts of Wonder: Enchanting Secularity – Walter De Maria, Diller + Scofidio, James Turrell, Andy Goldsworthy*. Chicago: University of Chicago Press.

Kratz, Corinne A.
2002 *The Ones That Are Wanted: Communication and the Politics of Representation in a Photographic Exhibition*. Berkeley: University of California Press.

Kreamer, Christine Mullen
2012 "African Cosmos: Performing the Moral Universe." In *African Cosmos: Stellar Arts*, edited by C. M. Kreamer, 114–53. Washington, D.C: Narional Museum of African Art, Smithsonian Insitution.

Kreamer, Christine Mullen, Mary Nooter Roberts, Elizabeth Hasrney, and Allyson Purpura
2007 *Inscribing Meaning: Writing and Graphic Systems in African Art*. Milan: 5 Continents for the Smithsonian Institution's National Museum of African Art, Washington, D.C.

Kriger, Colleen E.
1992 "Ironworking in 19th Century Central Africa." PhD diss., York University, Toronto.
1999 *Pride of Men: Ironworking in 19th Century West Central Africa*. Portsmouth, NH: Heinemann.
2009 "'Guinea Cloth': Production and Consumption of Cotton Textiles in West Africa Before and During the Atlantic Slave Trade " In *The Spinning World: A Global History of Cotton Textiles, 1200–1850*, edited by G. Riel o and P. Parthasaranthi, 105–26. New York: Oxford University Press.
2017 *Making Money: Life, Death, and Early Modern Trade on Africa's Guinea Coast*. Athens: Ohio University Press.

Krupp, Edwin C.
1991 *Beyond the Blue Horizon: Myths and Legends of the Sun, Moon, Stars, and Planets*. New York: Harper Collins.

Kubik, Gerhard
1999 "African and African American Lamellophones: History, Typology, Nomenclature, Performers, and Intracultural Concepts." In *Turn up the Volume: A Celebration of African Music*, edited by Jacqueline Cogdell Djedje, 20–57. Los Angeles: UCLA Fowler Museum of Cultural History.

Kunene, Masizi
(1979) 2017 *Emperor Shaka the Great: A Zulu Epic*. Durban: University of KwaZulu-Natal Press.

Kusimba, Chapurukha M.
1997 "The archaeology of the Swahili." In *Encyclopedia of Pre-colonial Africa*, edited by J. O. Vogel, 507–13. Walnut Creek, CA: AltaMira.

Kusimba, Chapurkha M., David J. Killick, and R. G. Cresswell
1994 "Indigenous and imported metals in Swahili sites on the Kenyan coast." In *Society, Culture and Technology in Africa*, edited by S. T. Childs, 63–77. MASCA Research Papers in Science and Archaeology supplement to vol. 11. Philadelphia: University of Pennsylvania Museum.

LaGamma, Alisa
2007 *Eternal Ancestors: The Art of the Central African Reliquary*. New York: The Metropolitan Museum of Art.
2015 *Kongo: Power and Majesty*. New York: The Metropolitan Museum of Art.

LaGamma, Alisa, with Ellen G. Howe and Adr ana Rizzo
2015 "Mangaaka." In *Kongo: Power and Majesty*, edited by A. LaGamma, 220–65. New York: The Metropolitan Museum of Art.

Lagercrantz, Sture
1950 *Contribution to The Ethnography of Africa*. Studia Ethnographica Upsaliensia series, no. 1. Lund, Sweden: Håkan Ohlssons Boktryckeri.

Lakoff, George, and Mark Johnson
1980 *Metaphors We Live By*. Chicago: University of Chicago Press.

Lambert, Nicole
1972 "Objets en cuivre et Néolithique de Mauritanie Occidentale." In *Proceedings of the 6th Panafrican Congress of Prehistory*, Dakar, 1967, 159–75. Chambéry: Les Imprimeries Réunies de Chambéry.
1983 "Nouvelle contribution à l'étude du Chalcolithique de Mauritanie." In *Métallurgies Africaines: Nouvelles Contributions*, edited by N. Echard, 63–88. Paris: Mémoire de la Société des Africanistes 9.

Landgraf, Walter, and Marc Banks
2008 *2007 Archaeological Investigation at the Richard Smith Forge Site, Colebrook, Connecticut*. Colebrook, CT: Farmington River Coordinating Committee, Barkhamsted and Colebrook Historical Societies and The Colebrock Land Conservancy.

Landor, A. Henry Savage
1907 *Across widest Africa: an account of the country and people of eastern, central and western Africa as seen during a twelve months' journey from Djibuti to Cape Verde*, vol. 2. London: Hurst and Blackett.

Langlands, Alexander
2018 *Cræft: An Inquiry into the Origins and True Meaning of Traditional Crafts*. New York: W. W. Norton.

Langlois, Olivier
1995 "Histoire du peuplement post-néolithique du Diamaré." PhD diss., Université de Paris I, Pantheon Sorbonne.
2010 "Aux Origines de l'Endogamie des Forgerons dans les Monts Mandara (Nord-Cameroun): Mythes, Hypothèses Historiques et Arguments Matériels." *Journal des Africanistes* 79, no. 2: 316–33.

Lane, J. Paul
2017 "Early agriculture in sub-Saharan Africa to c. 500 CE." In *The Cambridge World History*, vol. 2, edited by Graeme Barker and Candice Goucher, 472–98. Cambridge: Cambridge University Press.

Lanning, E. C.
1954 "Genital Symbolism on Smiths' Bellows in Uganda." *Man* 54: 262.

Larson, Leila M., Kamija S. Phiri, and Pasricha Sant-Rayn
2017 "Iron and Cognitive Development: What is the Evidence?" *Annals of Nutrition and Metabolism* 71: 25–38.

Larroque, Olivier, and Martine Nougarède
2010 *Vaca bruto du quotidien au sacré: Tauraux de l'Afrique Noire*. Paris: Editions du Musée des Cultures Taurines.

Latour, Bruno
1996 "On actor-network theory: A few clarifications." *Soziale Welt* 47, no. 4: 369–81.
1999 *Pandora's Hope: Essays on the Reality of Science Studies*. Cambridge, MA: Harvard University Press.

Laude, Jean
1994 "Le sens de la forme: une approche des arts dogon." In Falgayrettes-Leveau 1994, 169–220.

Launay, Robert, ed.
2016 *Islamic Education in Africa: Writing Boards and Blackboards*. Bloomington: Indiana University Press.

Lavachery, Philippe, Scott MacEachern, Tchago Bouimon, Bienvenu Gouem Gouem, Pierre Kinyock, Jean J. Mbairo, and Olivier Nkokonda
2005 "Komé to Ebomé: Archaeological Research for the Chad Export Project, 1999–2003." *Journal of African Archaeology* 3, no. 2: 175–93.

Lavachery, Philippe, Scott MacEachern, Tchago Bouimon, and Christophe Mbida Mindzie
2010 *Komé – Kribi: Rescue Archaeology Along the Chad-Cameroon Oil Pipeline, 1999–2004*. Journal of African Archaeology Monograph Series. Frankfurt: Africa Magna.

LaViolette, Adria
1987 "An Archaeological Ethnography of Blacksmiths, Potters and Masons in Jenné, Mali." PhD diss., Washington University, St. Louis.
2018 "Craft and Industry." In Wynne-Jones and LaViolette 2018, 319–34.

Law, Robin
2004 *Ouidah: The Social History of a West African Slaving 'Port', 1727–1892*. Athens: Ohio University Press.

Lawal, Babatunde
1970 "Yoruba Sango Sculpture in Historical Perspective." PhD diss., Indiana University.
1997 *The Gelede Spactacle: Art, Gender, and Social Harmony in an African Culture*. Seattle: University of Washington Press.

Layton, Robert
1991 *The Anthropology of Art*. Cambridge: Cambridge University Press.

Lebling, Robert
2010 *Legends of the Fire Spirits: Jinn and Genies from Arabia to Zanzibar*. Berkeley, CA: Counterpoint.

Lee, Stan
1963 *Tales of Suspense* 39. New York: Marvel Comics.

Lembezat, Bertrand
1952 *Mukuléhé, un Clan Montagnard du Nord-Cameroun*. Paris: Berger-Levrault.

Lemelle, Sidney J.
1992 "Ritual, Resistance and Social Reproduction: A Cultural Economy of Iron-Smelting in Colonial Tanzania 1890–1975." *Journal of Historical Sociology* 5, no. 2: 161–82.

Lewis, David Levering
1987 *The Race to Fashoda: Colonialism and African Resistance*. New York: Henry Holt.

Logan, William Bryant
1995 *Dirt: The Ecstatic Skin of the Earth*. New York: W. W. Norton.

Longfellow, Henry Wadsworth
(1840) 1992 "The Village Blacksmith." In *Favorite Poems*. Mineola, NY: Dover.

Levtzion, Nehemia
1973 *Ancient Ghana and Mali*. London: Methuen.

Lijadu, E. M.
1972 *Orunmila*. Ado-Ekiti: Omolayo Standard Press of Nigeria.

Livingston, James D.
1996 *Driving Force: The Natural Magic of Magnets*. Cambridge, MA: Harvard University Press.

Lorenz, Carol Ann
1982 "Lower Niger Bronze Bells: Form, Iconography, and Function." In *The Art of Metal in Africa*, edited by Marie-Therese Brincard, 52–60. New York: African American Institute.

Lukes, Steven
1978 "Power and Authority." In *A History of Sociological Analysi*, edited by T. Bottomore and R. Nisbet, 633–76. New York: Basic Books.

Luna, Kathryn de
2016 *Collecting Food, Cultivating People: Subsistence and Society in Central Africa*. New Haven: Yale University Press.

Lupo, Karen D., Dave N. Schmitt, Christopher A. Kiahtipes, Jean-Paul Ndanga, D. Craig Young, and Bernard Simiti
2015 "On Intensive Late Holocene Iron Mining and Production in the Northern Congo Basin and the Environmental Consequences Associated with Metallurgy in Central Africa." *PLOS ONE* 10, no. 7. https://doi.org/10.1371/journal.pone.0132632.

MacDonald, Kevin C.
1996 "Tichitt-Walata and the Middle Niger: Evidence for Cultural Contact in the Second Millennium BC." In Pwiti and Soper 1996, 429–40.

MacDonald, Kevin, Robert Vernet, Dorian Fuller, and James Woodhouse
2003 "New Light on the Tichitt Tradition: A Preliminary Report on Survey and Excavation at Dhar Nema." In *Researching Africa's Past: New Contributions from British Archaeologists*, edited by Peter Mitchell, Anne Haour, and John Hobart, 73–80. Oxford: Oxford University School of Archaeology.

MacDonald, Kevin C., Robert Vernet, Marcos Martinón-Torres, and Dorian Q. Fuller
2009 "Dhar Néma: From Early Agriculture to Metallurgy in Southeastern Mauritania." *Azania: Archaeological Research in Africa* 44: 1, 3–48.

MacEachern, Scott
1996 "Iron Age Beginnings North of the Mandara Mountains, Cameroon and Nigeria." In Pwiti and Soper 1996, 489–95.
2005 "Two Thousand Years of West African History." In Stahl 2005, 441–66.
2012 "The Prehistory and Early History of the Northern Mandara Mountains and Surrounding Plains." In David 2012, 29–67.
2013 "Time on the Timeless Continent:History and Archaeological Chronologies in the Southern Lake Chad Basin." In *Big Histories, Human Lives: Tackling Problems of Scale in Archaeology*, edited by Timothy R. Pauketat and John Robb, 123–44. Santa Fe: School for Advanced Research Press.
2016 "Globalization: Contact Between West Africa, North Africa and Europe During the European Medieval Period." In *Routledge Handbook of Archaeology and Globalization*, edited by Tamar Hodos, 90–103. London: Routledge.

MacEachern, Scott, David Scott, Jean-Marie Datouang Djoussou, and Molly O'Guinness Carlson
2013 "A Cache of Iron Artifacts from the DGB-1 Site, Northern Cameroon: Metallurgical Analysis and Ethnoarchaeological Analogies." *Journal of African Archaeology* 11, no. 1: 39–54.

MacGaffey, Wyatt
1986 *Religion and Society in Central Africa: The Bakongo of Lower Zaïre*. Chicago: University of Chicago Press.
1990 "The Personhood of Ritual Objects: Kongo 'Minkisi.'" *Etnofoor* 3, no. 1: 45–61.
1991 *Art and Healing of the Bakongo Commented by Themselves: Minkisi from the Laman Collection*. Stockholm: Folkens Museum – Etnografiska.
1993 "The Eyes of Understanding: Kongo Minkisi." In *Astonishment and Power*, by W. MacGaffey and Michael D. Harris, 18–103. Washington, D.C.: Smithsonian Institution.
1998 "Magic, or as We Usually Say, Art." In Schildkraut and Keim 1998, 217–35.
2000 *Kongo Political Culture: The Conceptual Challenge of the Particular*. Bloomington: Indiana University Press.
2014–15 "Franchising Minkisi in Loango: Questions of Form and Function." *Res: Anthropology and Aesthetics* 65–66: 148–57.

Mack, John
1990 *Emil Torday and the Art of the Congo, 1900–1909*. Seattle: British Museum Press.
2003 *The Museum of the Mind: Art and Memory in World Cultures*. London: British Museum Press.
2007 *The Art of Small Things*. Cambridge, MA: Harvard University Press.

Mackenzie, John M.
1975 "A pre-colonial industry: the Njanja and the iron trade." *Nada* 11, no. 2: 200–20.

MacRae, Suzanne H.
1995 "Yeelen: A Political Fable of the Komo Blacksmith/Sorcerers." *Research in African Literatures* 26, no. 3: 57–66.

Maes, Joseph
1922 "Armes de jet des populations du Conge belge." *Congo* 1, no. 2: 181–93.

Maggs, Tim
1991 "Metalwork from Iron Hoards as a Record of Metal Production in the Natal Region." *South African Archaeological Bulletin* 46, no. 154: 131–36.

Magnavita, Sonja
2010 "Sahelian Crossroads: Some Aspects on the Iron Age Sites of Kissi, West Africa." In Magnavita et al. 2010, 79–104.

Magnavita, Sonja, Lassina Koté, Peter Breunig, and Oumarous A. Idé, eds.
2010 *Crossroads/Carrrefour Sahel: Cultural and Technological Developments in First Millennium BC/AD West Africa*. Journal of African Archaeology Monograph Series. Frankfurt: Africa Magna.

Mahieu, Alfred
(1910) 1925 *Numismatique au Congo 1485–1925: instruments d'échange, valeurs monétaires, méreaux, médailles*. Brussels: Imprimerie médicale et scientifique.

Mahvunga, Clapperton Chakanestsa
2014 *Transient Workspaces: Technologies of Everyday Innovation in Zimbabwe*. Cambridge, MA: MIT Press.

Malcom, Corey
1998 "Trade Goods on the *Henrietta Marie* and the Price of Men in 1699–1700." Paper presented at the Society for Historical Archaeology Conference on Underwater Archaeology, Atlanta. http://www.melfisher.org/pdf/The%20Price%20of%20Men%20in%201699-1700_Updated-Version.pdf [updated October, 2003].

Manning, Patrick
1982 *Slavery, Colonialism, and Economic Growth in Dahomey, 1640–1960*. Cambridge: Cambridge University Press.

Manning, Katie, Ruth Pelling, Tom Higham, Jean-Luc Schwenniger, and Dorian Q. Fuller
2011 "4500-Year Old Domesticated Pearl Millet (Pennisetum Glaucum) from the Tilemsi Valley, Mali: New Insights into an Alternative Cereal Domestication Pathway." *Journal of Archaeological Science* 38, no. 2: 312–22.

Mapunda, Bertram
1995 "*An Archaeological View of the History and Variation of Iron Working in South-Western Tanzania*." PhD diss., University of Florida, Gainesville.
2010 *Contemplating the Fipa Ironworking*. Kampala: Fountain Publishers.
2011 "Jack of Two Trades, Master of Both: Smelting and Healing in Ufipa, Southwestern Tanzania." *The African Archaeological Review* 28, no. 3: 161–75.
2018 "Metals and metal-working along the Swahili coast." In Wynne-Jones and LaViolette 2018, 306–18.

Maquet, Emma
1965 *Outils de Forge du Congo, du Rwanda et du Burundi*. Tervuren, Belgium: Royal Museum for Central Africa.

Maret, Pierre de
1977 "Sanga: New Excavations, More Data, and Some Related Problems." *The Journal of African History* 18, no. 3:19.
1980 "Ceux qui jouent avec le feu: la place du forgeron en Afrique centrale." *Africa* 50, no. 3: 263–79.
1981 "Évolution monétaire du Shaba Central entre le 7e et le 18e siècle." *African Economic History* 10: 117–49.
1985 "The Smith's Myth and the Origin of Leadership in Central Africa." In Haaland and Shinnie 1985, 73–87. Oslo: Norwegian University Press.
1992 "Fouilles archéologiques dans la vallée du Haut-Lualaba, Zaïre, III, Kamilamba, Kikulu, et Malemba-Nkulu, 1975." In *Annales Sciences Humaines*, vol. 131. Tervuren, Belgium: Royal Museum for Central Africa.
1993 "Metallurgies, Metamorphoses, and Metaphors." Paper presented at "Iron, Master of Them All," Fifth Stanley Conference on African Art, University of Iowa, Iowa City.
1995 "Croisette Histories." In *Annales 145: Objects, Signs of Africa*, edited by Luc de Heusch, 133–45. Tervuren, Belgium: Royal Museum for Central Africa.
1999 "The Power of Symbols and the Symbols of Power Through Time: Probing the Luba Past." In *Beyond Chiefdoms: Pathways to Complexity in Africa*, edited by Susan Keech McIntosh, 151–65. Cambridge: Cambridge University Press.
2005 "L'oryctérope, un animal 'bon à penser' pour les Africains, est-il à l'origine du dieu égyptien Seth?" *Bulletin de l'Institut français d'archéologie orientale*: 105, 107–28.
2013 "Archaeologies of the Bantu Expansion." In Mitchell and Lane 2013, 628–43.

Marliac, Alain, Olivier L. Anglois, and Michelle Delneuf
2000 "Archéologie de la région Mandara-Diamaré." In *Atlas de la Province Extrême-Nord, Cameroun*, edited by Christian Seignobos and Olivier Iyebi-Mandjek, 71–76. Paris: Editions de l'IRD.

Marquart, J., J. D. E. Schmeltz, and J. P. B. de Josselin de Jong
1904–16 *Ethnographisch album van het stroomgebiel van den Congo*. Vol. 2. Leiden: Rijks Ethnographisch Museum.

Martin, Phyllis M.
1986 "Power, Cloth and Currency on the Loango Coast." *African Economic History* 15: 1–12.

Martínez-Ruiz, Bárbaro
2007 "Flying over Dikenga: The Circle of New Life." In Kreamer et al. 2007, 186–99.
2009 "Kongo Atlantic Body Language." *Les actes de colloques*. Paris: Musée du quai Branly. http://actesbranly.revues.org/462.
2012 *Escritura Gráfica Kongo y Otros Narrativas del Signo*. Mexico City: El Colegio de México. Published in English as *Kongo Graphic Writing and Other Narratives of the Sign* (Philadelphia: Temple University Press, 2013).

Masonen, Pekka, and Humphrey J. Fisher
1996 "Not Quite Venus from the Waves: The Almoravid Conquest of Ghana in the Modern Historiography of Western Africa." *History in Africa* 23: 197–232.

Matenga, Edward, and William J. Dewey
Forthcoming "Materiality of Artifacts, Great Zimbabwe and the Shona." In *The World of Great Zimbabwe*, edited by Innocent Pikirayi. New York: Routledge.

Maupoil, Bernard
1943 *La Géomancie à l'Ancienne Côte des Esclaves*. Institut d'Ethnologie 42. Paris: Musée de l'Homme.

Mauss, Marcel
(1907) 1969 *Oeuvres completes*. Vol. 2. Paris: Editions de Minuit.
(1935) 1973 "Techniques of the Body." Translated by B. Brewster. *Economy and Society* 2, no. 1: 70–77.
(1925) 2016 *The Gift, Expanded Edition*. Translated and edited by Jane Guyer. Chicago: HAU Books.

Mayor, Anne
2010a "Outils de potières au Mali: chaînes opératoires et traditions techniques." *Bulletin de la Société préhistorique française* 107, no. 4: 643–66.
2010b "Ceramic Traditions and Ethnic Identity in the Niger Bend, West Africa." *Ethnoarchaeology* 2, no. 1: 5–48.

Mbacké, Khadim
2005 *Sufism and Religious Brotherhoods in Senegal*. Princeton, NJ: Marcus Wiener.

McClusky, Pamela
2002 "The Fetish and the Imagination of Europe: Sacred Medicines of the Kongo." In *Art from Africa: Long Steps Never Broke a Back*, by Pamela McClusky and Robert Farris Thompson, 142–67. Seattle: Seattle Art Museum.

McCoy, Timothy. J.
2016 "Meteoritic Archaeologic Objects as a Key to Understanding Trade in Ancient Eastern North America." Seventy-Ninth Annual Meeting of the Meteoritical Society.

McCoy, Timothy J., A. E. Marquardt, E. P. Vicenzi, R. D. Ash, and J. T. Wasson
2008 "Meteoritic Metal Beads from the Havana, Illinois, Hopewell Mounds: A Source in Minnesota and Implications for Trade and Manufacture." *Lunar and Planetary Science* 39: 1984.

McIntosh, Roderick J.
1998 *People of the Middle Niger: Island of Gold*. Oxford: Blackwell.
2005 *Ancient Middle Niger: Urbanism and the Self-Organizing Landscape*. Cambridge: Cambridge University Press.

McIntosh, Susan Keech, and Brian Fagan
2017 "Re-dating the Ingombe Ilede burials." *Antiquity* 91, no. 358: 1069–77.

McNaughton, Patrick R.
1970 "The Throwing Knife in African History." *African Arts* 3, no. 2: 54–60, 89.
1971 "The African Throwing Knife." Paper for MA in the area of African Studies, UCLA.
1979a "Bamana Blacksmiths." *African Arts* 12, no. 2: 65–71, 92.
1979b *Secret Sculptures of Komo: Art and Power in Bamana (Bambara) Initiation Associations*. Philadelphia: Institute for the Study of Human Issues.
1982a "Language, Art, Secrecy and Power: The Semantics of *Dalilu*." *Anthropological Linguistics* 24: 487–505.
1982b "The Shirts that Mande Hunters Wear." *African Arts* 15, no. 3: 54–8, 91.
1987 "*Nyamakalaw*: The Mande Bards and Blacksmiths." *Word and Image* 3: 271–88.
1988 *The Mande Blacksmiths: Knowledge, Power, and Art in West Africa*. Bloomington: Indiana University Press.
1995 "The Semantics of *Jugu*: Blacksmiths, Lore and Who's 'Bad' in Mande." In Conrad and Frank 1995, 46–57.
2001a "The Power Associations." In Colleyn 2001, 167–74.
2001b "The *Komo*." In Colleyn 2001, 175–83.
2003 "The Lore of Smiths." In *Encyclopedia of African Folklore*, edited by Phillip Peek and Kwesi Yankah. London: Routledge.
2008 *A Bird Dance Near Saturday City: Sidi Ballo and the Art of West African Masquerade*. Bloomington: Indiana University Press.
2011 "The Smiths in Sunjata: What Epics and Oral Traditions Suggest About West African History." *Mande Studies* 13: 1–19.
2014 "A Little Metal Object of Provocative Potential: Re-considering the History of West African Metalurgy." In *Les Actes du Huitième Congrès International des Études Mandé/Proceedings of the Eighth International Conference on Mande Studies*. Bamako: La Sahélienne et Presses Universitaires du Sahel.

McSween, Harry Y., Jr.
1999 "Meteorites." In Beatty et al. 1999, 351–63.

Meek, C. K.
1931a *A Sudanese Kingdom: An Ethnographical Study of the Jukun-Speaking Peoples of Nigeria*. London: Kegan, Paul, Trench, Trubner.
1931b *Tribal Studies in Northern Nigeria*. 2 vols. London: Kegan, Paul, Trench, Trubner.
1931c "Illustration to Notes on Jukons of Gwana." Unpublished paper.

Meier, Prita, and Alyson Purpura, eds.
2018 *World on the Horizon: Swahili Arts across the Indian Ocean*. Urbana-Champaign: Krannert Art Museum, University of Illinois.

Meister, Conny
2010 "Remarks on Early Iron Age Burial Sites from Southern Cameroon." *African Archaeological Review* 27, no. 3: 237–49.

Meister, Conny, and Manfred K. H. Eggert
2008 "On the Early Iron Age in Southern Cameroon: The Sites of Akonétye." *Journal of African Archaeology* 6, no. 2: 20.

Messier, R. A., and James A. Miller
2015 *The Last Civilized Place: Sijilmasa and Its Saharan Destiny*. Austin: University of Texas Press.

Meyer, Piet
1981 *Kunst und Religion der Lobi*. Zurich: Museum Rietberg.

Michaud, Philippe-Alain
2007 *Aby Warburg and the Image in Motion*. New York: Zone Books.

Miers, Suzanne, and Igor Kopytoff
1977 "African 'Slavery' as an Institution of Marginality." In *Slavery in Africa: Historical and Anthropological Perspectives*, edited by Suzanne Miers and Igor Kopytoff, 3–84. Madison: University of Wisconsin Press.

Miller, Duncan
2002 "Smelter and smith: Iron Age metal fabrication technology in southern Africa." *Journal of Archaeological Science* 29, no. 10: 1083–131.

Miller, Joseph
1988 *Way of Death: Merchant Capitalism and the Angolan Slave Trade 1730–1830*. Madison: University of Wisconsin Press.

Mines, Diane P., and Brad Weiss
1997 "Materializations of Memory: The Substance of Remembering and Forgetting: Introduction." *Anthropological Quarterly* 70, no. 4: 161–63.

Mitchell, Peter, and Paul Lane, eds.
2013 *The Oxford Handbook of African Archaeology*. Oxford: Oxford University Press.

Moffett, Abigail Joy, Tim Maggs, and Johnny van Schalkwyk
2017 "Breaking Ground: Hoes in Precolonial South Africa—Typology, Medium of Exchange and Symbolic Value." *African Archaeological Review* 34, no. 1: 1–20.

Mollien, Gaspard T.
1822 *Voyage dans l'Intérieur de l'Afrique, aux Sources du Sénégal et de la Gambie*. Paris: l'Imprimerie de Mme. V. Courcier.

Moñino, Yves
1983 "Accoucher du fer: La métallurgie gbaya (Centrafrique)." In *Métallurgies africaines, nouvelles contributions*, edited by N. Échard, 281–309. Mémoires de la Société des Africanistes 9. Paris: Société des africanistes.

Monteil, Charles
1953 "La légende de Ouagadou et l'origine des Soninké." *Mélanges ethnologiques. Mémoires* IFAN 23: 359–408.

Monteil, Vincent
1965 Un visionnaire musulman sénégalais." *Archives de sociologie des religions* 10, no. 19: 69–98.

Montilus, Guerin
1972 "L'Homme dans la pensée traditionnelle fon." Mimeographed paper, Cotonou, Republic of Bènin.

Morgan, Andy
2011 "Kel Inedan—The Touareg Blacksmiths." *Andy Morgan Writes* (blog), January 12. http://www.andymorganwrites.com/kel-inedan-the-touareg-blacksmiths/.

Morgan, Jo-Ann
2014 "Huey P. Newton Enthroned—Iconic Image of Black Power." *Journal of American Culture* 37, no. 2: 129–48.

Morisseau, Jules
1910 *Sur le lac Moero (Encore le Katanga)*. Brussels: Charles Bulens.

Morrall, Michael T.
1862 *History and Description of Needle Making*. Manchester: Abel Morall.

Morris, Randall
1995 "The Style of His Hand: The Iron Art of Georges Liautaud." In *Sacred Arts of Haitian Vodou*, edited by Donald J. Cosentino, 383–95. Los Angeles: UCLA Fowler Museum of Cultural History.

Morris, William, ed.
1967 *The American Heritage Dictionary of the English Language*. Boston: Houghton Mifflin.

Mouchet, M.
1947a "Prospection Ethnologique Sommaire de Quelques Massifs du Mandara: Massif Hurza." *Bulletin de la Société d'Etudes Camerounaises* 17–18: 111–24.
1947b "Prospection Ethnologique Sommaire de Quelques Massifs du Mandara: Massif Mora." *Bulletin de la Société d'Etudes Camerounaises* 17–18: 124–39.
1947c "Prospection Ethnologique Sommaire de Quelques Massifs du Mandara: Massif Uldham." *Bulletin de la Société d'Etudes Camerounaises* 19–20: 93–103.

Mudenge, S. I. G.
1988 *A Political History of Munhumutapa c 1400–1902*. London: James Currey.

Mudimbe, V. Y.
1986 "African Art as a Question Mark." *African Studies Review* 29, no. 1: 3–4.
1988 *The Invention of Africa: Gnosis, Philosophy, and the Order of Knowledge*. Bloomington: Indiana University Press.
1994 *The Idea of Africa*. Bloomington: Indiana University Press.
1996 "Afterword: The Idea of Luba." In Roberts and Roberts 1996, 245–47.

Muhly, James D.
1995 "Mining and Metalwork in Ancient Western Asia." In *Civilization of the Ancient Near East*, vol. 3, edited by Jack M. Sasson, 1501–21. New York: Charles Scribner's Sons.

Muhly, Frank, Don Wescott, Peter R. Schmidt, and Peter O'Neill
1988 *The Tree of Iron*. Watertown, MA: Documentary Educational Resources. Video, 57 min.

Mulinda, Habi Buganza
1985/1987 "Le nkisi dans la tradition Woyo du Bas-Zaïre." *Systèmes de pensée en Afrique noire* 8: 201–20.

Muller, Jean-Claude
1974 "Métaux, chefs, et forgerons chez les Rukuba (Benue-Plateau State, Nigeria)." *Anthropologica* N.S. 16, no. 1: 97–121.
2001 "Le foyer des métaphores mal soudées. Forgerons et potières chez les Dìì de l'Adamaoua (Nord-Cameroun)." *Anthropologica* 43, no. 2: 209–20.

Müller-Kosack, Gerhard
2003 *The Way of the Beer: Ritual Re-enactment of History Among the Mafa, Terrace Farmers of the Mandara Mountains (North Cameroon)*. London: Mandaras Publishing.

Mumbanza mwa Bawele na Nyabakombi Ensobato
1989 "Les Forgerons de la Ngiri. Une Elite Artisanale parmi les Pêcheurs." *Enquêtes et Documents d'Histoire Africaine* 4: 114–32.

Park, Mungo
(1813) 1931 "Travels of Mungo Park." In *Documents Illustrative of the History of the Slave Trade to America*, vol. 2, edited by Elizabeth Donnan, 632–42. Washington, D.C.: Carnegie Institution of Washington.

Munson, Patrick J.
1968 "Recent Archaeological Research in the Dhar Tichitt Region of South-Central Mauretania." *The West African Archaeological Newsletter* 10: 6–13.
1976 "Archaeological Data on the Origins of Cultivation in the Southwestern Sahara and their Implications for West Africa." In *Origins of African Plant Domestication*, edited by J. R. Harlan, J. M. J. De Wet, and A. B. Stemler, 187–209. The Hague: Mouton.
1980 "Archaeology and the Prehistoric Origins of the Ghana Empire." *Journal of African History* 21: 457–66.
1989 "About: economie et société neolithique du Dhar Tichitt (Augustin Holl)." *Sahara* 2: 107–09.

Mupanda, Bertram
2010 *Contemplating the Fipa Iron Working*. Kampala: Fountain Publishers.

Murray-Kolb, Laura E.
2013 "Iron and Brain Functions." *Journal of Clinical Nutrition and Metabolic Care* 16, no. 6: 703–07.

Nachtigal, Gustav
(1879–89) 1971 *Sahara and Sudan, Vol. 4: Wadai and Darfur*, translated by Allan G. B Fisher and Humphrey J. Fisher. Reprint, London: C. Hurst and Company.
(1879–89) 1974 *Sahara and Sudan, Vol.1: Tripoli and Fezzan, Tibesti or Tu*, translated by Allan G. B. Fisher and Humphrey J. Fisher. Reprint, London: C. Hurst and Company.
(1879–89) 1980 *Sahara and Sudan, Vol. 2: Kawar, Bornu, Kanem, Borku, Ennedi*, translated by Allan G. B. Fisher and Humphrey J. Fisher. Reprint, London: C. Hurst and Company.
(1879–89) 1987 *Sahara und Sudan, Vol. 3: The Chad Basin and Bagirmi*, translated by Allan G. B. Fisher and Humphrey J. Fisher. Reprint, London: C. Hurst and Company.

Nalder, L.F.
1935 "Throwing Knives in the Sudan." *Sudan Notes and Records* 18, no. 2: 297–302.

NAK (National Archives, Kanuda)
1913–29 "Lala District. Miscellaneous Papers." G2.Q no. P, 147/13.
1926–28 "Ethnology Guduen, Gudu." 2710-H no. 225.

Napier, David
1992 *Foreign Bodies: Performance, Art, and Symbolic Anthropology. Berkeley*: University of California Press.

Nathanson, Rick
2012 "'Memory' Captured in 9/11 Steel." *Albuquerque Journal*, 11 September. https://www.abqjournal.com/129900/memory-captured-in-911-steel.html.

N'Diayé, Bokar
1970 *Les Castes au Mali*. Bamako: Éditions Populaire.

Ndiaye, Ken Kelountang
2011 "Ethnography Museums, Diasporas and Modernity: What's at Stake? " In Bouttiaux and Seiderer 2011, 28–39.

Ndoro, Weber
1994 "Natural draught furnaces south of the Zambezi River." *Zimbabwean Prehistory* 21: 29–32.

Nevadomsky, Joseph
1986 "The Benin Bronze Horseman as the Ata of Idah." *African Arts* 19, no. 4: 40–47.
1988 "Kemwin-Kemwin: The Apothecary Shop in Benin City." *African Arts* 22, no. 1: 72–83, 100.

Newbury, David
1991 *Kings and Clans: Ijwi Island and the Lake Kivu Rift, 1780–1840*. Madison: University of Wisconsin Press.
2009 *The Land beyond the Mists: Essays on Identity and Authority in Precolonial Congo and Rwanda*. Athens: Ohio University Press.

Ngom, Fallou
2016 *Muslims Beyond the Arab World: The Odyssey of 'Ajami and the Muridiyya*. New York: Oxford University Press.

Niemeyer, Hans Georg
2001 "Archaeological Evidence of Early Iron Technology at Carthage and Other Phoenician Settlements." *Mediterranean Archaeology* 14: 83–93.

Ngonadi, John Uche, and Pamela Eze-Uzomaka
2015 "Fipa Furnaces and Slag: Distinguishing Criteria Based on Physical Analysis." *Archaeological Discovery* 3: 32–39. DOI: 10.4236/ad.2015.31004.

Niane, D. T.
1984 "Mali and the Second Mandingo Expansion." In *General History of Africa 4: Africa from the Twelfth to the Sixteenth Century*, edited by D. T. Niane, 117–71. Paris: UNESCO; Berkeley: University of California Press.
2006 *Sundiata: An Epic of Old Mali*. Revised ed., translated by G.D. Pickett. Essex: Longman.

Nininger, Harvey H.
1952 *Out of the Sky: An Introduction to Meteoritics*. Denver: University of Denver Press.

Nooter, Mary H.
1990 Catalogue entries. In *Art of Central Africa: Masterpieces from the Berlin Museum für Völkerkunde*, edited by H.-J. Koloss. New York: The Metropolitan Museum of Art.
1991 "Luba Art and Polity: Creating Power in a Central African Kingdom." PhD diss., Columbia University Department of Art History and Archaeology.
1993 "The Visual Language of Secrecy." In Nooter 1993, 48–52.

Nooter, Mary H., ed.
1993 *Secrecy: African Art that Conceals and Reveals*. New York: Museum for African Art.

Ntole, Kazadi
1996 "Objects and People: Relationships and Transformations in the Culture of the Bambala." In Arnoldi et al. 1996, 130–41.

Odera Oruka, Henry
1990 "Sage Philosophy: The Basc Questions and Methodology." In *Sage Philosophy: Indigenous Tinkers and Modern Debate on African Philosophy*, edited by H. Odera Oruka, 27–40. Leiden: Brill.

O'Hern, Robin, Ellen Pearlstein, and Susan Elizabeth Gagliardi
2016 "Beyond the Surface: Where Cultural Contexts and Scientific Analyses Meet in Museum Conservation of West African Power Association Helmet Masks." *Museum Anthropology* 39, no. 1: 70–86.

Oguibe, Olu
2004 *The Culture Game*. Minneapolis: University of Minnesota Press.

Okafor, Edwin
1993 "New Evidence on Early Iron-smelting in Southeastern Nigeria." In Shaw et al. 1993, 432–48.

Okafor, Edwin Eme, and Patricia Phillips
1992 "New 14C Ages from Nsukka, Nigeria and the Origins of African Metallurgy." *Antiquity* 66, no. 252: 686.

Okediji, Moyosore
1991 "Yoruba Pidgin Chromacy." In *Oritameta: Proceedings of the 1990 International Conference on Yoruba Art*, 16–28. Ile-Ife, Nigeria: Obafemi Awolowo University-Department of Fine Arts.
2003 *The Shattered Gourd: Yoruba Forms in Twentieth-Century American Art*. Seattle: University of Washington Press.

Olderogge, D.
1934 "Survivals of the Throwing-Knife in Darfur." *Man* 34: 106–07.

Oliver, Roland Anthony and Brian Fagan
1975 *Africa in the Iron Age, c. 500 BC to AD 1400*. Cambridge: Cambridge University Press.

Oppenheim, A. Leo
1964 *Ancient Mesopotamia: Portrait of a Dead Civilization*. Chicago: The University of Chicago Press.
1978 "Man and Nature in Mesopotamian Civilization." In *Dictionary of Scientific Biography*, vol. 15, supplement 1, edited by Charles Coulson Gillespie, 634–66. New York: Charles Scribner's Sons.

Osborn, Emily L.
2009 "Casting Aluminum Cooking Pots: Labour, Migration, and Artisan Production in West Africa's Informal Sector, 1945–2005." *African Identities* 7, no. 3: 373–86.

Osório de Oliveira, José
1958 *Flagrantes da vida na Lunda*. Lisbon: Diamang, Dundo Museum.

Otsuka, K., and C. M. Wayman, eds.
1998 *Shape Memory Materials*. Cambridge: Cambridge University Press.

Ottenberg, Simon
1975 *Masked Rituals of Afikpo*: The Context of an African Art. Seattle: University of Washington Press.

Pâques, Viviana
1964 "Les bouffons sacrds de Bougouni." *Journal de la Société des Africanistes* 34, no. 1: 63–110.

Paris, Francois, Alan Person, Gerard Quéchon, and Jean-Francois Saliège
1992 "Débuts de la métallurgie au Niger septentrional: Air, Azawagh, Ighazer, Termit." *Journal des Africanistes* 62, no. 2: 55–68.

Park, Mungo
1799 *Travels in the Interior Districts of Africa: Performed Under the Patronage of the African Association, in the Years 1795, 1796, and 1797*. London: W. Bulmer.

Paudrat, Jean-Louis
1994 "Résonances mythiques de la statuaire du pays dogon." In Falgayrettes-Leveau 1994, 55–81.

Peek, Phillip and John Picton
2016 "The Resonance of Osun across a Millennium of Lower Niger History." *African Arts* 49, no. 1: 40–55.

Pemberton, John, III
1994 "Introduction: In Praise of Artistry." In Abiodun et al. 1994, 119–36.

Pemberton, John, III, ed.
2000 *Insight and Artistry: Crosscultural Study of Divination in West and Central Africa*. Washington, D.C.: Smithsonian Institution Press.

Perinbam, Marie B.
1996 "The Salt-Gold Alchemy in the Eighteenth and Nineteenth Century Mande World: If Men Are Its Salt, Women Are Its Gold." *History in Africa* 23: 257–78.

Pereira, Duarte Pacheco
1937 *Esmeraldo de Situ Orbis*. London: Printed for the Hakluyt Society.

Perrois, Louis
2001 "Art of the Kwele of Equatorial Africa: Ancestor Masks, Bush Spirit Masks." *The World of Tribal Arts* 6, no. 4: 80–113.

Petit, Pierre
1993 "Rites familiaux, rites royaux. Étude du système cérémoniel des Luba du Shaba (Zaïre)." PhD diss., Université Libre de Bruxelles.

Petrie, W. M. Flinders
1940 *Wisdom of the Egyptians*. Vol. 53. London: Bernard Quaritch.

Pévenasse, François
1920 "Les termitières du Haut-Katanga." *Journal de Pharmacie de Belgique*, 3–36: 44–46.

Phillips, Tracy
1926 "African Throwing Knives." *Man* 26: 80.

Phillipson, David
1968 "Cewa, Leya and Lala Iron-Smelting Furnaces." *The South African Archaeological Bulletin* 23, no. 91: 102–13.
2002 *Ancient Ethiopia, Aksum: Its Antecedents and Successors*. London: British Museum Press.
2005 *African Archaeology*. New York: Cambridge University Press.

Phillipson, David, and Brian Fagan
1969 "The Date of the Ingombe Ilede Burials." *The Journal of African History* 10, no. 2: 199–204.

Phillipson, David, Brian Fagan, and S. G. H. Daniels, eds.
1969 *Iron Age Cultures in Zambia*, vol. 2. London: Chatto and Windus.

Picton, John, and John Mack
1979 *African Textiles: Looms, Weaving and Design*. London: Westview.

Pietz, William
1985 "The Problem of the Fetish, I." *Res: Anthropology and Aesthetics* 9: 5–17
1987 "The Problem of the Fetish, II." *Res: Anthropology and Aesthetics* 13: 23–45.

Pwiti, Gilbert, and Robert Soper, eds.
1996 *Aspects of African Archaeology: Proceedings of the Tenth Pan-African Congress*. Harare: University of Zimbabwe Press.

Pole, Len
2010a "The Hammers of Mawu: Ironworking Traditions in the Togo Hills, Ghana." *African Archaeological Review* 27, no. 1, 43–78.
2010b "Recent Developments in Iron-working Research in West Africa." In *West African Archaeology: New Developments, New Perspectives*, edited by Philip Allsworth Jones, 53–65. Oxford: Archaeopress.

Pollan, Michael
2013 "The Intelligent Plant." *The New Yorker*, December 23 and 30, 92–105.

Poyner, Robin
2013 "Crucifixes." In Cooksey et al. 2013, 66–71.

Prasse, Karl G.
1995 *The Tuaregs: The Blue People*. Copenhagen, Denmark: Museum Tusculanum Press.

Prendergast, Martin D.
1975 "A new furnace type from the Darwendale Dam basin." *Journal of the Prehistory Society of Rhodesia* 7, no. 14: 6–20.

Pringle, Heather
2009 "Seeking Africa's First Iron Men." *Science*, January 9, 2009, 3.

Pwiti, Gilbert
1996 *Continuity and change: an archaeological study of farming communities in northern Zimbabwe AD 500–1700*. Uppsala: Societa Archaeologica Uppsaliensis.

Pye, David
(1978) 1982 *The Nature & Aesthetics of Design*. New York: Van Nostrand Reinhold Company.
(1968) 2015 *The Nature and Art of Workmanship*. New York: Bloomsbury Academic.

Quiggen, Alison Hingston
1949 *A Survey of Primitive Money: The Beginnings of Currency*. London: Methuen.

Rabaté, Jean-Michel
2018 *Rust*. Object Lesson series. New York: Bloomsbury Academic.

Radcliffe-Brown, Alfred R., and Daryll Ford, eds.
(1950) 2015 *African Systems of Kinship and Marriage*. New York: Routledge.

Radimilahy, Chantal
1988 *L'ancienne métallurgie du fer à Madagascar*. International Series. London: British Archaeological Reports.

Rasmussen, Susan
2013 *Neighbors Strangers, Witches, and Culture-Heroes: Ritual Powers of Smith/Artisans in Tuareg Society and Beyond*. New York: University Press of America.

Ravenhill, Philip L.
1996 "The Passive Object and the Tribal Paradigm: Colonial Museography in French West Africa." In Arnoldi et al. 1996, 265–82.

Read, F. W.
1902 "Iron-smelting and Native Blacksmithing in Ondulu Country, South-East Angola." *Journal of the African Society* 5: 44–49.

Redinha, José
1953 *Campanha Etnográfica ao Tchiboco*. Lisbon: Diamang, Dundo Museum.

Reed, Graham
1987 "Kaskara from Northern Darfur, Sudan." *Journal of the Arms and Armour Society* 12, no. 3: 165–201.

Reefe, Thomas
1981 *The Rainbow and the Kings: A History of the Luba Empire to 1891*. Berkeley: University of California Press.

Rehren, Thilo, Tamás Belgya, Albert Jambon, György Káli, Zsolt Kasztovsky, Zoltán Kis, Imre Kovács, Boglárka Maróti, Marcos Martinón-Torres, Gianluca Miniaci, Vincent C. Pigoyy, Miljana Radivojević, László Rosta, László Szentmiklósi, Zoltán Szökefalvi-Nagy
2001 "Meroe, Iron and Africa." *Der Antike Sudan* 12: 102–09.

Rehren, Thilo, Michael Charlton, Shadreck Chirikure, Jane Humphris, Akin Ige, and Harald Alexander Veldhuijzen
2007 "Decisions Set in Slag: The Human Factor in African Iron Smelting." In *Metals and Mines: Studies in Archaeometallurgy*, edited by Susan La Niece, Duncan R. Hook, and Paul T. Craddock, 211–18. London: Archetype.

Rehren, Thilo, Michael Charlton, Shadreck Chirikure, Jane Humphris, Akin Ige, and Harald Alexander Veldhuijzen
2013 "5,000 Years Old Egyptian Beads Made from Hammered Meteoritic Iron." *Journal of Archaeological Science* 40, no. 12: 4785–92.

Reid, Andrew
2013 "The Emergence of States in Great Lakes Africa." In Mitchell and Lane 2013, 887–99.

Reid, Andrew, and Rachel MacLean
1995 "Symbolism and the Social Contexts of Iron Production in Karagwe." *World Archaeology* 27, no. 1: 144–61.

Reid, J. A.
1936 "Reminiscences of the Sudan Mahdi: By His Personal Servant Mohammed el Mekki Ghuleib." *Journal of the Royal African Society* 35, no. 138: 71–75.

Reid, Richard J.
2012 *Warfare in African History*. New York: Cambridge University Press.

Reyna, Stephen P.
1990 *Wars without End: The Political Economy of a Precolonial African State*. Hanover, NH: University Press of New England.

Richards, Audrey
1935 "Bow Stand or Trident?" *Man* 35: 30–32.

Richards, Polly
2005 "Masques Dogons in a Changing World." *African Arts* 38, no. 4: 46–53, 93.

Rickard, Thomas Arthur
1941 "The Use of Meteoric Iron." *Journal of the Royal Anthropological Institute of Great Britain and Ireland* 71, nos. 1–2: 55–66.

Ringquist, John
2008 "Kongo Iron: Symbolic Power, Superior Technology and Slave Wisdom." *Newsletter of the African Diaspora Archaeology Network*. http://www.diaspora.uiuc.edu/news0908.html.

Robert, J. M.
1948 *Croyances et coutumes magico-religieuses des Wafipa païens*. Tabora, Tanzania: Tanganyika Mission Press.

Robert-Chaleix, Dennis
1994 "Métallurgie du fer dans la moyenne vallée du Sénégal: les bas fourneaux de Silla." *Journal des Africanistes* 64, no. 2: 113–27.

Roberts, Allen F.
1980 "Heroic Beasts, Beastly Heroes: Principles of Cosmology and Chiefship among the Lakeside BaTabwa of Zaïre." PhD diss., University of Chicago Department of Anthropology.
1981 "Passage Stellified: Speculation upon Archæoastronomy in Southeastern Zaïre." *Archæoastronomy* 4, no. 4: 27–37.
1982 "'Comets Importing Change of Times and States': Ephemerae and Process Among Tabwa of Zaïre." *American Ethnologist* 9, no. 4: 712–29.
1984 "'Fishers of Men': Religion and Political Economy among Colonized Tabwa." *Africa: Journal of the International African Institute*, 54, no. 2: 49–70.
1986 "The Serious Business of Dogon Cosmology." *Archaeoastronomy* 10: 148–53.
1987 "An Eye in the Sky, One Deep in the Earth: Elements of Zaose Religion." In *Ethnologiques: Hommages à Marcel Griaule*, edited by S. de Ganay, J.-P. Lebeuf, and D. Zahan, 291–306. Paris: Eds. Hermann.
1988a "Of Dogon Crooks and Thieves." *African Arts* 21, no. 4: 70–75, 90–92.
1988b "A Note of the Precolonial Iron Currency of the Laka of Southwestern Chad." *Journal des africanistes* 58, no. 1: 99–105.
1988c "Tabwa 'Tegumentary Inscription.'" In Rubin 1988, 41–56.
1989 "History, Ethnicity, and Change in the 'Christian Kingdom' of Southeastern Zaïre." In *The Creation of Tribalism in South and Central Africa: Studies in the Political Economy of Ideology*, edited by Leroy Vail, 193–214. Berkeley: University of California Press.
1991 "Where the King Is Coming From." In Jacobson-Widding 1991, 249–69.
1992 "The Uncommon Currency of Masks." Paper presented at Ninth Triennial Symposium on African Art, University of Iowa, Iowa City.
1993a "Smelting Ironies: The Performance of a Precolonial Technology." Paper presented at "Iron, Master of Them All," Fifth Stanley Conference on African Art, University of Iowa, Iowa City.
1993b "Insight, or NOT Seeing Is Believing." In Nooter 1993, 64–79.
1995 *Animals in African Art: From the Familiar to the Marvelous*. New York: Museum for African Art.
1996 "The Ironies of System D." In *Recycled, Re-seen: Folk Art from the Global Scrap Heap*. In Cerny and Seriff 1996, 82–101. New York: Harry N. Abrams; Santa Fe: Museum of International Folk Art.
2000 "Difficult Decisions, Perilous Acts: Producing Potent Histories with the Tabwa Boiling Water Oracle." In Pemberton 2000, 83–98.
2004 "When They Begin the Between." In *Intermedia: Enacting the Liminal*, edited by Hans Breder and Klaus-Peter Busse, 179–81. Dortmund, Germany: University of Dortmund Press.
2008 "Image Ethics in and about Africa." *African Arts* 41, no. 2: 1, 4–8.
2012 "Performing Cosmology: Harmonies of Land, Lake, Body, and Sky." In *African Cosmos: Stellar Arts*, edited by C. M. Kreamer, 184–203. New York: Monacelli Press for the National Museum of African Art, Smithsonian Institution.
2013a *A Dance of Assassins: Performing Early Colonial Hegemony in the Congo*. Bloomington: Indiana University Press.
2013b "Movement of Ideas and Forms between Central and Eastern Africa." In *Shangaa: Art of Tanzania*, edited by Gary Van Wyk, 196–215. New York: QCC Art Gallery, City University of New York.
2013c "Icons from the End of Days: Visual Hagiography among Layennes of Senegal." *World Art* 3, no. 2: 235–58.
2016 "Why the Hero Lost His Teeth: Reflections on the Great Luba Helmet Mask." In *La Danse du Masque / The Dancing Mask: A Liber Amicorum in Honour of Anne-Marie Bouttiaux*, edited by Anne Seiderer, 51–68. Tervuren, Belgium: Royal Museum for Central Africa.

Roberts, Allen F., and Evan M. Maurer, eds.
1985 *The Rising of a New Moon: A Century of Tabwa Art*. Seattle: University of Washington Press for University of Michigan Museum of Art.

Roberts, Andrew D.
1973 *A History of the Bemba: Political Growth and Change in North-Eastern Zambia Before 1900*. London: Longman.

Roberts, Avery, and Allen Roberts
1988 "What's Funny About Money?" *FACES: The Magazine About People* 4, no. 10: 4–7.

Roberts, Mary Nooter
1994 "Does an Object Have a Life?" In *EXHIBITION-ism: Museums and African Art*, by M. N. Roberts and Susan Vogel, 37–56. New York: Museum for African Art.
1996a "Ceremonial Spear." In *Masterpieces from Central Africa*, edited by Gustaaf Verswijver et al., 179. Tervuren, Belgium: Royal Museum for Central Africa.
1996b "Luba Memory Theater." In Roberts and Roberts 1996, 116–49.
2000 "Proofs and Promises: Setting Meaning Before the Eyes." In Pemberton 2000, 63–82.
2013 "The King Is a Woman: Shaping Power in Luba Royal Arts." *African Arts* 46, no. 3: 68–81.
2017 "Exhibition Review: *The Inner Eye: Vision and Transcendence in African Arts*." African Arts 50, no. 1: 60–79.

Roberts, Mary Nooter, and Alison Saar
2000 *Body Politics: The Female Image in Luba Art and the Sculpture of Alison Saar*. Los Angeles: UCLA Fowler Museum of Cultural History.

Roberts, Mary Nooter, and Allen F. Roberts
2007 *Visions of Africa*: Luba. Milan: 5 Continents.

Roberts, Mary Nooter, and Allen F. Roberts, eds.
1996 *Memory: Luba Art and the Making of History*. New York: Museum for African Art.
2003 *A Saint in the City: Sufi Arts of Urban Senegal*. Los Angeles: UCLA Fowler Museum of Cultural History.

Robertson, Eric
2007 *The Art of Money—African Metalwork and Currency*. Eatonville, FL: Zora Neale Hurston National Museum of Fine Arts.

Robinson, Arthur E.
1935 "Throwing-Swords and Knives from the Sudan." *Man* 35: 72–73.
1936 "Sudan Weapons." *Sudan Notes and Records* 19, no. 1: 189–90.

Robion-Brunner, Caroline
2010 *Forgerons et sidérurgie en pays dogon: vers une histoire de la production du fer sur le plateau de Bandiagara (Mali) durant les empires précoloniaux*. Journal of African Archaeology Monograph Series 3. Frankfurt: Africa Magna.

Robles De La Torre, Gabriel, and Vincent Hayward
2001 "Force Can Overcome Object Geometry in the Perception of Shape Through Active Touch." *Nature* 412: 445–48.

Ross, Doran
n.d. "Akan Leadership Arts." *Art & Life in Africa*, University of Iowa Stanley Museum of Art. https://africa.uima.uiowa.edu/topic-essays/show/8.

Roth, H. Ling
1968 *Great Benin: Its Customs, Art and Horrors*. London: Routledge & K. Paul.

Rowlands, Michael, and Jean-Pierre Warnier
1996 "Magical Iron Technology in the Cameroon Grassfields." In Arnoldi et al. 1996, 51–72.

Rubin, Arnold
1985 "Akuma: Carving the Masks at Takum." *African Arts* 18, no. 2: 60–62, 103.
1989 "Herders: Fulani." In *Art as Technology*, edited by Zena Pearlstone, 67–72. Beverly Hills, CA: Hillcrest.

Rubin, Arnold, ed.
1988 *Marks of Civilization: Artistic Transformations of the Human Body*. Los Angeles: Museum of Cultural History, UCLA.

Rupp, Nicole, James Ameje, and Peter Breunig
2005 "New Studies on the Nok Culture of Central Nigeria." *Journal of African Archaeology* 3, no. 2: 283–90.

Rupp, Nicole, Peter Breunig, and Stefanie Kahlheber
2008 "Exploring the Nok Enigma." *Antiquity* 82, no. 316.

Ruskin, John
1905 "The Work of Iron, in Nature, Art, and Policy." Lecture from 1858. In vol. 16 of *The Complete Works of John Ruskin, Library Edition*, edited by E. T. Cook and A. Wedderburn, 375–411. New York: Longmans, Green.

Rustad, John Austin
1980 "The Emergence of Iron Technology in West Africa, with a Special Emphasis on the Nok Culture of Nigeria." In *West African Culture Dynamics*, edited by B. K. Swartz and Raymond E. Dumett, 227–46. The Hague: Mouton.

Saggs, H. W. F.
1984 *The Might That Was Assyria*. London: Sidgwick & Jackson.

Santos, Boaventura de Sousa
2014 *Epistemologies of the South: Justice against Epistemicide*. New York: Routledge.

Sassoon, Hamo
1971 "The Collection of Metalwork from the Kingdom of Karagwe and its Relationship to the Insignia of Neighbouring Territories." Unpublished paper.
1983 "Kings, Cattle and Blacksmiths: Royal Insignia and Religious Symbolism in the Interlacustrine States." *Azania*, 18: 93–106.
1995 "Omusinga." In *Africa: The Art of a Continent*, edited by Tom Philips, 157. London: Royal Academy of Arts.

Sayce, A. H.
1911 "Second Interim Report on the Excavations at Meroe in Ethiopia II. The Historical Results." *Liverpool Annals of Archaeology and Anthropology* 4: 53–65.

Schechner, Sara
1997 *Comets, Popular Cultures, and the Birth of Modern Cosmology*. Princeton, NJ: Princeton University Press.

Scheel, Bernd
1989 *Egyptian Metalworking and Tools*. Aylesbury, U.K.: Shire Publications.

Scheymans, J.
1903 "Lettre à M. le Curé de Kinroy." *Missions d'Afrique* (Pères Blancs): 212–225.

Schildkrout, Enid, and Curtis A. Keim
1990 *African Reflections: Art From Northeastern Zaire*. New York: American Museum of Natural History.

Schildkrout, Enid, and Curtis A. Keim, eds.
1998 *The Scramble for Art in Central Africa*. London: Cambridge University Press.

Schmidt, A. M., and P. Westerdijk
2008 *The Cutting Edge: West Central African 19th Century Throwing Knives in the National Museum of Ethnology, Leiden*. Leiden: National Museum of Ethnology.

Schmidt, Peter R.
1996a "Reconfiguring the Barongo: Reproductive Symbolism and Reproduction among a Work Association of Iron Smelters." In Schmidt 1996b, 74–127.
1997 *Iron Technology in East Africa: Symbolism, Science, and Archaeology*. Bloomington: Indiana University Press.
2009 "Tropes, Materiality, and Ritual Embodiment of African Smelting Furnaces as Human Figures." *Journal of Archaeological Method and Theory* 16, no. 3: 262–82.
2013 "Bricolage, ritual performance, and Habitus [forgotten] in Barongo iron smelting." In Humphris and Rehren 2013, 66–72.

Schmidt, Peter R., ed.
1996b *The Culture and Technology of African Iron Production*. Gainesville: University Press of Florida.

Schmidt, Peter, and D. H. Avery
1978 "Complex Iron Smelting and Prehistoric Culture in Tanzania." *Science* 201: 1085–89.

Schmidt, Peter R., and S. Terry Childs
1985 "Innovation and industry during the early Iron Age in East Africa: The KM2 and KM3 sites of northwest Tanzania." *The African Archaeological Review* 3: 53–94.

Schmitz, Bruno
1903 "Les minerais de fer du Haut-Congo." *La Belgique coloniale* 8: 40.

Schweinfurth, Georg
1873 *The Heart of Africa: Three Years' Travels and Adventures in the Unexplored Regions of Central Africa from 1868 to 1871*. Translated by Ellen E. Frewer. 2 vols. London: S. Low, Marston, Low, and Searle.
1874 *Artes africanae: Abbildungen und beschreibungen von erzeugnissen des kunstfleisses centralafrikanischer völker*. Leipzig: F. A. Brockhaus; New York: Harper and Brothers.

Searcy, Kim
2011 *The Formation of the Sudanese Mahdist State, Ceremonies and Symbols of Authority: 1882–1898*. Boston: Brill.

Sennett, Richard
2008 *The Craftsman*. New Haven: Yale University Press.

Severi, Carlo
2015 *The Chimera Principle: An Anthropology of Memory and Imagination*. Chicago: HAU Books.

Severin, T., T. Rehren, and H. Schleicher
2011 "Early metal smelting in Aksum, Ethiopia: copper or iron?" *European Journal of Mineralogy* 23, no. 6: 981–92.

Shah, Tahir
2010 "Forward." In Lebling 2010, ix–xiv.

Sharify-Funk, Meena
n.d. "Geometry of the Spirit': Sufism, Calligraphy, and Letter Mysticism." *Society for the Arts in Religious and Theological Studies*. https://www.societyarts.org/geometry-of-the-spirit-sufism-calligraphy-and-letter-mysticism.html.

Shaw, Thurstan
1970 *Igbo-Ukwu: An Account of Archaeological Discoveries in Eastern Nigeria*. 2 vols. Evanston, IL: Northwestern University Press.
1993 "Further Light on Igbo-Ukwu, Including New Radiocarbon Dates." In *Proceedings of the 9th Congress of the Pan-African Association of Prehistory and Related Studies: Jos, 1983*, edited by Bassey Andah, Pierre de Maret, and Robert Soper, 79–83. Ibadan, Nigeria: Rex Charles Publications.

Shaw, Thurstan, Paul Sinclair Bassey Andah, and Alex Okpoko, eds.
1993 *The Archaeology of Africa: Food, Metals and Towns*. London and New York: Routledge.

Shinnie, Peter L., and Francois J. Kense
1989 *Archaeology of Gonja, Ghana: Excavations at Daboya*. Calgary: University of Calgary Press.

Shinnie, Peter L.
1985 "Iron working at Meroe." In Haaland and Shinnie 1985, 28–35.

Shusterman, Richard
2011 "Muscle Memory and the Somaesthetic Pathologies of Everyday Life." *Human Movement* 12, no. 1: 4–15.

Sidibé, Mamby
1959 "Les gens de caste ou nyamakala au Soudan français." *Notes africaines* 81: 13–17.

Siegmann, William
n.d. "Masquerades Among the Dan People." *Art & Life in Africa*, University of Iowa Stanley Museum of Art. https://africa.uima.uiowa.edu/topic-essays/show/29.

Silva, Sónya
2014 "Mind, Body, and Spirit in Basket Divination: An Integrative Way of Knowing." *Religions* 5, no. 4: 1175–87.

Simbao, Ruth
2014 "Cosmological Efficacy and the Politics of Sacred Place: Soli Rainmaking in Contemporary Zambia." *African Arts* 47, no. 3: 40–57.

Simmel, Georg
(1900) 1978 *The Philosophy of Money*. London: Routledge.

Simoons, Frederick
1958 "The Agricultural Implements and Cutting Tools of Begemder and Memyen, Ethiopia." *Southwestern Journal of Anthropology* 14, no. 4: 386–406.

Smith, Barbara Herrnstein
1988 *Contingencies of Value: Alternative Perspectives for Critical Theory*. Cambridge, MA: Harvard University Press.

Smith, Cyril Stanley
1961 "The Interaction of Science and Practice in the History of Metallurgy." *Technology and Culture* 2, no. 4: 357–67.
1968 "Matter versus Materials: A Historical View." *Science*, New Series 162: 637–44.

Smith, Robert Sydney
1989 *Warfare and Dilomacy in Precolonial West Africa*. 2nd ed. Madison: University of Wisconsin Press.

Sousa Santos, Boaventura de
2016 *Epistemologies of the South: Justice Against Epistemicide*. New York: Routledge.

Speke, John Hanning
1863 *Journal of the Discovery of the Nile*. Edinburgh: Blackwood.

Spring, Christopher
1993 *African Arms and Armor*. Washington, D.C.: Smithsonian Institution Press, for the Trustees of the British Museum.

Stahl, Ann Brower, ed.
2005 *African Archaeology: A Critical Introduction*. London: Blackwell.

Stamp, Daryl
2008 "Hamo Sassoon's Personal Archive. Comments on Material Recently Gifted to the BIEA." *African Archaeological Review* 25, nos. 3–4: 169–74.

Stanley, Henry Morton
1878 *Through the Dark Continent*. New York: Harper and Brothers.

Steel, R. H.
1975 "Ingot Casting and Wire Drawing in Iron Age Southern Africa." *Journal of the South African Institute of Mining and Metallurgy* 76, no. 4: 232–37.

Stefaniszyn, Bronislaw.
1954 "African Rincarnation Re-Examined." *African Studies* 13, nos. 3–4: 131–46.

Stewart, Susan
1993 *On Longing: Narratives of the Miniature, the Gigantic, the Souvenir, the Collection*. Durham, NC: Duke University Press.
1994 "Death and Life, in that Order, in the Works of Charles Wilson Peele." In *The Cultures of Collecting*, edited by J. Elsner and R. Cardinal, 204–33. Cambridge, MA: Harvard University Press.
2011 *The Poet's Freedom: A Notebook on Making*. Chicago: University of Chicago Press.

Stiansen, Endre, and Jane I. Guyer, eds.
1999 *Credit, Currencies and Culture: African Financial Institutions in Historical Perspective*. Stockholm: Nordiska Afrikainstitutet.

Storms, Emile
ca. 1884 "Voyage à Oudjidji." Unpublished manuscript. Archives of the Royal Museum for Central Africa, Tervuren, Belgium, F.S. 11, F-III.

Strom, Robert G.
1987 *Mercury: The Elusive Planet*. Washington, D.C.: Smithsonian Institution Press.

Summers, R.
1969 *Ancient mining in Rhodesia and adjacent areas*. Salisbury: Trustees of the National Museums of Rhodesia.

Surgy, Albert de
1985/1987 "Présentation." In *Fétiches: Objets enchantés, Mots réalisés. Systèmes de pensée en Afrique noire* 8: 1–11.

Stuer, P.
1903 "Chez les montagnards du Maroungou." *Missions d'Afrique* (Pères Blancs): 161–70.

Tamari, Tal
1991 "The Development of Caste Systems in West Africa." *The Journal of African History* 32, no. 2: 221–50.
1995 "Linguistic Evidence for the History of West African 'Castes.'" In Conrad and Frank 1995, 61–85.
1997 *Les castes de l'Afrique occidentale. Artisans et musiciens endogames*. Nanterre, France: Société d'Ethnologie.
2012 "De l'apparition et de l'expansion des groupes de spécialistes endogames en Afrique: essai d'explication." In *Métallurgie du fer et sociétés africaines. Bilans et nouveaux paradigmes dans la recherché anthropologique et archéologique*, edited by Caroline Robion-Brunner and Bruno Martinelli, 5–31. Oxford: British Archaeological Reports.

Taneja, Anand Vivek
2013 "Jinnealogy: Everyday Life and Islamic Theology in Post-Partition Delhi." *HAU: Journal of Ethnographic Theory* 3, no. 3: 139–65.

Terashima, Hideaki
1980 "Hunting Life of the Bambote: An Anthropological Study of Hunter-Gatherers in a Wooded Savanna." *Africa 2, SENRI Ethnological Studies* 6: 223–68.

Thomas, Ernest S.
1925 "The African Throwing Knife." *The Journal of the Royal Anthropological Institute of Great Britain and Ireland* 55: 129–45.

Thomas, Nicholas
1991 *Entangled Objects: Exchange, Material Culture, and Colonialism in the Pacific*. Cambridge, MA: Harvard University Press.

Thompson, Robert Farris
1973a *African Art in Motion*. Berkeley: University of California Press.
1973b "Yoruba Artistic Criticism." In *The Traditional Artist in African Societies*, edited by Warren d'Azevedo, 19–61. Bloomington: Indiana University Press.
1976 *Black Gods and Kings: Yoruba at UCLA*. Bloomington: Indiana University Press.
1978 "The Grand Detroit 'N'Kondi.'" *Bulletin of the Detroit Institute of Arts* 56, no. 4: 206–21.
1981 "Kongo Civilization and Kongo Art." In *The Four Moments of the Sun: Kongo Art in Two Worlds*, by Robert Farris Thompson and Joseph Cornet, 34–140. Washington, D.C: National Gallery of Art.
1984 *Flash of the Spirit: African and Afro-American Art and Philosophy*. New York: Vintage.
1993 *Face of the Gods: Art and Altars of Africa and the African Americas*. New York: Museum for African Art.

Thomson, Joseph
(1881) 1968 *To the Central African Lakes and Back*. 2 vols. London: Frank Cass.

Thornton, John
1992 "The Regalia of the Kingdom of the Kongo 1491–1895." In *The Kings of Africa—Art and Authority in Central Africa*, edited by E. Beumers and H.-J. Koloss, 57–63. Maastricht: Foundation Kings of Africa.
2000 *Warfare in Atlantic Africa*, 1500–1800. New York: Routledge.
2004 "Origin Traditions and History in Central Africa." *African Arts* 37, no. 1: 32–37, 93–94.
2008 "Introduction to the Araldi Manuscript." *Central African History* (blog), August 19. https://centralafricanhistory.blogspot.com/search?q=%E2%80%9CCentral+African+History%3A+The+Araldi+Manuscript+of+Giovanni+Antonio+Cavazzi%2C+1668.%E2%80%9D+.

Time-Life Books
1988 *Stars*. Voyage Through the Universe series. Alexandria, VA: Time-Life Books.

Togola, Téréba
1996 "Iron Age Occupation in the Méma Region, Mali." *African Archaeological Review* 13, no. 2: 91–110.
(1736) 2008 *Archaeological Investigations of Iron Age Sites in the Mema Region, Mali (West Africa)*. Oxford: Archaeopress.

Torday, Emil
1925 *On the Trail of the Bushongo; An Account of a Remarkable and Hitherto Unknown African People, their Origin, Art, High Social and Political Organization and Other Culture, Derived From the Author's Personal Experience Amongst Them*. London: Seeley, Service and Company.

Torday, Emil, and T. A. Joyce
1905 "Notes on the Ethnography of the Ba-Mbala." *Journal of the Anthropological Institute of Great Britain and Ireland* 35: 398–426.
1910 *Notes ethnographiques sur les peuples communément appelés Bakuba, ainsi que sur les peuplades apparentées, Les Bushongo*. Brussels: Ministère des colonies: En vente chez Falk fils.
1922 *Notes Ethnographiques sur Des Populations Habitant Les Bassins Du Kassi Et Du Kwango Oriental*. Tervuren, Belgium: Royal Museum for Central Africa.

Toropov, Andrey, and M. Grosso
1998 "Dynamic Calibration of Impact Test Instruments." *Journal of Testing and Evaluation* 26, no. 4: 315–19.

Traore, Dominique
1965 *Comment le noir se soigne-t-il? Ou Médecine et magie africaines*. Paris: Présence Africaine.

Travélé, M.
1929 "Le Komo ou Koma." *Outre Mer* 1: 127–50.

Trigger, B.
1969 "The myth of Meroe and the African Iron age." *African Historical Studies* 2: 23–50.

Trimble, Virginia
1996 "The Origin and Evolution of the Chemical Elements." In *The Origin and Evolution of the Universe*, edited by Ben Zuckerman and Matthew A. Malkan, 89–108. Sudbury, MA: Jones and Bartlett Publishers.

Tully, R. Brent, Hélène Courtois, Yehuda Hoffman, and Daniel Pomarède
2014 "The Laniakea Supercluster of Galaxies." *Nature* 513, no. 7516: 71–73.

Turner, Victor
1968 "Myth and Symbol." In *International Encyclopedia of the Social Sciences*, edited by David Sills and Robert K. Merton. New York: MacMillan Reference USA.
1970 *The Forest of Symbols: Aspects of Ndembu Ritual*. Ithaca, NY: Cornell University Press.

Tylecote, R.
1975a "The Origin of Iron Smelting in Africa." *West African Journal of Archaeology* 5: 1–9.
1975b "Iron Smelting at Taruga, Nigeria." *Journal of Historical Metallurgy* 9, no. 2: 49–56.

Van Avermaet, E., and Benoît Mbuya
1954 *Dictionnaire Kiluba-Français*. Sciences de l'Homme, Linguistique 7. Tervuren, Belgium: Annales du Musée Royal du Congo Belge.

Van Binsbergen, Wim
1981 *Religious Change in Zambia*. London: Kegan Paul for the African Studies Center, Leiden University.

Van der Merwe, Nikolaas, and Donald Avery
1987 "Science and Magic in African Technology: Traditional Iron Smelting in Malawi." *Africa* 57, no. 2: 143–72.

Van Grunderbeek, M.-C., Roche, E., and Doutrelepont, H.
1982 "L'Age du Fer Ancien au Rwanda et au Burundi: Archéologie et environnement." *Journal des Africanistes* 52: 5–58.
2001 "Un type de fourneau de fonte de fer associe´a` la culture Urewe (Age du Fer Ancien) au Rwanda et au Burundi." *Mediterranean Archaeology* 14: 291–97.

Van Noten, Francis L.
1972 *Les Tombes du Roi Cyrima Rujugira et de la Reine-Mere Nyirayuhi Kanjogera*. Tervuren, Belgium: Royal Museum for Central Africa.

Vanhee, Hein
2013 "The Power to Kill (with Proverbs)." In Cooksey et al. 2013, 112–13.

Vansina, Jan
1955 "Initiation Rituals of the Bushongo." *Africa* 25: 138–53.
1969 "The Bells of Kings." *Journal of African History* 10, no. 2: 187–97.
1973 *The Tio Kingdom of the Middle Congo, 1880–1892*. New York: Oxford University Press.
1978 *The Children of Woot: A History of the Kuba Peoples*. Madison: University of Wisconsin Press.
1969 "The Bells of Kings." *Journal of African History* 10, no. 2: 189–97.
1986 "Do Pygmies Have a History?" *SUGIA: Sprache und Geschichte in Afrika* 7, no. 1: 431–45.
1990 *Paths in the Rainforests: Towards a History of Political Tradition in Equatorial Africa*. Madison: University of Wisconsin Press.
2006 "Linguistic Evidence for the Introduction of Ironworking into Bantu-Speaking Africa." *History in Africa* 33: 321–61.

Van Wyk, Gary
2013 "The Metal Regalia of Karagwe," with a long caption by Silvia Dolz. In *Shangaa: Art of Tanzania*, edited by G. Van Wyk, 216–21. New York: QCC Art Gallery, City University of New York.

Vaughan, James H., Jr.
1973 "*Eŋkyagu* As Artists in Marghi Society." In *The Traditional Artist in African Societies*, edited by Warren L. d'Azavedo, 162–93. Bloomington: Indiana University Press.

Verger, Pierre Fatumbi
1981 *Orixas: Deuses Iorubas na Africa e no Novo Mundo*. Salvador, Bahia, Brazil: Corrupio.

Vermaas, Wim F. J.
2001 "Photosynthesis and Respiration in Cyanobacteria." In *Encyclopedia of Life Sciences*. London: Nature Publishing Group.

Vernet, Robert
1993 *Préhistoire de la Mauritanie*. Nouakchott, Mauritania: Centre culturel Français – Sépia.
2000 "Un habitat de l'âge du cuivre (2500 B.P.) de la région de Nouakchott (Mauritanie occidentale): Imbich-Est." *Sahara: History and Prehistory of the Sahara* 12: 83–90.

Vleugels, Joseph
1973 "Moeurs et coutumes principales des Batabwa." *Bulletin du Centre d'Etude des Problèmes Sociaux et Economiques* (Lubumbashi, DRC), I 98–99, II 3–71.

Vogel, Joseph O., ed.
2000 *Ancient African Metallurgy: The Socio-Cultural Context*. Walnut Creek, CA: Alta Mira Press.

Vogel, Susan
1997 *Baule: African Art, Western Eyes*. New Haven: Yale University Press.

Vogel, Susan, and Francine N'Diaye
1985 *African Masterpieces from the Musée de l'Homme*. New York: Center for African Art.

Volper, Julien
2011 *Ora pro nobis: Étude sur les crucifix bakongo*. Brussels: Impresor-Pauwels.

Volper, Julien
2016 *Du Jourdain au Congo: Art et christianisme en Afrique centrale*. Paris: Flammarion.

Walker, Roslyn A.
1994 "Anonymous Has a Name: Ọlọ́wẹ̀ of Ìsẹ̀." In Abiodun et al. 1994, 91–106.

Warburg, Aby
1923 "Memories of a Journey through the Pueblo Region." Unpublished notes from the Kreuzlingen Lecture on the Serpent Ritual. In *Aby Warburg and the Image in Motion*, by Philippe-Alain Michaud, 293–331. Chicago: University of Chicago Press, 2007.

Ward, Herbert
1910 *A voice from the Congo, comprising stories, anecdotes, and descriptive notes, by Herbert Ward; with illustrations from photographs, sculpture, and drawings by the author*. London: W. Heinemann.

Ward, Michael, and Peter Westerdijk
1988 *Symbols of Wealth: Abstractions in African Metalwork*. New York: Michael Ward Gallery.

Watson, John
1998 *Himba: Photographs by John Watson*. Cambridge, MA: Cultural Survival.

Webb, Virginia-Lee
2000 *Perfect Documents: Walker Evans and African Art, 1935*. New York: The Metropolitan Museum of Art.

Weghsteen, Joseph
1962 “Origine et dispersion des hommes d’après les legends tabwa.” *Annali del Pontificio Museo Missionario Ethnologico Lateranensi* 26: 213–19.

Weinryb, Ittai
2017 “Votives and Material Religion.” *Material Religion* 13, no. 1: 102–03.

Weiss, Brad
1997 “Forgetting Your Dead: Alienable and Inalienable Objects in Northwest Tanzania.” *Anthropological Quarterly* 70, no. 4: 164–72.

Wiessner, Polly
1983 “Style and Social Information in Kalahari San Projectile Points.” *American Antiquity* 48, no. 2: 253–76.

Wembah-Rachid, J. A. R.
1969 “Iron Workers of Ufipa.” *Bulletin of the International Committee of Urgent Anthropological Research*, 11: 65–72.

Westerdijk, Peter
1988 “The African Throwing Knife: A Style Analysis.” PhD diss., Utrecht University.

Whitehead, Colson
2016 *The Underground Railroad*. New York: Doubleday.

Westermarck, Edward A.
(1926) 2003 *Ritual and Belief in Morocco*. London: Trubner & Co.

Wilks, Ivor
1999 “The History of the Sunjata Epic: A Review of the Evidence.” In *In Search of Sunjata: the Mande Oral Epic as History, Literature, and Performance*, edited by Ralph A. Austen, 25–67. Bloomington: Indiana University Press.

Williams, Denis
1974 *Icon and Image: A Study of Sacred and Secular Forms of African Classical Art*. London: Allen Lane.

Willett, Frank
1971 “Nigeria.” In *The African Iron Age*, edited by Peter L. Shinnie, 1–35. Oxford: Oxford University Press.

Willis, Roy
1964 “Traditional History and Social Structure in Ufipa.” *Africa* 34, no. 4: 340–52.
1981 *A State in the Making: Myth, History and Transformation in Pre-colonial Ufipa*. Bloomington: Indiana University Press.

Wilson, Frank R.
1998 *The Hand: How Its Use Shapes the Brain, Language, and Human Culture*. New York: Vintage Books.

Wingate, Francis Reginald
1893 *The Years’ Captivity in the Mahdi’s Camp*, 1882–1892. 3rd ed. London: Sampson Low, Marston.

Wiredu, Kwasi
1997 “African Philosophy and Intercultural Dialogue.” *Quest: An African Journal of Philosophy* 11: 30–41.
2002 “Conceptual Decolonization as an Imperative in Contemporary African Philosophy.” *Rue Descartes* 36, no. 2: 53–64.

Wise, R.
1958a “Some Rituals of Iron-Making.” *Tanganyika Notes and Records* 51: 232–38.
1958b “Iron Smelting in Ufipa.” *Tanganyika Notes and Records* 50: 106–11.

Womersley, Harold
1975 *In the Glow of the Log Fire*. London: Peniel Press.

Wood, John
1999 “The Origin of the Solar System.” In Beatty et al. 1999, 13–22.

Woodhouse, James
1998 “Iron in Africa: Metal from Nowhere.” In *Transformations in Africa: Essays on Africa’s Later Past*, edited by Graham Connah, 160–85. London: Leicester University Press.

Wright, David K.
2017 “Accuracy vs. Precision: Understanding Potential Errors from Radiocarbon Dating on African Landscapes.” *African Archaeological Review* 34, no. 3: 303–19.

Wright, Donald R.
2015 *The World and a Very Small Place in Africa: A History of Globalization in Niumi, The Gambia*. 3rd ed. New York: Routledge.

Wright, Marcia
1985 “Iron and Regional History: Report on a Research Project in Southwestern Tanzania.” *African Economic History* 14: 147–65.
2002 “Life and Technology in Everyday Life: Reflections on the Career of Mzee Stefano, Master Smelter in Ufipa, Tanzania.” *Journal of African Cultural Studies* 15, no. 1: 17–34.

Wyckaert, R. P.
1914 “Forgerons paeins et forgerons chrétiens au Tanganyika.” *Anthropos* 9: 371–80.

Wynne-Jones, Stephanie and Adria LaViolette, eds.
2018 *The Swahili World*. Oxford: Routledge.

Yarranton, Andrew
1681 *England’s Improvement by Sea and Land. To Out-do the Dutch without Fighting, to Pay Debts without Moneys* [. . .]. London.

Zahan, Dominique
1960 *Sociétés d’initiation bambara: Le N’domo, le koré*. Paris: Mouton.
1988 “The Boat of the World as a Pendant.” *African Arts* 21, 4: 56–57.

Zangato, Etienne, and Augustin Holl
2010 “On the Iron Front: New Evidence from North-Central Africa.” *Journal of African Archaeology* 8, no. 1: 7–23.

Zhang, Zhongshi, Gilles Ramstein, Mathieu Schuster, Camille Li, Camille Contoux, and Qing Yan
2014 “The Aridification of the Sahara Desert caused by the Tethys Sea shrinkage during the Late Miocene.” *Nature: International Journal of Science* 513: 401–04.

Zirngibl, Manfred A.
1983 *Seltene Afrikanische Kurzwaffen*. Grafenau: Morsak-Verlag.

INDEX

CONTRIBUTORS

Rowland Abiodun is John C. Newton Professor of Art, the History of Art, and Black Studies at Amherst College. He is the author of *What Follows Six Is More than Seven: Understanding African Art* (1995) and *Yoruba Art and Language: Seeking the African in African Art* (2014); co-author of *Yoruba Art and Aesthetics* (with Henry J. Drewal and John Pemberton III; 1991) and *Cloth Only Wears to Shreds: Yoruba Textiles and Photographs from the Beier Collection* (with Ulli Beier and J. Pemberton III; 2004); and co-editor of *The Yoruba Artist: New Theoretical Perspectives on African Arts* (with H. J. Drewal, and J. Pemberton III; 1994) and *Ifá Divination, Knowledge, Power, and Performance* (with Jacob K. Olupona; 2016). Abiodun is a member of the Advisory Board of the National Museum of African Art at the Smithsonian Institution, Washington, D.C. In 2011, he received the Leadership Award of the Arts Council of the African Studies Association in recognition of his excellence, innovative contributions, and vision in the fields of African and Diasporic Arts

Since 2001, **Marla C. Berns** has served as the Shirley and Ralph Shapiro Director of the Fowler Museum at UCLA and as adjunct assistant professor in the UCLA department of art history. Her publishing and curatorial work has concentrated on women's arts of northeastern Nigeria, encompassing ceramic sculpture, decorated gourds, programs of body scarification, and issues of gender and identity. Berns was the lead curator of the major international traveling exhibition *Central Nigeria Unmasked: Arts of the Benue River Valley* (2011–13) and co-editor and author of its accompanying publication, which won the 2014 Arnold Rubin Outstanding Publication Award from the Arts Council of the African Studies Association (ACASA). To celebrate the Fowler's fiftieth anniversary, Berns wrote a history of the museum for the book *World Arts, Local Lives: The Collections of the Fowler Museum at UCLA* (2014). She has served on the editorial board of the journal *African Arts* since 2001, and in 2013 received the medal of chevalier of the Order of Arts and Letters of the French Republic.

At the University of Cape Town, **Shadreck Chirikure** is professor in the department of archaeology and directs the Archaeological Materials Laboratory. His research explores the political economies of state and nonstate formations in precolonial Africa over the last two thousand years, and engages with topics including technological advancements, trade and exchange, and inequality. In so doing, Chirikure's work fuses scientific technique with interpretive flavors based in anthropological theories and African-centered positions. He is the author of *Metals in Past Societies: A Global Perspective on Indigenous African Metallurgy* (2015) and a co-editor of *Managing Africa's Heritage: Who Cares* (with Webber Ndoro and Janette Deacon; 2016).

As an art historian and director of the African Studies Program at Penn State University, **William J. Dewey** has largely focused his field research on blacksmith-carvers of Zimbabwe and the Democratic Republic of the Congo. His publications include *Sleeping Beauties: The Jerome L. Joss Collection of Headrests* (1993); *Iron, Master of Them All* (co-authored with A. F. Roberts; 1993); "Forging Memory" (co-authored with S. Terry Childs), in *Memory: Luba Art and the Making of History* (edited by Mary Nooter Roberts and Allen F. Roberts; 1996); *Legacies of Stone, Zimbabwe: Past and Present* (co-edited with Els Palmenaer; 1997); "Iron Sculpture in Africa," in *Material Differences: Art and Identity in Africa* (edited by Frank Herreman; 2003); "Blacksmiths and Kings and Their Work in Iron," in *Fatal Beauty: Traditional Weapons of Central Africa* (edited by Jan Elsen and Marc Leo Felix; 2009). He also produced the video *Weapons for the Ancestors* (1990), on the iron smelting of Njanja, Shona peoples of Zimbabwe.

Isaie Dougnon is professor of anthropology at the University of Bamako, Mali, and assistant professor of French and humanitarian affairs at Fordham University in New York. His study of work and migration in Mali and Ghana resulted in his first book, *'Travail de Blanc' 'Travail de Noir' la migration des paysans dogons vers l'Office du Niger et au Ghana 1910–1980* (2007). He is currently finishing a second book, on life cycles, careers, and rites in Malian public service. Dougnon has received fellowships from the Fulbright Foundation, re:work Berlin, the Alexander von Humboldt Foundation, and the Nantes Institute for Advanced Study Foundation.

Apprenticeships with Yorùbá sculptors in Nigeria transformed **Henry J. Drewal**'s life, leading to interdisciplinary doctoral studies at Columbia University in African arts, histories, and cultures. His books, articles, films, and catalogues include *Introspectives: Contemporary Art by Americans and Brazilians of African Descent* (1989); *Yoruba: Nine Centuries of African Art and Thought* (co-authored with John Pemberton III and Rowland Abiodun; 1989); *Beads, Body, and Soul: Art and Light in the Yoruba Universe* (co-authored with John Mason; 1998); *Mami Wata: Arts for Water Spirits in Africa and its Diasporas* (2008); and *Dynasty and Divinity: Ife Art in Ancient Nigeria* (co-authored with Enid Schildkrout; 2010). At the University of Wisconsin-Madison, he is the Evjue-Bascom Professor of Art History and Afro-American Studies. Since 1990, he has been exploring, through his approach called *sensiotics*, the role of the senses and sense-abilities in shaping arts, persons, cultures, and histories.

Candice Goucher, Professor Emerita of history at Washington State University, has conducted historical and archaeological research on African metallurgy, gender, and food studies on both sides of the Atlantic. The award-winning author of *Congotay! Congotay! A Global History of Caribbean Food* (2014), Goucher has written and edited sixteen additional books, among them *World History: Journeys from Past to Present* (co-authored with Linda Walton; 2013). She has been a lead collaborator on twenty-seven films, including many for public television. Her research on African metallurgy has focused on lead isotope studies of Nigerian bronzes, gender and technology, and the impact of iron smelting on African environmental history.

Tom Joyce, a sculptor and MacArthur Fellow, is widely acknowledged as one of the foremost contemporary practitioners in the art and science of forging iron. As lead curator for *Striking Iron*, he conducted comparative research on over 7,000 iron artworks in 147 public and private collections in the U.S. and abroad, and organized six research trips to west Africa to interview, photograph, and film blacksmiths in the Republic of Bénin, Ghana, Togo, Burkina Faso, and Mali. In addition to lecturing on African forge work, Joyce has previously curated the exhibitions *Life Force at the Anvil: The Blacksmith's Art from Africa* (University of North Carolina Asheville, 1998), *Forged Iron Currency and the Art of African Blacksmiths* (Museum of International Folk Art, Santa Fe, 2003), and *Artful Exchanges: Sub-Saharan African Metal Trade Tokens* (Santa Fe Institute, 2015).

Colleen E. Kriger writes about west and central Africa, particularly before the twentieth century, and is professor of history at the University of North Carolina at Greensboro. Her research into African, European, and American museum collections focuses on social and cultural history, material culture and artisans, and commodity currencies and trade. Major publications include *Pride of Men: Ironworking in 19th century West Central Africa* (1999), *Cloth in West African History* (2006), and *Making Money: Life, Death, and Early Modern Trade on Africa's Guinea Coast* (2017). Her most recent work is "'Words and Things' and 'The Kuba Miracle'," published in the June 2018 issue of *History in Africa*.

Manuel Jordán is deputy director, chief curator, and curator for Africa at the Musical Instrument Museum in Phoenix, Arizona. He formerly served as Phyllis Wattis Curator for the Arts of Africa, Oceania, and the Americas at the Iris & B. Gerald Cantor Center for Visual Arts at Stanford University; prior to that, he was Curator for Arts of Africa and the Americas at the Birmingham Museum of Art. Jordán has conducted extensive field research on art and culture in Zambia among Chokwe, Lunda, Luvale, Luchazi, and Mbunda peoples. He is considered a leading authority on the arts of Chokwe and related peoples.

E. C. Krupp is an astronomer and director of the Griffith Observatory in Los Angeles. He attended Pomona College as an undergraduate, majoring in physics and astronomy, and received his MA and PhD in astronomy at UCLA, where the late George O. Abell was his advisor. Since 1973, Krupp has studied cultural dimensions of astronomy. He is the author and editor of five books on this subject and has personally visited, studied, and photographed more than twenty-one hundred ancient, historic, and prehistoric sites throughout the world. He has also written hundreds of articles for the general reader, dozens of research papers, and four children's books on astronomy.

Scott MacEachern, professor of archaeology and anthropology at Duke Kunshan University, has been involved in archaeological research projects in different parts of Africa since the early 1980s, especially around the Mandara Mountains of northern Cameroon and Nigeria. His main research interests are state formation processes in Africa, the archaeological study of ethnicity and social boundaries, African cultural heritage management, and archaeogenetics. He is the lead co-author (with Philippe Lavachery, Tchago Bouimon, and Christophe Mbida Mindzie) of *Komé – Kribi: rescue archaeology along the Chad-Cameroon Oil Pipeline, 1999–2004* (2010) and the author of *Searching for Boko Haram: a history of violence in Central Africa* (2018), in addition to having co-edited four volumes and published approximately sixty papers on African prehistory.

Retired Chancellor's Professor and Indiana University art history chair **Patrick McNaughton** has received Smithsonian, Guggenheim, and National Endowment for the Humanities fellowships. He has published numerous articles and a monograph on the Mande Kòmò Association. His books *The Mande Blacksmiths: Knowledge, Power, and Art in West Africa* (1988) and *A Bird Dance Near Saturday City: Sidi Ballo and the Art of West African Masquerade* (2008) were finalists for the Melville J. Herskovits Award. He also directed the CD-ROM *Five Windows into Africa* (2000). McNaughton has served on the African Studies Association Board of Directors, was an editor for *Africa Today*, is general editor of the Indiana University Press series *African Expressive Cultures*, and serves on the Oxford Research Encyclopedia of African History Advisory Board.

Philip M. Peek's focus has been southern Nigeria since his service in the Peace Corps in the mid-1960s, and he has published extensively on African visual and verbal arts, expressive behavior, and divination. He edited *African Divination Systems: Ways of Knowing* (1991) and *Twins in African and Diaspora Cultures* (2011), and co-edited *Ways of the Rivers: Arts and Environment in the Niger Delta* (with Martha Anderson; 2002), *African Folklore: An Encyclopedia* (with Kwesi Yankah; 2004), *Divination and Healing* (with Michael Winkelman; 2004), and *Reviewing Reality: Dynamics of African Divination* (with Walter van Beek; 2013). Peek is Professor Emeritus of anthropology at Drew University and research associate at the Smithsonian Institution.

Allen F. Roberts is professor of world arts and cultures and co-editor of the journal *African Arts* at UCLA. His most notable scholarly accomplishments, all with his late spouse Mary "Polly" Nooter Roberts (1959–2018), include the National Endowment for the Humanities–funded traveling exhibitions and award-winning books *Memory: Luba Art and the Making of History* (1996), which was based upon Polly's Congolese dissertation research, and *A Saint in the City: Sufi Arts of Urban Senegal* (2003), built from more than nine years of collaborative research. His award-winning monograph *A Dance of Assassins: Performing Early Colonial Hegemony in the Congo* (2013) is one of several hundred scholarly publications based upon his more than fifty years of research in and about sub-Saharan Africa.

A historian of the visual cultures of Africa, **Raymond Silverman** is professor of the history of art, African studies, and museum studies at the University of Michigan. His research and writing explore historical and contemporary visual practices in Ethiopia and Ghana, and museum and heritage discourse in Africa. Silverman's recent work focuses on the visual culture of the Ethiopian Orthodox Church in the twentieth and twenty-first centuries, as represented in his books *Painting Ethiopia: The Life and Work of Qes Adamu Tesfaw* (2005) and *Icons of Devotion/Icons of Trade: Contemporary Painting and the Orthodox Church in Ethiopia* (with Neal Sobania; forthcoming).

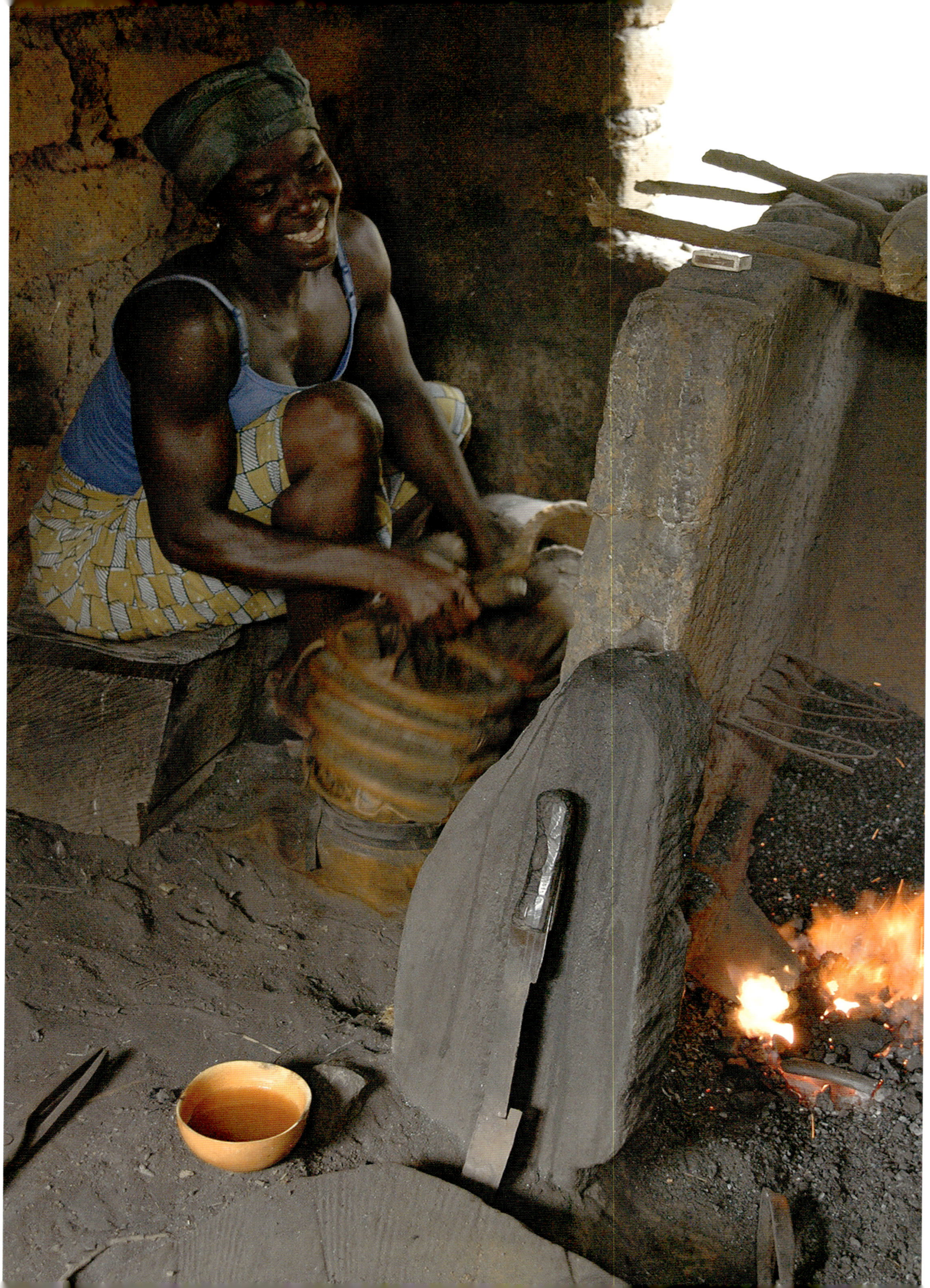

The Fowler Museum is part of UCLA's School of the Arts and Architecture

Fowler Museum at UCLA
Box 951549
Los Angeles, California 90095-1549

Striking Iron: The Art of African Ironsmiths

Deirdre O'Dwyer, *Managing Editor*
Danny Brauer, *Designer and Production Manager*
Don Cole, *Photographer*
Gassia Armenian and Sophia Livsey, *Editorial Assistants*
David Fuller, *Cartographer*

Printed in Hong Kong by Great Wall Printing Company, Limited.

Library of Congress Cataloging-in-Publication Data

Names: Roberts, Allen F., 1945- editor. | Joyce, Tom, 1956- editor. | Berns, Marla, editor. | Dewey, William Joseph, editor. | Drewal, Henry John, editor. | Goucher, Candice Lee, editor. | Abiodun, Rowland, writer of added commentary. | Fowler Museum at UCLA, issuing body, host institution.
Title: Striking iron : the art of African blacksmiths / edited by Allen F. Roberts, Tom Joyce, and Marla C. Berns, with William J. Dewey, Henry J. Drewal, and Candice Goucher ; with essays and contributions by Rowland Abiodun [and 17 others].
Description: Los Angeles, California : Fowler Museum at UCLA, 2019. | Published on the occasion of an exhibition held at the Fowler Museum at UCLA, the Smithsonian Museum's National Museum of African Art, and Musee du quai Branly-Jacques Chirac, Paris. | Includes bibliographical references and index.
Identifiers: LCCN 2019007323 | ISBN 9780990762669 (hardcover)
Subjects: LCSH: Blacksmiths—Africa—Exhibitions. | Artist-blacksmiths—Africa—Exhibitions. | Art metal-work—Africa—Exhibitions. | Ironwork—Africa—Exhibitions.
Classification: LCC NK8287 .S77 2019 | DDC 739.4096--dc23
LC record available at https://lccn.loc.gov/2019007323

(FRONT COVER) SEE FIG. 4.4, PG 107.
(PAGE 21) SEE FIG. 4.14B, PG 21, 120. PHOTOGRAPH BY TOM JOYCE, OUIDAH, REPUBLIC OF BÉNIN, 2010. © TOM JOYCE
(PAGE 508) SEE FIG. 4.29, PG 132. PHOTOGRAPH BY TOM JOYCE, TCHARÉ, TOGO, 2010. © TOM JOYCE
(BACK COVER): SEE FIG. 4.40, PG 142. PHOTOGRAPHS BY TOM JOYCE, TCHARÉ, TOGO, 2010. © TOM JOYCE